BURGER'S MEDICINAL CHEMISTRY, DRUG DISCOVERY AND DEVELOPMENT

BURGER'S MEDICINAL CHEMISTRY, DRUG DISCOVERY AND DEVELOPMENT

BURGER'S MEDICINAL CHEMISTRY, DRUG DISCOVERY AND DEVELOPMENT

Seventh Edition

Volume 1: Methods in Drug Discovery

Edited by

Donald J. Abraham
Virginia Commonwealth University

David P. Rotella
Wyeth Research

Burger's Medicinal Chemistry, Drug Discovery and Development
is available Online in full color at
http://mrw.interscience.wiley.com/emrw/9780471266945/home/

A JOHN WILEY & SONS, INC., PUBLICATION

Published by John Wiley & Sons, Inc., Hoboken, New Jersey
Published simultaneously in Canada

Library of Congress Cataloging-in-Publication Data:

Abraham, Donald J., 1936-
Burger's medicinal chemistry, drug discovery, and development/Donald J. Abraham, David P. Rotella. – 7th ed.
p. ; cm.
Other title: Medicinal chemistry, drug discovery, and development
Rev. ed. of: Burger's medicinal chemistry and drug discovery. 6th ed. / edited by Donald J. Abraham. c2003.
Includes bibliographical references and index.
ISBN 978-0-470-27815-4 (cloth)
1. Pharmaceutical chemistry. 2. Drug development. I. Rotella, David P. II. Burger, Alfred, 1905-2000. III. Burger's medicinal chemistry and drug discovery. IV. Title. V. Title: Medicinal chemistry, drug discovery, and development.
[DNLM: 1. Chemistry, Pharmaceutical–methods. 2. Biopharmaceutics–methods. 3. Drug Compounding–methods. QV 744 A105b 2010]
RS403.B8 2010
615'.19–dc22 2010010779

Printed in Singapore

10 9 8 7 6 5 4 3 2 1

CONTENTS

PREFACE

The seventh edition of Burger's Medicinal Chemistry resulted from a collaboration established between John Wiley & Sons, the editorial board, authors, and coeditors over the last 3 years. The editorial board for the seventh edition provided important advice to the editors on topics and contributors. Wiley staff effectively handled the complex tasks of manuscript production and editing and effectively tracked the process from beginning to end. Authors provided well-written, comprehensive summaries of their topics and responded to editorial requests in a timely manner. This edition, with 8 volumes and 116 chapters, like the previous editions, is a reflection of the expanding complexity of medicinal chemistry and associated disciplines. Separate volumes have been added on anti-infectives, cancer, and the process of drug development. In addition, the coeditors elected to expand coverage of cardiovascular and metabolic disorders, aspects of CNS-related medicinal chemistry, and computational drug discovery. This provided the opportunity to delve into many subjects in greater detail and resulted in specific chapters on important subjects such as biologics and protein drug discovery, HIV, new diabetes drug targets, amyloid-based targets for treatment of Alzheimer's disease, high-throughput and other screening methods, and the key role played by metabolism and other pharmacokinetic properties in drug development.

The following individuals merit special thanks for their contributions to this complex endeavor: Surlan Alexander of John Wiley & Sons for her organizational skills and attention to detail, Sanchari Sil of Thomson Digital for processing the galley proofs, Jonathan Mason of Lundbeck, Andrea Mozzarelli of the University of Parma, Alex Tropsha of the University of North Carolina, John Block of Oregon State University, Paul Reider of Princeton University, William (Rick) Ewing of Bristol-Myers Squibb, William Hagmann of Merck, John Primeau and Rob Bradbury of AstraZeneca, Bryan Norman of Eli Lilly, Al Robichaud of Wyeth, and John Lowe for their input on topics and potential authors. The many reviewers for these chapters deserve special thanks for the constructive comments they provided to authors. Finally, we must express gratitude to our lovely, devoted wives, Nancy and Mary Beth, for their tolerance as we spent time with this task, rather than with them.

As coeditors, we sincerely hope that this edition meets the high expectations of the scientific community. We assembled this edition with the guiding vision of its namesake in mind and would like to dedicate it to Professor H.C. Brown and Professor Donald T. Witiak. Don collaborated with Dr. Witiak in the early days of his research in sickle cell drug discovery. Professor Witiak was Dave's doctoral advisor at Ohio State University and provided essential guidance to a young scientist. Professor Brown, whose love for chemistry infected all organic graduate students at Purdue University, arranged for Don to become a medicinal chemist by securing a postdoctoral position for him with Professor Alfred Burger.

It has been a real pleasure to work with all concerned to assemble an outstanding and up-to-date edition in this series.

Donald J. Abraham
David P. Rotella

March 2010

CONTRIBUTORS

Jeffrey Aubé, University of Kansas, Lawrence, KS
Soumitra Basu, Kent State University, Kent, OH
J. Ellis Bell, University of Richmond, Richmond, VA
Jessica K. Bell, Virginia Commonwealth University, Richmond, VA
Francesc J. Corcho, Universitat Politecnica de Catalunya, Barcelona, Spain
Jeffrey H. Dahl, University of Illinois, Chicago, IL
Apurba Dutta, University of Kansas, Lawrence, KS
Jennifer E. Golden, University of Kansas, Lawrence, KS
Timothy J. Hagen, deCODE Chemistry, Inc., Woodridge, IL
Anke-Hilse Maitland-van der Zee, Utrecht University, Utrecht, The Netherlands
Yvonne C. Martin, Martin Consulting, Waukegan, IL
Jonathan S. Mason, Heptares Therapeutics, Welwyn Garden City, Hertfordshire, UK; Lundbeck Research, Valby, Denmark
Lester A. Mitscher, University of Kansas, Lawrence, KS
Richard Morphy, Schering-Plough Corporation, Newhouse, Lanarkshire, UK
Andrew G. Newsome, University of Illinois, Chicago, IL
Adegboyega Oyelere, Georgia Tech, Atlanta, GA
Juan J. Perez, Universitat Politecnica de Catalunya, Barcelona, Spain
Brian J. Puckett, Virginia Commonwealth University, Richmond, VA
Zoran Rankovic, Schering-Plough Corporation, Newhouse, Lanarkshire, UK
Jaime Rubio-Martinez, Universitat de Barcelona, Barcelona, Spain
Matthieu Schapira, University of Toronto, Toronto, Canada
Cynthia Selassie, Pomona College, Claremont, CA
Jasbir Singh, deCODE Chemistry, Inc., Woodridge, IL
Steven G. Terra, JHM Health Science Center, Gainesville, FL
Bernard Testa, University Hospital Centre, Lausanne, Switzerland
Alexander Tropsha, University of North Carolina, Chapel Hill, NC
Richard B. van Breemen, University of Illinois, Chicago, IL
Rajeshwar P. Verma, Pomona College, Claremont, CA
Joe Walker, Orchid BioSciences, Princeton, NJ
Camille G. Wermuth, Prestwick Chemical, Inc., Illkirch-Graffenstaden, France

HISTORY OF QUANTITATIVE STRUCTURE–ACTIVITY RELATIONSHIPS

Cynthia Selassie
Rajeshwar P. Verma
Chemistry Department, Pomona College, Claremont, CA

1. INTRODUCTION

Over the past 50 years, the quantitative structure–activity relationship (QSAR) paradigm has flourished and evolved to the point where it is many pronged and diverse as well as a critical element of computer-assisted molecular design (CAMD). The utilization of statistical methodologies is critical to the formulation and development of a QSAR model and the encoding of the relationship between the structure of a molecule and its ability to induce a measurable biological endpoint. This general approach provides an understanding of the interactions of various chemicals with biological macromolecules as determined by underlying intermolecular forces that could be hydrophobic, electrostatic, polar, and steric in nature. It delineates how the various physicochemical attributes of molecules affect their chemical reactivities and biological activities. The overall objective of QSAR is twofold: to be predictive in nature and to be conducive to mechanistic interpretation, although in recent years the former goal now appears to overshadow the latter. Developments in combinatorial chemistry and the subsequent availability of rapidly expanding, synthetic, and virtual libraries of compounds combined with an increase in the structural knowledge and number of clinical targets for screening and escalation in the proliferation of easily calculable descriptors foments a need for a rapid and organized approach to molecular design and development. In this chapter, we will take a few steps back and examine the history, critical milestones, and development of the QSAR paradigm.

1.1. Historical Development of QSAR

Forty-five years ago, although Hansch defined and crystallized the general approach to QSAR, the faint beginnings of structure–activity relationships go back to the time of the Renaissance when Paracelsus (1493–1541) the famous alchemist, physician, and astrologer argued about the importance of proper dosages in internal and external remedies [1]. In 1868, Crum-Brown and Fraser determined that the paralyzing property of a set of alkaloids including strychnine, morphine, codeine, and atropine was related to the nature of the quaternary head group [2]. They expressed their observations in the form of Equation 1, where Φ is a measure of physiological activity and C is a measure of chemical constitution.

$$\Phi = f(C) \tag{1}$$

Despite extensive studies, they found it to be less successful in applying their knowledge to the discovery of useful drugs; curare-type compound did not attain medical status for another 70 years, but clearly their mathematical definition of biological activity was of fundamental importance. A few years later, Richardson noted that the anesthetic potency of aliphatic alcohols generally paralleled their carbon chain length and molecular weight [3]. In 1893, Richet and his student Houdaille established that the toxicities of narcotics were inversely proportional to their water solubilities [4]. Soon, Meyer and Overton separately and independently noted that the anesthetic potencies of small organic molecules in tadpoles and small fish were not adequately predicted by water solubility, but could be assessed by using olive oil–water partition coefficients [5,6]. Both Meyer and Overton realized that olive oil by virtue of its properties, served as a surrogate for a hydrophobic site of action.

In the late 1930s, the focus shifted from similarities in groups or atom types to a more generalized emphasis on overall properties of the molecules. This approach was exemplified by the work of Albert et al. who examined the effects of ionization/electron distribution and steric access on the potencies of a large number of amino acridines [7]. They also established that not all drug receptors were on proteins but could also be present on nucleic acids. Around the same time, Bell and Roblin studied the correlation between p*K*a and the antibacterial activities of a series of sulfanilamides [8]. They showed that activity increased as the p*K*a decreased from 11 to 7, but decreased as the p*K*a values continued to decline. They concluded that activity was optimized at p*K*a = 7, but with a higher degree of ionization (p*K*a < 7) the charge distribution on the SO_2N moiety decreased because of the substituent. They suggested that the most significant factor governing the potencies of the sulfanilamides was the magnitude of the charge on the oxygen atom of the SO_2 group as opposed to the overall ionization, results later confirmed by others [9–11].

2. INDEPENDENT VARIABLES UTILIZED IN QSAR

Physicochemical parameters are of critical importance in describing the types of intermolecular forces that represent drug–receptor interactions. The three major groups of parameters that were initially utilized and that still hold sway are electronic, hydrophobic, and steric in nature [12,13]. Indicator or dummy variables are sometimes included in a QSAR to pinpoint a particular structural feature that may confer unusual activity or lack of it, to a series of molecules. An extensive list of common parameters/descriptors used in QSAR/QSPR studies is given in Table 1.

2.1. Electronic Parameters

Classical QSAR as it is known today was guided by the earlier work of many chemists and physicians, but it really vaulted into perspective with the critical contributions of Hammett who was highly influenced by Brönsted's early work in catalysis [14,15]. Brönsted suggested that catalytic rate constants would increase with increased base or acid strength and described this relationship by what is now known as the Brönsted catalysis law, the first example of a linear free-energy relationship where k_A represents the catalytic rate constants of individual acids with ionization constants, K_A [14].

$$\log k_A = \alpha K_A + \log \text{constant} \qquad (2)$$

These studies piqued Hammett's interest and led to a search for analogous log–log relationships for other types of reactions. Experimental work by Hammett and Pfluger and extensive literature analysis of logarithmic reaction relationships led to his critical publication in 1935 [16]. At the same time, G.N. Burkhardt was independently also examining the effects of substituents on organic reactions. He proposed the use of the ionization of substituted benzoic acids as a reference series [17].

In 1937, Hammett published his seminal study on the effects of *meta-* and *para-*substituents on the rate constants or equilibrium constants of side-chain reactions of benzene derivatives [18,19]. The Hammett equation is now a well-established empirical relationship for correlating changes in structures with proportional changes in the activation energy ($\Delta G^{\ddagger}$) for many such reactions and is a well-established example of an LFER. It has been widely applied in diverse areas of ground-state kinetics to delineate the outcomes of changes in reaction parameters. Thus, it can be used in reactions involving equilibrium constants (K_X/K_H) or for reaction rate constants (k_X/k_H). The measure of the electronic effect of each substituent on a reaction is relative to that of hydrogen and is described by its substituent constant, σ_X.

$$\sigma_X = \log K_X - \log K_H \qquad (3)$$

Hammett used the ionization of substituted benzoic acids in water at 25°C as a reference system to determine σ_X values (Fig. 1) and hence to reflect the effect of the substituent

COOH + H_2O ⇌ COO^- + H_3O^+

Figure 1. Ionization of X-benzoic acids in water.

Table 1. Common Parameters/Descriptors Used in QSAR/QSPR Studies

No.	Type of Descriptor	Name of Descriptor (Symbol)	Brief Description
1	Conformational	LowEne	Lowest energy conformer
		EPenalty	Conformational energy penalty
2	Constitutional/ Structural	D_{Atoms}	Total number of atoms
		$D_{Atoms,X}$	Number of individual types of atoms
		$D_{Rel,H}$	Relative number of H atoms
		D_{Bonds}	Total number of bonds
		NSB	Number of single bonds
		NDB	Number of double bonds
		NTB	Number of triple bonds
		D_{Rings}	Number of rings
		NRA	Count of ring atoms
		$2SP^2$	Number of sp^2-hybridized carbon atoms attached to two C atoms.
		$3SP^2$	Number of sp^2-hybridized carbon atoms attached to three C atoms.
		HBD	Hydrogen bond donor
		HBA	Hydrogen bond acceptor
		Rotlbonds	Number of rotatable bonds
		Chiral	Number of the chiral centers (R or S)
		MW	Molecular weight
		AAW	Average atomic weight
3	Electronic	Apol	Sum of atomic polarizabilities
		NVE	Number of valance electrons (polarizability)
		BDE	Bond dissociation energy
		Charge	Sum of partial charges
		Fcharge	Sum of formal charges
		μ	Dipole moment
		σ, σ^+, σ^-	Hammett electronic substituent constant
		pKa	Ionization constant
		δ^*	Taft polar constant
		δ_I	Taft inductive component
		δ_R	Taft resonance component
		HOMO	Highest occupied molecular orbital energy
		LUMO	Lowest unoccupied molecular orbital energy
		HOMO–LUMO	Difference in HOMO and LUMO energy
		Sr	Superdelocalizability
		IP	Ionization potential
		EA	Electron affinity
		EN	Electronegativity
		η	Absolute hardness
4	Electrostatic	CPSA	Charged partial surface area
		D_{PPSA}	Partial positive surface area
		D_{RPCG}	Relative positive charge
		D_{RNCG}	Relative negative charge
5	E-State	S_sCH_3	E-state value of carbon atom in the fragment $-CH_3$
		S_ssCH_2	E-state value of carbon atom in the fragment $-CH_2-$
		S_tCH	E-state value of carbon atom in the fragment $\equiv$CH
		S_dsCH	E-state value of carbon atom in the fragment =CH–
		S_tsC	E-state value of carbon atom in the fragment $\equiv$C–

(continued)

Table 1. *(Continued)*

No.	Type of Descriptor	Name of Descriptor (Symbol)	Brief Description
		S_dssC	E-state value of carbon atom in the fragment 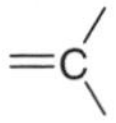
		S_sssCH	E-state value of carbon atom in the fragment 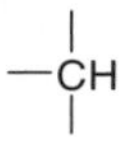
		S_ssssC	E-state value of carbon atom in the fragment 
		S_aaCH	E-state value of carbon atom in the fragment 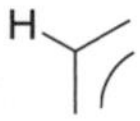
		S_aasC	E-state value of carbon atom in the fragment 
		S_tN	E-state value of nitrogen atom in the fragment ≡N
		S_sNH_2	E-state value of nitrogen atom in the fragment $-NH_2$
		S_ssNH	E-state value of nitrogen atom in the fragment 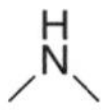
		S_sssN	E-state value of nitrogen atom in the fragment 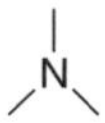
		S_ddsN	E-state value of nitrogen atom in the fragment 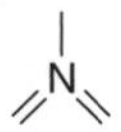
		S_sOH	E-state value of oxygen atom in the fragment −OH
		S_dO	E-state value of oxygen atom in the fragment =O
		S_ssO	E-state value of oxygen atom in the fragment −O−
		S_dS	E-state value of sulfur atom in the fragment =S
		S_ddssS	E-state value of sulfur atom in the fragment 
		S_sF	E-state value of fluorine atom in the fragment −F
		S_sCl	E-state value of chlorine atom in the fragment −Cl
		S_sBr	E-state value of bromine atom in the fragment −Br
		S_sI	E-state value of iodine atom in the fragment −I
6	Geometrical	SA	Molecular surface area

Table 1. *(Continued)*

No.	Type of Descriptor	Name of Descriptor (Symbol)	Brief Description
		V	Molecular volume
		D_I	Moment of inertia
		GI	Gravitation index
		PSA	Polar surface area
7	Hydrophobic	$\log P_{o/w}$	Log of partition coefficient of compound in 1-octanol/water system
		$\log D$	Log of partition coefficient of ionizable compound in 1-octanol/water system
		R_m	Hydrophobicity of compound obtained from TLC
		π	Hansch hydrophobic constant for the substituent
		$A\log P98$	Hydrophobicity index
		f	Rekker's fragment constant
8	Quantum-Chemical	Q_A	Net atomic charge on atom A
		Q_{min}, Q_{max}	Net charges of the most negative and most positive atoms
		Q_{AB}	Net group charge on atoms A, B
		$E_{HOMO,A}$	Fraction of HOMO energy arising from the atomic orbitals of the atom A
		$E_{LUMO,A}$	Fraction of LUMO energy arising from the atomic orbitals of the atom A
		$q_{A,\sigma}$, $q_{A,\pi}$	σ- and π-electron densities of the atom A
		$Q_{A,H}$, $Q_{A,L}$	HOMO/LUMO electron densities on the atom A
		α	Molecular polarizability
		Δ	Submolecular polarity parameter
		D	Local dipole index
		τ	Quadrupole moment tensor
		EHC	Effective hydrogen charge
9	Shape	DIFFV	Difference between the volume of the individual compound and the volume of the shape reference compound
		COSV	The common volume between each of the individual molecule and the reference molecule
		Fo	The common overlap steric volume (COSV) descriptor divided by the individual molecular volume
		NCOSV	Difference between the volume of the individual molecule and the COSV
		ShapeRMS	Root mean square deviation between the individual molecule and the shape reference compound
		SRVolume	The volume of the shape reference compound
10	Solubility	S	Molar solubility
		X	Mole fraction solubility
		$\log \gamma_w$	Activity coefficients
		δ_H	Hildebrand solubility parameter
11	Spatial	Density	Ratio of molecular weight to molecular volume
		PMI-mag	Principal moment of inertia about the principal axes of the molecule
		V_m	Molecular volume inside the contact surface
		Area	van der Waals area of the molecule
		Shadow indices	A set of descriptors helps to characterize the shape of the molecule (surface area projections descriptors)

(continued)

Table 1. *(Continued)*

No.	Type of Descriptor	Name of Descriptor (Symbol)	Brief Description
		Jurs descriptors	A set of descriptors combines shape and electronic information to characterize the molecules (Jurs charged partial surface areas descriptors)
		RadOf-Gyration	Radius of gyration
12	Steric	E_s	Taft steric parameter
		MR	Molar refractivity
		MR_X	Molar refractivity of the X-substituent
		P	Parachor
		υ	Charton steric parameter
		V_{vdw}	van der Waals volume
		L	Verloop's length sterimol parameter of the substituent
		B1	Verloop's minimum width sterimol parameter of the substituent
		B5	Verloop's maximum width sterimol parameter of the substituent
		CAA	Connolly accessible surface area
		CMA	Connolly molecular surface area
		CSEV	Connolly solvent-excluded volume
		EM	Exact mass
		MW	Molecular weight
		OVAL	Ovality
13	Thermodynamic	$D_{max,vib}$	Highest normal-modevibrational frequency
		D_{Srot}	Rotational energy (300 K)
		BP, MP	Boiling point, melting point
		CP	Critical pressure
		CT	Critical temperature
		ΔH	Heat of formation
		HLC	Henry's law constant
		IGTC	Ideal gas thermal capacity
		$\log P$	Logarithmic partition coefficient
		MR	Molar refractivity
		SGP	Standard Gibb's free energy
		VDW	van der Waals force
		STERG	Stretch energy
		STBERG	Stretch bend energy
		TORERG	Torsion energy
		TOTERG	Total energy
14	Topological	JX	Balaban's J index
		κ	Kappa shape indices
		D_{Randic}	Randic index
		$D_{Kier,flex}$	Kier flexibility index
		$D_{Kier,shape}$	Kier shape index
		χ	Molecular connectivity index
		W	Wiener index
		IC	Information content indices
		SC	Subgraph count index
		Zagreb	Sum of the squares of vertex valencies
		TPSA	Topological polar surface area
15	Miscellaneous	^{1}H, ^{13}C	Chemical shifts (δ_{ppm})
		ν	IR frequencies
		I	Indicator variable
		DCW(SMILES)	SMILES-based optimal descriptor

group on the free energy of ionization of the substituted benzoic acid as follows:

In an extension of this approach to a similar system, the ionization constants of substituted phenylacetic acids (PAs) were also measured and compared to those of the corresponding benzoic acids (BAs). It was quickly apparent that the effect of any substituent X, on the ionization of X-phenylacetic acid was proportional to its effect on the ionization of X-benzoic acid as represented in Equation 4.

$$\log\left(\frac{K'_{\text{X-(PA)}}}{K'_{\text{H-(PA)}}}\right) \propto \log\left(\frac{K_{\text{X-(BA)}}}{K_{\text{H-(BA)}}}\right) = \rho\sigma \qquad (4)$$

The susceptibility of a reaction to polar effects of a substituent is ascertained from its reaction constant, ρ. Thus, Hammett relationships for equilibria or rate data can be expressed by the following general equation:

$$\log\left(\frac{K_X}{K_H}\right) \text{ or } \log\left(\frac{k_X}{k_H}\right) = \rho\sigma \qquad (5)$$

The operational definition of σ is a measure of the size of the electronic effect of a substituent and represents the electronic charge distribution in the benzene nucleus. Electron-withdrawing substitutents are characterized by positive values while electron-releasing substituents have negative values. A positive value of the proportionality or reaction constant, ρ, signifies enhancement of a reaction by electron withdrawal at the reaction site while a negative value implies that a reaction is assisted by electron-releasing substituents.

Over the years, substituent constants and their applications have been determined, calculated, thoroughly analyzed and reviewed in great detail while reaction constants have come into their own, particularly in terms of mechanistic interpretation of organic reactions as well as biological interactions [20–22]. It is clearly evident that the Hammett approach, which is a quintessential representation of an LFER, helped lay the foundation for the study of correlation analysis and subsequently QSAR.

Several sets of modified Hammett parameters have been introduced to explain anomalous behavior of reactions that is inadequately addressed by the original Hammett constants. *Ortho*-substituted compounds present a vexing problem and their reactivity has to be resolved by separation into at least two components that represent steric and field effects transmitted through space. This limitation was first addressed by Charton and then by Fujita and Nishioka in an integrated approach that included *meta*- and *para*-substituted analogs [23,24]. See Equation 6.

$$\log K_X = \rho\sigma + \delta E_{S,\text{ortho}} + fF_{\text{ortho}} + C \qquad (6)$$

E_S is Taft's steric constant while F is defined as the Swain Lupton, field-inductive constant. In this approach (Eq. 5), activity is attributed to the sum of three components that reflect the standard Hammett polar effect ($\rho\sigma$), a primary steric effect ($\delta E_{S,\text{ ortho}}$) and a proximity polar effect (fF_{ortho}). In a recent kinetic study of a mononuclear heterocyclic rearrangement of six *ortho*-substituted *Z*-phenyl hydrazones to their corresponding triazoles, Frenna et al., and Consiglio et al., utilized the Fujita–Nishioka treatment to dissect the substituent effects into their various components that helped shed light on the different mechanisms at play in these reactions [25,26].

Despite the success of the Hammett equation and its extended applications, it had some limitations. The most significant one involved the influence of resonance or direct conjugation between the substituent and the reaction site, on reaction rates. Thus, reactivity was enhanced or reduced by occurrence of coupling between the substituent and reaction center via the pi-electron transmitting system. This led to the development of σ^+ (electrophilic substituents) and σ^- (nucleophilic substituents) constants for use in the following modified linear free-energy relationships [27–29]:

$$\log\left(\frac{K_X}{K_H}\right) = (\rho^+)(\sigma^+) \text{ or } (\rho^-)(\sigma^-) \qquad (7)$$

The Yukawa–Tsuno approach as described in Equation 8, corrects for this enhanced resonance contribution; the constant r reflects the magnitude of the resonance contribution to a particular reaction. When r approaches zero,

Equation 8 morphs back into the Hammett equation [30].

$$\log\left(\frac{K_X}{K_H}\right) = \rho\sigma + \rho(r)(\sigma^+ - \sigma) \qquad (8)$$

Other limitations of the Hammett approach included the lack of additivity of substituent constants in multisubstituted aromatic compounds due to proximity effects [31], changes in mechanism [32], rate-determining step or transition state [33–35], changes in reactivity depending on solvent polarity [36], and the variable behavior of highly charged substituents compared to their neutral counterparts [37]. Exner's studies with infrared data on a series of thiobenzanilides have shown that problems may arise when large sterically hindered groups with significant conformational freedom are included in a Hammett analysis [38]. Questions have also been raised about the applicability of Hammett relationship and its ground state constants to photochemical reactions, that is, reactions occurring in the excited electronic state [39]. Cordes et al. addressed this issue by examining the photochemical *Z*/*E* isomerization of a series of hemithioindigo derivatives and found it to be well correlated with σ^+ [40].

A refined treatment of substituent effects involved partitioning of the electronic effect into its inductive and resonance components based on earlier work done by Taft in aliphatic systems where resonance effects were absent; this factorization led to the formulation of the dual-substituent parameter (DSP) equation [41,42]. See Equation 9. The terms ρ_I and ρ_R reflect the sensitivity of a reaction to inductive and resonance effects, respectively.

$$\log\left(\frac{K_X}{K_H}\right) \text{ or } \log\left(\frac{k_X}{k_H}\right) = \rho_I\sigma_I + \rho_R\sigma_R \qquad (9)$$

However, this approach also has its limitations when dealing with constrained systems: the geometry of the molecules warrants attention [43] and the appropriate combination of a donor reaction center with acceptor substituents and vice versa have to be taken into consideration [44].

The success of the Hammett approach to the quantitation of substituent effects in aromatic systems and its failure in aliphatic systems lead to Taft's formulation of the aliphatic polar constant, σ^* that was defined as follows [45,46]:

$$\sigma^* = \frac{1}{2.48}\left[\log\left(\frac{k_X}{k_H}\right)_{\text{base}} - \log\left(\frac{k_X}{k_H}\right)_{\text{acid}}\right] \qquad (10)$$

$\log k_X$ and k_H represent the rate constants of base and acid catalyzed hydrolysis of XCH_2COOR and CH_3COOR, respectively. The acid hydrolysis term was introduced in Equation 9 since it was found to be dependent only on the steric effects of the substituent. The scaling factor 1/2.48 was needed to align σ^* values with Hammett constants. This Taft parameter eventually morphed into the closely related σ_I that was widely used but eventually overshadowed by σ_χ, based on electronegativity, σ_F (field), and σ_α (polarizability) parameters [47,48].

In the last 60 years, Hammett constants and subsequent extensions have been extremely effective at assessing the importance of substituents on chemical reactivity and biological activity. Today, they continue to help shed light on quantitative structure property relationships (QSPRs), QSAR, and comparative QSAR studies. Despite the proliferation of sigma constants for some well-utilized substituents, there are a substantial number of newer substituents whose sigma constants have not been measured and that are now being determined by extrapolation or calculation using quantum chemical approaches. The limitations of Hammett constants have already been addressed; their empirical nature contributes to some of their failures to correlate chemical reactivity. The speed and efficiency of computers and software facilitates the ease of calculation of atom charges, energy states, etc., using molecular orbital computations. Nevertheless, the Hammett equation still holds sway in QSAR modeling because of its ability to enhance mechanistic interpretation. It provides us with a general approach to understanding the effect of structural changes on organic molecules.

2.2. Hydrophobic Parameters

More than a 100 years ago, Meyer and Overton made their seminal discovery on the correlation between oil/water partition coefficients and the narcotic potencies of small organic molecules [5,6]. Ferguson extended this analysis by placing the relationship between depressant action and hydrophobicity in a thermodynamic context; the relative saturation of the depressant in the biophase was a critical determinant of its narcotic potency [49]. Mullins suggested that volume associated with shape and size was the determining factor for anesthetic potency. If an anesthetic of appropriate size could fit into the "holes" of the membrane structure, ionic channels would be blocked and the excitability of the cell would be reduced [50].

Meanwhile, Collander determined that there was a high degree of correlation between the $\log P$ values of a mixed set of solutes and their corresponding number of carbons in various solvent–water partitioning systems, for example, octanol–water, chloroform–water, etc. [51]. This work led to the birth of the Collander equation that is simple in form, but had laid the basis for most QSAR models.

$$\log P_2 = a \log P_1 + b \qquad (11)$$

In a comparison of partitioning systems, it has been noted that as the organic phase becomes less polar, the coefficient of the $\log P$ term (a) increases while the intercept, b, decreases. However, there are practical limitations to the validity of this equation. If the solvent systems vary considerably in hydrogen-bonding ability such as octanol versus ether, then Equation 11 will have different intercepts for solutes that are pure hydrogen-bond acceptors than for those with hydrogen-bond donor–acceptor attributes [52]. Around this time, the Hammett equation began to take hold and emphasis was place on correlating chemical reactivity with Hammett's constants and Taft's steric parameter, E_s. Hansen utilized this approach and obtained a fortuitous and moderate correlation of the toxicity of substituted benzoic acids with sigma constants [53]. Meanwhile, Hansch and Muir were busy synthesizing and testing substituted phenoxyacetic acids as plant growth regulators in oak sprouts. Hansch first employed a direct Hammett approach to determine the structural features pertinent to cell elongation but his efforts were to no avail although it was apparent that electron-withdrawing groups were more effective at promoting auxin activity than electron-releasing substituents. Meanwhile, T. Fujita who had prior experience working with plant growth regulators joined the Hansch group and made significant contributions to the pursuit of the parameters that favored cell elongation. Meanwhile, Veldstra had doggedly maintained that it was the hydrophilic–hydrophobic balance of the compounds that was of prime importance [54]. Armed with octanol as the partitioning solvent, Hansch and Fujita began the measurements of the partition coefficients of the phenoxyacetic acids. It turned out that a combination of both factors electronic and hydrophilic–hydrophobic balance was important as can be seen in the seminal equation(s) that laid the foundation for the development of the QSAR paradigm [55].

$$\log\left(\frac{1}{C}\right) = 4.08\,\pi - 2.14\,\pi^2 + 2.78\,\sigma_p + 3.36$$

$$n = 20 \qquad (12)$$

Expansion of this data set led to the formulation of Equation 13 with three outliers [56].

$$\log\left(\frac{1}{C}\right) = 3.24\,\pi - 1.97\,\pi^2 + 1.86\,\sigma_p + 4.16$$

$$n = 23,\ r^2 = 0.776 \qquad (13)$$

These studies not only laid the basis for the correlation of biological activity with a combination of electronic and hydrophobic parameters but also highlighted the importance of the parabolic dependence on the hydrophobicity of the compounds. This particular study underscored the critical importance of hydrophobicity in many areas of chemistry, pharmacology, toxicology, and environmental sciences just to name a few [57–59]. Over the past 50 years, it has generated more critical analysis, interest, and excitement than any other parameter or descriptor in QSAR.

Hydrophobicity is strongly implicated in a large number of biological interactions at the molecular and cellular level, but it is still not well characterized and is thus subject to extensive experimental and computational analysis. It would be difficult to overestimate the importance of hydrophobic interactions in biology; it has a central role in stabilizing biological structures from native conformations of proteins to cellular membranes, protein folding and aggregation and in ligand–receptor binding. Excellent reviews on this subject have been written by Ball, Blokzijl and Engerts, and Taylor [60–63]. The passive role of water in these interactions has come under renewed scrutiny. Ball has defined it as an active matrix that engages and interacts with biomolecules in complex, subtle and essential ways [60]. Despite the use of the term "hydrophobic bond," there is no evidence of the strong attractive force between two apolar surfaces in close contact as determined by Hildebrand who frowned on this notion [64]. Frank and Evans applied a thermodynamic treatment to the solvation of apolar molecules in water and invoked the "iceberg" model that suggested a more structured environment for the hydrated apolar solute [65]. The delineation of this model led Nemethy and Scherega to the development of the "flickering cluster" model that attested to the strong hydrogen-bonding network in water [66]. In his 1959 publication, Kauzmann defined the hydrophobic interaction as the tendency of nonpolar atoms to interact with each other rather than with water via van der Waals attractions and subsequent squeezing out of water molecules [67–69]. The driving force behind these interactions is attributed to the strong mutual attraction between water molecules, their capacity to maintain their hydrogen-bonded network and the fact that water "repulses" alkanes [60,70]. Recently, Chandler has suggested that water that is part of the lower liquid density film near monolayers of hydrophobic alkanes, evaporates out of this gap when large hydrophobic surfaces approach each other [70]. The imbalance in pressure then propels the two surfaces together. Thus, analysis of the definition of the term, "hydrophobic interaction" continues to move forward and undoubtedly more sophisticated experimental and theoretical studies will help to shed light on this complex and confounding phenomena. A great deal of attention was also focused on the development of partition coefficients as a quantitative representation of hydrophobicity.

Buoyed by Veldstra's suggestions that biological activity was related to distribution of chemicals between the fatty and aqueous phases, Maloney and then Fujita in the Hansch laboratory began the measurement of octanol–water partition coefficients in earnest. Octanol was a fortuitous and prescient choice as evidenced by the thousands of partition coefficients carried out in this system. Its advantages include its cost, nontoxic nature, non volatility, lack of reactivity, and ability to mimic a biological membrane with its polar and nonpolar constituent parts. The hydroxyl group has both hydrogen-bond acceptor and hydrogen-bond donor features capable of interacting with a large variety of polar groups. Despite its hydrophobic attributes, it is able to dissolve many more organic compounds than alkanes, cycloalkanes, or aromatic hydrocarbons. It is UV transparent over a large range and has a low enough vapor pressure that allows for reproducible measurements. It is also elevated enough to allow for its removal under mild conditions. Water, saturated with octanol only contains 10^{-3} M of octanol while water-saturated octanol contains 2.3 mol/L of water that enhances hydration of polar entities on transfer from the aqueous phase to octanol. Hydrophobic solutes are not measurably solvated by the miniscule amounts of octanol in the aqueous phase unless their intrinsic partition coefficients are more than 6.0. It was fortuitous that octanol was chosen as the solvent most likely to mimic biomembranes. Extensive experimental measurements of approximately 42,000 partition coefficients in 400 different solvent systems over the past 40 years have failed to dislodge octanol from its secure perch [71,72]. This suggests that octanol can indeed act as an excellent mimic for biomembranes since it shares the traits of amphiphilicity and hydrogen-bonding capability reminiscent of the fluid phospholipids bilayer and globular proteins found in cell membranes [73]. However, it must be noted that there are two critical differences between biomembranes and octanol: octanol is an isotropic phase while

membranes as represented by phospholipids are anisotropic phases and octanol is neutral while phospholipids are charged. Thus, differences in partitioning and failure of the octanol–water system are exacerbated when solutes are charged and/or when their biomembrane interactions are dependent on the topography of the solute–membrane interaction.

The choice of the octanol–water partitioning system as a standard reference for assessing the compartmental distribution of molecules of biological interest was investigated via molecular dynamics (MD) simulations, NMR, near-IR, and X-ray diffraction analyses [74–77]. All these studies indicate that the water present in water-saturated octanol plays an active role in the interactions with solutes, thus lending complexity to the "isotropic" system. In the MD simulations, it was determined that pure 1-octanol contains a mix of hydrogen-bonded "polymeric" species; mostly four-, five-, and six-membered ring clusters at 40°C. These small ring clusters form a central hydroxyl core from which their corresponding alkyl chains radiate outward. On the other hand, water-saturated octanol tends to form well-defined, inverted, micellar aggregates. Long hydrogen-bonded chains are absent and water molecules congregate around the octanol hydroxyls. "Hydrophilic channels" are formed by cylindrical formation of water and octanol hydroxyls with the alkyl chains extending outward. Thus, water-saturated octanol has centralized polar cores where polar solutes can localize. Hydrophobic solutes would migrate to the alkyl-rich regions [74]. NMR, near-IR, and XRD studies also indicate that water is not evenly dispersed, but clusters in groups of four that are then surrounded by approximately 16 octanol hydroxyl groups creating a polar region [75–77]. The simulation study also provides insight into the partitioning of benzene and phenol by analyzing the structure of the octanol/water solvation shell and delineating octanol's capability to serve as a surrogate for biomembranes.

The partition coefficient of a solute is conventionally defined as its concentration ratio between its organic phase and aqueous phase:

$$P = \frac{\{\text{Concentarion in organic phase}\}}{\{\text{Concentarion in aqueous phase}\}} \quad (14)$$

In QSAR studies, hydrophobicity is mostly associated with octanol–water partition coefficient, $\log P_{\text{oct}}$ [72].

$$\log P_{\text{oct}} = \log_{10} \frac{\{\text{Solute}\}_{\text{oct}}}{\{\text{Solute}\}_{\text{water}}} \quad (15)$$

Partition coefficients do not take into account the fact that many compounds ionize in aqueous solution. The degree of this ionization will have a substantial impact on the pharmacodynamic and pharmacokinetic properties of these compounds. The hydrophobicity of most neutral compounds in octanol is at least three units higher than their corresponding charged species. In a wide range of pH values, the contributions of a charged form are usually negligible. Distribution coefficients represent the ratio of the concentration of neutral and charged species of a solute between an organic solvent and the aqueous medium at a particular pH. This is not constant and will vary according to the nature of the solute. When the contribution of an ionized form to overall partitioning is miniscule, $\log D$ values can be estimated using the $\log P$ and $\mathrm{p}K_{\mathrm{a}}$ values for acids and bases. See Equations 16 and 17.

$$\log D = \log P - \log(1 + 10^{\mathrm{pH}-\mathrm{p}K\mathrm{a}}) \text{ for acids} \quad (16)$$

$$\log D = \log P - \log(1 + 10^{\mathrm{p}K\mathrm{a}-\mathrm{pH}}) \text{ for bases} \quad (17)$$

Although most partition coefficients have been determined in the octanol–water system, a substantial number have been determined in other partitioning systems. These partition coefficients can be compared and converted to the octanol–water scale in Collander-type equations as first envisioned by Collander [51]. He stressed the importance of the following linear relationship: $\log P_2 = a \log P_1 + b$. This type of relationship works well when the two solvents are both alkanols. However, when two solvent systems have varying hydrogen-bond donor and acceptor capabilities, the relationship tends to fray. A classical example involves the relationship between $\log P$ values in chloroform and octanol [78,79].

$$\log P_{\mathrm{CHCl_3}} = 1.012 \log P_{\mathrm{oct}} - 0.513 \quad n = 72, r^2 = 0.811, s = 0.733 \tag{18}$$

Only 81% of the variance in the data is explained by this equation. However, a separation of the various solutes into hydrogen-bond donors, acceptors, and neutrals helped account for 94% of the variance in the data. Generally, octanol gives the most consistent results for drugs absorbed in the gastrointestinal tract while cyclohexane gives better results for drugs crossing the blood–brain barrier. These limitations led Seiler to extend the Collander equation by incorporating a corrective term for H-bonding in the cyclohexane system [80]. Fujita generalized this approach and formulated Equation 19 [81].

$$\log P_2 = a \log P_1 + \sum b_{\mathrm{i}} \cdot \mathrm{HB_i} + C \tag{19}$$

P_1 is the reference solvent and $\mathrm{HB_i}$ is a hydrogen-bonding parameter. Leahy, Tailor, and Wait suggested that a more sophisticated approach incorporating four model systems would be needed to adequately address issues of hydrogen bonding in solute partitioning in membranes [82]. Thus, four distinct solvent types were chosen—apolar, amphiprotic, proton donor, and proton acceptor, and represented by cyclohexane, octanol, chloroform, and propyleneglycol dipelargonate (PGDP), respectively. The demands of measuring four partition coefficients for each solute have slowed progress in this particular area. Recent studies have shown that $\Delta \log P$ values have utility in a number of QSAR studies addressing cardioselectivity in oxypropanolamines [83], skin penetration [84], and blood–brain barrier penetration of antidepressant and antipsychotic molecules [85,86]. In the latter study, a series of D2/D3 antagonists showed excellent correlation between penetration through the blood–brain barrier and $\Delta \log P$ ($\log P_{\mathrm{oct}} - \log P_{\mathrm{cy}}$) [86].

$$\log \mathrm{BB} = -0.422 \Delta \log P + 1.772 \quad n = 12, r = 0.926, s = 0.200 \tag{20}$$

These results suggested that a $\Delta \log P < 2$ has a good chance of gaining access to the brain.

2.2.1. Measurement of Partition Coefficients

Shake-Flask Method and Modifications The "shake-flask" method is most commonly used to measure partition coefficients with great accuracy and precision and with a log P range that extends from -3 to $+6$ [87,88]. The procedure calls for the use of pure, distilled, and deionized water, high purity octanol, and pure solutes. At least three concentration levels of solute should be analyzed and the volumes of octanol and water should be varied according to a rough estimate of the log P value. Care should be exercised to ensure that the eventual *amounts* of the solute in each phase are approximately the same after equilibrium. Standard concentration curves utilizing three to four known concentrations in water saturated with octanol are usually established. Methods of quantitation of the solute include UV-based procedures, GC and HPLC [89].

Generally, 10 mL stopped centrifuge tubes or 200 mL centrifuge bottles are utilized. They are inverted gently for 2–3 min and then centrifuged at 1000–2000 g for 20 min before the phases are analyzed. Analysis of both phases is highly recommended to minimize errors incurred by adsorption to glass walls at low solute concentration. For highly hydrophobic compounds, the slow stirring procedure of de Bruijn and Hermens is recommended [90]. The filter-probe extractor system of Tomlinson et al. is a modified, automated, shake-flask method that is efficient, reliable, and flexible [91]. Despite the reliability of these methods, they are tedious, time consuming, and require appreciable amount of solute in some cases. Detailed reviews on difficulties associated with the shake-flask method and pointers on how to maximize efficiency have been described [92,93].

Potentiometric Methods When an organic compound is ionizable in aqueous solution, a potentiometric titration may be used to determine $\log P$. The basic $\mathrm{p}K_{\mathrm{a}}$ is determined by titration and after a back titration the new $\mathrm{p}K_{\mathrm{a(oct)}}$ is determined in the presence of octanol [94,95]. Partitioning by the compound (passing of the neutral species into octanol) will induce a change in the apparent $\mathrm{p}K_{\mathrm{a(oct)}}$ and from this data the $\log P$ can be ascertained. See Equation 21.

$$P = (10^{pKa(oct)\text{-}pKa} - 1)\frac{V_{water}}{V_{oct}} \quad \text{for acids} \quad (21)$$

This method has advantages in terms of automation, speed, convenience, and minute amounts of solute needed for these determinations.

Reversed-Phase Chromatography Reversed-phase chromatography that includes high-performance liquid chromatography (HPLC) and thin-layer chromatography (TLC) provides an alternate tool for the estimation of hydrophobicity parameters. R_m values derived from thin-layer chromatography provide a simple, rapid, and easy way to ascertain approximate values of hydrophobicity [96,97].

$$R_m = \log\left(\frac{1}{R_f - 1}\right) \quad (22)$$

Recent developments in chromatography techniques have led to the development of powerful tools that entail reduced handling and sample sizes and which rapidly, reproducibly, and accurately measure octanol–-water partition coefficients. Countercurrent chromatography is one of these methods. The stationary and mobile phases include two nonmiscible solvents (water and octanol) and the total volume of the liquid stationary phase is utilized for solute partitioning [98,99]. Thus, the retention volumes of solutes are directly proportional to their distribution coefficients K(D) in this biphasic liquid system. Octanol– water partition coefficients that fall in the range from -1 to 4 can be determined by this system. $\log P_{app}$ values of several diuretics including ionizable drugs have been measured at different pH values using countercurrent chromatography; the $\log P$ values ranged from -1.3 to 2.7 and were consistent with literature values [100].

An indirect, rapid and high-throughput method for the determination of partition coefficients using gradient reversed-phase high-performance liquid chromatography (RP-HPLC) has been developed and has found wide utility. It hinges on the linear relationship between $\log P_{oct}$ and retention factor, $\log k$. These isocratic retention factors are relatable to $\log P$ by a Collander-type equation.

$$\log P = a \log k + c \quad (23)$$

In most cases, the stationary phases include silanized silica gel although the polymer-based octadecyl-poly(vinyl alcohol) (ODP) stationary phase completely free of reactive silanol groups has seen some usage for hydrophobicity measurements [101,102]. The most widely used mobile phases are methanol, acetonitrile, and tetrahydrofuran. RP-HPLC is touted as a high-throughput hydrophobicity screen for combinatorial libraries [103,104]. A chromatography hydrophobicity index (CHI) was established for a diverse set of compounds. Acetonitrile was used as the modifier and 50 mm ammonium acetate as the mobile phase. A linear relationship was established between $C\log P$ and CHIN for neutral molecules.

$$C\log P = 0.057\,\text{CHIN} - 1.107$$
$$n = 52,\ r^2 = 0.724,\ s = 0.82,\ F = 131 \quad (24)$$

Partitioning into a liposome–water system has also been utilized for the determination of partition coefficients since it represents a biological system most similar to a membrane. The actual process is tedious and difficult since it entails equilibration of the solute with liposomes, separation via ultrafiltration, centrifugation, or equilibrium dialysis followed by quantitation of the solute in the lipid-free milieu [105,106].

A new alternative method for $\log P_{oct}$ determination is based on immobilized artificial membrane (IAM) approaches via HPLC or LC-MS [107–110]. It represents a close simulation of liposome–water partitioning and hence cell membrane partitioning, without the difficulties associated with it. IAM columns are frequently made with phospholipids attached to a propylamino-silica support and the mobile phase is usually phosphate buffered saline. It must be noted that there is a significant difference between HPLC and IAM approaches. The former utilizes alkyl-bonded phases that retain analytes based purely on hydrophobicity while the latter incorporates "phospholipophilicity" a combination of hydrophobicity, ion-paring, and hydrogen-bonding interactions that could wield an important role in membrane transport [111]. Despite the mediocre correlations between the IAM retention factors, k_{IAM} and $\log P$, phospholipophilicity parameters have been successfully utilized in various drug studies [112,113]. A number of extensive reviews on various chromatographic methods for determining partition coefficients may be found in the literature [114–116].

2.2.2. Calculation Methods for Partition Coefficients The measurement of partition coefficients can be a laborious process fraught with difficulties unless special attention is paid to the process. Factors that can influence the measurement include temperature, poor solubility of the analyte, pH of the solution, and subsequent ionization as well as the degree of tautomerization and micelle formation. With developments in combinatorial chemistry, there is a real need for rapid and accurate calculations of partition coefficients.

Partition coefficients are additive–constitutive, free-energy related properties. $\log P$ represents the overall hydrophobicity of a molecule that includes the sum of the hydrophobic contributions of the "parent" molecule and its substituent. Thus, the π value for a substituent may be defined as follows:

$$\pi_X = \log P_{R-X} - \log P_{R-H} \quad (25)$$

where π_H is set to zero. The π value for a cyano-substituent is thus calculated from the $\log P$ of benzonitrile and benzene. See Equation 26.

$$\pi_{CN} = \log P_{benzonitrile} - \log P_{benzene} = 1.56 - 2.13 = -0.57 \quad (26)$$

An extensive list of π values for aromatic substituents appears in Table 2. Aliphatic fragment values were developed a few years later. *Pi* values for side chains of amino acids in peptides have also been well characterized and are easily available [117–121].

For a more extensive list of substituent constants, refer to the extensive compilation by Hansch, Leo, and Hoekman [122]. Initially, the π system was applied only to substitution on aromatic rings and when the hydrogen undergoing replacement, was of innocuous character. However, it was quickly apparent that not all hydrogens on aromatic systems could be substituted without correction factors because of strong electronic interactions. It became necessary to determine π values in various electron-rich and -deficient systems, for example, X-phenols and X-nitrobenzenes. Correction factors were introduced for special features such as unsaturation, branching, and ring fusion. The proliferation of π scales made it difficult to ascertain which system was more appropriate for usage, particularly with complex structures.

The shortcomings of this approach provided the impetus for Nys and Rekker to design the fragmental method, a "reductionist" approach, which was based on the statistical analysis of a large number of measured partition coefficients and the subsequent assignment of appropriate values for particular molecular fragments [79,123]. Hansch and Leo took a "constructionist" approach and developed a fragmental system that included correction factors to account for various intramolecular interactions [124–126]. Labor-intensive efforts and inconsistency in manual calculations were eliminated with the debut of the automated system CLOGP and its powerful SMILES notation [127–129]. A recent analysis of the accuracy of CLOGP versus experimental values yielded equation 27 [130].

$$\text{Measured} \log P = 0.959\, \mathrm{C} \log P + 0.08$$
$$n = 12,107,\ r^2 = 0.973,\ s = 0.299 \quad (27)$$

The $C \log P$ values of 228 structures (1.8% of the data set) were not well predicted. It must be noted that Starlist (most accurate values in the database) contains almost 300 charged nitrogen solutes (ammonium, pyridinium, imidazolium, etc.) and more than 2200 in all, which amounts to 5% of Masterfile (database of measured values). CLOGP adequately handles these molecules within the 0.30 standard deviation limit. Most other programs make no attempt to calculate them. For more details on computing $\log P_{oct}$ from structures, see excellent reviews by Leo and Sangster and a recent one by Tetko [131–133].

Table 2. Substituent Constants for QSAR Analysis

No.	Substituent	PI	MR	*L*	*B*1	*B*5	S-P	S-M
1	$+N(CH_3)_3$	−5.96	1.94	4.02	2.57	3.11	0.82	0.88
2	$EtN(CH_3)_3+$	−5.44	2.87	5.58	1.52	4.53	0.13	0.16
3	$CH_2N(CH_3)_3+$	−4.57	2.40	4.83	1.52	4.08	0.44	0.40
4	CO_2-	−4.36	0.61	3.53	1.60	2.66	0.00	−0.10
5	$+NH_3$	−4.19	0.55	2.78	1.49	1.97	0.60	0.86
6	PR-$N(CH_3)_3+$	−4.15	3.33	6.88	1.52	5.49	−0.01	0.06
7	CH_2NH_3+	−4.09	1.01	4.02	1.52	3.05	0.29	0.32
8	IO_2	−3.46	6.35	4.25	2.15	3.66	0.78	0.68
9	$C(CN)_3$	−2.33	1.86	3.99	2.87	4.12	0.96	0.97
10	$NHNO_2$	−2.27	1.07	4.50	1.35	3.66	0.57	0.91
11	$C(NO_2)_3$	−2.01	2.27	4.59	2.55	3.72	0.82	0.72
12	$SO_2(NH_2)$	−1.82	1.23	4.02	2.04	3.05	0.60	0.53
13	$C(CN)=C(CN)_2$	−1.77	2.58	6.46	1.61	5.17	0.98	0.77
14	$CH_2C=O(NH_2)$	−1.68	1.44	4.58	1.52	4.37	0.07	0.06
15	$N(COCH_3)_2$	−1.68	2.48	4.45	1.35	4.33	0.33	0.35
16	SO_2CH_3	−1.63	1.35	4.11	2.03	3.17	0.72	0.60
17	$P(O)(OH)_2$	−1.59	1.26	4.22	2.12	2.88	0.42	0.36
18	$S=O(CH_3)$	−1.58	1.37	4.11	1.40	3.17	0.49	0.52
19	$N(SO_2CH_3)_2$	−1.51	3.12	4.83	1.36	3.72	0.49	0.47
20	$C=O(NH_2)$	−1.49	0.98	4.06	1.50	3.07	0.36	0.28
21	$CH(CN)_2$	−1.45	1.43	3.99	1.85	4.12	0.52	0.53
22	$CH_2NHCOCH_3$	−1.43	1.96	5.67	1.52	4.75	−0.05	0.05
23	$NHC=S(NH_2)$	−1.40	2.22	5.06	1.35	4.18	0.16	0.22
24	NH(OH)	−1.34	0.72	3.87	1.35	2.63	−0.34	−0.04
25	$CH=NNHCONHNH_2$	−1.32	2.42	7.57	1.60	4.55	0.16	0.22
26	$NHC=O(NH_2)$	−1.30	1.37	5.06	1.35	3.61	−0.24	−0.03
27	$C=O(NHCH_3)$	−1.27	1.46	5.00	1.54	3.16	0.36	0.35
28	2-Aziridinyl	−1.23	1.19	4.14	1.55	3.24	−0.10	−0.06
29	NH_2	−1.23	0.54	2.78	1.35	1.97	−0.66	−0.16
30	$NHSO_2CH_3$	−1.18	1.82	4.70	1.35	4.13	0.03	0.20
31	$P(O)(OCH_3)_2$	−1.18	2.19	5.04	2.42	3.25	0.53	0.42
32	$C(CH_3)(CN)_2$	−1.14	1.90	4.11	2.81	4.12	0.57	0.60
33	$N(CH_3)SO_2CH_3$	−1.11	2.34	4.83	1.35	3.72	0.24	0.21
34	SO_2Et	−1.10	1.81	4.92	2.03	3.49	0.77	0.66
35	CH_2NH_2	−1.04	0.91	4.02	1.52	3.05	−0.11	−0.03
36	1-Tetrazolyl	−1.04	1.83	5.28	1.71	3.12	0.50	0.52
37	CH_2OH	−1.03	0.72	3.97	1.52	2.70	0.00	0.00
38	$N(CH_3)COCH_3$	−1.02	1.96	4.77	1.35	3.71	0.26	0.31
39	NHCHO	−0.98	1.03	4.22	1.35	3.61	0.00	0.19
40	$NHC(=O)CH_3$	−0.97	1.49	5.09	1.35	3.61	0.00	0.21
41	$C(CH_3)(NO_2)_2$	−0.88	2.17	4.59	2.55	3.72	0.61	0.54
42	$NHNH_2$	−0.88	0.84	3.47	1.35	2.97	−0.55	−0.02
43	OSO_2CH_3	−0.88	1.70	4.66	1.35	4.10	0.36	0.39
44	$SO_2N(CH_3)_2$	−0.78	2.19	4.83	2.03	4.08	0.65	0.51
45	$NHC=S(NHC_2H_5)$	−0.71	3.17	7.22	1.45	4.38	0.07	0.30
46	$SO_2(CHF_2)$	−0.68	1.31	4.11	2.03	3.70	0.86	0.75
47	OH	−0.67	0.29	2.74	1.35	1.93	−0.37	0.12
48	CHO	−0.65	0.69	3.53	1.60	2.36	0.42	0.35
49	$CH_2CHOHCH_3$	−0.64	1.64	4.92	1.52	3.78	−0.17	−0.12
50	$CS(NH_2)$	−0.64	1.81	4.10	1.64	3.18	0.30	0.25
51	$OC=O(CH_3)$	−0.64	1.25	4.74	1.35	3.67	0.31	0.39
52	$SOCHF_2$	−0.63	1.33	4.70	1.40	3.70	0.58	0.54
53	4-Pyrimidinyl	−0.61	2.18	5.29	1.71	3.11	0.63	0.30

(continued)

Table 2. *(Continued)*

No.	Substituent	PI	MR	*L*	*B*1	*B*5	S-P	S-M
54	2-Pyrimidinyl	−0.61	2.18	6.28	1.71	3.11	0.53	0.23
55	$P(CF_3)_2$	−0.59	1.99	4.96	1.40	3.86	0.69	0.60
56	CH_2CN	−0.57	1.01	3.99	1.52	4.12	0.18	0.16
57	CN	−0.57	0.63	4.23	1.60	1.60	0.66	0.56
58	$COCH_3$	−0.55	1.12	4.06	1.60	3.13	0.50	0.38
59	$CH_2P{=}O(OEt)_2$	−0.54	3.58	7.10	1.52	5.73	0.06	0.12
60	$P{=}O(OEt)_2$	−0.52	3.12	6.26	2.52	5.58	0.60	0.55
61	NHCOOMe	−0.52	1.57	5.84	1.45	3.99	−0.17	−0.02
62	$NHC{=}O(NHC_2H_5)$	−0.50	2.32	7.29	1.45	3.98	−0.26	0.04
63	$NHC{=}O(CH_2Cl)$	−0.50	1.98	6.26	1.55	4.26	−0.03	0.17
64	$NHCH_3$	−0.47	1.03	3.53	1.35	3.08	−0.70	−0.21
65	$N(CH_3)COCF_3$	−0.46	1.95	5.20	1.56	3.96	0.39	0.41
66	$C{=}S(NHCH_3)$	−0.46	2.23	5.00	1.88	3.18	0.34	0.30
67	$NHC{=}S(CH_3)$	−0.42	2.34	5.09	1.45	4.38	0.12	0.24
68	$C(Et)(NO_2)_2$	−0.35	3.66	4.92	2.55	3.72	0.64	0.56
69	CO_2H	−0.32	0.69	3.91	1.60	2.66	0.45	0.37
70	$C(OH)(CH_3)_2$	−0.32	1.64	4.11	2.40	3.17	0.60	0.47
71	$EtCO_2H$	−0.29	1.65	5.97	1.52	3.31	−0.07	−0.03
72	NO_2	−0.28	0.74	3.44	1.70	2.44	0.78	0.71
73	$CH{=}NNHCSNH_2$	−0.27	2.96	7.16	1.60	5.41	0.40	0.45
74	NHCN	−0.26	1.01	3.90	1.35	4.05	0.06	0.21
75	$CH_2C(OH)(CH_3)_2$	−0.24	2.11	4.92	1.52	4.19	−0.17	−0.16
76	$CH{=}CHCHO$	−0.23	1.69	5.76	1.60	3.46	0.13	0.24
77	$NHCH_2CO_2Et$	−0.21	2.69	7.91	1.35	5.77	−0.68	−0.10
78	CH_2OCH_3	−0.21	1.21	4.78	1.52	3.40	0.01	0.08
79	$NHC{=}OCH(CH_3)_2$	−0.18	2.43	5.53	1.35	4.09	−0.10	0.11
80	$CH_2OC{=}O(CH_3)$	−0.17	1.65	5.46	1.52	4.46	0.05	0.04
81	$CH_2N(CH_3)_2$	−0.15	1.87	4.83	1.52	4.08	0.01	0.00
82	CH_2SCN	−0.14	1.81	6.63	1.52	3.41	0.14	0.12
83	1-Aziridinyl	−0.12	1.35	4.14	1.35	3.24	−0.22	−0.07
84	NO	−0.12	0.52	3.44	1.70	2.44	0.91	0.62
85	ONO_2	−0.12	0.85	4.46	1.35	3.62	0.70	0.55
86	$S{=}O(C_6H_5)$	−0.07	3.34	4.62	1.40	6.02	0.44	0.50
87	$CH_2SO_2C_6H_5$	−0.06	3.79	8.33	1.52	3.78	0.16	0.15
88	OCH_3	−0.02	0.79	3.98	1.35	3.07	−0.27	0.12
89	$C{=}O(OCH_3)$	−0.01	1.29	4.73	1.64	3.36	0.45	0.36
90	H	0.00	0.10	2.06	1.00	1.00	0.00	0.00
91	$C{=}O(CF_3)$	0.02	1.12	4.65	1.70	3.67	0.80	0.63
92	$CH{=}C(CN)_2$	0.05	1.97	6.46	1.60	5.17	0.84	0.66
93	$SO_2(F)$	0.05	0.87	3.33	2.01	2.70	0.91	0.80
94	COEt	0.06	1.58	4.87	1.63	3.45	0.48	0.38
95	$C(CF_3)_3$	0.07	2.08	4.11	3.13	3.64	0.55	0.55
96	NH-Et	0.08	1.50	4.83	1.35	3.42	−0.61	−0.24
97	$NHC{=}O(CF_3)$	0.08	1.43	5.62	1.79	3.61	0.12	0.30
98	$SC{=}O(CH_3)$	0.10	1.84	5.11	1.70	4.01	0.44	0.39
99	CF_3	0.10	0.50	3.30	1.99	2.61	0.54	0.43
100	OCH_2F	0.10	0.72	4.57	1.35	3.07	0.02	0.20
101	$CH{=}CHNO_2$ (TR)	0.11	1.64	4.29	1.60	4.78	0.26	0.32
102	CH_2F	0.13	0.54	3.30	1.52	2.61	0.11	0.12
103	F	0.14	0.09	2.65	1.35	1.35	0.06	0.34
104	$C(OMe)_3$	0.14	2.48	4.78	2.56	4.29	−0.04	−0.03
105	$SECF_3$	0.15	1.63	4.50	1.85	4.09	0.45	0.44
106	$NHC{=}O(OEt)$	0.17	2.12	7.25	1.35	3.92	−0.15	0.11
107	CH_2Cl	0.17	1.05	3.89	1.52	3.46	0.12	0.11

Table 2. *(Continued)*

No.	Substituent	PI	MR	*L*	*B*1	*B*5	S-P	S-M
108	$N(CH_3)_2$	0.18	1.56	3.53	1.35	3.08	−0.83	−0.16
109	CHF_2	0.21	0.52	3.30	1.71	2.61	0.32	0.29
110	$CCCF_3$	0.22	1.41	5.90	1.99	2.61	0.51	0.41
111	$SO_2C_6H_5$	0.27	3.32	5.86	2.03	6.02	0.68	0.62
112	$COCH(CH_3)_2$	0.29	1.98	4.84	1.99	4.08	0.47	0.38
113	$OCHF_2$	0.31	0.79	3.98	1.35	3.61	0.18	0.31
114	$CH_2SO_2CF_3$	0.33	1.75	5.35	1.52	4.07	0.31	0.29
115	$C(NO_2)(CH_3)_2$	0.33	2.06	4.59	2.58	3.72	0.20	0.18
116	$P(O)(OPR)_2$	0.35	4.05	7.07	2.52	6.90	0.50	0.38
117	$CH_2S{=}O(CF_3)$	0.37	1.90	5.35	1.52	4.07	0.24	0.25
118	OCH_2CH_3	0.38	1.25	4.80	1.35	3.36	−0.24	0.10
119	SH	0.39	0.92	3.47	1.70	2.33	0.15	0.25
120	$N{=}NCF_3$	0.40	1.39	5.45	1.70	3.48	0.68	0.56
121	CCH	0.40	0.96	4.66	1.60	1.60	0.23	0.21
122	$N{=}CCl_2$	0.41	1.84	5.65	1.70	4.54	0.13	0.21
123	SCCH	0.41	1.62	4.08	1.70	4.85	0.19	0.26
124	SCN	0.41	1.34	4.08	1.70	4.45	0.52	0.51
125	$P(CH_3)_2$	0.44	2.12	3.88	2.00	3.32	0.06	0.03
126	$NHSO_2C_6H_5$	0.45	3.79	8.24	1.35	3.72	0.01	0.16
127	$SO_2NHC_6H_5$	0.45	3.78	8.24	2.03	4.50	0.65	0.56
128	CH_2CF_3	0.45	0.97	4.70	1.52	3.70	0.09	0.12
129	NNN	0.46	1.02	4.62	1.50	4.18	0.08	0.37
130	NNN	0.46	1.02	4.62	1.50	4.18	0.08	0.37
131	4-Pyridyl	0.46	2.30	5.92	1.71	3.11	0.44	0.27
132	$N{=}NN(CH_3)_2$	0.46	2.09	5.68	1.77	3.90	0.44	0.27
133	$C{=}O(NHC_6H_5)$	0.49	3.54	8.24	1.63	4.85	−0.03	−0.05
134	2-pyridyl	0.50	2.30	6.28	1.71	3.11	0.41	0.23
135	$OCH_2CH{=}CH_2$	0.51	1.61	6.22	1.35	4.42	0.17	0.33
136	C=O(OEt)	0.51	1.75	5.95	1.64	4.41	−0.25	0.09
137	$S{=}O(CF_3)$	0.53	1.31	4.70	1.40	3.70	0.45	0.37
138	$CHOHC_6H_5$	0.54	3.15	4.62	1.73	6.02	0.69	0.63
139	OCH_2Cl	0.54	1.20	5.44	1.35	3.13	−0.03	0.00
140	$SO_2(CF_3)$	0.55	1.29	4.70	2.03	3.70	0.08	0.25
141	CH_3	0.56	0.57	2.87	1.52	2.04	0.96	0.83
142	SCH_3	0.61	1.38	4.30	1.70	3.26	−0.17	−0.07
143	$SC{=}O(CF_3)$	0.66	1.82	5.55	1.70	4.51	0.00	0.15
144	$COC(CH_3)_3$	0.69	2.44	4.87	1.87	4.42	0.46	0.48
145	$CH{=}NC_6H_5$	0.69	3.30	8.50	1.70	4.07	0.32	0.27
146	$P{=}O(C_6H_5)_2$	0.70	5.93	5.40	2.68	6.19	0.42	0.35
147	Cl	0.71	0.60	3.52	1.80	1.80	.530	.380
148	$N{=}CHC_6H_5$	0.72	3.30	8.40	1.70	4.65	0.23	0.37
149	$SeCH_3$	0.74	1.70	4.52	1.85	3.63	−0.55	−0.08
150	SCH_2F	0.74	1.34	4.89	1.70	3.41	0.00	0.10
151	$OCH{=}CH_2$	0.75	1.14	4.98	1.35	3.65	0.20	0.23
152	CH_2Br	0.79	1.34	4.09	1.52	3.75	−0.09	0.21
153	$CCCH_3$	0.81	1.41	5.47	1.60	2.04	0.14	0.12
154	$CH{=}CH_2$	0.82	1.10	4.29	1.60	3.09	0.03	0.21
155	Br	0.86	0.89	3.82	1.95	1.95	−0.16	−0.08
156	$NHSO_2CF_3$	0.93	1.75	5.26	1.35	4.00	0.23	0.39
157	$OSO_2C_6H_5$	0.93	3.67	8.20	1.35	3.64	0.39	0.44
158	1-Pyrryl	0.95	1.95	5.44	1.71	3.12	0.33	0.36
159	$N(CH_3)SO_2CF_3$	1.00	2.28	5.26	1.54	4.00	0.37	0.47
160	$SCHF_2$	1.02	1.38	4.30	1.70	3.94	0.44	0.46

(continued)

Table 2. *(Continued)*

No.	Substituent	PI	MR	*L*	*B*1	*B*5	S-P	S-M
161	CH_2CH_3	1.02	1.03	4.11	1.52	3.17	0.37	0.33
162	OCF_3	1.04	0.79	4.57	1.35	3.61	−0.15	−0.07
163	$OCH_2CH_2CH_3$	1.05	1.71	6.05	1.35	4.42	0.35	0.38
164	$C{=}O(C_6H_5)$	1.05	3.03	4.57	1.92	5.98	−0.25	0.10
165	$NHCO_2C_4H_9$	1.07	3.05	9.50	1.45	5.05	0.43	0.34
166	S-Et	1.07	1.84	5.16	1.70	3.97	−0.05	0.06
167	$N(CF_3)_2$	1.08	1.43	4.01	1.52	3.58	0.03	0.18
168	$CHCl_2$	1.09	1.53	3.89	1.88	3.46	0.53	0.40
169	$CH_2CH{=}CH_2$	1.10	1.45	5.11	1.52	3.78	0.32	0.31
170	CH_2I	1.10	1.86	4.36	1.52	4.15	−0.14	−0.11
171	NH-Bu	1.10	2.43	6.88	1.35	4.87	0.11	0.10
172	$CClF_2$	1.11	1.07	3.89	1.99	3.46	−0.51	−0.34
173	I	1.12	1.39	4.23	2.15	2.15	0.46	0.42
174	Cyclopropyl	1.14	1.35	4.14	1.55	3.24	0.18	0.35
175	$C(CH_3){=}CH_2$	1.14	1.56	4.29	1.73	3.11	−0.21	−0.07
176	NCS	1.15	1.72	4.29	1.50	4.24	0.05	0.09
177	$SCH_2CH{=}CH_2$	1.15	2.26	6.42	1.70	5.02	0.38	0.48
178	$N(Et)_2$	1.18	2.49	4.83	1.35	4.39	0.12	0.19
179	OSO_2CF_3	1.23	1.45	5.23	1.35	3.24	−0.72	−0.23
180	SF_5	1.23	0.99	4.65	2.47	2.92	0.53	0.56
181	$OCHCl_2$	1.26	1.69	3.98	1.35	4.41	0.68	0.61
182	CF_2CF_3	1.26	0.92	4.11	1.99	3.64	0.26	0.38
183	$C(OH)(CF_3)_2$	1.28	1.52	4.11	2.61	3.64	0.52	0.47
184	$SCH{=}CH_2$	1.29	1.77	5.33	1.70	4.23	0.30	0.29
185	NHC_6H_5	1.37	3.00	4.53	1.35	5.95	0.20	0.26
186	$SCH(CH_3)_2$	1.41	2.41	5.16	1.70	4.41	−0.56	−0.02
187	SCF_3	1.44	1.38	4.89	1.70	3.94	0.07	0.23
188	$OC{=}O(C_6H_5)$	1.46	3.23	8.15	1.64	4.40	0.50	0.40
189	$COOC_6H_5$	1.46	3.02	8.13	1.94	3.50	0.13	0.21
190	Cyclobutyl	1.51	1.79	4.77	1.77	3.82	0.44	0.37
191	O-Bu	1.52	2.17	6.86	1.35	4.79	−0.14	−0.05
192	$CH(CH_3)_2$	1.53	1.50	4.11	1.90	3.17	−0.32	0.10
193	$CHBr_2$	1.53	1.68	4.09	1.92	3.75	−0.15	−0.04
194	Pr	1.55	1.50	4.92	1.52	3.49	0.32	0.31
195	$C(F)(CF_3)_2$	1.56	1.34	4.11	2.45	3.64	−0.13	−0.06
196	$C_6H_4(NO_2)$-*p*	1.64	3.17	7.66	1.71	3.11	0.05	0.09
197	$CH_2OC_6H_5$	1.66	3.22	8.19	1.52	3.53	0.13	0.10
198	$N{=}NC_6H_5$	1.69	3.13	8.43	1.70	4.31	0.07	0.06
199	$SO_2CF_2CF_3$	1.73	1.97	5.35	2.03	4.07	0.39	0.32
200	$CF_2CF_2CF_2CF_3$	1.74	1.77	6.76	1.99	5.05	1.08	0.92
201	1-Cyclopentenyl	1.77	2.21	5.24	1.91	3.08	0.52	0.47
202	OCF_2CHF_2	1.79	1.08	5.23	1.35	3.94	−0.05	−0.06
203	$C_6H_4(OCH_3)$-*p*	1.82	3.17	7.71	1.80	3.11	0.25	0.34
204	CH_2SCF_3	1.83	1.76	5.82	1.52	4.10	−0.08	0.05
205	C_6H_5	1.96	2.54	6.28	1.71	3.11	0.04	0.01
206	$C(CH_3)_3$	1.98	1.96	4.11	2.60	3.17	−0.01	0.06
207	CCl_3	1.99	2.01	3.89	2.64	3.46	−0.20	−0.10
208	$CH_2Si(CH_3)_3$	2.00	2.96	5.39	1.52	4.75	0.46	0.40
209	$CH_2C_6H_5$	2.01	3.00	4.62	1.52	6.02	−0.21	−0.16
210	$CH(CH_3)(Et)$	2.04	1.96	4.92	1.90	3.49	−0.09	−0.08
211	C_6H_4F-*p*	2.04	2.53	6.87	1.71	3.11	−0.12	−0.08
212	OC_5H_{11}	2.05	2.63	8.11	1.35	5.81	0.06	0.12
213	$N(C_3H_7)_2$	2.08	3.24	6.07	1.35	5.50	−0.34	0.10
214	OC_6H_5	2.08	2.77	4.51	1.35	5.89	−0.93	−0.26

Table 2. *(Continued)*

No.	Substituent	PI	MR	*L*	*B*1	*B*5	S-P	S-M
215	$C_6H_4N(CH_3)_2$-*p*	2.10	3.99	7.75	1.79	3.11	−0.03	0.25
216	Bu	2.13	1.96	6.17	1.52	4.54	−0.56	−0.06
217	Cyclopentyl	2.14	2.20	4.90	1.90	4.09	0.28	0.35
218	CHI_2	2.15	3.15	4.36	1.95	4.15	−0.14	−0.05
219	SC_6H_5	2.32	3.43	4.57	1.70	6.42	0.26	0.26
220	1-Cyclohexenyl	2.33	2.67	6.16	2.23	3.30	0.07	0.23
221	$OCCl_3$	2.36	2.18	5.44	1.35	4.41	−0.08	−0.10
222	$C(Et)(CH_3)_2$	2.37	2.42	4.92	2.60	3.49	0.35	0.43
223	$CH_2C(CH_3)_3$	2.37	2.42	4.89	1.52	4.18	−0.18	−0.06
224	$SC_6H_4NO_2$-*p*	2.39	4.11	4.92	1.70	7.86	−0.17	−0.05
225	SCF_2CHF_2	2.43	1.84	5.60	1.70	4.55	0.24	0.32
226	C_6H_4Cl-*p*	2.61	3.04	7.74	1.80	3.11	−0.07	−0.04
227	C_6F_5	2.62	2.40	6.87	1.71	3.67	0.12	0.15
228	C_5H_{11}	2.63	2.42	6.97	1.52	4.94	0.27	0.26
229	CCC_6H_5	2.65	3.32	8.88	1.71	3.11	−0.15	−0.08
230	CBr_3	2.65	2.88	4.09	2.86	3.75	0.16	0.14
231	EtC_6H_5	2.66	3.47	8.33	1.52	3.58	0.29	0.28
232	$C_6H_4(CH_3)$-*p*	2.69	3.00	7.09	1.84	3.11	−0.12	−0.07
233	C_6H_4I-*p*	3.02	3.91	8.45	2.15	3.11	−0.03	0.06
234	C_6H_4I-*m*	3.02	3.91	6.72	1.84	5.15	0.12	0.15
235	1-Adamantyl	3.37	4.03	6.17	3.16	3.49	−0.15	−0.05
236	$C(Et)_3$	3.42	3.36	4.92	2.94	4.18	0.10	0.14
237	$CH(C_6H_5)_2$	3.52	5.43	5.15	2.01	6.02	0.06	0.13
238	$N(C_6H_5)_2$	3.61	5.50	5.77	1.35	5.95	0.01	0.08
239	Heptyl	3.69	3.36	9.03	1.52	6.39	−0.13	−0.12
240	$C(SCF_3)_3$	4.17	4.40	5.82	3.32	5.00	−0.20	−0.07
241	C_6Cl_5	4.96	4.95	7.74	1.81	4.48	−0.05	−0.03

The proliferation of methodologies and programs to calculate partition coefficients continues unabated. These programs are based on substructure approaches or whole molecule approaches [134,135]. Substructure methods are based on molecular fragments, atomic contributions or computer-identified fragments [125,136–140]. Variations of atomic-contribution methods initially developed by Klopman et al. were refined and extended by Broto et al., and Ghose et al. who computed the contributions for atoms in diverse topological environments [141–143]. G. Loew's group proposed atom-based parameterization for a conformationally dependent hydrophobic index [144]. Whole molecule approaches utilize molecular properties or spatial properties to predict $\log P$ values [145–148]. The associated programs run on different platforms (e.g., Mac, PC, Unix, VAX, etc.) and utilize different computational strategies. An excellent review by Mannhold and van de Waterbeemd addresses the advantages and limitations of the various approaches [135]. Statistical indices yield insight as to the effectiveness of such programs.

In the 1980s, attempts to compute more accurate $\log P$ values resulted in the development of specialized solvatochromic parameters, pioneered by Kamlet, Taft, et al. and that focused on molecular properties [149,150]. In its simplest form, this approach can be expressed as follows:

$$\log P_{\text{oct}} = aV + b\pi^* + c\,\beta_{\text{H}} + d\alpha_{\text{H}} + e \quad (28)$$

V is a solute volume term, π^* represents the solute polarizability, β_{H} and α_{H} are measures of hydrogen-bond acceptor strength and hydrogen-bond donor strength, respectively, and e is the intercept. An extension of this model has been formulated by Abraham and used by researchers to refine molecular descriptors and characterize hydrophobicity scales [151–154].

2.3. Steric Parameters

The quantitation of steric effects is complex at best and challenging in all other situations, particularly at the molecular level. An added level of confusion comes into play when attempts are made to delineate size and shape. Yet, sterics are of overwhelming importance in ligand–receptor interactions as well as in transport phenomena in cellular systems. The first steric parameter to be quantified and utilized in QSAR studies was Tafts E_S constant [155,156]. E_S is defined as follows:

$$E_S = \log\left(\frac{k_X}{k_H}\right)_A \quad (29)$$

Where, k_X and k_H represent the rates of acid hydrolysis of esters, XCH_2COOR and CH_3COOR, respectively. To correct for hyperconjugation in the α-hydrogens of the acetate moiety, Hancock devised a correction for E_S such that the following relationship holds [157]:

$$E_S^C = E_S + 0.306 \times (n-3) \quad (30)$$

In Equation 30, n represents the number of alpha-hydrogens and 0.306 is a constant derived from molecular orbital calculations [157]. Unfortunately, the limited availability of E_S and $E_S{}^C$ values for a great number of substituents precludes extensive usage in QSAR studies. Despite this limitation, Hansch et al. demonstrated that E_S played a prominent role in the glucuronidation of primary alcohols in rabbits [158]. Charton demonstrated a strong correlation between E_S and van der Waals radii ($E_S = \psi \times r_X + h$) which led to the development of the upsilon parameter, ν_X [159].

$$\nu_X = r_X - r_H = r_X - 1.20 \quad (31)$$

r_X and r_H are the minimum van der Waals radii of the substituent and hydrogen, respectively. Extension of this approach from symmetrical substituents to nonsymmetrical substituents must be handled with caution.

One of the most widely used steric parameters is molar refraction that has been aptly described as a "chameleon" parameter by Tute [160]. Although it is generally considered to be a crude measure of overall bulk, it does incorporate a polarizability component that may describe cohesion and is related to London dispersion forces as follows: $MR = 4\pi N\alpha/3$, where N is Avogadro's number and α is the polarizability of the molecule. It contains no information on the shape of the whole molecule. MR is also defined by the Lorentz–Lorenz equation:

$$MR = \left(\frac{n^2-1}{n^2+2}\right) \times \left(\frac{MW}{\text{Density}}\right) \quad (32)$$

MR is generally scaled by 0.1 and utilized in biological QSAR where intermolecular effects are of primary importance. The refractive index of the molecule is represented by n. With alkyl substituents, there exists a high degree of collinearity with hydrophobicity; hence, caution must be exercised in the QSAR analysis of data sets populated with such derivatives. The MR descriptor does not distinguish shape; thus, the MR value for amyl ($-CH_2CH_2CH_2$-CH_2CH_3) is the same as that for ($-C(Et)$-$(CH_3)_2$): 2.42. The coefficients with MR terms challenge interpretation although extensive experience with this parameter suggests that a negative coefficient implies steric hindrance at that site and a positive coefficient attests to either dipolar interactions in that vicinity or anchoring of a ligand in an opportune position for interaction with a critical residue in the binding site of a receptor [161].

The failure of the MR descriptor to adequately address three-dimensional shape issues led to Verloop's development of STERIMOL parameters [162]. STERIMOL parameters define the steric constraints of a given substituent along several fixed axis. Five parameters were deemed necessary to define shape: L, $B1$, $B2$, $B3$, and $B4$. L represents the length of a substituent along the axis of a bond between the parent molecule and the substituent. $B1$ to $B4$ represented four different width parameters. However, the high degree of collinearity between $B1$, $B2$, and $B3$ and the large number of training set members needed to establish the statistical validity of this group of parameters led to their demise in QSAR studies. Verloop subsequently established the adequacy of just three parameters for QSAR analysis—a slightly modified length L, a minimum width $B1$, and a maximum width $B5$, that is orthogonal to L [163]. The use of these insightful parameters has done much to enhance correlations with biological activities. Recent analysis in our laboratory has established that in many cases, $B1$ alone is superior to Tafts E_S and a combination of $B1$ and $B5$ can adequately replace E_S [164].

Molecular weight terms have also been utilized as descriptors, particularly in cellular systems, or in distribution/transport studies where diffusion is the mode of operation. According to the Einstein–Sutherland equation, molecular weight affects diffusion rate. The log MW term has been utilized extensively in some studies [165–167] and an example of such usage is given below. In correlating permeability (Perm) of nonelectrolytes through chara cells, Lien, et.al. obtained the following QSAR [168]:

$$\log \text{Perm} = 0.889 \log P^* - 1.544 \log \text{MW} - 0.144\, \text{H}_b + 4.653 \quad (33)$$

$$n = 30,\ r^2 = 0.899,\ s = 0.322,\ F = 77.39$$

In QSAR 33, $\log P^*$ represents the olive oil–water partition coefficient, MW is the molecular weight of the solute and defines its size, while H_b is a crude approximation of the total number of hydrogen bonds for each molecule. The molecular weight descriptor has also been an omnipresent variable in QSAR studies pertaining to cross-resistance to various chemotherapeutic agent in multidrug resistant cell lines [169]. $\sqrt[3]{\text{MW}}$ was used since it most closely approximates the size (radii) of the drugs involved in the study and their interactions with GP-170 (QSAR 34).

$$\log \text{CR} = 0.70\sqrt[3]{\text{MW}} - 1.01 \log (\beta' \times 10^{\sqrt[3]{\text{MW}}} + 1) - 0.10 \log P + 0.38\, \text{I} - 3.08 \quad (34)$$

$$n = 40,\ r^2 = 0.794,\ s = 0.344,\ \log\beta = -6.851,\ \text{optimum}\sqrt[3]{\text{MW}} = 7.21$$

2.4. Other Variables

Indicator variables I, are often used to highlight a structural feature present in some of the molecules in a data set that confers unusual activity or lack of it to these particular members. Usage of this variable could be beneficial in cases where the data set is heterogeneous and includes large numbers of members with an unusual feature that may or may not impact a biological response. QSAR for the inhibition of trypsin by X-benzamidines utilized indicator variables to denote the presence of unusual features such as positional isomers and vinyl/carbonyl containing substituents [170]. A recent study on the inhibition of lipoxygenase catalyzed production of leukotriene B4 and 5-hydroxyeicosatetraenoic from arachidonic acid in guinea pig leukocytes by X-vinyl catechols led to the development of the following QSAR [171]:

$$\log\left(\frac{1}{C}\right) = 0.49(\pm 0.11)\log P - 0.75(\pm 0.22) \times \log(\beta \times 10^{\log P} + 1) - 0.62(\pm 0.18) \times D2 - 1.13(\pm 0.20)D3 + 5.50(\pm 0.33) \quad (35)$$

$$n = 51,\ r^2 = 0.801,\ s = 0.269,\ \log P_0 = 4.61(\pm 0.49),\ \log\beta = -4.33$$

The indicator variables, D2 and D3 pertain to simple X-catechols ($D2 = 1$) and to X-naphthalene diols, ($D3 = 1$), respectively. The negative coefficients with both terms (D2 and D3) underscore the detrimental effects of these structural features on inhibitory activity. Thus, discontinuities in the structural features of the molecules of this data set are accounted for by the use of indicator variables. An indicator variable may be visualized graphically as a constant that adjusts two parallel lines so that they are superimposable. The use of indicator variables in QSAR analysis is also described in the following example. An analysis of a comprehensive set of nitroaromatic and heteroaromatic compounds that induced mutagenesis in TA 98 cells was conducted by Debnath et al., and QSAR 36 was formulated [172].

$$\log \text{TA98} = 0.65(\pm 0.16)\log P - 2.90(\pm 0.59) \times \log(\beta \times 10^{\log P} + 1) - 1.38(\pm 0.25)E_{\text{LUMO}} + 1.88(\pm 0.39)I_1 - 2.89(\pm 0.81)I_a - 4.15(\pm 0.58) \quad (36)$$

$$n = 188,\ r^2 = 0.810,\ s = 0.886,$$
$$\log P_0 = 4.93(\pm 0.35),\ \log \beta = -5.48$$

TA98 represents the number of revertants per nanomole of nitro compound. E_{LUMO} is the energy of the lowest unoccupied molecular orbital and I_a is an indicator variable that signifies the presence of an acenthrylene ring in the mutagens. I_1 is also an indicator variable that pertains to the number of fused rings in the data set. It acquires a value of 1 for all congeners containing 3 or more fused rings and a value of zero for those containing one or two fused rings (e.g., naphthalene, benzene). Model 36 indicates that the mutagenic potential of nitro congenors parallels the number of fused rings in the molecules. The E_{LUMO} term indicates that the lower the energy of the LUMO, the more potent the mutagen. In this QSAR, the combination of indicator variables affords a mixed blessing. One variable helps to enhance activity while the other leads to a decrease in mutagenicity of the acenthrylene congeners. In both of these QSARs, Kubinyi's bilinear model [173] is used. See Section 4.1.2 for a description of this approach.

The importance of hydrogen bonding in biological systems and molecular recognition is underscored by its critical role in defining the structure of biomacromolecules such as DNA and RNA as well as delineating protein–ligand interactions. Taft et al., were the first to devise a quantitative scale of hydrogen-bond strength of H-bond acceptors by borrowing from the $\log K$ values of hydrogen-bond complexation with 4-fluorophenol in carbon tetrachloride [174]. See Equation 37.

$$\log K = [m \times \alpha_2 \times \beta_2] + c \qquad (37)$$

In Equation 37, α_2 and β_2 denote the H-bond donor and H-bond acceptor ability, respectively. Other attempts were made to devise hydrogen-bonding parameters for use in QSAR by Seiler with his I_H values and Moriguchi with his E_R polarity terms [175,176]. Fujita utilized an indicator variable, HB to describe a possible hydrogen-bond presence [177]. Many attempts have been made to factor hydrogen-bonding ability into its component hydrogen-bond donor and hydrogen-bond acceptor portions; a few approaches to characterizing the explicit nature of the hydrogen bond have been undertaken by Charton and Charton, Dearden and Ghafourian, and Raevsky [178–180].

Charton and Charton used a simple but efficient method to represent the number of hydrogen bonds that a molecule could induge in—an amino group would score a 1 as a proton acceptor and 2 as proton donor in contrast to a hydroxyl group with its two pairs of unshared electrons and its proton acceptor score of 2 and its proton donor score of 1 [178]. Dearden and Ghafourian have used the atomic charge on the hydrogen-bonding hydrogen atom (Q_H) and E_{LUMO} energy to model hydrogen-bond donor ability and the charge on the most negatively charged atom (Q_{MN}) the energy of E_{HOMO} to delineate hydrogen-bond acceptor ability [179].

Raevsky has utilized empirical free energy and/or enthalpy values to calculated factor values for H-bond donors (C_d) and H-bond acceptors (C_a) using the HYBOT program. The well-worked relationship between tadpole narcosis and $\log P$ was examined from a new angle by using polarizability (Pol) and hydrogen-bond acceptor ability [180].

$$\log\left(\frac{1}{C}\right) = 0.49(\pm 0.20) + 0.23(\pm 0.02) \times \text{Pol} - 0.42(\pm 0.05) C_a \qquad (38)$$

$$n = 85 \quad r = 0.954 \quad s = 0.33 \quad F = 413.1$$

$$Rcv = 0.950$$

Extensive studies on the delineation of H-bond scales for H-bond donors and acceptors have approached this complex problem from a solvatochromic angle that is based on linear solvation energy relationships (LSERs) elaborated by Kamlet and coworkers, and Schuurmann [181,182]. The independent variables utilized to correlate activity include a volume or cavity term, polarizability and H-bond interaction terms. Wilson and Famini have developed a theoretical set of LSER descriptors that are more easily obtained via computation for multiple solutes and a single solvent [183,184]. In addition to the well-established LSER terms, electrostatic equivalents (q^-, q^+) associated with proton donors and acceptors, are added to the mix. Recently, Charton has derived 45 hydrogen-bonding parameters for hydrogen acceptor groups and 15 for hydrogen acceptor and donor groups of substituents bonded to sp^2 hybridized carbon atoms from octanol–water partition data [185]. It was emphasized that in aqueous solutions, a leveling effect occurs with H-bond donors and thus groups that have H-bond donor and H-bond acceptor capabilities function primarily as H-bond acceptors.

Quantum chemical methods and modeling techniques have facilitated the rapid and accurate computation of a large number of molecular attributes that define the chemical reactivities and biological activities of whole molecules or their substituents. Direct derivation of these descriptors can be obtained from their molecular wave functions [186]. Two different approaches may be used to provide quantitative description of molecular structures: *ab initio* and semiempirical quantum chemical methods. For a detailed description of methods and descriptors, see Ref. [186]. The following examples illustrate the utility of semiempirical methods to calculate molecular descriptors appropriate for QSAR studies.

Using MNDO, Shusterman et al. formulated the following QSAR 39 pertaining to the induction of mutagenicity of triazenes ($R{-}N{=}NN(CH_3)_2$) in *Salmonella typhimurium* TA92 [187].

$$\log\left(\frac{1}{C}\right) = 0.95(\pm 0.25)\log P + 2.22(\pm 0.88) \times E_{\mathrm{HOMO}} + 22.69(\pm 7.79) \qquad (39)$$

$n = 21 \quad r^2 = 0.845 \quad s = 0.631 \quad q^2 = 0.780$

The two-parameter Equation 39 indicates that hydrophobic and electron-releasing substituents increase the mutagenic potential of these aryl triazenes. In a study to assess the inhibitory effects of substituted benzalacetones on UV induced mutagenesis in *Escherichia coli* WP2uvrA, Yamagami et al. used semiempirical MO calculations (AM1 and PM3) to obtain the following QSAR that contrasts with Equation 40 [188].

$$\log\left(\frac{1}{C}\right) = 0.643(\pm 0.240)\mathrm{HB}_{2\mathrm{OH}} - 1.240 \times (\pm 0.238)E_{\mathrm{LUMO}} + 2.248(\pm 0.2130) \qquad (40)$$

$n = 28 \quad r = 0.912 \quad s = 0.154 \quad F = 61.8$

This two-parameter equation suggests that X-benzalacetones, with strong hydrogen-bonding capability and electron-withdrawing substituents that would increase susceptibility to nucleophilic attack, would make excellent antimutagenic agents. These few examples exemplify the applicability of quantum chemical descriptors in QSAR studies and their solid entrenchment in the area. However, critical analysis must still be exercised in order to minimize problems of collinearity, misinterpretation, and misuse.

2.5. Molecular Descriptors

2.5.1. Topological Descriptors Topological descriptors are an important class of whole molecule descriptors. They are truly structural descriptors since they are based only on the two-dimensional representation of a chemical structure. The most widely known descriptors are those that were originally proposed by Randic [189,190] and extensively developed by Kier and Hall [191,192]. Molecules are treated as topological entities such that atoms become vertices and bonds constitute the edges of a molecular graph. The strength of this approach is that the required information is embedded in the hydrogen-suppressed framework and thus no experimental measurements are needed to define molecular connectivity indices. For each bond, the C_K term is calculated. The summation of these terms then leads to the derivation of X, the molecular connectivity index for the molecule.

$$C_k = (\delta_i \delta_j)^{-0.5}, \text{ where } \delta = \sigma - h \qquad (41)$$

δ is the count of formally bonded carbons while h is the number of bonds to hydrogen atoms.

$$^1\mathrm{X} = \Sigma C_k = \sum (\delta_i \delta_j)_k^{-0.5} \qquad (42)$$

^{1}X is the first bond order since it considers only individual bonds. Higher molecular connectivity indices encode more complex attributes of molecular structure by considering longer paths. Thus, ^{2}X and ^{3}X account for all two-bond paths and three-bond paths, respectively, in a molecule. In order to correct for differences in valence, Kier and Hall proposed a valence delta (δ^v) term to calculate valence connectivity indices [193].

Molecular connectivity indices have been shown to be closely related to many physicochemical parameters such as boiling points, molar refraction, polarizability, and partition coefficients [191,194]. Ten years ago, the E-state index was developed to define an atom or group centered numerical code to represent molecular structure [195,196]. The E-state was established as a composite index encoding both electronic and steric properties of atoms in molecules. It reflects an atom's electronegativity, the electronegativity of proximal and distal atoms, and topological state. Extensions of this method include the HE-state, atom-type E-state and the polarity index Q. $\log P$ showed a strong correlation with the Q index of a small set ($n = 21$) of miscellaneous compounds [195]. Various models using electrotopological indices have been developed to delineate a variety of biological responses [197–199]. Some criticism has been leveled at this approach [200,201]. Chance correlations are always a problem when dealing with such a wide array of descriptors. The physicochemical interpretation of the meaning of these descriptors is not transparent, although attempts have been made to address this issue [202].

2.5.2. Molecular Polarizability The polarizability of a molecule, α, is a significant descriptor that is extensively used in QSAR and QSPR studies. It is defined as the proportionality constant between the strength of an applied electrical field (E) and the magnitude of the induced dipole moment (μ_{induced}).

$$\mu_{\text{induced}} = \alpha \times E \tag{43}$$

The polarizability of a molecule is actually influenced by the strength of the interaction between electrons and atomic nuclei. This means that any molecule contains only a few electrons will have lower polarizability than that of the molecule containing many electrons and a more diffuse electron distribution. The electron distribution can be easily distorted if LUMO energy lies close to the HOMO energy, that is, molecules with smaller HOMO–LUMO gap are typically associated with larger polarizabilities and vice versa [203]. Experimentally, polarizability can be determined from the Lorentz–Lorenz equation that was previously described in Equation 32. Thus, if MR is expressed in milliliters, and α in Å^3, polarizability, α, can be expressed as follows:

$$\alpha = 0.3964 \times \text{MR} \tag{44}$$

It has well established that molecular polarizability is an additive property, which is based on the fact that the molar refraction is an additive property, and is proportional to the molecular polarizability. Thus, the value for a given group or atom is fairly constant for a variety of molecules. The additive hypothesis has been extensively used by a number of researchers in the rapid calculation of molecular polarizability. Bosque and Sales [203] recently developed a simple 10-descriptor model using the atomic polarizabilities for 10 elements, which allowed for an accurate calculation of the molecular polarizability without considering any other structural parameters. The training set contained 340 compounds while the test set contained 86 compounds. The QSPR of the training set is as follows:

$$\begin{aligned}\alpha_{\text{cal}} = {} & 1.51N_{\text{C}} + 0.17N_{\text{H}} + 0.57N_{\text{O}} + 1.05N_{\text{N}} \\ & + 2.99N_{\text{S}} + 2.48N_{\text{P}} + 0.22N_{\text{F}} + 2.16N_{\text{Cl}} \\ & + 3.29N_{\text{Br}} + 5.45N_{\text{I}} + 0.32\end{aligned} \tag{45}$$

$$n = 340,\ r^2 = 0.994,\ q^2 = 0.994,\ s = 0.340,\ F = 5784$$

In Equation 45, N is the number of atoms of the corresponding element, n is the number of compounds in the set, r is the correlation coefficient, q^2 is the cross-validated r^2, s is the standard deviation, and F is the Fisher test. By using different methods, the average atomic polarizabilities (Å^3) of the elements C, H, O, N, S, P, F, Cl, Br, and I were determined to be 1.51, 0.17, 0.57, 1.05, 2.99, 2.48, 0.22, 2.16, 3.29, and 5.45, respectively. The above QSPR model was further modified by Zhokhova et al. [204] considering atomic polarizabilities of the elements as well as the structural features of the molecules resulting in the development of 14-descriptor model. The training set contained 552 compounds while the test set contained 61 compounds. The QSPR equation of the training set is as follows:

$$\alpha_{\mathrm{cal}} = 1.08f_1 + 0.38f_2 + 0.92f_3 + 0.61f_4 + 3.04f_5 + 2.18f_6 + 0.44f_7 + 2.34f_8 + 3.35f_9 + 5.49f_{10} + 0.38f_{11} + 0.15f_{12} + 0.34f_{13} + 0.36f_{14} - 0.04 \quad (46)$$

$$n = 552,\ r^2 = 0.997,\ s = 0.380,\ F = 10931$$

where $f_1, f_2, f_3, f_4, f_5, f_6, f_7, f_8, f_9, f_{10}$ represent the number of C, H, N, O, S, P, F, Cl, Br, and I atoms, respectively; f_{11} is the number of triple bonds in the molecule; f_{12} is the number of double bonds in the molecule; f_{13} is the number of aromatic bonds; and f_{14} is the number of atoms in the ring junctions in the aromatic system ($CAr(CAr)_2$).

An easily calculable polarizability parameter, NVE, was further developed by Hansch et al. [205], which has shown to be effective at delineating various chemicobiological interactions. It was computed by adding up the number of valence electrons (NVE) in a molecule such that H = 1, C = 4, Si = 4, N = 5, P = 5, O = 6, S = 6, and halogens = 7. NVE may also be represented as follows: NVE = $n_\sigma + n_\pi + n_\mathrm{n}$, where n_σ is the number of electrons in σ-orbital, n_π is the number of electrons in π-orbitals, and n_n is the number of lone pair electrons. NVE shows an excellent correlation with molecular polarizability, α ($n = 146$, $r^2 = 0.987$) as well as calculated molar refraction ($n = 146$, $r^2 = 0.995$) [206]. In a recent report, Verma, Kurup, and Hansch [207] successfully applied NVE as a polarizability parameter in an extensive QSAR study pertaining to various chemical–biological interactions. The general form of those QSAR models is represented by Equation 47.

$$\log\left(\frac{1}{C}\right) = K \times \mathrm{NVE} \pm \mathrm{constant} \quad (47)$$

2.5.3. Molecular Polar Surface Area Molecular polar surface area (PSA) is a descriptor that correlates well with passive molecular transport through membranes and, therefore, allows prediction of transport properties of drugs. The PSA of a molecule is defined as the area of its van der Waals surface that arises from heteroatoms and hydrogen atoms attached to the heteroatoms. Thus, it is related to a compound with their capacity to form hydrogen bonds [208,209]. It has already been established that a PSA value of over 140 Å^2 generally yields molecules that are poorly absorbed from stomach and gastrointestinal tract, whereas values below 60 Å^2 suggest potential penetration of the blood–brain barrier. Thus, the general rule developed for the predicted classification of molecules is: PSA $\leq$ 61 Å^2 for good, PSA $\geq$ 140 Å^2 for poor and 140 Å^2 > PSA > 61 Å^2 for OK [209,210]. A simple QSAR 48 for the prediction of log BB from a set of 55 diverse organic compounds was developed by Clark using two descriptors PSA and $C\log P$ [211].

$$\log \mathrm{BB} = -0.015(\pm 0.001)\mathrm{PSA} + 0.152 \times (\pm 0.036) C\log P + 0.139(\pm 0.073) \quad (48)$$

$$n = 55,\ r = 0.887,\ s = 0.354,\ F = 95.800$$

Indeed, it has been proven that PSA is one of the best predictive descriptors to build a QSAR model for drugs affecting the central nervous system (CNS). The calculation of PSA, however, is very time-consuming process because of the necessity to generate reasonable 3D molecular geometry and the calculation of the surface itself. A simple approach for the calculation of PSA is based on the summation of the surface contributions of polar fragments. This method is termed as topological PSA (TPSA), that provides practically identical results as that of 3D PSA (the correlation coefficient between 3D PSA and fragment-based TPSA for 34,810 molecules from the World Drug Index is 0.99). The correlation is demonstrated by Equation 49 [208].

$$3\mathrm{D\ PSA} = \sum_{i}^{n_{\mathrm{types}}} n_i \times c(\mathrm{Fragment}_i) \quad (49)$$

Where 3D PSA is the calculated PSA (based on 3D molecular structure), n_{types} is the number of types of polar fragments, $c(\text{Fragment}_i)$ is the coefficient to optimize (i.e., surface contribution of fragment i), and n_i is the frequency of fragment i in the molecule. TPSA is a simple and convenient measure of the polar surface area of the molecule associated with the oxygen, nitrogen, sulfur, and phosphorus atoms, including also their attached hydrogens. A QSAR 50 for the toxicity (LD_{50}) of a series of 19 alkaloids with the lycoctonine skeleton was developed using TPSA descriptor [212].

$$\log\left(\frac{1}{LD_{50}}\right) = 0.017(\pm 0.005)\text{TPSA} + 3.265(\pm 0.339) \quad (50)$$

$$n = 19,\ r = 0.880,\ s = 0.318,\ q^2 = 0.700,\ F = 56.000$$

2.5.4. van der Waals Volumes The van der Waals volume (V_{vdW}) in Å^3/molecule is another widely used descriptor in QSAR/QSPR modeling. However, its calculation is also a time-consuming process. A new, easy, and fast calculating method for van der Waals volume was recently developed by Zhao et al. [213], which is based on Bondi atomic volumes and only requires the atomic contributions and the number of atoms, bonds, and rings. It can thus be calculated from the following equation:

$$V_{vdW} = \sum \text{all atom contributions} - 5.92 N_B - 14.7 R_A - 3.8 R_{NA} \quad (51)$$

where N_B is the number of bonds (no matter whether they are single, double, or triple bonds, and no matter what elements comprise the bond, that is, CC, CO, CN, CS, NN, etc.), R_A is the number of aromatic rings, and R_{NA} is the number of nonaromatic rings. The number of bonds present in a molecule (N_B) can be calculated by using a simple equation, $N_B = N - 1 + R_g$, where N is the total number of atoms and R_g is the total number of ring structures ($R_A + R_{NA}$). Bondi atomic volumes of some frequently used atoms are as follows: atom/V_{vdW} (Å^3) = H/7.24, C/20.58, N/15.60, O/14.71, F/13.31, Cl/22.45, Br/26.52, I/32.52, P/24.43, S/24.43, As/26.52, B/40.48, Si/38.79, Se/28.73, and Te/36.62. A representative sample, that is, the van der Waals volume of naphthalene ($C_{10}H_8$) is calculated by using this approach: $V_{vdW} = (10 \times 20.58) + (8 \times 7.24) - (19 \times 5.92) - (2 \times 14.7) = 121.84$ Å^3. Results utilizing this method are equivalent to the TSAR and MacroModel computer-calculated van der Waals volumes.

The V_{vdW} was successfully utilized by Agrawal et al. [214] in the construction of a QSAR 52 for the anti-HIV activities of a series of HEPT analogs (**I**; Fig. 2):

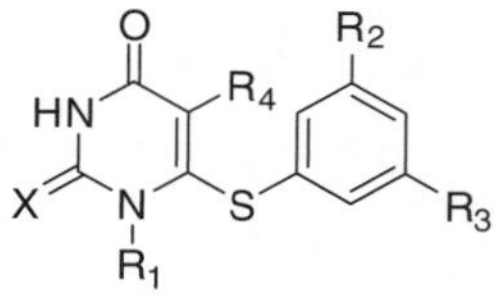

Figure 2. Structure of HEPT analogs (**I**).

$$pCC_{50} = 0.717(\pm 0.084) V_{vdW} - 0.426 \times (\pm 0.051) I_1 + 0.130(\pm 0.064) \times I_2 - 3.926 \quad (52)$$

$$n = 48,\ r = 0.875,\ s = 0.170,\ F = 47.912,\ Q = 5.147$$

Note that pCC_{50} is the cytotoxic activity of HIV-inhibitors (HEPT analogs). I_1 is an indicator variable, which takes the value as 1 when the $-CH_2CH_2OH$ moiety is present at R_1, otherwise it is zero. Similarly, I_2 takes the value 1 when halogen is present at R_2, otherwise it is zero. Since the V_{vdW} is a steric parameter, its positive coefficient suggests that pCC_{50} depends on the size of the substituents, that is, increase in the bulk of the compounds likewise increases pCC_{50}. The negative coefficient of the indicator variable (I_1) suggests that the presence of $-CH_2CH_2OH$ moiety at R_1 position is not helpful to the favorable cytotoxic activity (pCC_{50}). On the other hand, the presence of halogen at R_2 position is favorable for the cytotoxic activity as indicated by the positive coefficient with the indicator variable, I_2.

2.5.5. DCW (SMILES) The new SMILES-based optimal descriptor, DCW is also calculated by the additive scheme and was found to be useful in the development of QSAR/QSPR models. It is represented by Equation 53.

$$\mathrm{DCW(SMILES)} = \sum \mathrm{CW(SF}_k) \quad (53)$$

Where SF_k is a symbol of the SMILES fragment and $CW(SF_k)$ is the correlation weight of the SF_k. The SMILES fragment (SF_k) can be (i) a symbol of the SMILES notation of one character (e.g., "c," "C," "N," ")," "=," etc.); (ii) two symbols of the SMILES encoding a physicochemical image (e.g., "Cl," "Si," "Pb," etc.); (iii) fragments of three characters (e.g., "C=C," "C#C," "C#N," etc.); (iv) fragments of four characters (e.g., "[O$^-$]," "[N+]," "[S+]," "[Au]," etc.); and (v) all others. It must be noted that the symbols ")" and "]" have been replaced by "(" and "[" because these are indicators of the same molecular phenomenon, that is, branching and ions, respectively. The $CW(SF_k)$ values are calculated by using a Monte Carlo method, optimization procedure [215–217] that provides the values of the DCW descriptor. The descriptor DCW was successfully utilized by Toropov et al. [216] in the construction of QSPR model 54 for the octanol/water partition coefficient of organic compounds of different classes.

$$\log P = 9.343(\pm 0.013)\mathrm{DCW} - 9.393(\pm 0.020) \quad (54)$$

$$n = 69,\ r_2 = 0.987,\ s = 0.156,\ F = 5184$$

The following QSPR model was also developed by Toropov et al. [217] for the solubility of fullerene C_{60} in various organic solvents:

$$\log S = 104.428(\pm 0.348)\mathrm{DCW} - 111.006 \times (\pm 0.358) \quad (55)$$

$$n = 92,\ r^2 = 0.937,\ q^2 = 0.934,\ s = 0.270,\ F = 1342$$

DCW was also applied by Roy et al. [218] for QSAR modeling of peripheral benzodiazepine receptor (PBR) binding affinity (ovary and cortex) and PBR binding selectivity (peripheral versus central benzodiazepine receptor) of 2-phenylimidazo[1,2-a]pyridineacetamides. The statistics of the training sets were r^2 0.756 and $q^2 = 0.717$ (PBR cortex), $r^2 = 0.852$ and $q^2 = 0.836$ (PBR ovary), $r^2 = 0.784$ and $q^2 = 0.732$ (PBR cortex selectivity), and $r^2 = 0.845$ and $q^2 = 0.828$ (PBR ovary selectivity). The above results highlight the promising potential of the SMILES-based optimal descriptor, DCW in QSPR/QSAR modeling studies.

2.5.6. Effective Hydrogen Charge An attempt was made by Esteki and Khayamian [219] to develop QSAR models for the prediction of LD_{50} values of three important classes of psychotropic drugs such as phenothiazines, antidepressants and anxiolytics. In that study, the screening of a wide selection of descriptors, from topological to quantum chemical in nature, was carried out. Unfortunately, none of them were found to be adequate enough to describe the toxicity of the investigated drugs. Consequently, a new quantum chemical descriptor, that is, hydrogen charge (HC) descriptor was proposed; it was based on the oxidation mechanism of bioactivation of these drugs. The new HC descriptor is the sum of the hydrogen charges of amine moieties, carbons in the α-position of the heteroatoms (mainly N and O), carbonyl groups and unsaturated bonds. The results showed that a descriptor comprising the ratio of the HC over volume of the molecules, now specified as the effective hydrogen charge (EHC) was superior to a model with just the unadulterated HC descriptor. This indicates that the new EHC descriptor is sufficiently rich in chemical information and does encode the structural features contributing significantly to the toxicity of these drugs. The pertinent QSAR models with the new EHC descriptors are shown below:

Toxicity (LD_{50}) of phenothiazines

$$\mathrm{LD}_{50} = -774.400(\pm 46.200)\mathrm{EHC} + 1044.000(\pm 43.700) \quad (56)$$

$$n = 17,\ r^2 = 0.950,\ q^2 = 0.940,$$
$$\mathrm{SE} = 38.480,\ F = 280.800$$

Toxicity (LD_{50}) of antidepressants

$$\mathrm{LD}_{50} = -1018.700(\pm 91.300)\mathrm{EHC} + 1520.900(\pm 95.900) \quad (57)$$

$$n = 21,\ r^2 = 0.860,\ q^2 = 0.830,$$
$$\mathrm{SE} = 106.300,\ F = 124.400$$

Toxicity (LD_{50}) of anxiolytics

$$LD_{50} = -3013.700(\pm 246.500)EHC + 2930.800(\pm 159.600) \quad (58)$$

$$n = 18,\ r^2 = 0.910,\ q^2 = 0.870,\ SE = 137.000,\ F = 150.000$$

3. DEPENDENT VARIABLES IN QSAR

3.1. Role of Receptors

The development of the QSAR paradigm would not have proceeded so rapidly had it not been for concurrent breakthroughs in the structural elucidation of molecular receptors. A central theme of molecular pharmacology and the underlying basis of SAR studies are the elucidation of the structure and function of drug receptors. It is an endeavor that continues to proceed with unparalleled vigor, fueled by the rapid developments in genomics. It is generally widely accepted that endogenous and exogenous chemicals interact with a binding site on a specific macromolecular receptor according to the occupation theory first proposed by Clark and in keeping with the laws of mass action [220]. This interaction that is determined by intermolecular forces may or may not elicit a pharmacological response depending on its eventual site of action.

The idea that exogenous chemicals interacted with specific receptors originated with Langley, who studied the mutually antagonistic action of the alkaloids, pilocorpine, and atropine. He realized that both these chemicals interacted with some receptive substance in the nerve endings of the gland cells [221]. Paul Ehrlich defined the receptor as the "binding group of the protoplasmic molecule to which a foreign newly introduced group binds" [222]. In 1905, Langley's studies on the effects of curare on muscular contraction led to the first delineation of critical characteristics of a receptor—recognition capacity for certain ligands and an amplification component that result in a pharmacological response [223].

Receptors are mostly integral proteins embedded in the phospholipid bilayer of cell membranes. Rigorous treatment with detergents is needed to dissociate the proteins from the membrane that often results in loss of integrity and activity. Pure proteins such as enzymes also act as drug receptors. Their relative ease of isolation and amplification have made enzymes desirable targets in structure-based ligand design and QSAR studies. Nucleic acids comprise an important category of drug receptors. Nucleic acid receptors (aptamers), which interact with a diverse number of small organic molecules have been isolated via *in vitro* selection techniques and studied [224]. Recent binary complexes provide insight into the molecular recognition process in these biopolymers and also establish the importance of the architecture of tertiary motifs in nucleic acid folding [225]. Groove binding ligands such as lexitropsins hold promise as potential drugs, and are therefore, suitable subjects for focused QSAR studies [226].

Over the past 30 years, extensive QSAR studies on ligand–receptor interactions have been carried out with most of them focusing on enzymes. Two significant developments augmented QSAR studies and established an attractive approach to the elucidation of the mechanistic underpinnings of ligand–receptor interactions: the advent of molecular graphics, molecular modeling, and the ready availability of X-ray crystallography coordinates of various binary and ternary complexes of enzymes with diverse ligands and cofactors. Early studies with serine and thiol proteases (chymotrypsin, trypsin and papain), alcohol dehydrogenase and numerous dihydrofolate reductases (DHFR) not only established molecular modeling as a powerful tool but also helped clarify the extent of the role of hydrophobicity in enzyme–-ligand interactions [227–230]. Empirical evidence indicated that the coefficients with the hydrophobic term could be related to the degree of desolvation of the ligand by critical amino acid residues in the binding site of an enzyme. Total desolvation as characterized by binding in a deep crevice/pocket resulted in coefficients of approximately 1.0 (0.9–1.1) [230]. An extension of this agreement between the mathematical expression and structure as determined by X-ray crystallography led to the expectation that the binding of a set of substituents on the surface of an enzyme would yield a coefficient of approximately 0.5 (0.4–0.6) in the regression equation, indicative of partial desolvation.

Probing of various enzymes by different ligands also aided in dispelling the notion of Emil Fischer's rigid lock-and-key concept, in which the ligand (key) fits precisely into a receptor (lock). Thus, a "negative" impression of the substrate was considered to exist on the enzyme surface (geometric complementarity). Unfortunately, this rigid model fails to account for the effects of allosteric ligands, and this encouraged the evolution of the induced fit model. Thus, "deformable" lock-and-key models have gained acceptance on the basis of structural studies, especially NMR [231].

It is now possible to isolate and crystallize membrane-bound receptors, albeit with great difficulty as outlined by Parrill [232]. Some of the difficulties include low protein concentrations, lack of purification procedures that render functional receptors, conformational heterogeneity of the membrane-bound protein and the lack of polar surface areas to form crystollagraphic constacts. Nevertheless, great advances have been made in this arena, and the three-dimensional structures of some integral membrane-bound proteins such as rhodopsin and beta2-adrenergic receptor have recently been elucidated [233,234]. In order to gain an appreciation for mechanisms of ligand–receptor interactions, it is necessary to consider the intermolecular forces at play. Considering the low concentration of drugs and receptors in the human body, the law of mass action alone cannot account for the ability of a minute amount of a drug to elicit a pronounced pharmacological effect. The driving force for such an interaction may be attributed to the low energy state of the drug–receptor complex: K_D = [Drug] [Receptor]/[Drug–receptor complex]. Thus the biological activity of a drug is determined by its affinity for the receptor that is measured by its K_D, the dissociation constant at equilibrium. A smaller K_D implies a large concentration of the drug–receptor complex and hence a greater affinity of the drug for the receptor. The latter property is promoted and stabilized by mostly noncovalent interactions sometimes augmented by a few covalent bonds. The spontaneous formation of a bond between atoms results in a decrease in free energy; that is, ΔG is negative. The change in free energy, ΔG^0 is related to the equilibrium constant K_{eq} by the well-established relationship, where $\Delta G^0 = -RT \ln K_{eq}$. Thus, small changes in ΔG^0 can have a profound effect on equilibrium constants.

In the broadest sense, these "bonds" would include covalent, ionic, hydrogen, dipole–-dipole, van der Waals, and hydrophobic interactions. Most drug–receptor interactions comprise a combination of the bond types listed in Table 3, most of which are reversible under physiological conditions.

Covalent-bond interactions are not as important in drug–receptor binding as noncovalent interactions but they do occur in some specific cases. Alkylating agents in chemotherapy tend to react and form an immonium ion that then alkylates proteins, preventing their normal participation in cell divisions. B.R. Baker's concept of active site directed irreversible inhibitors was well established by covalent formation of Baker's antifolate and dihydrofolate reductase [235].

Ionic (electrostatic) interactions are formed between ions of opposite charge with energies that are nominal and which tend to fall off with distance. They are ubiquitous and since they act across long distances, they play a prominent role in the actions of ionizable drugs. The strength of an electrostatic force is directly dependent on the charge of each ion and inversely dependent on the dielectric constant of the solvent and the distance between the charges.

Hydrogen bonds are ubiquitous in nature: their muliple occurrences contribute to the stability of the α-helix and base pairing in DNA. Hydrogen bonding is based on an electrostatic interaction between the nonbonding electrons of a heteroatom (e.g., N, O, and S) and the electron deficient hydrogen atom of an −OH, −SH or −NH group. Hydrogen bonds are strongly directional, highly dependent on the net degree of solvation, and rather weak having energies ranging from 1–10 kcal/mol [236,237]. Bonds with this type of strength are of critical importance because they are stable enough to provide significant binding energy but weak enough to allow for quick dissociation. The greater electronegativity of atoms such as oxygen, nitrogen, sulfur, and

Table 3. Types of Intermolecular Forces

No.	Bond Type	Bond Strength (kcal/mol)	Example
1	Covalent	40–140	$CH_3CH_2O{-}H$
2	Ionic (electrostatic)	5	$R_4\overset{+}{N}\cdots\overset{-}{O}{-}C({=}O){-}$
3	Hydrogen	1–10	$-NH\cdots O(H){-}H$
4	Dipole–dipole	1	$R_3N\colon\cdots C{=}O$
5	van der Waals	0.5–1	$C\cdots C$
6	Hydrophobic	1	

halogen in comparison to carbon, causes bonds between these atoms to have an asymmetric distribution of electrons that results in the generation of electronic dipoles. Since so many functional groups have dipole moments, ion–dipole and dipole–dipole interactions are frequent. The energy of dipole–dipole interactions can be described by Equation 59, where μ is the dipole moment, θ is the angle between the two poles of the dipole, D is the dielectric constant of the medium, and r is the distance between the charges involved in the dipole.

$$E = \frac{2\mu_1\mu_2 \cos\theta_1 \cos\theta_2}{Dr3} \qquad (59)$$

While electrostatic interactions are generally restricted to polar molecules, there are also strong interactions between nonpolar molecules over small intermolecular distances. Dispersion or London/van der Waals forces are the universal attractive forces between atoms that hold nonpolar molecules together in the liquid phase. They are based on polarizability and these fluctuating dipoles or shifts in electron clouds of the atoms tend to induce opposite dipoles in adjacent molecules resulting in a net overall attraction. The energy of this interaction decreases very rapidly in proportion to $1/r^6$ where r is the distance separating the two molecules. These van der Waals forces operate at a distance of approximately 0.4–0.6 nm and exert an attraction force of less than 0.5 kcal/mol. Yet, while individual van der Waals forces make a low energy contribution to an event, they become significant and additive when summed up over a large area with close surface contact of the atoms.

Hydrophobicity refers to the tendency of nonpolar compounds to transfer from an aqueous phase to an organic phase [238,239]. When a nonpolar molecule is placed in water, it gets solvated by a "sweater" of water molecules ordered in a somewhat "ice-like" manner. This increased order in the water molecules surrounding the solute results in a loss of entropy. Association of hydrocarbon molecules leads to a "squeezing out" of the structured water molecules. The displaced water becomes bulk water, less ordered resulting in a gain in entropy, which provides the driving force for what has been referred to as a "hydrophobic bond." Although this is a generally accepted view of hydrophobicity, the hydration of apolar molecules and the non covalent interactions between these molecules in water are still poorly understood and thus the source of continued examination and frustration [63,240,241].

Since noncovalent interactions are generally weak, cooperativity by several types of interactions is essential for overall activity. Enthalpy terms will be additive, but once the first interaction occurs, translational entropy is lost. This results in a reduced entropy loss in the second interaction. The net result is that eventually several weak interactions combine to produce a strong interaction. One can safely state that it is the involvement of myriad interactions that contribute to the overall selectivity of drug–receptor interactions.

3.2. Biological Parameters

In QSAR analysis, it is imperative that the biological data be both accurate and precise in order to develop a meaningful model. It must be realized that any resulting QSAR model that is developed is only as valid statistically as the data that led to its development. The equilibrium constants and rate constants that are used extensively in physical organic chemistry and medicinal chemistry are related to free-energy values ΔG. Thus, for use in QSAR, standard biological equilibrium constants such as K_i or K_m should be used in QSAR studies. Likewise, only standard rate constants should be deemed appropriate for a QSAR analysis. Percent activities, for example, percent inhibition of growth at certain concentrations, are *not* appropriate biological endpoints due to the nonlinear characteristic of dose–response relationships. These types of endpoints may be transformed to equieffective molar doses. Only equilibrium and rate constants pass muster in terms of the free-energy relationships or influence on QSAR studies. Biological data is usually expressed on a logarithmic scale because of the linear relationship between response and log dose in the midregion of the log dose–response curve. Inverse logarithms for activity ($\log 1/C$) are utilized so that higher values are obtained for more effective analogs. Various types of biological data have been utilized in QSAR analysis. A few common endpoints are outlined in Table 4.

Biological data should pertain to an aspect of biological/biochemical function that can be measured. The events could be occurring in enzymes, isolated or bound receptors, in cellular systems or whole animals. Since there is considerable variation in biological responses, test samples should be run in duplicate or preferably triplicate, except in whole animal studies where assay conditions (e.g., plasma concentrations of a drug) preclude such measurements.

It is also important to design a set of molecules that will yield a range of values in terms of biological activities. It is understandable that most medicinal chemists are loathe to synthesize molecules with poor activity but these data points are important in developing a meaningful QSAR. Generally, the larger the range (>2 log units) in activity, the easier it is to generate a predictive QSAR. This kind of equation is more forgiving in terms of errors of measurement. A narrow range in biological activity is less forgiving in terms of accuracy of data. Another factor that merits consideration is the time structure. Should a particular reading be taken after 48 or 72 h? Knowledge of cell cycles in cellular systems or biorhythms in animals would be advantageous.

Table 4. Types of Biological Data Utilized in QSAR Analysis

Source of Activity	Biological Parameters	Source of Activity	Biological Parameters
1. Isolated receptors		**2. Cellular systems**	
Rate constants	$\log k_{cat}$; $\log k$	Inhibition constants	$\log(1/IC_{50})$
Michaelis–Menten	$\log(1/K_m)$	Cross-resistance	log CR
Inhibition constants	$\log(1/K_i)$	General biological data	$\log(1/C)$
Affinity data	pA_2; pA_1	Mutagenicity states	$\log TA_{98}$; $\log TA_{100}$
3. "*In vivo*" systems			
Biocencentration factor			log BCF
Reaction rates			$\log I$ (induction)
Pharmodynamic rates (e.g., clearance)			$\log T$ (total clearance)

Each single step of drug transport, binding and metabolism involves some form of partitioning between an aqueous compartment and a nonaqueous phase that could be a membrane, serum protein, receptor, or enzyme. In the case of isolated receptors, the endpoint is clear-cut and the critical step is evident. But in more complex systems, such as cellular systems or whole animals, many localized steps could be involved in the random walk process and the eventual interaction with a target. Usually, the observed biological activity is reflective of the slow step or the rate-determining step.

In order to determine a defined biological response, for example, IC_{50}, a dose response curve is first established. Usually six to eight concentrations are tested to yield percents of activity or inhibition between 20% and 80%, the linear portion of the curve. Using the curves, the dose responsible for an established effect can easily be determined. This procedure is meaningful if, at the time the response is measured, the system is at equilibrium, or at least under steady-state conditions.

Other approaches have been utilized to apply the additivity concept and ascertain the binding energy contributions of various substituent (R) groups. Fersht et al. [242] have measured the binding energies of various alkyl groups to aminoacyl-tRNA synthetases. Thus, the ΔG values for methyl, ethyl, isopropyl, and thio substituents were determined to be 3.2, 6.5, 9.6, and 5.4 kcal/mol, respectively.

An alternative generalized approach to determining the energies of various drug–receptor interactions was developed by Andrews et al. [243] who statistically examined the drug–receptor interactions of a diverse set of molecules in aqueous solution. Using Equation 60, a relationship was established between ΔG and E_X (intrinsic binding energy), E_{DOF} (energy of average entropy loss) and the $\Delta S_{r,t}$ (energy of rotational and translational entropy loss).

$$\Delta G = T\Delta S_{r,t} + n_{DOF}\, E_{DOF} + n_X E_X \qquad (60)$$

E_X denotes the sum of the intrinsic binding energy of each functional group of which n_X is present in each drug in the set. Using Equation 60, the average binding energies for various functional groups were calculated. These energies followed a particular trend with charged groups showing stronger interactions and nonpolar entities such as sp^2, sp^3 carbons, contributing very little. The applicability of this approach to specific drug–receptor interactions remains to be seen.

To obtain a statistically sound QSAR, it is important that certain caveats be kept in mind. One needs to be cognizant about collinearity between variables and chance correlations. Using a correlation matrix ensure that variables of significance and/or interest, are orthogonal to each other. With the rapid proliferation of parameters, caution must be exercised in amassing too many variables for a QSAR analysis. Topliss has elegantly demonstrated that there is a high risk of ending up with a chance correlation when too many variables are tested [244].

Outliers in QSAR model generation present their own problems. If they are badly fit by the model (off by more than two standard deviations), they should be dropped from the data set but their elimination should be noted and addressed. Their aberrant behavior may be attributed to inaccuracies in the testing procedure (usually dilution errors) or unusual behavior. They often provide valuable information in terms of the mechanistic interpretation of a QSAR model. They could be participating in some intermolecular interaction that is not available to other members of the data set or have a drastic change in mechanism. A later section will address these issues in greater detail.

3.3. Compound Selection

In setting up to run a QSAR analysis, compound selection is an important angle that needs to be addressed. One of the earliest manual methods was an approach devised by Craig, which involves 2D plots of important physicochemical properties. Care is taken to select substituents from all four quadrants of the plot [245]. Topliss developed a nonmathematical, nonstatistical, manual guide to the efficient usage of Hansch principles [246]. Later on he devised an operational scheme, which allowed one to start with two compounds and construct a decision tree that grows branches as the substituent set is expanded in a stepwise fashion [247]. The "tree" included certain substituents such as the 4-H, 3,4-Cl_2, 4-Cl, 4-CH_3 and 4-OCH_3 analogs that were tested in a stepwise fashion [247]. Other methods of manual substituent selection include the Fibonacci search method, sequential simplex strategy and parameter focusing by Magee [248–250].

One of the earliest computer-based and statistical selection methods; cluster analysis was devised by Hansch to accelerate the process and diversity of the substituents [125]. A combination of fractional factorial design in tandem with a principal property approach has also proven useful in QSAR [251]. Extensions of this approach using multivariate design have shown promise in environmental QSAR with nonspecific responses, where the clusters overlap and a cluster-based design approach has to be utilized [252]. With strongly clustered data containing several classes of compounds, a new strategy involving local multivariate designs within each cluster is described. The chosen compounds from the local designs are grouped together in the overall training set which is representative of all clusters [253].

4. QUANTITATIVE APPROACHES

4.1. Multilinear Regression Analysis (MLRA)

The correlation of biological activity with physicochemical properties is often termed an "extrathermodynamic relationship." Since it follows in the line of Hammett and Taft equations that correlate thermodynamic and related parameters, it is appropriately labeled. The Hammett equation represents relationships between the logarithms of rate or equilibrium constants and substituent constants. The linearity of many of these relationships led to their designation as linear free-energy relationships. The Hansch approach represents an extension of the Hammett equation from physical organic systems to a biological milieu. It should be noted that the simplicity of the approach belies the tremendous complexity of the intermolecular interactions at play in the overall biological response.

Biological systems are a complex mix of heterogeneous phases. Drug molecules usually traverse many of these phases to get from the site of administration to the eventual site of action. Along this random walk process, they perturb many other cellular components such as organelles, lipids, proteins, etc. These interactions are complex and vastly different from organic reactions in test tubes, even though the eventual interaction with a receptor may be chemical or physicochemical in nature. Thus, depending on the biological system involved—isolated receptor, cell or whole animal—one expects the response to be multifactorial and complex. The overall process particularly *in vitro* or *in vivo* studies a mix of equilibrium and rate processes, a situation that defies easy separation and delineation.

4.1.1. Linear Models Meyer and Overton were the first to attempt to get a grasp on biological responses by noting the relationship between oil/water partition coefficients and their narcotic activity. Ferguson recognized that equitoxic concentrations of small organic molecules were markedly influenced by their phase distribution between the biophase and exobiophase. This concept was generalized in the form of Equation 61 and extended by Fujita to Equation 62 [254,255].

$$C = kA^m \tag{61}$$

$$\log\left(\frac{1}{C}\right) = m\log\left(\frac{1}{A}\right) + \text{constant} \tag{62}$$

C represents the equipotent concentration, k and m are constants for a particular system and A is a physicochemical constant representative of phase distribution equilibria such as aqueous solubility, oil–water partition coefficient and vapor pressure. In examining a large and diverse number of biological systems, Hansch and coworkers defined the following relationship that expressed biological activity as a function of physicochemical parameters, for example, partition coefficients of organic molecules [71].

$$\log\left(\frac{1}{C}\right) = a\log P + b \tag{63}$$

Model systems have been devised to elucidate the mode of interactions of chemicals with biological entities. Examples of linear models pertaining to nonspecific toxicity are described. The effects of a series of alcohols (ROH) have been routinely studied in many model as well as biological systems. See QSAR 64–66.

Penetration of ROH into phosphatidylcholine monolayers [256]

$$\log\left(\frac{1}{C}\right) = 0.87(\pm0.01)\log P + 0.66(\pm0.01) \quad (64)$$

$$n = 4,\ r^2 = 0.998,\ s = 0.002$$

Changes in EPR signal of labeled ghost membranes by ROH [257]

$$\log\left(\frac{1}{C}\right) = 0.93(\pm0.09)\log P - 0.41(\pm0.16) \quad (65)$$

$$n = 6,\ r^2 = 0.996,\ s = 0.092$$

Inhibition of growth of Tetrahymena pyriformis by ROH [258,259]

$$\log\left(\frac{1}{C}\right) = 0.82(\pm0.04)C\log P + 0.89(\pm0.10) \quad (66)$$

$$n = 34,\ r^2 = 0.982,\ s = 0.173$$

In all cases, there is a strong dependence on $\log P_{oct}$ since all these processes involve transport of alcohols through membranes. The low intercepts speak to the nonspecific nature of the alcohol mediated toxic interaction. An equilibrium pseudo-equilibrium modeled by $\log P$ can be defined as shown in Fig. 3.

The Hammett type relationship for this conceptual idea of distribution is

$$\log P_{\text{bio}} = a \times \log P_{\text{octanol}} + b \quad (67)$$

This postulate assumes that steric, hydrophobic, electronic, and hydrogen-bonding factors that affect partitioning in the biophase are handled by the octanol–water system. Since the biological response (log 1/C) is proportional to $\log P_{\text{bio}}$, then it follows that

$$\log\left(\frac{1}{C}\right) = a \times \log \mathrm{P}_{octanol} + \text{constant} \quad (68)$$

Hansch and coworkers have amply demonstrated that Equation 68 applies to systems at or near phase distribution equilibrium as well as to systems removed from equilibrium [256,257].

4.1.2. Nonlinear Models Extensive studies on development of linear models led Hansch and coworkers to note that a breakdown in the linear relationship occurred when a greater range in hydrophobicity was assessed with particular emphasis placed on test molecules at extreme ends of the hydrophobicity spectrum. Thus, Hansch et al. suggested that the compounds could be involved in a "random walk" process: low hydrophobic molecules had a tendency to remain in the first aqueous compartment, while highly hydrophobic analogs sequestered in the first lipoidal phase that they encountered. This led to the formulation of a parabolic equation—relating biological activity and hydrophobicity [260].

$$\log\left(\frac{1}{C}\right) = -a(\log P)^2 + b(\log P) + \text{constant} \quad (69)$$

In the "random walk" process, the compounds partition in and out of various compartments and interact with myriad biological components in the process. In order to deal with this conundrum, Hansch proposed a general, comprehensive equation for QSAR 70 [261].

$$\log\left(\frac{1}{C}\right) = -a(\log P)^2 + b(\log P) + \rho\sigma + \delta E_S + \text{constant} \quad (70)$$

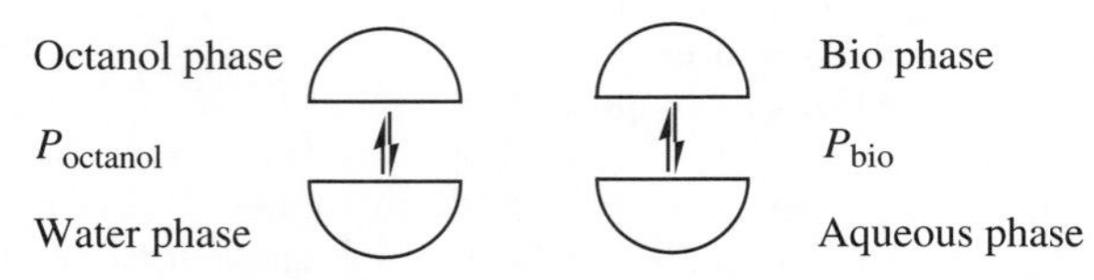

Figure 3. $\log P_{\text{octanol}}$ mirrors $\log P_{\text{bio}}$.

The optimum value of $\log P$ for a given system is $\log P_0$ and it is highly influenced by the number of hydrophobic barriers a drug encounters in its walk to its site of action. Hansch and Clayton formulated the following parabolic model to elucidate the narcotic action of alcohols on tadpoles [262].

Narcotic action of ROH on tadpoles

$$\log\left(\frac{1}{C}\right) = 1.38(\pm 0.34)\log P - 0.08(\pm 0.07)(\log P)^2 + 0.52(\pm 0.34) \quad (71)$$

$$n = 10,\ r^2 = 0.990,\ s = 0.210,$$

$$\log P_O = 8.69(5.78\text{–}43.43)$$

This is an example of nonspecific toxicity where the last step probably involved partitioning into a hydrophobic membrane. $\log P_0$ represents the optimal hydrophobicity (as defined by $\log P$) that elicits a maximal biological response.

Despite the success of the parabolic equation, there are a number of weaknesses in this approach. Firstly, it forces the data into a symmetrical parabola with the result that there are usually deviations between the experimental and parabola-calculated data. Secondly, the ascending slope is curved and inconsistent with the observed linear data. Finally, thus, the slope of a linear model cannot be compared to the curved slope of the parabola. In 1973, Franke devised a sophisticated, empirical model consisting of a linear ascending part and a parabolic section [263]. See Equations 72 and 73.

$$\log\left(\frac{1}{C}\right) = a \times \log P + c \quad (\text{if} \log P < \log P_X) \quad (72)$$

$$\log\left(\frac{1}{C}\right) = -a(\log P)^2 + b \times \log P + c \quad (\text{if} \log P > \log P_X) \quad (73)$$

The binding of drugs to proteins is linearly dependent on hydrophobicity up to a limited value, $\log P_X$ after which steric hindrance causes the linear dependence to alter to a nonlinear one. The major limitation of this approach involves the inclusion of highly hydrophobic congeners that tend to cause systematic deviations between experimental and predicted values.

Another cutoff model which deals with nonlinearity in biological systems, is one defined by McFarland [264]. It attempts to elucidate the dependence of drug transport on hydrophobicity in multicompartment models. McFarland addressed the probability of drug molecules traversing several aqueous lipid barriers from the first aqueous compartment to a distant, final aqueous compartment. The probability $P_{o,n}$ of a drug molecule to access the final compartment, n of a biological system was used to define the drug concentration in this compartment.

$$\log C_R = a \times \log P - 2a \times \log(P+1) + \text{constant} \quad (74)$$

The ascending and descending slopes are equal (= 1) and linear. However, a major drawback of this model is that it forces the activity curves to maximize at $\log P = 0$. These studies were extended and refined by Kubinyi who developed the elegant and powerful bilinear model that is superior to the parabolic model and is extensively utilized in QSAR studies [265].

$$\log\left(\frac{1}{C}\right) = a \times \log P - b \times \log(\beta \times P + 1) + \text{constant} \quad (75)$$

where β is the ratio of the volumes of the organic phase and the aqueous phase. An important feature of this model lies in the symmetry of the curves. For aqueous phases of this model system, symmetrical curves with linear ascending and descending sides (such as a teepee) and a limited parabolic section around the hydrophobicity optimum are generated. Unsymmetrical curves arise for the lipid phases. It is highly compatible with the linear model and allows for quick comparisons of the ascending slopes. It can also be utilized with other parameters such as MR and σ, where it appears to pinpoint a change in mechanism similar to the breaks in linearity of the Hammett equation. The following example of the bilinear model reveals the unsymmetrical, tepee-like nature of the curve. In a recent QSAR study of the toxicity relationships of dapsone derivatives with anti-inflammatory activity, the following bilinear model was formulated by Seydel et al. using an HPLC-capacity factor, k' in lieu of π hydrophobic constants [266]:

Percent production of methemoglobin (log%) by 100 μM of dapsone derivatives [266]

$$\log\% = 0.528(\pm 0.094)\log k' - 0.608(\pm 0.088) \times \log(\beta \times k' + 1) + 1.765(\pm 0.117) \quad (76)$$

$$n = 27,\ r^2 = 0.85,\ s = 0.12,\ \beta = 0.391(\pm 0.290),\ F = 31.6$$

A highly significant correlation is observed between methemoglobin formation and $\log k'$ and it also accurately predicts the activity of two derivatives not included in the original test set. The range in $\log k'$ was appropriate and extended from −1.53 to 4.25.

The bilinear model has been utilized to model biological interactions in isolated receptor systems and in adsorption, metabolism, elimination, and toxicity studies. It has a few limitations. These include the need for at least fifteen data points (because of the presence of the additional disposable parameter, β) and data points beyond optimum $\log P$ of $\log k'$. If the range in values for the dependent variable is limited, unreasonable slopes are obtained.

4.1.3. Free-Wilson and Other Approaches

The Free-Wilson approach is truly a structure–-activity based methodology, since it incorporates the contributions made by various structural fragments to the overall biological activity [267–269]. It is represented by Equation 77.

$$\mathrm{BA}_i = \sum_j a_j X_{ij} + \mu \quad (77)$$

Indicator variables are used to denote the presence or absence of a particular structure feature. Like classical QSAR, this *de novo* approach assumes that substituent effects are additive and constant. BA is the biological activity, X_j is the jth substituent that carries a value 1 if present, 0 if absent. The term a_j represents the contribution of the jth substituent to biological activity and μ is the overall, average activity. The summation of all activity contributions at each position must equal zero. The series of linear equations that are formulated are solved via linear regression analysis. It is necessary for each substituent to appear more than once at a position in different combinations with substituents at other positions.

There are certain advantages to the Free-Wilson method that have been addressed [268–270]. Any type of quantitative biological data can be subject to such analysis. There is no need for any physicochemical constants. The molecules of a series may be structurally dissected in any way and multiple sites of substitution are necessary and easily accommodated [271]. Limitations include the large number of molecules with varying substituent combinations that are needed for this analysis and the inability of the system to handle nonlinearity of the dependence of activity on substituent properties. Intramolecular interactions between the substituents are not handled very well although special treatments can be utilized to accommodate proximal effects. Extrapolation outside of the substituents used in the study is not feasible. Another problem inherent with this approach is that usually a large number of variables are required to describe a smaller number of compounds that creates a statistical "faux-pas." Fujita and Ban modified this approach in two important ways [272]. They expressed the biological activity on a logarithmic scale in order to bring it into line with the extrathermodynamic approach. See Equation 78.

$$\log X_C = \sum a_i X_i + \mu \quad (78)$$

This allowed the derived substituent constants to be compared with other free-energy-related parameters. The overall average intercept, μ took on a "new look" akin to an intercept in other QSAR analyses.

Analysis of a Free-Wilson type by Yalcin et al. have included the *in vitro* inhibitory activity of a series of heterocyclic compounds against *Klebsiella pneumoniae* [273]. Other applications of the Free-Wilson approach have included studies on the antimycobacterial activity of 4-alkylthiobenzanilides, the antibacterial activity of fluoronapthyridines and the benzodiazepine receptor binding ability of some nonbenzodiapzepine compounds such as 3-X-imidazo[1,2-b]pyridazines, 2-

phenylimidazo[1,2-α]pyridines, 2-(alkoxy-carbony) imidazo[2,1-p]benzothia-zoles and 2-arylquinolones [274–276].

The similarity in approaches of Hansch Analysis and Free-Wilson Analysis allows them to be used within the same framework. This is based on their theoretical consistency and the numerical equivalencies of activity contributions. This development has been called the mixed approach and can be represented by the following equation:

$$\log\left(\frac{1}{C}\right) = \sum a_i + \sum \mathrm{c_j ø_j} + \text{constant} \quad (79)$$

The term, a_i denotes the contribution for each ith substituent, while $ø_j$ is any physicochemical property of a substituent X_j. For a thorough review of the relationship between Hansch and Free-Wilson analysis, see the excellent review by Kubinyi [270]. A recent study of the P-glycoprotein inhibitory activity of 48 propafenone-type modulators of multidrug resistance, using a combined Hansch/Free-Wilson approach was deemed to have higher predictive ability than a stand-alone Free-Wilson analysis [277]. Molar refractivity that has a high collinearity with molecular weight was a significant determinant of modulating ability. It is of interest to note that molecular weight has been shown to be an omnipresent parameter in cross-resistance profiles in multidrug resistance phenomena [167].

4.2. Partial Least Squares Regression (PLSR)

PLSR is an established multivariate statistical method that relates data from a set of dependent variables (activities) to a set of independent variables [278]. It uses factors (latent variables) to describe the variance in both the independent and the dependent variables. The data used to construct the PLS model is termed the training set while the data used to validate the subsequent PLS model forms the test set. Leave-one-out cross-validation is often used to ascertain the optimum number of factors that could be used to establish a PLS model since the utilization of too many latent variables invariably leads to overfitting of the data. The advantage of PLSR is that it handles collinear descriptors well and thus results in the generation of better predictive models [279]. It has a limitation in that regression is performed on latent variables that do not have any physical meaning and are consequently less amenable to ease in interpretation than multilinear regression analyses.

In a recent study, Kulkarni's group used PLSR to develop a QSAR on a series of 5-aryl thiazolidinedione and oxazolidinedione derivatives acting as dual activators of peroxisome proliferator-activated receptors (PPARs), alpha and gamma [280]. The partitioning of the data set of 34 molecules into the training (27 molecules) and test (5 molecules) sets was accomplished by using random selection [281] and rotational selection via sphere exclusion [282]. Two molecules were excluded from the analysis as statistical outliers. PLSR resulted in an optimum four PLS components with excellent r^2 but poor q^2 for both PPARs. The descriptors with the higher contribution along with descriptors based on chemical structure were then used to develop statistically significant dual response and individual QSAR models for both receptors. The resulting models stressed the importance of hydrogen bonding, thermodynamics, electrostatic interaction, and shape/size of the molecules on PPAR activation.

4.3. Principal Component Regression (PCR)

As a result of the collinearity between predictor variables, MLRA is not able to extract useful information from structural data and thus there is a danger of overfitting the data. Improved predictive models can be obtained by orthogonalization of the variables by means of principal component analysis [283]. However, there has been minor criticism of the original approach to PCRA in QSAR [284,285]. One facet of PCRA pertains to the choice of appropriate principal components that comprise the best subset for predictive purposes. Usually, the simplest and most general route involves top–down variable selection where the factors are listed in the order of decreasing eigenvalues. The factor with the highest eigenvalue is included in the calibration model and factors are added sequentially till there is little or no improvement in the statistical output. The magnitude of an eigenvalue is not reflective of its significance to calibration. PC selection is clearly an optimization limitation and techniques involving factor correlation rankings and genetic search algorithms have been used to address this problem [286,287].

A recent study used PCRA and factor analysis (FA)-MLRA to examine the antimicrobial activity of a series of 3-hydroxypyridine-4-ones and 3-hydroxypyran-4-ones versus *Staphylococcus aureus* and *Candida albicans* [288]. Using FA-MLRA, the following equation was obtained:

$$\text{pMIC} = 4.786(\pm 0.484) + 0.196(\pm 0.063)\text{DMy} + 0.167(\pm 0.063)n\text{CONHR} - 0.130 \times(\pm 0.058)\text{PJI3} \quad (80)$$

$$n = 31,\ r^2 = 0.73,\ \text{SE} = 0.31,\ F = 11.41,\ q^2 = 0.68$$

The corresponding equation using PCRA is described as

$$\text{pMIC} = 3.756(\pm 0.036) + 0.400(\pm 0.036)f_3 \quad (81)$$

$$n = 31,\ r^2 = 0.81,\ \text{SE} = 0.19,\ F = 35.05,\ q^2 = 0.79$$

In Equation 80, the inhibitory activity is enhanced by the presence of amide groups and the quantum chemical index (DMy). With PCRA, there is a stronger correlation with the factor score (f_3) and the overall statistics are better (81% explained variance versus 73% explained variance for Equation 80). This improvement could be attributed to the usage of factor scores as opposed to descriptors; the former are more information rich and this is reflected in the quality of the regressions. The loading values of factor 3 underscore the importance of quantum and functional descriptors. Similar statistical patterns were seen with the data for *C. albicans*.

4.4. Artificial Neural Networks

The artificial neural networks (ANNs) approach has been shown to exhibit superior predictive power and flexibility in the development of QSAR models. It can be used for both the linear and the nonlinear correlations. One of the most important features of ANN is that "no prior knowledge" is required about the relation among variables. The rapid increase in the number of ANN-QSAR models is the best proof of a productive cooperation between artificial intelligence (AI) and chemistry. However, this technique has been labeled as a "black box" because it provides a very little explanatory insight into the relative influence of the independent variables in the predictive process. Thus, a little information on how and why compounds work can be obtained [289].

This viewpoint has evolved in recent years due to the development of ANN of minimal size to address this limitation, that is, networks having a small number of weights and/or descriptors, but possessing a reasonably high generalization capacity. There are various selection or pruning techniques that aim to eliminate redundant weights and/or descriptors, leaving only those offering a significant contribution to the model. Elimination of connections can also lead to better ANN generalizations. Pruning can be divided into two terms, that is, sensitivity and penalty, and can be implemented either manually or incorporated within the training algorithms [290,291]. An interpretation method focused on a pruning algorithm was well described by Hervás et al. [292]. Unfortunately, this method was not found applicable to those ANN models developed using alternative algorithms. Guha et al. [293] have described a method to interpret the weights and biases of an ANN-QSAR model that allows one to understand how an input descriptor is correlated to the predicted output by the network. This method has been divided into two parts. In the first part, the nonlinear transform for a given neuron is linearized that determines how a given neuron affects the downstream output. In the second part, a ranking scheme for neurons in a layer is developed that allows developing interpretations of an ANN model as similar to the partial least squares (PLS) interpretation method.

ANN with a layered structure is a mathematical tool that stimulates biological neural network, consisting of computing units (neurons) and connections between neurons (synapses) [294]. Input (independent) variables are considered as neurons of input layer. On the other hand, output (dependent) variables are considered as output neurons. Synapses connect input neurons to hidden neurons and hidden neurons to output neurons. The synapse strength from neuron i to neuron j is usually determined by a weight, W_{ij}. ANN-QSAR method is required to establish the possible relationship between the input descriptors and output bioactivity. Back propagation neural networks (BNNs) are the most often used techniques in analytical applications. The BNN receives a set of inputs, which is multiplied by each node and then a nonlinear transfer function is applied. The changes in the values of the weights between the layers in order to minimize the output errors can be obtained by using Equation 82:

$$\Delta W_{ij,n} = F_n + \alpha \Delta W_{ij,n-1} \quad (82)$$

where $\Delta W_{ij,n}$ is the change in the weight factor for each network node, α is the momentum factor, and F is a weight update function. Various types of algorithms have been used for most practical purposes including basic back propagation algorithm (BBP), conjugate gradient (CG) algorithm, Levenberg–Marquardt (L–M) algorithm, etc. [295].

The sample size (or number of patterns) in any ANN-QSAR model can be related to the number of connections by the parameter ρ, which is defined as the ratio of the former to the latter and given by Equation 83:

$$\rho = \frac{P}{(I+1) \times H + (H+1) \times O} \quad (83)$$

where P is the number of patterns, I is the number of input variables, H is the number of hidden neurons, and O is the number of output neurons. Equation 83 allows also for the bias connection to the input and hidden layers. The range $1.8 < \rho < 2.2$ is generally used as the guideline for an optimum ANN-QSAR models. It has been claimed that, for $\rho \ll 1.0$, the network simply memorizes the data, whereas for $\rho \gg 3.0$, the network loses its ability to generalize [291,296,297].

In a recent study, ANN-QSAR models were constructed to predict differential relative binding affinities of a series of structurally diverse compounds with estrogenic activity [298]. The models were developed with a data set of 93 compounds and tested with an additional 30 independent compounds. High correlations ($r^2 = 0.83$–0.91) for training sets were observed while validation results ($r^2 = 0.62$–0.86) for the external sets were encouraging. The models were used to identify structural features of phytoestrogens that are responsible for selective ligand binding to estrogen receptors (ERs) α and β.

In another study, a QSAR model was developed on binding affinity ($\log K'$hsa) for 94 different HSA drug/drug-like compounds by using the principal component-artificial neural network (PC-ANN) modeling technique, with application of eigenvalue ranking factor selection procedure. The results obtained by PC-ANN gave regression models with good predictive ability using a relatively low number of principal components (PCs). The coefficient of determination was 0.8497 with six extracted PCs [299].

4.5. Multidimensional QSAR

The recognition of the 3D structural importance of drug molecules to their biological activities, increasing 3D structural knowledge of biological macromolecules such as proteins or enzymes, and the importance of stereochemistry led to the development of quantitative structure–activity relationships at the three-dimensional level (3D QSAR), which is, in general, considered as an important tool in the field of computer-aided drug design (CADD). In 3D QSAR, the structures of the involved molecules are represented by three-dimensional entities, allows to quantifying steric/electrostatic forces, hydrogen-bond strengths, and hydrophobic interactions at the atomic level. 3D QSAR models typically represent a binding site surrogate with physicochemical properties mapped onto its surface or a grid surrounding the ligand molecules, superimposed in 3D space [300]. There are a number of available 3D QSAR approaches, such as CoMFA, CoMSIA, CoMMA, SOMFA, GRID, Apex-3D, VolSurf, and the HypoGen module of CATALYST.

There are five important caveats related to 3D QSAR methodologies discussed recently by Cramer and Wendt [301]: (a) A CoMFA model ability to predict binding affinities of ligands often exceeds the abilities of the method that may also consider the ligand's binding partner; (b) Comparing the activity prediction errors from 3D QSAR model using "topomer" alignments with those from receptor-based methods; (c) Although the biological activities that underlie all SAR models must often be influenced by differences in ligand transport, "alignment-averaged" ligand properties (e.g., $\log P$) are not explicitly addressed by (alignment-dependent) 3D QSAR descriptors. Does the omission of these ligand properties have an important impact on the predictive ability of 3D QSAR?; (d) What are the strengths and weaknesses of leave-one-out cross-validated r^2 (q^2) as the measure of accuracy for a 3D QSAR?; and (e) Are there any general rule among the fluctuations in r^2 and q^2 values and predictivity of 3D QSAR that are encountered when new results are added to a data set?

The 3D QSAR, comparative molecular field analysis (CoMFA) method, proposed by Cramer et al. in 1988 [302], is the most popular and extensively used method, in the present practice of drug design and discovery. An advantage of CoMFA is its ability to predict the biological activity of molecules by deriving a relation between steric/electrostatic properties and biological activities in the form of contour maps. The important steps in the CoMFA method are (a) determination of the bioactive conformation of each molecule in the training set; (b) introduction of the partial atomic charges so that an electrostatic field can be generated; (c) alignment of the molecules using either manual or automated methods in such a manner that the best alignment represents the interaction of the molecule with the target receptor; (d) creation of a cubic lattice of points around the molecules; (e) computing interaction energies using a probe (e.g., sp^3 with a 1 + charge) placed in all intersections of a regularly placed grid with a grid size of 1–2 Å. This generates a steric interaction energy based on a Lennard–Jones potential and an electrostatic interaction energy based on a Coulombic potential; (f) establishment of quantitative relationships to the biological activity and these interaction energies using the PLS technique and the cross-validation; and (g) predictions for a test set, and visualization of the results as contour plots on displays of the individual molecules in the set [303,304].

Another well-utilized 3D QSAR method involves comparative molecular similarity indices analysis (CoMSIA), which was introduced in 1994 by Klebe et al. [305]. The CoMSIA method makes use of similarity measures calculated between a common probe and each molecule in the set over a regularly spaced grid. The distance dependence of the similarities is computed using a Gaussian function, which avoids singularities near the atomic positions, unlike the Lennard–Jones and Coulomb potentials used in a standard CoMFA technique. This method also computes hydrophobic, hydrogen-bond donor (HBD), and hydrogen-bond acceptor (HBA) fields for each ligand in addition to steric and electrostatic fields. This analysis is not sensitive to changes in the orientation of the superimposed molecules in the lattice, and the correlation results obtained by CoMSIA can be graphically interpreted in terms of the field contribution maps allowing physicochemical properties relevant for binding to be easily mapped back to molecular structures [306,307]. In thermolysin inhibitors analysis using five different property fields, it was demonstrated that the CoMSIA results were easier to interpret than standard CoMFA [308].

There are a number of good examples for the application of 3D QSAR (CoMFA and CoMSIA) in the literature. Some of them are as follows:

(i) Kaur and Talele [309] conducted CoMFA and CoMSIA studies on a series of 1,3,4-benzotriazepine-based CCK2 receptor antagonists in order to gain helpful information for designing potent antagonists with novel structures. 3D QSAR models were derived from a training set of 46 compounds. By applying leave-one-out (LOO) cross-validation study, cross-validated (r^2_{cv}) values of 0.673 and 0.608 and non-cross-validated (r^2_{ncv}) values of 0.966 and 0.969 were obtained for the CoMFA and CoMSIA models, respectively. The predictive ability of the CoMFA and CoMSIA models was determined using a test set of 13 compounds, which gave predictive correlation coefficients (r^2_{pred}) of 0.793 and 0.786, respectively.

(ii) A training set of 42 structurally diverse carboline derivatives for their MAPKAPK2 (mitogen-activated protein kinase-activated protein kinase 2, a substrate of P38 MAPKs) inhibition activities was used by Nayana et al. [310] to construct 3D QSAR models with CoMFA and CoMSIA techniques. The CoMFA and CoMSIA models gave leave-one-out cross-validated coefficients (q^2) of 0.804 and 0.765, leave-five-out cross-validated coefficients (r^2_{LFO}) of 0.852 and 0.801, and regression coefficients (r^2) of 0.984 and 0.986, respectively. An external test set of 12 compounds have suggested that the predictive power of CoMFA model ($r^2_{pred} = 0.931$) was better than that of CoMSIA model ($r^2_{pred} = 0.921$).

(iii) The kinase insert domain receptor (KDR) is an attractive target for the development of novel anticancer agents. In a recent work, 3D QSAR was performed by Du et al. [311] using the CoMFA and CoMSIA methods on a series of 82 selective inhibitors of KDR (63 in the training set and 19 in the test set). The docked conformer-based alignment (DCBA) gave the best 3D QSAR models. The best CoMFA model gave the cross-validated q^2 of 0.546 with an optimal number of components (ONC) of 5 and a conventional correlation coefficient r^2 of 0.936 for the noncross-validated final model. The steric and electrostatic field contributions are 46.5% and 53.5%, respectively. The best CoMSIA model gave the cross-validated q^2 of 0.715 with an ONC of 6 and a noncross-validated coefficient r^2 of 0.961. The corresponding field contributions of steric, electrostatic, hydrophobic, HBD, and HBA are 7.4%, 25.0%, 28.4%, 20.4%, and 18.8%, respectively. The predictive power of the CoMFA and CoMSIA models was determined using a test set of 19 compounds, which gave predictive correlation coefficients (r^2_{pred}) of 0.673 and 0.797, respectively.

The quantitative structure–activity relationship at the four-dimentional level (4D QSAR) was introduced in 1997 by Hopfinger et al. [312]. The basic difference of this method from CoMFA is that it incorporates conformational and alignment freedom into the development of QSAR models by performing conformational ensemble sampling from a set of structures with biological activity. The descriptors in 4D QSAR analysis are Cartesian coordinates of each of the atom-types (such as steric, polar positive, polar negative, HBD, HBA, and aromatic atoms) of the ligands at each point of the 3D-grid. The molecular dynamics simulations (MDS) have been used to obtain the conformational ensemble of the molecules (the fourth dimension of this method). The 4D QSAR technique can be applied to both receptor-dependent (RD) and receptor-independent (RI) schemes. In the first scheme, the structure of the receptor (or molecular target) is explicitly used. In contrast, in the second scheme either the structure of the receptor is not available or it is neglected because of the uncertainty in the receptor structure and/or ligand binding mode [313]. The current methodology for the development of RI 4D QSAR consists of 10 operational steps [314] whereas RD 4D QSAR performs in 12 operational steps [315].

The 4D QSAR approach can play an important role in identifying the most probable tautomeric form, as even X-ray crystallography is often not able to unambiguously determine the protonation state [316,317]. This method can be interpreted as an extension of 3D QSAR methodology to address uncertainties during the alignment process. It can also have the fundamental biological relevance, when dealing with multimode binding targets. 4D QSAR can also account for different ligand configurations in a single simulation [317]. Some examples of 4D QSAR are described below.

A training set of 30 structurally diverse hydrazides that had been assessed for their minimum inhibitory concentrations against *Mycobacterium tuberculosis* var. *bovis* was used by Pasqualoto et al. [314] to develop the following receptor-independent 4D QSAR model:

$$\begin{aligned} \text{pMIC} = & -11.94\,\text{GC1(np)} + 11.37\,\text{GC2(any)} \\ & + 23.07\,\text{GC3(any)} + 1.57\,\text{GC4(any)} \\ & + 6.43\,\text{GC5(any)} + 30.76\,\text{GC6(np)} \\ & -0.23 \end{aligned} \quad (84)$$

$$n = 30,\ r^2 = 0.88,\ q^2 = 0.80,\ \text{LSE} = 0.16$$

It is important to note that this model is composed of only two classes of IPEs (interaction pharmacophore elements): np (nonpolar atoms) and "any" (no differentiation of all-atom occupancy) atoms. Moreover, the grid cell occupancy descriptor (GC1) responsible for predicted decreases in the biological activity corresponds to occupancy by a nonpolar IPE-type (np). This model was also externally validated by a test set of seven compounds (not included in the development of the above RI 4D QSAR model).

Receptor-dependent 4D QSAR models were generated by Santos-Filho and Hopfinger [315] using a series of 4-hydroxy-5,6-dihydropyrones as HIV-1 protease inhibitors. The optimized RD 4D QSAR models were statistically significant ($r^2 = 0.86 - 0.88$, q^2 $0.78 - 0.82$ for four- to six-term models). One example for four-term model is shown in Equation 85.

$$\begin{aligned}\mathrm{pIC_{50}(M)} = {} & 3.79\,\mathrm{GC1}(3,4,5,\mathrm{np}) - 1.36\,\mathrm{GC2} \\ & \times(0,-2,5,\mathrm{np}) - 3.20\,\mathrm{GC3} \\ & \times(0,5,3,\mathrm{np}) + 3.64\,\mathrm{GC4} \\ & \times(2,-3,3,\mathrm{any}) + 6.69 \end{aligned} \tag{85}$$

$$n = 39,\ r^2 = 0.86,\ q^2 = 0.80$$

In this equation, $\mathrm{pIC_{50}}$ (M) is the molar HIV-1 PR inhibition potency and GC$i(x, y, z,$ IPE) are the grid cell occupancy descriptors (GCODs), where $x, y,$ and z are the Cartesian coordinates of the reference grid cell and IPE is the specific interaction pharmacophore element of the GCOD.

The 4D QSAR approach significantly reduces bias with selecting a bioactive conformer, orientation, or protonation state, but a major unknown still persists: manifestation and magnitude of the induced fit—a ligand-induced adaptation of the binding site to the topology of the small molecule. The magnitude of induced fit cannot be estimated in the absence of the true biological receptor. Thus, 5D QSAR methodology was introduced in 2002 by Vedani and Dobler [318], which is not only able to evaluate simultaneous the different induced-fit scenarios but also allows for the induced-fit crossover and linear combination of the individual hypotheses. The model hypotheses can be generated using different scenarios for adapting the mean envelope to each molecule used in the QSAR analysis. The current 5D QSAR method defines up to six different protocols simulating the induced fit. The adaptation is based on (i) steric, (ii) electrostatic, (iii) H-bond field, (iv) lipophilic potential, (v) energy minimization, or (vi) linear scheme. The energetic cost associated with the adaptation of receptor-to-ligand may be estimated from the "mean envelope" → individual envelope" rms shift [319].

Vedani et al. [320] used this QSAR methodology (5D QSAR) for the design of novel compounds to inhibit the chemokine receptor-3 (CCR3) in the low-nanomolar range. The study was based on 141 compounds, representing four different substance classes. They built two receptor surrogates that yielded a cross-validated r^2 of 0.950 and 0.861 (106 training ligands) and predictive r^2 of 0.877 and 0.798 (for 35 test ligands), respectively. The model was then employed to predict the activity of 58 hypothetical compounds featuring two variation patterns: lipophilic substitutions and amphiphilic H-bond acceptors. The most potent molecule was predicted to bind with an $\mathrm{IC_{50}}$ of 0.3 nM.

A molecular modeling study was also carried out by Ducki et al. [321] to develop predictive 5D QSAR models for combretastatin-like analogs populating the colchicines binding site of β-tubulin. A series of compounds was selected that includes two aromatic groups linked by various moieties such as alkenes (stilbenes), enones (chalcones), or ethers. The 5D QSAR model was developed in stepwise manner. First a model was generated for the chalcone series (training/test set 14/5 compounds, $r^2 = 0.982$, $q^2 = 0.980$, and $r^2_{\mathrm{pred}} = 0.711$), then for the stilbene series (training/test set = 12/6 compounds, $r^2 = 0.964$, $q^2 = 0.960$, and $r^2_{\mathrm{pred}} = 0.510$), and finally for the combined data set (training/test set = 36/11 compounds, $r^2 = 0.920$, $q^2 = 0.917$, and $r^2_{\mathrm{pred}} = 0.460$). Although the models for the chalcone and stilbene series appeared slightly different when represented by QSAR colored surfaces, the combined model represents a highly predictive model for compounds that bind to the colchicine binding site of tubulin.

The most recent extension of the multidimensional QSAR concept is the six-dimensional (6D QSAR). This methodology was introduced in 2005 by Vedani et al. [322], which allows for the simultaneous consideration of different solvation scenarios. This can be achieved by mapping parts of the surface area with solvent properties whereby position and size are optimized by the genetic algorithm. The solvation terms (ligand desolvation and solvent stripping) can be scaled independently for each different model within the surrogate family, reflecting varying solvent accessibility of the binding pocket. In 6D QSAR, the fourth dimension represents each molecule by an ensemble of conformations, orientations, protonation states, and tautomers; the fifth dimension refers to an ensemble of different induced-fit models by adapting the mean van der Waals surface, generated about all ligands defining the training set; and the sixth dimension allows for the evaluation of different solvation models [313].

Vedani et al. [300] developed 6D QSAR model on a series of 106 diverse agonists (training/test set = 88/18 agonists) for their binding to the estrogen receptor. The model gave a cross-validated r^2 of 0.903 and a predictive r^2 of 0.885. Vedani et al. [323] also reported the 6D QSAR studies for peroxisome proliferator-activated receptor γ (PPAR γ). Its receptor surrogate was based on the experimental structure of the protein and 95 compounds (89 tyrosine-based compounds and 6 thiazolidinediones) with K_i values ranging from 6.9×10^{-10} to 2.0×10^{-6} M. The simulation reached a cross-validated $r^2 = 0.832$ (75 training ligands) and yielded a predictive $r^2 = 0.723$ (20 test compounds).

5. OUTLIERS IN QSAR

Outliers are those compounds that have unexpected biological activities and are unable to fit in a QSAR model. They are valuable in defining the limitations under which compounds act by a common molecular mechanism modeled by one or more descriptors, and also in defining the experimental limitations of the biological test data. Outliers, in the QSAR modeling, can be associated with one of the following reasons: (i) experimental errors in the primary data, (ii) molecules may act by different mechanisms, (iii) multiple binding modes of structurally similar ligands, (iv) special effects of some functional groups on structurally similar compounds, (v) steric hindrance toward the reactive centers, (vi) unexpected hydrogen bonding, (vii) ionization of the functional groups, (viii) members of the data set may have different rate of metabolism, (ix) used parameters may not be the best, and (x) intrinsic noise associated with either the original data or the methodological aspects involved in the development of a QSAR model. Thus, the outliers may provide new directions or opportunities in drug discovery research and need to be given special consideration [324–328].

There are mainly two types of outliers: leverage outliers and activity outliers. The leverage outliers are dissimilar from all other compounds in the data set. They can be found by using a sphere-exclusion algorithm. In contrast, the activity outliers are similar to other compounds in the data set, but their activities are significantly different from those of their nearest neighbors. These outliers are of particular interest in QSAR studies and can be addressed as "activity cliffs" in descriptor space [329]. An activity cliff is defined by the ratio of the difference in activities of two compounds to their "distance" of separation in a given chemical space. The activity cliffs can have the following implications in QSAR modeling: (a) linear models are unlikely to satisfactorily account for activity landscapes with significant numbers of cliffs, (b) outliers in the data set may reflect the presence of activity cliffs that may not be due to the statistical fluctuations or measurement errors, and (c) the presence of activity cliffs may requires the assay of additional compounds in the neighborhoods around these cliffs to further ensure that the activity landscapes are adequately represented. Thus, the detection of true outliers that can lead to dramatic changes in the nature of the activity cliffs is not to be underestimated [330].

There are several available methods for the detection of outliers in the development of QSAR models.

(i) *The Mahalanobis Distance* [331]: In this method, the distance between each test set compound and the centroid of the training set compounds is computed, and compared to a cutoff value of χ^2 distribution. Any compound with a distance larger than the value given by the appropriate χ^2 distribution is considered as an outlier.

(ii) *The Smallest Half-Volume Method* [332]: In the smallest half-volume (SHV) method, the distances between each pair of the observations in the multivariate space are considered. In each observation, the first $n/2$ smallest distances are summed and the $n/2$ observations with the smallest sum are taken as a clean subset. The distributions of the Mahalanobis distances for all n observations toward this clean subset are obtained. The compounds are considered to be outliers by comparing with their χ^2 distribution.

(iii) *X-Residual* [332]: This is a simple indication of an outlier, if the total residual standard deviation of the test set is much larger than that of the training set (three times larger as recommended).

(iv) *The Potential Functions* [331]: Each data point of the training set creates a "potential" in the space in such a way that the value of the potential becomes maximum at a particular location, and decreases continuously with distance from that point. By averaging all individual potentials from the training set, a global potential can be generated at any place in the space. Thus, at any location in the space, a global value of the potential induced by the training set compounds exists. If the potential value of a compound is 0, it means that this compound is "far" from the bulk of the training set, and therefore, that compound is an outlier.

(v) *The Uncertainty Method*: Høy et al. [333] developed Equation 86 to estimate the prediction uncertainty for an object i:

$$\hat{\sigma}^2_{yi,\text{pred}} = V_{y,\text{val}}\left(1-\frac{A+1}{I}\right) \times \left(h_{i,\text{pred}} + \frac{V_{xi,\text{pred}}}{V_{X\text{tot,val}}} + \frac{2}{I}\right) \quad (86)$$

where $\hat{\sigma}^2_{yi,\text{pred}}$ is the estimated variance of the predicted y_i-value, I is the number of compounds in the training set, $V_{xi,\text{pred}}$ is the X-residual variance of prediction object i, $V_{y,\text{val}}$ is the Y-residual variance in a validation data set, $V_{X\text{tot,val}}$ is the total residual of the validation data set, and $h_{i,\text{pred}}$ is the leverage of the prediction object i with respect to the PLS factor A. The high value of the estimated variance, $\hat{\sigma}^2_{yi,\text{pred}}$, suggests that the compound is not well predicted and it can be treated as an outlier.

(vi) *The Convex Hull Method* [334]: In this method, the convex hull is constructed in two dimensions by computing the most distant object from the centroid. The first face for the convex hull is defined by the line that joins this object with another object, chosen in such a way that the rest of the points are located on the same side of the line as the gravity center. This process is then repeated for all the extreme objects till the closing of the boundary. Objects outside the hull are considered to be outliers.

(vii) *The R-NN Curves* [335]: The first step of this method is to evaluate the R-NN curve for a given descriptor space that determines the maximum pair-wise distance in the data set (Dmax). In the next step, the first observation in the data set is considered to determine the number of neighbors that lie within a sphere of radius, R, centered on it. This step is then repeated for different R. Thus, a count of nearest neighbors will be generated for a given observation and within a set of specific radii. This procedure is repeated for all the molecules present in the data set. In general, a molecule will have few neighbors at lower radii, whereas for higher radii, the number of neighbors will always increase. The visual inspection of R-NN curves can be utilized to distinguish between molecules in dense and sparse regions of a descriptor space. However, for larger data sets visual inspection can become cumbersome. Thus, a numerical characterization of R-NN curves would be a more efficient approach to determine an outlier. The numerical characterization of R-NN curve for each molecule can be written as Equation 87:

$$N_N = a \times \frac{1 + me^{-R/\tau}}{1 + ne^{-R/\tau}} \qquad (87)$$

where N_N is the number of neighbors within a radius R and a, m, n, and τ are the parameters of the fit. Thus, we can avoid the actual nonlinear fitting step by simply determining the slope of Equation 87 at varying R. This can be simplified by considering the value of R for which the slope of the curve is maximal. This is further simplified since the slope can be calculated numerically by Equation 88:

$$S_R = N_{N,R} - N_{N,R-1} \qquad (88)$$

In this equation, S_R is the slope of the R-NN curve at a radius R and $N_{N,R}$ is the number of neighbors at that radius. Thus for each molecule, we can evaluate S_R for $2 \leq R \leq 100$ and determine the radius, $R_{\max(S)}$, at the maximum value of S_R. Plotting the values of $R_{\max(S)}$ will allow us the identification of outliers in a data set.

(viii) *The ODC Method* [336]: The ODC (outlier detection by distance toward training set compounds) method calculates the euclidean distance between each test set compound and its closest training set compound. In addition to this, all the nearest neighbor distances within the training set compounds are also computed. The 90th percentile of these distances is then used as a cutoff for defining outliers.

(ix) *The Difference Between r^2 and q^2* [337]: The difference between r^2 and q^2 ($r^2 - q^2$) will not exceed by 0.3. A substantially large difference indicates the presence of outliers in the data set.

(x) *The Standard Deviation (s) Test* [338]: Compounds are considered to be outliers on the basis of their deviation between observed and predicted activities from the equation (obsd − pred > $2s$)

After detection of outliers, it is necessary to eliminate those compounds for substantial improvement of prediction errors in the training set. There is a valid reason for outlier removal because it is normally assumed that all the compounds in a data set operate by a similar mechanism but the atypical behavior of an outlier raises questions about steric hindrance of reactive centers (or functional groups required for receptor binding), unexpected hydrogen bonding and/or ionization. Removal of significant outliers will allow for the development of rigorous and more significant QSAR models. To assess the effects of excluding outliers, QSAR models must be examined before and after the removal of outliers [339]. An example for the exclusion of statistical outliers was provided by Verma et al. [340] in the development of Equation 89 for the cytotoxicity of 4-X-thiophenols to L1210 cell line:

$$\log\left(\frac{1}{C}\right) = -0.93(\pm 0.18)\sigma^+ + 0.86(\pm 0.24)I_H + 3.99(\pm 0.13) \qquad (89)$$

$$n = 23 \; r^2 = 0.852 \; s = 0.168 \; q^2 = 0.793$$

$$\text{outliers}: X = 4\text{-NO}_2,\ 4\text{-CH(CH}_3)_2,\ \text{and } 4\text{-OC}_6\text{H}_5$$

In QSAR 89, C represents the ID_{50}, the concentration of the substituted thiophenol that induces 50% inhibition of growth in the L1210 cells after 48 h. σ^+ is the electronic parameter, and I_H is an indicator variable, which acquires a value of 1 when the substituent is a halogen (e.g., Cl, Br, I) or pseudo-halogen (e.g., CF3) and is set at 0 for all other substituents. Three data points were considered to be outliers and excluded from the analysis since their deviations were greater than twice the standard deviation of the regression line. The 4-isopropyl analog was twice as active as predicted. It may be due to the isopropyl group, which is particularly labile in terms of hydrogen abstraction and subsequent radical formation. The 4-nitro analog was nine times more active than predicted. This unusual behavior could be attributed to the formation of the active nitro anion radical and its subsequent reduction to a nitrosobenzene and/or phenylhydroxylamine. The other statistical outlier (4-phenoxy analog) was overpredicted by a factor of 10. Possible reasons for its unusually low activity are its bulk and/or geometry that may reduce coplanarity with the thiophenyl ring and minimize interactions with a critical receptor.

Another good example involves permeability coefficient (K_p) data for 114 compounds across excised human skin *in vitro*, reported initially by Kirchner et al. [341]. The complete data set of 114 compounds was used by Cronin et al. [342] in the development of QSAR 90, which is based on $\log P$ and molecular weight:

$$\log K_p = 0.612 \log P - 0.007\,\mathrm{MW} - 2.47 \quad (90)$$

$$n = 114,\ r = 0.816,\ s = 0.708,\ F = 110$$

Later on, the authors [342] improved the correlation by pinpointing seven significant outliers on the basis of their residuals. See QSAR 91:

$$\log K_p = 0.772 \log P - 0.010\,\mathrm{MW} - 2.33 \quad (91)$$

$$n = 107,\ r = 0.927,\ s = 0.394,\ F = 317$$

6. VALIDATION OF QSAR

QSAR model validation is essential in the development of statistically robust models, because the real utility of a QSAR model is in its ability to accurately predict the modeled property for new compounds not present in the data set. Validation is comprised of three steps: (a) statistical diagnostics, (b) internal validation, and (c) external validation.

(a) *Statistical Diagnostics*: It is extremely important to ascertain that a QSAR model is of sufficiently high quality, and is worth validating for practical purposes. A QSAR model may be acceptable for validation only if the following criteria are satisfied by the model under consideration:

(i) *n/p Ratio:* $n/p \geq 4$ or $n \geq 4p$, where n is the number of data points and p is the number of descriptors used in the QSAR model [343].

(ii) *Fraction of the Variance (r^2)*: It is important to note that a QSAR model must have to explain a sufficiently high fraction of the variance for any data set. The fraction of the variance is expressed by the coefficient of determination, r^2 (measure of the goodness of fit between model-predicted and experimental values). It is believed that the closer the value of r^2 to unity, the better the QSAR model. The value of r^2 may vary 0 to 1, where 1 reflects a perfect model explaining 100% of the variance in the data, and 0 means a model without any explanatory power at all. According to the accepted guidelines a QSAR model must have $r^2 > 0.6$ for its predictive ability and is then acceptable for validation [344–346].

(iii) *Cross-Validation Test (q^2)*: According to the literature, a QSAR model must have $q^2 > 0.5$ for its predictive ability and acceptability for validation [344–346]. q^2 is the cross-validated r^2 (a measure of the quality of the QSAR model), and is usually obtained by using a leave-one-out (LOO) procedure [347].

(iv) *$r^2 - q^2 < 0.3$*: This difference ought not to exceed 0.3. A substantially large difference indicates either of the following: (a) overfitting model, (b) presence of irrelevant descriptors, or (c) presence of outliers in the data [337].

(v) *Standard Deviation (s)*: The smaller the value of s, the better the QSAR model. The value of $s \leq 0.3$ is usually acceptable, although it may vary depending on the error in the biological endpoint.

(vi) *QUIK Rule*: The QUIK rule states that the only models with K_{XY} correlation among the [X + Y]—variables greater than the K_X correlation among the [X]—variables can be considered, or if $K_{XY} - K_X < \delta K$ the model will be rejected [348]. This rule has been demonstrated to be very effective in avoiding models with multicollinearity without predicting power.

(vii) *Quality Factor (Q)*: Q is the quality factor, for which $Q = r/s$ (where, r is the correlation coefficient and s is the standard deviation). Chance correlation, due to the excessive number of descriptors, can be detected by the examination of the Q value. High values of Q indicate not only the high predictive power of the QSAR models but also their lack of "overfitting" [349].

(viii) *Fischer Statistics (F)*: F represents the Fischer statistics, for which $F = fr^2/[(1 - r^2)m]$, where f is the number of degrees of freedom [$f = n - (m + 1)$], n is the number of data points, and m is the number of variables. Since the F value is actually the ratio between explained and unexplained variance for a given number of degrees of freedom, it must be comparable to the literature F value at 95% (or 99%) level for the QSAR model [350]. The model is considered to be optimal when Q reached a maximum together with F, even if slightly nonoptimal F values may also be accepted. A significant decrease in F with one additional variable could mean that the new descriptor is not as significant as expected; i.e., its introduction has endangered to the statistical quality of combination. However, the statistical quality could be improved by the addition of a more convincing descriptor [349].

(ix) $0.85 \leq k \leq 1.15$ or $0.85 \leq k' \leq 1.15$, where k and k' are the slopes of the regression lines forced through zero, relating observed versus predicted and predicted versus observed values, and calculated as follows: $k = \frac{\sum y_i \tilde{y}_i}{\sum \tilde{y}_i^2}$, and $k' = \frac{\sum y_i \tilde{y}_i}{\sum y_i^2}$, where y_i and $\tilde{y}_i$ are the observed and predicted activities, respectively [345].

(x) $\frac{r^2 - r_0^2}{r^2} < 0.1$ or $\frac{r^2 - r\prime_0^2}{r^2} < 0.1$, where r_0^2 and $r\prime_0^2$ are the coefficient of determination characterizing linear regression with Y-intercept set at zero, the first related to observed versus predicted values, the second associated with predicted versus observed values, and calculated as follows:

$$r_0^2 = 1 - \frac{\sum (\tilde{y}_i - y_i^{r0})^2}{\sum (\tilde{y}_i - \hat{y})^2} \quad \text{and}$$

$$r'^2_0 = 1 - \frac{\sum (y_i - \tilde{y}_i^{r0})^2}{\sum (y_i - \bar{y})^2} \tag{92}$$

$$y_i^{r0} = k\tilde{y}_i \quad \text{and} \quad \tilde{y}_i^{r0} = k' y_i \tag{93}$$

Where y_i and $\tilde{y}_i$ are the observed and predicted activities, respectively, and $\bar{y}$ and $\hat{y}$ are the average values of the observed and predicted activities, respectively [344,345].

(xi) $r_0^2 - r\prime_0^2 < 0.3$ as suggested by Zhang et al. [346].

(b) *Internal Validation*: There are a number of methods for internal validation of QSAR models. Some of the important ones are described herein.

(i) *Cross-Validated $r^2(q^2)$*: Cross-validations (CVs) are the most commonly used statistical techniques for internal validation, in which different proportions of compounds are removed from the training set used for model development. New models are then developed in order to verify internal predictive ability, for example, q^2_{LOO} (leave-one-out), q^2_{LFO} (leave-five-out), and q^2_{LMO}

(leave-many-out), etc. Cross-validated correlation coefficient r^2 (q^2_{LOO} or q^2_{LFO} or q^2_{LMO}) is calculated according to the following formula:

$$q^2 = 1 - \frac{\sum (Y_{obs} - Y_{pred})^2}{\sum (Y_{obs} - \hat{Y})^2} \quad (94)$$

where Y_{obs}, Y_{pred}, and $\hat{y}$ are, respectively, the observed, predicted, and averaged activities. The value of q^2 may be q^2_{LOO}, or q^2_{LFO} or q^2_{LMO}. In the case of leave-one-out cross-validation, each member of the data point in turn is removed, and the remaining n–1 members are used in model development. On the other hand, five members of the data point in turn are removed in the case of leave-five-out cross-validation, and the remaining n–5 members are used in model development. Similarly, a certain data points are removed in the case of leave-many-out cross-validation technique. Although a low value of q^2 in the LOO, LFO, or LMO test typically indicates low predictivity in a model, high q^2 does not necessarily imply high predictive power of the model. Nevertheless, the value of cross-validated r^2 (q^2_{LOO} or q^2_{LFO}or q^2_{LMO}) is frequently used as a criterion of both robustness and predictive ability of the model. Often, a high value of q^2 ($q^2 > 0.5$) is considered as an ultimate proof for the high predictive power of the model [344–346]. This method has also been criticized by a number of authors mainly due to its two major limitations: (a) if one of the compounds essentially determines one of the model parameters, then it probably cannot be predicted well from the other compounds, and so its squared prediction error is likely to be large, and (b) it does not sufficiently penalize overfitting [351]. Recently, it has been suggested that while high q^2 is the necessary condition for a QSAR model to have a high predictive power, it is not a sufficient condition [344].

(ii) *Bootstrapping*: It is another approach to internal validation where the data point are randomly selected from the data set. In this validation, K groups of n objects are generated by repeating random selection from the original data set. Some objects can be included several times in the same random sample, while other objects may never be selected. The new QSAR model is developed on the n randomly selected objects, which is then used to predict the target properties of the excluded sample. A high average q^2 in the bootstrap validation is an indication toward the robustness of the model [344,352].

(iii) *Asymptotic q^2rule*: Asymptotic q^2 (q^2_{ASYM}) is calculated according to the following formula:

$$q^2_{ASYM} = 1-(1-R^2)\left(\frac{n}{n-p}\right)^2 \quad (95)$$

where n is the number of data points and p is the number of descriptors. The asymptotic q^2 rule is based on the comparison of the asymptotic and the actual q^2 values of the model: if $q^2_{LOO} - q^2_{ASYM} < \delta q \rightarrow$ the model will be rejected. In this rule, it has been assumed that a model with an actual q^2_{LOO}value lower than the asymptotic value of a quantity δq does not guarantee its future predictive ability. The simplest threshold value of $\delta q = 0$, but a more conservative value could be –0.005, while a less conservative value could be 0.005 [353].

(iv) *Y-Randomization Test*: This is a widely used test method that ensures the robustness of a QSAR model. In this test, the dependent-variable vector is shuffled randomly, and a new QSAR model is developed using the original (unchanged) independent variable matrix. The process is repeated several times and is deemed successful when the resulting QSAR models emerge with low r^2 and q^2 values [354].

(c) *External Validation*: The external validation of QSAR models constitutes mainly of two steps: (i) Splitting of a parent data set into training and test sets, and (ii) predictive power of QSAR models.

(i) *Splitting of a Parent Data Set into Training and Test Sets*: In typical situations, locating new experimental data for validation purposes is usually difficult. Therefore, the available data set is sequestered into training and test sets, which are then used for establishing the QSAR model and external validation, respectively. There are a number of methods for splitting the original data set into training and test sets. Some important methods among them are as follows:

(I) *Random Selection*: This is a very simple approach that constitutes one of the most widely used methods for dividing a data set into training and test sets by mere random selection [355,356].

(II) *Kohonen's Self-Organizing Neural Network (KohNN)*: KohNN has the special property of creating an organized internal representation of various features of input signals and their abstractions [357]. In this method, the neurons are arranged in a two-dimensional (2D) array map to generate a 2D feature map so that the similarity is preserved in the data. In other words, if two input data vectors are similar, they will be mapped into the same neuron or closely together in the 2D map. The data set is then split into training and test sets with the generated feature map. The KohNN division is certainly more sophisticated and superior to random selection [358].

(III) *K-Means Clustering*: Cluster analysis is an alternative rational technique that offers more specific control by assigning every single structure to a group of compounds. *K*-means clustering is a method of splitting the data set where the final cluster member for each case is expressed. It must be supplied with the number of cluster (*K*) for which the data is grouped [359]. The main idea consists of clustering a compound series into several statistically representative classes of compounds and then the data is split between *K* clusters. In this method, one can examine the means for each cluster on each dimension to assess how distinct the *K* clusters are [359,360].

(IV) *Statistical Molecular Design (SMD)*: SMD is an efficient technique for selecting a representative and diverse training set of compounds in QSAR, combinatorial technologies and other areas of research depending on optimization of molecular properties. This method uses a large number of compounds for which the *Y* variable does not need to be measured. Molecular descriptors are calculated/measured for all the compounds and then principal component analysis (PCA) is performed. The PCs with combinations of molecular properties explain the variation among the molecules in an optimal way and refer to them as the principal properties (PP) of the data set. The subset selection of the compounds is then performed with respect to PP, which is the most efficient in spanning the compound space and thus it is the best selection method of the training set for a QSAR model [361,362].

(V) *D-Optimal Onion Designs (DOODs)*: The DOOD is a method for selecting a training set of reasonable size, which is representative for the chemical property space defined by the molecular structures. In this approach, the data set is divided into a number of subsets (shells or layers), and a D-optimal selection is performed from each shell. This makes it possible to select representative sets of molecular structures throughout any property space, for example, the physicochemical space, with reasonable design sizes. The total number of selected compounds is then easily controlled by varying (i) the number of shells and (ii) the model for which the design is based [363].

(VI) *Kennard– Stone(KS) Selection*: Kennard– Stone selection method takes the compounds into the training set based on their Euclidean distance from the already selected compounds. The KS method adopts a max–min approach to the selection. The distance between all samples is calculated from the descriptor values and the first two data points selected as the members of the training set are those that are furthest apart. The compound, which is selected as the next one to join the training set, is the one that has the largest minimal distance to all previously selected objects. The technique repeats the selection by comparing all distances of compounds to the already selected objects until a certain predefined number of objects have been found [355,364].

(VII) *Sphere Exclusion*: In this method, a compound with the highest activity is selected for the training set. A sphere with the center in the representative point of this compound with radius $R = c(V/N)^{1/K}$ is constructed, where K, c, N, and V are, respectively, the number of descriptors, the dissimilarity level, the total number of compounds in the data set, and the total volume occupied by the representative points of the compounds. Dissimilarity level is varied to construct different training and test sets. Compounds corresponding to the representative points within this sphere (except for the center of it) are included in the test set. All the points within this sphere from the initial set of compounds are excluded. Let n be the number of remaining compounds and if $n = 0$, the splitting process will be stopped [365].

(VIII) *Activity Ranking*: In this method, only activities of the compounds are used to divide a data set into training and test sets. A specified number of the most active compounds are included into the first group, and the same number of the next most active compounds into the second group, etc. Of course, the last group of compounds may be smaller than the specified size. Now, a number of compounds in each group are specified, which will be included into the training set. This number of most active compounds of each group is selected for the training set, and the remaining compounds for the test set [365].

(ii) *Predictive Power of QSAR Models*: The true predictive power of a QSAR model is estimated by comparing the predicted and observed activities of the test set compounds that are not used in the model development. The predictive power of a QSAR model can be estimated by their predictive value of R^2, R^2_{pred}, which is calculated by Equation 96:

$$R^2_{\text{pred}} = 1 - \frac{\sum (Y_{\text{pred(test)}} - Y_{\text{test}})^2}{\sum (Y_{\text{test}} - \bar{Y}_{\text{training}})^2} \tag{96}$$

where $Y_{\text{pred(test)}}$ and Y_{test} are the respective predicted and observed activities of the test set compounds and $\bar{Y}_{\text{training}}$ is the observed mean activity of the training set compounds [344,356].

7. APPLICATIONS OF QSAR

Over the past 40 years, the glut in scientific information has resulted in the development of thousands of equations pertaining to structure–activity relationships in biological systems. In its original definition, the Hansch equation was defined to model drug–receptor interactions involving electronic, steric and hydrophobic contributions. Nonlinear relationships helped to refine this approach in cellular systems and organisms where pharmacokinetic constraints had to be considered and tackled. They have also found increased utility in addressing the complex QSAR of some receptor–ligand interactions. In many cases the Kubinyi bilinear model has provided a sophisticated approach to delineation of steric effects in such interactions. Examples of ligand–receptor interactions will be drawn from receptors such as the much-studied DHFR, α-chymotrypsin and 5-α reductase [366–368].

7.1. Isolated Receptor Interactions

The critical role of DHFR in protein, purine, and pyrimidine synthesis, the availability of crystal structures of binary and ternary complexes of the enzyme and the advent of molecular graphics combined to make DHFR an attractive target for well-designed heterocyclic ligands generally incorporating a 2,4-diamino-1,3-diazapharmacophore [369]. The earliest study focused on the inhibition of DHFR by 4,6-diamino-1,2-dihydro-2,2-dimethyl-1*R*-*s*-triazines (**II**; Fig. 4) [366].

Inhibition of crude pigeon liver DHFR by triazines [370]

$$\log\left(\frac{1}{\text{IC}_{50}}\right) = 2.21(\pm 1.00)\pi - 0.28(\pm 0.17)\pi^2 + 0.84(\pm 0.76)D + 2.58(\pm 1.30)$$

$$n = 15,\ r^2 = 0.861,\ s = 0.553,\ \pi_o = 4(3.6-6.0) \tag{97}$$

In all equations, n is the number of data points, r^2 is the square of the correlation coefficient, s represents the standard deviation and the figures in parentheses are for construction of the 95% confidence intervals, π represents the hydrophobicity of the substituent R, and π_o is the optimum hydrophobic contribution of the R substituent. D is an indicator variable that acquires a value of 1.0 when a phenyl ring is present on the nitrogen

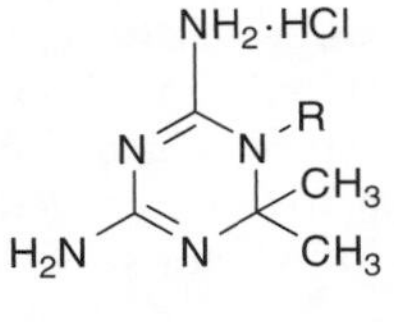

Figure 4. 4,6-Diamino-1,2-dihydro-2,2-dimethyl-1*R*-*s*-triazines (**II**).

and a value of zero for all other R. This is an example of a Hansch–Fujita–Ban analysis where the indicator variable D establishes the contribution and hence importance of a phenyl ring in DHFR inhibition. This equation has some limitations. Improper choice of N-substituents led to a high degree of collinearity between size and hydrophobicity and in terms of electronic contributions, spanned space was limited and thus inadequate. A subsequent study on the binding of these compounds to DHFR isolated from chicken liver was more revealing.

Inhibition of chicken liver DHFR by 3-X-triazines [371]

$$\log\left(\frac{1}{K_i}\right) = 1.01(\pm 0.14)\pi' - 1.16(\pm 0.19) \times \log(\beta \times 10^{\pi'} + 1) + 0.86(\pm 0.57) \times \sigma + 6.33(\pm 0.14) \quad (98)$$

$$n = 59,\ r^2 = 0.821,\ s = 0.906,$$

$$\pi'_{\mathrm{o}} = 1.89(\pm 0.36),\ \log\beta = -1.08$$

In this example, the R group on the 2-nitrogen was restricted to an (3-X-phenyl) aromatic ring. Accurate K_{i} values were obtained from highly purified DHFR isolated from chicken liver. In most cases, π' represented the hydrophobicity of the substituent except in certain instances where $X = -OR$ or $-CH_2ZC_6H_4$-Y. It was ascertained that alkoxy substituents were not making direct hydrophobic contact with the enzyme since their inhibitory activities were essentially constant from the methoxy to the nonyloxy substituent. In the bridged substituents where $Z = O$, NH, S, Se, the Y substituent again did not contact the enzyme surface. Variation in Y led to the same, constant biological activity. The coefficient with π', suggests that the substituent is engulfed in a hydrophobic pocket that has an optimal π'_{o} of 2. This value is consistent with that seen in the crude pigeon liver DHFR corrected for the presence of the phenyl group $(4.0 - 2.0 = 2)$. The 0.86 rho value (coefficient with σ) suggests that there could be a dipolar interaction between the electron deficient phenyl ring and a region of positively charged electrostatic potential in the enzyme—perhaps an arginine, lysine, or histidine residue. Hathaway et al. developed a QSAR for the inhibition of human DHFR by 3-X-triazines and obtained equation 99 [372].

Inhibition of human DHFR by 3-X-triazines [372]

$$\log\left(\frac{1}{K_i}\right) = 1.07(\pm 0.23)\pi' - 1.10(\pm 0.26) \times \log(\beta \times 10^{\pi'} + 1) + 0.50(\pm 0.19)I + 0.82(\pm 0.66)\sigma + 6.07(\pm 0.21) \quad (99)$$

$$n = 60,\ r^2 = 0.792,\ s = 0.308,$$

$$\pi'_{\mathrm{o}} = 2.0(\pm 0.87),$$

$$\log\beta = -0.577$$

The enhanced activity of the "bridged" substituents was corrected by the indicator variable I. Triazines bearing the bridge moieties $-CH_2NHC_6H_4Y$, $-CH_2OC_6H_4Y$, and $-CH_2SC_6H_4Y$ had unusually high enzyme binding activity. It was noted that the $-CH_2NHC_6H_5$ bridge was present in the endogenous substrate, folic acid.

The bilinear dependence on hydrophobicity of the substituents parallels that seen in the case of chicken liver DHFR. A similar QSAR 100 was obtained for DHFR isolated from L1210 murine leukemia cells [373].

Inhibition of L1210 DHFR by 3-X-triazines [373]

$$\log\left(\frac{1}{K_i}\right) = 0.98(\pm 0.14)\pi' - 1.14(\pm 0.20) \times \log(\beta \times 10^{\pi'} + 1) + 0.79(\pm 0.57)\sigma + 6.12(\pm 0.14) \quad (100)$$

$$n = 58,\ r^2 = 0.810,\ s = 0.264,$$

$$\pi'_{\mathrm{o}} = 1.76(\pm 0.28),\ \log\beta = -0.979$$

The consistency in these models, versus prokaryotic DHFR is established by the coefficient with the hydrophobic term, the optimum π' value and the rho value. These numerical coefficients can be contrasted sharply with those obtained from fungal and protozoal DHFR. Inhibition constants were determined for 3-X-triazines versus *Pneumocystis carinii* DHFR [374].

Inhibition of P. carinii DHFR by 3-X-triazines [374]

$$\log\left(\frac{1}{K_i}\right) = 0.73(\pm 0.12)\pi' - 1.36(\pm 0.35) \times \log(\beta \times 10^{\pi'} + 1) - 0.78(\pm 0.42)I_{OR} + 0.28(\pm 0.21)MR_Y + 6.48(\pm 0.23) \quad (101)$$

$$n = 43,\ r^2 = 0.840,\ s = 0.435,$$
$$\pi'_o = 3.99(\pm 0.68),\ \log\beta = -3.925$$

In Equation 101, I_{OR} is an indicator variable that assumes a value of 1 when an alkoxy substituent is present and 0 for all other substituents. It is of interest to note that the Y-substituent on the second phenyl ring now contributes to activity. The MR_Y term suggests that it most probably accesses a polar region of the active site of the enzyme. The positive coefficient with MR_Y suggests that an increase in bulk and/or polarizability enhances binding. The descending slope of the bilinear equation is much steeper ($1.36 - 0.73 = 0.63$) than that seen with the mammalian and avian enzymes.

A similar model 102 is obtained versus the bifunctional protozoal DHFR from *Leishmania major*, which is coupled to thymidylate synthase [375].

Inhibition of L. major DHFR by 3-X-triazines [375]

$$\log\left(\frac{1}{K_i}\right) = 0.65(\pm 0.08)\pi' - 1.22(\pm 0.29) \times \log(\beta \times 10^{\pi'} + 1) - 1.12(\pm 0.29)I_{OR} + 0.58(\pm 0.16)MR_Y + 5.05(\pm 0.16) \quad (102)$$

$$n = 41,\ r^2 = 0.931,\ s = 0.298,$$
$$\pi'_o = 4.54,\ \log\beta = -4.491$$

QSAR analysis on a limited set of 3-X-triazines assayed by Chio and Queener versus *Toxoplasmosis gondii* led to the formulation of Equation 103 [366,376].

Inhibition of T. gondii DHFR by 3-X-triazines

$$\log\left(\frac{1}{IC_{50}}\right) = 0.39(\pm 0.20)\pi' - 0.43(\pm 0.19)MR_Y + 6.65(\pm 0.30) \quad (103)$$

$$n = 17,\ r^2 = 0.810,\ s = 0.289$$

A quick comparison of QSAR 98–102 reveals the strong similarity between the avian and mammalian models. In fact, because of its increased stability, chicken liver DHFR has often been used as a surrogate for human DHFR in enzyme-inhibition studies. The intercepts, coefficients with π' and optimum π'_o for avian (6.33, 1.01, 1.9), human (6.07, 1.07, 2.0) and mouse leukemia (6.12, 0.98, 1.76) can be compared to the corresponding values for *P. carinii* (6.48, 0.73, 3.99) and *L. major* (5.05, 0.65, 4.54). QSAR 97 and 103 are not included in the comparison because crude pigeon enzyme was used in the former and the testing for QSAR 103 was conducted under different assay conditions; K_i values were not determined. A noteworthy difference between these models is the wide disparity in π_o values. The binding site of the protozoal and fungal species comprises an extensive hydrophobic surface unlike the constrained pockets in the mammalian and avian enzymes. The positive coefficients with the MR_Y terms suggest that added bulk on the bridged phenyl ring enhances inhibitory potency. The study versus *T. gondii* DHFR (QSAR 103) included a number of mostly small, polar substituents (NH_2, NO_2, $CONMe_2$) on the bridged phenyl and their activities were considerably lower than the unsubstituted analog. Comparative QSAR can be useful, particularly if the biological data is consistent (tested under the same assay conditions, excellent purity of enzymes, substrates, inhibitors, and buffers) and the choice of substituents is appropriate.

One of the major problems that crops up with some QSAR studies is extrapolation from beyond spanned space. Predictive ability is only sound when one has probed an adequate range in electronic, hydrophobic, and steric space. At the onset of the study, the training set should address these concerns. Lack of adequate attention to diversity in the training set can result in QSAR models that are misleading. When examined on its own, such a model may appear to withstand statistical rigor and apparent transparency but on being subjected to lateral validation, loopholes emerge. A brief but illustrative example is used to address what appear to be discrepancies between the four equations but is in rea-

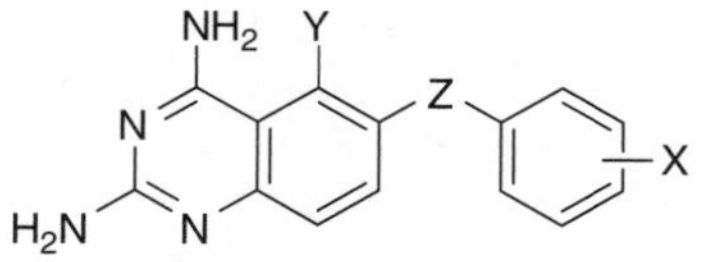

Figure 5. 2,4-Diamino, 5-Y, 6-Z-quinazolines (**III**).

lity quite consistent except for focusing on "different parts of the same elephant" and trying to determine what it is.

Four different QSAR were derived for the inhibition of DHFR from rat liver, human leukemia, mouse L1210 and bovine liver by 2,4-diamino, 5-Y, 6-Z-quinazolines (**III**; Fig. 5) [366,377–379]. A comparison of their QSAR presents an interesting study on the importance of spanned space in delineating enzyme–receptor interactions.

*Inhibition of rat liver DHFR by 2,4-diamino, 5-Y, 6-Z-quinazolines (**III**)* [377]

$$\log\left(\frac{1}{\mathrm{I}_{50}}\right) = 0.78(\pm 0.12)\pi_5 + 0.81(\pm 0.12) \times \mathrm{MR}_6 - 0.06(\pm 0.02)\mathrm{MR}_6^2 - 0.73(\pm 0.49) \times I_1 - 2.15(\pm 0.38)I_2 - 0.54(\pm 0.21)I_3 - 1.40 \times (\pm 0.41)I_4 + 0.78(\pm 0.37)I_6 - 0.20(\pm 0.12) \times \mathrm{MR}_6 \times I_1 + 4.92(\pm 0.23) \quad (104)$$

$n = 101,\ r^2 = 0.924,\ s = 0.441,$
$\mathrm{MR}_{6,\mathrm{o}} = 6.4(\pm 0.8)$

*Inhibition of human liver DHFR by 2,4-diamino, 5-Y, 6-Z-quinazolines (**III**)* [378]

$$\log\left(\frac{1}{K_i}\right) = -2.87(\pm 0.16)I_1 + 0.29(\pm 0.14) \times I_2 - 0.38(\pm 0.11)\mathrm{MR}_6 - 0.29(\pm 0.06)\pi_{\mathrm{R}} - 0.19(\pm 0.07)\mathrm{MR}_R + 10.12(\pm 0.45) \quad (105)$$

$n = 47,\ r^2 = 0.914,\ s = 0.420$

*Inhibition of murine L1210 DHFR by 2,4-diamino, 5-Y, 6-Z-quinazolines (**III**)* [378]

$$\log\left(\frac{1}{\mathrm{I}_{50}}\right) = 0.49(\pm 0.11)I_2 - 1.23(\pm 0.25) \times I_3 - 0.30(\pm 0.07)\mathrm{MR}_6 - 0.12(\pm 0.04)\pi_{\mathrm{R}} + 9.36(\pm 0.27) \quad (106)$$

$n = 24,\ r^2 = 0.817,\ s = 0.235$

*Inhibition of bovine liver DHFR by 2,4-diamino, 5-Y, 6-Z-quinazolines (**III**)* [379]

$$\log\left(\frac{1}{\mathrm{I}_{50}}\right) = 0.70(\pm 0.24)\mathrm{MR}_6 + 4.72(\pm 0.59) \quad (107)$$

$n = 11,\ r^2 = 0.823,\ s = 0.420$

These QSAR vary in size and species and the number of variables used to define inhibitory activity. A brief focus on the MR_6 term reveals that its coefficients vary remarkably in all four sets. QSAR 104 is a parabola with an optimum of 6.4. Since it is parabolic in nature, the coefficient of the ascending slope cannot be compared with the linear slopes in QSAR 105–107. Figure 6 illustrates the problems with QSAR 104–107, which failed to test analogs across the available space.

Figure 6 reveals that QSAR 105 and 106 were sampled in the suboptimal MR_6 range; hence, the negative dependence on MR_6. On the other hand, QSAR 107 was focused on the ascending portion of the curve and thus only molecules in the 0.1–3.4 MR ranges were tested. Thus with a limited data set, one gets an erroneous picture of the biological interactions. Selassie and Klein have described a more thorough comparative analysis of these QSAR [366].

Enzymatic reactions in nonaqueous solvents have generated a great deal of interest, fueled in part by the commercial application of enzymes as catalysts in speciality synthesis. The increasing demand for enantiopure pharmaceuticals has accelerated the study of enzymatic reactions in organic solvents containing little or no water [380]. In order to investigate the substrate specificity of α-chymotrypsin in pentanol, a series of X-phenyl esters of *N*-benzoyl-L-alanine (**IV**; Fig. 7) were synthesized and their binding constants were evaluated in buffer and in pentanol [367]. The following QSAR 108 and 109 were derived in phosphate buffer and pentanol.

*Binding of X-phenyl-N-benzoyl-L-alaninates (**IV**) to α-chymotrypsin in phosphate*

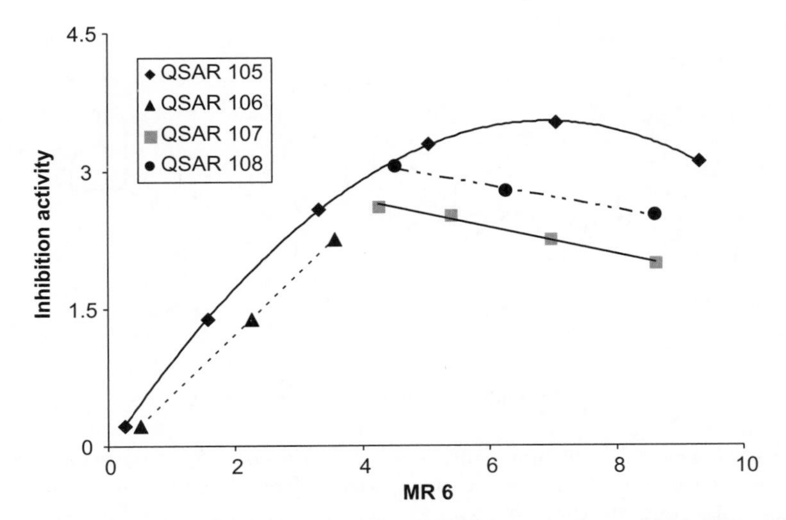

Figure 6. Gaps in spanned space of MR_6 for 2,4-diamino-quinazolines. (This figure is available in full color at http://mrw.interscience.wiley.com/emrw/9780471266945/home.)

buffer pH 7.4 [367]

$$\log\left(\frac{1}{K_M}\right) = 0.28(\pm0.11)\pi + 0.51(\pm0.24)\sigma^- + 0.38(\pm0.23)\text{MR} + 3.70(\pm0.24) \quad (108)$$

$$n = 16,\ r^2 = 0.834,\ s = 0.198$$

*Binding of X-phenyl- N-benzoyl-L-alaninates (**IV**) to α-chymotrypsin in pentanol* [367]

$$\log\left(\frac{1}{K_M}\right) = 0.25(\pm0.09)\pi + 0.24(\pm0.18)\sigma^- + 4.10(\pm0.09) \quad (109)$$

$$n = 17,\ r^2 = 0.762,\ s = 0.156$$

Outliers in QSAR 108 included the 4-*t*-butyl and 4-OH analogs, while the 4-$CONH_2$ analog was an outlier in QSAR 109. These results were reanalyzed by Kim [381,382] with respect to the role of enthalpic and entropic contributions to ligand binding with α-chymotrypsin. Usage of the Fujiwara hydrophobic enthalpy parameter π_H and the hydrophobic entropy parameter π_S, led to the development of QSAR 110 and 111 [383].

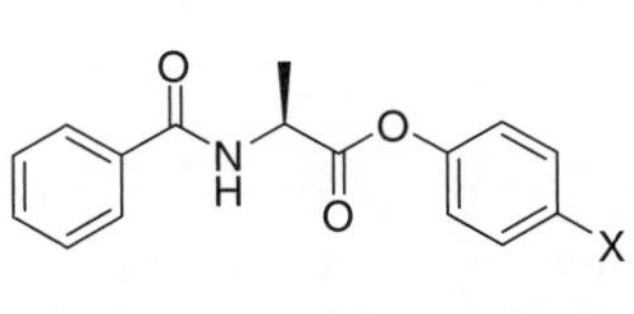

Figure 7. X-Phenyl, *N*-benzoyl-L-alaninates (**IV**).

*Binding of X-phenyl-N-benzoyl-L-alaninates (**IV**) in aqueous phosphate buffer* [382]

$$\log\left(\frac{1}{K_M}\right) = 0.38(\pm0.11)\pi_H + 0.19 \times(\pm0.07)\pi_S + 0.53(\pm0.11)\sigma^- + 0.26 \times(\pm0.10)\text{MR} + 3.77(\pm0.11) \quad (110)$$

$$n = 15,\ r^2 = 0.806,\ s = 0.200$$

*Binding of X-Phenyl-N-benzoyl-L-alaninates (**IV**) in pentanol* [382]

$$\log\left(\frac{1}{K_M}\right) = 0.21(\pm0.08)\pi_H + 0.31(\pm0.05)\pi_S + 0.20(\pm0.08)\sigma^- + 4.16(\pm0.04) \quad (111)$$

$$n = 15,\ r^2 = 0.787,\ s = 0.160$$

The disappearance of the MR term in QSAR 109 and 111 is significant. The MR term

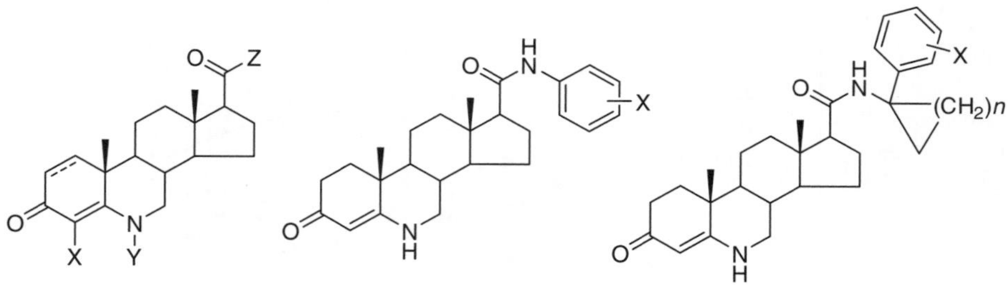

V VI VII

Figure 8. Steroidal inhibitors of 5α-reductase (**V–VII**).

usually relates to nonspecific, dispersive interactions in polar space. Thus, its presence in QSAR 108 and 110 suggests that substrates bearing polarizable substituents may displace the ordered-category-two, water molecules. In pentanol, the substrate may be faced with the task of displacing pentanol not water from the enzyme and thus the MR term is no longer of consequence. QSAR 110 also indicates that the enthalpy term π_H plays a more critical role in binding than the entropy term π_S. Note that these roles are reversed in QSAR 111, suggesting that binding in pentanol is largely an entropic driven process. Similar results were obtained by Compadre et al. in a study on the hydrolysis of X-phenyl-*N*-benzoyl-glycinates by cathepsin B in aqueous buffer and acetonitrile [384]. Kim's analysis provides an excellent example of a study that focuses on mechanistic interpretation and clearly demonstrates that a thermodynamic approach in QSAR can provide pertinent information about the energetics of the ligand binding process.

5α-Reductase, a critical enzyme in male sexual development, mediates the reduction of testosterone to dihydrotestosterone (DHT). Elevated levels of DHT in certain disease states such as benign prostatic hypertrophy and prostatic cancer drives the need for effective inhibitors of 5α-reductase. A recent QSAR study on inhibition of human 5α-reductase, type 1 by various steroid classes was carried out by Kurup et al. [368,385,386]. A few of the models will be examined to demonstrate the importance and power of lateral validation. The three classes of steroidal inhibitors (V, VI, and VII; Fig. 8) are depicted as follows:

Inhibition of 5-α-reductase by 4-X, N-Y-6-azaandrost-17-CO-Z-4-ene-3-ones, ***V***

$$\log\left(\frac{1}{K_i}\right) = 0.42(\pm 0.22)C\log P - 1.47(\pm 0.43)I - 0.32(\pm 0.30)L_Y + 6.88(\pm 0.13) \quad (112)$$

$n = 21,\ r^2 = 0.829,\ s = 0.406$

outliers : $X = Y = H,\ Z = NHCMe_3$; $X = Me,\ Y = H,\ Z = CH_2CHMe_2$

Inhibition of 5-α-reductase by 17β-(N-(X-phenyl)carbamoyl)-6-azaandrost-4-ene-3-ones, ***VI***

$$\log\left(\frac{1}{K_i}\right) = 0.35(\pm 0.09)C\log P + 0.26(\pm 0.11)B5_{ortho} + 5.08(\pm 0.58) \quad (113)$$

$n = 12,\ r^2 = 0.942,\ s = 0.154$

outlier : $2,5\text{-}(CF_3)_2$

Inhibition of 5-α-reductase by 17β(N-(1-X-phenyl-cycloalkyl)carbamoyl)-6-azaandrost-4-ene-3-ones, ***VII***

$$\log\left(\frac{1}{K_i}\right) = 0.32(\pm 0.17)C\log P + 6.34(\pm 1.15) \quad (114)$$

$n = 5,\ r^2 = 0.920,\ s = 0.090$

outlier : $n = 5$, X = 4-*t*-Bu

In all these equations, the coefficients with hydrophobicity as represented by $C\log P$, suggest that binding of these azaandrostene-ones occurs on the surface of the binding site where partial desolvation can occur. I is an indicator variable that pinpoints the negative effect of a double bond at C-1. A bulky substituent on N-6 is detrimental to activity while a large substituent in the *ortho*-position on the aromatic ring enhances activity (QSAR 113. Bulky *ortho*-substituents such as tertiary-butyl may destroy coplanarity with the amide bridge by twisting of the phenyl ring and enhancing its hydrophobic contact with the binding site on the enzyme.

7.2. Interactions at the Cellular Level

QSAR analysis of studies at the cellular level allows us to get a handle on the physicochemical parameters critical to pharmacokinetics processes—adsorption, distribution, metabolism, and excretion. Cell culture systems offer an ideal way to determine the optimum hydrophobicity of a system with enhanced complexity compared to an isolated receptor. Some of the first QSAR in cellular systems pertained to the assessment of the selective toxicity of 3-X-triazines in mammalian cell lines [373]. A comparison of the cytotoxicities of these analogs versus sensitive murine leukemia cells (L1210/S) and methotrexate resistant murine leukemia cells (L1210/R) reveals some startling differences.

Inhibition of growth of L1210/S by 3-X-triazines [373]

$$\log\left(\frac{1}{\mathrm{IC}_{50}}\right) = 1.13(\pm 0.18)\pi - 1.20(\pm 0.21) \times \log(\beta \times 10^{\pi} + 1) + 0.66(\pm 0.23)I_{\mathrm{R}} - 0.32 \times (\pm 0.17)I_{\mathrm{OR}} + 0.94(\pm 0.37)\sigma + 6.72(\pm 0.13) \quad (115)$$

$n = 61,\ r^2 = 0.792,\ s = 0.241,$

$\pi_{\mathrm{o}} = 1.45(\pm 0.93),\ \log\beta = -0.274$

Inhibition of growth of L1210/R by 3-X-triazines [373]

$$\log\left(\frac{1}{\mathrm{IC}_{50}}\right) = 0.42(\pm 0.05)\pi - 0.15 \times (\pm 0.05)\mathrm{MR} + 4.83(\pm 0.11) \quad (116)$$

$n = 62,\ r^2 = 0.885,\ s = 0.220$

There are substantive differences between these two QSARs. QSAR 115 is very similar to the one (QSAR 100) obtained versus the L1210 DHFR and it does suggest that the cytotoxicity in the sensitive cell line results from the inhibition of the enzyme. The intercepts indicate that slight interference with folate metabolism significantly affects growth. A comparison of the sensitive and resistant QSAR reveals a significant difference in the coefficients with π, the hydrophobic term. The absence of many variables in QSAR 116 and its overall simplicity suggests that inhibition of the enzyme is not the critical step, rather transport to the site of action in these resistant cells may be of utmost importance. This particular cell line is resistant to methotrexate by virtue of elevated levels of DHFR and also overexpression of glycoprotein, GP-170 [373]. Thus modified transport through the dysfunctional membrane would severely curtail the partitioning process resulting in a coefficient with π that is only half (0.42) of what is normally seen. The negative coefficient with the MR term indicates that size plays a role, albeit a negative one in passage through the GP-170-fortified membrane and to the site of action.

The QSAR paradigm has been shown to be particularly useful in environmental toxicology, especially in acute toxicity determinations of xenobiotics [387]. QSAR-based methods are not only useful for providing insight into the effects of structural changes of a compound on its biological activity but also provide a more holistic and global approach to the prediction of the effects of structurally unrelated compounds with different mechanisms of action on biological activity. There has recently been an emphasis on "transparent, mechanistically comprehensive QSAR for toxicity," a move that is welcomed by many researchers in the field [388–391]. Cronin and Schultz developed QSAR 117 to describe the polar, narcotic toxicity of a large set of substituted phenols. A number of phenols with ionizable or reactive groups, for example, $-COOH$, $-NO_2$, $-NO$, $-NH_2$, or $-NHCOCH_3$, were omitted from the final analysis [392].

Inhibition of growth of T. pyriformis (40 h)

$$\log\left(\frac{1}{C}\right) = 0.67(\pm 0.02)C\log P - 0.67 \times(\pm 0.55)E_{\mathrm{LUMO}} - 1.12 \quad (117)$$

$n = 120$, $r^2 = 0.893$, $s = 0.271$

Using Hammett σ constants, Garg et al. rederived QSAR 118 for the same set and QSAR 119 and 120 for the diverse set of multi, di, and mono phenols, which were sequestered into two subsets containing electron-releasing and electron-attracting substituents, respectively [393].

Inhibition of growth of T. pyriformis by all phenols (using σ) [393]

$$\log\left(\frac{1}{C}\right) = 0.64(\pm 0.04)C\log P + 0.61 \times(\pm 0.12)\Sigma\sigma + 1.84(\pm 0.13) \quad (118)$$

$n = 119$, $r^2 = 0.896$, $s = 0.265$

Inhibition of growth of T. pyriformis by electron-releasing phenols [393]

$$\log\left(\frac{1}{C}\right) = 0.66(\pm 0.05)C\log P + 1.63(\pm 0.15) \quad (119)$$

$n = 44$, $r^2 = 0.946$, $s = 0.182$

Inhibition of growth of T. pyriformis by electron-attracting phenols [393]

$$\log\left(\frac{1}{C}\right) = 0.63(\pm 0.07)C\log P + 0.59 \times(\pm 0.16)\Sigma\sigma + 1.92(\pm 0.18) \quad (120)$$

$n = 100$, $r^2 = 0.836$, $s = 0.327$

There is excellent agreement between QSARs 117–120, in terms of the importance of hydrophobicity and electron demand of the substituents: the coefficients with $C\log P$ are similar and there is a good correspondence between E_{LUMO} and σ. Nevertheless, a separation of the phenols into subsets based on their electronic attributes indicates that different mechanisms of toxicity might be operative in this organism—a phenomenon that has been duplicated in mammalian cells [394]. In a recent extension of toxicity studies on aromatics, Cronin and Schultz utilized a two-parameter or response-surface approach to define toxicity [395]. In addition, indicator variables and group counts were included to broaden the applicability of the approach. An excellent comparison of the different modeling approaches (MLR, PLS, and Bayesian regularized neural networks) in QSAR is also made [395].

Inhibition of growth of T. pyriformis by aromatic compounds [395]

$$\log\frac{1}{\mathrm{IGC}_{50}} = 0.633\log P - 0.526\,E_{\mathrm{LUMO}} + 0.721\,I_{2,4\mathrm{AP}} - 1.61\,I_{\text{strong acid}} + 0.314\sum\text{H-donor} - 1.39 \quad (121)$$

$n = 268$, $r^2 = 0.780$, $s = 0.393$

The indicator variables $I_{2,4\ \mathrm{AP}}$ and $I_{\text{strong acid}}$ suggest that 2 and 4-aminosubstituted phenols enhance toxicity, while strong acids decrease toxicity, respectively. The H-bond donor parameter may be correcting for the added potency of amino phenols. The low r^2 may be attributed to inherent variability in biological data and to the commingling of data from four different studies. The wide variety of compounds with different toxicity mechanisms, present in this combined study would also be a contributing factor to the low r^2. Overall, this regression-based approach shows adequate predictability and is transparent which aids in mechanistic interpretation.

In related studies on the toxicity of various chlorophenols to mouse connective tissue fibroblasts L929 cells *in vitro*, Liu et al. derived the following equation [396]:

Inhibition of growth of L929 cells by chlorophenols (48 h) [396]

$$\log \frac{1}{LC_{50}} = 0.480 \log K_{OW} + 1.968 \quad (122)$$

$$n = 10,\ r^2 = 0.873,\ SE = 0.191,\ F = 55.18$$

There is a significant dependence of toxicity on the sum of the electron-attracting capability of the chlorophenols but strong collinearity ($r = 0.980$) between hydrophobicity and sum of sigma contributions precludes inclusion in QSAR 122.

7.3. Interactions *In Vivo*

The paucity of extensive data in whole animal studies is understandable in terms of the costs, the heterogeneity of the biological data and the complexity of the results. Hence, the dearth of QSAR analyses that relies on *in vivo* data. Nevertheless, in the few studies that have been done, excellent QSAR have been obtained despite the small number of subjects in the data set. One particular example is insightful. The renal and nonrenal clearance rates of a series of 11 β-blockers, including bufuralol, tolamolol, propranolol, alprenolol, oxprenolol, acebutol, timolol, metoprolol, prindolol, atenolol, and nadolol were measured [397]. The following QSARs were formulated using that data [397,398].

Renal clearance of β-adrenoreceptor antagonists

$$\log k = -0.42(\pm 0.12) C\log P + 2.35(\pm 0.24) \quad (123)$$

$$n = 10,\ r^2 = 0.888,\ s = 0.185$$

Nonrenal clearance of β-adrenoreceptor antagonists

$$\log k = 1.94(\pm 0.61) C\log P - 2.00 \times (\pm 0.80) \log(\beta \times P + 1) + 1.29 \times (\pm 0.30) \quad (124)$$

$$n = 10,\ r^2 = 0.950,\ s = 0.168,$$

$$C\log P_o = 2.6 \pm 1.5 \log \beta = -0.813$$

outlier : oxprenolol

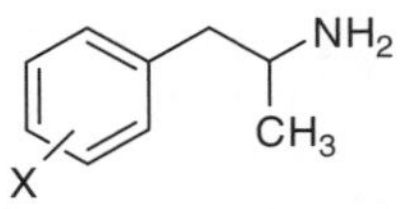

Figure 9. Structure of mescaline derivatives (VIII).

It is apparent from QSAR 123 and 124 that the hydrophobic requirements as represented by the signs of the coefficients with $C\log P$, vary considerably. As expected, renal clearance is enhanced in the case of hydrophilic drugs while nonrenal clearance shows a strong dependence on hydrophobicity. Note that QSAR 124 is stretching the limits of the bilinear model with only ten datapoints! The 95% confidence intervals are also large, but nevertheless, the equations serve to emphasize the difference in clearance mechanisms that are clearly linked to hydrophobicity.

In an interesting and audacious study to determine the hallucinogenic potential of a series of mescaline derivatives, Shulgin et al. measured the relative biological response (MU) of 26 analogs, relative to that of mescaline [399]. Using their data, the following QSAR was formulated using only 23 of the 25 1-(X-Phenyl)-2-propylamine derivatives (**VIII**; Fig. 9) [400]:

$$\log MU = 3.32(\pm 0.92) C\log P - 0.56(\pm 0.17) \times C\log P^2 - 0.34(\pm 0.28) I_{4\text{-}OR} + 0.30 \times (\pm 0.17) I_{n(OMe)} - 3.45(\pm 1.23) \quad (125)$$

$$n = 23,\ r^2 = 0.844,\ s = 0.275,$$

$$q^2 = 0.765,\ \log P_o = 2.94\{2.79\text{–}3.14\}$$

The MU indices in this particular study were not easy to ascertain but were doable because of Shulgin's extensive and longstanding experience working in the CNS field. Thus hydrophobic analogs with an optimum hydrophobicity around ~3.0 had a higher potency than mescaline in inducing a hallucinogenic haze! Two indicator variables were used in this derivation and they have opposing effects; the number of methoxy substituents on the phenyl ring ($I_{n(OMe)}$) contributed to potency while the presence of any alkoxy moiety at para position ($I_{4\text{-OR}}$) on the aromatic ring was deleterious to hallucinogenic potency. A QSAR study was conducted by Mracec et al. using electrotopological parameters but different parameters cannot be utilized in a comparative QSAR analysis. The combined data pooled from two sources by Mrarec et al. [401] and Clare [402], was subjected to QSAR 126 analysis [400].

$$\log \mathrm{MU} = 2.44(\pm 0.71) C\log P - 0.44 \times (\pm 0.14) C\log P^2 - 0.78(\pm 0.33) I_{4\text{-OR}} + 0.42(\pm 0.20) I_{n(\mathrm{OMe})} - 2.04(\pm 0.89) \quad (126)$$

$$n = 26,\ r^2 = 0.833,\ s = 0.319,\ q^2 = 0.779,\ \log P_O = 2.77\{2.58\text{–}3.05\}$$

QSAR 126 is very similar to QSAR 125 and underscores the importance of optimum hydrophobicity and certain structural features for hallucinogenic activity. Again only the 1-(X-phenyl), 2-propylamine derivatives were considered and compared for consistency. Despite the difficulty involved in measuring psycho mimetic activity of such compounds, these QSAR results are invaluable in providing direction for further optimization and design.

7.4. Anti-HIV Activity

Despite the approval of several drugs by the US FDA to reduce the morbidity and mortality caused by human immunodeficiency virus (HIV) infection, the success of these drugs is tempered by high costs, side effects, and/or development of resistance. Numerous QSAR models delineating the drug–receptor interactions of anti-HIV agents have helped in the design of more effective analogs. An extensive compilation of QSAR models pertaining to anti-HIV drugs has been published by Garg et al. [403].

QSAR 127 was formulated for a series of aryl-substituted phosphoramidate analogs (**IX**; Fig. 10) of the anti-HIV drug d4T that was synthesized and evaluated for their anti-HIV activities in cell culture by Siddiqui et al. [404]. The model revealed that the anti-HIV-1 activities of these compounds were mostly dependent on hydrophobicity in a parabolic fashion with an optimum hydrophobicity of 1.72.

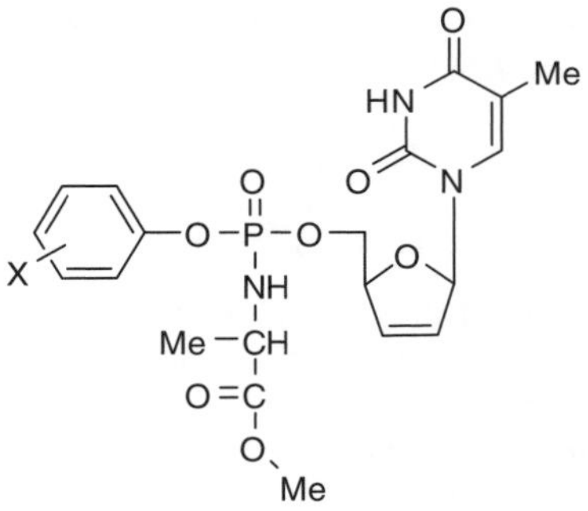

Figure 10. Structure of aryl-substituted phosphoramidate analogs (**IX**).

*Inhibition of viral (HIV-1) activity by **IX** in CEM cells*

$$\log \frac{1}{\mathrm{EC}_{50}} = 5.16(\pm 1.12)\log P - 1.50 \times (\pm 0.40)\log P^2 - 2.44 \quad (127)$$

$$n = 21,\ r^2 = 0.807,\ s = 0.200,\ q^2 = 0.745,\ F_{2,18} = 37.735,\ \log P_O = 1.72$$

In this QSAR, EC_{50} represents the concentration of each compound required to protect CEM cells against the cytopathicity of HIV-1 by 50% and $\log P$ is the experimental hydrophobicity in octanol/aqueous buffer (pH 7.0).

A similar QSAR study on a series of aryl sulfonamide HIV protease inhibitors (**X**; Fig. 11) [405] was reported by Ravichandran et al. in 2007 [406] using STATISTICA (SoftStat, Inc., Tulsa, OK) program. Multiple

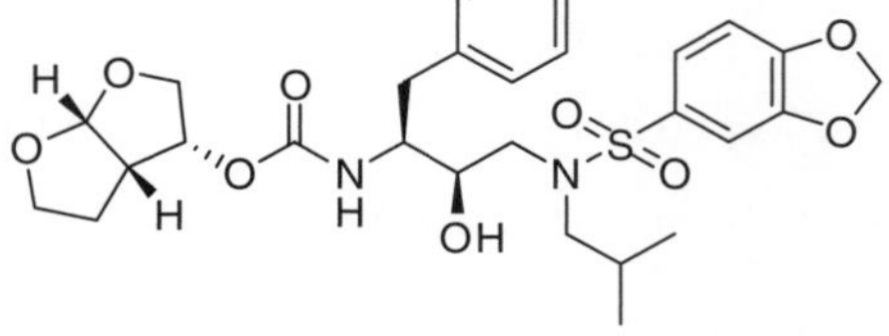

Figure 11. Structure of aryl sulfonamide derivatives (**X**).

linear regression analysis was performed to derive QSAR 128, which was further evaluated for statistical significance and predictive power by internal and external validation.

Inhibition of viral (HXB-2) activity by ***X*** *in MT4 cells*

$$\log\frac{1}{\mathrm{IC}_{50}} = 1.18(\pm 0.22)\log P - 0.03(\pm 0.01) \times \mathrm{MR} - 0.01\,\mathrm{HF} - 4.06(\pm 1.74) \qquad (128)$$

$$n = 18,\ r^2 = 0.808,\ s = 0.259,\ q^2 = 0.657,\ F_{3,14} = 19.640$$

IC_{50} represents the antiviral activities of compounds (**X**) against the wild-type HIV virus HXB2 (in an MT4 cell line). In this QSAR model, a high correlation was shown between thermodynamical descriptors such as $\log P$, molar refractivity (MR), and heat of formation (HF), and anti-HIV activities.

In a recent publication, 2D QSAR and 3D QSAR analyses of 42 thiourea derivatives (**XI-XIV**; Fig. 12) [407] were reported by Li et al. [408]. The best 2D QSAR model was obtained with four descriptors as shown in QSAR 129. The validation of this model was carried out by full cross-validation tests, randomization tests, and external test set prediction.

Induction of maximal protection against HIV-1 by ***XI-XIV****in MT-4 cells* [408]

$$\log\frac{1}{M_p} = 0.52\,\mathrm{LUMO} + 0.53\,\mathrm{HBA} - 2.86\,\mathrm{Shad\_XYfrac} - 0.02\,\mathrm{S\_aaCH} + 0.41 \qquad (129)$$

$$n = 31,\ r^2 = 0.897, s = 0.021,\ q^2 = 0.862,\ F_{4,26} = 56.347,\ r^2_{\mathrm{pred}} = 0.899$$

M_p is the maximum protection (%) against HIV-1 (IIIB) in MT-4 cells, LUMO is the lowest unoccupied molecular orbital energy, HBA is the hydrogen-bond acceptors, Shad_XYfrac is the fraction of the area of molecular shadow in the XY plane over the area of enclosing rectangle, and S_aaCH is an E-state descriptor that represents the fragment type, =CH– in the aromatic ring. The above QSAR model suggests that the bioactivity of thiourea derivatives (**XI–XIV**) is mainly controlled by its four descriptors, for example, LUMO, HBA, Shad_XYfrac, and S_aaCH. In a related molecular field analysis (MFA) coupled with genetic partial least squares (GPLS) approach for 130 thiourea derivatives (**XI–XIV**), electrostatic fields were shown to contribute significantly toward bioactivity.

In an elegant study, Debnath [409] using CoMFA examined a large data set (118 compounds) of diverse cyclic urea analogs as pro-

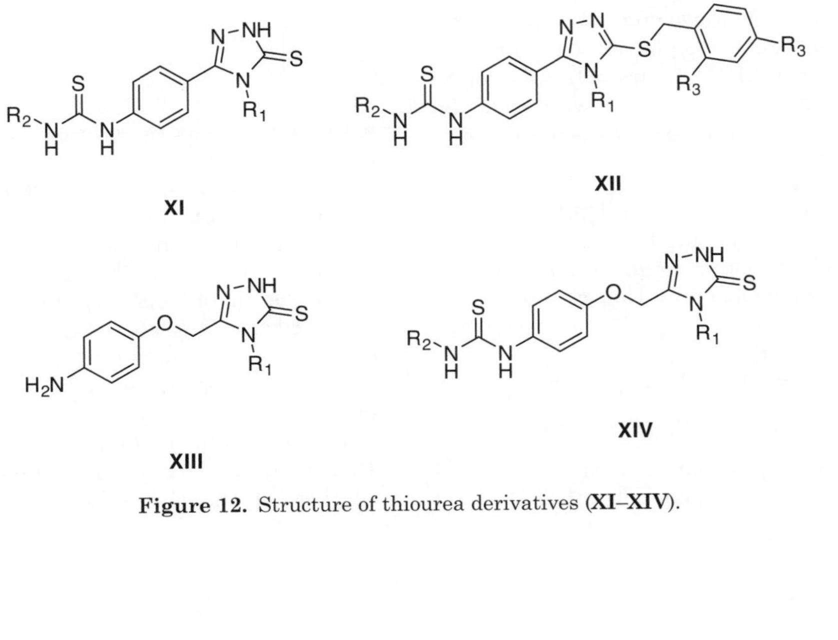

Figure 12. Structure of thiourea derivatives (**XI–XIV**).

tease inhibitors against HIV-1. The X-ray crystal structures of HIV-1 protease bound with this class of inhibitors were used to derive the bioactive conformations of the inhibitors. Multiple predictive models were developed by dividing the data set into training sets (93 compounds in each set) and corresponding test sets (25 compounds in each set). All models yielded high values of r^2 (0.965–0.973) and q^2 (0.699–0.727), and reasonably low values of s (0.239–0.265). Steric and electrostatic effects were found to exert approximately equal contributions, 45% and 55%, respectively, toward protease inhibition. An extensive application of 3D QSAR techniques in anti-HIV-1 drug design and discovery can be seen in an excellent overview published by the same author [410].

7.5. ADMET Properties

Although there has been a dramatic increase in the speed of synthesis and biological evaluation of new chemical entities each year, the increase in the number of marketed new drugs is not commensurate with these rapid developments. This discrepancy has mostly been attributed to undesired ADMET (absorption, distribution, metabolism, excretion, and toxicity) constraints. To circumvent these problems and increase the speed of marketed new drugs, the application of ADMET is now given a high priority by the pharmaceutical industry in the early stage of the drug discovery process. QSAR/QSPR has been shown to be a useful tool in the development of predictive ADMET models using pharmacokinetic data. Some common ADMET parameters as well as their specific contribution(s) with brief description and experimental tools are listed in Table 5 [411–413].

7.5.1. Solubility A simple QSPR 130 between the aqueous solubility of organic liquids (log S) and octanol–water partition coefficient (log P) was developed by Hansch and coworkers [414]

$$\log S = -1.339 \log P + 0.978 \quad (130)$$

$$n = 156,\ r^2 = 0.874,\ s = 0.472$$

In order to expand the applicability of this model to organic compounds either in liquid or solid form, Yalkowsky and Valvani [415] developed Equation 131 with two additional terms, that is, melting point (MP) and entropy of fusion (ΔS_f):

$$\log S = -1.00 \log P - \Delta S_f(\mathrm{MP}-25)1.11/1364 + 0.54$$

$$n = 167, \quad (131)$$

$$r^2 = 0.988,\ s = 0.242$$

In this model, calculated log P values were used while the entropies of fusion were a mix of experimental and calculated values. However, this model was limited by the need for experimental melting points. For early screening in drug discovery, models based on descriptors calculated from molecular structure are preferable. Thus in a later study in 1999, the aqueous solubility of solids and liquids (log S) was calculated by Abraham and Le with an amended solvation Equation 132 that incorporated an extra $\sum\alpha_2^H \times \sum\beta_2^H$ term, which represented hydrogen-bond interactions between acid sites and basic sites in the solid or liquid [416].

$$\log S = -1.004\, R_2 + 0.771\, \pi_2^H + 2.168 \sum\alpha_2^H + 4.238 \sum\beta_2^H - 3.362 \sum\alpha_2^H \times \sum\beta_2^H - 3.987 V_x + 0.518$$

$$n = 659, \quad (132)$$

$$r^2 = 0.920,\ s = 0.557,\ F = 1256$$

In this equation, R_2 describes excess molar refraction in units of (cm^3/mol)/10, π_2^H is the dipolarity/polarizability, $\sum\alpha_2^H$ is the overall or summation of hydrogen-bond acidity, $\sum\beta_2^H$ is the overall or summation of hydrogen-bond basicity, and V_x is the McGowan volume in units of (cm^3/mol)/100 of solutes. It has been concluded from the above equation that the hydrogen-bonding propensity of a compound leads to an increase in solubility, even though the $\sum\alpha_2^H \times \sum\beta_2^H$ term opposes solubility due to interactions in the liquid or solid. On the other hand, the increase in solute dipolarity/polarizability increases solubility, whereas an

Table 5. Common ADMET (Absorption, Distribution, Metabolism, Excretion, and Toxicity) Parameters with Their Brief Description, Experimental Tools, and Specific Contribution(s)

No.	ADMET Parameter(s)	Brief Description[a]	Some Experimental Tools[b]	Specific Contribution(s) in ADMET
1	Solubility	Drug ability to dissolve in water. It is also known as water solubility or aqueous solubility and represented by log S.	Turbidometry, shake-flask, dissolution, HTSA	A and D
2	Permeability	An *in vitro* measurement of intestinal drug absorption and transport. Commonly reported as permeability coefficient ($P_{app\text{-}Caco\text{-}2}$, $P_{Caco\text{-}2}$, $P_{app\text{-}pampa}$, P_m, P_{am}, $P_{pp\text{-}rf}$, P_{eff}, or F)	Caco-2 cells, MDCK cells, IAMC, PAMPA, PAMPA-PP-RF, BAMPA	A and D
3	Human intestinal absorption	An *in vivo* measurement of percent human intestinal absorption (percent oral absorption) or the ratio of the total absorbed mass divided by the drug dose (percent fractional absorption) (F_a)	Autoradiography, LC-MS/MS, Radio-HPLC, GC-MS, NMR	A, D, and M
4	BBB penetration	Experimentally measured ratio of the drug concentration in the brain to that in the blood. An *in vitro* measurement by PAMPA-BBB (P_e)	Bovine brain microvessel cells, IAMC, PAMPA-BBB	A, D, and M
5	Oral bioavailability	Percent drug availability to the general circulation or the site of pharmacological actions	GC-MS, LC-MS/MS	A and E
6	Volume of distribution	Relative partitioning of drug between the plasma and the tissues	Animal pharmacokinetics, LC-MS/MS, GC-MS,	D
7	Binding to plasma or serum proteins	Percent concentration of a drug bound to all plasma proteins	Ultrafiltration, equilibrium dialysis, HPLC	D
8	Metabolism	Metabolic stability and liability. An *in vitro* determination of half-life and intrinsic clearance	CYP450 metabolism ADME assays, supersome or hepatocyte assays, LC-MS/MS	M and E
9	Excretion	Time required by the body to eliminate (or break down) one-half of a dose of the drug	LC-MS/MS, Radio-HPLC, GC-MS, NMR	M and E

(*continued*)

Table 5. *(Continued)*

No.	ADMET Parameter(s)	Brief Description[a]	Some Experimental Tools[b]	Specific Contribution(s) in ADMET
10	Toxicity	An *in vivo* or *in vitro* determination of toxicological endpoints, for example, acute, aquatic, carcinogenicity, mutagenicity, polar narcotic, skin sensitization, skin irritation, eye irritation, etc.	Animal, Draize rabbit eye irritation test, *in vitro* skin corrosive test, Ames test	T

[a] $P_{app\text{-}pampa}$: apparent permeability coefficient measured in centimeters per second by PAMPA; $P_{Caco\text{-}2}$ or $P_{app\text{-}Caco\text{-}2}$: permeability coefficient measured in centimeters per second by Caco-2 assay; P_{am}: artificial membrane permeability coefficient determined by BAMPA in centimeters per second; $P_{pp\text{-}rf}$: permeability measured in centimeters per second by PAMPA-PP-RF; P_{eff}: human intestinal permeability measured in centimeters per second; F: permeation measured as percent flux by PAMPA; F_a: fraction of dose absorbed in humans; P_e: permeability measured in centimeters per second by PAMPA-BBB.

[b] HTSA: high-throughput solubility assay; MDCK cells: Madin–Darby canine kidney cells; IAMC: immobilized artificial membrane chromatography; PAMPA: parallel artificial membrane permeability assay; PAMPA-PP-RF: parallel artificial membrane permeability assay using paracellular pathway model based on the Renkin function; PAMPA-BBB: parallel artificial membrane permeability assay of blood–brain barrier; BAMPA: biomimetic artificial membrane permeation assay; LC-MS/MS: liquid chromatography-mass spectrometry/mass spectrometry; HPLC: high-performance liquid chromatography; GC-MS: gas chromatography-mass spectrometry; NMR: nuclear magnetic resonance; CYP450: cytochrome P450.

increase in solute excess molar refraction and especially, volume decreases solubility.

In a recent study, Gao and Cao [417] developed QSPR 133 for the aqueous solubility of 134 polychlorinated biphenyls (PCBs) using three parameters, for example, the volume (V), the HOMO energy (E_{HOMO}), and the LUMO energy (E_{LUMO}) of the solute:

$$\log S = -0.0298\,V + 11.1821\,E_{HOMO} + 32.7300\,E_{LUMO} \quad (133)$$

$n = 134,\ r^2 = 0.9485,\ q^2 = 0.9451,\ s = 0.260,\ F = 804$

Here, note the negative dependence on volume.

7.5.2. Permeability The experimental data of Caco-2 permeability coefficients ($P_{Caco\text{-}2}$) of a diverse set of drug molecules was compiled from the literature by Kulkarni et al. [418] who developed various MI-QSAR models, of which QSAR 134 was the best.

$$P_{Caco\text{-}2} = 0.65\,F(H_2O) + 0.06\,\Delta E_{TT}(hb) - 0.19\,E_{SS}(hb) + 0.10\,E_{TT}(14) - 0.03\,E_{TT}(tor) - 5.61\,\chi_3 - 40.50 \quad (134)$$

$n = 30,\ r^2 = 0.86,\ q^2 = 0.77$

In this model, $F(H_2O)$ is the aqueous solvation free energy, $\Delta E_{TT}(hb)$ is the change in the hydrogen-bonding energy, $E_{SS}(hb)$ is the intramolecular hydrogen-bonding energy, $E_{TT}(14)$ is the 1,4-van der Waals plus electrostatic interaction energy, $E_{TT}(tor)$ is the torsion energy, and χ_3 is the topological index measuring the size and shape of a molecule. The descriptors $\Delta E_{TT}(hb)$, $E_{SS}(hb)$, $E_{TT}(14)$, and $E_{TT}(tor)$ are for the entire membrane-solute system of the solute located at the position corresponding to the lowest solute–membrane interaction energy state of the model. In contrast, the same data set of Caco-2 permeability coefficients of drug molecules was further used by Hansch et al. [419] in the development of a parabolic QSAR 135:

$$\log P_{\text{Caco-2}} = 0.41(\pm 0.06) C\log P - 0.06 \times (\pm 0.02) C\log P^2 - 0.31(\pm 0.16) \times \text{MgVol} + 1.43(\pm 0.31) \quad (135)$$

$$n = 34, r^2 = 0.870, s = 0.205, q^2 = 0.820, \log P_O = 3.35$$

Fujikawa et al. [420] measured the permeability coefficients ($P_{\text{app-pampa}}$) of a diverse set of 97 compounds including several drugs, pesticides and peptides using the parallel artificial membrane permeation assay (PAMPA) and developed QSAR 136:

$$\log P_{\text{app-pampa}} = 0.53(\pm 0.10)\log P_{\text{app}} - 1.18(\pm 0.25)\log(\beta \times 10^{\log \text{Papp}} + 1) - 0.74(\pm 0.35)\text{SA}_{\text{HA}} - 1.13 \times (\pm 0.39)\text{SA}_{\text{HD}} - 5.00(\pm 0.24) \quad (136)$$

$$n = 97, r^2 = 0.720, s = 0.360, q^2 = 0.680, \log P_{\text{app(O)}} = 2.08, \log \beta = -2.17$$

In this equation, $\log P_{\text{app}}$ ($\log P_{\text{app}} = \log P_{\text{oct}} - 0.67 \mid pKa\text{–pH} \mid$) is the apparent hydrophobicity of compounds at particular pH when the partition of ion-pair complexes to 1-octanol cannot be neglected, while SA_{HA} and SA_{HD} are van der Waals surface areas ($\text{Å}^2 \times 1/100$) occupied by hydrogen-bond acceptor and donor atoms, respectively. A good correlation 137 was obtained between the permeability coefficients of 35 chemicals (20 commercial drugs, 7 peptide related compounds, and 8 chemicals/agrochemicals) measured from PAMPA and Caco-2 cell assays [420].

$$\log P_{\text{app-Caco-2}} = 1.03(\pm 0.21) \log P_{\text{app-pampa}} + 0.49(\pm 1.09) \quad (137)$$

$$n = 35, r^2 = 0.760, s = 0.350, q^2 = 0.730$$

7.5.3. Human Intestinal Absorption The percentage of human intestinal absorption (percent oral absorption) is an *in vivo* measurement, which is defined as the percent dose of an orally administered drug to reach the hepatic portal vein. It can also be defined in terms of percent fractional absorption, that is, the total mass absorbed divided by the drug dose [413]. The percent human intestinal absorption data (F_a) of a large diverset set of drugs [421] was used in the development of QSAR 138 [411]:

$$\log F_a = 0.28(\pm 0.05) C\log P - 0.69(\pm 0.22) I_{\text{COOH}} + 0.72 \times (\pm 0.36) I_{\text{N(CH3)2}} + 1.47(\pm 0.11) \quad (138)$$

$$n = 57, r^2 = 0.711, s = 0.328, q^2 = 0.654$$

The indicator variables I_{COOH} and $I_{\text{N(CH3)2}}$ acquire a value of one only for the presence of –COOH and –N(CH3)2 groups, respectively. This model illustrates the positive contribution of both the hydrophobicity and the presence of the dimethylamino function ($I_{\text{N(CH3)2}}$) as well as the detrimental effects of the carboxyl group (I_{COOH}).

7.5.4. Blood–Brain Barrier The interest of medicinal chemists in drug penetration of the blood–brain barrier (BBB) continues to escalate due to its critical importance in drug design targeting the CNS. The degree of BBB penetration is usually expressed as log BB or $\log (C_{\text{brain}}/C_{\text{blood}})$, where C_{brain} is the concentration of the test compound in the brain, and C_{blood} is the concentration of the test compound in blood. A simple QSAR 48 for the prediction of log BB from a large set of 55 diverse organic molecules was reported by Clark [211] using only two descriptors, for example, PSA and calculated $\log P$. Using data of artificial membrane permeability of miscellaneous drugs obtained by PAMPA-BBB assay [422], a bilinear QSAR 139 was developed with an optimum hydrophobicity of 3.67 [411].

$$\log P_e = 0.43(\pm 0.10) C\log P - 0.54(\pm 0.25)\log(\beta \times 10^{C\log P} + 1) - 0.09(\pm 0.19) \quad (139)$$

$$n = 23, r^2 = 0.842, s = 0.249, q^2 = 0.780, C\log P_O = 3.67, \log \beta = -3.087$$

7.5.5. Oral Bioavailability Human oral bioavailability is directly related to the percent of drug amount available in the systemic circulation to exert pharmacological and therapeutic effects. Recently, the known human oral bioavailability of 250 structurally diverse molecules (training set) was used by Moda et al. [423] in the development of a significant hologram quantitative structure–activity relationship (HQSAR) model with $r^2 = 0.93$ and $q^2 = 0.70$. The HQSAR model was obtained using fragment descriptors such as atoms, bonds, connections, and chirality. The predictive ability of this model was evaluated by an external test set of another 52 molecules.

7.5.6. Volume of Distribution Volume of distribution is a very complex combination of multiple chemical and biochemical phenomena. It is a measure of the relative partitioning of drug between the plasma and the tissues, which is important in drug dose calculations in therapeutics. The best example comes from the work of Nestorov et al. [424], who showed that there is a good correlation between the tissue-to-unbound plasma distribution ratios (K_{pu}) and $\log P$ using the examples of 13 rat tissues after intravenous administration of nine 5-*N*-alkyl-5-ethyl barbituric acids. These data of K_{pu} [424] were further analyzed by Hansch and coworkers [419] in the development of 13 QSAR models with calculated $\log P$. All these QSAR models for 13 different rat tissues can be represented by Equation 140 that had excellent statistical indices.

$$\log K_{pu} = A \times C\log P + C \tag{140}$$

The coefficient A for stomach, pancreas and adipose tissue ($n = 3$) was 0.72–0.82 while in muscle, spleen, heart, lung, testis, gut, liver, skin, brain, and kidney ($n = 10$) it was 0.44–0.63.

7.5.7. Binding of Plasma or Serum Protein Although the plasma protein binding (PPB) of drug molecules is an important pharmacokinetic property in ADME processes, its role is still not thoroughly understood. A drug with strong plasma–protein binding tends to show lower activity than predicted from *in vitro* assays, and/or cause several adverse effects including low clearance, low BBB penetration, drug–drug interaction, etc. In a recent study, 250 structurally diverse molecules with known PPB were used as a training set by Moda et al. [425] who developed a significant HQSAR model with good statistics ($r^2 = 0.91$, $q^2 = 0.72$). This model was then successfully used to predict the PPB of 62 test set compounds not included in the training set. QSAR models for the human serum protein binding activity of a large data set of 808 compounds (training set) were developed by Votano and coworkers [426] using four different modeling algorithms: multiple linear regression (MLR), ANN, k-nearest neighbors (kNN), and support vector machine (SVM). The best model with 33 descriptors was obtained from ANN technique ($r^2 = 0.90$, mean absolute error, i.e., MAE = 7.6). Prediction results from the validation set of 200 compounds yielded $r^2 = 0.70$ and MAE 4.1. A simple QSAR 141 for the binding of organic compounds with bovine serum albumin (BSA) was developed by Helmer et al. [427].

$$\log\left(\frac{1}{C}\right) = 0.75(\pm 0.07)\log P + 2.30(\pm 0.15) \tag{141}$$

$$n = 42,\ r^2 = 0.922,\ s = 0.159$$

In this equation, C is the molar concentration of compound necessary to produce a 1:1 complex with BSA.

7.5.8. Metabolism Rapid metabolism is one of the primary reasons for the inadequacy of drug systemic levels due to either phase I or phase II metabolism. Low metabolic stability generally results in a short half-life and high clearance of the drug molecule. Cytochromes P450 (CYPs) constitute the most important family of biotransformation enzymes and play a critical role in the disposition of drugs, their pharmacological and toxicological effects. *In vitro* CYP450 inhibition and induction are generally used to assess the potential of compound toward drug–drug interactions [412,428–430]. QSARs 142–145 represent some of the many faceted activities of CYP450.

Inhibition of cytochrome P450 by ROH [431,432]

$$\log\left(\frac{1}{K_i}\right) = 0.94(\pm 0.10)C\log P$$

$$-3.87(\pm 0.59)\log(\beta \times 10^{C\log P} + 1) + 1.25(\pm 0.21)$$

$$n = 26, \quad (142)$$

$r^2 = 0.951$, $s = 0.258$, $q^2 = 0.944$,

$$C\log P_O = 4.42 \log\beta = -4.915$$

K_i is the molar concentration of alcohol (ROH) needed to inhibit CYP450. Thus, inhibition of the enzyme increases as the hydrophobicity of the alcohols increase with inhibition maximizing ~4.4.

Induction of cytochrome P450 by barbiturates in chick embryo hepatocytes [431,433]

$$\log\left(\frac{1}{C}\right) = 1.01(\pm 0.16)\log P + 2.77(\pm 1.27) \quad (143)$$

$n = 9$, $r^2 = 0.969$, $s = 0.183$, $q^2 = 0.951$

Hydroxylation of halogenated anilines by P450 [431,434]

$$\log\left(\frac{k_{cat}}{K_m}\right) = 3.48(\pm 0.79)E_{HOMO} + 33.2(\pm 6.88)$$

$$n = 13, r^2 = 0.896, s = 0.158, q^2 = 0.863 \quad (144)$$

Binding affinity of azoles to CYP2B [435]

$$pK_d = 1.12(\pm 0.14)\log P - 1.26 \times (\pm 0.22)\log(\beta \times P + 1) + 0.30 \times (\pm 0.21)I_{azole} + 3.02(\pm 0.30) \quad (145)$$

$n = 18$, $r^2 = 0.962$, $s = 0.198$,

$$F = 118, \log P_O = 3.47 \log\beta = -2.57$$

In the above QSAR model, the indicator variable (I_{azole}) is unity for imidazole and zero for triazole. Once again, binding is clearly enhanced by the hydrophobicity of the azoles. In recent years, several reviews have been published on CYPs, explaining a large numbers of QSAR approaches and models [429,431,436–440].

7.5.9. Excretion Excretion is the process by which the body eliminates the waste products of metabolism and other nonuseful or harmful materials. The important routes of excretion, whether for a metabolite or the parent compound, are renal and urinary excretions. For the QSAR point of view, we shall focus only those examples where the excretion results in unmetabolized forms of drugs/compounds. A good example comes from the work of Cantelli-Forti et al. [441]. They studied the relationship between hydrophobicity and the urinary excretion of the unchanged form of a series of nitroimidazoles and nitrothiazoles. The unmetabolized forms of the drugs were detected in the urine by means of UV and HPLC procedures at different time intervals, that is, 18, 36, 54, and 72 h. There was excellent correlation ($n = 11$, $r = 0.985$) between the results obtained from these two methods. The best QSAR model developed by these authors [441] was Equation 146 for the data obtained after 36 h.

$$\log BR_{UV} = -0.325(\pm 0.065) \times \log P + 2.479(\pm 0.042) \quad (146)$$

$n = 26$, $r = 0.716$, $s = 0.229$, $F = 25.29$

In this model, there is an inverse relationship between $\log BR_{UV}$ and $\log P$. $\log BR_{UV}$ represents log (percent × 10) of the administered dose that is excreted after 36 h. Another example pertaining to the renal clearance (excretion) of β-adrenoreceptor antagonists can be seen in QSAR 123.

7.5.10. Toxicity Toxicity of a compound is quite different from the ADME properties as it involves different toxicological mechanisms and thus provides different types of toxicological responses. Common toxicological endpoints are acute, aquatic, carcinogenicity, mutagenicity, polar narcosis, skin sensitization, skin irritation, eye irritation, etc. Huang et al. [442] modeled the acute toxicity

(12 h – log 1/LC_{50} mol/L) of 51 benzene derivatives to *Rana japonica* tadpoles using three parameters:

Toxicity of benzene derivatives to R. japonica [442]

$$\log \frac{1}{LC_{50}} = 0.399 \log D_{ow} - 0.453$$

$$\times E_{LUMO} + 0.011\,\text{Vol.} + 1.394 \qquad (147)$$

$$n = 51,\ r^2 = 0.914,\ s = 0.175,$$

$$q^2 = 0.785,\ F = 167$$

In this QSAR, $D_{ow} = K_{ow}/(1 + 10^{pH-pKa})$, where K_{ow} is the 1-octanol–water partition coefficient, pH is the pH of the test medium, and pKa is the acidity dissociation constants in logarithmic form for benzoic acid derivatives. With regards to aquatic toxicity, QSAR 148 was recently published by Schultz et al. [443] for the toxicity of 384 aromatic compounds to *T. pyriformis* [log (IGC_{50} – 1)] using only two descriptors, log P (hydrophobicity) and maximum acceptor superdelocalizability (A_{max}):

Toxicity of aromatic compounds to T. pyriformis [443]

$$\log(IGC_{50} - 1) = 0.545(0.015)\log P$$

$$+ 16.2(0.62)A_{max} - 5.91(0.20) \qquad (148)$$

$$n = 384,\ r^2(\text{adj}) = 0.859,$$

$$r^2(\text{pred}) = 0.856,\ s = 0.275,\ F = 1163$$

A QSAR 149 for the carcinogenic potency of aromatic amines in the mouse was published by Benigni et al. [444]:

Carcinogenecity of aromatic amines to mice [444]

$$\text{BRM} = 0.56(\pm 0.18)\log P + 1.03$$

$$\times(\pm 0.74)E_{HOMO} - 1.19(\pm 0.58)$$

$$\times E_{LUMO} - 0.79(\pm 0.37)\sum \text{MR}_{2,6} - 0.93$$

$$\times(\pm 0.90)\text{MR}_3 - 0.22(\pm 0.19)E_s(\text{R})$$

$$+ 8.51(\pm 6.31) \qquad (149)$$

$$n = 37,\ r^2 = 0.714,\ s = 0.485,\ F = 12.5$$

In this QSAR, BRM = log $(MW/TD_{50})_{mouse}$; where MW is the molecular weight and TD_{50} is the daily dose rate (mg/kg/day) of a compound required to halve the probability of tumor formation in a mouse. Es(R) is Tafts steric parameter for the substituent on the amino nitrogen.

QSAR 36, 39, and 40 are examples for mutagenicity, while QSAR 117 is for polar narcotic toxicity. For the skin sensitization toxicity, QSAR 150 was developed by Roberts et al. [445] using a combined data set of 11 aliphatic aldehydes, 1 α-ketoester and 4 α,β-diketones (RCOR′):

Sensitization of various compounds to skin [445]

$$\text{pEC3} = 1.12(\pm 0.07)\sum \sigma^* + 0.42$$

$$(\pm 0.04)\log P - 0.62(0.13) \qquad (150)$$

$$n = 16,\ r^2 = 0.952,\ r^2(\text{adj}) = 0.945,$$

$$s = 0.120,\ F = 129.6$$

In the above QSAR model, pEC3 = log(MW/EC3); where MW is the molecular weight and EC3 is the dose (percent concentration by weight) of a compound giving stimulation index (SI) of 3. $\sum\sigma^*$ is the sum of Taft σ^* values for the two groups R and R′ in RCOR′. Compounds with EC3 < 0.1% are classified as extreme sensitizers, while those with EC3 ranging from 0.1–1%, 1–10%, 10–100%, and >100% are strong, moderate, weak, and nonsensitizers, respectively [446].

8. COMPARATIVE QSAR

8.1. Database Development

Due to the multifactorial nature of biological data as well as the large number of possible predictions that can be derived with many different QSAR models, it becomes imperative to develop a systematic QSAR database; the utility of such a database lies in powerful, rapid, and easy searching facility. The content of a QSAR database needs to be rigorously modeled using standard and controlled vocabulary in order to provide accurate information within a hypotheses-building environment. The practical purpose of a QSAR database varies widely from a source of QSAR data sets for modelers to a basis for "read-access" for regulators. Many tasks in the use of a database are closely related to the data mining; hence, database and data mining are the requisite technological pairs [447]. QSAR databases are likely to become a very powerful tool in upcoming years and will undoubtedly play a critical role in the chemoinformatics arena. Available QSAR databases as well as QSAR and molecular descriptors programs are listed in Table 6.

The C-QSAR database is one of a few databases for QSAR models [398,448]. It was designed to organize QSAR data on physical (PHYS) organic reactions as well as chemico-biological (BIO) interactions, in numerical terms in order to bring cohesion and understanding to mechanisms of chemical biodynamics. The two databases are organized on a similar format with the emphasis on reaction types in the PHYS database. The entries in the BIO database are sequestered into six main groups: macromolecules, enzymes, organelles, single celled organisms, organs/tissues, and multicellular organisms, for example, insects. The combined databases or the separate PHYS or BIO databases can be searched independently via string search or searching using the SMILES notation. A SMILES search can be approached in three ways: one can identify every QSAR that contains a specific molecule, one can utilize a MERLIN search that locates all derivatives of a given structure or one can search on single or multiple parameters. The net result of searching the QSAR database is to "mine" for models; one can call it "model mining!"

8.2. Database: Mining for Models

In order to enhance our understanding of ligand–receptor interactions and bring coherence to these relationships, there needs to be a concerted effort to not only develop high quality QSAR models but also formulate models that resonate with those drawn from mechanistic organic chemistry. A comprehensive, integrated database C-QSAR allows us to do so; it contains more than 22,000 examples drawn from all facets of chemistry and biology. An example on the toxicity of X-phenols and X-thiophenols will illustrate the usefulness of this database [340,398]. Recently, increasing numbers of QSAR for phenols have been based on Brown's σ^+ term; an electronic term that was first designed to rationalize polar electronic effects of substituents in electrophilic aromatic substitution, but has found increased utility in radical reactions as well. Recent studies in our laboratory have focused on the radical mediated toxicity of phenolic compounds and were extended to studies of butylated hydroxyanisole (BHA), butylated hydroxytoluene (BHT), and its derivatives (Fig. 13) in order to gauge the effects of large, bulky groups on the chemical and biological reactivity of the phenolic group [449].

Hydrogen Abstraction from 4-X-2,6-di-tert-butylphenols by styrene-peroxide at 65°C [[449]]

$$\log k = -1.15(\pm 0.22)\sigma^+ + 0.21(\pm 0.11) \tag{151}$$

$$n = 7,\ r^2 = 0.974,\ s = 0.114,\ q^2 = 0.952$$

The negative dependence on Brown's σ^+ constant indicates that electron-releasing substituents can stabilize the radical via spin delocalization [450]. Cytotoxicity studies of BHA/BHT derivatives resulted in the formulation of the following QSAR 128 [449].

Inhibition of growth of L1210 cells by 2-alkyl/2,6-dialkyl, 4-X-phenols [[449]]

$$\log\left(\frac{1}{C}\right) = 2.42(\pm 0.70)E_{\rm R} - 0.47(\pm 0.10)E_{\rm S-2} + 2.43(\pm 0.21) \tag{152}$$

$$n = 18,\ r^2 = 0.934,\ s = 0.136,\ q^2 = 0.904$$

$E_{\rm S\text{-}2}$, Taft's constant describes the larger of the two, *ortho*-substituents while $E_{\rm R}$ is Otsu's parameter that was specifically defined for radical reactions and was based on hydrogen abstraction from X-cumenes by a polystyryl radical. The strong dependence on $E_{\rm S\text{-}2}$ clearly establishes the critical role of steric factors on the ease of radicalization of hindered phenols. Even though different radical parameters are used to describe these reactions in different environments, they both accentuate the

Table 6. Available QSAR and Molecular Descriptor Programs, and QSAR Databases

No.	Name of the Programs/ Databases	Description in Brief	URL
1	C-QSAR	Development of QSAR model and a database of more than 22,000 QSARs of which 13,000 + pertain to biological systems and 8900 + are from mechanistic organic chemistry.	http://www.biobyte.com
2	Bio-Loom and QSAR Web database	To access BioByte's entire Masterfile database, which includes more than 60,000 measured log *P* and log *D* values (in many solvent systems), as well as 14,000 p*K*a values, including associated references. To access entire C-QSAR database and easy searching online. Hundreds of searchable biological activity types, both TARGETS (e.g., HmgCoA reductase inhibitors) and RESULTS (e.g., antihyperlipidemic) lead to thousands of active structures, many of which have been carefully studied via QSAR. Still calculating hydrophobic and molecular refractivity parameter via *C* log *P* and CMR, calculations.	http://www.biobyte.com
3	JRC QSAR model database	The JRC QSAR model database is freely accessible.	http://ecb.jrc.ec.europa.eu/qsar/qsar-tools/index.php?c=QRF
4	Danish QSAR database	A database for QSAR models.	http://www.mst.dk/English/Chemicals/Substances_and_materials/QSAR/
5	Molecular modeling Pro™ Plus	QSAR/QSPR model development and database storage	http://www.chemsw.com/13071.htm
6	Tsar	QSAR package for library design and lead optimization	http://accelrys.com/products/accord/desktop/tsar.html
7	TerraQSAR	**TerraQSAR**: QSAR development using personal computers with a windows operating system.	http://www.terrabase-inc.com
	TerraTox databases	**TerraTox – Databases**: Provides the search of compounds with structure/fragment-specific biological effects and properties.	
8	AMBIT	QSAR model development and chemical database	http://ambit.sourceforge.net
9	Cheminformatics and QSAR	QSAR/QSPR, Binary QSAR, ADME assessments, calculation of over 300 molecular descriptors, and combinatorial libraries	http://www.chemcomp.com/software-chem.htm
10	ChemSpider	A free source of structure-based chemistry information and the calculation of molecular descriptors	http://www.chemspider.com

Table 6. (*Continued*)

No.	Name of the Programs/ Databases	Description in Brief	URL
11	Partek QSAR solution	A comprehensive suite of advanced statistics, data mining, and interactive visualization. It automatically selects the most predictive chemical properties and constructs validated QSAR models using cross-validation.	http://www.partek.com
12	MCASE/ MC4PC	QSAR model development	http://www.multicase.com/products/prod01.htm
13	Strike	Strike (Statistical tool for revealing insight and knowledge): Software for statistical modeling and QSAR	http://www.schrodinger.com/ProductDescription.php?mID=6&sID=17
14	GQSAR	A fragment based approach to QSAR development	http://www.vlifesciences.com/products/QSARPro/gqsar.htm
15	Discovery Studio	QSAR modeling, includes molecular descriptors calculation, ADMET, pharmacophore modeling, simulation, and visualization	http://accelrys.com/products/discovery-studio/QSAR/
16	GOLPE	3D QSAR	http://www.miasrl.com/golpe.htm
17	Biograf^{3R}	**Quasar 5.2:** 6D QSAR **Raptor 2.0:** dual-shell 5D QSAR	http://www.biograf.ch/index.php
18	HQSAR, Almond, QSAR with CoMFA, VolSurf	**HQSAR:** It uses molecular holograms and PLS to generate fragment-based structure–activity relationships. **Almond:** Rapidly Generate, Validate, and Apply Pharmacodynamic QSAR Models without Alignment. **QSAR with CoMFA:** Builds statistical and graphical models that relate the properties of molecules (including biological activity) to their structures **VolSurf:** Calculate ADME Properties and Create Predictive ADME Models	http://www.tripos.com/index.php?family=modules,SimplePage,sybyl_ligand_based_design
19	DMax Chemistry Assistant™	QSAR/SAR development, compound screening data analysis, and virtual screening	http://www.pharmadm.com
20	CODESSA	Computes 600 + descriptors and development of QSAR models	http://www.semichem.com/codessa/default.php
21	ChemSAR	ChemSAR is a Chem3D Windows add-in for MS Excel with descriptive statistics and plots for SAR, and calculated some molecular descriptors e.g. log P, PSA etc.	http://chembionews.cambridgesoft.com/Articles/Default.aspx?articleID = 148
22	DRAGON	Calculation of 3200 + molecular descriptors	http://www.talete.mi.it
23	Spartan'06	Calculation of HOMO and LUMO Energies, Polar Surface Area, Electronegativity, Hardness, Dipole, etc.	http://www.wavefun.com
24	Hyperchem 7.5	Calculation of molecular descriptors	http://www.computational-chemistry.com/products/Professional/Features.htm

(*continued*)

Table 6. (*Continued*)

No.	Name of the Programs/ Databases	Description in Brief	URL
25	ACD/PhysChem Suite	Calculation of physicochemical properties of compounds (for example log *P*, log *D*, pKa, aqueous solubility, etc.)	http://www.acdlabs.com/products/phys_chem_lab/physchemsuite/
26	Molecule Evoluator and Polar Surface Area	**Molecule Evoluator:** Automatic calculation of physicochemical properties of molecules **Polar Surface Area:** A free software for the calculation of PSA of the molecules	http://www.cidrux.com
27	Mold2	A free software for fast-calculating molecular descriptors from a two-dimensional chemical structure and suitable for small and large data sets	http://www.fda.gov/nctr/science/centers/toxicoinformatics/Mold2/Mold.htm
28	PETRA	A free online service for calculating various physicochemical properties e.g. heats of formation, bond dissociation energies, sigma charge distribution, pi charge distribution, inductive effect, resonance effect, delocalization energies, polarizability, etc.	http://www2.chemie.unierlangen.de/services/petra/smiles.html
29	Molinspiration	Java based software for free online calculations of basic molecular properties/descriptors (for example, log *P*, polar surface area, number of hydrogen-bond donors and acceptors etc.)	http://www.molinspiration.com
30	Topological polar surface area (TPSA)	A free online calculation of TPSA for the organic molecules	http://www.daylight.com/meetings/emug00/Ertl/tpsa.html

importance of radicalization of the phenolic group and thus provide a consistent understanding of mechanistic behavior in chemical and biological systems—an excellent illustration of lateral validation.

In formulating QSAR, it is useful to utilize a well-designed series in order to optimize a particular biological activity. It is also important to insure that the ratio of compounds to parameter equal five, collinearity is minimized while spanned space is maximized. A normal distribution of biological data is necessary. A violation of these guidelines usually leads to statistically insignificant QSAR or models that defy predictability. One of our earliest works on the inhibition of *E. coli* DHFR by 2,4-diamino-5-X-benzyl pyrimidines led to the derivation of the following equa-

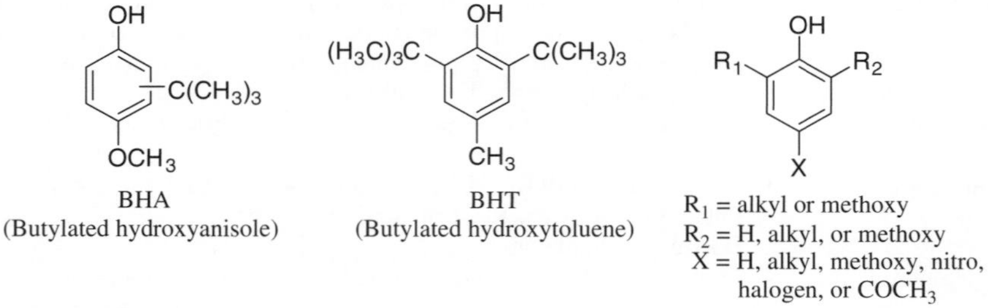

BHA
(Butylated hydroxyanisole)

BHT
(Butylated hydroxytoluene)

R_1 = alkyl or methoxy
R_2 = H, alkyl, or methoxy
X = H, alkyl, methoxy, nitro, halogen, or $COCH_3$

Figure 13. Structures of BHA, BHT, and BHT analogs.

tion [451]:

$$\log\left(\frac{1}{K_i}\right) = -1.13\,\sigma_R + 5.54 \tag{153}$$

$$n = 10,\ r^2 = 0.972,\ s = 0.182$$

Most of the variance in this data was explained by the Hammett through-resonance constant (σ_R). It implied that electron-releasing substituents enhanced inhibitory potency. Later, expanded and extensive studies on this system revealed that inhibition of the bacterial enzyme was related to mostly steric effects and there was little or no dependence on electronic terms. Careful analysis of the initial data revealed that it had a limited range in hydrophobicity and steric attributes. The lack of other QSAR to validate the findings in QSAR 153 made it statistically significant, at that time, but mechanistically weak as more elaborate studies were conducted.

8.3. Progress in QSAR

The past four decades has seen major changes in the QSAR paradigm. In tandem with developments in molecular modeling, NMR spectroscopy and X-ray crystallography, it has impacted drug design and development in numerous ways. It has also spawned 3D QSAR approaches such as CoMFA and CoMSIA that are routinely used in computer-assisted molecular design. In terms of ligand design, it shares center stage with other computational approaches such as structure-based ligand design including docking methods, molecular dynamics, pharmacophore searching and virtual screening, and genetic/evolutionary algorithms [452]. QSAR applications now cover a wide variety of fields and have been successfully employed for predictive, classification, and mechanistic purposes. Some of the biological phenomena that have been subjected to such analyses include toxicity prediction, drug resistance, apoptosis, drug absorption, metabolism, blood–brain barrier partitioning, drug "likeness" and prediction of physicochemical properties [167,393,411,453–457]. Success stories utilizing QSAR in new drug/pesticide development have been reviewed [458,459]. Bioactive compounds have emerged in agrochemistry, pesticide chemistry, and medicinal chemistry.

Bifenthrin, a pesticide resulted from a design strategy that utilized cluster analysis [458] (Fig. 14). Guided by QSAR analysis, the chemists at Kyorin Pharmaceutical Company designed and developed Norfloxacin, a 6-fluoro quinolone, which heralded the arrival of a new class of antibacterial agents [460] (Fig. 14). Two azole containing fungicides, metconazole and ipconazole were launched in 1994 in France and Japan, respectively [461]. Lomerizine, a 4-F-benzhydryl-4-(2,3,4-trimethoxy benzyl) piperazine, was introduced into the market in 1999 after extensive design strategies utilizing QSAR [462] (Fig. 15). Flobufen, an anti-inflammatory agent was designed by Kuchar et al. as a long acting agent without the usual gastric toxicity [463] (Fig. 15). Other examples of the commercial utility of QSAR include the development of metamitron and bromobutide [463]. In most of these examples, QSAR was used in combination with other rational drug design strategies, which is a useful and generally fruitful approach.

In addition to these commercial successes, the QSAR paradigm has steadily evolved into a

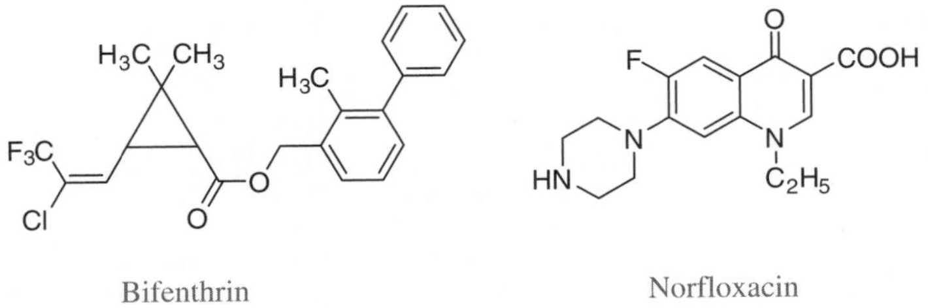

Bifenthrin

Norfloxacin

Figure 14. Structures of bifenthrin and norfloxacin. (This figure is available in full color at http://mrw.interscience.wiley.com/emrw/9780471266945/home.)

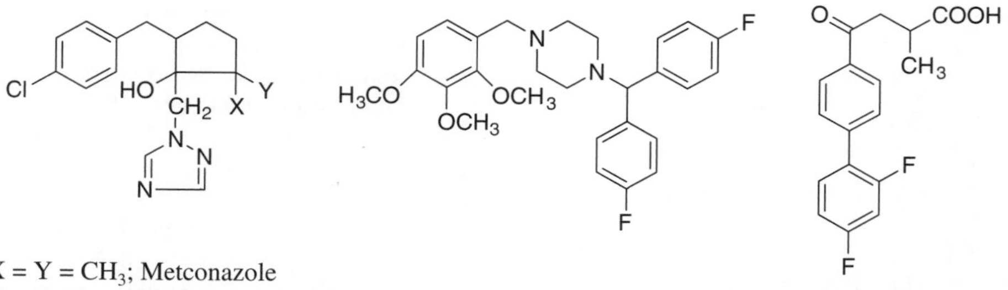

X = Y = CH_3; Metconazole
X = H, Y = $CH(CH_3)_2$; Ipconazolez

Lomerizine

Flobufen

Figure 15. Structures of metconazole, ipconazole, lomerizine, and flobufen. (This figure is available in full color at http://mrw.interscience.wiley.com/emrw/9780471266945/home.)

science. It is empirical in nature and it seeks to bring coherence and rigor to the QSAR models that are developed. By comparing models and using meaningful descriptors, one is able to more fully comprehend scientific phenomena with a "global" perspective; trends in patterns of reactivity or biological activity become self-evident. Recently in an elegant review, Guha has stressed the dire need for adequate interpretation of QSAR models and a careful analysis of the factors that impact the interpretability of the models [464].

9. SUCCESS OF QSAR

The scope of 2D QSAR also extends into the realm of not only optimizing biological activity but also making the jump to the development of potential commercial entities. It is difficult to exactly ascertain the role of this approach to the drug design and development process since it blends seamlessly with many other approaches such as 3D QSAR, quantum calculations and docking that are also used in these ventures. Thus, teasing out of the contribution of 2D QSAR is difficult at best. This limitation is compounded by the reality that pharmaceutical/biotechnology companies are unable to disclose results in this competitive arena due to confidentiality restrictions. However, there are examples of successful studies using this approach and they are listed in Table 7. Many thanks go to Toshio Fujita who has judiciously monitored commercial progress in this area for over the past 25 years.

Table 7. Chemical Entities Suggested by QSAR

No.	Agent	Class	Type of Approach	Reference(s)
1	Norfloxacin	Antibacterial	2D QSAR	[465,466]
	Ofloxacin	Antibacterial	2D QSAR, Bioisoterism	[467]
2	Bromobutide	Herbicide	2D QSAR	[465,468]
3	(*S*)-RS-8359 (derivative of RS-2232)	Antidepressant	2D QSAR	[469–471]
4	Zaprinast	Antiallergy	2D QSAR	[469,472]
5	KB-R9032	Na/H exchange inhibitor	2D QSAR	[473]
6	Lomerizine	Antimigraine	2D QSAR	[459,474]
7	Metconazole	Fungicide	2D QSAR	[459,461]
	Ipconazole	Fungicide	2D QSAR	
8	Flobufen	Antiflammatory	2D QSAR	[475,476]
9	Metamitron	Herbicide	2D QSAR	[477]
10	Muzolimine	Diuretic	Quantum chemical calculations	[478]
11	Amsacrine	Antineoplastic	2D QSAR	[479]

10. SUMMARY

The QSAR approach has been instrumental in enhancing our understanding of fundamental processes and phenomena in medicinal chemistry and drug design [480]. The concept of hydrophobicity and calculation of its key, representative descriptors has generated knowledge and extensive discussion as well as spawned a mini industry in this focused area. QSAR has also refined our thinking on selectivity at the molecular and cellular level. Hydrophobicity requirements vary considerably between sensitive tumor cells and resistant ones. Clear delineation of hydrophobic and steric needs has allowed us to impart more selectivity into the design of antifolate, antibacterial agents. QSAR studies in the pharmacokinetic arena have established different hydrophobic requirements for renal/nonrenal clearance, while the optimum hydrophobicity for CNS penetration has been determined in a classical study by Hansch, Bjorkroth and Leo [481]. QSAR has also shown some success in uncovering allosteric effects in the well-defined and complex hemoglobin system, topoisomerases and anthrax lethal factor [356,482,483]. QSAR has evolved and morphed into a multipronged paradigm. It has made great inroads in the fields of drug design and development, environmental science and toxicology [484–488]. The reader is referred to many excellent reviews and tomes on this subject [71,255,270,304].

Working with biological systems to obtain accurate and precise data is usually a complex process and in attempting to formulate a QSAR, one must always be cognizant of the biochemistry of the system analyzed and the limitations of the experimental systems. Today, modeling approaches are different and diverse, descriptors (the bedrock of QSAR) have multiplied, and outputs are elaborated in many forms including equations and visual representations of contour maps. Newer technologies in the forefront include the utility of QSAR in regulatory guidelines, in pharmacophoric "hopping" and in recognition of certain features associated with certain target classes.

Yet the basic tenets of QSAR remain the same. They encompass accurate and unambiguous biological endpoints, robust, statistical methodologies, appropriate application domains that are restricted to spanned space, internal and external validation of the developed models, and excellent predictivity and mechanistic interpretability of the resulting model. It is clear to most practitioners in the field that the applications and approaches to QSAR are growing in leaps and bounds. One area that continues to merit consideration and careful deliberation is the one associated with descriptors. We would not be overzealous if we suggested that descriptors warrant refinement in order to enhance transparency and interpretability.

REFERENCES

1. Rozman KK, Doull J. Paracelsus, Haber and Arndt. Toxicology 2001;160(1–3):191–196.
2. Crum-Brown A, Fraser TR. On the connection between chemical constitution and physiological action. Part 1. On the physiological action of the ammonium bases, derived from Strychia, Brucia, Thebaia, Codeia, Morphia and Nicotia Trans R Soc Edin 1868–1869;25: 151–203.
3. Richardson BW. Physiological research on alcohols. Med Times Gaz 1869;2:703–706.
4. Richet MC. Noté sur le rapport entre la toxicité et les propriétés physiques des corps. Compt Rend Soc Biol 1893;45:775–776.
5. Meyer H. Zur Theorie der Alkoholnarkose. I. Welche Eigenschaft der anaesthetika bedingt ihre narkotische Wirkung? Arch Exp Pathol Pharmakol 1899;42:109–118.
6. Overton CE. In: Lipnick RL, editor. Studies of Narcosis. London: Chapman and Hall; 1991. (English translation of Overton CE. *Studien uber die narkose*. Jena, Germany: Gustav Fischer; 1901).
7. Albert A, Goldacre R, Phillips J. The strength of heterocyclic bases. J Chem Soc 1948;70: 2240–2249.
8. Bell PH, Roblin RO Jr. Studies in chemotherapy. VII. A theory of the relation of structure to activity of sulfanilamide type compounds. J Am Chem Soc 1942;64(12):2905–2917.
9. Thijssen HHW. Relation between structure of sulfonamides and inhibition of H2-pteroate [7,8-dihydropteroic acid] synthesis in *Escherichia coli*. J Pharm Pharmacol 1974;26 (4):228–234.

10. Foernzler EC, Martin AN. Molecular orbital calculations on sulfonamide molecules. J Pharm Sci 1967;56(5):608–615.
11. Rastelli A, De Benedetti PG, Gavioli Battistuzzi G, Albasini A. Role of anionic, imidic, and amidic forms in structure–activity relations. Correlation of electronic indexes and bacteriostatic activity in sulfonamides. J Med Chem 1975;18(10):963–967.
12. Hansch C. Quantitative approach to biochemical structure–activity relationships. Acc Chem Res 1969;2(8):232–239.
13. Hansch C, Fujita T. ρ–σ–π Analysis; method for the correlation of biological activity and chemical structure. J Am Chem Soc 1964;86 (8):1616–1626.
14. Brönsted JN, Pedersen K. The catalytic decomposition of nitramide and its physicochemical applications. Z Physik Chem 1924; 108:185–235.
15. Brönsted JN. Acid and basic catalysis. Chem Rev 1928;5(3):231–338.
16. Hammett LP. Some relations between reaction rates and equilibrium constants. Chem Rev 1935;17(1):125–136.
17. Shorter J. The prehistory of the Hammett equation. Chem Listy 2000;94(4):210–214.
18. Hammett LP. The effect of structure upon the reactions of organic compounds. Benzene derivatives. J Am Chem Soc 1937;59(1):96–103.
19. Hammett LP. Physical Organic Chemistry, 2nd ed. New York: McGraw-Hill; 1970.
20. Jaffe HH. A reexamination of the Hammett equation. Chem Rev 1953;53(2):191–261.
21. Exner O. Correlation Analysis of Chemical Data, 2nd ed. New York: Plenum Publishing; 1988. p 55–128.
22. Shorter J. Hammett memorial lecture. Prog Phys Org Chem 1990;17:1–29.
23. Fujita T, Nishioka T. The analysis of the *ortho* effect. Prog Phys Org Chem 1976;12:49–89.
24. Charton M. Quantitative treatment of the *ortho* effect. Prog Phys Org Chem 1971;8: 235–317.
25. Frenna V, Macaluso G, Consiglio G, Cosimelli B, Spinelli D. Mononuclear heterocyclic rearrangements. Part 16. Kinetic study of the rearrangement of some *ortho*-substituted *Z*-phenylhydrazones of 3-benzoyl-5-phenyl-1,2,4-oxadiazole into 2-aryl-4-benzoylamino-5-phenyl-1,2,3-triazoles in dioxane-water and in benzene. Tetrahedron 1999;55(44):12885–12896.
26. Consiglio G, Noto R, Spinelli D, Arnone C. Kinetics of the reactions of 2-bromo-3,5-dinitrothiophene with *ortho*-substituted anilines in methanol. An application of the Fujita–Nishioka equation. J Chem Soc Perkin Trans 2 1979;(2):219–221.
27. Okamoto Y, Brown HC. A quantitative treatment of electrophilic reactions of aromatic derivatives. J Org Chem 1957;22(5):485–494.
28. Johnson CD. The Hammett Equation. London: Cambridge University Press; 1973. p 28–31.
29. Yoshioka M, Hamamoto K, Kubota T. Relation between acid dissociation constants of *N*-arylsulfanilamides and the Hammett equation. Bull Chem Soc Jpn 1962;35(10): 1723–1728.
30. Yukawa Y, Tsuno Y. Resonance effect in Hammett relation. III. The modified Hammett relation for electrophilic reactions. Bull Chem Soc Jpn 1959;32(9):971–981.
31. Kalfus K, Kroupa J, Vecera M, Exner O. Additivity of substituent effects in *m*- and *p*-substituted benzoic acids. Coll Czech Chem Commun 1975;40(10):3009–3019.
32. Swain CG, Langsdorf WP Jr. Concerted displacement reactions. VI. *m*- and *p*-substituent effects as evidence for a unity of mechanism in organic halide reactions. J Am Chem Soc 1951;73(6):2813–2819.
33. Hoffmann J, Klicnar J, Sterba V, Vecera M. Kinetics of hydrolysis of substituted sylicylideneanilides. Coll Czech Chem Commun 1970;35(5):1387–1398.
34. Hart H, Sedor EA. Mechanism of cyclodehydration of 2-phenyltriarylcarbinols. J Am Chem Soc 1967;89(10):2342–2347.
35. Schreck JO. Nonlinear Hammett relations. J Chem Ed 1971;48(2):103–107.
36. Unger SH, Hansch C. Model building in structure–activity relations. Reexamination of adrenergic blocking activity of β-halo-β-arylalkylamines. J Med Chem 1973;16(7): 745–749.
37. Hoefnagel AJ, Hoefnagel MA, Wepster BM. Substituent effects. 6. Charged groups: a simple extension of the Hammett equation. J Org Chem 1978;43(25):4720–4745.
38. Palát K Jr, Böhm S, Braunerová G, Waisser K, Exner O. Reaction series not obeying the Hammett equation: conformational equilibria of substituted thiobenzanilides. New J Chem 2002;26(7):861–866.
39. Zimmerman HE, Schuster DI. Photochemical rearrangement of 4,4-diphenylcyclo-hexadienone. I. A general theory of photochemical reactions. J Am Chem Soc 1961;83(21): 4486–4488.

40. Cordes T, Schadendorf T, Priewisch B, Rueck-Braun K, Zinth W. The Hammett relationship and reactions in the excited electronic state: Hemithioindigo Z/E-photoisomerization. J Phys Chem A 2008;112(4):581–588.
41. Wells PR, Ehrenson S, Taft RW. Substituent effects in the naphthalene series. An analysis of polar and pi delocalization effects. Prog Phys Org Chem 1968;6:147–322.
42. Ehrenson S, Brownlee RTC, Taft RW. Generalized treatment of substituent effects in the benzene series. Statistical analysis by the dual substituent parameter equation. I. Prog Phys Org Chem 1973;10:1–80.
43. Jovanovic BZ, Marinkovic AD, Vitnik Z, Juranic IO. Substituent and structural effects on the kinetics of the reaction of *N*-(substituted phenylmethylene)-*m*- and -*p*-aminobenzoic acids with diazodiphenylmethane. J Serb Chem Soc 2007;72(12):1191–1200.
44. Böhm S, Exner O. Revision of the dual-substituent-parameter treatment; reaction series with a donor reaction center. Coll Czech Chem Commun 2007;72(8):1158–1176.
45. Taft RW Jr. Linear free-energy relationships from rates of esterification and hydrolysis of aliphatic and *ortho*-substituted benzoate esters. J Am Chem Soc 1952;74(11):2729–2732.
46. Taft RW Jr. The general nature of the proportionality of polar effects of substituent groups in organic chemistry. J Am Chem Soc 1953;75 (17):4231–4238.
47. Hansch C, Leo A, Taft RW. A survey of Hammett substituent constants and resonance and field parameters. Chem Rev 1991;91(2):165–195.
48. Taft RW, Topsom RD. The nature and analysis of substituent effects. Prog Phys Org Chem 1987;16:1–83.
49. Ferguson J. Use of chemical potentials as indexes of toxicity. Proc R Soc B 1939;127: 387–404.
50. Mullins LJ. Some physical mechanisms in narcosis. Chem Rev 1954;54(2):289–323.
51. Collander R. Partition of organic compounds between higher alcohols and water. Acta Chem Scand 1951;5:774–780.
52. Leo AJ, Hansch C. Linear free energy relations between partitioning solvent systems. J Org Chem 1971;36(11):1539–1544.
53. Hansen OR. Hammett series with biological activity. Acta Chem Scand 1962;16:1593–1600.
54. Veldstra H. The reaction of chemical structure to biological activity in growth substances. Annu Rev Plant Physiol 1953;4:151–198.
55. Hansch C, Maloney PP, Fujita T, Muir RM. Correlation of biological activity of phenoxyacetic acids with Hammett substituent constants and partition coefficients. Nature 1962;194(4824):178–180.
56. Hansch C, Muir RM, Fujita T, Maloney PP, Geiger F, Streich M. The correlation of biological activity of plant growth regulators and chloromycetin derivatives with Hammett constants and partition coefficients. J Am Chem Soc 1963;85(18):2817–2824.
57. Hansch C. The QSAR paradigm in the design of less toxic molecules. Drug Metab Rev 1985;15 (7):1279–1294.
58. Cronin MTD. The role of hydrophobicity in toxicity prediction. Curr Comput Aided Drug Des 2006;2(4):405–413.
59. Fujita T, Nakagawa Y. SAR and QSAR analyses of substituted dibenzoylhydrazines for their mode of action as ecdysone agonists. In: Devillers J,editor. Endocrine Disruption Modeling. 2009. Boca Raton, FL: CRC Press, p 357–377.
60. Ball P. Water as a biomolecule. ChemPhysChem 2008;9(18):2677–2685.
61. Blokzijl W, Engberts JBFN. Hydrophobic effects: opinion and fact. Angew Chem 1993;105 (11):1610–1648.
62. Blokzijl W, Engberts JBFN. Hydrophobic effects: opinion and fact. Angew Chem Int Ed Engl 1993;32(11):1545–1579.
63. Taylor PJ. Hydrophobic properties of drugs. In: Ramsden CA, editor. Comprehensive Medicinal Chemistry: The Rational Design, Mechanistic Study and Therapeutic Application of Chemical Compounds, Vol. 4 (*Quantitative Drug Design*). Elmsford, NY: Pergamon; 1990. p 241–294.
64. Hildebrand JH. A criticism of the term "hydrophobic bond". J Phys Chem 1968;72 (5):1841–1842.
65. Frank HS, Evans MW. Free volume and entropy in condensed systems. III. Entropy in binary liquid mixtures; partial molal entropy in dilute solutions; structure and thermodynamics in aqueous electrolytes. J Chem Phys 1945;13(11):507–532.
66. Nemethy G, Scheraga HA. Structure of water and hydrophobic bonding in proteins. I. A model for the thermodynamic properties of liquid water. J Chem Phys 1962;36(12):3382–3400.
67. Kauzmann W. Some factors in the interpretation of protein denaturation. Adv Protein Chem 1959;14:1–63.

68. Tanford C. The hydrophobic effect and the organization of living matter. Science 1978;200(4345):1012–1018.
69. Tanford C. How protein chemists learned about the hydrophobic factor. Protein Sci 1997;6(6):1358–1366.
70. Chandler D. Hydrophobicity: two faces of water. Nature 2002;417(6888):491.
71. Hansch C, Leo A. Exploring QSAR. In: Heller SR, editor. Fundamentals and Applications in Chemistry and Biology. Washington, DC: American Chemical Society; 1995.
72. Smith RN, Hansch C, Ames MM. Selection of a reference partitioning system for drug design work. J Pharm Sci 1975;64(4):599–606.
73. Singer SJ, Nicolson GL. Fluid mosaic model of the structure of cell membranes. Science 1972;175(4023):720–731.
74. DeBolt SE, Kollman PA. Investigation of structure, dynamics, and solvation in 1-octanol and its water-saturated solution: molecular dynamics and free-energy perturbation studies. J Am Chem Soc 1995;117(19):5316–5340.
75. Iwahashi M, Hayashi Y, Hachiya N, Matsuzawa H, Kobayashi H. Self-association of octan-1-ol in the pure liquid state and in decane solutions as observed by viscosity, self-diffusion, nuclear magnetic resonance and near-infrared spectroscopy measurements. J Chem Soc Faraday Trans 1993;89(4):707–712.
76. Franks NP, Abraham MH, Lieb WR. Molecular organization of liquid *n*-octanol: an X-ray diffraction analysis. J Pharm Sci 1993;82(5):466–470.
77. MacCallum JL, Tieleman DP. Structures of neat and hydrated 1-octanol from computer simulations. J Am Chem Soc 2002;124 (50):15085–15093.
78. Leo A, Hansch C, Elkins D. Partition coefficients and their uses. Chem Rev 1971;71 (6):525–616.
79. Rekker RF. The hydrophobic fragmental constant: its derivation and application: a means of characterizing membrane systems. Amsterdam: Elsevier; 1977.
80. Seiler P. Interconversion of lipophilicites from hydrocarbon/water systems into the octanol/water system. Eur J Med Chem 1974;9 (5):473–479.
81. Fujita T, Nishioka T, Nakajima M. Hydrogen-bonding parameter and its significance in quantitative structure–activity studies. J Med Chem 1977;20(8):1071–1081.
82. Leahy DE, Taylor PJ, Wait AR. Model solvent systems for QSAR. Part I. Propylene glycol dipelargonate (PGDP). A new standard solvent for use in partition coefficient determination. Quant Struct Act Relat 1989;8(1):17–31.
83. Leahy DE, Morris JJ, Taylor PJ, Wait AR. Membranes and their models: towards a rational choice of partitioning system. In: Silipo C, Vittoria A, editors. QSAR: Rational Approaches to the Design of Bioactive Compounds. Amsterdam: Elsevier; 1991. p 75–81.
84. El Tayar N, Tsai RS, Testa B, Carrupt PA, Leo A. Partitioning of solutes in different solvent systems: the contribution of hydrogen-bonding capacity and polarity. J Pharm Sci 1991;80 (6):590–598.
85. Deák K. Physico-chemical profiling of centrally acting molecules for prediction of pharmacokinetic properties. Acta pharm Hung 2008;78 (3):110–120.
86. Deák K, Takács-Novák K, Kapás M, Vastag M, Tihanyi K, Noszal B. Physico-chemical characterization of a novel group of dopamine D3/D2 receptor ligands, potential atypical antipsychotic agents. J Pharm Biomed Anal 2008;48(3):678–684.
87. Leo AJ. Some advantages of calculating octanol-water partition coefficients. J Pharm Sci 1987;76(2):166–168.
88. Leo AJ. Hydrophobic parameter: measurement and calculation. Methods Enzymol 1991;202:544–591.
89. Grunewald GL, Pleiss MA, Gatchell CL, Pazhenchevsky R, Rafferty MF. Gas chromatographic quantitation of underivatized amines in the determination of their octanol-0.1 M sodium hydroxide partition coefficients by the shake-flask method. J Chromatogr 1984;292(2):319–331.
90. De Bruijn J, Hermens J. Relationships between octanol/water partition coefficients and total molecular surface area and total molecular volume of hydrophobic organic chemicals. Quant Struct Act Relat 1990;9(1):11–21.
91. Tomlinson E, David SS, Parr GD, James M, Farraj N, Kinkel JFM, Gaisser D, Wynn HJ. Partition Coefficient: Determination and Estimation. In: Dunn WJ III, Block JH, Perlman RS, editors. Oxford, UK: Pergamon; 1986. p 83–99.
92. Dearden JC, Bresnen GM. The measurement of partition coefficients. Quant Struct Act Relat 1988;7(3):133–144.
93. Sangster J. Octanol–Water Partition Coefficients: Fundamentals and Physical Chemistry. Chichester: John Wiley & Sons, Inc.; 1997. p 113–156.

94. Avdeef A. pH-metric log$P$. Part 1. Difference plots for determining ion-pair octanol–water partition coefficients of multiprotic substances. Quant Struct Act Relat 1992;11(4):510–517.

95. Franke U, Munk A, Wiese M. Ionization constants and distribution coefficients of phenothiazines and calcium channel antagonists determined by a pH-metric method and correlation with calculated partition coefficients. J Pharm Sci 1999;88(1):89–95.

96. Dearden JC, Patel AM, Tubby JH. Hydrophobic substituent constants from thin-layer chromatography on polyamide plates. J Pharm Pharmacol 1974;26(Suppl): 74P–75P.

97. Draber W, Buchel KH, Dickore K. Mode of action and structure–activity correlations of 1,2,4-triazinone herbicides. Pestic Chem Proc Int Congr Pestic Chem 2 1972;5:153–175.

98. Vallat P, El Tayar N, Testa B, Slacanin I, Marston A, Hostettmann K. Centrifugal counter-current chromatography, a promising means of measuring partition coefficients. J Chromatogr 1990;504(2):411–419.

99. Berthod A, Carda-Broch S. Determination of liquid-liquid partition coefficients by separation methods. J Chromatogr A 2004;1037 (1–2):3–14.

100. Berthod A, Carda-Broch S, Garcia-Alvarez-Coque MC. Hydrophobicity of ionizable compounds. A theoretical study and measurements of diuretic octanol–water partition coefficients by countercurrent chromatography. Anal Chem 1999;71(4):879–888.

101. Liu X, Tanaka H, Yamauchi A, Testa B, Chuman H. Lipophilicity measurement by reversed-phase high-performance liquid chromatography (RP-HPLC): a comparison of two stationary phases based on retention mechanisms. Helv Chim Acta 2004;87(11):2866–2876.

102. Donovan SF, Pescatore MC. Method for measuring the logarithm of the octanol–water partition coefficient by using short octadecyl-poly (vinyl alcohol) high-performance liquid chromatography columns. J Chromatogr A 2002; 952(1–2):47–61.

103. Valko K, Bevan C, Reynolds D. Chromatographic hydrophobicity index by fast-gradient RP-HPLC: a high-throughput alternative to log P/log D. Anal Chem 1997;69(11):2022–2029.

104. Valko K, Du CM, Bevan C, Reynolds DP, Abraham MH. Rapid method for the estimation of octanol/water partition coefficient (Log P_{oct}) from gradient RP-HPLC retention and a hydrogen bond acidity term ($\sum\alpha$ 2H). Curr Med Chem 2001;8(9):1137–1146.

105. Austin RP, Davis AM, Manners CN. Partitioning of Ionizing Molecules between Aqueous Buffers and Phospholipid Vesicles. J Pharm Sci 1995;84(10):1180–1183.

106. Ottiger C, Wunderli-Allenspach H. Partition behavior of acids and bases in a phosphatidylcholine liposome–buffer equilibrium dialysis system. Eur J Pharm Sci 1997;5(4):223–231.

107. Taillardat-Bertschinger A, Carrupt PA, Barbato F, Testa B. Immobilized artificial membrane HPLC in drug research. J Med Chem 2003;46(5):655–665.

108. Faller B, Grimm HP, Loeuillet-Ritzler F, Arnold S, Briand X. High-throughput lipophilicity measurement with immobilized artificial membranes. J Med Chem 2005;48(7):2571– 2576.

109. Barbato F. The use of immobilized artificial membrane (IAM) chromatography for determination of lipophilicity. Curr Comp Aided Drug Des 2006;2(4):341–352.

110. Ong S, Cai SJ, Bernal C, Rhee D, Qiu X, Pidgeon C. Phospholipid immobilization on solid surfaces. Anal Chem 1994;66(6):782–792.

111. Ward RS, Davies J, Hodges G, Roberts DW. Applications of immobilised artificial membrane chromatography to quaternary alkylammonium sulfobetaines and comparison of chromatographic methods for estimating the octanol–water partition coefficient. J Chromatogr A 2003;1007(1–2):67–75.

112. Kaliszan R. Chromatography and capillary electrophoresis in modeling the basic processes of drug action. Trends Anal Chem 1999;18(6):400–410.

113. Abraham MH, Chadha HS, Leitao RAE, Mitchell RC, Lambert WJ, Kaliszan R, Nasal A, Haber P. Determination of solute lipophilicity, as log P(octanol) and log P(alkane) using poly(styrene-divinylbenzene) and immobilized artificial membrane stationary phases in reversed-phase high-performance liquid chromatography. J Chromatogr A 1997;766 (1–2):35–47.

114. Kaliszan R. High performance liquid chromatographic methods and procedures of hydrophobicity determination. Quant Struct Act Relat 1990;9(2):83–87.

115. Nasal A, Kaliszan R. Progress in the use of HPLC for evaluation of lipophilicity. Curr Comp Aided Drug Des 2006;2(4):327–340.

116. Giaginis C, Tsantili-Kakoulidou A. Alternative measures of lipophilicity: from octanol–water partitioning to immobilized artificial membrane retention. J Pharm Sci 2008;97 (8):2984–3004.

117. Fauchère JL, Pliska V. Hydrophobic parameters π of amino acid side chains from the partitioning of *N*-acetyl-amino acid amides. Eur J Med Chem 1983;18(4): 369–375.
118. Fauchère JL. Elements for the rational design of peptide drugs. In: Testa B, editor. Advances in Drug Research, Vol. 15, London: Academic Press; 1986. p 29.
119. Abraham DJ, Leo AJ. Extension of the fragment method to calculate amino acid zwitterion and side chain partition coefficients. Proteins 1987;2(2):130–152.
120. Akamatsu M, Yoshida Y, Nakamura H, Asao M, Iwamura H, Fujita T. Hydrophobicity of di- and tripeptides having unionizable side chains and correlation with substituent and structural parameters. Quant Struct Act Relat 1989;8 (3):195–203.
121. Sotomatsu-Niwa T, Ogino A. Evaluation of the hydrophobic parameters of the amino acid side chains of peptides and their application in QSAR and conformational studies. J Mol Struct 1997;392:43–54.
122. Hansch C, Leo A, Hoekman D. In: Heller SR, editor. Exploring QSAR, Hydrophobic, Electronic and Steric Constants, Vol. 2. Washington, DC: American Chemical Society; 1995.
123. Nys GG, Rekker RF. Statistical analysis of a series of partition coefficients with special reference to the predictability of folding of drug molecules. Introduction of hydrophobic fragmental constants (*f* values). Chim Ther 1973;8 (5):521–535.
124. Leo A, Jow PYC, Silipo C, Hansch C. Calculation of hydrophobic constant (log *P*) from π and *f* constants. J Med Chem 1975;18(9):865–868.
125. Hansch C, Leo A. Substituent Constants for Correlation Analysis in Chemistry and Biology. New York: Wiley; 1979.
126. Fujita T, Nishimura K, Takayama C, Yoshida M, Uchida M. Hydrophobicity as a key physicochemical parameter of environmental toxicology of pesticides. In: Krieger R, editor. Handbook of Pesticide Toxicology, Vol. 1.London: Academic Press; 2001. p 649–670.
127. Weininger D. SMILES, a chemical language and information system. 1. Introduction to methodology and encoding rules. J Chem Inf Comput Sci 1988;28(1):31–36.
128. Weininger D, Weininger A, Weininger JL. SMILES. 2. Algorithm for generation of unique SMILES notation. J Chem Inf Comput Sci 1989;29(2):97–101.
129. Leo A. Methods of calculating partition coefficients. In: Ramsden CA, editor. Comprehensive Medicinal Chemistry: The Rational Design, Mechanistic Study and Therapeutic Application of Chemical Compounds, Vol. 4 (*Quantitative Drug Design*) Elmsford, NY: Pergamon; 1990. p 295–319.
130. Leo A, Hoekman D,personal communication.
131. Leo AJ. Calculating $\log P_{\mathrm{oct}}$ from structures. Chem Rev 1993;93(4):1281–1306.
132. Leo AJ, Hoekman D. Calculating log *P*(oct) with no missing fragments; the problem of estimating new interaction parameters. Perspect Drug Discov Des 2000;18(*Hydrophobicity and Solvation in Drug Design*, Pt II): 19–38.
133. Tetko IV, Poda GI. Prediction of log *P* with property-based methods. Methods and Principles in Medicinal Chemistry 2008;37(*Molecular Drug Properties*): 381–406.
134. van de Waterbeemd H, Mannhold R. Programs and methods for calculation of log *P*-values. Quant Struct Act Relat 1996;15(5):410–412.
135. Mannhold R, van de Waterbeemd H. Substructure and whole molecule approaches for calculating log *P*. J Comput Aided Mol Des 2001;15 (4):337–354.
136. Leo AJ, Hansch C. Role of hydrophobic effects in mechanistic QSAR. Perspect Drug Discov Des 1999;17(*Hydrophobicity and Solvation in Drug Design*, Pt I): 1–25.
137. Rekker RF, De Kort HM. The hydrophobic fragmental constant; an extension to a 1000 data point set. Eur J Med Chem 1979;14 (6):479–488.
138. Klopman G, Li JY, Wang S, Dimayuga M. Computer automated log *P* calculations based on an extended group contribution approach. J Chem Inf Comput Sci 1994;34(4):752–781.
139. Ghose AK, Crippen GM. Use of physicochemical parameters in distance geometry and related three-dimensional quantitative structure–activity relationships: a demonstration using *Escherichia coli* dihydrofolate reductase inhibitors. J Med Chem 1985;28(3):333–346.
140. Suzuki T, Kudo Y. Automatic log *P* estimation based on combined additive modeling methods. J Comput Aided Mol Des 1990;4 (2):155–198.
141. Klopman G, Iroff LD. Calculation of partition coefficients by the charge density method. J Comput Chem 1981;2(2):157–160.
142. Broto P, Moreau G, Vandycke C. Molecular structures: perception, autocorrelation de-

scriptor and SAR studies. System of atomic contributions for the calculation of the n-octanol/water partition coefficients. Eur J Med Chem 1984;19(1):71–78.

143. Ghose AK, Pritchett A, Crippen GM. Atomic physicochemical parameters for three dimensional structure directed quantitative structure–activity relationships III: modeling hydrophobic interactions. J Comput Chem 1988;9(1):80–90.

144. Kantola A, Villar HO, Loew GH. Atom based parametrization for a conformationally dependent hydrophobic index. J Comput Chem 1991;12(6):681–689.

145. Moriguchi I, Hirono S, Liu Q, Nakagome I, Matsushita Y. Simple method of calculating octanol/water partition coefficient. Chem Pharm Bull 1992;40(1):127–130.

146. Kellogg GE, Joshi GS, Abraham DJ. New tools for modeling and understanding hydrophobicity and hydrophobic interactions. Med Chem Res 1992;1:444–453.

147. Devillers J, Domine D, Guillon C, Karcher W. Simulating lipophilicity of organic molecules with a back-propagation neural network. J Pharm Sci 1998;87(9):1086–1090.

148. Bodor N, Gabanyi Z, Wong CK. A new method for the estimation of partition coefficient. J Am Chem Soc 1989;111(11):3783–3786.

149. Kamlet MJ, Carr PW, Taft RW, Abraham MH. Linear solvation energy relationships. 13. Relationship between the Hildebrand solubility parameter, δH, and the solvatochromic parameter, π^*. J Am Chem Soc 1981;103 (20):6062–6066.

150. Kamlet MJ, Abboud JLM, Abraham MH, Taft RW. Linear solvation energy relationships. 23. A comprehensive collection of the solvatochromic parameters, π^*, α, and β, and some methods for simplifying the generalized solvatochromic equation. J Org Chem 1983;48 (17):2877–2887.

151. Platts JA, Butina D, Abraham MH, Hersey A. Estimation of molecular linear free energy relation descriptors using a group contribution approach. J Chem Inf Comput Sci 1999;39 (5):835–845.

152. Ishihama Y, Asakawa N. Characterization of lipophilicity scales using vectors from solvation energy descriptors. J Pharm Sci 1999;88 (12):1305–1312.

153. Platts JA, Abraham MH, Butina D, Hersey A. Estimation of molecular linear free energy relationship descriptors by a group contribution approach. 2. Prediction of partition coefficients. J Chem Inf Comput Sci 2000;40 (1):71–80.

154. Leo AJ. Evaluating hydrogen-bond donor strength. J Pharm Sci 2000;89(12):1567–1578.

155. Taft RW. Steric Effects in Organic Chemistry, In: Newman MS, editor. New York: Wiley; 1956. p 556.

156. Gallo R. Treatment of steric effects. Prog Phys Org Chem 1983;14:115–163.

157. Hancock CK, Meyers EA, Yager BJ. Quantitative separation of hyperconjugation effects from stearic substituent constants. J Am Chem Soc 1961;83(20):4211–4213.

158. Hansch C, Lien EJ, Helmer F. Structure–activity correlations in the metabolism of drugs. Arch Biochem Biophys 1968;128(2):319–330.

159. Charton M. Steric Effects in Drug Design. In: Charton MM, Motoc I, editors. Berlin: Springer; 1983; p 57.

160. Tute MS. History and objectives of quantitative drug design. In: Ramsden CA, editor. Comprehensive Medicinal Chemistry: The Rational Design, Mechanistic Study and Therapeutic Application of Chemical Compounds. Vol. 4 (*Quantitative Drug Design*). Elmsford, NY: Pergamon; 1990. p 1–31.

161. Hansch C, Klein TE. Molecular graphics and QSAR in the study of enzyme–ligand interactions. On the definition of bioreceptors. Acc Chem Res 1986;19(12):392–400.

162. Verloop A, Hoogenstraaten W, Tipker J. In: Ariens EJ, editor. Drug Design, Vol. VII. New York: Academic Press; 1976. p 165.

163. Verloop A. The STERIMOL Approach to Drug Design. New York: Marcel Dekker; 1987.

164. Hansch C, Hoekman D, Leo A, Weininger D, Selassie CD. Unpublished results.

165. Levin VA. Relationship of octanol/water partition coefficient and molecular weight to rat brain capillary permeability. J Med Chem 1980;23(6):682–684.

166. Lien EJ, Wang PH. Lipophilicity, molecular weight, and drug action: reexamination of parabolic and bilinear models. J Pharm Sci 1980;69(6):648–650.

167. Selassie CD, Hansch C, Khwaja TA. Structure–activity relationships of antineoplastic agents in multidrug resistance. J Med Chem 1990;33(7):1914–1919.

168. Lien EJ, Lien LL, and Gao H. In: Sanz F, Guiraldo J, Manaut F, editors. QSAR and Molecular Modelling: Concepts, Computational Tools and Biological Applications. Barcelona: Prous Science; 1995. p 94.

169. Selassie C. Unpublished results.

170. Recanatini M, Klein T, Yang CZ, McClarin J, Langridge R, Hansch C. Quantitative structure–activity relationships and molecular graphics in ligand receptor interactions: amidine inhibition of trypsin. Mol Pharmacol 1986;29 (4):436–446.

171. Naito Y, Sugiura M, Yamaura Y, Fukaya C, Yokoyama K, Nakagawa Y, Ikeda T, Senda M, Fujita T. Quantitative structure–activity relationship of catechol derivatives inhibiting 5-lipoxygenase. Chem Pharm Bull 1991;39 (7):1736–1745.

172. Debnath AK, Lopez de Compadre RL, Debnath G, Shusterman AJ, Hansch C. Structure–activity relationship of mutagenic aromatic and heteroaromatic nitro compounds. Correlation with molecular orbital energies and hydrophobicity. J Med Chem 1991;34(2):786–797.

173. Kubinyi H. Quantitative structure–activity relations. IV. Non-linear dependence of biological activity on hydrophobic character: a new model. Arzneim Forsch 1976;26(11): 1991–1997.

174. Abraham MH, Grellier PL, Prior DV, Taft RW, Morris JJ, Taylor PJ, Laurence C, Berthelot M, Doherty RM, Kamlet MJ, Abbound J-LM, Sraidi K, Guihëneuf G. A general treatment of hydrogen bond complexation constants in tetrachloromethane. J Am Chem Soc 1988; 110(25):8534–8536.

175. Seiler P. Interconversion of lipophilicites from hydrocarbon/water systems into the octanol/water system. Eur J Med Chem 1974;9 (5):473–479.

176. Moriguchi I. Quantitative structure–activity studies. I. Parameters relating to hydrophobicity. Chem Pharm Bull 1975;23(2):247–257.

177. Fujita T, Nishioka T, Nakajima M. Hydrogen-bonding parameter and its significance in quantitative structure–activity studies. J Med Chem 1977;20(8):1071–1081.

178. Charton M, Charton BI. The structural dependence of amino acid hydrophobicity parameters. J Theoret Biol 1982;99(4): 629–644.

179. Dearden JC, Ghafourion T. In: Sanz F, Guiraldo J, Manaut F, editors. QSAR and Molecular Modelling: Concepts, Computational Tools and Biological Applications. Barcelona: Prous Science; 1995. p 117.

180. Raevsky OA. Quantification of non-covalent interactions on the basis of the thermodynamic hydrogen bond parameters. J Phys Org Chem 1997;10(5):405–413.

181. Kamlet MJ, Doherty RM, Abboud JLM, Abraham MH, Taft RW. Linear solvation energy relationships. 36. Molecular properties governing solubilities of organic nonelectrolytes in water. J Pharm Sci 1986;75(4): 338–349.

182. Schüürmann G. Quantitative structure–property relationships for the polarizability, solvatochromic parameters and lipophilicity. Quant Struct Act Relat 1990;9:326–333.

183. Wilson LY, Famini GR. Using theoretical descriptors in quantitative structure–activity relationships: some toxicological indices. J Med Chem 1991;34(5):1668–1674.

184. Famini GR, Aguiar D, Payne MA, Rodriquez R, Wilson LY. Using the theoretical linear energy solvation energy relationship to correlate and predict nasal pungency thresholds. J Mol Graph Model 2002;20(4):277–280.

185. Charton M. Hydrogen bonding contribution to lipophilicity parameters. Hydrogen acceptor and hydrogen acceptor–donor parameters for substituents bonded to sp^2-hybridized carbon atoms. Curr Comp Aided Drug Des 2006;2 (4):353–368.

186. Karelson M, Lobanov VS, Katritzky AR. Quantum-chemical descriptors in QSAR/QSPR studies. Chem Rev 1996;96(3):1027–1043.

187. Shusterman AJ, Debnath AK, Hansch C, Horn GW, Fronczek FR, Greene AC, Watkins SF. Mutagenicity of dimethyl heteroaromatic triazenes in the Ames test: the role of hydrophobicity and electronic effects. Mol Pharmacol 1989;36(6):939–944.

188. Yamagami C, Motohashi N, Akamatsu M. Quantum chemical- and 3-D-QSAR (CoMFA) studies of benzalacetones and 1,1,1-trifluoro-4-phenyl-3-buten-2-ones. Bioorg Med Chem Lett 2002;12(17):2281–2285.

189. Randic M. Characterization of molecular branching. J Am Chem Soc 1975;97 (23):6609–6615.

190. Randic M. On computation of optimal parameters for multivariate analysis of structure–property relationship. J Comput Chem 1991;12(8):970–980.

191. Kier LB, Hall LH. Molecular Connectivity in Chemistry and Drug Research. New York: Academic Press; 1976.

192. Hall LH, Kier LB. Electrotopological state indices for atom types: a novel combination of electronic, topological, and valence state information. J Chem Inf Comput Sci 1995;35 (6):1039–1045.

193. Kier LB, Hall LH. General definition of valence delta-values for molecular connectivity. J Pharm Sci 1983;72(10):1170–1173.

194. Murray WJ, Hall LH, Kier LB. Molecular connectivity. III. Relation to partition coefficients. J Pharm Sci 1975;64(12):1978–1981.

195. Kier LB, Hall LH. Molecular Structure Descriptors, The Electrotopological State. San Diego, CA: Academic Press; 1999.

196. Hall LH, Mohney B, Kier LB. The electrotopological state: an atom index for QSAR. Quant Struct Act Relat 1991;10(1):43–51.

197. Buolamwini JK, Raghavan K, Fesen MR, Pommier Y, Kohn KW, Weinstein JN. Application of the electrotopological state index to QSAR analysis of flavone derivatives as HIV-1 integrase inhibitors. Pharm Res 1996;13(12): 1892–1895.

198. Mukherjee S, Mukherjee A, Saha A. QSAR studies with E-state index: predicting pharmacophore signals for estrogen receptor binding affinity of triphenylacrylonitriles. Biol Pharm Bull 2005;28(1):154–157.

199. Cash GG, Anderson B, Mayo K, Bogaczyk S, Tunkel J. Predicting genotoxicity of aromatic and heteroaromatic amines using electrotopological state indices. Mutat Res 2005;585 (1–2):170–183.

200. Kubinyi H. The physicochemical significance of topological parameters. A rebuttal. Quant Struct Act Relat 1995;14(2):149–150.

201. Lopez de Compadre RL, Compadre CM, Castillo R, Dunn WJ III On the use of connectivity indexes in quantitative structure–activity studies. Eur J Med Chem 1983;18(6): 569–571.

202. Hall LH, Kier LB. Structure–activity studies using valence molecular connectivity. J Pharm Sci 1977;66(5):642–644.

203. Bosque R, Sales J. Polarizabilities of Solvents from the Chemical Composition. J Chem Inf Comput Sci 2002;42(5):1154–1163.

204. Zhokhova NI, Baskin II, Palyulin VA, Zefirov AN, Zefirov NS. Fragmental descriptors in QSPR: application to molecular polarizability calculations. Russ Chem Bull Int Ed 2003;52 (5):1061–1065.

205. Hansch C, Steinmetz WE, Leo AJ, Mekapati SB, Kurup A, Hoekman D. On the role of polarizability in chemical–biological interactions. J Chem Inf Comput Sci 2003;43 (1):120–125.

206. Verma RP, Hansch C. A comparison between two polarizability parameters in chemical–biological interactions. Bioorg Med Chem 2005;13 (7):2355–2372.

207. Verma RP, Kurup A, Hansch C. On the role of polarizability in QSAR. Bioorg Med Chem 2005;13(1):237–255.

208. Ertl P, Rohde B, Selzer P. Fast Calculation of molecular polar surface area as a sum of fragment-based contributions and its application to the prediction of drug transport properties. J Med Chem 2000;43(20):3714–3717.

209. Kelder J, Grootenhuis PDJ, Bayada DM, Delbressine LPC, Ploemen J-P. Polar molecular surface as a dominating determinant for oral absorption and brain penetration of drugs. Pharm Res 1999;16(10):1514–1519.

210. Clark DE. Rapid calculation of polar molecular surface area and its application to the prediction of transport phenomena. 1. Prediction of intestinal absorption. J Pharm Sci 1999;88 (8):807–814.

211. Clark DE. Rapid calculation of polar molecular surface area and its application to the prediction of transport phenomena. 2. Prediction of blood–brain barrier penetration. J Pharm Sci 1999;88(8):815–821.

212. Turabekova MA, Rasulev BF. A QSAR toxicity study of a series of alkaloids with the lycoctonine skeleton. Molecules 2004;9(12): 1194–1207.

213. Zhao YH, Abraham MH, Zissimos AM. Fast calculation of van der Waals volume as a sum of atomic and bond contributions and its application to drug compounds. J Org Chem 2003;68(19):7368–7373.

214. Agrawal VK, Mishra K, Sharma R, Khadikar PV. Topological estimation of cytotoxic activity of some anti-HIV agents: HEPT analogues. J Chem Sci 2004;116(2):93–99.

215. Toropov AA, Benfenati E. SMILES as an alternative to the graph in QSAR modelling of bee toxicity. Comput Biol Chem 2007;31(1):57–60.

216. Toropov AA, Toropova AP, Raska I Jr. QSPR modeling of octanol/water partition coefficient for vitamins by optimal descriptors calculated with SMILES. Eur J Med Chem 2008;43 (4):714–740.

217. Toropov AA, Rasulev BF, Leszczynska D, Leszczynski J. Multiplicative SMILES-based optimal descriptors: QSPR modeling of fullerene C60 solubility in organic solvents. Chem Phys Lett 2008;457(4–6):332–336.

218. Roy K, Toropov A, Raska I Jr. QSAR modeling of peripheral versus central benzodiazepine receptor binding affinity of 2-phenylimidazo

[1,2-a]pyridineacetamides using optimal descriptors calculated with SMILES. QSAR Comb Sci 2007;26(4):460–468.

219. Esteki M, Khayamian T. Mechanistic-based descriptors for QSAR study of psychotropic drug toxicity. Chem Biol Drug Des 2008;72 (5):409–435.

220. Perez DM, Karnik SS. The Handbook of Experimental Pharmacology. 148 (*Pharmacology of Functional, Biochemical and Recombinant Receptor System*). Springer-Verlag; 2000. p 283.

221. Langley JN. On the Physiology of the Salivary Secretion. Part II. On the mutual antagonism of atropin and pilocarpin, having especial reference to their relations in the sub-maxillary gland of the Cat. J Physiol 1878;1(4–5):339–369.

222. Ehrlich P. Die Wertbemessung des Diphtherieheilserums und deren theoretische Grundlagen. Klin Jahrb 1897;6:299–333.

223. Langley JN. Over the reaction of the cells and the nerve ending to certain poisons, in particular in consideration of the reaction of the transverse striped muscle to nicotine and to curare. [machine translation]. J Physiol 1905;33:374–413.

224. Famulok M. Oligonucleotide aptamers that recognize small molecules. Curr Opin Struct Biol 1999;9(3):324–329.

225. Wang KY, Swaminathan S, Bolton PH. Tertiary structure motif of oxytricha telomere DNA. Biochemistry 1994;33(24):7517–7527.

226. Lown JW. Molecular Aspects of Anticancer Drug-DNA Interactions. In: Neidle S, Waring M-J, editors. Basinstoke, UK: Macmillan; 1993. p 322.

227. Smith RN, Hansch C, Kim KH, Omiya B, Fukumura G, Selassie CD, Jow PYC, Blaney JM, Langridge R. The use of crystallography, graphics, and quantitative structure–activity relationships in the analysis of the papain hydrolysis of X-phenyl hippurates. Arch Biochem Biophys 1982;215(1):319–328.

228. Hansch C, Klein T, McClarin J, Langridge R, Cornell NW. A quantitative structure–activity relationship and molecular graphics analysis of hydrophobic effects in the interactions of inhibitors with alcohol dehydrogenase. J Med Chem 1986;29(5):615–620.

229. Selassie CD, Fang ZX, Li RL, Hansch C, Klein T, Langridge R, Kaufman BT. Inhibition of chicken liver dihydrofolate reductase by 5-(substituted benzyl)-2,4-diaminopyrimidines. A QSAR and graphics analysis. J Med Chem 1986;29(5):621–626.

230. Blaney JM, Hansch C. Application of molecular graphics to the analysis of macromolecular structures. In: Ramsden CA, editor. Comprehensive Medicinal Chemistry: The Rational Design, Mechanistic Study and Therapeutic Application of Chemical Compounds. Vol. 4 (*Quantitative Drug Design*) 1990. Elmsford, NY: Pergamon. p 459–496.

231. Roberts GCK. Flexible keys and deformable locks: ligand binding to dihydrofolate reductase. Pharmacochem Libr 1983;6 (*Quantitaive Approaches to Drug Design*): 91–98.

232. Parrill AL. Crystal structures of a second G protein-coupled receptor: triumphs and implications. ChemMedChem 2008;3(7): 1021–1023.

233. Palczewski K, Kumasaka T, Hori T, Behnke CA, Motoshima H, Fox BA, Le Trong I, Teller DC, Okada T, Stenkamp RE, Yamamoto M, Miyano M. Crystal structure of rhodopsin: a G protein-coupled receptor. Science 2000;289 (5480):739–745.

234. Rasmussen SGF, Choi H-J, Rosenbaum DM, Kobilka TS, Thian FS, Edwards PC, Burghammer M, Ratnala VRP, Sanishvili R, Fischetti RF, Schertler GFX, Weis WI, Kobilka BK. Crystal structure of the human $\beta 2$ adrenergic G-protein-coupled receptor. Nature 2007;450 (7168):383–387.

235. Kumar AA, Mangum JH, Blankenship DT, Freisheim JH. Affinity labeling of chicken liver dihydrofolate reductase by a substituted 4,6-diaminodihydrotriazine bearing a terminal sulfonyl fluoride. J Biol Chem 1981;256 (17):8970–8976.

236. Rose GD, Wolfenden R. Hydrogen bonding, hydrophobicity, packing, and protein folding. Annu Rev Biophys Biomol Struct 1993;22: 381–415.

237. Hagler AT, Dauber P, Lifson S. Consistent force field studies of intermolecular forces in hydrogen-bonded crystals. 3. The C:O···H–O hydrogen bond and the analysis of the energetics and packing of carboxylic acids. J Am Chem Soc 1979;101(18):5131–5141.

238. Kauzmann W. Some factors in the interpretation of protein denaturation. Adv Protein Chem 1959;14:1–63.

239. Ben-Naim A. Solvation and solubility of globular proteins. Pure Appl Chem 1997;9 (11):2239–2243.

240. Muller N. Search for a realistic view of hydrophobic effects. Acc Chem Res 1990;23 (1):23–28.

241. Eisenhaber F. Hydrophobic regions on protein surfaces. Perspect Drug Discov Des 1999;17

(*Hydrophobicity and Solvation in Drug Design*, Part I): 27–42.

242. Fersht AR, Shindler JS, Tsui W-C. Probing the limits of protein-amino acid side chain recognition with the aminoacyl-tRNA synthetases. Discrimination against phenylalanine by tyrosyl-tRNA synthetases. Biochemistry 1980;19 (24):5520–5524.
243. Andrews PR, Craik DJ, Martin JL. Functional group contributions to drug–receptor interactions. J Med Chem 1984;27(12):1648–1657.
244. Topliss JG, Edwards RP. Chance factors in studies of quantitative structure–activity relationships. J Med Chem 1979;22(10):1238–1244.
245. Craig PN. Interdependence between physical parameters and selection of substituent groups for correlation studies. J Med Chem 1971;14(8):680–684.
246. Topliss JG. Utilization of operational schemes for analog synthesis in drug design. J Med Chem 1972;15(10):1006–1011.
247. Topliss JG. A manual method for applying the Hansch approach to drug design. J Med Chem 1977;20(4):463–469.
248. Bustard TM. Optimization of alkyl modifications by Fibonacci search. J Med Chem 1974;17 (7):777–778.
249. Darvas F. Application of the sequential simplex method in designing drug analogs. J Med Chem 1974;17(8):799–804.
250. Magee PS. Pesticide chemistry: human welfare and environment. In: Miyamoto J. and Kearney PC, editors. Proceedings of the International Congress on Pesticide Chemistry, Vol. 1. Oxford, UK: Pergamon; 1983. p 251.
251. Sjöström M, Eriksson L. Applications of statistical experimental design and PLS modelling in QSAR. In: van de Waterbeemd H, editor. Chemometric Methods in Molecular Design. Weinheim, Germany: VCH; 1995. p 63–90.
252. Eriksson L, Johansson E, Mueller M, Wold S. Cluster-based design in environmental QSAR. Quant Struct Act Relat 1997;16(5):383–390.
253. Eriksson L, Johansson E, Muller M, Wold S. On the selection of the training set in environmental QSAR analysis when compounds are clustered. J Chemom 2000;14(5–6):599–616.
254. Janssen PAJ, Eddy NB. Compounds related to pethidine. IV. General chemical methods of increasing the analgesic activity of pethidine. J Med Pharm Chem 1960;2(1):31–45.
255. Fujita T. The extrathermodynamic approach to drug design. In: Ramsden CA, editor. Comprehensive Medicinal Chemistry: The Rational Design, Mechanistic Study and Therapeutic Application of Chemical Compounds. Vol. 4 (*Quantitative Drug Design*). Elmsford, NY: Pergamon; 1990. p 503.
256. Hansch C, Kim D, Leo AJ, Novellino E, Silipo C, Vittoria A. Toward a quantitative comparative toxicology of organic compounds. Crit Rev Toxicol 1989;19(3):185–226.
257. Hansch C, Dunn WJ III. Linear relations between lipophilic character and biological activity of drugs. J Pharm Sci 1972;61(1):1–19.
258. Schultz TW, Tichy M. Structure–toxicity relationships for unsaturated alcohols to *Tetrahymena pyriformis*: C5 and C6 analogs and primary propargylic alcohols. Bull Environ Contam Toxicol 1993;51(5):681–688.
259. C-QSAR Database, BioByte Corp., Claremont, CA.
260. Penniston JT, Beckett L, Bentley DL, Hansch C. Passive permeation of organic compounds through biological tissue; a non-steady-state theory. Mol Pharmacol 1969;5(4):333–341.
261. Hansch C. Computerized approach to quantitative biochemical structure–activity relations. Adv Chem Ser 1972;114 (*Biological Correlation: the Hansch Approach*, Symposium): 20–40.
262. Hansch C, Clayton JM. Lipophilic character and biological activity of drugs. II. Parabolic case. J Pharm Sci 1973;62(1):1–21.
263. Franke R, Schmidt W. Parabolic relations between biological activity and hydrophobicity in congenergic series. Transport or protein binding. Acta Biol Med Germ 1973;31(2):273–287.
264. McFarland JW. Parabolic relation between drug potency and hydrophobicity. J Med Chem 1970;13(6):1192–1196.
265. Kubinyi H, Kehrhahn OH. Quantitative structure–activity relationships. VI. Non-linear dependence of biological activity on hydrophobic character: calculation procedures for the bilinear model. Arzneim Forsch 1978;28 (4):598–601.
266. Kurup A, Garg R, Carini DJ, Hansch C. Comparative QSAR: angiotensin II antagonists. Chem Rev 2001;101(9):2727–2750.
267. Free SM Jr, Wilson JW. A mathematical contribution to structure–activity studies. J Med Chem 1964;7(4):395–399.
268. Kubinyi H. Lipophilicity and biological activity. Drug transport and drug distribution in model systems and in biological systems. Arzneim Forsch 1979;29(8):1067–1080.

269. Franke R. In: Nauta WTh, Rekker RF, editors. Theoretical drug design methods. New York: Elsevier; 1984. p 256.

270. Kubinyi H. The Free-Wilson method and its relationship to the extrathermodynamic approach. In: Ramsden CA, editor. Comprehensive Medicinal Chemistry: The Rational Design, Mechanistic Study and Therapeutic Application of Chemical Compounds. Vol. 4 (*Quantitative Drug Design*). Elmsford, NY: Pergamon; 1990. p 589–643.

271. Blankley CJ. In: Topliss JG, editor. Quantitative Structure Activity Relationships of Drugs. New York: Academic Press; 1983. p 5.

272. Fujita T, Ban T. Structure–activity relation. 3. Structure–activity study of phenethylamines as substrates of biosynthetic enzymes of sympathetic transmitters. J Med Chem 1971;14 (2):148–152.

273. Yalcin E, Sener SE, Owren I, Temiz O.In: Sanz E, Giraldo J, Manaut F, editors. QSAR and Molecular Modelling: Concepts, Computational Tools and Biological Applications. Barcelona: Prous Science; 1995. p 147.

274. Kunes J, Jachym J, Jirasko P, Odlerova Z, Waisser K. Relationships between the chemical structure of substances and their antimycobacterial activity to a typical strains. XII. Combination of the Topliss approach with the Free-Wilson analysis in the study of antimycobacterial activity of 4-alkylthiobenzanilides. Collect Czech Chem Commun 1997;62 (9):1503–1510.

275. Terada Y, Nanya K. Free-Wilson analysis of the antibacterial activity of fluoronaphthyridines against various microbes. A new application of indicator variables. Die Pharmazie 2000;55(2):133–135.

276. Gupta SP, Paleti A. Quantitative structure–activity relationship studies on some nonbenzodiazepine series of compounds acting at the benzodiazepine receptor. Bioorg Med Chem 1998;6(11):2213–2218.

277. Tmej C, Chiba P, Huber M, Richter E, Hitzler M, Schaper K-J, Ecker G. A combined Hansch/Free-Wilson approach as predictive tool in QSAR studies on propafenone-type modulators of multidrug resistance. Arch der Pharmazie 1998;331(7–8):233–240.

278. Wold S. PLS for multivariate linear modeling. In: van de Waterbeemed H, editor. Chemometric Methods in Molecular Design. 1995. Weinheim, Germany: VCH. p 195.

279. Deeb O, Hemmateenejad B, Jaber A, Garduno-Juarez R, Miri R. Effect of the electronic and physicochemical parameters on the carcinogenesis activity of some sulfa drugs using QSAR analysis based on genetic-MLR and genetic-PLS. Chemosphere 2007;67(11):2122–2130.

280. Ajmani S, Kulkarni SA. A dual-response partial least squares regression QSAR model and its application in design of dual activators of PPARα and PPARγ. QSAR Comb Sci 2008;27 (11–12):1291–1304.

281. Khanna S, Sobhia ME, Bharatam PV. Additivity of molecular fields: CoMFA study on dual activators of PPARα and PPARγ. J Med Chem 2005;48(8):3015–3025.

282. Golbraikh A, Tropsha A. QSAR modeling using chirality descriptors derived from molecular topology. J Chem Inf Comput Sci 2003;43 (1):144–154.

283. Hawkins DM. On the investigation of alternative regressions by principal component analysis. Appl Stat 1973;22:275–286.

284. Sutter JM, Kalivas JH, Lang PM. Which principal components to utilize for principal component regression. J Chemometr 1992;6 (4):217–225.

285. Hemmateenejad B. Optimal QSAR analysis of the carcinogenic activity of drugs by correlation ranking and genetic algorithm based PCR. J Chemometr 2004;18:475–485.

286. Verduj-Andres J, Massart DL. Comparison of prediction- and correlation-based methods to select the best subset of principal components for principal component regression and detect outlying objects. Appl Spectrosc 1998;52 (11):1425–1434.

287. Barros AS, Rutledge DN. Genetic algorithm applied to the selection of principal components. Chemom Intell Lab Syst 1998;40 (1):65–81.

288. Sabet R, Fassihi A. QSAR study of antimicrobial 3-hydroxypyridine-4-one and 3-hydroxypyran-4-one derivatives using different chemometric tools. Int J Mol Sci 2008;9 (12):2407–2423.

289. Yang L, Wang P, Jiang Y, Chen J. Studying the explanatory capacity of artificial neural networks for understanding environmental chemical quantitative structure–activity relationship models. J Chem Inf Model 2005;45 (6):1804–1811.

290. Tetko IV, Villa AEP, Livingstone DJ. Neural network studies. 2. Variable selection. J Chem Inf Comput Sci 1996;36(4):794–803.

291. Turner JV, Cutler DJ, Spence I, Maddalena DJ. Selective descriptor pruning for QSAR/

QSPR studies using artificial neural networks. J Comput Chem 2003;24(7):891–897.

292. Hervás C, Silva M, Serrano JM, Orejuela E. Heuristic extraction of rules in pruned artificial neural networks models used for quantifying highly overlapping chromatographic peaks. J Chem Inf Comput Sci 2004;44(5):1576–1584.

293. Guha R, Stanton DT, Jurs PC. Interpreting computational neural network quantitative structure–activity relationship models: a detailed interpretation of the weights and biases. J Chem Inf Model 2005;45(4):1109–1121.

294. Zupan J, Gasteiger J. Neural networks: a new method for solving chemical problems or just a passing phase? Anal Chim Acta 1991;248 (1):1–30.

295. Jalali-Heravi M, Asadollahi-Baboli M, Shahbazikhah P. QSAR study of heparanase inhibitors activity using artificial neural networks and Levenberg–Marquardt algorithm. Eur J Med Chem 2008;43(3):548–556.

296. Andrea TA, Kalayeh H. Applications of neural networks in quantitative structure–activity relationships of dihydrofolate reductase inhibitors. J Med Chem 1991;34(9):2824–2836.

297. So SS, Richards WG. Application of neural networks: quantitative structure–activity relationships of the derivatives of 2,4-diamino-5-(substituted-benzyl)pyrimidines as DHFR inhibitors. J Med Chem 1992;35(17):3201–3207.

298. Agatonovic-Kustrin S, Turner JV, Glass BD. Molecular structural characteristics as determinants of estrogen receptor selectivity. J Pharm Biomed Anal 2008;48(2):369–375.

299. Deeb O, Hemmateenejad B. ANN-QSAR model of drug-binding to human serum albumin. Chem Biol Drug Des 2007;70(1):19–29.

300. Lill MA, Dobler M, Vedani A. Multi-dimensional QSAR in drug discovery: probing ligand alignment and induced fit—application to GPCRs and nuclear receptors. Curr Comput Aided Drug Des 2005;1(3):307–324.

301. Cramer RD, Wendt B. Pushing the boundaries of 3D-QSAR. J Comput Aided Mol Des 2007;21 (1–3):23–32.

302. Cramer RD III, Patterson DE, Bunce JD. Comparative molecular field analysis (CoMFA). 1. Effect of shape on binding of steroids to carrier proteins. J Am Chem Soc 1988;110 (18):5959–5967.

303. Livingstone DJ. The characterization of chemical structures using molecular properties. a survey. J Chem Inf Comput Sci 2000;40 (2):195–209.

304. Debnath AK. Quantitative structure–activity relationship (QSAR) paradigm: Hansch era to new millennium. Mini Rev Med Chem 2001;1 (2):187–195.

305. Klebe G, Abraham U, Mietzner T. Molecular similarity indices in a comparative analysis (CoMSIA) of drug molecules to correlate and predict their biological activity. J Med Chem 1994;37(24):4130–4146.

306. Klebe G. Comparative molecular similarity indices analysis. CoMSIA. Perspect Drug Discov Des 1998;12/13/14(*3D QSAR in Drug Design: Recent Advances*): 87–104.

307. Boehm M, Stuerzebecher J, Klebe G. Three-dimensional quantitative structure-activity relationship analyses using comparative molecular field analysis and comparative molecular similarity indices analysis to elucidate selectivity differences of inhibitors binding to trypsin, thrombin, and factor Xa. J Med Chem 1999;42(3):458–477.

308. Klebe G, Abraham U. Comparative molecular similarity index analysis (CoMSIA) to study hydrogen-bonding properties and to score combinatorial libraries. J Comput Aided Mol Des 1999;13(1):1–10.

309. Kaur K, Talele TT. 3D QSAR studies of 1,3,4-benzotriazepine derivatives as CCK2 receptor antagonists. J Mol Graph Model 2008;27 (4):409–420.

310. Nayana RS, Bommisetty SK, Singh K, Bairy SK, Nunna S, Pramod A, Muttineni R. Structural analysis of carboline derivatives as inhibitors of MAPKAP K2 using 3D QSAR and docking studies. J Chem Inf Model 2009;49 (1):53–67.

311. Du J, Lei B, Qin J, Liu H, Yao X. Molecular modeling studies of vascular endothelial growth factor receptor tyrosine kinase inhibitors using QSAR and docking. J Mol Graph Model 2009;27(5):642–654.

312. Hopfinger AJ, Wang S, Tokarski JS, Jin B, Albuquerque M, Madhav PJ, Duraiswami C. Construction of 3D-QSAR models using the 4D-QSAR analysis formalism. J Am Chem Soc 1997;119(43):10509–10524.

313. Santos-Filho OA, Cherkasov A. Using molecular docking, 3D-QSAR, and cluster analysis for screening structurally diverse data sets of pharmacological interest. J Chem Inf Model 2008;48(10):2054–2065.

314. Pasqualoto KFM, Ferreira EI, Santos-Filho OA, Hopfinger AJ. Rational design of new antituberculosis agents: receptor-Independent four-dimensional quantitative-structure

activity relationship analysis of a set of isoniazid derivatives. J Med Chem 2004;47 (15):3755–3764.

315. Santos-Filho OA, Hopfinger AJ. Structure-based QSAR analysis of a set of 4-hydroxy-5,6-dihydropyrones as inhibitors of HIV-1 protease: an application of the receptor-dependent (RD) 4D-QSAR formalism. J Chem Inf Model 2006;46(1):345–354.
316. Vedani A, Briem H, Dobler M, Dollinger H, McMasters DR. Multiple-conformation and protonation-state representation in 4D-QSAR: the neurokinin-1 receptor system. J Med Chem 2000;43(23):4416–4427.
317. Lill MA. Multi-dimensional QSAR in drug discovery. Drug Discov Today 2007;12 (23–24):1013–1017.
318. Vedani A, Dobler M. 5D-QSAR: The key for simulating induced fit? J Med Chem 2002;45 (11):2139–2149.
319. Vedani A, Dobler M. Multidimensional QSAR: moving from three- to five-dimensional concepts. Quant Struct Act Relat 2002;21 (4):382–390.
320. Vedani A, Dobler M, Dollinger H, Hasselbach K-M, Birke F, Lill MA. Novel ligands for the chemokine receptor-3 (CCR3): a receptor-modeling study based on 5D-QSAR. J Med Chem 2005;48(5):1515–1527.
321. Ducki S, Mackenzie G, Lawrence NJ, Snyder JP. Quantitative structure–activity relationship (5D-QSAR) study of combretastatin-like analogues as inhibitors of tubulin assembly. J Med Chem 2005;48(2):457–465.
322. Vedani A, Dobler M, Lill MA. Combining protein modeling and 6D-QSAR. Simulating the binding of structurally diverse ligands to the estrogen receptor. J Med Chem 2005;48 (11):3700–3703.
323. Vedani A, Descloux A-V, Spreafico M, Ernst B. Predicting the toxic potential of drugs and chemicals *in silico*: a model for the peroxisome proliferator-activated receptor γ (PPAR γ). Toxicol Lett 2007;173(1):17–23.
324. Verma RP, Hansch C. Camptothecins: a SAR/QSAR study. Chem Rev 2009;109(1):213–235.
325. Verma RP, Hansch C. An approach toward the problem of outliers in QSAR. Bioorg Med Chem 2005;13(15):4597–4621.
326. Kim KH. Outliers in SAR and QSAR: is unusual binding mode a possible source of outliers? J Comput Aided Mol Des 2007;21(1–3):63–86.
327. Selassie CD, Garg R, Kapur S, Kurup A, Verma RP, Mekapati SB, Hansch C. Comparative QSAR and the radical toxicity of various functional groups. Chem Rev 2002;102 (7):2585–2605.
328. Polanski J, Bak A, Gieleciak R, Magdziarz T. Modeling robust QSAR. J Chem Inf Model 2006;46(6):2310–2318.
329. Golbraikh A, Zhu H, Ye L, Wang-Bell M, Tang H, Tropsha A. Automatic detection of outliers prior to QSAR studies. 235th ACS National Meeting, New Orleans, LA; April 6–10, 2008. COMP-237.
330. Maggiora GM. On outliers and activity cliffs-why QSAR often disappoints. J Chem Inf Model 2006;46(4):1535.
331. Jouan-Rimbaud D, Bouveresse E, Massart DL, de Noord OE. Detection of prediction outliers and inliers in multivariate calibration. Anal Chim Acta 1999;388(3):283–301.
332. Fernández Pierna JA, Wahl F, de Noord OE, Massart DL. Methods for outlier detection in prediction. Chemom Intell Lab Syst 2002;63 (1):27–39.
333. Høy M, Steen K, Martens H. Review of partially least squares regression prediction error in Unscrambler. Chemom Intell Lab Syst 1998;44(1–2):123–133.
334. Fernández Pierna JA, Jin L, Daszykowski M, Wahl F, Massart DL. A methodology to detect outliers/inliers in prediction with PLS. Chemom Intell Lab Syst 2003;68(1–2):17–28.
335. Guha R, Dutta D, Jurs PC, Chen T. R-NN curves: an intuitive approach to outlier detection using a distance based method. J Chem Inf Model 2006;46(4):1713–1722.
336. Busemann M, Baumann K. Detection of prediction outliers in QSAR analysis using descriptor data only. In: Aki-Sener E, Yalcin I, editors. QSAR and Molecular Modelling in Rational Design of Bioactive Molecules. Istanbul, Turkey: EuroQSAR 2004 Proceedings; 2004. p 144–146.
337. Eriksson L, Jaworska J, Worth AP, Cronin MTD, McDowell RM, Gramatica P. Methods for reliability and uncertainty assessment and for applicability evaluations of classification- and regression-based QSARs. Environ Health Perspect 2003;111(10):1361–1375.
338. Selassie CD, Kapur S, Verma RP, Rosario M. Cellular apoptosis and cytotoxicity of phenolic compounds: a quantitative structure–activity relationship study. J Med Chem 2005;48 (23):7234–7242.
339. Cronin MTD, Schultz TW. Pitfalls in QSAR. J Mol Struct (Theochem) 2003;622(1–2):39–51.

340. Verma RP, Kapur S, Barberena O, Shusterman A, Hansch CH, Selassie CD. Synthesis, cytotoxicity, and QSAR analysis of X-thiophenols in rapidly dividing cells. Chem Res Toxicol 2003;16(3):276–284.

341. Kirchner LA, Moody RP, Doyle E, Bose R, Jeffery J, Chu I. The prediction of skin permeability using physicochemical data. ATLA 1997;25:359–370.

342. Cronin MTD, Dearden JC, Moss GP, Murray-Dickson G. Investigation of the mechanism of flux across human skin *in vitro* by quantitative structure–permeability relationships. Eur J Pharm Sci 1999;7(4):325–330.

343. Hansch C, Verma RP. Understanding tubulin/microtubule-taxane interactions: a quantitative structure–activity relationship study. Mol Pharm 2008;5(1):151–161.

344. Tropsha A, Gramatica P, Gombar VK. The importance of being earnest: validation is the absolute essential for successful application and interpretation of QSPR models. QSAR Comb Sci 2003;22(1):69–77.

345. Golbraikh A, Tropsha A. Beware of q2! J Mol Graph Model 2002;20:269–276.

346. Zhang S, Wei L, Bastow K, Zheng W, Brossi A, Lee K-H, Tropsha A. Antitumor Agents 252. Application of validated QSAR models to database mining: Discovery of novel tylophorine derivatives as potential anticancer agents. J Comput Aided Mol Des 2007;21 (1–3):97–112.

347. Cramer RD III Bunce JD, Patterson DE, Frank IE. Cross validation, bootstrapping and partial least squares compared with multiple regression in conventional QSAR studies. Quant Struct Act Relat 1988;7:18–25.

348. Todeschini R, Consonni V, Maiocchi A. The K correlation index: theory development and its application in chemometrics. Chemom Intell Lab Syst 1999;46(1):13–29.

349. Pogliani L. From molecular connectivity indices to semiempirical connectivity terms: recent trends in graph theoretical descriptors. Chem Rev 2000;100(10):3827–3858.

350. Bennett CA, Franklin NL. Statistical Analysis in Chemistry and the Chemical Industry. New York: John Wiley & Sons; 1967. p 708–709.

351. Hawkins DM, Basak SC, Mills D. Assessing model fit by cross-validation. J Chem Inf Comput Sci 2003;43(2):579–586.

352. Wehrens R, Putter H, Buydens LMC. The bootstrap: a tutorial. Chemom Intell Lab Syst 2000;54(1):35–52.

353. Todeschini R, Consonni V, Mauri A, Pavan M. Detecting "bad" regression models: multicriteria fitness functions in regression analysis. Anal Chim Acta 2004;515(1):199–208.

354. Wold S, Eriksson L. Statistical validation of QSAR results, validation tools. In: van de Waterbeemd H, editor. Chemometrics Methods in Molecular Design. Weinheim, Germany: VCH; 1995. p 309–318.

355. Wu W, Walczak B, Massart DL, Heuerding S, Erni F, Last IR, Prebble KA. Artificial neural networks in classification of NIR spectral data: design of the training set. Chemom Intell Lab Syst 1996;33(1):35–46.

356. Verma RP, Hansch C. Combating the threat of anthrax: a quantitative structure–activity relationship approach. Mol Pharm 2008;5 (5):745–759.

357. Zupan J, Gasteiger J. Neural Networks in Chemistry and Drug Design, 2nd ed. Weinheim, Germany: Wiley-VCH; 1999.

358. Yan A, Wang Z, Cai Z. Prediction of human intestinal absorption by GA feature selection and support vector machine regression. Int J Mol Sci 2008;9(10):1961–1976.

359. Everitt BS, Landau S, Leese M. Cluster Analysis. London: Edward Arnold; 2001.

360. Kowalski RB, Wold S. Handbook of Statistics. Amsterdam; Holland Publishing Company; 1982.

361. Eriksson L, Arnhold T, Beck B, Fox T, Johansson E, Kriegl JM. Onion design and its application to a pharmaceutical QSAR problem. J Chemometr 2004;18(3–4):188–202.

362. Roy K. On some aspects of validation of predictive quantitative structure–activity relationship models. Expert Opin Drug Discov 2007;2(12):1567–1577.

363. Olsson I-M, Gottfries J, Wold S. D-optimal onion designs in statistical molecular design. Chemom Intell Lab Syst 2004;73(1):37–46.

364. Kennard RW, Stone LA. Computer aided design of experiments. Technometrics 1969;11: 137–148.

365. Golbraikh A, Tropsha A. Predictive QSAR modeling based on diversity sampling of experimental datasets for the training and test set selection. J Comput Aided Mol Des 2002;16 (5–6):357–369.

366. Selassie C, Klein TE. Comparative QSAR. In: Devillers J, editor. Washington DC: Taylor & Francis; 1998. p 235.

367. Selassie CD, Gan WX, Fung M, Shortle R. QSAR and molecular modelling: concepts,

computational tools and biological applications. In: Sanz F, Giraldo J, Manaut F, editors. Barcelona: Prous Science; 1995. p 128.

368. Kurup A, Garg R, Hansch C. Comparative QSAR analysis of 5α-reductase inhibitors. Chem Rev 2000;100(3):909–924.
369. Blaney JM, Hansch C, Silipo C, Vittoria A. Structure–activity relationships of dihydrofolated reductase inhibitors. Chem Rev 1984;84 (4):333–407.
370. Hansch C. Folate inhibitors. Structure–activity analysis using linear modeling. Ann NY Acad Sci 1971;186:235–247.
371. Hansch C, Hathaway BA, Guo Z, Selassie CD, Dietrich SW, Blaney JM, Langridge R, Volz KW, Kaufman BT. Crystallography, quantitative structure–activity relationships, (QSAR) and molecular graphics in a comparative analysis of the inhibition of dihydrofolate reductase from chicken liver and *Lactobacillus casei* by 4,6-diamino-1,2-dihydro-2,2-dimethyl-1-(substituted-phenyl)-*s*-triazines. J Med Chem 1984;27(2):129–143.
372. Hathaway BA, Guo ZR, Hansch C, Delcamp TJ, Susten SS, Freisheim JH. Inhibition of human dihydrofolate reductase by 4,6-diamino-1,2-dihydro-2,2-dimethyl-1-(substituted-phenyl)-*s*-triazines. A quantitative structure–activity relationship analysis. J Med Chem 1984;27(2):144–149.
373. Selassie CD, Strong CD, Hansch C, Delcamp TJ, Freisheim JH, Khwaja TA. Comparison of triazines as inhibitors of L1210 dihydrofolate reductase and of L1210 cells sensitive and resistant to methotrexate. Cancer Res 1986;46(2):744–756.
374. Marlowe CK, Selassie CD, Santi DV. Quantitative structure–activity relationships of the inhibition of *Pneumocystis carinii* dihydrofolate reductase by 4,6-diamino-1,2-dihydro-2,2-dimethyl-1-(X-phenyl)-*s*-triazines. J Med Chem 1995; 38(6):967–972.
375. Booth RG, Selassie CD, Hansch C, Santi DV. Quantitative structure–activity relationship of triazine-antifolate inhibition of *Leishmania* dihydrofolate reductase and cell growth. J Med Chem 1987;30(7):1218–1224.
376. Chio LC, Queener SF. Identification of highly potent and selective inhibitors of *Toxoplasma gondii* dihydrofolate reductase. Antimicrob Agents Chemother 1993;37(9):1914–1923.
377. Fukunaga JY, Hansch C, Steller EE. Inhibition of dihydrofolate reductase. Structure–activity correlations of quinazolines. J Med Chem 1976;19(5):605–611.
378. Chen B-K, Horvath C, Bertino JR. Multivariate analysis and quantitative structure–activity relationships. Inhibition of dihydrofolate reductase and thymidylate synthetase by quinazolines. J Med Chem 1979;22(5):483–491.
379. Harris NV, Smith C, Bowden K. Antifolate and antibacterial activities of 6-substituted 2,4-diaminoquinazolines. Eur J Med Chem 1992;27(1):7–18.
380. Klibanov AM. Improving enzymes by using them in organic solvents. Nature 2001;409 (6817):241–246.
381. Kim KH. Thermodynamic aspects of hydrophobicity and biological QSAR. J Comput Aided Mol Des 2001;15(4):367–380.
382. Kim KH. Thermodynamic quantitative structure–activity relationship analysis for enzyme–ligand interactions in aqueous phosphate buffer and organic solvent. Bioorg Med Chem 2001;9(8):1951–1955.
383. Nakamura K, Hayashi K, Ueda I, Fujiwara H. Micelle/water partition properties of phenols determined by liquid chromatographic method. Proposal for versatile measure of hydrophobicity. Chem Pharm Bull 1995;43 (3):369–373.
384. Compadre CM, Sanchez RJ, Bhuraneswaran C, Compadre RL, Plunkett D, Novick SG. In: Wermuth CG, editor. Trends in QSAR and Molecular Modelling. Strasbourg, France: Escom; 1993. p 112.
385. Frye SV, Haffner CD, Maloney PR, Mook RA Jr, Dorsey GF, Hiner RN, Cribbs CM, Wheeler TN, Ray JA, Andrews RC, Batchelor KW, Branson HN, Stuart JD, Schwiker SL, van Arnold J, Croom S, Bickett DM, Moss ML, Tian G, Unwalla RJ, Lee FW, Tippin TK, James MK, Grizzle MK, Long JE, Schuster SV. 6-Azasteroids: structure–activity relationships for inhibition of type 1 and 2 human 5α-reductase and human adrenal 3β-hydroxy-Δ5-steroid dehydrogenase/3-Keto-Δ5-steroid isomerase. J Med Chem 1994;37(15):2352–2360.
386. Frye SV, Haffner CD, Maloney PR, Hiner RN, Dorsey GF, Noe RA, Unwalla RJ, Batchelor KW, Bramson HN, Stuart JD, Schwiker SL, van Arnold J, Bickett DM, Moss ML, Tian G, Lee FW, Tippin TK, James MK, Grizzle MK, Long JE, Croom DK. Structure–activity relationships for inhibition of type 1 and 2 human 5α-reductase and human adrenal 3β-hydroxy-Δ5-steroid dehydrogenase/3-keto-Δ5-steroid isomerase by 6-azaandrost-4-en-3-ones: optimization of the C17 substituent. J Med Chem 1995;38(14):2621–2627.

387. Kaiser KLE. Evolution of the international workshops on quantitative structure–activity relationships (QSARs) in environmental toxicology. SAR QSAR Environ Res 2007;18 (1–2):3–20.

388. Cronin MTD, Dearden JC. QSAR in toxicology. 4. Prediction of non-lethal mammalian toxicological endpoints, and expert systems for toxicity prediction. Quant Struct Act Relat 1995;14 (6):518–523.

389. Cronin MTD, Gregory BW, Schultz TW. Quantitative structure–activity analyses of nitrobenzene toxicity to *Tetrahymena pyriformis*. Chem Res Toxicol 1998;11(8):902–908.

390. Enoch SJ, Cronin MTD, Schultz TW, Madden JC. An evaluation of global QSAR models for the prediction of the toxicity of phenols to *Tetrahymena pyriformis*. Chemosphere 2008; 71(7):1225–1232.

391. Schultz TW. Structure-toxicity relationships for benzenes evaluated with *Tetrahymena pyriformis*. Chem Res Toxicol 1999;12 (12):1262–1267.

392. Cronin MTD, Schultz TW. Structure–toxicity relationships for phenols to *Tetrahymena pyriformis*. Chemosphere 1996;32(8):1453–1468.

393. Garg R, Kurup A, Hansch C. Comparative QSAR: on the toxicology of the phenolic OH moiety. Crit Rev Toxicol 2001;31(2):223–245.

394. Selassie CD, DeSoyza TV, Rosario M, Gao H, Hansch C. Phenol toxicity in leukemia cells: a radical process? Chem Biol Interact 1998;113 (3):175–190.

395. Cronin MTD, Schultz TW. Development of quantitative structure–activity relationships for the toxicity of aromatic compounds to *Tetrahymena pyriformis*: comparative assessment of the methodologies. Chem Res Toxicol 2001;14(9):1284–1295.

396. Liu X, Chen J, Yu H, Zhao J, Giesy JP, Wang X. Quantitative structure activity relationship (QSAR) for toxicity of chlorophenols on L929 cells *in vitro*. Chemosphere 2006;64(10): 1619–1626.

397. Hinderling PH, Schmidlin O, Seydel JK. Quantitative relationships between structure and pharmacokinetics of β-adrenoceptor blocking agents in man. J Pharmacokinet Biopharm 1984;12(3):263–287.

398. Hansch C, Hoekman D, Leo A, Weininger D, Selassie CD. Chem-bioinformatics: comparative QSAR at the interface between chemistry and biology. Chem Rev 2002;102 (3):783–812.

399. Shulgin AT, Sargent T, Naranjo C. Structure–activity relations of one-ring psychotomimetics. Nature 1969;221(5180):537–541.

400. Verma RP, Selassie C.Unpublished results.

401. Mracec M, Muresan S, Mracec M, Simon Z, Naray-Szabo G. QSARs with orthogonal descriptors on psychotomimetic phenylakylamines. Quant Struct Act Relat 1997;16 (6):459–464.

402. Clare BW. Structure–activity correlations for psychotomimetics. 1. Phenylalkylamines: electronic, volume, and hydrophobicity parameters. J Med Chem 1990;33(2):687–702.

403. Garg R, Gupta SP, Gao H, Babu MS, Debnath AK, Hansch C. Comparative quantitative structure–activity relationship studies on anti-HIV drugs. Chem Rev 1999;99 (12):3525–3601.

404. Siddiqui AQ, McGuigan C, Ballatore C, Zuccotto F, Gilbert IH, Clercq ED, Balzarini J. Design and synthesis of lipophilic phosphoramidate d4T-MP prodrugs expressing high potency against HIV in cell culture: structural determinants for *in vitro* activity and QSAR. J Med Chem 1999;42(20):4122–4128.

405. Miller JF, Brieger M, Furfine ES, Hazen RJ, Kaldor I, Reynolds D, Sherrill RG, Spaltenstein A. Novel P1 chain-extended HIV protease inhibitors possessing potent anti-HIV activity and remarkable inverse antiviral resistance profiles. Bioorg Med Chem Lett 2005;15 (15):3496–3500.

406. Ravichandran V, Jain PK, Mourya VK, Agrawal RK. QSAR study on some arylsulfonamides as anti-HIV agents. Med Chem Res 2007;16(7–9):342–351.

407. Küçükgüzel Í, Tatar E, Küçükgüzel ŞG, Rollas S, Clercq ED. Synthesis of some novel thiourea derivatives obtained from 5-[(4-aminophenoxy)methyl]-4-alkyl/aryl-2,4-dihydro-3H-1,2,4-triazole-3-thiones and evaluation as antiviral/anti-HIV and anti-tuberculosis agents. Eur J Med Chem 2008;43 (2):381–392.

408. Li Z-G, Chen K-X, Xie H-Y, Gao J-R. Quantitative structure–activity relationship analysis of some thiourea derivatives with activities against HIV-1 (IIIB). QSAR Comb Sci 2009;28(1):89–97.

409. Debnath AK. Three-Dimensional quantitative structure–activity relationship study on cyclic urea derivatives as HIV-1 protease inhibitors: application of comparative molecular field analysis. J Med Chem 1999;42(2):249–259.

410. Debnath AK. Application of 3D-QSAR techniques in Anti-HIV-1 drug design: an overview. Curr Pharm Des 2005;11(24):3091–3110.
411. Verma RP, Hansch C, Selassie CD. Comparative QSAR studies on PAMPA/modified PAMPA for high throughput profiling of drug absorption potential with respect to Caco-2 cells and human intestinal absorption. J Comput Aided Mol Des 2007;21(1–3):3–22.
412. Van de Watetbeemd H, Gifford E. ADMET *in silico* modelling: towards prediction paradise? Nat Rev Drug Discov 2003;2(3):192–204.
413. Wishart DS. Improving early drug discovery through ADME modelling an overview. Drugs R D 2007;8(6):349–362.
414. Hansch C, Quinlan JE, Lawrence GL. The linear free energy relationship between partition coefficients and the aqueous solubility of organic liquids. J Org Chem 1968;33 (1):347–350.
415. Yalkowsky SH, Valvani SC. Solubility and partitioning. I. Solubility of nonelectrolytes in water. J Pharm Sci 1980;69(8):912–922.
416. Abraham MH, Le J. The correlation and prediction of the solubility of compounds in water using an amended solvation energy relationship. J Pharm Sci 1999;88(9):868–880.
417. Gao S, Cao C. A new approach on estimation of solubility and *n*-octanol/water partition coefficient for organohalogen compounds. Int J Mol Sci 2008;9(6):962–977.
418. Kulkarni A, Han Y, Hopfinger AJ. Predicting Caco-2 cell permeation coefficients of organic molecules using membrane-interaction QSAR analysis. J Chem Inf Comput Sci 2002;42 (2):331–342.
419. Hansch C, Leo A, Mekapati SB, Kurup A. QSAR and ADME. Bioorg Med Chem 2004;12(12):3391–3400.
420. Fujikawa M, Nakao K, Shimizub R, Akamatsu M. QSAR study on permeability of hydrophobic compounds with artificial membranes. Bioorg Med Chem 2007;15(11):3756–3767.
421. Sugano K, Takata N, Machida M, Saitoh K, Terada K. Prediction of passive intestinal absorption using bio-mimetic artificial membrane permeation assay and the paracellular pathway model. Int J Pharm 2002;241(2):241–251.
422. Di L, Kerns EH, Fan K, McConnell OJ, Carter GT. High throughput artificial membrane permeability assay for blood–brain barrier. Eur J Med Chem 2003;38(3):223–232.
423. Moda TL, Carlos A, Montanari C.A. Andricopulo A.D. Hologram QSAR model for the prediction of humanv oral bioavailability. Bioorg Med Chem 2007;15(24):7738–7745.
424. Nestorov I, Aarons L, Rowland M. Quantitative structure–pharmacokinetics relationships. II. A mechanistically based model to evaluate the relationship between tissue distribution parameters and compound lipophilicity. J Pharmacokinet Biopharm 1998;26(5): 521–545.
425. Moda TL, Montanari CA, Andricopulo AD. *In silico* prediction of human plasma protein binding using hologram QSAR. Lett Drug Des Discov 2007;4(7):502–509.
426. Votano JR, Parham M, Hall LM, Hall LH, Kier LB, Oloff S, Tropsha A. QSAR modeling of human serum protein binding with several modeling techniques utilizing structure–information representation. J Med Chem 2006;49 (24):7169–7181.
427. Helmer F, Kiehs K, Hansch C. The linear free-energy relationship between partition coefficients and the binding and conformational perturbation of macromolecules by small organic compounds. Biochemistry 1968;7 (8):2858–2863.
428. Pajouhesh H, Lenz GR. Medicinal chemical properties of successful central nervous system drugs. NeuroRx® 2005;2(4):541–553.
429. de Graaf C, Vermeulen NPE, Feenstra KA. Cytochrome P450 *in Silico*: an integrative modeling approach. J Med Chem 2005;48 (8):2725–2755.
430. Singh SS. Preclinical pharmacokinetics: an approach towards safer and efficacious drugs. Curr Drug Metabol 2006;7(2):165–182.
431. Hansch C, Mekapati SB, Kurup A, Verma RP. QSAR of cytochrome P450. Drug Metabol Rev 2004;36(1):105–156.
432. LaBella FS, Chen QM, Stein CD, Queen G. The site of general anesthesia and cytochrome P450 oxygenases: similarities defined by straight chain and cyclic alcohols. Br J Pharm 1997;120(6):1158–1164.
433. Hansch C, Sinclair JF, Sinclair PR. Induction of cytochrome P450 by barbiturates in chick embryo hepatocytes: a quantitative structure analysis. Quant Struct Act Relat 1990;9 (3):223–226.
434. Cnubben NHP, Peelen S, Borst JW, Vervoort J, Veeger C, Rietjens IMCM. Molecular orbital-based quantitative structure–activity relationship for the P450-catalyzed 4-hydroxylation of halogenated anilines. Chem Res Toxicol 1994;7(5):590–598.

435. Itokawa D, Nishioka T, Fukushima J, Yasuda T, Yamauchi A, Chuman H. Quantitative structure–activity relationship study of binding affinity of azole compounds with CYP2B and CYP3A. QSAR Comb Sci 2007;26 (7):828–836.
436. Lewis DFV. Quantitative structure–activity relationships (QSARs) within the cytochrome P450 system: QSARs describing substrate binding, inhibition and induction of P450s. Infammopharmacology 2003;11(1):43–73.
437. Locuson CW, Wahlstrom JL. Three-dimensional quantitative structure–activity relationship analysis of cytochromes P450: effect of incorporating higheraffinity ligands and potential new applications. Drug Metab Dispos 2005;33(7):873–878.
438. Yang GF, Huang X. Development of quantitative structure–activity relationships and its application in rational drug design. Curr Pharm Des 2006;12(35):4601–4611.
439. Chohan KK, Paine SW, Waters NJ. Quantitative structure–activity relationships in drug metabolism. Curr Top Med Chem 2006;6 (15):1569–1578.
440. Li H, Sun J, Fan X, Sui X, Zhang L, Wang Y, He Z. Considerations and recent advances in QSAR models for cytochrome P450-mediated drug metabolism prediction. J Comput Aided Mol Des 2008;22(11):843–855.
441. Cantelli-Forti G, Guerra MC, Barbaro AM, Hrelia P, Biagi GL, Borea PA. Relationship between lipophilic character and urinary excretion of nitroimidazoles nitrothiazoles in rats. J Med Chem 1986;29(4):555–561.
442. Huang H, Wang X, Ou W, Jinsong Zhao J, Shao Y, Wang L. Acute toxicity of benzene derivatives to the tadpoles (*Rana japonica*) and QSAR analyses. Chemosphere 2003;53 (8):963–970.
443. Schultz TW, Hewitt M, Netzeva TI, Cronin MTD. Assessing applicability domains of toxicological QSARs: definition, confidence in predicted values, and the role of mechanisms of action. QSAR Comb Sci 2007;26(2):238–254.
444. Benigni R, Giuliani A, Franke R, Gruska A. Quantitative structure–activity relationships of mutagenic and carcinogenic aromatic amines. Chem Rev 2000;100(10):3697–3714.
445. Roberts DW, Aptula AO, Patlewicz G. Mechanistic applicability domains for non-animal based prediction of toxicological endpoints. QSAR analysis of the Schiff base applicability domain for skin sensitization. Chem Res Toxicol 2006;19(9):1228–1233.
446. Kimber I, Basketter DA, Butler M, Gamer A, Garrigue JL, Gerberick GF, Newsome C, Steiling W, Vohr HW. Classification of contact allergens according to potency: proposals. Food Chem Toxicol 2003;41(12):1799–1809.
447. Yang C, Richard AM, Cross KP. The art of data mining the minefields of toxicity databases to link chemistry to biology. Curr Comput Aided Drug Des 2006;2(2):135–150.
448. Hansch C, Gao H, Hoekman D. Comparative QSAR. In: Devillers J, editor. Washington, DC: Taylor & Francis; 1998. p 285.
449. Selassie CD, Verma RP, Kapur S, Shusterman AJ, Hansch C. QSAR for the cytotoxicity of 2-alkyl or 2,6-dialkyl, 4-X-phenols: the nature of the radical reaction. J Chem Soc Perkin Trans 2 2002;(6):1112–1117.
450. Bordwell FG, Zhang X-M, Satish AV, Cheng J-P. Assessment of the importance of changes in ground-state energies on the bond dissociation enthalpies of the O−H bonds in phenols and the S−H bonds in thiophenols. J Am Chem Soc 1994;116(15):6605–6610.
451. Selassie C, Klein TE. In: Kubinyi H, editor. 3D QSAR in Drug Design. Theory, Methods and Applications. Leiden, The Netherlands: Escom Science; 1993. p 257.
452. Boyd DB. Is rational design good for anything? In: Parrill AL, Rami-Reddy M, editors. Rational Drug Design, ACS Symposium Series 719. Washington, DC: American Chemical Society. 1999. p 346–355.
453. Benigni R. Structure–activity relationship studies of chemical mutagens and carcinogens: mechanistic investigations and prediction approaches. Chem Rev 2005;105(5):1767–1800.
454. Hansch C, Gao H. Comparative QSAR: radical reactions of benzene derivatives in chemistry and biology. Chem Rev 1997;97(8):2995–3059.
455. Ajay Walters WP, Murcko MA. Can we learn to distinguish between "drug-like" and "nondrug-like" molecules? J Med Chem 1998;41(18):3314–3324.
456. Lewis DFV. Structural characteristics of human P450s involved in drug metabolism: QSARs and lipophilicity profiles. Toxicology 2000;144(1–3):197–203.
457. Selassie CD, Kapur S, Verma RP, Rosario M. Cellular apoptosis and cytotoxicity of phenolic compounds: a quantitative structure–activity relationship study. J Med Chem 2005;48 (23):7234–7242.
458. Plummer EL. Successful application of the QSAR paradigm in discovery programs. In:

Hansch C, Fujita T, editors. Classical and Three-Dimensional QSAR in Agrochemistry, ACS Symposium Series 606. Washington, DC: American Chemical Society; 1995. p 240–253.

459. Fujita T. Recent success stories leading to commercializable bioactive compounds with the aid of traditional QSAR procedures. Quant Struct Act Relat 1997;16(2):107–112.

460. Koga H, Itoh A, Murayama S, Suzue S, Irikura T. Structure-activity relationships of antibacterial 6,7- and 7,8-disubstituted 1-alkyl-1,4-dihydro-4-oxoquinoline-3-carboxylic acids. J Med Chem 1980;23(12):1358–1363.

461. Chuman H, Ito A, Shaishoji T, Kumazawa S. QSAR and three-dimensional shape studies of fungicidal azolylmethylcyclopentanols. Molecular design of novel fungicides metconazole and ipconazole. In: Hansch C, Fujita T, editors. Classical and Three-Dimensional QSAR in Agrochemistry, ACS Symposium Series 606. Washington, DC: American Chemical Society; 1995. p 171–185.

462. Ohtaka H, Tsukamoto G. Benzylpiperazine derivatives. V. Quantitative structure–activity relationships of 1-benzyl-4-diphenylmethylpiperazine derivatives for cerebral vasodilating activity. Chem Pharm Bull 1987;35 (10):4117–4123.

463. Kvasnicková E, Szotáková B, Wsól V, Trejtnar F, Skálová L, Hais IM, Kuchar M, Poppová M. Metabolic pathways of flobufen. A new antirheumatic and antiarthritic drug. Interspecies comparison. Exp Toxicol Pathol 1999;51 (4–5):352–356.

464. Guha R. On the interpretation and interpretability of quantitative structure–activity relationship models. J Comput Aided Mol Des 2008;22(12):857–871.

465. Fujita T. The role of QSAR in drug design. In: Jolles G, Woolridge KRH, editors. Drug Design: Fact or Fantasy? London: Academic Press; 1984. pp. 19–33.

466. Koga H, Itoh A, Murayama S, Suzue S, Irikura T. Structure–activity relationships of antibacterial 6,7- and 7,8-disubstituted 1-alkyl-1,4-dihydro-4-oxoquinoline-3-carboxylic acids. J Med Chem 1980;23(12):1358–1363.

467. Hayakawa I, Hiramitsu T, Tanaka Y. Synthesis and antibacterial activities of substituted 7-oxo-2,3-dihydro-7*H*-pyrido[1,2,3-de][1,4] benzoxazine-6-carboxylic acids. Chem Pharm Bull 1984;32(12):4907–4913.

468. Kirino O, Furuzawa K, Matsumoto H, Hino N, Mine A. *N*-Benzylbutanamides as new potential herbicides. Agric Biol Chem 1981;45 (11):2669–2670.

469. Fujita, T. The extrathermodynamic approach to drug design. In: Ramsden CA, editor. *Comprehensive Medicinal Chemistry: The Rational Design, Mechanistic Study and Therapeutic Application of Chemical Compounds*. Vol. 4 (*Quantitative Drug Design*). Elmsford, NY: Pergamon; 1990. pp. 497–560.

470. Yokoyama T, Iwata N, Kobayashi T. RS-2232, a compound with a reversible and specific type-A monoamine oxidase inhibiting property in mouse brain. Jpn J Pharmacol 1987;44 (4):421–427.

471. Iwata N, Puchler K, Plenker A. Pharmacology of the new reversible inhibitor of monoamine oxidase A, RS-8359. Intl Clin Psychopharmacol 1997;12(Suppl 5): S3–S10.

472. Wooldridge, KRH. 1980. Antiallergic purinones: a successful application of QSAR. *Drugs Affecting Respiratory Systems*. ACS Symposium Series 118. Washington, DC: American Chemical Society; 1980. p 117–123.

473. Yamamoto T, Hori M, Watanabe I, Harada K, Ikeda S, Ohtaka H. Quantitative structure–activity relationship study of *N*-(3-oxo-3,4-dihydro-2H-benzo[1,4]thiazine-6-carbonyl)guanidines as potent Na/H exchange inhibitors. Chem Pharm Bull 2000;48(6):843–849.

474. Hara H, Morita T, Sukamoto T, Culter FM. Lomerazine (KB-2796), A new anti-migraine drug. CNS Drug Rev 1995;1:204–226.

475. Kuchar M, Maturova E, Brunova B, Grimova J, Tomkova H, Holubek J. Quantitative relations between structure and antiinflammatory activity of aryloxoalkanoic acids. Coll Czech Chem Commun 1988;53(8):1862–1872.

476. Kuchar M, Poppova M, Zunova H, Knezova E, Vosatka V, Prihoda M. 4-(2′,4′-Difluorobiphenyl-4-yl)-2-methyl-4-oxobutanoic acid and its derivatives. Coll Czech Chem Commun 1994;59(12):2705–2713.

477. Schmidt RR, Draber W, Eue L, Timmler H. Herbicidal activity and selectivity of new 3-alkyl-4-amino-6-aryl-1,2,4-triazin-5-ones. Pest Sci 1975;6(3):239–244.

478. Moeller E, Horstmann H, Meng K. The chemistry of muzolimine (Bay g 2821), a new non-sulfonamide diuretic. Pharmatherapeutica 1977;1(8):540–545.

479. Denny WA, Cain BF, Atwell GJ, Hansch C, Panthananickal A, Leo A. Potential antitumor agents. 36. Quantitative relationships between experimental antitumor activity, toxicity, and structure for the general class of 9-

anilinoacridine antitumor agents. J Med Chem 1982;25(3):276–315.

480. Topliss JG. Some observations on classical QSAR. Persp Drug Discov Des 1993;1 (2):253–268.
481. Hansch C, Bjoerkroth JP, Leo A. Hydrophobicity and central nervous system agents: on the principle of minimal hydrophobicity in drug design. J Pharm Sci 1987;76(9):663–687.
482. Mekapati SB, Kurup A, Verma RP, Hansch C. The role of hydrophobic properties of chemicals in promoting allosteric reactions. Bioorg Med Chem 2005;13(11):3737–3762.
483. Verma RP. Understanding topoisomerase I and II in terms of QSAR. Bioorg Med Chem 2005;13(4):1059–1067.
484. Martin YC. What works and what does not: lessons from experience in a pharmaceutical company. QSAR Comb Sci 2006;25 (12):1192–1200.
485. Winkler DA. The role of quantitative structure–activity relationships (QSAR) in biomolecular discovery. Brief Bioinformatics 2002;3 (1):73–86.
486. Yang GF, Huang X. Development of quantitative structure–activity relationships and its application in rational drug design. Curr Pharm Des 2006;12(35):4601–4611.
487. Chen JW, Li XH, Yu HY, Wang YN, Qiao XL. Progress and perspectives of quantitative structure–activity relationships used for ecological risk assessment of toxic organic compounds. Sci China Ser B Chem 2008;51 (7):593–606.
488. Benigni R. Structure–activity relationship studies of chemical mutagens and carcinogens: mechanistic investigations and prediction approaches. Chem Rev 2005;105(5):1767–1800.

MASS SPECTROMETRY AND DRUG DISCOVERY

Richard B. van Breemen
Andrew G. Newsome
Jeffrey H. Dahl
Department of Medicinal Chemistry and Pharmacognosy, University of Illinois, Chicago, IL

1. INTRODUCTION

At the beginning of the twentieth century, mass spectrometers were invented to help physicists and physical chemists to prove the existence of isotopes of the elements. As radioactivity and nuclear physics were explored, specialized mass spectrometers were used to characterize the fission products of radioactive elements as they were created or discovered. In addition, mass spectrometers were used for the measurement of isotopic enrichment of radioactive elements, their inorganic derivatives, and even the isotopic purification of radioactive elements as inorganic compounds. As this era of mass spectrometry reached maturity by the 1940s, some physicists announced that there would no longer be any need for mass spectrometry since virtually all of the elements had been discovered and characterized. Of course, these prognosticators were wrong because the entire field of organic mass spectrometry was about to begin.

1.1. Electron Impact (EI)

While mass spectrometers were being used for the purification of fissionable material for atomic weapons as part of the Manhattan Project of World War II, organic mass spectrometry was being invented for the analysis and quality control of petroleum distillates and petroleum-based fuels. By 1945, the application of mass spectrometry to organic chemistry had emerged as a productive new area of research and discovery. Commercial production of organic mass spectrometers began during the 1940s, and petroleum companies became the first customers for these new analytical instruments. Early commercial mass spectrometers used electron impact ionization (see Eqs 1 and 2) to generate ions from gas-phase molecules that were separated by acceleration through an electromagnetic field provided by either a fixed magnet or an electromagnet. After separation, the ions were detected using a simple impact detector such as a Faraday cup. This basic design is still in use today for the identification and quantitative analysis of volatile organic compounds.

$$M + e^{-}(70\,\text{eV}) \rightarrow M^{+\bullet} + 2e^{-} \quad \text{formation of positive molecular ions using EI ionization} \tag{1}$$

$$M + e^{-}(2\text{-}10\,\text{eV}) \rightarrow M^{-\bullet} \quad \text{electron capture EI ionization} \tag{2}$$

Toward the late 1950s, organic mass spectrometers began to be used for the analysis of a wider variety of organic molecules and eventually became a fundamental analytical tool for the characterization of synthetic organic compounds. Today, mass spectrometers are used routinely to confirm the molecular masses of organic compounds, to determine elemental compositions and to verify their structures based on fragmentation patterns. Fragmentation results from the cleavage of chemical bonds within an ion resulting in the formation of a product ion of lower mass and one or more neutral products. Of course, only the fragment ions and not the neutral species are detected in a mass spectrometer because this instrument measures the mass-to-charge ratio (*m/z*) of ions in the gas phase. The energy for fragmentation is either the result of excess energy imparted to the molecular ion during ionization or during a process known as collision-induced dissociation (CID) which will be discussed along with tandem mass spectrometry (MS–MS) below. Since the fragmentation pattern reflects the relative strengths of chemical bonds in a compound, mass spectra (a plot of ion relative abundance versus *m/z*) provide structurally significant fragment ions for compound identification. Rules for structure elucidation of chemical structures through the interpretation of mass spectra have been developed. (For a review of EI and ion fragmentation pathways, see McLafferty et al., 1993 provided in Section 4.)

In many cases, EI imparts so much excess energy into a molecule that only fragment ions and no molecular ions are observed. Therefore, "softer" ionization techniques were developed to enhance molecular mass information. The first of these ionization methods was chemical ionization (CI). Developed by researchers in the petroleum industry [1], CI became another standard ionization technique for organic mass spectrometry. During CI, high-energy electrons (as in EI) are used to ionize a gas called a reagent gas at a constant pressure (usually ~1 mbar) in the mass spectrometer ionization source. The reagent gas in turn ionizes the sample molecules through ion-molecule reactions that usually involve the exchange of protons. Less frequently, sample molecule ionization might involve a charge exchange. Two of the most common ionization mechanisms in CI are summarized in Equations 3 and 4.

$$M + RH^{+} \rightarrow MH^{+} + RCI \quad \text{through proton transfer, } R = \text{reagent gas} \tag{3}$$

$$M + R^{+\bullet} \rightarrow M^{+\bullet} + R \quad \text{CI through charge exchange} \tag{4}$$

1.2. Types of Mass Analyzers

During the 1960s, high-resolution double-focusing magnetic sector instruments became standard tools for the determination of elemental compositions using a type of analysis called accurate mass measurement. In mass spectrometry, resolving power is defined as $M/\Delta M$ (resolution is the inverse of this term, $\Delta M/M$), where M is the m/z value of a singly charged ion, and ΔM is the difference (measured in m/z) between M and the next highest ion. Alternatively, ΔM may be defined in terms of the width of the peak. High resolution is typically regarded as a resolving power of at least 10,000 such that the molecular ions of most drug-like molecules (that is, compounds with molecular masses less than ~500) can be resolved from each other. After resolving a sample ion from others in a mass spectrum, an accurate mass measurement may be carried out by comparing the m/z value of the unknown to that of a calibration standard.

Since the 1960s, other types of mass spectrometers capable of high-resolution accurate mass measurements have become available as commercial products including Fourier transform ion cyclotron resonance (FTICR) mass spectrometers, reflectron TOF instruments, quadrupole time-of-flight hybrid (QqTOF) mass spectrometers, and recently, ion trap-TOF hybrid mass spectrometers (see Table 1 for a listing of types of organic mass spectrometers and a comparison of their performance characteristics). By the early 2000s, FTICR and QqTOF instruments became more popular than magnetic sector mass spectrometers for accurate mass measurements, high-resolution measurements, and drug discovery applications. As will be discussed below, accurate mass measurements are essential to many types of mass spectrometry-based screening and drug discovery today.

Table 1. Types of Mass Spectrometers and Tandem Mass Spectrometers

Instrument	Resolving Power	*m/z* Range	Tandem MS
Magnetic sector	100,000	12,000	Low resolution
Quadrupole	<4,000	4,000	None
Triple quadrupole	<4,000	4,000	Low resolution
Time-of-flight (TOF)	15,000	>200,000	None
FTICR	>200,000	<10,000	MS^n, high resolution
Orbitrap	<200,000	6,000	MS^n, high resolution
Ion trap	<4,000	<10,000	MS^n, low resolution
QqTOF	>14,000	4,000	High resolution
Ion trap-TOF	>10,000	4,000	High resolution
TOF–TOF	15,000	>10,000	High resolution

1.3. Gas Chromatography-Mass Spectrometry

Biomedical applications of mass spectrometry began during the 1960s both at academic institutions and at pharmaceutical companies. These applications depended upon the volatilization (usually by heating) of pharmaceutical compounds and biochemicals prior to their gas-phase ionization using EI or CI. In order to increase the thermal stability and volatility of these compounds, a variety of derivatization methods were developed to mask polar functional groups and reduce hydrogen bonding between molecules. These methods were particularly useful for use with gas chromatography-mass spectrometry (GC-MS) which was introduced during the 1960s as a practical and powerful tool for qualitative and quantitative analysis of compounds in mixtures. Both EI and CI were immediately useful for GC-MS, since both of these ionization methods require that the analytes be in the gas phase. When capillary GC was incorporated into GC-MS, this technique reached maturity. GC-MS may be used to select, identify and quantify organic compounds in complex mixtures at the femtomole level. The speed of GC-MS is determined by the chromatography step, which typically requires several minutes up to one hour per analysis. By the 1970s, some organic chemists were announcing that organic mass spectrometry had reached maturity and that no new applications were possible. Like the physicists and physical chemists who had pronounced the end of mass spectrometry a generation earlier, this group would soon be proved wrong.

Although GC-MS remains important for the analysis of many organic compounds, this technique is limited to volatile and thermally stable compounds that comprise only a small fraction of all organic compounds and even fewer biomedically important molecules. Therefore, thermally unstable compounds including many pharmaceutical compounds such as nucleic acid analogs and biomolecules such as proteins, carbohydrates, and nucleic acids cannot be analyzed in their native forms using GC-MS. (For more details regarding GC-MS and its applications, see Watson, 2007 provided in Section 4) Although derivatization facilitates the GC-MS analysis of many of these compounds, alternative ionization techniques were needed for the analysis of the vast majority of polar and nonvolatile compounds of interest to drug discovery.

1.4. Desorption Ionization Techniques

During the 1970s and the early 1980s, desorption ionization techniques such as field desorption (FD), desorption EI, desorption CI (DCI), and laser desorption were developed to extend the utility of mass spectrometry toward the analysis of more polar and less volatile compounds (See Watson, 2007 provided in Section 4) for more information regarding desorption ionization techniques including DCI and FD). Although these techniques helped extend the mass range of mass spectrometry beyond a general limit of *m/z* 1000 and toward ions of *m/z* 5000, a breakthrough in the analysis of polar, nonvolatile compounds occurred in 1982 with the invention of fast atom bombardment (FAB) [2]. FAB and its counterpart liquid secondary ion mass spectrometry (LSIMS) facilitate the formation of abundant molecular ions, protonated molecules, and deprotonated molecules of nonvolatile and thermally labile compounds such as peptides, chlorophylls, and complex lipids up to approximately *m/z* 12,000. FAB and LSIMS use energetic particle bombardment (fast atoms or ions from 3–30,000 V of energy) to ionize compounds dissolved in nonvolatile matrices such as glycerol or 3-nitrobenzyl alcohol and desorb them from this condensed phase into the gas phase for mass spectrometric analysis. Protonated or deprotonated molecules are usually abundant and fragmentation is minimal. Although still in use, FAB and LSIMS have been replaced by electrospray and matrix-assisted laser desorption ionization (MALDI) for most biomedical applications.

Introduced in the late 1980s, MALDI has helped solve the mass limit barriers of laser desorption mass spectrometry so that singly charged ions may be obtained up to *m/z* 500,000 and sometimes higher [3]. For most commercially available MALDI mass spectrometers, ions up to *m/z* 200,000 are readily obtained. Like FAB and LSIMS, MALDI samples are mixed with a matrix to form a solution that is loaded onto the sample stage for analysis. Unlike the other matrix-mediated techniques, the solvent is evaporated prior to MALDI analysis leaving sample molecules trapped in crystals of solid-phase matrix. The MALDI matrix is selected to absorb the pulse of laser light directed at the sample. Most MALDI mass spectrometers are equipped with a pulsed UV laser, although IR lasers are available as an option on some commercial instruments. Therefore, matrices are often substituted benzenes or benzoic acids with strong UV absorption properties. During MALDI, the energy of the short but intense UV laser pulse obliterates the matrix and in the process desorbs and ionizes the sample (see Fig. 1). Like FAB and LSIMS, MALDI typically produces abundant protonated or deprotonated molecules with little fragmentation.

1.5. Liquid Chromatography-Mass Spectrometry (LC-MS)

By the time that GC-MS had become a standard technique in the late 1960s, LC-MS was still in the developmental stages. Producing gas-phase sample ions for analysis in a vacuum system while removing the HPLC mobile phase proved to be a challenging task. Early LC-MS techniques included a moving belt interface to desolvate and transport the HPLC eluate into a CI or EI ion source or a direct inlet system in which the eluate was pumped at a low flow rate (1–3 μL/min) into a CI source. However, neither of these systems was robust enough or suitable for a broad enough range of samples to gain widespread acceptance. Since FAB and LSIMS require that the analyte be dissolved in a liquid matrix, this ionization technique was easily adapted for use with HPLC in an approach known as continuous-flow FAB [4]. Because the matrix often interfered with HPLC and required postcolumn addition and because the LC-MS interface required frequent maintenance and cleaning, continuous-flow FAB is rarely used today.

Like continuous-flow FAB, the popularity of particle beam interfaces has diminished. During particle beam LC-MS, the HPLC eluate is sprayed into a heated chamber connected to a vacuum pump. As the droplets evaporate, aggregates of analyte (particles) form and pass through a momentum

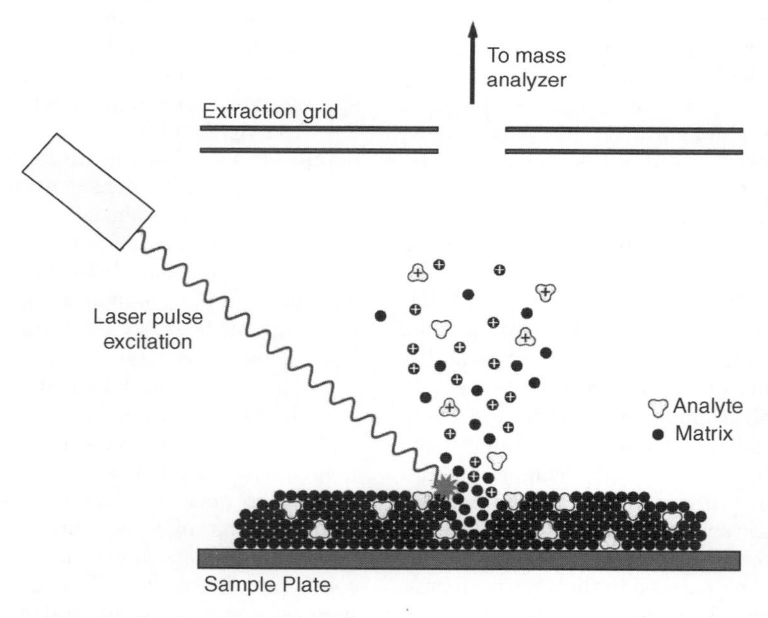

Figure 1. Scheme showing the process of MALDI from a solid matrix absorbing strongly at the wavelength of the incident laser pulse. (This figure is available in full color at http://mrw.interscience.wiley.com/emrw/9780471266945/home.)

separator that removes the lower molecular mass solvent molecules. Finally, the particle beam enters the mass spectrometer ion source where the aggregates strike a heated plate from which the analyte molecules evaporate and are ionized using conventional EI or CI. Particle beam LC-MS is limited to the analysis of volatile and thermally stable compounds that are amenable to flash evaporation and EI or CI mass spectrometry.

Thermospray became the first widely utilized LC-MS technique (during the late 1970s and the early 1980s). During thermospray, the HPLC eluate is sprayed through a heated capillary into a heated desolvation chamber at reduced pressure. Gas-phase ions remaining after desolvation of the droplets are extracted through a skimmer into the mass spectrometer for analysis. The sensitivity of thermospray is poor since there is no mechanism or driving force to enhance the number of sample ions entering the gas phase from the spray during desolvation. Also, thermally labile compounds tend to decompose in the heated source. These problems were solved when thermospray was replaced by electrospray.

Electrospray and atmospheric pressure chemical ionization (APCI) have become the most widely used ionization sources and HPLC-interfaces for drug discovery using mass spectrometry. Unlike thermospray, particle beam or continuous-flow FAB, electrospray and APCI interfaces operate at atmospheric pressure, do not depend upon vacuum pumps to remove solvent vapor, and are compatible with a wider range of HPLC flow rates. Also, no matrix is required. Both APCI and electrospray are compatible with a wide range of HPLC columns and solvent systems. Like all LC-MS systems, the solvent system should contain only volatile solvents, buffers or ion pair agents to reduce fouling of the mass spectrometer ion source. In general, APCI and electrospray form abundant molecular ion species. When fragment ions are formed, they are usually more abundant in APCI than electrospray mass spectra.

The APCI ion source and HPLC interface (see Fig. 2) uses a heated nebulizer to form a fine spray of the HPLC eluate, which is much finer than the particle beam system but similar to that formed during thermospray. Heated nitrogen gas is used to facilitate the

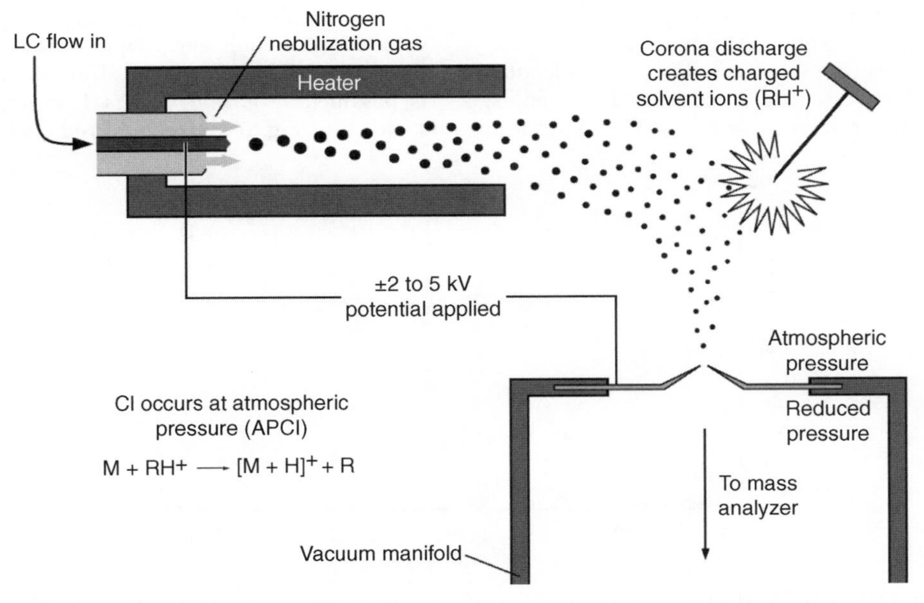

Figure 2. During APCI, eluate from a HPLC system is sprayed through a heated capillary and desolvated using heated nitrogen. Functioning as a CI reagent gas, solvent molecules are ionized by a corona discharge and then ionize sample molecules through proton transfer or charge exchange.

evaporation of solvent from the droplets. The resulting gas-phase sample molecules are ionized by collisions with solvents ions, which are formed by a corona discharge in the atmospheric pressure chamber. Molecular ions, $M^{+\cdot}$ or $M^{-\cdot}$, and/or protonated or deprotonated molecules can be formed. The relative abundance of each type of ion depends upon the sample itself, the HPLC solvent, and the ion source parameters. Next, ions are drawn into the mass spectrometer analyzer for measurement through a narrow opening or skimmer that helps the vacuum pumps to maintain very low pressure inside the analyzer while the APCI source remains at atmospheric pressure. For example, the positive ion APCI mass spectrum of lycopene is shown in Fig. 3. The carotenoid lycopene is the red pigment of ripe tomatoes and is under clinical investigation for the prevention of prostate cancer [5].

During electrospray, the HPLC eluate is sprayed through a capillary electrode at high potential (usually 2000–7000 V) to form a fine mist of charged droplets at atmospheric pressure. As the charged droplets migrate toward the opening of the mass spectrometer due to electrostatic attraction, they encounter a cross-flow of heated nitrogen that increases solvent evaporation and prevents most of the solvent molecules from entering the mass spectrometer (see Fig. 4). Molecular ions, protonated or deprotonated molecules, and cationized species such as $[M + Na]^+$ and $[M + K]^+$ can be formed. (For additional information on electrospray ionization, see Cole, 1997 provided in Section 4). In addition to singly charged ions, electrospray is unique as an ionization technique in that multiply charged species are common and often constitute the majority of the sample ion abundance. The relative abundance of each of these species depends upon the chemistry of the analyte, the pH, the presence of proton donating or accepting species, and the levels of trace amounts of sodium or potassium salts in the mobile phase. In contrast, APCI, APPI, MALDI, EI, CI, and FAB/LSIMS usually produce singly charged species. A consequence of forming multiply charged ions is that they are detected at lower *m/z* values (i.e., $z > 1$) than the corresponding singly charged species. This has the benefit of allowing mass spectrometers with modest *m/z* ranges to detect and measure ions of molecules with very high masses. For example, electrospray has been used to measure ions with molecular masses of hundreds of thousands or even millions of daltons on mass spectrometers with *m/z* ranges of only a few thousands. (For a review of LC-MS techniques, see Niessen, 2006 provided in Section 4.)

An example of the C_{18} reversed phase HPLC-negative ion electrospray mass spectrometric (LC-MS) analysis of an extract of the botanical *Trifolium pratense* L. (red clover) is shown in Fig. 5. Extracts of red clover are

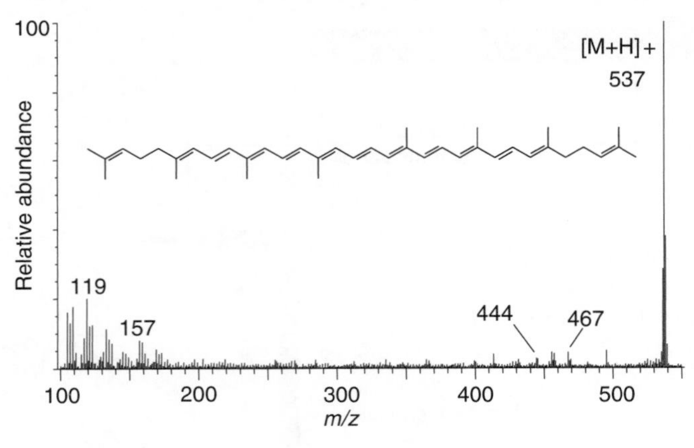

Figure 3. Positive ion APCI mass spectrum of the red carotenoid lycopene in a solution of methanol and *tert*-butyl methyl ether (1 : 1; v/v). In this analysis, lycopene formed a protonated molecule detected at *m/z* 537 as the base peak of the mass spectrum instead of a molecular ion, $M^{+\bullet}$.

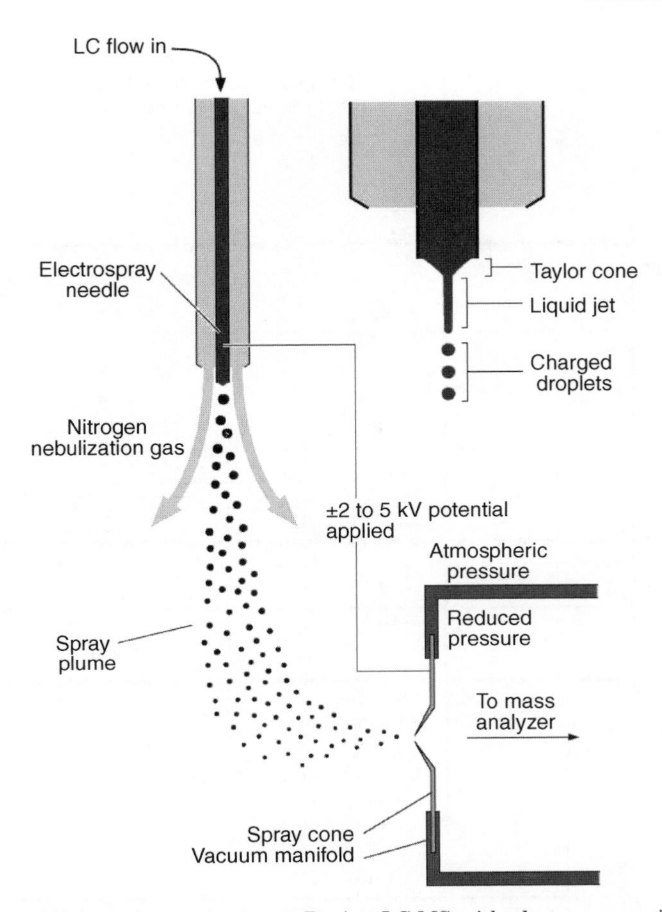

Figure 4. Schematic of an electrospray source. During LC-MS with electrospray, nitrogen is used as a nebulization and drying gas to facilitate the formation of small droplets that become charged as they emerge from a Taylor cone and electrospray needle at high potential. As the solvent evaporates, analytes becomes gas-phase ions that are accelerated into the mass spectrometer through a potential gradient. In some designs, heated nitrogen is directed across the entrance to the mass spectrometer as a curtain gas to provide additional desolvation and to prevent the solvent from entering the mass spectrometer.

used as dietary supplements by menopausal and postmenopausal women and are under investigation as alternatives to estrogen replacement therapy [6]. The two-dimensional map illustrates the amount of information that may be acquired using hyphenated techniques such as LC-MS. In the time dimension, chromatograms are obtained, and a sample computer-reconstructed mass chromatogram is shown for the signal at *m/z* 269. An intense chromatographic peak was detected eluting at 12.4 min. In the *m/z* dimension, the negative ion electrospray mass spectrum recorded at 12.4 min shows a base peak at *m/z* 269. Based on comparison to authentic standards (data not shown), the ion of *m/z* 269 was found to correspond to the deprotonated molecule of genistein that is an estrogenic isoflavone [6]. Since almost no fragmentation of the genistein ion was observed, additional characterization would require CID and MS–MS as discussed in the next section.

When analyzing complex mixtures such as the botanical extract shown in Fig. 5, the use of chromatographic separation prior to mass spectrometric ionization and analysis is

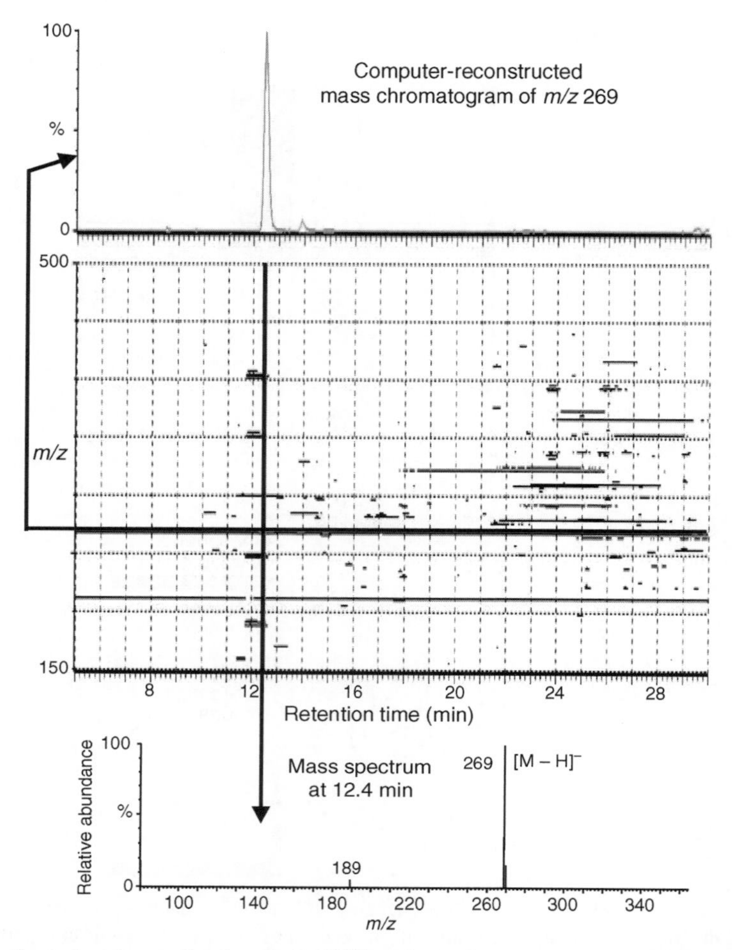

Figure 5. Two-dimensional map showing the LC-MS analysis of an extract of red clover under investigation for the management of menopause. Reversed phase separation was carried out using a C_{18} HPLC column in the time dimension and negative ion electrospray mass spectrometry was used for compound detection and molecular mass determination in the second dimension.

essential in order to distinguish between isomeric compounds. Even simple mixtures of synthetic compounds might contain isomers that would require LC-MS for adequate characterization. Another problem overcome by utilizing a chromatography step prior to mass spectrometric analysis is ion suppression. No matter what ionization technique is used, the presence of multiple compounds in the ion source might enhance the ionization of one compound while suppressing the ionization of another. Usually, only some of the compounds in a complex mixture can be detected by mass spectrometry without chromatographic separation. The presence of salts and buffers in a sample can also suppress sample ionization. Therefore, LC-MS has become a powerful tool for analyzing natural products, synthetic organic compounds, and pharmaceutical agents and their metabolites.

In general, APCI facilitates the ionization of nonpolar and low molecular mass species and electrospray is more useful for the ionization of polar and high molecular mass compounds. In this sense, APCI and electrospray are often complementary ionization techniques. However, during the analysis of large or diverse combinatorial libraries, both polar

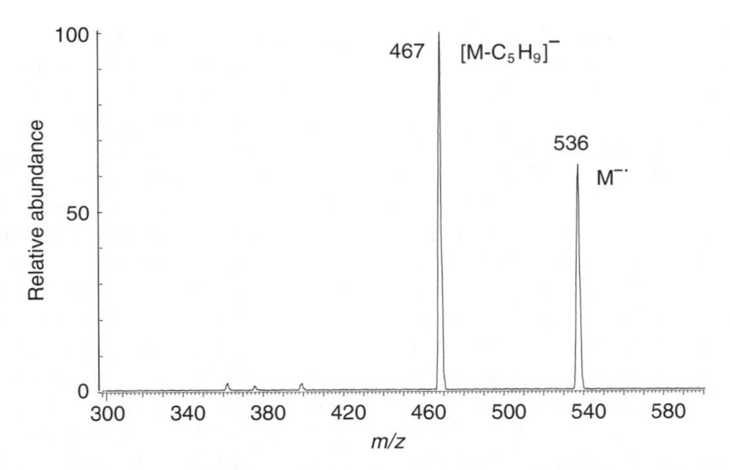

Figure 6. Negative ion electrospray tandem mass spectrum of lycopene. CID was used to induce fragmentation of the molecular ion of *m/z* 536. As a result, the fragment ion of *m/z* 467 was formed by the loss of a terminal isoprene unit. This fragment ion may be used to distinguish lycopene from isomeric α-carotene and β-carotene that lack terminal isoprene groups.

and nonpolar compounds are usually present. As a result, no one set of ionization conditions using APCI or electrospray is adequate to detect all the compounds contained the library of compounds. Therefore, a UV ionization technique called atmospheric pressure photoionization (APPI) has been developed for use with combinatorial libraries and LC-MS [7]. During APPI, a liquid solution or HPLC eluate is sprayed at atmospheric pressure as in APCI. Instead of using a corona discharge as in APCI, ionization occurs during APPI due to irradiation of the analyte molecules by an intense UV light source. Obviously, the carrier solvent must not absorb UV light at the same wavelengths or interference would prevent sample ionization and detection.

1.6. Tandem Mass Spectrometry (MS–MS)

Desorption ionization techniques such as FAB, MALDI, and electrospray facilitate the molecular mass determination of a wide range of polar, nonpolar, low, and high molecular mass compounds including drugs and drug targets such as proteins. However, the "soft" ionization character of these techniques means that most of the ion current is concentrated in molecular ions and few structurally significant fragment ions are formed. In order to enhance the amount of structural information in these mass spectra, CID may be used to produce more abundant fragment ions from molecular ion precursors formed and isolated during the first stage of mass spectrometry. Then, a second mass spectrometric analysis may be used to characterize the resulting product ions. This process is called tandem mass spectrometry or MS–MS and is illustrated in Fig. 6.

Another advantage of the use of tandem mass spectrometry is the ability to isolate a particular ion such as the molecular ion of the analyte of interest during the first mass spectrometry stage. This precursor ion is essentially purified in the gas phase and free of impurities such as solvent ions, matrix ions, or other analytes. Finally, the selected ion is fragmented using CID and analyzed using a second mass spectrometry stage. In this manner, the resulting tandem spectrum contains exclusively analyte ions without impurities that might interfere with the interpretation of the fragmentation patterns. In summary, CID may be used with LC-MS-MS or desorption ionization and MS-MS to obtain structural information such as amino acid sequences of peptides, sites of alkylation of nucleic acids, or to distinguish structural isomers such as β-carotene and lycopene.

The types of tandem mass spectrometers used for biomedical research are shown in

Table 1 and include triple quadrupole instruments (ideal for quantitative analysis), ion trap and FTICR (high resolution) mass spectrometers which are capable of multiple stages of tandem mass spectrometry, and hybrid instruments such as QqTOF and ion trap-TOF mass spectrometers (capable of high-resolution tandem mass spectrometry). Except for the TOF–TOF mass spectrometers, all the instruments listed in Table 1 are compatible with electrospray, APPI, and APCI and may be interfaced with HPLC systems. The TOF–TOF mass spectrometer is ideally suited for use with a pulsed ionization technique such as MALDI.

Over the course of the last century, mass spectrometry has become an essential analytical tool for a wide variety of biomedical applications including drug discovery and development. By combining mass spectrometry with chromatography as in LC-MS or by adding another stage of mass spectrometry as in MS–MS, the selectivity of the technique increases considerably. As a result, mass spectrometry offers all of the analytical elements that are essential to modern drug discovery namely speed, sensitivity, and selectivity.

2. CURRENT TRENDS AND RECENT DEVELOPMENTS

Since the early 1990s, drug discovery research has focused on combinatorial chemistry [8,9] and high-throughput screening [10] in an effort to accelerate the pace of drug discovery. The goal has been to produce in a short time large numbers of synthetic organic compounds representing a great diversity of chemical structures through a process called combinatorial chemistry and then quickly screen them *in vitro* against pharmacologically significant targets such as enzymes or receptors. The "hits" identified through these high-throughput screens may then be optimized by quickly and efficiently synthesizing and then screening large numbers of analogs called targeted or directed libraries. As a result, lead compounds might emerge from such combinatorial chemistry drug discovery programs in a few weeks instead of several years. Furthermore, a single organic chemist using combinatorial synthetic methods might synthesize thousands of compounds or more in a single week instead of less than five in the same time using conventional techniques, and a single medicinal chemist might identify hundreds of lead compounds per month instead of just one or two in the same period of time.

Accompanying this new drug discovery paradigm, new scientific journals have been established such as *Combinatorial Chemistry & High Throughput Screening, Journal of Combinatorial Chemistry, Journal of Biomolecular Screening*, and *Molecular Diversity* (see the list of journal Web sites provided in Section 4). The variety of topics published in these journals reflects the multidisciplinary nature of the current drug discovery process and ranges from organic chemistry, medicinal chemistry, molecular modeling, molecular biology, and pharmacology to analytical chemistry. As described below, the most significant analytical component of drug discovery has become mass spectrometry. Only mass spectrometry has become an essential element at all stages of the drug discovery and development process.

Although a variety of spectroscopic and chromatographic techniques including infrared spectroscopy, nuclear magnetic resonance spectroscopy, fluorescence spectroscopy, gas chromatography, high-performance liquid chromatography (HPLC) and mass spectrometry are being used to support drug discovery in various capacities, some of them such as gas chromatography and fluorescence spectroscopy are not applicable to most new chemical entities, some are not specific enough for chemical identification (e.g., infrared spectroscopy), and other techniques suffer from low throughput (e.g., nuclear magnetic resonance spectroscopy). Unlike gas chromatography, HPLC is compatible with virtually all drug-like molecules without the need for chemical derivatization to increase thermal stability or volatility. In addition, mass spectrometry provides a universal means to characterize and distinguish drugs based on both molecular mass and structural features while at the same time providing high throughput. With the development of routine LC-MS interfaces and ionization techniques such as electrospray and APCI, mass spectrometry has also become an ideal HPLC detector for the analysis of combinatorial libraries [11], and LC-MS, MS-MS, and LC-MS-MS have become fundamental tools in the analysis of combinatorial libraries and subsequent drug development studies [12–14].

The application of combinatorial chemistry and high-throughput screening to drug discovery has altered the traditional serial process of lead identification and optimization that previously required years of human effort. Consequently, neither the synthesis of new chemical entities nor their screening is limiting the pace of drug discovery. Instead, a new bottleneck is the verification of the structure and purity of each compound in a combinatorial library or of each lead compound obtained from an uncharacterized library using high-throughput screening. Since the number of lead compounds entering the drug development process has increased in part because compounds are entering development at earlier stages than in the past, the traditional drug development investigations concerning absorption, distribution, metabolism, and excretion (ADME) and even toxicology evaluations of new drug entities have become additional bottlenecks. As a solution to the drug development bottlenecks, high-throughput assays to assess the metabolism, bioavailability, and toxicity of lead compounds are being developed and applied earlier than ever during the drug discovery process so that only those compounds most likely to become successful drugs enter the more expensive and slower preclinical pharmacology and toxicology studies. In support of these new combinatorial chemistry synthetic programs and new high-throughput assays, mass spectrometry has emerged as the only analytical technique with sufficient throughput, sensitivity, selectivity, and robustness to address all of these bottlenecks.

2.1. LC-MS Purification of Combinatorial Libraries

Although combinatorial libraries were originally synthesized as mixtures, today most libraries are prepared in parallel as discrete compounds and then screened individually in microtiter plates of 96-well, 384-well, or 1536-well formats. In order to facilitate subsequent structure–activity analyses and to assure the validity of the screening results, many laboratories verify the structure and purity of each compound prior to high-throughput screening. Semipreparative HPLC has become the most popular technique for the purification of combinatorial libraries on the milligram scale because of high throughput and the ease of automation. Typically during semipreparative HPLC, fraction collection is initiated whenever a UV signal is observed above a predetermined threshold. This procedure usually results in the collection of several fractions per analysis and hence creates additional issues such as the need for large fraction collector beds and the need for secondary analysis using flow injection mass spectrometry, LC-MS, or LC-MS-MS to identify the appropriate fractions. When purification of large numbers of combinatorial libraries is required, this approach can become prohibitively time consuming and expensive.

To enhance the efficiently of this purification procedure, the steps of HPLC purification and mass spectrometric analysis may be combined into automated mass-directed fractionation [15–17]. Any size HPLC column may be used, and only a small fraction of the eluant (~μL/min) is diverted to the mass spectrometer equipped for APCI or electrospray ionization. Since all of the components including autosampler, injector, HPLC, switching valve, mass spectrometer, and fraction collector are controlled by computer, the procedure may be fully automated. For greatest efficiency, the system may be programmed to collect only those peaks displaying the desired molecular ions or alternatively, all peaks displaying abundant ions within a specified mass range. An example of the MS-guided purification of a compound synthesized during the parallel synthesis of a combinatorial library of discrete compounds is shown in Fig. 7. Although the crude yield of the reaction product was only 30% (Fig. 7a), the desired product was detected based on its molecular ion (Fig. 7b). After MS-guided fractionation, reanalysis using LC-MS showed that the desired product was >90% pure (Fig. 7c).

The use of MS-guided purification of combinatorial libraries provides a means for

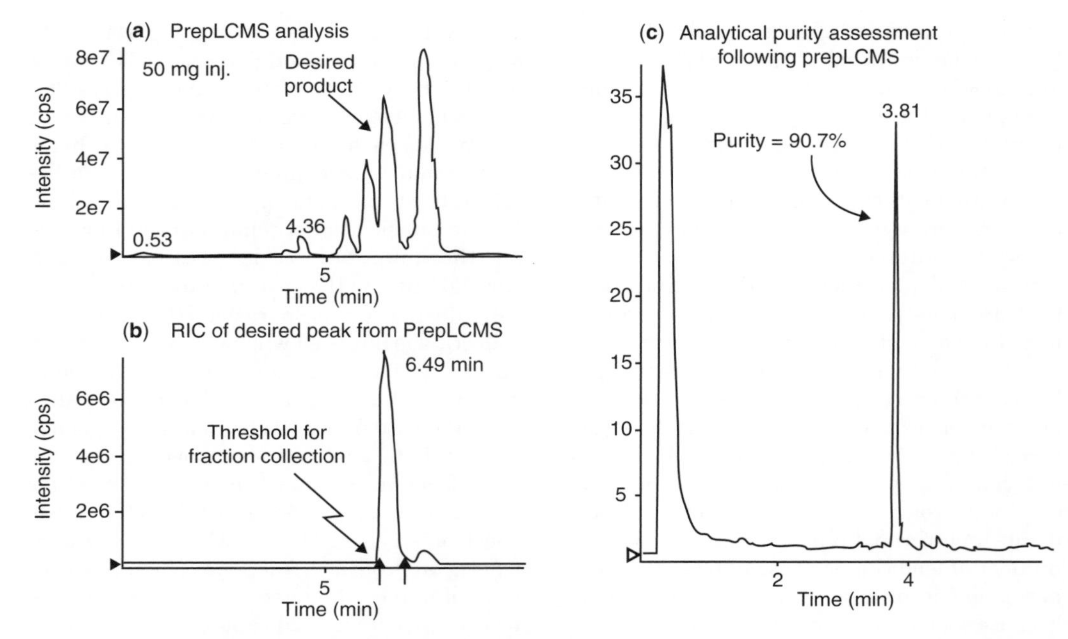

Figure 7. Mass-directed purification of a combinatorial library. Chromatographic separation was carried out using gradient elution of 10–90% acetonitrile in water for 7 min following an initial hold at 10% acetonitrile for 1 min. (a) Total ion chromatogram showing desired product and impurities. (b) Computer-reconstructed ion chromatogram (RIC) corresponding to the expected product. (c) Postpurification analysis of the isolated component with a purity >90%. (Reproduced from Ref. [15] by permission of Elsevier Science.)

reducing the number of HPLC fractions collected per sample and eliminates the need for postpurification analysis to further characterize and identify each compound as would be necessary when using UV-based fractionation. The ionization technique (i.e., electrospray, APCI, or APPI), and ionization mode (positive or negative) must be suitable for the combinatorial compound, so that molecular ion species are formed. Also, a suitable mobile phase and HPLC column must be selected. As an alternative to HPLC, supercritical fluid chromatography-mass spectrometry (SFC-MS) has been used for the high-throughput analysis of combinatorial libraries [18,19]. The advantages of SFC-MS relative to conventional LC-MS for the purification of combinatorial libraries of compounds are the lower viscosities and higher diffusivities of condensed CO_2 compared to HPLC mobile phases, and the ease of solvent removal and disposal after analysis. However, SFC instrumentation remains more expensive and less widely available than conventional HPLC systems. Furthermore, many drug-like molecules are poorly soluble in condensed CO_2.

2.2. Confirmation of Structure and Purity of Combinatorial Compounds

The determination of molecular masses, elemental compositions, and structures of compounds used for high-throughput screening, whether discrete compounds or combinatorial library mixtures, is typically carried out using mass spectrometry, since traditional spectroscopic and gravimetric techniques are too slow to keep pace with combinatorial chemical synthesis. In addition, mass spectrometry may be used to assess the purity of compounds being used for high-throughput screening. The highest throughput technique for confirming molecular masses and structures of drug candidates is flow injection analysis of sample solutions using electrospray, APCI, or APPI mass spectrometry. Typically, no sample preparation is necessary.

Although any organic mass spectrometer may be used to confirm the molecular mass of a compound, tandem mass spectrometers provide additional structural information through the use of CID to produce fragment ions. As discussed above (see also Table 1), tandem mass spectrometers include triple quadrupole instruments, QqTOF mass spectrometers, ion trap-TOF mass spectrometers, ion trap mass spectrometers, multiple sector magnetic sector instruments, FTICR instruments, and TOF–TOF mass spectrometers, and the new orbitrap mass spectrometers. For most medicinal chemistry applications, APCI or electrospray ionization is used.

In addition to molecular mass and fragmentation patterns, high-precision and high-resolution mass spectrometers such as QqTOF instruments, ion trap-TOF mass spectrometers, reflectron TOF mass spectrometers, double focusing magnetic sector mass spectrometers, and FTICR instruments are necessary for the measurement of exact masses of drugs and drug candidates for the determination of elemental compositions. The combination of high-resolution and high-precision is especially useful for determining the elemental compositions of compounds in combinatorial library mixtures without having to isolate each compound using chromatography or some other separation technique. Since FTICR instruments, orbitrap mass spectrometers, and hybrid ion trap-TOF and QqTOF mass spectrometers are capable of accurate mass measurements at high resolving power of both molecular ions and fragment ions generated during MS–MS, these instruments are becoming extremely popular within drug discovery programs.

Although accurate mass measurements of compounds in combinatorial libraries can usually be measured using electrospray or APCI mass spectrometry with infusion, online HPLC separation is sometimes required to overcome ion suppression problems due to the presence of buffer, contaminants or reaction by-products. However, LC-MS is a relatively slow process due to the slow chromatographic separation step. Several approaches have emerged to increase the throughput of this technique; parallel LC-MS, fast LC-MS and ultrahigh pressure liquid chromatography (UHPLC)-MS. One approach to increasing throughput of the rate-limiting chromatographic separation has been to simultaneously interface multiple HPLC columns to a single mass spectrometer. This approach is called parallel LC-MS. Commercial parallel electrospray interfaces and HPLC systems are now available that can accommodate up to 8 HPLC columns simultaneously [20–22]. Although the multiple sprays are introduced to the ion source simultaneously, these streams may be sampled in a time-dependent manner to minimize cross-contamination between channels.

Another solution to increasing the throughput of LC-MS has been to minimize the time required for HPLC separation through an approach called fast HPLC. HPLC separations are accelerated by using shorter columns and higher mobile phase flow rates. Since coelution of some species is likely to occur during fast chromatographic separations, the selectivity of the mass spectrometer is essential for the characterization and/or quantitative analysis of the target compound. However, samples prepared using combinatorial chemistry are usually simple mixtures of reagents, by-products, and product that require only partial chromatographic purification to prevent ion suppression effects during mass spectrometric analysis. A variety of HPLC columns are used for fast LC-MS that include narrow bore (2 mm) and analytical bore (4.6 mm) columns with length typically from 0.5–5 cm. The mobile phase flow rate for these fast LC-MS analyses is usually from 1.5–5 mL/min.

Similar stationary phase chemistries are used for UHPLC, but the particle size of the packing material is reduced to $<2\,\mu M$ for greater column efficiency. If the UHPLC columns are shortened compared to conventional columns, similar separation efficiencies may be obtained while the analysis times and solvent consumption can be reduced to two- to threefold. Alternatively, UHPLC may be operated at higher flow rates to obtain good separations in just seconds instead of minutes. However, the small particle size of UHPLC columns produce high back pressures of up to 800 bar, which is twice the maximum pressure of conventional HPLC systems. Therefore,

UHPLC systems are engineered to operate at pressures up to at least 800 bar.

In addition to molecular mass determination using conventional MS or high-resolution accurate mass measurement and structural confirmation using MS–MS, fast LC-MS, and UHPLC-MS are also used to assess the purity and yield of combinatorial products [15,23]. Prior to high-throughput screening, many researchers analyze combinatorial libraries for both purity and structural identity using mass spectrometry so as to assure the validity of structure–activity relationships that might be derived from the screening data. Fast LC-MS, UHPLC-MS, and LC-MS–MS may be carried out to satisfy this requirement using gradients (usually a step gradient with a reverse phase HPLC or UHPLC column) with a total cycle time of <3 min [24] for fast HPLC or <1 min per analysis using UHPLC or isocratic systems.

2.3. Encoding and Identification of Compounds in Combinatorial Libraries and Natural Product Extracts

The utility of mass spectrometric identification in combinatorial chemistry is limited not only to the analysis of synthetic products as a means of quality control but also for the identification of active compounds or "hits" during high-throughput screening. Although the synthesis and screening of discrete compounds [25] enables them to be followed through the entire process by using partial encoding or bar coding, it is sometimes advantageous to screen libraries prepared as mixtures [26] and use a technique such as mass spectrometry to rapidly identify the hit(s) in the mixture. One approach to the rapid deconvolution of combinatorial library mixtures is to prepare libraries containing compounds of unique molecular mass, and then identify them using mass spectrometry. However, such libraries are necessarily small since the molecular mass of most drug-like molecules is between 150–400 Da. Because of the molecular mass degeneracy of larger combinatorial libraries, several encoding strategies have been devised to rapidly identify active compounds in these mixtures [27–29].

Since most combinatorial libraries contain compounds with degenerate molecular masses, various tagging strategies have been devised to uniquely identify library compounds bound to beads. Most of these tagging approaches are based on the synthesis of encoding molecules. For example, peptide [30] or oligonucleotide [31] labels have been synthesized on the beads in parallel to the target molecules and then sequenced for bead decoding. Alternatively, haloarene tags have been incorporated during synthesis and then identified with high sensitivity using electron-capture gas chromatography detection [32]. In addition to the increased time and cost for the synthesis of a library containing tagging moieties, the tagging groups themselves might interfere with screening giving false-positive or -negative results.

For peptide libraries, one solution to this problem uses MALDI mass spectrometry to directly desorb and identify peptides from beads that were screened and found to be hits [33]. This technique is called the termination synthesis approach. Since the peptide library compounds are analyzed directly, products with amino acid deletions or substitutions, side-reaction products, or incomplete deprotection are readily observed. Also, since there are no extra molecules used for chemical tagging, this source of interference is avoided. However, this approach is specific to peptide libraries and not necessarily applicable to other types of combinatorial libraries.

Another approach that eliminates possible interference from the chemical tags, "ratio encoding," has been developed for the mass spectrometric identification of bioactive leads using stable isotopes incorporated into the library compounds [29,34]. Within the ligand itself, the code might be a single labeled atom that is conveniently inserted whenever a common reagent transfers at least one atom to the target compound or ligand. The code consists of an isotopic mixture having one of the many predetermined ratios of stable isotopes and can be incorporated in the linker or added through a reagent used during the synthesis. The mass spectrum of the compound shows a molecular ion with a unique isotope ratio that codes for a particular library compound. For example, Wagner et al. [29] used isotope ratio encoding during the synthesis of 1000 compound peptoid library and was able to identify uniquely all the components based on their isotopic patterns and molecular masses. Since isotope ratio codes are contained within each combinatorial compound, a chemical tag is not required. The speed of MS-based decoding outperforms most other decoding technologies, which are time consuming and decode a restricted set of active compounds.

Although combinatorial synthesis provides rapid access to large numbers of compounds for screening during drug discovery and lead optimization, these libraries are usually based on a small number of common structures or scaffolds. There is a constant need for increasing the molecular diversity of combinatorial libraries and finding new scaffolds, and natural products have always been a rich source of chemical diversity for drug discovery. The traditional approach to screening natural products for drug leads utilizes bioassays to test organic solvent extracts for activity. If strong activity is detected, then activity-guided fractionation of the crude extract is used to isolate the active compound(s), which are identified using mass spectrometry (including tandem mass spectrometry and accurate mass measurements), IR, UV/VIS spectrometry, and NMR. Recently, a variety of mass spectrometry-based affinity screening methods have been developed to streamline the tedious process of activity-guided fractionation. These approaches are discussed in Section 2.4.

Whether lead compounds in natural product extracts are isolated using bioassay-guided fractionation or mass spectrometry-based screening, there is a high probability that the structure of the active compound(s) has already been reported in the natural product literature. In such cases, the tedious process of complete structure elucidation using a battery of spectrometric tools should be unnecessary. Instead, mass spectrometry alone may be used to quickly "dereplicate" or identify the known compounds based on molecular mass, fragmentation patterns and elemental composition in combination with natural product database searching [35–39]. Commercially available natural products databases include NAPRALERT [40], Scientific & Technical information Network (STN) [41], and the Dictionary of Natural Products [42]. Since some of these databases also contain UV–vis absorbance data, it is also advantageous to use a photodiode array detector between the HPLC and mass spectrometer to obtain additional spectrometric data during LC-UV-MS dereplication [36,37].

2.4. Mass Spectrometry-Based Screening

The earliest approaches to combinatorial synthesis used portioning and mixing [26] and enabled the synthesis of combinatorial libraries containing hundreds of thousands to millions of compounds. However, efficient screening techniques did not exist to rapidly identify the "hits" within large combinatorial mixtures. Therefore, chemists were motivated to develop ways to prepare large numbers of discreet compounds using massively parallel synthesis, which could be assayed quickly for pharmacological activity using high-throughput screening one compound at a time. Although high-throughput screening of discreet compounds has become the standard approach to drug discovery in medicinal chemistry, there is a shortage of chemically diverse novel structures to serve as lead compounds for combinatorial synthesis and lead optimization. A traditional source of molecular diversity for drug discovery, natural products are receiving renewed interest in drug discovery programs. However, the complexity of natural products sources such as botanical or microbial extracts requires screening techniques that are suitable for mixtures. To address this requirement, mass spectrometry-based screening assays have been developed that are suitable for screening complex mixtures including natural product extracts. All of the mass spectrometry-based screening methods use receptor binding of ligands as the basis for identification of lead compounds.

2.4.1. Affinity Chromatography-Mass Spectrometry Since the introduction of affinity chromatography approximately 40 years ago, this technique has become a standard biochemical tool for the isolation and identification of new binding partners to specific target molecules. Therefore, the coupling of affinity chromatography to mass spectrometry is a logical extension of this approach, and the application of affinity LC-MS to the screening of combinatorial libraries has been demonstrated by several groups [43,44]. During affinity LC-MS screening, a receptor molecule such as a binding protein or enzyme is immobilized on a solid support within a chromatography column. The library mixture is pumped through the affinity column in a suitable binding buffer so that any ligands in the mixture with affinity for the receptor would be able to bind. Then, unbound material is washed away. Finally, the specifically bound ligands are eluted using a destabilizing mobile phase and identified using mass spectrometry. This affinity-column LC-MS assay is summarized in Fig. 8.

In some applications [43], ligands are eluted from the affinity column and then trapped on a second column such as a reverse phase HPLC column. LC-MS or LC-MS–MS identification of the ligands (hits) is then carried out using the trapping column. In other systems, ligands are identified directly from the affinity column using mass spectrometry. For example, Kelly et al. [44] prepared an affinity column containing immobilized phosphatidylinositol-3-kinase and used it for direct LC-MS screening of a 361-component peptide library. Electrospray mass spectrometry and tandem mass spectrometry were used to identify the ligands released from the affinity column using pH gradient elution.

Advantages of affinity chromatography-mass spectrometry for screening during drug discovery include versatility and reuse of the column. Both combinatorial libraries and natural product extracts can be screened using this approach, and a wide range of binding buffers may be used. Mass spectrometry-compatible mobile phases are only required during the final LC-MS detection step. Furthermore, a single column may be used multiple times to screen different samples for ligands unless the destabilization solution irreversibly denatures, releases, or inhibits the receptor.

Despite these advantages, affinity chromatography has numerous drawbacks that have prompted the development of alternative mass spectrometer screening tools. For example, immobilization of the receptor might change its affinity characteristics causing false-negative or false-positive hits. This is particularly problematic for receptors that are solution-phase in their native state. Also, developing and then implementing an immobilization scheme is often a slow, tedious, and even expensive process, which is unique for each new receptor. Finally, false-positive hits are often obtained when screening large, molecularly diverse libraries, since there are usually compounds in such mixtures that have affinity for the stationary phase or linker molecule instead of the receptor.

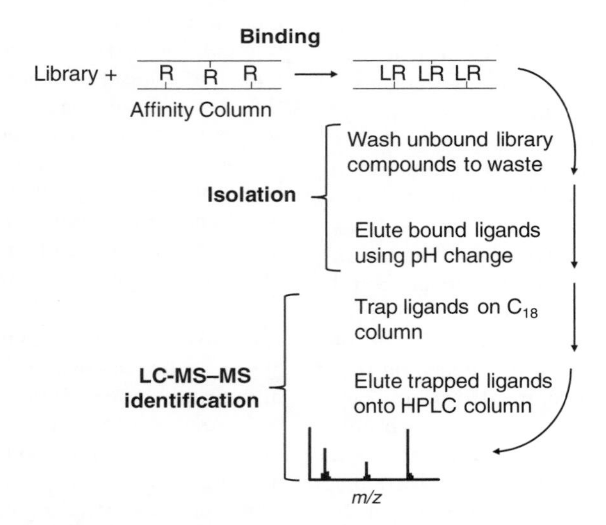

Figure 8. Affinity chromatography combined with LC-MS–MS for screening combinatorial library mixtures.

2.4.2. Gel Permeation Chromatography-Mass Spectrometry Another type of chromatography that has been combined with mass spectrometry as a screening system for drug discovery is gel permeation chromatography (GPC) [45,46]. Also called size exclusion chromatography, GPC separates molecules according to size as they pass through a stationary phase containing particles with a defined pore size. During GPC-based screening, a library mixture is preincubated with a macromolecular receptor to allow any ligands in the library to bind, and then GPC is used to separate the large receptor–ligand complexes from the unbound low molecular mass compounds in the mixture. Finally, ligands are released from the receptor during reversed phase HPLC and identified either on-line or off-line using tandem mass spectrometry. This screening method is illustrated in Fig. 9.

During the preincubation and GPC steps, any binding buffer may be used, since the binding buffer will be removed during reverse phase LC-MS analysis. However, the GPC separation step must be carried out quickly, since ligands begin to dissociate from the receptor immediately and can become lost into the size exclusion gel. Despite this disadvantage, this approach allows both receptor and ligand to be screened in solution, which avoids some of the problems associated with the use of affinity columns for screening. The GPC LC-MS-MS screening method is suitable for screening natural product extracts as well as combinatorial library mixtures.

2.4.3. Affinity Capillary Electrophoresis-Mass Spectrometry Affinity capillary electrophoresis was originally used for the determination of the binding constants of small molecules to proteins [47–49]. This solution-based technique is rapid and requires only small amounts of ligands. Affinity constants are measured based on the mobility change of the ligand upon interaction with the receptor present in the electrophoretic buffer [50]. By combining affinity capillary electrophoresis with on-line mass spectrometric detection and identification, affinity constants for multiple compounds can be measured in a single analysis [51]. Recognizing that on-line mass spectrometric detection was helpful for the identification of each ligand, Chu et al. [52] extended this approach to include the screening of combinatorial libraries as a means of drug discovery. The data in Fig. 10 show the results of screening a 100-tetrapeptide library for affinity to vancomycin using affinity capillary electrophoresis-mass spectrometry. Without vancomycin in the electrophoresis buffer, all the peptides eluted within 3 min. When vancomycin was present, the peptides eluted in

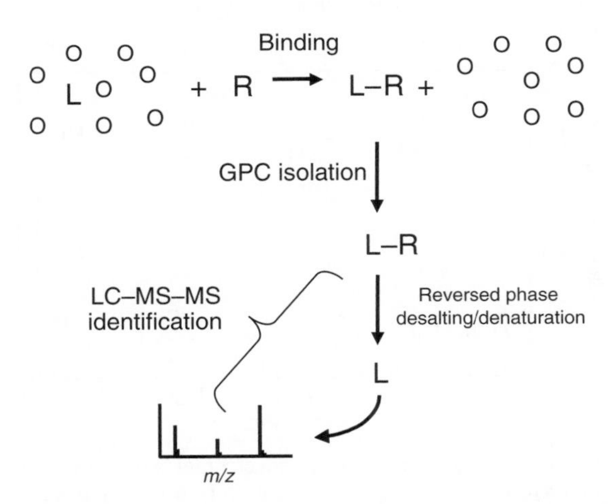

Figure 9. GPC followed by LC-MS–MS for screening mixtures of combinatorial libraries. After incubation of a receptor with a library of compounds, the ligand–receptor complexes (L–R) are separated from the low molecular mass unbound library compounds using GPC. Next, the L–R complexes are denatured during reversed phase HPLC to release the ligands for MS–MS identification.

order of affinity with the highest affinity compounds being detected between 4.5 and 5 min. Positive ion electrospray tandem mass spectrometry was used to identify the highest affinity ligands (see Fig. 10B).

Note that the some peptide ligands such as Fmoc-DDFA were detected as adducts with Tris that was used in the electrophoresis buffer. Although the identification of this peptide was not prevented by the formation of this adduct, some buffers used during electrophoresis might interfere with mass spectrometric ionization and detection. Also, the types of libraries that have been screened using this approach have contained modest numbers of synthetic analogs such as peptides. Libraries exceeding 400 members required preliminary purification using affinity chromatography to reduce the number of compounds [52]. As a result, this approach is probably not ideal for screening libraries containing molecularly diverse compounds or for screening natural product extracts. However, affinity capillary electrophoresis-mass spectrometry is fast with each analysis requiring less than 10 min. Also, it may be used to measure affinity constants for ligand–receptor interactions.

2.4.4. Frontal Affinity Chromatography-Mass Spectrometry Like affinity chromatography-mass spectrometric screening (see Section 2.4.1), frontal affinity chromatography utilizes an affinity column containing immobilized receptor molecules [53]. The difference between the two screening methods is that the ligands are continuously infused into the column during frontal affinity chromatography and detected using mass spectrometry. Compounds with no affinity for the immobilized receptor elute immediately in the void volume,

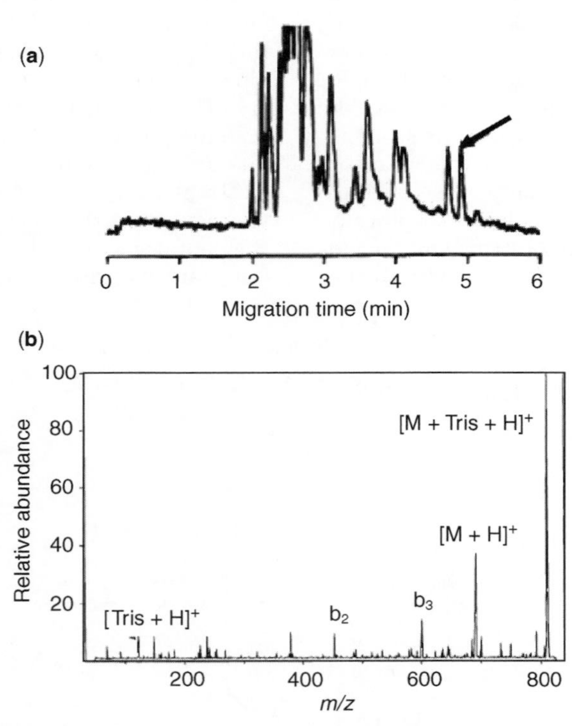

Figure 10. Affinity capillary electrophoresis-UV-mass spectrometry of a 100-tetrapeptide library screened for binding to vancomycin (104 μM in the electrophoresis buffer). (a) The elution of peptides was monitored with UV absorbance during capillary electrophoresis, and the elution time increased with increasing affinity for vancomycin. (b) Positive ion electrospray mass spectrum with CID of the Tris adduct of the protonated peptide detected at ~5 min in the electropherogram shown in A. (Reproduced from Ref. [52] by permission of the American Chemical Society.)

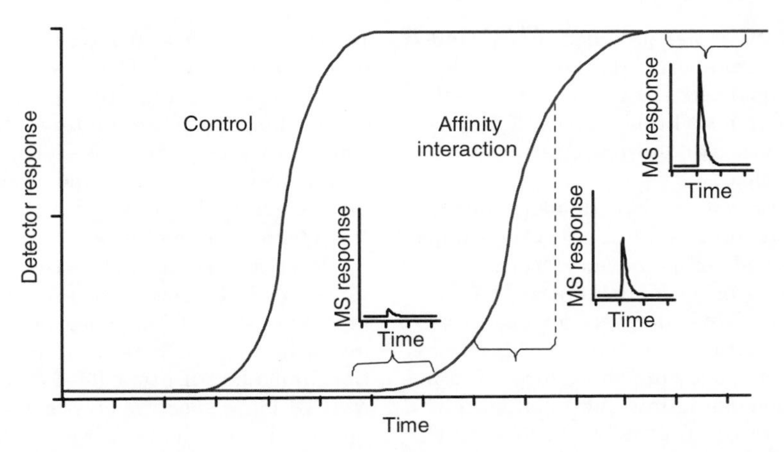

Figure 11. Frontal affinity chromatography screening using mass spectrometry. The control represents the void volume of the column and the elution profile of compounds with no affinity for the immobilized receptor. Compounds eluting later than those in the control elute in increasing order of affinity for the immobilized receptor. When the binding buffer is not compatible with mass spectrometry, fractions can be collected and analyzed off-line using LC-MS to determine the relative amounts of each potential ligand eluting from the column. Based on Ref. [55]. (This figure is available in full color at http://mrw.interscience.wiley.com/emrw/9780471266945/home.)

but the elution of the ligands is delayed. As compounds compete for binding sites on the affinity column, these sites become saturated until ligands begin to elute from the column at their infusion concentration (see Fig. 11). In this manner, frontal affinity chromatography may be used to measure affinity constants for ligands, and by using a mass spectrometer for on-line identification of ligands, this technique becomes a screening method [54,55].

During frontal affinity chromatography-mass spectrometry, signals for compounds eluting from the affinity column are recorded on-line by a mass spectrometer, and the last compounds to elute at their infusion concentrations represent the highest affinity compounds or "hits." Since frontal affinity chromatography utilizes a conventional affinity column, this technique would be most accessible to investigators already using affinity-mass spectrometry (see Section 2.4.1). However, the same limitations and disadvantages of using immobilized receptors still apply such as nonspecific binding to the stationary phase, the development time and cost of preparing the affinity columns, and the possibility that immobilizing the receptor might alter its binding characteristics and specificity. In addition, mass spectrometric detection creates some additional limitations. Since all library compounds must be monitored simultaneously, the compounds must be selected so that they have unique molecular masses. Also, one compound in the mixture should not suppress the ionization of another. Therefore, this approach is probably restricted to the screening of small combinatorial libraries that are similar in chemical structure and ionization efficiencies. Finally, the binding buffer used for affinity chromatography must be compatible with on-line APCI or electrospray mass spectrometry. This means that the mobile phase must be volatile and usually of low ionic strength (i.e., typically <40 mM for electrospray ionization). However, the mobile phase limitations may be overcome by collecting fractions from the affinity column and the analyzing them off-line using LC-MS as indicated in Fig. 11 [55].

2.4.5. Solid-Phase Mass Spectrometric Screening Since drugs are usually in a soluble form in order to be transported to the active sites in cells and tissues, it is logical that most mass spectrometry-based screening methods utilize solution-phase analysis of these compounds, and it is no surprise that most successful mass spectrometry screening assays use electrospray ionization or APCI. However, solid-phase ionization techniques such as MALDI might be effective also provided that ligand–receptor interactions are allowed to take place in an environment similar to *in vivo* conditions and provided that a suitable separation step is carried out prior to the preparation of the MALDI sample.

In order to utilize MALDI mass spectrometry for screening, several research groups have developed immobilized receptors on MALDI targets or on solid supports that can be placed on a MALDI target for use in the affinity purification of potential drugs from test solutions. Following procedures originally developed for affinity chromatography, the preparation of affinity surfaces for MALDI mass spectrometry has been achieved quite easily. However, the utility of these affinity MALDI chips for screening mixtures of small molecules during drug discovery has been unproductive. One of the problems has been the high background noise at low *m/z* values caused by the matrix used for MALDI. This problem may be mitigated by eliminating the matrix or using alternative sample stages such as porous silicon chips [56,57]. However, noise persists due to the affinity support and immobilized receptor molecules. Another problem to be overcome is to eliminate the high background noise caused by nonspecific binding of test compounds to the affinity target. Although this problem is similar to the false-positive results and nonspecific binding that occurs during affinity chromatography-mass spectrometry (see Section 2.4.1), the signals for nonspecific binding is magnified by the fact that the actual affinity surface is being irradiated and sampled by the MALDI laser beam. As a result, affinity-based screening coupled with MALDI mass spectrometry has not been a successful drug discovery approach.

A recent method termed self-assembled monolayers for MALDI (SAMDI) offers a solution to the problems associated with MALDI screening [58]. In an anthrax lethal factor inhibition assay, self-assembled monolayers of an anthrax protein were constructed against a background of triethylene glycol groups on a glass plate. The self-assembled monolayers provided well-defined arrays of exposed receptor sites for ligand interaction, and the background noise and nonspecific binding problems pervasive in previous MALDI screening appear to have been largely subdued. Next, MALDI-TOF mass spectrometry was used successfully to screen a library of 10,000 molecules for inhibitors of the receptor. The SAMDI technique was even used to derive quantitative values (IC_{50} and Z-factor) in good agreement with those obtained by conventional screening methods.

Progress is being made in the use of affinity probes for the capture of proteins and other macromolecules from biological solutions followed by MALDI mass spectrometric detection and identification [59,60]. One affinity MALDI mass spectrometry method has been paired with affinity probes using surface plasmon resonance systems [61]. These affinity-based MALDI mass spectrometry screening assays are promising approaches for testing blood or other biological fluids for the presence of specific proteins or other macromolecules. As a result, these have the potential to become clinical diagnostic tools or might even lead to the identification of new therapeutic targets. Recently, MALDI mass spectrometry has been used for molecular imaging of drug compounds, metabolites, biomarkers, proteins, and other chemicals in tissue sections [62].

2.4.6. Pulsed Ultrafiltration-Mass Spectrometry

A versatile approach to screening solution-phase combinatorial libraries and natural product extracts is pulsed ultrafltration-mass spectrometry [63,64], which utilizes a standard LC-MS system with an ultrafiltration chamber substituted for the HPLC column. The principle of pulsed ultrafiltration screening of combinatorial libraries is shown in Fig. 12. During pulsed ultrafiltration, ligand–receptor complexes remain in solution in the ultrafiltration chamber while unbound library compounds and buffers are washed away. After unbound compounds are removed, the hits from the library are eluted from the chamber by destabilizing the ligand–receptor complex using an organic solvent, a pH change, or a combination of both. The released ligands are identified on-line using APCI or electrospray mass spectrometry [63] or collected and analyzed off-line using mass spectrometry, LC-MS, or LC-MS–MS [65].

An example of pulsed ultrafiltration mass spectrometry for the screening of a library of

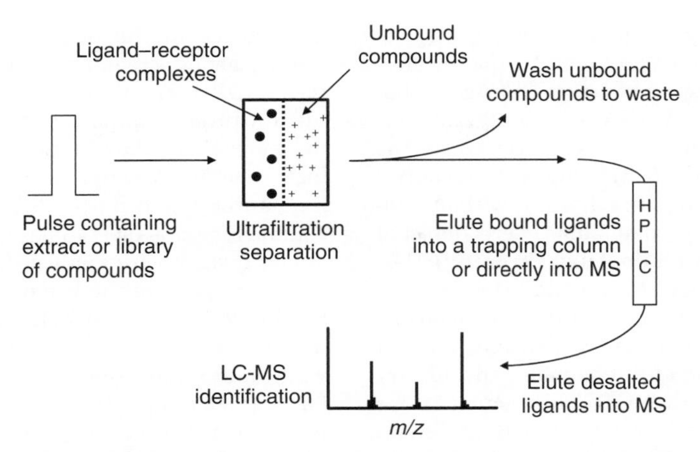

Figure 12. Combinatorial library or natural product extract screening using pulsed ultrafiltration mass spectrometry. During the loading step (left), ligands are bound to the receptor either on-line (top) using a flow-through approach or off-line (bottom two incubations). Unbound compounds and binding buffer, cofactors, etc. are washed out of the ultrafiltration chamber to waste during a separation step (middle). Bound ligands are dissociated from the receptor molecules and eluted from the chamber by introducing a destabilizing solution such as methanol, pH change, etc. Finally, released ligands are identified using mass spectrometry, tandem mass spectrometry, or LC-MS (right). (Reproduced from Ref. [66] by permission of John Wiley & Sons.)

20 adenosine analogs for ligands to adenosine deaminase is shown in Fig. 13. After a 15 min preincubation of the library compounds (17.5 μM each except for EHNA, which was present at 1.75 μM) with 2.1 μM adenosine deaminase in 50 mM phosphate buffer, an aliquot containing 420 pmol of the receptor was injected into the ultrafiltration and washed for 8 min at 50 μL/min with water to remove the phosphate buffer and unbound or weakly binding library compounds. Methanol was introduced into the mobile phase to dis-

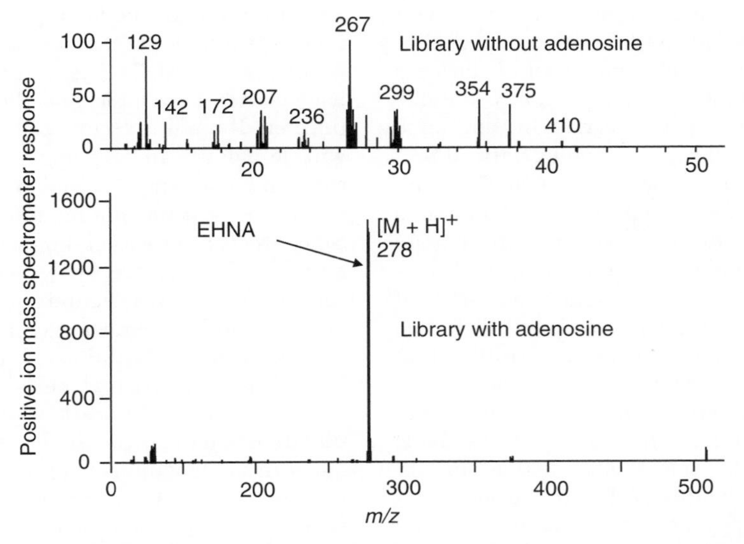

Figure 13. Identification of EHNA as the highest affinity ligand for adenosine deaminase in a combinatorial library of 20 adenosine analogs using ultrafiltration electrospray mass spectrometry. (Reproduced from Ref. [63] by permission of the American Chemical Society.)

sociate the enzyme–ligand complex and release bound ligands for identification by electrospray mass spectrometry. During methanol elution, only EHNA (erythro-9-(2-hydroxy-3-nonyl) adenine) was detected as the $[M + H]^+$ ion of *m/z* 278 (Fig. 13). In control experiments using the library without enzyme, no library compounds were detected during methanol elution (Fig. 13 control). Despite being present at a 10-fold lower concentration than the natural substrate adenosine analogs, EHNA was easily identified since it had the highest affinity among the library compounds ($K_d = 1.9$ nM). This demonstrates the utility of ultrafiltration electrospray mass spectrometry for identifying a high-affinity ligand among a set of analogs that bind to a specific receptor. In a follow-up lead optimization study using pulsed ultrafiltration mass spectrometry, a synthetic combinatorial library of EHNA analogs was screened for binding to adenosine deaminase, and structure–activity relationships for EHNA binding were identified [67].

As an illustration of the versatility of pulsed ultrafiltration-mass spectrometry, binding assays for a variety of receptors have been reported including dihydrofolate reductase [65], cyclooxygenase-2 [64], serum albumin [68,69] and estrogen receptors [70]. Pulsed ultrafiltration is not only useful for identifying ligands to different receptors but also a wide range of combinatorial libraries and natural product extracts in any suitable binding buffer may be screened. In addition to combinatorial libraries, complex natural product extracts have been screened [70], and neither plant nor fermentation broth matrices were found to interfere with screening [64]. As another example of the flexibility of this screening system, a centrifuge tube equipped with an ultrafiltration membrane [71] has been used instead of an on-line ultrafiltration chamber. Other applications of pulsed ultrafiltration-mass spectrometry include screening drugs and drug candidates for metabolic stability [72], metabolic activation to reactive metabolites [73] and the measurement of affinity constants for ligand–receptor interactions [68,69].

Metabolism and toxicity screening applications of pulsed ultrafiltration use hepatic microsomes in the ultrafiltration chamber. For metabolic screening, drugs and the cofactor NADPH are flow injected through the ultrafiltration chamber (oxygen is dissolved in the mobile phase), and the metabolites formed by microsomal cytochrome P450 and any unreacted compounds flow out of the chamber for mass spectrometric identification and/or quantitative analysis [72]. On-line applications require the use of volatile buffers, but LC-MS and LC-MS-MS may be used off-line to analyze the ultrafiltrate no matter what buffer had been used. Screening drugs for metabolic activation using pulsed ultrafiltration-mass spectrometry is carried out in a similar manner, except that glutathione is coinjected along with NADPH and the drug substrate [73]. MS-MS may be used on-line or LC-MS-MS used off-line to screen for glutathione adducts as an indication that the drug was metabolized to a reactive intermediate(s) that was trapped by reaction with glutathione. Finally, pulsed ultrafiltration may be used with UV or mass spectrometric detection to measure affinity constants of individual compounds [68].

In order to measure affinity constants and other physicochemical properties of binding such as the number of binding sites, two pulsed ultrafiltration measurements are carried out. First, an aliquot or pulse of a liquid is injected through the chamber, and the elution profile is recorded. Then, the chamber is loaded with a receptor, and the ligand is re-injected. If binding occurs, the elution profile will be delayed in proportion to the affinity constant. The control injection is used to control for nonspecific binding to the apparatus. Since the concentration of receptor and total amount of liquid are known, and since the concentration of free ligand is measured as it elutes from the chamber over a wide range of concentrations, the affinity constant and other binding parameters may be calculated.

In most of the applications of pulsed ultrafiltration to date, serial analyses were carried out with a throughput of approximately one or two assays per hour. Since the purpose of these assays was to screen complex mixtures or to obtain metabolism data for new drug entities, the throughput of these analyses was acceptable but not high throughput. The rate limiting step in these analyses is the ultrafiltration separation and not the mass spectrometric detection. Several solutions have been reported to increase the throughput of pulsed ultrafiltration mass spectrometry. van Breemen et al. [72] used a multiplex ultrafiltration system with ultrafiltration chambers arranged in parallel and interfaced to a single mass spectrometer. An off-line variation of this approach is to use centrifuge tubes fitted with ultrafiltration membranes to obtain ultrafiltrates of multiple samples simultaneously [74]. Then, each ultrafiltrate can be analyzed using LC-MS. Another solution to increasing the throughput of pulsed ultrafiltration mass spectrometry has been to miniaturize the ultrafiltration chamber volume while maintaining the flow rate and chamber pressure. Since the ultrafiltration membrane cannot withstand high pressure without rupturing, the ultrafiltration process cannot be accelerated simply by increasing the flow rate through the chamber. For example, Beverly et al. [75] fabricated a 35 μL ultrafiltration chamber which was approximately threefold lower in volume than previously reported versions [63]. As a result, ultrafiltration mass spectrometric analyses could be carried out at the rate of at least three per hour, which corresponded to a threefold enhancement of throughput. This study suggests that chip-based ultrafiltration mass spectrometry would have the potential to result in a truly high-throughput system.

The advantages of pulsed ultrafiltration-mass spectrometry include the variety of different applications that may be carried out, the convenience of on-line screening, solution-phase screening, the ability to screen combinatorial libraries and natural product extracts, the diversity of receptors that may be screened, and the freedom to use either volatile or nonvolatile binding buffers. For metabolic and toxicity screening, flow injection analyses have the additional advantages that product feed-back inhibition is prevented so that the metabolic profile more closely approximates the *in vivo* system [72]. Finally, the disadvantages of pulsed ultrafiltration screening for drug discovery include the washing step, during which dissociation and loss of weakly bound ligands might occur, and the slow speed of each experiment, which can take up to one hour.

2.4.7. Drug Development Assays Based on Mass Spectrometry Often, lead compounds identified during high-throughput screening have undesirable properties that preclude their use as drugs, such as poor oral bioavailability, unacceptably rapid metabolism, or the formation of toxic metabolites [76]. Identification of these problems early in the drug development process can save time and reduce the costs of drug development by guiding medicinal chemists toward lead compounds more likely to become drugs. The application of mass spectrometry to drug development assays such as the high-throughput estimation of lipophilicity, metabolic stability, and the detection of reactive metabolites are just a few examples of how LC-MS and LC-MS-MS are being used to expedite drug development. Many drug development applications such as screening for metabolic stability, *in vitro* metabolism and the formation of reactive metabolites can be carried out using pulsed ultrafiltation mass spectrometry as described in Section 2.4.6. Some other drug development assays utilizing mass spectrometry are described below.

Determination of log P Using LC-MS The lipophilicity of a drug candidate is usually estimated by its log P, where P is the partition coefficient between water and octanol. Because most compounds contain ionizable groups that significantly affect solubility, the partition coefficient is often measured at a specific pH in which case the value is known as the log D. Traditionally, log P and log D measurements have been carried out using the shake-flask method, in which the test compound is added to water and octanol, agitated and the phases allowed to separate. The concentration of the analyte is then measured in both solvents by a suitable technique such as LC-UV or LC-MS. The shake-flask method is the most accurate method for log P or log D estimation; however, it is the most costly and time consuming. Faster and less expensive alternatives to the shake-flask method include computer-based *in silico* prediction, LC-MS retention time-based estimation, and a recent 96-well plate format coupled to

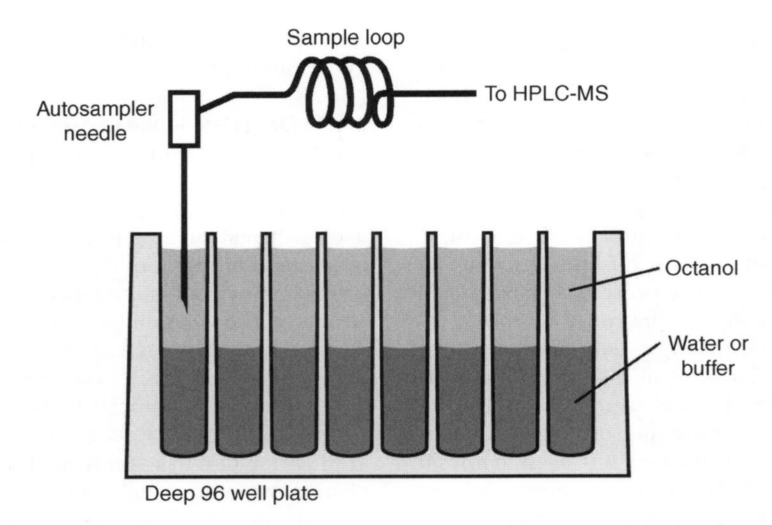

Figure 14. Determination of log D and log D using LC-MS in a 96-well plate format.

LC-MS [77]. In the microtiter plate method, a deep 96-well plate is loaded with aqueous and organic phases at the appropriate pH along with the compound of interest, and then vortex mixed. The phases are separated by centrifugation, and the plate is loaded into an autosampler for quantitative analysis using LC-MS. The needle depth of the autosampler can be controlled to sample from either the aqueous or organic layers (see Fig. 14). A needle wash is performed to eliminate carryover. Quantitation of the analytes in each phase is carried out using LC-MS. This approach facilitates the rapid and accurate determination of log P or log D while eliminating tedious liquid handling operations.

Drug Metabolism and Metabolic Activation Metabolism studies of lead compounds are usually initiated *in vitro* by incubating the lead compound with liver microsomes and enzymatic cofactors such as NADPH or hepatocytes. The products are analyzed by using LC-MS and compared to control reactions to find new peaks, which correspond to drug metabolites. To detect metabolites of lead compounds formed *in vivo*, laboratory animals are administered the investigational compound followed by LC-MS analysis of body fluids and tissues. To enhance the throughput of metabolite identification and quantification, several LC-MS and LC-MS–MS approaches have been developed [78].

The use of LC-MS with electrospray or APCI facilitates the detection and characterization of metabolites resulting from common biotransformations, such as metabolic oxidation or dealkylation. Accurate mass measurements recorded using high-resolution mass spectrometers such as TOF, QqTOF, orbitrap, or IT-TOF mass spectrometers (see Table 1) facilitate the discrimination of drug metabolites from endogenous compounds by providing molecular formulas of metabolites. Data-dependent product ion tandem mass spectrometry combined with accurate mass measurements on QqTOF, orbitrap, or IT-TOF mass spectrometers further enhance metabolite identification by providing structural information as well as elemental composition data. Commercially available software is available from several vendors, which enables automatic data analysis and metabolite detection based on elemental composition, mass defect filtering and isotope pattern-matching functions [79]. Mass defect filtering relies on the unique difference between the nominal mass of a compound (an integer value) and its exact mass (value including the fractional mass), which is specific to its elemental composition. The mass defects and isotope patterns of drug metabolites are usually similar to those of the precursor drug. For example, Bateman et al. [79] investigated the metabolism of the antiretroviral agent indinavir by using

UHPLC-MS on a QqTOF mass spectrometer. By applying a mass defect filter, the background was dramatically reduced and several new metabolites were detected that were not otherwise apparent.

Electrophilic metabolites of lead compounds that might cause toxicity may be detected by trapping the metabolite with glutathione (GSH) or other water-soluble nucleophile and then detecting the GSH conjugate using LC-MS–MS. Since GSH is an endogenous nucleophile, LC-MS–MS screening may be used to detect GSH conjugates formed *in vitro* or *in vivo*. Glutathione conjugates formed *in vivo* are either excreted in bile or as mercapturic acids in the urine. The potential of LC-MS–MS to be used as a screening tool to detect GSH conjugates with selectivity was reported in the 1990s by Nikolic et al. [73]. Initially described as an application of pulsed ultrafiltration LC-MS–MS, screening for reactive drug metabolites trapped as GSH conjugates using tandem mass spectrometry has become widely accepted, although sample cleanup may involve protein precipitation or solid-phase extraction instead of ultrafiltration.

During LC-MS–MS with CID, GSH conjugates form characteristic GSH fragment ions that are useful for their selective detection. For example, during positive ion electrospray tandem mass spectrometry, GSH conjugates fragment to eliminate a neutral molecule of pyroglutamic acid (129 mass units) [80]. The tandem mass spectrometry technique of constant neutral loss scanning may be used on instruments such as a triple quadrupole mass spectrometer to detect GSH conjugates due to their characteristic loss of 129 units. Since some GSH conjugates do form this type of fragment ion in high abundance, negative ion electrospray may be used instead to detect another characteristic GSH ion of *m/z* 272 that is formed by cleavage at the sulfur linkage between the drug metabolite and the GSH moiety [81]. Whether used for the characterization of drug metabolites or the detection of reactive metabolites after trapping with nucleophiles such as GSH, LC-MS, and LC-MS-MS are unsurpassed for their sensitivity, speed and selectivity. By applying LC-MS, LC-MS-MS and drug metabolite analysis software, information about the types of metabolites, their structures, and rates of formation may be obtained and used to select the best lead compounds to be advanced in the drug discovery and development process.

Other Drug Development Assays High-throughput LC-MS assays to measure other important pharmacokinetic parameters such as serum protein binding have been reported. Serum protein binding can be measured by online ultrafiltration MS [68] or with a 96-well filtrate assembly [82]. Intestinal permeability measurements are usually carried out using the Caco-2 cell model in a multi-well format that can be analyzed by LC-MS [83]. It should be noted that LC-MS or LC-MS–MS quantitative analysis to support drug discovery and drug development assays may be accelerated using UHPLC. These UHPLC-MS systems reduce the runtime per sample to below 5 min while maintaining or even improving analytical sensitivity and reproducibility [84].

3. THINGS TO COME

Mass spectrometry has become an essential analytical tool at every stage of the drug discovery and development process. In this chapter, the various applications of mass spectrometry to combinatorial chemistry, drug discovery, and lead optimization have been highlighted. In the past decade, advances in instrument technology have progressed rapidly. In particular, UHPLC, high-resolution mass spectrometers, and tandem mass spectrometers are becoming more widely available and are being used to make drug discovery and development more rapid than ever before.

Intense competition between instrument vendors and the demands of biomedical mass spectrometrists (the consumers) continue to drive technological innovation. For example, innovations in ion sources and ion optics improve detection limits by an order of magnitude with each generation instruments, although this trend may not be able to continue indefinitely. New combinations of ion traps with TOF and orbitrap mass spectrometers make it possible to explore MS^n fragmentation patterns with accurate mass measurements without the high cost of FT-ICR mass spectrometers. A full exploration of the applications of these hybrid instruments has only begun.

Among the most promising of the new mass spectrometry technologies just becoming commercially available is ion mobility spectrometry (IMS). Specialized IMS instruments are already in use by security teams for the detection of drugs, explosives, and toxins in the field. Recently, IMS has been interfaced with MS to add a new dimension of separation in an instrument termed ion mobility mass spectrometry (IMMS). These instruments direct ions along a high-pressure path constrained by an electric field. The ion packet undergoes multiple collisions with a drift tube gas, which retards the motion of molecules with larger cross-sectional areas. Distinct from both gas chromatography and time-of-flight, ion mobility spectrometry separates ions based on their shape and charge within a few tens of milliseconds. After exiting the drift tube, the ions can be separated and detected by conventional mass spectrometer analyzers. IMMS is ideal for the separation and analysis of large biomolecules in the gas phase, such as studies of protein conformation and binding. Small molecule separations, particularly for complex natural products extracts, might also benefit from the ability of IMMS to separate and characterize isomers or other compounds with similar structures within just a few microseconds instead of the minutes required for HPLC or even the seconds needed for UHPLC.

To match the increased analytical performance of LC-MS technology, data analysis software must be developed that is both user-friendly and highly functional. Tools such as automatic feature picking, mass defect or isotope pattern filtering, and metabolite identification are only just becoming available. These tools will become more sophisticated, and will become better integrated into the data analysis workflow. Even more powerful will be software that allows rapid profiling of large sample sets such as is needed for metabolomics. As the quality and quantity of mass spectrometry data increases, robust strategies to manage and archive the data will need to be implemented to allow efficient data mining and interchange of data between vendor platforms.

In conclusion, mass spectrometry provides rapid, reliable, sensitive, and selective analysis of combinatorial libraries for structure confirmation, purity analysis, and library deconvolution. In addition, mass spectrometric screening methods have been developed and applied to drug discovery and development. High-throughput LC-MS methods have become routine for solubility and bioavailability testing and metabolite screening to ensure that only high-quality lead compounds are advanced to more expensive clinical testing. In the case of natural products, mass spectrometry facilitates the rapid and accurate activity screening, dereplication, and characterization of complex extracts.

At different times during the last 100 years, first physicists and physical chemists and then organic chemists pronounced that mass spectrometry had run out of new applications and had no future. Gladly, they were wrong. Today, medicinal chemists recognize the potential of mass spectrometry to contribute to all facets of drug discovery and development. Mass spectrometry has become a fundamental analytical tool for drug discovery, and this role should continue to grow in the future.

4. WEB SITE ADDRESSES AND RECOMMENDED READING FOR FURTHER INFORMATION

http://www.asms.org Homepage of the American Society for Mass Spectrometry.

This web site contains additional information about mass spectrometry and links to a variety of eference materials regarding biomedical mass spectrometry.

http://benthamscience.com\cchts *Combinatorial Chemistry & High Throughput Screening*

http://pubs.acs.org/journals/jcchff/ *Journal of Combinatorial Chemistry*

http://www.5z.com/moldiv/ *Molecular Diversity*

http://www.liebertpub.com/BSC/default1.asp *Journal of Biomolecular Screening*

R. B. Cole, editor. *Elecrospray Ionization Mass Spectrometry*. New York: John Wiley & Sons; 1997.

F. W. McLafferty, F. Turecek. *Interpretation of Mass Spectra*, 4th ed. Mill Valley, CA: University Science Books; 1993.

W. M. A. Niessen. *Liquid Chromatography-Mass Spectrometry*. 3rd ed. Boca Raton, FL: CRC Press; 2006.

J. T. Watson. *Introduction to Mass Spectrometry*, 4th ed. Philadelphia, PA: Lippincott-Raven; 2007.

ACKNOWLEDGMENTS

The authors would like to thank Young Geun Shin, Benjamin Johnson, and Jennifer Mosel for assistance with the preparation of the manuscript. In addition, the authors acknowledge the support from NIH grants S10 RR10485, R01 CA101052, and P01 CA48112.

REFERENCES

1. Field F. J Am Soc Mass Spectrom 1990;1:277–283.
2. Barber M, Bordoli RS, Elliott GJ, Sedgwick RD, Tyler AN. Anal Chem 1982;54:645A–657A.
3. Hillenkamp F, Karas M, Beavis RC, Chait BT. Anal Chem 1991;63:1193A–1203A.
4. Ito Y, Takeuchi T, Ishii D, Goto M. J Chromatogr 1985;346:161–166.
5. Chen L, Stacewicz-Sapuntzakis M, Duncan C, Sharifi R, Ghosh L, van Breemen R, Ashton D, Bowen PE. J Natl Cancer Inst 2001;93:1872–1879.
6. Liu J, Burdette JE, Xu H, Gu C, van Breemen RB, Bhat KPL, Booth N, Constantinou AI, Pezzuto JM, Fong HHS, Farnsworth NR, Bolton JL. J Agric Food Chem 2001;49:2472–2479.
7. Syage JA, Evans MD. Spectroscopy 2001;16:14–21.
8. Gordon EM, Gallop MA, Patel DV. Acc Chem Res 1996;29:144–154.
9. Thompson LA, Ellman JA. Chem Rev 1996;29:132–143.
10. Loo JA. Eur Mass Spectrom 1997;3:93–104.
11. Kyranos JN, Hogan JC. Anal Chem 1998;70:389A–395A.
12. Dunayevskiy Y, Vouros P, Carell T, Wintner EA, Rebek J. Anal Chem 1995;67:2906–2915.
13. Dunayevskiy Y, Lyubarskaya YV, Chu YH, Vouros P, Karger BL. J Med Chem 1998;41:1201–1204.
14. Demirev PA, Zubarev RA. Anal Chem 1997;69:2893–2900.
15. Zeng L, Burton L, Yung K, Shushan B, Kassel DB. J Chromatogr A 1998;794:3–13.
16. Zeng L, Kassel DB. Anal Chem 1998;70:4380–4388.
17. Kiplinger JP, Cole RO, Robinson S, Roskamp EJ, Ware RS, O'Connell HJ, Brailsford A. J. Batt, Rap Commun Mass Spectrom 1998;12:658–664.
18. Ventura MC, Farrell WP, Aurigemma CM, Greig MJ. Anal Chem 1999;71:2410–2416.
19. Ventura MC, Farrell WP, Aurigemma CM, Greig MJ. Anal Chem 1999;71:4223–4231.
20. de Biasi V, Haskins N, Organ A, Bateman R, Giles K, Jarvis S. Rap Commun Mass Spectrom 1999;13:1165–1168.
21. Wang T, Zeng L, Cohen J, Kassel DB. Comb Chem High Throughput Screen 1999;2:327–334.
22. Yang L, Wu N, Clement RP, Rudewicz PJ. In: Proceedings of the 48th ASMS Conference Mass Spectrometry and Allied Topics. 2000; p 861–862.
23. Weller HN, Young MG, Michalczyk SJ, Reitnauer GH, Cooley RS, Rahn PC, Loyd DJ, Fiore D, Fischman SJ. Mol Diversity 1997;3:61–70.
24. Fang AS, Vouros V, Stacey CC, CC Kruppa GH, Laukien FH, Wintner EA, Carell T, Rebek J Jr. Comb Chem High Throughput Screen 1998;1:23–33.
25. Powers DG, Coffen DL. Drug Discov Today 1999;4:377–383.
26. Furka A, Bennett WD. Comb Chem High Throughput Screen 1999;2:105–122.
27. Janda KD. Proc Natl Acad Sci USA 1994;91:10779–10785.
28. Czarnik AW. Proc Natl Acad Sci USA 1997;94:12738–12739.
29. Wagner DS, Markworth CJ, Wagner CD, Schoenen FJ, Rewerts CE, Kay BK, Geysen HM. Comb Chem High Throughput Screen 1998;1:143–153.
30. Kerr JM, Banville SC, Zuckermann RN. J Am Chem Soc 1993;115:2529–2531.
31. Brenner S, Lerner RA. Proc Natl Acad Sci USA 1992;89:5381–5383.
32. Ohlmeyer MHJ, Swanson RN, Dillard LW, Reader JC, Asouline G, Kobayashi R, Wigler

M, Still WC. Proc Natl Acad Sci USA 1993;90:10922–10926.

33. Youngquist RS, Fuentes GR, Lacey MP, Keough T. Rap Commun Mass Spectrom 1994;8:77–81.
34. Karet G. Drug Discov Develop 1999;1:32–38.
35. Potterat O, Wagner K, Haag H. J Chromatogr A 2000;872:85–90.
36. Cordell GA, Shin YG. Pure Appl Chem 1999;71:1089–1094.
37. Shin YG, Cordell GA, Dong Y, Pezzuto JM, Rao AVNA, Ramesh M, Kumar BR, Radhakishan M. Phytochem Anal 1999;10:208–212.
38. Zani CL, Alves TMA, Queiroz R, Chaves MAL, Fontes ES, Shin YG, Cordell GA. Phytochemistry 2000;53:877–880.
39. Constant HL, HL Beecher CWW. Nat Prod Lett 1995;6:193–196.
40. Corley DG, Durley RC, RC J Nat Prod 1994;57:1484–1490.
41. Stinson S. Chem Eng News 1994;72:18.
42. Running WE. J Chem Info Comp Sci 1993;33:934–935.
43. Nedved ML, Habibi-Goudarzi S, Ganem B, Henion JD. Anal Chem 1996;68:4228–4236.
44. Kelly MA, Liang HB, Sytwu II, Vlattas I, Lyons NL, Bowen BR, Wennogle LP. Biochemistry 1996;35:11747–11755.
45. Kaur S, McGuire L, Tang D, Dollinger G, Huebner V. J Protein Chem 1997;16:505–511.
46. Siegel MM, Tabei K, Bebernitz GA, Baum EZ. J Mass Spectrom 1998;33:264–273.
47. Chu YH, Avila LZ, Biebuyck HA, Whitesides GM. J Med Chem 1992;35:2915–2917.
48. Chu YH, Whitesides GM. J Org Chem 1992;57:3524–3525.
49. Rundlett KL, Armstrong DW. J Chromatogr 1996;721:173–186.
50. Avila LZ, Chu YH, Blossey EC, Whitesides GM. J Med Chem 1993;36:126–133.
51. Karger BL. J Med Chem 1998;41:1201–1204.
52. Chu YH, Dunayevskiy YM, Kirby DP, Vouros P, Karger BL. J Am Chem Soc 1996;118:7827–7835.
53. Kasai K, Oda Y. J Chromatogr 1986;376:33–47.
54. Schriemer DC, Bundle DR, Li L, Hindsgaul O. Angew Chem Int Ed 1998;37:3383–3387.
55. Ng ESM, Yang F, Kameyama A, Palcic MM, Hindsgaul O, Schriemer DC. Anal Chem 2005;77:6125–6133.
56. Shen Z, Thomas JJ, Averbuj C, Broo KM, Engelhard M, Crowell JE, Finn MG, Siuzdak G. Anal Chem 2001;73:612–619.
57. Wei J, Buriak JM, Siuzdak G. Nature 1999;399:243–246.
58. Mrksich M, Min DH, Tang W. Nat Biotechnol 2004;22:717–723.
59. Nelson RW. Mass Spectrom Rev 1997;16:353–376.
60. Nelson RW, Nedelkov D, Tubbs KA. Anal Chem 2000;72:404A–411A.
61. Nelson RW, Krone JR. J Mol Recognit 1999;12:77–93.
62. McDonnell LA, Heeren RM. Mass Spectrom Rev 2007;26:60643.
63. van Breemen RB, Huang CR, Nikolic D, Woodbury CP, Zhao YZ, Venton DL. Pulsed ultrafiltration mass spectrometry: a new method for screening combinatorial libraries. Anal Chem 1997;69:2159–2164.
64. Nikolic D, Habibi-Goudarzi S, Corley DG, Gafner S, Pezzuto JM, van Breemen RB. Anal Chem 2000;72:3853–3859.
65. Nikolic D, van Breemen RB. Comb Chem High Throughput Screen 1998;1:47–55.
66. Shin YG, van Breemen RB. Biopharm Drug Dispos 2001;22:353–372.
67. Zhao YZ, van Breemen RB, Nikolic D, Huang CR, Woodbury CP, Schilling A, Venton DL. J Med Chem 1997;40:4006–4012.
68. Gu C, Nikolic D, Lai J, Xu X, van Breemen RB. Comb Chem High Throughput Screen 1999;2:353–359.
69. van Breemen RB, Woodbury CP, Venton DL. In: Larsen BS, McEwen CN, editors. Mass Spectrometry of Biological Materials. New York: Marcel Dekker; 1998; p 99–113.
70. Liu J, Burdette JE, Xu H, Gu C, van Breemen RB, Bhat KP, Booth N, Constantinou AI, Pezzuto JM, Fong HH, Farnsworth NR, Bolton JL. J Agric Food Chem 2001;49: 2472–2479.
71. Wieboldt R, Zweigenbaum J, Henion J. Anal Chem 1997;69:1683–1691.
72. van Breemen RB, Nikolic D, Bolton JL. Drug Metab Dispos 1998;26:85–90.
73. Nikolic D, Fan PW, Bolton JL, van Breemen RB. Comb Chem High Throughput Screen 1999;2:165–175.
74. Liu D, Guo J, Luo Y, Broderick DJ, Schimerlik MI, Pezzuto JM, van Breemen RB. Anal Chem 2007;79:9398–9402.
75. Beverly MB, West P, Julian RK. Comb Chem High Throughput Screen 2002;5:65–73.
76. van de Waterbeemd H, Gifford E. Nat Rev Drug Discov 2003;2:192–204.

77. Chiang P, Hu Y. Comb Chem High Throughput Screen 2009;12:250–257.
78. Prakash C, Shaffer C, Nedderman A. Mass Spectrom Rev 2007;26:340–369.
79. Bateman K, Castro-Perez J, Wrona M, Shockcor J, Yu K, Oballa R, Nicoll-Griffith D. Rap Commun Mass Spectrom 2007;21:1485–1496.
80. Ma L, Wen B, Ruan Q, Zhu M. Chem Res Toxicol 2008;21:1477–1483.
81. Mahajan M, Evans C. Rap Commun Mass Spectrom 2008;22:1032.
82. Fung E, Chen Y, Lau Y. J Chromatogr B 2003;795:187–194.
83. van Breemen RB, Li Y. Expert Opin Drug Metab Toxicol 2005;1:175–185.
84. Xu R, Fan L, Rieser M, El-Shourbagy T. J Pharm Biomed Anal 2007;44:342–355.

CHIRALITY AND BIOLOGICAL ACTIVITY

Jasbir Singh
Timothy J. Hagen
deCODE Chemistry, Inc., Woodridge IL

1. INTRODUCTION

1.1. Definition of Chirality

Chirality is the property of any molecule that is nonsuperimposable on its mirror image. The definition of chirality, its measurement, examples, and implications are described in detail in a number of different texts [1–3]. In most cases, chirality results from the three-dimensional orientation of four different substituents around a (sp^3) carbon atom that form the chiral center. In addition to carbon, the orientation of atoms or groups around a sulfur, phosphorus, and nitrogen atom can sometimes form a chiral center. There are many examples of chiral drugs; naproxen (**1**), cyclophosphoramide (**2**), and esomeprazole (**3**) are shown below as examples of chirality at carbon, phosphorus, and sulfur, respectively (Fig. 1).

1.2. Atropisomer

A special case of stereoisomerism occurs when rotation about a single bond is sufficiently restricted due to steric or other factors such that the different conformers can be separated. Biaryl ring systems represent one of the most important classes of atropisomers. Diphenic acid, a derivative of biphenyl that has a set of four ortho substituents (Fig. 2a), and dimers of naphthalene derivatives such as 1,1′-bi(2-naphthol) (Fig. 2b) represent examples of atropisomers. Aliphatic ring systems like cyclohexanes that are linked through a single bond may also display atropisomerism if bulky substituents are present. An example of atropisomerism has been reported for 1,2,4,5-tetrasubstituted cyclohexane derivatives [4]. Examples of naturally occurring, biologically active atropisomers include those from the stegane series (e.g., steganone, Fig. 2c), gossypol (Fig. 2d), and the antibacterial drug vancomycin (Fig. 2e).

Unlike stereogenic isomers derived from a chiral center, which are practically unable to interconvert into each other [5], atropisomers possess axial chirality and may interconvert between enantiomeric forms depending on the rotational energy barrier. The enantiomers are resolved at room temperature when the interconversion is slow and it has a half-life of at least 1000 s, as defined by Oki [6].

2. CHIRAL COMPOUNDS AND MOLECULAR TARGETS

2.1. Interactions of a Chiral Drug with a Generalized Receptor

Biological systems are chiral. DNA is composed of right-handed helices and is made of 2-deoxyribose monomer of the D- (or *R*-)-configuration, while proteins, including serum and transport proteins and drug metabolizing enzymes, are composed of L- or usually *S*-monomers, amino acids. Thus, these biological molecules can differentiate the two enantiomers of a racemate as two distinct substances. This differentiation arises through differences in the strengths and orientations of the intermolecular interactions between the two enantiomers of a racemate and the chiral biological target. This property is extensively utilized for a wide class of enzymes (proteases, esterases, lipases, etc.) for chiral resolution of racemic compounds (amino acid derivatives, hydrolysis of esters, for example, using chymotrypsin, lipases, etc). There are numerous examples of these in the published literature [7]. These different enantiomers of a chiral drug can differ in their interactions with biological systems such as enzymes, proteins, and receptors and these differences can lead to differentiation in activities such as efficacy, pharmacokinetics, or toxicity, and these are addressed below in Section 16.5. The reason for the differing biological activities is chiral recognition by drug receptors. Easson and Stedman proposed a three-point interaction of the drug with the receptor site to explain these differences [8–11]. In this model, when three groups (*A*, *B*, and *C*) of the tetrahedral carbon atom bind to a protein surface at specific sites *A*′, *B*′, and *C*′, it is impossible to bind the equivalent groups *A*, *B*, and *C* of its mirror image

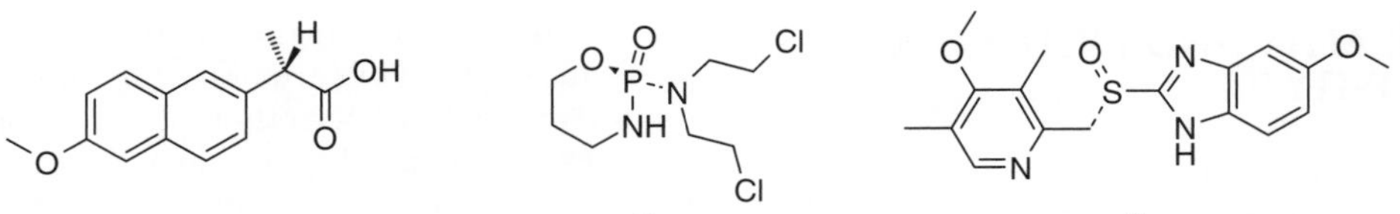

Figure 1. Naproxen, cyclophosphoramide, and esomeprazole

(enantiomer) at the same three sites (Fig. 3a). This model was effective in explaining the differences in biological activity between enantiomers for decades; however, more recently this model has been modified. In particular, sterics is now considered at least as important as binding interactions (three-point model). This results in the creation of a "four-location model" that is closer to reality and has been verified with crystallographic data [12]. Mesecar and Koshland [13] found that the traditional three-point model did not always hold, and they used the enzyme isocitrate dehydrogenase (IDH) as an example to show that the three-point model needed to be revised to provide a more general mechanism for stereospecificity. According to Mesecar and Koshland, the three-point model works only if it is assumed that the ligand can approach a flat protein surface only from the one direction (e.g., top). A fourth location is essential to distinguish between enantiomers in an actual

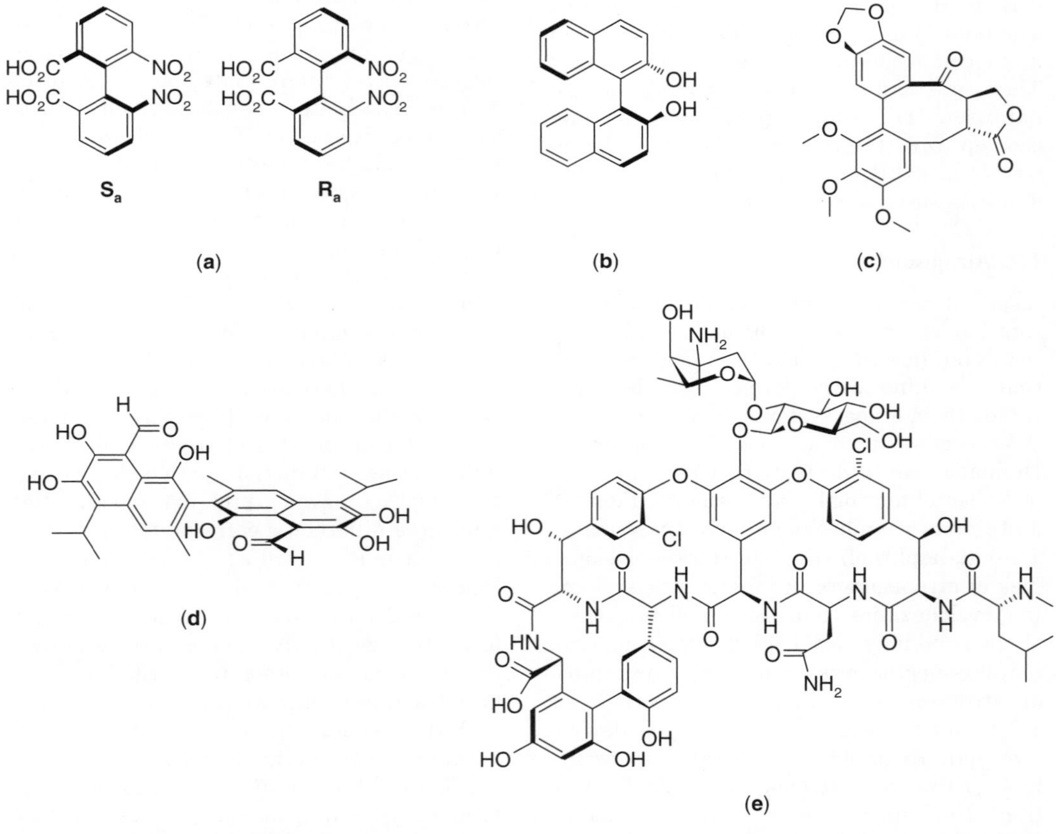

Figure 2. Atropisomers of (a) dinitrodiphenic acid, and atropisomer examples (b) 1,1'-bi-2-naphthol, (c) steganone, (d) gossypol, and (e) vancomycin.

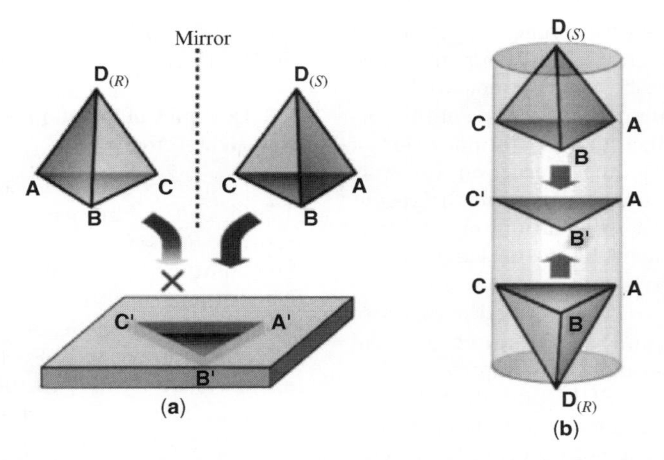

Figure 3. (a) Three-point attachment model for binding on a surface. (b) Four-point location model for stereoselectivity of a protein, showing how a protein might provide two sites (D_R and D_S) in either of the two locations for interaction with group D on a chiral carbon atom: D_R would bind one enantiomer and D_S would bind its mirror image. (Figure reproduced from Ref. 14. With permission from Wiley.)

protein structure—especially if the binding sites on the protein are in a cleft or on protruding residues. A subsequent report from Kato et al. [14] accounts that a four-location model (Fig. 3b) is necessary for supporting the enantioresolution data with host frameworks with channels where aliphatic alcohols are accommodated. Inclusion complexes formed from nordeoxycholic acid (3α,12α-dihydroxy-5β,23-norcholan-23-oic acid), a series of secondary alcohols formed three kinds of host frameworks, and the 3-methyl-2-pentanol formed an inclusion complex with 47% ee. The essence of this is shown in the Fig. 3b, where directionality is important for elucidating enantioresolution in a channel.

2.2. Pfeiffer's Rule

In the 1950s, Carl Pfeiffer formulated a theoretical description of interaction between an enantiomer and a biological system. He observed that there was a correlation between the dose of the racemate used clinically and the relative potencies of the enantiomers. The difference between the clinical dose and the relative potency of the enantiomer was greatest with most active compounds. This theory became known as "Pfeiffer's rule." [15] Simply stated, Pfeiffer's rule is that stereoselectivity increases with increasing drug potency.

The relative potency is defined by the term eudismic ratio (ER), which is the quotient of the potencies of the more potent and less potent enantiomers (eutomer and distomer, respectively). A quantitative structure–activity relation (QSAR) for the ER can therefore be of great help to the medicinal chemist in making an informed decision whether to develop a single enantiomer drug or a racemate. A rule governing the potency of chiral drugs was first suggested by Pfeiffer who claimed that the better the drug–receptor match, the greater the drug potency and the higher the ER. This rule was demonstrated using 14 randomly selected drugs by drawing a linear correlation ($r = 0.9$) between the logarithm of the ER and the logarithm of the average human dose [16]. Pfeiffer did not restrict his examples to a specific therapeutic activity, and this generality has been used as the main reason for much debate about this rule.

Barlow points out that there is a corollary to the rule, that the activity of the weaker enantiomer is determined by the activity of the more potent one, which is intuitively unlikely [16]. The ideas behind Pfeiffer's rule overlook differences in molecular flexibility and multiple receptor states. There is a need to revise Pfeiffer's model so that it considers entropy differences. The model also overlooked differences in pharmacokinetics

(PK). [17] Pfeiffer's rule holds for chemical series that are highly homologous and have a common target; however, it does not hold for divergent structures. The slope of the correlation can, in fact, give some indication of the nature of the binding interactions (primarily H-bonding or lipophilic). With the recent developments in the elucidations of the structural biology, one is likely to see renewed interest in Pfeiffer's rule.

Bunnage et al. [18] described a series of novel sildenafil analogs containing a chiral center that exhibit markedly improved selectivity for PDE_5 over PDE_6 inhibition. In this series, the (*R*)-enantiomer was found to be the eutomer providing a 2–3 log order selectivity for PDE_5 versus PDE_6. A series of analogs were subjected to a eudismic analysis and the plot (Fig. 4) of eudismic index (EI) *versus* pIC_{50} of the eutomer showed a strong linear correlation ($r^2 = 0.98$), with the eudismic affinity quotient (EAQ) of 1. This analysis provides a clear illustration of Pfeiffer's rule for a congeneric series.

A high level of enantioselectivity favoring the conjugation of the (*R*)-enantiomers over their respective (*S*)-enantiomers at eudismic ratios up to 256 has been reported for glucuronidation of secondary alcohols. [19] In this study, 28 enantiomers (14 enantiomeric pairs) comprising of rigid and flexible secondary alcohols were subjected to enzymatic glucuronidation assays employing the human QUDP-glucuronosyltransferases (UGTs) 2B7 and 2B17. The diastereoselectivity with EAQ of 0.83 ± 0.14 was observed exhibiting good agreement with Pfeiffer's rule for the UGT2B17-catalyzed reaction.

2.3. Examples of Enantiomers Exhibiting Biological Activity

Molecular complexity coded by chirality should lead to improved selectivity under *in vitro* and/or *in vivo* environment. Based on the four-point model described above, interaction of enantiomers with the respective molecular target may show modest to significantly large differences in activity. Furthermore, the differential activity may result from interaction of enantiomers with (a) the same target, (for example, one enantiomer may possess activity while the other enantiomer is weakly or essentially inactive), (b) enantiomers displaying activity for diverse targets, and (c) one enantiomer behaving as agonist while the other as antagonist (GPCRs, ion channels, etc.).

2.3.1. Enantiomers Exhibiting Differing Activity on the Same Target

The mitochondrial ATP-sensitive potassium (mito-K_{ATP}) channel is considered a major effector in anti-ischemic cardiac protection. Activation of the mito-K_{ATP} channel gives myocardiocytes an increased resistance against ischemia-induced cell damage during myocardial ischemia, thereby resulting in a marked reduction of the infarct size. This makes mito-K_{ATP} channel an attractive target. Spirocyclic benzopyran-based derivatives have recently been reported as novel anti-ischemic activators of the mito-K_{ATP} channel. [20] Recently, individual enan-

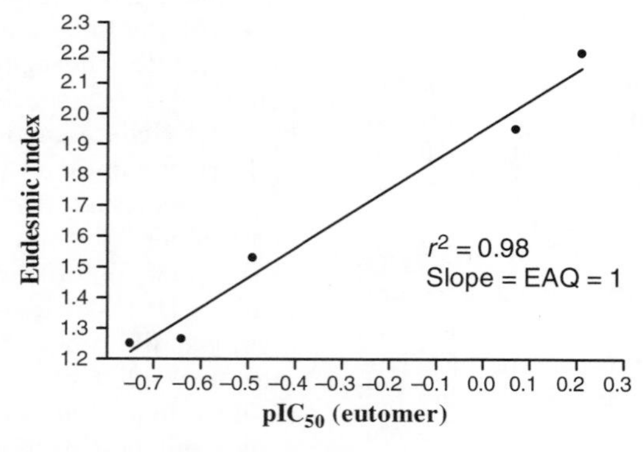

Figure 4. Plot of eudismic index versus pIC_{50} supporting Pfeiffer's rule. (Figure reproduced from Ref. 18. With permission from Elsevier.)

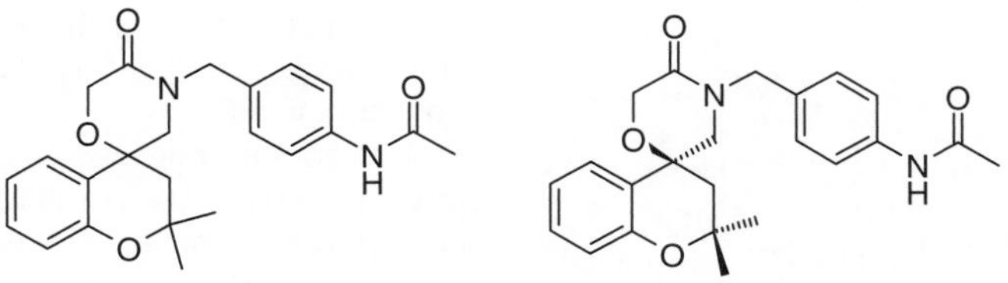

4 (Racemate)
(a)

5 (S)-Enantiomer
(b)

Figure 5. Structures of (a) racemic active spirochromane derivatives **4** and (b) the active (*S*)-enantiomer **5**.

tiomers of 4-spiro-chromane derivatives **4** and **5** showed a high degree of cardioprotection in a Langendorff-perfused rat heart model. The enantiomers were separated by chiral HPLC. Rapposelli et al. demonstrated that this intracellular target is highly enantioselective as the cardioprotective properties reside in the levorotatory enantiomer **5**, while the dextrorotatory enantiomer is devoid of any activity. Single crystal X-ray diffraction and circular dichroism were employed to assign the absolute configuration as (*S*) for the levorotatory enantiomer **5**.

2.3.2. Enantiomers Exhibiting Activity on Different Targets Racemic zilpaterol, a β-adrenergic agonist, has been approved for use in cattle. It is used to increase the rate of weight gain, improve feed efficiency, and increase carcass leanness in cattle. However, it has not been approved for human use. Recently, individual enantiomers of zilpaterol (Fig. 6) were prepared and their biological activity evaluated. It was observed that the (–)-[6*R*,7*R*]-enantiomer of zilpaterol (**6b**) showed potent activity in radioligand displacement assay using recombinant human β2-adrenergic receptor ($K_i = 320$ nM), whereas the (+)-enantiomer was essentially inactive, showing only 3% displacement of the radioligand at 10 μM. [21] Interestingly, the (+)-enantiomers, when evaluated using recombinant human μ-opioid receptor assay, provided K_i value of 7 μM versus 130 μM for the (–)-enantiomer. These activities have been confirmed in follow-up functional assays. Therefore, the two enantiomers of zilpaterol exhibit affinity for two distinct receptors: (–)-enantiomer **6a** for human β2-adrenergic receptor and the (+)-enantiomer **6b** for human μ-opioid receptor (Fig. 6).

Several other β-adrenergic agonists have been evaluated as enantiomers. For example, levalbuterol (*R*-enantiomer) is used as a bronchodilator for the treatment of asthma, including in pediatric population [22].

2.3.3. Enantiomers Exhibiting Agonist Versus Antagonist Characteristics This phenomenon of different enantiomers displaying opposite functional activities has been reported for several GPCR receptor ligands and ion channels. Below are some representative examples.

6a

(–)-[6*S*,7*S*]-Zilpaterol
human β_2-adrenergic receptor

6b

(+)-[6R,7R]-Zilpaterol
humanfor μ-opioid receptor

Figure 6. Structures of (–) and (+)-enantiomers of zilpaterol.

2.4. GPCR Ligands

Cholecystokinin-2 receptor (CCK-2R, previously referred to as the "CCK-B gastrin" receptor) is a G-protein-coupled 7TM domain protein that has subnanomolar affinity for two endogenous peptide ligands, CCK octapeptide (CCK-8) and gastrin. The CCK-2R is expressed in the stomach and the CNS. Stimulation of CCR-2R with either CCK-8 or gastrin triggers activation of phospholipase C and leads to the production of inositol phosphates (IPs). Production of IPs, in turn, has been postulated to modulate a wide spectrum of physiologic functions including gastric acid secretion, differentiation of the gastric mucosa, and perception of pain. Among these small molecule modulators of CCK-2R, the 1,4-benzodiazepine derivatives have been identified as a potent series of ligands [23]. The binding affinity of each compound was assessed at the human wild-type receptor using ^{125}I-CCK for a series of the analogs. All *R*-enantiomers of this series demonstrated higher affinity (IC_{50} 0.25–3.3 nM), for the CCK-2R than the corresponding *S*-enantiomers (IC_{50} ranged from 6.2 to 48 nM). In addition to the binding affinity, the pairs of enantiomers were further explored by measuring ligand-induced production of IP in COS-7 cells expressing the recombinant human CCK-2R. Two pairs of enantiomers (Table 1), (*S*)-**7** and (*S*)-**8**, had activities of 74% and 77%, respectively, when normalized to the hormone-induced maximum, while the (*R*)-**7** and (*R*)-**8** enantiomers provided <10% response. These data clearly show agonistic behavior, and the ability to trigger receptor activation is highly dependent on compound chirality for this series of small molecule ligands for the CCK-2R. This pattern was also observed with another pair of enantiomers reported by Kopin et al. [23]

Additional examples of contrasting functional activities between enantiomers have also been reported within the CCK receptor literature. The *S,S* and *R,R* enantiomers of L-740093 (Fig. 7) were also demonstrated to have weak agonist and inverse agonist function, respectively, for the CCK-2R [24]. The CCK-1 receptor ligand PD-149164 (Fig. 8) is an agonist, whereas the enantiomer of this compound (PD-151932) is an antagonist [25].

Enantiomers of medetomidine have opposite effects on signal transduction through the endogenous α_2-adrenoceptors. Dexmedetomidine acts as an agonist by elevating intracellular Ca^{2+} and inhibits cAMP production, while levomedetomidine acts as an inverse agonist by reducing $[Ca^{2+}]_i$ and enhances cAMP production [26].

The two enantiomers A and B of the constrained N1-arylsulfonyltryptamine analog shown in Table 2 are very potent in the 5-HT6 binding assay. However, only one enantiomer shows essentially 100% agonist activity, while the other enantiomer is devoid of activity in the inhibition of cAMP production functional assay in HeLa cells stably transfected with human 5-HT6 receptors [27].

2.5. Ion Channel Ligands

The 1,4-dihydropyridines that interact with the L-type calcium channels provide an example where pairs of enantiomers of BayK-8644 (Fig. 10) have opposite effects on voltage-dependent calcium channels with the (−)-form being an activator and the (+)-form acting as an inhibitor [28–30].

Jinag et al. [31] have prepared all four stereoisomers of the 3-amino-4-methyl piperidine derived analog **9** and have reported that all isomers retained some level of binding to Jak3 and Jak2. However, additional studies showed that only the 3*R*,4*R* enantiomer **10** (Fig. 11) is capable of blocking Jak3-dependent Stat5 phosphorylation. The pure enantiomer **10** (CP-690550) is currently in phase III clinical evaluation as an orally active immunosuppressant for autoimmune disease and transplant patients.

An interesting report examining the muscarinic receptor-subtype selectivity for structurally related analogs and their enantiomers [32] would suggest that a routine evaluation of enantiomers for their impact on biological activity is usually warranted even with slight modification of ligand structure (see structure and data for analogs shown in Fig. 12 and Table 3).

Table 1. Comparison of 1,4-Benzodiazepine-Based Enantiomeric Ligands and Corresponding Affinities at the Wild-Type Human CCK-2 Receptor

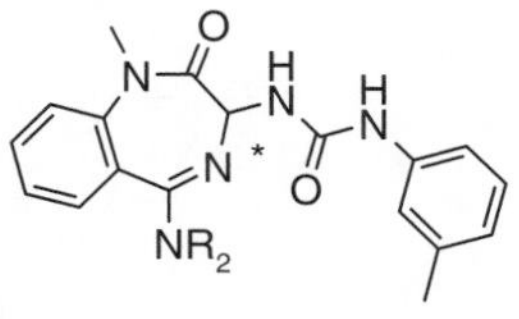

(NR_2)	Chirality*	IC_{50} at Wild-Type Human CCK-2 Receptor
N-ethyl-N-cyclopentyl	(*S*)-**7** (*R*)-**7**	7.6 ± 1.2 0.85 ± 0.25
2,4,6-trimethylpiperidin-1-yl	(*S*)-**8** (*R*)-**8**	48 ± 3.6 0.99 ± 0.25

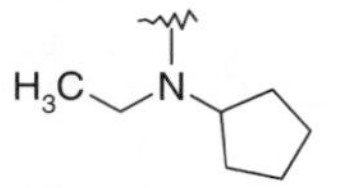

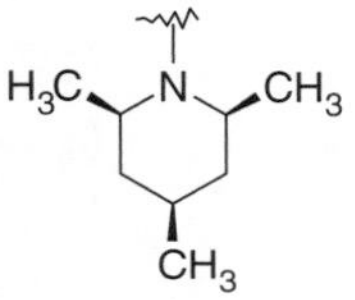

*Denote chiral center.

3. EXAMPLES OF ATROPISOMERS

Several drug and drug candidates that display atropisomerism have been reported in the literature. The classic example is the antiobiotic vancomycin (Fig. 2) that exhibits atropisomerism and is clinically used to treat life-threatening infections [33]. The antipsoriatic drug candidate SCH40120 (15) [34] and the benzodiazepine anxiolytic agent etizolam (**16**) [35] both possess a high rotational energy barrier (e.g., $\Delta G^{\#} = 21.6$ kcal/mol for **15**). Etizolam is an antianxiety agent that is marketed as Depas in Japan. The AMPA antagonist piriqualone (**17**) is atropisomeric due to a sterically crowded environment around N3 and one of its analogs was taken into development to treat neurodegenerative conditions [36]. The endothelin receptor antagonist BMS-207940 (**18**) exists as a pair of atropisomers that were shown to interconvert in aqueous solution [37]. The NK_1 antagonist ZM374979 (**19**) is a tertiary amide that displays atropisomerism due to hindered rotation about the amide bond [38,39]. The anticancer Bcl-2 inhibitor, UCB-1350883 (**20**), exists as diastereomers due to restricted rotation about the aryl–CO bond [40]. A potent and orally active gonadotropin-releasing hormone receptor antagonist, NBI 42902, (**21**, Fig. 13), has been clinically evaluated to treat endometriosis and uterine fibroids [41]. Compound **21** has one conventional chiral center that is derived from phenylglycine. Tucci and coworkers unexpectedly found by NMR analysis that diastereomers were present.

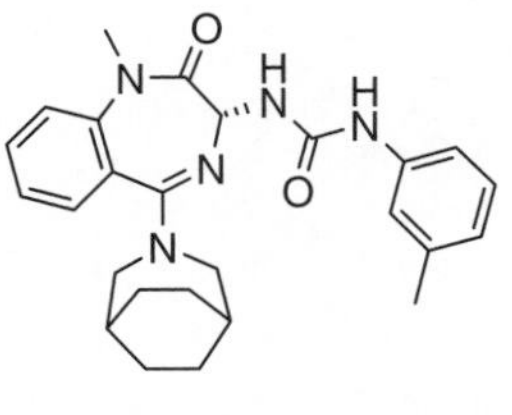

L-740093

Figure 7. Structure of L-740093.

Vrudhula and coworkers described a series of 4-arylquinolin-2-ones (Fig. 14) that are maxi-K ion channel openers [42]. Reasons for unpredictable activity in electrophysiological assays for a series of 4-arylquinolin-2-ones

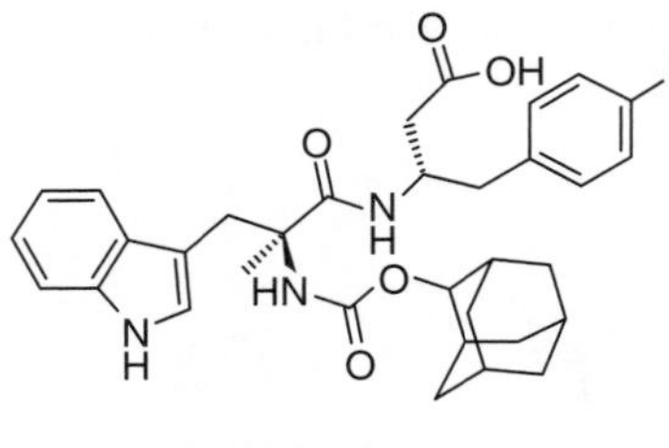

PD-149164

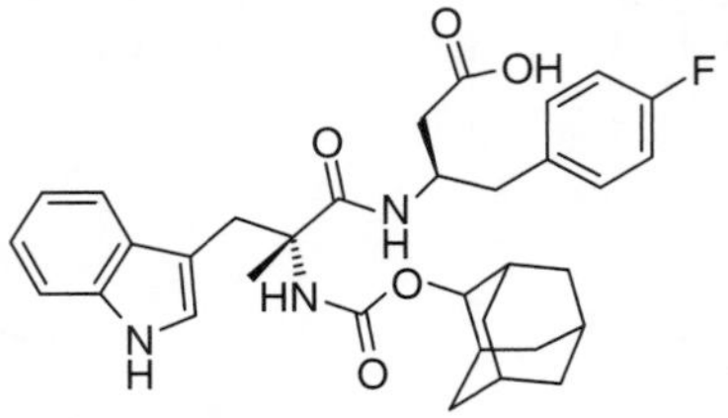

PD-151932

Figure 8. Structure of PD-149164 versus PD-151932.

Table 2. N1-Arylsulfonyltryptamine Enantiomers

		^{3}H-LSD Binding Assay	HeLa Cell (cAMP Assay)	
		(IC_{50})	(IC_{50})	E_{max}
	Enantiomer-A (*)	1 nM	36 nM	100%
	Enantiomer-B (*)	3 nM	0	0

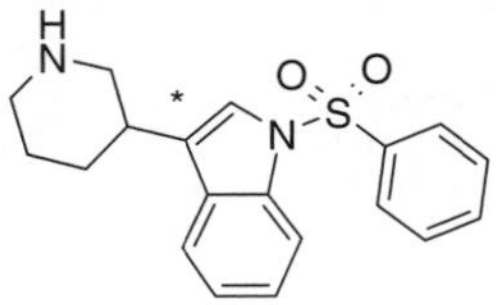

*Denote chiral center.

were investigated. The researchers found that these compounds demonstrated atropisomeric properties, as determined by chiral HPLC analysis. The isomers of (±)-**22** had a calculated energy barrier to rotation of **22** of 31 kcal/mol around the aryl–aryl single bond. The enantiomers were separated by preparative chiral HPLC. Furthermore, the structures of (+)-**22** and (+)-**23** were determined by X-ray crystallography and are shown in Figure 15. The atropisomers of (±)-**22** and (±)-**23** were then tested for their effect on maxi-K-mediated outward ion current in *hSlo* injected into *X. laevis* oocytes. The (−)-isomer in each case was found to be more active than the corresponding (+)-isomer and these data suggested that the ion channel exhibited enantioselective activation. The conversion of (−)-**22** to (+)-**22** in *n*-butanol at 80 °C showed a 19.6% conversion to (+)-**22** over a 72 h period. Racemization of (−)-**22** in human serum at 37 °C was not observed over a 30 h period. This demonstrates that if there is a possibility of atropisomers, it should be thoroughly investigated early during discovery and development and the possibility of dealing with atropisomers needs to be addressed.

Ustiloxins A–F are antimitotic cyclic peptides containing a 13-membered macrocyclic core and an unusual chiral tertiary alkyl ether

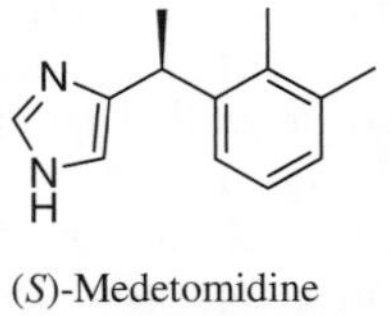

(*S*)-Medetomidine (dexmedetomidine)

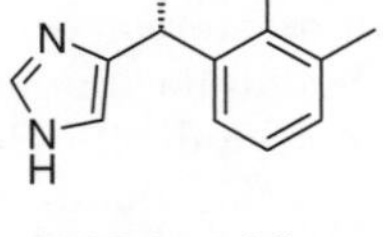

(*R*)-Medetomidine

Figure 9. Structure of medetomidine enantiomers.

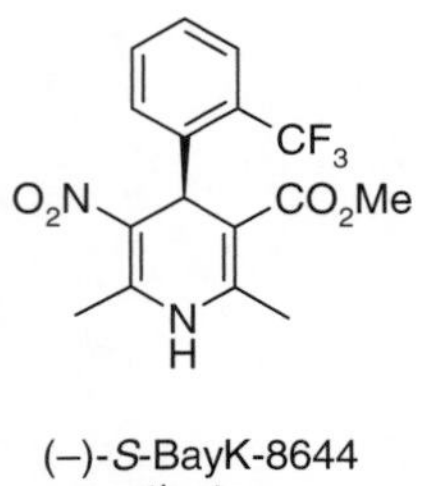

(−)-*S*-BayK-8644 activator

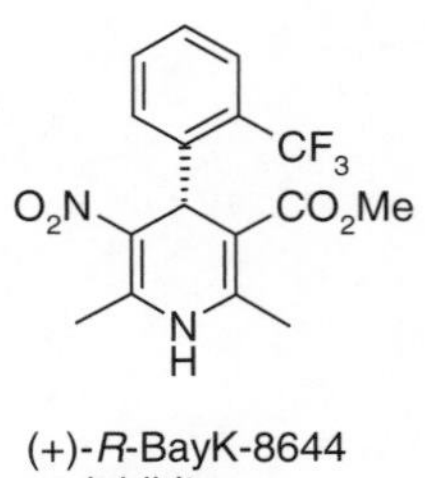

(+)-*R*-BayK-8644 inhibitor

Figure 10. Structure of BayK-8644 enantiomers (activator and inhibitor).

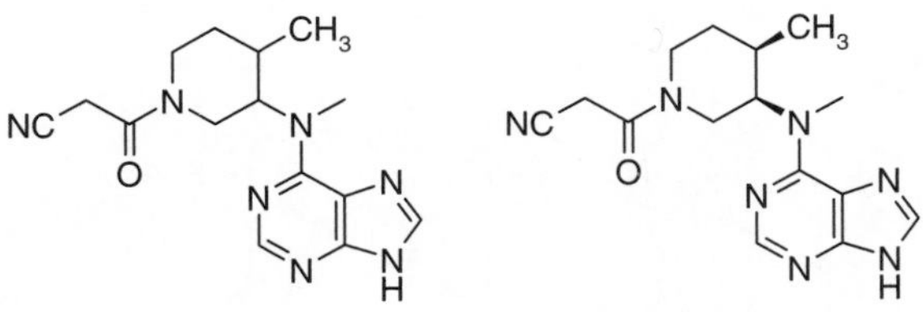

9 10 [CP-690550 (3*R*,4*R*)]

Figure 11. Structure of 3-amino-4-methyl piperidine derived Jak inhibitors and CP-690550.

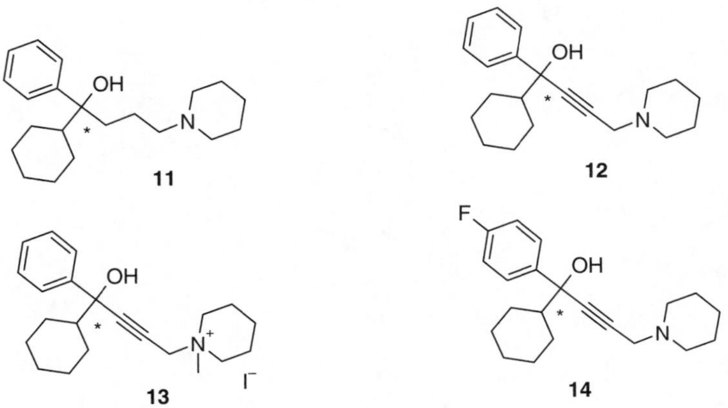

11 12 13 14

Figure 12. Structures of chiral hexahydrodifenidol and acetylenic analogs. *Denote chiral center.

linkage that is challenging from a synthetic perspective. Li and coworkers described the first total synthesis of ustiloxin D that was achieved in 31 linear steps [43]. Like other macrocyclic natural products, such as vancomycin, the ustiloxin D (Fig. 16) may exist as two atropisomers. A two-dimensional $^{1}H-^{1}H$ NOESY experiment was performed to determine if atropisomers were present. Ustiloxin D exhibited diagnostic NOE cross-peaks between protons on C16 and C10, C12 and C9, and C16 and C21 (Fig. 16a). No NOE cross-peaks were observed between protons on C12 and C10, C16 and C9, or C12 and C21 [43]. The NOE data confirmed that the ustiloxin D existed as a single atropisomer and not as a mixture. Li et al. generated the three-dimensional structure that is shown in Figure 16b, which is based on the information from the NOE experiment [43].

Table 3. Muscarinic Receptor Subtype Selectivity and Stereoselectivity

		Receptor Selectivity			Stereoselectivity		
Compound	Chirality[a]	M1/M2	M3/M1	M3/M2	M1	M2	M3
11	R	48	0.44	21	550	17	191
	S	1.5	1.3	1.9			
12	R	10	1.0	1.9	107	8.5	44
	S	0.81	2.5	2.0			
13	R	6.5	0.26	1.7	40	17	39
	S	2.7	0.27	0.72			
14	R	13	2.6	34	34	1.7	8.5
	S	0.53	10.5	6.6			

[a]The values shown represent the antilogs of the differences between corresponding mean pA2 values determined at M1 receptors (rabbit vas deferens), and at atrial M2 and ileal M3 receptors of guinea pigs.

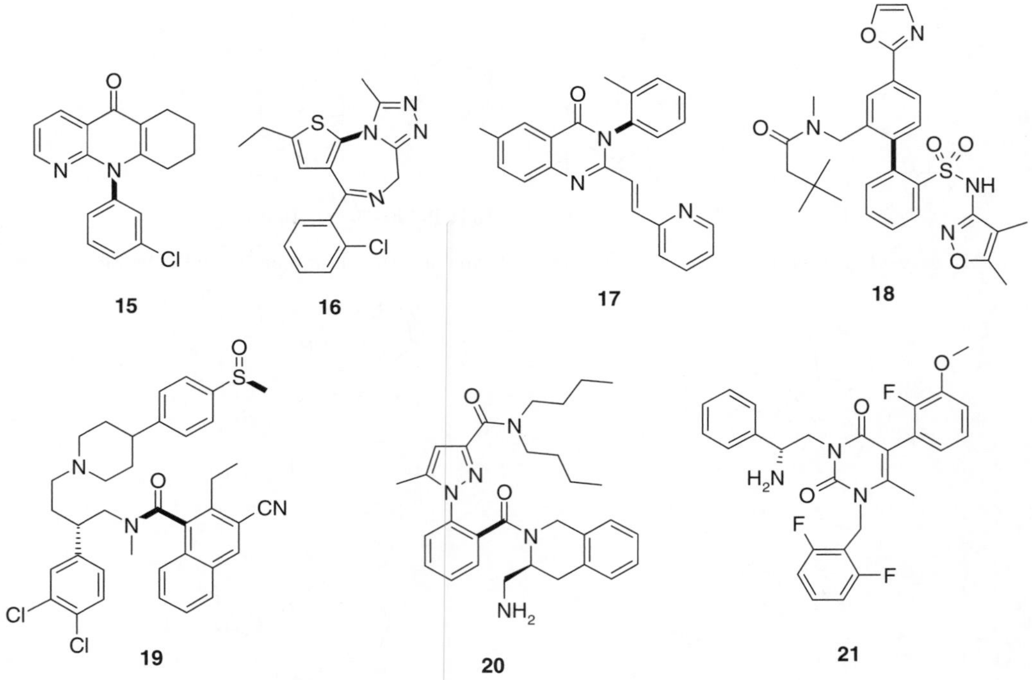

Figure 13. Atropisomeric drugs or drug candidates.

3.1. Conformers of Single Molecule with Different Biological Response

It is commonly stated that a target receptor or enzyme recognizes a "bioactive" conformation and that a molecule that could not adopt this bioactive conformation is inactive or may exhibit weak activity or response. This concept is usually used in the literature when discussing structure–activity studies of biologically active substances that have some degree of conformational flexibility. The phytotoxin tentoxin, [cyclo(-L-MeAla-L-Leu-MePhe[Z(Δ)-Gly] (**24**), a cyclic tetrapeptide natural product, is a species-specific inhibitor of photophosphorylation and chloroplast coupling factor 1 (CF1) [44]. Synthetic analog [D-MeAla] tentoxin was prepared and it was shown that this analog exists in multiple conformations in solution (Fig. 17). In fact the conformers were sufficiently stable and were separated by thin-layer chromatography (TLC) on silica gel at 4 °C into two fractions designated "U" and "L" based on TLC mobility (U: upper R_f and L: lower R_f). After determining the relative rate of interconversion of the two conformation

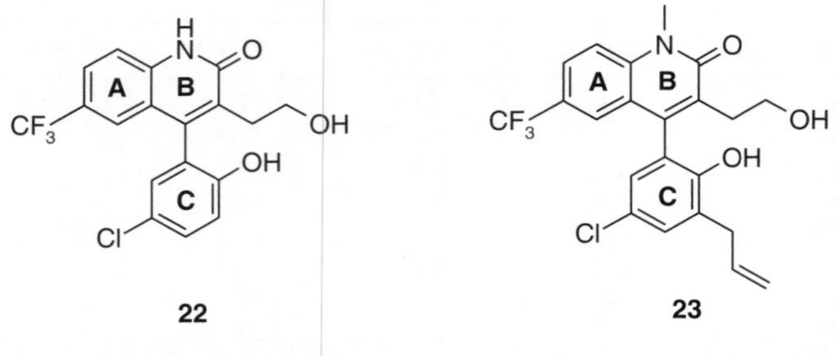

Figure 14. Potassium channel openers that are atropisomers [42].

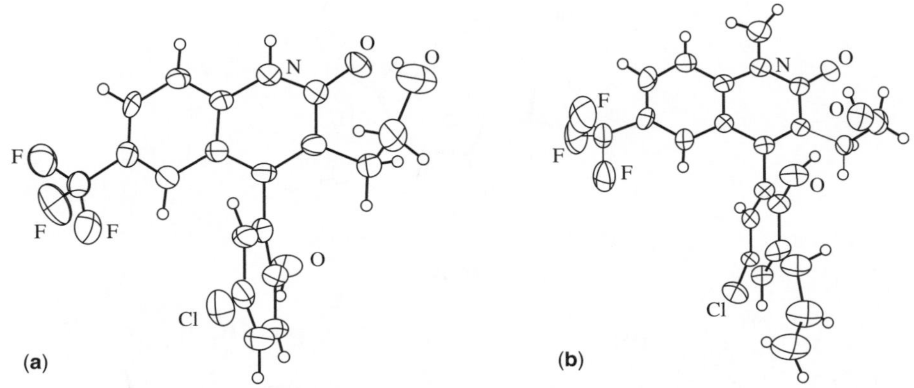

Figure 15. ORTEP drawings of (a) (+)-**22** and (b) (+)-**23** with thermal ellipsoids at 30% probability for non-H atoms and open circles for H-atoms. Carbon and hydrogen atoms are not labelled [42]. (Figure reproduced from Ref. [42]. With permission from The American Chemical Society.)

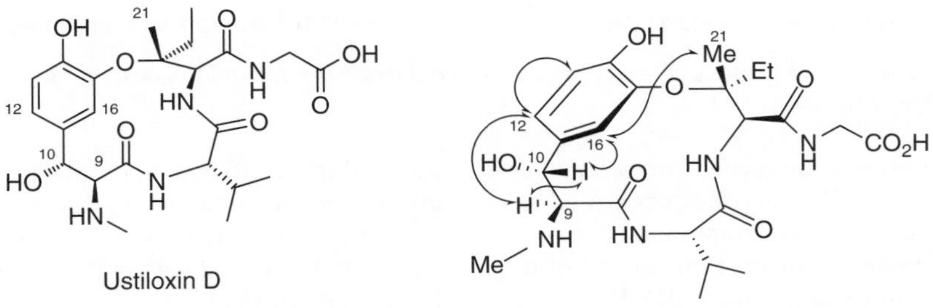

Figure 16. (a) Structure of ustiloxin D. (b) Assignment of atropstereochemistry of ustiloxin D by $^{1}H-^{1}H$ NOESY (500 MHz, CD_3OD) study [43]. (Figure reproduced from Ref. [43]. With permission from The American Chemical Society.)

populations under the bioassay conditions, it was established that the two isolable conformational populations of [D-MeAla]tentoxin differ in their biochemical properties and that the "L" (**26**) is a substantially more potent inhibitor than "U" (**25**) for the inhibition of coupled electron transport in lettuce chloroplasts (Table 4). This is the first demonstration that separate amide conformers of a single molecule can have differing biological activities.

4. CHIRALITY AND ADME PROPERTIES

The first observation of biological enantioselectivity was made by Louis Pasteur. He found, in 1858, that when solutions of racemic ammonium tartrate were fortified with "organic matter" (i.e., a source of microorganisms) and allowed to stand, the solution "fermented" and (+)-tartaric acid was consumed rapidly while (–)-tartaric acid was left behind unreacted [45]. Eventually the (–)-enantiomer was also metabolized, but considerably more slowly than (+)-tartrate [46]. In later experiments, Pasteur showed that the common mold *Penicillium glaucum* metabolized (+)-tartaric acid with high enantioselectivity [47]. He correctly theorized that the enantioselective destruction of tartaric acid by microorganisms involves selective interaction of the tartrate enantiomers with a key chiral molecule within the microorganism. Toward the end of the 19th century, the role of chirality in biological activity began to receive serious attention. Two lines of investiga-

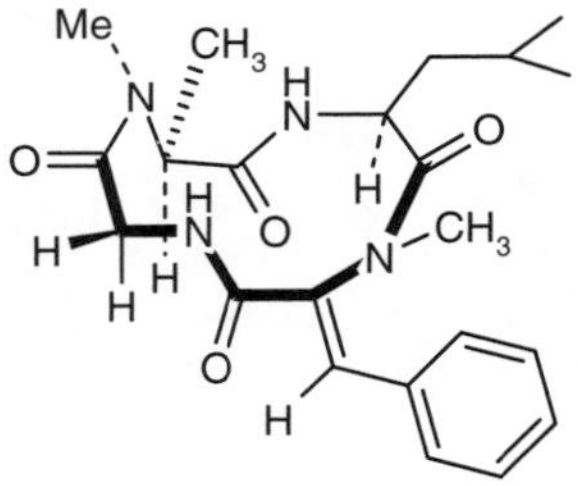

Tentoxin (**24**)

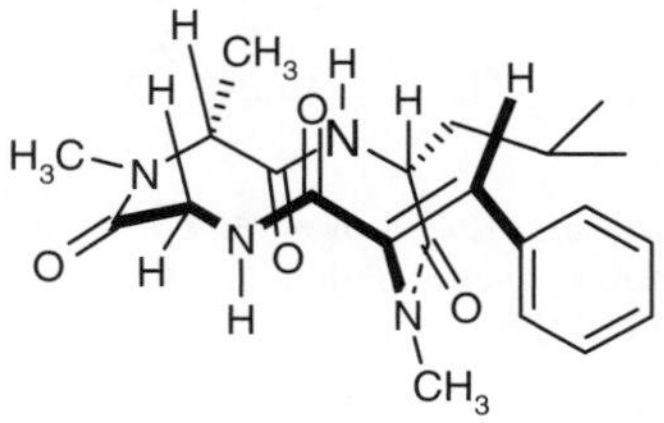

(**25**) D-MeAla[1]-tentoxin (U)

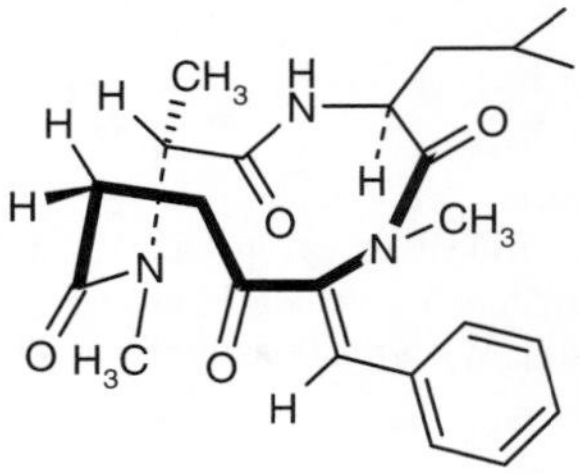

(**26**) D-MeAla[1]-tentoxin (L) (s-*cis*)

Figure 17. Structures of tentoxin (24); upper R_f spot conformer (U) **25** and major (80%) aqueous solution conformer—lower R_f spot (L) **26**.

tion were pursued: one focused on pharmacological activity while the other examined the metabolic fate of chiral compounds.

The processes of absorption, distribution, metabolism, and elimination (ADME) are crucial determining factors of drug action and can be equally relevant to the actual biological activity. It is prudent to study and obtain ADME data on the enantiomers so that the differences can be analyzed. There are many examples of chiral drugs where enantioselective behavior has been observed in PK and metabolic profiles [48,49]. It is noteworthy that the relative dynamic range for enantioselective differences for these two environments (pharmacology and ADME/PK) differs greatly, where the range in activity (at the molecular target) could potentially be of much greater magnitude (up to several log units) than the range for PK being by definition within maximum 2 log (ca. bioavailability defined as 0–100%).

4.1. Chirality and Absorption/Permeability

Absorption of orally administered drugs from the gastrointestinal (GI) tract is essentially the first step that determines its entry into the systemic circulation and eventually its disposition. *In vitro* assays, such as Caco-2 and MDCK representing the epithelial cells lines, are widely used for determining the permeability of bioactive compounds. One would expect the transport of chiral drugs across the

Table 4. Activity of Tentoxin and its Derivatives

Compound	R_f (TCL)	ED50 (μM) Inhibition of Coupled Electron Transport in Lettuce Chloroplasts
Tentoxin (**24**)	0.45	0.4
[D-MeAla]tentoxin "U" (**25**)	0.45	6[a]
[D-MeAla]tentoxin "L" (**26**)	0.30	1.4

Data in table adopted from Ref. 44. With permission from Elsevier.

[a]Best estimate from linear extrapolation of the data from 0 to 4 μM out to about 12 μM. In the absence of saturation effects, this may be an artificially low value.

epithelial cell monolayer to be different. Although the liver is considered to be the key metabolizing organ, gut metabolism, particularly for compounds metabolized by CYP3A4, plays an important role in governing overall absorption of drugs from GI tract. The enantioselective metabolism of chiral drugs is discussed in Section 5.3.

The entry of topically administered drugs, on the other hand, following absorption through the skin determines its entry into systemic circulation. Siccardi and coworkers have investigated the potential epithelial permeability of a series of pharmacologically active aryloxy phosphoramidate derivatives of the anti-HIV agent d4T (Fig. 18) using the Caco-2 and MDCK cell lines [50]. Since each compound is prepared from L-alanine, the equimolar mixture of two diastereoisomers is produced due to chirality at the phosphorous atom of the phosphoramidate. The phosphoramidate diastereoisomers of each prodrug nucleotide were separated and quantified by reversed-phase HPLC. The enantiomers were classified as FE (fast eluting; less lipophilic) and SE (slow eluting; more lipophilic) based on the relative retention time for each diastereomeric sample. Interestingly, a clear statistical difference in Caco-2 permeability between the corresponding FE and SE diastereoisomers for the more hydrophobic derivatives (*p*-Cl, *p*-Br, and *p*-I) was observed. The more lipophilic (SE) diastereoisomer showed greater permeability, which was approximately 1.6-fold, 3-fold, and 1.9-fold greater for the *p*-Cl, *p*-Br and *p*-I derivatives, respectively. In contrast, such selective permeability for less hydrophobic diastereoisomer was not observed for this series.

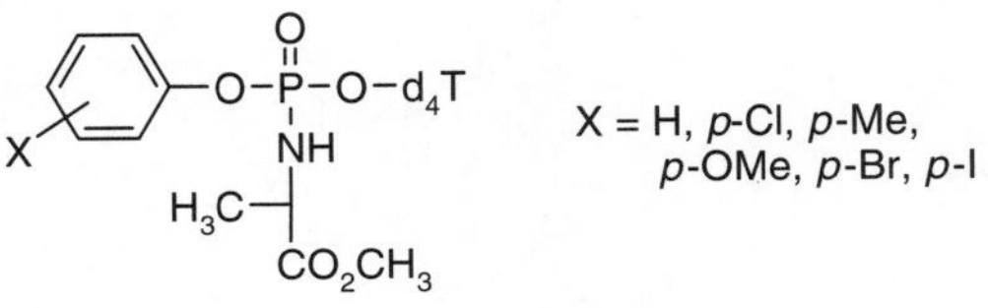

X = H, *p*-Cl, *p*-Me, *p*-OMe, *p*-Br, *p*-I

Figure 18. Structures of anti-HIV agent d4T (2′,3′-didehydro-2′,3′-dideoxythymidine).

The permeability of the (*S*)- and (*R*)-enantiomers of esmolol was determined using Caco-2 cell monolayers and samples were quantified by enantioselective reversed-phase HPLC [51]. Statistically significant differences in P_{app} values between (*S*)- and (*R*)-esmolol at concentrations ranging from 5 to 100 μM were observed for both apical-to-basolateral (absorptive) and basolateral-to-apical (secretory) directions of transport. In the absorptive direction, the permeability of (*S*)-esmolol was less than (*R*)-esmolol, while in the secretory direction, P_{app} of (*S*)-esmolol was greater than that for (*R*)-esmolol. Therefore, distribution of esmolol showed enantioselective behavior in the concentration range evaluated. The overall analysis showed that (*R*)-esmolol was transported mainly by passive diffusion, while a carrier-mediated mechanism was involved in the transport of (*S*)-esmolol.

The (*S*)-enantiomer of metoprolol (β1-selective adrenergic blocking agent) is the biological active component, while the (*R*)-enantiomer is therapeutically inactive. Due to extensive first-pass effect, metoprolol has ~50% oral bioavailability. A transdermal delivery of racemic metoprolol has recently been reported to provide therapeutic levels for 24 h in rats, with improved absolute bioavailability. Absorption of active agents has been enhanced by additives. A chiral additive could potentially alter the permeability of another chiral compound as enantiomers may exhibit different behavior in chiral environments. The

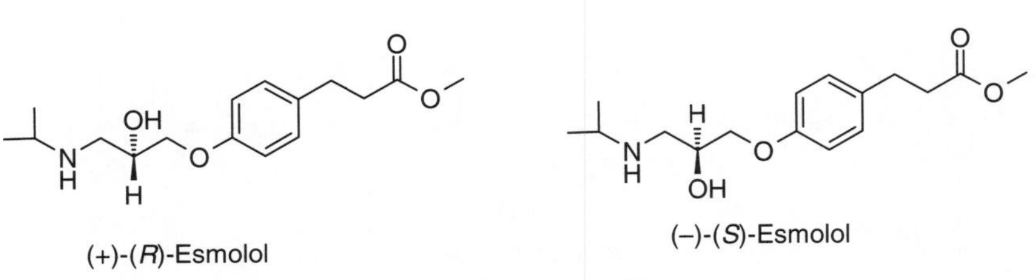

(+)-(*R*)-Esmolol (−)-(*S*)-Esmolol

Figure 19. Esmolol enantiomers

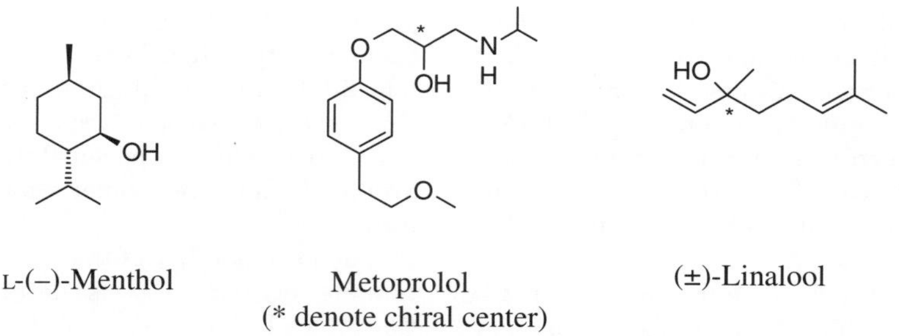

L-(−)-Menthol Metoprolol (* denote chiral center) (±)-Linalool

Figure 20. Structures of L-(−)-menthol, metoprolol, and (±)-linalool.

permeation characteristics of individual enantiomers of metoprolol free base (MB) were investigated using hairless mouse skin and the influence of chiral permeation enhancers, L-menthol and (±)-linalool (Fig. 20) was also investigated [52]. In the absence of the additive, the permeation profiles of (*R*)-MB and (*S*)-MB from donor solutions containing either (*R*,*S*)-MB or pure enantiomers were comparable. Whereas, in presence of L-menthol, when donor solution contained pure enantiomers, the permeation enhancing effect of L-menthol on (*S*)-MB was significantly higher (by 25%) than on (*R*)-MB. In addition, the flux of (*S*)-MB from donor solution containing pure (*S*)-MB was 35% higher than the flux of (*R*,*S*)-MB (racemate). Moreover, similar effects were not observed when (±)-linalool was used as additive. The researchers believed that the observed enantioselective permeability effects are due to the stereospecific interactions between ceramides in stratum corneum and L-menthol/MB enantiomer.

4.2. Chirality and Distribution

Many antiarrhythmic drugs are marketed as racemates (Fig. 21) such as disopyramide, encainide, flecainide, mexiletine, propafenone, tocainide, and so on [53]. The absorption of chiral antiarrhythmic drugs appears to be nonstereoselective. However, their distribution, metabolism, and renal excretion usually favor one enantiomer versus the other [48].

In terms of distribution, plasma protein binding is stereoselective for most of these

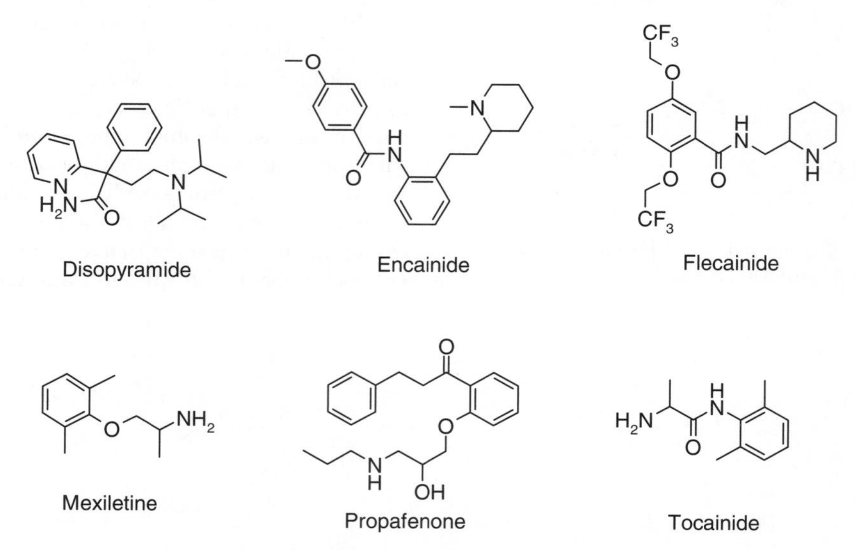

Figure 21. Selected marketed racemic antiarrhythmic drugs.

drugs, resulting in up to twofold differences between the enantiomers in their unbound fractions in plasma and volume of distribution. Hepatic metabolism plays a significant role in the elimination of these antiarrhythmics. In addition, in most cases, significant enantioselectivity is observed in different pathways of metabolism of these drugs. Therefore, it is not surprising that a wide interindividual variability exists in the metabolism of these drugs. Overall, substantial enantioselectivity has been observed in both the pharmacokinetics and pharmacodynamics of chiral antiarrhythmic agents. Because the effects of these drugs are related to their plasma concentrations, this information is of special clinical relevance [48].

As reported by Stoschitzky et al., [54] there are marked pharmacokinetic differences between the D- and L-enantiomers of most β-blockers, particularly under exercise and when extensive and poor metabolizers are compared. Plasma concentrations of these D- and L-enantiomers usually differ significantly and in wide ranges when the racemic mixture is administered orally or intravenously. Mehvar et al. [55] also reported that the β-blockers are quite diverse in pharmacokinetic profile, as they display a high range of values in plasma protein binding, in percentage of drug eliminated by metabolism or unchanged in the urine, and in hepatic extraction ratio. With respect to plasma concentrations attained after oral or intravenous dosing, in most cases, the enantiomers of the β-blockers show only a modest degree of enantioselectivity. However, the relative magnitude of the concentrations of the enantiomers in plasma is not constant in all situations and varies from drug to drug. Furthermore, various factors related to the drug (e.g., dosing rate or enantiomer–enantiomer interaction) or the patient (e.g., racial background, cardiovascular function, or the patient metabolic phenotype) may affect the enantiospecific pharmacokinetics and pharmacodynamics of β-blockers [50].

4.3. Chirality and Metabolism

The extensive metabolites of propranolol (Inderal™) have been reported to be mostly in the urine [56]. There are three major routes of metabolism for propranolol: aromatic hydroxylation at position-4, *N*-dealkylation followed by further side-chain oxidation, and direct glucuronidation (Fig. 22). The estimated percentage contributions of these metabolic routes are 42%, 41%, and 17%, respectively [56]. The identified major metabolites are propranolol glucuronide, naphthyloxylactic acid, and glucuronic acid and sulfate conjugates of 4-hydroxy propranolol. The interindividual variability and the pharmacokinetic profile for propranolol enantiomers and metabolites in humans have been confirmed [46].

Racemic methadone is an analgesic drug that is used to treat opiate addiction. The *R*-(+)-enantiomer primarily dictates the pharmacological activity. There is a considerable variation in human serum and urine levels due to the interindividual differences in the pharmacokinetics of the methadone enantiomers; the ratios of *R*/*S*-methadone levels

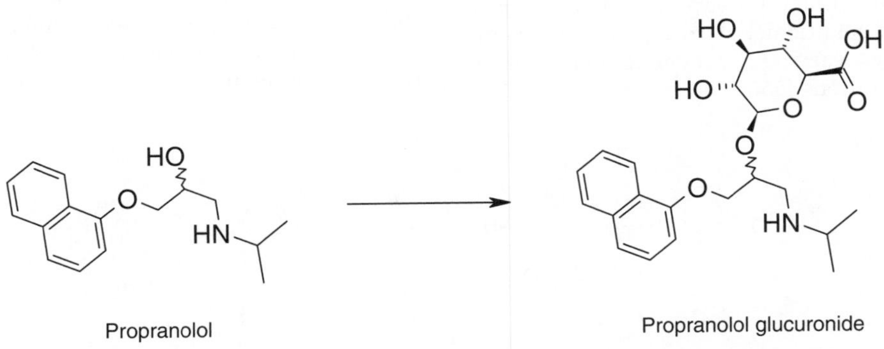

Propranolol

Propranolol glucuronide

Figure 22. Propranolol and glucuronidation in phase II metabolism.

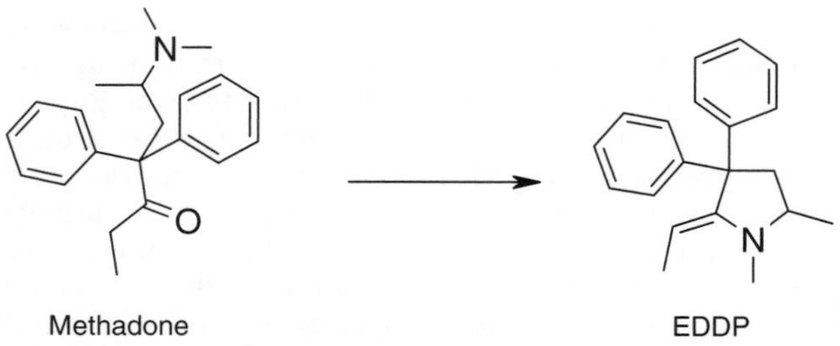

Methadone

EDDP

Figure 23. (*R*/*S*)-Methadone and its metabolite.

range from 0.6 to 2.0 in serum and 1.2–2.0 in urine [57–60]. The major route of metabolism for methadone utilizes hepatic cytochrome P_{450} 3A4 and P_{450} 2D6 to form a major methadone metabolite EDDP (2-ethylidene-1,5-dimethyl-3,3-diphenylpyrrolidine) (Fig. 23). Due to its polarity, EDDP levels were always 1.5–5-fold higher than those of the parent molecule in human urine. On the other hand, its concentrations were undetectable in human saliva and always lower than those of methadone in human serum [61].

The chiral inversion of 3-hydroxy-benzodiazepines has been documented *in vitro*. [62] Pham-Huy and coworkers were the first to demonstrate that the enantiomers of oxazepam displayed different serum concentrations [63]. When rabbits were treated with pharmacological and toxic doses, the *S*(+)-form of oxazepam was dominant. When rabbits were treated with the antagonist flumazenil and oxazepam, the levels of the *R*(−)-isomer of oxazepam were higher. The differences were attributed to a lack of racemization and the high plasma protein binding (97%) for the oxazepam enantiomers. The authors speculated that differences in binding affinity for each enantiomer of oxazepam can account for the varying *R*/*S*-enantiomer serum concentrations [55].

Originally terfenadine was marketed as a racemate. It was discovered that one enantiomer of terfenadine actually caused a potentially fatal cardiac arrhythmia in some patients [64]. Subsequently, it was shown that the racemic terfenadine is preferentially oxidized in rats to form a carboxylic acid metabolite, fexofenadine. The active metabolite fexofenadine is currently marketed as Allegra. An understanding of the enantiospecific pharmacokinetics of chiral drugs may help clinicians to interpret and predict differences in pharmacologic responses among patients when racemic drugs are prescribed.

4.4. Chirality and Excretion

The racemates and individual enantiomers of hydroxyl urea-based 5-LO inhibitors (**27** and **28**, Fig. 24) were evaluated for their relative rate of glucuronidation using monkey liver microsomes *in vitro* and the compound's activity was measured in the whole-blood assay *ex vivo*. Although a small difference in 5-LO inhibiting activity was observed (IC_{50} value 80 nM for (*S*)-**27** versus 50 nM for (*R*)-**28**), in the whole-blood assay, fivefold difference in glucuronidation rate was observed between the pair of enantiomers, with the (*R*)-**27**

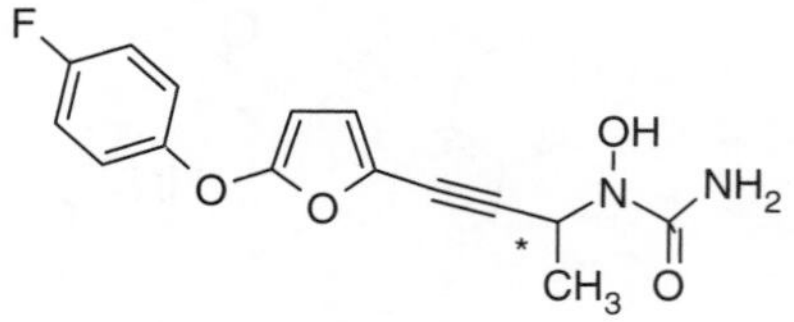

27

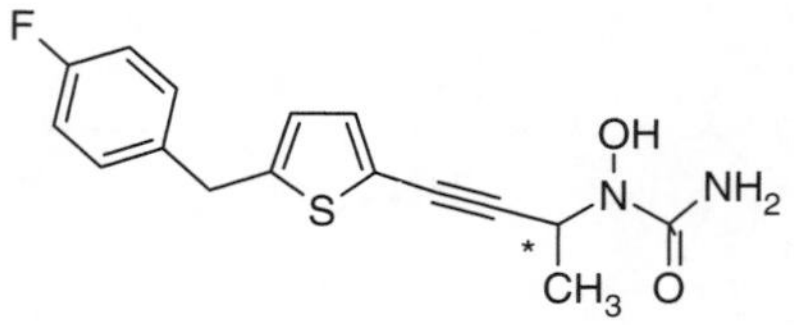

28

Figure 24. *N*-hydroxy urea-based 5-LO inhibitors, enantiomers exhibiting different rates of glucuronidation. *Denote chiral center.

enantiomer showing reduced glucuronidation [65]. This difference in phase II metabolism rate translated into a much longer $t_{1/2}$ for the enantiomer in monkey following i.v. dosing, (*R*)-**27** $t_{1/2} = 9$ h versus 1.8 h for the (*S*)-**27** enantiomer. A similar analysis for the second analog **28** was even more encouraging with very little glucuronidation detected for the (*R*)-**28** enantiomer *in vitro*. The monkey i.v. PK analysis revealed an exceptionally long $t_{1/2}$ of 16 h for the (*R*)-**28** enantiomer (ABT-761). The improved duration of action ($t_{1/2} = 18$ h and >95% inhibition of *ex vivo* stimulated LTB_4 formation in human blood samples up to 18 h post dose with ABT-761) was observed during phase I clinical studies [65].

The sections above highlight the impact of chirality on individual aspects of ADME. However, these various aspects produce a net effect on total drug disposition. The following examples are meant to highlight this.

Enantiomeric discrimination in absorption and disposition has been noted for some drugs that are administered as racemic mixtures. Since pharmacokinetics is the sum total of such ADME processes, the resultant effects on the general pharmacokinetic properties of such drugs are obvious. In a study of nilvadipine, a dihydropyridine calcium antagonist, administered as a racemic mixture to rats and dogs, Tokuma et al. observed that pharmacokinetic properties such as C_{max} and AUC were different for the enantiomers following p.o. dosing, whereas the AUC, V_{dss} and $t_{1/2}$ (beta) values were similar following i.v. dosing in the dog [66].

The route of administration also affects the pharmacokinetic properties of verapamil when administered as racemic mixtures in dogs and humans. The racemates are stereoselective with respect to such properties as volume of distribution, terminal elimination half-life, and systemic clearance, based on the dosing route [67].

Acebutolol shows stereoselective first-pass metabolism; therefore, it is likely that a differential saturation of the primary metabolic pathway for any enantiomer would affect the resultant pharmacokinetic profile of that enantiomer. Such a situation applies only to oral dosing but not i.v. dose administration of acebutolol [68].

In contrast to carbon-based chiral compounds described above, another example of a phosphorus derived chiral drug is exemplified in Section 7.3.

5. CHIRALITY AND TOXICITY

Since the reason(s) a compound may exhibit toxicity in *in vitro* or *in vivo* environment may vary widely, a few selected areas where a clear role of chirality has been reported are provided.

5.1. hERG and Chirality

QT prolongation is a risk that must be managed whenever a clinical candidate is advanced. Drugs that do not have a cardiovascular indication can often prolong the action potential duration (APD) and QT interval. This event can lead to Torsade de Pointes (TdP) and sudden death. TdP is a very rare event and is correlated to QT prolongation, which can be monitored clinically. The risk of prolonged QT is associated with hERG channel activity. The hERG activity is often measured using patch clamp techniques in canine Purkinje fiber. The activity in the hERG assay can be evaluated by calculating the safety margin from the ratio of the IC_{50} for inhibition of hERG current, measured by patch clamp, divided by the effective therapeutic plasma concentration of the drug. The hERG liability should be determined during the lead optimization process and the safety margin can be maximized. There exist numerous examples in which chirality affects hERG activity; a few examples are cited below.

Methadone can cause a prolonged QT interval and has been reported to inhibit the cardiac potassium channel hERG. As shown in Fig. 25, methadone is chiral and its therapeutic activity is mainly due to its *R*-enantiomer. Methadone is metabolized predominately by CYP3A4, CYP2B6, and CYP2D6. The enantiomers of methadone are metabolized differently by CYP2B6. In whole-cell patch clamp, expressing hERG, (*S*)-methadone was shown to block the hERG current 3.5-fold more potently than (*R*)-methadone with IC_{50} values of 2 and 7 μM, respectively [69]. Eap and coworkers reported a study in which 179 patients

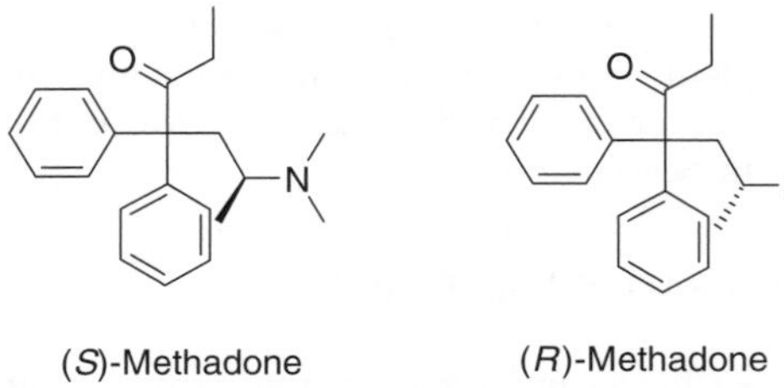

(*S*)-Methadone (*R*)-Methadone

Figure 25. Enantiomers of methadone.

received (*R*,*S*)-methadone; differences in genotypes for CYP2B6, electrocardiograms, and (*R*)- and (*S*)-methadone plasma concentrations were observed. It was found that the genotypic differences for CYP2B6 correlated to higher levels of (*S*)-methadone and led to an increased cardiovascular risk. This report by Eap et al. was the first study for a genetic factor implicated in methadone metabolism that may increase the risk of cardiac arrhythmias and sudden death [69].

Selective inhibitors of the slow component of the cardiac delayed rectifier K^+ current (IKs) such as chromanol 293B are of interest as novel class III antiarrhythmic agents. Synthesis of the enantiomers of chromanol 293B allowed the compounds to be evaluated for their relative potency to block recombinant human K^+ channels [70]. Using whole-cell patch-clamp technique, these researchers found that the physicochemical properties and relative potency of the enantiomers differed. The (−)-isomer was determined to be approximately sevenfold more potent than (+)-isomer for the block of a subtype IK (KvLQT1 + minK). Thus, the (−)-[3*R*,4*S*]-enantiomer of chromanol 293B was an enantioselective inhibitor of KvLQT1 + minK and therefore could serve as a useful tool for studying IKs [70].

Aryl sulfonamidoindane analogs were evaluated as Kv1.5 inhibitors with potential to develop atrium-selective agents. For this series, the (1*R*,2*R*)-enantiomer (**29**, Fig. 27) with $IC_{50} = 33$ nM was shown to be approximately 25 times more potent than (1*S*,2*S*)-enantiomer ($IC_{50} = 760$ nM). In fact, the (1*R*,2*R*)-enantiomer was also shown to be at least 300-fold selective compared to other ion channels including human cardiac potassium channel (hERG), a human cardiac voltage-dependent sodium channel (hSCN5A), and calcium channels found in a rat pituitary cell (GH3) [71].

For the bicyclo[3.1.0]hexane derived melanin-concentrating hormone receptor R1 (MCHR1) antagonists [72] chirality at the bicyclo[3.1.0]hexane core but not the side chain (Fig. 28) had displayed a significant impact on antagonist activity (**30B** being ~19 times more active than **30A**, and **31B** being ~28 times more active than **31A**, whereas analogs **30B** and **31B** gave K_i values of 6.1 and 4.0 nM, respectively). Although all enantiomers had reduced liability in *in vitro* hERG assessment, the side-chain chirality had a greater differentiation in Rb efflux assay, where **30A/30B** displayed a weaker activity compared to **31A/31B.**

5.2. Metal Toxicity: Effect of Chirality for Amine-Based Platinum Complexes

Platinum complexes containing enantiomeric ligands pose an intriguing arena to investigate structure–pharmacological activity relationships of these complexes, especially as DNA is the molecular target for this class of compounds. Even though complexes derived from the (*R*,*R*)- and (*S*,*S*)-chiral diamines produce the same type of intra- and interstrand cross-links, the two sets of enantiomers exhibit different biological activity [73]. Analysis

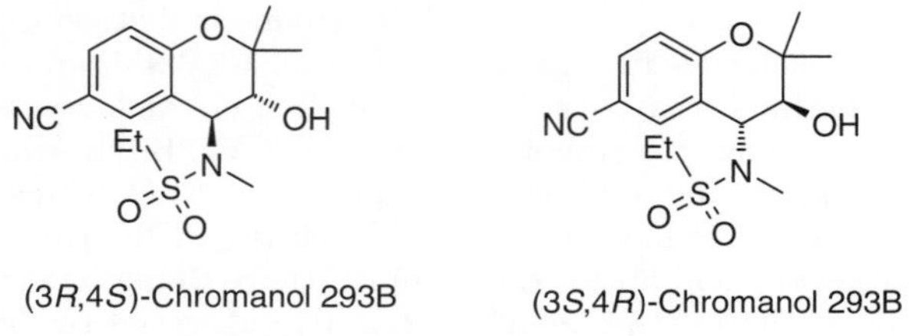

(3*R*,4*S*)-Chromanol 293B (3*S*,4*R*)-Chromanol 293B

Figure 26. Enantiomers of chromanol 293B.

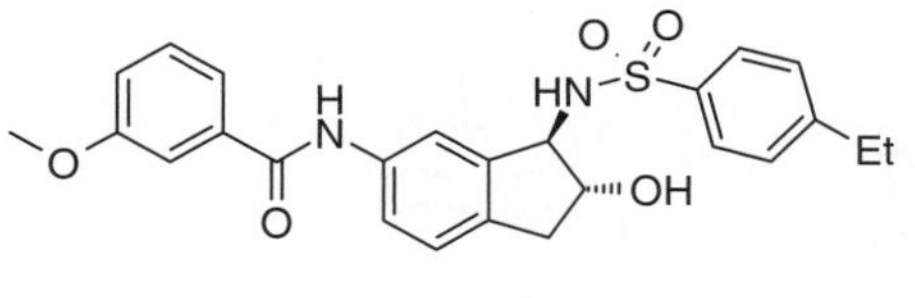

29

Figure 27. Structure of Kv1.5 inhibitor, chiral aryl sulfonamidoindane.

of several chiral amines show that the (*R,R*)-enantiomer exhibits higher antitumor activity and lower mutagenicity than the (*S,S*)-isomer. Evaluation of the *in vitro*, mutagenic activity, which is strictly related to the interaction of the drug with DNA, revealed an eudismic ratio of ~10. Consequently, only the (*R,R*)-enantiomer of [Pt(DACH)(oxalato)] (oxaliplatin) (Fig. 29) has been approved for clinical use. Thus, impact of chirality extends to all types of biological macromolecules and therapeutics.

5.3. Chirality and Carcinogenicity

(*S*)-NNAL [4-(methylnitrosamino)-1-(3-pyridyl)-1-butanol] (**33**) has been shown to be equivalent to NNK [4-(methylnitrosamino)-1-(3-pyridyl)-1-butanone] (**32**) in carcinogenic potency and significantly more potent than (*R*)-NNAL. (*R*)-NNAL (**34**), the main metabolite of NNK (Fig. 30), is produced by enantioselective and reversible reduction of NNK. It was hypothesized and subsequently shown that stereoselective differences in metabolism and/or tissue distribution contributed to the difference in carcinogenicity between the enantiomers [74]. (*R*)/(*S*)-NNAL metabolites were quantified by HPLC analysis from bile, urine, blood, and tissue samples collected over 24 h following equal i.v. dose of the individual NNAL enantiomers to bile duct cannulated male Fischer F344 rats. NNAL was also collected from the HPLC and silylated, and the two NNAL enantiomers were separated by chiral GC-TEA. It was determined that the (*S*)-NNAL had a much larger tissue distribution V_{ss} 1792 mL) than (*R*)-NNAL (V_{ss} 645 mL) and that (*S*)-NNAL appeared to be stereoselectively retained in lung [74].

This study also showed that the metabolic profiles of the two enantiomers were significantly different. When the (*R*)-enantiomer was dosed, almost 40% of the dose was excreted in the bile as the (*R*)-NNAL-Gluc. (*R*)-NNAL-Gluc excretion in bile was predominant over (*S*)-NNAL-Gluc even when the

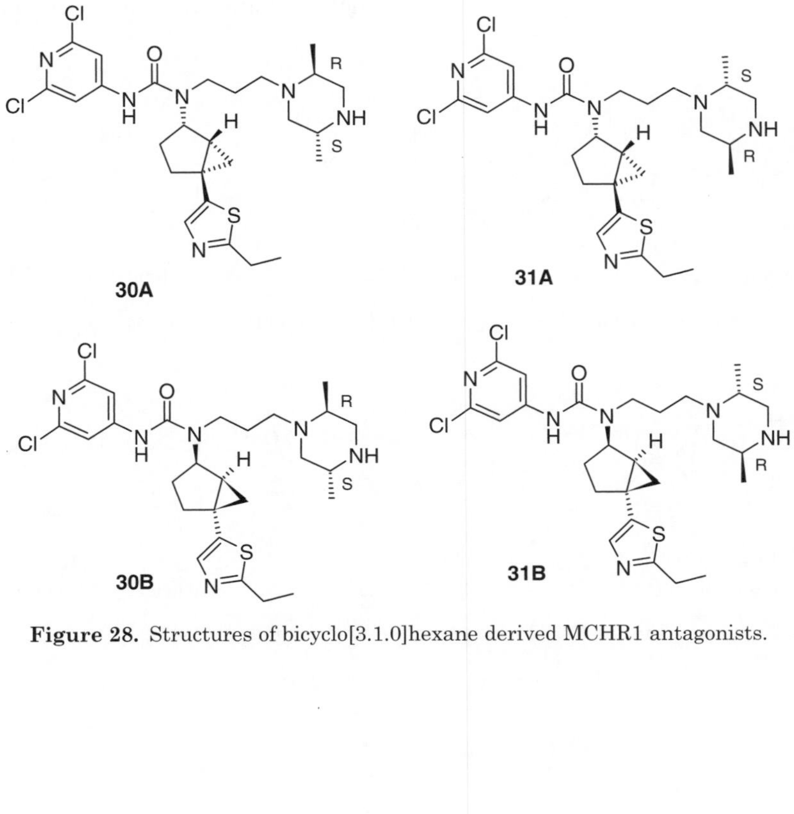

30A

31A

30B

31B

Figure 28. Structures of bicyclo[3.1.0]hexane derived MCHR1 antagonists.

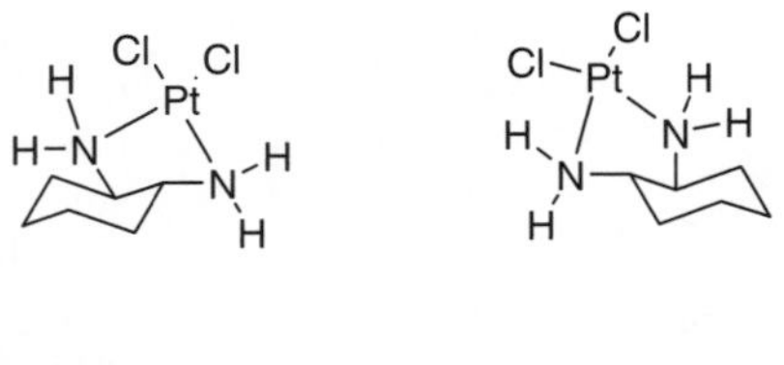

Configuration at carbon (R,R) (S,S)

Figure 29. Structures of platinum complex with chiral diamine [$PtCl_2$(1,2-DCHA)].

(*S*)-enantiomer was dosed. The (*S*)-enantiomer was preferentially converted back to NNK, while the (*R*)-NNAL tended to enter detoxification pathways, particularly glucuronidation. In human studies, despite nicotine and nicotine metabolites reaching background levels (within 7 days) in people who had stopped smoking or stopped using smokeless tobacco, NNAL and NNAL-Gluc in urine had elimination half-lives of 40–45 days. These authors report that "more importantly, the enantiomeric ratios of (*S*)-NNAL/(*R*)-NNAL and (*S*)-NNAL-Gluc/(*R*)-NNAL-Gluc in urine were significantly (3.1–5.7 times) higher 7 days after cessation than at baseline in both smokeless tobacco users and smokers, indicating enantioselective retention of (*S*)-NNAL in humans." [74]

5.4. Chirality and Clinical Chemistry

The malabsorption of carbohydrates in the gut leads to large amounts of D- and L-lactate, which is produced by intestinal flora. In most clinical laboratories, routine analysis is only performed to determine the level of L-lactate. A high level of D-lactate is a very rare metabolite in healthy individuals. If undetected, the unrecognized D-lactic acidosis could lead to severe neurological consequences. Patients with short bowel syndrome (SBS) are prone to periods of acute life-threatening metabolic acidosis that could be attributable to the accumulation of D-lactate that may be caused by bacterial overgrowth. In a recent case, a child who had surgical correction for a small intestinal volvulus with SBS was admitted, and a routine qualitative urinary organic acid ana-

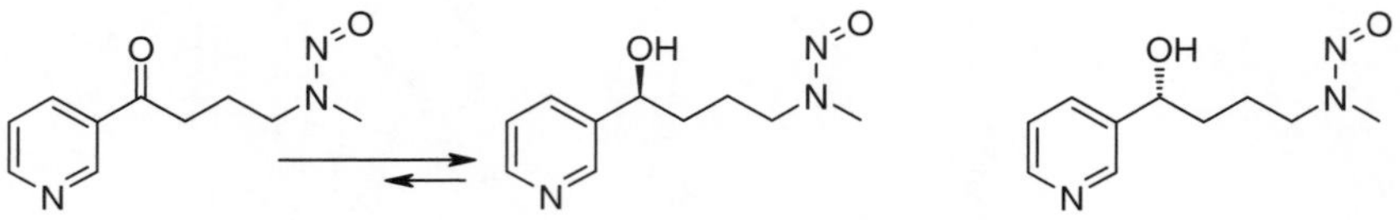

NNK **32** (*S*)-NNAL **33** (*R*)-NNAL **34**

Figure 30. NNK (**32**), (*S*)-NNAL (**33**), and (*R*)-NNAL (**34**).

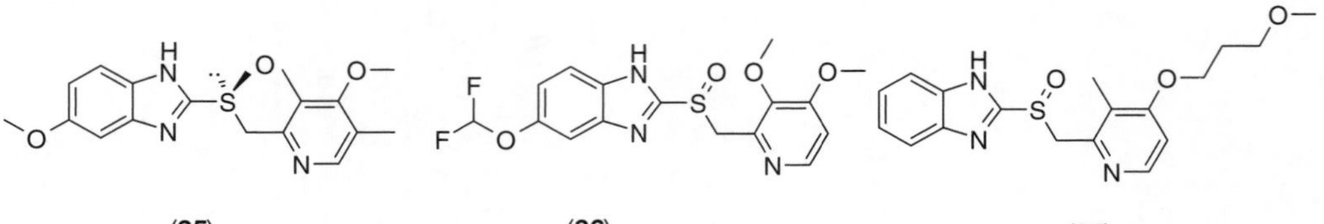

(35) Nexium™ (esomeprazole) **(36)** Pantoprazole **(37)** Aciphex™ (rabeprazole)

Figure 31. Marketed proton pump inhibitors (**35–37**).

lysis established a pattern characteristic of bacterial overgrowth syndrome [75]. In the follow-up tests, enantiomeric analysis of urinary organic acids using enantio-selective multidimensional capillary gas chromatography–mass spectrometry (enantio-MDGC–MS) was performed that provided a D:L-lactate ratio of 95:5, further indicating D-lactic acidosis [73]. Subsequent antibiotics treatment lead to a positive outcome. Follow-up enantio-MDGC–MS urine analysis, when the patient was clinically well, also showed a D:L-lactate ratio of 50:50. This enantio-MDGC–MS method has been employed for studies of inborn errors of metabolism where enantiodifferentiation is important [75].

5.5. Chirality and Environment Toxicity

Organic compounds that contain a chiral center with significant environmental exposure (air, water, soil) are being recommended to be treated as individual enantiomers from an environmental perspective. In one such recent study, metabolism of linear alkyl benzenes (LAB) was evaluated as these are reported to be present in natural waters, coastal sediments, and sewage sludges. The microbial degradation of this class of compounds, LAB, starts with ω-oxidation of the alkyl side chain. However, as the chain cuts back to four or five carbon atoms from the point of attachment to the benzene ring, further shortening by β-oxidation seems to become hindered. The microbial enantioselective metabolism of chiral 3-phenylbutyric acid, an intermediate of linear alkylbenzene degradation, was examined by Simoni and coworkers. In this study, degradation of individual enantiomers and racemic 3-phenylbutyric acid was evaluated and was shown that only one enantiomer was utilized as the growth substrate by the Gram-positive bacterium *Rhodococcus rhodochrous* PB1. The data presented are consistent with the view that the enantioselective step discriminating between the (*R*)- and (*S*)-enantiomers of 3-phenylbutyric acid is located at the beginning of the metabolic sequence, as *R. rhodochrous* PB1 was able to grow with the (*R*)-enantiomer as the sole carbon and energy source [76]. This example highlights the importance of evaluating the fate of compounds containing chiral center(s) by using enantio-selective techniques for their environmental impact.

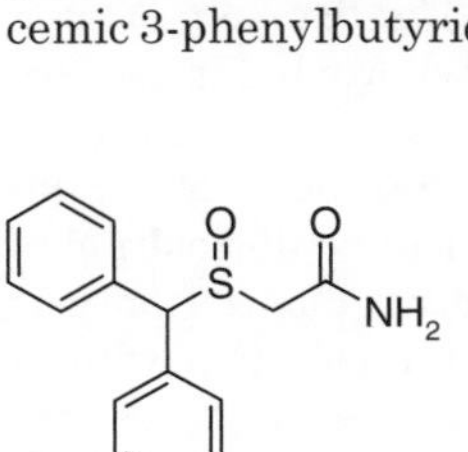

38

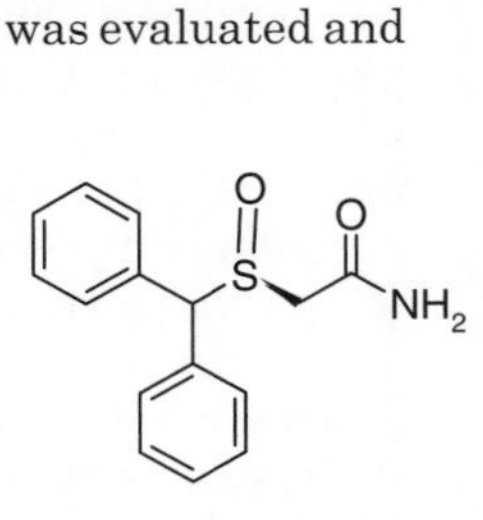

39

Figure 32. Structures of modafinil (**38**) and armodafinil (**39**).

6. CHIRALITY AT NONCARBON ATOM

6.1. Chirality at Sulfur

The sulfoxide functionality and the sulfonium ion are nonplanar, displaying a pyramidal geometry with a lone pair of electrons. These pyramidal structures can undergo an atomic inversion proceeding through a near-planar or planar transition state. However, this inversion requires overcoming a significant energy barrier, typically in the range 25–30 kcal/mol [77]. Listed below are some examples of chiral sulfoxide derived compounds and the impact of chirality at sulfur on biological activity.

One of the more prominent chiral sulfoxide drugs is Nexium™ (esomeprazole), and it is marketed as a proton pump inhibitor for the treatment of gastric acid-related problems such as erosive esophagitis and gastroesophageal reflux disease (GERD). Nexium has gained high popularity over the marketed racemic mixture, Prilosec (omeprazole), due to its superior clinical properties. Interestingly, the Nexium precursor sulfide is converted to the chiral sulfoxide, Nexium (**35**), utilizing a titanium tetraisopropoxide and (*S*,*S*)-(2)-diethyl tartrate-mediated chemical oxidation at industrial scale. Pantoprazole (**36**) and rabeprazole (Aciphex™), **37**, represent the second- and third-generation drug for this disease target.

Similar to the differential ADME behavior described above for enantiomers derived from chirality at carbon centers, enantiomers involving chirality at sulfur would be expected to show preferential absorption, metabolism, and clearance properties. Examples of drugs showing more rapid clearance of one sulfoxide enantiomer, or promoting chiral inversion, include modafinil and pantoprazole. Modafinil (Provigil™), **38**, has been marketed as a racemate in Europe and the United States for the treatment of excessive daytime sleepiness (EDS) associated with narcolepsy. Recently, its *R*-enantiomer **39** has been approved and is being marketed as armodafinil (Nuvigil™) for EDS associated with narcolepsy and shift work sleep disorder [78]. Armodafinil has been reported to have a long duration of action [78].

6.2. Chirality at Nitrogen

A quaternary ammonium salt derived from tetrahydroberberine, CPU86017 (Fig. 33), bears two chiral centers: 7N and 13aC (see atom labels, Fig. 33). It has potential for effective antiarrhythmic activity. However, in addition to the blockade on *I*Kr, *I*Ks, and calcium influx, the racemic CPU86017 causes adverse hypotension following i.v. infusion due to α-adrenoceptor blockade activity. In order to improve activity at ion channels, the four stereoisomers were separated, and each of these four enantiomers was evaluated for effects on (i) *I*Kr and *I*Ks (using the whole-cell patch-clamp assay), (ii) calcium channels (voltage-dependent calcium channels (VDC) in vascular smooth muscle), and (iii) α-adrenoceptors activity, compared with the effects of the racemate [79]. The *I*Kr and *I*Ks inhibition IC_{50} values for the enantiomers are shown in Table 5. The (7*S*,13a*R*)-isomer showed balanced blockade of *I*Kr and *I*Ks that was associated with a loss of α-adrenoceptor antagonism but enhanced VDC blockade.

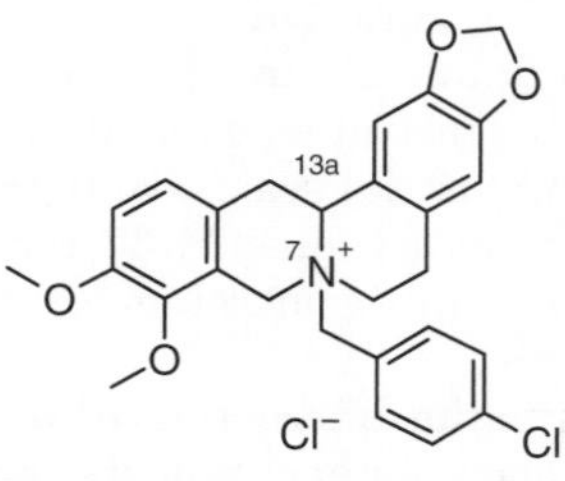

Figure 33. Structure of CPU86017.

Table 5. *In Vitro* Activity for the CPU86017 Enantiomers Versus the Racemate

Compound	*I*Kr Tail Inhibition, IC50 nM	*I*Ks tail inhibition, IC50 nM
7*S*, 13a*S*	2.86 ± 1.20	16.9 ± 4.0
7*S*, 13a*R*	39.4 ± 8.5	20.0 ± 2.1
7*R*, 13a*R*	3.48 ± 0.80	99.1 ± 5.9
7*R*, 13a*S*	7.65 ± 1.50	160 ± 81
Racemate	12.5 ± 7.8	65.0 ± 4.7

Analysis showed that configuration of 13aC essentially dictates *I*Kr blockade and the Ca^{2+} antagonism, whereas the configuration of the 7N position only affects the IKs blocking activity of CPU86017. The (7*S*,13a*R*)-enantiomer that displayed less α-adrenoceptor antagonism compared to the racemate, exhibited mild blockade of *I*kr, and moderately enhanced blockade of *I*ks and Ca^{2+} influx, may prove to be a promising antiarrhythmic agent.

6.3. Chirality at Phosphorous

Cyclophosphamide [(*R*,*S*)-CP], **2** (Fig. 1), contains an oxazaphosphorine ring and a chiral center at phosphorous atom with a nitrogen mustard side chain. The racemate form of CP is used clinically as an anticancer drug. The commercial process used for the synthesis provides a racemic product. Recently, a new HPLC assay suitable for CP enantiomers in plasma to support PK studies, involving analyses of multiple samples, was developed. This method involves ethyl acetate extraction of CP enantiomers from plasma followed by precolumn derivatization to produce diastereomers in a two-step process utilizing chloral and (+)-naproxen acid chloride. The two-step derivatization approach used is shown in Fig. 34. In a PK study, rabbits were used to compare the influence of i.p., i.v., and oral routes of administration on the stereoselective disposition of (*R*,*S*)-CP. No stereoselective disposition was observed following i.v. administration, while a preference for (*R*)-enantiomer was seen in rabbits following i.p. dosing (the ratios of (*R*/*S*)-enantiomers were 1.1–2.0 (i.v.) and 1.4–2.5 (i.p.)). The (*S*)-CP was cleared faster than (*R*)-CP following i.p. administration, and only (*R*)-CP was detectable in plasma following oral

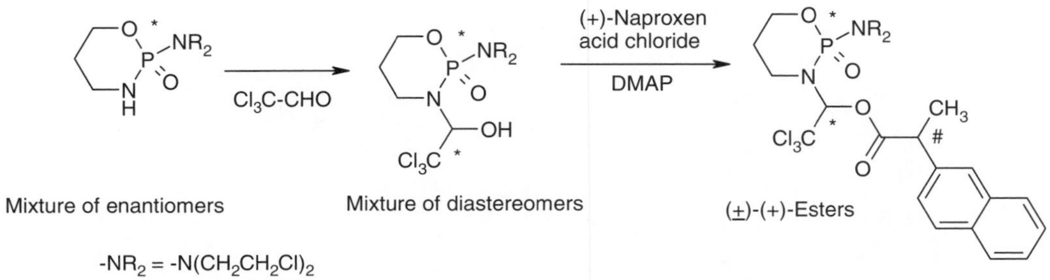

Figure 34. Two-step derivatization approach for separation of diastereomers via HPLC. Approach has been applied to analysis of plasma derived samples[1]. *Denote chiral center.

administration. A marked enantioselective metabolism of the (*S*)-CP enantiomer depending on the route of administration of the drug was observed. In this study, the i.p. and oral routes provided the greatest differences due to first-pass metabolism as reported by Holm et al. [80a] In *in vitro* studies, incubation of rabbit liver microsomes with (*R*,*S*)-CP was demonstrated to exhibit marked enantioselectivity in the CP metabolism. For the *in vitro* incubation studies, the ratio of (*R*)-CP:(*S*)-CP increased with time, from 1 : 1 (at initial timepoints) to 4.5 : 1 after 60 min incubation. The *in vivo* rabbit studies indicate that first-pass metabolism is highly enantioselective for this drug and this is supported by *in vitro* incubation studies using liver microsome incubation studies. Nine cancer patients received (*R*,*S*)-CP, dosed intravenously. While there was significant intersubject variability in the levels of the two enantiomers at various timepoints evaluated, the ratio of enantiomers was effectively 1 : 1, thus reconfirming no enantioselectivity, for the elimination of the enantiomers in human, similar to a previous report [80b]. The lack of enantioselectivity in human is similar to what was observed in rabbit following the i.v. route. It will be interesting to see if enantioselective first-pass metabolism of this drug also occurs in humans following oral dosing, and if so, for which enantiomer.

7. CHIRALITY IN DRUG DESIGN

During lead optimization, typically medicinal chemists undertake diverse structural variations to improve selectivity, enhance PK properties, and minimize off-target activity and toxicity. These variations typically result in analogs with varying molecular weight (<400–500 Da) and molecular properties (clog *P*, clog *D*, pK_a, PSA, and aqueous solubility). Chirality provides a source of molecular complexity that is molecular weight independent, and increasing molecular complexity can drive selectivity [81]. Chirality is increasingly being introduced into a compound during the lead optimization stage of drug discovery to improve potency and selectivity, and there are numerous examples in the literature [82]. This section provides some examples from recent drug discovery stage literature. Chirality has also been of key importance for development candidates and this is addressed in the Section 10, where the rationale for incorporation of chirality in the target molecule is more than simply gaining selectivity or improvements in *in vitro* activity. Chiral switch is a procedure used for the development of an old racemic drug into its single active enantiomer. This new enantiomeric drug, developed by a pharmaceutical manufacturer, will receive additional patent protection [83] and a new generic name. This is addressed in Section 9.

Bunnage et al. looked to introduce a chiral substituent into the sildenafil template as a new tactic to differentiate activity of PDE_5 versus PDE_6 [18]. The selective SAR observed

[1]Retention times for (*R*)-CP = 18.8 min and (*S*)-CP = 22.9 for analysis of samples using the two-step derivatization approach, as described in Ref. 78.

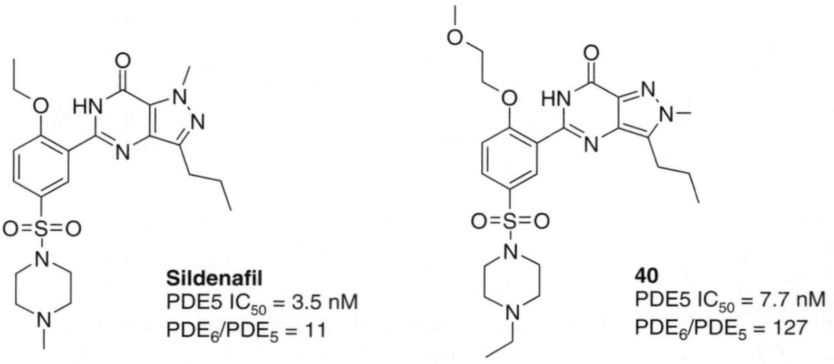

Sildenafil
PDE5 IC_{50} = 3.5 nM
PDE_6/PDE_5 = 11

40
PDE5 IC_{50} = 7.7 nM
PDE_6/PDE_5 = 127

Figure 35. Structure of sildenafil and lead compound (**40**) for introduction of chirality.

with the alkoxyarene domain, coupled with synthetic expediency, led them to target a small panel of compounds with chiral alkoxy substituents at the 20-position of lead compound **40** (Fig. 35). Although PDE_5 and PDE_6 have high binding site homology, the authors were intrigued to see if these novel chiral PDE_5 inhibitors were able to exploit subtle differences in the PDE_5 and PDE_6 binding site cavities and deliver greater selectivity. As shown in Table 6, this was indeed found to be the case: all the analogs showed impressive levels of selectivity and some analogs showed selectivity in excess of 1000-fold (which was previously unprecedented in the sildenafil scaffold). These data further illustrate the potential of utilizing chirality as a tactic to drive selectivity between closely related targets.

In HMG-CoA reductase inhibitors, the lactone moiety can be opened to a (3*R*,5*R*)-dihydroxyacid moiety. It was found that the carboxylate anion with two hydroxyl groups in an erythro relationship is needed for potency and

Table 6. In Vitro Inhibition of PDE5 Exhibited by Chiral Compounds (41a–e) and Their Corresponding (*S*)-Enantiomers, and Selectivity of (41a–e) over PDE6

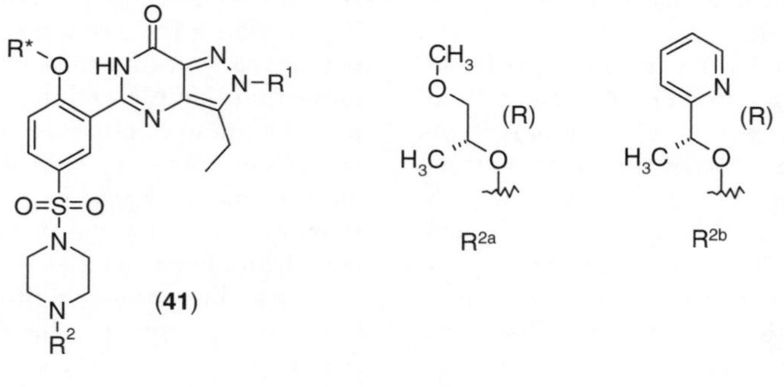

(**41**) R*O, R[1], R[2]; R^{2a} (R); R^{2b} (R)

Compound	R*O	R^1	R^2	PDE5 IC50 (nM)	PDE_6 IC_{50} (nM)	Selectivity PDE_6/PDE_5
41a	R^{2a}	Me	Et	4.2 [(*S*)41a = 77]	1978	471
41b	R^{2a}	Me	pr	5.6 [(*S*)41b = 100]	1249	223
41c	R^{2a}	$-CH_2$(2-Pyr)	Et	0.62 [(*S*)41c = 100]	1539	2482
41d	R^{2a}	$-CH_2-O-CH_3$	Et	0.89 [(*S*)41d = 80]	734	825
41e	R^{2b}	Me	Et	3.0 [(*S*)41e = 100]	4323	1441

See Ref. 18 and citations listed therein for details. Data excerpted from Ref. 18. With permission from Elsevier.

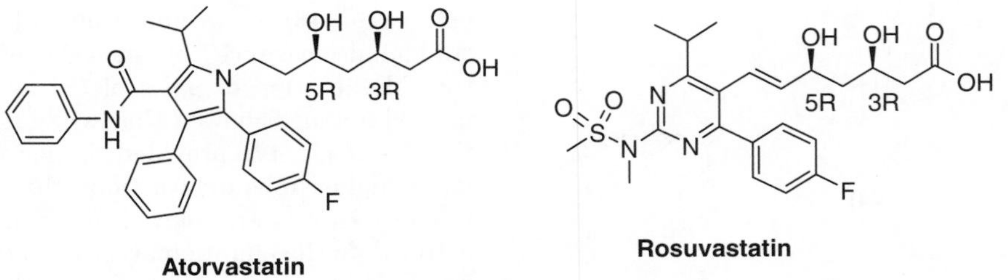

Atorvastatin

Rosuvastatin

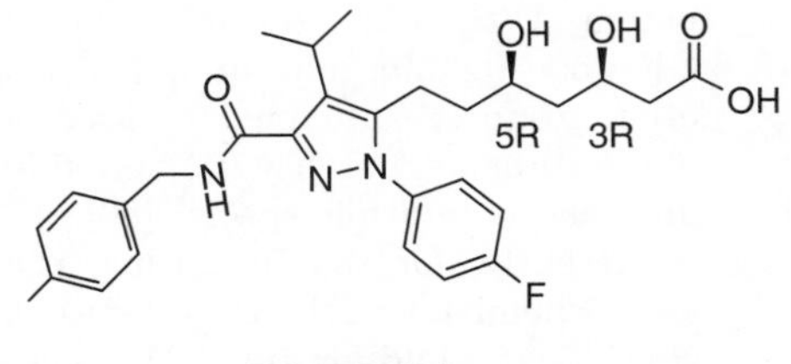

PF-3052334

Figure 36. HMG-CoA reductase inhibitors with chiral (3*R*,5*R*)-dihydroxy acid moieties that are conserved for potency and selectivity.

selectivity, as this chirality mimics the structures of the endogenous substrate/product (3-hydroxy-methylglutaryl-CoA/mevalonic acid) in the enzyme's active site. The same stereochemical array is also present in the naturally occurring mevinic acids (known HMG-CoA reductase inhibitors) and has been shown to be essential for bioactivity. Consequently, this chirality has been conserved in marketed and clinical HMG-CoA reductase inhibitors (Fig. 36) [84].

In structure–activity studies targeting CCR2 antagonism, the initial hit **42** (IC_{50} value 80 nM) showed that substitution of methyl groups at 2- and 4-positions reduced activity while at 3-position it abolished activity (Fig. 37). Subsequently, a systematic and rational approach directed at designing conformationally constrained analogs resulted in the identification of a 3-amino-1-cyclopentanecarboxamide as the optimal scaffold for CCR2 antagonism [85]. The 1,3-disubstituted cyclopentane **43a** had an IC_{50} value of 65 nM on the hCCR2 receptor as a mixture of four isomers. Binding results for the four isomers, listed in Table 7, revealed the *cis*-(1*S*,3*R*) **43b** as the only active isomer, showing 36 nM potency in the human monocyte binding assay. Having established 1,3-disubstituted cyclo-

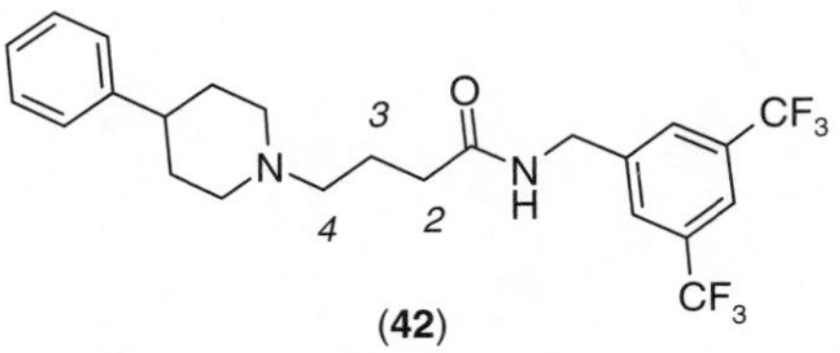

(42)

Figure 37. Structure of linear analogs, initial CCR2 antagonist.

Table 7. Binding Affinity of Enantiomers 43a–e to Human Monocyte

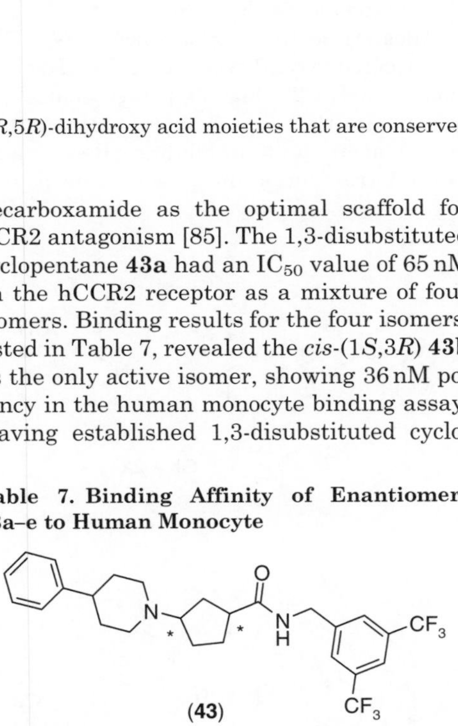

(43)

Compound	Stereochemistry	IC_{50}
43a	Mixture of four	65 nM
43b	*cis*-(1*S*,3*R*)	36 nM
43c	*cis*-(1*R*,3*S*)	30% @ 1 μM
43d	*trans*-(1*R*,3*S*)	44% @ 1 μM
43e	*trans*-(1*R*,3*S*)	25% @ 1 μM

Data excerpted from Ref. 85. With permission from The American Chemical Society.

*Denote chiral center.

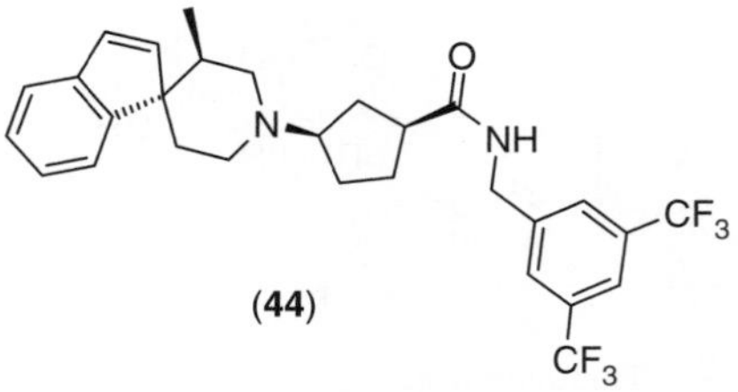

(44)

Figure 38. Binding affinity to human monocyte.

pentane as the optimal lead, the researchers utilized this chiral core to further increase potency through incorporation of additional chiral substituent resulting in an optimized lead compound **44** (Fig. 38).

Allosteric activators of glucokinase (GK) are targeted to treat type-2 diabetes. The lead molecule GKA-22 (Fig. 39) possessed oral PK parameters that were nonoptimal as it had high plasma protein binding. Researchers from Astra-Zeneca introduced α-branching (**45** versus **46**) as an approach to lower the "unbound clearance" by an order of magnitude. Evaluation of pairs of (*R*) and (*S*) α-methyl analogs showed that while the enantiomers of **47a/47b** provided similar and relatively higher unbound fractions, the (*S*)-enantiomer **47a** was an order of magnitude more active than the enantiomer **47b**. On the other hand, both enantiomers of α-methylbenzyl ether **48a** and **48b** were significantly more potent. Ultimately, the enantiopure analog GKA-50 (Fig. 39), containing two α-branched ether groups, resulted in increased selectivity and oral bioavailability (99%) and low clearance across multiple species [86].

Recently, for the first time, the role of stereochemistry in the C2-substituent on the *S*-DABO scaffold for anti-HIV-1 activity was investigated. The *trans*-double bond was replaced by an isosteric cyclopropyl moiety. This change resulted in the introduction of two new chiral centers (**50**). The enantiopure cyclopro-

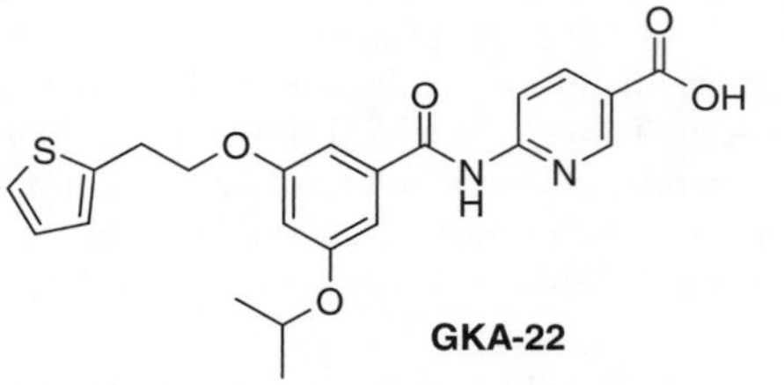

GKA-22

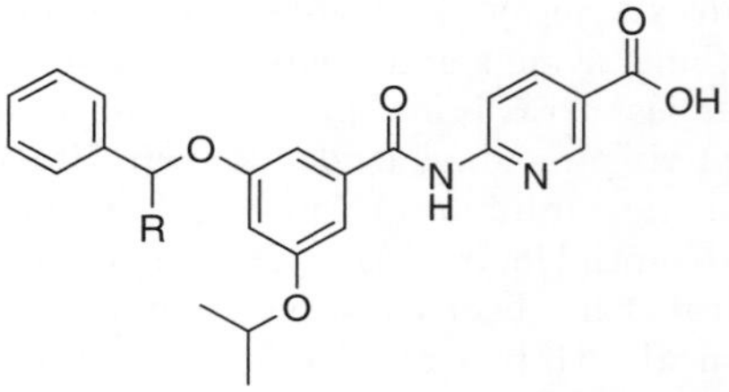

	R	Unbound clearance[a]
45	H	1850
46	(*S*)-CH3	171

(a): defined as CL/*f*, (clearance/fraction unbound)

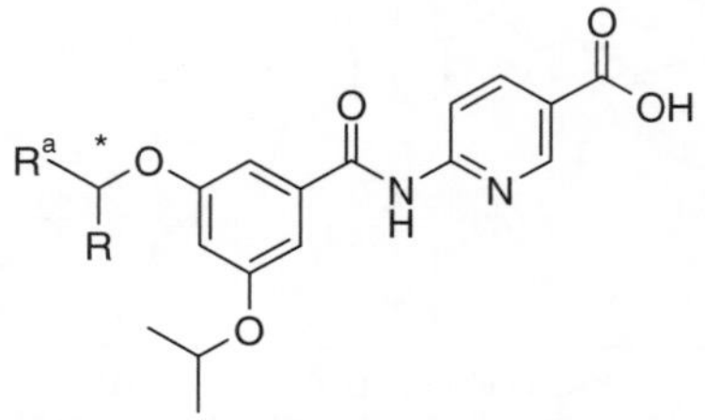

Compound	Chirality (*)	R[a]	% Free (rat)	EC_{50} (nM)
47a	(S)	-CH2OCH3	5.3	610
47b	(R)	-CH2OCH3	6.2	5510
48a	(S)	-Ph	0.23	20
48b	(R)	-Ph	0.23	90

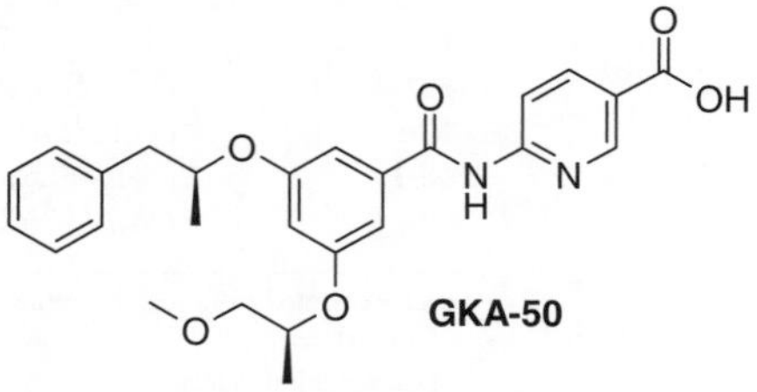

GKA-50

Figure 39. Structures of GK activators.

Table 8. Anti-HIV-1 Activity of Enantiomers of 49 and 50 in Cell-Free and Cellular Assays

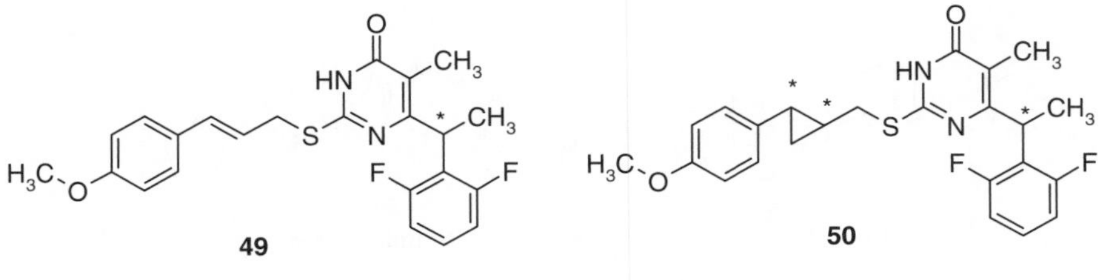

Compound	Cell-Free Assay, ID_{50} nM[a]		Cellular Assay, EC_{50} μM[b]	
	Wt	K103N	NL4-3 wt	K103N
(*S*)-**49**	0.60	7.01	NR	NR
(*R*)-**49**	0.01	0.47	NR	NR
(*R*,*R*,*R*)-**50**	0.02	0.04	0.00007	0.036
(*R*,*R*,*S*)-**50**	0.30	0.60	0.0049	2.55
(*S*,*S*,*S*)-**50**	0.12	0.83	0.09	5.51
(*S*,*S*,*R*)-**50**	0.04	0.4	0.01	3.71

ID, inhibition dose; EC, effective concentration; NR, not reported.
Data in table excerpted from Ref. 87. With permission from The American Chemical Society.
*Denote chiral center.

pyl analogs retaining the more active C6-(*R*)-enantiomer were prepared and these new enantiomers were evaluated for bioactivity. As highlighted from data in Table 8, the "RRR" enantiomer not only provided pM activity against wild-type reverse transcriptase (RT), but it also resulted in nM activity against clinically relevant drug-resistant mutants, especially the K103N mutant [87].

Incorporation of chiral center(s) during lead optimization has led to several reported examples of potent, selective, and highly efficacious compounds. In general, the incorporation of chiral center(s) is expected to result in increased cost for the production of the API. Judicious use of chirality can result in enhanced *in vivo* profile while achieving the desired cost objective. The following two recent reports serve as suitable examples.

Application of structure-based design (SBD) approaches for the elaboration of hits identified from fragment-based screening resulted in the incorporation of chirality to provide a lead compound (**51**). Subsequent structure–activity relations (SAR) studies lead to an optimized potent LTA4H inhibitor (**52**) with excellent PK profile. Finally, both enantiomers of the optimized lead compound were prepared and evaluated. These studies led to the identification of DG-051 (Fig. 40), an analog derived from (*S*)-proline (a more cost-effective raw material), as the clinical candidate [88]. DG-051 and its enantiomers exhibited essentially identical activity and similar PK profile. DG-051 is under phase II clinical evaluations.

Often chirality is introduced to increase potency and selectivity, as illustrated by examples cited above. However, the increased structure complexity due to presence of chiral center(s) would need to result in some advantage over the nonchiral compound(s). In a recent example, where the chirality was distal to the pharmacophore, a number of chiral centers were eliminated while retaining bioactivity. For this example, a potent analog

Table 9. SMN2 Promoter Assay Results for C5-Substituted 2,4-Diaminoquinazoline Analog

Compound	Chirality[a]	SMN2 Promoter EC_{50} (nM)[a]
58	NA	78
59	*RS* (racemate)	21
59	*R*	32
59	*S*	2

See Ref. [91] for details.
[a]Data in table excerpted from Ref. 91. With permission from The American Chemical Society.

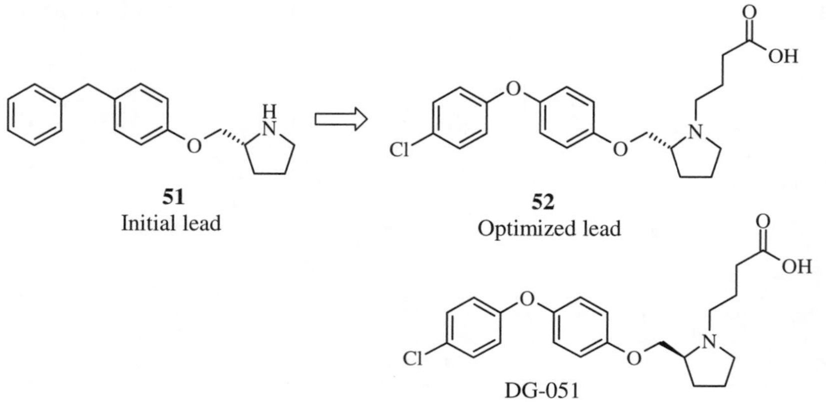

Figure 40. Structure of LTA4H inhibitors: initial lead (**51**), optimized lead (**52**), and clinical candidates (DG-051).

of the antitumor spliceosome inhibitor FR901464, Fig. 41, was shown to possess some stereocenters that were superfluous for activity. Lagisetti and coworkers [89] were able to make modifications (conversion of the pyran ring oxygen to carbon, removal of two methyl groups, and removal of a hydroxyl oxygen) that eliminated six chiral centers when compared to FR901464, which made the synthesis readily achievable. Molecule **53** (Fig. 41) displays a similar bioactivity profile as FR901464.

In addition to the design and interaction of chiral small molecules with protein derived biological targets, incorporation of chirality for interaction with DNA derived therapeutic targets has also been investigated. The helical groove of DNA is considered to be the most pronounced determinant of enantioselectivity between a chiral pair. Naphthalimide analogs bearing tertiary amine side chains are well-known antitumor agents against a variety of murine and human tumor cells, and the photo-cleaving ability for this class of compounds is well known. Considering that DNA is a chiral molecule with right-handed helical configuration, in order to determine the impact of chirality for interaction of DNA with naphthalimides, analogs bearing chiral (tertiary amine) functionality were prepared [90]. A diverse set of heterocyclic-fused naphthalimides **54**, **55**, **56**, **57** (Fig. 42) bearing side chains with (*S*) and (*R*) chirality were prepared and evaluated for their DNA photocleavage activities. The order of DNA cleavage efficiency by these chiral compounds (*S*-enantiomers) was **55** > **56** **54**. The Scatchard binding constants for calf thymus DNA (CT-DNA) were monitored by fluorescence spectroscopy (in 30 mM Tris–HCl buffer, pH 7.0). The Scatchard binding constants of (*S*)-enantiomers were higher than those of the corresponding (*R*)-enantiomers.

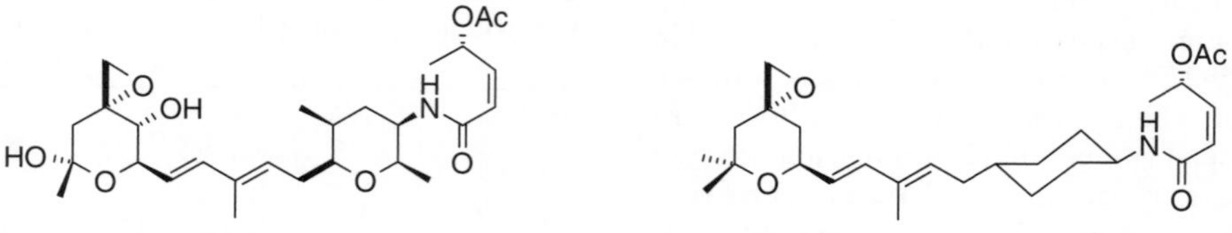

Figure 41. Removal of superfluous stereocenters with retention of activity.

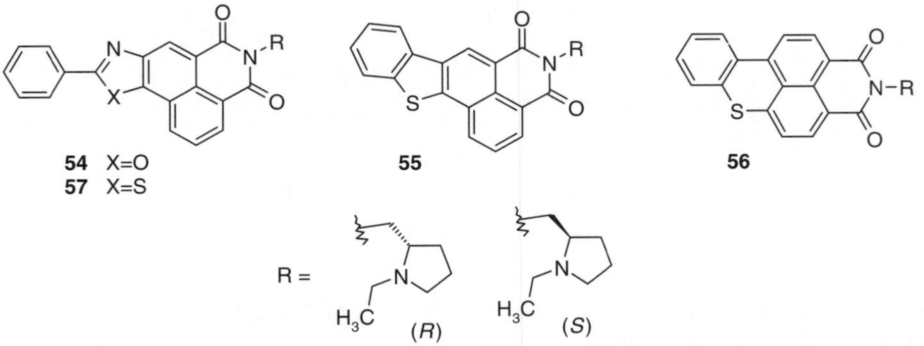

54 X=O
57 X=S
55
56
R = (R) (S)

Figure 42. Structures of chiral naphthalimides with photocleaving activities.

The order of the Scatchard binding constants for the (*S*)-enantiomers was determined to be **55** > **56** > **57** > **54**. These analogs were also evaluated for cytotoxicity and it was observed that (*R*/*S*)-enantiomers behaved quite differently in terms of cytotoxicity against different cell lines. Even though there were no direct relationships between the DNA binding affinities and cytotoxicities for these series of analogs, the *in vitro* trend and high enantioselectivity for DNA binding affinity, DNA photocleavage activity was very clear (for *S*-enantiomers). Future explorations incorporating chirality should be continued, as it would be expected to lead to the development of novel antitumor agents.

Finally, incorporation of chirality has been utilized for impacting gene induction. During the lead optimization process for the identification of small molecule activators of the SMN2 promoter for the potential treatment of spinal muscular atrophy, Thurmond et al. [91] incorporated a chiral center at the benzyl methylene in the initial lead compounds (**58**) to facilitate ligand preorganization. Individual enantiomers of **59** were synthesized and evaluated in a β-lactamase reporter gene assay in NSC-34 (a mouse motor neuron hybrid) cell line [92]. The (*S*)-enantiomer (*S*)-**59**) resulted in ~40-fold enhancement in EC_{50} compared to the achiral analog **58**. The chiral analogs also provided significantly improved PK profile as well [91]. In follow-up studies, scavenger decapping enzyme, DcpS, was identified as the molecular target modulating gene expression for this series of small molecules [93].

8. REGULATORY CONSIDERATIONS

In 1992, the Food and Drug Administration (FDA) in the United States published a policy statement for the development of new stereoisomeric drugs, which was closely followed by European guidelines in 1993, and it came into force in 1994 [94,95]. FDA invited discussions with sponsors concerning whether to pursue the development of the racemate or the individual enantiomer. The sponsor of a drug candidate needs to make the decision as to develop a racemate or single enantiomer as guided by regulatory agencies. It is critical to develop quantitative analytical assays early in the process for the individual enantiomers so that the appropriate data can be obtained [96]. All information developed by the sponsor or available from the literature that is relevant to the chemistry, pharmacology, toxicology, or clinical actions of the enantiomers should be included in the IND and NDA submissions [96]. The corresponding agencies will evaluate the rationale of the sponsor.

When considering different scenarios that could arise from a racemate consisting of two enantiomers, there could be different scenarios with respect to the PK and pharmacodynamics (PD) of the two enantiomers: for example, same PK and PD (efficacy and toxicity) profiles, same PK but different PD profile. Carvedilol, a nonselective β-adrenergic receptor antagonist and α-adrenoceptor blocker, is administered as a racemate for the treatment of hypertension and congestive heart failure [68,97]. Although both *S*(−) and *R*(+)-

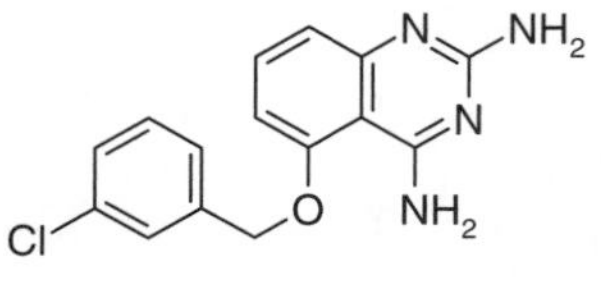

58

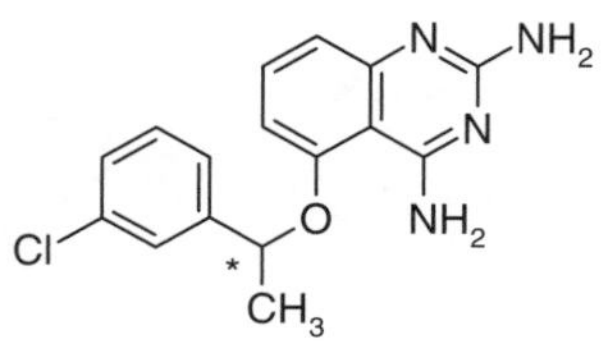

59

Figure 43. SMN2 promoter activators [91].

enantiomers have similar PK properties, they differ in their PD properties on the two receptors of interest; the *S*(−)-enantiomer is more potent on the β-adrenergic receptor. This is an example of enantiomers having similar PK but different PD effects, especially in consideration of the β-adrenergic receptor. On the other hand, both enantiomers are equally effective in blocking the α-adrenergic receptor, indicating similar PK and PD effects on that receptor site [54].

The number of single enantiomer drugs that have been launched since 1985 has steadily increased. Between 1985 and 1988, one-third of all NCEs that contained chiral centers were launched as a single enantiomer [98]. Majority of chiral drugs launched in 2003 were launched as single enantiomers (14 of 17) [94]. Reasons for this change to develop the single enantiomers are numerous and include a higher selectivity and potent pharmacodynamic profile, less complex pharmacokinetic profile, and an increased safety profile.

9. CHIRAL SWITCH

The term racemic or chiral switch refers to developing a single enantiomer of an already marketed racemic drug (Table 10 and Fig. 44) [95]. There are several reasons for this strategy. The potential advantages of the single enantiomer are (1) a more selective and potent pharmacodynamic profile, (2) potential for an improved therapeutic index, (3) a less complex pharmacokinetic profile, (4) reduced drug–drug interactions, and (5) less complex relationship between plasma concentrations and effect. There is also the potential for patent coverage and increased revenue generation.

The arguably most successful example of a chiral switch is the well known "purple pill". [99] Nexium™ or esomeprazole is the *S*-enantiomer of omeprazole. Omeparazole is the active ingredient in the heartburn medication Prilosec™ by Astra-Zeneca. The patent for Prilosec™ expired in 2002 and it had worldwide sales of $5.6 billion in 2001. Astra-Zeneca was able to develop and market Nexium™ and

Table 10. Examples of Successful Chiral Switches

Racemate	Single Enantiomer	Indication/Mechanism
Ofloxacin	Levofloxacin	Antibacterial
Ibuprofen	Dexibuprofen	Anti-inflammatory
Salbutamol	Levalbuterol	β2-agonist
Ketoprofen	Dexketoprofen	Anti-inflammatory
Citalopram	Escitalopram	Antidepressant
Ketamine	(*S*)-Keatmine	Anesthetic
Bupivacaine	Levobupivacaine	Anesthetic
Methylphenidate	(*R,R*)-Methylphenidate	Attention deficit disorder
Cetirizine	Levocertirizine	Antihistamine
Atracurium besylate	Cisatracurium	Muscle relaxant
Omeprazole	Esomeprazole	Proton pump inhibitor
Zopiclone	Eszopiclone	Hypnotic
Formoterol	Arformoterol tartrate	Beta-adrenoceptor agonists

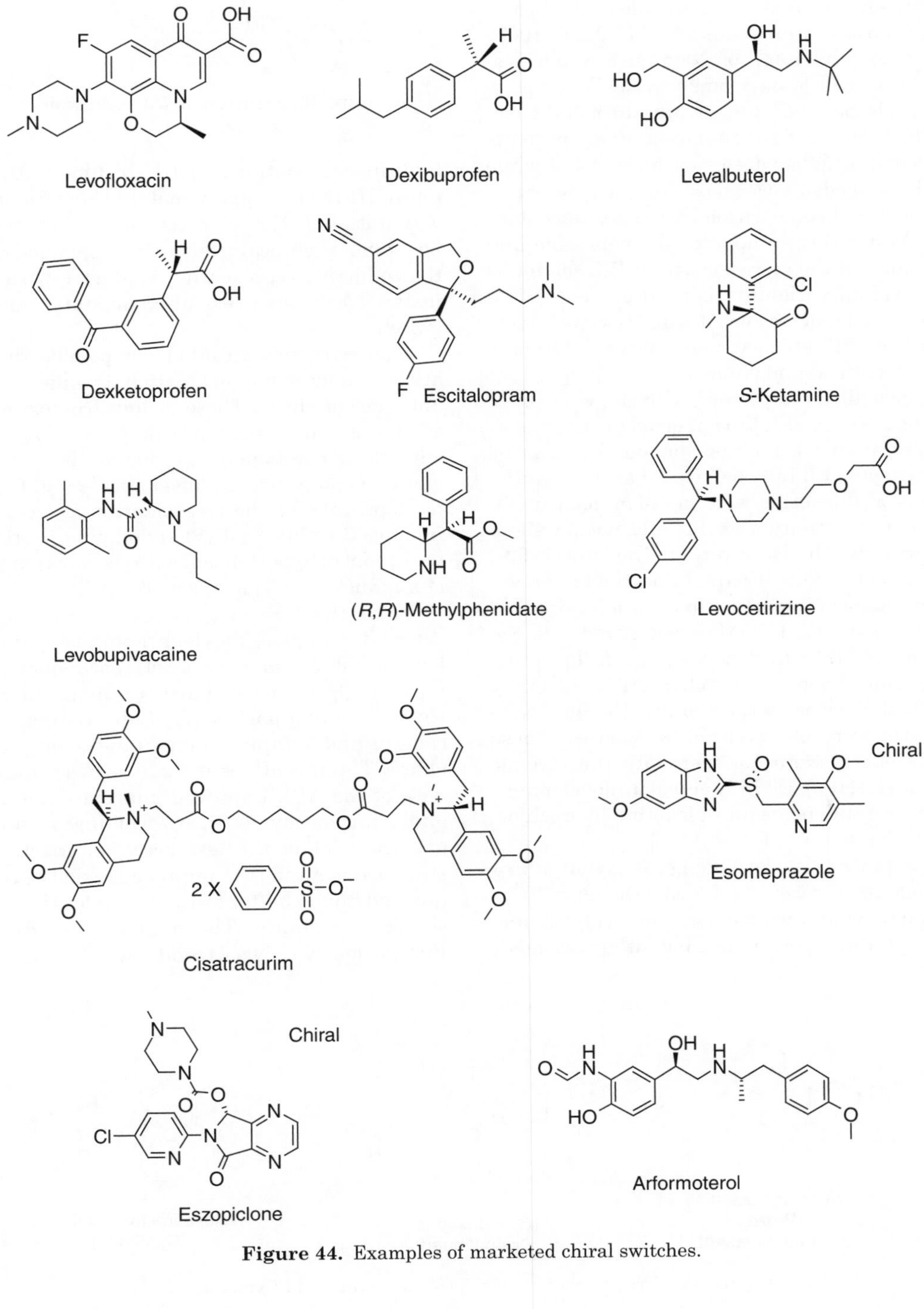

Levofloxacin

Dexibuprofen

Levalbuterol

Dexketoprofen

Escitalopram

S-Ketamine

Levobupivacaine

(*R*,*R*)-Methylphenidate

Levocetirizine

Cisatracurim

Esomeprazole

Eszopiclone

Arformoterol

Figure 44. Examples of marketed chiral switches.

maintain worldwide sales for both drugs of $6.6 billion in 2002. In 2007, the U.S. sales of Nexium™ were $4.4 billion while the U.S. sales of Prilosec™ were a mere $0.17 billion [100].

One advantage of Nexium™ (esomeprazole) over Prilosec™ (omeprazole) is that omeprazole has variable PK due to metabolism differences in the population. Some patients needed a higher dose to achieve relief while others needed lower doses due to slower metabolism. Esomeprazole has more consistent bioavailability compared to omeprazole and because of the more consistent PK, the inter-individual variability is reduced. There is a solid body of science behind esomeprazole such as PK, efficacy, and species differences in preclinical and clinical species [99].

Not all attempts of a chiral switch have been effective. Eli Lilly aborted an attempt at a chiral switch for the compound fluoxetine (Prozac™). Eli Lilly believed that the enantiomers of fluoxetine were not significantly different in activity and they did not pursue a chiral switch. However, in the mid 1990s, Sepracor obtained patents on the single enantiomers of fluoxetine to treat migraine and depression. Eli Lilly then partnered with Sepracor in 1998 to co-develop the *R*-fluoxetine for depression. In October 2000, Eli Lilly halted the deal with Sepracor because side effects were observed at high doses. These side effects were not seen with the racemic material [100]. Eli Lilly ended up developing a better antidepressant duloxetine hydrochloride (Cymbalta™, Fig. 45) [78].

Sepracor was developing (*S*)-oxybutynin, a single-isomer version of Alza's Ditropan™ (racemic oxybutynin), a muscarinic acetylcholine receptor antagonist, as a potential treatment for urinary incontinence [101]. In June 2001, a phase III trial of approximately 850 patients was initiated. However, as per a recent report [78], its clinical trials were suspended by the company, representing another reported instance of an unsuccessful example of a chiral switch.

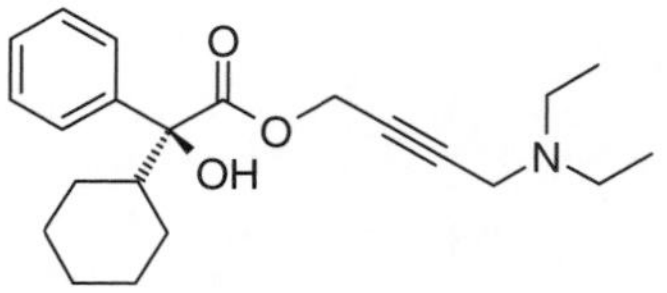

O
O
OH
N

Figure 46. Structure of (*S*)-oxybutynin.

There are several other compounds that are still undergoing evaluation as single enantiomer products. These include (*S*)-doxazosin for use in benign prostatic hyperplasia with the advantage of a reduction in orthostatic hypotension, (*S*)-lansoprazole and (−)-pantoprazole for the treatment of gastro-oesophageal reflux, and (*S*)-amlodipine for the treatment of hypertension with the advantage of a reduction in side effects [76,93].

From the foregoing discussion and the selected examples, it is clear that the impetus for the development of stereochemically pure, single enantiomer drugs is simply more than extending patent life of an existing racemate and retaining or gaining a market share. The incentives include reduced total dose of the API, improved pharmacokinetic and pharmacodynamic, overcoming off-target activities of the less active or inactive stereoisomer and thus improved toxicity profile, and finally better prospects for less inter-subject variability. The data presented in Fig. 47 clearly shows a trend toward enantio-

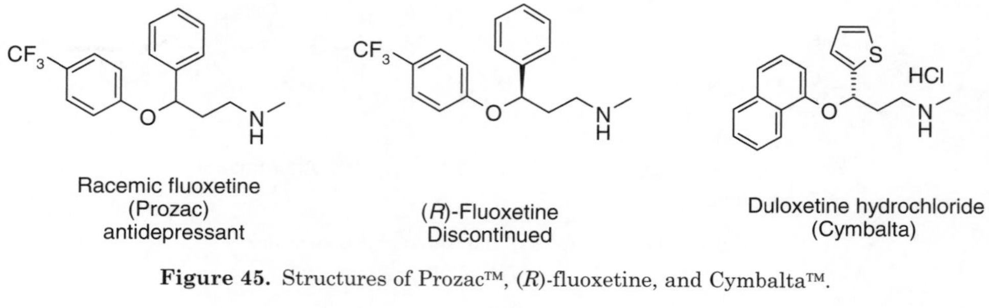

CF_3
O
N
H
Racemic fluoxetine
(Prozac)
antidepressant

CF_3
O
N
H
(*R*)-Fluoxetine
Discontinued

S
HCl
O
N
H
Duloxetine hydrochloride
(Cymbalta)

Figure 45. Structures of Prozac™, (*R*)-fluoxetine, and Cymbalta™.

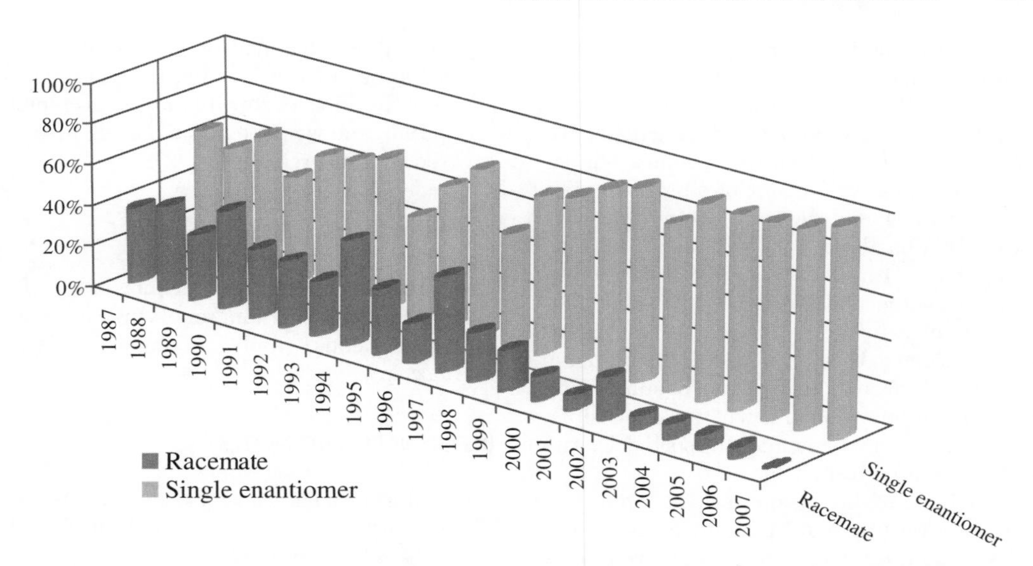

Figure 47. Annual distribution of approved racemic versus single enantiomeric NCEs (1987–2007).

pure new chemical entities (NCEs) launched over the past two decades, with >95% of the chiral NCEs launched as single enantiomers since 2003 [102]. Figure 48 shows the data for all NCEs, including achiral drugs launched from 1989 to 2007.

10. SUMMARY AND FUTURE CONSIDERATIONS

Biological systems are chiral and differences in efficacy, metabolism, adsorption, excretion, and toxicity have been observed for chiral

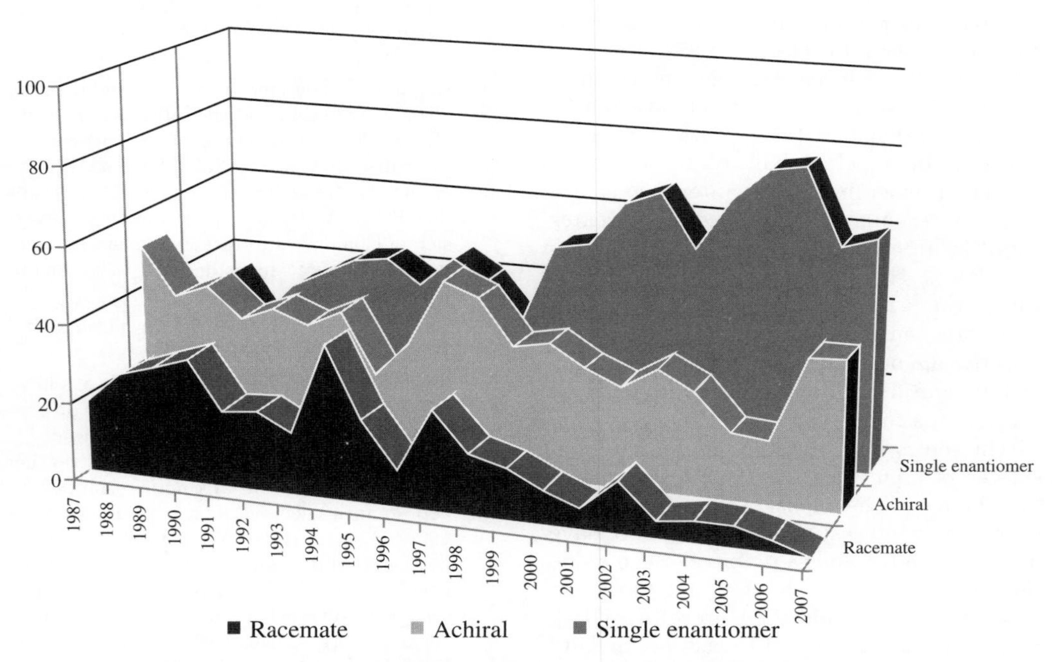

Figure 48. Annual world distribution of all approved NCEs (1987–2007).

drugs. Introduction of chirality as a mean to improve selectivity (on-target and off-target) enhances PK characteristics, and safety is being increasingly employed during lead optimization. The use of chiral switches has focused on the development of single enantiomers of marketed racemic drugs. In some cases, this has been very successful, such as omeprazole. Other attempts at chiral switches have not succeeded. The FDA and other regulatory agencies do not require single enantiomer drugs to be submitted but the industry has moved toward the direction of submitting only single enantiomers, whenever possible. This trend is evident and supported by the data presented in Fig. 47.

The focus of discussion in this chapter has been on profiling and highlighting representative examples with enantioselective effects of small molecules on biological activities at a wide range of molecular targets under *in vitro* and *in vivo* environments. However, chiral differentiation extends well into biological targets itself. It is well known that the complex biological effects are exerted by macromolecular–macromolecular (protein–protein, protein–DNA, RNA, etc.) interactions, which often involves conformation changes. These in turn have the potential to create a new environment for enantioselective recognition, or alternatively conformational changes could be induced by stereoselectivity interaction(s) with a chiral ligand [103]. With the recent advances in structure biology, these macromolecular–macromolecular interactions are being increasingly well defined in a greater detail at an atomic level (high-resolution X-ray data). These changes in environment within the active (and allosteric) site of proteins are being exploited for the design of selective ligands and these environments may include chiral differentiation and should be covered in a future chapter.

The topics and subtopics chosen are considered of high interest to practicing medicinal chemists. The impact of chirality on plant biology and interest to wildlife habitat have not been covered and were considered outside the scope of current chapter, but these are described in the published literature [104]. Similarly, environmental impact of chirality has been touched upon in this chapter and this topic could be expanded in a future chapter. There has been increasing numbers of reports showing the effect of chirality at the proteome level [105], and we hope this topic will also be covered in a future chapter.

ACKNOWLEDGMENTS

The authors gratefully acknowledge Dr Wayne Zeller and Mr. Emanuel Onua for helpful discussions and proofreading this manuscript.

REFERENCES AND NOTES

1. Smith M, March J. March's Advanced Organic Chemistry. 5th ed. New York: John Wiley & Sons; 2001. p 125–217.
2. Carey FA, Sundberg RJ. Advanced Organic Chemistry. 4th ed. New York: Plenum; 2000. p 75–122.
3. (a) Eliel EL, Wilen SH, Mander LN. Stereochemistry of Organic Compounds. New York: Wiley; 1994;(b) Eliel EL, Wilen SH, Doyle MP. Basic Organic Stereochemistry. New York: Wiley; 2001.
4. Wolfe S, Campbell JR. Atropisomerism of cyclohexane compounds. Chem. Commun. (London) 1967; 874–876.
5. (a) Eriksson T, Bjorkman S, Roth B, Fyge A, Hoglund P. The enantiomers of thalidomide: blood distribution and the influence of serum albumin on chiral inversion and hydrolysis. Chirality 1998;10:223–228.(b) Reist M, Roy-deVos M, Montseny J-P, Mayer JM, Carrupt P-A, Berger Y, Testa B. Very slow chiral inversion of clopidogrel in rats: a pharmacokinetic and mechanistic investigation. Drug Metab Dispos 2000;28:1405–1410.
6. Oki M. Recent advances in atropisomerism. Top Stereochem 1998;14:1–81.
7. Several recent review articles and books have covered this area. Some selected and diverse examples related to this topic from recent literature include (a) Rakels J, Wolff A, Straathof A, Heijnen J. Sequential kinetic resolution by two enantioselective enzymes. Biocatalysis 1994;9:31–47, (1994). (b) Kielbasinski P, Zwanenburg B, Damen T, Wieczorek M, Majzner W, Bujacz G. Kinetic resolution of racemic cyclic sulfoxides using hydrolytic enzymes. European J Org Chem 1999;10:2573–2578. (c) Nagaoka H. Chiral resolution function with

immobilized food proteins. Biotechnol Prog 19,1149–1155, (2003). (d) Gardimalla H, Mandal D, Stevens P, Yen M, Gao Y. Superparamagnetic nanoparticle-supported enzymatic resolution of racemic carboxylates. Chem Commun 2005;37(3):4432–4434 (e) Dlugy C, Wolfson A. Lipase catalyse glycerolysis for kinetic resolution of racemates. Bioprocess Biosyst Eng 2007;30(5):327–330 (f) Overmeyer A, Schrader-Lippelt S, Kasche V, Brunner G. Lipase-catalysed kinetic resolution of racemates at temperatures from 40 °C to 160 °C in supercritical CO_2. Biotechnol Lett 1992;21(1): 65–69. (g) Andrade LH, Barcellos T. Lipase-catalyzed highly enantioselective kinetic resolution of boron-containing chiral alcohols. Org Lett 2009;11(14):3052–3055. (h) Ogorevc M, Faber K. Biocatalytic resolution of sterically hindered alcohols, carboxylic acids and esters containing fully substituted chiral centers by hydrolytic enzymes. J Mol Cat B Enzym 2000;10(4):357–376. *Some recent books* are (i) Asymmetric Organic Synthesis with Enzymes, Editors: Gotor V.; Alfonso I.; García-Urdiales E, eds 2008 Wiley-VCH, Weinheim(j) Immobilization of Enzymes and Cells, Methods in Biotechnology. Editor: Guisan J.M., Humana Press 2006. (k) Multi-Step Enzyme Catalysis: Biotransformations and Chemoenzymatic Synthesis. Garcia-Junceda E.; ed., 2008 Wiley-VCH, Weinheim. (l) Industrial Enzymes: Structure, Function and Applications. Polaina J, MacCabe A.P.; eds, 2007 Springer, Netherlands.

8. Easson LH, Stedman E. Studies on the relationship between chemical constitution and physiological action. Biochem J 1933; 27:1257–66.
9. Burke D, Henderson DJ. Chirality: a blueprint for the future. Br J Anesth 2002; 88:563–576.
10. McConalthy J, Owens MJ. Stereochemistry in drug action. J Clin Psychiatry—Prim Care Companion. 2003;5:70–73.
11. Bentley R, Chemical methods for the investigation of stereochemical problems in biology. In: Stereochemistry. 49–112. Gamm CT,Ed., 1982 Elsevier Biomedical Press, New York.
12. Sundaresa V., Abrol R., Protein Sci 2002;11:1330–1339.
13. Mesecar AD, Koshland DE. A new model for protein stereospecificity. Nature 2000; 403:614–615.
14. Kato K, Aoki Y, Sugahara M, Tohnai N, Sada K, Miyata M. Interpretation of enantioresolution in nordeoxycholic acid channels based on the four-location model. Chirality 2003;15:53–59.
15. Pfeiffer CC. Optical isomerism and pharmacological action, a generalization. Science 1956;124:29–31.
16. Barlow R. Enantiomers: how valid is Pfeiffer's rule? Trends Pharmacol Sci 1990;11:148–50.
17. Gualtieri F. Pfeiffer's rule OK?. Trends Pharmacol Sci 1990;11:315–316.
18. Bunnage ME, Mathias JP, Wood A, Miller D, Street S. Highly potent and selective chiral inhibitors of PDE5: an illustration of Pfeiffer's rule. Bioorg Med Chem Lett 2008; 18:6033–6036.
19. Bichlmaier I, Siiskonen A, Finel M, Yli-Kauhaluoma J. Stereochemical sensitivity of the human UDP-glucuronosyltransferases 2B7 and 2B17. J Med Chem 2006;49:1818–1827.
20. Rapposelli S, Calderone V, Cirilli R, Digiacomo M, Faggi C, Torre F, Manganaro M, Martelli A, Testai L. Enantioselectivity in cardioprotection induced by (*S*)-(−)-2,2-dimethyl-*N*-(4′-acetamido-benzyl)-4-spiromorpholone-chromane. J Med Chem 2009; 5:1477–1480.
21. Kern C, Meyer T, Droux S, Schollmeyer D, Miculka C. Synthesis and pharmacological characterization of 2-adrenergic agonist enantiomers: zilpaterol. J Med Chem 2009; 5:1773–1777.
22. Carl JC, Myers TR, Kirchner HL, Kercsmar CM. Comparison of racemic albuterol and levalbuterol for treatment of acute asthma. J Pediatr 2003;143:731–736.
23. Kopin AS, McBride EW, Chen C, Freidinger RF, Chen D, Zhao CM, Beinborn M. Identification of a series of CCK-2 receptor nonpeptide agonists: sensitivity to stereochemistry and a receptor point mutation. Proc Natl Acad Sci USA 2003;100:5525–5530.
24. Beinborn M, Quinn SM, Kopin AS. Minor modifications of a cholecystokinin-B/gastrin receptor non-peptide antagonist confer a broad spectrum of functional properties. J Biol Chem 1998;273:14146–14151.
25. Hughes J, Dockray GJ, Hill D, Garcia L, Pritchard MC, Forster E, Toescu E, Woodruff G, Horwell DC. Characterization of novel peptoid agonists for the CCK-A receptor. Regul Pept 1996;65:15–21.
26. Jansson CC, Kukkonen JP, Nasman J, Huifang G, Wurster S, Virtanen R, Savola JM, Cockcroft V, Akerman KE. Protean agonism

at α_{2A}-adrenoceptors. Mol Pharmacol 1998; 53:963–968.

27. Cole DC, Lennox WJ, Stock JR, Ellingboe JW, Mazandarani H, Smith DL, Zhang G, Tawac GJ, Schechter LE. Conformationally constrained N1-arylsulfonyltryptamine derivatives as 5-HT6 receptor antagonists. Bioorg Med Chem Lett 2005;15:4780–4785.
28. Artigas P, Ferreira G, Reyes N, Brum G, Pizarro G. Effects of the enantiomers of BayK 8644 on the charge movement of L-type Ca channels in guinea-pig ventricular myocytes. J Membr Biol 2003;193:215–227.
29. Gerhard F, Martin B, Matthias S, Guenter T. The optical isomers of the 1,4-dihydropyridine BAY K 8644 show opposite effects on calcium channels. Eur J Pharmacol 1985; 114:223–6.
30. Ravens U, Schoepper H. Opposite cardiac actions of the enantiomers of Bay K 8644 at different membrane potentials in guinea pig papillary muscles. Naunyn Schmiedebergs Arch Pharmacol 1990;341:232–239.
31. Jiang JJ, Ghoreschi K, Deflorian F, Chen Z, Perreira M, Pesu M, Smith J, Nguyen D, Liu EH, Leister W, Costanzi S, Shea J, Thomas CJ. Examining the chirality, conformation and selective kinase inhibition of 3-((3R,4R)-4-methyl-3-(methyl(7H-pyrrolo[2,3-d]pyrimidin-4-yl)amino)piperidin-1-yl)-3-oxopropanenitrile (CP-690,550). J Med Chem 2008; 51:8012–8018.
32. Feifel R, Waner-Roder M, Strohmann C, Tacke R, Waelbroeck M, Christophe J, Mutschler E, Lambrecht G. Stereoselective inhibition of muscarinic receptor subtypes by the enantiomers of hexahydro-difenidol and acetylenic analogues. Br J Pharmacol 1990; 99:455–460.
33. Lloyd-Williams P, Giralt E. Atropisomerism, biphenyls and the Suzuki coupling: peptide antibiotics. Chem Soc Rev 2001;30:145–157.
34. Zbaida S, Du Y, Shannon D, Laudicina D, Thonoor CM, Ng K, Blumenkrantz N, Patrick JE, Cayen MN, Friary R, Seidl V, Chan TM, Pramanik B, Spangler M, McPhail AT. *In vitro* metabolism of 10-(3-chlorophenyl)-6,8,9,10-tetrahydrobenzo[*b*][1,8]naphthyridin-5(7*H*)-one, a topical antipsoriatic agent. Use of precision-cut rat, dog, monkey and human liver slices, and chemical synthesis of metabolites. Biopharm Drug Dispos 1998; 19:315–332.
35. (a) Sanna E, Pau D, Tuveri F, Massa F, Maciocco E, Acquas C, Floris C, Fontana SN, Maira G, Biggio G. Molecular and neurochemical evaluation of the effects of etizolam on GABAA receptors under normal and stress conditions. Arzneimittelforschung 1999, 49:88–95. (b) Marubayashi N, Ogawa T, Moriwaki M, Haratake M. Atropisomerism in 4-(2-thienyl)-4*H*-1,2,4-triazole derivatives. Chem Pharm Bull 1992,40:220–223.
36. Welch WM, Ewing F, Huang J, Menniti F, Pagnozzi MJ, Kelly K, Seymour PA, Guanowsky V, Guhan S, Guinn MR, Critchett D, Lazzaro J, Ganong AH, DeVries KM, Staigers TL, Chenard BL. Atropisomeric quinazolin-4-one derivatives are potent noncompetitive α-amino-3-hydroxy-5-methyl-4-isoxazolepropionic acid (AMPA) receptor antagonists. Bioorg Med Chem Lett 2001;11:177–181.
37. Zhou YS, Tay LK, Hughes D, Donahue S. Simulation of the impact of atropisomer interconversion on plasma exposure of atropisomers of an endothelin receptor antagonist. J Clin Pharmacol 2004;44:680–688.
38. Parker JS, Smith NA, Welham MJ, Moss WO. A new approach to the rapid parallel development of four neurokinin antagonists. Part 5. Preparation of ZM374979 cyanoacid and selective crystallization of ZM374979 atropisomers. Org Process Res Dev 2004;8:45–50.
39. Albert JS, Ohnmacht C, Bernstein PR, Rumsey WL, Aharony D, Alelyunas Y, Russell DJ, Potts W, Sherwood SA, Shen L, Dedinas RF, Palmer WE, Russell K. Structural analysis and optimization of NK1 receptor antagonists through modulation of atropisomer interconversion properties. J Med Chem 2004; 47:519–529.
40. (a) Porter J, Payne A, de Candole B, Ford D, Hutchinson B, Trevitt G, Turner J, Edwards C, Watkins C, Whitcombe I, Davis J, Stubberfield C. Tetrahydroisoquinoline amide substituted phenyl pyrazoles as selective Bcl-2 inhibitors. Bioorg Med Chem Lett 19:230–233 2009. (b) Payne A, Porter J, Watkins I, de Candole B, Ford D, Hutchinson B, Tavernier J, Davis J, Stubberfield C. *In vitro* profiling of UCB1350883, a novel, highly selective antagonist of Bcl-2. American Association for Cancer Research NCI-EORTC, International Congress (Abs B297), October 24, 2007.
41. (a) Tucci FC, Zhu Y-F, Struthers RS, Guo Z, Gross TD, Rowbottom MR, Acevedo O, Gao G, Saunders J, Xie Q, Reinhart GJ, Liu X-J, Ling N, Bonneville AKL, Chen TK, Bozigian H, Chen C. 3-[(2*R*)-Amino-2-phenylethyl]-1-(2,6-difluorobenzyl)-5-(2-fluoro-3-methoxyphenyl)-6-methylpyrimidin-2,4-dione (NBI 42902) as a

potent and orally active antagonist of the human gonadotropin-releasing hormone receptor—design, synthesis and *in vitro* and *in vivo* characterization. J Med Chem 2005;48:1169–1178. (b) Tucci FC, Hu T, Mesleh MF, Bokser A, Allsopp E, Gross TD, Guo Z, Zhu Y-F, Struthers S, Ling N, Chen C. Atropisomeric property of 1-(2,6-difluorobenzyl)-3-[(2*R*)-amino-2-phenethyl]-5-(2-fluoro-3-methoxyphenyl)-6-methyluracil. Chirality 2005;17: 559–564.

42. Vrudhula VM, Dasgupta B, Qian-Cutrone J, Kozlowski ES, Boissard CG, Dworetzky SI, Wu D, Gao Q, Kimura R, Gribkoff VK, Starrett JE, J Med Chem 2007;50:1050–1057.
43. Li P, Evans CD, Wu Y, Cao B, Hamel E, Joulli M. Evolution of the total syntheses of ustiloxin natural products and their analogues. J Am Chem Soc 2008;130:2351–2364.
44. Rich DH, Bhatnager PK, Jasensky RD, Steele JA, Uchytil TF, Durbin RD. Two conformations of the cyclic tetrapeptide. [D-MeAla'l]-tentoxin have different biological activities. Bioorg Chem 1978;7:207–214.
45. Gal J. Chirality in Drug Research. Edited by E. Francotte and W. Lindner Wiley-VCH Verlag GmbH & Co. KGaA, Weinheim, 2006.
46. Pham-Huy C, Sahui-Gnassi A, Saada V, Gramond JP, Galons H, Ellouk-Achard S, Levresse V, Fompeydie D, Claude JR. Microassay of propranolol enantiomers and conjugates in human plasma and urine by HPLC after chiral derivatization for pharmacokinetic study. J Pharm Biomed Anal 1994;12:1189–1198.
47. Pasteur L, Compt Rend Acad Sci 1860; 51:298–299.
48. Lin JH, Lu AYH. Role of pharmacokinetics and metabolism in drug discovery and development. Pharmacol Rev 1997;49:403–449.
49. Somogyi A, Bochner F, Foster D. Inside the isomers: the tale of chiral switches. Aust Prescr 2004;27:47–49.
50. Siccardi D, Kandalaft L, Gumbleton M, Mcguigan C. Stereoselective and concentration-dependent polarized epithelial permeability of a series of phosphoramidate triester prodrugs of d4T: an *in vitro* study in Caco-2 and Madin-Darby canine kidney cell monolayers. J. Pharamcol Exp Ther 2003; 307:1112–1119.
51. He Y, Zeng S. Determination of the stereoselectivity of chiral drug transport across Caco-2 cell monolayers. Chirality 2006;18:64–69.
52. Kommmuru TR, Khan MA, Reddy IK. Effect of chiral enhancers on the permeability of optically active and racemic metoprolol across hairless mouse skin. Chirality 1999; 11:536–540.
53. Mehvar R, Brocks DR, Vakily M. Impact of stereoselectivity on the pharmacokinetics and pharmacodynamics of antiarrhymic drugs. Clin Pharmacokinet 2002;41:533–558.
54. Stoschitzky K, Lindner W, Zernig G. Racemic beta-blockers—fixed combinations of different drugs. J Clin Basic Cardiol 1998;1:15–19.
55. Mehvar R, Brocks DR, Stereospecific pharmacokinetics and pharmacodynamics: cardiovascular drugs In: Chirality in Drug Design and Development. Reddy IK, and Mehvar R,Editors, New York, Marcel Dekker, 2004.
56. http://www.wyeth.com prescribing information for Inderal.
57. Rentsch KM. The importance of stereoselective determination of drugs in the clinical laboratory. J Biochem Biophys Methods 2002; 54:1–9.
58. Pham-Huy C, Chikhi-Chorfi N, Galons H, Sadeg N. Enantioselective HPLC determination of methadone enantiomers and its major metabolite in human biologic fluids using a new derivatized cyclodextrin-bonded phase. J Chromatogr B Biomed Sci Appl 1997; 700:155–163.
59. Chikhi-Chorfi N, Pham-Huy C, Sandouk P, Sadeg N, Pecquart C, Massicot F, Galons H. Scherrmann JM, J. M. Warnet JM, Claude JR. Pharmacokinetic profiles of methadone enantiomers and its major metabolite in serum, saliva and urine of treated patients. Toxicol Lett 1998;Supl.1(95):103.
60. Eap CB, Finkbeiner T, Gastpar M, Scherbaum N. Replacement of (*R*)-methadone by a double dose of (*R*,*S*)-methadone in addicts: interindividual variability of the (*R*)/(*S*) ratios and evidence of adaptive changes in methadone pharmacokinetics. Eur J Clin Pharmacol 1996;50:385–389.
61. Chikhi-Chorfi N, Pham-Huy C, Galons H, Manuel N. Rapid determination of methadone and its major metabolite in biological fluids by gas–liquid chromatography with thermoionic detection for maintenance treatment of opiate addicts. J Chromat B 1998;718:278–284.
62. Nguyen LA, He H, Pham-Huy C. Chiral drugs: an overview. Int J Biomed Sci 2006; 2:85–100.
63. Pham-Huy C, Vilain-Pautet G, He H, Chikhi-Chorfi N, Galons H, Thevenin M, Claude JR, Warnet JM. Separation of oxazepam, lorazepam and temazepam enantiomers by HPLC on a derivatized cyclodextrin-bonded phase: application to the determination in plasma. J Biochem Biophys Methods 2002;54:287–299.

64. Fura A, Yue-Zhong S, Zhu M, Hanson RL, Vikram Roongta V, Griffith Humphreys W. Discovering drugs through biological transformation: role of pharmacologically active metabolites in drug discovery. J Med Chem 2004;47:4339–4350.
65. Brooks C, Stewart A, Basha A, Bhatia P, Ratajczyk JD, Martin JG, Craig R, Kolasa T, Bouska J, Lanni C, Harris R, Malo P, Carter G, Bell R. (*R*)-(1)-*N*-[3-[5-[(4-Fluorophenyl) methyl]-2-thienyl]-1-methyl-2-propynyl]-*N*-hydroxyurea (ABT-761), a second generation 5-lipoxygenase inhibitor. J Med Chem 1995; 38:4768–4775.
66. Tokuma Y, Fujiwara T, Niwa T, Hashimoto T, Noguchi H. Stereoselective disposition of nilvadipine, a new dihydropyridine calcium antagonist, in the rat and dog. Res Commun Chem Pathol Pharmacol 1989;63 (2): 249–262.
67. (a) Bhatti MM, Foster RT. Pharmacokinetics of the enantiomers of verapamil after intravenous and oral administration of racemic verapamil in a rat model. Biopharm Drug Dispos 1997,18:387–396. (b) Bai SA, Lankford SM, Johnson LM. Pharmacokinetics of the enantiomers of verapamil in the dog. Chirality 1993,5:436–442
68. Mehvar R, Brocks DR. Stereospecific pharmaceutics and pharmacodynamics of beta-adrenergic blockers in humans. J Pharm Pharmaceut Sci 2001;4(2): 185–200.
69. Eap CB, Crettol S, Rougier JS, Schläpfer J, Grilo S, Déglon JJ, Besson J, Croquette-Krokar M, Carrupt A, Abriel H. Stereoselective Block of hERG channel by (*S*)-methadone and QT interval prolongation in CYP2B6 slow metabolizers. Clin Pharmacol Ther 2007; 81:719–728.
70. Yang CH, Scherz MW, Anthony B, Bennett PB, Murray KT. Stereoselective interactions of the enantiomers of chromanol 293B with human voltage-gated potassium channels. J. Pharmacol Exp Ther 2000;294:955–962.
71. Gross MF, Beaudoin S, McNaughton-Smith G, Amato GS, Castle NA, Huang C, Zoua A, Yua W. Aryl sulfonamido indane inhibitors of the Kv1.5 ion channel. Bioorg Med Chem Lett 2007;17:2849–2853.
72. Su J, Tang H, McKittrick BA, Gu H, Guo T, Qian G, Burnett DA, Clader JW, Greenlee WJ, Hawes BE, O'Neill K, Spar B, Weig B, Kowalskic T, Sorotad S. Synthesis of novel bicyclo[4.1.0]heptane and bicyclo[3.1.0]hexane derivatives as melanin-concentrating hormone receptor R1 antagonists. Bioorg Med Chem Lett 2007;17:4845–4850.
73. Benedetti M, Malina J, Kasparkova J, Brabec V, Natile G. Chiral discrimination in platinum anticancer drugs. Environ Health Perspect 2002;110:779–782. and references cited therein
74. Zimmerman CL, Wu Z, Upadhyaya P, Hecht SS. Stereoselective metabolism and tissue retention in rats of the individual enantiomers of 4-(methylnitrosamino)-1-(3-pyridyl)-1-butanol (NNAL), metabolites of the tobacco-specific nitrosamine, 4-(methylnitrosamino)-1-(3-pyridyl)-1-butanone (NNK). Carcinogenesis 2004; 25:1237–1242.
75. Sewell AC, Heil M, Blieke A, Böhles M. Rapid enantiomeric differentiation of urinary metabolites in a patient with bacterial overgrowth syndrome. Clin Chem 2002;46(9): 1444–1445. and references cited therein
76. Simoni S, Klinke S, Zipper C, Angst W, Kohler H. Enantioselective metabolism of chiral 3-phenylbutyric acid, an intermediate of linear alkylbenzene degradation, by *Rhodococcus rhodochrous* PB1. Appl Environ Microbiol 1996;62(3): 749–755.
77. Bentley R. Role of sulfur chirality in the chemical processes of biology. Chem Soc Rev 2005; 34:609–624.
78. Investigational Drug Report, IDdb[3].
79. Li N, Yang L, Dai D-Z, Wang Q-J, Dai Y. Chiral separation of racemate CPU86017, an antiarrhythmic agent, produces stereoimers possessing favourable ion channel blocker and less α-adrenoceptor antagonism. Clin Exp Pharamcol Physiol 2008;35:643–650.
80. (a) Holm KA, Kindberg CG, Stobaugh JF, Slavik M, Riley CM. Stereoselective pharmacokinetics and metabolism of the enantiomers of cyclophosphamide: preliminary results in humans and rabbits. Biochem Pharmacol 1990;39:1375–1384. (b) See Ref. 13 in Holm et.al.
81. Michael MH, Andrew RL, Gavin H. Molecular complexity and its impact on the probability of finding leads for drug discovery. J Chem Inf Comput Sci 2001;41:856–864.
82. (a) Beroza P, Suto MJ. Designing chiral libraries for drug discovery. Drug Discov Today 2000,5:364–372.(b) Crossley. R. Chirality and the Biological Activity of Drugs. Boca Raton, FL: CRC Press, Inc. 1995.
83. Darrow JJ, The patentability of enantiomers: implications for the pharmaceutical industry. Stan. Tech. L. Rev. 2007.

84. Pfefferkorn JA, Choi C, Larsen SD, Auerbach B, Hutchings R, Park W, Askew V, Dillon L, Hanselman JC, Lin Z, Lu GH, Robertson A, Sekerke C, Harris MS, Pavlovsky A, Bainbridge G, Caspers N, Kowala M, Tait BD. Substituted pyrazoles as hepatoselective HMG-CoA reductase inhibitors: discovery of (3*R*,5*R*)-7-[2-(4-fluoro-phenyl)-4-isopropyl-5-(4-methyl-benzylcarbamoyl)-2*H*-pyrazol-3-yl]-3,5-dihydroxyheptanoic acid (PF-3052334) as a candidate for the treatment of hypercholesterolemia. J Med Chem 2008;51:31–45.

85. Yang L, Butora G, Jiao R, Pasternak A, Zhou C, Parsons W, Mills S, Vicario P, Ayala J, Cascieri M, MacCoss M. Discovery of 3-piperidinyl-1-cyclopentanecarboxamide as a novel scaffold for highly potent CC chemokine receptor 2 antagonists. J Med Chem 2007;50:2609–2611.

86. McKerrecher D, Allen J, Caulkett P, Donald C, Fenwick M, Grange E, Johnson K, Johnstone C, Jones C, Pike K, Rayner J, Walker R. Design of a potent, soluble glucokinase activator with excellent *in vivo* efficacy. Bioorg Med Chem Lett 2006; 2705–2709.

87. Radi M, Maga G, Alongi M, Angeli L, Samuele A, Zanoli S, Bellucci L, Tafi A, Casaluce G, Giorgi G, Armand-Ugon M, Gonzalez E, Este JA, Baltzinger M, Bec G, Dumas P, Ennifar E, Botta M. Discovery of chiral cyclopropyl dihydro-alkylthio-benzyl-oxopyrimidine (*S*-DABO) derivatives as potent HIV-1 reverse transcriptase inhibitors with high activity against clinically relevant mutants. J Med Chem 2009;52:840–851.

88. Kiselyov AS. Fragment-based discovery of a prolinol derivative (DG-051) as a novel leukotriene A4 hydrolase inhibitor for the prevention of myocardial infarction. Cambridge Healthtech Institute conference: Fragment-Based Techniques–Challenges and Solutions, San Diego, CA, April 7–9, 2009.

89. Chandraiah L, Alan P, Qin J, Xiaoli C, Tinopiwa G, Stephan WM, Thomas RW. Antitumor compounds based on a natural product consensus pharmacophore. J Med Chem 2008;51: 6220–6224.

90. Yang Q, Yang P, Qian X, Tong L. Naphthalimide intercalators with chiral amino side chains: effects of chirality on DNA binding, photodamage and antitumor cytotoxicity. Bioorg Med Chem Lett 2008;18:6210–6213.

91. Thurmond J, Butchbach M, Palomo M, Pease B, Rao M, Bedell L, Keyvan M, Pai G, Mishra R, Haraldsson M, Andresson T, Bragason G, Thosteinsdottir M, Bjornsson JM, Coovert DD, Burghes AHM, Gurney ME, Singh J. Synthesis and biological evaluation of novel 2,4-diaminoquinazoline derivatives as SMN2 promoter activators for the potential treatment of spinal muscular atrophy. J Med Chem 2008;51: 449–469.

92. Jarecki J, Chen X, Bernardino A, Coovert DD, Whitney M, Burghes A, Stack J, Pollok BA. Diverse small-molecule modulators of SMN expression found by high-throughput compound screening early leads towards a therapeutic for spinal muscular atrophy. Hum Mol Genet 2005;14:2003–2018.

93. Singh J, Salcius M, Liu S, Staker B, Mishra R, Thurmond J, Michaud G, Mattoon D, Printen J, Christensen J, Bjornsson JM, Pollok B, Kiledjian M, Stewart L, Jarecki J, Gurney ME. DcpS as a therapeutic target for spinal muscular atrophy. ACS Chem Biol 2009;3: 711–722.

94. http://www.fda.gov/cder/guidance/stereo.htm.

95. Hutt AJ, Valentova J. The chiral switch: the development of single enantiomer drugs from racemates. Acta Facult Pharm Univ Comenianae 2003;50:7–23.

96. Sahajwalla C. Regulatory considerations in drug development of stereoisomers.. Chirality in Drug Design and Development. Marcel Dekker; New York: 2004; 419–432.

97. Phuong NT, Lee BJ, Choi JK, Kang JS, Kwon K. Enantioselective pharmacokinetics of carvedilol in human volunteers. Arch Pharm Res 2004;27(9): 973–977.

98. Murakami H, Topics in Current Chemistry 2007; 273–299.

99. Rouhi AM. Chirality at work. Chem Eng News 2003;81:56–61.

100. http://www.drugs.com/top200.html.

101. Dmochowski R, Current Opin Investig Drugs 2002;3:1508–1511.

102. (a) The data for 1989-2007 was compiled from To Market, To Market in Annual Reports in Medicinal Chemistry Vols (25 to 43). (b) The data for 1989-2000 was also reported in Figure 3: Agranat I, Hava-Caner H, Caldwell J, Putting chirality to work: the strategy of chiral switches. Nat Rev Drug Discov 2002, 1:753–768.

103. Inaba Y, Yoshimoto N, Sakamaki Y, Nakabayashi M, Ikura T, Tamamura H, Ito N, Shimizu M, Yamamoto K. A new class of vitamin D analogues that induce structural rearrangement of the ligand-binding pocket of the receptor. J Med Chem 2009;52:1438–1449.

104. (a) Liu W, Ye J, Jin M. Enantioselective phytoeffects of chiral pesticides. J. Agric Food Chem 2009,57:2087–95. (b) Cai X, Liu W, Sheng G. Enantioselective degradation and ecotoxicity of the chiral herbicide diclofop in three freshwater alga cultures. J Agric Food Chem 2008,56:2139–46. (c) Liao YW, Li HX. Enantioselective synthesis and antifungal activity of optically active econazole and miconazole. Acta Pharm Sin 1993,28:22–27 (d) Norin T. Chiral chemodiversity and its role for biological activity. Some observations from studies on insect/insect and insect/plant relationships. Pure Appl Chem 1996,68:2043–2049. (e) Silversteine RM. Chirality in insect communication. J Chem Ecol 1988,14:11. (f) Hoekstra PF, Braune BM, Wong CS, Williamson M, Elkin B, Muir DCG. Profile of persistent chlorinated contaminants, including selected chiral compounds, in wolverine (*Gulo gulo*) livers from the Canadian Arctic. Chemosphere 2003,53: 551–560.

105. Some recent reports on effect of chirality and proteomics are (a) Sui J, Zhang J, Tan TL, Ching CB, Chen WN. Comparative proteomics analysis of vascular smooth muscle cells incubated with *S*- and *R*-enantiomers of atenolol using iTRAQ-coupled two-dimensional LC-MS/MS Mol Cell Proteomics 2008,7: 1007–1018. (b) Sui J, Tan TL, Zhang J, Ching CB, Chen WN. iTRAQ-coupled 2D LC-MS/MS analysis on protein profile in vascular smooth muscle cells incubated with *S*- and *R*-enantiomers of propranolol: possible role of metabolic enzymes involved in cellular anabolism and antioxidant activity. J Proteome Res 2007,6: 1643–1651. (c) Zhang J, Sui J, Ching CB, Chen WN. Protein profile in neuroblastoma cells incubated with *S*- and *R*-enantiomers of ibuprofen by iTRAQ-coupled 2-D LC-MS/MS analysis: possible action of induced proteins on Alzheimer's disease. Proteomics 2008; 8:1595–607.

ANALOG DESIGN

CAMILLE G. WERMUTH
Prestwick Chemical, Inc.,
Illkirch-Graffenstaden,
France

1. INTRODUCTION

Analog design is usually defined as the modification of a drug molecule or of any bioactive compound in order to prepare a new molecule showing chemical and biological similarity with the original model compound. Excellent descriptions of the various possibilities of bioisosteric replacement are found in Cannon's chapter "Analog Design" published in the previous edition of this book. As a rule, it is expected that the analog exhibits some improvements over the original drug molecule, but sometimes financial motivations are predominant and justify the decision to prepare an analog in order to share market parts. Analog design is a fruitful procedure, easy to practice, and very popularly employed in pharmaceutical research from the beginning. Particularly, from the second half of the twentieth century, the production of very sophisticated molecules such as steroids, prostaglandins, anticancer drugs, and antibiotics became available and considerable advances could be made in medicinal chemistry. Analog design represents two-third of all small molecule sales. Among the 29 new drugs launched in 2000, 24 were copies [1].

This continued preference of analog-based approaches has been criticized as being mainly restricted to the synthesis of "me-too compounds" [2]. In fact, the situation is more complicated as one has to take into account the chemical originality as well as the optimal ADME and toxicity profiles. Moreover, a distinction needs to be established between three different categories of analogs (see below). Finally, it has to be noticed that analog design deals with the production of *new chemical entities* while drug-repositioning targets *new uses* for old drugs. The SOSA approach occupies an intermediate situation insofar as it generates *new leads* from old drugs [3–5].

2. THE THREE CATEGORIES OF ANALOGS

By definition, two analogs should show similar structural and functional properties. Applied to drugs, this definition implies that the analog of an existing drug molecule shares chemical and therapeutic similarities with the original compound. This represents the usual objective for medicinal chemists. However, two chemical compounds having the same structure can have totally different pharmacological profiles and, vice versa, compounds having similar pharmacological activities can differ in their structure. Thus, three categories of analogs can be foreseen:

(1) Analogs possessing chemical and pharmacological similarities,
(2) Analogs possessing only chemical similarities,
(3) Compounds chemically different, but displaying similar pharmacological properties.

The first class of analogs, simultaneously having chemical and pharmacological similarities, can be considered as "*direct*" *analogs* [5,6]. These analogs correspond to the category of drugs often referred to as "me-too drugs." Usually, they are an advantage to the consumer. In the case of drugs, this means that they provide an advance in therapeutic benefit.

The second class, made of "*structural analogs*," contains compounds originally prepared as close and patentable analogs of a novel lead, but for which the biological assays revealed totally unexpected pharmacological properties. Observation of the new activity can be purely fortuitous but can also result from a planned systematic investigation.

A historical example of the emergence of such a new activity is provided by the discovery of the antibacterial sulfonylureas prepared by Rhône-Poulenc scientists for use in typhoid cases were found in the clinic as potent hypoglycemic agents and opened the route to orally active antidiabetic drugs such as carbutamide, tolbutamide, chlorpropamide, and glybenclamide. Similarly, the antidepressant imipramine was originally designed as an

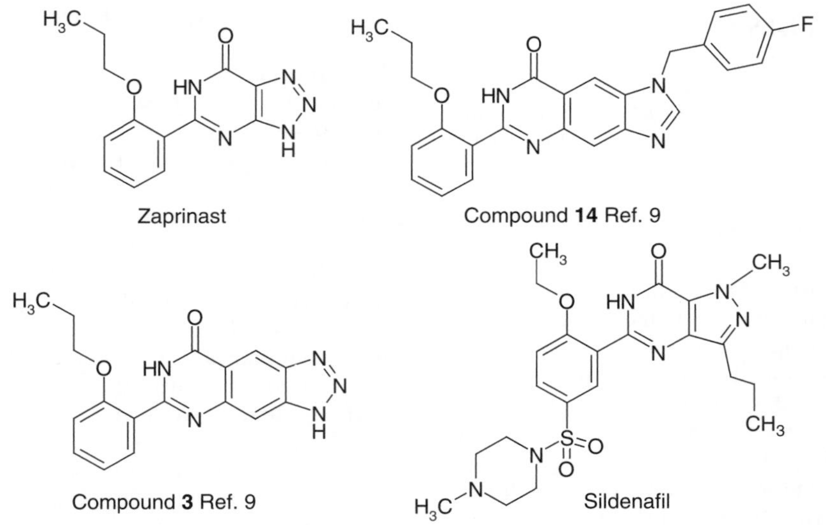

Zaprinast

Compound **14** Ref. 9

Compound **3** Ref. 9

Sildenafil

Figure 1. The activity of sildenafil on male erectile dysfunction was discovered by chance. The benzologs 3 and 14 can be considered as either direct analogs (similar structure and similar functionality) or functional analogs (similar functionality but different structure).

analog of the potent neuroleptic drug chlorpromazine. A more recent example is found in the PDE5 inhibitor sildenafil (Fig. 1).

In continuation of earlier observations with phosphodiesterase inhibitors such as erythro-9-(2-hydroxy-3-nonyl)adenine (EHNA) [7] and zaprinast [8], sildenafil was designed as a hypotensive drug [9]. Surprisingly, it exhibited activity on male erectile dysfunction during the clinical studies. Using the prototypical PDE5 inhibitor zaprinast (Fig. 1) as a template-directed screening identified "compound **3**," the benzolog of zaprinast, as a moderately active but nonselective lead (PDE5 IC_{50} sildenafil 1.6 nM, compound **3** = 44 nM) [9]. Further modifications of compound **3** yielded the potent and selective benzolog **14** (PDE5 $IC_{50} = 0.48$ nM) that can be considered as a direct analog of sildenafil (similar structure and similar functionality), or as a functional analog (similar functionality but different structure).

For the third class of analogs, chemical similarity is not observed; however, they share common biological properties. We propose the term "*functional analogs*" for such compounds. Examples are the neuroleptics chlorpromazine and haloperidol, the tranquilizers diazepam and zopiclone, and the dopamine agonists dopamine and pramipexole. Despite having totally different chemical structures, they show similar affinities for the dopamine and the benzodiazepine receptors, respectively. The design of such drugs is presently facilitated thanks to virtual screening of large libraries of diverse structures.

3. INSPIRATION SOURCES FOR ANALOG DESIGN

The synthesis of new analogs can originate from *natural sources* such as plants and animals. A second inspiration source considers already existing drugs as models for the synthesis of improved followers. Sometimes drug metabolites are excellent starting compounds.

3.1. Drugs from Natural Sources

Excellent inspiration source investigated were the *natural compounds*, particularly the alkaloids. An example is given by the acetylcholinesterase inhibitors neostigmine and rivastigmine that were clearly derived from the natural alkaloid physostigmine (Fig. 2).

Similarly, natural compounds such as cocaine, morphine, or quinine gave rise to the production of an impressive number of analogs. In general, the synthetic copies have

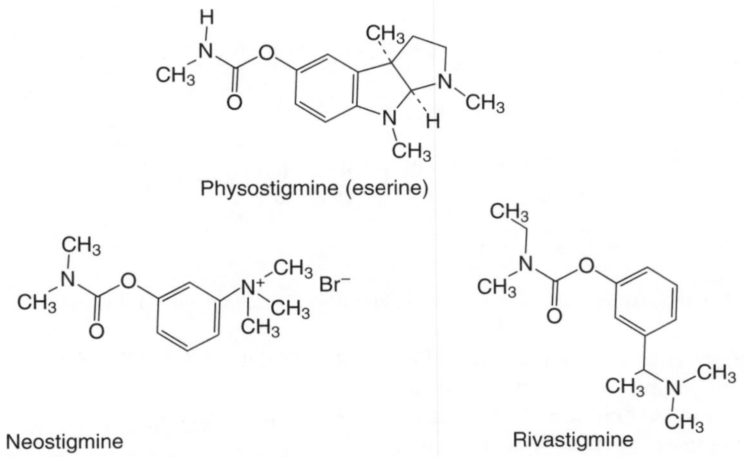

Physostigmine (eserine)

Neostigmine

Rivastigmine

Figure 2. The critical phenylcarbamic structure present in the natural compound physostigmine is conserved in the synthetic analogs neostigmine and rivastigmine.

simpler chemical structures than the original compound. Often, the main active component of a plant extract is accompanied by several satellite structures, which may also be of interest. Opium poppy, for example, contains over 40 opium alkaloids, including codeine (about 1%), morphine (up to 20%), narcotine (about 5%), and papaverine (about 1%).

Endogenous substances such as neurotransmitters, hormones, second messengers, and enzyme cofactors represent another source of model compounds (see Fig. 3). Many

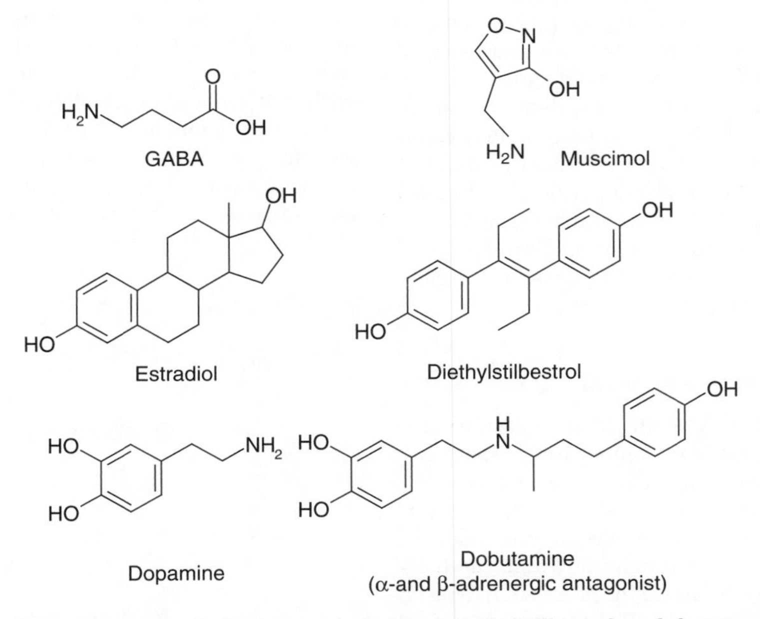

Figure 3. The isoxazole muscimol, the estrogenic diphenol diethylstilbestrol, and the α- and β-adrenergic antagonist dobutamine derive, respectively, from the endogenous compounds GABA, estradiol, and dopamine.

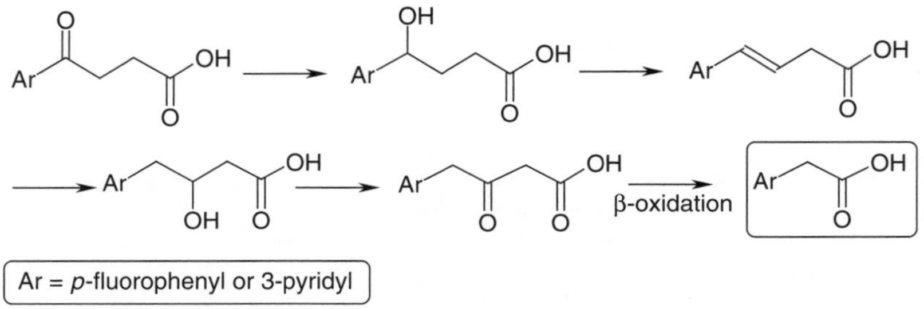

Ar = *p*-fluorophenyl or 3-pyridyl

Figure 4. Progressive metabolic degradation of β-aroylpropionic acids into arylacetic acids [10].

of these substances are very potent but suffer from insufficient chemical or metabolic stability. As a consequence, their administration by the oral route is useless. Analog design aims then to overcome these drawbacks.

3.2. Analogs of Metabolites

A particular aspect related to natural products consists of the study of their different metabolites. Besides drug inactivation, drug metabolism can generate pharmacologically active metabolites. These compounds may have superior ADME and safety profiles compared to their respective parent molecules. Current metabolic transformations are hydroxylation, glucuronidation, sulfation, and *O*- and *N*-dealkylation. Examples of marketed drug metabolites are etaminophen, oxyphenbutazone, oxazepam, cetirizine, and fexofenadine. In some cases, the metabolite serves itself as lead compound to produce further metabolites as illustrated for nicotine and haloperidol [10] (Fig. 4). For the first time, nicotine and haloperidol are oxidized into the corresponding β-aroylpropionic acid, which undergo an oxidoreductive cascade yielding finally arylacetic acids.

3.3. Analogs of Existing Drugs

Original and successful drug molecules attract the attention of competitors, especially when the molecule is of the "blockbuster" type, in other words when it realizes billion dollar sales. "Fast-followers" are compounds developed during the prelaunch years and "me-too" drugs are introduced when the original drug is already on the market. The histamine receptor H_2 antagonist ranitidine is a typical me-too copy of cimetidine (Fig. 5).

4. PRODUCTION OF ANALOGS

4.1. Direct Analogs

The design of *direct analogs* starting from a prototype drug (a "pioneer" drug [11]) is a current practice in the pharmaceutical industry. Direct analogs are also the most wanted drug candidates. Among the several classes that reached the market, the "triptans" (Fig. 6) represent a recent example of direct analog design [12].

Generally, the design of direct analogs involves simple molecular modifications relevant to bioisosterism such as the synthesis of homologs, vinylogs, isosteres, positional isomers, optical isomers, modified ring systems, and twin drugs (homodimers) [13,14]. In some cases, bioisosteric changes generate analogs with reversed pharmacological profile (agonists → antagonists). Over the time, many wordings such as isosterism, bioisosterism, classical isosterism, ring equivalents, and so on, were used in relationship to the isosterism

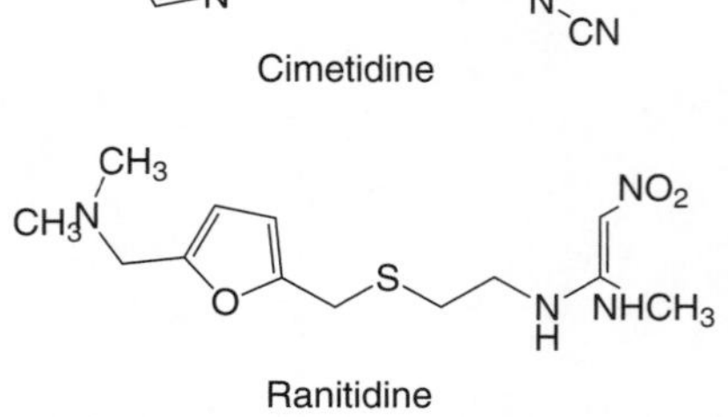

Cimetidine

Ranitidine

Figure 5. Ranitidine represents a typical "me-too" copy of cimetidine.

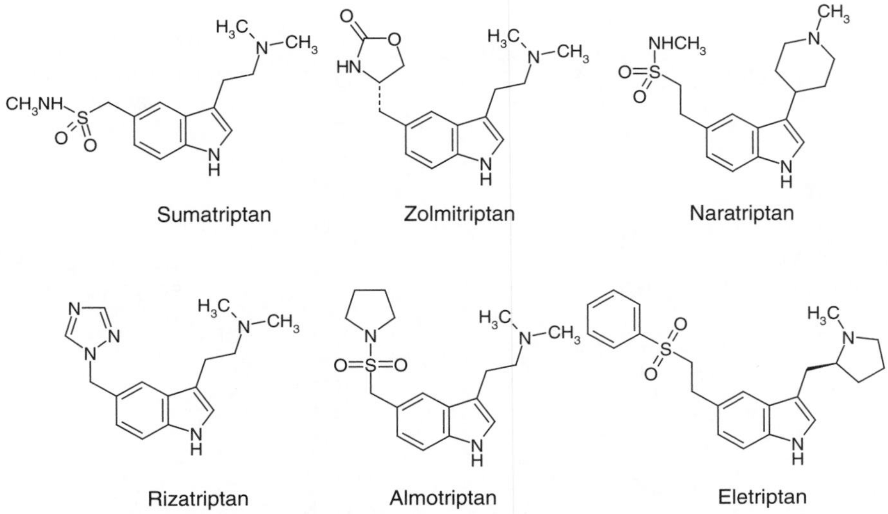

Sumatriptan Zolmitriptan Naratriptan

Rizatriptan Almotriptan Eletriptan

Figure 6. The triptans show chemical and pharmacological similarities. Thus, they can be considered direct analogs.

concept; we propose a very general definition of the term bioisostere:

> A bioisostere is the compound that results from the replacement in an active molecule of an atom, a group of atoms, or a scaffold by another one, with conservation of the initial biological affinity for a given target (Fig. 7).

This definition is sufficiently broad to include all molecules relevant to the isosterism concept and at the same time it distinguishes the three classes of bioisosteres.

The comparison of bioisosteres and analogs in general shows that the word analog covers clearly a more general concept: all bioisosteres are analogs (see Fig. 8), but many analogs are not bioisosteres (analogs produced by combinatory chemistry, fragment libraries, fermentation, etc.)

As a rule, the basic scaffold is conserved or only slightly modified. Fine-tuning can be achieved by means of substituent effects (see Fig. 9) possibly monitored by means of QSAR approaches.

Drug-like properties are not a major concern in direct drug design, because the analogs prepared stem from a starting lead that is already a recognized drug molecule. However, if needed, some filter methods are used thereby eliminating reactive or toxic groups or accounting for central bioavailability or other ADME properties [15].

4.1.1. Bioisosterism: A Practical Example The passage from the angiotensin II receptor antagonist losartan (see Fig. 10) to its direct analog valsartan represents a typical example of analog design [16]. Apparently, the tetrazolyl-diphenyl part of the molecule is essential and does not allow various changes. The substituted pyrazole moiety of losartan bears four characteristic features that match nicely: (1) the *n*-butyl chain, common in both molecules, probably interacts with some lipophilic residues of the receptor, (2) the partial charge of one of the pyrazolic nitrogens corresponds with the partial charge of the amide carbonyl of valsartan, (3) the lipophilic chlorine was replaced by an methyl group, and (4) the primary alcoholic function of losartan is probably metabolized to the corresponding carboxylic acid and presumably losartan functions as a prodrug; valsartan already contains the carboxylic function. Taken together losartan and valsartan both present four similar interaction features.

4.2. Structural Analogs

Despite their similar chemical structure, *structural analogs* exhibit different pharma-

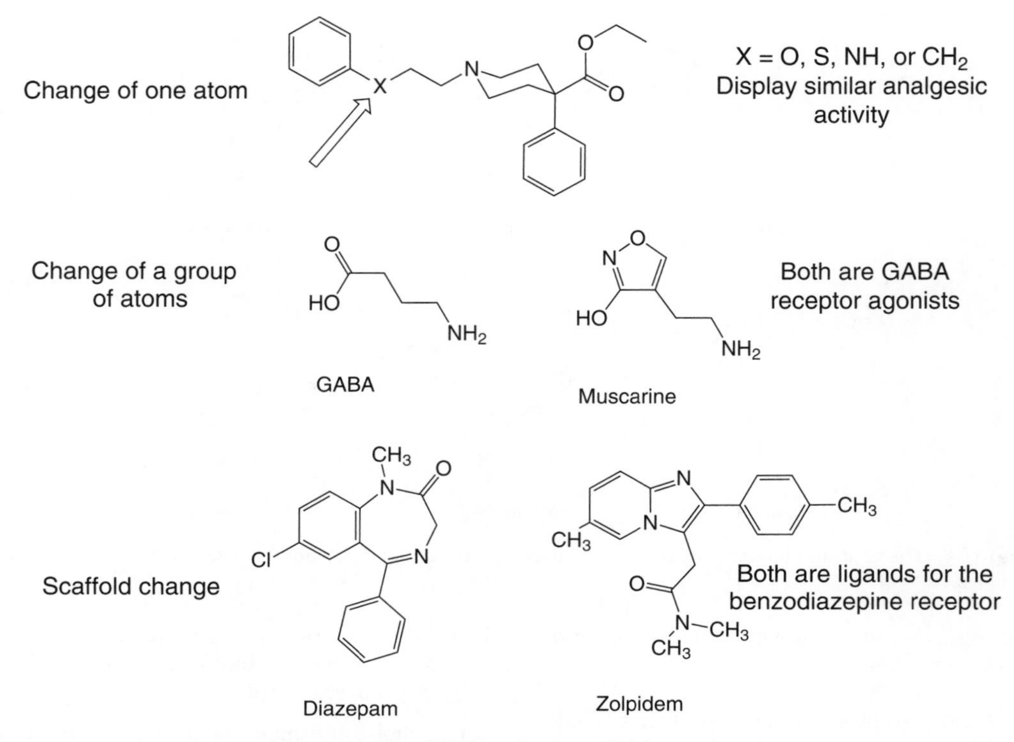

Figure 7. The different degrees of bioisosterism.

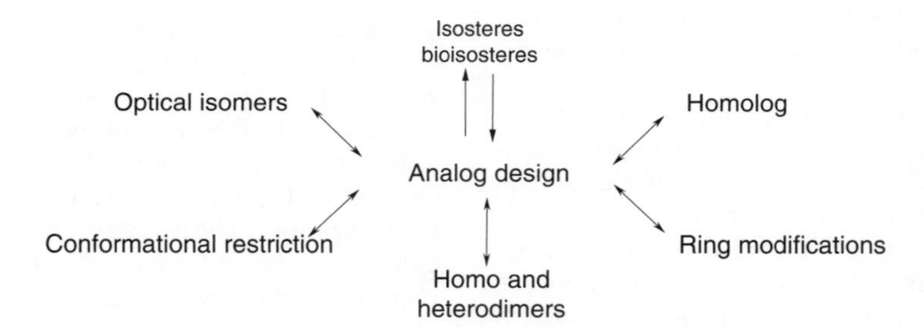

Figure 8. Bioisosterism is a privileged provider of analogs.

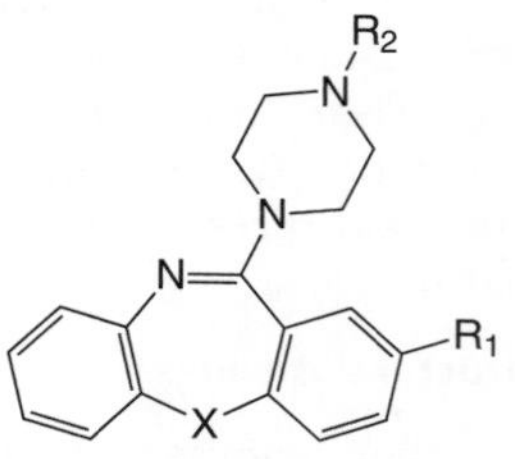

X	R_1	R_2	Compound
S	Cl	CH_3	Clotiapine
O	Cl	CH_3	Loxapine
O	Cl	H	Amoxapine
CH_2	H	CH_3	Perlapine

Figure 9. Analogs derived from clotiapine.

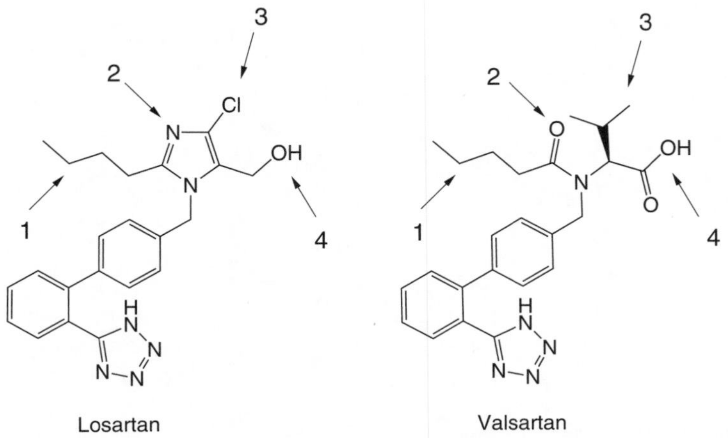

Figure 10. The rational design of valsartan as an analog of losartan.

cological profiles. Typical examples of this type are the steroid hormones estradiol and testosterone (Fig. 11a); the antimalarial quinacrine and the antipsychotic chlorpromazine; (Fig. 11b), and the antidepressant minaprine and its cyano analog, SR 95191, a selective inhibitor of type A monoamine oxidase (see Fig. 11c) [17].

As mentioned previously, most of the structural analogs originate from those serendipitous discoveries that often happen during clinical investigations. For a long time, no

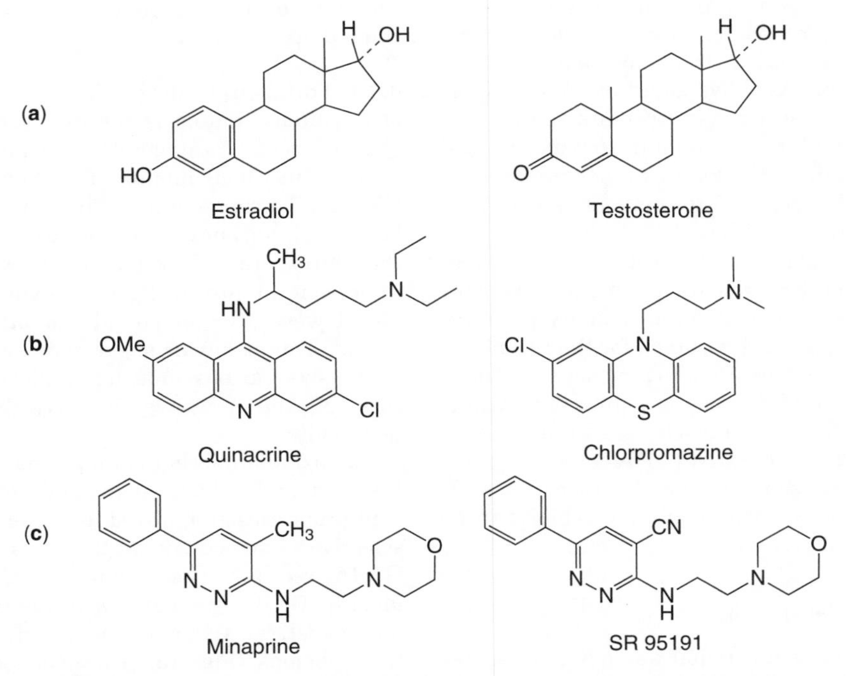

Figure 11. Examples of structural analogs. Despite their structural analogy, these compounds present different pharmacological activities.

K_i (nM)		
α2A	α2C	5-HT_t
0.9	1.7	16
2.4	**0.1**	8.9
>1000	>1000	**22**

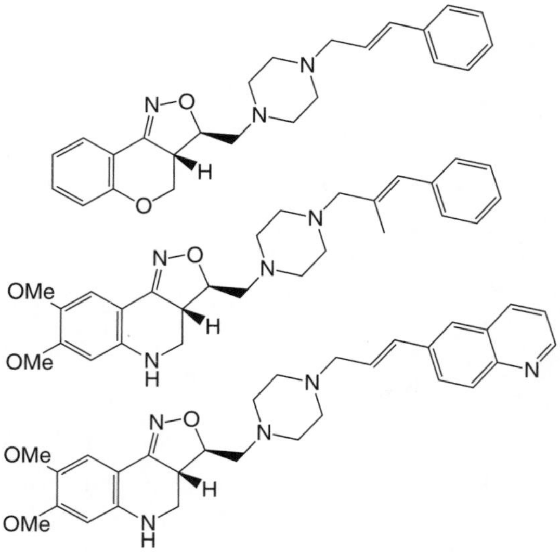

Figure 12. Similar structures exhibit different affinities for pharmacological targets [18].

straightforward methodologies leading to structural analogs were available. However, presently we can systematically apply multitarget screening to large series of structurally similar compounds. An example is a series of closely related tricyclic isoxazole derivatives in which scientists from Johnson & Johnson [18] observed diverse affinity profiles for the α_{2A}, α_{2C} adrenergic receptors and for the inhibition of the serotonin transporter (5-HT_T). Figure 12 illustrates how relatively modest changes lead to compounds with different activity profiles.

A particular case of structural analogs is found in optical isomerism when the two enantiomers present different activity profiles. Such a situation is observed for the racemate tetramisole where S(–)-levamisole has nematocidal and immunostimulant properties, whereas R(+)-dexamisole is an antidepressant [19,20]. Similarly, (–)-3-methoxy-cyproheptadine has anticholinergic properties, and (+)-3-methoxy-cyproheptadine exhibits antiserotonin activity [18].

4.3. Functional Analogs

The design of functional analogs represents the most original method of analog design. For a long time, discoveries of these analogs were the unpredictable result of fortuitous observations. Presently, the search for functional analogs can be planned either by using systematic screening or by applying virtual screening or related approaches.

4.3.1. Fortuitous Findings A typical example of fortuitous finding is the discovery of the $GABA_A$ receptor antagonist gabazine [21–23] as the functional analog of (+)-bicuculline (Fig. 13). During a study of minaprine metabolism [21,23], one of its putative metabolites, the hydroxyethylamino metabolite, was found to present an interesting anticonvulsant profile. However, the compound was rather short acting. This behavior suggested the need to synthesize various prodrugs and bioprecursors of the putative hydroxyethylamino metabolite.

Gabazine and other *endo*-pyridazinyl-GABAs lost the antidepressant and anticonvulsant properties and proved to be competitive, selective, and potent antagonists for the $GABA_A$ receptor. The resemblance of the pyridazinyl-GABA, gabazine, with the reference antagonist, the alkaloid (+)-bicuculline, is far from obvious (Fig. 13). However, computer modeling studies of several $GABA_A$ antago-

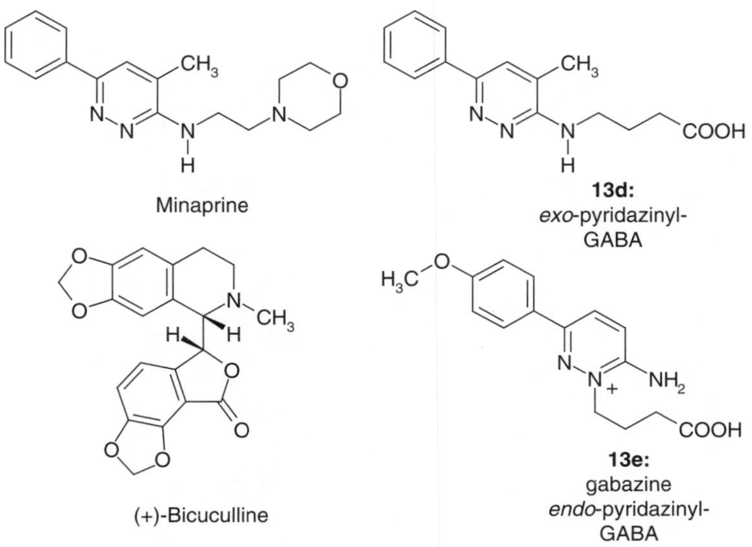

Minaprine

13d: *exo*-pyridazinyl-GABA

(+)-Bicuculline

13e: gabazine *endo*-pyridazinyl-GABA

Figure 13. The surprising change from the antidepressant minaprine to the GABA-A receptor antagonist gabazine (*endo*-pyridazinyl-GABA) [23,24].

nists revealed the isofunctional nature of both gabazine and (+)-bicuculline [23].

4.3.2. Systematic Screening Functional analogs can be produced deliberately as the result of a screening campaign. Thus, after the discovery of the anxiolytic and tranquilizing activity of the benzodiazepines, scientists from Rhône Poulenc initiated, in the early 1970s, a systematic search of other molecules showing similar binding and *in vivo* pharmacological profiles. The result was the discovery of zopiclone [25], the first nonbenzodiazepine showing a pharmacological profile similar to chlordiazepoxide [26] (Fig. 14).

Another screening search, performed in the early 1980s by Synthelabo scientists [27], led to the imidazopyridine, zolpidem, which is presently one of the most popular sleep inducers.

4.3.3. Virtual Screening and Scaffold Hopping Systematic screening, despite being a fruitful method for identifying functional analogs, needs to be applied to real libraries made of thousands of compounds. This is not the case for virtual screening, an approach based on computer-aided detection in a database of 2D or 3D chemical structures and/or chemical feature-based pharmacophore models. One can consider that the passage from the original drug molecule to its analog consists of a kind of hopping from one given platform to another. The goal of scaffold hopping is to design structurally new compounds by a significant modification of the platform of a known active drug (template). Other terms like "leapfrogging," "lead hopping," and "structure morphing" have been used to point out the discovery of novel molecules with biological activities comparable to a template and with a significantly different architecture. The concept is based on the assumption that structurally and chemically diverse structures can interact with the same receptor eliciting similar biological activity. This appears to contradict the usual approach to the drug discovery process. In fact, from time immemorial medicinal chemists have relied on the principle that structurally similar molecules have similar biological activity [28] to modify the structure of biologically active compounds in order to get drug candidates.

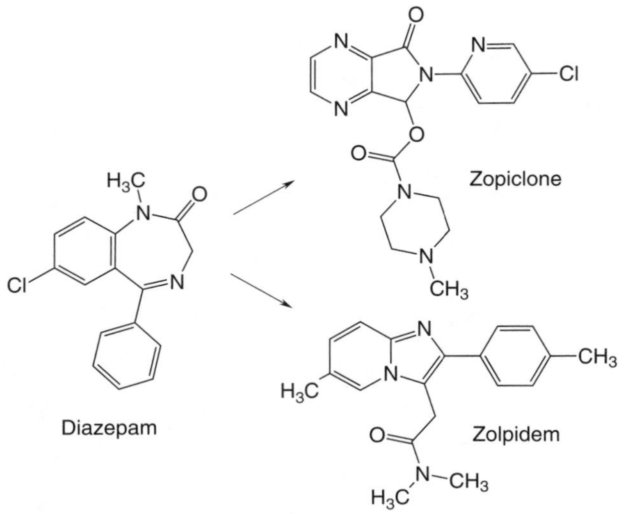

Figure 14. Zopiclone and zolpidem are selective benzodiazepine receptor agonists not related chemically to benzodiazepines [27].

In a prospective test, for example [29], a program called CATS (chemically advanced template search) was applied to the prediction of novel cardiac T-type Ca^{2+}-channel-blocking agents by exploring the Roche in-house compound depository. Mibefradil, a known T-channel-blocking agent ($IC_{50} = 1.7\,\mu M$) (see Fig. 15), was used as the seed structure for CATS. The 12 highest-ranking molecules were tested, using a cell culture assay for their ability to inhibit cellular Ca^{2+} influx. Nine compounds (75%) showed significant activity ($IC_{50} < 10\,\mu M$), and one of these, clopomozide, had an $IC_{50} < 1\,\mu M$. The IC_{50} values of the next best structures (not shown in this chapter) were 1.7, 2.2, 3.2, and 3.5 μM.

These hits have structural scaffolds that differ significantly from the query structure mibefradil. However, essential function-determining points are preserved, forming the basis of a relevant pharmacophoric pattern.

In another publication [30], the protein ERG2, the emopamil binding protein (EBP), and the σ1 receptor were investigated; these compounds are involved in sterol metabolism and share high affinity for various and struc-

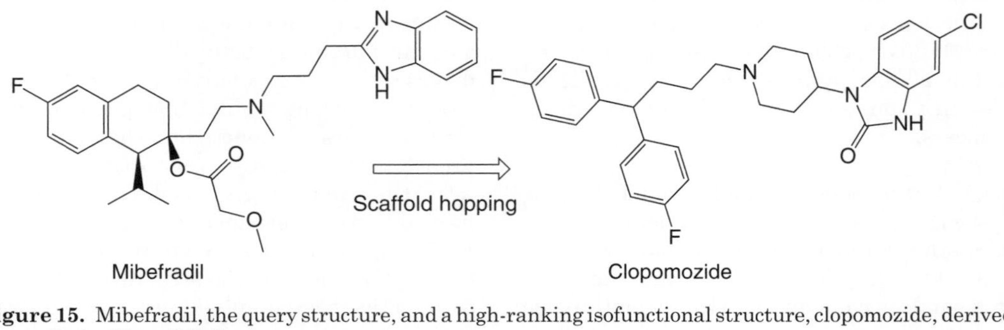

Figure 15. Mibefradil, the query structure, and a high-ranking isofunctional structure, clopomozide, derived from mibefradil by CATS.

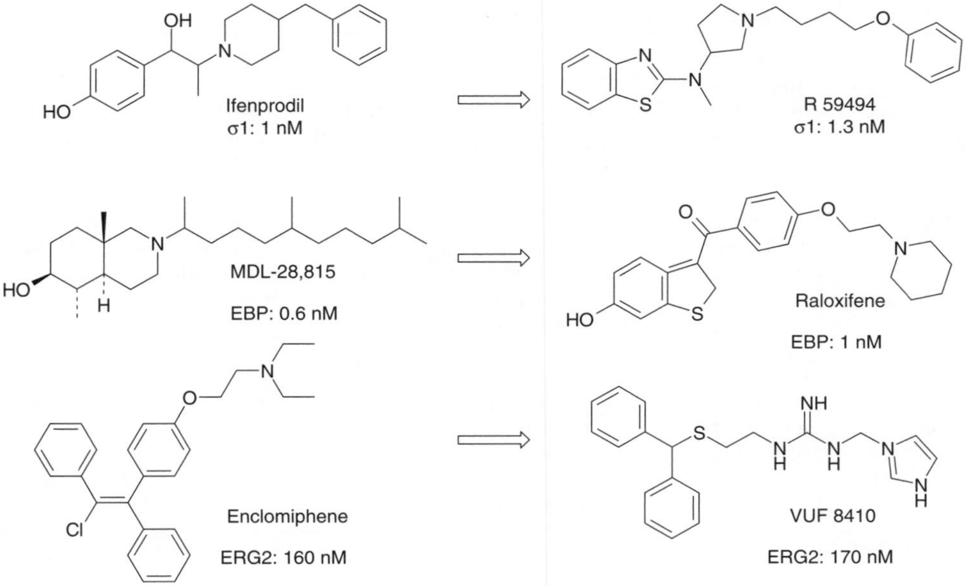

Figure 16. Examples of original structures in the training set (left column) and the corresponding newly discovered bioactive compounds.

turally diverse proteins. To discover novel high-affinity ligands, pharmacophore models were built with catalyst based on a series of 23 structurally diverse chemicals exhibiting K_i *values* from 10 pM to 100 μM for all three proteins.

In virtual screening experiments, ligands were identified, which were previously reported to bind to one or several of these proteins (see Fig. 16). Some of them were also included into a set of compounds that were tested experimentally for their affinity toward the three proteins. The authors obtained 11 hits, of which three, including raloxifene, had affinities for σ1 or EBP that were less than 60 nM. When searched a database of 3525 biochemicals of intermediary metabolism, a slightly modified ERG2 pharmacophore model successfully retrieved 10 substrate candidates among the top 28 hits. These findings indicate that inhibitor-based pharmacophore models for σ1, ERG2, and EBP can be used to screen drug and metabolite databases for chemically diverse compounds and putative endogenous ligands.

5. CONCLUSION

Improvement of an existing active molecule, natural or synthetic—that is, practicing analog design—is one of the major approaches of medicinal chemistry. The objective of this chapter is to clarify the terminology of analog design by establishing a clear distinction among three kinds of analogs (Table 1): *Direct* analogs, that is, analogs possessing structural and functional similarities; *structural analogs*, that is, analogs possessing only chemical similarities but presenting different pharmacological profiles; and *functional analogs*, that is, chemically different compounds displaying similar pharmacological properties.

In addition to the *full*, the *structural*, or the *functional* nature of the designed analogs, it appears worthwhile to consider also the *degree of relationship* between the copy and the original. As discussed earlier, very diverse chemical modifications can be used to design analogs. Among them, the synthesis of bioisosteres is the most fruitful and can be considered as main provider for analog design.

Table 1. The Three Categories of Drug Analogs

Nature of the Analog	Degree of Relationship with the Original Structure	Production Design	Originality Patentability	Result of the Design	Likelihood of Drug Success
S + F:	Isosterism	Molecular variations			
Direct analog	Functional exchange	Parallel synthesis	Low	"Me-too"	Medium to high
S: Structural analog	Bioisosterism	Traditional medicinal chemistry	Medium	Patentable copy	Low
	Fragment exchange				
F: Functional analog	Pharmacophore identification scaffold hopping	Fortuitous discovery screening, Computer-aided design	High	Original back-up	Variable

Redrawn from Ref. [6].

ACKNOWLEDGMENTS

I wish to express my gratitude to Sharon Bryant for her constructive and critical contribution to the manuscript, particularly, to the section "Virtual Screening and Scaffold Hopping."

REFERENCES

1. Proudfoot JR. Drugs, leads, and drug-likeness: an analysis of some recently launched drugs. Bioorg Med Chem Lett 2002;12:1647–1650.
2. Angell M. The Truth About the Drug Companies: How They Deceive Us and What to do About It. New York: Random House; 2004. p 305.
3. Wermuth CG. The "SOSA" approach: an alternative to high-throughput screening. Med Chem Res 2001;10:431–439.
4. Wermuth CG. Selective optimization of side activities: another way for drug discovery. J Med Chem 2004;47:1303–1314.
5. Wermuth CG. Analogues as means of discovering new drugs. In: Fischer J, editors. Analog Based Drug Discovery. Weinheim: John Wiley - VCH; Vol. 1:2006. p 3–23.
6. Wermuth CG. Similarity in drugs: reflections on analogue design. Drug Discov Today 2006;11: 348–354.
7. Duncan GS, Wolberg G, Schmitged CJ, Deeprose RD, Zimmerman TP. Inhibition of lymphocytes-mediated-cytolysis and cyclic AMP phosphodiesterase by erythro-9-(2-hdroxy-3-nonyl)adenine. J Immunopharmacol 1982;4:79–100.
8. Lugnier C, Schoefter P, Le Bec A, Strouthou E, Stoclet JC. Selective inhibition of cyclic nucleotide phosphodiesterases of human, bovine and rat aorta. Biochem Pharmacol 1986;35:1743–1751.
9. Rotella DP, Sun Z, Zhu Y, Krupinski J, Pongrac R, Seliger L, Normandin D, Macor JE. *N*-3-Substituted imidazo-quinazo quinazolinones; potent and selective PDE5 inhibitors as potential agents for treatment of erectile dysfunction. J Med Chem 2000;43:1257–1263.
10. Testa B, Jenner P, Drug Metabolism, Chemical and Biochemical Aspects. New York: Marcel Dekker; 1976.
11. Fischer J, Gere A. Timing of analog research in the medicinal chemistry. Pharmazie 2001; 56:675–682.
12. Proudfoot JR. Drug likeness and analog based drug discovery. In: Fischer J,Editor. Analog Based Drug Discovery. Weinheim: John Wiley -VCH; 2006. p 25–52.
13. Wermuth CG. The Practice of Medicinal Chemistry. 2nd ed.; San Diego: Academic Press; 2003. p 768.
14. Wermuth CG. Analogues as means of discovering new drugs. In: Fischer J, Ganellin CR,Edi-

tors. Analog Based Drug Discovery. Weinheim: John Wiley-VCH; 2006. p 3–23.

15. Bajorath J. Integration of virtual and high-throughput screening. Nat Rev Drug Discov 2002;1:882–894.
16. Buehlmayer P, Furet P, Cricione L, de Gasparo M, Whitebread S, Schmidlin T, Lattman R. Wood J Valsartan, a potent, orally active angiotensin II antagonist developed from the structurally new amino acid series. Bioorg Med Chem Lett 1994;4:29–34.
17. Worms P, Khan JP, Wermuth CG, Roncucci R, Bizière K. SR 95191 a selective inhibitor of type A monoamine oxidase with dopaminergic properties. I. Psychopharmacological profile in rodents. J Pharmacol Exp Ther 1987;240:241–250.
18. Andrès JI, Alcazar J, Alonzo JM, Alvarez RM, Bakker MH, Biesmans I, Cid JM, De Lucas AI, Fernandez J, Font LM, Hens KA, Iturrino L, Lenaerts I, Martinez S, Megens AA, Pastor J, Vermote PCM, Steckler T. Discovery of a new series of centrally active tricyclic isoxazoles combining serotonin (5-HT) reuptake inhibition with α_2-adrenoceptor blocking activity. J Med Chem 2005;48:2054–2071.
19. Bullock MW, Hand JJ, Waletzky E. Resolution and racemization of dl-tetramisole, dl-6-phenyl-2,3,5,6-tetrahydroimidazo-[2,1-b]thiazole. J Med Chem 1968;11:169–171.
20. Schnieden H. Levamisole: a general pharmacological perspective. Int J Immunopharmacol 1981;3:9–13.
21. Chambon J-P, Feltz P, Heaulme M, Restle S, Schlichter R, Biziere K, Wermuth CG. An arylaminopyridazine derivative of γ-aminobutyric acid (GABA) is a selective and competitive antagonist at the $GABA_A$ receptor site. Proc Natl Acad Sci USA 1985;82:1832–1836.
22. Wermuth CG, Bizière K. Pyridazinyl-GABA derivatives: a new class of synthetic $GABA_A$ antagonists. Trends Pharmacol Sci 1986;7:421–424.
23. Wermuth CG, Bourguignon J-J, Schlewer G, Gies J-P, Schoenfelder A, Melikian A, Bouchet M-J, Chantreux D, Molimard J-C, Heaulme M, Chambon J-P, Bizière K. Synthesis and structure–activity relationships of a series of aminopyridazine derivatives of γ-aminobutyric acid acting as selective $GABA_A$ antagonists. J Med Chem 1987;30:239–249.
24. Wermuth CG. The development of pyridazinyl-GABA's, a new class of synthetic GABA antagonists In: Ravina E,Editor. 1st Spanish-Portuguese Congress of Therapeutical Chemistry, Santiago de Compostela, 1988, 1987; Universidade de Santiago de Compostela: Santiago de Compostela 1987; 107–140.
25. Jeanmart C, Cotrel C. Synthèse de (chloro-5 pyridyl-2) (méthyl-4 piperazinyl-1) carbonyloxy-5 *oxo*-7 dihydro-6, 7 5H-pyrrolo [3-4b] pyrazine. Compt Rend Acad Sci Série C 1978;287:377–380.
26. Doble A, Canto T, Piot O, Zundel J-L, Stutzmann J-M, Cotrel C, Blanchard J-C. The pharmacology of cyclopyrrolone derivatives acting at the $GABA_A$/benzodiazepine receptor. Adv Biochem Psychopharmacol 1992;47:407–418.
27. Arbilla S, Depoortere H, George P, Langer SZ. Pharmacological profile of the imidazopyridine zolpidem at benzodiazepine receptors and electrocorticogram in rats. Naunyn Schmiedebergs Arch Pharmacol 1985;330:248–251.
28. Martin YC, Kofron JL, Traphagen LM. Do structurally similar molecules have similar biological activity? J Med Chem 45:19 2002; 4350–4358.
29. Schneider G, Giller T, Neidhart W, Schmid G. "Scaffold-hopping" by topological pharmacophore search: a contribution to virtual screening. Angew Chem Int Ed 1999;38:2894–2896.
30. Laggner C, Schieferer C, Fiechtner B, Poles G, Hoffmann RD, Glossmann H, Langer T, Moebius FF. Discovery of high-affinity ligands of $\sigma 1$ receptor, ERG2, and emopamil binding protein by pharmacophore modeling and virtual screening. J Med Chem 2005;48:4754–4764.

SNPS: SINGLE NUCLEOTIDE POLYMORPHISMS AND PHARMACOGENOMICS: INDIVIDUALLY DESIGNED DRUG THERAPY

ANKE-HILSE MAITLAND-VAN DER ZEE[1]
BRIAN J. PUCKETT[2]
STEVEN G. TERRA[3]
JOE WALKER[4]
[1]Pharmaceutical Sciences, Utrecht University, Utrecht, The Netherlands
[2]Virginia Commonwealth University, Richmond, VA
[3]JHM Health Science Center, Gainesville, FL
[4]Orchid BioSciences, Princeton, NJ

1. INTRODUCTION

1.1. History

1.1.1. Watson and Crick to the Human Genome Project It seems to have all started at the turn of the twentieth century. Mendel's laws came back into favor within the scientific community, leading to discoveries in the cellular basis for heredity. But it was not until 1953 when Watson and Crick first described and built the elegant model of the structure of DNA that the field of genetics became not just a science, but an obsession. Over the past 50 years, scientists have been feverishly trying to unlock and decode the secrets of the human genome. Then, a monumental accomplishment was achieved when the first draft of the entire human genome sequence was published in February 2001 as a result of the Human Genome Project [1]. This has sparked a renewed fervor for research in genomics that will certainly carry us through and beyond the twenty-first century, the postgenomic era.

1.1.2. Drug Response and Toxicity Variability It is an incontrovertible fact that large interpatient variability exists in response to medications. Variation in response has existed as long as medications have been used for the prevention and treatment of disease. In many ways, the field of pharmacogenomics began serendipitously in the 1950s after seminal observations describing variability in response to medications. Examples included peripheral neuropathy from isoniazid among slow acetylators [2], prolonged apnea from succinylcholine caused by pseudocholinesterase deficiency [3], and severe hypotension from debrisoquine among cytochrome P450 (CYP) 2D6 poor metabolizers [4]. For the next 40 years, pharmacogenetic studies focused almost exclusively on the etiologies of altered variability in pharmacokinetic responses to medications.

As we entered the 1990s, pharmacogenomic studies began to include studies that examined pharmacodynamic variability in drug response. Now, instead of examining only differences in drug metabolizing enzymes, scientists began to focus on genes that encode drug transporters, drug targets, and ion channels. The goal of this chapter is to provide a primer on pharmacogenomics and describe how the human genome and molecular biology techniques are transforming medicine to create an era of personalized therapeutics.

1.2. Genetic/Genomic Definitions

1.2.1. Basics The human genome is made up of DNA, which is organized into 23 chromosomes. DNA is a double helical structure that is composed of sugar, phosphate, and a nitrogenous base. The DNA strands are held together by hydrogen bonds. The four nitrogenous bases that make up DNA are adenine (A), guanine (G), cytosine (C), and thymine (T). The base pairing is consistent in that adenine always binds to thymine and cytosine always binds to guanine. A combination of three base pairs makes up a codon, and each codon specifies an amino acid that will be incorporated into a protein. The process of making a protein begins when an RNA polymerase attaches to a region of DNA known as the promoter region. This single-stranded chain now serves as a template to synthesize a single-stranded RNA molecule. Once RNA has been formed in the nucleus of the cell, it is transported to the ribosomes in the cytoplasm of the cell where translation will occur. However, before translation occurs, the RNA is processed and introns, noncoding regions of the DNA, are removed. The removal of noncoding regions is termed splicing. Exons, coding regions of DNA, constitute only 5% of the human genome. Once this has occurred, the RNA molecule is translated into amino acids and proteins. Because there are four nucleotides, a total of 64 different codons are possible. However, there are only 20 amino acids, hence several different codons may specify the same amino acid.

1.2.2. Pharmacogenetics or Pharmacogenomics?

Pharmacogenetics is literally a combination of pharmacology and genetics. It thus stands to reason that pharmacogenetics is the study of how an individual's genetic makeup may influence response to medications. The field of pharmacogenetics has been around in one form or another for over 50 years. Historically, differences in drug response were observed and documented. Based on these phenotypic (i.e., the outward physical effect of a genotype) observations, scientists explored heredity and gene lines to pharmacogenetically explain these differences in response. Scientists further explored and concentrated efforts on the pharmacokinetics (i.e., absorption, distribution, metabolism, and excretion) of these medications in search of genetic differences in, for example, how fast or slow a particular drug is metabolized. Then came the Human Genome Project and birth of the term pharmacogenomics.

There has been considerable debate about whether the terms pharmacogenetics and pharmacogenomics mean the same thing or whether they truly are different sciences. Some state that pharmacogenomics is just the new en vogue terminology of late. Some state that pharmacogenetics looks at a single gene whereas pharmacogenomics looks at multiple interacting genes. Others state that there is a fundamental difference in the way the research problem or hypothesis is approached. Few would argue that pharmacogenomics has its roots in pharmacogenetics, but it is the recent technology in molecular biology along with the Human Genome Project that has shaped and defined the field of pharmacogenomics. Pharmacogenetics historically relied heavily on phenotypic observations to drive hypotheses about genetic differences [5]. But, with today's technology, one can take a genomewide approach to *a priori* hypothesize about differences in drug response and/or search for novel drug targets. This is the essence of pharmacogenomics. Furthermore, pharmacogenetics typically concentrated its research in the area of pharmacokinetics thus looking at drug metabolizing enzyme polymorphisms [6]. However, pharmacogenomics opens a genomic Pandora's box, allowing the ability to explore not only drug metabolizing polymorphisms but also drug target polymorphisms, drug transporter polymorphisms, and disease progression polymorphisms. If the goal of pharmacogenetics is to explain the interindividual differences in drug response based on genetic information, then it is the promise of pharmacogenomics to truly individualize pharmacotherapy [5].

2. SINGLE NUCLEOTIDE POLYMORPHISMS

2.1. Definition

Single nucleotide polymorphisms (SNPs, pronounced "snips") are single base pair substitutions that occur in a sequence of DNA. It has been estimated that SNPs occur at a frequency of approximately 1 in 1000 bp. Because the human genome contains approximately 3 billion base pairs, there should be approximately 30,000,000 SNPs in the genome [7]. However, the number of SNPs in coding regions (i.e., the regions that actually code for proteins) of the genome has been estimated at 500,000. Recently, a "SNP map" of the human genome was published containing 1.42 million SNPs [1]. The researchers found an average density of one SNP every 1.9 kb, or approximately 1 in 1900 bp. They also estimated that of the 1.42 million SNPs identified, only 60,000 SNPs actually fell in coding regions. However, it is important to consider that although a SNP may fall outside the coding region it may be in linkage disequilibrium with a coding SNP, thus indirectly affecting a biologic or pharmacologic response. Irregardless, in order to be classified as a polymorphism, it must have a frequency of at least 1% in the population. However, to be of routine clinical use, the frequency of polymorphisms may need to be much higher.

2.2. Classification of SNPs

As alluded to above, there are different types of SNPs. Coding SNPs are those polymorphisms that are located within the coding block of the gene, whereas noncoding SNPs occur outside of the coding block. Coding SNPs can further be classified as either synonymous or nonsynonymous. A synonymous SNP is a polymorphism in which, although the codon has been changed, both the wild type and the polymorphic variant code for the exact same amino acid. Thus, a nonsynonymous SNP is one in which a different amino acid is coded and is therefore the most common SNP described in the literature. A nonsynonymous SNP can further be classified as conservative or nonconservative. A conservative nonsynonymous SNP confers an amino acid substitution of similar size and charge to that of the original amino acid, whereas a nonconservative nonsynonymous SNP substitutes an amino acid that is very different in size and/or charge, which may greatly impact protein folding. It thus stands to reason that a nonconservative nonsynonymous SNP may potentially cause the most obvious genetic variations.

It is becoming increasingly important given the dogma of pharmacogenomics to evaluate multiple SNPs. A haplotype is a collection of SNPs on a particular locus. The locus could be multiple genes, one entire gene, or merely a segment of a gene. By examining haplotypes, the interaction between SNPs can be further addressed and elucidated. For example, SNPs can often travel together because of linkage disequilibrium, thus making it very important to study the haplotype. It therefore may be potentially misleading to only look at the individual SNPs without considering their interplay. This has indeed been proven to be the case in a seminal paper looking at β_2-adrenergic receptor polymorphisms in asthma [8]. The details of this case are described in more detail in section "β_2-Adrenergic Receptor Haplotypes."

It is also very important to note that SNPs are not the only polymorphisms out there that affect drug response. Up until now, we have been discussing what happens when a single nucleotide base pair is substituted. However, there are also polymorphisms causing insertion or deletion of a segment of DNA, splice site mutations resulting in exon skipping, microsatellite nucleotide repeats, gene duplication, point mutations resulting in early stop codons, and complete gene deletions. This makes for an incredibly complex variety of SNPs that are potentially responsible for interindividual response differences observed with certain medications.

3. PHARMACOGENOMICS

3.1. Drug Metabolizing Enzymes

More than 40 years ago, the deficiency of glucose-6-phosphate dehydrogenase (G6PD) and arylamine *N*-acetyltransferase type 2 (NAT2) were the first examples that revealed that hereditary variants of drug metabolizing enzymes could be responsible for side effects and interindividual variability in response to drugs [9,10]. Since then, significant progress has been made in the field of pharmacogenetics. Much of that work involves investigation into the clinical relevance of genetic variability in the enzymes responsible for the metabolism of both endogenous and exogenous substrates (Fig. 1).

These drug metabolizing enzymes, such as the cytochrome P450s (CYP), are responsible for the metabolic elimination of most of the drugs currently used in medicine. Genetically determined variability in the function of these enzymes can have a profound effect on drug safety and efficacy.

There are many molecular mechanisms of variability or inactivation of drug metabolizing enzymes. These include splice site mutations resulting in exon skipping (CYP2C19), microsatellite nucleotide repeats (CYP2D6), gene duplication (CYP2D6), point mutations resulting in early stop codons (CYP2D6), amino acid substitutions that alter protein stability or catalytic activity (e.g., TPMT, NAT2, CYP2D6, CYP2C19, and CYP2C9), or complete gene deletions (CYP2D6).

3.1.1. Polymorphisms in the Cytochrome P450 System Cytochrome P450 (CYP) enzymes, a very large gene family comprising numerous isoforms, oxidatively metabolize xenobiotics, including many drugs. Specializing in the removal of lipophilic foreign chemicals, these enzymes rank among the most abundant proteins in the liver. Table 1 lists selected medications that are metabolized by CYP enzymes.

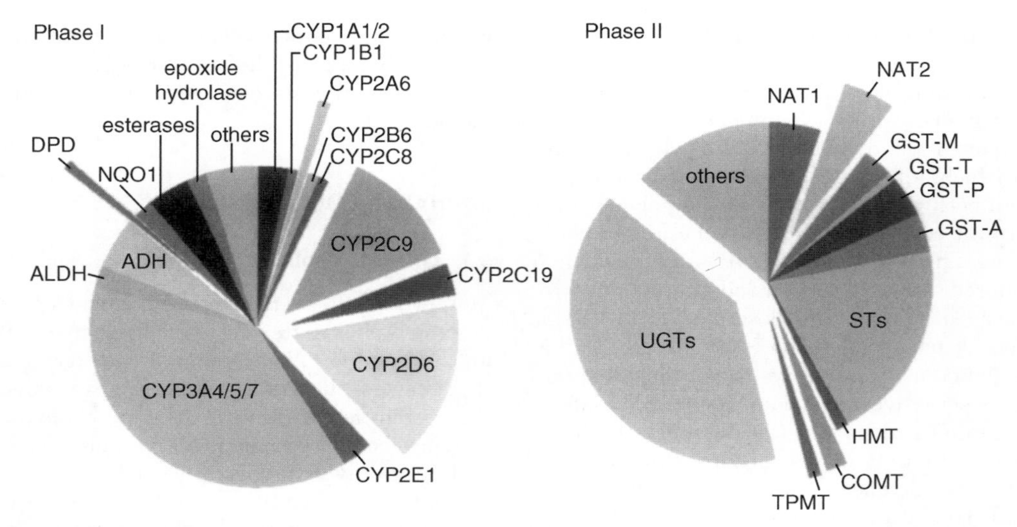

Figure 1. Drug metabolizing enzymes. Those drug metabolizing enzyme polymorphisms that have already been associated with altered drug effects are separated from the pie chart. For each of the corresponding phases, the size of the pie slice approximates the drug metabolizing enzyme's contribution to the overall metabolism of drugs. ADH, alcohol dehydrogenase; ALDH, aldehyde dehydrogenase; CYP, cytochrome P450; DPD, dihydropyrimidine dehydrogenase; NQO1, NADPH:quinine oxidoreductase or DT diaphorase; COMT, catechol *O*-methyltransferase; GST, glutathione *S*-transferase; HMT, histamine methyltransferase; NAT, *N*-acetyltransferase; STs, sulfotransferases; TPMT, thiopurine methyltransferase; UGTs, uridine 5′-triphosphate glucuronosyltransferases. Reproduced with permission from Ref. [6].

When CYP mutations result in null alleles (inactivation), a complete lack of active enzyme and a severely compromised ability to metabolize drugs results. Drugs may reach toxic plasma concentrations if given in regular doses to these "poor metabolizers." For example, mutations in the gene encoding cytochrome P450 CYP2C9, which metabolizes warfarin, affect patients' response to the drug and their dose requirements [11].

CYP2D6 CYP2D6, also known as debrisoquine/sparteine hydroxylase, is highly polymorphic and is inactive in about 8% of Caucasian Americans and 2–5% of African Americans [12]. It is involved in the metabolism of approximately 30–40 commonly used drugs. Millions of patients with compromised metabolism are thus at risk of adverse drug reactions when prescribed drugs that are CYP2D6 substrates. Many such drugs are used for treating psychiatric (such as antidepressants and antipsychotics) and cardiovascular diseases (such as β-blockers and antiarrhythmics), where the therapeutic window can be narrow and side effects common.

More than 70 variant alleles of the CYP2D6 locus have been described, of which at least 15 encode nonfunctional gene products. These alleles as well as several functional allelic variants of CYP2D6 have been described that occur at variable frequencies in racially diverse populations [13]. Most of the null alleles have interrupted open reading frames because of splice-site mutations, single base deletions, nonsense mutations, or deletion of the entire gene. Alleles encoding nonfunctional full-length proteins have also been described. Based on genetic diagnosis, it is now possible to identify individuals with poor metabolizer (PM) phenotype as carriers of two null alleles with over 99% certainty [14–16].

Whereas using a molecular diagnostic to identify CYP2D6 PMs has become much easier, it has remained much more difficult to predict the metabolic capacity of extensive metabolizers (EM, i.e., individuals carrying one or more functional gene copies) [16]. Even

Table 1. Selected Substrates of Polymorphic Drug Metabolizing Enzymes

CYP2D6	CYP2C9	CYP2C19	CYP3A4/5/7
Propranolol	*S*-Warfarin	Omeprazole	Clarithromycin
Metoprolol	Glipizide	Lansoprazole	Erythromycin
Timolol	Glimepiride	Carisoprodol	Midazolam
Amitriptyline	Tolbutamide	Diazepam	Triazolam
Nortriptyline	Phenytoin	Pantoprazole	Cyclosporine
Imipramine	Celecoxib	Citalopram	Tacrolimus
Desipramine	Ibuprofen	Clomipramine	Indinavir
Paroxetine	Losartan	Hexobarbital	Nelfinavir
Fluoxetine	Irbesartan		Ritonavir
Venlafaxine	Torsemide		Saquinavir
Codeine			Cisapride
Dextromethorphan			Astemizole
Tolterodine			Estradiol
Propafenone			Hydrocortisone
Mexiletine			Progesterone
Tamoxifen			Testosterone
			Sildenafil
			Trazodone
			Vincristine
			Zaleplon
			Zolpidem
			Amlodipine
			Diltiazem
			Nifedipine
			Verapamil

in extensive metabolizers, CYP2D6 activity is known to vary greatly. For example, the CYP2D6 activity represented by the metabolic ratio (MR) values of debrisoquin and desipramine has been reported to show more than a 70-fold variation among extensive metabolizers in Korean and white populations [17,18].

In poor metabolizers, the genes often contain inactivating mutations, which result in a complete lack of active enzyme and a severely compromised ability to metabolize drugs. Thus, poor metabolizers of CYP2D6 are potentially at risk for increased plasma concentrations of drugs given at conventional doses. For example, the metabolism of the antidepressant venlafaxine is controlled by genetic polymorphism. Poor metabolizers of CYP2D6 have significantly reduced venlafaxine clearance and an increased risk of cardiovascular toxicity [19].

Conversely, ultrarapid metabolizers (UM) often do not reach therapeutic concentrations when given standard doses. In some cases, these individuals inherit up to 13 copies of the CYP2D6 gene, arranged in tandem [20]. This amplification polymorphism results in affected people metabolizing drugs that are CYP2D6 substrates so quickly that a therapeutic effect cannot be obtained at conventional doses. For example, it has been estimated that whereas a daily dose of 10–20 mg nortriptyline would be sufficient for a patient who is a CYP2D6 poor metabolizer, an UM inheriting multiple copies of the gene could require as much as 500 mg/day [21]. Ultrarapid metabolizers are found in 1–10% of Caucasians and 2–3% of African Americans. Among Ethiopian and Saudi Arabian populations, there is a very high frequency (20–30%) of the UM phenotype [22].

In addition to detoxifying and eliminating drugs and metabolites, CYP2D6 is required for activation of prodrugs. For example, codeine must be converted to its active metabolite, morphine, by CYP2D6, rendering the 2–8% of the population who are homozygous for nonfunctional CYP2D6 alleles resistant to the analgesic effects of this commonly used medication. Thus, this common polymorphism explains at least some of the interindividual variability in pain relief from standard doses of codeine [6].

Another important example is the pivotal role of CYP2D6 in the metabolism of tamoxifen. Tamoxifen is an estrogen receptor modulator that is widely used for the endocrine treatment of hormone receptor positive breast cancer. Although many other CYP enzymes are involved in the metabolism of tamoxifen, CYP2D6 is responsible for transforming tamoxifen in the active metabolite endoxifen that has approximately 100 times greater affinity for the estrogen receptor. Women with nonfunctional CYP2D6 alleles would therefore achieve much lower plasma levels of this more active metabolite [23]. This should be considered when treating these slow metabolizers for hormone receptor positive breast cancer.

CYP2C9 CYP2C9 is one of the most abundant cytochromes P450 in the human liver and has been shown to metabolize a large number of drugs, including *S*-warfarin, losartan, glipizide, tetrahydrocannabinol, phenytoin, torsemide, celecoxib, and various nonsteroidal anti-inflammatory drugs [24–27].

There have been six different CYP2C9 alleles described. The two most common variant alleles (CYP2C9*2 and CYP2C9*3) differ from the wild type allele (CYP2C9*1) by a single point mutation: CYP2C9*2 is characterized by an Arg144Cys amino acid substitution, whereas CYP2C9*3 has an Ile359Leu substitution. CYP2C9*2 and *3 have been reported to occur at a frequency of 8% and 6%, respectively, in the Caucasian population. Both of these variants are much less common in African American (1% and 0.5%, respectively) and Asian populations (0% and 2–3%, respectively) [28]. Both allelic variants are associated with reduced catalytic activity compared with the wild type. They are reported to show approximately 12% (*2) and less than 5% (*3) of wild type enzyme activity [29].

The most important and widely replicated clinical implication of the CYP2C9 polymorphism is related to oral anticoagulant treatment. Warfarin is the most widely prescribed oral anticoagulant for reducing thromboembolic events; however in Europe, acenocoumarol and phenprocoumon are also used. Oral anticoagulation can be difficult to dose, because of the narrow therapeutic window and the large inter- and intraindividual variations in the dosage needed to achieve the desired therapeutic effect. The effect of CYP2C9 polymorphisms on the pharmacokinetics of the anticoagulant is most pronounced in warfarin and least in phenprocoumon [30]. A systematic review and a meta-analysis of 39 studies comprising 7907 patients showed that, compared to the Cyp2C9*1/*1 genotype, dose requirement was lower for 19.6% in CYP2C9*1/*2, 33.5% in CYP2C9*1/*3, 36% in CYP2C9*2/*2, 56.7% in CYP2C9*2/*3, and 78,1% in CYP2C9*3/*3 [31]. To further explain the variability in anticoagulant dose requirements, the VKORC1 genotype should be determined. The two genes and environmental factors together can predict 50–60% of the variability in warfarin dose requirement [32].

Patients carrying at least one of the CYP2C9 variants have also been shown to require 30% less phenytoin to achieve therapeutic phenytoin concentrations [29].

Much less is known about the less common CYP2C9 variants. CYP2C9*4 results in an Ile359Thr substitution. It has been reported to be extremely rare [33]. CYP2C9*5 is reported to lead to an Asp360Glu substitution. The CYP2C9*5 variant has only been observed in African Americans, such that approximately 3% of this population carries the CYP2C9*5 allele. *In vitro* intrinsic clearances for CYP2C9*5, calculated as the ratio of V_{max}/K_m, ranged from 8% to 18% of CYP2C9*1 values in the initial report [34]. The CYP2C9*6 variant characterized by deletion of adenine in nucleotide 818 that results in a frameshift mutation causing a null allele was only found among African Americans with a frequency of 0.6% [35].

CYP2C19 CYP2C19, also known as mephenytoin hydroxylase, was first described in 1993 [36]. Since then, eight alleles have been identified. Each of the alleles other than CYP2C19*1 has been associated with almost complete absence of gene expression. Most of the alleles (CYP2C19*2, *3, *4, *5, *6, *7, and *8) occur infrequently (approximately 3–5% in total) in random Caucasian and African American populations. In all racial groups studied, CYP2C19*2 is the allele most commonly associated with an inactive gene product. Within Asian populations, the higher frequency of CYP2C19*2 and CYP2C19*3 alleles accounts for the higher prevalence in this racial group (approximately 35% PM) [37,38].

CYP2C19 metabolizes many clinically important drugs (Table 1). Subjects with the CYP2C19 PM phenotype have an area under the curve (AUC) of omeprazole that is more than sixfold higher than efficient metabolizers (EM), and the drug has a severely prolonged half-life in PM individuals [39]. To reach similar plasma levels, PMs of CYP2C19 would take about 1–2 mg of omeprazole instead of the recommended dose of 20 mg [40].

Regarding proton pump inhibitors, the effect of CYP2C19 PM status is not limited to pharmacokinetic alterations. The difference in the pharmacokinetics has been shown to influence the outcome of *H. Pylori* eradication therapy. Furuta et al. showed that in patients with confirmed *H. Pylori* infection treated with omeprazole or lansoprazole plus clarithromycin and amoxicillin, CYP2C19 PMs had an eradication rate of 97.8% compared with a rate of 72.7% ($p < 0.001$) for CYP2C19 EMs [41]. These findings have been confirmed in a meta-analysis [42].

In addition to the proton pump inhibitors, CYP2C19 genotype has also been shown to be associated with reduced elimination of diazepam [43], proguanil [44], imipramine [45], citalopram [46], carisoprodol [47], and hexobarbital [48].

CYP3A4/5/7 The CYP3A family consists of CYP3A4, CYP3A5, CYP3A7, and CYP3A43. The CYP3A members are the most abundant CYPs in the human liver and small intestine. Substantial interindividual differences in CYP3A expression, exceeding 30-fold in some populations, contribute greatly to variation in oral bioavailability and systemic clearance of CYP3A substrates [49]. One factor contributing to this large variability in 3A expression is the presence or absence of CYP3A5. CYP3A5 was previously detected in livers and small intestines of only some adult individuals, but the basis for this "polymorphic" expression was unknown [50,51].

Recently, two important SNPs in CYP3A5 (CYP3A5*3 and *6) were described. Relative to the wild type (CYP3A5*1), these mutations have been shown to cause alternative splicing and protein truncation, which results in the absence of CYP3A5. Only carriers of at least one CYP3A5*1 allele have been shown to express large amounts of CYP3A5. The ethnic distribution of the CYP3A5*1 allele indicates that relatively high levels of CYP3A5 are expressed by an estimated 30% of Caucasians, 30% of Japanese, 30% of Mexicans, 40% of Chinese, and more than 50% of African Americans, Southeast Asians, Pacific Islanders, and Southwestern American Indians [49].

For most Caucasians and African Americans who carry the CYP3A5*1 allele, CYP3A5 accounts for at least 50% of the total CYP3A content. Because most CYP3A4 substrates are also substrates for CYP3A5, this CYP3A5 polymorphism influences overall CYP3A activity in humans [52]. Thus, the presence or absence of CYP3A5 should contribute substantially to the total metabolic clearance of the many CYP3A substrates. Indeed, those heterozygous or homozygous for CYP3A5*1 should have the highest clearance and lowest oral bioavailability of CYP3A substrates. Moreover, these people might be more likely to encounter a lack of efficacy from standard doses [49]. The CYP3A5 polymorphism seems to be of importance for treatment with immunosuppressive drugs because they have a narrow therapeutic window and there is a large variability between patients in pharmacokinetics and pharmacodynamics. The influence of CYP3A5*1 and CYP3A5*3 on the pharmacokinetics of tacrolimus is well established. However, data on cyclosporine A and sirolimus disposition are not conclusive [53].

Important variation in other clinically relevant CYP enzymes such as CYP1A2, CYP2A6, and CYP2E1 has been demonstrated and reviewed in detail elsewhere [40].

3.1.2. Polymorphisms in Other Important Drug Metabolizing Enzymes

Dihydropyrimidine Dehydrogenase Dihydropyrimidine dehydrogenase (DPD) is the initial and rate-limiting enzyme in the catabolism of the chemotherapeutic agent 5-fluorouracil (5-FU). Diasio et al. conducted a familial study that suggested an autosomal recessive pattern of inheritance of deficiency of DPD.

While it seems that a complete absence of DPD activity is extremely rare, even partial enzyme inactivity might result in severe toxicity from 5-FU [54]. Prospectively evaluating the gene encoding DPD is a typical example of what could be a useful pharmacogenomic approach to preventing toxicity from a very effective drug that has a high level of toxicity [55].

N-Acetyltransferase-2 The *N*-acetyltransferase-2 (NAT2) polymorphism is one of the most common polymorphisms known in human populations. While more than 50% of Caucasians are NAT2 slow acetylator phenotype, there is a tremendous amount of interethnic variation in the frequency of the slow acetylator polymorphism [56]. For instance, the slow acetylator phenotype is much more frequent in Egyptians but is much less frequent in Asians [57].

Grant et al. first demonstrated that the classical isoniazid slow acetylator phenotype is due, at least in part, to reduction of the expression of NAT2 protein [58]. Seven SNPs in NAT2 form the most important haplotypes. Five of these lead to amino acid changes: 191G A (Arg63Glu), this one occurs mostly in African Blacks; 341T > C (Ile114Thr); 590G > A (Arg197Gln); 803A > G (Lys268Arg); and 857G A (Gly286Glu). The remaining two SNPs (282C > T and 481C > T) are silent mutations. Next to these SNPs, further less common SNPs exist. No variation has been reported in regulatory or untranslated regions. In 2008, an update on a new nomenclature on NAT in mammalians was published [59]. The current known haplotypes can be found on the Internet (http://louisville.edu/medschool/pharmacology/consensus-human-arylamine-n-acetyltransferase-gene-nomenclature/).

This polymorphism (NAT2) was discovered almost 50 years ago after differences were observed to isoniazid toxicity in tuberculosis patients [60]. Subsequently, the differences in isoniazid toxicity were attributed to genetic variability in NAT2, a cytosolic Phase II conjugation enzyme primarily responsible for deactivation of isoniazid [61]. Indeed, the polymorphism was termed the "isoniazid acetylation polymorphism" for many years until the importance of the polymorphism in the metabolism and disposition of other drugs and chemical carcinogens was fully appreciated [57].

Since these first observations, a wealth of clinical evidence has shown that the disposition of a variety of drugs (including sulfonamides, dapsone, hydralazine, procainamide, and caffeine) possessing primary aromatic amino or hydrazine functional groups is affected by the same genetic defect [62]. In addition to metabolizing drugs, NAT2 is also known to catalyze both *N*-acetylation (usually deactivation) and *O*-acetylation (usually activation) of aromatic and heterocyclic amine carcinogens. Epidemiological studies suggest that the NAT2 acetylation polymorphisms modify risk of developing urinary bladder, colorectal, breast, head and neck, lung, and possibly prostate cancers. Associations between slow NAT2 acetylator genotypes and urinary bladder cancer and between rapid NAT2 acetylator genotypes and colorectal cancer are the most consistently reported [57]. The importance of the NAT2 polymorphisms in clinical pharmacology and toxicology has been extensively reviewed [62,63]. Consideration of NAT2 genotype would possibly be beneficial for patients that start treatment with *N*-acetylated drugs; however, the results of large prospective studies are lacking.

Thiopurine Methyltransferase Azathioprine, thioguanine, and 6-mercaptopurine are thiopurine drugs that are used to treat acute lymphoblastic leukemia, autoimmune disorders, inflammatory bowel disease, and organ transplant recipients. These drugs are metabolized by the genetically polymorphic enzyme thiopurine methyltransferase (TPMT). Thiopurines are very useful drugs, but they have a relatively narrow therapeutic index, with life-threatening myelosuppression as a major toxicity [64].

Population studies have found that approximately 11% of Caucasians are heterozygous and 0.3% homozygous for TPMT deficiency [65]. For the TPMT polymorphism, all patients who inherit two nonfunctional TPMT alleles will develop dose-limiting hematopoietic toxicity. Patients deficient in this drug metabolizing enzyme can require up to a 15-fold reduction in mercaptopurine to prevent fatal hematotoxicity [66–69].

TPMT genotyping is recommended by the FDA prior to treatment with irinotecan.

TPMT poor metabolizers have a higher change of side effects. The clinical importance was shown in pediatric acute lymphoblastic leukemia patients. The study showed that TPMT poor metabolizers are at elevated risk of severe azathioprine side effects. Therefore, it is necessary to adjust the dose in these children [70].

3.2. Polymorphisms in Drug Transporter Genes

3.2.1. SLCO1B1 Organic anion transporters (OATPs) are classified within the solute carrier class subfamily referred to as solute carrier organic anion transporter family (SLCO). OATP1B transporters are expressed in the sinusoidal membrane of hepatocytes [71] and have a very broad substrate specificity. These substrates include many clinically relevant drugs. The SLCO1B1 gene has been extensively studied in the pharmacokinetics and dynamics of statins. Statins are cholesterol lowering drugs that inhibit 3-hydroxy-3-methylglutaryl coenzyme A (HMG-CoA) reductase. The pharmacokinetic different profile of *SLCO1B1* genotypes has also translated in reports of *SLCO1B1* genotype associated increased risk of myopathy. The SEARCH Collaborative Group study conducted a genomewide association study (GWAS) in 85 myopathy cases and 90 controls that were all taking 80 mg of simvastatin on a daily basis [72]. Only a noncoding SNP (rs4363657) within the *SLCO1B1* gene showed a strong association. This SNP is in strong linkage disequilibrium with the nonsynonymous rs4149056 SNP, which had previously been associated with altered statin pharmacokinetics [73,74]. For each copy of the variant allele, there was an approximate four times higher risk of myopathy. Importantly, they successfully replicated their findings in a trial with subjects treated with 40 mg of simvastatin daily [72]. The STRENGTH study investigated the genetics of four *CYP* genes and the *SLCO1B1* gene in relation to statin-induced myopathy (simva-, ator-, and pravastatin were included). Not only did they confirm the findings from the SEARCH study, but they also reported an association between the *SLCO1B1* risk allele and myalgia symptoms without CK elevation for simvastatin and atorvastatin (weaker), but not for pravastatin treatment [75]. An additional potentially important but very rare SNP in the *SLCO1B1* gene is T1628G. This novel mutation was discovered by a Japanese group in a patient with pravastatin-induced myopathy [76], and was shown to reduce transporter activity [77].

Various other drugs have been studied for their interactions with SNPs in the SLCO1B1 gene (e.g., torasemide, fexofenadine, and repaglinide). However, changes in drug efficacy or in frequency or severity of adverse events have not been described.

3.2.2. ABCB1 P-glycoprotein (P-gp/ABCB1) is an ATP-dependent drug efflux pump. It is the best characterized ABC-transporter and has been demonstrated to facilitate the export of a variety of compounds including several drugs in clinical use. In humans, P-gp is encoded by the multidrug resistance gene (MDR-1) that is located on the long arm of chromosome 7. Overexpression of P-gp in neoplastic cells is associated with the phenomenon of multidrug resistance to chemotherapeutic agents by promoting efflux of chemotherapy [78]. P-glycoprotein is also expressed in normal cells including the intestinal epithelium, renal proximal tubule, liver, adrenal cortex, placenta, testes, and blood–brain barrier. Recently, P-gp has been implicated as the causative factor in numerous pharmacokinetic interactions [79]. For example, amiodarone and quinidine therapy increases serum digoxin concentrations through inhibition of P-gp in the intestine and renal tubule, which increases digoxin absorption and decreases total body clearance. Table 2 provides a partial list of substrates, inhibitors, or inducers of P-gp. Considerable substrate overlap and tissue location exist between P-gp and the CYP 3A4 isoenzyme.

Many SNPs have been identified within the ABCB1 gene, and influence of these SNPs on mRNA and protein expression have been studied. [81,82]. Data concerning the influence of SNPs are not always consistent. It seems that functional findings might be more a consequence of haplotypes than that of isolated SNPs. Haplotype frequency significantly differs between ethnicities [83]. Digoxin is probably the most studied drug in association with

Table 2. Selected Substrates, Inducers, and Inhibitors of P-Glycoprotein [80]

P-Glycoprotein Substrates	P-Glycoprotein Inhibitors	P-Glycoprotein Inducers
Amiodarone	Amiodarone	Dexamethasone
Anthracyclines	Atorvastatin	Phenobarbital
Cisplatin	Clarithromycin	Phenytoin
Cyclosporine	Cyclosporine	Rifampin
Cytarabine	Diltiazem	St. John's Wort
Dactinomycin	Erythromycin	
Daunorubicin	Itraconazole	
Dexamethasone	Ketoconazole	
Digoxin	Quinidine	
Docetaxel	Quinine	
Doxorubicin	Ritonavir	
Etoposide	Tacrolimus	
Fexofenadine	Tamoxifen	
Fluorouracil	Verapamil	
Glucocorticoids		
Indinavir		
Loperamide		
Losartan		
Methotrexate		
Mitoxantrone		
Nelfinavir		
Paclitaxel		
Ritonavir		
Saquinavir		
Sirolimus		
Tacrolimus		
Topotecan		
Vinblastine		
Vincristine		
Vindesine		
Vinorelbine		

ABCB1 polymorphisms. This cardiac glycoside has a very narrow therapeutic window. The 3435C>T variant has been associated with enhanced digoxin plasma levels [84], however this could not be replicated by other studies [85]. The most likely explanation for the differences in results is that the study populations have different haplotypes. Several studies have shown associations with haplotypes of ABCB1 and digoxin plasma levels [86,87].

Other drugs (such as the antihistaminic agent fexofenadine and the immunosuppressant cyclosporine A) have been studied for interactions with the ABCB1 SNPs. Analog to the example of digoxin, the results of the studies were conflicting [88–90]. Therefore, it seems unlikely that ABCB1 plays a pivotal role in the pharmacogenetics of these drugs.

3.3. Polymorphisms in Drug Target Genes and Clinical Efficacy

The biomedical literature contains a multitude of publications that attempt to correlate genetic variation underlying differential response to medications. It is obviously not feasible to review all of the examples from the literature; however, Table 3 summarizes several of these polymorphisms that are associated with altered drug response. Our goal is to provide several illustrative examples of how polymorphisms in drug targets are being used to establish a personalized medicine platform.

Table 3. Selected Polymorphisms Associated with Altered Drug Response

Gene/Gene Product	Medication	Effect Associated with Polymorphism	Reference
ALOX5	Lipoxygenase inhibitors	Improvement in FEV_1	[91]
Angiotensin 1 receptor	Losartan	Reduction in mean arterial pressure	[92]
β_1-Adrenergic receptor	Metoprolol	Reduction in blood pressure	[93]
β_2-Adrenergic receptor	Albuterol	Improvement in FEV_1	[94,95]
Bradykinin B2 receptor	ACE inhibitors	ACE inhibitor induced cough	[96]
Dopamine D2 receptor	Dopamine antagonists	Anxiolytic/antidepressive effects of neuroleptics	[97]
Estrogen receptor	Equine estrogen	Increased bone mineral density	[98]
Gs α	β-Blockers	Reduction in blood pressure	[99]
Platelet Fc	Heparin	Heparin-induced thrombocytopenia	[100]
Serotonin transporter	SSRIs	Antidepressant response	[101,102]

3.3.1. ACE Insertion/Deletion Polymorphism One of the most studied polymorphisms occurs in the gene encoding angiotensin-converting enzyme. This polymorphism occurring in intron 16 is not a SNP. Rather it is an insertion/deletion (I/D) of a 287 bp product. Numerous studies have demonstrated that the D allele is associated with higher concentrations of the hormone, angiotensin II [103]. The functional consequences of these findings are apparent and have been supported by several examples from the literature demonstrating that the D allele is associated with increased risk of hypertension, myocardial infarction, and ventricular arrhythmias [104–106]. Presence of the D allele is also associated with a poorer prognosis among patients with heart failure [107]. Among 328 patients with heart failure followed at a cardiomyopathy clinic, after 2 years, the percentage of patients with transplant-free survival was 78% for the II genotype, 65% ID, and 60% DD ($p = 0.044$). Interestingly, these investigators also examined the impact of the ACE I/D polymorphism with β-blocker therapy in this cohort of patients. In the 208 patients who were not receiving β-blocker therapy, the 2-year transplant-free survival was 81% II, 61% ID, and 48% DD. However, a provocative finding was that β-blocker therapy obviated the influence of the D allele with poor prognosis. Among patients receiving β-blockers, the 2-year transplant-free survival was 70% for II genotype, 71% for ID, and 77% DD (p = NS).

More than 20 studies have examined the impact of the insertion/deletion polymorphism on clinical response to ACE inhibitor therapy [108–115]. Review of these studies reveals numerous conflicting results. Uncertainty about the expected biological mechanism-based direction of the association together with many studies showing opposite or no association directions of the association suggests that the *ACE* I/D polymorphism is not likely to be a strong modifier of blood pressure response to treatment with ACE inhibitors. Also stratification of results by study design, study population, or type of ACE inhibitor cannot explain the conflicting results.

However, perhaps the largest factor resides in the limitation of examining a single SNP. Medications act with numerous transporters and receptors to illicit a therapeutic response. Consequently, it is more plausible that variability in drug efficacy will be caused by polymorphisms in multiple genes involved in the drug response pathway. Thus, the findings with the ACE I/D polymorphism provide a rationale for future studies to eschew the simplicity of examining a single SNP and instead incorporate a more genomic approach.

3.3.2. 5-Lipoxygenase Polymorphisms Leukotrienes mediate airway inflammation and play an integral role in the pathophysiology of asthma. Zileuton, montelukast, and zafirlukast are leukotriene antagonists and interfere with leukotriene synthesis by inhibition of the enzyme 5-lipoxygenase, reducing the formation of leukotrienes and thus improves the symptoms of asthma. A polymorphism exists in the gene encoding the 5-lipoxygenase (ALOX5) promoter region. This polymorphism contains three to six tandem repeats of GGGCGG. Several studies have been performed and showed that individuals that do not carry the five tandem repeats allele show a decreased response (measured by change in FEV_1 or PEF and exacerbation rate) to therapy with leukotriene antagonists. [116].

3.3.3. β_1-Adrenergic Receptor Polymorphisms

The β_1-adrenergic receptor (AR) is a G-protein-coupled receptor expressed in a number of cell types including the heart and kidneys. The gene that codes for the β_1-AR is intronless and is located on chromosome 10q21. There are two common SNPs within the β_1-AR gene at codon 49 and codon 389 [117].

Codon 49 is located in the extracellular tail of the amino terminus end of the receptor, a potentially important region for receptor binding, regulation, and expression [118]. A nonsynonymous SNP produces a glycine (Gly) for serine (Ser) substitution at codon 49 (Ser49Gly). Although there are no data *in vivo* associating this polymorphism with drug response, a recent site-directed *in vitro* mutagenesis study suggests that agonist-promoted downregulation of the receptor is amplified with the Gly49 variant [119].

Codon 389 is located in the intracellular tail of the carboxy terminus end of the receptor, a potentially important region for G-protein coupling [118]. A nonsynonymous SNP produces a Gly for arginine (Arg) substitution at codon 389 (Arg389Gly). This polymorphism has been shown to vary by race, with African Americans possessing an allele frequency of 42% for the Gly389 variant while Caucasians possess a frequency of only 27% [120]. Furthermore, *in vitro* mutagenesis studies have shown a functional difference with this polymorphism [121]. In this study, those cells carrying the Arg389 variant had a nearly twofold greater resting activity rate, as measured by adenylyl cyclase levels, and an almost fourfold greater activity when stimulated with β-agonist, thus suggesting the Gly389 is a less active or perhaps less reactive receptor form. This theory has been recently put to test in a prospective study of patients with hypertension, and indeed, this polymorphism may be an important determinant of the antihypertensive response to β-blocker therapy [93]. Further studies are needed to evaluate whether this polymorphism confers the racial differences observed in response to β-blocker therapy in both hypertensive and heart failure patients.

3.3.4. β_2-Adrenergic Receptor Polymorphisms

β_2-Adrenergic Receptor SNPs The β_2-adrenergic receptor is a G-protein-coupled receptor that interacts with endogenous catecholamines and various pharmacologic agents. The mainstay of therapy for acute bronchoconstriction is the administration of β_2-AR agonists such as albuterol.

Several variations have been identified in *ADRB2*, the gene coding for the β_2-receptor. Nine single base substitutions have been identified in the ADRB2 coding region. Five of these substitutions are synonymous and therefore unlikely to be functionally important. Of the remaining four polymorphisms, three substitutions (an arginine substitution for cystine at codon 16, glutamine substitution for glutamic acid at codon 27, and a threonine substitution for isoleucine at codon 164) have functional effects. One rare polymorphism (valine substitution for methionine at codon 34) seems to have no influence on receptor function [122]. The polymorphisms at codon 16 (Arg16Gly) and 27 (Gln27Glu) are quite common (allele frequency 40% and 45%, respectively) and have been validated in several studies [123].

Patients who are homozygous carriers of arginine at position 16 showed improved therapeutic response after treatment with short-acting $(_2$-agonists when compared to patients homozygous for glycine at position 16 [124]. However, subsequent pharmacogenetic studies found opposite results (improved treatment response for homozygous glycine carriers) or no association [125]. Patients homozygous for glutamine at position 27 showed reduced treatment response compared to patients homozygous for glutamic acid [126].

A nonsynonymous SNP in the β_2-AR gene produces a Gly for Arg substitution at codon 16 (Arg16Gly). *In vitro* studies indicate that the Gly16 form of the receptor undergoes enhanced agonist-mediated downregulation compared with the Arg allele [127]. These findings have been supported by several clinical studies. One study demonstrated that subjects with the Arg16 homozygous genotype were 5.3 times more likely to have a positive bronchodilator response to a single dose of albuterol compared with Gly16 homozygotes [94]. Another study enrolled 16 patients with asthma and measured FEV_1 response to an 8 mg oral dose of albuterol [95]. The study population was divided into two groups: Arg16 homozygotes and Gly16 homozygotes. Patients who were homozygous for Arg16 had a fourfold greater FEV_1 response compared with Gly16 homozygotes despite nearly identical plasma albuterol concentrations. Moreover, the codon 16 genotype may also influence long-term response to albuterol. One study of 107 patients with mild-to-moderate asthma demonstrated that Arg16 homozygous patients receiving regularly scheduled albuterol had nearly double the number of asthma exacerbations per year compared with placebo. Furthermore, the rate of asthma exacerbations was significantly greater among Arg16 homozygotes during treatment with albuterol compared with heterozygotes and Gly16 homozygotes [128]. A separate study found that Arg16 homozygotes receiving regularly scheduled albuterol experienced a decrease in morning peak expiratory flow rate (PEFR). In contrast, Gly16 homozygous patients who received regularly scheduled albuterol did not experience a decline in PEFR [129]. These results suggest that the Arg16 homozygous genotype is associated with deleterious effects from regularly scheduled albuterol therapy and that patients with this genotype should only receive albuterol for breakthrough symptoms.

β_2-Adrenergic Receptor Haplotypes In all, 13 polymorphisms occur in the β_2-AR gene. Thus, if these polymorphisms occurred completely randomly, one would expect a total of 2^{13} variations (haplotypes) in the β_2-AR gene. However, only 12 haplotypes occur, and 5 haplotypes describe 88% of the population [8]. Thus, the polymorphisms in the β_2-AR are in strong linkage disequilibrium. A seminal paper has investigated the impact of haplotypes, rather than individual SNPs in predicting response to albuterol among patients with asthma. Importantly, there was no association between an individual SNP and response to albuterol. However, haplotype pair was significantly related to improvements in FEV_1 from albuterol [8]. Therefore, examination of multiple SNPs in a receptor that is physiologically linked to drug response resulted in the best prediction of therapeutic efficacy. Focusing on multiple genes and/or multiple SNPs to determine disease associations or drug response is analogous to the multiple factors that a clinician must consider when dosing. For example, when prescribing digoxin, the likelihood of prescribing the appropriate dose is increased when a clinician considers multiple factors such as patient age, body size and weight, renal function, and concomitant drug therapy.

3.4. Single Gene Pharmacogenetic Studies in Genes Influencing Disease Progression

Once again, it would not be feasible to review all the examples in the biomedical literature of polymorphisms influencing disease severity or progression. However, Table 4 summarizes many of these polymorphisms that influence disease severity and associated drug response. We will discuss two of these disease states in detail.

3.4.1. Alzheimer's Disease Alzheimer's disease is a progressive complex disorder, where genetic predisposition interacts with environmental factors [145]. One of the most well-studied disease association genes is the apolipoprotein E (APOE) gene located on chromosome 19; this is the gene related with the risk of developing Alzheimer's disease. Three allelic variants exist for APOE. The frequency for E3, E4, and E2 are 77%, 15%, and 8%, respectively. Several studies have determined an association between the presence of the E4 allele with late-onset (greater than 60 years of age) Alzheimer's disease [146,147]. Furthermore, a gene-dose response exists for the E4 allele and the risk of late-onset Alzheimer's.

Table 4. Selected Polymorphisms Influencing Disease Severity and Associated Drug Response

Gene	Altered Disease Severity	Impact of Polymorphism on Drug Response	References
ACE	D allele ↑ risk of death or need for heart transplant	β-Blockers abolished poorer prognosis of patients with DD genotype	[107]
APOE	E4 allele ↑ in Alzheimer's disease	Presence of E4 allele associated with poor response to tacrine	[130]
APOE	Smokers with an E4 allele had a threefold ↑ risk of CHD event	Unknown	[112]
β_1	Heart failure patients with the Ser49Ser genotype had a worse 5-year prognosis compared with patients with a Gly variant	Unknown	[113]
β_2	Reduced survival rate among heart failure patients with an Ile164 allele	Unknown	[131]
	Reduced exercise capacity among heart failure patients with an Ile164 allele	Unknown	[132]
Calcitonin	Association between heterozygosity and ↓ fracture risk among postmenopausal women	Unknown	[133]
Cystathionine beta synthase	Risk of coronary artery disease	Response to homocysteine lowering from folic acid	[134]
Endothelin A	↑ Frequency of TT homozygotes in idiopathic dilated cardiomyopathy	Unknown	[135]
Factor V	Risk of venous thrombosis	↑ Risk of venous thrombosis from oral contraceptives	[136]
Factor VII	Patients with the ArgArg genotype have a threefold ↑ risk of complications after PCI	Unknown	[137]
	Odds ratio of MI among patients with ArgGln genotype = 0.47 (0.27–0.81) compared with ArgArg genotype	Unknown	[138]
GP IIIa	Association between Pl^{A2} allele and acute coronary thrombosis	Unknown	[139]
G-protein	Increased BMI among TT homozygous primiparous women	Unknown	[140]
HERG, KvLQT1, MiRP1	Long QT syndrome	↑ Risk of drug-induced Torsade de pointes	[141,142]
P-selectin	Risk of myocardial infarction	Unknown	[143]
Prothrombin	Risk of venous thrombosis	↑ Risk of venous thrombosis with oral contraceptives	[144]

CHD: coronary heart disease; PCI: percutaneous coronary intervention; MI: myocardial infarction; GP: glycoprotein; BMI: body mass index.

In one case-control study, the odds ratio for developing Alzheimer's disease was 3.9 for the E3/E4 genotype and 15.6 for the E4/E4 genotype. Interestingly, the allelic variant of APOE also influences treatment response with the cholinesterase inhibitor, tacrine [130]. In one study, 83% of non-APOE4 carrying patients had improvements in cognition when given tacrine. In contrast, 60% of patients with an E4 allele were unchanged or declined after tacrine administration. Despite these data, single SNP studies of the APOE gene would not justify withholding therapy because 40% of patients with the E4 allele had a positive response to tacrine [130]. The combination of the APOE genotype with the presenilin 1 (PS1) and the presenilin 2 (PS2) genotypes in haplotypes has been studied [148]. Subjects with a defective PS2 exon 5 show less therapeutic response compared with subjects without this defective exon. Furthermore, 15% of the Caucasian population with Alzheimer's disease are reported to slow metabolizers for CYP2D6. This can also be a reason why some patients do not react well to cholinesterase inhibitors [149].

3.4.2. Hypercholesterolemia Statins primarily reduce the risk on coronary artery disease (CAD) by lowering blood cholesterol through inhibition of the HMG-CoA reductase enzyme. Although large clinical trials found a 27% average relative risk reduction of major coronary events [150], there is large variability in benefits from statin therapy. Many genes involved in the pharmacodynamic pathway of statins have been part of pharmacogenetic research in patients with hypercholesterolemia, with an emphasis on genes involved in the cholesterol pathway, although genes involved with possible pleiotropic effects of statins gain more and more interest [151]. The enormous amount of candidate genes that have been associated with statin efficacy highlights the complexity of the mechanism by which statins exert their beneficial effects.

In 1998, Kuivenhoven et al. published a well-cited paper on the importance of a common SNP (Taq1B) in the CETP (cholesteryl ester transfer protein) in predicting the efficacy of statin therapy in reducing intima media thickness [152]. The CETP enzyme plays a central role in transport of cholesterol from peripheral tissues back to the liver. Kuivenhoven et al. showed that pravastatin therapy slowed the progression of coronary atherosclerosis in B1B1 carriers, whereas B2B2 carriers did not benefit from pravastatin therapy although higher HDL levels were observed. More recently, a 10-year follow-up analysis in the same REGRESS cohort was published, showing similar results: more benefit from statin therapy for B1B1 carriers on cardiovascular clinical outcomes and all cause mortality despite higher HDL levels observed in *B2* carriers [153]. Several other studies investigated the *CETP* polymorphism as well but could not find an association of the Taq1B polymorphism with altered efficacy of statins in preventing cardiovascular diseases [154–156] or found the opposite [157]. A large meta-analysis including over 13,000 patients found no gene treatment interaction between statins and the Taq1B polymorphism of the *CETP* gene [158]. The conclusion after this meta-analysis has to be that the polymorphism in the CETP gene does not play an important role in predicting efficacy of statins.

HMG-CoA reductase, encoded by the *HMGCR* gene, is responsible for the conversion of HMG-CoA to mevalonic acid, an intermediate in the cholesterol synthesis. The minor alleles of two SNPs, SNP12 and SNP29 (rs17238540), jointly uniquely define haplotype 7 of the of the *HMGCR* gene [159]. The minor alleles for both the SNPs were associated with less pronounced total cholesterol (TC) and low-density lipoprotein (LDL) cholesterol reduction in response to pravastatin treatment (PRINCE study) [159]. Again replication studies showed conflicting results [160–162]. Both SNP 12 and SNP 29 are located in a noncoding region and further research should determine whether there is a molecular explanation for the results that were found in several studies. These SNPs might explain a small part of the variability in response to statins due to genetics.

Cholesterol is transported throughout the body by apolipoproteins. The apolipoprotein E is a major component of very low-density lipoproteins (VLDLs) and ligand for the LDL receptor. Moreover, apolipoprotein E is involved in intestinal cholesterol absorption and reverse cholesterol transport (RCT). Apolipoprotein E is a genetically polymorphic protein defined by three alleles, *ApoE2*, *E3*, and *E4*, encoding proteins with increasing affinity for the LDL receptor. Consequently, lipoproteins carrying the E4 isoform are cleared most efficiently from the circulation and cholesterol synthesis and thereby HMG-CoA reductase levels are lower. As a result, *E4* carriers could benefit less from statin therapy. Ordovas et al. reported that carriers of the *E2* genotype experience greatest LDL reduction in response to statin therapy in comparison to *E3* and *E4* carriers [163]. Many similar studies were conducted trying to clarify the role of the ApoE polymorphism in the pharmacogenetics of statins. There is a reasonable body of evidence supporting the findings from Ordovas et al.; however, studies that could not show similar results have also been reported. *ApoE2* carriers seem to benefit most from statin therapy regarding lipid profile improvement. However, contrary to these findings, a substudy of the Scandinavian Simvastatin Survival Study (S4) and the GISSI-Prevenzione study (both multicentre, double-blind, randomized trials) reported subjects carrying the ApoE4 genotype to have the largest risk reduction of mortality [164,165]. This would mean that patients with the least effect on cholesterol have the best protection against mortality. Statins not only exert their effects via cholesterol lowering but also have so-called pleiotropic effects. These effects might be more pronounced in E4 carriers. If this is true, measuring cholesterol might not be the best way to assess the efficacy of statins in the individual patient.

At this time, it is premature to withhold statin therapy from any patient meeting criteria for treatment based on national consensus guidelines. Future prospective genetic epidemiology studies may further delineate the role of genetic variants on the development and progression of coronary artery disease and response to treatment.

3.5. Clinical Relevance

For many decades, we have known that patients respond differently to drugs. The contribution of genetic variation to interindividual response to isoniazid, an antimycobacterial drug, was described as early as in 1954 [166]. Although technological advances have provided us with relatively easy and cheap methods for genotyping high expectations about personalized medicine have not yet been met. However, recently several clinical relevant pharmacogenetic interactions have been proven and implemented into clinical practice. A good recent example in prevention of adverse reactions is hypersensitivity to abacavir, an antiviral drug. Prospective HLA-B*5701 screening greatly reduces the risk of abacavir hypersensitivity [167].

At this time, expectations of pharmacogenetics are becoming more realistic. Pharmacogenetics is no longer a panacea, but a useful tool for a subset of drugs. It might be expected that pharmacogenetics will play an important role in clinical practice within the next decade.

4. RESEARCH AND DEVELOPMENT

The numbers are staggering. It costs the pharmaceutical industry approximately $880 million and 15 years to go from target identification through regulatory approval for a novel drug. One-half of this cost and time occur during Phase II and Phase III clinical trials. Contributing to the prodigious cost and time are the many inefficiencies of drug discovery, development, and clinical trials. Seventy-five percent of the costs of drug development are incurred from late-stage clinical trials. Incorporation of pharmacogenomic data may lead to a dramatic change in the way clinical trials are designed and conducted. Pharmacogenomics may result in more efficient trials that would be associated with a lower cost to bring a new chemical entity to market. Moreover, this could result in shorter time to drug approval, longer patent protection, and more importantly, increased time of market exclusivity. A recent report concluded that incorporation of genomic technologies currently available could result in savings of up to $300 million per novel drug and cut 2 years from the drug development process (http://www.bcg.com). This section will focus on specific examples of how pharmacogenomic data can be incorporated into clinical trials to identify responders and exclude nonresponders to reduce unnecessary adverse events.

4.1. Influence of Pharmacogenomics on Clinical Trials

In recent years, the complexity and cost of clinical trials has increased. It is not uncommon for Phase III drug studies to involve thousands of patients with several years of follow-up. However, the results of these trials only provide information on the average treatment effect in a population, not for an individual patient. In addition, numerous drugs advance all the way to Phase III trials only then failing to demonstrate any treatment benefit or having an unacceptable adverse event profile that prevents regulatory approval. Drug safety continues to be an enduring problem. High profile drug withdrawals are still a common practice, as recently illustrated with respect to the COX-2 inhibitors, three of which have been taken off the market for different adverse reactions (rofecoxib—cardiovascular events; valdecoxib—blistering skin reactions; and lumiracoxib—hepatotoxicity). Furthermore, serious adverse reactions of relatively high frequency have occurred in recent years. Regulatory action has been required in order to manage and minimize risk. For example, new prescribing indications have been initiated for rosiglitazone, which has been associated with cardiovascular events.

4.2. Use of Pharmacogenomic Data in the Clinical Trial Process

Assume that a company has a novel drug to treat chronic heart failure in Phase II trials. Data from this study demonstrate that 35% of the study population receiving active drug achieves a therapeutic response compared with 15% of placebo-treated patients. Do these data warrant further study? If this hypothetical Phase II trial had incorporated pharmacogenomic data, the potential exists that a genetic subgroup of responders could have been identified. Suppose that 60% of patients with the hypothetical AA genotype respond to this drug while patients with genotypes Aa and aa have a response rate similar to placebo. A homogeneous patient population (AA genotype) likely to respond to the medication could now be selected for a confirmatory Phase III trial. The sample size needed for a Phase III trial would be reduced because patients likely to respond to the medication have been enrolled. In fact, the majority of clinical trials that use pharmacogenomic information will require smaller sample sizes compared with trials in which no genotypic information is collected [168].

Alternatively, assume a company has developed an investigational drug to treat type 2 diabetes mellitus. Phase II studies demonstrate that 75% patients reduce their glycosylated hemoglobin (HbA_{1c}) levels by 1.5%. However, the drug is associated with a serious adverse event in 5% of patients. What impact will this serious adverse drug event have on the approval and use of this agent? If DNA samples had been collected from all study participants, a population of patients predisposed to developing this toxicity may have been identified. The susceptible patient population could then be excluded from Phase III trials, and assuming that the agent is eventually approved, a bedside molecular diagnostic test could be used to prospectively identify patients predisposed to the serious adverse event. In this aforementioned example, everyone benefits, including the drug manufacturer, patients, and the healthcare system by avoiding the costs of a serious adverse drug event. Furthermore, pharmacogenomic data could also be used in postmarketing surveillance, which would dramatically improve our current surveillance system [169].

Four pieces of information are important before planning a clinical trial using pharmacogenomic data [168]. First, the allele frequency for the polymorphism of interest must be known. Most trials with a nominal outcome variable would require sample sizes of greater than 1000 patients for allele frequencies of less than 10%. For these rare SNPs, association with efficacy and/or toxicity might only be discerned after the drug has been approved. However, studies of polymorphisms with an allele frequency of 30–50% would require fewer subjects compared with a trial that does not use pharmacogenomic information [168].

Second, the gene action of the SNP must be understood. Does the SNP behave in a dominant, additive, or recessive fashion? A dominant action means that the genotype Aa displays the same phenotype as genotype AA. This example would require fewer study participants compared with alleles displaying additive or recessive action. Third, the investigator should have knowledge of genotype relative risk (GRR). This is potentially the most difficult of the factors to have *a priori* knowledge, and may require assumptions, as are typically done to estimate the sample size of most clinical trials. As with any trial, the smaller the genotype relative risk, the larger the number of subjects required. The GRR is likely to be small given the multitude of factors influencing drug response. Fourth, as the number of alleles tested increases, the sample size also must increase. For example, studying 10 loci would require a sample size 1.5 times larger than studying a single locus. Studies of 100 loci would require sample sizes twice as large compared with a single locus. Studying 100,000 loci necessitates a sample size three times larger compared with a single locus [168].

4.3. Examples from the Literature

A study prospectively genotyped all patients for the CYP2D6 gene and excluded poor metabolizers to enhance patient safety [170]. The study was a randomized, double-blinded comparison of lamotrigine with desipramine in patients with unipolar depression. Desipramine is a substrate for CYP2D6 and poor metabolizers of this enzyme have serum desipramine concentrations markedly higher than extensive metabolizers [17]. In all, 6.1% of subjects screened were excluded from the trial because they were identified as poor metabolizers. Clearly, using genotyping as an inclusion or exclusion criterion requires that the genotype(s) of interest can be determined in a rapid fashion. For trials that involve acute treatment, a point of care or "bedside" test would need to be developed to potentially exclude patients at an increased risk of adverse events.

It is understandable why the pharmaceutical industry may have some trepidation of using pharmacogenomics in the clinical trial process. The basic tenet of most of the pharmaceutical industry has been to develop drugs in a one-size-fits-all approach with the hope that they will become blockbusters, generating greater than $1 billion in annual sales. There has been some concern that pharmacogenomic data will reduce the market share of a drug by identifying the subgroups of patients who actually derive therapeutic benefit from a drug. However, the example of abacavir proved that even though the group of patients that are HLA B5701 positive are excluded from the use of abacavir, the market share of the drug grew after the implementation of the pharmacogenetic test in clinical practice. Physicians who did not dare to prescribe abacavir because of the risk of developing severe side effects did start to prescribe abacavir again to patients that have a negative test result [171]. Several major unanswered questions exist as to how the FDA will react to new drug applications that contain pharmacogenomic data. Companies are not forced to include pharmacogenomic data in their drug dossier, but can do that on a facultative basis. It is more and more common that a pharmacogenomic dossier is submitted. On the Web site of the FDA, pharmacogenomic recommendations that have been added to drug labels can be found. Clearly, drugs approved for genetic subgroups will require molecular diagnostic testing and thus raise several compelling questions [172]. First, how will this affect off-label prescribing? What impact will this have on liability? Second, what happens if a patient refuses to have a molecular diagnostic test performed? Are they now limiting themselves to new therapeutic agents? What happens to the low socioeconomic populations who have no insurance and cannot afford the cost of the genetic test? Is this population now excluded from receiving potentially safe and effective pharmacotherapy? How will pharmacogenomics influence orphan drug status? In the United States, drugs developed for conditions affecting less than 200,000 individuals get 7 years of market exclusivity, unless an alternative medication is proven superior. Will this be the financial incentive that some companies require? Compelling questions indeed.

5. OBSTACLES FACING THE FIELD OF PHARMACOGENOMICS

5.1. Complexity and Cost

Current methods to sequence DNA are too costly and laborious to allow implementation into routine clinical practice. However, methods are rapidly inproving and this might soon change. Advances in high-throughput technology and increased competition have recuded the cost of genotyping SNPs to $0.001/SNP, thus making it viable to evaluate up to millions of SNPs per patient. Another factor contributing to cost involves the initial investment in the equipment and technology required to fully integrate pharmacogenomics into all aspects of drug discovery and development. However, these upfront costs may result in significant cost savings by increasing efficiency of clinical trials with the identification and termination of drugs with little potential of eventual regulatory approval. The deluge of the data generated from pharmacogenomic studies also requires advances in bioinformatics and data mining strategies. The extent to which pharmacogenomics pervades clinical practice is largely dependent on the ability of bioinformatics to transform the prodigious amounts of data into knowledge. Consequently, the bioinformatics budgets of some pharmaceutical companies have increased 20–60%—a sure indication of the desperate need for this technology.

5.2. Will Pharmacogenomics Improve Medical Care?

It is still unclear if prospectively genotyping patients for many of the genetic variants described within this chapter improves medial care and whether it is cost effective. Genotyping patients for the presence of TPMT deficiency to prevent life-threatening hematological toxicity from azathioprine, mercaptopurine, or thioguanine provides an equivocal advantage compared with empirical dosing. However, in other examples, genotyping may not be an advantage over the current best medical care. For example, several studies have demonstrated that individuals carrying mutant alleles for CYP2C9 have decreased metabolism or elimination of warfarin [11,173–175]. One retrospective trial also showed that these individuals have a higher risk for both minor and major bleeding [11]. However, it remains to be elucidated whether prospectively genotyping patients receiving warfarin reduces bleeding complications when compared with the best available standard of care. This question can only be answered through a randomized controlled trial comparing a genomic approach with a traditional dosing approach. Both in the US (the COAG trial) and in Europe (the EU-PACT trial), trials have started to answer this question [176].

5.3. Paradigm Shift in Healthcare

Transitioning from the one-size-fits-all approach to personalized medicine will create a paradigm shift for both healthcare providers and the pharmaceutical industry. There is some precedence for drugs specifically targeted for a subset of disease. Trastuzumab (Herceptin) is approved only for the 25–30% women whose breast cancer overexpresses the HER-2/neu protein. Trastuzumab is marketed along with a molecular diagnostic, the DAKO Hercep Test. This test is a semiquantitative assay for testing breast tumor tissue that overexpresses the HER-2/neu protein. In its first year on the market, trastuzumab generated $188.4 million in sales. Thus, an agent that probably would not have received FDA approval is now an effective alternative for a specific subgroup of women with breast cancer.

Physicians will also need to adjust to the shift from the art of medicine to the science of medicine. Rather than selecting a drug based on experience, the selection may be based on the analysis from a computer program. Rather than making a diagnosis based on phenotypic symptoms, a disease may be diagnosed years before it manifests, based on an individual's genetic makeup.

5.4. Ethical Considerations

Pharmacogenomics has created a new lexicon that all healthcare providers must familiarize themselves, and thus precise language is fundamental when dealing with pharmacogenomics. It is imperative that pharmacogenomics be distinguished from genetic predisposition testing. All investigators in the field must convey this concept to the public, members of the healthcare team, and the insurance sector. In the majority of cases, identification of SNPs to predict drug response carries no prognostic information for diseases. Determination of SNPs to predict drug response is analogous to obtaining culture and sensitivity data to guide antimicrobial therapy. However, in other cases the potential for discrimination exists. One example involves the apolipoprotein E4 allele. Genetic variation in the apolipoprotein E4 allele may be examined to explain the variability among statin therapy [177]. However, these results could have significant implications in predicting the risk of Alzheimer's disease later in life [146,147]. Clearly, an enormous potential for discrimination exists if an insurance carrier discovers the results of this diagnostic test. As an example, the New Jersey Genetic Privacy Act, passed in 1996, prevents employers and insurance carriers from discriminating based on genetic tests. Indeed, further state and federal legislation like New Jersey's would help to allay some of these ethical concerns and public fear of genetic testing.

Undeniably, genomics and pharmacogenomics opens up a whole host of legal, ethical, and societal issues that will have implications in patient confidentiality, discrimination, malpractice, and informed consent [174,175,178]. However, as clinicians, scientists, and healthcare practitioners, we must always remember the unquestionable power of an individual's right to choose, and prospectively fight to ensure patients' rights and prevent genomic discrimination.

6. CONCLUSION

The publication of the draft of the Human Genome Project represents an acmatic scientific breakthrough. One of the first discernible benefits from the Human Genome Project will be advances in pharmacogenomics. Pharmacogenomics is likely to have a major role in the daily practice of medicine in the near future. In many ways, a proof of principle is required. Although trastuzumab is an effective agent, its use is based on differences in protein expression and not genetic variation. Whatever our future holds, we must always remember the ultimate goal of pharmacogenomics: to develop or use truly individualized pharmacotherapy that will produce the most benefit and the least harm thus extending and enhancing human life.

REFERENCES

1. Anonymous. Nature 2001;409:813–958.
2. Evans DA. Ann NY Acad Sci 1968;151: 723–733.
3. Theodore J, Millen JE, Murdaugh HV, et al. Am Rev Respir Dis 1967;96:508–511.
4. Mahgoub A, Idle JR, Dring LG, et al. Lancet 1977;2:584–586.
5. Kalow W, Meyer UA, Tyndale R. Pharmacogenomics. New York: Marcel Dekker; 2001.
6. Evans WE, Relling MV. Science 1999;286: 487–491.
7. Stephens JC. Mol Diagn 1999;4:309–317.
8. Drysdale CM, McGraw DW, Stack CB, et al. Proc Natl Acad Sci USA 2000;97:10483–10488.
9. Carson PE, et al. Science 1956;124:484–485.
10. Kalow W. Lancet 1956;211:576.
11. Aithal GP, Day CP, Kesteven PJ, et al. Lancet 1999;353:717–719.
12. Meyer UA. Lancet. 2000;356:1667–1671.
13. Meyer UA, Zanger UM. Annu Rev Pharmacol Toxicol 1997;37:269–296.
14. Griese EU, Zanger UM, Brudermanns U, et al. Pharmacogenetics 1998;8:15–26.
15. Sachse C, Brockmoller J, Bauer S, et al. Am J Hum Genet 1997;60:284–295.
16. Zanger UM, Fischer J, Raimundo S, et al. Pharmacogenetics 2001;11:573–585.
17. Dahl ML, Johansson I, Palmertz MP, et al. Clin Pharmacol Ther 1992;51:12–17.
18. Roh HK, Dahl ML, Johansson I, et al. Pharmacogenetics 1996;6:441–447.
19. Lessard E, Yessine MA, Hamelin BA, et al. Pharmacogenetics 1999;9:435–443.

20. Johansson I, Lundqvist E, Bertilsson L, et al. Proc Natl Acad Sci USA 1993;90: 11825–11829.
21. Bertilsson L, Dahl ML, Sjoqvist F, et al. Lancet 1993;341:63.
22. Wolf CR, Smith G, Smith RL. BMJ 2000;320: 987–990.
23. Lim HS, et al. Clin Oncol 2007;25:3837–3845.
24. Bourrie M, Meunier V, Berger Y, et al. Drug Metab Dispos 1999;27:288–296.
25. Goldstein JA, De Morais SM. Pharmacogenetics 1994;4:285–299.
26. Kidd RS, Straughn AB, Meyer MC, et al. Pharmacogenetics 1999;9:71–80.
27. McCrea JB, Cribb A, Rushmore T, et al. Clin Pharmacol Ther 1999;65:348–352.
28. Sullivan-Klose TH, Ghanayem BI, Bell DA, et al. Pharmacogenetics 1996;6:341–349.
29. van der WJ, Steijns LS, van Weelden MJ, et al. Pharmacogenetics 2001;11:287–291.
30. Beinema M, et al. Thromb Haemost 2008;100; 1052–1057.
31. Lindh JD. Eur J Clin Pharmacol. 2009;65: 365–375.
32. Wadelius M. Blood 2009,113,784–792.
33. Imai J, Ieiri I, Mamiya K, et al. Pharmacogenetics 2000;10:85–89.
34. Dickmann LJ, Rettie AE, Kneller MB, et al. Mol Pharmacol 2001;60:382–387.
35. Kidd RS. Pharmacogenetics 2001,11, 803–808.
36. Wrighton SA, Stevens JC, Becker GW, et al. Arch Biochem Biophys 1993;306:240–245.
37. Goldstein JA, Ishizaki T, Chiba K, et al. Pharmacogenetics 1997;7:59–64.
38. Wedlund PJ. Pharmacology 2000;61:174–183.
39. Andersson T, Regardh CG, Lou YC, et al. Pharmacogenetics 1992;2:25–31.
40. Ingelman-Sundberg M. Drug Metab Dispos 2001;29:570–573.
41. Furuta T, Shirai N, Takashima M, et al. Clin Pharmacol Ther 2001;69:158–168.
42. Zhao F, et al. Helicobacter 2008,13,532–541.
43. Qin XP, Xie HG, Wang W, et al. Clin Pharmacol Ther 1999;66:642–646.
44. Setiabudy R, Kusaka M, Chiba K, et al. Br J Clin Pharmacol 1995;39:297–303.
45. Skjelbo E, Brosen K, Hallas J, et al. Clin Pharmacol Ther 1991;49:18–23.
46. Sindrup SH, Brosen K, Hansen MG, et al. Ther Drug Monit 1993;15:11–17.
47. Dalen P, Alvan G, Wakelkamp M, et al. Pharmacogenetics 1996;6:387–394.
48. Adedoyin A, Prakash C, O'Shea D, et al. Pharmacogenetics 1994;4:27–38.
49. Kuehl P, Zhang J, Lin Y, et al. Nat Genet 2001;27:383–391.
50. Schuetz JD, Molowa DT, Guzelian PS. Arch Biochem Biophys 1989;274:355–365.
51. Wrighton SA, Ring BJ, Watkins PB, et al. Mol Pharmacol 1989;36:97–105.
52. Wrighton SA, Schuetz EG, Thummel KE, et al. Drug Metab Rev 2000;32:339–361.
53. Wang J. Expert Rev Mol Diagn 2009;9; 383–390.
54. Diasio RB, Beavers TL, Carpenter JT. J Clin Invest 1988;81:47–51.
55. Rioux PP. Am J Health Syst Pharm 2000;57: 887–898.
56. Grant DM, Hughes NC, Janezic SA, et al. Mutat Res 1997;376:61–70.
57. Hein DW, Doll MA, Fretland AJ, et al. Cancer Epidemiol Biomarkers Prev 2000;9:29–42.
58. Grant DM, Morike K, Eichelbaum M, et al. J Clin Invest 1990;85:968–972.
59. Hein DW, et al. Pharmacogenet Genomics 2008;18(4):367–368.
60. Hughes HB, Biehl JP, Jones AP, et al. Am Rev Respir Dis 1954;70:266–273.
61. Weber WW, Hein DW. Clin Pharmacokinet 1979;4:401–422.
62. Grant DM, Goodfellow GH, Sugamori K, et al. Pharmacology 2000;61:204–211.
63. Weber WW, The Acetylator Genes and Drug Response. New York: Oxford University Press; 1987.
64. Weinshilboum R. Drug Metab Dispos 2001;29: 601–605.
65. Lennard L, Van Loon JA, Weinshilboum RM. Clin Pharmacol Ther 1989;46:149–154.
66. Evans WE, Horner M, Chu YQ, et al. J Pediatr 1991;119:985–989.
67. Iyer L, Ratain MJ. Eur J Cancer 1998;34: 1493–1499.
68. McLeod HL, Krynetski EY, Relling MV, et al. Leukemia 2000;14:567–572.
69. Weinshilboum RM, Otterness DM, Szumlanski CL. Annu Rev Pharmacol Toxicol 1999;39: 19–52.
70. Stanulla M. JAMA 2005;293(12):1485–1489.
71. Konig J. Am J Physiol Gastrointest Liver Physiol 2000;278(1):G156–G164.

72. Link E, Parish S, Armitage J, Bowman L, Heath S, Matsuda F, et al. N Engl J Med 2008;359(8):789–799.
73. Niemi M. Pharmacogenomics. 2007;8(7): 787–802.
74. Pasanen MK, Neuvonen M, Neuvonen PJ, Niemi M. Pharmacogenet Genomics. 2006;16 (12):873–879.
75. Voora D, Shah SH, Spasojevic I, Ali S, Reed CR, Salisbury BA, et al. J Am Coll Cardiol 2009;54 (17):1609–1616.
76. Morimoto K, Oishi T, Ueda S, Ueda M, Hosokawa M, Chiba K. Drug Metab Pharmacokinet. 2004;19(6):453–455.
77. Furihata T, Satoh N, Ohishi T, Ugajin M, Kameyama Y, Morimoto K, et al. Pharmacogenomics J 2009;9(3):185–193.
78. Lehne G. Curr Drug Targets 2000;1:85–99.
79. Fromm MF, Kim RB, Stein CM, et al. Circulation 1999;99:552–557.
80. Matheny CJ, Lamb MW Brouwer KR, et al. Pharmacotherapy 2001;21:778–796.
81. Cascorbi I, Pharmacol Ther 2006;112(2); 457–473.
82. Leschziner GD. Pharmacogenomics J 2007;7 (3):154–179.
83. Kim RB. Clin Pharmacol Ther 2001;70(2); 189–199.
84. Verstuyft C. Eur J Clin Pharmacol 2003;58 (12):809–812.
85. Gerloff T. Br J Clin Pharmacol 2002;54 (6):610–616.
86. Aarnoudse AJ, Pharmacogenet Genomics 2008;18(4):299–305.
87. Johne A. Clin Pharmacol Ther 2002;72: 584–594.
88. Drescher S. Br J Clin Pharmacol 2002;53 (5):526–534.
89. Yi SY. Clin Pharmacol Ther 2004;76(5) 418–427.
90. Jiang ZP. Basic Clin Pharmacol Toxicol 2008; 103(5):433–444.
91. Drazen JM, Yandava CN, Dube L, et al. Nat Genet 1999;22:168–170.
92. Miller JA, Thai K, Scholey JW. Kidney Int 1999;56:2173–2180.
93. Puckett BJ, Pauly DF, Zineh I, et al. Clin Pharmacol Ther 2002;71:P2.
94. Martinez FD, Graves PE, Baldini M, et al. J Clin Invest 1997;100:3184–3188.
95. Lima JJ, Thomason DB, Mohamed MH, et al. Clin Pharmacol Ther 1999;65:519–525.
96. Mukae S, Aoki S, Itoh S, et al. Hypertension 2000;36:127–131.
97. Suzuki A, Kondo T, Mihara K, et al. Pharmacogenetics 2001;11:545–550.
98. Ongphiphadhanakul B, Chanprasertyothin S, Payatikul P, et al. Clin Endocrinol 2000;52: 581–585.
99. Jia H, Hingorani AD, Sharma P, et al. Hypertension 1999;34:8–14.
100. Brandt JT, Isenhart CE, Osborne JM, et al. Thromb Haemost 1995;74:1564–1572.
101. Smeraldi E, Zanardi R, Benedetti F, et al. Mol Psychiatry 1998;3:508–511.
102. Zanardi R, Benedetti F, Di Bella D, et al. J Clin Psychopharmacol 2000;20:105–107.
103. Tiret L, Rigat B, Visvikis S, et al. Am J Hum Genet 1992;51:197–205.
104. Cambien F, Poirier O, Lecerf L, et al. Nature 1992;359:641–644.
105. Evans AE, Poirier O, Kee F, et al. Q J Med 1994;87:211–214.
106. Zee RY, Lou YK, Griffiths LR, et al. Biochem Biophys Res Commun 1992;184:9–15.
107. McNamara DM, Holubkov R, Janosko K, et al. Circulation 2001;103:1644–1648.
108. Penno G, Chaturvedi N, Talmud PJ, et al. Diabetes 1998;47:1507–1511.
109. Perna A, Ruggenenti P, Testa A, et al. Kidney Int 2000;57:274–281.
110. Stavroulakis GA, Makris TK, Krespi PG, et al. Cardiovasc Drugs Ther 2000;14: 427–432.
111. Ohmichi N, Iwai N, Uchida Y, et al. Am J Hypertens 1997;10:951–955.
112. Humphries SE, Talmud PJ, Hawe E, et al. Lancet 2001;358:115–119.
113. Borjesson M, Magnusson Y, Hjalmarson A, et al. Eur Heart J 2000;21:1853–1858.
114. Arnett DK, Davis BR, Ford CE, et al. Circulation. 2005;111(25):3374–3383.
115. Harrap SB, Tzourio C, Cambien F, et al. Hypertension 2003;42(3):297–303.
116. Drazen JM. Nat Genet 1999,22 (2),Telleria JJ. Respir Med 2008,102,Klotsman M Pharmacogenet Genomics 2007,17 (3),Silverman E, Clin Exp Allergy 1998, 28.
117. Maqbool A, Hall AS, Ball SG, et al. Lancet 1999;353:897.
118. Podlowski S, Wenzel K, Luther HP, et al. J Mol Med 2000;78:87–93.
119. Rathz DA, Brown KM, Kramer LA, et al. J Cardiovasc Pharmacol 2002;39:155–160.

120. Moore JD, Mason DA, Green SA, et al. Hum Mutat 1999;14:271.
121. Mason DA, Moore JD, Green SA, et al. J Biol Chem 1999;274:12670–12674.
122. Brodde OE. Pharmacogenet Genomics 2005, 15 (5).
123. Bleecker EL. Lancet 2007, 370,Hall IP Lancet 2006,368,Green SA, Pulm Pharmcol 1995,8 (1), Migita O. Int Arch Allergy Immunol 2004,134 (2), Brodde OE, Pharmacogenet Genomics 2005,15 (5).
124. Cho SH. Clin Exp Allergy 200525 (9),Martinez FD, JACI, 1997,100 (12).
125. Wechsler ME. Am Respir Crit Care Med 2006, 163;Tan S. Lancet 1997, 350;Israel E, Int Arch Allergy Immunol 2001, 124; Hawkins GA, Am J Respir Crit Care Med 2006, 174; Drysdale CM, Proc Natl Acad Sci USA 2000, 97.
126. Cho SH. Clin Exp Allergy 2005,35 (9); Tan S. Lancet 1997,350 (9083); Drysdale CM. Proc Natl Acad Sci USA 2000, 97.
127. Green SA, Turki J, Innis M, et al. Biochemistry 1994;33:9414–9419.
128. Taylor DR, Drazen JM, Herbison GP, et al. Thorax 2000;55:762–767.
129. Israel E, Drazen JM, Liggett SB, et al. Am J Respir Crit Care Med 2000;162:75–80.
130. Poirier J, Delisle MC, Quirion R, et al. Proc Natl Acad Sci USA 1995;92:12260–12264.
131. Liggett SB, Wagoner LE Craft LL, et al. J Clin Invest 1998;102:1534–1539.
132. Wagoner LE, Craft LL, Singh B, et al. Circ Res 2000;86:834–840.
133. Taboulet J, Frenkian M, Frendo JL, et al. Hum Mol Genet 1998;7:2129–2133.
134. Kruger WD, Evans AA, Wang L, et al. Mol Genet Metab 2000;70:53–60.
135. Charron P, Tesson F Poirier O, et al. Eur Heart J 1999;20:1587–1591.
136. Vandenbroucke JP, Koster T, Briet E, et al. Lancet 1994;344:1453–1457.
137. Mrozikiewicz PM, Cascorbi I, Ziemer S, et al. J Am Coll Cardiol 2000;36:1520–1525.
138. Girelli D, Russo C, Ferraresi P, et al. N Engl J Med 2000;343:774–780.
139. Weiss EJ, Bray PF, Tayback M, et al. N Engl J Med 1996;334:1090–1094.
140. Gutersohn A, Naber C, Muller N, et al. Lancet 2000;355:1240–1241.
141. Abbott GW, Sesti F, Splawski I, et al. Cell 1999;97:175–187.
142. Napolitano C, Schwartz PJ Brown AM, et al. J Cardiovasc Electrophysiol 2000;11: 691–696.
143. Kee F, Morrison C, Evans AE, et al. Heart 2000;84:548–552.
144. Martinelli I, Sacchi E, Landi G, et al. N Engl J Med 1998;338:1793–1797.
145. Gupta, Pharmacogenomics 2008;9(7):895–903.
146. Corder EH, Saunders AM, Risch NJ, et al. Nat Genet 1994;7:180–184.
147. Saunders AM, Strittmatter WJ, Schmechel D, et al. Neurology 1993;43:1467–1472.
148. Cacabelos R. Mini Rev Med Chem 2002;2,59–84.
149. Cacabelos R. Mol Diagn Ther 2007;11, 385–405.
150. Cheung BM, Lauder IJ, Lau CP, Kumana CR. Br J Clin Pharmacol 2004;57(5):640–651.
151. Peters BJM, Klungel OH, de Boer A, Maitland-van der Zee A-H. Expert Rev Cardiovasc Ther 2009;7(8):977–983.
152. Kuivenhoven JA, Jukema JW, Zwinderman AH, et al. N Engl J Med 1998;338(2):86–93.
153. Regieli JJ, Jukema JW, Grobbee DE, et al. Eur Heart J 2008;29(22):2792–2799.
154. de Grooth GJ, Zerba KE, Huang SP, et al. J Am Coll Cardiol 2004;43(5):854–857.
155. Freeman DJ, Samani NJ, Wilson V, et al. Eur Heart J 2003;24(20):1833–1842.
156. Marschang P, Sandhofer A, Ritsch A, Fiser I, Kvas E, Patsch JR. J Int Med 2006;260 (2):151–159.
157. Carlquist JF, Muhlestein JB, Horne BD, et al. Am Heart J 2003;146(6):1007–1014.
158. Boekholdt SM, Sacks FM, Jukema JW, et al. Circulation 2005;111(3):278–287.
159. Chasman DI, Posada D, Subrahmanyan L, Cook NR, Stanton VP, Jr., Ridker PM. JAMA. 2004;291(23):2821–2827.
160. Thompson JF, Man M, Johnson KJ, et al. Pharmacogenomics J 2005;5(6):352–358.
161. Donnelly LA, Doney AS, Dannfald J, et al. Pharmacogenet Genomics 2008;18(12):1021–1026.
162. Polisecki E, Muallem H, Maeda N, et al. Atherosclerosis 2008;200(1):109–114.
163. Ordovas JM, Lopez-Miranda J, Perez-Jimenez F, et al. Atherosclerosis 1995;113(2):157–166.
164. Gerdes LU, Gerdes C, Kervinen K, et al. Circulation 2000;101:1366–1371.
165. Chiodini BD, Franzosi MG, Barlera S, et al. Eur Heart J 2007;28(16):1977–1983.
166. Hughes HB, Biehl JP, Jones AP, Schmidt LH. Am Rev Tuberc 1954;70(2):266–273.

167. Mallal S, Phillips E, Carosi G, Molina JM, Workman C, Tomazic J, et al. N Engl J Med 2008;358(6):568–579.
168. Cardon LR, Idury RM, Harris TJ, et al. Pharmacogenetics 2000;10:503–510.
169. Roses AD. Lancet 2000;355:1358–1361.
170. Murphy MP, Beaman ME, Clark LS, et al. Pharmacogenetics 2000;10:583–590.
171. Raaijmakers JAM, Koster E, Maitland-van der Zee A-H. Curr Pharm Des 2010; 16:238–244.
172. Robertson JA Nat Genet 2001;28,207–209.
173. Taube J, Halsall D Baglin T. Blood 2000;96: 1816–1819.
174. Annas GJ. JAMA 2001;286:2326–2328.
175. Beskow LM, Burke W Merz JF, et al. JAMA 2001;286:2315–2321.
176. van Schie RM, Wadelius MI, Kamali F, Daly AK, Manolopoulos VG, de Boer A, Barallon R, Verhoef TI, Kirchheiner J, Haschke-Becher E, Briz M, Rosendaal FR, Redekop WK, Pirmohamed M, Maitland-van der Zee A-H. Pharmacogenomics. 2009;10(10):1687–1695.
177. Gerdes LU, Gerdes C, Kervinen K, et al. Circulation 2000;101(12):1366–1371.
178. Reilly PR. Nat Med 2001;7:268–271.

DESIGN OF PEPTIDOMIMETICS

JUAN J. PEREZ[1]
FRANCESC J. CORCHO[1]
JAIME RUBIO-MARTINEZ[2]
[1] Universitat Politecnica de Catalunya, Departament d'Enginyeria Quimica, Barcelona, Spain
[2] Universitat de Barcelona and the Institut de Recerca en Química Teòrica i Computacional (IQTCUB), Departament de Quimica Fisica, Barcelona, Spain

1. INTRODUCTION

Peptides are important mediators in the regulation of most of the physiological processes either through the endocrine signaling pathway or sometimes through the local paracrine and autocrine signaling pathways. Peptides elicit a plethora of functions being identified to serve as hormones, neurotransmitters, enzyme substrates or inhibitors, neuromodulating agents, and immunomodulators [1]. Representative examples include angiotensin II, bombesin, bradykinin, cholecystokinin, endothelins, enkephalins, oxytocin, secretin, somatostatin, or tachykinins. The wide spectrum of signaling activities exerted makes them very appealing molecules for pharmaceutical intervention, aimed at playing roles as agonists, antagonists, or inhibitors. In addition, the repertoire of possible roles for therapeutical intervention rendered by peptides can be further expanded if we take into consideration that fragments of proteins may also be used to mediate or inhibit biological actions, playing the same role as those protein epitopes where they are originally found [2]. This is currently an area of an enormous activity, mostly focused on the inhibition of protein–protein interactions.

Although peptides are important signal mediators, they exhibit a low profile as therapeutic agents: poor oral bioavailability, low absorption, are easily metabolized by peptidases, and are immunogenic. About 15 years ago, it was thought that biotechnology could cope with some of the pharmacokinetic problems of peptides. However, in spite of the advances in delivery techniques in the past 10 years [3–5], peptides cannot compete with small-molecule compounds as therapeutic agents. At present, there are only a few peptides approved for therapeutic intervention, including oxytocin, the segment 17–36 of the glucagon-like peptide 1 receptor agonist, the parathyroid hormone, or calcitonin. All these peptides are about 30 residues long and can only be delivered intracerebroventricularly or through nasal administration. Interestingly after the human genome sequencing, it has been estimated that among the 5000–10,000 potential drug targets available, two thirds could be amenable to traditional small-molecule drugs whereas the nearly one-third left to biopharmaceuticals [6]. Thus, there may still be specific treatments that could only be amenable using peptides. Accordingly, in today's view, peptides should be considered as valuable molecules for therapeutic intervention where important research efforts in achieving more efficient delivery procedures should be spent and, on the other hand, as valuable source of inspiration for the design of small-molecule compounds.

The term peptidomimetic concerns the development of small-molecule mimics of peptides. However, it has been used in different ways in the past 30 years. Historically, the term was coined and used in the medicinal chemistry arena, aimed at finding new leads for therapeutical intervention, although with time the term has been widened and used to describe molecules designed to emulate the structural features of peptides. Throughout this review, we will specifically use the term peptidomimetic to describe small molecules aimed at therapeutical intervention, while leaving the term peptide surrogate or pseudopeptide for those molecules inspired by the structural features (either primary or secondary) of a peptide that still keep a peptide structural appearance. The former view is the focus of this review in line with the scope of this book.

The concept of peptidomimetic can be considered to be coined with the discovery of the enkephalins, the endogenous peptide analgesics discovered in 1975. Isolated in 1804 from the *Papaver somniferum* and synthesized in 1952, morphine had long been known to elicit its analgesic action through the opiate receptor with important adverse effects.

The endogenous ligand had been sought for a long time, in order to get a better understanding of its analgesic action and with the hope that an endogenous opiate would not exhibit the same adverse effects. Interestingly enough, the enkephalin sequence exhibits moieties that are already present in morphine [8]. Moreover, a more elaborated analysis could be argued that enkephalins adopt a type I β-turn conformation when bound to the receptor, mimicking well the 3D structure of morphine [9]. The perfect match between enkephalins and morphine marked the starting point and the paradigm of peptidomimetic design. However, it is known nowadays from site-directed mutagenesis studies that peptides and small molecules do not necessarily interact in the same manner with the receptor, but share some receptor points at recognition, with peptides usually exhibiting a larger number of anchoring points [10].

In simple terms, a peptidomimetic may be considered as a small molecule that mimics the pharmacological action of a peptide. However, most of the natural peptides mediate their actions in the signaling process as agonists, and a systematic replacement of the residues of a peptide sequence may lead to discover analogs with an antagonistic profile, inhibitors of enzymes, or inhibitors of other proteins. A definition of a peptidomimetic should be broadened to include these considerations. Accordingly, we will consider a peptidomimetic as a small molecule designed to exhibit the same pharmacodynamical profile of a set of peptide analogs and to elicit a similar pharmacological response mediated through the same receptor. Alternatively, a peptidomimetic can be considered at the molecular level as a small molecule that mimics some of the molecular recognition interactions of a peptide with its receptor, mediating the same pharmaceutical action on the therapeutic target.

Excellent reviews on peptidomimetics were written in the early 1990s pointing their views on medicinal chemistry aspects. In these reviews, several definitions of peptidomimetic were provided. Thus, Giannis and Kolter [11] defined a peptidomimetic as "a compound that, as the ligand of a receptor, can imitate or block the biological effect of a peptide at the receptor level." Wiley and Rich [12] defined peptidomimetics as "chemical structures designed to convert the information contained in peptides into small nonpeptide structures." A year later, Gante [13] defined a peptidomimetic as "a substance having a secondary structure as well as other structural features analogous to that of the original peptide, which allows to displace the original peptide from receptors or enzymes." More recently, there is still a strong interest in designing peptidomimetics as orally active compounds inspired by the structures of endogenous peptides and peptide epitopes [14], although with a strong tendency to offer a broadened definition of peptidomimetics that includes the design of new architectures inspired by the structure of peptides, oriented to the design of nanomaterials [15,16].

2. PEPTIDE FEATURES

Peptides as well as proteins are polymers of amino acids linked through peptide bonds. Traditionally, peptides have been considered polymers with lengths of up to 30 residues, leaving the term protein for larger polymers. However, recently the term evolved to specifically consider proteins as those amino acid polymers that exhibit a defined tertiary structure, whereas the term peptides was left to those amino acid polymers that are unstructured in solution. Indeed, this definition covers well the traditional one, but is more specific since short peptide segments with precise 3D structures have already been reported [17].

Unstructured in solution implies that peptides are flexible molecules in a constant interchange among different conformations (states). This is a consequence of the rough topology of the potential energy surface, with many low-energy conformations accessible at room temperature. Since the amino acid sequence contains all the information necessary for a polypeptide to adopt its three-dimensional structure, it can be inferred that the topology of the potential energy surface that defines its conformational profile is a consequence of its own sequence. In the first approximation, it can be considered that the conformational profile of the peptide is the result of a combination of the conformational features of each of the residues constituting the peptide chain. This approach, known as the rotational isomeric model, assumes that residues are independent beads of a chain and that the final conformations are the superposition of the different contributions of each of them. Although very simplistic, this model provides a rough estimate of the number of conformations attainable by the peptide. Specifically, considering that if a standard residue can attain typically three different low-energy conformations, a peptide of N residues will adopt around $3^N{\sim}10^{N/2}$ different conformations. More interestingly, the model also provides a qualitative description of the density of states [18]. However, residues are not interdependent to each other and peptides may exhibit conformations that result from cooperative interactions among them. On the one hand, intramolecular interactions may provide new conformations not featured in the rotational isomeric model, and on the other, they may also discard other structures as unattainable due to steric hindrance. The balance between new and unattainable conformations may be the explanation of the bimodal distribution of states observed in real characterizations of the conformational space [19].

The conformational profile of a peptide is also modulated by the immediate environment. Depending on its nature, the solvent may enhance or damp certain conformations in solution. Thus, in water peptides tend to exhibit a more flexible behavior than in other solvents. On the other hand, there are structuring solvents such as trifluoroethanol that are capable of enhancing the structure of a peptide in solution. This is also found in crystal structures where depending on the mother liquor the peptide may exhibit different conformations [20]. There are two solvent features that may affect the attainability to the accessible states of peptides in solution: on the one hand, the solvent dielectric constant, affecting long-range interactions, and on the other, the capability of the solvent to elicit direct solute–solvent hydrogen bonding interactions. This is particularly evident in shorter peptides that do not have a sufficient number of intramolecular interactions. Accordingly, it can be said that peptides are expected to be flexible molecules due to their rough conformational landscape that gives rise to consider in general their observable structure as an average image of different structures attainable according to their sequence and subject to the immediate environment.

3. PEPTIDOMIMETIC DESIGN PROCESS

In order to design a small molecule that exhibits a similar pharmacodynamic profile of the original peptide with improved pharmacokinetics, it is necessary to have a reasonable hypothesis of the chemical moieties of the peptide responsible for the interaction with the therapeutic target as well as knowledge of their spatial distribution. This is referred as construction of a pharmacophore hypothesis.

The very first step in the design process is to establish structure–activity relationships from a set of analogs derived from the original peptide. The goal of this first step is twofold: to establish the shortest peptide segment that retains activity and to identify those residues that are important for function. The former information is established through the measurement of the activity and affinity of systematically shorter segments of the original peptide, whereas the latter information is normally obtained through the synthesis of analogs obtained by replacement of the residues derived from the original sequence. In this process, it is very useful to perform an alanine scan, where different analogs are synthesized from a systematic replacement with alanine at each position of the sequence. Another strategy involves conservative substitutions of the residues through the replacement of residues with similar chemical properties, or less conservative substitutions that will provide information about the necessity of specific interactions in certain positions. Also, the substitution of L- for D-amino acids may provide information about the bioactive conformation. It may be the case that in the process of residue substitutions an analog with antagonistic profile may be produced. For this reason, it is necessary to characterize both the binding and the biological activity in regard to the reference peptide along the process.

In the second step, additional structure–activity relationships need to be worked out to understand the conformational features of the bioactive conformation. Accordingly, all efforts need to be devoted to develop a pharmacophoric hypothesis of the interaction of the peptide with the receptor. Obviously, the best scenario is to have an experimental structure of the peptide bound to its receptor (bioactive conformation). In this way, we know the bound conformation and the spatial distribution of the side chain moieties or backbone elements involved in the recognition with the receptor. This is possible nowadays for the design of enzyme inhibitors, with great success as will be shown later. This procedure can also be successfully used in the design of protein–protein inhibitors. In contrast, this is not the case for peptides that mediate their function through the interaction with membrane proteins such as G-protein-coupled receptor (GPCR) for which only the structures of rhodopsin and α- and β-adrenergic receptors are available.

In the absence of an experimental structure of the complex, a pharmacophoric hypothesis of the peptide-receptor recognition needs to be developed. This represents a combined effort involving peptide synthesis, biophysical methods, computational methods, pharmacology, and molecular biology [20]. In the case of GPCRs, in contrast to previous models, aimed at providing an explanation to the basal activity shown by certain GPCRs and the existence of reverse agonists, the accepted model for the activation is the so-called two-state model. In this model, antagonists are supposed to bind to the inactive conformation whereas agonists are suited to bind to the active conformation. Although most of the GPCR-mediated peptidomimetics designed are antagonists, agonism will represent to find a different bound conformation and consequently the same methodology might be applied.

Figure 1 shows pictorially a general scheme regarding the process of peptidomimetic design. At the top of the left-hand side, a new sequence has been identified as important biological mediator. A process of analog synthesis is followed in order to understand the key residues involved in the interaction and the minimum length required. The next step involves the synthesis of conformationally constrained analogs including cyclic analogs and peptidomimetics that still incorporate much of the peptide chain. This results in the establishment of the chemical requirements needed for recognition as well as their spatial arrangement. On the other hand, different information is accumulated on the right-hand side of Fig. 1. Screening of different sources can identify new hits providing information about the chemical structures of binders, augmenting our knowledge about the diversity of scaffolds possible. On the other hand, the use of molecular modeling techniques and biophysical methods can provide a wealth of information regarding the hypothesis of ligand–target interaction through pharmacophore development or the characteristics of the binding through structural analysis. The information provided by these methods together with the information provided from the first-generation peptidomimetics and the new hits can be put together to design second-generation peptidomimetics ready for optimization.

4. LOCAL CONSTRAINTS ON PEPTIDE CONFORMATION

Design of analogs in this category involves modifications of the peptide backbone, such as methylation of the amide nitrogen or the α-carbon, modifications of the peptide bond, modifications of the side chains, or the synthesis of cyclic analogs. These modifications are aimed at modifying the conformational features of the amino acids, basically restricting their conformational freedom or impeding the formation of secondary structures by the elimination of typical hydrogen bonding sites [14].

Modifications of the backbone include methylation of the hydrogen amide giving rise to *N*-methylated residues (structure **1** in Fig. 2) [21] or substitution of the alpha hydrogen by an alkyl chain giving rise to the $C^{\alpha,\alpha}$ disubstituted amino acids (structure **2** in Fig. 2) [22]. Modifications of the amino acidic nature of the backbone are referred to as

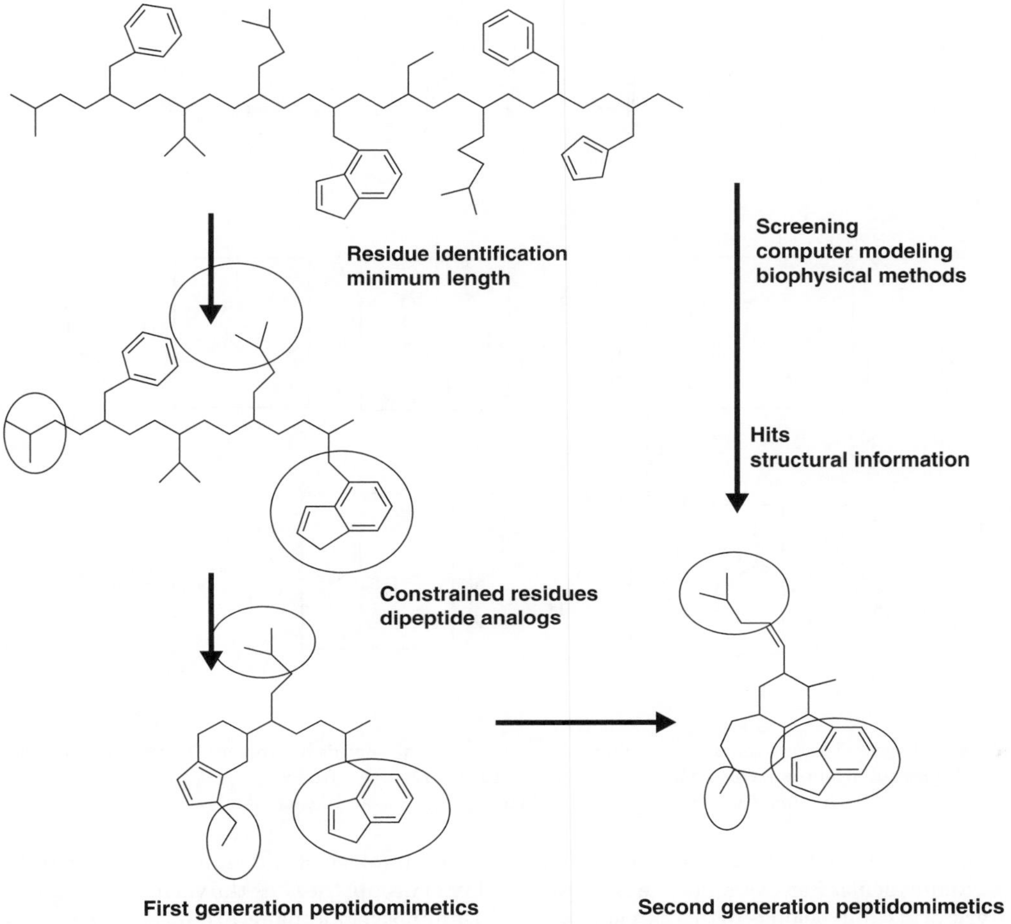

Residue identification
minimum length

Screening
computer modeling
biophysical methods

Hits
structural information

Constrained residues
dipeptide analogs

First generation peptidomimetics

Second generation peptidomimetics

Figure 1. Scheme of the process followed to design peptidomimetics.

isosteric or isoelectronic exchange of the constituting units. These modifications change the capacity of the backbone to elicit hydrogen bond interactions and may also change its conformational degrees of freedom. For example, the amide nitrogen can be substituted by oxygen or sulfur atoms giving rise to the corresponding depsi derivatives (structure **3** in Fig. 2) found in nature [23]; substitution of the C^{α} group by a nitrogen atom gives rise to the so-called aza derivatives (structure **4** in Fig. 2) [24]; reduction of the peptide bond represents another interesting backbone modification (structure **5** in Fig. 2) [25]. Larger modifications of the backbone include the retro-inverso peptides (structure **6** in Fig. 2), where the configuration of the C^{α} is changed and the direction of peptide bond is reversed [26], or peptoids (structure **7** in Fig. 2), polymers constituted by amino acids where the side chain is linked to the backbone amide nitrogen [27,28]. Alternatively, the number of carbon atoms of the residue can be enlarged by attaching the amino group in an atom different from that in α- in regard to the carboxylic moiety, giving rise to ω-amino acids (structure **8** in Fig. 2). In particular, β- [29] and γ-residues [30] have been used extensively. These modifications provide a plethora of possibilities that can be used to generate new analogs to help understand the role of backbone atoms in the recognition process.

Modifications of the side chains can be done with the aim to complement the repertoire of

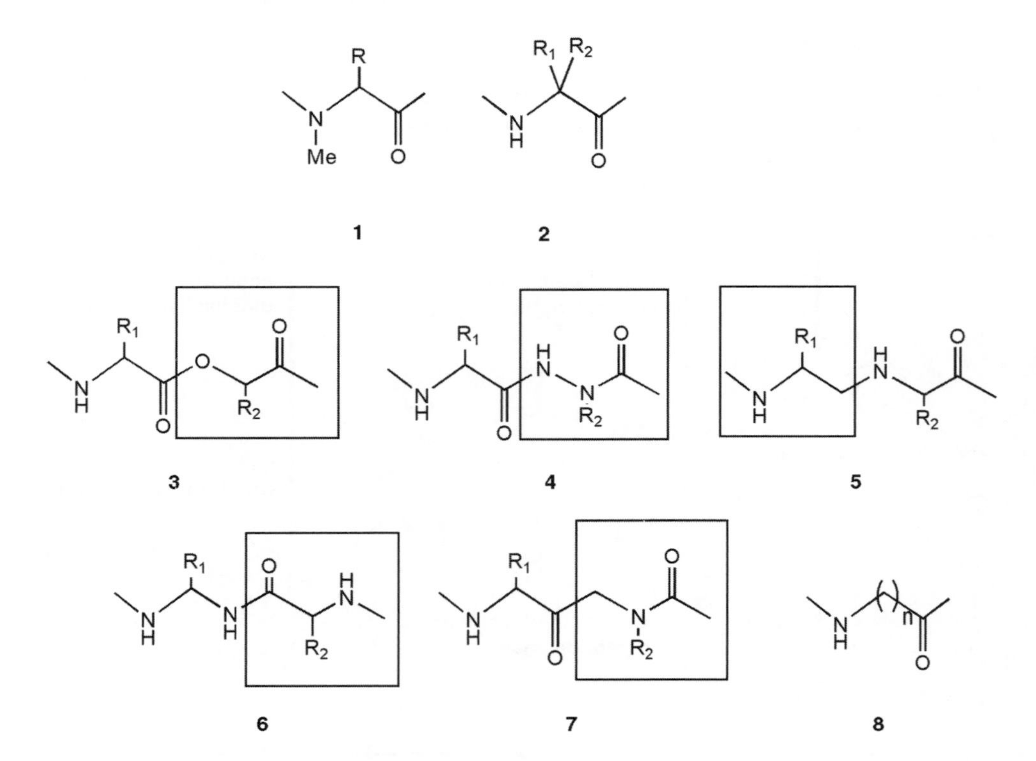

Figure 2. Diverse isostere replacements and substitutions Sof the peptide backbone: (1) a regular amino acid that is represented for comparison purposes; (2) C^{α} substituted residue; (3) depsi derivative; (4) aza derivative; (5) peptide bond reduction; (6) retro-inverso peptide; (7) peptoid; ω-amino acids.

chemical moieties available from the 20 natural amino acids. For example, the α-aminobutyric acid (Abu) belongs to this category and can be considered as a threonine lacking the hydroxyl group (structure **9** in Fig. 3); or different aromatic moieties, for example, nitrophenylalanine (Phe(NO_2)) (structure **10** in Fig. 3) or naphthalenealanine (Nap) (structure **11** in Fig. 3).

More interesting are those amino acids with side chains that limit the conformational freedom of the amino acid [22,31]. There is an important group of residues generated by replacement of the alpha hydrogen by an alkyl chain giving rise to the so-called $C^{\alpha,\alpha}$ disubstituted amino acids. The substitution can be carried out with another linear chain or both the C^{α} and the C^{β} can be part of a cyclic structure. The simplest representative of the former group is the α-aminoisobutyric acid (Aib) (structure **12** in Fig. 3), which can be considered as an alanine with a methyl group in place of the alpha hydrogen. In the latter group, the C^{α} is the vertex of a cycle, the 1-aminocyclopropanecarboxylic acid (c_3c) (structure **13** in Fig. 3) being its simplest representative. Another example is the 2-aminoindane-2-carboxylic acid (Aic), shown as structure **14** in Fig. 3. Also in this category dehydro-amino acids exhibiting a double bond between the C^{α} and the C^{β} can be included. As example of this kind of residues, structure **15** in Fig. 3 depicts pictorially dehydrophenylalanine (ΔPhe) [32]. Another class of residues exhibits cyclic structures involving two atoms of the backbone like in proline, for example, the pipecolic acid (structure **16** in Fig. 3) or more complex residues such as the tetrahydroisoquinoline-3-carboxylic acid (Tic) shown as structure **17** in Fig. 3. All these residues, in addition to offering a great deal of side chain diversity, also constrain the conformational freedom of the residue, resulting in useful tools to get

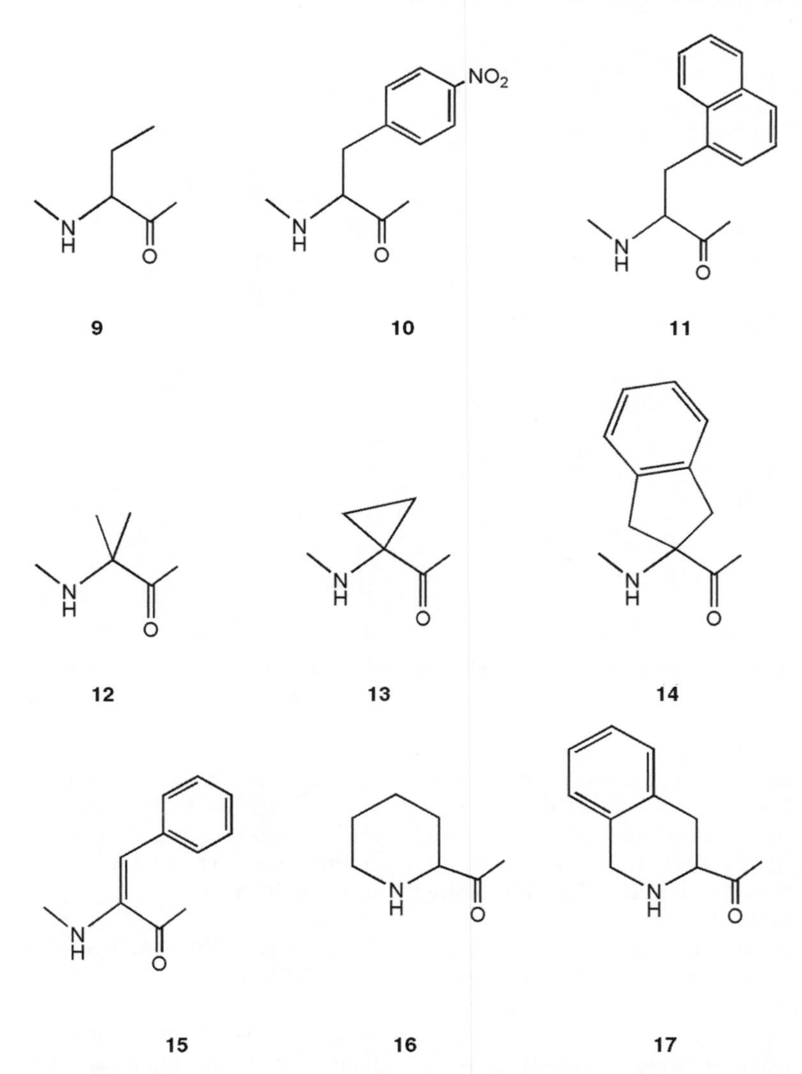

Figure 3. Side chain modifications: (**9**) α-aminobutyric acid (Abu); (**10**) nitrophenylalanine; (**11**) naphthylalanine; (**12**) α-aminoisobutyric acid (Aib); (**13**) 1-aminocyclopropanecarboxylic acid (c_3c); (**14**) the 2-aminoindane-2-carboxylic acid (Aic); (**15**) dehydrophenylalanine; (**16**) pipecolic acid; (**17**) tetrahydroisoquinoline-3-carboxylic acid (Tic).

insight into the features of the bound conformation of the peptide.

An alternative procedure to the synthesis of constrained analogs using unnatural amino acids consists of designing dipeptide analogs by bridging atoms of the peptide backbone, backbone-side chains, or even side chains generating heterocycles of different lengths [11,13,33,34]. Molecules produced in this way are half peptide and half small molecule. These compounds in addition to being less vulnerable to peptidase degradation, normally exhibit a bent structure that may help to produce analogs closer to the bioactive conformation.

This has been published in the literature in different ways. For example, bridging through backbone atoms or through the backbone amide nitrogen and side atoms typically yields piperazinones (structures **18** and **19** in

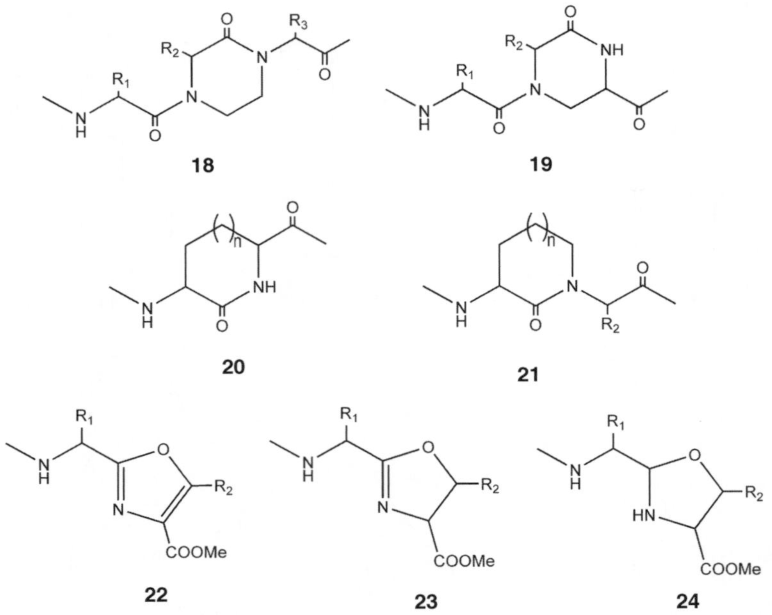

18 19 20 21 22 23 24

Figure 4. Structures of dipeptide isostere analogs to constraint the peptide backbone conformation. See text for details.

Figure 4, respectively). Whereas when cyclization occurs between the side chains of two consecutive residues, it yields lactams with different ring size (structures **20** and **21** in Figure 4, respectively). Similarly bridging between the side chain of cysteine, threonine or serine and the amide nitrogen yields oxazolones oxazolines and oxazolidines (structures **22–24** in Figure 3) or the corresponding thio-derivatives [35].

An alterative strategy to produce local geometrical restrictions is to use an organic scaffold where different peptide chains are attached in a suitable orientation. This approach has originated a series of interesting contributions in which peptidomimetics were designed in this way. Different scaffolds have been used including cyclohexane or steroids although monosaccharides have been extensively used for this purpose [13,36]. However, these analogs have the disadvantage to be designed without using a clear model of the secondary structure. There is a long list of dipeptide mimetics that work as scaffolds with the purpose of providing a specific secondary structure to the peptide chain that will be discussed later. Scaffold peptidomimetic design offers the great advantage that can be used in combinatorial chemistry to produce peptidomimetic libraries [37].

5. GLOBAL CONSTRAINTS ON PEPTIDE CONFORMATION

Whereas substitution of different residues of a peptide chain by constrained amino acids provides local bents to the structure of the peptide, cyclization represents a procedure to globally restrain its conformational space [38]. Cyclic peptides have shown to (1) increase agonist and antagonist potency; (2) withstand proteolytic degradation; (3) increase receptor selectivity; (4) enhance bioavailability; and (5) provide conformational insight for receptor binding [39]. Cyclization can be done through condensation of the N- and C-termini, by connecting functionalized side chains, through a linkage between the N-terminus and a side chain or two backbone atoms. There are different examples of cyclic peptides in nature [40]: for example, oxytocin and vasopressin are connected through a disulfur bridge, whereas in cyclosporin the cyclization is performed by condensation. However, it should be stressed that for designing purposes cyclic peptides may not necessarily provide good results, since cyclization may force moieties important for ligand recognition to adopt a conformation facing the interior of the structure [41].

6. ASSESSMENT OF THE BIOACTIVE CONFORMATION

The synthesis and evaluation of a series of peptide analogs may provide enough information to generate a hypothesis about the important moieties involved in recognition and about the three-dimensional arrangement. Depending on how well defined is this hypothesis, it could be used to find hits using an *in silico* approach [42–44]. These results will improve the knowledge of the necessary requirements for binding and eventually activation of the receptor.

Theoretical procedures concern the assessment of the bound conformation of the peptide using indirect methods through the analysis of the conformational profile of a set of analogs including binders and nonbinders. Once their conformational profile is established, the bound conformation may be assessed using the paradigm that active compounds share in common a conformation (the bound conformation) that is not found in the conformational profile of the nonbinders. A critical step is the selection of the analogs to perform the analysis, since the reason for an analog to be a nonbinder can be that it does not exhibit the necessary chemical moieties or because it cannot adopt the right conformation to have them properly arranged. There are a number of issues to be considered for the computational approach including the representation of the peptide, the inclusion of the solvent, and even the strategy to search the conformational space. The reader is referred to previous reviews in this regard [44].

7. FIRST-GENERATION PEPTIDOMIMETICS

If there is not enough information to build a pharmacophoric hypothesis, medicinal chemistry can help to provide a step further in the design of a first generation of peptidomimetics. One strategy is to suppose the sequence exhibits either a β- or a γ-turn and substitute the dipeptide responsible with a rigid organic scaffold [45–48]. γ- or β-turns change the direction of the peptide chain and are consequently assumed to be on the surface, and therefore they may be important epitopes in the ligand-receptor recognition. Figure 5 depicts pictorially the most relevant turns found in proteins and peptides. For designing purposes, it is useful to learn that a β-turn forms 10-membered cycles, whereas a γ-turn renders a 7-membered cycle.

Dipeptide mimicking is carried out through the substitution of the turn cycle by a heterocycle framework that restricts the geometry of the peptide backbone in a similar fashion. Specifically, in the case of β-turns the heterocycle mimics the features of residues $i + 1$ and $i + 2$. Several scaffolds have been suggested in the literature for this purpose [49], including a nine-membered ring lactam [50] (structure **25** in Fig. 6); other cycles include the 4-(2-aminoethyl)-6-dibenzofuranpropionic acid (structure **26** in Fig. 6) or the 6,6′-bis(acylamino)-2,2′-bipyridine-based amino acid (structure **27** in Fig. 5) [13,51]. Other scaffolds are the *N*-substituted benzimidazoles (structure **28** in Fig. 5 [52]; structure **29** in Fig. 6 [53]). Benzodiazepines have also been widely used to mimic β-turns with success (structure **30** in Fig. 6) [54]. The 1-azaoxobicycloalkane skeleton has demonstrated to exhibit a high degree of versatility that gives rise to a range of different bicycles including various ring sizes and heteroatoms embedded. Figure 6 depicts pictorially different structures available (structures **31–33**) [55,56].

A great deal of interest has also been devoted to design cyclic structure mimetics of γ-turns. According to the features of this secondary structure motif (see Fig. 5), seven-membered rings have been widely chosen as template scaffolds [13] including azepinones (structure **34** in Fig. 7) [57,58], diazepinones (structure **35** in Fig. 7) [59], and even six-membered ring lactams have also been used satisfactorily [60] (structure **36** in Fig. 7).

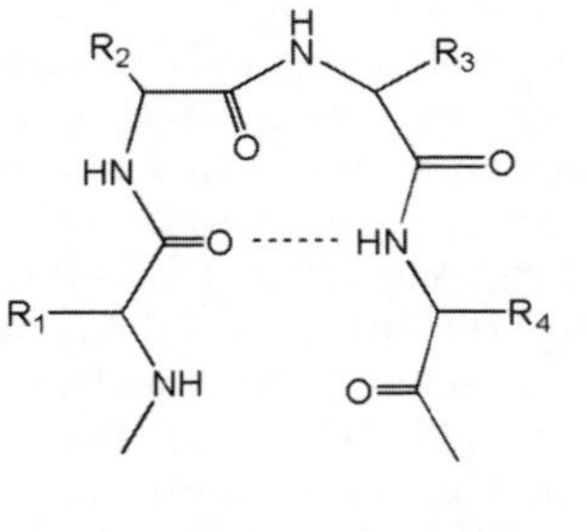

Type I β-turn

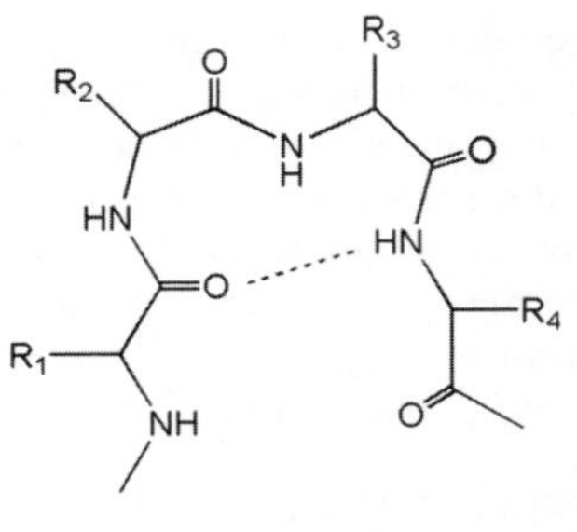

Type II β-turn

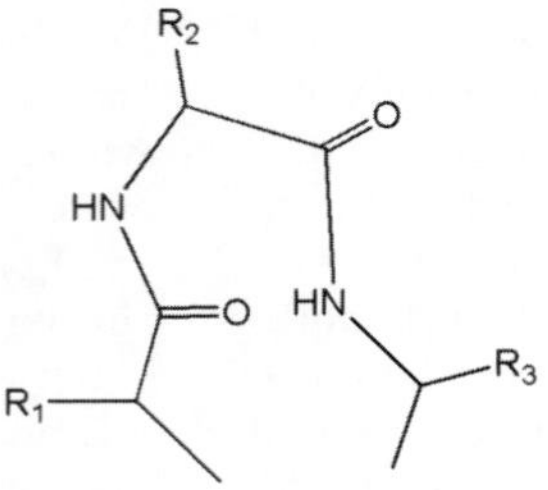

γ-turn

Figure 5. Schematic representation of the interaction involved in a β-turn (types I and II) and a γ-turn

25

26

27

28

29

30

31

32

33

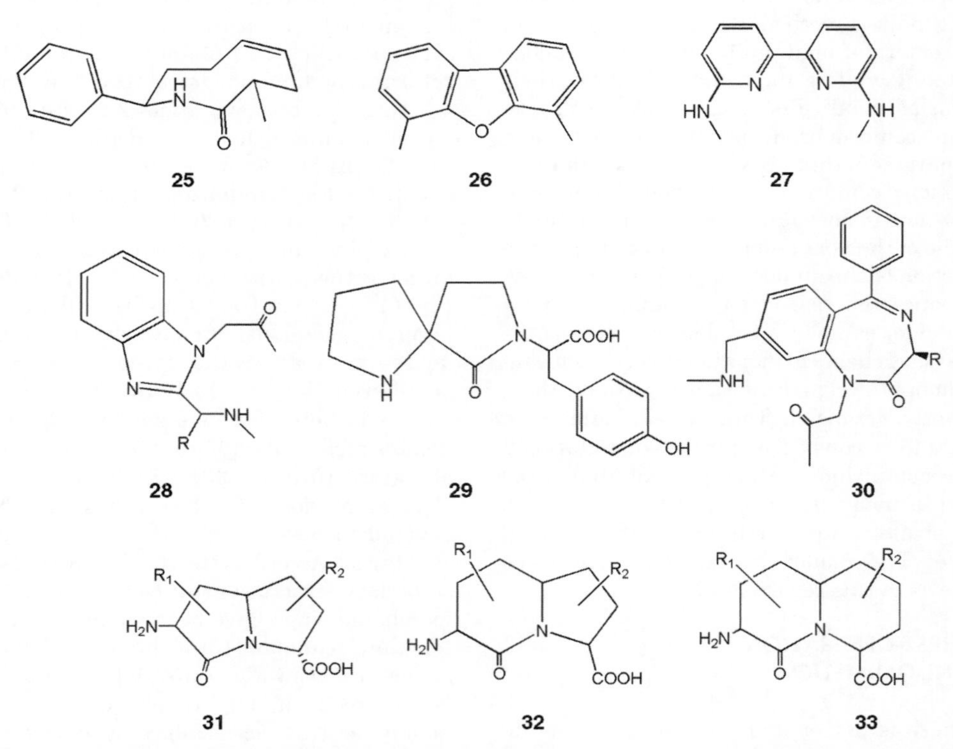

Figure 6. Structures of different dipeptide isosteres designed to provide a β-turn. See text for details.

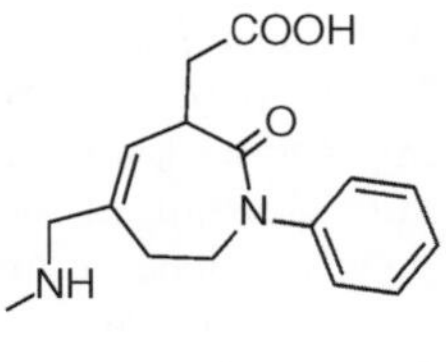

Figure 7. Structures of different isosteres designed to provide a γ-turn. See text for details.

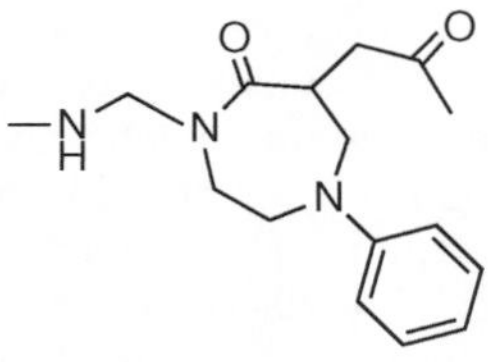

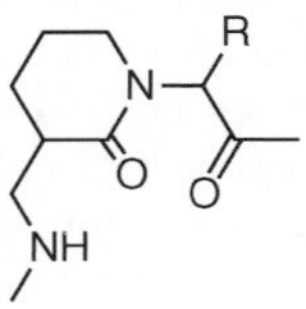

8. SECOND-GENERATION PEPTIDOMIMETICS

First-generation peptidomimetics may be considered molecules halfway between peptides and small molecules. Although they may not yet exhibit a good pharmacokinetic profile, they can provide useful information to understand the spatial arrangement of the necessary moieties involved in the ligand–peptide interaction and can be considered as a logical step to refine the pharmacophore of the interaction. A note of caution should be raised regarding definition of a peptide–receptor pharmacophore since the interaction surface between a peptide and its receptor is normally large and a competitive inhibitor or an antagonist may be blocking only part of this larger area. In this sense, it should be also taken into account that different families of compounds, although competing with the same peptide ligand, may not actually bind at the same spot and, consequently, may not obey the same pharmacophores.

Second-generation peptidomimetics should be considered more drug-like molecules, whose design is based on all the information gathered from the previous structure–activity studies. These molecules in a general sense can be thought as scaffolds either based or not on conformational restrictions and on the structure of possible hits obtained from screening either through high-throughput screening or *in silico*, with the proper moieties arranged in a suitable manner.

9. PEPTIDOMIMETIC DESIGN CASE STUDIES

There is an abundant literature describing interesting stories about the improvement and discovery of new peptidomimetics in different areas and using different approaches. Instead of providing an exhaustive list of examples, we have decided for the present review to describe a few example case studies selected on different kinds of peptide targets, suggesting the reader to browse on the different reviews on peptidomimetics quoted in the reference list for further information. Specifically, we briefly describe a few examples covering three broad groups of peptidomimetics: enzyme inhibitors, antagonists of GPCRs, and protein–protein inhibitors.

10. ENZYME INHIBITORS

Enzymes are involved in the regulation of key biological processes. Precise regulation of their activity is critical for organism survival, as deviations from their activity can be harmful. Accordingly, enzyme inhibition can be considered in some cases as a good strategy for therapeutic intervention. Inhibitors have specifically been designed in viral infections, as antimalarial agents, as regulators of blood clot formation, as anti-inflammatory agents, or as anticancer agents, among others.

From a structural point of view, the capacity of enzymes to catalyze biochemical reactions is due to their ability to stabilize transition-state species [61]. The use of this hypothesis to design inhibitors was successfully used for the first time in the early 1970s [62], and this paradigm still represents a valid approach to design first-generation enzyme inhibitors. Transition-state analog inhibitors are peptide analogs that exhibit a structure similar to the substrate, but with the scissile amide bond of the substrate sequence replaced by an amino acid containing groups such as hydroxymethylene, hydroxyethylene, or hydroxymethylamine or with a chelating moiety in the case of metalloproteases that mimic the transition state [63]. From the information gathered, subsequent structural studies involving the enzyme–inhibitor interaction of these first-generation peptidomimetics provide the opportunity to develop hypothesis of the important features involved in the enzyme–ligand interaction, which in turn allows the design of a new generation of inhibitors with improved characteristics [64].

10.1. Inhibitors of the HIV-1 Protease [65,66]

Among the different groups of enzymes described in the literature, proteases represent one of the largest. These enzymes are responsible for the cleavage of a peptide bond of a peptide sequence with high specificity and efficacy. Proteases can be classified into six different groups depending on the key residues involved in the active site, including serine, threonine, cysteine, aspartic acid, glutamic acid, and metalloproteases. The mechanism of action of the former three involves a nucleophilic attack of a serine, a threonine, or a cysteine to the peptide bond, respectively, whereas the latter three require the nucleophilic attack of a water molecule through a tetrahedral intermediate state. In order for the different proteases to exhibit substrate selectivity, the active site of a protease typically consists on a cleft capable of accommodating three to nine residues [67].

The HIV-1 aspartic protease is responsible for processing the viral polypeptide chain necessary for the organization of structural proteins and the release of viral enzymes during infection. Thus, inhibition of the protease results in immature and noninfectious virions. Stories of the design of potent inhibitors have been recently reviewed [65,66,68]. HIV-1 protease was shown to cleave Tyr-Pro or Phe-Pro sequences in the viral polypeptide. This information led to the design of an initial collection of inhibitors of HIV-1 designed as transition-state analogs based on previous experience of other proteases [69]. After some efforts, the first two inhibitor–protease structures with bound MVT-101 (structure **37** in Fig. 8) and JG-365 (structure **38** in Fig. 8), respectively, gave structural clues that were used to design more potent compounds.

The structure of the HIV protease is a dimer consisting of two identical 99 amino acid chains, each exhibiting a four-stranded β-sheet formed by N- and C-terminal β-strands. The enzyme active site is located at the interface of the two subunits and contains the catalytic triad (Asp_{25}-Thr_{26}-Gly_{27}) responsible for the protease activity. Interestingly, the 3D structure reveals that each monomer exhibits a flap whose conformation changes significantly when an inhibitor is bound. The binding cleft contains hydrophobic residues and is long enough to accommodate peptides of approximately six to eight amino acids. Inhibitor residues are numbered toward the C-terminus of the polypeptide chain in regard to the scissile bond as P'_1, P'_2, and P'_3, whereas they are called P_1, P_2, and P_3 toward the N-terminus. Similarly, the corresponding subsites of the protease binding cleft accommodating the different inhibitor side chains are termed S_1–S_3 and S'_1–S'_3, respectively.

The crystallographic structures of the complexes with MVT-101 and JG-365 were used to design improved inhibitors. These new molecules were developed simultaneously in different laboratories following a similar strategy: the design of different *in silico* analogs and the assessment of their binding affinities using different methods. Interestingly, the predicted binding affinities correlated well in all the cases with the IC_{50} obtained experimentally [70–72]. These studies led to the first-generation inhibitors of the protease presently in use, including the first one approved by the FDA: saquinavir (**39**), ritonavir (**40**), indinavir (**41**), or nelfinavir (**42**) in Fig. 9.

Rational design of second-generation nonpeptide inhibitors began with the discovery of the conserved water molecule mediating contacts between the carbonyl oxygen atoms of the transition-state analog inhibitors and the amide group of Ile50 of the enzyme pocket. This tetrahedrally coordinated water is reported in practically all the HIV protease structures, and its displacement had been postulated as a possible way of obtaining highly specific protease inhibitors with increased affinity. Chenera et al. [73] identified core structures containing *trans*-1,4-cyclohexanediol or hydroquinone by using molecular

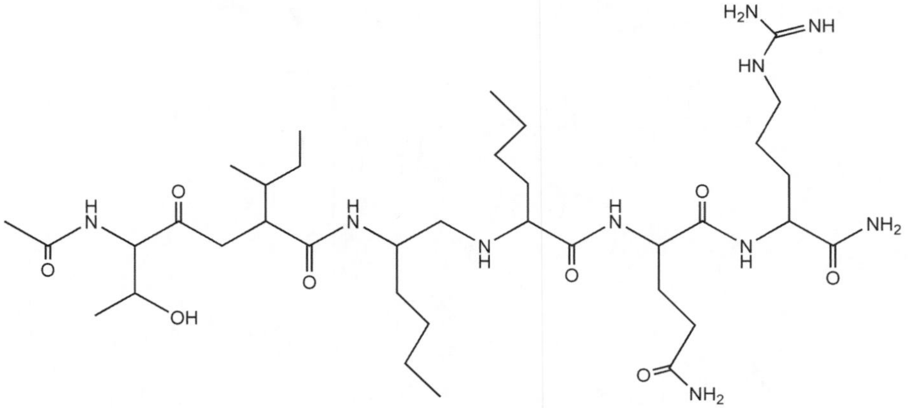

37

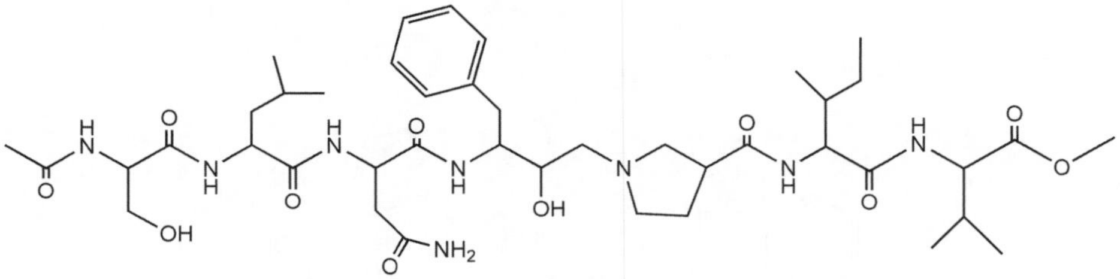

38

Figure 8. Structures of different first-generation HIV protease inhibitors. See text for details.

modeling and showed that six-membered rings with two oxygen atoms in *para* mimic the water function. Adequate substitutions in the scaffold led to the discovery of inhibitors with affinities as low as 7 μM. However, the best candidate obtained in this series lacked hydrophobic moieties to fill the S2 subsite. A more successful approach was carried out at Dupont Merck [74] that led to the design of very potent inhibitors. The process started with the generation of pharmacophores derived from 3D models of a previously designed diol transition-state analog bound to the active site of HIV protease. The search of this pharmacophore in different chemical databases provided Amprenavir (**43**, Fig. 10) that, in addition to fit the initial search criteria, also exhibited a methoxy phenyl ring mimicking the structural water molecule. Due to the fact that the phenyl ring does not adequately orient all substituents of the inhibitor, a cyclohexanone ring containing the water mimic was proposed instead. Subsequent enlargement of this ring to a seven-membered ring allowed the incorporation of a diol moiety and was shown to bind the receptor with high affinity. Further modeling studies were performed to determine the optimal stereochemistry and conformation required for an optimal interaction with the active site. Conformational analysis of designed cyclic ureas established the preferred conformations of the ring when the urea nitrogen is substituted. The preferred conformation subsequently confirmed by small-molecule crystallography studies was the 4*R*,5*S*,6*S*,7*R* stereoisomer. These new scaffolds were considered as good potential HIV protease inhibitors because they were able to properly place their substituents into the different subsites of the binding cleft, to displace the structural water 301, to interact with both Asp25, and finally, they were synthetically feasible. Substitutions in the ring predicted to fill S1 and S2 sites led to

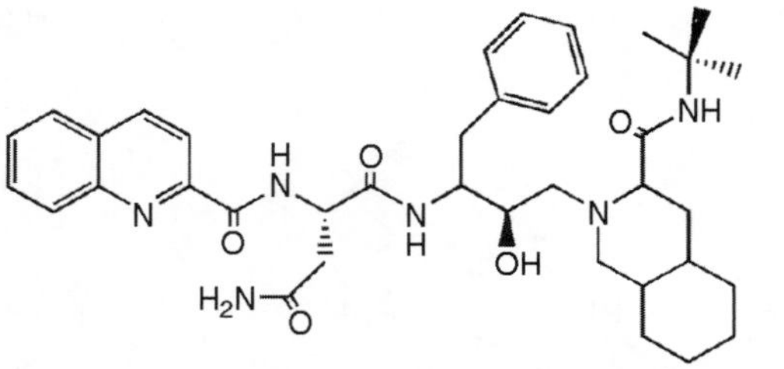

39

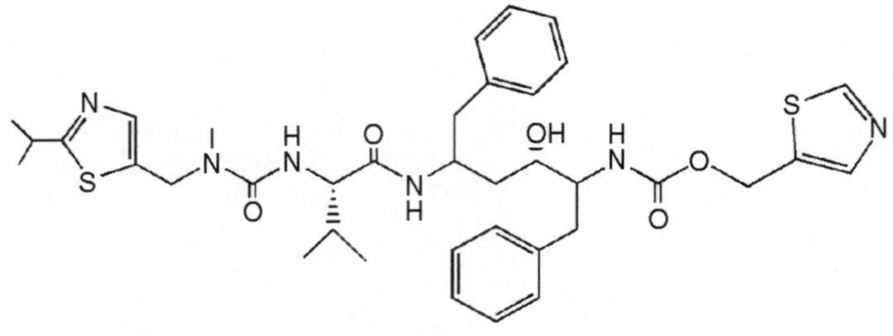

40

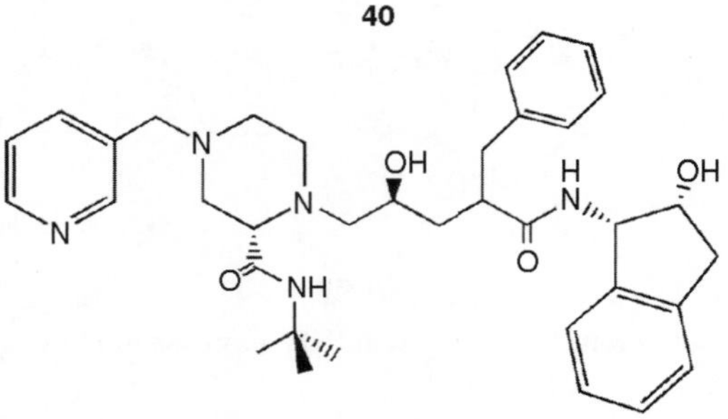

41

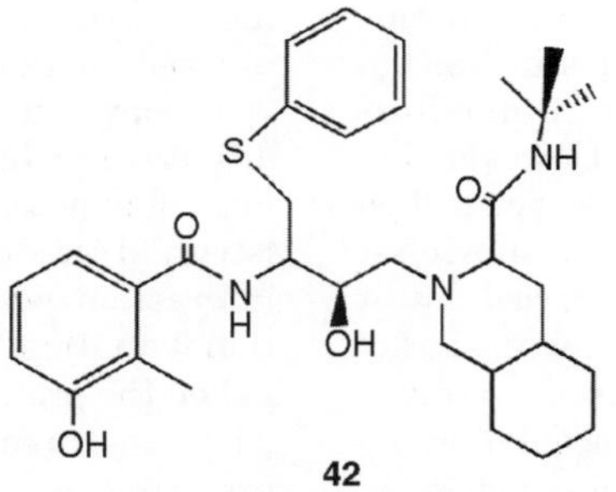

42

Figure 9. Structures of diverse second-generation HIV protease inhibitors. See text for details.

subnanomolar inhibitors such as DMP-450 (**44**) and DMP-323 (**45**) (see Fig. 10). The predicted interactions of these inhibitors were later confirmed by X-ray studies. The development of DMP-450 is now halted after early clinical trials.

10.2. Inhibitors of Ras Farnesyl Transferase [75,76]

In the previous case study, the inhibition of a protease was shown to be a useful antiviral strategy. Enzyme inhibition can also be used

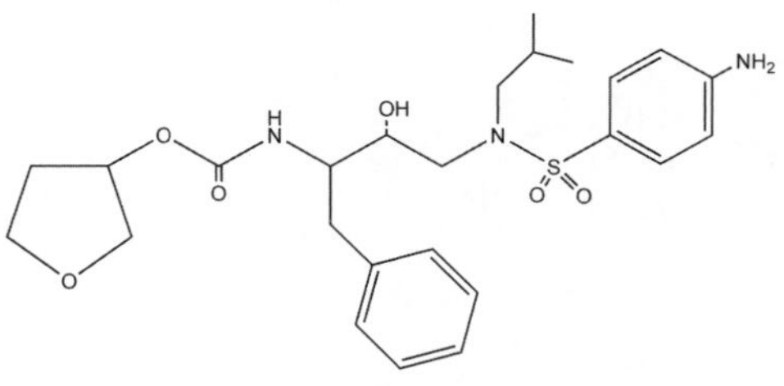

43

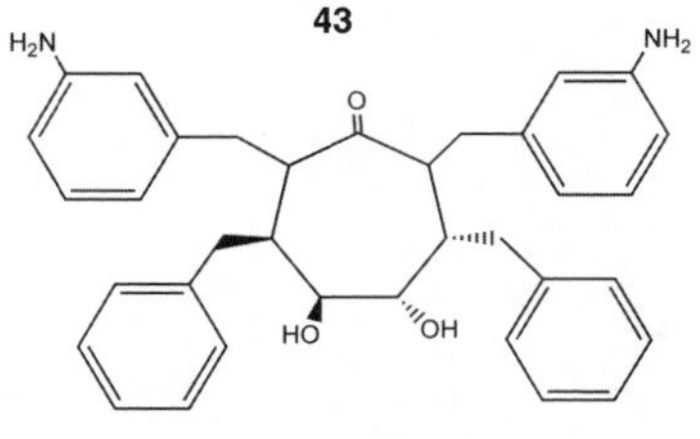

44

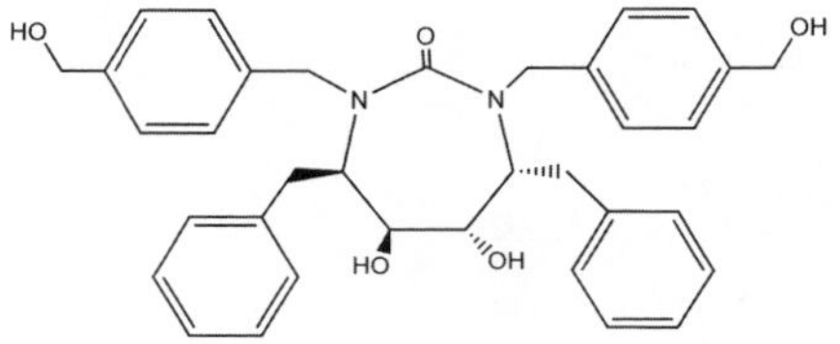

45

Figure 10. Structures of diverse second-generation HIV protease inhibitors. See text for details.

to avoid a harmful protein function. Specifically, there are many proteins that are not active without certain posttranslational modifications that are catalyzed by enzymes. Ras proteins are involved in the signal transduction process and require a farnesyl group for being fully active. Inhibition of this posttranslational modification was proposed as an interesting mechanism for the treatment of certain kinds of cancer [77]. Unfortunately, recent results have demonstrated lower efficacy for these drugs than expected [78]. At present, tipifarnib is a Ras inhibitor in clinical trials for the treatment of certain types of hematological and solid tumors. Interestingly, some of the inhibitors developed have been demonstrated effective in parasites and are presently being studied for the treatment of infectious diseases such as malaria [79].

Ras proteins are plasma membrane-bound GTP binding proteins involved in mitotic signal transduction [80]. Mutations that constitutively activate Ras result in uncontrolled cell growth and play a critical role in malignant transformations [81]. As mentioned before, Ras requires farnesylation and subsequent plasma membrane association for its transforming activity. This posttranslational modification is mediated by farnesyltransferase (FT), which transfers a farnesyl moiety from farnesylpyrophosphate to the cysteine of the CA1A2X motif (C = cysteine; A1 and A2 = isoleucine, leucine, or valine; X = methionine or serine) C-terminal tetrapeptide of Ras. Peptide analogs with the sequence CAAX, such as Cys-Val-Phe-Met (CVFM), show inhibition of Ras farnesylation.

Different conformationally constrained inhibitors of FT have been described in the past. They are modified tetrapeptides with the sequence CAAX where the middle two residues have been replaced by conformationally constrained amino acids, by an organic scaffold, or by N-alkylated glycine derivatives.

Compound BZA-5B (**46**, Fig. 11) was the best inhibitor of a series of compounds based on the replacement of the aliphatic portion of the CAAX tetrapeptide by a 3-amino-1-carboxymethyl-5-phenyl-benzodiazepin-2-one (BZA) [82,83]. A different approach used the hydrophobic spacer 4-aminobenzoic acid (4ABA) to link Cys and Met [84,85], Cys-4ABA-Met (**47**, Fig. 11) being the most potent analog. The rationale used in this work was to suppose that the N-terminal cysteine of the CAAX peptide binds to the zinc ion through a thiol with the subsequent looping around the zinc ion in order to allow additional coordination to the metal through the terminal carboxylate. Following the same hypothesis, a chemical series of compounds using an imidazole group prompted a series of quinolones that lead to tipifarnib (**48**, Fig. 11) (R115777) presently in clinical trials for the treatment of solid cancers [86].

A different group of inhibitors was discovered using alternative procedures. Thus, from a random screening of an extensive chemical collection of antihistamines, the 8-clorobenzocycloheptapyridine derivative SCH 44342 (**49,** Fig. 11) was found to be a FT inhibitor, and constituted an early lead compound (IC_{50} = 250 nM) [87]. From this compound, a series of derivatives was developed with the aim of optimizing its affinity and selectivity. These studies resulted in molecules with FT inhibition constants in the low nanomolar range and

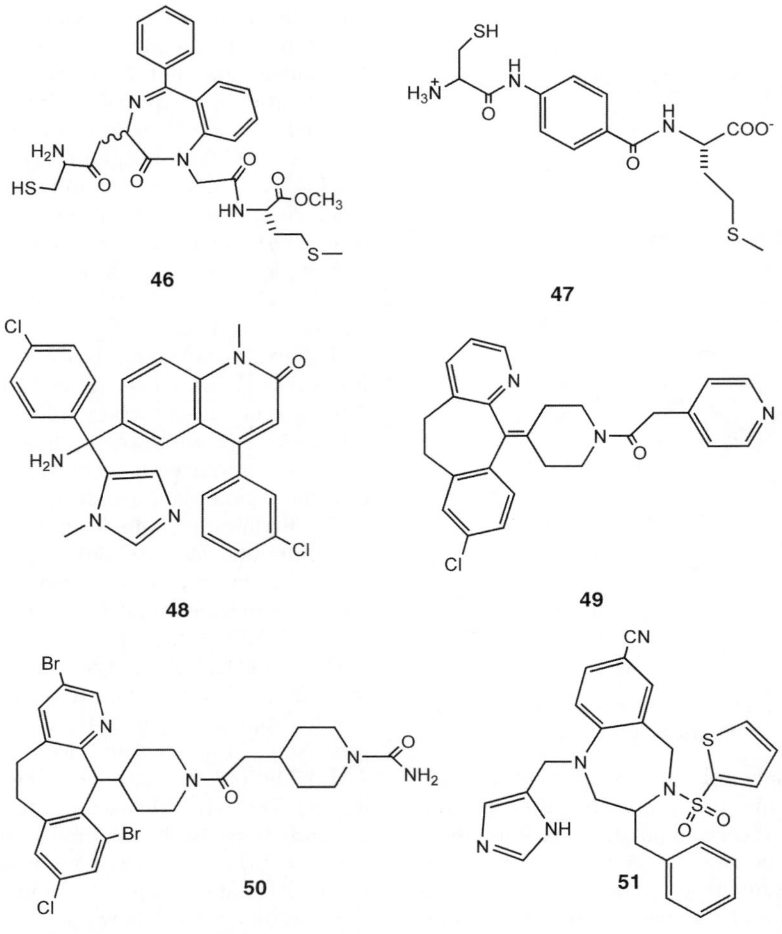

Figure 11. Structures of different Ras farnesyltransferase inhibitors. See text for details.

about 10,000-fold more selective for FT than for the structurally related geranylgeranyltransferase (GGT). All these compounds are characterized by a tricycle core composed of two six-membered aromatic rings, fused to a central seven-membered ring. One example of this series is the tricyclic compound Lona-farnib (SCH66336) (**50**, Fig. 11), which is presently undergoing clinical trials as anticancer agent [88,89]. Attempts made to define the bound conformation of this class of inhibitors by using molecular modeling techniques were unsatisfactory. This can be due to the size of the FT active site cavity that binds farnesylpyrophosphate and the carboxyl terminal residues of the Ras protein that is much larger than a single tricycle inhibitor. Another family of disclosed FT inhibitors is the 2,3,4,5-tetrahydro-1-(imidazol-4-ylalkyl)-1,4-benzodiazepines (BMS-214662) (**51**, Fig. 11) [90]. Low nanomolar inhibitors were identified that are able to revert the transformed phenotype of Ras transformed cells at submicromolar concentrations and that can prevent the anchorage-independent growth in soft agar of Ras transformed cells at concentrations as low as 160 nM.

11. GPCR LIGANDS

The problem of designing peptidomimetics associated to GPCR ligands has been hampered in the past by the lack of an experimental structure of the ligand–receptor complex. At present, there are only three GPCR structures available and there is an interesting discussion about the analysis of their similarities [91,92]. Accordingly, all the information gathered needs to be retrieved from the different structure–activity studies available. There are many stories regarding this kind of ligands, mostly related to the design of antagonists.

11.1. Design of Bradykinin B2 Antagonists

Bradykinin (BK) is a linear nonapeptide hormone of sequence Arg^1-Pro^2-Pro^3-Gly^4-Phe^5-Ser^6-Pro^7-Phe^8-Arg^9, produced from its precursor kininogen by the action of a group of proteases called kallikreins. The peptide hormone is released in response to inflammation, trauma, burns, shock, allergy, and some cardiovascular diseases, influencing vascular tone and permeability and decreasing blood pressure [93]. BK actions are mediated through two G-protein-coupled receptors: the B_1 and B_2 kinin receptors. B_1 has a low level of expression in normal tissues, being expressed during trauma or inflammation [94]. On the other hand, the B_2 kinin receptor is constitutively expressed by different cell types [93]. B_2 receptor agonists have been sought in the past for the treatment and prevention of various cardiovascular diseases, such as hypertension, ischemic heart disease, congestive heart failure, as well as diabetic disorders. B_2 antagonists are also sought to reduce the effects mediated by BK, such as pain and inflammation in several diseases (asthma, rhinitis, and septic shock) [95].

The discovery of the antagonistic profile of the [D-Phe^7]bradykinin analog led, in the late 1980s, to the characterization of the first-generation B_2 bradykinin antagonists, including NPC349, NPC567, and NPC17731 [96]. Analysis of the results obtained from biophysical studies [97] suggested that the peptide exhibits a β-turn involving the last four residues of the peptide. These studies led to the synthesis of HOE140 (icatibant) (**52**, Fig. 12), a peptide analog of sequence D-Arg^0-Arg^1-Pro^2-Hyp^3-Gly^4-Thi^5-Ser^6-D-Tic^7-Oic^8-Arg^9 (where Hyp stands for hydroxyproline; Thi, β-(2-thienyl)-alanine; Tic, 1,2,3,4-tetrahydroisoquinoline-3-carboxylic acid; Oic, (2*S*,3a*S*,7a*S*)-octahydroindole-2-carboxylic acid) that contains several conformational constrained residues to enhance the secondary motif at the C-terminus [97]. The pharmacology of this compound has been extensively studied and presently has been approved for the treatment of acute hereditary angioedema [98].

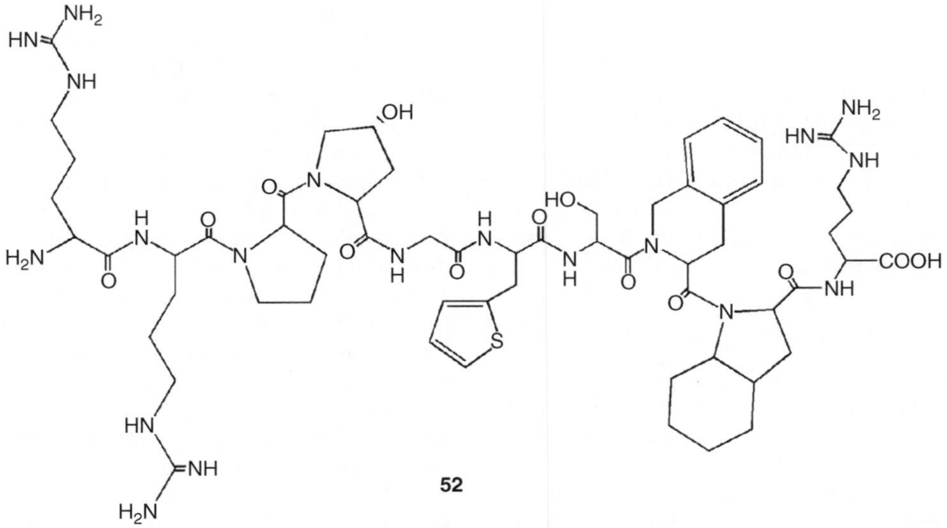

52

Figure 12. Structure of HOE140.

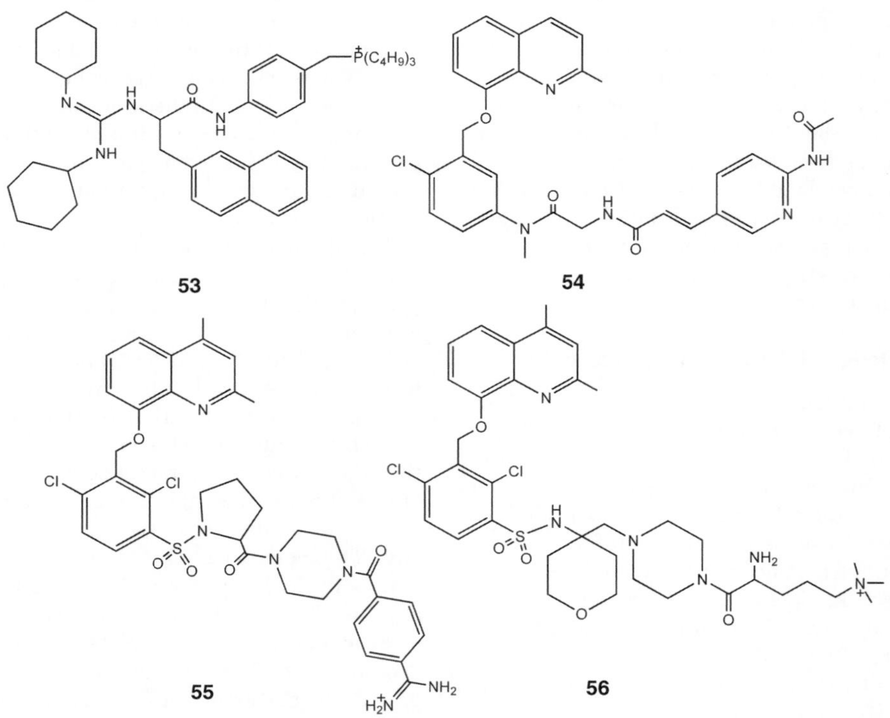

Figure 13. Structures of different B_2 bradykinin antagonists.

The first nonpeptide antagonist described was WIN64338 (**53**, Fig. 13) [99] based on a low-profile pharmacophore. Although this molecule exhibits a reasonable binding affinity for the B_2 receptor in rat tissue models, the compound exhibits a low affinity for the human receptor. A few years later, the first orally bioavailable B_2 antagonists were described, such as FR173657 (FK3657) (**54**, Fig. 13) [100] or LF-16-0335C (**55**, Fig. 13) [101], and more recently the compound MEN16132 (**56**, Fig. 13) [102] about 30 times more effective B_2 antagonist than HOE140. These compounds are presently in clinical trials.

11.2. Design of Endothelin Antagonists [103]

Endothelins (ET) are 21-residue peptides partially constrained by two disulfide bonds with potent vasoconstrictor properties. There are three isoforms described, known as ET-1, ET-2, and ET-3 [104,105], that mediate their action through at least two different G-protein-coupled receptors: ET_A and ET_B [106]. The former receptor is selective for ET-1 and ET-2, whereas the latter binds ET-1, ET-2, and ET-3 with equal affinity [107]. Upon binding ET initiate a complex signal transduction cascade including activation of various kinase-mediated pathways involved in mitogenic responses [108]. ET affects cell proliferation in various types of cells; they also modulate apoptosis induced by serum starvation and chemical treatment and act as survival factors for fibroblasts, endothelial, and smooth muscle cells [109]. ET-1 may play a pivotal role in the pathogenesis of cell growth disorders such as cancer, restenosis, and benign prostatic hyperplasia. Great efforts have been devoted in the past to block the endothelin system, leading to the discovery of a large number of peptide and nonpeptide receptor antagonists. Most preclinical and clinical pharmacological studies reported on endothelin receptor antagonism are focused on their therapeutic potential in relation to pulmonary arterial hypertension (PAH) and heart failure.

Early efforts to design ET antagonists produced two results simultaneously. On the one hand, the identification of the cyclic peptide BE-18257B (**57**, Fig. 14) as an ET_A antagonist, found as a fermentation by-product of *Streptomyces misakiensis* [110], and on the other, the peptide analog [Dpr^1,Asp^{15}]ET-1 (where Dpr is diaminopropionic acid), designed by replacement of the Cys^1–Cys^{15} disulfide bond of ET-1 by an amide bond between the side chains of Dpr^1 and Asp^{15} [111]. Shortly after, chemical modification of BE-18257B yielded BQ-123 (**58**, Fig. 14), an antagonist in the nanomolar range [112]. The same authors, in the course of a structure–activity relationships study of BQ-123, observed that some structural alterations intensified ET_B rather than ET_A affinity of some of the analogs. These results led to the development of BQ-788 (**59**, Fig. 15), a selective ET_B antagonist [113].

In parallel, structure–activity relationship studies carried out on the peptide established that the C-terminal hexapeptide His-Leu-Asp-Ile-Ile-Trp or (16–20)ET-1 was the minimum fragment preserving biological activity in some, but not all the tissues responding to ET [114,115]. These results together with the information provided by different reports aimed at understanding the conformational features of the peptides in solution [116–118] permitted scientists at Parke Davis to design the pseudopeptide analogs PD 145065 (**60**,

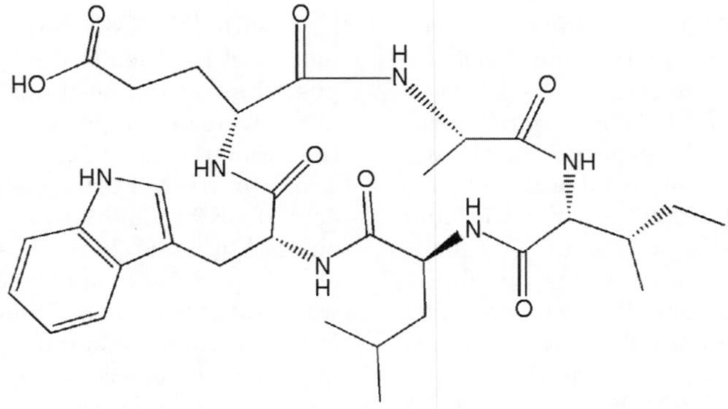

57

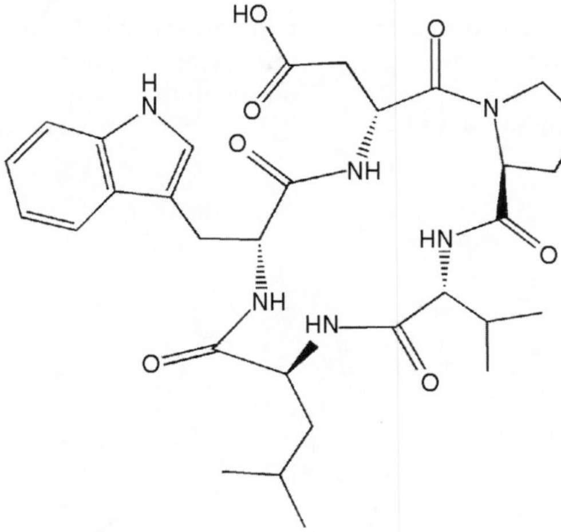

58

Figure 14. Chemical structure of BE-18257B (**57**) and BQ-123 (**58**), two cyclic peptide antagonists of endothelin.

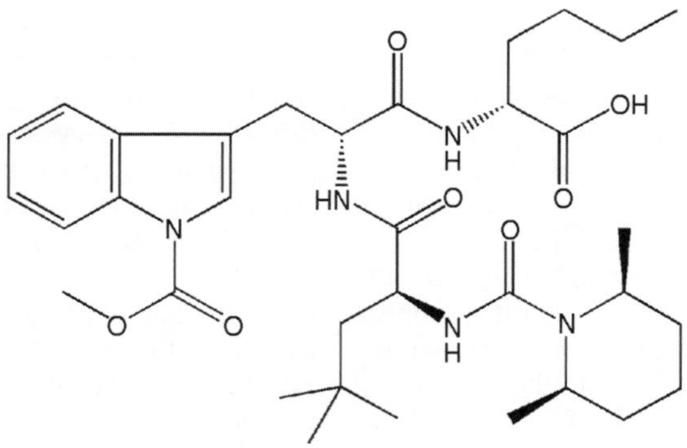

59

Figure 15. Structure of BQ-788, a small-molecule selective antagonist of ET_B.

Fig. 16) and PD 156252 (a modified version of PD 145065 where an *N*-methyl amino acid was added at position 20 to stabilize from proteolytic degradation) [119]. An alternative strategy to design endothelin peptidomimetics was also followed in the same laboratories. The lead compound PD 012527 (**61**, Fig. 17) found through library screening was used to generate a series of compounds. Using the *Topliss Decision Tree* approach for lead optimization [120,121], different ET_A-selective antagonists were found, the best representatives being PD 155080 (**62**, Fig. 17) and PD 156707 (**63**, Fig. 17) [122,123].

Design of ET antagonists at SmithKline Beecham began by identification of compound SK&F 66861 (**64**, Fig. 18) by screening a G-protein-coupled receptor ligand database [124]. Subsequently, comparison of the lead with ^{1}H NMR-derived conformational models of ET-1 led to the hypothesis that the phenyl groups in positions 1 and 3 in SK&F 66861 were mimicking a combination of two of the aromatic side chains of Tyr^{13}, Phe^{14}, and Trp^{21} in ET-1. Furthermore, the carboxylic acid of SK&F 66861 was proposed to mimic either Asp^{18} side chain or C-terminal carboxyl of ET-1. Subsequent medicinal chemistry work substituting the indene scaffold by an indane to increase stability together with the incorporation of a second carboxylic acid moiety to mimic the C-terminal carboxyl of ET-1 yielded compound SB 209670 (**65**, Fig. 18), an ET_A partially selective compound in the nanomolar range [125]. Further studies aimed at improving its poor bioavailability, intestinal permeability screening yielded the compound SB 217242 (Enrasentan) (**66**, Fig. 18), unfor-

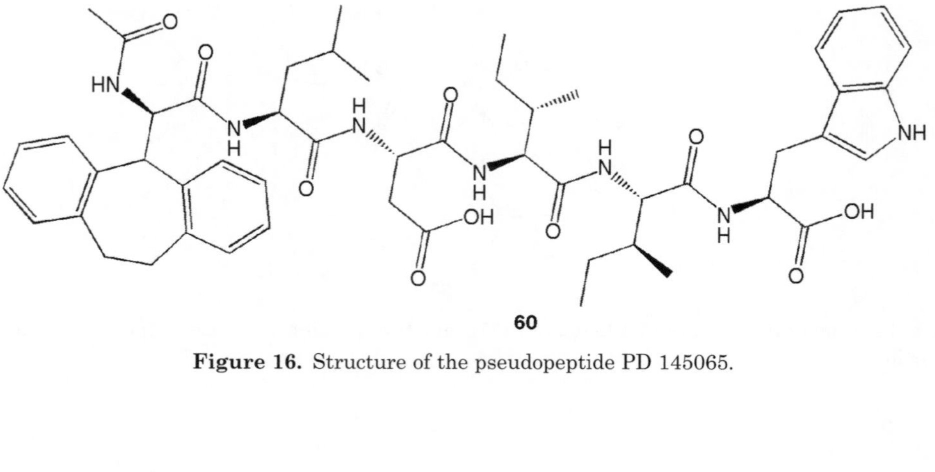

60

Figure 16. Structure of the pseudopeptide PD 145065.

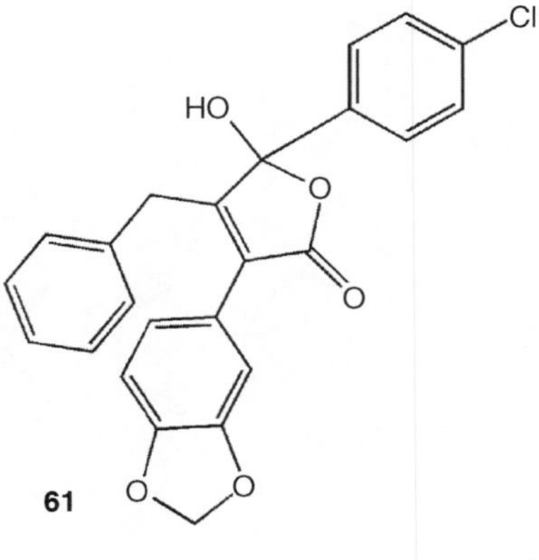

Figure 17. Structures of PD 012527 (**61**), PD 155080 (**62**), and PD 156707 (**63**).

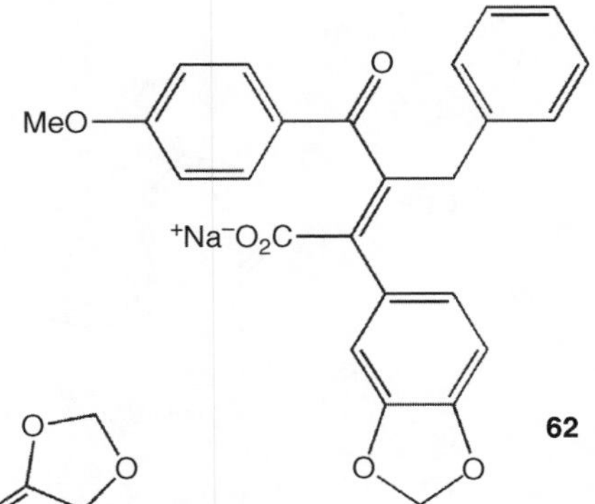

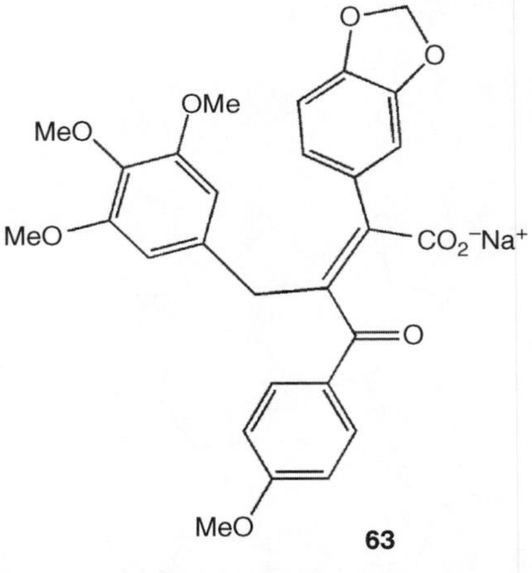

tunately with poor results for the treatment of chronic heart failure as found in clinical studies [126]. Using SB209670 as a template and following the Kohonen neural network method, scientists at Merck identified the benzothiadiazole moiety as a surrogate of the methylendioxyphenyl group of the model compound [127]. These studies yielded the compound EMD 122801 with an IC_{50} for ET_A of 0.30 nM (**67**, Fig. 18) [128].

The first orally active endothelin ET_A/ET_B antagonists came after optimization of Ro 46-2005 (**68**, Fig. 19), found in a library screening [129]. Optimization of the compound yielded Ro 47-0203 (Bosentan) (**69**, Fig. 19) [130] that was the first competitive antagonist at ET_A and ET_B receptors that passed clinical phases and reached the market in 2002 for the treatment of pulmonary arterial hypertension and chronic heart

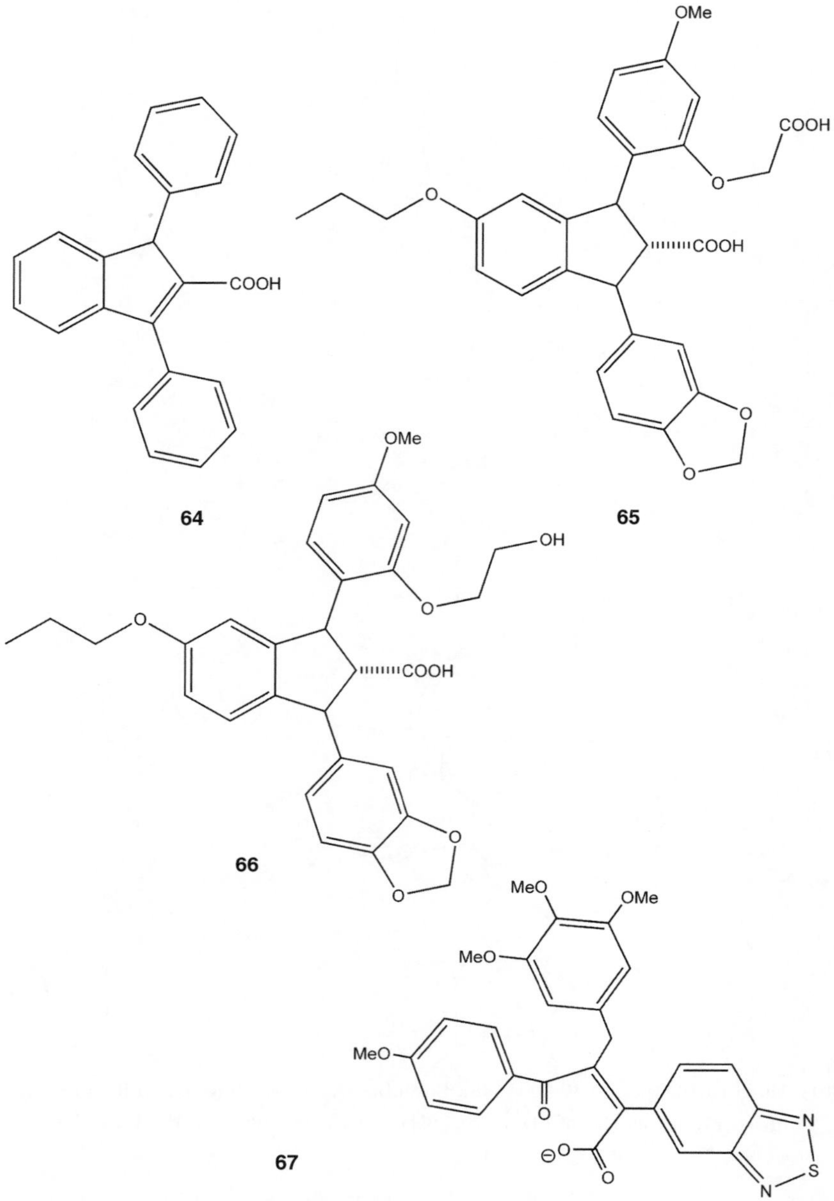

Figure 18. Structures of SK&F 66861 (**64**), SB 209670 (**65**), Enrasentan (SB 217242) (**66**), and EMD 122801 (**67**).

failure. Scientists at the same laboratory designed Ro 46-8443 (**70**, Fig. 19), the first nonpeptide ET_B-selective antagonist of the endothelin receptor [131]. This compound has the same structure as Bosentan, where the second pyrimidine and the hydroxyethoxy side chain were replaced by a *p*-methoxyphenyl moiety and a glycerol residue, respec-

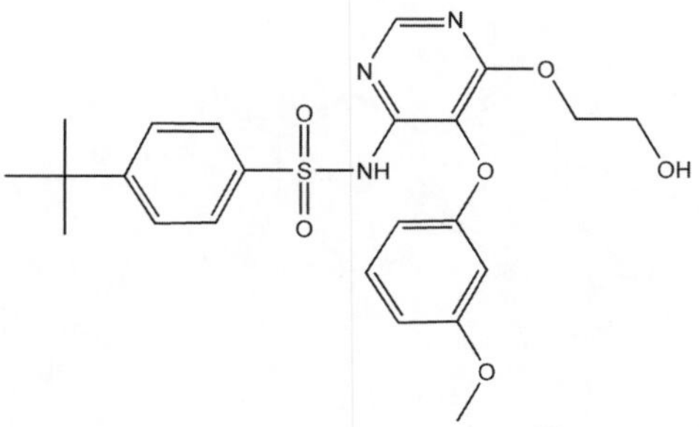

Figure 19. Structures of Ro 46-2005 (**68**), Bosentan (Ro 47-0203) (**69**), and Ro 46-8443 (**70**).

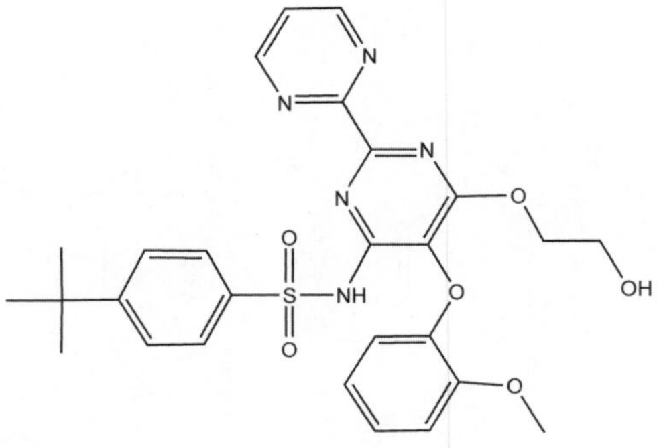

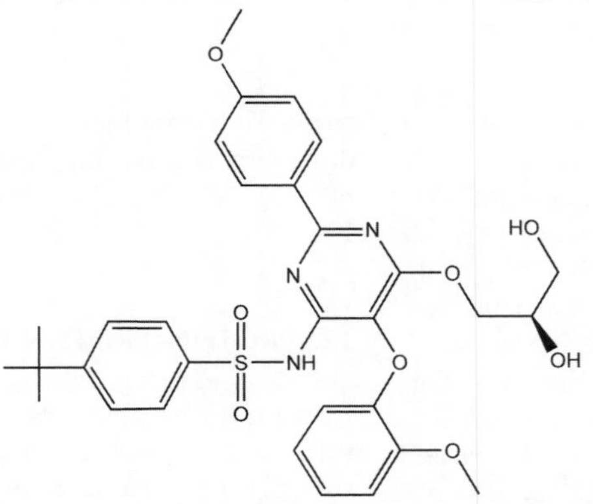

tively. Wu et al. [132] investigated further the use of amide sulfonamides as scaffold to deign ET_A-selective endothelin receptor antagonists aiming at improving their oral bioavailability and half-life. A series of compounds were produced by replacement of the amide bond with various linker groups including, urethane, imide, ureido, and other extended amide linkages. They found carbonyl group to be the optimal spacer, leading to the discovery of TBC-11251 (**71**, Fig. 20), a potent ET_A antagonist with good bioavailability and half-life.

At Bristol-Myers Squibb, Stein et al. [133] used a benzene sulfonamide scaffold to design endothelin antagonists leading to the potent, orally bioavailable, highly selective ET_A receptor antagonist BMS-182874

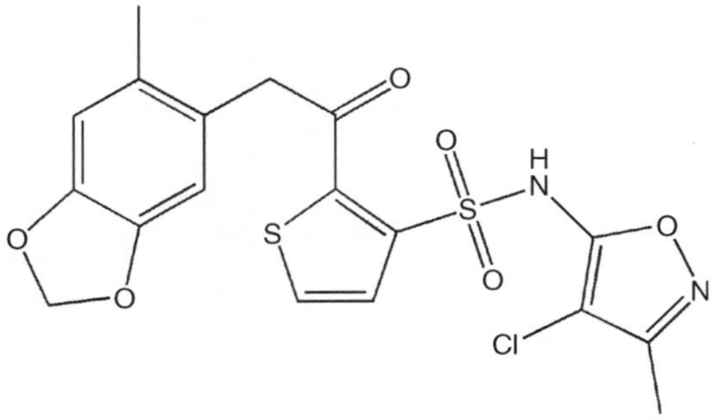

71

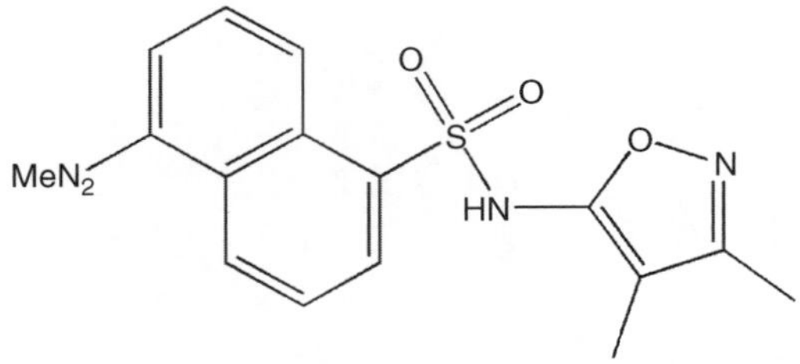

72

Figure 20. Structures of TBC-11251 (**71**) and BMS-182874 (**72**).

(**72**, Fig. 20). At Abbot Laboratories, Wu-Wong et al. [134] reported a series of compounds derived from the same scaffold that showed different selectivities for ET_A and ET_B. Thus, ABT-627 (**73**, Fig. 21) and ABT-546 (**74**, Fig. 21) were ET_A selective, A-182086 (**75**, Fig. 21) was nonselective, and A-192621 was ET_B selective (**76**, Fig. 21). Of these compounds, ABT-627, also known as Atrasentan, is undergoing clinical trials for the treatment of different types of cancers [135].

Lead compounds acting as ET_A/ET_B antagonists have been identified from the search in 3D chemical databases using different pharmacophores. Thus, Funk et al. [136] used HypoGen and HipHop algorithms implemented in the Catalyst molecular modeling software, to create two pharmacophores of five and six features, respectively. These pharmacophores were validated for their activity prediction capability and were used to screen the Maybridge 3D database of chemical compounds. As a result, two new potential lead compounds were identified that displaced ET-1 from its receptor out of six compounds tested with IC_{50} values of 220 nM (**77**, Fig. 22) and 6200 nM (**78**, Fig. 22).

12. PROTEIN–PROTEIN INHIBITORS

Protein–protein interactions are involved in numerous signaling cascades. Disruption of this process can become a useful process for therapeutic intervention [137]. Many cases have been reported recently where short stretches of peptides are capable of mimicking protein epitopes and act as inhibitors [138]. In this section, we briefly describe two stories related to the discovery of small-molecule inhibitors for the treatment of cancer. On the one hand, small-molecule mimics of the BH3 domain of the proapoptotic members of the Bcl2 family of proteins and, on the other, small-molecule inhibitors of a fragment of the Smac protein involved in inhibiting the inhibitor of apoptosis proteins (IAPs).

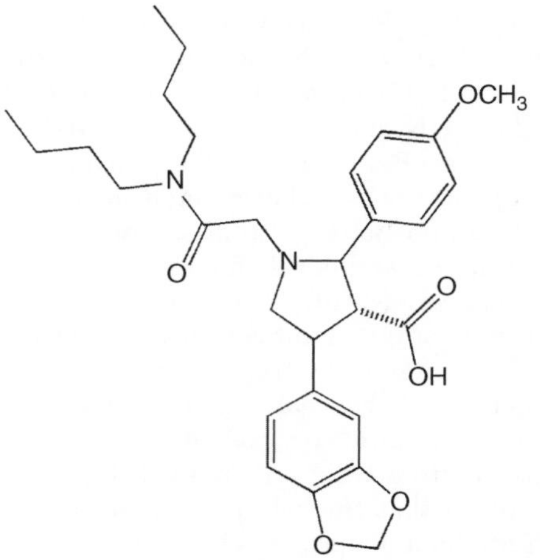

73 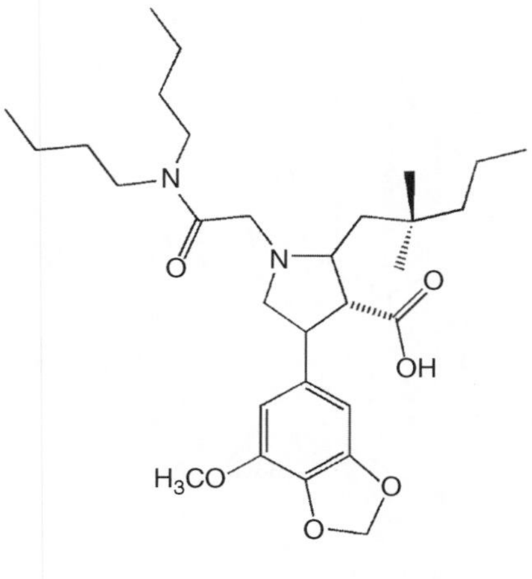 74

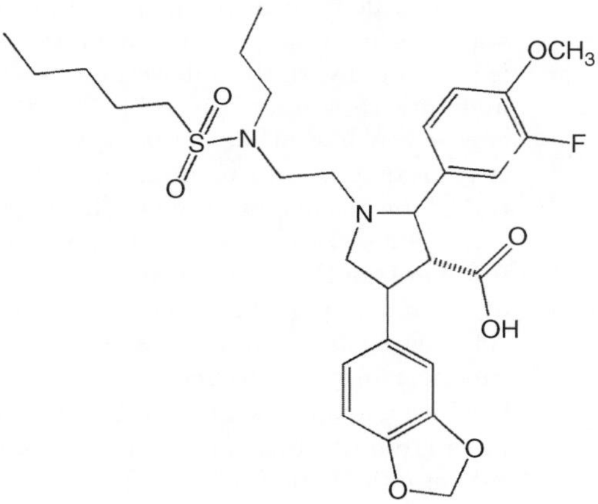

75 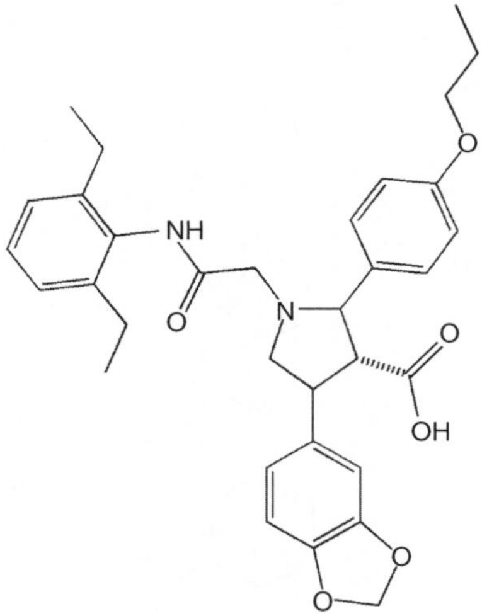 76

Figure 21. Structures of ABT-627 (**73**), ABT-546 (**74**), A-182086 (**75**), and A-192621 (**76**).

12.1. Design of Bcl-2 Inhibitors [139–141]

Apoptosis or *programmed cell death* is an essential process designed for cells to commit suicide under certain conditions, providing a mechanism for a normal development of an organism. Deregulation of this process can result in an undesirable increase of cell numbers as in cancer, or in cell loss as found in autoimmune or neurodegenerative diseases [142]. Apoptosis is executed by a family of cysteine aspartyl proteases called caspases [143–145], which are activated by two major interconnected pathways [146,147]: the *extrinsic* pathway activated by cell surface receptors and the *intrinsic* pathway that is activated by numerous signals in which cytochrome *c* is released from the mitochondria. The intrinsic pathway, and in some cases the extrinsic one, is regulated by the B-cell lymphoma-2 (Bcl-2) family of proteins [148–152].

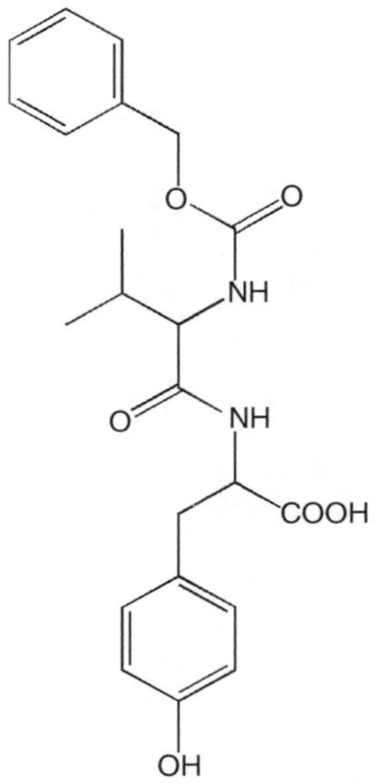

Figure 22. Structures of two hits obtained from database search. See text for details.

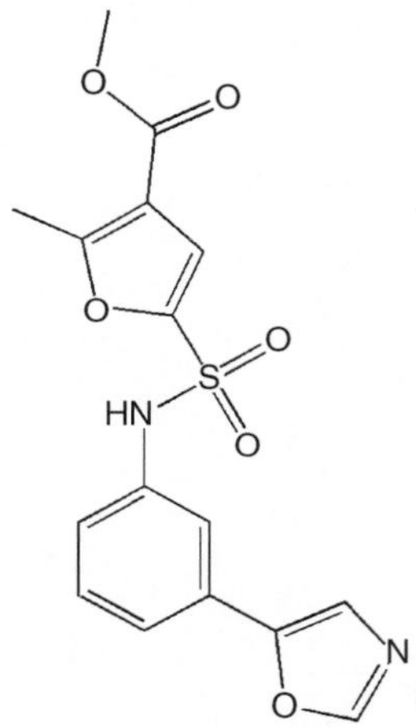

More than 20 members of this family have been identified so far; some of them exhibit proapoptotic activity whereas the rest exhibit an antiapoptotic profile. Sequence analysis of these proteins shows that all members share at least one out of four homology domains (BH1–BH4), and there is a relation between the function of the proteins and their sequence. Specifically, the antiapoptotic members, such as Bcl-2, Bcl-x_L, Bcl-w, or Mcl-1 are characterized for exhibiting the four homology domains, whereas the proapoptotic members are characterized by exhibiting either the BH3 domain only, such as Bad, Bim, Bid, or Noxa, or domains from BH1 to BH3 such as Bax, Bak, or Bok. Several studies support the hypothesis that overexpression of Bcl-2 and Bcl-x_L proteins is connected to the initiation and development of different types of cancer, as well as to an elevated resistance to conventional anticancer therapies.

In spite of the tremendous progress made in the past few years, the mechanism of apoptosis regulation by the Bcl-2 family of proteins is far from being understood. However, it is known that the relative amount of pro- and antiapoptotic proteins in a cell, together with the capability of these proteins to form heterodimers, determines cell susceptibility to undergo apoptosis [153–155]. Specifically, when the BH3 domain of proapoptotic proteins is suppressed, binding to antiapoptotic proteins is prevented with the subsequent loss of their apoptotic capability [156,157]. Accordingly, it can be considered that proapoptotic proteins trigger apoptosis through an inhibitory function of the antiapoptotic ones. This hypothesis was demonstrated when different peptide fragments of this domain were shown to bind at the nanomolar level and induce apoptosis in cell-free systems and HeLa cells [158,159]. Furthermore, cell-permeable BH3 peptides produce apoptosis in different cellular lines [160] demonstrating the pharmacological potential benefit of these peptides. Efforts have been devoted to improve the natural sequences in order to achieve a consensus sequence for the BH3–antiapoptotic protein interaction [161,162]. Furthermore, the structural analysis of complexes of Bcl-x_L with short peptide fragments of the BH3 domain of Bak [163] and Bad [164] reveals that these peptides adopt an α-helix when bound to the antiapoptotic protein, with their hydrophobic side facing to a hydrophobic cleft formed by the BH1, BH2, and BH3 domains. Based on this information, different strategies have been worked out.

One strategy to augment the pharmacological profile of the BH3 peptides focused on improving the stability of the helical structure. One approach is to enhance the helical content through a lactam cross-link (**79**,

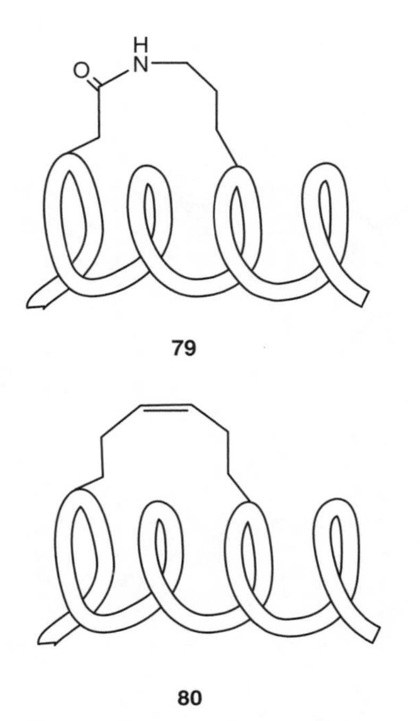

Figure 23. Structures of two peptidomimetics designed to stabilize the helical structure.

Fig. 23) [165], unfortunately although the helical contents increase, this kind of cross-link does not solve some of peptide shortcomings to act as a drug. A more successful approach is to add a hydrocarbon staple (**80**) [166] as shown in Fig. 23. These results can be improved if in addition some of the residues is replaced by α,α-disubstituted nonnatural amino acids that reinforce the tendency of the sequence to adopt a helical structure. This process has demonstrated to augment the helical content of the peptide and improve their pharmacokinetical profile by decreasing their resistance to proteolytic cleavage, their binding affinity, and permeability.

Other approaches include the construction of synthetic scaffolds that have attached the corresponding side chains of the BH3 motifs. Terphenyl, terpyridine, or terephthalamine have been used for this purpose [167,168]. This strategy puts the right-hand side chains exposed for interaction while preserving the amino acid skeleton from protease hydrolysis. Other strategies include the use of foldamers, chimerical peptides that juxtapose native BH3 domain sequence with alternating α- and β-peptide foldamer sequence yielding compounds with nanomolar affinities [169].

A different approach focused on finding small-molecule mimics of the BH3 domains. Different groups have reported small-molecule inhibitors of Bcl-2/Bcl-x_L using diverse strategies. Some inhibitors have been found from high-throughput screening of libraries. These compounds exhibit inhibitory constants in the micromolar range. One of the early compounds described is Antimycin A3 (**81**) [170], or BH3I-1 (**82**) and BH3I-2 (**83**) shown in Fig. 24 [171]. Natural products such as chelerythrine (**84**) [172] or gossypol (**85**) [173] shown in Fig. 25 have also been identified. Natural polyphenols isolated from tea extracts, such as epigallocatechin gallate (**86**) (from green tea) or theaflavinin (**87**) (from black tea), shown in Fig. 25, compete with the BH3 domain at the submicromolar range [174]. Obatoclax (GX15-070) (**88**, Fig. 26) is a derivative from a hit identified by screening of a natural product library that competitively binds to Bcl-xL, Bcl-w, and Mcl-1 in the high nanomolar range [175].

Virtual screening provided several hits in the micromolar range including HA14-1 (**89**) [176], compound 6 (**90**) [177], YC137 (**91**) [178], or the recently discovered NMNM (**92**) [179] shown in Fig. 26. Structure–activity relationships using NMR (fragment screening) have also provided small-molecule inhibitors of Bcl-2. This is the case of ABT-737 (**93,** Fig. 26) [180], a small-molecule inhibitor of Bcl-2, Bcl-xL, and Bcl-w in the subnanomolar range, with demonstrated antitumor activity *in vitro* and *in vivo*. Other small-molecule leads using similar fragments have been described using this methodology [181,182].

12.2. Design of XIAP Inhibitors [183]

The second mitochondrial-derived activator of caspases (Smac) is a protein released from mitochondria into the cytosol in response to an apoptotic stimulus. The protein is an endogenous inhibitor of different IAPs, its action

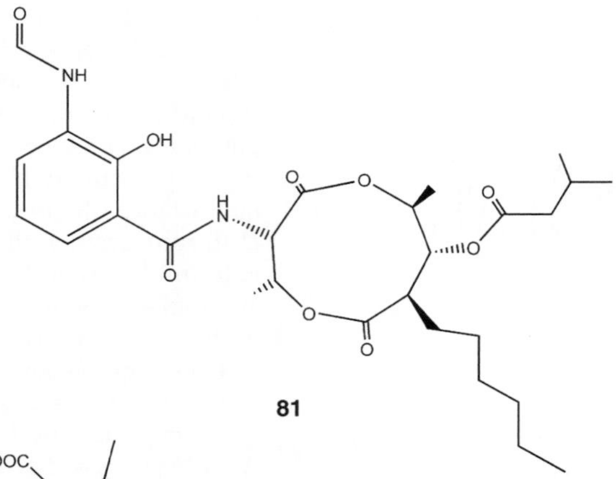

81

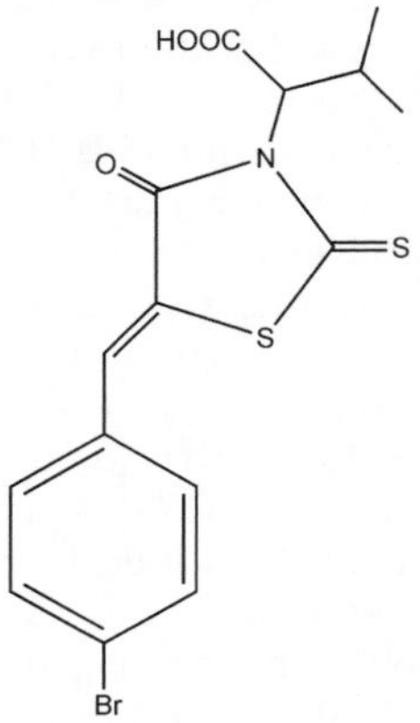

82

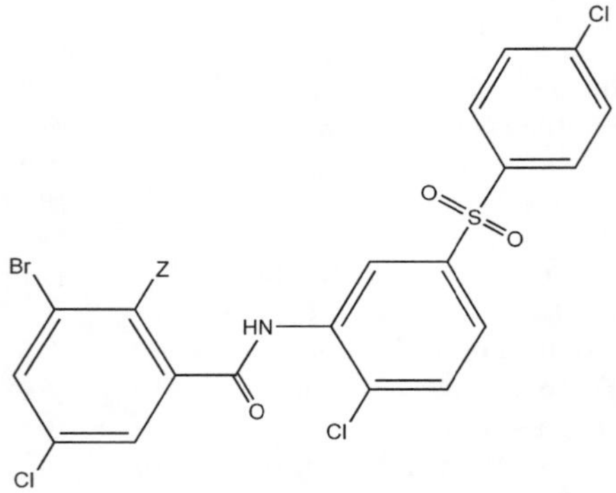

83

Figure 24. Structures of Antimycin A3 (**81**), BH3I-1 (**82**), and BH3I-2 (**83**)

being to prevent the binding of IAPs to caspase-3, caspase-7, and caspase-9, critical actors of the apoptotic process. IAPs are found highly expressed in tumor cells, playing an important role in the resistance of tumor cells to current chemotherapeutic agents [184]. IAP inhibition by Smac induces the cell to initiate the apoptotic process. Accordingly, mimics of the protein could be used as therapeutic agents for the treatment of cancer.

The X-linked inhibitor of apoptosis (XIAP) is one of the most potent IAPs. Structural analysis of the complex XIAP/Smac reveals that the two proteins interact via the N-terminal peptide segment of Smac: Ala-Val-Pro-Ile [185,186] (**94**, Fig. 27), which binds to the so-called baculovirus IAP repeat (BIR3) domain of XIAP. Moreover, the isolated peptide inhibits the protein with a $K_i = 290\,nM$ [187]. Accordingly, mimetics of the Smac N-terminus tetrapeptide could be transformed into useful therapeutic agents.

Early structure–activity relationship studies demonstrated the importance of Ala

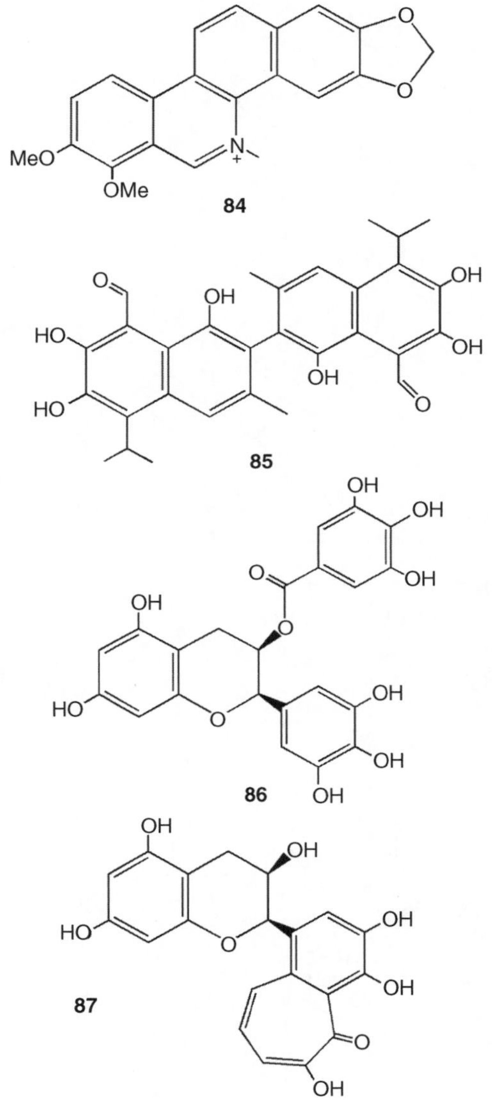

Figure 25. Structures of chelerythrine (**84**), gossypol (**85**), epigallocatechin gallate (**86**), and theaflavinin (**87**)

in position 1, although the ethyl side chain (α-aminobutyric acid) improves affinity. Similarly, it was shown that the binding affinity diminishes when Pro is replaced in position 3. In contrast, there is tolerance for positions 2 and 4. Specifically, most of the amino acids are admitted in position 2, with exception of Gly, Asp, and Pro. On the other hand, residues with bulky hydrophobic side chains such as Trp or Phe fit well in position 4. N-methylation of the amide moiety of residue 4 was tolerated in contrast to the N-methylation carried out on the other residues [187]. These analogs exhibit single-digit nanomolar affinity and promote cell death in different human cancer cell lines [188]. In a further step, the side chain of valine was substituted with several five-membered heterocyclic moieties while maintaining the Ala amino acid achieving submicromolar activity [189]. Li et al. using combinatorial synthesis and trying to mimic the last residue of the tetrapeptide by a heterocycle described a series of hybrid mimetics of the peptide [190], the most potent being the oxazoline containing pseudopeptide (**95**, Fig. 27). Further work in this direction has been pursued by other authors [191].

Several studies have been published reporting the design of conformationally constrained peptidomimetics of the tetrapeptide. One of the first attempts came from the analysis of the crystal structure of the Smac/XIAP complex. Inspection of the structure reveals that the isopropyl side chain of residue 2 is solvent exposed, whereas the proline ring sits on a hydrophobic pocket. These features suggested the design of a peptidomimetic fusing the side chains of valine and proline through a bicyclic scaffold. Although the first analog synthesized (**96**) (Fig. 28) exhibits a slight lower affinity than the original peptide, after subsequent optimization, low nanomolar compounds were obtained [192]. Compound (**97**) in Fig. 28 can be used as a proof of principle. Unfortunately, these series of compounds showed low *ex vitro* activity due to a low cell permeability, forcing to develop further a series of new analogs with improved cell penetration properties. As a result of an extensive study, several compounds were designed, compound **98** being an example of an analog that exhibits good activity *ex vivo* (Fig. 28) [193].

Other peptidomimetics based on bicyclic scaffolds were reported. Thus, Genetech described inhibitors using a bicyclic [7,5]-lactam scaffold (**99**, Fig. 29). These compounds exhibit both single-agent cell killing and addi-

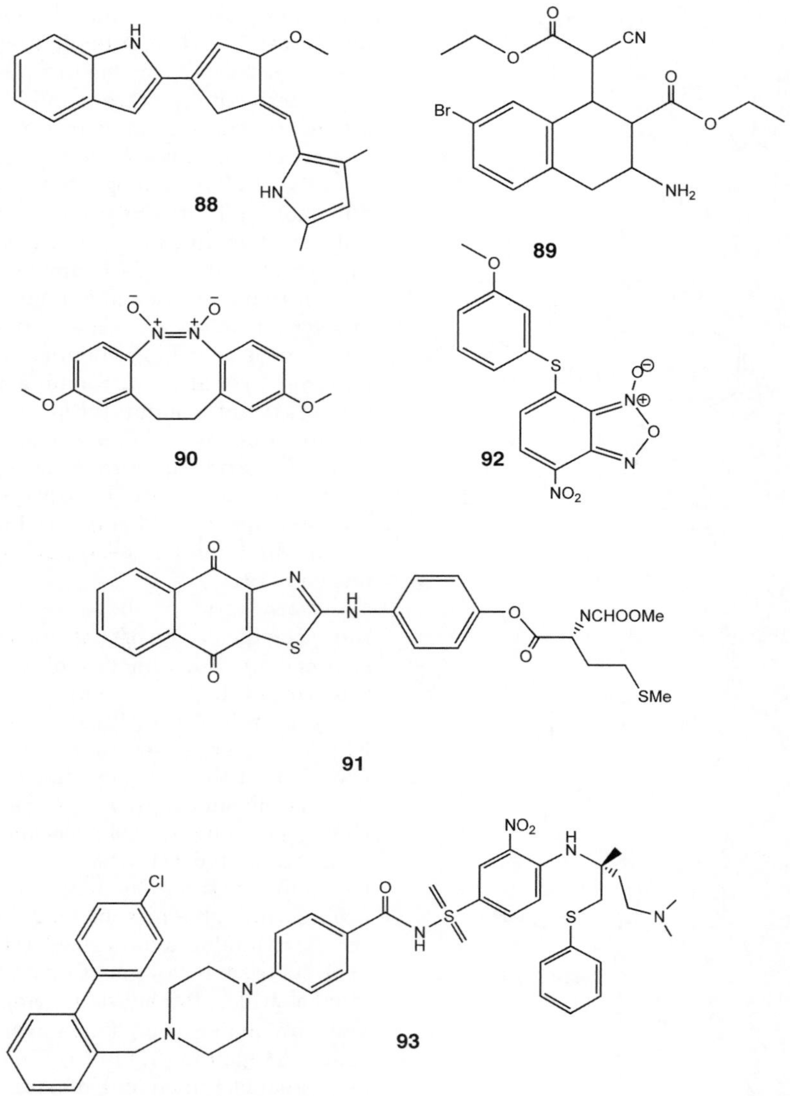

Figure 26. Structure of obatoclax (GX15-070) (**88**), HA14-1 (**89**), compound 6 (**90**), YC137 (**91**), NMNM (**92**), and ABT-737 (**93**).

tivity with doxorubicin in cancer cell lines [194]. More recently, the use of azabicyclooctane as scaffold has yielded orally bioavailable inhibitors (**100**, Fig. 29) [195]. Interestingly, this compound decreased the viability of cancer cell lines without affecting nontumor cells. Novartis also reported the design of peptidomimetic inhibitors of XIAP based on a pyrido[3,4*b*]pyrrole scaffold, their representative compound being LBW242 (**101**, Fig. 29) [196]. Similarly, taking as starting point the compound SM-130 described by Sun et al. [197], Bolognesi and colleagues synthesized a library of 4-subtituted azabicyclo[5.3.0]alkane compounds displaying high affinities (**102**, Fig. 29). X-ray crystallography

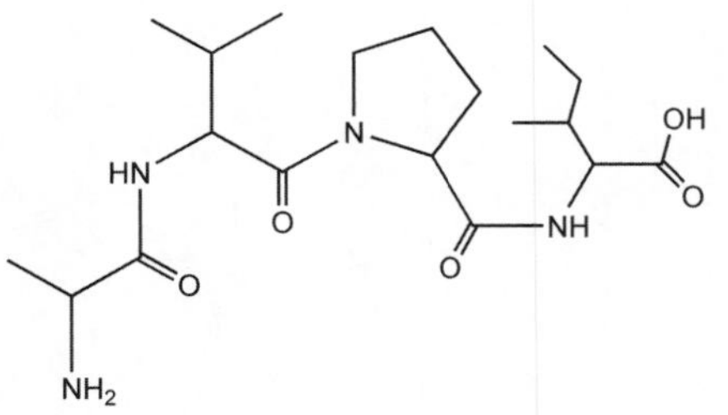

94

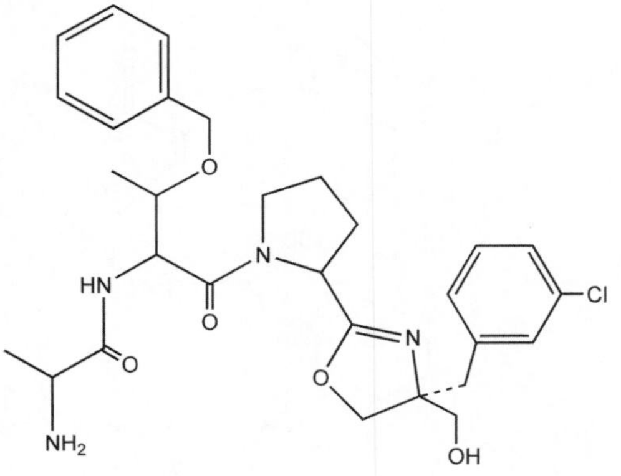

95

Figure 27. Structure of the peptide Ala-Val-Pro-Ile (**94**) and peptidomimetic **95**.

and molecular modeling simulations were used to properly analyze the structural features of XIAP domain–Smac mimetic complexes [198,199].

Other approaches have also been used to design peptidomimetics of the tetrapeptide. An early discovery was the discovery of the inhibition profile of embelin (**103**, Fig. 30), an active compound found by screening *in silico* of a traditional herbal medicine database. The compound was found to inhibit XIAP with an affinity in the low micromolar concentration [200]. The work was pursued beyond with the synthesis of different analogs with improved affinities [201]. Very recently, a new set of lead compounds obtained by fragment screening combined with NMR spectroscopy was disclosed. Of these compounds, molecule **104** (Fig. 30) exhibits cellular activity in *ex vivo* experiments [202].

In the early studies carried out by Harran et al. [189], the authors found that dimers of the analogs were more effective than the monomers. The reason for this was based on the fact that XIAP contains three BIR

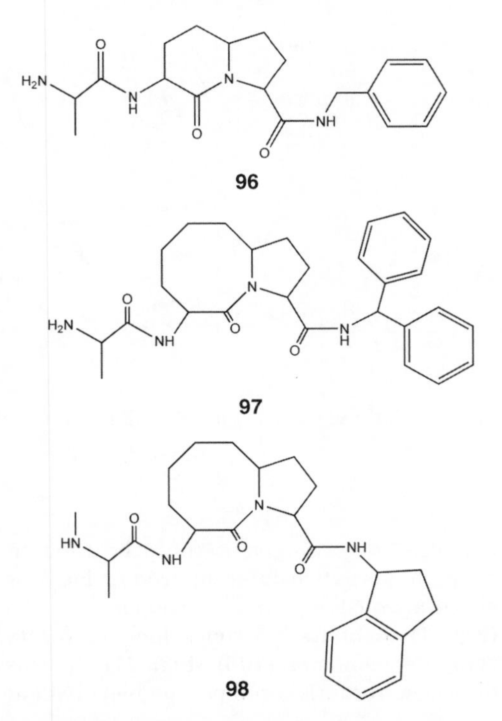

Figure 28. Structure of different bicyclic compounds peptidomimetics of **94**.

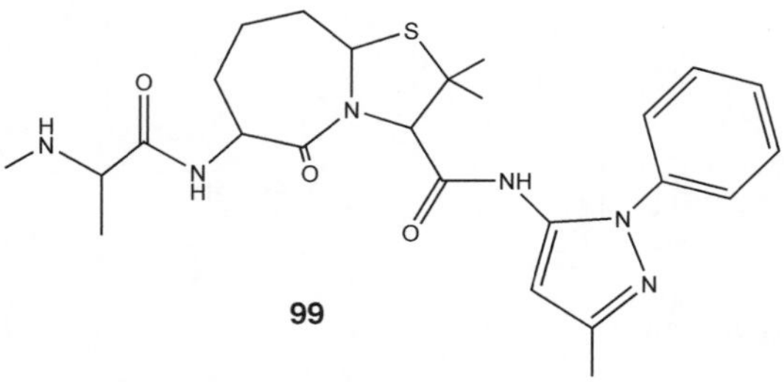

Figure 29. Structure of different bicyclic compounds peptidomimetics of **94** (cont).

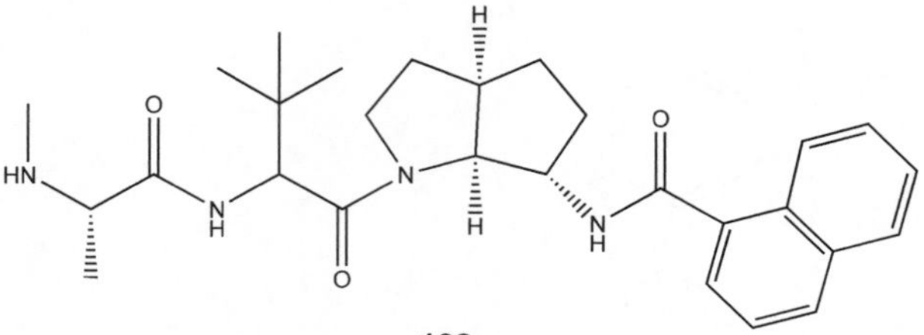

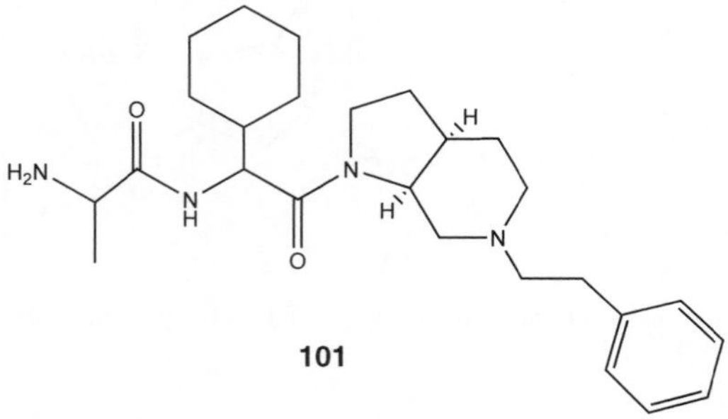

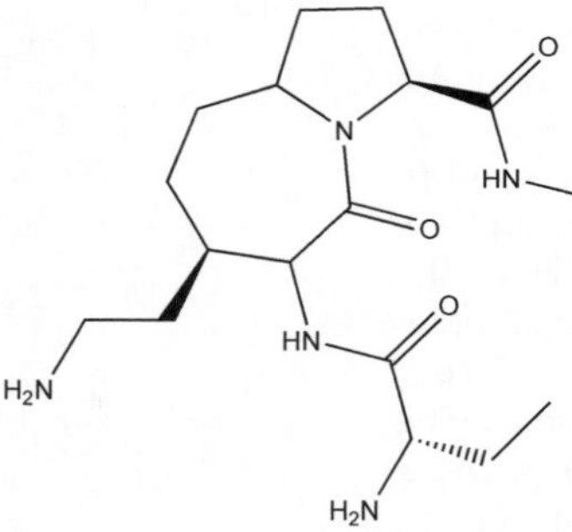

domains. Accordingly, it could be thought that it binds to two different protein domains simultaneously. Thus, compound **105** (Fig. 31) exhibits 1.5 times higher affinity than its monomer (**106**) (Fig. 31). In this direction, recently a cell permeable, bivalent Smac mimetic (**107**, Fig. 32) [203] and a cell permeable, bivalent and cyclic mimetic (**108**, Fig. 32) [204] that concurrently target both BIR2 and BIR3 domains of the XIAP protein with activity at the nanomolar range have been reported.

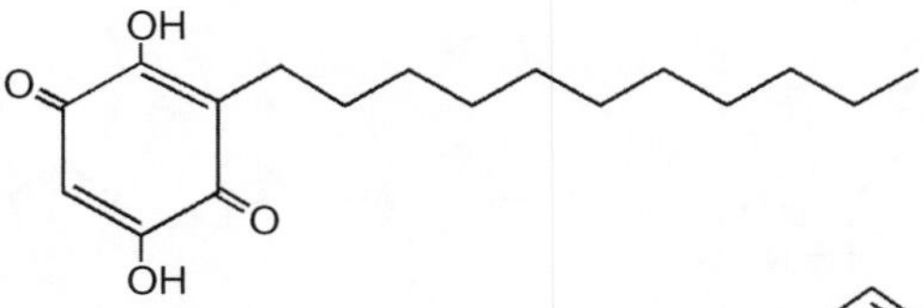

103

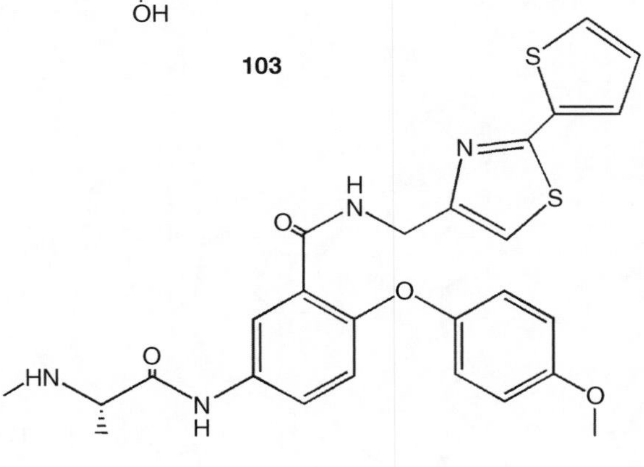

104

Figure 30. Structure of embelin (**103**) and compound **104**.

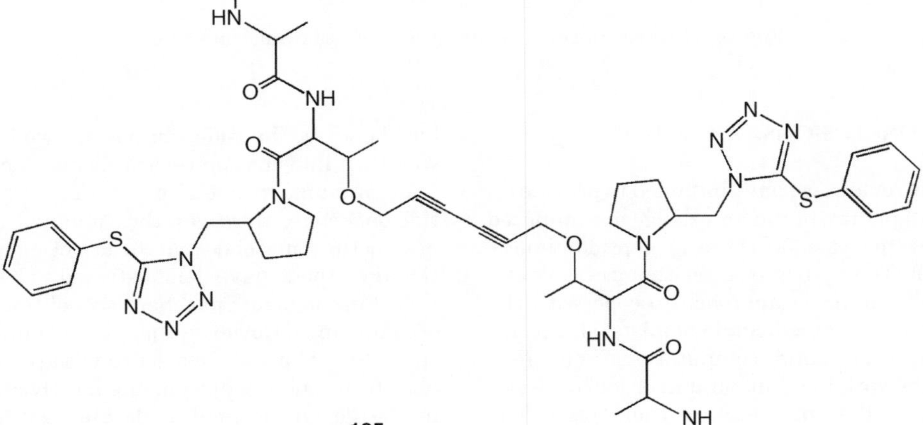

105

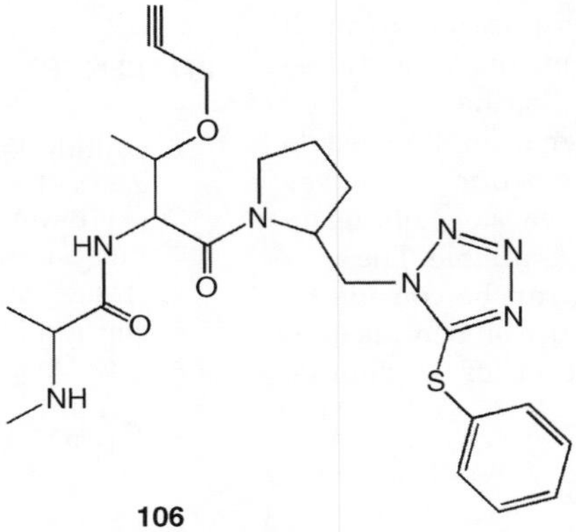

106

Figure 31. Structure of the inhibitor **105** and its monomer (**106**).

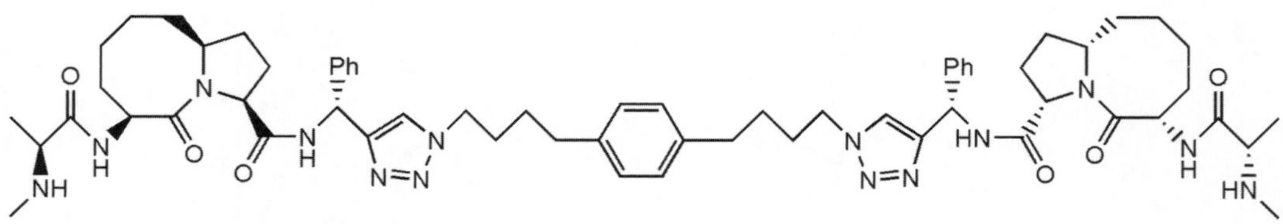

107

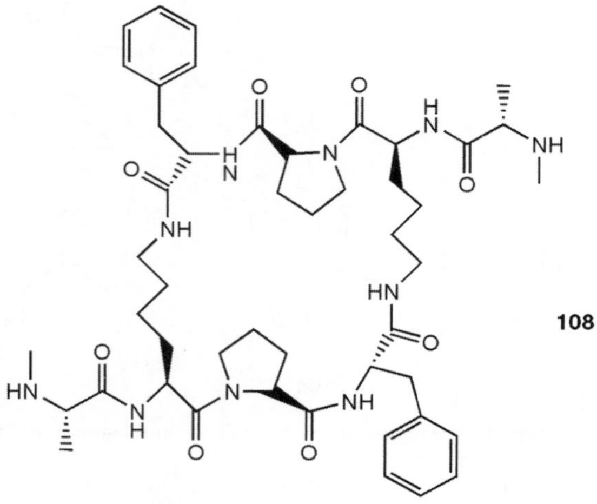

108

Figure 32. Structure of bivalent linear and cyclic Smac mimetics.

13. CONCLUSIONS

In this review, we have intended to provide an overall picture of the knowledge accumulated during the past 30 years in peptidomimetic design. The review stresses the necessity of a multidisciplinary approach to cope with the design, where medicinal chemists, structural biophysicists, and computational chemists need to work hand in hand in order to assemble the different pieces of the puzzle. The original idea of mimicking the stereospecific arrangement of the key moieties of a peptide at its recognition site has provided a much deeper insight into the features of peptide–-protein interactions. We also have outlined that the strategies to design a small molecule mimicking the features of a peptide as well as its pharmacological profile vary depending on the structural information available. The area of peptidomimetic design can be considered mature, with a vast literature on peptide analogs designed as modifications of the peptide backbone or the side chains, as well as a wealth of information regarding organic scaffolds to mimic geometrical features of a peptide backbone. Moreover, the combined action of structural methods and the use of computational methods have demonstrated to be very useful in the medicinal chemistry of this kind of compounds. However, there is still an ample space to explore. We expect to witness in the near future new achievements mostly related to the design of small-molecule ligands for GPCRs or protein–protein inhibitors.

REFERENCES

1. Griffiths EC. In: Krogsgaard-Larsend P, Bundgaard H, editors. A Textbook of Drug Design and Development. Chur: Hardwood Academic Publishers; 1991.
2. Hruby VJ, Ahn J-M, Liao S. Synthesis of oligopeptide and peptidomimetic libraries. Curr Opin Chem Biol 1997;1:114–119.

3. Niu C-H, Chiu Y-Y. FDA perspective on peptide formulation and stability issues. J Pharm Sci 1998;87:1331–1334.
4. Edwards CMB, Cohen MA, Bloom SR. Peptides as drugs. Q J Med 1999;92:1–4.
5. Schechter Y, Mironchick M, Saul A, Gerschonov E, Precido-Pratt L, Sasson K, Tsubery H, Mester B, Kapitkovsky A, Rubinraut S, Vachutinski Y, Fridkin G, Fidkin M. New technologies to prolong lifetime of peptide and protein drugs *in vivo*. Int J Pept Res Ther 2006;13:105–117.
6. Davies K. Cracking the Druggable Genome, Bio-IT World, http://www.bioitworld.com/archive/100902/firstbase.html, 2002.
7. Brownstein MJ. A brief history of opiates, opioid peptides, and opioid receptors. Proc Natl Acad Sci USA 1993;90:5391–5393.
8. Gu X, Ying J, Agnes RS, Naveatilova E, Davis P, Stahl G, Porreca F, Yamamura HI, Hruby VJ. Novel design of bicyclic β-turn dipeptides on solid-phase supports and synthesis of [3.3.0]-bicyclo[2,3]-leu-enkephalin. Org Lett 2004;6:3285–3288.
9. Gether U. Uncovering molecular mechanisms involved in activation of G protein-coupled receptors. Endocr Rev 2000;21:90–113.
10. Conklin BR, Hsiao EC, Claeysen S, Dumuis A, Srinivasan S, Forsayeth JR, Guettier J-M, Chang WC, Pei Y, McCarthy KD, Nissenson RA, Wess J, Bockaert J, Roth BL. Engineering GPCR signaling pathways with RASSLs. Nat Methods 2008;5:673–678.
11. Giannis A, Kolter T. Peptidomimetics for receptor ligands: discovery, development and medical perspectives. Angew Chem Int Ed Engl 1993;32:1244–1267.
12. Wiley RA, Rich DH. Peptidomimetics derived from natural products. Med Res Rev 1993;13:327–384.
13. Gante J. Peptidomimetics: tailored enzyme inhibitors. Angew Chem Int Ed Engl 1994;33:1699–1720.
14. Gentilucci L, Tolomelli A, Squassabia F. Peptides and peptidomimetics in medicine, surgery and biotechnology. Curr Med Chem 2006;13:2449–2466.
15. Wu Y-D, Gellman S. Peptidomimetics. Acc Chem Res 2008;41:1231–1232.
16. Wennemers H, Raines RT. Peptides and peptidomimetics as prototypes. Curr Opin Chem Biol 2008;12:690–691.
17. Honda S, Yamasaki K, Sawada Y, Morii H. 10-Residue folded peptide designed by segment statistics. Structure 2004;12:1507–1518.
18. Perez JJ, Villar HO, Arteca GA. Distribution of conformational energy minima in molecules with multiple torsional degrees of freedom. J Phys Chem 1994;98:2318–2324.
19. Corcho FJ, Filizola F, Perez JJ. Evaluation of the iterative simulated annealing technique in conformational search of peptides. Chem Phys Lett 2000;319:65–70.
20. Abdali S, Jensen MO, Bohr H. Energy levels and quantum states of [Leu]enkephalin conformations based on theoretical and experimental investigations. J Phys Condens Matter 2004;15:S1853–S1860.
21. Chatterjee J, Gilon C, Hoffman A, Kessler H. N-methylation of peptides: a new perspective in medicinal chemistry. Acc Chem Res 2008;41:1331–1342.
22. Cowell SM, Lee YS, Cain JP, Hruby VJ. Exploring Ramachandran and Chi space: conformationally constrained amino acids and peptides in the design of bioactive polypeptide ligands. Curr Med Chem 2004;11:2785–2789.
23. Moore RE. Cyclic peptides and depsipeptides from cyanobacteria: a review. J Ind Microbiol 1996;16:134–143.
24. Zega A. Azapeptides as pharmacological agents. Curr Med Chem 2005;12:589–597.
25. Doulut S, Rodriguez M, Lugrin D, Vecchini F, Kitabgi P, Aumelas A, Martinez J. Reduced peptide bond pseudopeptide analogues of neurotensin. Pept Res 1992;5:30–38.
26. Fletcher MD, Campbell MM. Partially modified retro-inverso peptides: development, synthesis and conformational behaviour. Chem Rev 1998;98:763–795.
27. Simon RJ, Robert SK, Zuckermann RN, Huebner VD, Jewell DA, Banville S, Ng S, Wang L, Rosenberg S, Marlowe CK, Spellmeyer DC, Tan R, Frankel AD, Santi DV, Cohen FE, Barlett PA. Peptoids: a modular approach to drug discovery. Proc Natl Acad Sci USA 1992;89:9367–9371.
28. Kessler H. Peptoids: a new approach to the development of pharmaceuticals. Angew Chem Int Ed Engl 1993;32:543–544.
29. Lelais G, Seebach D. β-Amino acids: syntheses, occurrence in natural products, and components of β-peptides. Biopolymers 2004; 76:206–243.
30. Trabocchi A, Guarna F, Guarna A. γ- and δ-amino acids: synthetic strategies and relevant applications. Curr Org Chem 2005;9: 1127–1153.

31. Balaram P. Non-standard amino acids in peptide design and protein engineering. Curr Opin Struct Biol 1992;2:845–851.
32. Latajka R, Jewginski M, Makowski M, Paweiczak M, Huber T, Sewald N, Kafarski P. Pentapeptides containing two dehydrophenylalanine residues: synthesis, structural studies and evaluation of their activity towards cathepsin C. J Pept Sci 2008;14:1084–1095.
33. Giannis A, Rübsam F. Peptidomimetics in drug design. Adv Drug Design 1997;29:1–80.
34. Freidinger RM. Design and synthesis of novel bioactive peptides and peptidomimetics. J Med Chem 2003;46:5553–566.
35. Falorni M, Giacomelli G, Porcheddu A, Dettori G. New oxazole-based conformationally restricted peptidomimetics: design and synthesis of pseudopeptides. Eur J Org Chem 2000; 3217–2322.
36. Von Roedern EG, Kessler H. A sugar amino acid as novel peptidomimetic. Angew Chem Int Ed Engl 1994;33:687–689.
37. Zuckermann RN. The chemicals synthesis of peptidomimetics libraries. Curr Opin Struct Biol 1993;3:580–584.
38. Kessler H. Peptide conformations. 19. Conformation and biological activity of cyclic peptides. Angew Chem Int Ed Engl 1982;21: 512–523.
39. Fung S, Hruby VJ. Design of cyclic and other templates for potent and selective peptide α-MSH analogues. Curr Opin Chem Biol 2005;9: 352–358.
40. Ovchinnikov YA, Ivanov VT. Conformational states and biological activity of cyclic peptides. Tetrahedron 1975;31:2177–2209.
41. Rizo J, Gierasch LM. Constrained peptides: models of bioactive peptides and protein substructures. Annu Rev Biochem 1992;61: 387–418.
42. Nikiforovich V. Computational molecular modelling in peptide drug design. Int J Pept Protein Res 1994;44:513–531.
43. Damewood JR, Jr. Peptide mimetic design with aid of computational chemistry. In: Lipkowitz KB, Boyd DB, editors. *Reviews of Computational Chemistry*, Vol. 9 New York, NY 1996. pp 1–66.
44. Perez JJ, Corcho F, Llorens O. Molecular modeling in the design of peptidomimetics and peptide surrogates. Curr Med Chem 2002;9: 209–2229.
45. Kahn M. Peptide secondary structure mimetics: recent advances and future challenges. Synlett 1993; 821.
46. Gillespie P, Cicariello J, Olson GL. Conformational analysis of dipeptide mimetics. Pept Sci 1997;43:191–217.
47. Trabocchi A, Scarpi D, Guarna A. Structural diversity of bicyclic amino acids. Amino Acids 2008;34:1–24.
48. Vagner J, Qu H, Hruby V. Peptidomimetics, asynthetic tool of drug discovery. Curr Opin Chem Biol 2008;12:292–296.
49. Robinson JA. The design, synthesis and conformation of some new β-hairpin mimetics: novel reagents for drug and vaccine discovery. Synlett 2000; 429–441.
50. Olson GL, Voss ME, Hill DE, Kahn M, Madison VS, Cook CM. Design and synthesis of a protein β-turn mimetic. J Am Chem Soc 1990;112:323–33.
51. Schneider JP, Kelly JW. Templates that induce α-helical, β-sheet, and loop conformations. Chem Rev 1995;95:2169–2187.
52. Lesma G, Sacchetti A, Silvani A. Synthesis and conformational analysis of benzimidazole-based reverse turns. Tetrahedron Lett 2008; 49:1293–1296.
53. Hinds MG, Richards NGJ, Robinson JA. Design and synthesis of a novel peptide β-turn mimetic. J Chem Soc Chem Commun 1988; 1447–1449.
54. James GL, Goldstein JL, Brown MS, Rawson TE, Somers TC, McDowell RS, Crowley CW, Lucas BK, Levinson AD, Marsters JC, Jr. Benzodiazepine peptidomimetics: potent inhibitors of ras farnesylation in animal cells. Science 1993;260:1937–1942.
55. Hanessian S, McNaughton-Smith G, Lombart H-G, Lubell WD. Design and synthesis of conformationally constrained amino acids as versatile scaffolds and peptide mimetics. Tetrahedron 1997;53:12789.
56. Cluzeau J, Lubell WD. Design, synthesis and application of azabicyclo [*X.Y.*0]alkanone amino acids as constrained surrogates and peptide mimics. Biopolymers 2005;80:98–150.
57. Callahan JF, Bean JW, Burges JL, Eggleston DS, Hwang SM, Kopple KD, Koster PF, Nichols A, Peishoff CE, Samanen JM, Vasko JA, Wong A, Huffmantand WF. Design and synthesis of a C7 mimetic for the predicted γ-turn conformation found in several constrained RGD antagonists. J Med Chem 1992;35: 3970–3972.
58. Schmidt B, Lindman S, Tong W, Lindeberg G, Gogoll A, Lai Z, Thörnwall M, Synnergren B, Nilsson A, Welch CJ, Sohtell M, Westerlund C, Nyberg F, Karlen A, Hallberg A. Design, synth-

esis and biological activities of four angiotensin II receptor ligands with γ-turn mimetics replacing amino acid residues 3–5. J Med Chem 1997;40:903–919.

59. Ramanathan SK, Keeler J, Lee H-L, Reddy DS, Lushington G, Aubé J. Modular synthesis of cyclic peptidomimetics inspired by γ-turns. Org Lett 2005;7:1059–1062.
60. Sato M, Lee JYH, Nakanishi H, Johnson ME, Chrusciel RA, Kahn M. Design, synthesis and conformational analysis of gamma-turn peptide mimetics of bradykinin. Biochem Biophys Res Commun 1992;187:999–1006.
61. Shramm VL. Enzymatic transition states and transition state analogs. Annu Rev Biochem 1998;67:693–720.
62. Byers LD, Wolfenden R. Binding of by-product analog benzylsuccinic acid by carboxypeptidase A. Biochemistry 1973;12:2070–2078.
63. Vacca JP. Design of tight-binding human-immunodeficiency-virus type-1 protease inhibitors. Methods Enzymol 1994;241:311–334.
64. Kurinov IV, Harrison RW. Prediction of new serine protease inhibitors. Struct Biol 1994;1:735–743.
65. Brik A, Wong C-H. HIV-1 protease: mechanism and drug discovery. Org Biomol Chem 2003;1:5–14.
66. Tsantrizos Y. Peptidomimetic therapeutic agents targeting the protease enzyme of the human immunodeficiency virus and hepatitis C virus. Acc Chem Res 2008;41:1252–1263.
67. Estiarte MA, Rich DH. Peptidomimetics for drug design. Burger's Medicinal Chemistry and Drug Discovery, 6th ed. New York: John Wiley & Sons; 2003.
68. Wlodawer A, Vondrasek J. Inhibitors of HIV-1 protease: a major success of structure-assisted drug design. Annu Rev Biophys Biomol Struct 1998;27:249–284.
69. Roberts NA, Martin JA, Kinchington D, Broadhurst AV, Craig JC, et al. Rational design of peptide-based HIV protease inhibitors. Science 1990;248:358–361.
70. Holloway MK, Wai JM, Halgren TA, Fitzgerald PM, Vacca JP, Dorsey BD, Levin RB, Thompson WJ, Chen LJ, deSolms SJ, Gaffin N, Ghosh AK, Giuliani EA, Graham SL, Guare JP, Hungate RW, Lyle TA, Sanders WM, Tucker TJ, Wiggins M, Wiscount CM, Woltersdoft OW, Young SD, Drake PL, Zugay JA. *A priori* prediction of activity for HIV-1 protease inhibitors employing energy minimization in the active site. J Med Chem 1995;38:305–317.
71. Tossi A, Bonin I, Antcheva N, Norbedo S, Benedetti F, Miertus S, Nair AC, Maliar T, Bello FD, Palù G, Romeo D. Aspartic protease inhibitors: an integrated approach for the design and synthesis of diaminodiol-based peptidomimetics. Eur J Biochem 2000;267: 1715–1722.
72. Reddy MR, Varney MD, Kalish V, Viswanadhan VN, Appelt K. Calculation of relative differences in the binding free energies of HIV-1 protease inhibitors: a thermodynamic cycle perturbation approach. J Med Chem 1994; 37:1145–1152.
73. Chenera B, DesJarlais RL, Finkelstein JA, Eggleston DS, Meek TD, Tomaszek TA, Dreyer GB. Nonpeptide HIV protease inhibitors designed to replace a bound water. Bioorg Med Chem Lett 1993;3:2717–2722.
74. Lam PY, Jadhav PK, Eyermann CJ, Hodge CN, Ru Y, Bacheler LT, Meek JL, Otto MJ, Rayner MM, Wong YN, Chang CH, Weber PC, Jackson DA, Sharpe TR, Erickson-Vitanen S. Rational design of potent, bioavailable, nonpeptide cyclic ureas as HIV protease inhibitors. Science 1994;263:380–384.
75. Caponigro F, Casale M, Bryce J. Farnesyl transferase inhibitors in clinical development. Expert Opin Investig Drugs 2003;12:943–954.
76. Sousa SF, Fernandes PA, Ramos MJ. Farnesyltransferase inhibitors: a detailed chemical view on an elusive biological problem. Curr Med Chem 2008;15:1478–1492.
77. Gibbs JB, Oliff A. The potential of farnesyltransferase inhibitors as cancer chemotherapeutics. Annu Rev Pharmacol Toxicol 1997;37:143–166.
78. Downward J. Targeting the RAS signalling pathway in cancer therapy. Nat Rev Cancer 2003;3:11–22.
79. Eastman RT, Buckner FS, Yokoyama K, Gelb MH, Van Voorhis WC. Fighting parasitic disease by blocking protein farnesylation. J Lipid Res 2006;47:233–240.
80. Lowy DR, Willumsen BM. Function and regulation of ras. Annu Rev Biochem 1993;62: 851–891.
81. Barbacid M. Ras genes. Annu Rev Biochem 1987;56:779–827.
82. Kohl NE, Mosser SD, deSolms SJ, Giuliani EA, Pompliano DL, Graham SL, Smith RL, Scolnick EM, Oliff A, Gibbs JB. Selective-inhibition of ras-dependent transformation by a farnesyltransferase inhibitor. Science 1993;260: 1934–1937.

83. James GL, Goldstein JL, Brown MS, Rawson TE, Somers TC, McDowell RS, Crowley CW, Lucas BK, Levinson AD, Marsters JC, Jr. Benzodiazepine peptidomimetics: potent inhibitors of ras farnesylation in animal cells. Science 1993;260:1937–1942.

84. Nigam M, Seong CM, Qian Y, Hamilton AD, Sebti SM. Potent inhibition of human tumor p21(ras) farnesyltransferase by A1A2-lacking p21(ras) CA1A2X peptidomimetics. J Biol Chem 1993;268:20695–20698.

85. Qian YM, Blaskovich MA, Saleem M, Seong CM, Wathen SP, Hamilton AD, Sebti SM. Design and structural requirements of potent peptidomimetic inhibitors of p21(ras) farnesyltransferase. J Biol Chem 1994;269:12410–12413.

86. Angibaud P, Bourdrez X, End DW, Freyne E, Janicot M, Lezouret P, Ligny Y, Mannens G, Damsch S, Mevellec L, Meyer C, Muller P, Pilatte I, Poncelet V, Roux B, Smets G, Van Dun P, Van Remoortere Venet M, Wouters W. Substituted azoloquinolines and -quinazolines as new potent farnesyl protein transferase inhibitors. Bioorg Med Chem Lett 2003;13: 4365–4369.

87. Bishop WR, Bond R, Petrin J, Wang L, Patton R, Doll R, Njoroge G, Catino J, Schwartz J, Windsor W, Syto R, Schwartz J, Carr D, James L, Krichmeier P. Novel tricyclic inhibitors of farnesyl protein transferase: biochemical characterization and inhibition of Ras modification in transfected COS cells. J Biol Chem 1995;270:30611–30618.

88. Njoroge FG, Taveras AG, Kelly J, Remiszewski S, Mallams AK, Wolin R, Afonso A, Cooper AB, Rane DF, Liu YT, Wong J, Vibulbhan B, Pinto P, Deskus J, Alvarez CS, del Rosario J, Connolly M, Wang J, Desai J, Rossman RR, Bishop WR, Patton R, Wang L, Kirschmeier P, Ganguly AK, Bryant MS, Nomeir AA, Lin C-C, Liu M, McPhail AT, Doll RJ, Girijavallabhan VM. (+)-4-[2-[4-(8-Chloro-3,10-dibromo-6,11-dihydro-5*H*-benzo[5,6]cyclohepta[1,2-*b*]-pyridin-11(*R*)-yl)-1-piperidinyl]-2-oxo-ethyl]-1-piperidinecarboxamide (SCH-66336): a very potent farnesyl protein transferase inhibitor as a novel antitumor agent. J Med Chem 1998;41:4890–4902.

89. Liu M, Bryant MS, Chen J, Lee S, Yaremko B, Lipari P, Malkowski M, Ferrari E, Nielsen L, Prioli N, Dell J, Sinha D, Syed J, Korfmacher WA, Nomeir AA, Lin CC, Wang L, Taveras AG, Doll RJ, Njoroge FG, Mallams AK, Remiszewski S, Catino JJ, Girijavallabhan VM, Bishop WR. Antitumor activity of SCH 66336, an orally bioavailable tricyclic inhibitor of farnesyl protein transferase, in human tumour xenograft models and wap-ras transgenic mice. Cancer Res 1998;58:4947–4956.

90. Hunt JT, Lee VG, Leftheris K, Seizinger B, Carboni J, Mabus J, Ricca C, Yan N, Manne V. Potent, cell active, non-thiol tetrapeptide inhibitors of farnesyltransferase. J Med Chem 1996;39:353–358.

91. Audet M, Bouvier M. Insights into signalling from β2-adrenergic receptor structure. Nat Chem Biol 2008;4:397–401.

92. Langerström MC, Schiöth HB. Structural diversity of G-protein coupled receptors and significance for drug discovery. Nat Rev Drug Discov 2008;7:339–357.

93. Regoli D, Barabé J. Pharmacology of bradykinin and related kinins. Pharmacol Rev 1980; 32:1–46.

94. Marceau F, Hess JF, Bachvarov DR. The B-1 receptors for kinins. Pharmacol Rev 1998; 50:357–386.

95. Abraham WM, Scuri M, Farmer SG. Peptide and non-peptide bradykinin receptor antagonists: role in allergic airway disease. Eur J Pharmacol 2006;533:215–221.

96. Stewart JM. Bradykinin antagonists: discovery and development. Peptides 2004;25:527–532.

97. Hock FJ, Wirth K, Albus U, Linz W, Gerhards HJ, Wiemer G, Henke S, Breipohl G, Konig W, Knolle J, Scholkens BA. HOE-140, a new potent and long-acting bradykinin antagonist: *in vitro* studies. Br J Pharmacol 1991; 102: 769–773.

98. Bork K, Yasothan U, Kirkpatrick P. Icatibant. Nat Rev Drug Discov 2008;7:801–802.

99. Salvino JM, Seoane PR, Douty BD, Awad MMA, Dolle RE, Houck WT. Design of potent non-peptide competitive antagonist of the human bradykinin B_2 receptor. J Med Chem 1993;36:2583–2584.

100. Asano M, Inamura N, Hatori C, Sawai H, Fujiwara T, Katayama A, Kayakiri H, Satoh S, Abe Y, Inoue T, Sawada Y, Nakahara K, Oku T, Okuhara M. The identification of an orally active, nonpeptide bradykinin B_2 receptor antagonist, FR173657. Br J Pharmacol 1997;120: 617–624.

101. Pruneau D, Luccarini JM, Fouchet C, Defrene E, Franck RM, Loillier B, Duclos H, Robert C, Cremers B, Belichard P, Paquet JL. *In vitro* and *in vivo* effects of the new nonpeptide bradykinin B-2 receptor antagonist, LF 16-0335C, on guinea-pig and rat kinin receptors. Fundam Clin Pharmacol 1999;13:75–83.

102. Valenti C, Cialdai C, Giuliani S, Lecci A, Tramontana M, Meini S, Quartara L, Maggi CA. MEN16132, a novel potent and selective nonpeptide kinin B_2 receptor antagonist: *in vivo* activity on bradykinin-induced bronchoconstriction and nasal mucosa microvascular leakage in anesthetized guinea pigs. J Pharmacol Exp Ther 2005;316:616–623.
103. Iqbal J, Sanghia R, Das SK. Endothelin receptor antagonists: an overview of their synthesis and structure–activity relationship. Mini Rev Med Chem 2005;5:381–408.
104. Yanagisawa M, Kurihara H, Kimura S, Tomobe Y, Kobayashi M, Mitsui Y, et al. A novel potent vasoconstrictor peptide produced by vascular endothelial cells. Nature 1988;332:411–415.
105. Inoue A, Yanagisawa M, Kimura S, Kasuya Y, Miyauchi T, Goto K, et al. The human endothelin family: three structurally and pharmacologically distinct isopeptides predicted by three separate genes. Proc Natl Acad Sci USA 1989;86:2863–2867.
106. Douglas SA, Ohlstein EH. Signal transduction mechanisms mediating the vascular actions of endothelin. J Vasc Res 1997; 34:152–164.
107. Motte S, McEntee K, Naeije R. Endothelin receptor antagonists. Pharmacol Ther 2006; 110:386–414.
108. Wu-Wong JR, Opgenorth TJ. Endothelin and isoproterenol counter-regulate cAMP and mitogen-activated protein kinases. J Cardiovasc Pharmacol 1998;31(Suppl 1): S185–S191.
109. Wu-Wong JR, Chiou WJ, Wang J. Extracellular signal-regulated kinases are involved in the antiapoptotic effect of endothelin-1. J Pharmacol Exp Ther 2000;293:514–521.
110. Ihara M, Fukuroda T, Saeki T, Nishikibe M, Kojiri K, Suda H, et al. An endothelin receptor (ETA) antagonist isolated from *Streptomyces misakiensis*. Biochem Biophys Res Commun 1991;178:132–137.
111. Spinella MJ, Malik AB, Everitt J, Andersen TT. Design and synthesis of a specific endothelin 1 antagonist: effects on pulmonary vasoconstriction. Proc Natl Acad Sci USA 1991;88:7443–7446.
112. Ihara M, Noguchi K, Saeki T, Fukuroda T, Tsuchida S, Kimura S, et al. Biological profiles of highly potent novel endothelin antagonists selective for the ETA receptor. Life Sci 1992;50:247–255.
113. Okada M, Nishikibe M. BQ-788, a selective endothelin ET(B) receptor antagonist. Cardiovasc Drug Rev 2002;20:53–66.
114. Maggi CA, Giuliani S, Patacchini R, Santicioli P, Rovero P, Giachetti A, et al. The C-terminal hexapeptide, endothelin-(16–21), discriminates between different endothelin receptors. Eur J Pharmacol 1989;166:121–122.
115. Rovero P, Patacchini R, Maggi CA. Structure–activity studies on endothelin (16–21), the C-terminal hexapeptide of the endothelins, in the guinea-pig bronchus. Br J Pharmacol 1990;101:232–234.
116. Saudek V, Hoflack J, Pelton JT. ^{1}H-NMR study of endothelin, sequence-specific assignment of the spectrum and a solution structure. FEBS Lett 1989;257:145–148.
117. Endo S, Inooka H, Ishibashi Y, Kitada C, Mizuta E, Fujino M. Solution conformation of endothelin determined by nuclear magnetic resonance and distance geometry. FEBS Lett 1989;257:149–154.
118. Munro S, Craik D, McConville C, Hall J, Searle M, Bicknell W, et al. Solution conformation of endothelin, a potent vaso-constricting bicyclic peptide. A combined use of ^{1}H NMR spectroscopy and distance geometry calculations. FEBS Lett 1991;278:9–13.
119. Cody WL, He JX, Reily MD, Haleen SJ, Walker DM, Reyner EL, et al. Design of a potent combined pseudopeptide endothelin-A/endothelin-B receptor antagonist, Ac-DBhg16-Leu-Asp-Ile-[NMe]Ile-Trp21 (PD 156252): examination of its pharmacokinetic and spectral properties. J Med Chem 1997;40: 2228–2240.
120. Topliss JG. Utilization of operational schemes for analog synthesis in drug design. J Med Chem 1972;15:1006–1011.
121. Topliss JG. A manual method for applying the Hansch approach to drug design. J Med Chem 1977;20:463–469.
122. Doherty AM, Patt WC, Edmunds JJ, Berryman KA, Reisdorph BR, Plummer MS, et al. Discovery of a novel series of orally active nonpeptide endothelin-A (ETA) receptor-selective antagonists. J Med Chem 1995;38:1259–1263.
123. Reynolds EE, Keiser JA, Haleen SJ, Walker DM, Olszewski B, Schroeder RL, et al. Pharmacological characterization of PD 156707, an orally active ETA receptor antagonist. J Pharmacol Exp Ther 1995;273:1410–1417.
124. Elliott J, Ohlstein E, Peishoff C, Ellens H, Lago M. Endothelin receptor antagonists. In: *Integration of Pharmaceutical Discovery and Development*. 2002. p. 113–129 (cited 2009, May 13). Available from http://dx.doi.org/10.1007/0-306-47384-4_6.

125. Ohlstein EH, Nambi P, Douglas SA, Edwards RM, Gellai M, Lago A, et al. SB 209670, a rationally designed potent nonpeptide endothelin receptor antagonist. Proc Natl Acad Sci USA 1994;91:8052–8056.

126. Kelland NF, Webb DJ. Clinical trials of endothelin antagonists in heart failure: a question of dose? Exp Biol Med 2006;231:696–699.

127. Anzali S, Mederski WW, Osswald M, Dorsch D. Endothelin antagonists: search for surrogates of methylendioxyphenyl by means of a Kohonen neural network. Bioorg Med Chem Lett 1998;8:11–16.

128. Mederski WW, Osswald M, Dorsch D, Anzali S, Christadler M, Schmitges CJ, et al. Endothelin antagonists: evaluation of 2,1,3-benzothiadiazole as a methylendioxyphenyl bioisoster. Bioorg Med Chem Lett 1998;8:17–22.

129. Clozel M, Breu V, Burri K, Cassal JM, Fischli W, Gray GA, et al. Pathophysiological role of endothelin revealed by the first orally active endothelin receptor antagonist. Nature 1993;365:759–761.

130. Clozel M, Breu V, Gray GA, Kalina B, Löffler BM, Burri K, et al. Pharmacological characterization of bosentan, a new potent orally active nonpeptide endothelin receptor antagonist. J Pharmacol Exp Ther 1994;270:228–235.

131. Breu V, Clozel M, Burri K, Hirth G, Neidhart W, Ramuz H. *In vitro* characterisation of Ro 46-8443, the first non-peptide antagonist selective for the endothelin ETB receptor. FEBS Lett 1996;383:37–41.

132. Wu C, Chan MF, Stavros F, Raju B, Okun I, Mong S, et al. Discovery of TBC11251, a potent, long acting, orally active endothelin receptor-A selective antagonist. J Med Chem 1997;40:1690–1697.

133. Stein PD, Hunt JT, Floyd DM, Moreland S, Dickinson KE, Mitchell C, et al. The discovery of sulfonamide endothelin antagonists and the development of the orally active ETA antagonist 5-(dimethylamino)-*N*-(3,4-dimethyl-5-isoxazolyl)-1-naphthalenesulfonamide. J Med Chem 1994;37:329–331.

134. Wu-Wong JR, Dixon DB, Chiou WJ, Sorensen BK, Liu G, Jae H, et al. Pharmacology of endothelin receptor antagonists ABT-627, ABT-546, A-182086 and A-192621: *in vitro* studies. Clin Sci 2002;103(Suppl): 48107S–48111S.

135. Jimeno A, Carducci M. Atrasentan: a novel and rationally designed therapeutic alternative in the management of cancer. Expert Rev Anticancer Ther 2005;5:419–427.

136. Funk OF, Kettmann V, Drimal J, Langer T. Chemical function based pharmacophore generation of endothelin-A selective receptor antagonists. J Med Chem 2004;47: 2750–2760.

137. Wells JA, McClendon L. Reaching for high-hanging fruit in drug discovery at protein–protein interfaces. Nature 2007;450:1001–1009.

138. Robinson JA, DeMarco S, Gombert F, Moehle K, Obrecht D. The design, structures and therapeutic potential of protein epitope mimetics. Drug Discov Today 2008;13:944–951.

139. Walensky LD. Bcl-2 in the crosshairs: tipping the balance of life and death. Cell Death Differ 2006;13:1339–1350.

140. Zhang L, Ming L, Yu J. BH3 mimetics to improve cancer therapy: mechanisms and examples. Drug Resist Update 2007;10: 207–217.

141. Marzo I, Naval J. Bcl-2 family members as molecular targets in cancer therapy. Biochem Pharmacol 2008;76:939–946.

142. Thompson CB. Apoptosis in the pathogenesis and treatment of disease. Science 1995;267: 1456–1462.

143. Thornberry NA, Lazebnik Y. Caspases: enemies within. Science 1998;281:1312–1316.

144. Chang HY, Yang X. Proteases for cell suicide: functions and regulation of caspases. Microbiol Mol Biol Rev 2000;64:821–846.

145. Grütter MG. Caspases: key players in programmed cell death. Curr Opin Struct Biol 2000;10:649–655.

146. Ashkenazi A, Dixit VM. Death receptors: signalling and modulation. Science 1998;282: 1305–1308.

147. Green DR, Reed JC. Mitochondria and apoptosis. Science 1998;281:1309–1312.

148. Chao DT, Korsmeyer SJ. Bcl-2 family: regulators of cell death. Annu Rev Immunol 1998; 16:395–419.

149. Adams JM, Cory S. The Bcl-2 protein family: arbiters of cell survival. Science 1998;281:1322–1326.

150. Reed JC. Bcl-2 family proteins. Oncogene 1998;17:3225–3236.

151. Kuwana T, Newmeyer DD. Bcl-2 family proteins and the role of mitochondria in apoptosis. Curr Opin Cell Biol 2003;15:691–699.

152. Youle RJ, Strasser A. The BCL-2 protein family: opposing activities that mediate cell death. Nat Rev Mol Cell Biol 2008; 9:47–59.

153. Oltvai ZN, Milliman CL, Korsmeyer SJ. Bcl-2 heterodimerizes *in vivo* with a conserved

homolog, Bax, that accelerates programmed cell death. Cell 1993;74:609–619.

154. Chinnaiyan AM, O'Rourke K, Lane BR, Dixit VM. Interaction of CED-4 with CED-3 and CED-9: a molecular framework for cell death. Science 1997;275:1122–1126.

155. Ottilie S, Wang Y, Banks S, Chang J, Vigna NJ, Weeks S, Armstrong RC, Fritz LC, Osterdorf T. Mutational analysis of the interacting cell death regulators CED-9 and CED-4. Cell Death Differ 1997;4:526–533.

156. Kelekar A, Thompson CB. Bcl-2-family proteins: the role of the BH3 domain in apoptosis. Trends Cell Biol 1998;8:324–330.

157. Chittenden T. BH3 domains: intracellular death-ligands critical for initiating apoptosis. Cancer Cell 2002;2:165–166.

158. Cosulich SC, Worral V, Hedge PJ, Green S, Clarke PR. Regulation of apoptosis by BH3 domains in a cell-free system. Curr Biol 1997;7:913–920.

159. Holinger EP, Chittenden T, Lutz RJ. Bak BH3 peptides antagonize Bcl-xL function and induce apoptosis through cytochrome *c*-independent activation of caspases. J Biol Chem 1999;274:13298–13304.

160. Wang J-L, Zhang ZJ, Choksi S, Shan S, Lu Z, Croce CM, Alnemri ES, Korngold R, Huang Z. Cell permeable Bcl-2 binding peptides: a chemical approach to apoptosis induction in tumor cells. Cancer Res 2000;60: 1498–1502.

161. Gemperli AC, Rutledge SE, Maranda A, Shepartz A. Paralog-selective ligands for Bcl-2 proteins. J Am Chem Soc 2005;127:1596–1597.

162. Lugovskoy AA, Degterev AI, Fahmy AF, Zhou P, Gross JD, Yuan J, Wagner G. A novel approach for characterizing protein ligand complexes: molecular basis for specificity of small-molecule Bcl-2 inhibitor. J Am Chem Soc 2002;124:1234–1240.

163. Sattler M, Liang H, Nettesheim D, Meadows RP, Harlan JE, Eberstadt M, Yoon HS, Shuker SB, Chang BS, Minn AJ, Thompson CB, Fesik SW. Structure of Bcl-x_L-Bak peptide complex: recognition between regulators of apoptosis. Science 1997;275:983–986.

164. Petros AM, Nettesheim DG, Wang Y, Olejniczak ET, Meadows RP, Mack J, Swift K, Matayoshi ED, Zhang H, Thompson CB, Fesik SW. Rationale for Bcl-x_L/Bad peptide complex formation from structure, mutagenesis, and biophysical studies. Protein Sci 2000;9: 2528–2534.

165. Yang B, Liu D, Huang Z. Synthesis and helical structure of lactam bridged BH3 peptides derived from pro-apoptotic Bcl-2 family proteins. Bioorg Med Chem Lett 2004;14:1403–1406.

166. Walensky LD, Kung AL, Escher I, Malia TJ, Barbuto S, Wright R, Wagner G, Verdine GL, Korsmeyer SJ. Activation of apoptosis *in vivo* by a hydrocarbon stapled BH3 helix. Science 2004;305:1466–1470.

167. Kutzki O, Park HS, Ernst JT, Orner BP, Yin H, Hamilton AD. Development of a potent Bcl-xL antagonist based on alpha-helix mimicry. J Am Chem Soc 2002;124:11838–11839.

168. Yin H, Lee GI, Sedey KA, Rodriguez JM, Wang HG, Sebti SM, Hamilton AD. Terephthalamide derivatives as mimetics of helical peptides: disruption of the Bcl-xL Bak interaction. J Am Chem Soc 2005; 127: 5463–5468.

169. Sadowsky JD, Schmitt MA, Lee HS, Umezawa N, Wand S, Tomita Y, Gellman SH. Chimeric (alpha/beta + alpha)-peptide ligands for the BH3-recognition cleft of Bcl-xL: critical role of the molecular scaffold in protein surface recognition. J Am Chem Soc 2005;127:11966–11968.

170. Tzung SP, Kim KM, Basanez G, Giedt CD, Simon J, Zimmerberg J, Zhang KY, Hochenbery DM. Antimycin A mimics a cell-death-inducing Bcl-2 homology domain 3. Nat Cell Biol 2001;3:183–191.

171. Degterev A, Lugovskoy A, Cardone M, Mulley B, Wagner G, Mitchison T, Yuan J. Identification of small-molecule inhibitors of interaction between the BH3 domain and Bcl-xL. Nat Cell Biol 2001;3:173–182.

172. Chan S-L, Lee MC, Yang L-K, Lee ASY, Flotow H, Fu NY, Butler MS, Soejarto DD, Buss AD, Yu VC. Identification of chelerythrine as an inhibitor of Bcl-x_L functions. J Biol Chem 2003;278:20453–20456.

173. Kitada S, Leone M, Sareth S, Zhai D, Reed JC, Pellecchia M. Discovery, characterization, and structure–activity relationships studies of proapoptotic polyphenols targeting B-cell lymphocyte/leukemia-2 proteins. J Med Chem 2003;46:4259–4264.

174. Leone M, Zhai D, Sareth S, Kitada S, Reed JC, Pellecchia M. Cancer prevention by tea polyphenols is linked to their direct inhibition of antiapoptotic Bcl-2-family proteins. Cancer Res 2003;63:8118–8121.

175. Shoemaker AR, Oleksijew A, Bauch J, Belli BA, Borre T, Bruncko M, Deckwirth T, Frost DJ, Jarvis K, Joseph MK, Marsh K, McClellan

W, Nellans H, Ng S, Nimmer P, O'Connor JM, Oltersdorf T, Qing W, Shen W, Stavropoulos J, Tahir SK, Wang B, Warner R, Zhang H, Fesik SW, Rosenberg SH, Elmore SW. A small-molecule inhibitor of Bcl-XL potentiates the activity of cytotoxic drugs *in vitro* and *in vivo*. Cancer Res 2006;66:8731–8739.

176. Wang J-L, Liu D, Zhang Z-J, Shan S, Han X, Srinivasa SM. Croce CM,. Almeri ES, Huang Z. Structure-based discovery of an organic compound that binds Bcl-2 protein and induces apoptosis of tumour cells. Proc Natl Acad Sci USA 2000;97:7124–7129.
177. Enyedy I, Ling Y, Nacro K, Tomita Y, Wu X, Cao Y, Guo R, Li B, Zhu X, Huang Y, Long Y-Q, Roller P, Yang D, Wang SJ. Discovery of small-molecule inhibitors of Bcl-2 through structure-based computer screening. J Med Chem 2001;44:4313–4324.
178. Real PJ, Cao YY, Wang RX, Nikolovska-Coleska Z, Sanz-Ortiz J, Wang SM, Fernandez-Luna JL. Breast cancer cells can evade apoptosis-mediated selective killing by novel small molecule inhibitor of Bcl-2. Cancer Res 2004;64:7947–7953.
179. Zhang M, Ling Y, Yang CY, Liu H, Wang R, Wu X, Ding K, Zhu F, Griffith BN, Mohammad RM, Wang S, Yang D. A novel Bcl-2 small molecule inhibitor 4-(3-methoxy-phenylsulfannyl)-7-nitro-benzofurazan-3-oxide (MNB)-induced apoptosis in leukemia cells. Ann Hematol 2007;86(7): 471–81.
180. OltersdofT, Elmore SW, Shoemaker AR, Armstrong RC, David J, Augeri DJ, Belli BA, Bruncko M, Deckwerth TL, Dinges J, Hajduk PJ, Joseph MK, Kitada S, Korsmeyer SJ, Kunzer AR, Letai A, Li L, Mitten MJ, Nettesheim DG, Ng S, Nimmer PM, O'Connor JM, Oleksijew A, Petros AM, Reed JC, Shen W, Tahir SK, Thompson CB, Tomaselli KJ, Wang B, Wendt MD, Zhang H, Fesik SW, Rosenberg SH. An inhibitor of Bcl-2 family proteins induces regression of solid tumours. Nature 2005;435: 677–681.
181. Petros AM, Dinges J, Augeri DJ, Baumeister SA, Betebenner DA, Bures MG, Elmore SW, Hajduk PJ, Joseph MK, Landis SK, Nettesheim DG, Rosenberg SH, Shen W, Thomas S, Wang X, Zanze I, Zhang H, Fesik SW. Discovery of a potent inhibitor of the antiapoptotic protein Bcl-xL from NMR and parallel synthesis. J Med Chem 2006;49:656–663.
182. Wendt MD, Shen W, Kunzer A, McClellan WJ, Bruncko M, Oost TK, Ding H, Joseph MK, Zhang H, Nimmer PM, Ng SC, Shoemaker AR, Petros AM, Oleksijew A, Marsh K, Bauch J, Oltersdorf T, Belli BA, Martineau D, Fesik SW, Rosenberg SH, Elmore SW. Discovery and structure–activity relationship of antagonists of B-cell lymphoma 2 family proteins with chemopotentiation activity *in vitro* and *in vivo*. J Med Chem 2006;49:1165–1181.
183. Sun H, Nikolovska-Coleska Z, Yang C-Y, Qian D, Lu J, Qiu S, Bai L, Peng Y, Cai Q, Wang S. Design of small-molecule peptidic and nonpeptidic Smac mimetics. Acc Chem Res 2008;41:1264–1277.
184. Salvensen GS, Duckett CS. Apoptosis: IAP proteins: blocking the road to death's door. Nat Rev Mol Cell Biol 2002;3:401–410.
185. Liu Z, Sun C, Olejniczak ET, Meadows R, Betz S, Oost T, Hermann J, Wu J, Fesik S. Structural basis for binding of Smac/DIABLO to the XIAR BIR3 domain. Nature 2000;408:1004–1008.
186. Wu G, Chai J, Suber TL, Wu JW, Du C, Wang S, Shi Y. Structural basis of IAP recognition by Smac/DIABLO. Nature 2000;408:1008–1012.
187. Kipp RA, Case MA, Wist AD, Cresson CM, Carrell M, Griner E, Wiita A, Albiniak PA, Chai J, Shi Y, Semmelhack MF, McLendon GL. Molecular targeting of inhibitor of apoptosis proteins based on small molecule mimics of natural binding partners. Biochemistry 2002;41:7344–7349.
188. Oost TK, Armstrong RC, Al-Assad A, Betz SF, Deckweth TL, Ding H, Elmore SW, Meadows RP, Olejniczak ET, Oleksijew A, Oltersdorf T, Rosenberg SH, Shoemaker AR, Tomaselli KJ, Zou H, Fesik SW. Discovery of potent antagonists of the antiapoptotic protein XIAP for the treatment of cancer. J Med Chem 2004;47:4417–4426.
189. Park CM, Sun C, Olejniczak ET, Wilson AE, Meadows RP, Betz SF, Elmore SW, Fesik SW. Non-peptidic small molecule inhibitors of XIAP. Bioorg Med Chem Lett 2005;15:771–775.
190. Li L, Thomas RM, Suzuki H, De Brabaner JK, Wang X, Harran PG. A small molecule Smac mimic potentiates TRAIL- and TNFα-mediated cell death. Science 2004;305: 1471–1474.
191. Wist AD, Gu L, Riedl SJ, Shi Y, McLendon GL. Structure–activity based study of the Smac-binding pocket within the BIR3 domain of XIAP. Bioorg Med Chem 2007;15:2935–2943.
192. Sun H, Nikolovska-Coleska Z, Yang CY, Xu L, Tomita Y, Krajewski K, Roller PP, Wang S. Structure-based design, synthesis, and evaluation of conformationally constrained

mimetics of the second mitochondrial-derived activator of caspase that target the X-linked inhibitor of apoptosis protein/caspase 9 interaction site. J Med Chem 2004;47:4147–4150.

193. Sun H, Stuckey JA, Nikolovska-Coleska Z., Qin D, Meagher JL, Qiu S, Lu J, Yang C-Y, Saito NG, Wang, S. Structure-Based Design, Synthesis, Evaluation and Crystallographic Studies of Conformationally Constrained Smac Mimetics as Inhibitors of the X-linked Inhibitor of Apoptosis Protein (XIAP), J Med Chem 2008; 51:7169–7180.

194. Zobel K, Wang L, Varfolomeev E, Franklin CM, Elliot LO, Wallweber HJA, Okawa DC, Flygare JA, Vucic D, Fairbrother WJ, Deshayes K. Design, synthesis and biological activity of a potent Smac mimetic that sensitizes cancer cells to apoptosis by antagonizing IAPs. ACS Chem Biol 2006; 1:525–534.

195. Coen F, Alicke B, Elliot LO, Flygare JA, Goncharov T, Keteltas SF, Franklin MC, Frankovitz S, Stephan JP, Tsui V, Vucic D, Wong H, Fairbrother WJ. Orally bioavailable antagonist of inhibitor of apoptosis proteins based on an azabicyclooctane scaffold. J Med Chem 2009;52:1723–1730.

196. Gaither A, Porter D, Yao Y, Borawski J, Yang G, Donovan J, Sage D, Slisz J, Tran M, Straub C, Ramsey T, Iourgenko V, Huang A, Chen Y, Schlegel R, Labow M, Fawell S, Sellers WR, Zawel L. A Smac mimetic rescue screen reveals roles for inhibitor of apoptosis proteins in tumor necrosis factor-alpha signaling. Cancer Res 2007;67:11493–11498.

197. Sun H, Nikolovska-Coleska Z, Yang CY, Xu L, Liu M, Tomita Y, Pan H, Yoshioka Y, Krajewski K, Roller PP, Wang S. Structure-based design of potent, conformationally constrained Smac mimetics. J Am Chem Soc 2004;126: 16686–16687.

198. Mastrangelo E, Cossu F, Milani M, Sorrentino G, Lecis D, Delia D, Manzoni L, Drago C, Senici P, Scolastico C, Rizzo V, Bolognesi M. Targeting the X-linked inhibitor of apoptosis protein through 4-substituted azabicyclo[5.3.0] alkane Smac mimetics. Structure, activity and recognition principles. J Mol Biol 2008; 384:673–639.

199. Cossu F, Mastrangelo E, Milani M, Sorrentino G, Lecis D, Delia D, Manzoni L, Senici P, Scolastico C, Bolognesi M. Targeting the X-linked inhibitor of apoptosis protein through 4-substituted azabicyclo[5.3.0]alkane Smac mimetics. Structure, activity and recognition principles. Biochem Biophys Res Commun 2009;378:162–167.

200. Nikolovska-Coleska Z, Xu L, Hu Z, Tomita Y, Li P, Roller P, Wang R, Fang X, Guo R, Zhang M, Lippman M, Yang D, Wang S. Discovery of embelin as a cell-permeable, small-molecular weight inhibitor of XIAP trough structure-bases computational screening of a traditional herbal medicine three-dimensional structure database. J Med Chem 2004;47: 2430–2440.

201. Chen J, Nikolovska-Coleska Z, Wang G, Qiu S, Wang S. Design, synthesis, and characterization of new embelin derivatives as potent inhibitors of X-linked inhibitors of apoptosis protein. Bioorg Med Chem Lett 2006;16: 5805–5808.

202. Huang JW, Zhang Z, Wu B, Cellitti JF, Zhang X, Dahl R, Shiau CW, Wels K, Emdadi A, Stebbins JL, Reed JC, Pellecchia M. Fragment-based design of small molecule X-linked inhibitor of apoptosis protein inhibitors. J Med Chem 2008;51:7111–7118.

203. Sun H, Nikolovska-Coleska Z, Lu J, Meagher JL, Yang CY, Qiu S, Tomita Y, Ueda Y, Jiang S, Krajewski K, Roller PP, Stuckey JA, Wang S. Design, synthesis, and characterization of a potent, nonpeptide, cell-permeable, bivalent Smac mimetic that concurrently targets both the BIR2 and BIR3 domains in XIAP. J Am Chem Soc 2007;129:15279–15294.

204. Nikolovska-Coleska Z, Meagher JL, Jiang S, Yang CY, Qiu S, Roller PP, Stuckey JA, Wang S. Interaction of a cyclic, bivalent Smac mimetic with the X-linked inhibitor of apoptosis protein. Biochemistry 2008;47:9811–9824.

MEDICINAL CHEMISTRY APPROACHES FOR MULTITARGET DRUGS

Richard Morphy
Zoran Rankovic
Chemistry Department, Schering-Plough Corporation, Newhouse, Lanarkshire, UK

Over the last few decades, drug discovery has been driven primarily by a "one-target-one-disease" philosophy, increasingly dominated by *in vitro* high-throughput screening (HTS) technologies. Many successful drugs, that are selective for a single target, have emerged from this strategy, but despite the best efforts of drug discoverers, many diseases remain inadequately treated by such an approach. It has been suggested that intrinsic redundancy and robustness of biological networks are to blame for the failure of highly selective drugs to produce the desired therapeutic effect. Since agents that modulate multiple targets simultaneously have the potential to enhance efficacy or improve safety relative to drugs that address only a single target, it is not surprising that this area is attracting the attention of increasing numbers of drug discoverers [1–3].

There are three distinctly different approaches to multitarget drug discovery (MTDD): polypharmacy, fixed dose combinations (FDC), and multiple ligands. Traditional polypharmacy involves the treatment of unresponsive patients with cocktails of drugs that exploit different therapeutic mechanisms. Most frequently the cocktail is administered in the form of two (or more) individual tablets [4,5]. However, the benefits of this approach are often compromised by poor patient compliance, particularly for treating asymptomatic diseases, such as hypertension [6]. Recently, there has been a move toward fixed dose combination drugs, whereby the two (or more) agents are coformulated in a single tablet to make dosing regimes simpler and thereby improve patient compliance [7,8]. However, complications due to highly complex PK/PD relationships and the potential for drug–drug interactions could have a significant impact on the risks and costs of developing FDCs [9]. An alternative strategy with a different risk-benefit profile is to develop a single chemical entity able to modulate multiple biological targets simultaneously [1]. A lower risk of drug–drug interactions in comparison to cocktails or FDCs is a clear advantage of this strategy. Although the development of such multiple ligands can be challenging due to the increased complexity in the design and optimization of such ligands, these difficulties are associated with an early and therefore less expensive stage of the drug discovery process. The risks and costs of developing multiple ligands in the clinic are in principle no different to the development of any other single entity.

A number of clinically used drugs have been found to have activity at more than one target, which in some cases is associated with increased efficacy and in others with side effects. In most cases these are historical drugs for which the multiple activity profile was not designed but serendipitously discovered. The rational design of ligands that act selectively on specific multiple targets of therapeutic interest, termed designed multiple ligands (DMLs), is a more recent trend [1].

In a number of disease areas, drug discoverers have followed a three-stage evolutionary journey, from a nonselective drug with undesirable side effects, to a target-selective ligand with a safer profile, and onward toward a *selectively nonselective* DML that attempts to provide a more optimal balance of efficacy and safety. An example of a nonselective ligand is the atypical antipsychotic drug, clozapine, which shows antagonist activity at multiple aminergic G-protein coupled receptors (GPCRs). To circumvent the side effects of clozapine, a number of ligands that are selective for single receptors targeted by clozapine were developed, such as dopamine D_4 and serotonin 5-HT_{2a} antagonists, but these lacked sufficient efficacy in the clinic [10]. Research then shifted toward DMLs, such as the dual D_2/5-HT_{2a} antagonists [11,12]. Nonselective tricyclic antidepressants such as Amitryptyline were superseded by selective serotonin (5-HT) transporter inhibitors (SSRIs) that increased safety, but had a slow onset of action and lacked efficacy in some patients. Dual serotonin and norepinephrine (NA) reuptake inhibitors (SNRIs) were subsequently developed with the hope of addressing these deficiencies [13]. The same trend is observed in the area of nonsteroidal anti-inflammatory drugs (NSAIDs), starting from nonselective agents such as aspirin, to selective cyclooxygenase-2 (COX-2) inhibitors and then to dual COX-2/5-lipoxygenase (5-LOX) inhibitors [14]. Similarly, for the treatment of asthma, nonselective adrenergic agonists (e.g., epinephrine) have been replaced by selective β_2-adrenoceptor agonists such as salbutamol, with a significantly improved therapeutic window. Most recently, dual M3 antagonist/β_2 agonist and D_2/β_2 agonist have been developed [15,16].

1. HOW ARE NEW LEADS DISCOVERED?

Conceptually, there are two quite different methods of generating lead compounds, *screening* approaches that rely largely upon serendipity and *knowledge-based* approaches that exploit information either from the general literature or proprietary information from within an organization (Fig. 1 and 2).

1.1. Screening Approaches

The screening of either diverse or focused compound libraries can deliver a single molecule that has at least minimal activity at each of the targets of interest. In focused screening, compound classes that are already known to be active against one of the targets of interest are screened against another target. This helps to simplify the logistics of screening against multiple targets and improves screening hit rates. For example, DMLs for kinase targets are usually discovered serendipitously through the cross-screening of ligands from selective kinase programs against other kinases.

To date, there have not been many reported examples of DMLs derived via the HTS approach. This could be due to the fact that HTS is a relatively new lead discovery paradigm and there is an inevitable time lag to publication. Other factors could be the logistical complications of screening against multiple targets in parallel or to an inherently low probability of detecting a compound with a multiple profile of therapeutic interest from screening compounds at random. Due to the large number of compounds typically involved in diversity-based screening, they will usually be screened first at one target of interest and any actives will then be filtered on the basis of activity at the other target(s). Even if activity is observed for the second target, usually the balance of affinities is nonoptimal so the activity ratio must be adjusted during optimization.

Although DML lead compounds produced by either of these screening approaches would

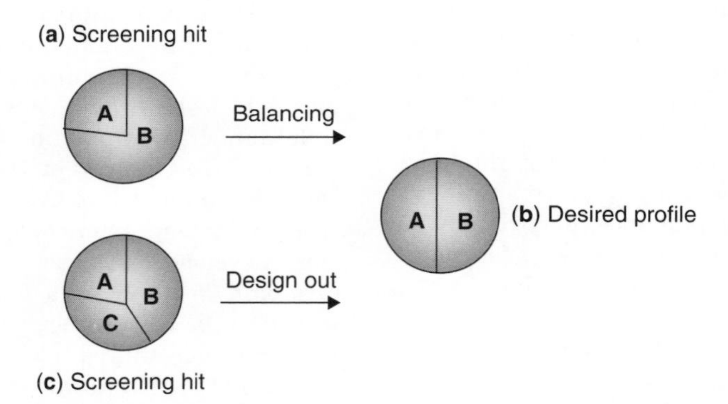

Figure 1. Screening approach: The screening of diverse or focused libraries can deliver a compound that has at least minimal activity at each target of interest, A and B. However it is unlikely that the hit compound has the optimal affinity for all targets so the profile must be balanced during optimization. Alternatively screening might deliver a compound that in addition to the desired activities (A and B) has undesired activity (C) and this must be designed out during optimization.

normally have all desired biological activities, it is highly unlikely they would have the desired pharmacological profile, too. Leads would often require "*balancing*," since one of the biological activities would need a greater improvement during the optimization in order to achieve the desired DML profile (Fig. 1a and b). In addition to the desired activities, screening hits frequently bind to other targets. To minimize the risk of side effects these undesired activities will need to be "designed out" (Fig. 1c).

1.2. Knowledge-Based Approaches

After focused screening, the second most common lead discovery strategy reported in the literature is a knowledge-based approach known as *framework combination*. It starts with two compounds, one of which binds with high selectively to one of the targets and the other with high selectively to the other target [1,17]. In this case, the first goal is to "design in" both activities into a single lead molecule by combining the frameworks (and the underlying pharmacophores) of the two selective molecules (Fig. 2). The intellectual elegance of the framework combination stems from the fact that often a wealth of SAR knowledge is on hand from previous selective ligand projects that can be used to guide the optimization process.

The resulting DMLs are termed *linked*, *fused*, or *merged*, depending upon the extent to which frameworks of the selective ligands have been integrated (Fig. 2). In linked DMLs (conjugates), the molecular frameworks are not at all integrated and there is a distinct linker group between the two components that is not found in either of the selective ligands. This linker is usually intended to be metabolically stable so that the single compound is capable of interacting with both targets, albeit different ends of the molecule may be responsible for the activity at the different targets. Some linked DMLs contain a cleavable linker that is designed to be metabolized to release two ligands that interact independently with each target. This could be seen as a half-way scenario between a true DML and a fixed dose combination.

If the frameworks are essentially touching, so there is neither a discernable linker nor any framework overlap, the DML can be viewed as fused. In the most common and most sought after type of DML, the frameworks are merged together by taking advantage of commonalities in the structures of the starting compounds. Medicinal chemists generally aspire to maximize the degree of framework overlap

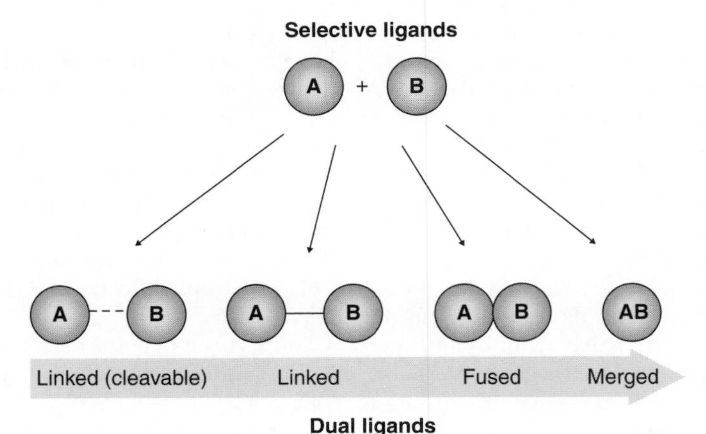

Figure 2. Knowledge-based approaches: framework combination is a knowledge-based approach to generating DMLs. There is a continuum in the degree of merger of the frameworks of the target-selective starting ligands. In linked DMLs, the frameworks are connected via a definable linker, which in some cases is designed to be cleaved *in vivo* to release two independently acting drugs. In fused DMLs, the frameworks are directly attached and in the commonest form of DML, the frameworks are merged together.

in order to produce smaller and simpler molecules with favorable physicochemical properties. Hence, the most common and most sought after are *merged* DMLs, where the frameworks are integrated by taking advantage of commonalities in the structures of the starting compounds. In reality, the degree of framework combination for the examples reported in the literature forms a continuum, with high molecular weight (MW) DMLs with lengthy linker groups at one extreme, and small DMLs with highly merged frameworks at the other.

1.3. Which Approach is Better?

The screening and knowledge-based approaches can be viewed as complementary strategies, and employing both approaches is often a sensible strategy to adopt in order to improve overall chance of success. One of the main advantages of the framework combination approach is a potentially rapid access to a DML starting point, which can be greatly assisted by leveraging the in-depth SAR knowledge from historical selective ligand projects. This is particularly true for conjugates that could be useful as intravenously (i.v.) administered drugs or as pharmacological tools. Furthermore, as the targets of a desired combination become increasingly dissimilar one might expect that the chance of screening success rapidly diminishes, in which case the framework combination could be a valuable alternative. However, rationally designing an additional activity into a selective ligand, while still maintaining the primary activity and wider selectivity, can prove to be a very challenging task. A major advantage of the screening approach is that you start from a compound that already has multiple activities built in, albeit these may be quite weak. Screening can add particular value if there is a lack of selective ligands for the targets of interest or little of the SAR information required for a knowledge-driven approach. Screening can deliver novel and unexpected chemotypes, sometimes providing hits for unusual target combinations that span unrelated receptor families. Since the framework combination strategy almost invariably produces dual ligands, discovering ligands that bind to more than two targets, usually demands that a screening approach is followed. Screening can also provide ligands with improved physiochemical and pharmacokinetic (PK) properties compared to framework combination (see the section 2.9 on physicochemical properties).

2. DISEASES OF HISTORICAL FOCUS IN DML DISCOVERY

Historically, the most common therapeutic areas for DML projects have been psychiatry, neurodegeneration, oncology, as well as metabolic, cardiovascular, and allergic disease. A common theme in the DML literature (often called a "plus") is to take a target that has been clinically validated for a given disease, and then add one or more additional activities that have been postulated might enhance efficacy or improve safety. For example, there are many examples of adding secondary activities to serotonin transporter inhibitors for depression ("SERT-plus"), to dopamine antagonists for schizophrenia ("D2-plus"), and so on (Fig. 3).

2.1. SERT-Plus for Depression

Depression is associated with reduced levels of serotonin (5-HT) in the brain, so serotonin transporter (SERT) blocking drugs (SSRIs) such as fluoxetine that increase 5HT levels have become widely used for the treatment of depression. However many patients still fail to respond and for the others the onset time of the effect is slow. In order to address these deficiencies of SSRIs, there have been many reports of the design of compounds with activity at an additional target, such as serotonin $5HT_{1A}$, $5HT_{1D}$, or adrenaline alpha2 receptors, or norepinephrine (NET) and dopamine (DAT) transporters.

It has been hypothesized that activation of $5\text{-}HT_{1A}$ autoreceptors is the cause of the prolonged onset time for SSRIs. Therefore $5\text{-}HT_{1A}$ receptor antagonists may accelerate the onset time. The following examples of dual $5\text{-}HT_{1A}$/SERT blockers illustrate how three different lead generation methods (framework combination, focused screening, HTS) can be employed. Using the framework combination

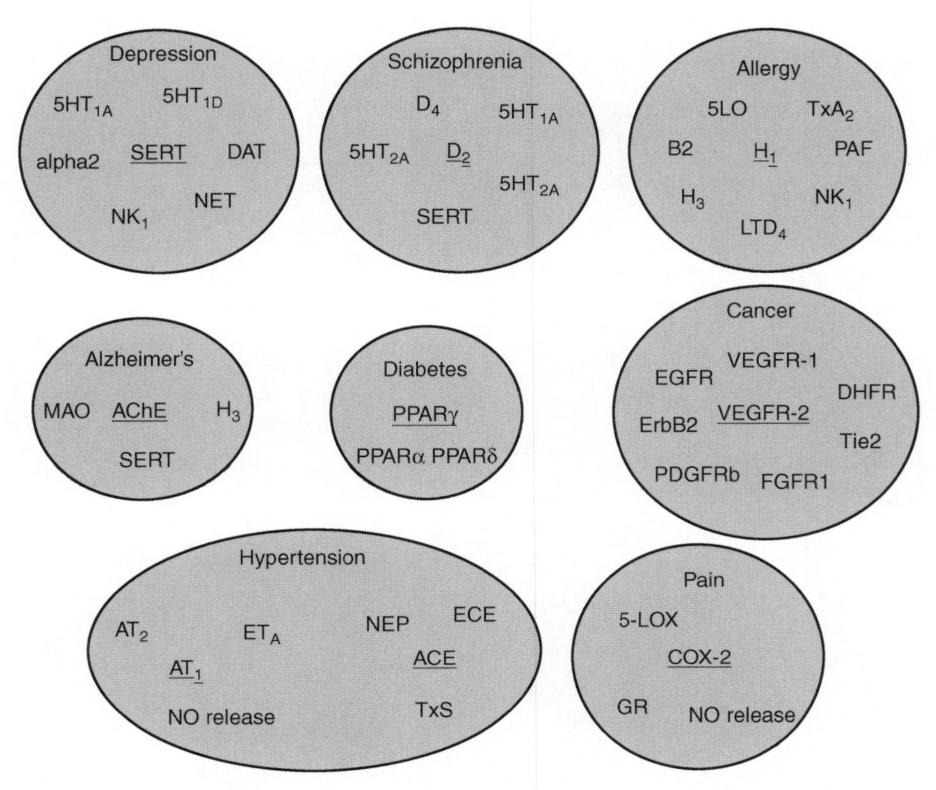

Figure 3. Secondary activities that have been added to a clinically validated primary target (underlined) in an effort to enhance efficacy and reduce side effects. Abbreviations: $5HT_{1A}$, $5HT_{1A}$ receptor; $5HT_{1D}$, $5HT_{1D}$ receptor; $5HT_{2A}$, $5HT_{2A}$ receptor; 5-LOX, 5-lipoxygenase; alpha2, alpha2 adrenergic receptor; ACE, angiotensin-converting enzyme; AChE, acetylcholinesterase; AT_1, angiotensin-1 receptor; AT_2, angiotensin-2 receptor; B2, bradykinin-2 receptor; COX-2, cyclooxygenase-2; D_2, dopamine-2 receptor; D_4, dopamine-4 receptor; DAT, dopamine transporter; DHFR, dihydrofolate reductase; ECE, endothelin-converting enzyme; EGFR, epidermal growth factor receptor; ET_A, endothelin-A receptor; FGFR1, fibroblast growth factor receptor 1; GR, glucocorticoid receptor; H_1, histamine-1 receptor; H_3, histamine-3 receptor; LTD4, leukotriene D4 receptor; MAO, monoamine oxidase; NEP, neutral endopeptidase; NET, norepinephrine transporter; NK_1, neurokinin-1 receptor; NO, nitric oxide; PAF, platelet-activating factor receptor; PDGFRb, platelet-derived growth factor receptor beta; PPAR, peroxisome proliferator-activated receptor; SERT, serotonin transporter; TxA_2, thromboxane-A2 receptor; TxS, thromboxane-A2 synthase; VEGFR-1, vascular endothelial growth factor receptor 1; VEGFR-2, vascular endothelial growth factor receptor 2. (This figure is available in full color at http://mrw.interscience.wiley.com/emrw/9780471266945/home.)

approach, Mewshaw et al. combined structural features of compounds **1** and **2**, which are SERT and 5-HT_{1A} ligands, respectively (Fig. 4) [18]. A basic nitrogen was the common pharmacophoric feature that allowed the two frameworks to be merged to give **3**. The high degree of framework integration helped to produce a DML with a relatively low molecular weight of 358 Da. In the second example, Van Niel et al. designed a focused screening library based upon the 3-aryloxy-2-propanolamine scaffold found in the 5-HT_{1A} antagonist pinadol **4** (Fig. 3) [19]. The variations at the amine and phenol positions included privileged structures and fragments reported to have affinity for either 5-HT_{1A} or SERT. The SAR around the indole region was reasonably tolerant for both targets but the only amine group that provided reasonable SERT inhibition was a spiro-piperidine **5**. This compound

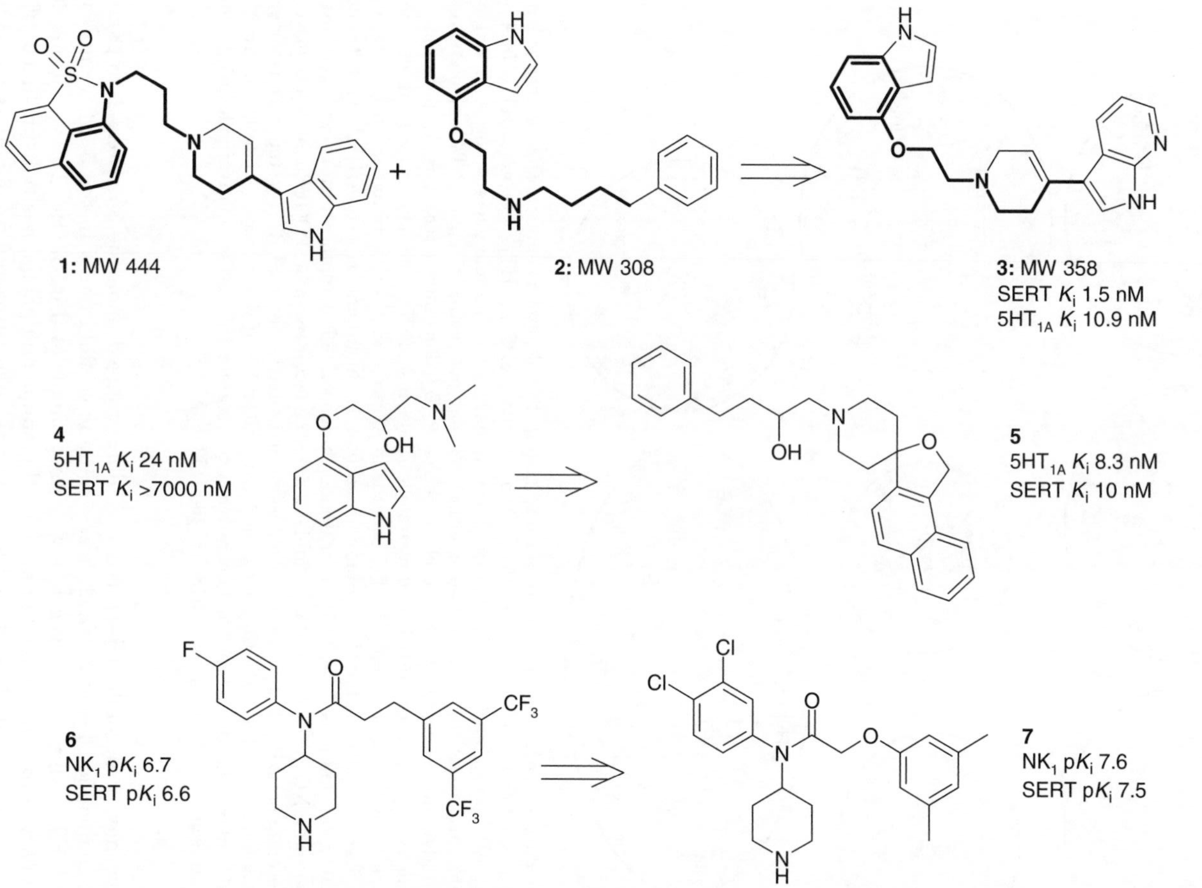

1: MW 444

2: MW 308

3: MW 358
SERT K_i 1.5 nM
$5HT_{1A}$ K_i 10.9 nM

4
$5HT_{1A}$ K_i 24 nM
SERT K_i >7000 nM

5
$5HT_{1A}$ K_i 8.3 nM
SERT K_i 10 nM

6
NK_1 pK_i 6.7
SERT pK_i 6.6

7
NK_1 pK_i 7.6
SERT pK_i 7.5

Figure 4. SERT-plus DMLs for depression.

provided balanced inhibition as well as good oral exposure ($F = 65\%$) and brain penetration in the rat.

Ryckmans et al. used HTS to identify a dual SERT/NK_1 receptor blockers that displayed an unusual combination of activities at a peptide GPCR and a monoamine transporter [20]. Although **6** had only modest activity, systematic optimization of each aromatic moiety in turn provided a more potent compound with a balanced activity at both targets **7** while maintaining drug-like properties (Fig. 4).

2.2. Dopamine D_2-Plus for Schizophrenia

D_2-selective antagonists are only efficacious against the positive symptoms of schizophrenia, such as hallucinations and delusions. To address the negative symptoms (such as social withdrawal), atypical antipsychotic drugs with complex multitarget profiles, such as clozapine **8**, are required. However, clozapine has disadvantages such as weight gain. One of a number of possible explanations for clozapine's atypical profile is its higher antagonist affinity for the 5-HT_2 receptor than for the D_2 receptor. This observation lead to the so-called "D_2/5-HT_2 ratio" hypothesis whereby agents with >10-fold selectivity for 5-HT_2 over D_2 were sought.

Using a framework combination approach, the endogenous agonist for the D_2 receptor, dopamine **9**, was fused with a large lipophilic group from a 5-HT ligand **10** (Fig. 5) [21]. This transformed the D_2-agonist activity of the endogenous ligand into an antagonist. Fused DMLs of this type (here only one nitrogen atom overlaps) can have an undesirably high molecular weight if the starting compounds are already quite large. In this case the starting ligands are much smaller, so the resulting DML has a relatively low MW of 371 Da. Various heterocyclic groups were selected containing hydrogen bonding groups that might mimick the phenolic interaction, such as the oxindole found in **11**. The naphthyl group was replaced by a 1,2-benzisothiazole group **12**, which provided a desirable clozapine-like D_2/5-HT_2 ratio of 11 [22]. The higher selectivity of **12** over the alpha1 receptor, relative to clozapine, should also provide a lower propensity to cause orthostatic hypotension. The D_2/5-HT_2 ratio hypothesis was validated in the clinic and **12** (Ziprasidone) was launched in 2001 by Pfizer for the treatment of schizophrenia. It has also been hypothesized that the unique profile of clozapine in treating psychosis might be due to its higher affinity at D_4 than D_2, although the approach still needs clinical validation [23].

To maximize the efficacy and safety profile of an antipsychotic drug, much evidence now suggests that it is necessary to address more than two receptors. Using a screening approach, Garzya et al. molecule discovered a molecule **13** that had five activities regarded as being critical for an antipsychotic drug, blocking the D_2, D_3, $5HT_{2A}$, $5HT_{2C}$, and $5HT_6$ receptors [24]. A DML **14** was discovered with the optimal balance of desired affinities. The vast majority of published work in the DML area is directed toward compounds with dual activity. This example is noteworthy since it represents a rational approach to a compound that has multiple activities, a goal that greatly complicates the task for the medicinal chemist. In addition, the medicinal chemist will aim to avoid activities associated with the side effects of many atypical drugs such as H_1 (weight gain and sedation) and alpha-1 antagonism (hypotension).

Clozapine binds strongly to the H_1 receptor but it lacks significant binding at the H_3 receptor. This is unfortunate since H_1 blockade has been associated with side effects in schizophrenia patients whereas H_3 has been proposed as an attractive target for CNS disorders. In a recently published example, van Coburg et al. introduced a H_3 pharmacophore element into clozapine **8** that had the benefit of increasing H_3 affinity, maintaining D_2/D_3 affinity while simultaneously reducing H_1 affinity, thereby potentially increasing the safety margin for weight gain. However, compound **15** (Fig. 5) proved to be inactive *in vivo*, possibly due to pharmacokinetic problems.

2.3. ACE-Plus and AT_1-Plus for Hypertension

Angiotensin-converting enzyme (ACE) inhibitors and angiotensin-1 receptor (AT_1) antagonists have gained widespread acceptance for the treatment of hypertension and congestive heart failure (Fig. 6). Neutral endopeptidase

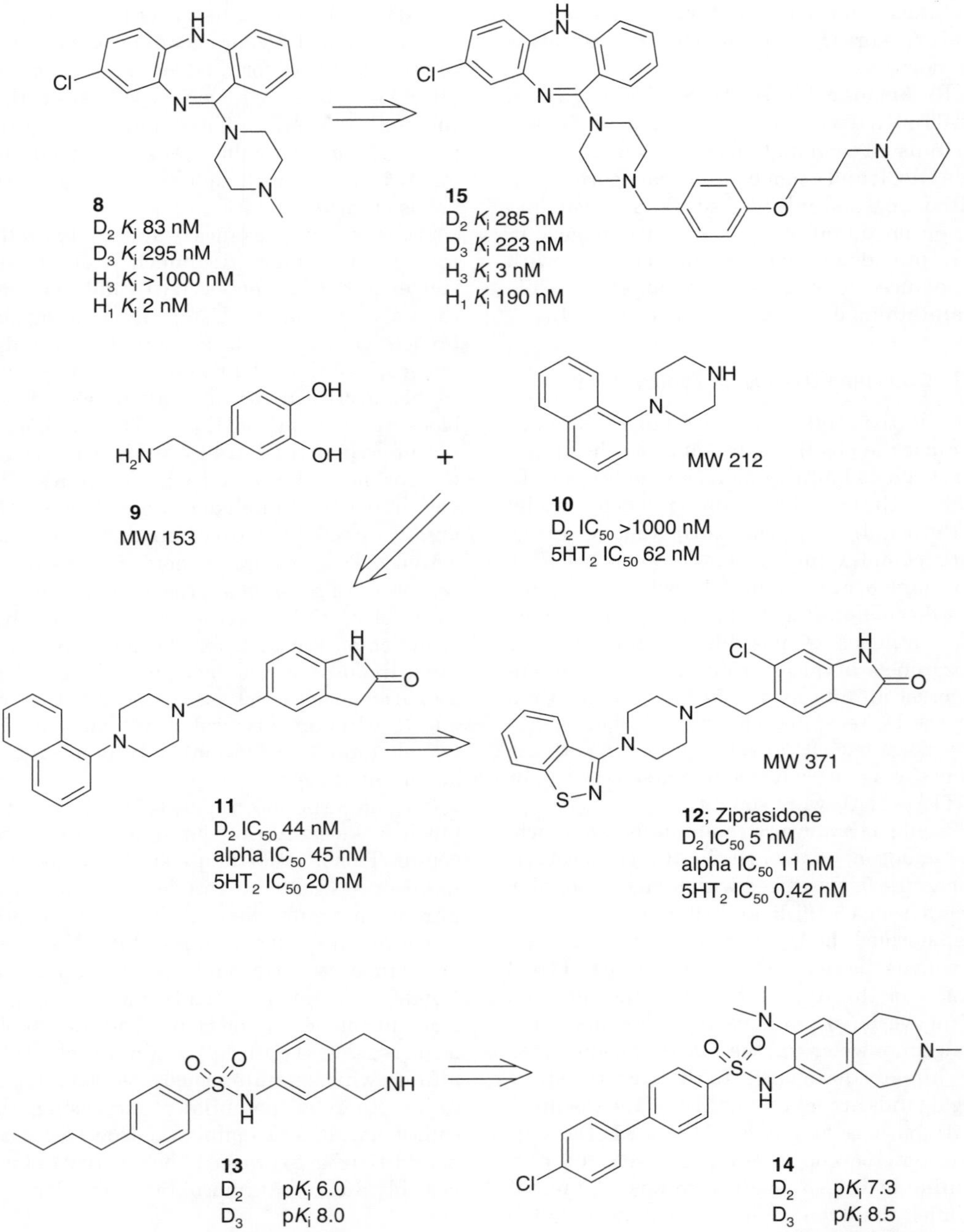

8
D_2 K_i 83 nM
D_3 K_i 295 nM
H_3 K_i >1000 nM
H_1 K_i 2 nM

15
D_2 K_i 285 nM
D_3 K_i 223 nM
H_3 K_i 3 nM
H_1 K_i 190 nM

9
MW 153

10
MW 212
D_2 IC_{50} >1000 nM
$5HT_2$ IC_{50} 62 nM

11
D_2 IC_{50} 44 nM
alpha IC_{50} 45 nM
$5HT_2$ IC_{50} 20 nM

12; Ziprasidone
MW 371
D_2 IC_{50} 5 nM
alpha IC_{50} 11 nM
$5HT_2$ IC_{50} 0.42 nM

13
D_2 pK_i 6.0
D_3 pK_i 8.0
$5HT_{2A}$ pK_i 7.5
$5HT_{2C}$ pK_i 7.9
$5HT_6$ pK_i 7.6

14
D_2 pK_i 7.3
D_3 pK_i 8.5
$5HT_{2A}$ pK_i 8.8
$5HT_{2C}$ pK_i 8.3
$5HT_6$ pK_i 8.1

Figure 5. Dopamine D_2-plus DMLs for schizophrenia.

(NEP) is another enzyme that has attracted interest in recent years as a potential target for treating hypertension and it has been postulated that dual ACE/NEP inhibition may have a beneficial synergistic effect in the management of the condition.

A good understanding of the pharmacophore requirements for ACE and NEP in the prior art facilitated the design of multiple ligands. NEP favors a benzyl moiety in its hydrophobic S1′ pocket, as in the NEP selective inhibitor **8**. ACE is more tolerant in this region, but strongly favors a proline residue at P2′, as in ACE selective inhibitor captopril **16**. The particularly tight SAR for NEP was fortuitously offset by a remarkably flexible SAR for ACE and this was the critical enabling feature of the optimization program. Robl et al. to designed **17**, one of the first dual ACE/NEP inhibitors (Fig. 6) [25]. By further exploiting the historical SAR from the selective inhibitors, a range of constrained analogs were produced in order to improve the *in vitro* and *in vivo* potency of dipeptide **18**. Ultimately this work led to discovery of omapatrilat **19**, a potent ACE/NEP inhibitor displaying a good pharmacokinetic profile and efficacy *in vivo* [26].

The interest in dual AT_1 and endothelin-A receptor (ET_A) antagonists stemmed from clinical studies suggesting that a combination of the AT_1 selective antagonist, Losartan, and the ET_A/ET_B selective antagonist, SB-290670, produced an additive reduction in blood pressure compared to either drug alone. Taking a framework combination approach, Murugesan et al. selected as the starting points, selective AT_1 and ET_A antagonists with common structural features, **20** and **21**, respectively (Fig. 6) [27]. Both selective ligands contained a biaryl core, a well-known privileged scaffold for GPCRs. Fortuitously, the heterocycle in the 4′-position of the biaryl, required for AT_1 activity, was tolerated by ET_A. A pyrolidinone moiety in the C2′ position of the merged DML was found to be optimal for the balanced dual activity at AT_1 and ET_A. As a consequence of the large size of the selective starting ligands and despite the extensive structural overlap, framework combination resulted in a DML **22** with a high molecular weight of 660 Da. Nonetheless, good oral bioavailability was observed for this compound in the rat ($F = 38\%$).

2.4. Histamine H_1-Plus for Allergies

Selective H_1 antagonists have found widespread utility in the treatment of hay fever and other allergic reactions, but have been largely ineffective for the treatment of asthma [28]. Almost all the H_1-antagonists that show some efficacy against asthma are reported to possess additional activities, suggesting that other chemical mediators are also involved in its pathogenesis. Several H_1-plus DMLs have been reported with a range of additional activities.

The thromboxane-A2 (TxA_2) receptor has also been linked to allergic disease so dual H_1 and TxA_2 receptor antagonists have been of interest. Although both these receptors are GPCRs, they might be expected to possess very different binding sites given that the former has a small polar amine **23** as its natural ligand and the latter has a lipophilic acid **24** (Fig. 7). While "designing in" activity for targets with highly dissimilar natural might be construed as near to impossible, a number of recent examples show that this need not necessarily be a barrier to the discovery of a DML. Ohshima et al. found that molecules with a common benzoxepine scaffold, the selective H_1 antagonist, **25**, and the TxA_2R antagonist **26**, bound to both targets [29]. The tertiary amine group in **27** successfully mimicked the benzimidazole moiety that was known to be crucial for the TxA_2 activity of **26**. Compound **27** was active at both GPCRs as well as being selective over related GPCRs.

Perhaps an even more striking example is provided by the dual H_1 antagonist/5-lipoxygenase (5-LOX) inhibitor **31** since this compound binds to two very different targets from unrelated proteomic families. The compound inhibits an enzyme that oxidizes highly lipophilic arachidonic acid **28**, while also antagonizing a GPCR that binds highly polar histamine **23**. The starting points for framework combination were the selective H_1 antagonist **29** and the 5-LOX inhibitor **30** (Fig. 7) [30]. The strategy took advantage of a tolerant SAR around the basic nitrogen of the H_1 antagonist

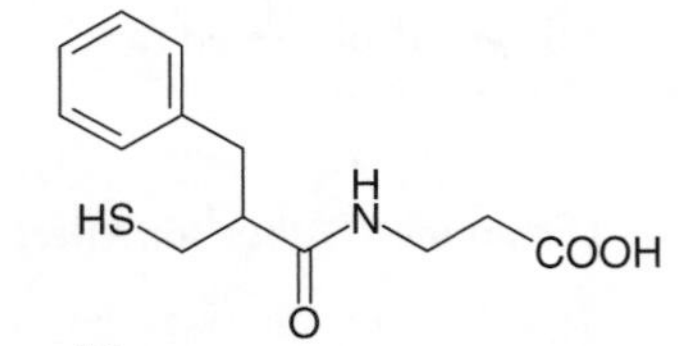

16
ACE IC_{50} 32,000 nM
NEP IC_{50} 9.4 nM

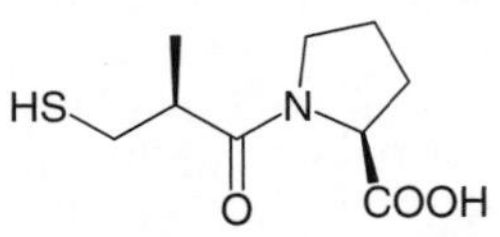

17; Captopril
ACE IC_{50} 23 nM
NEP IC_{50} 830,000 nM

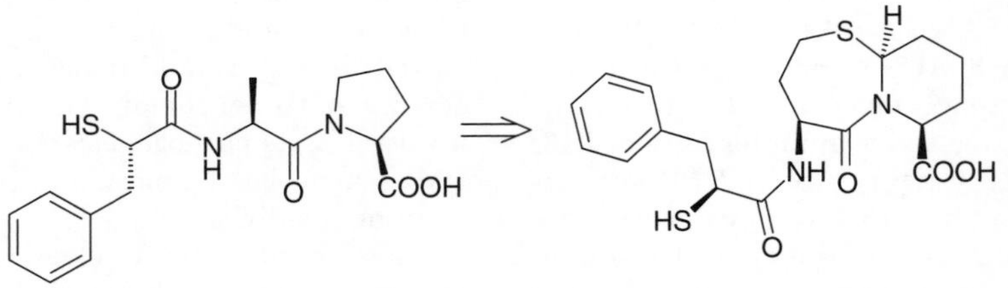

18
ACE IC_{50} 30 nM
NEP IC_{50} 400 nM

19; Omapatrilat
ACE IC_{50} 5 nM
NEP IC_{50} 8 nM

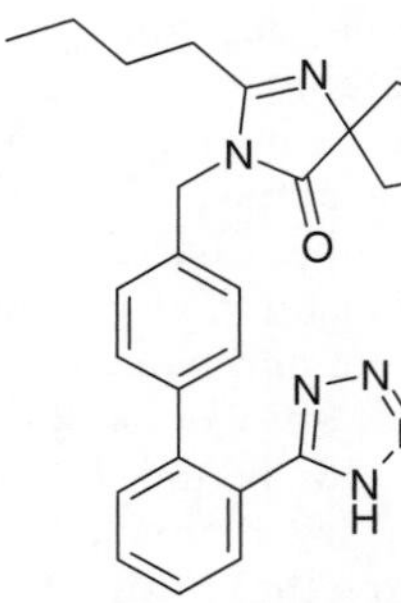

20; Irbesartan
AT_1 K_i = 0.8 nM
ET_A K_i > 10 μM
MW 429

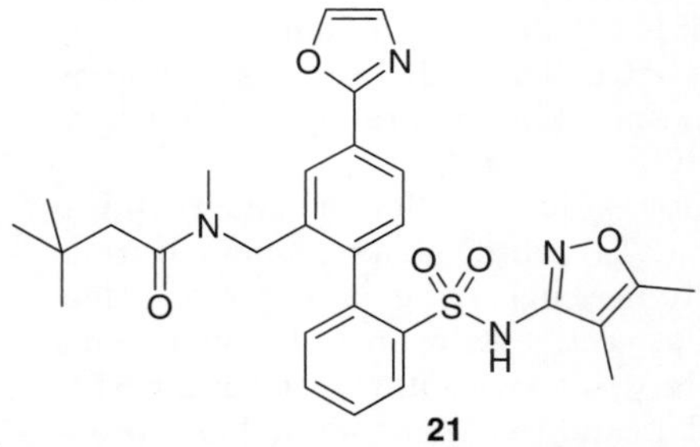

21
AT_1 K_i > 10 μM
ET_A K_i = 0.01 nM
MW 537

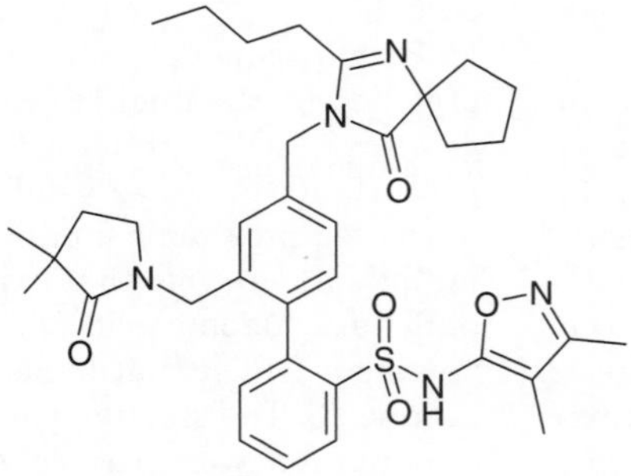

22
AT_1 K_i = 10 nM
ET_A K_i = 1.9 nM
AT_2, ET_B > 10,000 nM
MW 660

Figure 6. DMLs targeting the angiotensin system for hypertension.

to introduce a warhead required for 5-LOX inhibition, the butynyl-hydroxyurea group of **31**.

2.5. AChE-Plus for Alzheimer's Disease

Acetylcholinesterase (AChE) inhibitors have been used for the treatment of Alzheimer's disease to enhance cholinergic activity in the central nervous system, although such treatment is not particularly efficacious in many patients. In an attempt to increase efficacy, AChE inhibition has been combined with other activities.

Kogen and coworkers described a DML **35** that crosses two different proteomic families, inhibiting acetylcholinesterase and the 5-HT transporter (SERT) [31]. A model of the active site of AChE was used to guide the combination of the frameworks of the starting compounds. The AChE-selective inhibitor, Rivastigmine **32**, possessed only three elements of the proposed AChE pharmacophore, lacking a fourth hydrophobic binding site (Fig. 8). If the phenoxyethyl motif from the SERT blocker, Fluoxetine **33**, could provide this hydrophobic interaction, potency should be improved relative to Rivastigmine. Hybridizing the two inhibitors, followed by optimization of the carbamate and phenoxy substituents, provided a dual inhibitor **34**. Conformational constraint using an azepine ring gave a compound **35**, with potent and balanced inhibition at the two diverse targets. There are few examples so far of structure-based drug design (SBDD) being used in the DML area, and this elegant work from Kogen and coworkers illustrates how potentially difficult target combinations can be facilitated by the use of biostructural information.

Youdim and coworkers described dual AChE/MAO inhibitors for Alzheimer's disease [32]. In this case Rivastigmine **32** was hybridized with a selective MAO-B inhibitor Rasagiline **36**, yielding a dual inhibitor ladostigil **37** (Fig. 8). The carbamate and propargylamine groups are key pharmacophoric elements responsible for the AChE and brain MAO inhibition, respectively. The efficacy of Ladostigil in cognition and neuroprotection models in the rhesus monkey was demonstrated [33].

2.6. PPAR-Plus for Metabolic Diseases

The fibrate and glitazone classes of drugs are used to treat dyslipidemia and type-2 diabetes, respectively. They exert their effects through activation of peroxisome proliferators activated receptor-α and -γ (PPARα and PPARγ respectively. This stimulated interest in developing dual PPARα and PPARγ agonists to treat both conditions [34].

In one example from Xu et al., screening resulted in the identification of carboxylic acid **38** containing a bulky lipophilic group in the α-position as a weakly active dual PPARα/γ agonist (Fig. 9) [35]. Compound **38** activated both targets despite lacking the lipophilic "tail" characteristic of classical PPAR agonists. This suggested that addition of the α-benzyl group of **38** to another weak dual agonist **39** might improve its binding affinity. The α-benzyl derivative **40** indeed showed improved activity at both PPARα and PPARγ. Transposing the ether oxygen in **40** to the alternative benzylic position then provided a significantly more potent dual agonist **41**.

2.7. Multikinase Inhibitors for Treating Cancer

Compounds that inhibit several protein kinases, so-called multikinase inhibitors (MKIs), are of great interest in the fight against various forms of cancer. The most common approach to discovering MKIs is focused cross-screening of compounds originating from other kinase projects. Often the result is a nonselective inhibitor that hits both desired and undesired kinases and then the medicinal chemist attempts to "design out" the undesired activities.

The first kinase inhibitor to be developed for clinical use was Imatinib **42** a Bcr-Abl kinase inhibitor, first marketed in 2001 for chronic myelogenous leukemia (CML). Imatinib also possesses other kinase activities and this has lead to its exploitation in other cancer types, showing significant clinical activity against gastrointestinal stromal tumors via c-Kit inhibition and dermatofibroma sarcoma protuberans via PDGFR inhibition. Resistance to Imatinib in CML can become a pro-

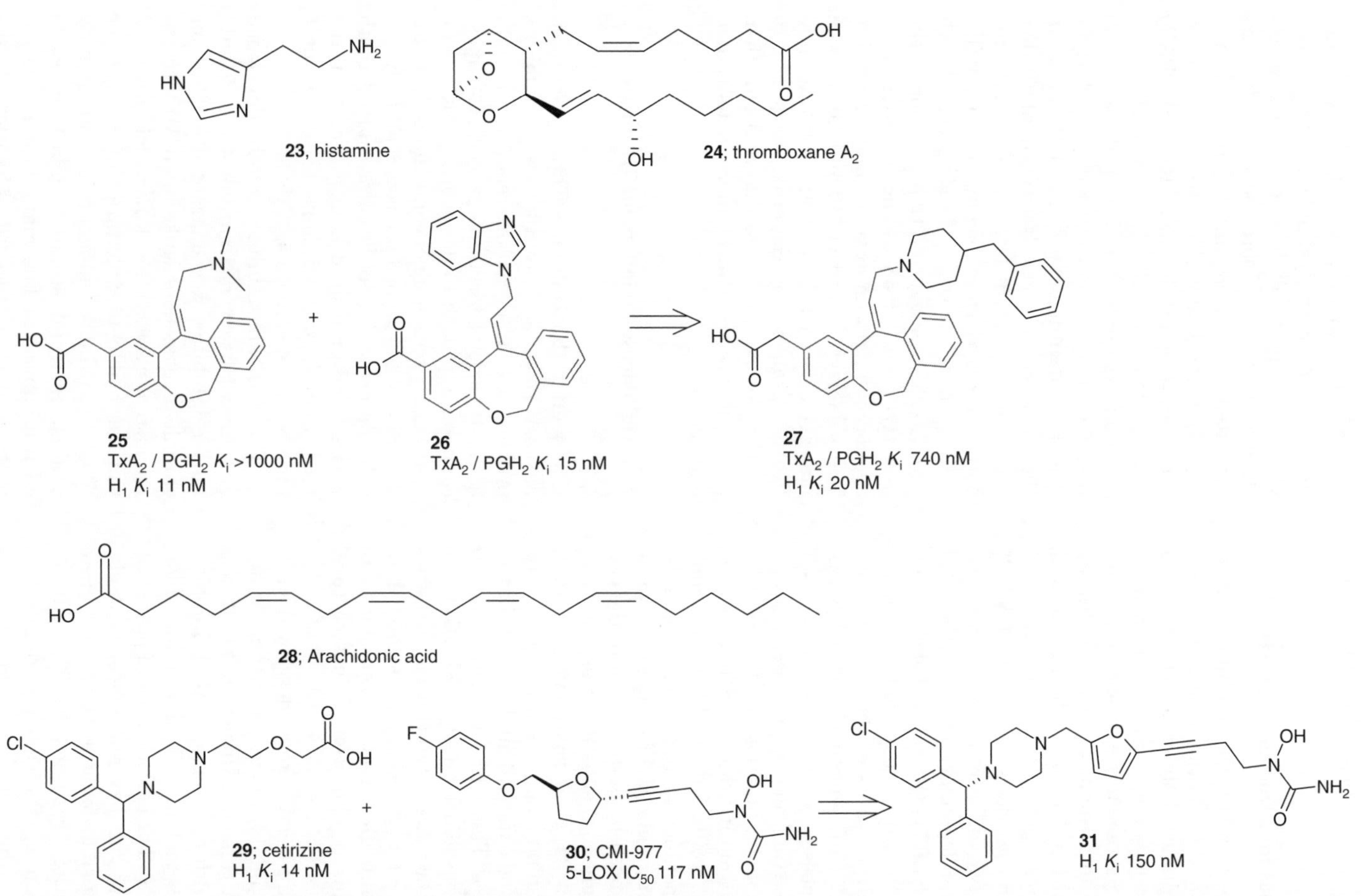

23, histamine

24; thromboxane A_2

25
TxA_2 / PGH_2 K_i >1000 nM
H_1 K_i 11 nM

+

26
TxA_2 / PGH_2 K_i 15 nM

27
TxA_2 / PGH_2 K_i 740 nM
H_1 K_i 20 nM

28; Arachidonic acid

29; cetirizine
H_1 K_i 14 nM

+

30; CMI-977
5-LOX IC_{50} 117 nM

31
H_1 K_i 150 nM

Figure 7. Histamine H_1-plus DMLs for allergies.

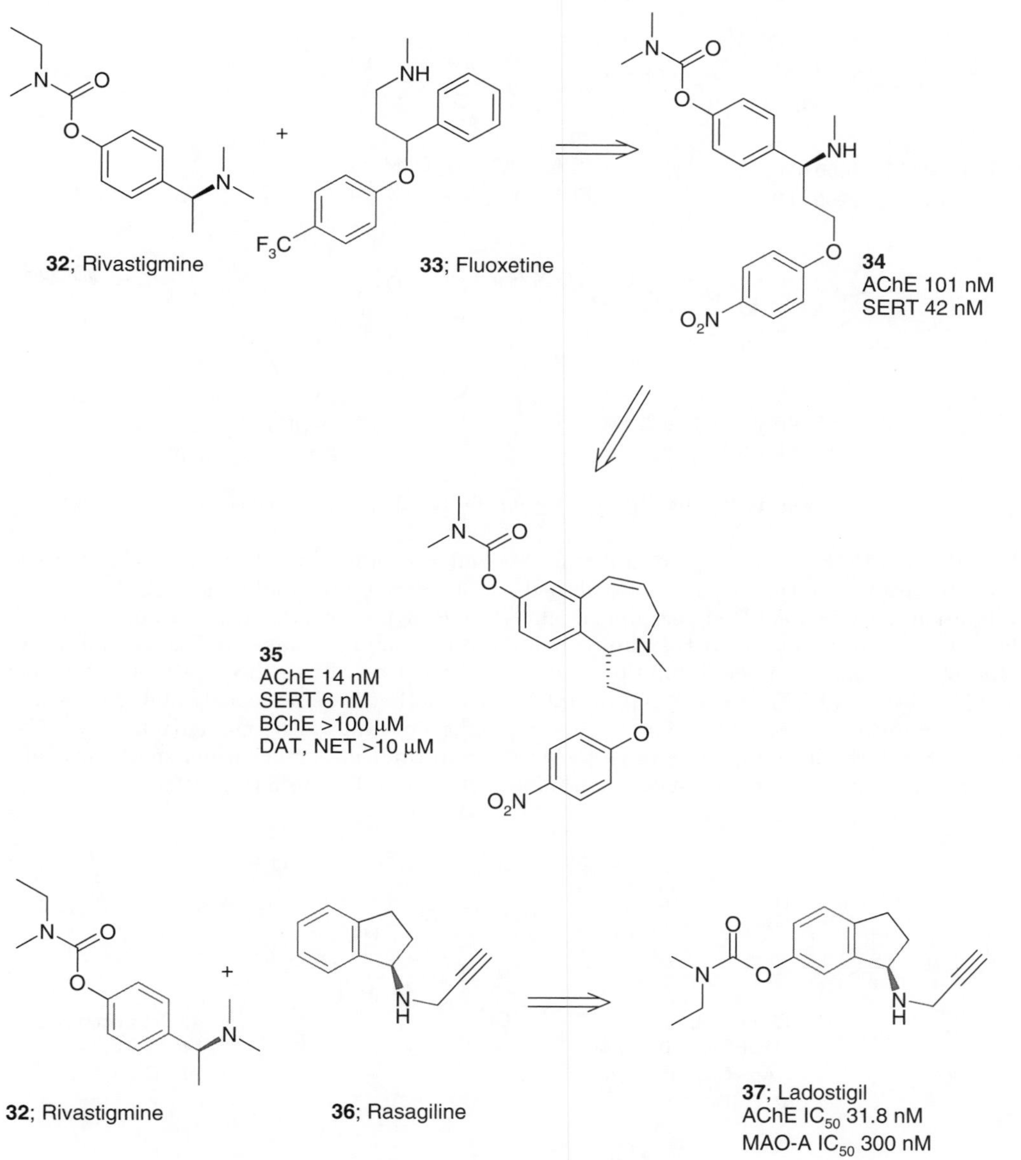

32; Rivastigmine

33; Fluoxetine

34
AChE 101 nM
SERT 42 nM

35
AChE 14 nM
SERT 6 nM
BChE >100 μM
DAT, NET >10 μM

32; Rivastigmine

36; Rasagiline

37; Ladostigil
AChE IC_{50} 31.8 nM
MAO-A IC_{50} 300 nM

Figure 8. AChE-based DMLs for Alzheimer's disease.

blem due to mutations in the Bcr-Abl gene [36]. Dual Src/Abl inhibitors are currently of interest for the treatment of CML in patients who are resistant to Imatinib. Whereas Imatinib itself has no measurable activity against Src, Boschelli et al. use a focused screening approach to identify an inhibitor **43** (Bosutinib) with dual Src/Abl activity (Fig. 10) [37]. They found a very close correlation between the Src and Abl SARs, reflecting the close homology of these kinases. Bosutinib is now in phase III clinical trials, and phase II studies have shown good activity in patients resistant to Imatinib.

The signaling pathways generated by receptor tyrosine kinases that are activated by

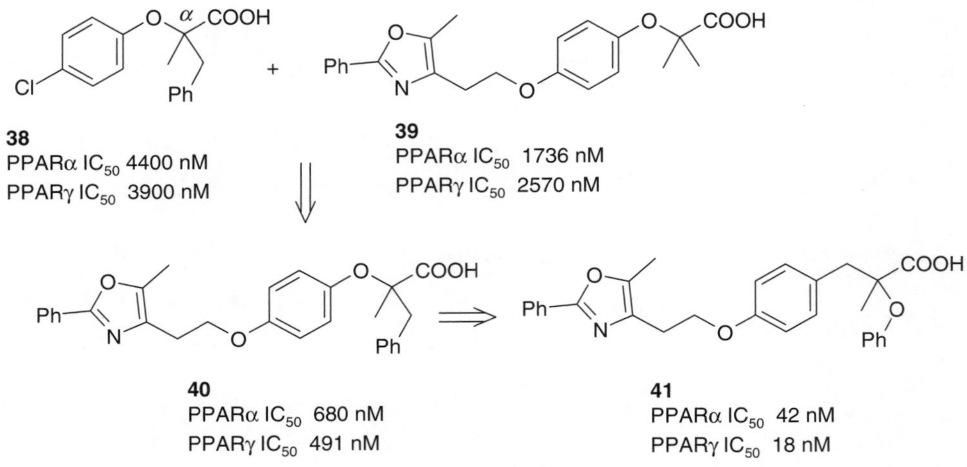

38
PPARα IC_{50} 4400 nM
PPARγ IC_{50} 3900 nM

39
PPARα IC_{50} 1736 nM
PPARγ IC_{50} 2570 nM

40
PPARα IC_{50} 680 nM
PPARγ IC_{50} 491 nM

41
PPARα IC_{50} 42 nM
PPARγ IC_{50} 18 nM

Figure 9. Dual PPARα/γ agonist for treating metabolic disease.

VEGF and PDGF collectively control tumor growth, survival, and angiogenesis. Combined inhibition of VEGF and PDGF receptors might be expected to result in broader antitumor efficacy than single kinase inhibition and a small molecule VEGFR-2 and PDGFRβ inhibitor, Sunitinib **44**, was approved in 2006 [38,39]. Sunitinib appears to be generally well tolerated with the most common side effects being generally manageable, for example, gastrointestinal toxicities, skin reaction, and hypertension. This demonstrates that broad-acting kinase inhibitors can provide clinical benefit in the treatment of solid tumors that were previously highly resistant to therapy and with manageable side effects. Sunitinib has a much wider spectrum of activities than Imatinib (Fig. 10).

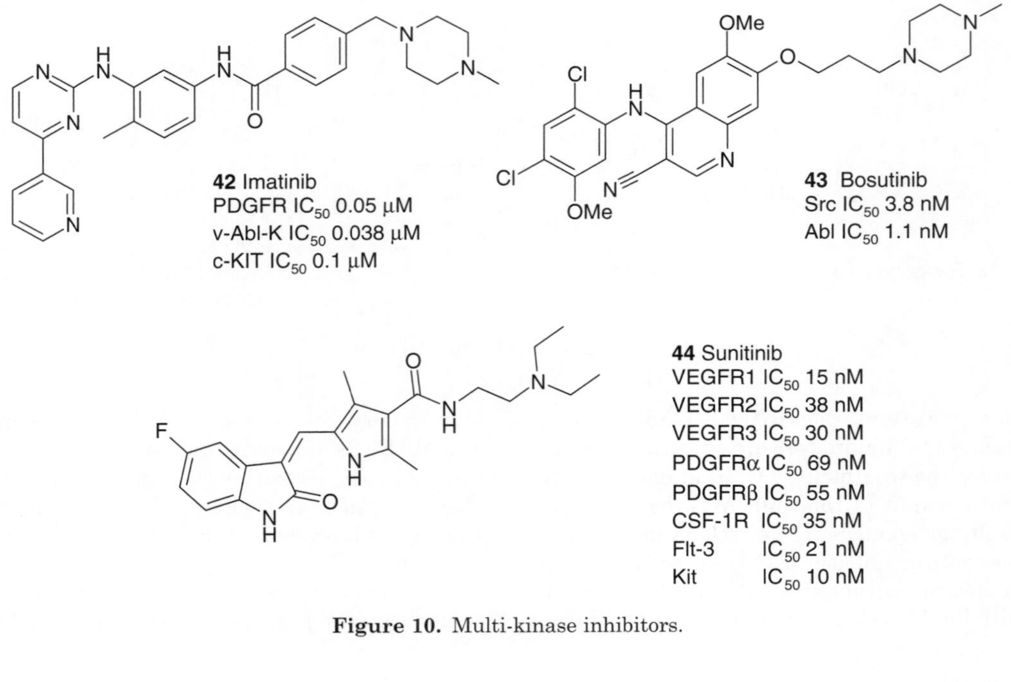

42 Imatinib
PDGFR IC_{50} 0.05 μM
v-Abl-K IC_{50} 0.038 μM
c-KIT IC_{50} 0.1 μM

43 Bosutinib
Src IC_{50} 3.8 nM
Abl IC_{50} 1.1 nM

44 Sunitinib
VEGFR1 IC_{50} 15 nM
VEGFR2 IC_{50} 38 nM
VEGFR3 IC_{50} 30 nM
PDGFRα IC_{50} 69 nM
PDGFRβ IC_{50} 55 nM
CSF-1R IC_{50} 35 nM
Flt-3 IC_{50} 21 nM
Kit IC_{50} 10 nM

Figure 10. Multi-kinase inhibitors.

The ability of the above kinase drugs to hit multiple targets of therapeutic significance was a fortuitous finding rather than an intentional design. Thus it is more likely that they may hit other, perhaps unknown, kinases associated with undue host toxicity. The first designed multikinase inhibitor to reach the market was Lapatinib **45** (Tykerb), where the advantage of a dual EGFR/ErbB2 inhibitor was rationalized prospectively and the optimization was conducted in a way to balance the desired activities and exclude undesired side activities [40]. The starting point was a series of 4-anilinoquinazolines with a structure similar to the selective EGFR inhibitor, gefitinib **46** (Fig. 11). It was found that increasing the size of the group on the aniline, by replacing the 4′-fluorine with a benzyloxy group, introduced potent erbB2 activity. Two noteworthy features of Lapatinib are its unusually high selectivity with no activity in a panel of 119 kinases and its long residence time on the target [40]. Both these features are thought to be related to its binding to an unusual unactivated conformation of the kinase.

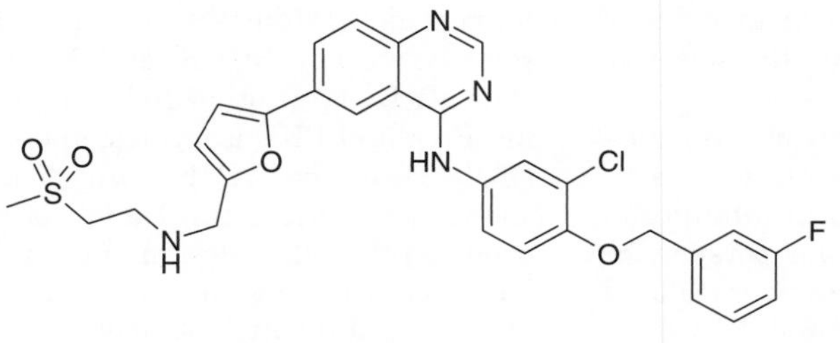

45 Lapatinib
erbB2 IC_{50} 10 nM
EGFR IC_{50} 10 nM

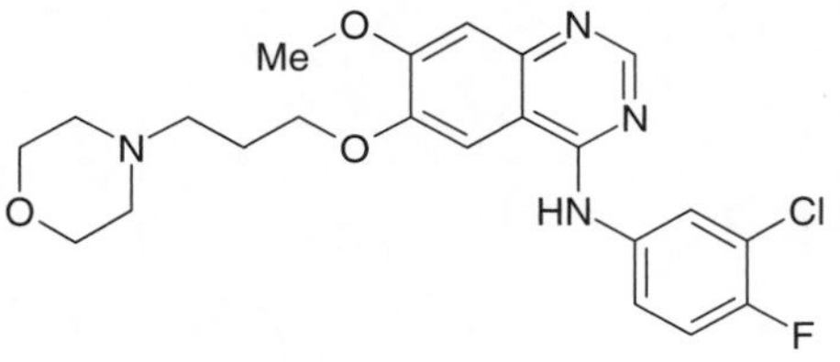

46 Gefitinib
EGFR IC_{50} 1 nM
erbB2 IC_{50} 240 nM

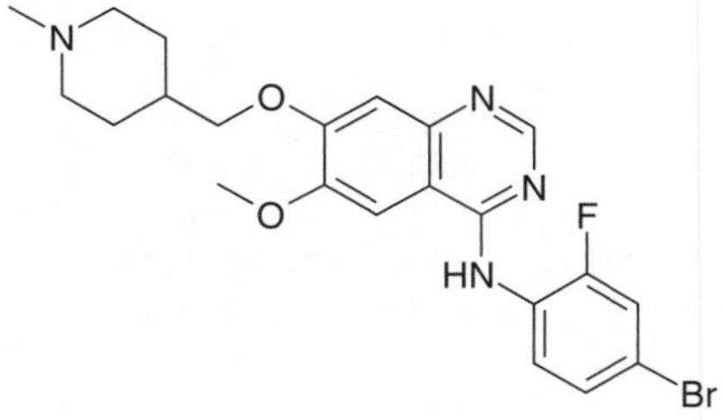

47 Vandetanib

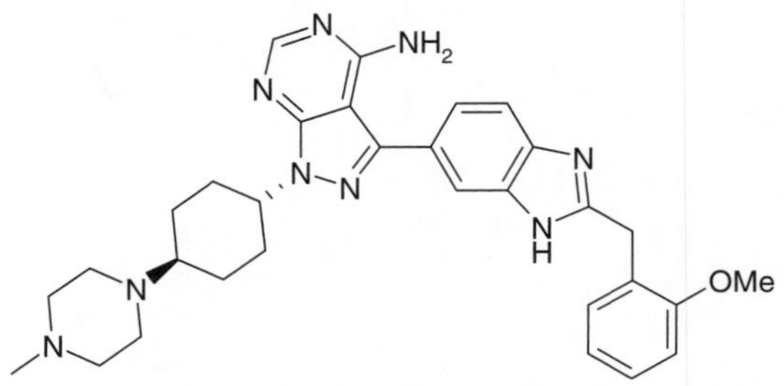

48
IGF-IR IC_{50} 81 nM
EGFRa IC_{50} 58 nM
ErbB-2 IC_{50} 54 nM

Figure 11. EGFR family kinase inhibitors.

The development of Lapatinib has stimulated much activity in the EGFR family area and there has been increasing interest in combining activity as EGFR family receptors with activity at other kinases. Vandetanib **47** (Zactima™) inhibits both the EGFR and the VEGFR pathways with triple inhibition of EGFR, ErbB2, and VEGFR-2 (Fig. 11) [41]. A group at Abbott developed a triple inhibitor **49** with balanced inhibition of insulin-like growth factor receptor 1 (IGF-1R), EGFR, and ErbB-2 (Fig. 11) following the discovery that a combination of selective EGFR and IGF1R inhibitors afforded a synergistic decrease in cellular proliferation across a diverse set of cancer cell lines compared to selective agents [42].

Heerding et al. described the discovery of a pan-AKT inhibitor [43]. The starting compound **49** (Fig. 12) was a modest inhibitor of the AKTs, with only micromolar potency for AKT2 and poor selectivity over the related AGC family kinases, MSK1 and ROCK. A significant feature of this work is that a docking model of **49** at the AKT2 active site was used to guide potency improvements at the AKTs and, at the same time, reduce activity at ROCK and RSK1. A 2-methyl-3-butyn-2-ol group was used to extend the compound through a narrow opening into the back pocket of AKT2, a change that was not well tolerated by ROCK and MSK1 due to differences in the residues lining the pocket. Compound **50** was cocrystallized with AKT2 that confirmed the binding mode predicted by the docking study with a key H-bond between the N5 of the oxadiazole and Alanine-232 in the AKT2 hinge region. While this compound shows good selectivity over ROCK and MSK1, it also shows activity at other AGC kinases, including PKA and PKC isozymes, and AMPK and DAPK3 from the CAMK family. This example shows how biostructural information can be used to rationally "design out" undesired kinase activities, but at the same time illustrates the difficulty medicinal chemists can

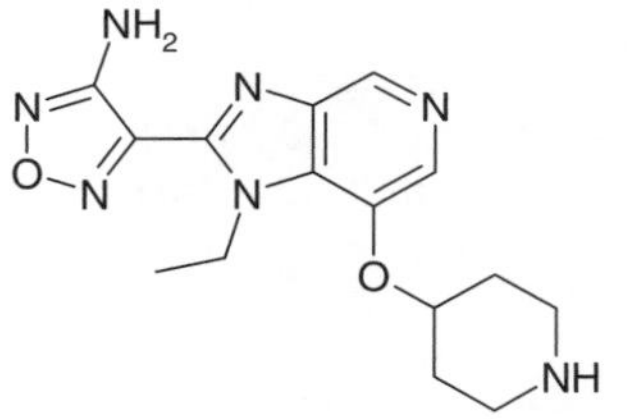

49

AKT_1 IC_{50} 79 nM
AKT_2 IC_{50} 1000 nM
AKT_3 IC_{50} 398 nM
MSK_1 IC_{50} 21 nM
$ROCK_1$ IC_{50} 8 nM

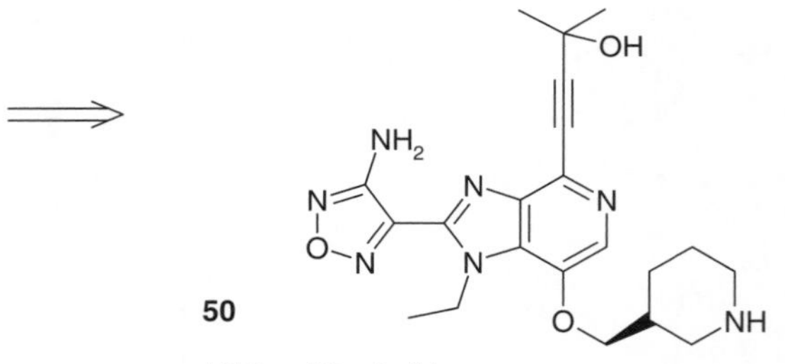

50

AKT_1 IC_{50} 2 nM
AKT_2 IC_{50} 13 nM
AKT_3 IC_{50} 9 nM
MSK_1 IC_{50} 8000 nM
$ROCK_1$ IC_{50} 890 nM

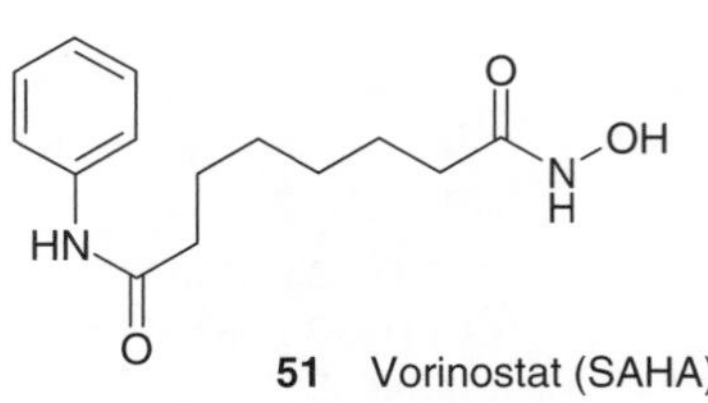

51 Vorinostat (SAHA)

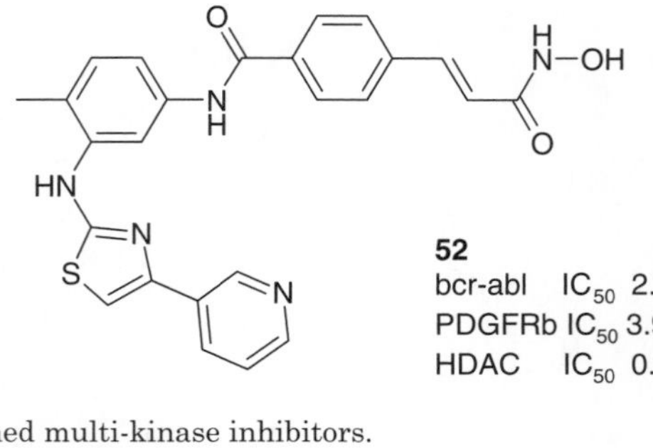

52
bcr-abl IC_{50} 2.7 μM
PDGFRb IC_{50} 3.9 μM
HDAC IC_{50} 0.080

Figure 12. Designed multi-kinase inhibitors.

face when trying to improve the selectivity over multiple off-targets.

The histone deacetylase (HDAC) inhibitor, Vorinostat (SAHA **51**), was approved for cutaneous T-cell lymphoma in 2006 and there has since been interest in inhibiting HDAC and kinases simultaneously (Fig. 12). A triple inhibitor **52** of BCR-ABL, PDGFR, and HDAC was designed by adding to the structure of Imatinib **42** a hydroxamic acid warhead that complexes the Zn^{2+} ion in the active site of HDAC [44].

2.8. COX-Plus for Inflammatory Pain

Nonsteroidal anti-inflammatory drugs counter inflammatory pain by inhibiting cyclooxygenases-1 and -2 (COX-1 and COX-2), key enzymes in prostaglandin (PG) biosynthesis from arachidonic acid [45]. Side effects often limit their use, in particular gastrointestinal ulcerogenic activity and renal toxicity [46]. A "selective drug" strategy adopted to address some of these issues resulted in the development of COX-2 inhibitors like Celecoxib **53** (Fig. 15). A number of alternative DML approaches, targeting multiple key proteins involved in the arachidonic acid biosynthesis, have also been reported, including COX/5-lipoxygenase (5-LOX), 5-LOX/TxA_2 and TxA_2/TxA_2 synthase (TxS).

The dual 5-LOX/COX-2 inhibition profile has attracted a particular attention in recent years [47]. Henichart et al reported a dual COX-2/5-LOX inhibitor designed by fusing the tricyclic moiety present in Celecoxib **53** with an aryltetrahydropyran moiety from the 5-LOX inhibitor, ZD-23138 **54** (Fig. 15) [48]. The knowledge of SAR around selective ligands was critical to identify key pharmacophoric features and most optimal position for connection of the two frameworks. Both starting compounds were completely inactive at the second target, but the resulting DML **55** possessed nanomolar potencies for both enzymes.

An in-depth understanding of the SAR for selective ligands was critical in another example of the framework combination approach, aiming to design dual TxA_2/TxS inhibitors [49]. The essential structural features of TxS inhibitors like Isbogrel **56** are a pyridine nitrogen and carboxylic group (Fig. 15). Since TxS belongs to cytochrome P450 enzyme family, it was postulated that the pyridine moiety forms a complex with the heme group of the enzyme catalytic site. Key features of TxA_2 receptor antagonists like Daltroban **57** are a carboxylic acid and a benzenesulfonamide group. Integration of the TxS and TxA_2 features produced compounds such as Samixogrel **58**, which showed high potency at both targets [50].

Nitric oxide (NO) exerts protective effects on the stomach, possibly due to it acting as a vasodilator and consequently increasing mucosal blood flow and mucus fluid secretion by the gastric epithelial cells. These findings provided rationale for "NO-NSAID" conjugates, based upon incorporation of a NO-releasing moiety into the structure of an established NSAID. One example is NO-Aspirin **59** (NCX-4016), which contains a cleavable ester linker to a nitric oxide-releasing moiety (Fig. 15) [51].

2.9. What are the Main Challenges of Optimizing DML Leads to Drugs?

The three main challenges presented to medicinal chemists working on a DML project are (1) balancing the desired activities, (2) removing any undesired activities, and (3) attaining the physicochemical property/pharmacokinetic profile required for an oral drug.

(1) *Balancing the desired activities*: The optimal ratio of desired activities needs to be established in order to maximize the efficacy and safety profile, but this is often a very difficult task. Most DML projects historically aimed to obtain the same degree of *in vitro* activity for each target, with the assumption that this will also lead to similar levels of target modulation *in vivo*. For example, this trend could be seen in the antidepressants field where agents evolved from the mostly selective serotonin transporter inhibitor fluoxetine **60** toward the dual SERT–NET inhibitor venlafaxine **61** and most recently duloxetine **62** (Fig. 13). Venlafaxine, despite being classified as a dual SERT/NET blocker

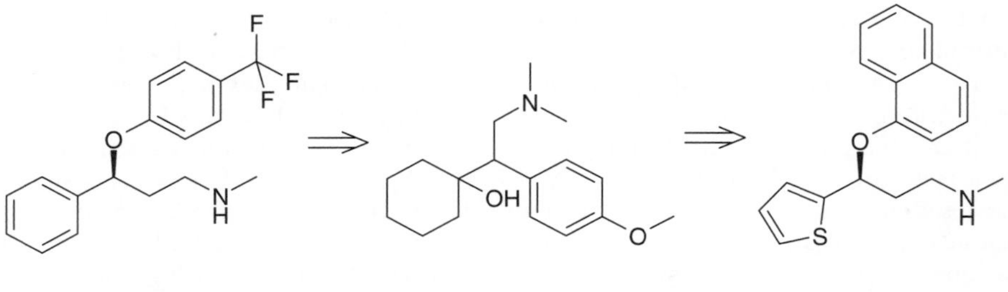

60; Fluoxetine
SERT K_i 0.8 nM
NET K_i 240 nM
NET/SERT ratio 300

61; Venlafaxine
SERT K_i 82 nM
NET K_i 2480 nM
NET/SERT ratio 30

62; Duloxetine
SERT K_i 0.8 nM
NET K_i 7.5 nM
NET/SERT ratio 9.4

Figure 13. The trend from SSRI to SNRI.

(SNRI), is 30-fold selective for SERT suggesting that it behaves as a multiple ligand *in vivo* only at high doses [52], whereas duloxetine has a more potent and balanced *in vitro* profile [53]. In this case, knowledge generated during clinical studies with venlafaxine-helped researchers to optimize the DML profile, but clearly this knowledge will not be available for novel mechanisms of action.

The vast majority of reported DMLs are dual ligands. As the number of targets in a profile increases, the complexity of the task of balancing the activities increases exponentially. Some more complex multiple activity profiles have been achieved, particularly for targets from families with conserved binding sites, such as monoamine transporters, monoamine GPCRs, proteases or kinases. Many efficacious antipsychotic drugs block more than two monoamine GPCR receptors. As described above, Garzya et al molecule discovered a molecule **15** that had five activities regarded as being critical, blocking the D_2, D_3, $5HT_{2A}$, $5HT_{2C}$, and $5HT_6$ receptors.

(2) *Removing any undesired activities*: In addition to adjusting the ratio of desired activities, optimizing wider selectivity against a broad panel of targets is often required. Bonnert et al. reported a successful example of "designing out" adrenergic α_1 receptor activities from a dual dopamine D_2/ adrenergic beta-2 (β_2) agonist [54]. Similarly, Atkinson et al. discussed how undesired adrenergic receptor β_2 activity was effectively removed from a 5-HT_{1A}/SERT ligand **63** by using knowledge of the antitarget pharmacophore (Fig. 14) [55].

Some of the examples described earlier give encouragement to the medicinal chemist that surprising activity and selectivity profiles can sometimes be achieved. The dual acetylcholines-

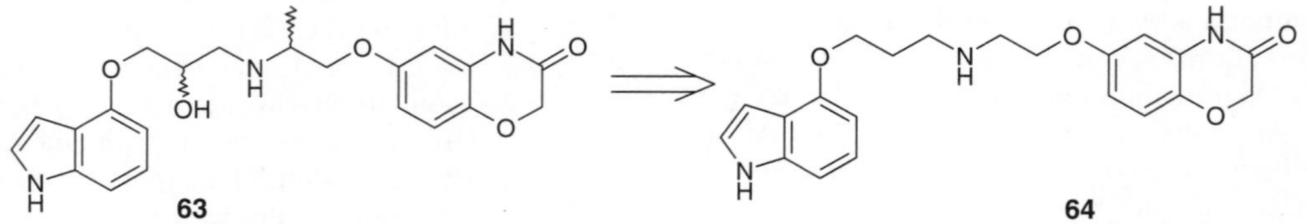

63
$5HT_{1A}$ pK_i 9.1
SERT pK_i 7.3
β2 pK_i 9.2

64
$5HT_{1A}$ pK_i 9.5
SERT pK_i 8.2
β2 pK_i <6.3

Figure 14. Optimization of DMLs wider selectivity.

terase AChE/SERT blocker **35** possessed high selectivity over several homologous targets, including butyrylcholinesterase and the norepinephrine/dopamine transporters (NET/DAT) (Fig. 8). Similarly the COX-2/5-LOX inhibitor **55** (Fig. 15) possessed surprisingly high selectivity over COX-1 and the AT_1/ET_A antagonist **22** was selective over the closely related AT_2 and ET_B receptors (Fig. 6).

(3) *Attaining the physicochemical property/pharmacokinetic profile required for an oral drug*: The fact that DMLs tend to be larger and more lipophilic than marketed drugs or preclinical compounds in general is a critical issue since it has been well documented that such molecules are often associated with poor oral absorption (Fig. 16) [56,57]. The generally less favorable physicochemical properties of DMLs can be explained by the historical popularity of the framework combination strategy whereby the molecular frameworks from two selective ligands are combined. Since the selective ligands are often already drug-like in size, and the extent of the framework integration is generally low, this approach frequently leads to large property increases that compromise oral bioavailability.

This is well illustrated by an example shown in Fig. 17. Here the framework of a selective gastrin receptor antagonist **65** was combined with that of a histamine H_2 ligand **66** [58] resulting in a typical "fused" DML **67** with a single carbon atom overlap. The incompatibility of the hydrophobic gastrin pharmacophore with the hydrophilic H_2 pharmacophore produces regions that are only relevant for binding at one of the targets, resulting in a molecule with high molecular weight (MW 744) and compromised oral absorption.

Despite these drawbacks, the framework combination approach has delivered oral drugs to the market as in the case of Ziprasidone **12** that was discovered by combining two small ligands, dopamine **9** and the 5-HT receptor ligand **10** (Fig. 5). To be successful, it is important that the size and complexity of the selective ligands is minimized while the degree of the framework overlap is maximized.

In comparison to framework combination, screening-derived DMLs are frequently smaller molecules with more attractive physicochemical properties (Fig. 18). This is related to the fact that starting compounds obtained via screening already possess multi-target activity, so that achieving the desired DML profile during lead optimization involves addition of only modestly sized groups, which in turn produces a smaller effect on the overall molecular size and physicochemical properties than the combination of two drug-like frameworks.

It has become apparent over recent years that the physicochemical properties of ligands are greatly influenced by the target gene family for which they were designed [56]. This renders some targets less amenable to drug discovery than others. A similar target family related effect has also been reported for DMLs [57]. The target family that has consistently given the highest property values for both preclinical compounds in general and DMLs is the peptide GPCRs (Fig. 19), whereas the ligands for transporters, monoamine GPCRs and oxidases generally possess favorable physicochemical properties, hence the feasibility of such targets for DML projects will be relatively high.

There are encouraging examples in the literature suggesting that even "difficult" target family combinations such as peptide GPCRs can sometimes be successfully addressed with a creative framework combination approach. One of such examples is the dual AT_1/ET_A antagonist **22** (Fig. 6) reported by Murugesan et al. to have a good oral bioavailability. A strong

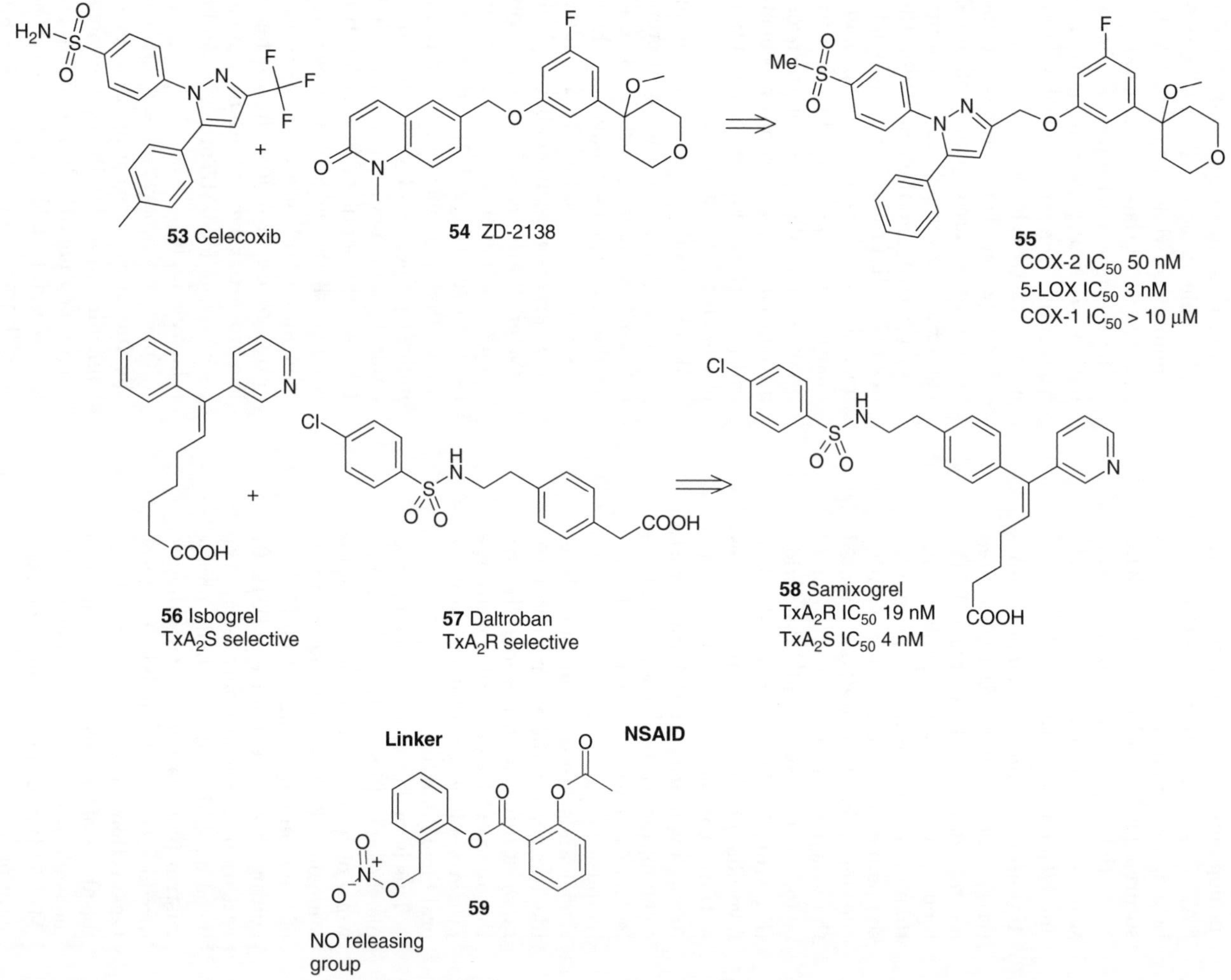

53 Celecoxib

54 ZD-2138

55
COX-2 IC_{50} 50 nM
5-LOX IC_{50} 3 nM
COX-1 IC_{50} > 10 μM

56 Isbogrel
TxA_2S selective

57 Daltroban
TxA_2R selective

58 Samixogrel
TxA_2R IC_{50} 19 nM
TxA_2S IC_{50} 4 nM

Linker

NSAID

59

NO releasing group

Figure 15. DMLs targeting the arachidonic acid cascade.

450
422
344
DMLs
SCOPE
Drugs
MW
4.4
4
2.3
DMLs
SCOPE
Drugs
clog P

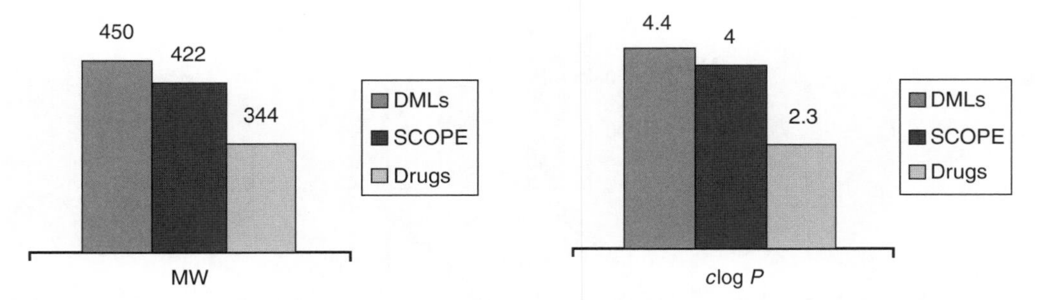

Figure 16. The median molecular weight and clog *P* values for designed multiple ligands are higher than those for oral drugs or a general set of preclinical compounds from Organon's SCOPE database [56,57].

focus during optimization on structural simplification of the DML lead is the main strategic commonality among these examples.

While the framework combination strategy tends to produce large molecules, this is less of an issue when the goal is the discovery of pharmacological tools. Indeed, developing high-quality pharmacological tools to explore the potential therapeutic value of novel target combinations is an important goal for future research in this field. For this purpose, oral exposure is not critical and optimization efforts can focus on exploring the optimal ratio of activities and wider selectivity profile. Portoghese et al. reported heterodimeric conjugates containing delta-antagonist (naltrindole) and kappa-agonist (ICI-199,441) pharmacophores tethered by variable length

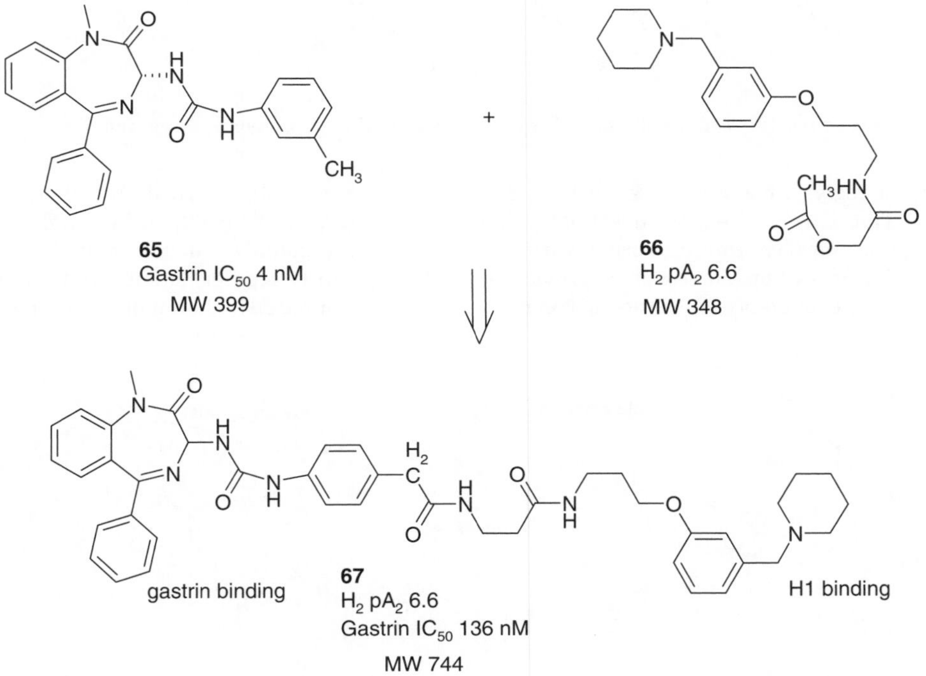

65
Gastrin IC50 4 nM
MW 399
66
H2 pA2 6.6
MW 348
gastrin binding
67
H2 pA2 6.6
Gastrin IC50 136 nM
MW 744
H1 binding

Figure 17. An example of a "fused" DML with nonoverlapping pharmacophores, a high MW, and low oral absorption.

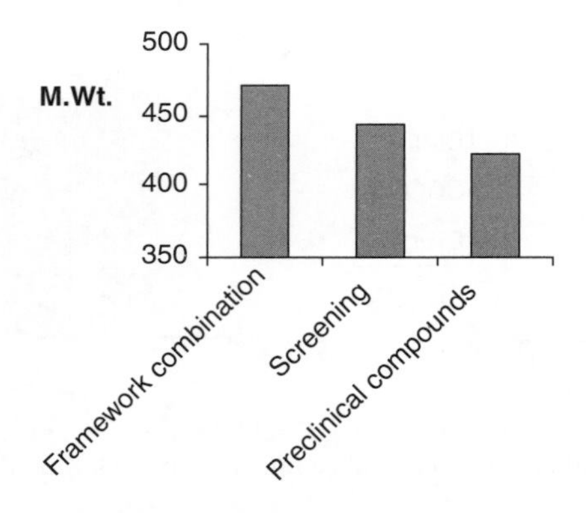

Figure 18. Median MW of DMLs derived via framework combination and screening compared to a general set of preclinical compounds.

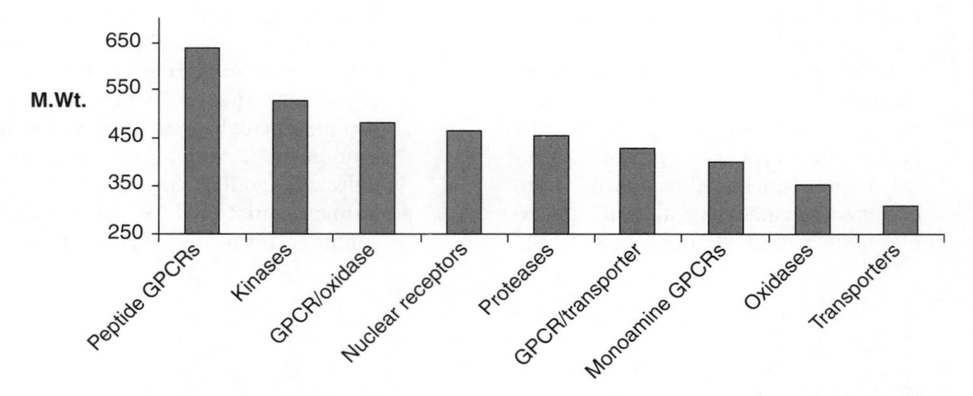

Figure 19. Median MW of DMLs classified according to proteomic target family.

oligoglycyl-based linkers, such as **68**. These were used to investigate the pharmacodynamic and organizational features of opioid receptors providing further evidence for the opioid receptor hetero-oligomerization phenomenon (Fig. 20) [59,60]. Additionally, these compounds demonstrated significantly greater potency and selectivity compared to their monomer congeners.

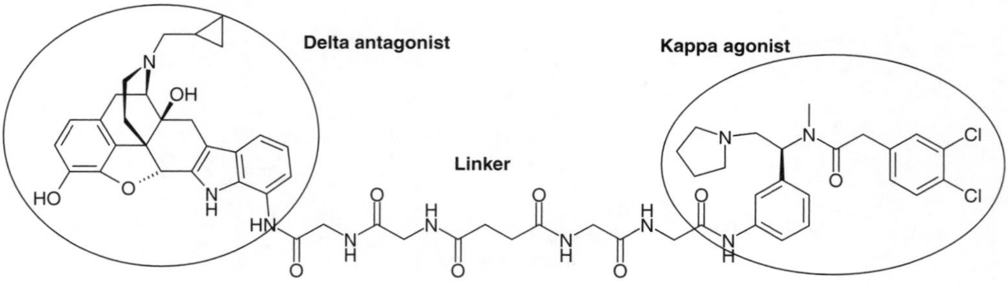

Figure 20. Linked DMLs derived via a framework combination approach can make useful biochemical tools.

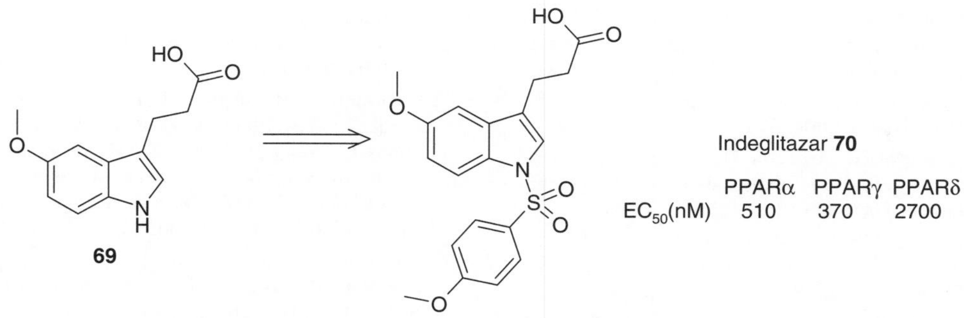

69

Indeglitazar **70**

	PPARα	PPARγ	PPARδ
EC_{50}(nM)	510	370	2700

Figure 21. PPAR agonist discovered using a FBDD approach.

Fragment-based drug discovery (FBDD) is potentially one way of discovering DMLs with improved physicochemical properties and higher "ligand efficiency" compared to framework combination [61]. A recent report describes the use of a FBDD approach to pan-PPAR inhibitors, in which a fragment hit **69** was optimized to indeglitazar **70** (Fig. 21) [62]. Despite the superficial similarity between the PPAR nuclear receptors it has not been trivial to design pan PPAR inhibitors even with the availability of crystallographic information. So this work is one of the first indications of the potential applicability of FBBD in MTDD.

3. CONCLUDING REMARKS

Increasing interest in the design of multiple ligands as a means of discovering drugs with superior efficacy and safety profiles has lead to a rapid evolution of the field in recent years. The discovery of DMLs is following an increasingly rational path compared to historical multitarget drugs that were mostly discovered serendipitously. The three principal challenges presented to medicinal chemists working on DML projects relate to the need to balance the desired activities, obtain wider selectivity over undesired targets, and often most challenging of all, optimize the oral pharmacokinetic profile. Looking to the future, new approaches are certainly needed and one possible way of generating leads is using the fragment-based approach [61]. The growing field of network pharmacology has the potential to provide many novel target combinations that are compelling from a disease-based perspective [63]. However far fewer will be attainable from an oral "druggability" angle, so new computational methods are required to define the true "opportunity space" for DMLs. [64] Despite the many challenges, the opportunity to discover novel and superior medicines should continue to motivate drug researchers in the area of multitarget drug discovery.

REFERENCES

1. Morphy R, Kay C, Rankovic Z. From magic bullets to designed multiple ligands. Drug Discov Today 2004;9:641–651.
2. Millan MJ. Multi-target strategies for the improved treatment of depressive states: conceptual foundations and neuronal substrates, drug discovery and therapeutic application. Pharmacol Ther 2006;110:135–370.
3. Keith CT, Borisy AA, Stockwell BR. Multicomponent therapeutics for networked systems. Nat Rev Drug Discov 2005;4:1–7.
4. Law MR, Wald J, Morris JK, Jordan RE. Value of low dose combination treatment with blood pressure lowering drugs: analysis of 354 randomized trials. Br Med J 2003;326:1427–1431.
5. Larder BA, Kemp SD, Harrigan PR. Potential mechanism for sustained antiretroviral efficacy of AZT-3TC combination therapy. Science 1995;269:696–699.
6. Eisen SA, Miller DK, Woodward RS, Spitznagel E, Przybeck TR. The effect of prescribed daily dose frequency on patient medical compliance. Arch Intern Med 1990;150:1881–1884.

7. Skolnik NS, Beck JD, Clark M. Combination antihypertensive drugs: recommendations for use. Am Fam Physician 2000;61:3049–3056.
8. Glass G. Cardiovascular combinations. Nat Rev Drug Discov 2004;3:731–732.
9. Edwards IR, Aronson JK. Adverse drug reactions: definitions, diagnosis and management. Lancet 2000;356:1255–1259.
10. Kramer MS, Last B, Getson A, Reines SA. The effects of selective D_4 dopamine receptor antagonist (L-745, 870) in acutely psychotic inpatients with schizophrenia. Arch Gen Psychiatry 1997;54:567–572.
11. Campiani G, Butini S, Fattorusso C, Catalanotti B, Gemma S, Nacci V, Morelli E, Cagnotto A, Mereghetti I, Mennini T, Carli M, Minetti P, Di Cesare MA, Mastroianni D, Scafetta N, Galletti B, Stasi MA, Castorina M, Pacifici L, Vertechy M, Di Serio S, Ghirardi O, Tinti O, Carminati P. Pyrrolo[1,3]benzothiazepine-based serotonin and dopamine receptor antagonists. Molecular modeling, further structure–activity relationship studies, and identification of novel atypical antipsychotic agents. J Med Chem 2004;47:143–157.
12. Jones HM, Pilowsky LS. New targets for antipsychotics. Exp Rev Neurother 2002;2:61–68.
13. Stahl SM, Entsuah R, Rudolph RL. Comparative efficacy between venlafaxine and SSRIs: a pooled analysis of patients with depression. Biol Psychiatry 2002;202:1166–1174.
14. Charlier C, Michaux C. Dual inhibition of cyclooxygenase-2 (COX-2) and 5-lipoxygenase (5-LOX) as a new strategy to provide safer non-steroidal anti-inflammatory drugs. Eur J Med Chem 2003;38:645–659.
15. Chao R, Rapta M, Colson PJ, Lee J. US Patent 0,182,092. 2005.
16. Investigational Drugs Database, 590916.
17. Contreras J, Bourguignon J. Identical and non-identical twin drugs. In: Wermuth CG, editor. The Practice of Medicinal Chemistry. New York: Academic Press; 2003. p 251.
18. Mewshaw R, Meagher K, Zhou P, Zhou D, Shi X, Scerni R, Smith D, Schechter L, Andree T. Studies toward the discovery of the next generation of antidepressants. Part 2: incorporating a 5-HT1A antagonist component into a class of serotonin reuptake inhibitors. Bioorg Med Chem Lett 2002;12:307–310.
19. Van Niel M, Beer M, Castro J, Cheng S, Evans D, Heald A, Hitzel L, Hunt P, Mortishire-Smith R, O'Connor D, Watt A, MacLeod A. Parallel synthesis of 3-aryloxy-2-propanolamines and evaluation as dual affinity 5-HT_{1A} and 5-HT re-uptake ligands. Bioorg Med Chem Lett 1999;9:3243–3248.
20. Ryckmans T, Balançon L, Berton O, Genicot C, Lamberty Y, Lallemand B, Pasau P, Pirlot N, Quéré L, Talaga P. First dual NK_1 antagonists–serotonin reuptake inhibitors: synthesis and SAR of a new class of potential antidepressants. Bioorg Med Chem Lett 2002;12:261–264.
21. Lowe J, Seeger T, Nagel A, Howard H, Seymour P, Heym J, Ewing F, Newman M, Schmidt A. 1-Naphthylpiperazine derivatives as potential atypical antipsychotic agents. J Med Chem 1991;34:1860–1866.
22. Howard H, Lowe J, Seeger T, Seymour P, Zorn S, Maloney P, Ewing F, Newman M, Schmidt A, Furman J, Robinson G, Jackson E, Johnson C, Morrone J. 3-Benzisothiazolylpiperazine derivatives as potential atypical antipsychotic agents. J Med Chem 1996;39:143–148.
23. Zhao H, He X, Thurkauf A, Hoffman D, Kieltyka A, Brodbeck R, Primus R, Wasley J. Indoline and piperazine containing derivatives as a novel class of mixed D_2/D_4 receptor antagonists. Part 2: asymmetric synthesis and biological evaluation. Bioorg Med Chem Lett 2002;12:3111–3115.
24. Garzy V, Forbes IT, Gribble AD, Hadley MS, Lightfoot AP, Payne AH, Smith AB, Douglas SE, Cooper DG, Stansfield IG, Meeson M, Dodds EE, Jones DNC, Wood M, Reavill C, Scorer CA, Worby A, Riley G, Eddershaw P, Ioannou C, Donati D, Hagan JJ, Ratti EA. Studies towards the identification of a new generation of atypical antipsychotic agents. Bioorg Med Chem Lett 2007;17:400–405.
25. Robl J, Sieber-McMaster E, Asaad MM, Bird JE, Delaney NG, Barrish JC, Neubeck R, Natarajan S, Cohen M, Rovnyak GC, Huber G, Murugesan N, Girotra R, Cheung HS, Waldron T, Petrillo EW. Mercaptoacyl dipeptides as dual inhibitors of angiotensin converting enzyme and neutral peptidase. Preliminary structure–activity studies. Bioorg Med Chem 1994;4:1783–1789.
26. Robl JA, Sun C-Q, Stevenson J, Ryono DE, Simpkins LM, Cimarusti MP, Dejneka T, Slusarchuk WA, Chao S, Stratton L, Misra RN, Bednarz MS, Asaad MM, Cheung HS, Abboa-Offei BE, Smith PL, Mathers PD, Fox M, Schaeffer TR, Seymour AA, Trippodo NC. Dual metalloprotease inhibitors: mercaptoacetyl-based fused heterocyclic dipeptide mimetics as inhibitors of angiotensin-converting enzyme and neutral endopeptidase. J Med Chem 1997;40:1570–1577.
27. Murugesan N, Tellew J, Gu Z, Kunst B, Fadnis L, Cornelius L, Baska R, Yang Y, Beyer S,

Monshizadegan H, Dickinson K, Panchal B, Valentine M, Chong S, Morrison R, Carlson K, Powell J, Moreland S, Barrish J, Kowala M, Macor J. Discovery of *N*-isoxazolyl biphenylsulfonamides as potent dual angiotensin II and endothelin A receptor antagonists. J Med Chem 2002;45:3829–3835.

28. Eiser N. Histamine antagonists and asthma. Pharmacol Ther 1982;17:239–250.
29. Ohshima E, Takami H, Harakawa H, Sato H, Obase H, Miki I, Ishii A, Ishii H, Sasaki Y. Dibenz[b,e]oxepin derivatives: novel antiallergic agents possessing thromboxane A_2 and histamine H_1 dual antagonizing activity. 1. J Med Chem 1993;36:417–420.
30. Lewis T, Bayless L, Eckman J, Ellis J, Grewal G, Libertine L, Nicolas J, Scannell R, Wels B, Wenberg K, Wypij D. 5-Lipoxygenase inhibitors with histamine H1 receptor antagonist activity. Bioorg Med Chem Lett 2004;14:2265–2268.
31. Toda N, Tago K, Marumoto S, Takami K, Ori M, Yamada N, Koyama K, Naruto S, Abe K, Yamazaki R, Hara T, Aoyagi A, Abe Y, Kaneko T, Kogen H. A conformational restriction approach to the development of dual inhibitors of acetylcholinesterase and serotonin transporter as potential agents for Alzheimer's disease. Bioorg Med Chem 2003;11:4389–4415.
32. Sterling J, Herzig Y, Goren T, Finkelstein N, Lerner D, Goldenberg W, Miskolczi I, Molnar S, Rantal F, Tamas T, Toth G, Zagyva A, Zekany A, Lavian G, Gross A, Friedman R, Razin M, Huang W, Krais B, Chorev M, Youdim MB, Weinstock M. Novel dual inhibitors of AChE and MAO derived from hydroxy aminoindan and phenethylamine as potential treatment for Alzheimer's disease. J Med Chem 2002;45:5260–5270.
33. Sagi Y, Drigues N, Youdim MB. The neurochemical and behavioral effects of the novel cholinesterase-monoamine oxidase inhibitor, ladostigil, in response to L-dopa and L-tryptophan, in rats. Br J Pharmacol 2005;146:543–552.
34. Henke B. Peroxisome proliferator-activated receptor α/γ dual agonists for the treatment of type 2 diabetes. J Med Chem 2004;47:4118–4127.
35. Xu Y, Rito CJ, Etgen GJ, Ardecky RJ, Bean SJ, Bensch WR, Bosley JR, Broderick CL, Brooks DA, Dominianni SJ, Hahn PJ, Dale SL, Mais DE, Montrose-Rafizadeh C, Ogilvie KM, Oldham BA, Peters M, Rungta DK, Shuker AJ, Stephenson GA, Tripp AE, Wilson SB, Winneroski LL, Zink R, Kauffman RK, McCarthy JR. Design and synthesis of α-aryloxy-α-methylhydrocinnamic acids: a novel class of dual peroxisome proliferator-activated receptor α/γ agonists. J Med Chem 2004;47:2422–2425.
36. Daub H, Specht K, Ullrich A. Strategies to overcome resistance to targeted protein kinase inhibitors. Nat Rev Drug Discov 2004;12:1001–1010.
37. Boschelli D, Wang Y, Johnson S, Wu B, Ye F, Sosa A, Golas J, Boschelli F. 7-Alkoxy-4-phenylamino-3-quinolinecarbonitriles as dual inhibitors of Src and Abl kinases. J Med Chem 2004;47:1599–1601.
38. Boyer SJ. Curr Top Med Chem 2002;2:973.
39. Sun L, Liang C, Shirazian S, Zhou Y, Miller T, Cui J, Fukuda JY, Chu J-Y, Nematalla A, Wang X, Chen H, Sistla A, Luu TC, Tang F, Wei J, Tang C. Discovery of 5-[5-fluoro-2-oxo-1,2-dihydroindol-(3*Z*)-ylidenemethyl]-2,4-dimethyl-1*H*-pyrrole-3-carboxylic acid (2-diethylaminoethyl) amide, a novel tyrosine kinase inhibitor targeting vascular endothelial and platelet-derived growth factor receptor tyrosine kinase. J Med Chem 2003;46:1116–1119.
40. Lackey KE. Lessons from the drug discovery of lapatinib, a dual ErbB1/2 tyrosine kinase inhibitor. Curr Top Med Chem 2006;6:435–460.
41. LoRusso PM, Eder JP. Therapeutic potential of novel selective-spectrum kinase inhibitors in oncology. Exp Opin Invest Drugs 2008;17:1013–1028.
42. Hubbard RD, Bamaung NY, Fidanze SD, Erickson SA, Palazzo F, Wilsbacher JL, Zhang Q, Tucker LA, Hu X, Kovar P, Osterling DJ, Johnson EF, Bouska J, Wang J, Davidsen SK, Bell RL, Sheppard GS. Development of multitargeted inhibitors of both the insulin-like growth factor receptor (IGF-IR) and members of the epidermal growth factor family of receptor tyrosine kinases. Bioorg Med Chem Lett 2009;19:1718–1721.
43. Heerding DA, Rhodes N, Leber JD, Clark TJ, Keenan RM, Lafrance LV, Li M, Safonov IG, Takata DT, Venslavsky JW, Yamashita DS, Choudhry AE, Copeland RA, Lai Z, Schaber MD, Tummino PJ, Strum SL, Wood ER, Duckett DR, Eberwein D, Knick VB, Lansing TJ, McConnell RT, Zhang S, Minthorn EA, Concha NO, Warren GL, Kumar R. Identification of 4-(2-(4-amino-1,2,5-oxadiazol-3-yl)-1-ethyl-7-{[(3*S*)-3-piperidinylmethyl]oxy}-1*H*-imidazo[4,5-c]pyridin-4-yl)-2-methyl-3-butyn-2-ol (GSK690693), a novel inhibitor of AKT Kinase. J Med Chem 2008;51:5663–5679.
44. Mahboobi S, Dove S, Sellmer A, Winkler M, Eichhorn E, Pongratz H, Ciossek T, Baer T, Maier T,

Beckers T. Design of chimeric histone deacetylase- and tyrosine kinase-inhibitors: a series of imatinib hybrids as potent inhibitors of wild-type and mutant BCR-ABL, PDGF-Rβ, and histone deacetylases. J Med Chem 2009;52:2265–2279.

45. Vane JR. Nat New Biol 1971;232–235; Dannhard G, Kiefer W. Cyclooxigenase inhibitors-current status and future prospects. Eur J Med Chem 2001;36:109–126.
46. Fosslien E. Adverse effects of nonsteroidal anti-inflammatory drugs on the gastrointestinal system. Ann Clin Lab Sci 1998;28:67–81.
47. Goossens L, Pommery N, Henichart P. COX-2/5-LOX. Dual acting anti-inflamatory drugs in cancer chemotherapy. Curr Top Med Chem 2007;7:283–296.
48. Barbey S, Goossens L, Taverne T, Cornet J, Choesmel V, Rouaud C, Gimeno G, Yannic-Arnoult S, Michaux C, Charlier C, Houssin R, Henichart JP. Synthesis and activity of a new methoxytetrahydropyran derivative as dual cyclooxygenase-2/5-lipoxygenase inhibitor. Bioorg Med Chem Lett 2002;12:779–782.
49. Kato K, Ohkawa S, Terao S, Terashita ZI, Nishikava K. Thromboxane synthetase inhibitors (TXSI). Design, synthesis, and evaluation of a novel series of omega-pyridylalkenoic acids. J Med Chem 1985;28:287–294.
50. Soyka R, Heckel A, Nickl J, Eisert W, Muller TH, Weisenberger H. 6,6-Disubstituted hex-5-enoic acid derivatives as combined thromboxane A2 receptor and synthetase inhibitors. J Med Chem 1994;37:26–39.
51. Di Napoli M, Papa F. NCX-4016 NicOX. Curr Opp Investig Drugs 2003;4:1126–1139.
52. Stahl SM. Comparative efficacy between venlafaxine and SSRIs: a pooled analysis of patients with depression. Biol Psychiatry 2002;52:1166–1174.
53. Bymaster FP, Beedle EE, Findlay J, Gallagher PT, Krushinski JH, Mitchell S, Robertson DW, Thompson DC, Wallace L, Wong DT. Duloxetine (Cymbalta TM), as dual inhibitor of serotonin and norepinephrine reuptake. Bioorg Med Chem Lett 2003;13:4477–4480.
54. Bonnert RV, Brown RC, Chapman D, Cheshire DR, Dixon J, Ince F, Kinchin EC, Lyons AJ, Davis AM, Hallam C, Harper ST, Unitt JF, Dougall IG, Jackson DM, McKechnie K, Young A, Simpson WT. Dual D_2-receptor and b_2-adrenoreceptor agonists for the treatment of airway diseases. 1. Discovery and biological evaluation of some 7-(2-aminoethyl)-4-hydroxybenzothiazol-2(3*H*)-one analogues. J Med Chem 1998;41:4915–4917.
55. Atkinson P, Bromidge S, Duxon M, Laramie M, Gaster L, Hadley M, Hammond B, Johnson C, Middlemiss D, North S, Price G, Rami H, Riley J, Scott C, Shaw T, Starr K, Stemp G, Thewlis K, Thomas D, Thompson M, Vong A, Watson J. 3,4-Dihydro-2*H*-benzoxazinones are 5-HT1A receptor antagonists with potent 5-HT reuptake inhibitory activity. Bioorg Med Chem Lett 2005;15:737–741.
56. Morphy JR. The influence of target family and functional activity on the physicochemical properties of pre-clinical compounds. J Med Chem 2006;49:2969–2978.
57. Morphy JR, Rankovic Z. The physicochemical challenges of designing multiple ligands. J Med Chem 2006;49:4961–4970.
58. Kawanishi Y, Ishihara S, Tsushim T, Seno K, Miyagoshi M, Hagishita S, Ishikawa M, Shima N, Shimamura M, Ishihar Y. Synthesis and pharmacological evaluation of highly potent dual histamine H_2 and gastrin receptor antagonists. Bioorg Med Chem Lett 1996;6:1427–1430.
59. Daniels DJ, Kulkarni A, Xie Z, Bhushan RG, Portoghese PS. A bivalent ligand (KDAN-18) containing δ-antagonist and κ-agonist pharmacophores bridges $\delta 2$ and $\kappa 1$ opioid receptor phenotypes. J Med Chem 2005;48:1713–1716.
60. Portoghese PS. From models to molecules: opioid receptor dimers, bivalent ligands, and selective opioid receptor probes. J Med Chem 2001;44:2259–2269.
61. Morphy R, Rankovic Z. Fragments, network biology and designing multiple ligands. Drug Discov Today 2007;12:156–160.
62. Artis DR, Lin JJ, Zhang C, Wang W, Mehra U, Perreault M, Erbe D, Krupka HI, England BP, Arnold J, Plotnikov AN, Marimuthu A, Nguyen H, Will S, Signaevsky M, Kral J, Cantwell J, Settachatgull C, Yan DS, Fong D, Oh A, Shi S, Womack P, Powell B, Habets G, West BL, Zhang KY, Milburn MV, Vlasuk GP, Hirth KP, Nolop K, Bollag G, Ibrahim PN, Tobin JF. Scaffold-based discovery of indeglitazar, a PPAR pan-active anti-diabetic agent. Proc Natl Acad Sci USA 2009;106:262–267.
63. Hopkins AL. Network pharmacology: the next paradigm in drug discovery. Nat Chem Biol 2008;4:682–690.
64. Bender A. When are multitarget drugs a feasible concept? J Chem Inf Model 2006;46:2445.

COMBINATORIAL CHEMISTRY AND MULTIPLE PARALLEL SYNTHESIS

Lester A. Mitscher[1],
Jeffrey Aubé[1]
Apurba Dutta[1]
Jennifer E. Golden[2]
[1] Department of Medicinal Chemistry, University of Kansas, Lawrence, KS
[2] Kansas University Specialized Chemistry Center, University of Kansas, Lawrence, KS

1. INTRODUCTION

In the nearly 5 years that have passed since medicinal chemical applications of combinatorial chemistry were last reviewed in this series [1], the field has evolved significantly. At first the ability to make large chemical libraries containing mostly informational macromolecules for speculative screening purposes was quite popular. Recently this attitude has changed with the realization that only a subset of the organic molecules capable of synthesis has a realistic chance of becoming oral drugs.

Combinatorial chemistry grew out of peptide chemistry performed on resin beads and initially served the needs of biochemists and the subset of medicinal chemists who specialized in peptide science. Many will agree that the path leading to the present state of combinatorial chemistry essentially started with the solid-phase synthetic experiments by Bruce Merrifeld in 1962 [2,3]. This work had immediate impact facilitated in large part because the reactions are essentially iterative, could be forced to completion by use of reagents in excess, are susceptibile to automation, and detritus can easily be removed from the products by simple washing and filtration.

With the later development of high-throughput screening methods for drug discovery the backlog of existing synthetic and natural molecules was worked through creating a greater demand for novel materials than chemists could supply. Combinatorial chemistry addressed this need. Peptides, however, while vital injectable therapeutic agents, are too metabolically unstable to make good oral drugs. The first library of small, potentially orally available drug-like molecules was that of Bunin and Ellman in 1992 who synthesized a group of benzodiazepine analogs using resin-bound solid-phase organic synthesis methods [4]. This caught the attention of a wider audience and combinatorial chemistry and multiple parallel syntheses now permeate virtually every corner of organic and medicinal chemistry. Initially it was hoped that a flood of new drugs would rapidly emerge as a consequence. This hope coincided with the discovery of many potential targets for chemotherapy arising out of genetic engineering advances. Faith in design of drugs from first principles using computers was also fading with the realization that a host of properties had to be gotten right in addition to novelty, receptor fit, potency, and selectivity and that these additional properties were difficult to add *ex post facto*. In addition, there was a general desire in management circles to shift the drug discovery paradigm in big pharma circles from the traditional cottage mode (crafting individual compounds in optimal yields and purity) to an industrial revolution mode (preparing large numbers of products as rapidly as possible with the minimal expenditure of human resources and as automated as possible, with lesser regard for absolute purity and cleverness of synthesis [5–8]). A major investment in personnel and equipment followed. Alas, intrinsic chemical diversity is more extensive than pharmacological utility and the hoped for increase in new drug introductions did not happen.

To deal with this problem, combinatorial methodology has been retained but emphasis has now shifted to the production of smaller, more realistic, compound libraries containing mostly drug-like molecules.

The primary benefit that combinatorial and multiple parallel chemistry bring to drug synthesis is speed. As with most other human endeavors, uncontrolled speed may be exhilarating but is not particularly useful. Rapid construction of compounds that have no chance of becoming drugs is of little value to the medicinal chemist. Following an initial euphoric period when many investigators thought that any novel compound had a realistic chance of becoming a drug, realism has now returned and libraries are being constructed that reflect the accumulated wisdom of the field of medicinal chemistry. If the requisite care had been exercised from the outset it is possible that a significant rate of emergence of new drugs would have taken place. Instead, a very large number of new compounds were made but the rate of introduction of new drugs into the marketplace has remained about the same as it had been before. A return has been made to understanding the kinds of characteristics that molecules need in order to have a chance of becoming orally active drugs. Intelligent applications of the Lipinski [9,10] and Veber and coworkers [11] rules are now common and are employed as "filters" for molecules proposed for inclusion in combinatorial libraries. Intensive work on understanding the structural barriers to good absorption, distribution, metabolism and excretion characteristics, once the principal reasons for drug failures in the clinic, has greatly improved drug design. Toxicity and side-effect profiles are the current big barriers to further success and these, hopefully, will also be overcome by intensive research. When combined with combinatorial chemical techniques, a new beginning has been made in compound library design that will, hopefully, lead to the desired increase in useful productivity.

With so many factors to balance, combinatorial chemistry research increasingly now involves smaller, focused libraries that serve as the inspiration for succeeding libraries until the drug seeking exercise has reached its objectives [12]. Most industrial concerns and many academic laboratories now regard combinatorial methods as an integral part of their activities and use them routinely in drug seeking [13].

Big pharma possesses legacy compound libraries containing individual compounds numbering in the millions. Maintaining these in shape for repeated sampling and screening, filling in diversity gaps by directed synthesis of interesting but missing chemotypes, removing "junk" compounds, determining stability, purity, identity, prevention of "crashing out" of solution, preventing unwanted molecular aggregation, keeping track of the data, selective retrieval of specific compounds for formation of screening sublibraries, the costs of screening, and the like are major problems yet to be completely mastered.

A number of boutique firms have been organized that lean heavily upon combinatorial chemistry in order to serve the needs of screening programs. These supplement the activities of in-house combinatorial laboratories in big pharma. A number of other firms have been developed in order to supply specialized equipment for the production, automation, separation, handling, and analysis of combinatorial efforts.

Despite the enormous labor that has gone into the enterprise, it is believed that at present only one compound has made it all the way from inception to FDA approval using combinatorial chemistry. Sunitinib (Fig. 1), a vascular endothelial growth factor (VEGF-R2) inhibitor and a platelet derived growth factor (PDGF-Rβ) tyrosine kinase inhibitor, was approved for treatment of renal carcinoma in 2006 having progressed from high-throughput screening (HTS) of a combinatorial chemical compound library followed by optimization of the resulting hits [14,15]. It is perhaps unfair to regard this meager showing as a pejorative argument against combinatorial chemistry since it still takes more than a decade to produce a marketed drug from inception to finish. Perhaps now that a successful beginning has been made, a flood of products will soon emerge, but it will be difficult to demonstrate this given that the mode of operation now is to integrate combinatorial methods into appropriate phases of drug seeking rather than to depend entirely on it. Thus extracting successful examples like this from the literature will be increasingly difficult. Few synthetic drugs now result without combinatorial methodologies having played a role at some stage in the process.

A richly specialized literature has developed about combinatorial chemistry. Specialized journals (for example, *The Journal of Combinatorial Chemistry, Molecular Diversity*, and *The Journal of Combinatorial Chemistry and High Throughput Screening*) are devoted to the topic whereas the usual drug discovery oriented and organic chemical journals

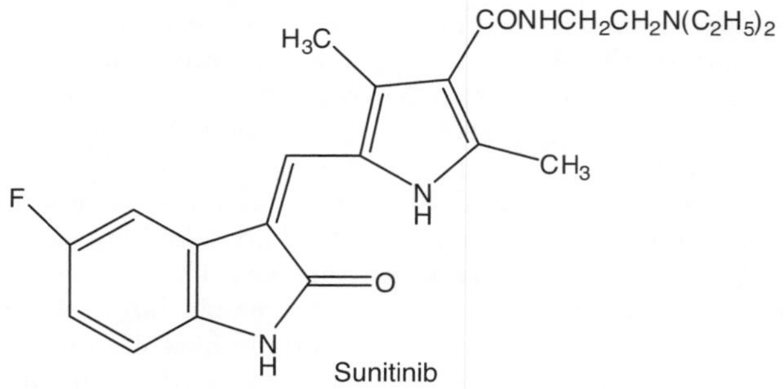

Figure 1. The structure of sunitinib.

frequently publish articles describing advances. In addition to these, a great many books on or about the topic have appeared [16–44]. In terms of reviews, Dolle and coworkers have undertaken the heroic task of reviewing and categorizing the combinatorial libraries that have appeared each year [45–56].

With respect to specific methodologies, the move toward production of smaller, more focused compound sets has taken away some of the advantages of solid-phase chemical synthesis. Bead-based chemistry has evolved in a rather different direction. Use of solvent soluble and solvent insoluble polymers has solved some of the previous problems inherent in scale up and reaction kinetics. Use of bead-bound reagents and removal of undesirable by-products and surplus reactants by resin capture have greatly diminished the bottleneck represented by purification needs. Thus "phase trafficking" has become popular. Polymers that are soluble under a given set of reaction conditions and insoluble in others have further enhanced the ability to prepare pure products rapidly. Mass-directed automated chromatographies have simplified the task in cases where classical chromatography or even chromofiltration are too labor intensive.

A few lengthy semiautomated syntheses of complex natural products have appeared forecasting the time when many molecules will be prepared by automated methods requiring chemist's attention primarily in the design and analysis phases. This will not be inexpensive but will relieve chemists able to afford this of much tedium in producing molecules with desirable characteristics.

The economic impact of combinatorial chemistry on drug seeking has been less than desired due to unrealistic expectations. When a clear structural prototype is lacking, the elapsed time from initial synthesis to marketing is estimated to lie on average between 10 and 15 years and to require the preparation and evaluation of a few thousand analogs. The costs are estimated to lie in the hundreds of millions of dollars per agent. Most large firms now target the introduction of from 1 to 3 novel drugs per year and target sales at a billion dollars or more from each [57]. Thus each day of delay in the drug seeking process not only deprives patients of the putative benefits of the new drug but also represents the loss of a million dollars or more of sales for the firm! While there are significant exceptions (such as atorvastatin (Lipitor®), first to market in an unserved therapeutic area can return a great profit if a sufficient number of sufferers exist that have access to the funds to pay for their treatment. The next two entries competing with this agent can also be expected to do well. After this, success is rather more problematic as the market grows more and more fragmented. Being first to finish the race, therefore, conveys very real survival value. From an economic standpoint, it is estimated that less than 10% of products introduced repay their development costs. Those few that do must return a sufficient surplus to amortize the costs of the losers, sufficient additional funds to cover the costs of future projects, and to gratify the shareholders. These imperatives have placed a premium on speed of discovery and development. The portion of this time devoted to synthesis and screening in the drug seeking campaign is usually about 3–5 years and these are not the most costly years in the process when compared with the costs of clinical trials. The enhanced speed of construction can be expected to decrease the time to market by perhaps as much as a year in favorable cases. While this potential economic reward is less than was originally hoped for when these methods were introduced, it is not trivial.

2. SOME DEFINITIONS

Combinatorial chemistry is now so multifaceted that it is somewhat hard to define precisely but generally speaking it is a collection of methods that allow the simultaneous chemical synthesis of large numbers of compounds utilizing a variety of starting materials and reagents. The number of chemical compounds covered by this definition is generally accepted to begin at 25 members. The resulting compound library can contain all of the possible chemical structures that can be produced in this manner. Many scientists prefer to reserve use of the term combinatorial chemistry for methods producing libraries that contain mixtures of compounds.

Multiple parallel synthesis is a related group of methodologies used to prepare a selected smaller, more focused, and less diverse collection of compounds than those constructed with combinatorial technology. Generally parallel synthesis techniques produce library members as single compounds.

Discretes is a term used to indicate that each compound in a chemical library is pure and a single component.

Cassette refers to a mixture of carefully chosen molecules assembled from discretes.

A *centroid* or *scaffold* is that constant portion of a molecule to which various substituents have been covalently attached.

Orthogonality literally refers to groups projecting at right angles from one another. In combinatorial chemistry it is loosely used to indicate that the chemistry to be used in constructing a compound library involves reactions that do not transform other functional groups present in the molecule.

Solid-phase organic synthesis(SPOS) generally refers to the synthesis of a target molecule in which the starting material is bound to a resin and remains so through successive reagent treatments, often until the last step.

Phase trafficking takes place when a substance or group of substances is transferred from one phase to another. This can involve liquid to solid, liquid to liquid, and so on, transfers. These procedures combine the advantages of solid-phase organic synthesis and solution-phase organic synthesis.

Fluorous phase chemistry takes advantage of the finding that polyflourinated ligands are often soluble in neither organic or aqueous phases allowing for convenient separation of materials from other substances (phase trafficking).

ROMP (ring-opening metathesis polymerization) resins are either saturated or unsaturated norbornene-based resins especially useful for applications in combinatorial chemistry in cases where polystyrene-based resins are insufficiently inert to electrophilic reagents [58–60]. Polymers generated in this manner are also high loading, of high purity, and are easy to prepare.

Safety catch linkers are those which allow attachment of an organic molecule through a relatively inert chemical linkage allowing many reactions to take place orthogonally yet which linkers can be chemoselectively modified to become labile for a product release step [61].

Kan Reactors™ are miniaturized devices that contain a funtionalized resin and a unique tag identifier. One discrete compound is synthesized in each microreactor. Reagents flow through the outer mesh walls of the microreactors which are made of high-grade polypropylene with polypropylene mesh sidewalls and cap or are made of a Teflon derivative.

3. SOLID-PHASE ORGANIC SYNTHESIS OF INFORMATIONAL MACROMOLECULES OF INTEREST TO MEDICINAL CHEMISTS

3.1. Peptide Arrays

At first the extremely useful solid-phase technology for peptide synthesis introduced by Merrifield [2,3] was employed in a linear fashion for construction of individual peptides. A. Furka, in Hungary, a decade or so later realized that the methodology could lead to simultaneous synthesis of large collections of peptides and conceived of the mix and split methods [62]. This work had an enormous impact at the time and colored subsequent events. Iterative synthesis and testing of subpools of compounds subsequently became popular [63–65]. M.H. Geysen made the synthetic process technically simpler in 1984 by producing compound collections of peptides attached to the ends of polyethylene rods [66] and R.A. Houghten introduced "tea bag" methodologies in 1985 in which porous bags containing resins were suspended in reagents [67]. This further simplified synthesis of arrays of peptides and their subsequent separation. The use of photolithographic techniques for constructing spatially separated individual peptides, the so-called "compound library on a credit card", was introduced by S.P.A. Fodor in 1991 [68] and was extended subsequently to an addressable masking technique [69]. The method is also applicable to reagents applied as spots when lower density arrays are satisfactory [70]. It is also possible to remove selectively terminal t-BOC groups from individual peptide sequences on such a compound chip by generating a suitable acid electrochemically [71]. Testing for active sequences can then proceed. Alternatively identification of products by individually tagging of chemical library contents by a variety of synthetic techniques developed. For example, radiofrequency tagging by K.C. Nicolaou [72], as well as laser activated coding methods were developed [73,74].

The peptide linkage has notable advantages for combinatorial chemistry being relatively chemically stable, nonchiral, and constructible by iterative processes amenable to automation. Peptides are rarely branched, possess a variety of interesting biological properties, and can be constructed in great variety from readily available materials. The counterbalancing defects of these compounds are that they are not easily delivered orally unless they are end capped and must be rather small in molecular weight, are readily destroyed by enzymatic action and generally fail to penetrate into cells. The physiological reason for this is readily understood. They, and other informational macromolecules, function in the body to create specific structures, to generate signals for cells to respond to according to their sequence and architecture, and to catalyze reactions. It would be dangerous if they were absorbed intact from ingestion of other life forms and subsequently incorporated in places where they are not intended to be. To prevent cellular disruption they are first digested in the GI tract, absorbed as monomers, and then reassembled following our own genetic pattern so that they join or supplement those already present without causing disruption in cellular architecture or function. Despite intensive study spanning several decades by some of the best minds of this generation, generally applicable means of delivering intact therapeutically significant blood levels of peptides through the oral route remain elusive.

For these reasons peptides are prominent among the compounds of interest to biochemists but less so to many medicinal chemists. Peptide libraries are valuable in lead seeking, in basic studies on cellular processes, or for the preparation of parenteral medications for replacement therapy. An added feature to bear in mind is that the preparation of certain medically important polypeptide drugs, such as human insulin and growth hormone, through genetic engineering methodologies, is well developed and convenient lessening the need for synthetic chemistry other than for the preparation of unnatural analogs.

The number of peptides that could in principle be made in combinatorial mode is immense. For example, given that there are approximately 20 common amino acids and, allowing for posttranslational modifications and the existence of many wholly synthetic amino acids, the available synthons are in excess of the number of letters in the Western alphabet. Considering the number of languages that have been generated using these letters, the potential number of peptides that could be made is clearly astronomical. It would require an incredible effort to make even a library containing only one molecule of each and Furka has estimated that the mass of materials needed for such a library would exceed the mass of the universe by more than 200 orders of magnitude! [62] In any event construction of peptide libraries with over a

Table 1. The Number of Possible Peptide Products as a Function of the Number of Naturally Occurring Amino Acids Utilized

Dipeptides	(20 × 20)	= 400
Tripeptides	(20 × 20 × 20)	8,000
Tetrapeptides	(20 × 20 × 20 × 20)	160,000
Pentapeptides	(20 × 20 × 20 × 20 × 20)	3,200,000
Hexapeptides	(20 × 20 × 20 × 20 × 20 × 20)	64,000,000
Heptapeptides	(20 × 20 × 20 × 20 × 20 × 20 × 20)	1,280,000,000

million components is not uncommon. For example, Furka was able to produce a library of more than one million different peptides in equimolar quantities by the positioning-mixing method in less than a week of experimental work! [75]

Were one to use just the common amino acids, the progression of peptides possible is shown in Table 1.

Large libraries of peptides continue to be made [76–90]. The ability to make such large libraries raised the necessity of identifying the active substances contained. Several ingenious methods soon emerged to deal with this problem.

By way of illustration, in Fig. 2, there is presented a hypothetical combinatorial compound library of dipeptides produced by reaction of every possible combination of just five amino acids. This produces a library containing 25 (5 × 5) discrete products. Before combinatorial chemistry this array would have been constructed using 25 individual reactions with each product being produced separate from all of the others. This would have taken some time. It could also have been constructed less laboriously by running all the reactions simultaneously so that a single mixture of all 25 substances would be obtained. If the experiment were performed in this way it would be ideal if all the reactions proceeded quantitatively and at the same rate so that each compound would be present in the final compound collection in equal molar concentration. This ideal is unlikely to be achieved. Individual amino acids differ greatly in their reactivity so if one simply placed all of the potential reactants in a flask under bond forming conditions, they would not react at the same rate so that complex mixtures of difficultly separable peptides would result. With each reaction the disparity between readily formed and poorly formed bonds would widen. One way to deal with this problem is to use less than fully equivalent amounts of each component and to allow the reactions sufficient time for all to go to completion. For example, if two components are employed, each should be added in about half molar quantities. At the end of the reaction one should have equimolar amounts of both products. This would allow the subsequent testing results to be quantitatively comparable [75]. If the whole library were tested instead as a mixture, then it would be seen that it contained an active component but one would not know which one it was. Even for this to happen it is also necessary that the components not interfere with one another so false positives and false negatives are not seen. The process would be even more complex if several active compounds were present. Things are rarely as simple as one would like so these complexities have to be dealt with.

	A	B	C	D	E
A	AA	BA	CA	DA	EA
B	AB	BB	CB	DB	EB
C	AC	BC*	CC	DC	EC
D	AD	BD	CD	DD	ED
E	AE	BE	CE	DE	EE

Figure 2. A combinatorial library constructed from five reacting components. The biologically active component is starred (*).

	A	B	C	D	E
C	AC	BC*	CC	DC	EC

Figure 3. A multiple parallel synthesis library constructed from five participants.

Many clever means of finding the active component expeditiously in collections of compounds have been developed. In a perfect drug seeking campaign only synthesis of component BC in Fig. 2 would be required but this level of efficiency is almost never achieved by any contemporary medicinal chemical method. The real problem is to construct a compound library that is sufficiently diverse and sufficiently large that there is a high possibility that at least one interesting component will be active in the chosen test system. The example assumes that this has been done. The reader will readily see that if instead of testing all of the compounds in the library in Fig. 2 simultaneously, if one prepared and tested the 10 individual mixtures resulting from combining separately the five products in each column and each row, then BC would reliably emerge as the active component as only the row B mixture and the column C mixture would be active and the active component must be the one where the rows and columns intersect.

In Fig. 3, a smaller multiple parallel synthesis library is illustrated that contains putatively active component BC. This library is much smaller and starts with the assumption that one knows or suspects that the best compound will have component C in the second position. In this case C is combined individually with all of the possible partners and BC is rapidly found by bioassay to be the best of five products.

The greater efficiency of this compound library is obvious. In case all of the peptides contained in Fig. 2 terminating in amino acid C have some activity, quantitative testing data of each provides useful structure–activity data.

A combinatorial equivalent of biological evolution is now commonly practiced that utilizes these strategies in an iterative manner. One identifies the best of a "training set" of peptides in a comparatively small library. One then creates a second library with further variants based upon the best compound in the first library and identifies the best compound in the second array. This can be continued until no further improvement is accomplished or until a sufficient level of activity and selectivity is obtained. A great many examples could be cited of which one is sufficient to make this point [91].

Clearly the smaller the library is that succeeds in solving the problem, the more efficient and effective the process will be. Whereas the number of reactions goes up arithmetically, the products increase logarithmically. The primary advantage is speed as the products are prepared simultaneously at each step. The efficiency is also enhanced when the condensation steps involve the same conditions so can be automated.

The utility of the library depends upon the specific structures included and the purposes for which the library is to be tested. If one has a general idea of the type of structure that would be useful, the library will have many promising compounds but contain fewer components. If one has no idea of the type of structure that would give satisfaction, then a successful library must have a larger and more diverse number of compounds in it.

The simplest and least ambiguous method for constructing and deconvoluting peptide libraries is the spatially separate or spatially addressed method. Here a single peptide is constructed on a single type of resin and the resin/products are kept separate. No decoding sequence needs to be attached to the beads in this kind of library. This method was introduced by Geysen in 1984 [66]. To make a suitable array of peptides simultaneously and to keep track of the products and facilitate their screening, the reactions were run on resins attached to the ends of individual polyethylene pins so arranged that they fit into individual wells of microtiter plates (Fig. 4). For the intended purpose (identifying the epitope region of an antigenic protein) biotesting of the array did not require release of the peptides from the rods. Thus only very small quantities of compound needed to be produced.

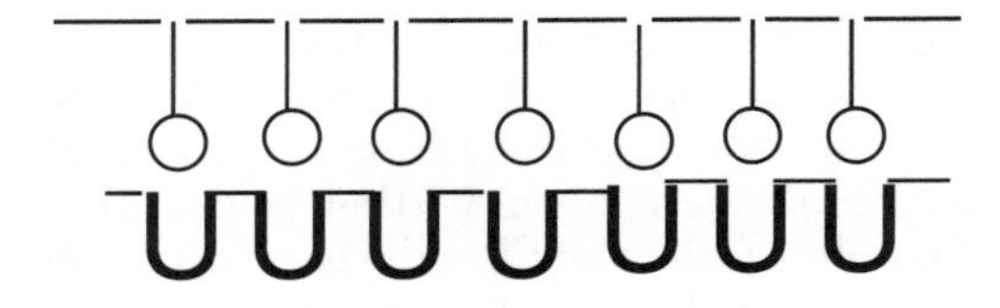

Figure 4. Geysen pins and wells in which individual peptides are prepared by dipping suitable supports on pins into wells containing suitable reagents.

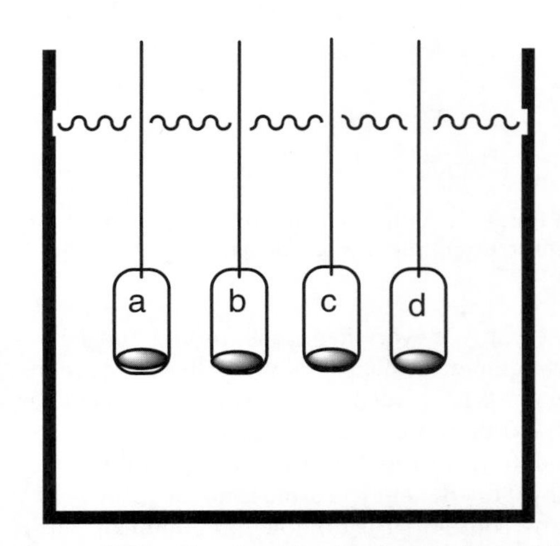

Figure 5. The Houghton "tea bag" method in which resins are enclosed in porous bags and then suspended in a solution of suitable reactants and reagents. The result is spatially separated peptides.

A convenient variation increasing the power of split and mix libraries was developed by R.A. Houghten for parallel synthesis in which beads were contained in porous polypropylene bags dipped into reagent solutions (Fig. 5). These are called "tea bags." The synthetic process utilized a divide-couple-recombine design in which individual residues were constructed in the individual bags which were spatially separated and so each could be tested without requiring tedious separation techniques. In this way the active sequence(s) could be systematically identified until the optimal sequence became clear. The identity of the peptide or peptides contained is recorded on the attached label. This convenient process overcame previous limitations by allowing the release of products into solution so that they could interact with a wide variety of test systems [67]. Using an iterative synthesis and testing scheme, a systematic identification of the best antigenic determinant was made from a chemical library of 34 million (!) hexapeptides. Scanning the same compound library identified antimicrobial peptides also.

At about the same time, Fodor [68] developed a diverse library (1024 peptides) on silicon wafers using photolithographic chemistry for protecting, deprotecting, forming, and releasing the peptides and at the same time controlling the specific place along an X/Y axis where each peptide would be located through use of variously configured masks (Fig. 6). Photolytic protecting groups were employed followed by coupling the newly revealed hot spots" with a suitable reactant. Following this the masks were moved as often as desired and the process repeated. In principle this method could produce thousands of individual peptides on a credit card-like surface and is not

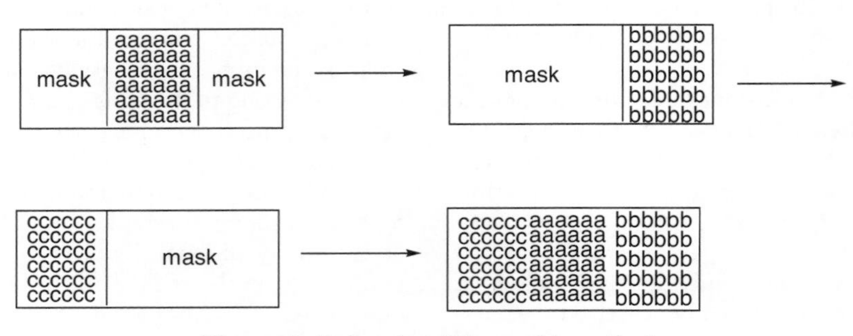

Figure 6. Fodor photolithographic method.

limited to peptides [92]. Each library component occupies a particular xy grid location. Although somewhat laborious, the synthesis can readily be automated. The method requires photosensitive protecting groups and testing methodologies compatible with support-bound products. These techniques are now widely employed for gene array amplification and identification experiments. The method, however, is not very scalable.

A biological method that enables the synthesis of very large arrays of peptides in solution has been developed by applying recombinant DNA techniques [93,94]. Different peptides are expressed on the surface of bacteriophages by inserting randomly synthesized oligonucleotide sequences into the genome. Each resulting clone encodes for a different peptide. Only natural amino acids can be employed in this method. The active compounds are in suspension and the receptor selects its ligand by panning.

In split and pool synthesis (Fig. 7), an initial reaction is run on a bead support to attach an amino acid and the resulting beads are then split into n equal portions (three in the figure) and each of these n groups of beads is deprotected and reacted with one of a group of second amino acids to form dipeptides. These are deprotected, pooled again, and thoroughly mixed. This pool is again separated into equal portions and each is reacted with another amino acid to produce a group of tripeptide mixtures. This sequence of operations is continued until satisfied and the last group of resin piles is often not mixed. Although each individual bead will contain a single peptide, the final products from this methodology consist of groups of related peptides all containing the same last amino acid if carried out in the illustrated manner. The last step is detachment from the beads for testing.

Since one rarely knows which amino acid one should start with, one could prepare a

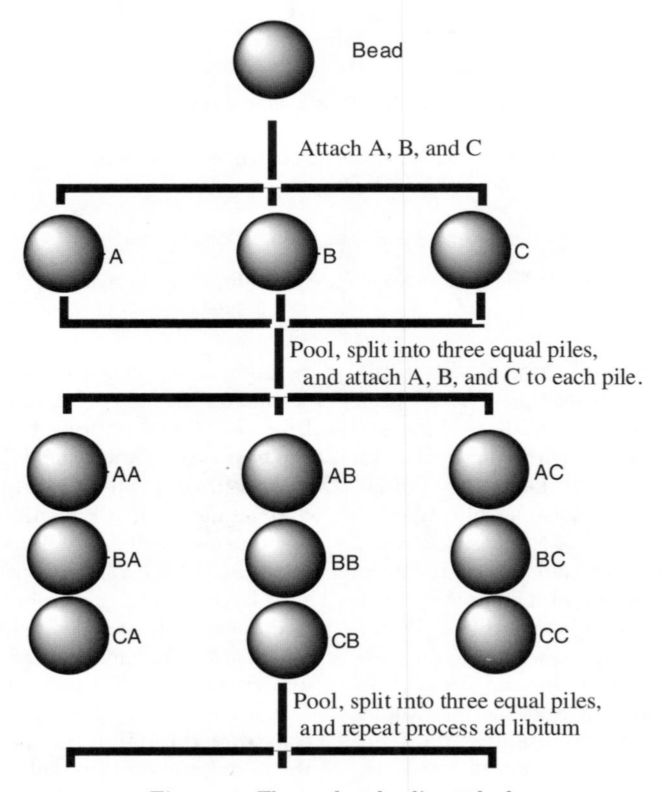

Figure 7. The pool and split method.

Step1. Cleave and test each pool separately.

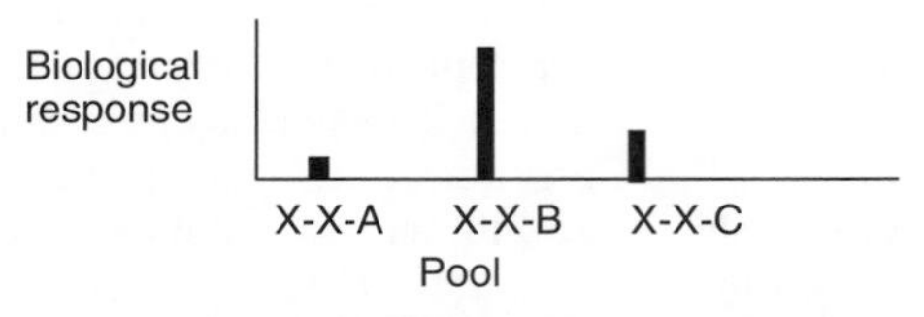

Identify Pool X-X-B as the more potent

Step2. Repeat sequence with Pool X-X-B however varying the second position.

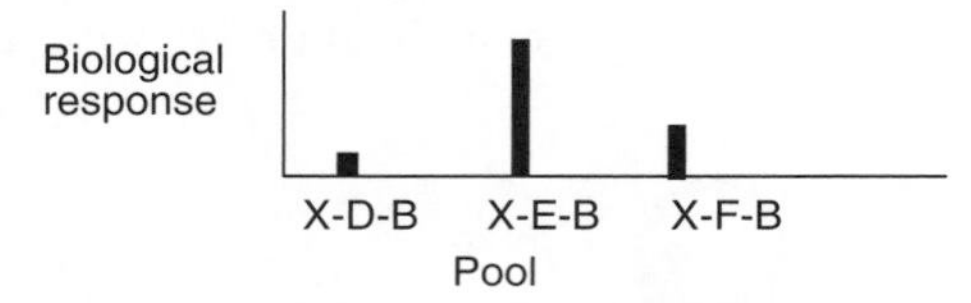

Identify Pool X-E-B as the more potent

Step3. Repeat the sequence with pool X-E-B however varying the first position.

Thus, identify G-E-B as the most potent analog.

Figure 8. The split and mix method with positional scanning.

group of resins each one starting with a different amino acid. Pooling these would produce a single pile of resins with 20 different amino acids attached. Separating this into pools and reacting each of these with 1 of 20 amino acids would produce a collection of dipeptides containing all of the possible combinations of the 20 amino acids. In a third step, the piles of beads could be reacted with on of the possible 20 amino acids to produce piles of beads each pile terminating with a given amino acid. If one did not combine the beads after the last step, detaching and testing would reveal which was the best amino acid to end with (XXB in the tripeptide example shown in Fig. 8). Keeping this one constant in the next reaction series, repeating the process with the second amino acid being individually selected from the 20 possibilities, testing would reveal the best second amino acid (XEB in the example). A third iteration would lead to the optimal sequence at all positions and produce a structure–activity relationship demonstrating that G-E-B was the optimal sequence in this hypothetical example. No assumptions had to be made in arriving at this experimental result. This method of deconvolution is known as positional scanning.

Clearly, a great diversity of peptides can be constructed in mixtures in short order using these methods. Many sub variations can be envisioned. In general it is often found that

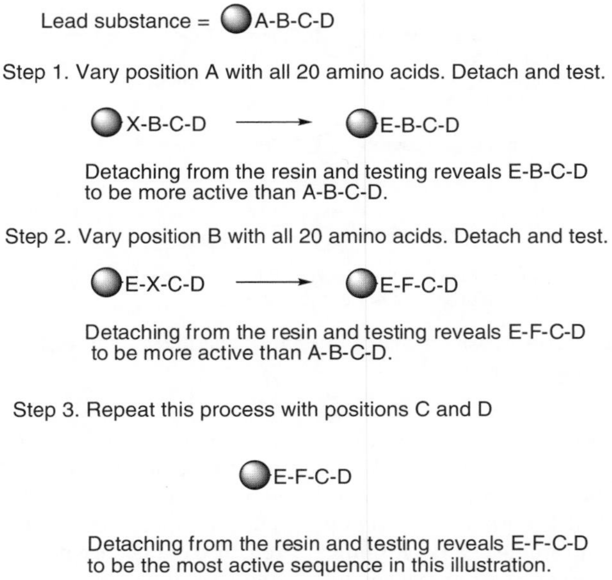

Lead substance = ● A-B-C-D

Step 1. Vary position A with all 20 amino acids. Detach and test.

● X-B-C-D ⟶ ● E-B-C-D

Detaching from the resin and testing reveals E-B-C-D to be more active than A-B-C-D.

Step 2. Vary position B with all 20 amino acids. Detach and test.

● E-X-C-D ⟶ ● E-F-C-D

Detaching from the resin and testing reveals E-F-C-D to be more active than A-B-C-D.

Step 3. Repeat this process with positions C and D

● E-F-C-D

Detaching from the resin and testing reveals E-F-C-D to be the most active sequence in this illustration.

Figure 9. The substitution/omission method.

more than one active peptide is obtained, making the process in practice somewhat more complex than the example. Synthesis of all of these actives as individual pure chemicals (discretes) will allow the development of a structure–activity relationship.

If one has, however, a lead peptide already and wishes to define the contributions of the amino acids or of a region in a lead peptide, then one can perform a limited and more focused study by systematically varying all of the individual positions in the lead. This process is illustrated in Fig. 9. Here one starts with a bead sequence whose activity is known and replaces sequentially each amino acid with all possible 20 analogs keeping the remainder the same. This is similar to the divide-pool method but evaluates only a single specific residue at a time. Testing the 20 libraries reveals which amino acid is optimal at a given position. This is repeated until all of the amino acids in the sequence being examined have been evaluated and an optimized sequence is at hand. In the illustration, one starts with known sequence A-B-C-D and discovers the optimal sequence to be E-F-C-D. A specific example demonstrates the power of this technique [95].

Although this method is laborious, it has no preconceptions and the residual sequences not of interest in this test series can be archived and examined in future against different assays. Another advantage of the method is that the final products are not tethered so are able to assume the solution conformation dictated by their sequence or by the receptor interaction and also to interact with insoluble receptors.

In the omission method, one deletes one of the amino acids from a given position (most often at the end). This is repeated until activity is lost. In this way one can determine the optimal length of a peptide lead. Alanine scans involve replacing an amino acid residue at any chosen position by an alanine and determining the effect that this has on bioactivity. Use of alanine preserves as well as possible the conformation of the resulting analog. This is done at each position to determine the relative contribution of each amino acid side chain.

Devolution by positional scanning is facilitated through testing groups of resins arranged in checkerboard rows and columns as illustrated earlier in Fig. 2. Subsequent libraries are much smaller so the process becomes progressively less laborious.

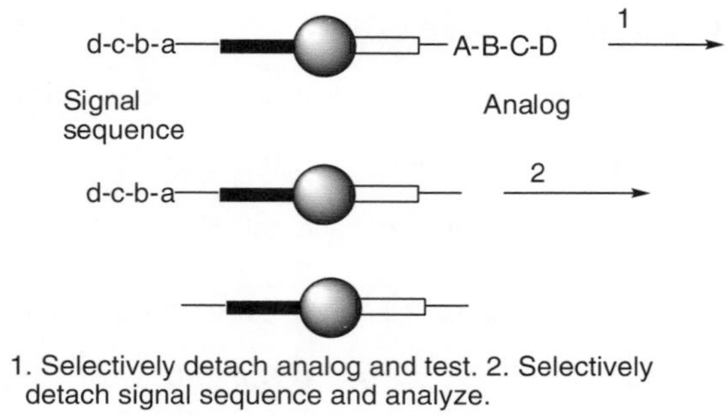

d-c-b-a — A-B-C-D 1

Signal sequence Analog

d-c-b-a — 2

1. Selectively detach analog and test. 2. Selectively detach signal sequence and analyze.

Figure 10. Illustration of a resin with arms containing an analog and a signal sequence each of which can be removed independently from the resin.

In addition to bioassay deconvolution, direct chemical methods of sequence analysis are also available. Mass spectrometry is popular for this as are NMR methods (involving magic angle methods on single beads). Edman degradation of peptides can also be performed.

Another popular method of identifying specific residues of interest is to attach a nonpeptide signaling molecule to a separate arm (Fig. 10) each time a new unit is added to the growing molecule. The signaling or coding molecule needs to be attached to its arm using chemistry that does not interfere with the growing peptide on the other arm and whose chemistry does not detach either molecule prematurely.

There are a variety of strategies employed in detaching the sequences from the arms (Fig. 11). Only the active analogs need be decoded. Whether the product is displaced from the bead before or after testing depends upon the requirements of the test protocol.

Detachment strategy (a) represents an internal displacement reaction and leaves no trace behind in the product of the original point of attachment. Strategy (b) uses hydrogenation or hydrolysis to release the product. This completes the product structure and often does not leave a linker trace. Strategy c is a displacement strategy that may or may not leave the attacking group attached to the peptide or to the resin.

When present the signal sequence carries the history of the bead and therefore codes for the history of the steps involved in the synthesis and thus for the identity of the peptide that one believes one has attached to the bead. The ease of identification of oligonucleotide sequences (by PCR methods) has made these popular for such coding [96,97]. Various halogenated aromatic residues have also been used in the form of a binary code for this purpose [98]. Electron capture mass spectrometry is sufficiently sensitive to detect the code on such beads. Encoding using dansylated secondary amines can also be done [99]. The use of stable isotopes in particular ratios has also been employed [100]. Another popu-

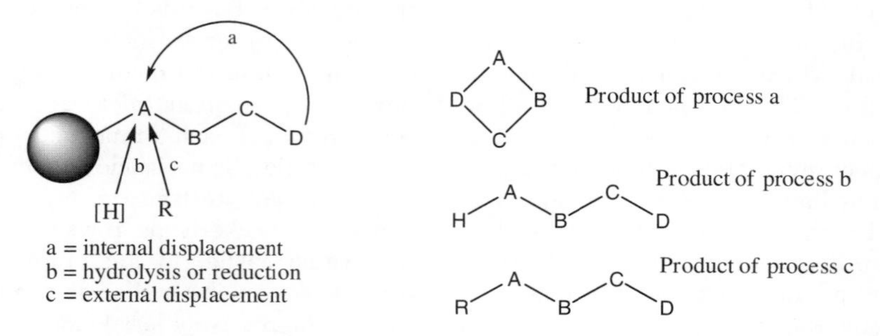

Figure 11. Some strategies for detachment of components from resins and the products produced.

lar method is to embed an rf generator tuned to individual frequencies in the resin itself so that the substance on the bead can be identified by tuning to the proper frequency [72]. This method requires significant instrumental resources. Color coding using oil-based organic dyes has also been used to label one-bead, one-compound chemical libraries [101]. The dyes are applied to the beads just before testing for cell-surface receptors which testing can then be done conveniently in mixtures for deconvolution.

Decoding of individual dipeptides present in 384-well microplate wells has also been done by a combination of mass spectroscopy and microflow NMR [102].

It is also possible to decode split and mix peptide arrays by amino acid analysis of the substances on the beads by incorporating an amino acid into two variable positions in the sequence grouped in a unique pair depending on the length of the peptide. This is limited to focused libraries but has been demonstrated in a compound library of 65,536 octapeptides [103].

A low tech method of devolution inspired by the classic work of Pasteur has been published. Here the resins are poured onto a solid surface previously seeded with a lawn of microorganisms or a substrate that generates a color. Those resins that contain an active material give a zone of inhibition or a color response. Both of these end points can be detected visually and the active resins taken off of the surface with tweezers. The active component can then be detached for analysis and identification.

Employment of these various methods simplifies the construction and purification work of the chemist but place a significantly greater burden on the biologist.

Clearly with very large libraries it is technically not possible to analyze each product so one must resort to statistical sampling instead or take it as an article of faith that each compound has been successfully prepared. Much work has been devoted to dealing with this problem but a comprehensive treatment of this complex topic is unfortunately too vast to cover here. Clearly careful rehearsal and fanatic attention to detail in the construction of the library helps but the analysis is conjecture not science. There are relatively few instances in the literature where a careful census has been performed from which to form an opinion of the reliability of this topic. In one important recent example, a statistical sampling of 7.5% of the contents of a library of 25,200 statin-containing pseudo-peptides showed that 85% had the anticipated structure. This is reasonably good for such complex work but leaves one with a sense of unease in that about 3800 wrong substances were probably produced. The wrong structures were not statistically distributed among all of the targeted compounds but rather showed a bias toward certain wrong structures. This is not particularly surprising but underscores the importance of caution when interpreting the results of screening such librries [104].

Initially, combinatorial libraries of peptides consisted primarily of products made from linear combinations of naturally occurring amino acids. Subsequently just about every maneuver employed in ordinary peptide work has been applied to combinatorial studies leading to a persistent evolutionary drift away from collections of natural peptides. For example, to enhance the metabolic stability of such libraries, end capping of the amino end or the carboxy end and also cyclization have been employed in order to stabilize these substances. Before long, the incorporation of unusual amino acids also began (Fig. 12). These substances are called pseudo-peptides [105–131]. More specifically, such residues as benzamidines [132,133], phosphinates [134–136], hydroxyethyleneketones [137], fluoroamides [138], hydroxamates [134,139,140], coumarins [141], oxazoles [142], nitriles [143,144], aldehydes or ketones [145], thiomethyleneketones [146], sulfonamides [147], phosphonates [148], and the like progressively appeared. Inclusion of these unusual residues is of special value when the substitution takes place at a site where the processing enzyme must act and employs mechanism-based inhibitory mechanisms. More recently libraries have appeared in which the overall conformation of the peptide has been replaced by a heterocycle so that the resulting product resembles topographically a beta-turn, for example [149,150]. The

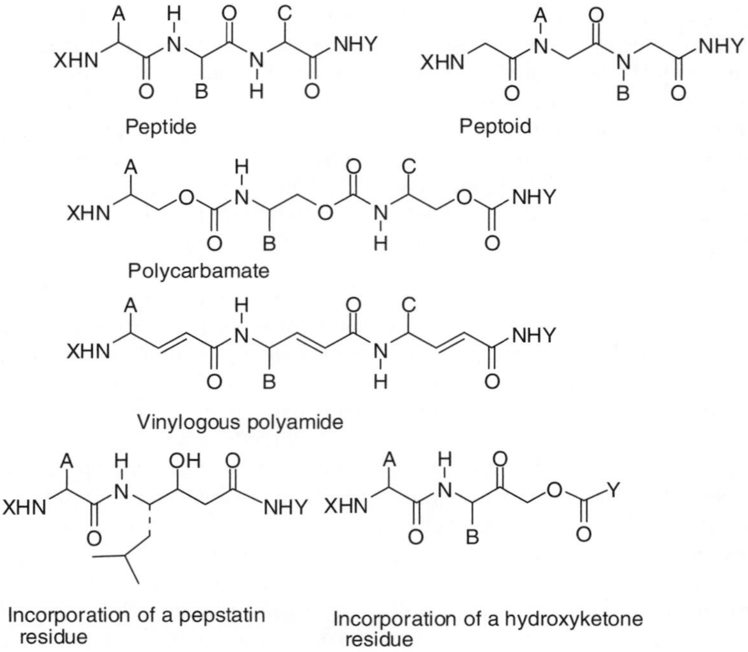

Peptide

Peptoid

Polycarbamate

Vinylogous polyamide

Incorporation of a pepstatin residue

Incorporation of a hydroxyketone residue

Figure 12. Some peptide surrogates employed in compound libraries.

peptidomimetic group need not incorporate any common amino acid components [123,151]. This is at present a very active subfield of medicinal chemistry.

Later, the peptide linkage itself began to be modified (Fig. 13). For example, the classical mode of stabilization against peptidase cleavage by conversion of the peptide bond NH into *N*-methyl subsequently evolved into the preparation of peptoids (polyglycine chains with each NH replaced by a variety of *N*-alkyl groups of the type that resemble the side

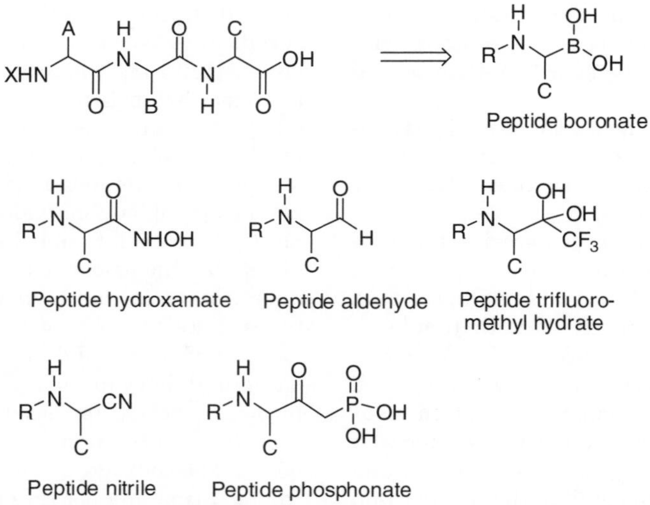

Peptide boronate

Peptide hydroxamate

Peptide aldehyde

Peptide trifluoromethyl hydrate

Peptide nitrile

Peptide phosphonate

Figure 13. Some peptides substituted with unusual carboxyl surrogate residues.

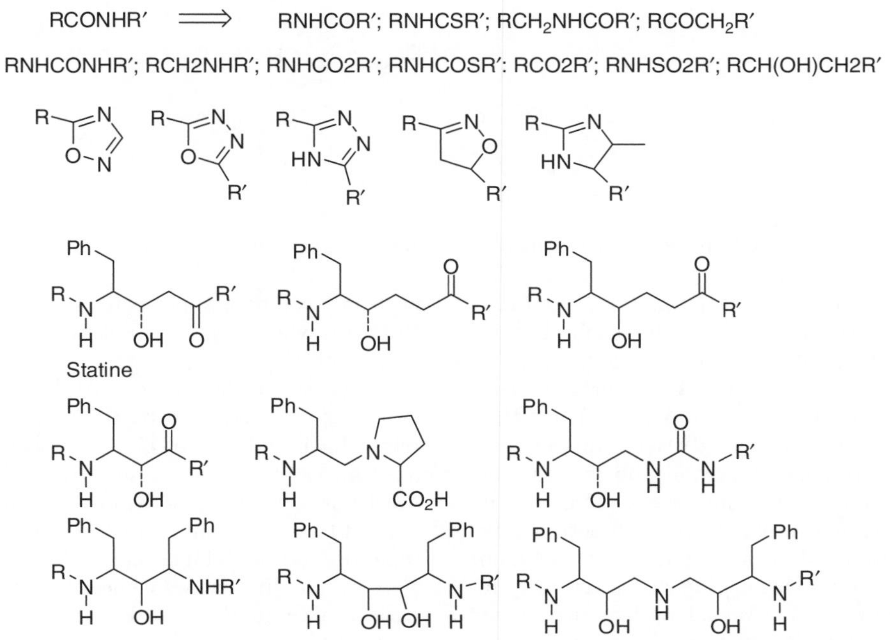

RCONHR′ ⟹ RNHCOR′; RNHCSR′; $RCH_2NHCOR′$; $RCOCH_2R′$

RNHCONHR′; RCH2NHR′; RNHCO2R′; RNHCOSR′: RCO2R′; RNHSO2R′; RCH(OH)CH2R′

Figure 14. Some bioisosteric replacements for peptide bonds that are used in combinatorial chemical compound libraries.

chains found in normal amino acids). Libraries of these compounds were very popular for a while but interest has decreased as time passes [152,153].

Alteration of peptide libraries by reaction of the peptide array with various reagents to transform the components into new substances is a powerful means of expanding the size and scope of the collection. This is known as "libraries from libraries" and the concept is not restricted to peptides [89,154–161].

Flirtation with other substitutes for normal peptide bonds includes preparation of libraries of polycarbamates [153], vinylogous amides, incorporation of a pepstatin residue [162], and ureas [153] as well (Fig. 14). In these libraries the side-chains project from each fourth rather than each third atom in the chain so these are not close models of amino acids. Such libraries, not surprisingly, more often lead to antagonists rather than agonists. Several clinically useful inhibitors of AIDS protease have emerged from studies of this type.

These peptide and pseudo-peptide libraries have been replaced progressively by collections containing smaller and more drug-like molecules. These will be covered in the respective sections below.

In summary, the ability to construct and test libraries containing peptides and peptidomimetics of modest length with known primary structure numbering in the millions is well in hand. This is a spectacular achievement. Some problems remain to be solved including constructing small to medium sized protein analogs of known chemical sequences with nonnatural amino acid residues that fold in dictated ways, and translation of primary and higher order structural features of peptides into smaller, orally active, drug-like molecules. Much effort has been expended in this area but as yet no generally successful stratagem has emerged.

3.2. Oligosaccharide Arrays

Carbohydrates play essential roles in many biological processes such as in immunology, cellular communication, cellular recognition, and adhesion. This alone is sufficient justification to construct their compound libraries.

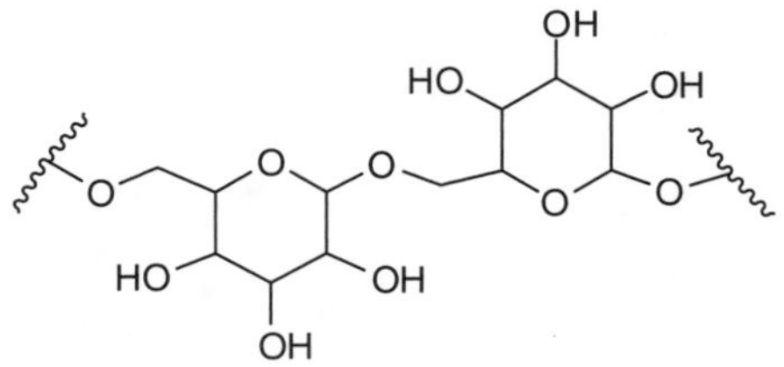

Figure 15. Carbohydrate compound libraries illustrated with normal strings of hexoses.

However, comparatively few synthetic drugs belong to this chemical family (excluding glycosides), reducing some of the motivation. Construction and analysis of diverse oligosaccharide libraries is much more difficult than making peptides. The glycoside linkage contains a stereogenic carbon and is relatively hard to control, the bonds are acid fragile, and there are many potentially competing functional groups that can be points of attachment (Fig. 15). When branching chains are present, their synthesis is often a lengthy and involved undertaking. The affinity of oligosaccharides to their receptors is also comparatively low so exploration of more tightly binding glycopeptides is more common [163]. Despite these complicating factors such libraries are appearing increasingly [163–183].

The early compound libraries of oligosaccharides were comparatively small, were generally not separated into individual compounds and were not well characterized, particularly with respect to their anomeric ratios. The reactants generally consisted of protected glycyl donors with protected or unprotected glycyl acceptors. The unprotected glycyl acceptor libraries consisted of relatively random glycosidation products so were fairly complex mixtures that were very difficult to separate. A seminal library of this type was that of Hindsgaul and coworkers [184,185] that consisted of a mixture of six α-linked positional diglycoside isomers prepared by solution-phase chemistry. An interesting feature was the incorporation of a lipophilic anomeric ether on the glycyl acceptor moiety to facilitate separation of the products. Further, the reactions were stopped at about 30% completion in order to minimize formation of multiple glycosidation products.

An illustrative library of 20 mixed trisaccharide α,β-anomers prepared in solution with each component protected so as to dictate the glycosylation site is that of Boons, Heskamp and Hout [186].

Oligosaccharide libraries are also prepared on beads as illustrated by the 12-membered trisaccharide library of Zhu and Boons [187] and the 1300-membered mixture di- and trisaccharide libraries of Kahne prepared by the split and mix method using halogen isotope encoded beads for decoding (Fig. 15) [165]. The Kahne library was successfully tested with the components still attached to the beads and the active beads were picked out by the Pasteur-like technique.

Convenient automated glycan synthesis, despite these advances, remains elusive. Recent progress has been briefed in Chemical and Engineering News [188].

3.3. Nucleoside Arrays

There are five commonly occurring nucleic acid bases and two sugars. Thus a fully realized compound library would be considerably smaller than those of peptides. The popularity of analogs with unnatural bases or sugars expands the possibilities considerably.

Nucleoside building blocks and oligonucleotides thereof have gained much attention by virtue of their significant biological profiles, including their use in antiviral, anticancer, and antisense, research. This awareness of the important therapeutic utility of nucleoside analogs has stimulated an intensive effort in the search for newer generations of structurally modified analogs, designed to improve their potency and circumvent various physical and biological limitations [189–193]. Traditionally, most nucleoside library construction efforts involved derivatization of the nucleobase component of preformed nucleosides. Employing the above strategy in both

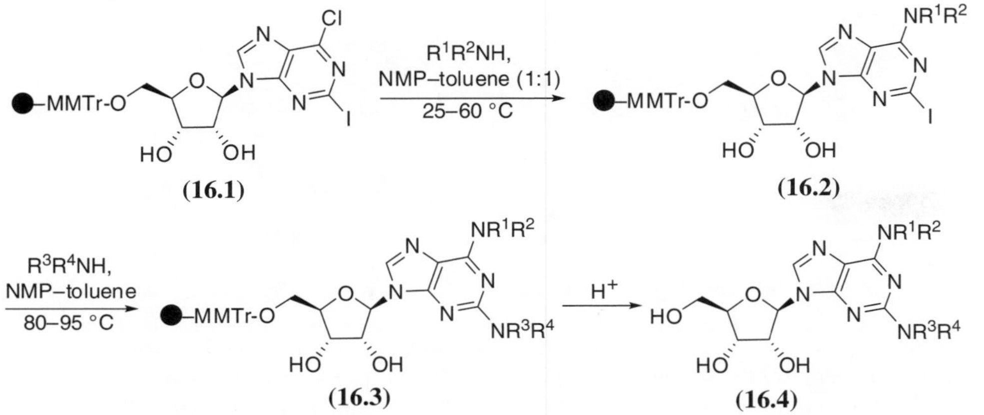

Figure 16. Combinatorial synthesis of N^2,N^6-disubstituted diaminopurine nucleosides.

solution, and solid-phase synthesis, large numbers of nucleoside libraries have been constructed. However, the above approach is somewhat limited in terms of diversification involving the core nucleobase and the carbohydrate framework.

In a representative example, solid-phase synthesis of a combinatorial library containing more than 1300 N^2,N^6-disubstituted purine ribonucleosides, was constructed starting from a polystyrene-methoxytrityl resin (PS-MMTr) bound 2-iodo-6-chloro-9-(β-D-ribofuranosyl)purine, as shown in Fig. 16 [194]. Sequential, regioselective displacement of the 6-chloro substituent with a variety of primary and secondary amines, followed by displacement of the 2-iodo group with a large pool of primary amines and final cleavage from the solid support resulted in the desired nucleosides.

In a strategically different solution-phase approach, involving microwave-accelerated reaction of various natural and nonnatural nucleobases (amines) with a suitably functionalized ribofuranosyl derivative under Vorbrüggen condition led to the rapid preparation of a structurally diverse nucleoside library (Fig. 17) [195]. In comparison to standard glycosidation conditions, use of microwave heating in the above reaction resulted in a significant reduction in reaction time. Quenching of the excess Lewis acidic reagent with triethanol amine in aq. acetonitrile allowed for direct high-throughput MPLC purification of the nucleoside adduct containing reaction mixture. Following the above method, combination of six different acylated ribofuranosyl donors with a variety of amine nucleophiles resulted in the formation of a library of more than 400 nucleosides.

3.4. Lipid Arrays

Lipid libraries have largely been neglected. For saturated fatty acids, the construction of carbon–carbon bonds and ester linkages is not difficult but this area of combinatorial chemistry is underdeveloped. Mixed triglycerides

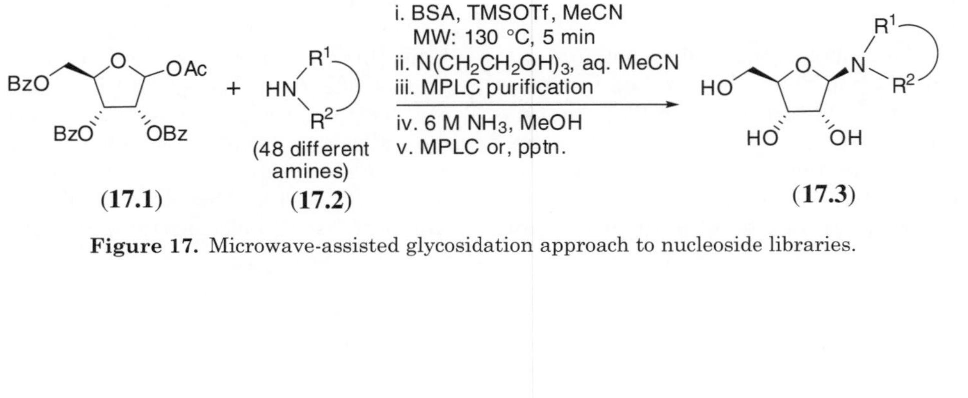

Figure 17. Microwave-assisted glycosidation approach to nucleoside libraries.

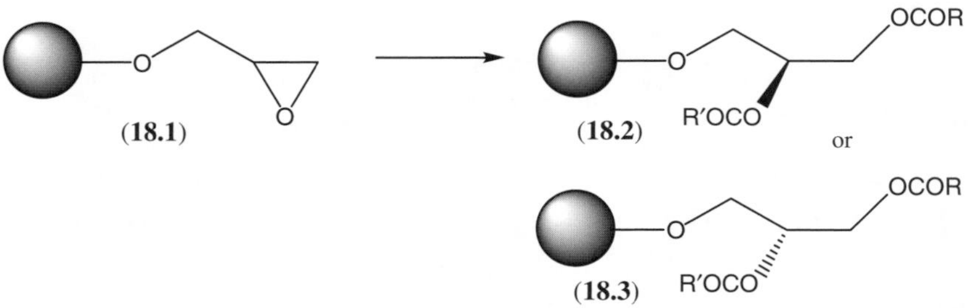

Figure 18. Combinatorial library synthesis of chiral lipids.

are reasonably accessible, particularly starting with oxiranemethanol to solve the regioselectivity problems [196]. Fig. 18 illustrates the preparation of chiral triglycerides by use of carboxylic acid-promoted ring-opening reactions of a glycidyl-bound resin (**18.1**). This leads to a polymer-bound monoacylglycerol that can be further converted to a chiral resin-bound diacylglycerol (**18.2** or **18.3**) by reaction with a second carboxylic acid moiety under configuration retaining or inversion conditions. The free diacylglycerol is released by BCl_3 and the remaining hydroxyl is available for further substitution. Synthesis of compound libraries containing phosphoglyceride, cephalin, sphingolipid, and lecithin residues would be more complex but surely would be doable.

Steroids [197–199] and other polyisoprenoids [200] have been the centroids for combinatorial libraries however they suffer from having significant molecular weights so minimal diversity can be accomplished with them with the intention of achieving oral activity. They are also fairly flat molecules so have a diminished capacity to project functional groups into three dimensional space.

4. SOLID- AND SOLUTION-PHASE LIBRARIES OF SMALL, DRUGGABLE MOLECULES

As has been shown, the field of combinatorial chemistry and multiple parallel synthesis started with libraries of peptides. In time unusual residues crept into the products and other informational macromolecule types were included in libraries. While this evolution is still ongoing, despite a major investment in equipment and personnel, few drugs of this type have emerged so the primary effort by medicinal chemists is now devoted to production of libraries of small, drug-like molecules instead. Many of the methods used for large molecules can carry over but the largely noniterative nature of small molecule synthesis is a significant complication and preparation and decoding is mostly done rather differently.

4.1. The Mythical Universal Screening Compound Library

The ideal drug-seeking library need contain only one substance. This would not really be a library in any case and no one knows how to do this. At the other extreme, a universal screening compound library would be capable of generating a pursuable lead in any screening program and might consist of a great many individual compounds. No such libraries exist. There are a great many approaches to making realistic compound libraries. In many cases compound filters are employed to generate special purpose screening libraries as subsets of massive legacy collections. An ideal practical screening compound library should contain only as many components as would produce a pursuable lead using available high-throughput screening methods. This is also a tall order.

The current "gold standard" in small molecule drug seeking is oral activity accompanied by one-a-day dosing. This is a high hurdle. The majority of passively transported molecules that have passed the filtering process have molecular weights of about 500 or less [9,10]. It has been calculated that the number of small molecules that would fall into this category is approximately 10^{62}![201] Clearly preparing all of these in reasonable time is beyond the capacity of the entire population of the earth even if they worked tirelessly. The number of compounds that can become satisfactory drugs encompassed in this impossible collection is probably in the range of a few thousand so most of the effort of producing such a compound library would be wasted. A further consideration is the estimate that druggable targets in the genome consist of something like 2000–3000 genes and gene products [202]. Hence medicinal chemical skills in mating up these hosts and guests are still at a premium [13].

Initial nonpeptide drug seeking programs utilizing combinatorial/multiple parallel synthesis coupled with high-throughput synthesis followed the traditional mode of finding a low micromolar potency or better lead whose structure appeared to be tractable. Disappointing productivity from this approach has been addressed in a number of ways. One of the more popular has come to be known as fragment-based drug discovery [203]. In this approach, one starts with smaller molecules (maximum molecular weight 300 or less) and accepts potencies rather weaker than would be pursued classically (often in the millimolar range). This requires very sensitive methods of detection so NMR, X-ray, or surface plasmon resonance (SPR) methodologies have been employed for this purpose. These techniques ideally also reveal some of the molecular features in the neighborhood of the binding site of this fragment to its receptor. In order to enhance potency and selectivity one then builds out from this fragment in an attempt to occupy binding sites in the vicinity that are not utilized in the initial host–guest relationship. This by itself is not particularly different from classical medicinal chemistry. In ideal cases one finds two small fragments that occupy sites near one another and then links them together by a suitable arm. This should, after further optimization, then bring the potency down into the conventional micro- or nanomolar range. This approach is often being taken along with rather than instead of classical screening and appears to be particularly suitable for receptors about which little structural information is available in advance and for which lead substances are not known. Thus it is particularly attractive for exploiting orphan receptors arising from genomic approaches.

As molecular weight increases the number of structural possibilities goes up rapidly. Use of fragment-based drug discovery techniques enables the medicinal chemist to utilize smaller screening compound libraries containing quite small molecules and thereby to explore a larger percentage of the potentially available biodiversity. It is thus particularly attractive to laboratories not possessing large legacy libraries.

Specific examples of successful drug seeking using fragment-based drug design will be presented later in this account.

4.2. The Molecular Diversity Versus Druggability Conundrum

A clear lesson from the poor results achieved in the early days of combinatorial drug seeking is that molecular diversity alone is insufficient for medicinal chemical purposes. Since molecular diversity greatly exceeds druggability, structural filters have become popular in order to deal with this.

4.3. Drug-Like Characteristics and Privileged Structures

The most frequently employed guidelines for drug-like character are the Lipinski rules of five [9,10] and the Veber rule of low polar surface area/rotatable bonds [204]. A significant modification of the Lipinski rules of five for screening hits taking into account the almost irresistible urge to add molecular weight and functional groups in lead evolution leads to the Congreve rule of three [205].

Examination of the structures of successful orally active drugs demonstrates that the molecular weight distribution is roughly Gaussian with the average range of about 300–399 being most densely populated. Larger molecules diffuse more slowly than smaller ones possibly rationalizing this observation. The calculated $\log P$ leads to values centering in a Gaussian distribution peaking at roughly 2 (100-fold more lipophilic than hydrophilic). Since drugs must pass from an aqueous environment into and through a lipid cell wall environment, this also appears to be logical. Since water solubility requires hydration and passage into or through lipid membranes requires dehydration, the rules that a molecule should have not more than 5 hydrogen bonding functional groups nor more than 10 hydrogen bond accepting functional groups is also understandable as this is a rough measure of hydrogen bonding and dehydration should not require excessive energy. Since most biological fluids in the body have a pH of about 7.4, molecules that are not ionized under these conditions are also favored for uptake by passive diffusion. The Lipinski and Congreve rules codify these considerations. It is also understandable that comparatively compact molecules without extensive surface distribution of charged moieties would also be favored so the Veber rules are also readily rationalized.

It is important to emphasize that these rules apply almost exclusively to potential drugs that are absorbed by passive diffusion. Some significant drugs enter cells by active processes instead. The interaction of such molecules with their macromolecular transporters requires strict structural compatibility so the range of permissible structural variation is much more limited. The Lipinski, Congreve, and Veber rules rarely apply to these substances.

Another method for prejudging the potential utility of a prospective compound library for screening purposes involves privileged structures [11,206]. The term refers to structural units that are found to be ligands to a variety of apparently different receptors and, in addition, to be reasonably able to pass through lipid membranes. Benzodiazepines serve as a well-known example. Using privileged structures as centroids for liganding is often found to produce attractive hits for further elaboration.

Fortunately chemists are well aware of the contributions to molecular properties various substituents are likely to produce so the potential quality of libraries can often be anticipated successfully by "eyeballing" the structures.

Slavish adherence to these guidelines will likely result in missing some useful starting compounds still they present ample latitude for production of large arrays of compounds for testing. By their intrinsic nature, screens have holes in them. One cannot therefore despair overmuch about possibly missing some compounds but rather must rejoice at the great ones one does find.

4.4. Structural Features Largely to be Avoided

Since drugs must survive for a reasonable period of time in a largely aqueous milieu containing a significant number of other molecules, it is obvious that significantly reactive functional groups should be avoided. Examples of bad actors are α-halo ketones and acyl and sulfonyl halides. Aldehydes are often excluded because of their tendency to react under mild conditions with primary amines to form Schiff's bases and Mannich products.

Toxicophores are groupings that are often associated with serious side effects. Consequently to be avoided are such groups as aromatic nitro groups (whose reduction products are often toxic) and anilines (whose oxidation products are likewise often toxic). Also compounds containing Michael acceptors and thiols that bind tightly to proteins and nucleic acids are mostly to be avoided.

4.5. Synthesis of Large Arrays Versus Pulsed-Iterative Compound Libraries

It is all too easy to "turn the crank" and produce large numbers of related molecules once a synthetic process has been developed. After a series of molecules has been found that has attractive if not privileged properties this can be very useful in defending one's discovery through comprehensive patents. Mostly, however, this is a seductive practice that can lead to a significant waste of time. A great many iterations of a worthless lead is a waste of time and resources.

Most laboratories now work in a pulsed mode preparing repeated series of smaller libraries. Each successive iteration is guided by the biological properties of the preceding one(s). This is a form of chemical evolution that enables experienced investigators to find the most attractive embodiments of a lead molecule rapidly and with minimal effort and expense.

4.6. Scaffolds, Centroids, and Adornments

The scaffold or centroid is the repeating part of molecules to be included in a combinatorial library. Adornments are attached to it in order to produce the various analogs. A centroid dictates the nature of the chemistry that can be employed and largely determines the diversity points that can be employed and so the ultimate success or failure of the compound library is significantly dependent upon a judicious choice of centroid [207,208]. A centroid should have a molecular weight as small as possible so that a wide range of adornments can be added to it without producing molecules of excessive molecular weight. Ideally it should also contain functional groups or reveal them as the library develops so that there is a reasonable possibility that its structure will contribute to receptor binding during testing. In this way its molecular weight contribution is not a total loss.

It is also important that the centroid project functionalizable arms into as many quadrants or octants about it as possible so that productive interactions with functional groups on the receptor be enabled.

The examples provided in Section 4.10 illustrate the application of these important concepts in drug seeking.

4.7. Stereochemical Aspects

Receptors are usually chiral so one might suppose that the molecules in a drug-seeking library should be complementarily chiral as well. Not so, at least in the early stages. It is common to find that one enantiomer (the eutomer) possesses stronger affinity and therefore potency than the other (the distomer). Resolution of an enantiomeric mixture can only result in doubling the activity if only one chiral center is present. In hit seeking, the potency loss suffered from using racemates is well worth it considering that if one's library contained only the distomer, one might miss the activity of the molecule entirely. When more than one chiral center is present, one may incur an even greater risk of missing the activity. In this case, as many epimers as possible should be present in the library. Chirality influences can readily be determined after one has gotten a hit.

4.8. Natural Products and Natural Product Mimetics

Mother Nature has a much greater structural imagination than most synthetic chemists have. The molecules of nature are generally more complex than those commonly designed for inclusion in combinatorial libraries. Just as is the case with synthetic libraries, the structural diversity of natural products is much greater than is their known pharmacological utility. Nevertheless it is clear that biota expend considerable energy and resources in their elaboration and those that have useful biological activity have often been coevolutionarily optimized over the centuries. Cytotoxic agents such as taxol come to mind in this context. These may be regarded as defensive or even aggressive substances of value to the producing organism. Clearly, however, the producing organism has different needs than do humans. Thus optimization of pharmacokinetic and toxicological properties of such agents is a worthwhile endeavor.

It is also instructive to consider the extremely remote likelihood of a molecule such as taxol being included in a drug-seeking compound library prepared entirely by chemical synthesis. On this basis, ideal screening libraries should contain at least a sprinkling of natural products as well as entirely synthetic substances.

Despite the comparatively large number of natural products that have become drugs or that have inspired the creation of drugs [206–219], combinatorial chemical enthusiasts initially believed that a flood of new products would emerge by multiple parallel synthesis and a number of firms became convinced that this would happen and disbanded their natural product efforts. In the light of subsequent experience this is now being rethought [13].

A number of structure–activity studies have resulted from the construction of compound libraries from fairly abundant natural products. These include as examples libraries from andrographolide [209], beauveriolide [210], santonin [211], taxol [212–214], vancomycin [215,216], flavonoids [217], shikimic acid [218], quinolizinones [219], usnic acid [220], indole alkaloids [221], and yohimbine [222].

A number of laboratories have constructed natural product-like libraries [223–228]. An early example exemplifies this approach wherein this concept has been married to privileged structures and to libraries from libraries inspired by the common appearance of 2,2-dimethylbenzopyrane moieties in nature [224–226]. It is too early as yet to judge the comparative success rate of combinatorial approaches utilizing synthesis of chemist-inspired compounds, natural product containing or inspired or mixed compound libraries.

Combinatorial biosynthesis is a related research area in which knowledge of biosynthetic pathways coupled with genetic engineering allows the production of unnatural analogs of natural products. These man made molecules have yet to be found in nature and indeed may never be found as natural products. Polyketides, such as the macrolide antibiotics, are commercially valuable substances. Their total chemical synthesis has been accomplished but this is a lengthy and expensive process of no known present commercial value. Much more practical is the combinatorial biosynthesis in which the gene cassettes normally present in microbes are partially disabled and then new gene segments not normally present in the pathway are introduced. This can provide novel molecules that are difficult to access using partial or totally synthetic means [229–237].

4.9. Reverse Chemical Genetics and the NIH Roadmap

A process that involves finding probes for proteins that have been uncovered by genomic mining but whose biological function is unknown is called the reverse chemical genetic approach [227,228,238–247]. Here newly synthesized chemical libraries containing diverse structures are screened for their ability to activate or inactivate various biological pathways. The identity of their protein partner is then ascertained. This process is likely to be particularly useful in uncovering the function of proteins newly uncovered through genome mapping. Many of these small molecule ligands have been patterned loosely after natural products but are densely functionalized and lack oral drug-like characteristics but they provide plausible starting points for determining the suitability of the partner protein as a drug target, and for subsequent drug seeking campaigns through analoging.

This technology has been a significant part of the inspiration for a major initiative by the National Institutes of Health, United States, in the form of the NIH Roadmap [248]. A recent summary of this program can be accessed electronically via http://nihroadmap.nih.gov/overview.asp and/or http://nihroadmap.nih.gov. A major investment has been made in this in an attempt to accelerate medical research by pursuing new avenues toward drugs by organizing interdisciplinary academic teams that would otherwise be working independently. These widely share progress as soon as it has been made, and do clinical research in a different way. Relevant to this review is that centers for innovative small molecule synthesis methods development, with an emphasis on combinatorial chemical methods, and for screening have been established in academia. Also part of the initiative is the formation of Molecular Libraries Screening Centers, the NIH Molecular Libraries Screening Center Network, the NIH Chemical Genomics Center, and the NIH Molecular Libraries Small Molecule Repository. The compounds in the latter are available to multiple NIH Screening centers. Associated with this is a publically available source of information in which the structures and their biological activities are placed into PubChem (pubchem.ncbi.nlm.nih.gov). The compound collections also include legacy materials obtained both from governmental and private sources.

This para departmental mode of integrative research is familiar to industrial scientists but less so to academics. This melding of basically antithetic cultures is rather novel. The central thrust of classic academia is individual investigator initiated work with an imperative to be creative and to share the results widely as soon as they are clearly realized. The industrial mode is to organize teams of experts, set a central theme, and to keep the progress and results secret (for patent purposes) until the material is ready for commercial exploitation or has been abandoned. A marriage between these two disparate styles requires significant accommodations but is clearly consistent with the intent of the roadmap. The central question is whether the pace of useful discovery in medicine can be forced by altering the organization and practice of drug seeking in this way.

4.10. Examples of Drug-Like Small Molecule Libraries

Obviously construction one at a time in the usual iterative or noniterative fashion will result in single molecules of a defined nature. Reaction of A and B produces a single product A-B. Reaction of this product with another substance produces product A-B-C. In each case, a single reaction produces a single product. The change brought about by combinatorial or multiple parallel synthesis methods is that the reactions are usually linear but the products can be logarithmic. For example, reacting A with 10 different Bs produces 10 products ($A\text{-}B_{1-10}$) and the reaction of these with 10 different Cs results in 100 products ($A\text{-}B_{1-10}\text{-}C_{1-10}$) either in mixtures or as discrete compounds. Rather large compound collections can be assembled quickly using this scheme.

A starting material (often called a centroid) possessing a number of different functional groups (preferably with different degrees of reactivity—known as orthogonality) can be reacted with a variety of substituents (often called adornments) to produce a large number of analogs. Reaction with two, three or four such functional groups and given the number of possible variants, this can lead to a very large library of analogs in a brief time (two functional groups with 10 variant adornments quickly results in 100 analogs, three in 1000 and four in 10,000). If the reaction conditions allow, these variations can be run in mixture or in parallel resulting in a very significant time savings. It can, however, put a significant strain on purification analysis, record keeping, budgets, and so on. Much work has been expended in addressing these potential limitations.

With 500 or so the normal practical upper molecular weight limit for oral activity, it can be seen that centroid A should be chosen to have the smallest practical molecular weight so that a variety of groups can be added to it. It is also helpful if, when fully adorned, it has exposed functional groups that can interact productively with a receptor so the weight devoted to this part of the molecule is not a net loss. The molecular weight of the centroid places practical limits on the net weight that the adornments can collectively have. If one adornment is rather large, then this requires one or more of the other adornments to be made compensatingly smaller.

It is particularly helpful if the adornments project into space into octants that fit precisely the needs of the receptor. Alternately, if one is hit seeking, they should project into various octants about the centroid so as to allow a fruitful exploration of potential receptor needs. In hit seeking, one often wants reasonable molecular flexibility whereas in lead optimization progressive rigidification is often more effective since reduced entropy of association often favors greater potency with rigid molecules.

In addition, the adornments must have the usual medicinal chemical characteristics [13]. They should not be chemically reactive, convey toxicity, or be inordinately polar or nonpolar.

Whether the centroid should be tethered to a solid support or be free in solution must be considered carefully. If tethered, it is important to consider whether the point of attachment will remain in the final analog and, if so, what affect this may have on its biological properties. One should consider whether this attachment will prevent the use of one of the potential adornment points. The beads must be inert to the reaction conditions to be employed as must the attachment arm. For the presently popular iterative small molecule libraries, the extra steps required for bead processes (attachment and detachment and also the possible constraints involved in decoding) are often a disincentive. Balanced against this is comparative ease of purifica-

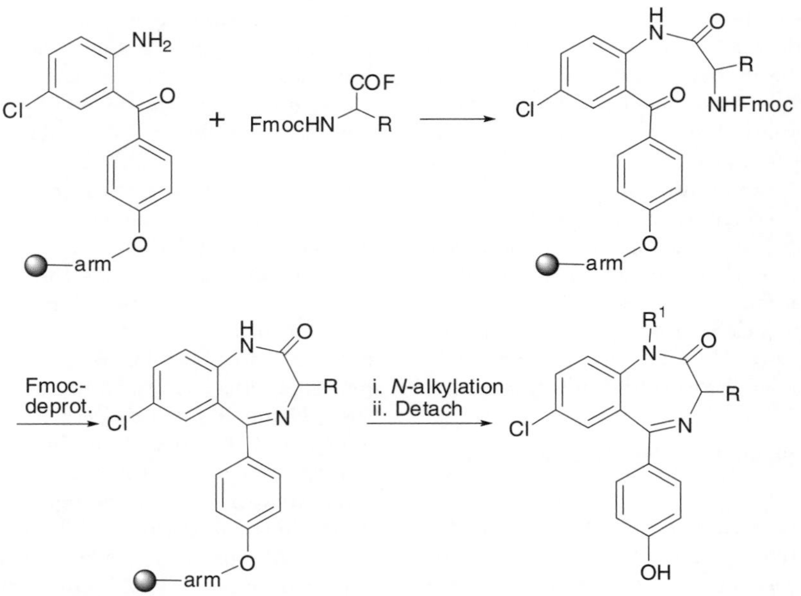

Figure 19. The Bunin and Ellman 1992 synthesis of 1,4-benzodiazepine-2-ones leaving a traceable linker attachment point.

tion. On the other hand, much progress has been made in mixed applications that combine the advantages of solution chemistry and bead processes without retaining most of their individual disadvantages.

Just as is the case with one at a time synthesis, linear combinatorial syntheses are the most risky and produce the lowest yields. Converging methodologies address these limitations successfully and, in combinatorial work, Ugi (four-component) and Passerini (three-component) reactions are very flexible and popular. Generally one has less control over the specific products being produced by such reactions but this is largely compensated for by the rapidity of the process and molecular diversity available in this way.

Clearly a great deal of thought should go into library design before the work begins. Many thousands of compound libraries have now been published embodying these concepts. It is not possible or desirable to treat them all so the remainder of this review contains a representative sampling illustrating the above concepts and revealing contemporarily productive approaches in a variety of pharmacologically promising test systems.

4.10.1. Benzodiazepines The first compound libraries containing heterocycles recognizable as being like orally active drugs were the 1,4-benzodiazepirie-2-ones prepared on resins by Bunin and Ellman in 1992 (Fig. 19) [4]. A notable feature of this reaction sequence is the use of amino acid fluorides to drive the amide formation to completion. Varying the point of attachment of the hydroxyl group would lead to multiple additional analogs. Indeed, a variant of this process resulted in a traceable linker in the other aromatic ring (Fig. 20). In yet another variation the use of stannanes and palladium acylations (Stille coupling) was employed successfully (Fig. 21) [249].

The choice of benzodiazepines was inspired because of the medicinal importance of these materials and their resemblance to peptides. The benzodiazepine scaffold mimics important aspects of peptide β-turns and have served as potentially useful ligands for GABA, cholecystokinin, opioids, vasopressin, *Ras*-farnesyltransferase, angiotensin-converting enzyme, fibrinogen, and so on. In the Bunin–Ellman work, the library was constructed by a sequential combination of three components

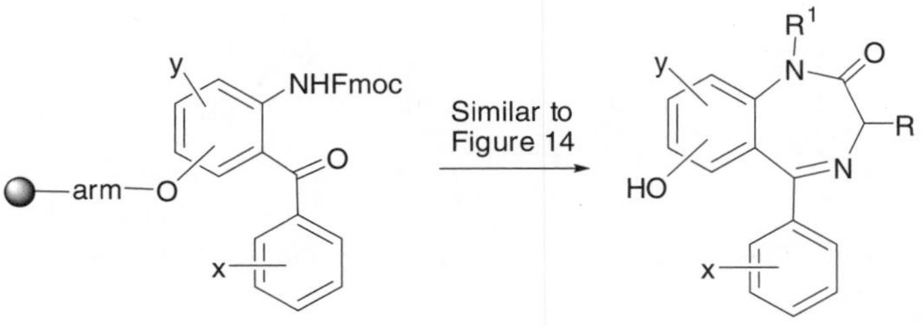

Figure 20. An alternate synthesis of 1,4-benzodiazepine-2-ones leaving a traceable linker attachment point.

so in principle very large arrays could have been assembled in short order. Being selective in the variations incorporated, a focused library of reasonable size was made that answered specific pharmacological questions. In this library the attached groups project into space at widely separated compass points around the molecule allowing a systematic exploration of receptor requirements. The centroid has a molecular weight of 160 when all of the available substitution points are occupied by hydrogen atoms. If one accepts an upper molecular weight limit of approximately 500, then four variations can occupy 340 atomic mass units (500 − 160) so each adornment can have an average of about 85 amu, if the weight is evenly distributed. This gives significant latitude for substitution. When chosen with care to convey drug-like properties and not to exceed collectively the guidelines that Lipinski has developed, the library can contain primarily substances that have a chance to be drug. They also can be so chosen that they allow for substitution independently of each other and also to be installed without premature separation from the beads. Centroids derived from molecular series that are known to be associated with good pharmacokinetics are often referred to as privileged molecules. Thus the choice of benzodiazepines to demonstrate the potential power of combinatorial chemistry and multiple parallel synthesis for medicinal chemical purposes involving drug-like molecules was excellent.

There is however in retrospect a defect in these pioneering libraries in that the final products were attached to the bead support through a phenolic hydroxyl group that remained as such in the products prior to testing. These initial libraries were pharmacologically less than completely satisfying because agents intended to penetrate well into the CNS should not usually contain such a polar substituent. Furthermore, phenolic hydroxyls are not optimal in drug seeking in that conjugative metabolism in the form of glucuronidation or sulfation often decreases the bioavailability of agents containing this function. Despite this, several components in these libraries were bioactive and the work drew widespread attention to the promise of the methodology and was soon followed by a flood of applications to the preparation of drug-like molecules.

From a medicinal chemistry viewpoint a somewhat more versatile synthesis of the schemes employed in Figs. 19–21 remedied that potential liability of leaving a polar group behind in the products. In this sequence, attachment to the resin arm through an aryl silicon bond led to benzodiazepine moieties upon acid cleavage from the resin (Fig. 22) that did not leave a polar substituent behind at the point of attachment [250]. The

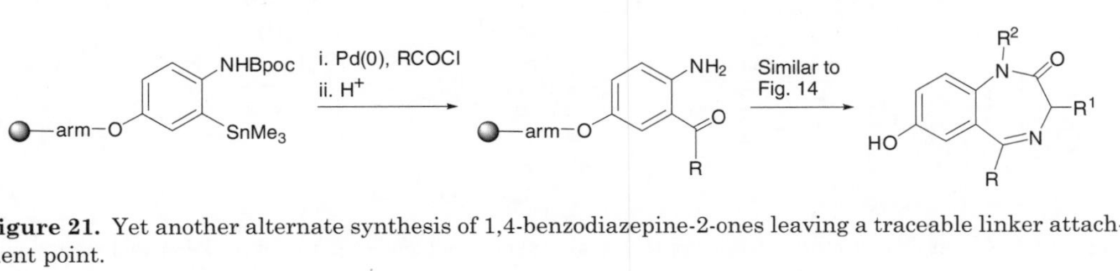

Figure 21. Yet another alternate synthesis of 1,4-benzodiazepine-2-ones leaving a traceable linker attachment point.

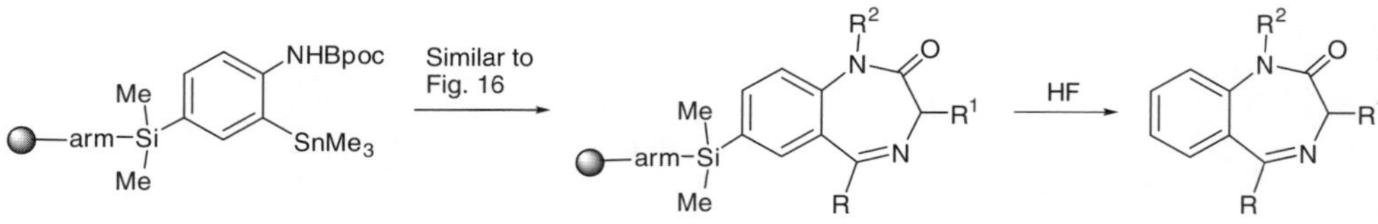

Figure 22. A synthesis of 1,4-benzodiazepine-2-ones that does not leave a traceable linker attachment point.

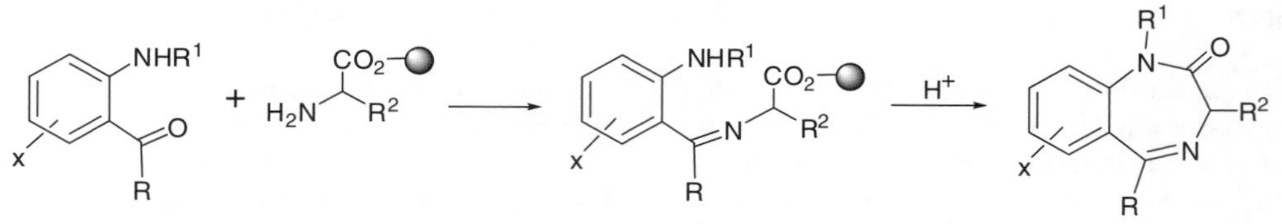

Figure 23. An alternate synthesis of 1,4-benzodiazepine-2-ones that does not leave a traceable linker attachment point.

"traceless linker" technology so introduced has now become standard. Furthermore, the products were now indistinguishable from benzodiazepines prepared by usual speed analoging methods.

Ellman's group also developed a traceless linker sequence of a different type based upon HF release of an aryl silicon link to the resin (Fig. 23) [251].

As it happens somewhat gratifyingly, testing of these agents revealed no structure–activity surprises. The intense study of the benzodiazepines in the empirical earlier years had apparently not missed much of significance. Nonetheless, these studies resulted in convincing proof that combinatorial chemical methods would be of dramatic use in preparing agents likely to become orally active drugs.

Benzodiazepine libraries [252] and close analogs such as 1,4-benzodiazepin-2,5-diones [253,254] continue to be popular. For example, one such library is prepared by a sequence involving a Borch reduction and an internal ester-amide exchange to form the seven-numbered ring followed by cleavage from the resin. The reaction conditions are mild enough to preserve the optical activity in this library (Fig. 24) [255]. Use of peptoid starting materials and reductive traceless linker technology are features of the approach taken by Zuckermanri et al. (Fig. 25) [254].

4.10.2. Dihydropyridine Analogs Representative of another important class of drugs is the 1,4-dihydropyridines. Hantzsch methodology (without the oxidative step) works efficiently on resins for this purpose. The conditions are mild in this synthesis and the yields are good (Fig. 26) [256].

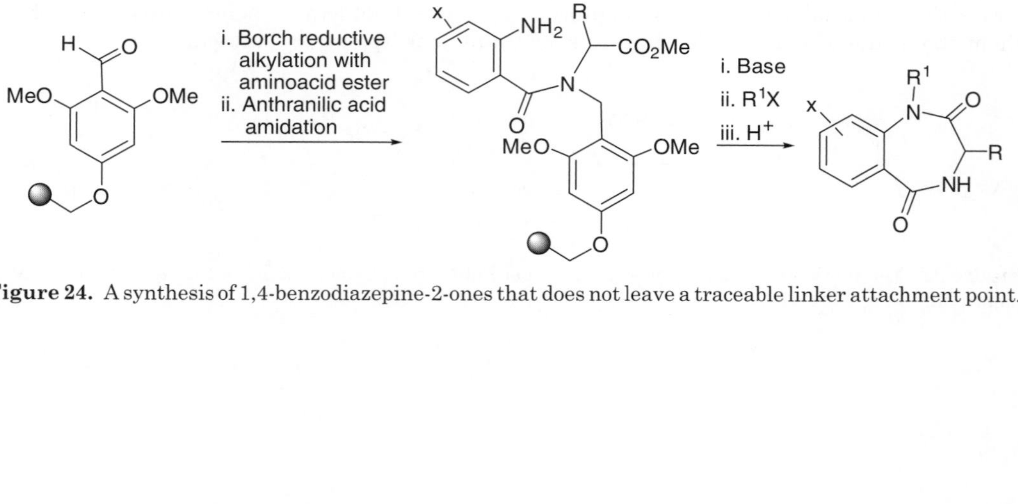

Figure 24. A synthesis of 1,4-benzodiazepine-2-ones that does not leave a traceable linker attachment point.

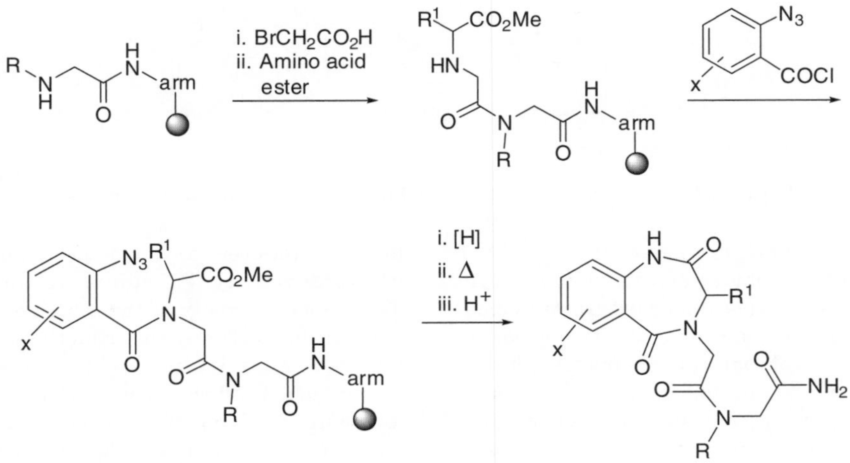

Figure 25. A Zuckermann synthesis of 1,4-benzodiazepine-2,5-diones that does not leave a traceable linker attachment point.

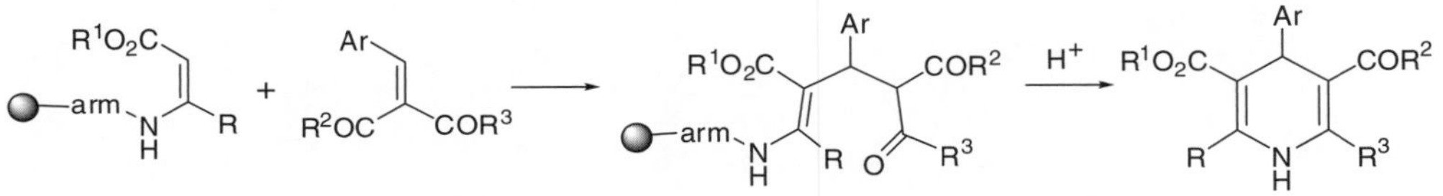

Figure 26. A combinatorial synthesis of dihydropyridine analogs.

4.10.3. Pseudo-Peptide Analogs Related to Captopril Angiotensin-converting enzyme inhibitors are million dollar molecules. An interesting library of captopril analogs was prepared on resin using a split-mix iterative resynthesis deconvolution procedure. From a collection of about 500 analogs, an analog emerged that was threefold more potent than captopril itself and possessed a K_i of 160 pM (Fig. 27)! [257]

4.10.4. Fluoroquinolone Anti-Infectives Another important class of contemporary drugs that have been made in library form, in this case both in solution and in solid state, are the fluoroquinolone antimicrobial agents (Fig. 28) [258,259]. The solution-based yields are superior to those obtained on the resins.

4.10.5. Oxazolidinone Anti-Infectives Recently, focused libraries have explored the SAR properties of series of contemporary interest where no drugs have yet emerged or where only the first of a promising series has been marketed. An example of this is the oxazolidinones. One example of this class,

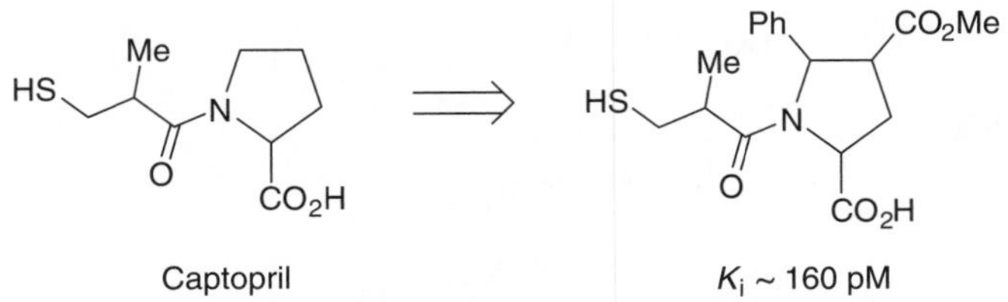

Figure 27. Synthesis of a compound library inspired by captopril.

Figure 28. Synthesis of a combinatorial library of fluoroquinolone analogs.

linezolid, has been marketed as an orally effective antimicrobial agent and most large firms have extensive analog programs in place in attempts to improve on its properties (Fig. 29). Combinatorial chemistry plays a significant role in this work. A Pharmacia group has shown that alteration of the morpholine function to a methylated pyrazole moiety produces a broad-spectrum analog with oral activity. Palladium coupling of the iodoaromatic moiety of the starting material allows construction of the trimethylsilylacetylene side chain. Hydrolysis of this group with formic acid produces a methyl ketone that, following a mixed aldol reaction to the dimethylenamine, reacts with methylhydrazine

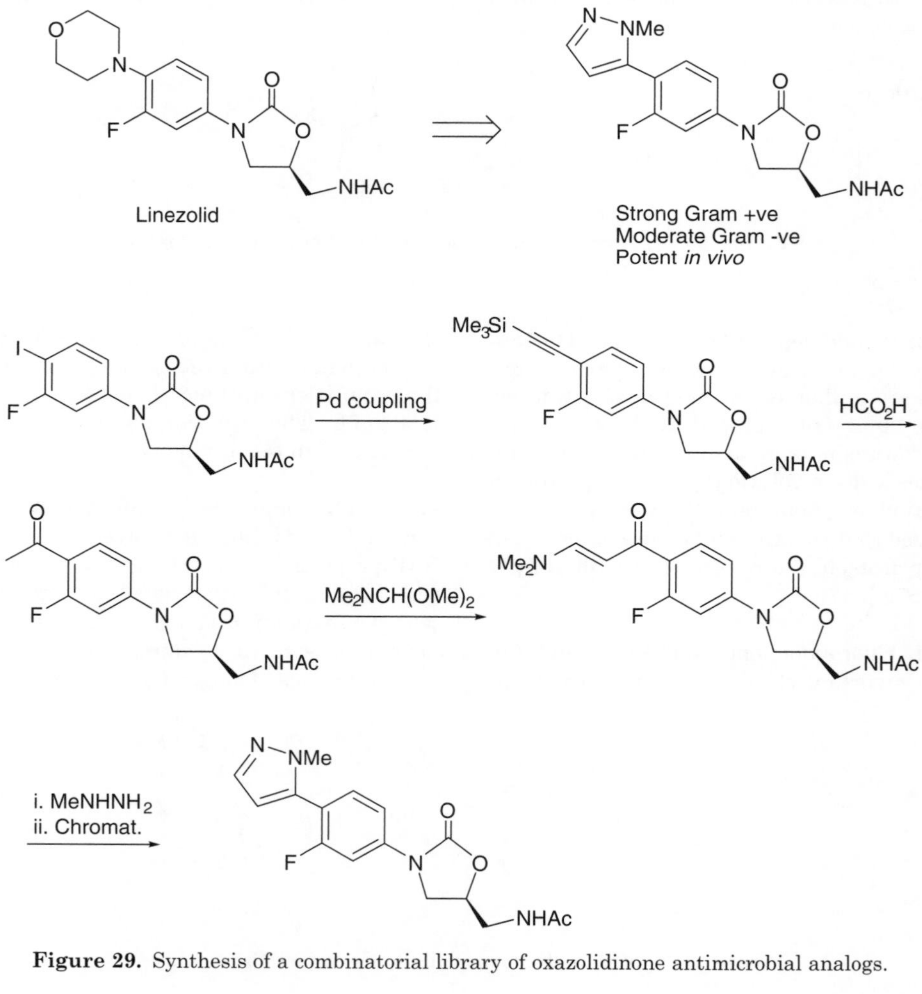

Figure 29. Synthesis of a combinatorial library of oxazolidinone antimicrobial analogs.

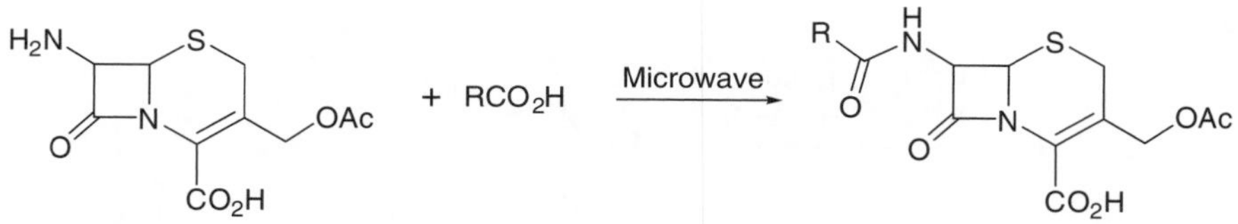

Figure 30. Microwave-assisted synthesis of a library of cephalosporin antibiotics.

to produce a mixture of the two possible methylated pyrazole isomers. These are separated by chromatography to produce the best of a large series of analogs produced by these and other reaction sequences. The product illustrated in Fig. 29 has the best combination of *in vitro* and *in vivo* properties in this grouping [260]. Reports of related studies have also appeared [261,262].

4.10.6. Cephalosporins A library of cephalosporin antibiotic analogs was made on a solid support (basic alumina) without requiring protection–deprotection. The compounds were prepared in high yield (82–93%) and purity in about 2 min with the aid of microwave irradiation (Fig. 30) [263,264].

4.10.7. Hit Seeking Libraries The small molecule libraries just exemplified belong to the class called focused. That is, in each case a discrete molecular target was available at the outset and chemical routes were generally available. Following some adaptation to the needs of the method and rehearsal of the chemistry libraries could be generated relatively quickly. Many analogs were then available by comparatively simple variations in the reactants employed. Clearly, in drug seeking one can operate in much the same manner following identification of a suitable hit molecule.

The strategy required in hit seeking, however, is often rather different. Here the initial libraries are usually bigger and more diverse. After the library is screened and useful molecules are uncovered, subsequent refining libraries are employed that are progressively smaller and more focused. Each succeeding library benefits from the information gained in the previous work so this can be considered the chemist's equivalent of biological evolution. As the work progresses the need for quantities of material for evaluation increase so the work usually proceeds back into the classical larger scale one-at-a-time mode.

Some examples illustrate the approach commonly followed in this manner. First, consider the discovery and progression of OC 144-093, an orally active modulator of P-glycoprotein-mediated multiple drug resistance that has entered clinical studies. Initially, a 500-membered library of variously substituted imidazoles was prepared on a mixture of aldehyde and amine beads (Fig. 31). The choice of materials was based upon prior knowledge of the structures of other F-glycoprotein modulators. Screening this library in whole cells led to the identification of two main leads, **31.1**, possessing an IC_{50} of 600 nM, and **31.2**, possessing an IC_{50} of 80 nM. In addition, B possessed an oral bioavailability in dogs of about 35%. These results were very encouraging.

The third stage involved making a solution-based library based upon the structures of **31.1** and **31.2**. Screening produced leads **31.3**, possessing an IC_{50} of 300 nM, and **31.4**, with an IC_{50} of 150 nM. Interestingly, **31.4** was found as an unexpected by-product. In addition to reasonable potency, **31.4** showed enhanced metabolic stability so it was chosen as the lead for the next phase. Analoging around structure **31.4** led ultimately to OC 144-093 with an IC_{50} of 50 nM and an estimated 60% bioavailability following oral administration to man [264,265].

Later biological studies *in vitro* and *in vivo* have shown that OC 144-093 enhances the activity of paclitaxel by interfering with its export by P-glycoprotein. It is not a substrate for CYP3A and does not interfere with paclitaxel metabolism unless at comparatively high doses. Following IV administration, OC 144-093 does not interfere with paclitaxel's pharmacokinetic profile but elevates its area under the curve when given orally. The results

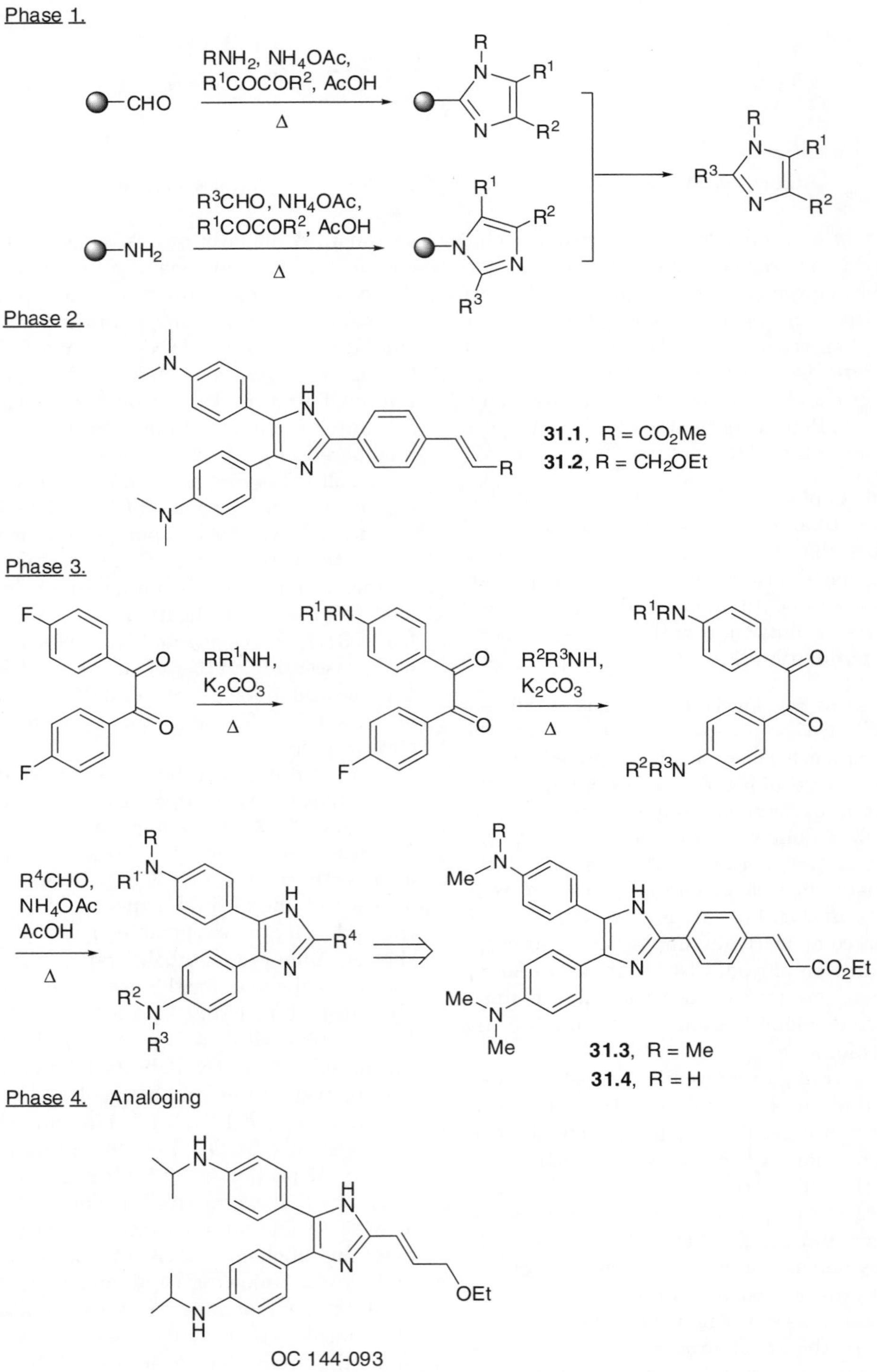

Figure 31. A compound library campaign leading to OC 144-093.

are interpreted as being due to interference with gut P-glycoprotein. Further studies are in progress and it is hoped that a marketed anticancer adjunct will emerge in due course [266].

In a different study, a search through a company compound collection was made in an attempt to find an inhibitor of the Em family of methyltransferases. These bacterial enzymes produce resistance to the widely used macrolide-lincosaminide-streptogramin B family of antibiotics by catalyzing *S*-adenosylmethionine-based methylation of a specific adenine residue in 23S bacterial ribosomes. This interferes with the binding of the antibiotics and conveys resistance to them. Using NMR (SAR by NMR) screening, a series of compounds including, 1,3-diamino-5-thiomethyltriazine, were found to bind to the active site of the enzyme, albeit weakly (1.0 mM for the triazine named) (Fig. 32). Analogs were retrieved from the compound collection and **32.1**, **32.2**, and **32.3** were identified as promising for further work. A solution-phase parallel synthesis study was performed from which compound **32.4** emerged as being significantly potent. Next a 232-compound library was prepared in order to discover the best R group on the left side of compound **32.4**. From this, compounds **32.5** and **32.6** emerged. These were now potent in the single digit micromolar range. The left side of analog **32.5** was fixed and the right side was investigated through a 411-membered library. From this, **32.7** emerged as the best substance with a K_i of 4 μM against Erm-AM and 10 μM against ErmC. Thus, starting with a very weak lead with a malleable structure, successive libraries produced analogs with quite significant potency for further exploration [267].

It has just been a decade after small molecule drug seeking via combinatorial chemistry began in earnest yet already most of the common drug series and hundreds of different heterocyclic classes have already been prepared in library form. Originally the emphasis had been upon bead-based chemistry and this actually slowed general acceptance of the method because few organic and medicinal chemists were familiar at that time with the techniques needed to make small molecules on beads through noniterative methods and, indeed, much of the needed technology had yet to be developed and disseminated. These problems have largely been overcome and today the choice of beads or not is partly a matter of taste and of the size of the libraries to be made and the length of the reaction sequences required.

Motilin Receptor Antagonists A continuing problem complicating the study of novel inhibitors of peptides is the difficulty of predicting the conformation these ligands adopt in the presence of their receptors. Conformational restriction is one of the successful, though chancy and laborious, means of dealing with this. Joining the ends of a peptide hormone together by inserting a spacer unit between them has been successful in a number of cases. Joining the ends together without a spacer unit is unlikely to work as the extended form of the peptide will probably be poorly mimicked by this. The proper choice of the linker insertion is therefore vital. An example of a compound library utilizing this approach involves motilin analogs.

Motilin is a 22-mer peptide hormone secreted by the M cells of the small intestine. It stimulates the rhythmic motions of the gut that propel GI contents along to the exit. It also stimulates the production of pepsin. Hypermotility of the GI track is uncomfortable leading to cramps or even diarrhea so inhibitors of the interaction between motilin and its receptors (in the antrum, duodenum, and proximal small intestine) are of interest. This problem has been addressed using combinatorial technology (see Fig. 33) [268]. Screening of a large compound library in a whole cell assay resulted in identification of a hit containing an 18-membered ring that possessed three amino acid residues in the ring that were joined by a tether unit (**33.1**) To optimize the action of this lead, a variety of protected tripeptides were affixed to a resin through a thiol bond giving **33.2**. A tether unit was added to the protected terminal amino acid residue by a Fukuyama-Mitsunobu sequence leading to **33.3**. The amino group on the tether unit was deprotected and the resulting intermediate was cyclized with the silver salt of trifluoroacetic acid releasing the product from the resin and forming the ring. Deprotection completed the process forming **33.4**. A modification of

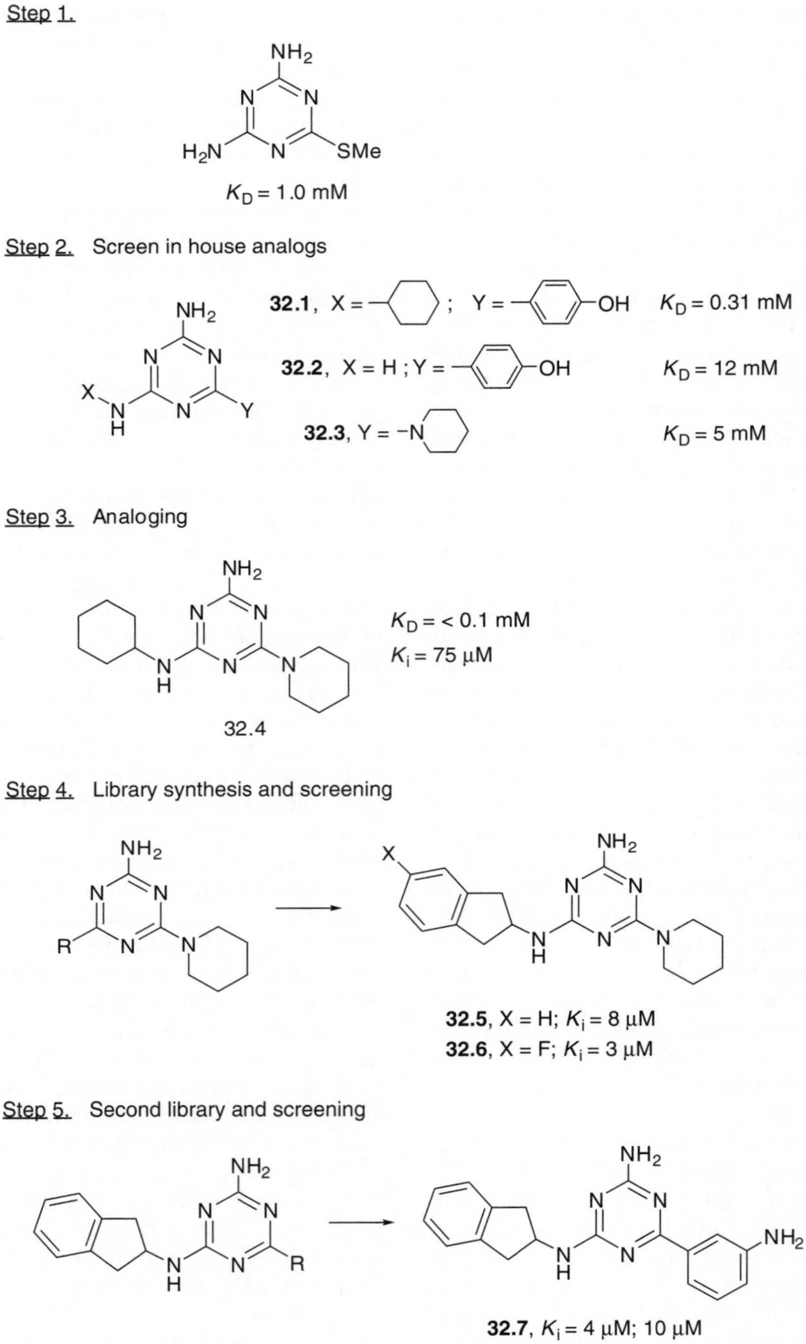

Step 1.

K_D = 1.0 mM

Step 2. Screen in house analogs

32.1, X = cyclohexyl; Y = 4-hydroxyphenyl K_D = 0.31 mM

32.2, X = H; Y = 4-hydroxyphenyl K_D = 12 mM

32.3, Y = –N(piperidinyl) K_D = 5 mM

Step 3. Analoging

K_D = < 0.1 mM

K_i = 75 μM

32.4

Step 4. Library synthesis and screening

32.5, X = H; K_i = 8 μM

32.6, X = F; K_i = 3 μM

Step 5. Second library and screening

32.7, K_i = 4 μM; 10 μM

Figure 32. A combinatorial library based search for inhibitors of Em family methyltransferases.

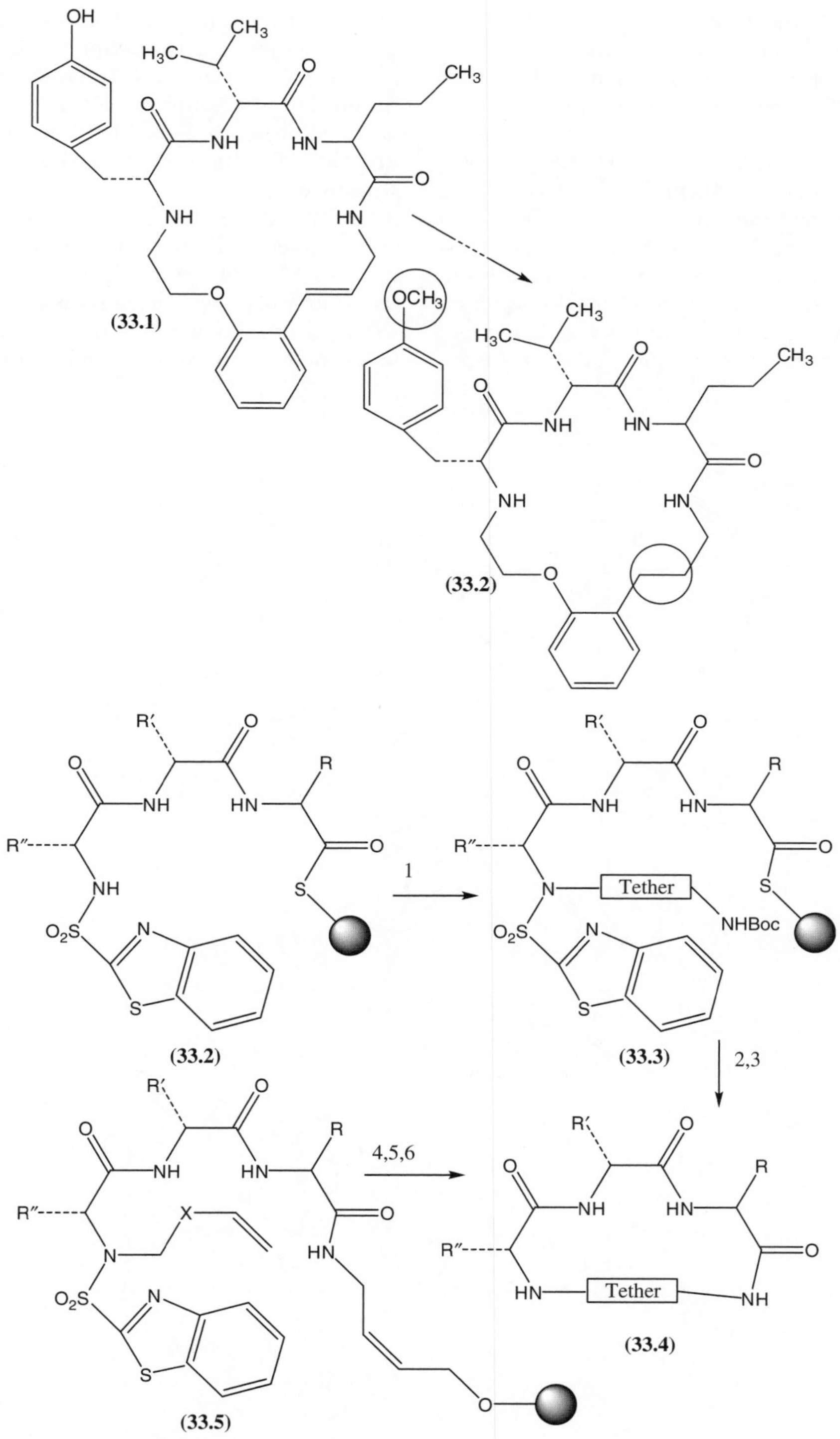

Reagents: 1, PPh_3, DIAD; 2, AgTFA, DIPEA; 3, deprotection; 4, RCM; 5, deprotection, 6, optionally, hydrogenaton.

Figure 33. Synthesis of a compound library of motilin receptor antagonists.

this process involved a different mode of attachment to the resin allowing cyclization-detachment by a ring forming metathesis reaction (**33.5**). Systematic modification of the amino acid units (including their absolute configuration) and of the interspersed tether finally led to an 8 nM inhibitor (**33.4**).

HIV-1 Protease Inhibitors In the AIDS virus infection cycle, the viral protein components are produced as an inactive larger protein. This must subsequently be cleaved into functional smaller peptide units before being assembled into complete, infectious, viral products. Of the various approaches to disabling the virus and thus extending the life of HIV sufferers, one of the most successful has been finding specific inhibitors of this symmetrical aspartyl protease. Several of these inhibitors are in clinical use by AIDS sufferers worldwide.

In an attempt to find structurally novel HIV protease inhibitors by combinatorial means (Fig. 34), a preliminary structure-based approach was attempted starting with *in silico* docking studies using amprenavir (**34.1**) as a comparator. Based on the results

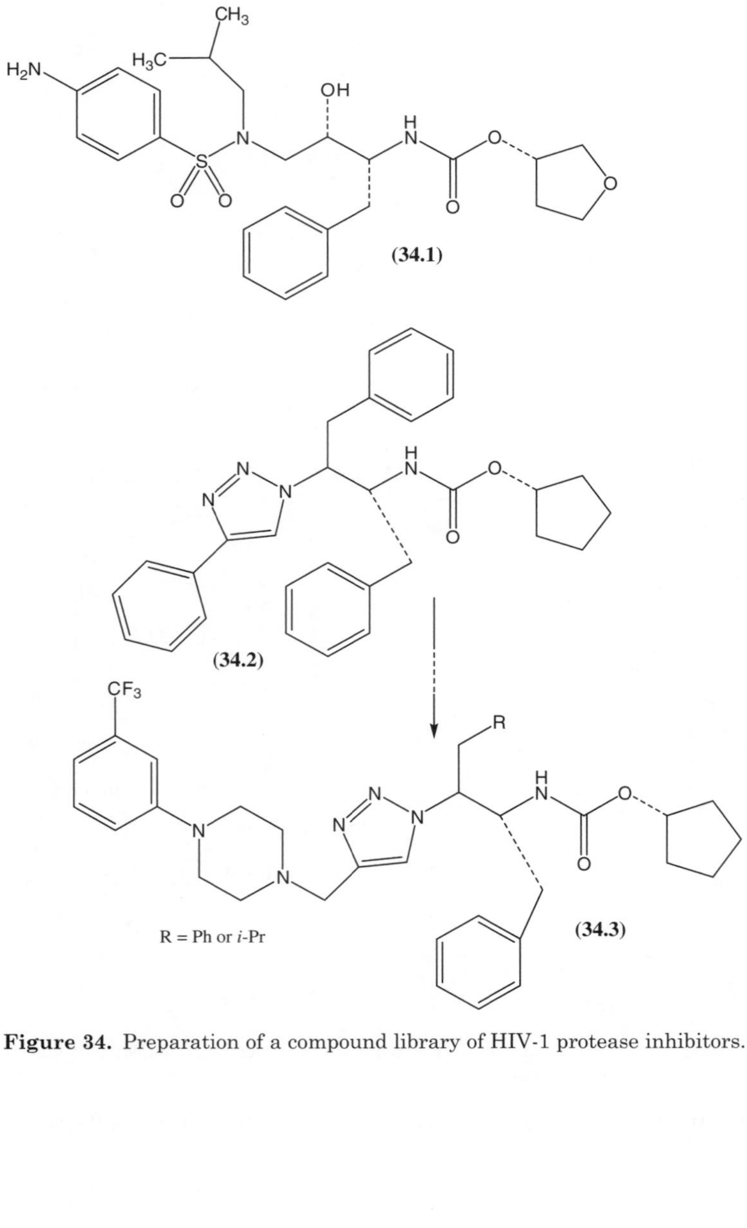

H_2N CH_3 H_3C OH H N O S O O N O O

(34.1)

N N N H N O O

(34.2)

CF_3 R H N N N N O N N O

R = Ph or *i*-Pr

(34.3)

Figure 34. Preparation of a compound library of HIV-1 protease inhibitors.

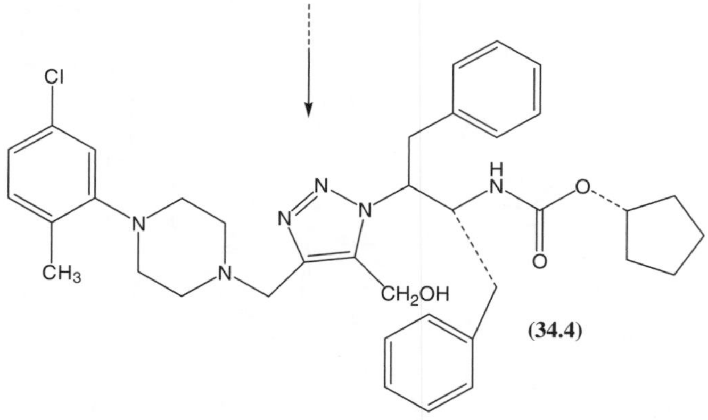

(34.4)

Figure 34. (*Continued*)

obtained from a preliminary set of azides, compound **34.2** was judged to be a promising starting point for further elaboration [269]. Consequently, a focused compound library [270] was prepared from which compounds **34.3** emerged as most potent. Further elaboration of the attachments to the triazole ring produced analogs with low nanomolar potency of which **34.4** is representative. Whereas the classical HIV protease inhibitors involve a chiral hydroxyl as a key pharmacophoric moiety, in these libraries the triazole moiety instead binds in the active site of the enzyme. The incorporation of the tetrazole moiety in place of the secondary hydroxyl present in most of the precedent HIV protease inhibitors has the advantage that inactivation by P450-mediated oxidation of the OH cannot take place.

Antagonists of 5-Hydroxytryptamine 2A Receptors These receptors are potentially important mediators of the central and peripheral responses to serotonin so antagonists are of potential value in the treatment of depression, schizophrenia, insomnia, and Parkinsonism. Among the class 2 serotonin receptors, the more abundant 2A receptors are presently of greatest interest. Using a FLIPR-based assay, a group of legacy library small molecules were identified and evaluated by computer means (Fig. 35) leading to compound **35.1** as a suitable starting point for drug seeking (35 nM IC_{50}) [271]. Parallel synthesis ultimately led to **35.2** that had good selectivity in a receptor panel and 35-fold enhanced potency against HT receptor 2A.

One route to these interesting analogs started with arylisoxazoles (**35.3**) that were alkylated to (**35.4**). Next ring opening with ethylenediamine (**35.5**) was followed by ring formation producing **35.6** containing three diversity points. A second route started with **35.7** that was converted to the desired iminoethers.

Carbonic Anhydrase II Inhibitors Carbonic anhydrase is a family of zinc containing isoenzymes that convert water and carbon dioxide to bicarbonate and protons. This serves to maintain the proper balance of acids and bases in body fluids and to clear carbon dioxide from tissues. They also are involved in generation of HCl in parietal cells for use in maintaining gastric acidity. These enzymes are extremely rapid in their forward action, the reaction being limited by the diffusion rate of its substrates, but the reverse reaction is comparatively slow. Inhibitors of carbonic anhydrase have a number of clinical applications including hypertension (by diuresis), altitude sickness, glaucoma, obesity, and so on. Sulfonamides are prominent elements in a number of inhibitors of carbonic anhydrases. A combinatorial approach to potential inhibitors involved application of a fragment-based methodology with alkene cross metathesis as a central design element (Fig. 36) [272]. Fragments **36.1** and **36.2** were combined in a variety of iterations using an immobilized heterogeneous Grubbs first-generation catalyst to

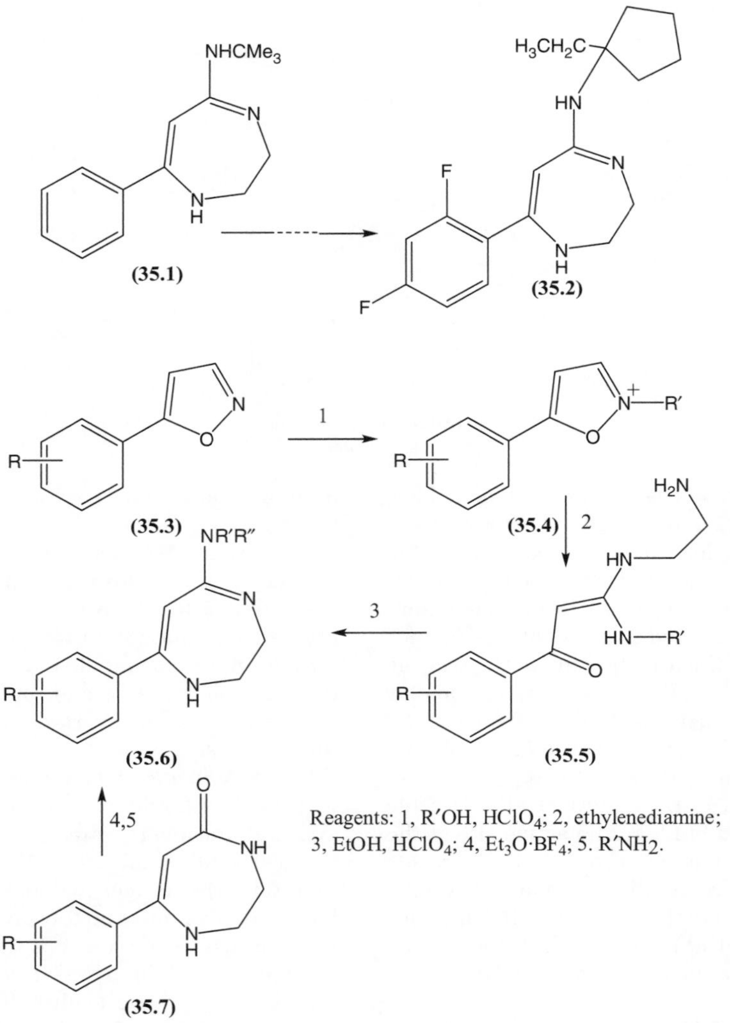

Reagents: 1, R′OH, $HClO_4$; 2, ethylenediamine; 3, EtOH, $HClO_4$; 4, $Et_3O{\cdot}BF_4$; 5. $R'NH_2$.

Figure 35. Preparation of a compound library of %-HT 2A antagonists.

give **36.3**. Ultimately 4.9 nM inhibitor **36.4** emerged from this study.

B-Cell Lymphocyte/Leukemia-2 (Bcl-2) Inhibitors Cells are intended to die by apoptosis when they become unrepairable or live as long as intended. The Bcl family of proteins regulates the apoptotic pathway and are over expressed in most human cancers leading to undesired cell "immortalization." Consequently they have become popular targets for chemotherapy. A balance of pro- and antiapoptotic interactions is characteristic of mature cells. The antiapoptotic proteins designated Bcl-2 are typical in that they have a BH3 dimerization domain. Small molecule inhibitors of this protein–protein interaction should, then, return cells to normal lifetimes and allow apoptosis to take place.

Interference by small molecules with protein–protein interactions is difficult to bring about. A combinatorial approach to this difficult task has been published applying a fragment-based strategy using SAR by NMR coupled with parallel synthesis as the central

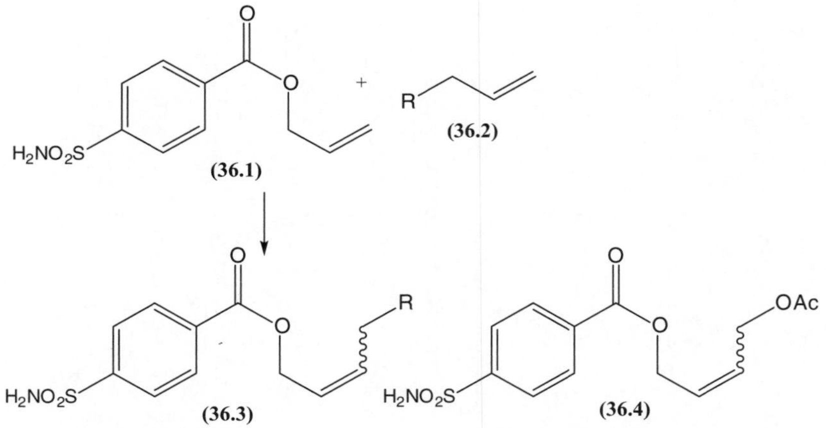

O
O
+
R
H_2NO_2S
(36.1)
(36.2)
O
R
O
H_2NO_2S
(36.3)
O
OAc
O
H_2NO_2S
(36.4)

Figure 36. Preparation of a compound library of carbonic anhydrase II inhibitors.

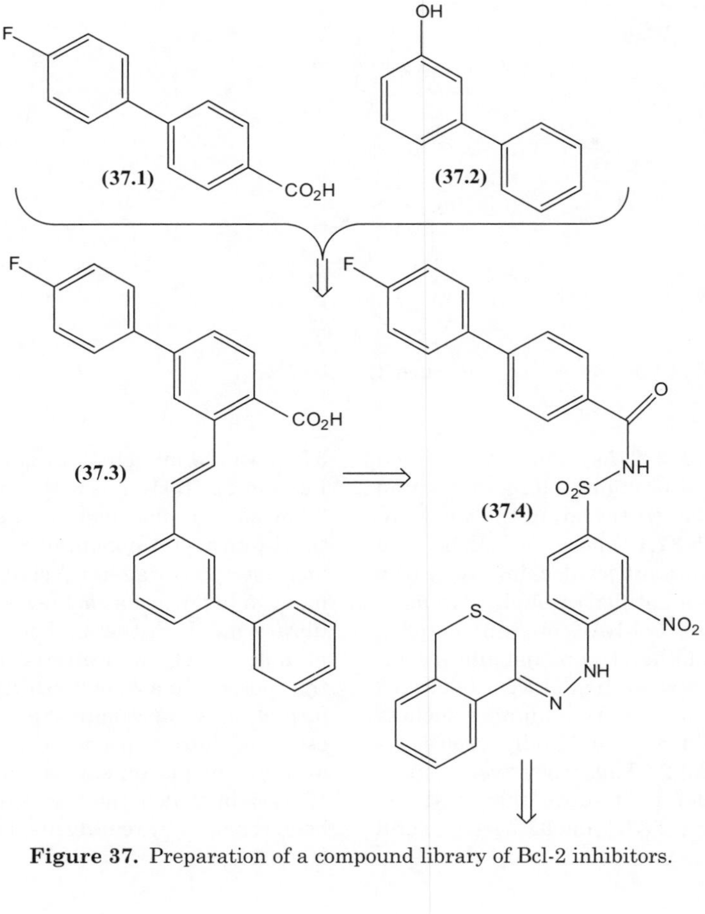

F
OH
(37.1)
CO_2H
(37.2)
F
F
CO_2H
O
(37.3)
NH
O_2S
(37.4)
S
NO_2
NH
N

Figure 37. Preparation of a compound library of Bcl-2 inhibitors.

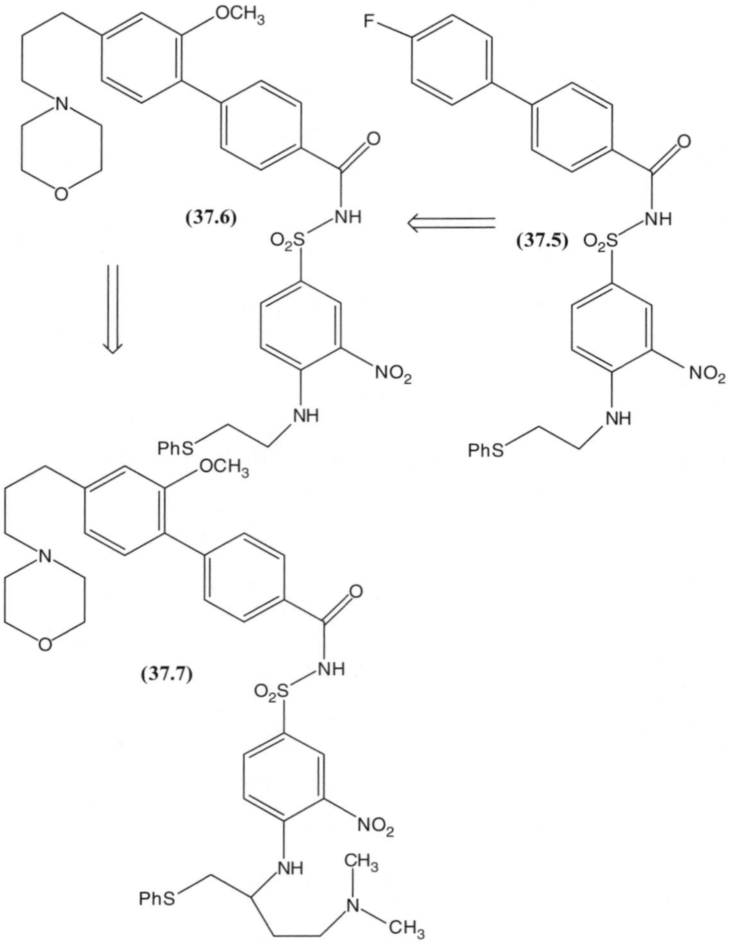

(37.6) (37.5) (37.7)

Figure 37. *(Continued)*

tool (Fig. 37) [273–275]. Using a modified form of the target proteins, a large compound library was screened revealing the weak (K_i 300 μM) ligand **37.1** whose carboxylic acid moiety was important for docking. A nearby binding site was identified through NMR measurements of the Ccl-BAK protein complex. Screening of another compound library revealed **37.2** as a weak ligand ($K_i = 5000$ μM) for this second site. Weak inhibitors such as these two would be individually useless as inhibitors of Bcl-2. Thus they were joined together in order to improve their activity. Synthetic linking of **37.1** and **37.2** led to olefin **37.3** as a low micromolar ligand ($K_i = 1.4$ μM). Further NMR studies with this complex were followed by replacement of the carboxyl function by an aryl sulfonamide moiety (a common bioisosteric stratagem). A significant improvement in potency was achieved by preparing a library in which the appendage to the sulfonamide moiety was altered leading to **37.4** that possessed a K_i of 245 nM. Further transformation was accomplished through a modest sized library prepared by a nucleophilic aromatic displacement reaction on an analog of **37.5** in which the hydrazine moiety had been replaced by readily displaceable chlorine substituent. Of this library, **37.6** emerged possessing a $K_i = 2.5\,\mu M$. Unfortunately, however, **37.6** also bound strongly to human serum albumen-III. To separate these, another compound library was generated with the finding that **37.7** (and an analog not illustrated here) possessed low μM potency against Bcl and this was not dramatically raised in the presence of human serum albumen. Adding an arm terminating in a tertiary amino moiety was responsible for further increased potency and decreased serum protein binding.

Compound (**37.7**) was tested against a human nonsmall cell lung carcinoma cell line known to over express Bcl with the result that cytotoxicity was induced and the substance potentiated the effect of taxol when the two were given simultaneously. More excitingly, the compound showed enhanced activity when used along with taxol in a murine A549 xenograft tumor model.

Caspase 3 Inhibitors Another approach to modulating apoptotic pathways involves finding selective inhibitors of the cysteinyl-aspartate proteases called caspases. Caspase 3 is the most prominent of the more than a dozen known caspases and in particular it is important in proteolysis of cytoskeleton proteins, kinases, and proteins involved in repair of DNA. Early approaches to inhibitors involved incorporation of aldehyde and nitrile moieties at the cleavage site in order to inhibit the enzyme. However, aldehydes usually do not make useful drugs and the resulting potency and selectivity was poor. A particular tetrapeptide segment helped with selectivity but did not produce inhibitors with good PK properties. Subsequently (Fig. 38) it was found that a shorter segment (**38.1**) was helpful and then that this could be reduced even further by incorporating a heterocyclic moiety. This subsequently served as the inspiration for a combinatorial library based upon a key aminopyrazinone segment (**38.2**) that mimicked a dipeptide segment in molecular modeling studies [276]. This was attached to resin-bound synthon **38.3** to give **38.4**. Detachment of **38.4** from the resin followed by deprotection produced library components **38.5**. Use of a furazan terminus to interact with a pocket in the enzyme produced a series of low nanomolar IC_{50} inhibitors. Solution-phase focused libraries further optimized interactions and led to selective caspase inhibitor M867 (**38.6**) ($K_i = 0.1\,nM$) against human caspase 3 and having good whole cell activity.

Dopamine D3 Receptor Ligands The action of ligands at the dopamine D3 receptors are of potential value in the treatment of Parkinsonism, schizophrenia, and substance abuse. One lead, BP 897 (**39.1**), is particularly interesting since it inhibits cocaine-seeking behavior without any apparent abuse potential itself. Further examination of this agent via resin-based combinatorial methodologies (Fig. 39) led to **39.2** possessing an order of magnitude enhanced potency and greater selectivity than BP 897 itself [277]. A complex aldehyde resin **39.6** was assembled by a click chemistry reaction by reacting indole aldehyde **39.3** with resin-bound azide **39.4** leading to resin-bound aldehyde synthon **39.5**. A primary amine terminus containing much of the linker arm of BP 897 was then reductively attached leading to **39.6** and this intermediate was removed form the resin and then acylated with a variety of activated aryl carboxylic acids to give **39.7**. *N*-Deprotection was followed by *N*-arylation using Buchwald chemistry. The best of a library constructed in this manner was **39.2**.

Methionine Aminopeptidase-2 (MetAP-2) Inhibitors Interest in inhibition of MetAP-2 metallopeptidases derives from their role in angiogenesis. This was heightened when the natural product fumagillin was shown to be an irreversible inhibitor of the enzyme and to play a significant role in inhibiting tumor growth *in vivo*. Reversible inhibitors of the enzyme (and MetAP-1) include bestatin and amastatin. Tumor masses require a vascularization in order to serve the metabolic needs of tumor masses, especially when they become large. As a consequence considerable interest has developed in finding selective antiangiogenic agents that would likely possess anticancer properties. MetAP-2 promotes protein maturation by removing N-terminal methionine residues and inhibitors should have useful activity.

X-ray study of MetAP-2 showed that significant room was available for ligands about the P1′ pocket and that exploiting this might

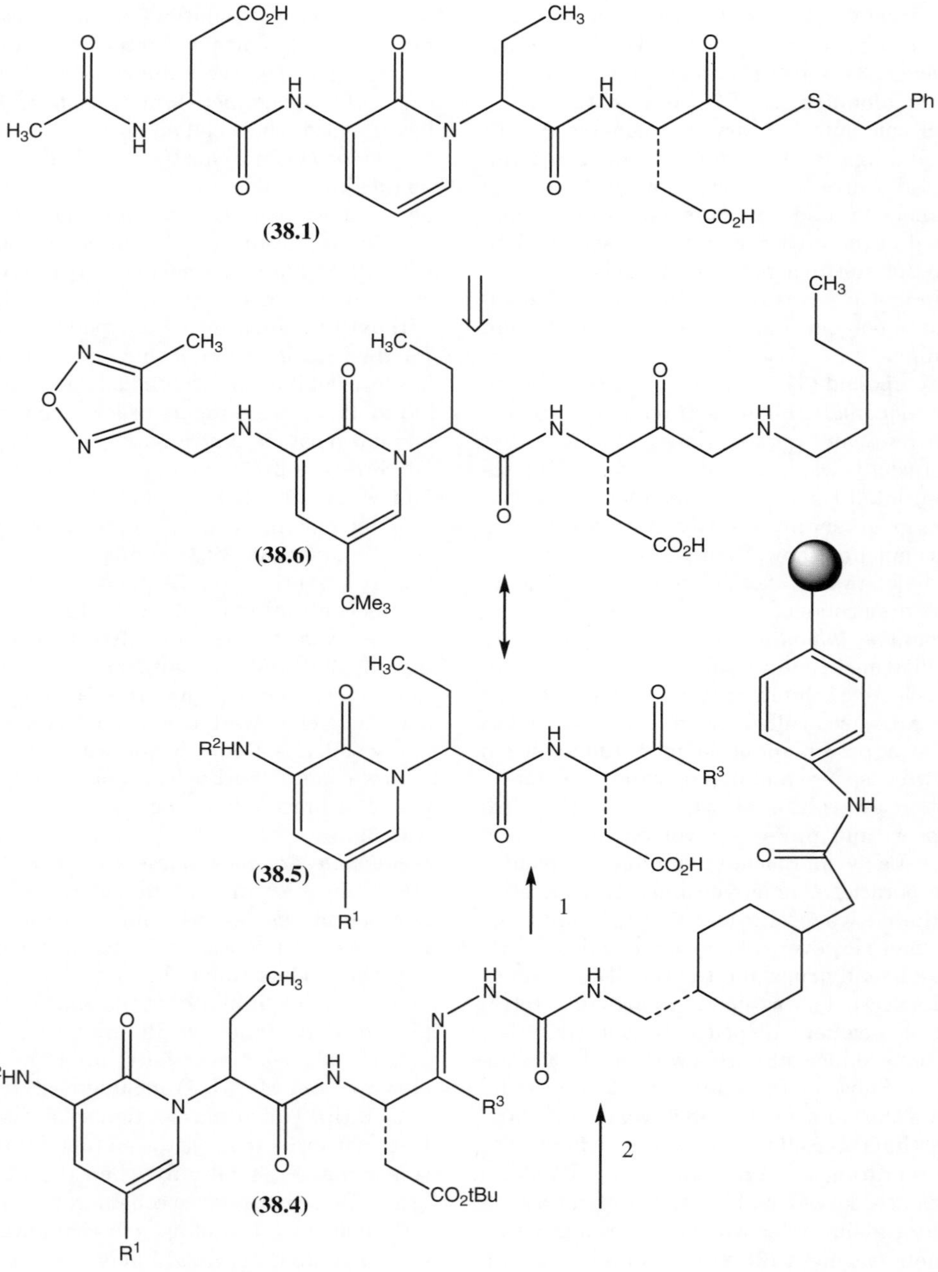

Figure 38. Preparation of a compound library of caspase-3 inhibitors.

also increase selectivity away from MetAP-1 [278–280]. A solution-phase combinatorial library was prepared for this purpose starting with weak inhibitor **40.1**. Considerable enhancement was found on preparation of the more lipophilic analog **40.2**. Another compound library was generated in order to optimize the amino acid moiety of **40.2** leading

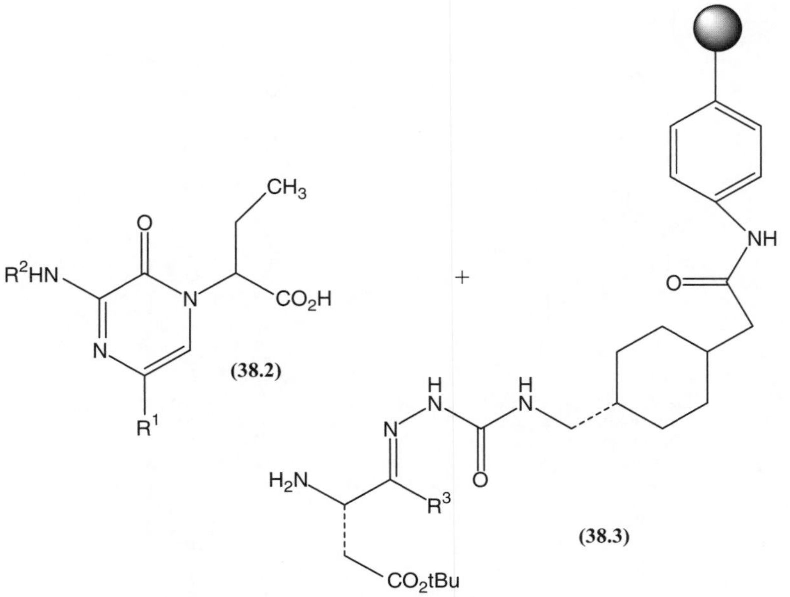

(38.2)

(38.3)

Reagents: 1, TFA; 2, HATU.

Figure 38. (*Continued*)

ultimately to naphthyl-containing lead **40.3**. Another compound library was then explored to structural requirements of the methionine substituted portion of **40.1** leading to **40.4**. This was tweaked with further compound generation leading finally to **40.5**, a lead compound with good potency and substantial selectivity. Success in this study was complicated by the flexibility of the target enzyme. This resulted in formation of different enzyme–ligand complexes at various stages of the analog program making the interpretation of SAR tables along the way puzzling (Fig. 40).

Dihydrofolic Acid Reductase Inhibitors The inhibition of bacterial dihydrofolic acid reductase (DHFR) has been shown to be a clinically effective means of curing bacterial infections of humans. Trimethoprim embodies this phenomenon. To identify structurally novel inhibitors of this enzyme, a novel approach was taken that allowed the simultaneous screening of large collections of compounds by mixing them with, in this case, *Escherichia coli* DHFR, allowing time for liganding to occur, then, using size-exclusion chromatography at low temperatures, to separate the newly formed enzyme–inhibitor complexes. Separation of the ligand from the enzyme takes place on another column and the eluted ligand is identified by electrospray high resolution mass spectrometry. Hits are verified by independent synthesis [281].

Using this powerful technique, a suitable lead was identified and then optimized in a follow-up compound library resulting in micro molar inhibitor **41.1** whose *S*-enantiomer displayed 15 μM potency and whose *R*-enantiomer displayed 40 μM potency against the enzyme. The eudesmic ratio is not impressive among these analogs (Fig. 41).

Success with this technique requires no foreknowledge of the mode of action or of any liganding requirements. With automation, screening and evaluation of compound collections can be done in remarkably short time and the compound size requirements even for the macromolecule are comfortably small.

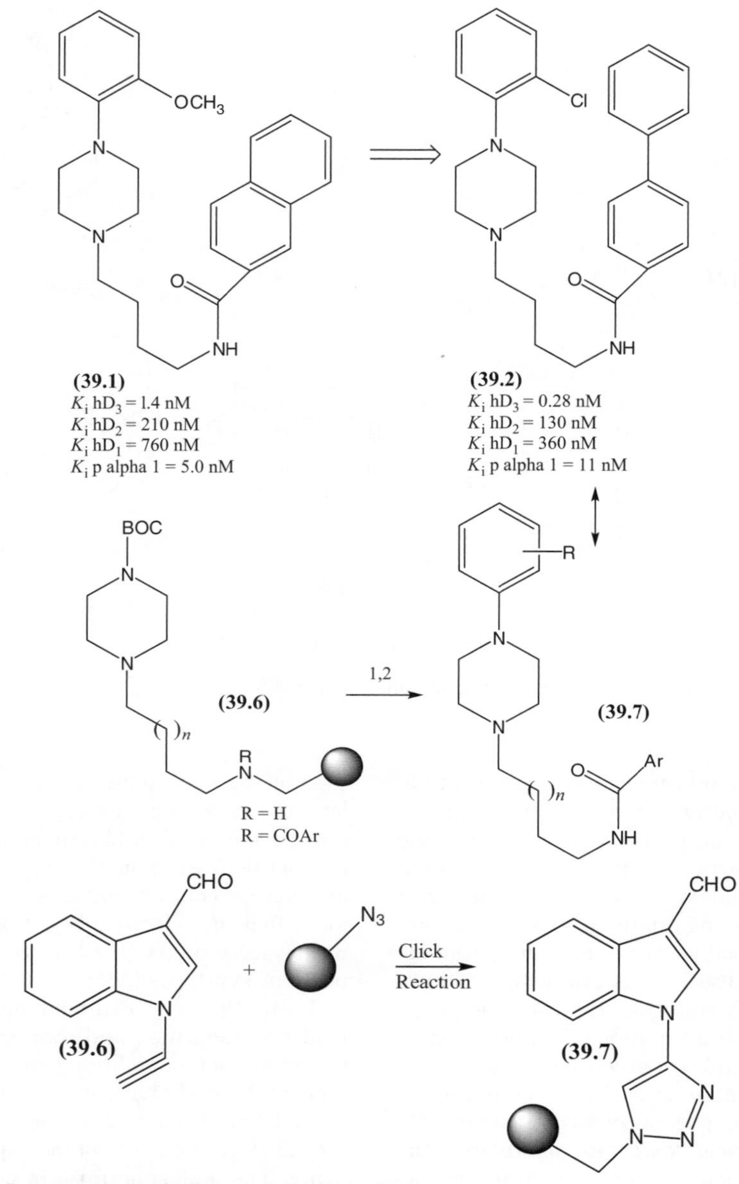

(39.1)
K_i hD_3 = 1.4 nM
K_i hD_2 = 210 nM
K_i hD_1 = 760 nM
K_i p alpha 1 = 5.0 nM

(39.2)
K_i hD_3 = 0.28 nM
K_i hD_2 = 130 nM
K_i hD_1 = 360 nM
K_i p alpha 1 = 11 nM

(39.6) R = H, R = COAr → 1,2 → (39.7)

(39.6) + bead-N_3 → Click Reaction → (39.7)

Reagents: 1. $NaCNBH_3$; 2, ArCOCl; 3, TFA.

Figure 39. Preparation of a compound library of dopamine D3 ligands.

Metabotropic Glutamate Receptor 5 (mGluR5)
A decrease in the functioning level of NMDA receptors is associated with schizophrenia and other psychoses. The activity of the NMDA receptors is modulated by the activity level of metabotropic glutamate receptors. Compounds that enhance the binding of glutamate to its receptors should consequently enhance

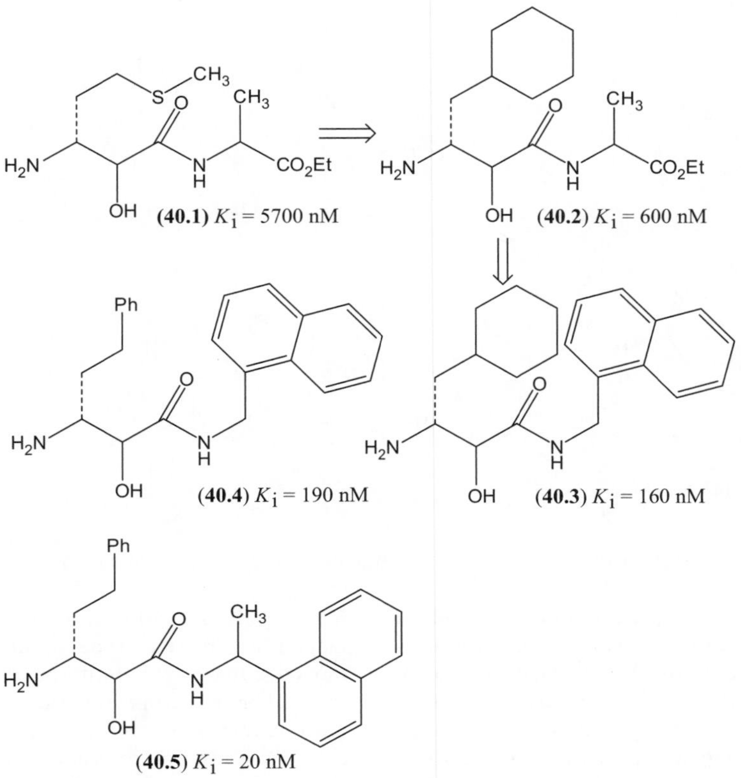

(40.1) $K_i = 5700$ nM

(40.2) $K_i = 600$ nM

(40.4) $K_i = 190$ nM

(40.3) $K_i = 160$ nM

(40.5) $K_i = 20$ nM

Figure 40. Preparation of a combinatorial library of MetAP-2 inhibitors.

the function of NMDA receptors and may in this indirect way lead to useful therapy for psychoses. Screening for such agents using a FLIPR-based end point led to a phthalimide containing lead substance **42.1** with fairly weak affinity [282]. The potency of **42.1** was not found in the absence of glutamate leading to the conclusion that the binding site was allosteric. Enhancing the activity of **42.1** was approached by the production and evaluation of iterative compound libraries ultimately leading by way of bioisostere **42.2** to heterocycle **42.3** and thence to low nM allosteric enhancing agent **42.4** (Fig. 42).

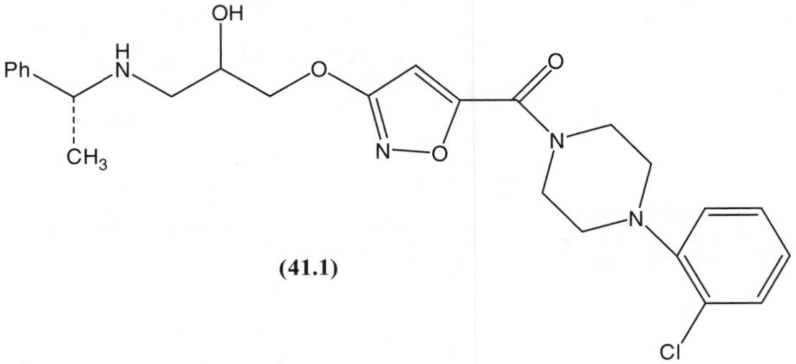

(41.1)

Figure 41. DHFR inhibitor **41.1**.

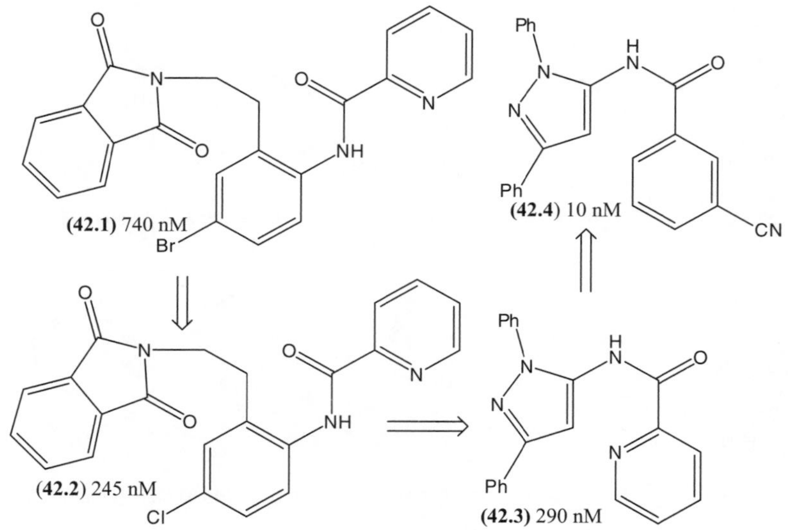

Figure 42. Preparation of a compound library of mGluR5 ligands.

RasGap and ERK-1 Dual Inhibitors Based on known kinase inhibitor scaffolds as structural leads, a combinatorial approach has been developed for the synthesis of diverse heterocyclic compound libraries [283]. The general synthetic strategy involved initial reaction of a variety of strategically substituted dichloroheterocyclic aromatic compounds with a resin-bound amine (**43.2**, itself derived from reductive amination with **43.1**), displacing one of the chloro groups with the nucleophilic amine to give **43.3** thus leading to attachment of the heterocyclic scaffold onto the solid support. Depending on the type of heterocycle employed, subsequent additional alkylation, acylation, or coupling reaction involving the amine functionality of the heterocycle (producing **43.4**) can add another diversity element in the synthesis. The remaining chloro substituent on the heterocycle is then displaced with various nucleophiles (such as **43.5**) followed by cleavage from the solid support thus resulting in the desired library of compounds (**43.6**). A representative example of the above

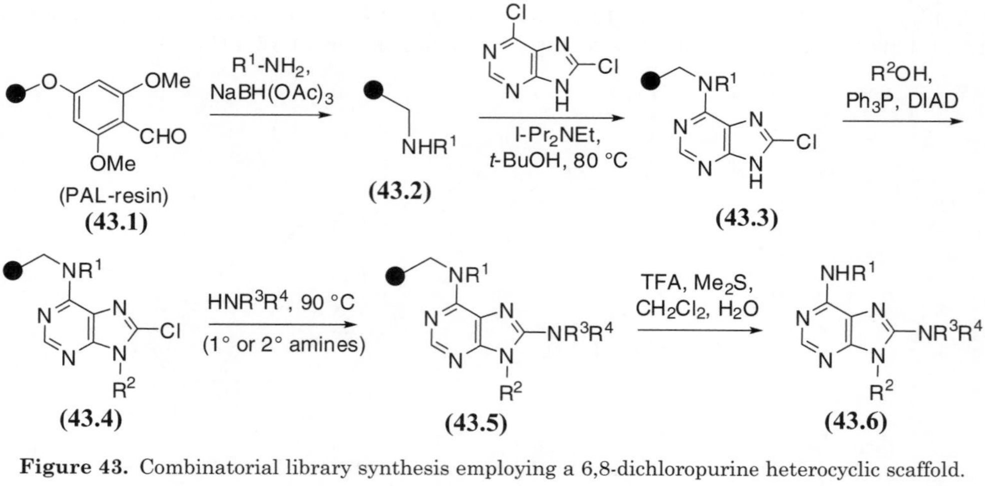

Figure 43. Combinatorial library synthesis employing a 6,8-dichloropurine heterocyclic scaffold.

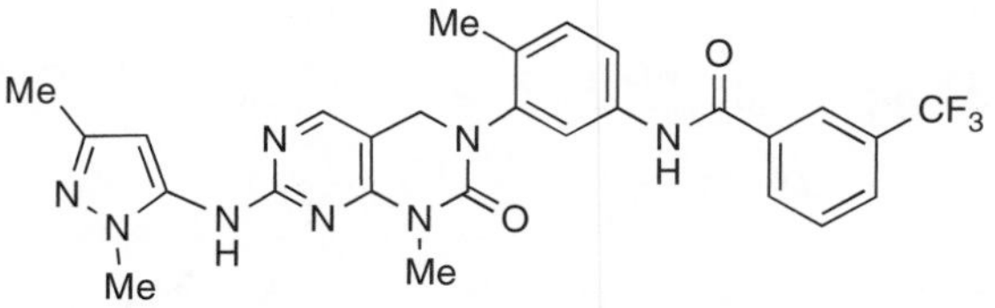

Figure 44. Structure of pluripotin (SC1)

combinatorial scaffold approach is shown below (Fig. 43).

Utilizing various combinations of dichloroheterocyclic scaffolds and appropriate nucleophilic counterparts, libraries consisting of more than 45,000 discrete and diverse heterocyclic small molecules were synthesized. The compound libraries obtained were subjected to biological evaluation in a variety of cell and protein-based assays. One of these studies involved a cell-based high-throughput assay directed at identification of small molecules that can control the self-renewal of embryonic stem (ES) cells [284]. An established reporter mES cell line, derived from transgenic OG2 mice was used for this assay. More than 20 hits were obtained from primary screening. From this set, compounds containing the 3,4-dihydropyrimido[4,5-*d*]pyrimidine structural framework were found to maintain the expression of multiple mES cell-specific markers in a dose-dependent manner. Further structure–activity relationship studies of a more focused, second-generation 3,4-dihydropyrimido[4,5-*d*]pyrimidine library ultimately led to the identification of pluripotin (SC1) (Fig. 44), a compound exhibiting potent activity in the ESC-SR assay ($EC_{50} = 1\,\mu m$) and with relatively low cellular toxicity ($>30\,\mu m$). Subsequent biochemical and cellular experiments revealed pluripotin's mechanism of action to be inhibition of ERK-1 and RasGap-dependent signaling pathways. This unique dual inhibition of a protein kinase and a GTPase-activating mechanism renders pluripotin a useful biological tool in understanding the mechanisms that control stem cell fate and its potential application in therapy.

This work was reviewed briefly with a somewhat different emphasis in a subsequent section of this chapter (see the kinase treatment on page 337).

Epoxide Hydrolase Inhibitors Epoxide hydrolases are a class of enzyme that catalyze the hydration of epoxides to their corresponding vicinal diol products. In mammalian species, there are several distinct epoxide hydrolases, including soluble, and microsomal epoxide hydrolase. The three main biological functions of these enzymes are regulation of signaling molecules, detoxification, and catabolism. For example, the soluble epoxide hydrolase (sEH) is involved in the selective metabolism of endogenous fatty acid epoxides, such as those derived from arachidonic acid, linoleic acid, and so on. The epoxyicosatrienoic acids (EETs), endogenous epoxidized products of arachidonic acid, were shown to exhibit antihypertensive, anti-inflammatory, and analgesic action. As the EETs are substrates for soluble epoxide hydrolase-catalyzed inactivation, inhibition of sEH is considered to be a promising target for the treatment of hypertension and inflammatory diseases [285].

In a study exploring optimization of the structural features of a previously discovered sEH inhibitor adamantyl urea derivative, Hammock and coworkers employed a solid-phase combinatorial approach to construct a focused library of urea derivatives [286]. Thus, using the IRORI AccuTag™ Combinatorial Chemistry System, and employing an acid-labile 3-formylindole-linked resin as the solid-support, a library of 192 urea derivatives was synthesized. The reaction steps included initial attachment of four different primary amines to the resin (**45.1**) via standard reductive amination protocol, reaction of the resulting secondary amines (**45.2**) with 48 different isocyanates forming the urea derivatives (**45.3**), and final cleavage from the solid support gave **45.4**. On high-throughput screening against recombinant human sEH, several of the library members thus synthesized

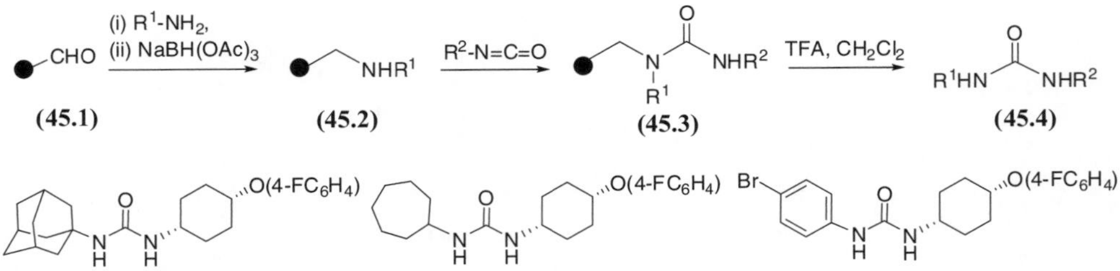

(45.1) (45.2) (45.3) (45.4)

(45.5) (Initial hit: IC_{50} = 0.5 nm) (45.6) (IC_{50} = 0.5 nm) (45.7) (IC_{50} = 0.6 nm)

Figure 45. Combinatorial synthesis of sEH inhibitor urea derivatives.

(**45.5–45.7**) showed potent inhibition of the target enzyme. Structures of some of the inhibitors thus identified are shown in Fig. 45.

Farnesyltransferase Inhibitors *Ras* proteins play a key role in signal transduction pathway, cell growth and cell proliferation. The *ras* oncogene is among the most frequently activated mutated genes in human tumors. A critical process in the activation of the oncogenic *ras* proteins involves a farnesylation step by *farnesyltransferase* (FTase). Consequently, inhibition of this enzyme is considered to be an attractive therapeutic target for the development of potential anticancer agents. Studies directed at the identification of FTase inhibitors (FTIs) have resulted in several promising candidates for further investigation. One such agent is the trihalogenated benzocycloheptapyridine derivative SCH 66336 (Sarasar® Fig. 46). Presently in clinical trial, sarasar has shown impressive activity against a number of solid tumor types and leukemia [287]. Interestingly, sarasar has also been found to be synergistic with the taxane class of anticancer drugs.

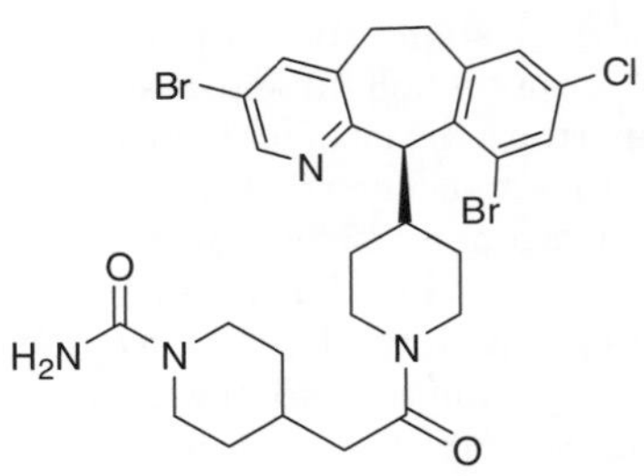

SCH 66336 (Sarasar®)
FTase IC_{50} = 1–2 nm

Figure 46. The chemical structure of Sarasar®

In a structure–activity relationship study directed at possible refinement of the sarasar substitution pattern, a focused library of analogs were prepared employing a solution-phase combinatorial approach [288]. The major structural components of this library of compounds included a dihalogenated benzocycloheptapyridine and a 1,2,4-trisubstituted piperazine core (**47.5**). Starting from an orthogonally protected piperazine carboxylic acid derivative (**47.1**), the synthesis involved initial formation of variously substituted amides at the C-2 position (**47.2**), followed by selective unmasking of the amine at *N*-1 position and its subsequent coupling with carboxylic acids or isocyanates to form the corresponding amide/urea derivatives (**47.3**). Further removal of the Boc-protection at *N*-4 (**47.4**) and amine alkylation with the appropriate benzocycloheptapyridine donor to give **47.5** then provided the desired library of compounds of which components **47.6–47.7** are representatives (Fig. 47).

Inhibitors of Human Papillomavirus 6 (HPV6) E1 Helicase Human papillomaviruses (HPVs) infect cutaneous and mucosal epithelia and are responsible for causing benign and sometimes malignant lesions. To address the lack of any antiviral drugs for the treatment of HPV-induced lesions and with the aim to identify anti-HPV compounds as potential drug development candidates, high-throughput screening of a compound collection was undertaken. This assay resulted in the identification of a biphenylsulfonacetic acid (**48.1**) as a potent inhibitor of the inhibition of the ATPase activity of recombinant HPV6 E1 helicase [289,290]. Further SAR studies on this initial hit revealed the substituted sulfonyl acetic acid component to be essential for activity, while modifications in the 4-phenyl ring resulted in analogs (**48.2**)

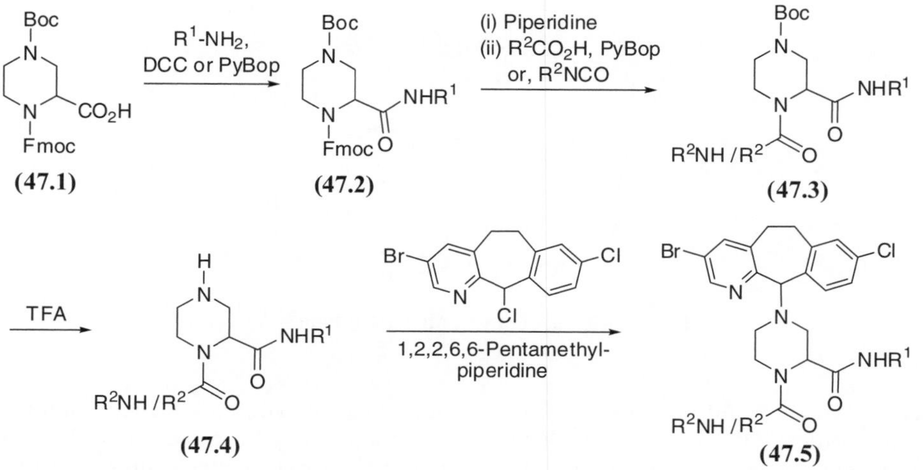

Boc
N
N CO2H
Fmoc
(47.1)
R1-NH2,
DCC or PyBop
Boc
N
NHR1
N
Fmoc O
(47.2)
(i) Piperidine
(ii) R2CO2H, PyBop
or, R2NCO
Boc
N
NHR1
N
O
R2NH / R2 O
(47.3)
TFA
H
N
NHR1
N
O
R2NH / R2 O
(47.4)
Br Cl
N
Cl
1,2,2,6,6-Pentamethyl-
piperidine
Br Cl
N
N
NHR1
N
O
R2NH / R2 O
(47.5)

Representative compounds:

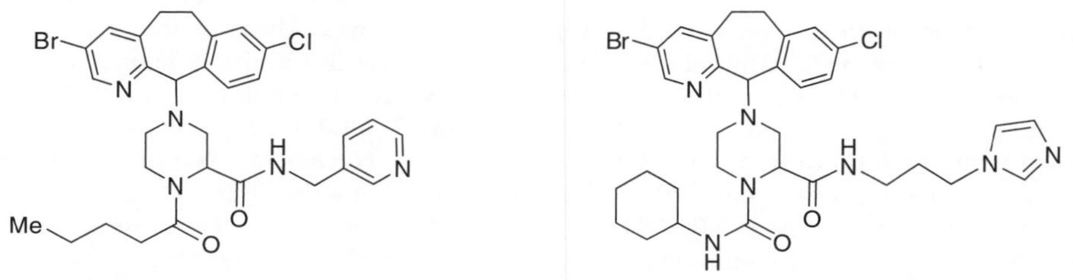

(47.6) FTase IC_{50} = 5.7 nm

(47.7) FTase IC_{50} = 4.5 nm

Figure 47. Combinatorial synthesis of a focused library of Sarasar® analogs.

with improved activity. Based on this SAR information, and employing a parallel synthesis approach, focused libraries of the corresponding meta- or para-substituted amides were constructed. On biological screening of these libraries, the most potent compounds were found to be the meta-substituted primary amides containing a hydrophobic aromatic appendage (**48.3–48.4**). Unfortunately, lack of cellular activity, and the instability of the sul-

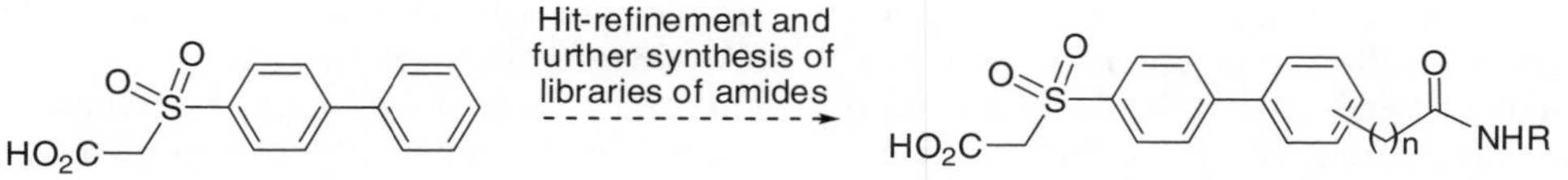

Initial screening hit
(48.1) (IC_{50} = 2 mM)

Hit-refinement and further synthesis of libraries of amides

Libraries of *m*-/*p*-substituted carboxamides
(48.2)

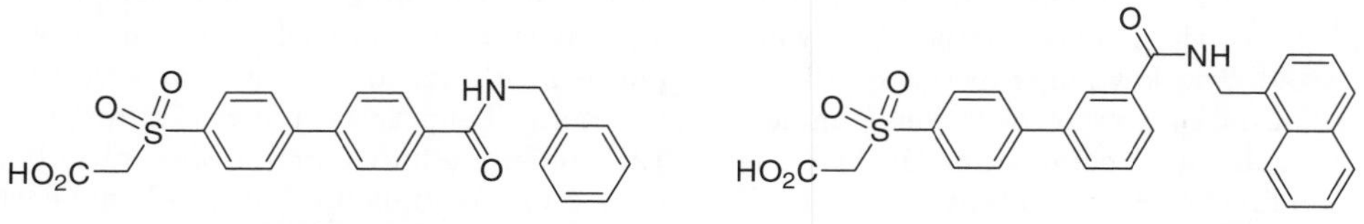

(48.3) (IC_{50} = 510 nm)

(48.4) (IC_{50} = 4.3 nm)

Figure 48. Inhibitors of HPV6 E1 helicase.

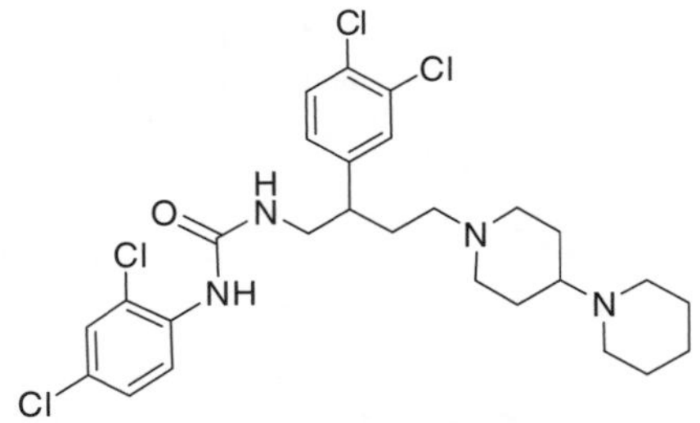

MCHR1 $K_i = 4.1$ μm

Figure 49. Structure of MCHB1 inhibitor.

fonylacetic acid functionality toward decarboxylative formation of inactive methylsulfone derivatives, continues to be a roadblock in the further development of these classes of compounds (Fig. 48).

Melanin-Concentrating Hormone 1 Receptor (MCHR1) Antagonists Melanin-concentrating hormone (MCH) is a cyclic peptide hormone, primarily expressed in the lateral hypothalamus region in the mammalian brain. MCH is the endogenous ligand for two distinct G-protein-coupled receptors, MCHR1 (Fig. 49) and MCHR2. While rodents express only MCHR1, both the MCHR1 and the MCHR2 are present in higher mammals including humans. On the basis of extensive murine *in vivo* studies, inhibition of MCHR1 was found to have a profound effect in controlling food intake and body weight. Accordingly, in recent years, small molecule inhibitors of MCHR1 are being actively pursued as agents for the potential treatment of obesity [291,292].

Starting from a 4-amino-2-arylbutylbenzamide derivative, a micromolar MCHR1 antagonist identified from previous screening, researchers conducted a lead optimization study in search of more potent compounds [293]. Employing solid-phase parallel synthesis approaches and exploring the SAR of the screening hit, different libraries, numbering more than 500 compounds were synthesized. The key steps in the syntheses involved initial conversion of various commercially available arylacetonitriles (**50.1**) to the corresponding 2-aryl-4-pentenylamine derivatives (**50.2**), followed by attachment of these amines to the aldehyde functionality of an acid labile solid support via a reductive amination protocol to give **50.3**. Subsequent diversification at the three strategic sites of the above scaffold culminated in the desired libraries of compounds (**48.4–48.6**). These hit-to-lead refinement studies resulted in the identification of several 4-amino-2-biarylbutylurea derivatives as nanomolar range MCHR1 antagonists (**50.7–50.9**) (Fig. 50) [293].

$P2X_7$ Nucleotide Receptor Antagonists The ATP-sensitive $P2X_7$ receptor belongs to the P2X family of ionotropic receptors. Present on cell types such as, mast cells, macrophages, lymphocytes, and glial cells in the CNS, activation of the $P2X_7$ receptor is thought to play an important role in inflammation and autoimmune processes [294]. Consequently, $P2X_7$ receptor antagonists are of current interest toward the development of new agents for inflammatory disorders [295,296].

Based on a high-throughput screening-derived 2,4,5-trisubstituted imidazoline derivative as a lead compound with potent $P2X_7$ antagonistic activity, researchers initiated an SAR investigation of the above compound. Failure in the initial attempts to prepare the desired imidazolines by one-step direct condensation between 1,2-diamines and carboxylic acid derivatives under various reaction conditions prompted the development of an alternative approach, combining solid-phase synthesis and microwave-accelerated reactions. Thus, in a stepwise strategy, *cis*-1,2-diphenyl ethanediamine was initially attached to a resin-bound *p*-nitrophenyl carbonate to form the corresponding carbamate derivative **51.1**. Subsequent EDC-mediated

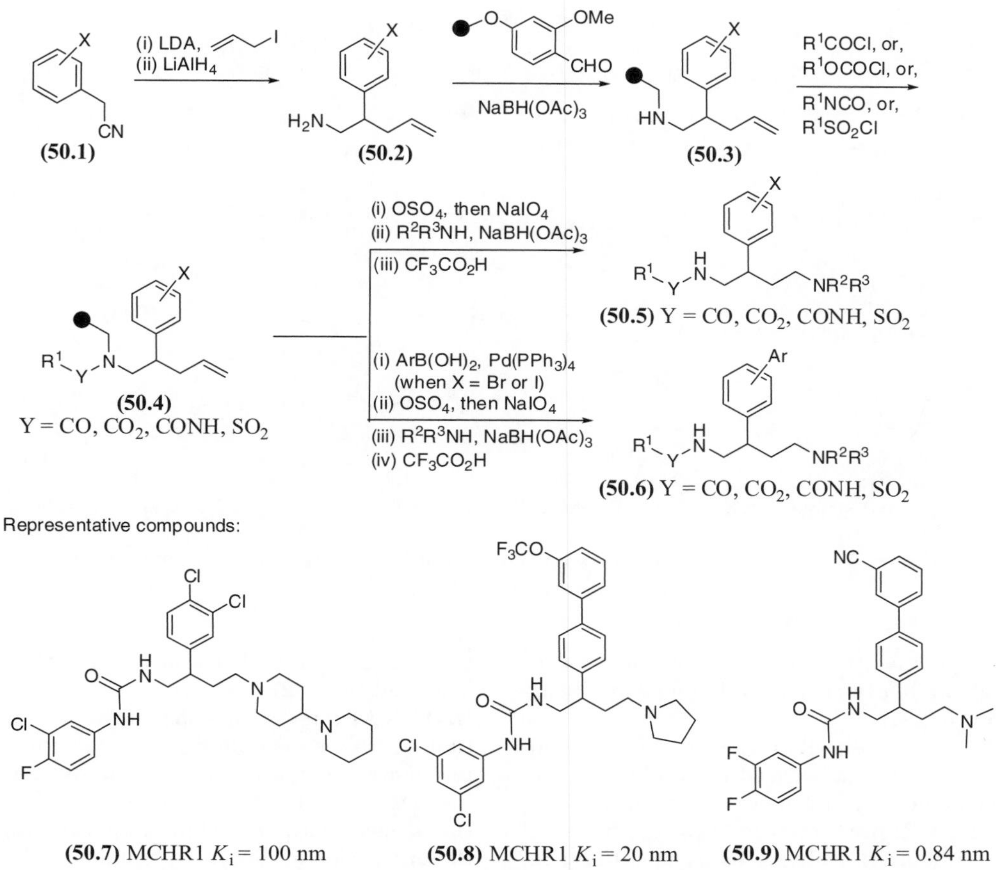

Figure 50. SAR directed solid-phase combinatorial synthesis of MCHR1 inhibitors.

coupling of the free amine with a variety of strategically chosen carboxylic acids provided the corresponding amides (**51.2**). At this stage, attempts to perform an acid-assisted one-pot resin cleavage and cyclization to form the desired imidazolines were unfortunately not successful, instead providing only the resin-cleaved acyclic amine derivative (**51.3**). However, when a mixture of the above free amines with trimethylsilyl polyphosphate (TMS-PP) were exposed to microwave irradiation at 140 °C for a short period of time, the corresponding imidazoline derivatives (**51.4**) were obtained in good yields. Employing the above approach, focused libraries of 2,4,5-trisubstituited imidazolines, differing in the substituents at the 2-position (**51.5–51.9**), were constructed. The SAR information obtained via subsequent evaluation of the above compounds in $P2X_7$ antagonist assay is expected to provide useful guidelines for future design of more potent analogs [296].

This work was reviewed briefly with a somewhat different emphasis in a subsequent section of this chapter (see the ion channels treatment on p. 336) (Fig. 51).

Metalloprotease Inhibitors The membrane-bound metalloprotease, endothelin-converting enzyme (ECE), is responsible for the generation of vasoactive endothelin (ET-1), via proteolytic processing of its precursor big endothelin (big-ET). Due to the potent vasoconstrictive action of endothelin, inhibition of ECE toward modulating ET-1 production is

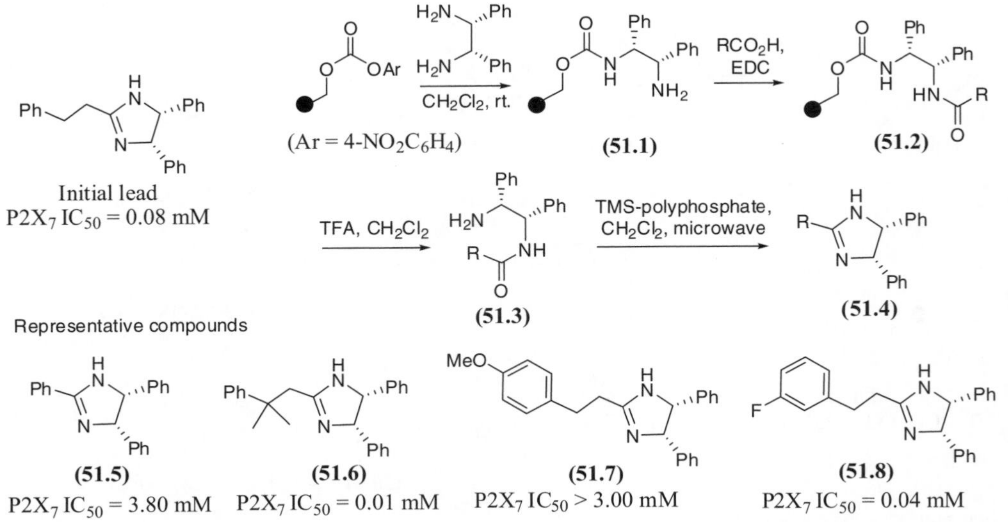

Initial lead
$P2X_7$ IC_{50} = 0.08 mM

$(Ar = 4\text{-}NO_2C_6H_4)$

H_2N, CH_2Cl_2, rt. (51.1) RCO_2H, EDC (51.2)

TFA, CH_2Cl_2 (51.3) TMS-polyphosphate, CH_2Cl_2, microwave (51.4)

Representative compounds

(51.5) $P2X_7$ IC_{50} = 3.80 mM

(51.6) $P2X_7$ IC_{50} = 0.01 mM

(51.7) $P2X_7$ IC_{50} > 3.00 mM

(51.8) $P2X_7$ IC_{50} = 0.04 mM

Figure 51. Microwave-accelerated solid-phase synthesis of $P2X_7$ antagonist imidazolines.

an attractive therapeutic target for diseases such as hypertension and congestive heart failure [297].

To facilitate SAR investigation and optimization of an indole-based ECE inhibitor lead compound, researchers employed a split-and-pool IRORI Kan® method toward synthesis of a combinatorial library of trisubstituted indoles [298]. Starting from a 2-carbomethoxy-5-amino substituted indole, linked to a acid labile resin at the 5-position (**52.1**), the first step in the synthesis involved acylation at the 5-amino position with various carboxylic acid chlorides to give **52.2**. Subsequent base-mediated alkylation of the indole ring nitrogen followed by ester hydrolysis provided the corresponding indole carboxylic acid derivatives (**52.3**). Derivatization of the carboxylic acid functionality to carboxamides via reaction with a variety of anilines/aminophenols, and resin cleavage with TFA resulted in the library of the desired indole derivatives (**52.4**). Utilizing the three sites of diversification in the starting indole scaffold, a 1332 member compound library could be obtained from the above endeavor (**52.5**). ECE inhibition studies of the above compounds indicated the aryl component of the C-2 amide chain as the only site allowing beneficial structural modifications compared to the initial lead (Fig. 52).

HDM2–p53 Interaction Antagonists The p53 tumor suppressor protein plays a critical role in cell cycle regulation. In its active form p53 induces growth arrest, apoptosis and cell senescence, with important implications in the prevention of cancer. In humans, HDM2, the human homolog of murine mdm2, is overproduced in many tumor cells. Acting as a regulator, binding of HDM2 to p53 inhibits the antitumor activity of p53. The disruption of the HDM2–p53 protein–protein interaction in cancer cells, and consequent up-regulation of p53 activity, is therefore considered to be a potential target for cancer therapy [299].

In search for small molecule inhibitor of the HDM2–p53 interaction, researchers have reported the discovery of a series of 1,4-benzodiazepin-2,5-diones as potent antagonists of the HDM2–p53 interaction. A library of 22,000 benzodiazepinones (**53.1**) were synthesized for the above study, employing initial coupling of various anthranilic acids, amines, aldehydes, and 1-isocyanocyclohexene under the Ugi four component reaction protocol (Fig. 53), followed by acid-catalyzed cyclization of the adducts [300]. A high-throughput screening of the above compounds by a

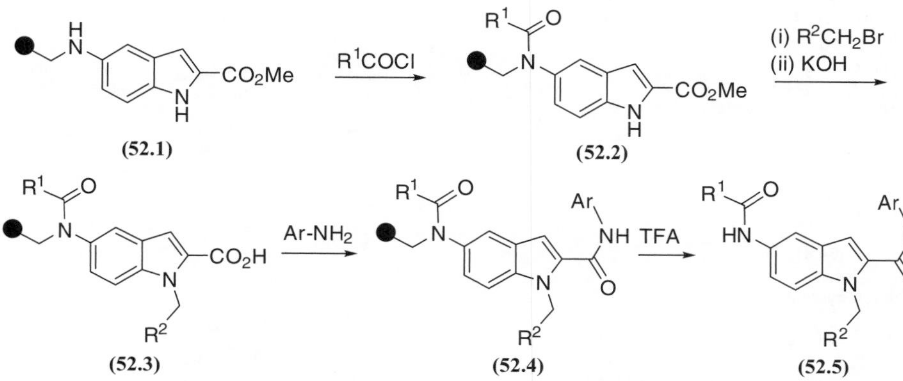

Figure 52. Combinatorial library of trisubstituted indoles as ECE inhibitors.

proprietary ThermoFluor® screening method [301] resulted in the identification of several potent antagonists of the HDM2–p53 interaction (**53.2**) [300,302].

Beta-Secretase Inhibitors Alzheimer's disease is a neurodegenerative disorder marked by the aggregation of β-amyloid proteins in the brain. β-Secretase, also known as BACE1 (β-site of amyloid precursor protein (APP) cleaving enzyme), is an aspartyl protease whose activity has been directly correlated with increased amyloid plaque accumulation that may result in nerve cell death and abnormal brain function over time. One in 10 people over age 65, and almost 50% of those over 85 are likely to be afflicted. This, coupled with the absence of treatments that halt the progression of the disease and a growing aged population, has prompted many researchers to investigate BACE1 inhibition as a therapeutic option [303].

The nature of this target requires that an effective BACE1 inhibitor must be adequately absorbed and penetrate the blood–brain barrier in sufficient concentration. Thus, many potent peptide-based inhibitors fail to demonstrate acceptable *in vivo* efficacy. Consequently, intense research has grown around the development of nonpeptide like small molecules as potential BACE1 inhibitors in an attempt to remedy some of the poor ADME characteristics suffered by their peptide-like counterparts. One approach involves using crystal structure data to define key binding elements of peptides that potently inhibit exogenous BACE1 enzyme and then incorporating these excised pieces into the design of novel small molecule libraries.

This tactic is exemplified by the extraction of the N-terminus Leu-Ala hydroxyethylene (HE) C-terminus unit from a known potent peptidomimetic BACE1 inhibitor, OM99-2. This particular segment of OM99-2 is intended as a nonhydrolyzable transition-state analog of the natural substrate, APP. This team chose the resin supported, orthogonally C- and N-terminus protected HE scaffold (**54.1**) as a template on which SAR optimization of the terminal ends was performed using standard amino acid coupling and sequential protection/deprotection procedures (Fig. 54).

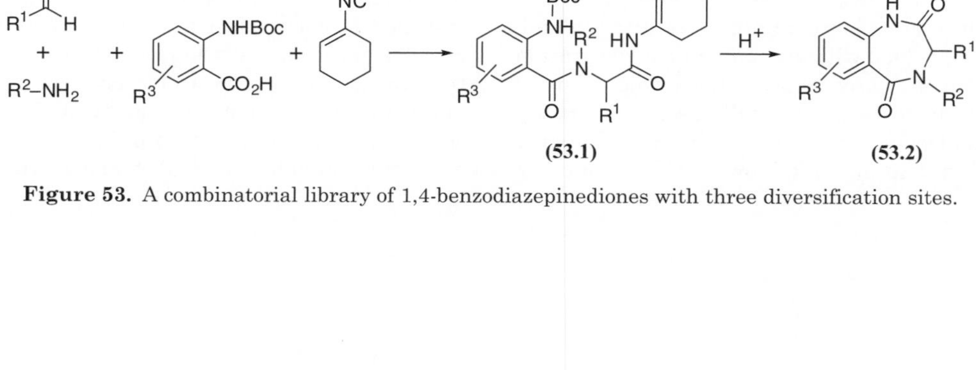

Figure 53. A combinatorial library of 1,4-benzodiazepinediones with three diversification sites.

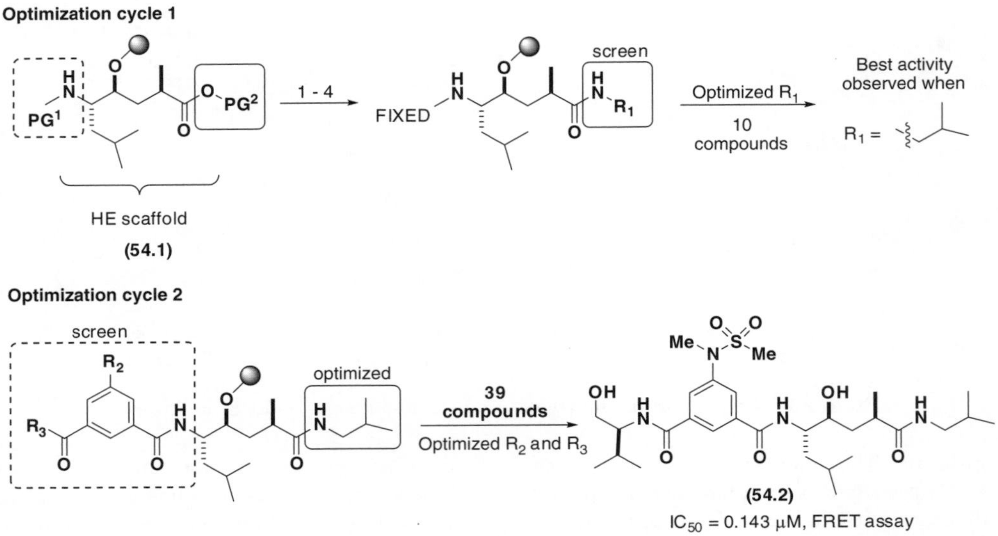

Reagents: 1. N-deprotection; 2. HOBt acid-coupling; 3. carboxylate deprotection; 4. HOBt acid-coupling.

Figure 54. Optimization of nonpeptidic BACE1 inhibitors.

Guided by enzyme inhibition data and the iterative modeling of designed compounds overlapped with the crystal structure of potent peptide OM99-2 complexed with BACE1, interactions were optimized with lipophilic binding pockets. Using a combinatorial approach, a small library of <50 members was generated that surveyed changes at the terminal ends of the HE scaffold. Of these, two compounds were prepared that possess double digit nanomolar activity in the *in vitro* assay which rivals results obtained with peptide inhibitors such as OM99-2. Specifically, these efforts resulted in nonpeptidic compound (**54.2**) that inhibits BACE1 with an IC_{50} of 143 nM in a FRET assay, and demonstrated 47% inhibition of β-amyloid production in CHO2B7 cells at a concentration of 10^{-7} M [304].

FSH Agonists Follicle-stimulating hormone (FSH) is a 38 kDa heteromeric peptide hormone secreted by the anterior pituitary gland. It is involved with ovarian follicle maturation in women; accordingly, the development of FSH agonists is potentially useful as fertility enhancement agents. This is a challenging task due to the well-known difficulty of using small molecules (desired here for oral bioavailability and the encouragement of patient compliance) to mimic a much larger peptide.

Two groups addressed these challenges by beginning with very broad screening activities, finding low micromolar hits. These hits, featuring a biphenyl unit [305] or a thiazolidinone [306], were derived from large libraries containing 2 million or 40,000 compounds, respectively. Both groups followed up on this initial discovery by synthesizing follow-up libraries that focused on these functional groups (Fig. 55).

The first of these libraries utilized cross-coupling chemistry to afford decorated biaryl groups such as those shown in the scheme. Carried out on Tentagel beads, the first follow-up library contained over 31,000 compounds as shown. Since the assay was carried out directly on the beads using pool and split techniques, MS-addressable tags were also installed in each synthetic step to allow for hit identification. This library entailed first the preparation of a series of polymer-bound amines on beads that were acetylated by a series of iodobenzene-containing carboxylic acids, thus providing a handle for downstream

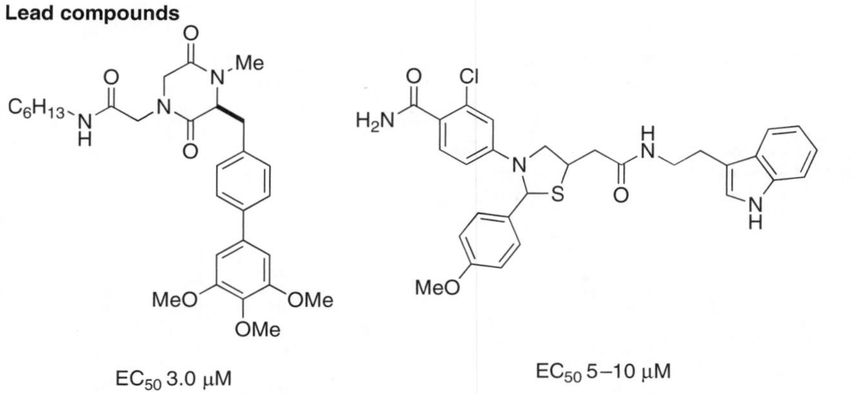

Figure 55. Screening lead compounds for combinatorial elaboration of follicle stimulating hormone agonists.

Suzuki reaction modification and ultimately affording one major library subset. A second library could be accessed by using aldehyde-containing aromatic substrates in this Suzuki step followed by straightforward reductive amination of the following compounds. An example of a hit from Library 1 is shown. It contains the familiar Freidinger lactam-type subunit in the linker section and a urea side chain. This compound was the most active hit obtained in the series and as such the investigators were able to pick up ca. 10-fold activity by carrying out the follow-up screen (Fig. 56).

Cyclin Dependent Kinase 5/p25 (cdk5/p25) Inhibitors Cyclin dependent kinase 5 is a serine/threonine kinase that recruits a transient, regulatory cofactor p35 to phosphorylate cytoskeletal stabilizing protein *tau*. Cleavage of cofactor p35 leads to a more stable p25 that accumulates in the brain. The discovery of extensive cytotoxic neurofibrillary *tau* protein tangles in the brains of Alzheimer's patients has led to the hypothesis that the unregulated presence of p25 may lead to uncontrolled activation of cdk5 and promiscuous hyperphosphorylation of cytoskeletal proteins. Inhibitors of the cdk5/p25 complex are envisioned to treat neurodegenerative disorders by preventing the formation of these neurofibrillary tangles [307].

Kinase selectivity is often difficult to attain, as many inhibitors bind in a highly conserved ATP binding cavity common across the kinase family of enzymes. For example, there are only two of 29 amino residues lining the ATP binding pocket that differ between cdk5 and cdk2, a cancer related target. Given that these two residues occupy a region of the binding site that are somewhat removed from

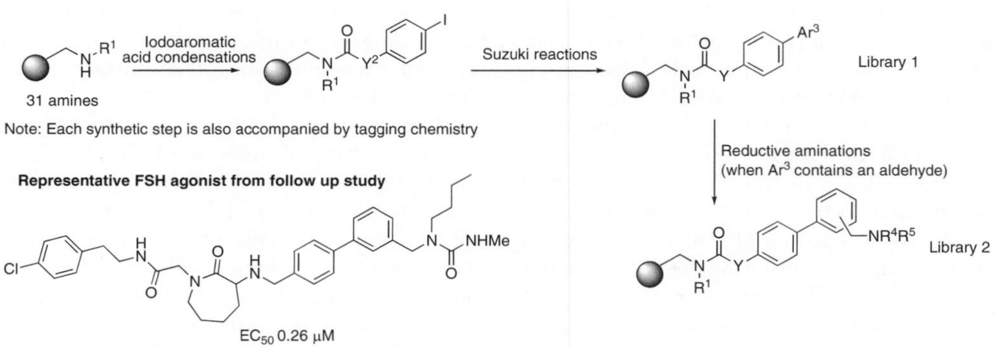

Figure 56. Solid-phase synthesis of libraries seeking nonpeptidic follicle stimulating hormone agonists.

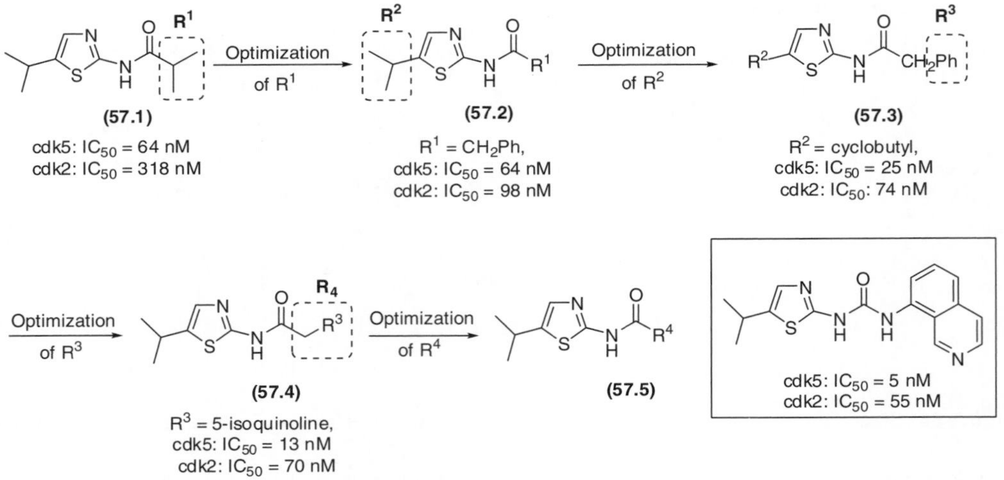

Figure 57. Hit optimization of amidothiazoles with Cdk5/p25 activity.

interacting with a ligand that resides there, achieving selectivity between these two kinases is challenging [318,319].

A high-throughput screening effort at Pfizer aimed at finding cdk5/p25 inhibitors revealed a low molecular weight amidothiazole **57.1** with good potency versus cdk5/p25 complex ($IC_{50} = 321$ nM). Amidothiazole **57.1** showed a high level of selectivity in a broad kinase panel; however, screening against other cyclin dependent kinases revealed an equipotent inhibition of cdk2/cylin E ($IC_{50} = 318$ nM). Competitive binding studies confirmed that the hit compound binds in the highly conserved ATP binding site of cdk5 (Fig. 57).

With the intent of improving potency and selectivity, optimization of the amide functionality was undertaken in the first round. In a parallel synthesis approach, 5-isopropyl-2-aminothiazole was derivatized to produce a library of amides and ureas, yielding amidothiazole **57.2** with a fivefold improvement in potency ($IC_{50} = 64$ nM). The next round of modifications focused on surveying the influence of the 5-substituent by acylating a series of 5-substituted 2-aminothiazoles with phenylacetyl chloride. This endeavor revealed that bulkier or more polar substitutions at the 5-position decreased potency, but that a 5-cyclobutyl-2-amidothiazole **57.3** has an improved potency and selectivity profile ($IC_{50} = 25$ nM; for cdk2/cyclin E, $IC_{50} = 74$ nM). A survey of the terminal benzyl moiety also revealed that amino-linked fused ring systems yielded more potent analogs. Ultimately, the process produced urea **57.4** with a 60-fold increase in potency for cdk5 compared to hit **57.1** and with a 10-fold improved selectivity ($IC_{50} = 5$ nM versus cdk2, $IC_{50} = 55$ nM).

Cysteine Proteases Proteases are responsible for the enzymatic cleavage of specific peptides. Cysteine proteases are thus named due to the feature of a nucleophilic cysteine thiol occupying the catalytic cavity in which the substrate peptide is fragmented. Cathepsin S is one such member of this group that has received attention from the research community due to its involvement in inflammatory and immune response mechanisms and potential as a therapeutic target [320].

Researchers at Boehringer Ingelheim initiated a study of dipeptide-based nitriles as potential inhibitors of cathepsin S using a combination of parallel solution and solid-phase synthesis. The interest here is due to the role of the enzyme in regulating immunological proteins that, once truncated, are localized on the cell surface as antigenic markers. Therefore, inhibitors of cathepsin S

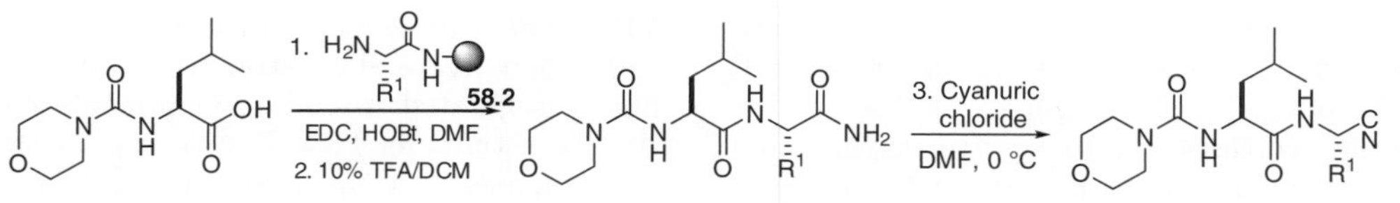

Figure 58. Synthesis of nitrile-containing library for cathepsin S screening.

would potentially serve as treatments in autoimmune disorders by short circuiting the presentation of antigens to T cells and eliciting an elaborate immune response.

Three methods were utilized to prepare nitrile-containing analogs whose structures explored three binding pockets of the enzyme. One of these methods employed a solid-phase component to elucidate the necessary binding elements of the P1 pocket of the enzyme. Coupling of the leucine derived morpholino urea **58.1** with Sieber resin supported amino acids **58.2**, after cleavage with TFA, afforded a series of amides **58.3** that were further dehydrated with cyanuric chloride to provide a library of nitriles **58.4** (Fig. 58).

Changes at R^1 demonstrated that as chain lengths increased (**59.1** → **59.2**), K_d values improved, culminating in picomolar active inhibitor **59.2**. Cocrystallization of **59.2** with cathepsin S clearly illustrated the formation of a covalent, reversible thioimidate bond between the substrate and the active site cysteine residue, thus confirming the necessity and type of interaction with the nitrile warhead. Additional structural permutations were analyzed that corresponded to P2 and P3 binding. These optimizations provided a series of equipotent analogs that were further

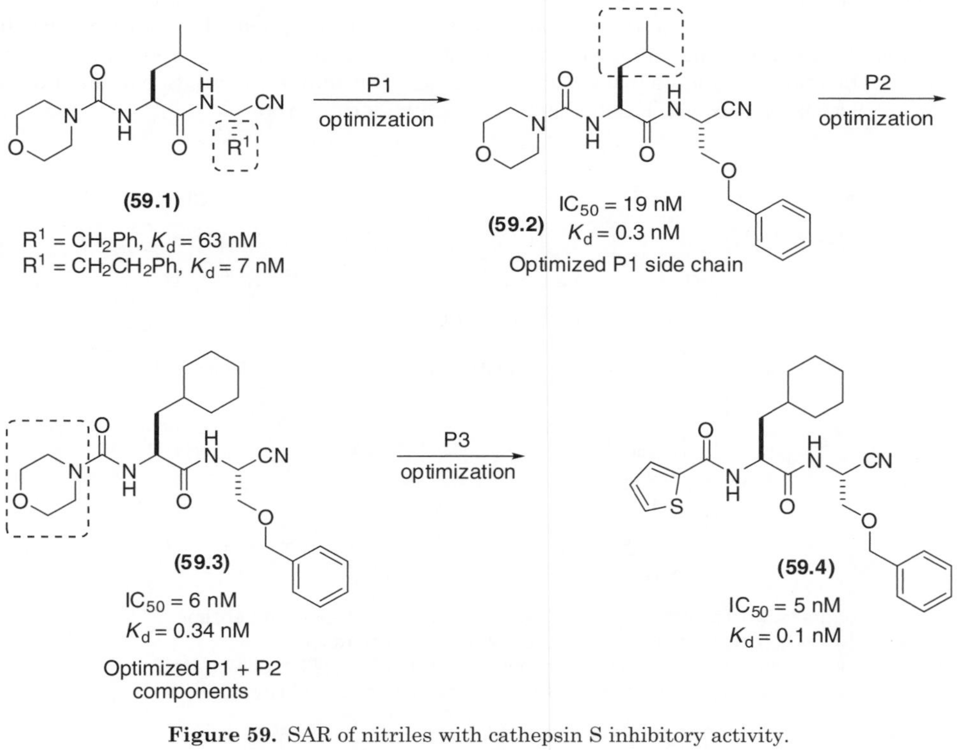

Figure 59. SAR of nitriles with cathepsin S inhibitory activity.

assessed using a human B cell assay, the most potent of these having MICs of 100 nM (Fig. 59).

Serine Proteases Proteases are enzymes that are responsible for hydrolyzing amide bonds of proteins. These proteolytic enzymes are classified into four major groups according to the actions that they perform and the makeup of residues in their catalytic site. Serine proteases utilize the hydroxyl group of a serine residue in concert with other neighboring amino acids in the active site to act as a nucleophile and cleave the protein substrate. These enzymes serve critical functions in digestion, inflammation, immunity, and in blood coagulation. The development of serine protease inhibitors to intervene in coagulation cascades has been an active area of research as thromboembolic diseases and conditions stemming from vascular injury are major causes of mortality throughout the world [308].

The factors responsible for blood clotting are generally serine proteases, though other enzyme classes are also involved. The key role of Factor VII, a circulating serine protease, is to initiate coagulation by complexing with tissue factor (TF) located on the external surface of blood vessels. Vascular injury permits exposure of the two components to each other to form an activated species, Tissue Factor VIIa (TF/VIIa), which functions proteolytically to activate downstream targets in a highly regulated cascade and promote clotting. Selectivity for TF/VIIa versus other plasma proteases was considered critical to avoid bleeding side effects that would be associated with antithrombotic efficacy [322].

As part of a program aimed at elucidating novel antithrombotic agents, researchers at Pharmacia (now Pfizer) utilized structure-based drug design and polymer-assisted solution-phase parallel synthesis to develop a series of orally bioavailable pyrazinone antithrombotics that selectively inhibit the TF/VIIa complex.

Armed with crystal structures of other proteins (thrombin and Factor Xa) that are involved in the coagulation cascade and TF/VIIa complexed with a peptidic substrate, the first set of pyrazinone inhibitors were designed. Preparation was done using a modified Strecker reaction with glycine benzyl ester, followed by cyclization to make the prototypic scaffold (Fig. 60). The diversification points were installed systematically and the need for chromatographic purification and intermediate isolation was avoided using polymer-assisted solution-phase chemistry. Several hundred compounds were prepared and purified using this amine-installation and resin-based sequestering protocol.

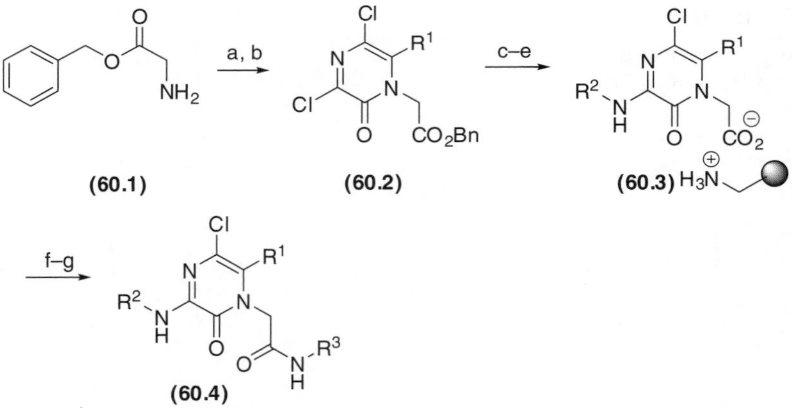

(60.1) (60.2) (60.3) (60.4)

Reagents: (a) TMSCN, R^1CHO, CH_2Cl_2; (b) $(COCl)_2$, 1,2-diclorobenzene; (c) R^2NH_2, AcCN; (d) H_2, Pd/C; (e) DCM/water, amine resin; (f) wash, then 4 N HLc/dioxane, then wash; (g) R^3NH_2, HOBt, CH_2Cl_2/DMF, polymer-supported amine for sequestering product and purification.

Figure 60. Synthesis of pyrazinone inhibitors of TF/VIIa complex.

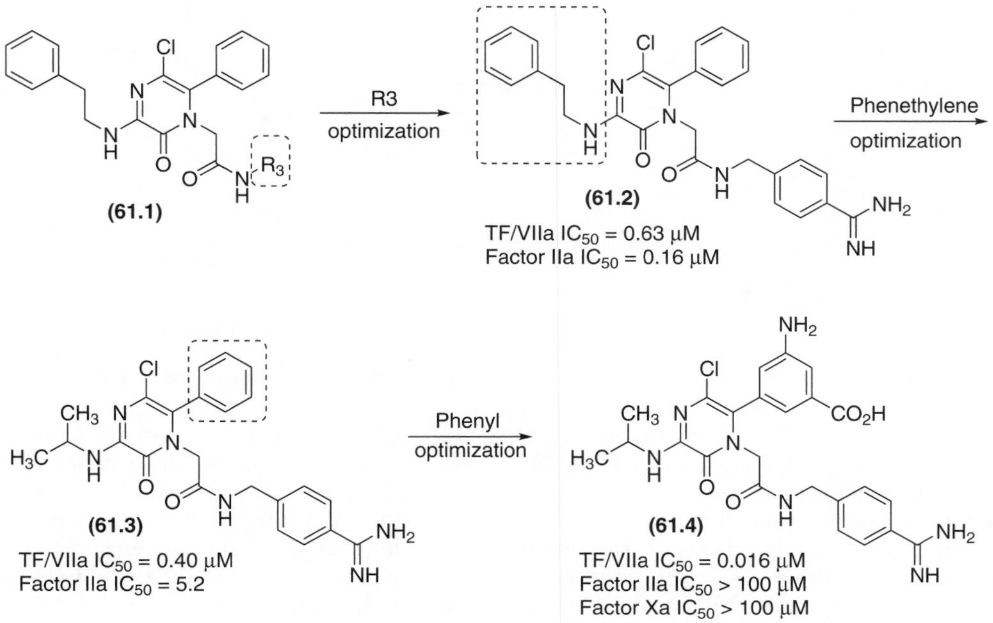

Figure 61. SAR and optimization of TF/VIIa inhibitors.

Early on, it was determined that a *p*-benzamidine moiety at R^3 was critical to maintain potency for TF/VIIa (**61.2**). This structural motif was preserved while modifications were surveyed at the 3-amino (R^2) and 6-substituent (R^1) positions. Substitution of the phenethylamino group with a small branched amino alkyl group delivered compounds like **61.3** with improved potency and selectivity for TF/VIIa. Further screening of functionalized aryl groups at C6 substantially refined the profile of these compounds and produced compound **61.3** with >6200-fold selectivity for TF/VIIa versus other coagulation cascade effectors, thrombin and Factor Xa (Fig. 61).

This program was able to demonstrate the productive use of structure-based drug design and solid-supported parallel synthesis as a powerful combination of techniques to generate very potent and selective TF/VIIa inhibitors. As a result of the enhanced profiles these compounds possess, they were advanced into preclinical, intravenous proof-of-concept studies.

Growth Hormone Secretagogue Receptor Antagonists The growth hormone secretagogue receptor (GHS-R) is a G-protein-coupled receptor that plays an increasingly appreciated role in the regulation of growth complementary to that ascribed to the better-known growth hormone releasing hormone receptor. The endogenous ligand for the GSH-R is ghrelin, a 28-amino acid peptide that increases food intake and is able to influence body weight, among other effects seen in both animal models and in humans. Taken together, these effects strongly suggest that an antagonist to the GHS-R could make a useful anti-obesity agent [309]. However, answering the question of whether a small molecule would be therapeutically useful in this context requires access to appropriate candidate probes to begin with—a notoriously difficult task when the only known ligand is an endogenous peptide. This is exactly the kind of question that a screening/parallel synthesis approach is well positioned to address.

The approach to a new class of GHS-R antagonists was initially informed by the existence of a lead isoxazoline-based compound with good potency. However, this structure suffered from poor pharmacokinetic properties, which led the investigators to seek alternative scaffolds for further structure–activity

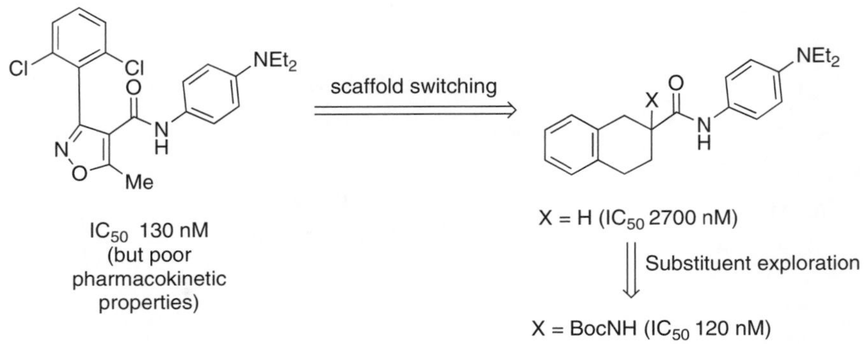

scaffold switching

IC_{50} 130 nM (but poor pharmacokinetic properties)

X = H (IC_{50} 2700 nM)

Substituent exploration

X = BocNH (IC_{50} 120 nM)

Figure 62. Scaffold hoping approach to inhibitors of growth hormone secretagogue receptors.

relationship studies. They first engaged in the process of "scaffold hopping," in which the *p*-diethylaminobenzamide substituent was held constant while a series of heterocyclic and carbocyclic substituents adjacent to the amide carbonyl were examined. This led to a series of weakly potent, but chemically attractive tetralin analogs (X = H in Fig. 62 below). Further systematic evaluation of ca. 10 series of compounds containing a substituent adjacent to the tetralin amide group afforded an *N*-Boc carbamate derivative in which all of the potency had been regained. These compounds were deemed appropriate for a systematic parallel synthesis approach [310].

The identification of a suitable carbamate-containing scaffold allowed for a systematic variation of both the tetralin substituents and the carbamate nitrogen through library synthesis. This process allowed for the identification of a very nicely potent compound (IC_{50} 16 nM) (Fig. 63). More importantly, this new lead had not only good binding to the receptor of interest but was demonstrated to be a functional antagonist with reasonable rat oral bioavailability and an acceptable pharmacokinetic/receptor selectivity profile for further studies. The synthesis of this compound, depicted below, featured the introduction of the crucial α-amido nitrogen atom through the application of a Curtius rearrangement reaction. Unfortunately, these compounds were eventually found not to have the appropriate antifeeding effects when tested *in vivo* and were ultimately abandoned in favor of additional scaffolds that were identified through later high-throughput screening campaigns.

Nuclear Hormone Receptors Nuclear receptors are cytosolic transcription factors that regulate gene expression involved with embryonic development, homeostasis, immune response, and metabolism. When activated by the appropriate ligand, these proteins translocate to the nucleus where they recruit other proteins to bind DNA directly and upregulate gene expression, thus initiating the production of proteins key to modifying cell function. A few examples of endogenous ligands that bind nuclear receptors include cortisol, testosterone, calcitriol, and 17β-estradiol. Drugs that interact with nuclear receptors have wide therapeutic utility in areas such as inflammation, cancer, and diabetes. As such, robust efforts continue to attempt to develop selective

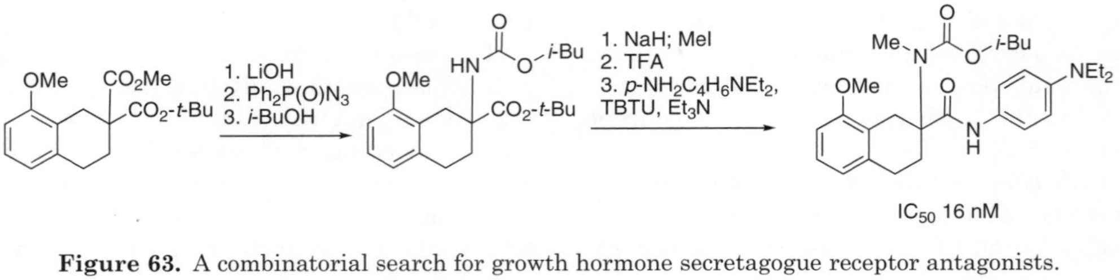

1. LiOH 2. $Ph_2P(O)N_3$ 3. *i*-BuOH

1. NaH; MeI 2. TFA 3. *p*-$NH_2C_4H_6NEt_2$, TBTU, Et_3N

IC_{50} 16 nM

Figure 63. A combinatorial search for growth hormone secretagogue receptor antagonists.

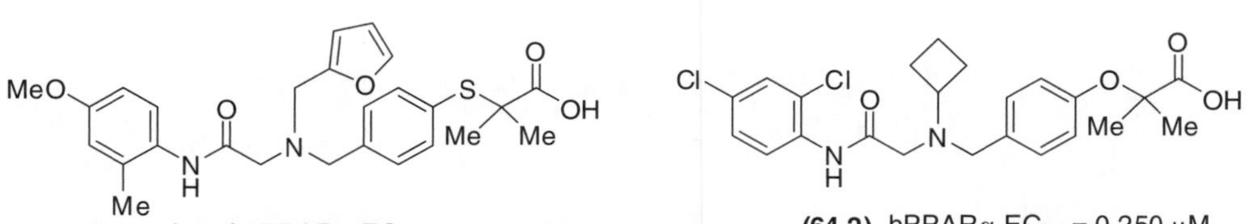

(64.1) hPPARα EC_{50} = 0.001 μM
hPPARγ EC_{50} = 0.100 μM
hPPARδ EC50 =1.500 μM

(64.2) hPPARα EC_{50} = 0.250 μM
hPPARγ EC_{50} = 1.00 μM
hPPARδ EC50 =0.080 μM

Figure 64. Lead structure **64.1** from PPARα program and starting point **64.2** for engineering PPARδ selectivity.

nuclear receptor modulators as therapeutic agents.

The three known subtypes of *peroxisome proliferator-activated receptors* (PPARα PPARγ and PPARδ belong to the nuclear hormone receptor superfamily and have been widely investigated as potential modulators of diabetes and other metabolic related disorders due to their implication in the metabolism and storage of fatty acids. A PPAR program within Bayer aimed at achieving potent selectivity among the PPAR family produced lead structures selectively effecting PPARα **64.1**). Using compounds from that effort that showed selectivity toward PPARδ as the starting point (**64.2**), a combination of solution-phase chemistry and solid-phase parallel synthesis and split-and-pool techniques were employed to generate libraries targeting PPARδ [311] (Fig. 64).

The general synthetic approach to the PPARδ selective library is outlined in Fig. 65. A Wang resin supported $\tilde{\alpha}$-bromoacetic acid ester was treated with variously substituted hydroxybenzaldehydes under Williamson ether synthesis conditions to provide a diversified core **65.1**. Reductive amination, followed by amino acid extension chemistry incorporated two additional variables, R^4 and R^5. Cleavage of these intermediates from the resin afforded a library of structurally related phenoxyacetic acid compounds from which structure–activity relationships were derived.

In this investigation, R^1 and R^2 were initially held constant as methyl groups, and 2,4-disubstituted aryl groups at R^5 were found to be advantageous. The initial optimization process surveyed alkyl alterations made at R^4 while toggling between a methyl or hydrogen substitution at R^3 (Fig. 66).

This effort revealed that n-alkyl or $\tilde{\beta}$-branched cycloalkyl substituents at R^4 did not enhance δ-selectivity; however, the presence of a cyclopentyl or cyclohexyl group at that

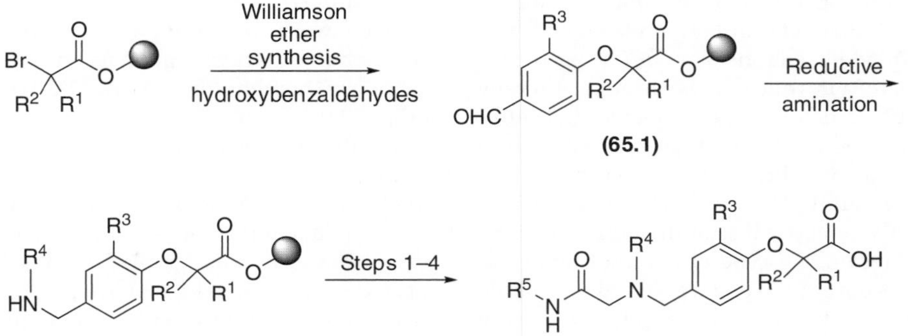

Williamson ether synthesis
hydroxybenzaldehydes
Reductive amination
(65.1)
Steps 1–4

Reagents: 1. TMS-bromoacetate; 2. TBAF; 3. amide coupling; 4. resin cleavage.

Figure 65. Synthetic approach to a PPARδ selective library.

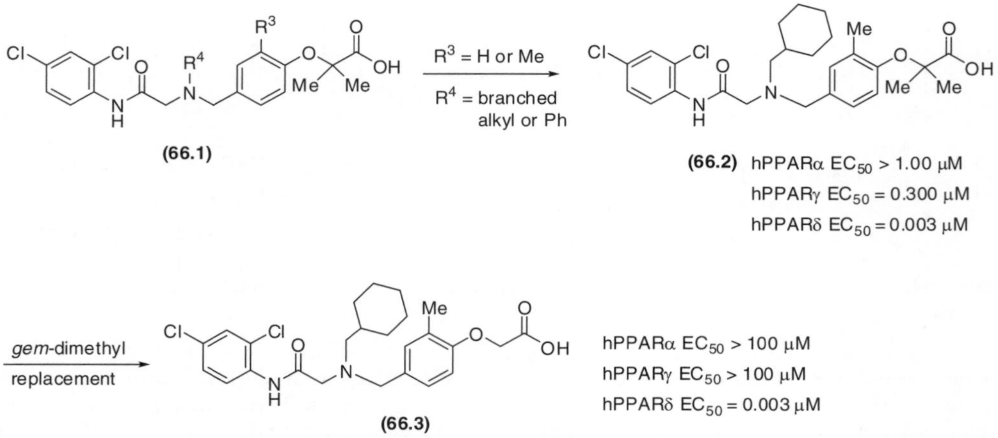

(66.1) → R^3 = H or Me; R^4 = branched alkyl or Ph → **(66.2)** hPPARα EC_{50} > 1.00 μM; hPPARγ EC_{50} = 0.300 μM; hPPARδ EC_{50} = 0.003 μM

gem-dimethyl replacement → **(66.3)** hPPARα EC_{50} > 100 μM; hPPARγ EC_{50} > 100 μM; hPPARδ EC_{50} = 0.003 μM

Figure 66. SAR evolution in PPARδ selective inhibitors.

position resulted in >25-fold boost in potency at the δ-receptor while decreasing activity at the remaining two PPAR receptors. Additionally, greater potency for PPARδ was observed with the introduction of the methyl group at R^3 versus hydrogen. Finally, the *gem*-dimethyl moiety comprising R^1 and R^2 was replaced with a methylene unit (R^1, $R^2 = H$). These combined changes resulted in compound **66.3** with an EC_{50} at the δ-receptor of 3 nM and >100 μM at both PPARα and PPARγ The pharmacological profiles of the lead compounds are under further investigation.

Ion Channels The regulated movement of ions across a biological membrane is necessary for many cellular functions. Transport of charged ions across a formidable lipophilic barrier is made possible through the formation of pore-forming proteins that assemble in the lipid bilayer of cellular membranes. These water-filled ion channels are often specific for the type of ion that passes through, thereby permitting fine control of ion-governed cellular processes and the voltage gradient across the cellular membrane. Passage through the pore is modulated by various effectors such as chemical or electrical stimuli or temperature, depending on the type of channel involved. These receptors have attracted considerable attention as therapeutic targets due to the processes with which they are associated. For instance, channels function prominently in the conductance of nerve impulses in the central nervous system as well as in cardiac muscle contraction, pancreatic cell insulin release and immunological T-cell activation [312].

The $P2X_7$ receptor, an ATP-activated, ligand-gated ion channel present on mast and other immune-response cells, is one such target that researchers have been investigating as a possible means of addressing various inflammatory and autoimmune disorders. Research has demonstrated that activation of this particular receptor initiates cytokine release and proinflammatory cascades; therefore, a program at Sanofi Aventis aimed at the design of a $P2X_7$ receptor antagonist was undertaken. A high-throughput screening effort delivered a potent $P2X_7$ lead structure **67.1** on which further development was focused [307].

This work was reviewed briefly with a somewhat different emphasis in a previous section of this chapter (see the $P2X_7$ treatment on p. 324) (Fig. 67).

Analog synthesis was completed by combining solid-phase and microwave techniques and allowed for rapid SAR exploration around the 4,5-diarylimidazoline scaffold. The parent ethanediamine was coupled to a solid-supported *p*-nitrophenylcarbonate to afford a resin-bound carbamate with a pendant amine (Fig. 68). The amine was then treated with various carboxylic acids to provide a diversified set of resin-bound amides quantitatively.

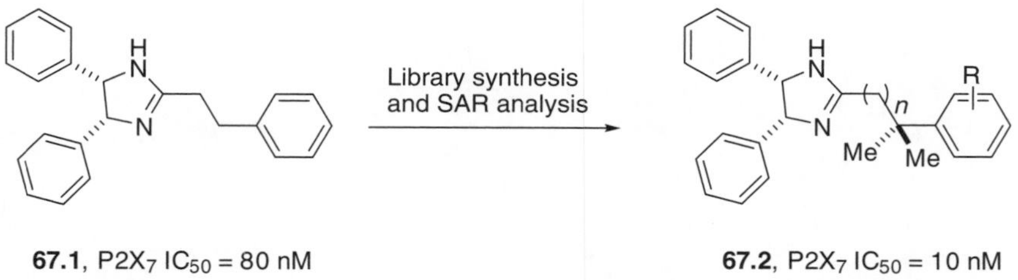

67.1, $P2X_7$ IC_{50} = 80 nM

Library synthesis and SAR analysis

67.2, $P2X_7$ IC_{50} = 10 nM
R = H, n = 1

Figure 67. Lead structure **67.1** and optimizations for potency affording **67.2**.

A streamlined approach in which heat and acid-catalyzed cyclization to the imidazoline products would occur concomitantly with resin cleavage failed to afford desired ring-formed materials; however, the imidazoline library was prepared once the intermediates were released from the resin and treated with trimethylsilylpolyphosphate in dichloromethane under microwave irradiation.

Various structural components were surveyed on the imidazoline scaffold, including appropriate chain length, branching, and constraining elements on the 2-alkylphenyl (R) substituent. The library was assessed for potency using a cell-based fluorescence assay employing U373 cells stably expressing human $P2X_7$.

This SAR effort revealed that alkyl linkers of 1–4 carbon atoms between the imidazoline and phenyl ring were tolerated; however, a two-carbon linker appeared to be ideal in terms of optimal potency. Complete removal of the linker dramatically attenuated potency. Incorporation of methyl or *gem*-dimethyl substituents on the linker were beneficial, the latter of which provided an analog with an IC_{50} of 10 nM. Another set of compounds were prepared that surveyed the effect of substitution of the alkyl-linked phenyl ring on potency. Unlike electron withdrawing groups, which enhanced activity, generally, electron donating substituents deteriorated potency. The study revealed a number of double-digit nanomolar active compounds for $P2X_7$ and highlighted an expeditious route to the imidazoline core that was amenable for SAR analysis by library synthesis.

Kinases "Kinase" is a general term for enzymes that add a phosphate group to a peptide or protein substrate, an act that is generally associated with the activation of cell signaling pathways. Typical sites for phosphorylation are tyrosine, serine, and threonine. Accordingly, the "kinome" comprises a dizzyingly broad array of targets of great potential medical value. Given this situation, it is not surprising that kinases inhibitors have been broadly targeted as one of the most prevalent targets in millennial medicinal chemistry. Accordingly, it is impossible to due justice to this

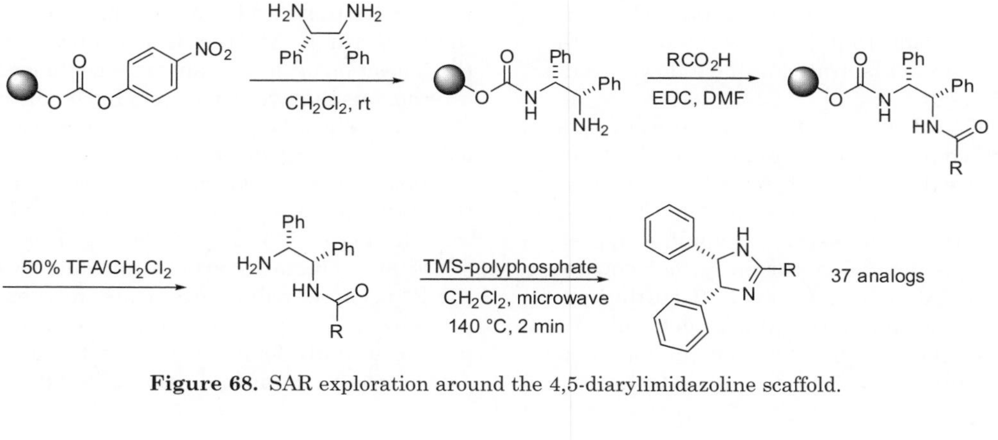

H_2N NH_2 Ph Ph; CH_2Cl_2, rt

RCO_2H; EDC, DMF

50% TFA/CH_2Cl_2

TMS-polyphosphate; CH_2Cl_2, microwave; 140 °C, 2 min

37 analogs

Figure 68. SAR exploration around the 4,5-diarylimidazoline scaffold.

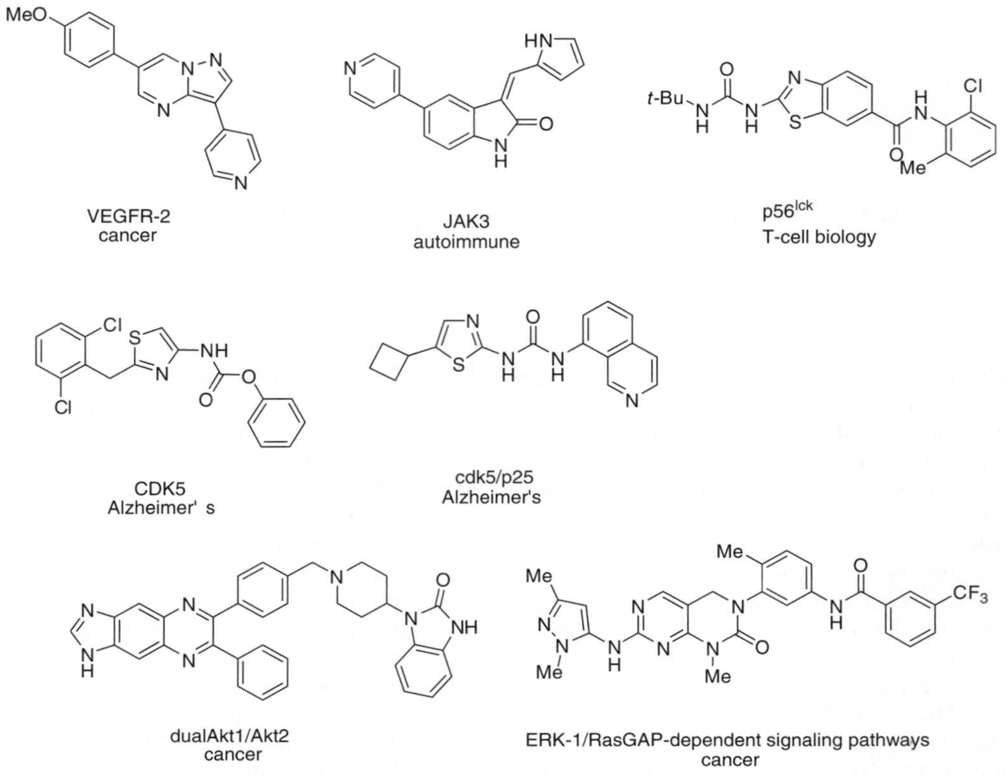

VEGFR-2
cancer

JAK3
autoimmune

p56[lck]
T-cell biology

CDK5
Alzheimer' s

cdk5/p25
Alzheimer's

dualAkt1/Akt2
cancer

ERK-1/RasGAP-dependent signaling pathways
cancer

Figure 69. Examples of developed kinase inhibitors for various therapeutic targets [328–335].

field in an entire chapter much less a section of one and so the coverage here will be brief and anecdotal. Interestingly, many kinase inhibitors share a certain level of structural similarity, possibly because of structural similarities in the enzyme class itself. General strategies toward libraries directed at kinase families have been developed and reviewed [313]. A few representative examples of scaffolds that have resulted from these efforts are illustrated in the following gallery (Fig. 69).

One of these scaffolds was the subject of a search for an inhibitor of the protein tyrosine kinase (PTK) Janus Kinase 3 (JAK 3) at Aventis (now Sanofi-Aventis) [336]. The observation that some severely immune-suppressed patient groups concurrently had downregulated JAK 3 activity contributed to the hypothesis that inhibitors of this target could potentially serve as important modulators in autoimmune disorders such as arthritis or diabetes. A focused high-throughput screening campaign produced hit compound **70.1** with activity of 900 nM. Further optimizations were designed using a JAK3 homology model derived from the crystal structure of cyclic adenosine monophosphate dependent protein kinase complex. Docking of the hit was accomplished by aligning the donor–acceptor elements with that of ATP in the binding cavity, thus indicating that structural modification of the 5-position of the oxindole ring could make interactions that were not exploited by the hit structure. The SAR of the 5-position was varied by preparing diversely 5-functionalized oxindole starting materials and condensing them with an assorted set of aldehydes, affording a library of 700 compounds (Fig. 70).

Despite screening various functionality in the R^2 position, no improvements in potency could be made beyond that observed with the pyrrole initially found in the hit structure. Alternatively, the introduction of bulky aryl

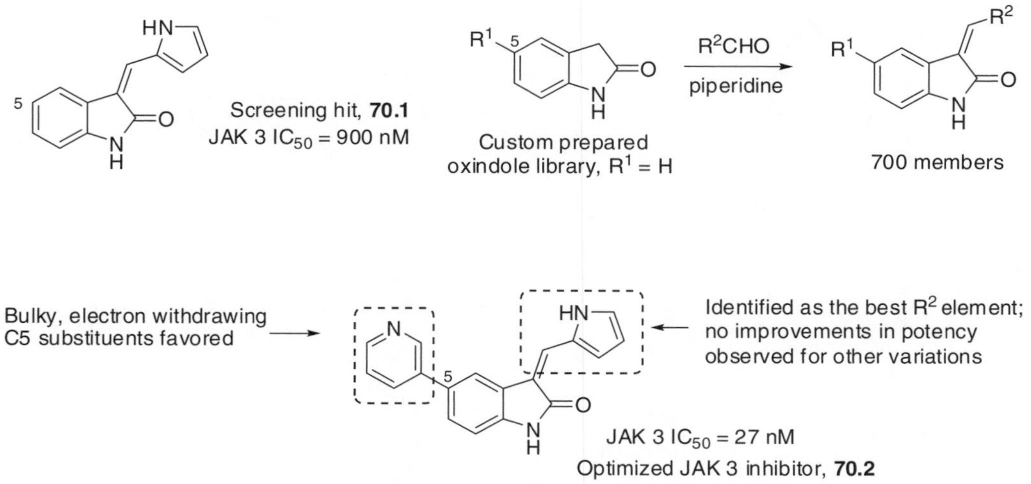

Screening hit, **70.1**
JAK 3 IC_{50} = 900 nM

R^2CHO, piperidine

Custom prepared oxindole library, R^1 = H

700 members

Bulky, electron withdrawing C5 substituents favored

Identified as the best R^2 element; no improvements in potency observed for other variations

JAK 3 IC_{50} = 27 nM
Optimized JAK 3 inhibitor, **70.2**

Figure 70. Screening hit for JAK 3 kinase and SAR development.

electron withdrawing substituents at R^1 enhanced affinity, producing inhibitor **70.2** with a JAK3 IC_{50} of 27 nM. Docking of **70.2** in the model showed the pyridyl C5 substituent advantageously occupying a previously unexploited region of the active site. Bioavailability was enhanced via formation of the methansulfonate salt, and a 200 μg topical dose of **70.2** demonstrated efficacy in reducing ear weight comparable to a 10 μg dose of dexamethasone in the ear oedema model of inflammation.

In a separate endeavor, researchers at the Scripps Research Institute undertook a combinatorial approach to generating biased heterocyclic libraries geared toward possessing kinase activity [283]. The method involved selection of a dichlorinated purine scaffold **71.2**, followed by orthogonal modification by mounting the template on an aminated PAL resin **71.1** and using various coupling protocols to diversify the remaining chloro substituent of **71.3** (Fig. 71).

This strategy was applied to other heterocycles such as pyrimidines, pyridazines, quinazolines, quinoxalines, phthalazines, and 1*H*-benzo[d]imidazoles and parlayed into multiple 140 membered heterocyclic libraries [54]. The collective set of heterocycles was then surveyed in a high-throughput, cell-based assay to find compounds that affected self-renewal and differentiation of embryonic stem cells [295]. Compounds that hit in the assay would serve as biological probes to aid in the understanding of the differentiation mechanisms at play. A thorough comprehension of these processes would guide therapeutic intervention when these same systems malfunction, resulting in cancer or other diseases. Members from the 3,4-dihydropyrimido[4,5-d] pyrimidine class of compounds were gleaned as hits from the screen of 45,000 compounds derived from the combinatorial approach described above. An SAR effort culminated in the identification of pluripotin (**72.1**, SC1) with an EC_{50} of 1 μM in the self-renewal assay and low cellular toxicity. Further study of the mode of action of **72.1** was discerned to be a result of inhibiting ERK1 and RasGap dependent signaling pathways. As such, this compound represents a useful new probe molecule that may elucidate novel therapeutic targets associated with stem cell self-renewal and differentiation.

This work was reviewed briefly with a somewhat different emphasis in a previous section of this chapter (see the nuclear receptor treatment on p. 334) (Fig. 72).

Protein Tyrosine Phosphatases The phosphorylation state of tyrosine residues of various proteins integral to cellular signaling cascades is regulated by opposing actions of protein tyrosine kinases (PTKs, promote phosphorylation) and protein tyrosine phosphatases (PTPs,

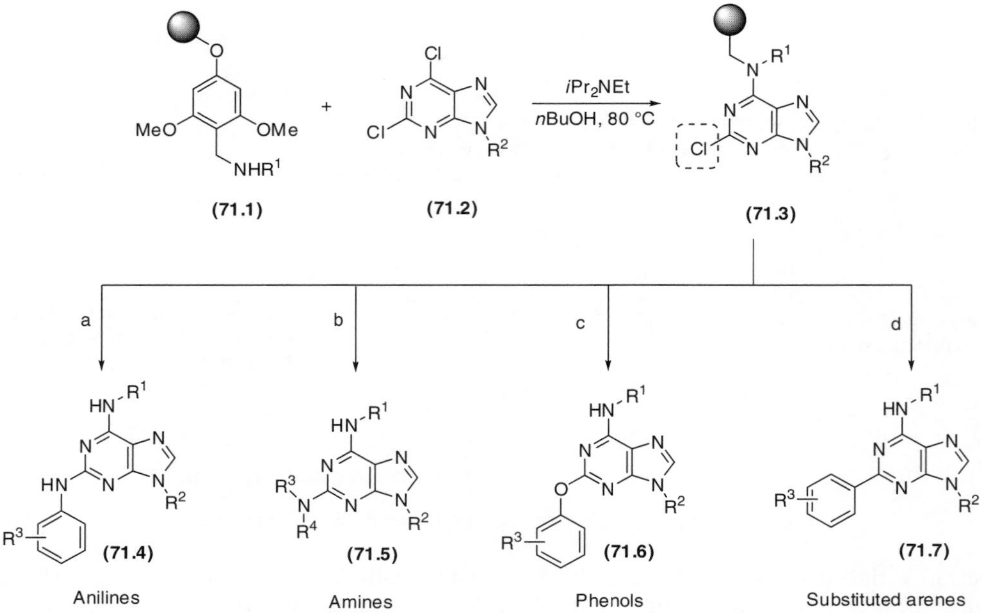

(71.1) (71.2) (71.3)

$i Pr_2NEt$, nBuOH, 80 °C

a b c d

(71.4) Anilines

(71.5) Amines

(71.6) Phenols

(71.7) Substituted arenes

Reagents: (a) $Pd_2(dba)_3$, carbene ligand, aniline, KOtBu, dioxane, 90 °C, then TFA, Me_2S, DCM, H_2O; (b) 1° or 2° amine, 90 °C, then TFA, Me_2S, DCM, H_2O; (c) $Pd_2(dba)_3$, phosphine ligand, phenol, K_3PO_4, toluene, 90 °C, then TFA, Me_2S, DCM, H_2O; (d) $Pd_2(dba)_3$, carbene ligand, boronic acid, Cs_2CO_3, dioxane, 90 °C, then TFA, Me_2S, DCM, H_2O.

Figure 71. Differentiation of a dichlorinated purine scaffold for library generation.

promote dephosphorylation). Extensive study of PTKs has led to clinical significance; however, PTPs have received far less attention until more recently when the roles that the defective variants play in disease has become better understood. For instance, some pathogenic bacteria possess evolved mechanisms of evading destruction by secreting their own PTPs and interfering with host-driven immune response signaling. *Mycobacterium tuberculosis*, the deadly infectious agent responsible for tuberculosis (TB), targets human macrophages by secreting PtpB protein tyrosine phosphatase. PtpB is an attractive target since it is secreted and operates outside of the protective barrier that the bacterium uses to

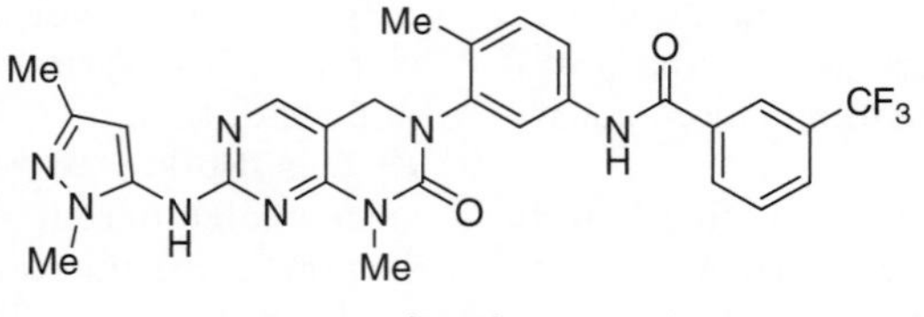

(72.1)
Pluripotin, SC1
Self renewal assay: EC_{50} = 1 μM
ERK1 K_d = 98 nM
RasGap K_d = 212 nM

Figure 72. Dual ERK-1/RasGAP inhibitor discovered from a kinase-directed combinatorial library.

render itself more robustly impervious to conventional antibiotics. Novel drugs are needed as TB remains one of the world's leading causes of illness and death. Current remedies require a 6–12 months course of treatment and the number of multidrug resistant strains continues to increase [337,338].

In an effort to find submicromolar *Mtb* PtpB inhibitors, Ellman and coworkers at the University of California Berkeley developed a fragment-based approach, described as substrate activity screening (SAS) [314]. A library of 140 *O*-aryl phosphates, prepared via solution-phase parallel synthesis from a diverse set of substituted phenols, was screened against PtpB using a fast and sensitive spectrophotometric coupled assay. The initial screen of the 140 member library revealed several weakly binding fragment hits that possessed improved K_M values relative to the control. A focused library of 45 members was then constructed around biphenyl phosphate fragment **73.1** due to the drug-like properties it possessed. This exercise, which surveyed substitutions on both phenyl components, provided tractable SAR and produced fragment **73.2** with a K_M of 19 μM. With suitable enzyme substrates now in hand, the team set out to find a suitable replacement of the phosphate moiety to convert the substrates to potential inhibitors. The choice of phosphate isostere was limited to the use of isothiazolidinone and isoxazole carboxylic acid in an effort to engineer acceptable physical properties to facilitate cell permeability and maintain activity. Replacement of the phosphate group with the nonhydrolyzable, monoacidic phosphate isostere generated PtpB inhibitor **73.3** with improved micromolar potency. Additional optimization of the inhibitor scaffold was undertaken with the isoxazole carboxylic acid in place to produce the most potent PtpB inhibitor **73.4** demonstrated in the literature with a K_i of 220 nM. Further studies revealed that compound **73.4** possesses good selectivity against a panel of mycobacterial (PtpA) and human PTPs (For PtPA, VHR, and TC-Ptp, $K_i > 50$ μM; for CD45, $K_i = 7.70$ mM; for LAR, $K_i = 21.4$ μM) (Fig. 73).

Integrins, Selectins, and Protein–Protein Interactions Integrins are a series of dimeric peptides expressed on cell surfaces that interact with various receptors, typically mediate cell–cell interactions, and are relevant to a broad selection of biological processes. Examples of integrins of interest to medicinal chemists include a_4b_1 (also known as VLA-4, which stands for very late antigen 4 and mediates various aspects of inflammation including cell adhesion and activation), a_vb_3 (angiogenesis), and $a_{IIb}b_{III}$ (platelet aggregation). Early approaches to using combinatorial methods for the identification of integrin antagonists have been reviewed [315].

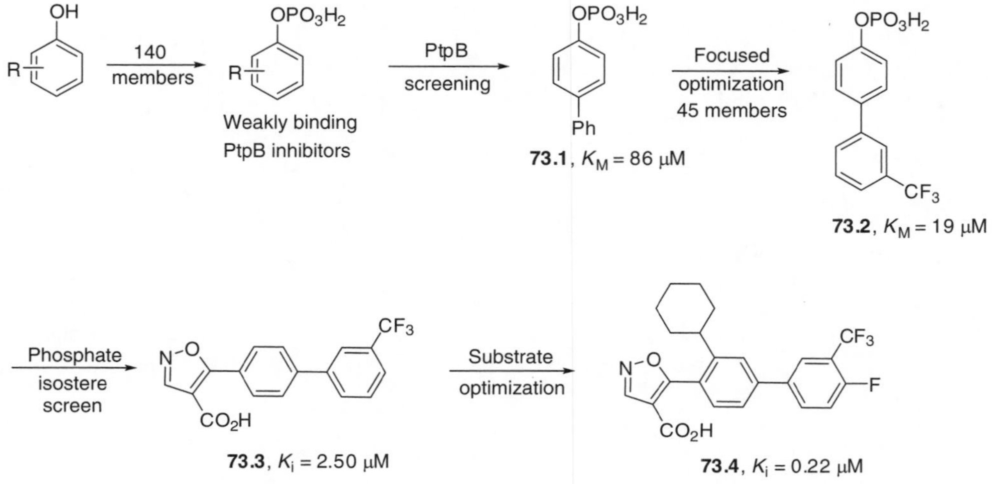

Figure 73. Development of a potent PtpB inhibitor from substituted phenols.

Several approaches to the design of inhibitors of VLA-4 were published nearly simultaneously by groups at Roche, Merck, and Celltech [340–342]. Interestingly, each of these groups began their work with similar leads, independently uncovered through high-throughput screening work. These compounds all featured a phenylalanine core decorated with aromatic rings and featuring several additional amide bonds. In hopes of addressing pharmacokinetic problems attributed to the amide substituent attached to the Phe a-amino group, each group carried out syntheses of additional library members bearing different groups at this position. Of these, the diversification chemistry carried out by the Celltech group is illustrated. Rather than utilize standard amine couplings, this group relied on the precedented use of a squarene core to serve as an amino acid bioisostere. Coupling of the methoxy-containing precursor to the left-hand part of the molecule identified initially afforded a central synthetic intermediate suitable for coupling with a variety of amines. The resulting compounds provided additional compounds of high binding affinity although pharmacokinetic problems persisted (Fig. 74).

Selectins are cell-surface glycopeptides that mediate cell adhesion, a key step in inflammation or metastasis; one possible use of a selectin antagonist would be for the treatment of asthma [316]. Since the key interaction in selectin binding to a call involves carbohydrate interactions, one approach to the development of specific selectin antagonists focuses on the synthesis of a series of novel carbohydrate–amino-acid-like conjugates via the Ugi four-component coupling reaction. Although the compounds prepared were only modestly active, optimization of the basic scaffold resulted in a second-generation version that was active against selectin (IC_{50} of 4 mM) (Fig. 75).

Both integrin and selectin antagonists fall into the general category of molecules that inhibit protein–protein interactions [317].

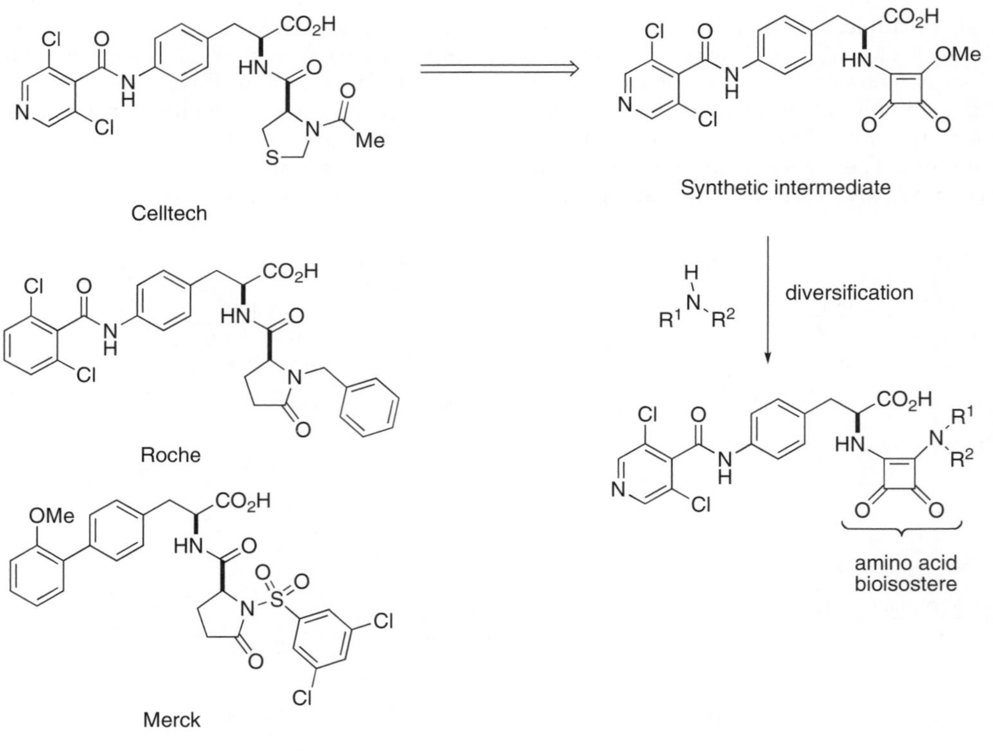

Figure 74. A search for integrin inhibitors.

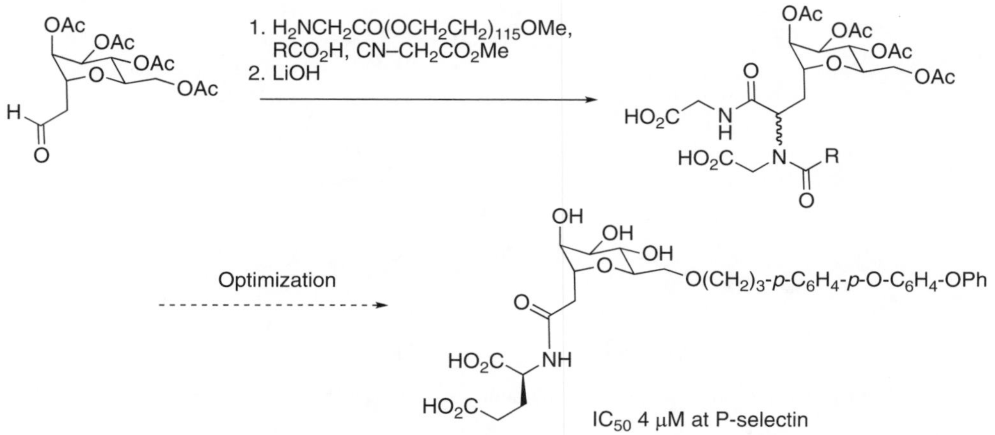

Figure 75. A search for some selectin inhibitors.

This is a knotty problem due to the often large surface area that must be covered in order to successfully compete against a target's natural binding partner. Besides computational approaches [345], some success has been obtained through the examination of peptoid libraries that are able to mimic the size and flexibility of bona fide protein interaction sites (Fig. 76). The compound shown was found to inhibit apoptosome-dependent activation of procaspase-9 [318].

The remainder of this chapter deals with selected examples that illustrate particular concepts and methodologies.

4.11. Resins and Solid Supports

A great variety of resins and solid supports are available for combinatorial work [18,149, 319–321]. The Aldrich Handbook of Fine Chemicals, for example, has an extensive listing of resins for a wide variety of purposes that are useful for synthesis by combinatorial chemistry. Gel-type supports are popular and consist of flexible polymeric matrixes to which are attached functional groups capable of binding small molecules. The particular advantage of this inert support is that the whole volume of the gel is available for use rather than just the surface. Generally these consist of cross-linked polystyrene or polyacrylamide resins, polyethylene glycol (PEG) grafted resins, and PEG-based resins. Surface functionalized supports have a lower loading capacity and many types are available. These include cellulose fibers, sintered polyethylene, glass, and silica gels. Composite gels are also used. These include treated Teflon membranes, kieselguhr and the like. Brush polymers consist of

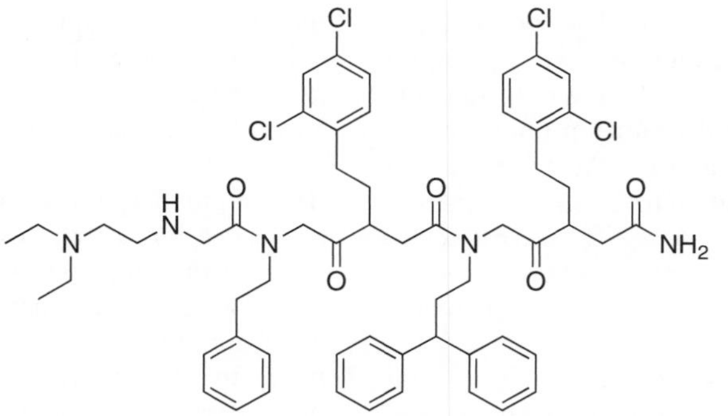

Figure 76. An inhibitor of procaspase-9.

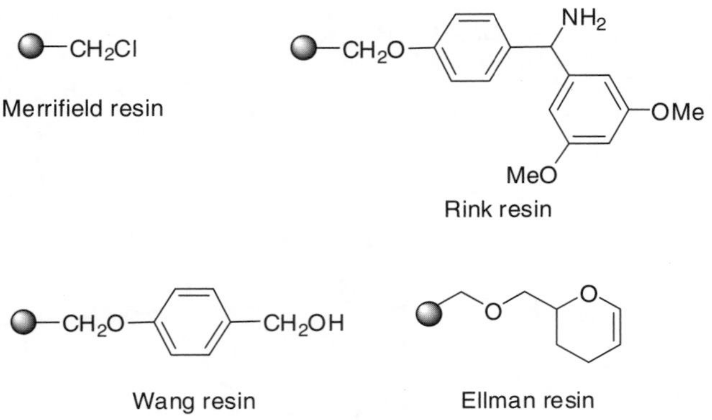

Merrifield resin

Rink resin

Wang resin

Ellman resin

Figure 77. Some resin types commonly used in combinatorial chemistry.

polystyrene or the like grafted onto a polyethylene film or tube.

The linking functionality varies. Commonly employed resins are the Merrifield, Wang, Rink, and Ellman types. These are illustrated in Fig. 77 and the reader can readily appreciate the kinds of chemistry that they allow.

Dendritic linker arms attached to resins have enlarged capacities and the physicochemical nature of the central resin support has a diminished influence on reactions and makes interaction of attached molecules less likely [322–324].

Solution-phase chemistry has been greatly facilitated by development of a wide range of reagents supported on resins. These are often attached to polystyrene which swells acceptably in organic solvents. Many other supports (polyacrylamide, polyureas, etc.) have also been developed for various purposes. Their special advantage is that they can easily be separated by filtration when the reaction has finished even when used in excess to promote reaction. They also generate less reaction detritus in comparison with solution-phase methods. Such reagent trash can be tenacious to remove. Further, they can often be regenerated and/or reused in order to control costs. They do, however, react more slowly. This can be compensated for by application of microwave technology.

Another important use of resins is scavenging excess reactants and or undesirable byproducts followed by filtration. Another use is to capture a desired product, remove it from the other materials present in the reaction, and then release it from the resin. This is a selective and efficient method of purification of compounds away from reaction junk.

Thus polymer-assisted solution-phase synthesis has been developed utilizing fully automated flow-through methods. Libraries can be made from libraries and the process leads to diverse products in high purity without the need for chromatography [325]. Some instructive examples of these uses are given below.

4.12. Solution-Phase Synthesis for Lead Optimization

Solution-phase synthesis for lead optimization now predominates in combinatorial work. Not only is this chemistry more readily scalable but it involves chemical operations that are more readily mastered by organic chemists who have mostly been trained to use solution methodologies from their earliest days. Furthermore a greater range of reagents, temperatures and solvents are possible in solution-phase chemistry. Solution phase does introduce complexities in purification of the products. These problems are, however, managed well by the use of phase trafficking methods and the use of resin-bound reagents. The use of resins also requires attention to complexities introduced by the need to choose attachment arms that do not interfere with subsequent steps and do not release the increasingly complex products prematurely. Attachment–detachment also lengthens the overall synthesis sequence by at least two steps. These problems are avoided when solution chemistry is employed.

4.13. Linear, Convergent, and Multicomponent Reactions

Linear reaction sequences are quite acceptable in combinatorial libraries for drug seeking provided that they are not more than about five reactions in sequence. Generally each stage requires purification otherwise undesirable reagents and by-products pile up making the preparation of a single pure end product rather difficult. Furthermore, efficiency of lead optimization is maximized when a common platform can be carried along in quantity through a synthetic sequence and then the diversity be added at a late stage. That is, the truly combinatorial portion of the process is best performed as late as possible in the construction phase.

Convergence in synthetic design has long been known to produce higher yields in multistep reaction sequences due to minimization of yield attrition as a function the decreased number of linear reaction steps. The ultimate in convergent design and reaction efficiency is multicomponent condensations. The Ugi, Passerini, and analogous name reactions are favored in this context. Initial problems encountered due to the limited number of isonitriles largely have been overcome. The remaining problem is caused by the necessity of the component condensations proceeding sequentially. When different reaction components are employed, occasionally the reactions proceed in a different sequence because of differing intrinsic reaction rates. If this occurs, then one looses control of the reaction and unexpected products can result. The products may well still be interesting but they are often not what one planned for. This can lead to difficultly separable mixtures.

A recent examination of a multicomponent synthetic design comparatively free of these complexities involved copper-catalyzed catalytic assembly of three components coupled with subsequent ring forming Diels–Alder chemistry rapidly leading to a complex series of isoindolones [326]. Presumably many other kinds of combinatorial reactions are possible using this general strategy.

4.14. Mixed Solid-Phase Drug Synthesis and Solution-Based Drug Synthesis

A great deal of progress has been made in developing resin-bound reagents that transform organic molecules efficiently in solution and which can then be removed by simple filtration. The use of capture resins to remove components that are not wanted in the final products or to abstract selectively a desired product from reaction detritus is also now an advanced art [327]. A major objective in designing practical combinatorial processes is the avoidance of chromatography. These solid-based reagents and the use of phase trafficking by capture resins is an exciting technology that largely owes its development to the necessity to reduce the labor required for producing solution-phase combinatorial compound libraries containing a significant number of components.

4.15. Purification

Clearly designing and constructing appropriate molecules to include in purpose built compound libraries is central to the success of combinatorial chemistry. In communicating their results chemists explain the design of the route and discuss the relative strengths and weaknesses of alternatives [328–330]. Separation and purification of the products in an efficient and timely manner is also essential but is less celebrated. Satisfactory isolation protocols are especially demanding in solution-phase combinatorial work. Little is gained if one saves much time in construction only to have to give this back by tedious and repetitious purification schemes. Increasingly attention is being paid to isolation and purification so that they do not become crippling bottlenecks [331,332].

Separation is comparatively uncomplicated when dealing with focused multiple parallel synthesis libraries if the compounds are rather similar in properties and the number of desired products made in parallel is comparatively small. Chromofiltration works well here especially when a manifold filtration devise is used [259,331,333]. Disposable or reusable cartridges prepacked with a variety of ion-exchange or solid-phase extraction supports are particularly convenient if cost is not a major factor [331]. Capture and subsequent release of desired products using bead protocols is also useful [319,320,327,331,334–344]. Use of perflourinated derivitizing reagents in order to direct products to a separated phase are also very useful [345–353]. It is even possible to separate products by solid-phase extraction in a 96 well plate format through gravity filtration [354]. Gravity filtration is made possible by development of large particle sized materials. Cartridges packed with solid-phase fluorous materials are also available for separation of reaction products under automation control [355]. Use of scavenger resins has become particularly popular in solution-phase chemistry and has obvious applications in purifying parallel synthesis compound library contents.

Even with use of these devices purification of large compound libraries requires automation to be truly efficient and a number of commercial firms have introduced equipment to meet these requirements [356–359]. In big pharma specialization is common. To assist medicinal chemists with their work it is often found to be helpful to set up separate "separation factories" manned by separation experts capable of delivering purified reaction products of an estimated rate of thousands of compounds per week [360].

Performing chemistry on beads addresses separation and purification needs in that simple filtration and washing often suffices although removal of excess reagents puts greater pressure on the purification techniques [327]. Separation from solution in solid form is very convenient when this is possible but from a druggability standpoint there is a danger in this. Compounds that separate readily from solvents are often of very low water solubility and present difficulties in biotesting and formulating. A number of commercially available combinatorial screening libraries historically, particularly from certain Eastern European sources, have been peppered with substances possessing solubility characteristics similar to that of ground up paving stones.

Column chromatography using reverse phase gradient HPLC has become very widely used as a generic purification procedure. Since the mobile phase is aqueous and may be acid-doped, simple evaporation of eluents can lead to contamination of the separated products due to the separation process itself. This can be minimized by use of in line trapping cartridges for collecting the desired components. Further purification can be accomplished then by washing techniques [361]. These various chromatographic separations are powerful but are usually labor intensive and solvent consuming. Separation of hundreds of analogs by column chromatography can be a logistical nightmare. To deal with this, column chromatography monitored by mass-directed techniques is automatable and efficient [362–364]. The essence of this is programming the instrumentation to collect peaks that display the desired mass. Use of standard compound mixtures prepared for this purpose raises the comfort level of chemists using this methodolgy [365]. It is important to design the experiment carefully so that each desired component has a unique mass [366]. When carried out appropriately this technique is more efficient than split-mix, tagging–detagging or tagging–coding methods, or preparation of one compound–one bead libraries. Indeed, separations can be automated to the extent that round the clock operation not requiring constant human supervision is now available. This chemistry also can be comparatively "green" in that solvent needs can be greatly reduced and disposal of unwanted materials is simplified. The application of mass-directed preparative supercritical fluid chromatography offers similar advantages [367].

An alternative to LC-MS-directed fractionation is the use of ultraviolet absorption for column monitoring purposes [368]. This method triggers collection when a particular UV absorbing peak is eluting. This is cheaper but less precise than LC-MS.

Due to the impact of these newer methods, today one rarely sees a separatory funnel in a combichem laboratory. This is not to say that liquid–liquid extraction is not performed. It is facilitating liquid–liquid processing, automated procedures have been developed. One work station recently described can extract up to 96 reactions in parallel avoiding much tedious labor [369]. Alternative methods of automated liquid–liquid extraction are available as well [370].

Solvent removal from multiple purification processes is ordinarily handled using vacuum centrifuges in order to prevent sample loss or cross contamination due to bumping or boiling over [371].

Clearly automated purification techniques greatly enhance the productivity of combinatorial libraries. It is important to note that the degree of purity that should be achieved is lesser in the earlier "hit seeking" phases. This allows useful leeway in designing equipment and processes. The degree of purity required is dramatically higher in the lead optimization stages as the data are required to be quite reliable and quantitative in assembling data to be used in devising the best forward path to producing clinical candidates [358].

4.15.1. Mixtures for Testing Natural product mixtures are historically screened as mixtures and are deconvoluted when active in a given assay but there is significant reluctance to employ this process in testing small molecule libraries. Nonetheless, this is not an impossible challenge and there are economies in the high-throughput laboratory that can be realized in this manner. Finding which component in a mixture is responsible for the activity can be assisted by assembling modest sized groups of molecules with different properties or diversity point differences followed by positional scanning deconvolution or by component removal methods, for example [372].

4.15.2. "Pure" Compounds for Testing Early on in a drug seeking campaign one is often satisfied to assemble a group of hits that reach a given level of activity in a screen. In screening large compound libraries a screen that generates much more than about 0.1% hits is usually too permissive. Too many fewer than this is usually too restrictive. The criteria are loose enough that the individual materials being screened need not be more than about 80% pure. When hits are progressing to lead and candidate compound stages, then the compounds need to be more than 95% pure so that the quantitative data produced are meaningful enough to guide further analoging.

Actionable quantitative biological data are best obtained from pure samples and uncertainties in selecting the most active constituents to pursue further are increasingly introduced by assaying less pure material. Three grades of products can be distinguished. Pure usually means greater than 95% in combinatorial work. A lower but generally acceptable grade of purity is arbitrarily chosen at about 80% and such compounds can be labeled as "practical grade". Less than this level of purity is generally unacceptable and is sometimes, disparagingly, called "practically". In very large libraries where purity analysis of each component is rarely available inevitably, one has some components present in this poor state of purity. Indeed, chemists occasionally report anecdotally finding that an active component in a library has none of the intended compound in it at all. This is more commonly found in legacy libraries where the compounds have been in storage for a long time and/or have been thawed and refrozen several times. This complicates analoging but is more satisfactory than basing SAR-based design on negative activity data wherein one can be significantly mislead.

In any case, the quality of the compounds in a compound library must be known or one can have no confidence in the biological test results obtained therefrom [373]. In this context it is important to bear in mind that a survey of the contents of such a library demonstrated that examples selected more or less at random often contained 10–20% of impurities that were not detected in LC-UV analysis [373]. In early screening for hit detection this is probably tolerable but in later hit to lead work more precise quantitative data are required to guide subsequent experiments. Proton NMR scanning of compounds in a collection is believed to reflect more accurately the actual compound content of weighed samples for biotesting.

4.16. Synthetic Success

There are comparatively few papers in the literature in which investigators report the success versus failure rate in constructing synthetic libraries. Everyone who does synthesis of any degree of complexity is well aware that molecules often behave in perverse ways and even excellently planned syntheses fail to work in every desired instance and often produce significant by-products as well. This clearly plays a role in determining the outcome of synthetic library construction. It is reasonable, therefore, to ask what is a reasonable success rate and what level of purity is it reasonable to expect in routine cases.

In small compound libraries it is possible to analyze each product and identify the specific success rate and degree of purity [259,374]. When making very large libraries, one can only estimate the success rate by statistical sampling.

Only a relatively few reports specifically assessing the success rate of combinatorial library construction are available [156,259,374] but the general consensus is that achieving an 85% success rate is fine.

One is disturbed to note that in a few cases where a census has been taken of very large libraries, the wrong or missing structures are not statistically distributed [139]. Thus such libraries have a structural bias. For example, the chemistry may selectively favor production of more lipophilic substances so that hydrophilic examples are underrepresented. It is hard to see how to get around this conveniently.

4.17. Automation

Shortening the time devoted to synthesis is a desirable characteristic in many drug seeking campaigns. When syntheses are iterative, such as with peptides, the process can be automated readily to achieve this. In many medicinal chemistry laboratories, however, the focus is on a variety of small molecule chemotypes making automation of the overall process rather more difficult. Various component parts of the overall work however can be automated successfully if sufficient fiscal resources are available. Not surprisingly, industrial laboratories have made the most progress in this direction and in some companies integration of most of the necessary phases has been accomplished following substantial expendatures [5–8,325,375]. Commonly the initial chemistry is developed in individual laboratories in the form of "rehearsal libraries." Significant efficiencies are achieved when this chemistry is developed with a view to its suitability for combinatorial expansion using the equipment and processes available in a centrally automated laboratory. Here generally the developing compound libraries are passed from one work station to the next until the final library is realized. Special care has to be taken to anticipate and identify the various potential bottle necks to avoid unacceptable delays.

In addition to the equipment developed to facilitate this process many laboratories have found it to be advantageous to prepare in advance sub libraries of frequently used synthons and reagents. These are preweighed or weighed automatically as needed and their utilization can also be computer slaved with special ease when the reagents and reactants can be present in solutions of defined concentrations and then delivered by fluidics [376].

Reactions intrinsically less suited for this kind of integrated automation are those in which gasses and solids are to be added or are generated during the course of the work. On the other hand, many reactions commonly used in drug seeking such as amide formation, alkylations, Mitsunobu reactions, opening of epoxides, nucleophilic aromatic substitution reactions, multicomponent condensation reactions, Suzuki couplings, Buchwald–Hartwig reactions, and the like can readily be performed [5]. Naturally, difficulties provide the inspiration for chemists to find solutions. For example, instruments suitable for automated flow in combinatorial library have recently been published [377,378]. These useful developments have been facilitated by the development of the H-cube technology in which hydrogen is generated electrolytically from deionized water [379,380].

When perfected to the extent practically feasible it is estimated that platform queuing can be limited to less than a month [6,7].

4.18. Microwave Accelerations

Molecules possessing a permanent dipole align themselves in a microwave apparatus and oscillate as the field oscillates. This rapid motion generates intense homogeneous internal heat greatly facilitating organic reactions, especially in the solid state [381–384]. For example, heat demanding Diels–Alder reactions can take days on solid support hours in solution and only minutes under microwave. This has been adapted to combinatorial methods and is even compatible with a 96 well plate format [158,384–389]. As a consequence, many publications have appeared combining these techniques [381]. Especially valuable has been the development of equipment allowing use of sealed reaction containers, pressure control, and minimal temperature variation between individual wells [390–393].

The outcome of multicomponent reactions are often intrinsically slow since they often proceed by a sequence involving irreversible trapping of a component produced by an equilibrium step. Furthermore, the necessary handling steps to monitor success also retard completion of experiments. Microwave acceleration coupled to a continuous flow process can be shown to have particular value in minimizing these limitations [394].

4.19. Analytical Considerations

Analysis of the degree of completeness of combinatorial experiments and the identity of the products is simpler in solution-phase chemistry than that seen in solid-phase work. This closely parallels general experience in the precombichem days with the exception that the work load is greatly magnified. Automation is called for and HPLC/TOF mass spectrometry is of particular value [360,395–398]. Even so, with very large libraries one is usually restricted by necessity to statistical sampling and compound identification rarely goes beyond ascertaining whether the product has the correct molecular weight. If activity is found then more detailed examination takes place [398].

With solid-state libraries, the problem is much more complex. An enormous effort has been put into working out analysis of single beads with mass spectrometry, Raman spectroscopy, magic angle NMR, and chemiluminescence techniques being particularly popular [399–402].

4.20. Sample Handling and Storage

Much of the information on this topic remains proprietary. Nonetheless this is especially important to those laboratories that maintain very large legacy compound libraries. All too often such libraries contain significant numbers of products that are mislabeled or that have degraded because of storage or subsequent handling conditions. A particularly pernicious problem is instability when components are stored frozen in DMSO solution and subjected to repeated freeze-thaw cycles. Screening "bad" samples can be misleading and is in any case expensive. It is also costly to examine the identity of samples so as to verify what is in them since in big pharma these libraries can contain in excess of a million individual compounds. Legacy compounds are often found to be in formats incompatible with the format needed for HTS necessitating the development of efficient means of integrating these substances. Physical property-based screening of such collections, however, is usually decided to be superior to risking false leads and follow-up structure identification of leads that are found to be inactive on re-synthesis and retesting [403]. In at least one firm, LC-MS proved to be the best method for the task given time and cost constraints. It is interesting to note that about one-third of the legacy samples proved to be of dubious quality. The time and effort required in going through a very large library in this manner will vary with the criteria set and the equipment brought to bear, but one laboratory relates a throughput rate of >400 samples per day [360].

Most laboratories prefer to maintain their sample collections in "hotels" containing samples preserved frozen in solution in well plates. The underlying bulk samples are generally maintained in solid state from which new working plates are made as needed. Automated weighing simplifies the task of subdividing for analysis and this is usually coupled with fluidics transfer. These crucial practical decisions are, of course, individual.

5. INFORMATICS AND DATA HANDLING

Currently the highly automated big pharma compound libraries consist of in excess of 1–2 million compounds. This number expands rapidly each year due to the application of combinatorial synthesis technology and external acquisitions. It is estimated that compound collections of this magnitude can be screened for bioactivity in less than a week [373]. This is truly an astonishing statistic.

Combinatorial synthesis and high-throughput screening generate an enormous amount of data [404–406]. Keeping track of this is a job for high speed computers [407–409]. Many firms have developed their own programs for the data handling and there are commercial packages that may be useful as well. The best of these have structure drawing capacity also.

Initially data handling consisted of relating structures and test data essentially as a book keeping exercise. Experienced medicinal chemists with good intuitive abilities could scan these compilations and decide upon a reasonable path to take toward significant improvements. As the quantity and nature of information grew and grew, even this became insufficient. In the transition, algorithms were developed as plausible alternatives to human intervention [410]. These basically depended upon increasingly complex pattern recognition methods used for platform selections [411–414]. These evolved into methods of fragment assembly as fragment-based drug design techniques became increasingly popular [415,416]. Increasingly these have evolved into considerations of the likely targets and how hits might interact in three dimensions with them.

Subsequent programs have assessed measures of molecular diversity to determine how much chemical space is represented in compound libraries [417,418]. Inevitably this progressed quickly into the design and construction of molecules to expand the coverage aiming at a realistic useful screening library and avoiding re-exploration of diversity space already plowed [419].

In time virtual screening assembled the fruits of several of these efforts so that prospective hit mining of large libraries could be done selectively and at more reasonable time and expense [420]. The inclusion of drug-like character filters, toxicity and side-effect profiles, and assembly of data from various species as leads progressed toward development are characteristics. The data involved in this last is truly mammoth. Doing this manually is beyond the capacity of all but a few.

6. PATENTS

Patent considerations are complex in combinatorial chemistry. The mass of potential data is hard to compress into a suitable format for patenting. Commonly, patenting takes place comparatively late in a drug seeking campaign and so patenting the results of combinatorial drug seeking differs little from traditional patenting. One notes however that the comparative speed and ease of molecule construction makes it possible to reduce to practice rather more examples that would have been possible in the one at a time days.

Rather more disturbing is the increasing tendency of patenting various means of making and evaluating libraries rather than focusing on their content. The fundamental purpose of patenting is to promote the useful arts and to provide protection for innovative discoveries for a period and then to share them with society in general. Patenting of means of producing libraries, if carried to an extreme would have a dampening effect on the development of the field and so would inhibit the development of the useful arts. This should be guarded against.

7. SUMMARY AND CONCLUSIONS

Combinatorial chemistry and multiple parallel syntheses have transformed the field of medicinal chemistry for the better. The last decade has seen revitalization and much dramatically useful technology have been discovered. No laboratory seriously involved in the search for new therapeutic agents can afford not to employ this technology.

From the vantage point of 2009, one can now look back at what has been done in the comparatively short time that this technique has been widely explored and one can see some things more clearly now and use the methodology more cunningly.

In the heady early days of combinatorial chemistry one frequently heard the opinion that existing drugs were only those to which nature or good fortune had laid a clear path. Some believed that there were large numbers of underexplored structural types that could be drugs if only they were prepared and screened. Combichem promised to make this a reality. It would be nice, indeed, if this had turned out to be true! It cannot be denied that there is some justice in this belief since speculative synthesis continues to reveal important drugs. Nonetheless, the cruel restraints that ADME and toxicity considerations place upon our chemical imagination [421] have ruined this dream of easy and unlimited progress. The present wedding of combichem with medicinal chemical knowledge is extremely powerful and we no longer in the main waste time on collections of molecules that have no chance of becoming drugs. Clearly space for chemical diversity is larger than space for medicinal diversity.

Before combichem (BC) there was little motivation for developing means of enhanced speed of synthesis. Generally synthesis could be accomplished much more quickly than screening and evaluation of the products. Enhanced speed of construction simply produced a greater backlog of work to be done in biological laboratories. The advent of high-throughput screening in the 1980s changed all this. The backlogs emptied rapidly and there was a demand for more compounds. In addition, new firms were founded to take advantage of newer technologies. These firms had no retained chemical libraries to screen and larger firms were reluctant to allow their libraries to be screened by outsiders. In significant part these needs were met by the methods in this chapter. With synthesis and screening back in phase, the next choke point in the pipeline has become pharmacokinetics, toxicity, solubility, and penetrability problems. These factors are presently under intensive examination in attempts to elucidate these properties in a similarly rapid fashion or to predict them so that favorable characteristics can be designed into chemical library members from the outset and thus largely avoid having to deal with them. The ADME choke point is no longer as serious as before but toxicity remains a problem. It can readily be seen that this and further choke points lie distally in the pipeline and these will have to be dealt with in turn. Some time can be saved by speeding things along the way and also by dealing with the remaining constrictions in parallel rather than simultaneously, but it is difficult to see how they can all be resolved in a rapid manner.

Trying to convert drug seeking from essentially a preindustrial revolution cottage industry to an assembly line model is a rather complex and difficult enterprise.

Part of the difficulty is that certain biological phenomena cannot be hurried. For example, no matter how much money and effort one is willing to throw at the problem, producing a baby requires essentially 9 months from conception. Hiring nine women will not result in producing a baby in 1 month. The problem in shortening drug seeking is further compounded in that the problem is not akin to laying bricks. To produce a brick wall of a given dimensions more quickly is largely a matter of buying enough bricks and hiring and motivating enough skilled labor. Drug seeking is not like this. In drug seeking one has to design the bricks first, develop the technology to produce and assemble them, and then to evaluate the final construction. Combichem does speed the process along but does not remove the elements of uncertainty that must be overcome. Given the strictures placed on clinical studies and their solidification in law and custom, it is hard to see how this phase of the drug seeking sequence can be shortened through chemical effort.

High-throughput screening can be likened to hastening the process of finding a needle in a haystack. Combinatorial chemistry can be likened to the preparation of needles. Ideally one should strive to make a few more useful needles embedded in progressively smaller haystacks. This involves mating as well as is possible productive chemical characteristics with productive biological properties. Medicinal chemistry reinvents itself every decade under the everlasting hope that there will be found a universally applicable methodology that will routinely produce satisfactory results. This may be likened to dealing with the needle-haystack problem by seeking a suitable match! For the present, combinatorial chemistry and multiple parallel synthesis in the hands of the skillful and lucky chemist rapidly zeros in on the best combination of atoms for a given purpose and the use of combinatorial chemical technologies is a very useful adjunct. This chemist receives approbation for his/her efforts. Those who consistently came up with useless compounds will eventually be encouraged to find other work.

There has been a dramatic increase in investment in drug discovery during the last decade (estimated at 10% annually). Unfortunately this has yet to result in a burst of new introductions. Certainly emphasis on chemical novelty has largely given way to the search for promises of utility. Diversity no longer rules. This is perhaps the combinatorial chemist's equivalent of paraphrasing the famous mantra of Peter Drucker that whereas efficiency is doing things properly, effectiveness is doing proper things. As with much of the points being discussed, achieving a proper balance is essential for optimum results.

It is interesting to note also that a 100-fold increase in screening activity has not yet resulted in a corresponding increase in the introduction of new pharmaceuticals. Part of the explanation for this is that ease of synthesis does not necessarily equate to equivalent value of the products. If each compound in chemical libraries was carefully designed and the resulting data carefully analyzed, then the disparity would be smaller than the present experience produces. Another exculpatory factor is that much of the low hanging fruit has already been harvested and the remaining diseases are more chronic than acute and are much more complex in their etiology.

Despite these considerations drug seeking is an exciting enterprise calling for the best of our talents and the appropriate use of high speed synthetic methods gives us a powerful new tool to use. A tool is a means to producing a product and not a product in itself.

REFERENCES

1. Mitscher LA, Dutta A. Combinatorial chemistry. In: Abraham D, editor. Burger's Medicinal Chemistry. 6th ed., Vol. 2. New York: John Wiley & Sons, Inc.; 2003. p 2–36.
2. Merrifeld RB. Solid phase peptide synthesis. I. The synthesis of a tetrapeptide. J Am Chem Soc 1963;85:2149–2154.
3. Merrifield RB. Solid phase synthesis. Science 1986;232:341–347.
4. Bunin BA, Ellman JA. A general and expedient method for the solid-phase synthesis of 1,4-benzodiazepine derivatives. J Am Chem Soc 1992;114:10997–10998.
5. Weber A, von Roedern E, Stilz HU. SynCar: an approach to automated synthesis. J Comb Chem 2005;7(2):178–84.
6. Weller HN, Nirschl DS, Petrillo EW, Poss MA, Andres CJ, Cavallaro CL, Echols MM, Grant-Young KA, Houston JG, Miller AV, Swann RT. Application of lean manufacturing concepts to drug discovery: rapid analogue library synthesis. J Comb Chem 2006;8(5):664–669.
7. Koppitz M. Maximizing efficiency in the production of compound libraries. J Comb Chem 2008;10(4):573–579.
8. Koppitz M, Eis K. Automated medicinal chemistry. Drug Discov Today 2006;11(11–12): 561–568.
9. Lipinski CA, Lombardo F, Dominy BW, Feeney PJ. Experimental computational approaches to estimate solubility and permeability in drug discovery and development settings. Adv Drug Deliv Rev 1997;23:3–25.
10. Lipinski CA, Lombardo F, Dominy BW, Feeney PJ. Experimenal computational approachs to estimate solubility and permeability in drug discovery and development settings. Adv Drug Deliv Rev 2001;46:3–26.
11. Evans BE, Rittle KE, Bock MG, DiPardo RM, Freidinger RM, Whitter WL, Lundell GF, Veber DF, Anderson PS, Chang RSL, Lotti VJ, Cerino DJ, Chen TB, Kling PJ, Kunkel KA, Springer JP, Hirshfield J. Methods for drug discovery: development of potent, selective, orally effective cholecystokinin antagonists. J Med Chem 1988;31:2235–2246.
12. Chen G, Zheng S, Luo X, Shen J, Zhu W, Liu H, Gui C, Zhang J, Zheng M, Puah CM, Chen K, Jiang H. Focused combinatorial library design based on structural diversity, druglikeness and binding affinity score. J Comb Chem 2005;7(3):398–406.
13. Kennedy JP, Williams L, Bridges TM, Daniels RN, Weaver D, Lindsley CW. Application of combinatorial chemistry science on modern drug discovery. J Comb Chem 2008;10(3):345–354.
14. Ojima I. Modern natural products chemistry and drug discovery. J Med Chem 2008;51 (9):2587–2588.
15. Sun L, Liang C, Shirazian S, Zhou Y, Miller T, Cui J, Fukuda JY, Chu JY, Nematalla A, Wang X, Chen H, Sistla A, Luu TC, Tang F, Wei J, Tang C. Discovery of 5-[5-fluoro-2-oxo-1,2-dihydroindol-(3*Z*)-ylidenemethyl]-2,4-dimethyl-1*H*-pyrrole-3-carboxylic acid (2-diethylaminoethyl)amide, a novel tyrosine kinase

inhibitor targeting vascular endothelial and platelet-derived growth factor receptor tyrosine kinase. J Med Chem 2003;46(7):1116–1119.

16. Miertus S, Fassina G. Combinatorial Chemistry. New York: Marcel Dekker; 1999. p 435.
17. Obrecht D, Villalgordo JM. Solid-Supported Combinatorial and Parallel Synthesis of Small-Molecular-Weight Compound Libraries. New York: Pergamon; 1998. p 339.
18. Bunin BA. The Combinatorial Index. San Diego: Academic Press; 1998. p 322.
19. Terrett NK. Combinatorial Chemistry. New York: Oxford; 1998. p 186.
20. Kerwin JF. Jr, Gordon EM. Combinatorial Chemistry and Molecular Diversity in Drug Discovery. New York: Wiley-Liss; 1998. p 516.
21. Cabilly S. Combinatorial Peptide Library Protocols. Totawa, NJ: Humana Press; 1998. p 313.
22. Czarnik AW, DeWitt SH. A Practical Guide to Combinatorial Chemistry. Washington, DC: American Chemical Society; 1997. p 450.
23. Wilson SR, Czarnik AW. Combinatorial Chemistry: Synthesis and Application. New York: Wiley; 1997. p 269.
24. Jung G. Combinatorial Peptide and Nonpeptide Libraries: A Handbook. New York: VCH; 1996. p 544.
25. Chaiken IW, Janda KD. Molecular Diversity and Combinatorial Chemistry: Libraries and Drug Discovery. Washington, DC: The American Chemical Society; 1996. p 328.
26. Yan B, Czarnik AW. Optimization of Solid-Phase Combinatorial Synthesis. New York: Marcel Dekker; 2002. p 385.
27. Beck-Sickinger A, Weber P. Combinatorial Strategies in Biology and Chemistry. New York: Wiley; 2002. p 179.
28. Seeberger PH. Solid Support Oligosaccharide Synthesis and Combinatorial Carbohydrate Libraries. New York: Wiley; 2001. p 308.
29. Ghose AP, Viswanadhan VN. Combinatorial Library Design and Evaluation: Principles, Software Tools, and Applications in Drug Discovery. New York: Marcel Dekker; 2001. p 631.
30. Bannwarth W, Felder E. Combinatorial Chemistry: A Practical Approach. Vol 9:New York: Wiley-VCH; 2000. p 430.
31. Fenniri H. Combinatorial Chemistry: A Practical Approach. Vol. 233:New York: Oxford University Press; 2000. p 476.
32. el-Basil S. Combinatorial Organic Chemistry: An Educational Approach. New York: Nova Science Publishers; 1999. p 234.
33. Sucholeiki I. High-Throughput Synthesis: Principles and Practices. New York: Marcel Dekker; 2001. p 366.
34. Seneci P. Solid Phase Synthesis and Combinatorial Technologies. New York: John Wiley & Sons Inc.; 2000. p 637.
35. Schwartz M. Analytical Techniques in Combinatorial Chemistry. New York: Marcel Dekker; 2000. p 301.
36. Jung G. Combinatorial Chemistry: Synthesis, Analysis Screening. Weinheim: Wiley-VCH; 1999. p 601.
37. Pirrung MC. Molecular Diversity and Combinatorial Chemistry: Principles and Applications. Oxford: Elsevier; 2004. p 173.
38. Yang B. Analysis and Purificaiton Methods in Combinatorial Chemistry. Hoboken, NJ: Wiley-Interscience; 2004. p 466.
39. English LB. Combinatorial Library Methods and Protocols. Vol. 201. Totawa, NJ: Humana Press; 2002. p 383.
40. Nicolaou KC, Hanko R, Hartwig W. Handbook of Combinatorial Chemistry: Drugs, Catalysts Materials. Weinheim: Wiley-VCH; 2002. p 1114.
41. Boldi AM. Combinatorial Synthesis of Natural Product-Based Libraries. Boca Raton, FL: CRC Press/Taylor & Francis; 2006. p 347.
42. Bannwarth W, Hinzen B. Combinatorial Chemistry: From Theory to Application. 2nd ed.; Vol. 26:Weinheim: Wiley-VCH; 2006. p 672.
43. Fassina G, Miertus S. Combinatorial Chemistry and Technology. Boca Raton: Taylor & Francis; 2004. p 597.
44. Kshirsagar T. High-Throughput Lead Optimizaiton in Drug Discovery. Boca Raton, FL: Taylor & Francis/CRC Press; 2008. p viii + 248.
45. Dolle RE. Comprehensive survey of chemical libraries yielding enzyme inhibitors, receptor agonists and antagonists, and other biologically active agents: 1992 through 1997. Mol Divers 1998;3(4):199–233.
46. Dolle RE. Comprehensive survey of combinatorial libraries with undisclosed biological activity: 1992–1997. Mol Divers 1998;4 (4):233–256.
47. Dolle RE, Nelson KH Jr. Comprehensive survey of combinatorial library synthesis: 1998. J Comb Chem 1999;1(4):235–282.
48. Dolle RE. Comprehensive survey of combinatorial library synthesis: 1999. J Comb Chem 2000;2(5):383–433.

49. Dolle RE. Comprehensive survey of combinatorial library synthesis: 2000. J Comb Chem 2001;3(6):477–517.
50. Dolle RE. Comprehensive survey of combinatorial library synthesis: 2001. J Comb Chem 2002;4(5):369–418.
51. Dolle RE, Comprehensive survey of combinatorial library synthesis: 2002 J Comb Chem 2003;5(6):693–753.
52. Dolle RE. Comprehensive survey of combinatorial library synthesis: 2003. J Comb Chem 2004;6(5):623–679.
53. Dolle RE. Comprehensive survey of combinatorial library synthesis: 2004. J Comb Chem 2005;7(6):739–798.
54. Dolle RE, Le Bourdonnec B, Morales GA, Moriarty KJ, Salvino JM. Comprehensive survey of combinatorial library synthesis: 2005. J Comb Chem 2006;8(5):597–635.
55. Dolle RE, Le Bourdonnec B, Goodman AJ, Morales GA, Salvino JM, Zhang W. Comprehensive survey of chemical libraries for drug discovery and chemical biology: 2006. J Comb Chem 2007;9(6):855–902.
56. Dolle RE, Le Bourdonnec B, Goodman AJ, Morales GA, Thomas CJ, Zhang W. Comprehensive survey of chemical libraries for drug discovery and chemical biology: 2007. J Comb Chem 2008;10(6):753–802.
57. DiMasi JA, Hansen RW, Grabowski HG. The price of innovation: new estimates of drug development costs. J Health Econ 2003;22 (2):151–185.
58. Lee BY, Mahajan S, Clapham B, Janda KD. Suspension ring-opening metathesis polymerization: the preparation of norbornene-based resins for application in organic synthesis. J Org Chem 2004;69:3319–3329.
59. Roberts RS. ROMPgel beads in IRORI format: acylations revisited. J. Comb Chem 2005;7:21–32.
60. Vedantham P, Zhang M, Gor PJ, Huang M, Georg GI, Lushington GH, Mitscher LA, Ye Q-Z, Hanson PR. Studies towards the synthesis of methionine aminopeptidase inhibitors: diversification utilizing a ROMP-derived coupling reagent. J. Comb Chem 2008;10: 195–203.
61. Patek M, Lebl M. Safety-catch and multiply cleavable linkers in solid-phase synthesis. Biopolymers (Peptide Sci) 1998;47:353–363.
62. Furka A. Combinatorial chemistry: 20 years on. Drug Discov Today 2002;7:1–4.
63. Konings DA, Wyatt JR, Ecker DJ, Freier SM. Deconvolution of combinatorial libraries for drug discovery: theoretical comparison of pooling strategies. J Med Chem 1996;39 (14):2710–2719.
64. Konings DA, Wyatt JR, Ecker DJ, Freier SM. Strategies for rapid deconvolution of combinational libraries: comparative evaluation using a model system. J Med Chem 1997;40 (26):4386–4395.
65. Freier SM, Konings DA, Wyatt JR, Ecker DJ. Deconvolution of combinatorial libraries for drug discovery: a model system. J Med Chem 1995;38(2):344–352.
66. Geysen MH, Meloen RH, Barteling SJ. Use of peptide synthesis to probe viral antigens for epitopes to a resolution of a single amino acid. Proc Natl Acad Sci USA. 1984;81:3998–4002.
67. Houghten RA, Pinilla C, Blondelle SE, Appel JR, Dooley CT, Cuervo JH. Generation and use of synthetic peptide combinatorial libraries for basic research and drug discovery. Nature 1990;354:84–85.
68. Fodor SPA, Read JL, Pirrung MC, Stryer L, Lu AT, Solas D. Light-directed, spatially addressable parallel chemical synthesis. Science 1991;251:767–773.
69. Pellois JP, Wang W, Gao X. Peptide synthesis based on *t*-Boc chemistry and solution photogenerated acids. J Combinat Chem 2000;2:355–360.
70. Min DH, Mrksich M. Peptide arrays: towards routine implementation. Curr Opin Chem Biol 2004;8(5):554–558.
71. Maurer K, McShea A, Strathmann M, Dill K. The removal of the *t*-BOC group by electrochemically generated acid and use of an addressable electrode array for peptide synthesis. J Comb Chem 2005;7(5):637–640.
72. Nicolaou KC, Xiao X-Y, Parandoosh Z, Senyei Y, Nova MP. Radiofrequency encoded combinatorial chemistry. Angew Chem Int Ed, Engl 1995;34:2289–2291.
73. Barnes C, Balasubramanian S. Recent developments in the encoding and deconvolution of combinatorial libraries. Curr Opin Chem Biol 2000;4(3):346–350.
74. Ede NJ, Wu Z. Beyond Rf tagging. Curr Opin Chem Biol 2003;7(3):374–379.
75. Sebestyen F, Dibo G, Kovacs A, Furka A. Chemical synthesis of peptide libraries. Bioorg Med Chem Lett 1993;3:413–418.
76. Reed JD, Edwards DL, Gonzalez CF. Synthetic peptide combinatorial libraries: a method for the identification of bioactive peptides against phytopathogenic fungi. Mol Plant Microbe Interact 1997;10(5):537–549.

77. Koppel G, Dodds C, Houchins B, Hunden D, Johnson D, Owens R, Chaney M, Usdin T, Hoffman B, Brownstein M. Use of peptide combinatorial libraries in drug design: the identification of a potent serotonin reuptake inhibitor derived from a tripeptide cassette library. Chem Biol 1995;2(7):483–7.
78. Muller D, Zeltser I, Bitan G, Gilon C. Building units for N-backbone cyclic peptides. 3. Synthesis of protected *N*(alpha)-(omega-aminoalkyl) amino acids and *N*(alpha)-(omega-carboxyalkyl)amino acids. J Org Chem 1997;62 (2):411–416.
79. Smith AB, 3rd Knight SD, Sprengeler PA, Hirschmann R. The design and synthesis of 2,5-linked pyrrolinones. A potential non-peptide peptidomimetic scaffold. Bioorg Med Chem 1996;4(7):1021–1034.
80. Wallace A, Koblan KS, Hamilton K, Marquis-Omer DJ, Miller PJ, Mosser SD, Omer CA, Schaber MD, Cortese R, Oliff A, Gibbs JB, Pessi A. Selection of potent inhibitors of farnesyl-protein transferase from a synthetic tetrapeptide combinatorial library. J Biol Chem 1996;271(49):31306–31311.
81. Blake J, Johnston JV, Hellstrom KE, Marquardt H, Chen L. Use of combinatorial peptide libraries to construct functional mimics of tumor epitopes recognized by MHC class I-restricted cytolytic T lymphocytes. J Exp Med 1996;184(1):121–130.
82. Eichler J, Houghten RA. Generation and utilization of synthetic combinatorial libraries. Mol Med Today 1995;1(4):174–180.
83. Eichler J, Houghten RA. Identification of substrate-analog trypsin inhibitors through the screening of synthetic peptide combinatorial libraries. Biochemistry 1993;32(41): 11035–11041.
84. Houghten RA, Pinilla C, Appel JR, Blondelle SE, Dooley CT, Eichler J, Nefzi A, Ostresh JM. Mixture-based synthetic combinatorial libraries. J Med Chem 1999;42(19):3743–3778.
85. Camarero JA, Ayers B, Muir TW. Studying receptor–ligand interactions using encoded amino acid scanning. Biochemistry 1998;37 (20):7487–7495.
86. Zysk JR, Baumbach WR. Homogeneous pharmacologic and cell-based screens provide diverse strategies in drug discovery: somatostatin antagonists as a case study. Comb Chem High Throughput Screen 1998;1(4):171–183.
87. Brinker A, Weber E, Stoll D, Voigt J, Muller A, Sewald N, Jung G, Wiesmuller KH, Bohley P. Highly potent inhibitors of human cathepsin L identified by screening combinatorial pentapeptide amide collections. Eur J Biochem 2000;267(16):5085–5092.
88. Blondelle SE, Lohner K. Combinatorial libraries: a tool to design antimicrobial and antifungal peptide analogues having lytic specificities for structure–activity relationship studies. Biopolymers 2000;55(1):74–87.
89. Houghten RA, Blondelle SE, Dooley CT, Dorner B, Eichler J, Ostresh JM. Libraries from libraries: generation and comparison of screening profiles. Mol Divers 1996;2(1–2):41–45.
90. Eichler J, Lucka AW, Houghten RA. Cyclic peptide template combinatorial libraries: synthesis and identification of chymotrypsin inhibitors. Pept Res 1994;7(6):300–307.
91. El Oualid F, van den Elst H, Leroy IM, Pieterman E, Cohen LH, Burm BE, Overkleeft HS, van der Marel GA, Overhand M. A combinatorial approach toward the generation of ambiphilic peptide-based inhibitors of protein: geranylgeranyl transferase-1. J Comb Chem 2005;7(5):703–713.
92. Xu Y, Lu H, Kennedy JP, Yan X, McAllister LA, Yamamoto N, Moss JA, Boldt GE, Jiang S, Janda KD. Evaluation of "credit card" libraries for inhibition of HIV-1 gp41 fusogenic core formation. J Comb Chem 2006;8(4):531–539.
93. Scott JK, Smith GP. Searching for peptide ligands with an epitope library. Science 1990;249(4967):386–390.
94. Ellington AD, Szostak JW. Selection *in vitro* of single-stranded DNA molecules that fold into specific ligand-binding structures. Nature 1992;355(6363):850–852.
95. Houghten RA, Appel JR, Blondelle SE, Cuervo JH, Dooley CT, Pinilla C. The use of synthetic peptide combinatorial libraries for the identification of bioactive peptides. Biotechniques 1992;13(3):412–421.
96. Brenner S, Lerner RA. Encoded combinatorial chemistry. Proc Natl Acad Sci USA 1992;89 (12):5381–5383.
97. Nielsen J, Brenner S, Janda KD. Synthetic methods for the implementation of encoded combinatorial chemistry. J Am Chem Soc 1993;115:9812–9813.
98. Ohlmeyer MH, Swanson RN, Dillard LW, Reader JC, Asouline G, Kobayashi R, Wigler M, Still WC. Complex synthetic chemical libraries indexed with molecular tags. Proc Natl Acad Sci USA 1993;90(23):10922–10926.
99. Ni ZJ, Maclean D, Holmes CP, Murphy MM, Ruhland B, Jacobs JW, Gordon EM, Gallop

MA. Versatile approach to encoding combinatorial organic syntheses using chemically robust secondary amine tags. J Med Chem 1996;39(8):1601–8.

100. Geysen HM, Wagner CD, Bodnar WM, Markworth CJ, Parke GJ, Schoenen FJ, Wagner DS, Kinder DS. Isotope or mass encoding of combinatorial libraries. Chem Biol 1996;3(8):679–688.

101. Luo J, Zhang H, Xiao W, Kumaresan PR, Shi C, Pan CX, Aina OH, Lam KS. Rainbow beads: a color coding method to facilitate high-throughput screening and optimization of one-bead one-compound combinatorial libraries. J Comb Chem 2008;10(4):599–604.

102. Simon RA, Schuresko L, Dendukuri N, Goers E, Murphy B, Lokey RS. One-bead-one-compound library of end-capped dipeptides and deconvolution by microflow NMR. J Comb Chem 2005;7(5):697–702.

103. Kofoed J, Reymond JL. A general method for designing combinatorial peptide libraries decodable by amino acid analysis. J Comb Chem 2007;9(6):1046–1052.

104. Dolle RE, Guo J, O'Brien L, Jin Y, Piznik M, Bowman KJ, Li W, Egan WJ, Cavallaro CL, Roughton AL, Zhao Q, Reader JC, Orlowski M, Jacob-Samuel B, Carroll CD. A statistical-based approach to assessing the fidelity of combinatorial libraries encoded with electrophoric molecular tags. Development and application of tag decode-assisted single bead LC/MS analysis. J Comb Chem 2000;2(6):716–731.

105. Boger DL, Lee JK, Goldberg J, Jin Q. Two comparisons of the performance of positional scanning and deletion synthesis for the identification of active constituents in mixture combinatorial libraries. J Org Chem 2000;65 (5):1467–1474.

106. Andrus MB, Turner TM, Sauna ZE, Ambudkar SV. The synthesis and evaluation of a solution phase indexed combinatorial library of non-natural polyenes for reversal of P-glycoprotein mediated multidrug resistance. J Org Chem 2000;65(16):4973–4983.

107. Gopalsamy A, Yang H. Combinatorial synthesis of heterocycles: solid-phase synthesis of 2-amino-4(1*H*)-quinazolinone derivatives. J Comb Chem 2000;2(4):378–381.

108. Lazo JS, Wipf P. Combinatorial chemistry and contemporary pharmacology. J Pharmacol Exp Ther 2000;293(3):705–709.

109. Walker B, Lynas JF, Meighan MA, Bromme D. Evaluation of dipeptide alpha-keto-beta-aldehydes as new inhibitors of cathepsin S. Biochem Biophys Res Commun 2000;275(2):401–405.

110. Ho KC, Sun CM. Liquid phase parallel synthesis of guanidines. Bioorg Med Chem Lett 1999;9(11):1517–1520.

111. Gong Y, Becker M, Choi-Sledeski YM, Davis RS, Salvino JM, Chu V, Brown KD, Pauls HW. Solid-phase parallel synthesis of azarene pyrrolidinones as factor Xa inhibitors. Bioorg Med Chem Lett 2000;10(10):1033–1036.

112. Illgen K, Enderle T, Broger C, Weber L. Simulated molecular evolution in a full combinatorial library. Chem Biol 2000;7(6):433–441.

113. Ducruet AP, Rice RL, Tamura K, Yokokawa F, Yokokawa S, Wipf P, Lazo JS. Identification of new Cdc25 dual specificity phosphatase inhibitors in a targeted small molecule array. Bioorg Med Chem 2000;8(6):1451–1466.

114. Rabinowitz M, Seneci P, Rossi T, Dal Cin M, Deal M, Terstappen G. Solid-phase/solution-phase combinatorial synthesis of neuroimmunophilin ligands. Bioorg Med Chem Lett 2000;10(10):1007–1010.

115. Wei Y, Yi T, Huntington KM, Chaudhury C, Pei D. Identification of a potent peptide deformylase inhibitor from a rationally designed combinatorial library. J Comb Chem 2000;2 (6):650–657.

116. Grabowska U, Rizzo A, Farnell K, Quibell M. 5-(hydroxymethyl)oxazoles: versatile scaffolds for combinatorial solid-phase synthesis of 5-substituted oxazoles. J Comb Chem 2000;2 (5):475–490.

117. Mohler DL, Shen G, Dotse AK. Solution- and solid-phase synthesis of peptide-substituted thiazolidinediones as potential PPAR ligands. Bioorg Med Chem Lett 2000;10(20):2239–2242.

118. Bergnes G, Gilliam CL, Boisclair MD, Blanchard JL, Blake KV, Epstein DM, Pal K. Generation of an Ugi library of phosphate mimic-containing compounds and identification of novel dual specific phosphatase inhibitors. Bioorg Med Chem Lett 1999;9(19):2849–2854.

119. Henry KJ Jr, Wasicak J, Tasker AS, Cohen J, Ewing P, Mitten M, Larsen JJ, Kalvin DM, Swenson R, Ng SC, Saeed B, Cherian S, Sham H, Rosenberg SH. Discovery of a series of cyclohexylethylamine-containing protein farnesyltransferase inhibitors exhibiting potent cellular activity. J Med Chem 1999;42(23):4844–4852.

120. Lumma WC Jr, Witherup KM, Tucker TJ, Brady SF, Sisko JT, Naylor-Olsen AM, Lewis SD, Lucas BJ, Vacca JP. Design of novel, potent, noncovalent inhibitors of thrombin with nonbasic P-1 substructures: rapid structure–activity studies by solid-phase synthesis. J Med Chem 1998;41(7):1011–1013.

121. Rohrer SP, Birzin ET, Mosley RT, Berk SC, Hutchins SM, Shen DM, Xiong Y, Hayes EC, Parmar RM, Foor F, Mitra SW, Degrado SJ, Shu M, Klopp JM, Cai SJ, Blake A, Chan WW, Pasternak A, Yang L, Patchett AA, Smith RG, Chapman KT, Schaeffer JM. Rapid identification of subtype-selective agonists of the somatostatin receptor through combinatorial chemistry. Science 1998;282(5389):737–740.
122. Warmus JS, Ryder TR, Hodges JC, Kennedy RM, Brady KD. Rapid optimization of an ICE inhibitor synthesis using multiple reaction conditions in a parallel array. Bioorg Med Chem Lett 1998;8(17):2309–2314.
123. Ogbu CO, Qabar MN, Boatman PD, Urban J, Meara JP, Ferguson MD, Tulinsky J, Lum C, Babu S, Blaskovich MA, Nakanishi H, Ruan F, Cao B, Minarik R, Little T, Nelson S, Nguyen M, Gall A, Kahn M. Highly efficient and versatile synthesis of libraries of constrained beta-strand mimetics. Bioorg Med Chem Lett 1998;8(17):2321–2326.
124. Kundu B, Bauser M, Betschinger J, Kraas W, Jung G. Identification of a potent analogue of Nazumamide A through iteration of combinatorial tetrapeptide libraries. Bioorg Med Chem Lett 1998;8(13):1669–1672.
125. Apletalina E, Appel J, Lamango NS, Houghten RA, Lindberg I. Identification of inhibitors of prohormone convertases 1 and 2 using a peptide combinatorial library. J Biol Chem 1998;273(41):26589–26595.
126. Bhandari A, Jones DG, Schullek JR, Vo K, Schunk CA, Tamanaha LL, Chen D, Yuan Z, Needels MC, Gallop MA. Exploring structure–activity relationships around the phosphomannose isomerase inhibitor AF14049 via combinatorial synthesis. Bioorg Med Chem Lett 1998;8(17):2303–2308.
127. Roychoudhury S, Blondelle SE, Collins SM, Davis MC, McKeever HD, Houghten RA, Parker CN. Use of combinatorial library screening to identify inhibitors of a bacterial two-component signal transduction kinase. Mol Divers 1998;4(3):173–182.
128. Bianco A, Brock C, Zabel C, Walk T, Walden P, Jung G. New synthetic non-peptide ligands for classical major histocompatibility complex class I molecules. J Biol Chem 1998;273(44):28759–28765.
129. Ellman J, Stoddard B, Wells J. Combinatorial thinking in chemistry and biology. Proc Natl Acad Sci USA 1997;94(7):2779–2782.
130. Schullek JR, Butler JH, Ni ZJ, Chen D, Yuan Z. A high-density screening format for encoded combinatorial libraries: assay miniaturization and its application to enzymatic reactions. Anal Biochem 1997;246(1):20–29.
131. Harada K, Martin SS, Tan R, Frankel AD. Molding a peptide into an RNA site by *in vivo* peptide evolution. Proc Natl Acad Sci USA 1997;94(22):11887–11892.
132. Kim SW, Shin YS, Ro S. Solution and solid phase combinatorial synthesis of peptidomimetic library containing diversified alpha-methylated amino acids. Bioorg Med Chem Lett 1998;8(13):1665–1668.
133. Ostrem JA, al-Obeidi F, Safar P, Safarova A, Stringer SK, Patek M, Cross MT, Spoonamore J, LoCascio JC, Kasireddy P, Thorpe DS, Sepetov N, Lebl M, Wildgoose P, Strop P. Discovery of a novel, potent, and specific family of factor Xa inhibitors via combinatorial chemistry. Biochemistry 1998;37(4):1053–1059.
134. Jiracek J, Yiotakis A, Vincent B, Checler F, Dive V. Development of the first potent and selective inhibitor of the zinc endopeptidase neurolysin using a systematic approach based on combinatorial chemistry of phosphinic peptides. J Biol Chem 1996;271(32):19606–19611.
135. Jiracek J, Yiotakis A, Vincent B, Lecoq A, Nicolaou A, Checler F, Dive V. Development of highly potent and selective phosphinic peptide inhibitors of zinc endopeptidase 24-15 using combinatorial chemistry. J Biol Chem 1995;270(37):21701–21706.
136. Dive V, Cotton J, Yiotakis A, Michaud A, Vassiliou S, Jiracek J, Vazeux G, Chauvet MT, Cuniasse P, Corvol P. RXP 407, a phosphinic peptide, is a potent inhibitor of angiotensin I converting enzyme able to differentiate between its two active sites. Proc Natl Acad Sci USA 1999;96(8):4330–4335.
137. Coats SJ, Schulz MJ, Hlasta DJ. Method for the parallel preparation of the aspartic protease isostere: hydroxyethylamino amides. J Comb Chem 2004;6(5):688–691.
138. Bastos M, Maeji NJ, Abeles RH. Inhibitors of human heart chymase based on a peptide library. Proc Natl Acad Sci USA 1995;92 (15):6738–6742.
139. Salvino JM, Mathew R, Kiesow T, Narensingh R, Mason HJ, Dodd A, Groneberg R, Burns CJ, McGeehan G, Kline J, Orton E, Tang SY, Morrisette M, Labaudininiere R. Solid-phase synthesis of an arylsulfone hydroxamate library. Bioorg Med Chem Lett 2000;10 (15):1637–1640.
140. Poreddy AR, Schall OF, Osiek TA, Wheatley JR, Beusen DD, Marshall GR, Slomczynska U.

Hydroxamate-based iron chelators: combinatorial syntheses of desferrioxamine B analogues and evaluation of binding affinities. J Comb Chem 2004;6(2):239–254.

141. Rano TA, Timkey T, Peterson EP, Rotonda J, Nicholson DW, Becker JW, Chapman KT, Thornberry NA. A combinatorial approach for determining protease specificities: application to interleukin-1beta converting enzyme (ICE). Chem Biol 1997;4(2):149–155.

142. Wipf P, Cunningham A, Rice RL, Lazo JS. Combinatorial synthesis and biological evaluation of library of small-molecule Ser/Thr-protein phosphatase inhibitors. Bioorg Med Chem 1997;5(1):165–177.

143. Ward YD, Thomson DS, Frye LL, Cywin CL, Morwick T, Emmanuel MJ, Zindell R, McNeil D, Bekkali Y, Girardot M, Hrapchak M, DeTuri M, Crane K, White D, Pav S, Wang Y, Hao MH, Grygon CA, Labadia ME, Freeman DM, Davidson W, Hopkins JL, Brown ML, Spero DM. Design and synthesis of dipeptide nitriles as reversible and potent Cathepsin S inhibitors. J Med Chem 2002;45(25):5471–5482.

144. Bondebjerg J, Fuglsang H, Valeur KR, Pedersen J, Naerum L. Dipeptidyl nitriles as human dipeptidyl peptidase I inhibitors. Bioorg Med Chem Lett 2006;16(13):3614–3617.

145. Fenwick AE, Garnier B, Gribble AD, Ife RJ, Rawlings AD, Witherington J. Solid-phase synthesis of cyclic alkoxyketones, inhibitors of the cysteine protease cathepsin K. Bioorg Med Chem Lett 2001;11(2):195–198.

146. Han Y, Giroux A, Grimm EL, Aspiotis R, Francoeur S, Bayly CI, McKay DJ, Roy S, Xanthoudakis S, Vaillancourt JP, Rasper DM, Tam J, Tawa P, Thornberry NA, Paterson EP, Garcia-Calvo M, Becker JW, Rotonda J, Nicholson DW, Zamboni RJ. Discovery of novel aspartyl ketone dipeptides as potent and selective caspase-3 inhibitors. Bioorg Med Chem Lett 2004;14(3):805–808.

147. Prokai L, Prokai-Tatrai K, Zharikova A, Li X, Rocca JR. Combinatorial lead optimization of a neuropeptide FF antagonist. J Med Chem 2001;44(10):1623–1626.

148. Makaritis A, Georgiadis D, Dive V, Yiotakis A. Diastereoselective solution and multipin-based combinatorial array synthesis of a novel class of potent phosphinic metalloprotease inhibitors. Chemistry 2003;9(9):2079–2094.

149. Fenster E, Rayabarapu DK, Zhang M, Mukherjee S, Hill D, Neuenswander B, Schoenen F, Hanson PR, Aube J. Three-component synthesis of 1,4-diazepin-5-ones and the construction of gamma-turn-like peptidomimetic libraries. J Comb Chem 2008;10(2):230–234.

150. Im I, Webb TR, Gong YD, Kim JI, Kim YC. Solid-phase synthesis of tetrahydro-1,4-benzodiazepine-2-one derivatives as a beta-turn peptidomimetic library. J Comb Chem 2004; 6(2):207–213.

151. Haskell-Luevano C, Rosenquist A, Souers A, Khong KC, Ellman JA, Cone RD. Compounds that activate the mouse melanocortin-1 receptor identified by screening a small molecule library based upon the beta-turn. J Med Chem 1999;42(21):4380–4387.

152. Wagner DS, Markworth CJ, Wagner CD, Schoenen FJ, Rewerts CE, Kay BK, Geysen HM. Ratio encoding combinatorial libraries with stable isotopes and their utility in pharmaceutical research. Comb Chem High Throughput Screen 1998;1(3):143–153.

153. Revesz L, Bonne F, Manning U, Zuber JF. Solid phase synthesis of a biased mini tetrapeptoid-library for the discovery of monodentate ITAM mimics as ZAP-70 inhibitors. Bioorg Med Chem Lett 1998;8(5):405–408.

154. Ostresh JM, Husar GM, Blondelle SE, Dorner B, Weber PA, Houghten RA. "Libraries from libraries": chemical transformation of combinatorial libraries to extend the range and repertoire of chemical diversity. Proc Natl Acad Sci USA 1994;91(23):11138–11142.

155. Ostresh JM, Dorner B, Houghten R.A., Peralkylation. "Libraries from libraries": chemical transformation of synthetic combinatorial libraries. Methods Mol Biol 1998;87:41–49.

156. Eichler J, Appel JR, Blondelle SE, Dooley CT, Dorner B, Ostresh JM, Perez-Paya E, Pinilla C, Houghten RA. Peptide, peptidomimetic, and organic synthetic combinatorial libraries. Med Res Rev 1995;15(6):481–496.

157. Dorner B, Husar GM, Ostresh JM, Houghten RA. The synthesis of peptidomimetic combinatorial libraries through successive amide alkylations. Bioorg Med Chem 1996;4 (5):709–715.

158. Nefzi A, Ostresh JM, Meyer J-P, Houghten RA. Solid phase synthesis of heterocyclic compounds from linear peptides: cyclic ureas and thioureas. Tetrahedron Lett 1997;38:931–934.

159. Ostresh JM, Schoner CC, Hamashin VT, Nefzi A, Meyer J-P, Houghten RA. Solid-phase synthesis of trisubstituted bicyclic guanidines via cyclization of reduced *N*-acylated dipeptides. J Org Chem 1998;63:8622–8623.

160. Houghten RA, Pinilla C, Blondelle SE, Appel JR, Dooley CT, Cuervo JH. Generation and use

of synthetic peptide combinatorial libraries for basic research and drug discovery. Nature 1991;354(6348):84–86.

161. Cuervo JH, Weitl F, Ostresh JM, Hamashin VT, Hannah AL, Houghten RA.In: Maria HLS, editor. Libraries from libraries, Peptides 94. Proceedings of the 23rd European Peptide Symposium, Leiden; 1995. Leiden: ESCOM; 1995. p 465–466.

162. Rano TA, Cheng Y, Huening TT, Zhang F, Schleif WA, Gabryelski L, Olsen DB, Kuo LC, Lin JH, Xu X, Olah TV, McLoughlin DA, King R, Chapman KT, Tata JR. Combinatorial diversification of indinavir: *in vivo* mixture dosing of an HIV protease inhibitor library. Bioorg Med Chem Lett 2000;10(14):1527–1530.

163. St Hilaire PM, Meldal M. Glycopeptide and oligosaccharide libraries. Angew Chem Int Ed 2000;39:1163–1179.

164. Kahne D. Combinatorial approaches to carbohydrates. Curr Opin Chem Biol 1997;1 (1):130–135.

165. Liang R, Yan L, Loebach J, Ge M, Uozumi Y, Sekanina K, Horan N, Gildersleeve J, Thompson C, Smith A, Biswas K, Still WC, Kahne D. Parallel synthesis and screening of a solid phase carbohydrate library. Science 1996;274(5292):1520–1522.

166. Ako T, Daikoku S, Ohtsuka I, Kato R, Kanie O. A method of orthogonal oligosaccharide synthesis leading to a combinatorial library based on stationary solid-phase reaction. Chem—Asian J 2006;1:798–813.

167. Wang Y, Zhang L, Ye X. Oligosaccharide synthesis and library assembly by one-pot sequential glycosylation strategy. Comb Chem High Throughput Screen 2006;9:63–75.

168. Yamago S, Yamada T, Ito H, Hara O, Mino Y, Yoshida J. Combinatorial synthesis of an oligosaccharide library by using beta-bromo-glycoside-mediated iterative glycosylation of seneno-glycosides: rapid expansion of molecular diversity with simple building blocks. Chem 11:6159–6174.

169. Clique B, Ironmonger A, Whittaker B, Colley J, Titchmarsh J, Stockley P, Nelson A. Synthesis of a library of stereo- and regio-chemically diverse amino-glycoside derivatives. Org Biomol Chem 2005;3:2776–2785.

170. Tanaka H, Matoba N, Tsukamoto H, Takimoto H, Yamada H, Takahashi T. Automated parallel synthesis of a protected oligosaccharide library based upon the structure of dimeric Lewis X by one-pot sequential glycosylation. Synlett 2005; 824–828.

171. Yamago S, Yamada T, Maruyama T, Yoshida J. Iterative glycosylation of 2-deoxy-2-aminothioglycosides and its application to the combinatorial synthesis of linear oligo- glucosamines. Angew Chem Int Ed 2004;43: 2145–2148.

172. Grathwohl M, Drinnan N, Broadhurst M, West ML, Meutermans W. Solid-phase oligosaccharide chemistry and its application to library synthesis. Methods Enzymol 2003;369B: 248–267.

173. Amaya T, Tanaka H, Takahashi T. Combinatorial synthesis of carbohydrate cluster on tree-type linker with orthogonally cleavable parts. Synlett 2004; 497–502.

174. Baytas S, Linhardt RJ. Combinatorial carbohydrate synthesis. Mini-Rev Org Chem 2004;1:27–39.

175. Fukase K. Combinatorial and solid-phase methods of oligosaccharide synthesis. Glycoscience 2001;2:1621–1660.

176. Kanemitsu T, Kanie O. Recent developments in oligosaccharide synthesis: tactics, solid-phase synthesis and library synthesis. Comb Chem High Throughput Screen 2002;5: 339–360.

177. St Hilaire PM, Meldal M. Glycopeptide and oligosaccharide libraries. Comb Chem 1999; 291–318.

178. Seeberger PH, Haase W-C. Solid-phase oligosaccharide synthesis and combinatorial carbohydrate libraries. Chem Rev 2000;100: 4349–4393.

179. Haase W-C, Seeberger PH. Recent progress in polymer-supported synthesis of oligosaccharides and carbohydrate libraries. Curr Org Chem 2000;4:481–511.

180. Taylor, CM. Strategies for combinatorial libraries of oligosaccharides. Comb Chem, 1997;207–224.

181. Izumi M, Ichikawa Y. Combinatorial synthesis of oligosaccharide library of 2,6-dideoxysugars. Tetrahedron Lett 1998;39:2079–2082.

182. Rademann J, Schmidt RR. A new method for the solid phase synthesis of oligosaccharides. Tetrahedron Lett 1996;37:3989–3990.

183. Krepinsky JJ, Douglas SP, In: Ernst B, Hart GW, Sinay P, editors. Polymer-Supported Synthesis of Oligosaccharides, Carbohydrates in Chemistry and Biology. Weinheim: Wiley-VCH Verlag GmbH; 2000. p 239–265.

184. Ding Y, Kanie O, Labbe J, Palcic MM, Ernst B, Hindsgaul O. Synthesis and biological activity of oligosaccharide libraries. Adv Exp Med Biol 1995;376:261–269.

185. Ding Y, Kanie O, Labbe J, Palcic MM, Ernst O, Hindsgaul O. Synthesis and biological activity of oligosaccharide libraries. In: Alivi A, Axford JS, editor. Glycoimmunology. New York: Plenum Press; 1995. p 261–269.

186. Boons G-J, Heskamp B, HoutF Vinyl glycosides in oligosaccharide synthesis: a strategy for the preparation of trisaccharide libraries based on latent-active glycosylation. Angew Chem Int Ed Engl 1996;106:3053–3056.

187. Zhu T, Boons G-J. A two-directional approach for the solid-phase synthesis of trisaccharide libraries. Angew Chem Int Ed, Engl 1898;37: 1898–1900.

188. Borman S. Automated glycan synthesis. Chem Eng News 2009; 39.

189. Aucagne V, Berteina-Raboin S, Guenot P, Agrofoglio LA. Palladium-catalyzed synthesis of uridines on polystyrene-based solid supports. J Comb Chem 2004;6(5):717–723.

190. Golisade A, Wiesner J, Herforth C, Jomaa H, Link A. Anti-malarial activity of *N*(6)-substituted adenosine derivatives. Part I. Bioorg Med Chem 2002;10(3):769–777.

191. Sun D, Jones V, Carson EI, Lee RE, Scherman MS, McNeil MR, Lee RE. Solid-phase synthesis and biological evaluation of a uridinyl branched peptide urea library. Bioorg Med Chem Lett 2007;17(24):6899–6904.

192. Oliviero G, Amato J, D'Errico S, Borbone N, Piccialli G, Mayol L. Solid phase synthesis of nucleobase and ribose modified inosine nucleoside analogues. Nucleos Nucleot Nucleic Acids 2007;26(10–12):1649–1652.

193. Poon KW, Liang N, Datta A. De novo synthetic route to a combinatorial library of peptidyl nucleosides. Nucleos Nucleot Nucleic Acids 2008;27(4):389–407.

194. Gunic E, Amador R, Rong F, Abt JW, An H, Hong Z, Girardet JL. Synthesis of nucleoside libraries on solid support. I. *N*2,*N*6-disubstituted diaminopurine nucleosides. Nucleos Nucleot Nucleic Acids 2004;23(1–2): 495–499.

195. Bookser BC, Raffaele NB. High-throughput five minute microwave accelerated glycosylation approach to the synthesis of nucleoside libraries. J Org Chem 2007;72(1):173–179.

196. Seo JS, Yoon CM, Gong YD. Solid-phase synthesis of sn-1,2- and sn-2,3-diacylglycerols via ring-opening of the glycidyl-bound resin. J Comb Chem 2007;9(3):366–369.

197. Roy J, DeRoy P, Poirier D. 2beta-(*N*-substituted piperazino)-5alpha-androstane-3alpha, 17beta-diols: parallel solid-phase synthesis and antiproliferative activity on human leukemia HL-60 cells. J Comb Chem 2007;9 (3):347–358.

198. Maltais R, Tremblay MR, Ciobanu LC, Poirier D. Steroids and combinatorial chemistry. J Comb Chem 2004;6(4):443–456.

199. Maltais R, Tremblay MR, Poirier D. Solid-phase synthesis of hydroxysteroid derivatives using the diethylsilyloxy linker. J Comb Chem 2000;2(6):604–614.

200. Yu X, Wang S, Chen F. Solid-phase synthesis of solanesol. J Comb Chem 2008;10(4):605–610.

201. Bohacek RS, McMartin C. Exploring the universe of molecules for new drugs. Nat Med 1995;1(2):177–178.

202. Russ AP, Lampel S. The druggable genome: an update. Drug Discov Today 2005; 101607–1610.

203. Congreve M, Chessari G, Tisi D, Woodhead AJ. Recent developments in fragment-based drug discovery. J Med Chem 2008;51 (13):3661–3680.

204. Veber DF, Johnson SR, Cheng HY, Smith BR, Ward KW, Kopple KD. Molecular properties that influence the oral bioavailability of drug candidates. J Med Chem 2002;45 (12):2615–2623.

205. Congreve M, Carr R, Murray C, Jhoti H. A 'rule of three' for fragment-based lead discovery? Drug Discov Today 2003;8(19):876–877.

206. Severinsen R, Bourne GT, Tran TT, Ankersen M, Begtrup M, Smythe MI. Library of biphenyl priveleged substructures using a safety-catch linker approach. J Comb Chem 2008;10: 557–566.

207. Zhao H. Scaffold selection and scaffold hopping in lead generation: a medicinal chemistry perspective. Drug Discov Today 2007;12 (3–4):149–155.

208. Bemis GW, Murcko MA. The properties of known drugs. 1. Molecular frameworks. J Med Chem 1996;39(15):2887–2893.

209. Mang C, Jakupovic S, Schunk S, Ambrosi HD, Schwarz O, Jakupovic J. Natural products in combinatorial chemistry: an andrographolide-based library. J Comb Chem 2006;8 (2):268–274.

210. Nagai K, Doi T, Sekiguchi T, Namatame I, Sunazuka T, Tomoda H, Omura S, Takahashi T. Synthesis and biological evaluation of a beauveriolide analogue library. J Comb Chem 2006;8(1):103–109.

211. Schwarz O, Jakupovic S, Ambrosi HD, Haustedt LO, Mang C, Muller-Kuhrt L. Natural

products in parallel chemistry—novel 5-lipoxygenase inhibitors from BIOS-based libraries starting from alpha-santonin. J Comb Chem 2007;9(6):1104–1113.

212. Bhat L, Liu Y, Victory SF, Himes RH, Georg GI. Synthesis and evaluation of paclitaxel C7 derivatives: solution phase synthesis of combinatorial libraries. Bioorg Med Chem Lett 1998;8(22):3181–3186.

213. Liu Y, Ali SM, Boge TC, Georg GI, Victory S, Zygmunt J, Marquez RT, Himes RH. A systematic SAR study of C10 modified paclitaxel analogues using a combinatorial approach. Comb Chem High Throughput Screen 2002;5 (1):39–48.

214. Xiao X-Y, Parandoosh Z, Nova MP. Design and synthesis of a taxoid library using radiofrequency encoded combinatorial chemistry. J Org Chem 1997;62:6029–6033.

215. Nicolaou KC, Hughes R, Cho SY, Winssinger N, Labischinski H, Endermann R. Synthesis and biological evaluation of vancomycin dimers with potent activity against vancomycin-resistant bacteria: target-accelerated combinatorial synthesis. Chemistry 2001;7 (17):3824–3843.

216. Xu R, Greiveldinger G, Marenus LE, Cooper A, Ellman JA. Combinatorial library approach for the identification of synthetic receptors targeting vancomycin resistant bacteria. J Am Chem Soc 1999;121:4898–4899.

217. Yao N, Song A, Wang X, Dixon S, Lam KS. Synthesis of flavonoid analogues as scaffolds for natural product-based combinatorial libraries. J Comb Chem 2007;9(4):668–676.

218. Miao H, Tallarico JA, Hayakawa H, Munger K, Duffner JL, Koehler AN, Schreiber SL, Lewis TA. Ring-opening and ring-closing reactions of a shikimic acid-derived substrate leading to diverse small molecules. J Comb Chem 2007;9 (2):245–253.

219. Liu JF, Kaselj M, Isome Y, Ye P, Sargent K, Sprague K, Cherrak D, Wilson CJ, Si Y, Yohannes D, Ng SC. Design and synthesis of a quinazolinone natural product-templated library with cytotoxic activity. J Comb Chem 2006;8(1):7–10.

220. Tomasi S, Picard S, Laine C, Babonneau V, Goujeon A, Boustie J, Uriac P. Solid-phase synthesis of polyfunctionalized natural products: application to usnic acid, a bioactive lichen compound. J Comb Chem 2006;8 (1):11–14.

221. Lee SC, Park SB. Solid-phase parallel synthesis of natural product-like diaza-bridged heterocycles through Pictet-Spengler intramolecular cyclization. J Comb Chem 2006;8 (1):50–57.

222. Atuegbu A, Maclean D, Nguyen C, Gordon EM, Jacobs JW. Combinatorial modification of natural products: preparation of unencoded and encoded libraries of Rauwolfia alkaloids. Bioorg Med Chem 1996;4(7):1097–1106.

223. Spring DR, Krishnan S, Blackwell HE, Schreiber SL. Diversity-oriented synthesis of biaryl-containing medium rings using a one bead/one stock solution platform. J Am Chem Soc 2002; 124(7):1354–1363.

224. Nicolaou KC, Pfefferkorn JA, Roecker AJ, Cao G-Q, Barluenga S, Mitchell HJ. Natural product-like combinatorial libraries based on privileged structures. 1. General principles and solid-phase synthesis of benzopyrans. J Am Chem Soc 2000;122:9939–9953.

225. Nicolaou KC, Pfefferkorn JA, Mitchell HJ, Roecker AJ, Barluenga S, Cao G-Q, Affleck RL, Lillig JE. Natural product-like combinatorial libraries based on privileged structures. 2. Construction of a 10,000-membered benzopyran library by directed split-and-pool chemistry using NanoKans and optical encoding. J Am Chem Soc 2000;122: 9954–9967.

226. Nicolaou KC, Pfferkorn JA, Barulenga S, Mitchell HJ, Roecker AJ, Cao G-Q. Natural product=like combinatorial libraries based on privileged structures. 3. The "libraries from libraries" principle for diversity enhancement of benzopyran libraries. J Am Chem Soc 2000;122:9968–9976.

227. Schreiber SL. Chemical genetics resulting from a passion for synthetic organic chemistry. Bioorg Med Chem 1998;6(8):1127–1152.

228. Tan DS, Foley MA, Stockwell BR, Shair MD, Schreiber SL. Synthesis and preliminary evaluation of a library of polycyclic small molecules for use in chemical genetic assays. J Am Chem Soc 1999;121:9073–9087.

229. Zhang W, Tang Y. Combinatorial biosynthesis of natural products. J Med Chem 2008;51 (9):2629–2633.

230. Katz L, Khosla C. Antibiotic production from the ground up. Nat Biotechnol 2007;25 (4):428–429.

231. Ward SL, Desai RP, Hu Z, Gramajo H, Katz L. Precursor-directed biosynthesis of 6-deoxyerythronolide B analogues is improved by removal of the initial catalytic sites of the polyketide synthase. J Ind Microbiol Biotechnol 2007;34(1):9–15.

232. Tang L, Chung L, Carney JR, Starks CM, Licari P, Katz L. Generation of new epothilones by genetic engineering of a polyketide synthase in *Myxococcus xanthus*. J Antibiot 2005;58(3):178–184.

233. Katz L, Ashley GW. Translation and protein synthesis: macrolides. Chem Rev 2005;105 (2):499–528.

234. Wilkinson B, Micklefield J. Mining and engineering natural-product biosynthetic pathways. Nat Chem Biol 2007;3(7):379–386.

235. Khosla C, Keasling JD. Metabolic engineering for drug discovery and development. Nat Rev Drug Discov 2003;2(12):1019–1025.

236. Baltz RH. Molecular engineering approaches to peptide, polyketide and other antibiotics. Nat Biotechnol 2006;24(12):1533–1540.

237. Baltz RH. Biosynthesis and genetic engineering of lipopeptide antibiotics related to daptomycin. Curr Top Med Chem 2008;8 (8):618–638.

238. Nielsen TE, Schreiber SL. Towards the optimal screening collection: a synthesis strategy. Angew Chem Int Ed 2008;47(1):48–56.

239. Tolliday N, Clemons PA, Ferraiolo P, Koehler AN, Lewis TA, Li X, Schreiber SL, Gerhard DS, Eliasof S. Small molecules, big players: the National Cancer Institute's Initiative for Chemical Genetics. Cancer Res 2006;66 (18):8935–8942.

240. Schreiber SL. Small molecules: the missing link in the central dogma. Nat Chem Biol 2005;1(2):64–66.

241. Butcher RA, Schreiber SL. Using genome-wide transcriptional profiling to elucidate small-molecule mechanism. Curr Opin Chem Biol 2005;9(1):25–30.

242. Haggarty SJ, Clemons PA, Wong JC, Schreiber SL. Mapping chemical space using molecular descriptors and chemical genetics: deacetylase inhibitors. Comb Chem High Throughput Screen 2004;7(7):669–676.

243. Burke MD, Berger EM, Schreiber SL. A synthesis strategy yielding skeletally diverse small molecules combinatorially. J Am Chem Soc 2004;126(43):14095–14104.

244. Haggarty SJ, Clemons PA, Schreiber SL. Chemical genomic profiling of biological networks using graph theory and combinations of small molecule perturbations. J Am Chem Soc 2003; 125(35):10543–10545.

245. Koeller KM, Haggarty SJ, Perkins BD, Leykin I, Wong JC, Kao MC, Schreiber SL. Chemical genetic modifier screens: small molecule trichostatin suppressors as probes of intracellular histone and tubulin acetylation. Chem Biol 2003;10(5):397–410.

246. Haggarty SJ, Koeller KM, Wong JC, Butcher RA, Schreiber SL. Multidimensional chemical genetic analysis of diversity-oriented synthesis-derived deacetylase inhibitors using cell-based assays. Chem Biol 2003;10(5):383–396.

247. Strausberg RL, Schreiber SL. From knowing to controlling: a path from genomics to drugs using small molecule probes. Science 2003;300 (5617):294–295.

248. Zerhouni E, Medicine The NIH roadmap. Science 2003;302:63–72.

249. Bunin BA, Plunkett MJ, Ellman JA. The combinatorial synthesis and chemical and biological evaluation of a 1,4-benzodiazepine library. Proc Natl Acad Sci USA 1994;91 (11):4708–4712.

250. DeWitt SH, Kiely JS, Stankovic CJ, Schroeder MC, Cody DM, Pavia MR. "Diversomers": an approach to nonpeptide, nonoligomeric chemical diversity. Proc Natl Acad Sci USA 1993;90 (15):6909–6913.

251. Plunkett MJ, Ellman JA. Germanium and silicon linking strategies for traceless solid-phase synthesis. J Org Chem 1997;62 (9):2885–2893.

252. Herpin TF, Van Kirk KG, Salvino JM, Yu ST, Labaudiniere RF. Synthesis of a 10,000 member 1,5-benzodiazepine-2-one library by the directed sorting method. J Comb Chem 2000;2(5):513–521.

253. Boojamra CG, Burow KM, Ellman JA. An expedient and high-yielding method for the solid-phase synthesis of diverse 1,4-benzodiazepine-2,5-diones. J. Org. Chem. 1995;60:5742–5743.

254. Goff DA, Zuckermann RN. Solid-phase synthesis of defined 1,4-benzodiazepine-2,5-dione mixtures. J. Org. Chem. 1995;60:5744–5745.

255. Landi JJ. Jr, Ramig K. Regioselective preparation of 4-formyl-3,5-dimethoxyphenol, an intermediate in the synthesis of the PAL solid-phase peptide synthesis handle. Synth Commun 1991; 21.

256. Gordeev MF, Patel DV, England BP, Jonnalagadda S, Combs JD, Gordon EM. Combinatorial synthesis and screening of a chemical library of 1,4-dihydropyridine calcium channel blockers. Bioorg Med Chem 1998;6 (7):883–889.

257. Look GC, Murphy MM, Campbell DA, Gallop MA. Trimethylorthoformate: a mild and effective dehydrating reagent for solution and solid

phase imine formation. Teterahedron Lett 1995;36:2937–2940.

258. MacDonald AA, DeWitt SH, Ramage R. Consideration of solid-phase synthesis with reference to quinolone antibiotics. Chimia 1996;50 (6):266–270.
259. Frank KE, Devasthale PV, Gentry EJ, Ravikumar VT, Keschavarz-Shokri A, Mitscher LA, Nilius A, Shen LL, Shawar R, Baker WR. A simple, inexpensive apparatus for performance of preparative scale solution phase multiple parallel synthesis of drug analogs. II. Biological evaluation of a retrospective library of quinolone antiinfective agents. Comb Chem High Throughput Screen 1998;1(2):89–99.
260. Lee CS, Allwine DA, Barbachyn MR, Grega KC, Dolak LA, Ford CW, Jensen RM, Seest EP, Hamel JC, Schaadt RD, Stapert D, Yagi BH, Zurenko GE, Genin MJ. Carbon-carbon-linked (pyrazolylphenyl)oxazolidinones with antibacterial activity against multiple drug resistant Gram-positive and fastidious Gram-negative bacteria. Bioorg Med Chem 2001;9 (12):3243–3253.
261. Gordeev MF. Combinatorial lead discovery and optimization of antimicrobial oxazolidinones. Curr Opin Drug Discov Dev 2001;4 (4):450–461.
262. Bergmeier SC, Katz SJ. A method for the parallel synthesis of multiply substituted oxazolidinones. J Comb Chem 2002;4 (2):162–166.
263. Kidwai M, Misra P, Bhushan KR, Saxena RK, Singh M. Microwave-assisted solid-phase synthesis of cephalosporin derivatives with antibacterial activity. Monatshr Chem 2000; 131:937–943.
264. Zhang C, Sarshar S, Moran EJ, Krane S, Rodarte JC, Benbatoul KD, Dixon R, Mjalli AM. 2,4,5-Trisubstituted imidazoles: novel nontoxic modulators of P-glycoprotein mediated multidrug resistance. Part 2. Bioorg Med Chem Lett 2000;10(23):2603–2605.
265. Sarshar S, Zhang C, Moran EJ, Krane S, Rodarte JC, Benbatoul KD, Dixon R, Mjalli AM. 2,4,5-Trisubstituted imidazoles: novel nontoxic modulators of P-glycoprotein mediated multidrug resistance. Part 1. Bioorg Med Chem Lett 2000;10(23):2599–2601.
266. Guns ES, Denyssevych T, Dixon R, Bally MB, Mayer L. Drug interaction studies between paclitaxel (Taxol) and OC144-093—a new modulator of MDR in cancer chemotherapy. Eur J Drug Metab Pharmacokinet 2002;27 (2):119–126.
267. Hajduk PJ, Dinges J, Schkeryantz JM, Janowick D, Kaminski M, Tufano M, Augeri DJ, Petros A, Nienaber V, Zhong P, Hammond R, Coen M, Beutel B, Katz L, Fesik SW. Novel inhibitors of Erm methyltransferases from NMR and parallel synthesis. J Med Chem 1999;42(19):3852–3859.
268. Marsault E, Hoveyda HR, Peterson ML, Saint-Louis C, Landry A, Vezina M, Ouellet L, Wang Z, Ramaseshan M, Beaubien S, Benakli K, Beauchemin S, Deziel R, Peeters T, Fraser GL. Discovery of a new class of macrocyclic antagonists to the human motilin receptor. J Med Chem 2006;49(24):7190–7197.
269. Whiting M, Tripp JC, Lin YC, Lindstrom W, Olson AJ, Elder JH, Sharpless KB, Fokin VV. Rapid discovery and structure–activity profiling of novel inhibitors of human immunodeficiency virus type 1 protease enabled by the copper(I)-catalyzed synthesis of 1,2,3-triazoles and their further functionalization. J Med Chem 2006;49(26):7697–7710.
270. Rodriguez-Borges JE, Goncalves S, do Vale ML, Garcia-Mera X, Coelho A, Sotelo E. Click chemistry approach to assembly proline mimetic libraries containing 1,4-substituted 1,2,3-triazoles. J Comb Chem 2008;10 (3):372–375.
271. Swain CJ, Teran A, Maroto M, Cabello A. Identification and optimisation of 5-amino-7-aryldihydro-1,4-diazepines as 5-HT2A ligands. Bioorg Med Chem Lett 2006;16 (23):6058–6062.
272. Poulsen SA, Bornaghi LF. Fragment-based drug discovery of carbonic anhydrase II inhibitors by dynamic combinatorial chemistry utilizing alkene cross metathesis. Bioorg Med Chem 2006;14(10):3275–3284.
273. Oltersdorf T, Elmore SW, Shoemaker AR, Armstrong RC, Augeri DJ, Belli BA, Bruncko M, Deckwerth TL, Dinges J, Hajduk PJ, Joseph MK, Kitada S, Korsmeyer SJ, Kunzer AR, Letai A, Li C, Mitten MJ, Nettesheim DG, Ng SC, Nimmer PM, O'Connor JM, Oleksijew A, Petros AM, Reed JC, Shen W, Tahir SK, Thompson CB, Tomaselli KJ, Wang B, Wendt MD, Zhang H, Fesik SW, Rosenberg SH. An inhibitor of Bcl-2 family proteins induces regression of solid tumours. Nature 2005;435 (7042):677–681.
274. Petros AM, Dinges J, Augeri DJ, Baumeister SA, Betebenner DA, Bures MG, Elmore SW, Hajduk PJ, Joseph MK, Landis SK, Nettesheim DG, Rosenberg SH, Shen W, Thomas S, Wang X, Zanze I, Zhang H, Fesik SW. Dis-

covery of a potent inhibitor of the antiapoptotic protein Bcl-xL from NMR and parallel synthesis. J Med Chem 2006;49(2):656–663.

275. Wendt MD, Shen W, Kunzer A, McClellan WJ, Bruncko M, Oost TK, Ding H, Joseph MK, Zhang H, Nimmer PM, Ng SC, Shoemaker AR, Petros AM, Oleksijew A, Marsh K, Bauch J, Oltersdorf T, Belli BA, Martineau D, Fesik SW, Rosenberg SH, Elmore SW. Discovery and structure–activity relationship of antagonists of B-cell lymphoma 2 family proteins with chemopotentiation activity in vitro and *in vivo*. J Med Chem 2006;49(3):1165–1181.

276. Han Y, Giroux A, Colucci J, Bayly CI, McKay DJ, Roy S, Xanthoudakis S, Vaillancourt J, Rasper DM, Tam J, Tawa P, Nicholson DW, Zamboni RJ. Novel pyrazinone mono-amides as potent and reversible caspase-3 inhibitors. Bioorg Med Chem Lett 2005;15(4):1173–1180.

277. Bettinetti L, Lober S, Hubner H, Gmeiner P. Parallel synthesis and biological screening of dopamine receptor ligands taking advantage of a click chemistry based BAL linker. J Comb Chem 2005;7(2):309–316.

278. Sheppard GS, Wang J, Kawai M, BaMaung NY, Craig RA, Erickson SA, Lynch L, Patel J, Yang F, Searle XB, Lou P, Park C, Kim KH, Henkin J, Lesniewski R. 3-Amino-2-hydroxyamides and related compounds as inhibitors of methionine aminopeptidase-2. Bioorg Med Chem Lett 2004;14(4):865–8.

279. Sheppard GS, Wang J, Kawai M, Fidanze SD, BaMaung NY, Erickson SA, Barnes DM, Tedrow JS, Kolaczkowski L, Vasudevan A, Park DC, Wang GT, Sanders WJ, Mantei RA, Palazzo F, Tucker-Garcia L, Lou P, Zhang Q, Park CH, Kim KH, Petros A, Olejniczak E, Nettesheim D, Hajduk P, Henkin J, Lesniewski R, Davidsen SK, Bell RL. Discovery and optimization of anthranilic acid sulfonamides as inhibitors of methionine aminopeptidase-2: a structural basis for the reduction of albumin binding. J Med Chem 2006;49(13):3832–49.

280. Kawai M, BaMaung NY, Fidanze SD, Erickson SA, Tedrow JS, Sanders WJ, Vasudevan A, Park C, Hutchins C, Comess KM, Kalvin D, Wang J, Zhang Q, Lou P, Tucker-Garcia L, Bouska J, Bell RL, Lesniewski R, Henkin J, Sheppard GS. Development of sulfonamide compounds as potent methionine aminopeptidase type II inhibitors with antiproliferative properties. Bioorg Med Chem Lett 2006;16 (13):3574–7.

281. Annis DA, Athanasopoulos J, Curran PJ, Felsch JS, Kalghatgi K, Lee WH, Nash HM, Orminati J-P, Rosner KE, Shipps GW Jr, Thaddupathy G.R.A., Tyler AN, Vilenchik L, Watgner CR, Wintner EA. An affinity selection-mass spectrometry method for the identification of small molecule ligands from self-encoded combinatorial libraries. Discovery of a novel antagonist of *E. coli* dihydrofolate reductase. Int. J. Mass Spectrom. 2004;238:77–83.

282. Lindsley CW, Wisnoski DD, Leister WH, O'Brien JA, Lemaire W, Williams DL Jr, Burno M., Sur C, Kinney GG, Pettibone DJ, Tiller PR, Smith S, Duggan ME, Hartman GD, Conn PJ, Huff JR. Discovery of positive allosteric modulators for the metabotropic glutamate receptor subtype 5 from a series of N-(1,3-diphenyl-1H- pyrazol-5-yl)benzamides that potentiate receptor function *in vivo*. J Med Chem 2004;47(24):5825–8.

283. Ding S, Gray NS, Wu X, Ding Q, Schultz PG. A combinatorial scaffold approach toward kinase-directed heterocycle libraries. J Am Chem Soc 2002;124(8):1594–1596.

284. Chen S, Do JT, Zhang Q, Yao S, Yan F, Peters EC, Scholer HR, Schultz PG, Ding S. Self-renewal of embryonic stem cells by a small molecule. Proc Natl Acad Sci USA 2006;103 (46):17266–17271.

285. Chiamvimonvat N, Ho CM, Tsai HJ, Hammock BD. The soluble epoxide hydrolase as a pharmaceutical target for hypertension. J Cardiovasc Pharmacol 2007;50(3):225–237.

286. Hwang SH, Morisseau C, Do Z, Hammock BD. Solid-phase combinatorial approach for the optimization of soluble epoxide hydrolase inhibitors. Bioorg Med Chem Lett 2006;16 (22):5773–5777.

287. Taveras AG, Kirschmeier P, Baum CM. Sch-66336 (sarasar) and other benzocycloheptapyridyl farnesyl protein transferase inhibitors: discovery, biology and clinical observations. Curr Top Med Chem 2003;3(10):1103–1114.

288. Huang CY, Stauffer TM, Strickland CL, Reader JC, Huang H, Li G, Cooper AB, Doll RJ, Ganguly AK, Baldwin JJ, Rokosz LL. Guiding farnesyltransferase inhibitors from an ECLiPS library to the catalytic zinc. Bioorg Med Chem Lett 2006;16(3):507–511.

289. Faucher AM, White PW, Brochu C, Grand-Maitre C, Rancourt J, Fazal G. Discovery of small-molecule inhibitors of the ATPase activity of human papillomavirus E1 helicase. J Med Chem 2004;47(1):18–21.

290. White PW, Faucher AM, Massariol MJ, Welchner E, Rancourt J, Cartier M, Archambault J. Biphenylsulfonacetic acid inhibitors of the hu-

man papillomavirus type 6 E1 helicase inhibit ATP hydrolysis by an allosteric mechanism involving tyrosine 486. Antimicrob Agents Chemother 2005;49(12):4834–4842.

291. Erickson SD, Banner B, Berthel S, Conde-Knape K, Falcioni F, Hakimi I, Hennessy B, Kester RF, Kim K, Ma C, McComas W, Mennona F, Mischke S, Orzechowski L, Qian Y, Salari H, Tengi J, Thakkar K, Taub R, Tilley JW, Wang H. Potent, selective MCH-1 receptor antagonists. Bioorg Med Chem Lett 2008;18 (4):1402–1406.

292. Su J, Tang H, McKittrick BA, Gu H, Guo T, Qian G, Burnett DA, Clader JW, Greenlee WJ, Hawes BE, O'Neill K, Spar B, Weig B, Kowalski T, Sorota S. Synthesis of novel bicyclo[4.1. 0] heptane and bicyclo[3. 1. 0]hexane derivatives as melanin-concentrating hormone receptor R1 antagonists. Bioorg Med Chem Lett 2007;17(17):4845–4850.

293. Guo T, Hunter RC, Gu H, Rokosz LL, Stauffer TM, Hobbs DW. Discovery and SAR of 4-amino-2-biarylbutylurea MCH 1 receptor antagonists through solid-phase parallel synthesis. Bioorg Med Chem Lett 2005;15 (16):3691–3695.

294. Donnelly-Roberts DL, Jarvis MF. Discovery of P2X7 receptor-selective antagonists offers new insights into P2X7 receptor function and indicates a role in chronic pain states. Br J Pharmacol 2007;151(5):571–579.

295. Nelson DW, Sarris K, Kalvin DM, Namovic MT, Grayson G, Donnelly-Roberts DL, Harris R, Honore P, Jarvis MF, Faltynek CR, Carroll WA. Structure–activity relationship studies on *N'*-aryl carbohydrazide P2X7 antagonists. J Med Chem 2008;51(10):3030–3034.

296. Merriman GH, Ma L, Shum P, McGarry D, Volz F, Sabol JS, Gross A, Zhao Z, Rampe D, Wang L, Wirtz-Brugger F, Harris BA, Macdonald D. Synthesis and SAR of novel 4,5-diarylimidazolines as potent P2X7 receptor antagonists. Bioorg Med Chem Lett 2005;15 (2):435–438.

297. Masaki T. Historical review: endothelin. Trends Pharmacol Sci 2004;25(4):219–224.

298. Brands M, Erguden JK, Hashimoto K, Heimbach D, Schroder C, Siegel S, Stasch JP, Weigand S. Novel, selective indole-based ECE inhibitors: lead optimization via solid-phase and classical synthesis. Bioorg Med Chem Lett 2005;15(19):4201–4205.

299. Chene P. Inhibiting the p53-MDM2 interaction: an important target for cancer therapy. Nat Rev Cancer 2003;3(2):102–109.

300. Parks DJ, Lafrance LV, Calvo RR, Milkiewicz KL, Gupta V, Lattanze J, Ramachandren K, Carver TE, Petrella EC, Cummings MD, Maguire D, Grasberger BL, Lu T. 1,4-Benzodiazepine-2 5-diones as small molecule antagonists of the HDM2-p53 interaction: discovery and SAR. Bioorg Med Chem Lett 2005;15(3):765–770.

301. Pantoliano MW, Petrella EC, Kwasnoski JD, Lobanov VS, Myslik J, Graf E, Carver T, Asel E, Springer BA, Lane P, Salemme FR. High-density miniaturized thermal shift assays as a general strategy for drug discovery. J Biomol Screen 2001;6(6):429–440.

302. Grasberger BL, Lu T, Schubert C, Parks DJ, Carver TE, Koblish HK, Cummings MD, LaFrance LV, Milkiewicz KL, Calvo RR, Maguire D, Lattanze J, Franks CF, Zhao S, Ramachandren K, Bylebyl GR, Zhang M, Manthey CL, Petrella EC, Pantoliano MW, Deckman IC, Spurlino JC, Maroney AC, Tomczuk BE, Molloy CJ, Bone RF. Discovery and cocrystal structure of benzodiazepinedione HDM2 antagonists that activate p53 in cells. J Med Chem 2005;48(4):909–912.

303. Mattson MP. Pathways towards and away from Alzheimer's disease. Nature 2004;430 (7000):631–639.

304. Xiao K, Li X, Li J, Ma L, Hu B, Yu H, Fu Y, Wang R, Ma Z, Qiu B, Li J, Hu D, Wang X, Shen J. Design, synthesis, and evaluation of Leu*Ala hydroxyethylene-based non-peptide beta-secretase (BACE) inhibitors. Bioorg Med Chem 2006;14(13):4535–4551.

305. Guo T, Adang AEP, Dolle RE, Dong G, Fitzpatrick D, Geng P, Ho K-K, Kultgen SG, Liu R, McDonald E, McGuinness BF, Saionz KW, Valenzano KJ, van Straten NCR, Xie D, Webb ML. Small molecule biaryl FSH receptor agonists. Part 1: lead discovery via encoded combinatorial synthesis. Bioorg Med Chem Lett 2004;14:1713–1716.

306. Maclean D, Holden F, Davis AM, Scheuerman RA, Yanofsky S, Holmes CP, Fitch WL, Tsutsui K, Barrett RW, Gallop MA. Agonists of the follicle stimulating hormone receptor from an encoded thiazolidinone library. J. Comb Chem 2004;6:196–206.

307. Helal CJ, Sanner MA, Cooper CB, Gant T, Adam M, Lucas JC, Kang Z, Kupchinsky S, Ahlijanian MK, Tate B, Menniti FS, Kelly K, Peterson M. Discovery and SAR of 2-aminothiazole inhibitors of cyclin-dependent kinase 5/p25 as a potential treatment for Alzheimer's disease. Bioorg Med Chem Lett 2004;14(22):5521–5525.

308. Coughlin SR. Thrombin signalling and protease-activated receptors. Nature 2000;407 (6801):258–264.

309. Zhao H, Liu G. Growth hormone secretagogue receptor antagonists as anti-obesity therapies? Still an open question. Curr Opin Drug Discov Dev 2006;9(4):509–515.

310. Zhao H, Xin Z, Liu G, Schaefer VG, Falls HD, Kaszubska W, Collins CA, Sham HL. Discovery of tetralin carboxamide growth hormone secretagogue receptor antagonists via scaffold manipulation. J Med Chem 2004;47(27): 6655–6657.

311. Weigand S, Bischoff H, Dittrich-Wengenroth E, Heckroth H, Lang D, Vaupel A, Woltering M. Minor structural modifications convert a selective PPAR[alpha] agonist into a potent, highly selective PPAR[delta] agonist. Bioorg Med Chem Lett 2005;15(20):4619–4623.

312. Tai K-K. Ion Channels as a Target for Drug Design. Curr Pharm Des 2007;13(31):1381–6128.

313. Lowrie JF, Delisle RK, Hobbs DW, Diller DJ, Comb Chem High Throughput Screen 2004;7:495–510.

314. Soellner MB, Rawls KA, Grundner C, Alber T, Ellman JA. Fragment-based substrate activity screening method for the identification of potent inhibitors of the *Mycobacterium tuberculosis* phosphatase PtpB. J Am Chem Soc 2007;129(31):9613–9615.

315. Hoekstra WJ, Poulter BL. Combinatorial chemistry techniques applied to nonpeptide integrin antagonists. Curr Med Chem 1998;5 (3):195–204.

316. Romano SJ. Selectin antagonists: therapeutic potential in asthma and COPD. Treat Respir Med 2005;4(2):85–94.

317. Vicent MJ, Perez-Paya E, Orzaez M. Discovery of inhibitors of protein–protein interactions from combinatorial libraries. Curr Top Med Chem 2007;7(1):83–95.

318. Malet G, Martin AG, Orzaez M, Vicent MJ, Masip I, Sanclimens G, Ferrer-Montiel A, Mingarro I, Messeguer A, Fearnhead HO, Perez-Paya E. Small molecule inhibitors of Apaf-1-related caspase-3//-9 activation that control mitochondrial-dependent apoptosis. Cell Death Differ 2005;13(9):1523–1532.

319. Ley SV, Baxendale IR, Brusotti G, Caldarelli M, Massi A, Nesi M. Solid-supported reagents for multi-step organic synthesis: preparation and application. Farmaco 2002;57(4):321–330.

320. Ley SV, Baxendale IR, Bream RN, Jackson PS, Leach AG, Longbottom DA, Nesi M, Scott JS, Storer RI, Taylor SJ. Multi-step organic synthesis using solid-supported reagents and scavengers: a new paradigm in chemical library generation. J Chem Soc, Perkin Trans 2000;1:3815–4195.

321. Kates SA, Albericio F. Solid-Phase Synthesis. New York: Marcel Dekker, Inc; 2000. p 826.

322. Kim RM, Manna M, Hutchins SM, Griffin PR, Yates NA, Bernick AM, Chapman KT. Dendrimer-supported combinatorial chemistry. Proc Natl Acad Sci USA 1996;93(19):10012–10017.

323. Swali V, Well NJ, Langley GJ, Bradley M. Solid-phase dendrimer synthesis and the generation of super-high-loading resin beads for combinatorial chemistry. J Org Chem 1997; 62:4902–4903.

324. Klein Gebbink RJM, Kruithof CA, van Klink GPM, Van Koten G. Dendritic supports in organic synthesis. Rev Mol Biotechnol 2002;90:183–193.

325. Griffiths-Jones CM, Hopkin MD, Jonsson D, Ley SV, Tapolczay DJ, Vickerstaffe E, Ladlow M. Fully automated flow-through synthesis of secondary sulfonamides in a binary reactor system. J Comb Chem 2007;9(3):422–430.

326. Zhang L, Lushington GH, Neuenswander B, Hershberger JC, Malinakova HC. Solution-phase parallel synthesis of hexahydro-1*H*-isoindolone libraries via tactical combination of Cu-catalyzed three-component coupling and Diels–Alder reactions. J Comb Chem 2008; 10(2):285–302.

327. Flynn DL. Phase-trafficking reagents and phase-switching strategies for parallel synthesis. Med Res Rev 1999;19(5):408–431.

328. Winger BE, Campana JE. Characterization of combinatorial peptide libraries by electrospray ionization Fourier transform mass spectrometry. Rapid Commun Mass Spectrom 1996;10(14):1811–1813.

329. Nawrocki JP, Wigger M, Watson CH, Hayes TW, Senko MW, Benner SA, Eyler JR. Analysis of combinatorial libraries using electrospray Fourier transform ion cyclotron resonance mass spectrometry. Rapid Commun Mass Spectrom 1996;10(14):1860–1864.

330. Dunayevskiy YM, Vouros P, Wintner EA, Shipps GW, Carell T, Rebek J Jr. Application of capillary electrophoresis-electrospray ionization mass spectometry in the determination of molecular diversity. Proc Natl Acad Sci USA 1996;93(12):6152–6157.

331. Cork D, Hird N. Work-up strategies for high-throughput solution synthesis. Drug Discov Today 2002;7:56–63.

332. Ripka WC, Barker G, Krakover J. High-throughput purification of compound libraries. Drug Discov Today 2001;6:471–477.

333. Frank KE, Jung M, Mitscher LA. A simple, inexpensive apparatus for performance of preparative scale solution phase multiple parallel synthesis of drug analogs. I. Preparation of a retrospective library of quinolone antiinfective agents. Comb Chem High Throughput Screen 1998;1(2):73–87.

334. Flynn DL, Berk SC, Makara GM. Recent advances in chemical library synthesis methodology. Curr Opin Drug Discov Devel 2002;5 (4):580–593.

335. Harned AM, Mukherjee S, Flynn DL, Hanson PR. Ring-opening metathesis phase-trafficking (ROMpt) synthesis: multistep synthesis on soluble ROM supports. Org Lett 2003;5(1):15–18.

336. Kaldor SW, Siegel MG. Combinatorial chemistry using polymer-supported reagents. Curr Opin Chem Biol 1997;1(1):101–106.

337. Kaldor SW, Siegel MG, Fritz JE, Dressman BA, Hahn PJ. Use of solid supported nucleophiles and electrophiles for the purification of non-peptide small molecule libraries. Tetrahedron Lett 1996;37:7193–7196.

338. Flynn DL, Crich JZ, Devraj RV, Hockerman SL, Parlow JJ, South MS, Woodard S. Chemical library purification stratagies based on principles of complementary molecular reactivity and molecular recognition. J Am Chem Soc 1997;119:4874–4881.

339. Booth RJ, Hodges JC. Polymer supported quenching reagents for parallel purification. J Am Chem Soc 1997;119:4882–4886.

340. Baxendale IR, Ley SV. Polymer-supported reagents for multi-step organic synthesis: application to the synthesis of sildenafil. Bioorg Med Chem Lett 2000;10(17): 1983–1986.

341. Baxendale IR, Ley SV. Solid supported reagents in multi-step flow synthesis. Ernst Schering Found Symp Proc 2006; (3):151–185.

342. Storer RI, Takemoto T, Jackson PS, Brown DS, Baxendale IR, Ley SV. Multi-step application of immobilized reagents and scavengers: a total synthesis of epothilone C. Chemistry 2004;10(10):2529–2547.

343. Ley SV, Baxendale IR. Organic synthesis in a changing world. Chem Rec 2002;2(6):377–388.

344. Ley SV, Baxendale IR. New tools and concepts for modern organic synthesis. Nat Rev Drug Discov 2002;1(8):573–586.

345. Curran DP. Fluorous synthesis: an alternative to organic synthesis and solid phase synthesis for the preparation of small organic molecules. Cancer J Sci Am 1998;4(Suppl 1):S73–S76

346. Curran DP, Hadida S, Studer A, He M, Kim S-Y, Luo Z, Larhed M, Hallberg A, Linclau B. Experimental techniques in fluorous synthesis: a users guide. In: Fenniri H, editor. Combinatorial Chemistry: A practical Approach. Vol. 2. New York: Oxford University Press; 2000. p. 327–352.

347. Curran DP. Fluorous techniques. In: Stoddard FR, Shibasaki M, editor. Stimulating Concepts in Chemistry. New York: Wiley-VCH; 2000; 25–37.

348. Curran DP. Strategy-level separations in organic synthesis: from planning to practice. Angew Chem Int Ed, Engl 1998;37:1175–1196.

349. Zhang W. Fluorous technologies for solution-phase high-throughput organc synthesis. Tetrahedron 2003;59:4475–4489.

350. Gladysz JA, Curran DP. Fluorous chemistry: from biphasic catalysis to a parallel chemical universe and beyond. Tetrahedron 2002;58: 3823–3825.

351. Lebl M. Solid-phase synthesis of combinatorial libraries. Curr Opin Drug Discov Dev 1999;2: 385–395.

352. Ortega A, Erra M, Navarro E, Roberts RS, Fernandez-Forner D. Perfluorous solid-phase organic synthesis (PF-SPOS) and IRORI technology for combinatorial chemistry. QSAR Comb Sci 2006;25:598–604.

353. Zhang W, Luo Z, Hiu-Tung C, Durran DP. Solution-phase preparation of a 560-compound library of individual pure mappicine analogs by fluorous mixture synthesis. J Am Chem Soc 2002;124:10443–10450.

354. Zhang W, Lu Y. 96-well plate-to-plate gravity fluorous solid-phase extraction (F-SPE) for solution-phase library purification. J Comb Chem 2007;9(5):836–843.

355. Zhang W, Lu Y. Automation of fluorous solid-phase extraction for parallel synthesis. J Comb Chem 2006;8(6):890–896.

356. Williams JP, Lavrador K. A solution-phase combinatorial synthesis of selective dopamine D4 ligands. Comb Chem High Throughput Screen 2000;3:43–50.

357. Hachenberg H. Automated chromatography in research—a link between laboratory and process analysis. Anal Chim Acta 1986;190: 107–117.

358. Isbell J. Changing requirements of purification as drug discovery programs evolve from hit discovery. J. Comb Chem 2008;10:150–157.

359. Reader JC. Automation in medicinal chemistry. Curr Top Med Chem 2004;4:671–686.

360. Espada A, Molina-Martin M, Dage J, Kuo M-S. Application of LC/MS and related techniques to high-throughput drug discovery. Drug Discov Today 2008;13:417–423.

361. Boughtflower B, Lane S, Mutton I, Stasica P. Generic compound isolation using solid-phase trapping as part of the chromatographic purification process. Part 1. Proof of generic trapping concept. J Comb Chem 2006;8(4): 441–454.

362. Xu R, Wang T, Isabell J, Cai Z, Sykes C, Brailsford A, Kassel DB. High-throughput mass-directed parallel purificaiton incorporating a multiplexed single quadrupole mass spectrometer. Anal Chem 2002;74:3055–3062.

363. Koppitz M, Brailsford A, Wenz M. Maximizing automation in LC/MS high-throughput analysis and purification. J Comb Chem 2005;7 (5):714–20.

364. Schaffrath M, von Roedern E, Hamley P, Stilz HU. High-throughput purification of single compounds and libraries. J Comb Chem 2005;7(4):546–553.

365. Li S, Julien L, Tidswell P, Goetzinger W. Enhanced performance test mix for high-throughput LC/MS analysis of pharmaceutical compounds. J Comb Chem 2006;8(6):820–828.

366. Hughes I. Design of self-coded combinatorial libraries to facilitate direct analysis of ligands by mass spectrometry. J Med Chem 1998;41 (20):3804–3811.

367. Zhang X, Towle MH, Felice CE, Flament JH, Goetzinger WK. Development of a mass-directed preparative supercritical fluid chromatography purification system. J Comb Chem 2006;8(5):705–714.

368. Karancsi T, Godorhazy L, Szalay D, Darvas F. UV-triggered main-component fraction collection method and its application for high-throughput chromatographic purification of combinatorial libraries. J Comb Chem 2005;7(1):58–62.

369. Carpintero M, Cifuentes M, Ferritto R, Haro R, Toledo MA. Automated liquid–liquid extraction workstation for library synthesis and its use in the parallel and chromatography-free synthesis of 2-alkyl-3-alkyl-4-(3H)-quinazolinones. J Comb Chem 2007;9(5):818–822.

370. Kuroda N, Hird N, Cork DG. Further development of a robust workup process for solution-phase high-throughput library synthesis to address environmental and sample tracking issues. J Comb Chem 2006;8(4):505–512.

371. Gelb MH, Holm BP. A simple and inexpensive device for removal of solvent from a large collection of sample tubes. J Comb Chem 2006;8(1):15–17.

372. Houghton RA, Pinilla C, Giulianotti MA, Appel JR, Dooley CT, Nefzi A, Ostresh JM, Yu Y, Maggiora GM, Medina-Franco JL, Brunner D, Schneider J. Strategies for the use of mixture-based synthetic combinatorial libraries: scaffold ranking, direct testing *in vivo* and enhanced deconvolution by computational methods. J. Comb Chem 2008;10:3–19.

373. Letot E, Koch G, Falchetto R, Bovermann G, Oberer L, Roth HJ. Quality control in combinatorial chemistry: determinations of amounts and comparison of the "purity" of LC-MS-purified samples by NMR, LC-UV and CLND. J Comb Chem 2005;7(3):364–371.

374. Laursen JB, Nielsen J, Haack T, Pusuluri S, David S, Balakrishna R, Zeng Y, Ma Z, Doyle TB, Mitscher LA. Further exploration of antimicrobial ketodihydronicotinic acid derivatives by multiple parallel syntheses. Comb Chem High Throughput Screen 2006;9 (9):663–681.

375. Koppitz M. Maximizing efficiency in the production of compound libraries. J Comb Chem 2008;10(4):573–579.

376. Avramova LV, Desai J, Weaver S, Friedman AM, Bailey-Kellogg C. Robotic hierarchical mixing for the production of combinatorial libraries of proteins and small molecules. J. Comb Chem 2008;10:63–68.

377. Clapham B, Wilson NS, Michmerhuizen MJ, Blanchard DP, Dingle DM, Nemcek TA, Pan JY, Sauer DR. Construction and validation of an automated flow hydrogenation instrument for application in high-throughput organic chemistry. J. Comb Chem 2008;10:88–93.

378. Desai B, Kappe CO. Heterogeneous hydrogenation reactions using a continuous flow high pressure device. J Comb Chem 2005;7 (5):641–643.

379. Clapham B, Wilson NS, Michmerhuizen MJ, Blanchard DP, Dingle DM, Nemcek TA, Pan JY, Sauer DR. Construction and validation of an automated flow hydrogenation instrument for application in high-throughput organic chemistry. J Comb Chem 2008;10(1): 88–93.

380. Jones RV, Godorhazy L, Varga N, Szalay D, Urge L, Darvas F. Continuous-flow high pressure hydrogenation reactor for optimization and high-throughput synthesis. J Comb Chem 2006;8(1):110–116.

381. Mavandadi F, Pilotti A. The impact of microwave-assisted organic synthesis in drug discovery. Drug Discov Today 2006;11:165–174.
382. Mavandadi F, Lidström P. Microwave assisted chemistry in drug discovery. Curr Top Med Chem 2004;4:773–792.
383. Kappe CO. Speeding up solid-phase chemistry by microwave irradiation, a tool for high-throughput synthesis. Am Lab 2001;33: 13–19.
384. Lew A, Krutzik PO, Hart ME, Chamberlin AR. Increasing rates of reaction: microwave-assisted organic synthesis for combinatorial chemistry. J. Comb Chem 2002;4:95–105.
385. Lidstrom P, Westman J, Lewis A. Enhancement of combinatorial chemistry by microwave-assisted organic synthesis. Comb Chem High Throughput Screen 2002;5:441–458.
386. Kappe CO, Stadler A. Microwaves in Organic and Medicinal Chemistry. Vol. 25. Weinheim: Wiley-VCH Verlag GmbH; 2005. p 409.
387. Bogdal D. Microwave-Assisted Organic Synthesis: One Hundred Reaction Procedures. Vol. 25. Amsterdam: Elsevier; 2005. p 202.
388. de la Hoz A, Díaz-Ortiz A, Moreno A, Sanchez-Migallon A, Prieto P, Carrillo JR, Vazquez E, Gomez MV, Herrero MA. Microwave-assisted reactions in heterocyclic compounds with applications in medicinal and supramolecular chemistry. Comb Chem High Throughput Screen 2007;10:877–902.
389. Martinez-Palou R. Advances in microwave-assisted combinatorial chemistry without polymer-supported reagents. Mol Divers 2006;10: 435–462.
390. Kremsner JM, Stadler A, Kappe CO. High-throughput microwave-assisted organic synthesis: moving from automated sequential to parallel library-generation formats in silicon carbide microtiter plates. J Comb Chem 2007;9(2):285–291.
391. Alcazar J. Reproducibility across microwave instruments: preparation of a set of 24 compounds on a multiwell plate under temperature-controlled conditions. J Comb Chem 2005;7(3):353–355.
392. Alcazar J, Diels G, Schoentjes B. Microwave assisted medicinal chemistry. Mini Rev Med Chem 2007;7(4):345–369.
393. Díaz-Ortiz A, de la Hoz A, Alcazar J, Carrillo JR, Antonia Herrero M, Fontana A, de Mata Munoz J. Reproducibility and scalability of solvent-free microwave-assisted reactions: from domestic ovens to controllable parallel applications. Comb Chem High Throughput Screen 2007;10(3):163–169.
394. Bremner WS, Organ MG. Multicomponent reactions to form heterocycles by microwave-assisted continuous flow organic synthesis. J Comb Chem 2007;9(1):14–16.
395. Whitehurst CE, Annis DA. Affinity selection-mass spectrometry and its emerging application to the high throughput screening of G protein-coupled receptors. Comb Chem High Throughput Screen 2008;11(6):427–438.
396. Markert C, Rösel P, Pfaltz A. Combinatorial ligand development based on mass spectrometric screening and a double mass-labeling strategy. J Am Chem Soc 2008;130:3234–3235.
397. Jerkovic V, Nguyen F, Nizet S, Collin S. Combinatorial synthesis, reversed-phase and normal-phase high-performance liquid chromatography elution data and liquid chromatography/positive atmospheric pressure chemical ionization tandem mass spectra of methoxylated and glycosylated resveratrol analogues. Rapid Commun Mass Spectrom 2007;21: 2456–2466.
398. Bowes S, Sun D, Kaffashan A, Zeng C, Chuaqui C, Hronowski X, Buko A, Zhang X, Josiah S. Quality assessment and analysis of Biogen Idec compound library. J Biomol Screen 2006;11:828–835.
399. Pivonka DE, Sparks RB. Implementation of Raman spectroscopy as an analytical tool throughout the synthesis of solid-phase scaffolds. Appl Spectros 2000;54:1584–1590.
400. Hochlowski J, Whittern D, Pan J, Swenson R. Applications of Raman spectroscopy to combinatorial chemistry. Drugs Future 1999;24: 539–554.
401. Lippens G, Warrass R, Wieruszeski J-M, Rousselot-Pailley P, Chessari G. High resolution magic angle spinning NMR in combinatorial chemistry. Comb Chem High Throughput Screen 2001;4:333–351.
402. Roda A, Guardigli M, Pasini P, Mirasoli M. Bioluminescence and chemiluminescence in drug screening. Anal Bioanal Chem 2003; 377:826–833.
403. Lane SJ, Eggleston DS, Brinded KA, Hollerton JC, Taylor NL, Readshaw SA. Defining and maintaining a high quality screening collection: the GSK experience. Drug Discov Today 2006;11:267–272.
404. Calvert S, Stewart FP, Swarna K, Wiseman JS. The use of informatics and automation to remove bottlenecks in drug discovery. Curr Opin Drug Discov Dev 1999;2:234–238.

405. Suh CR, Krishna V, Brandon M, Narasimhan B, Mallapragada SK. Informatics methods for combinatorial materials science. Comb Mater Sci 2007; 109–111.
406. Sehgal A. Anticancer drug discovery using chemical genomics. Curr Med Chem 2003;10: 749–755.
407. Parker CN, Shamu CE, Kraybill B, Austin CP, Bajorath J. Measure, mine, model, and manipulate: the future for HTS and chemoinformatics? Drug Discov Today 2006;11:863–865.
408. Howe TJ, Mahieu G, Marichal P, Tabruyn T, Vugts P. Data reduction and representation in drug discovery. Drug Discov Today 2007;12: 45–54.
409. Fay N. The role of the informatics framework in early lead discovery. Drug Discov Today 2006;11:1075–1084.
410. Holbrook JD, Sanseau P. Drug discovery and computational evolutionary analysis. Drug Discov Today 2007;12:826–832.
411. Schnecke V, Boström J. Computational chemistry-drive decision making in lead generation. Drug Discovery Today 2006;11:43–50.
412. Ghose AP, Herbertz T, Salvino JM, Mallamo JP. Knowledge-based chemoinformatic approaches to drug discovery. Drug Discovery Today 2006;11:1107–1114.
413. Wolber G, Seidel T, Bendix B, Langer T. Molecule-pharmacophore superpositioning and pattern matching in computational drug design. Drug Discov Today 2008;13:23–29.
414. Manly CJ, Chandrasekhar J, Ochterski JW, Hammer JD, Warfield BB. Strategies and tactics for optimizing the hit-to-lead process and beyond—a computational chemistry perspective. Drug Discov Today 2008;13:99–109.
415. Orry AJ, Abagyan RA, Cavasotto CN. Structure-based development of target-specific compound libraries. Drug Discov Today 2006;11: 261–266.
416. Edwards P. Fragment-based drug discovery of carbonic anhydrase II inhibitors by dynamic combinatorial chemistry. Drug Discov Today 2007;12:497–498.
417. Eckert H, Bajorath J. Molecular similarity analysis in virtual screening: foundations, limitations and novel approaches. Drug Discov Today 2007;12:225–233.
418. Fitzgerald SH, Sabat M, Geysen HM. Survey of the diversity space coverage of reported combinatorial libraries. J Comb Chem 2007;9 (4):724–734.
419. Boldt GE, Dickerson TJ, Janda KD. Emerging chemical and biological approaches for the preparation of discovery libraries. Drug Discov Today 2006;11:143–148.
420. Waszkowycz B. Towards improving compound selection in structure-based virtual screening. Drug Discov Today 2008;13:219–226.
421. Kerns EH, Di L. Drug-like Properties: Concepts, Structure, Design, and Methods New York: Academic Press; 2008; xix, 526.

ALLOSTERIC PROTEINS AND DRUG DISCOVERY

J. Ellis Bell[1]
Jessica K. Bell[2]
[1] Department of Chemistry, Gottwald Science Center, University of Richmond, Richmond, VA
[2] Department of Biochemistry, Virginia Commonwealth University, Richmond, VA

1. INTRODUCTION

The concept of allostery for regulation of protein function has been known for about a half a century. Medicinal chemists have only recently been successful in developing allosteric effectors as therapeutic agents. Allosteric modifiers do not act as agonists or antagonists; instead, they regulate substrate binding or release, or catalysis. Allosteric proteins regulate some of the most important biochemical pathways in living systems. Allosteric effectors represent a new mechanistic approach in drug therapy. Pharmacologists have explored similar and closely related areas using different terminology such as inverse agonists. The purpose of this chapter is to integrate the underpinnings grounded in structural biology and mechanistic biochemistry with drug discovery. As an increasing number of human illnesses are now known to result from mutations affecting allosteric regulation of various enzymes [1], it is safe to assume that a number of new drug entities will emerge in the near future that modulate allosteric interactions. The major barrier to more rapid discovery of allosteric effectors has been the lack of detailed structural information of the different allosteric states that an allosteric protein assumes. Until recently, hemoglobin represented the only well-defined allosteric protein with plethora of structural and mechanistic studies. However, those involved in unraveling the mechanistic details on the allosteric transition still argue over what triggers and what is the order of tertiary and quaternary conformational changes.

In part as a result of using allosteric proteins as potential drug targets, in recent years there has been renewed interest in allosteric properties. Most allosteric proteins are homooligomers and exhibit internal symmetry. It has been suggested that this symmetry result in properties such as greater folding efficiency, reduced aggregation, greater adaptability, and the possibility of allosteric regulation. Since there is an entropic cost for oligomerization, evolutionary optimization suggests that this must have been overcome by the interaction energy between monomers that must contribute to fitness of the organism [2].

The various manifestations of cooperativity that impact both biology and drug discovery as they relate to the different levels of complexity have been discussed and further highlight the importantance of allostery [3].

Allosteric regulation requires a mechanism by which ligand binding or catalysis at one site on a protein structure causes an effect at another site. Ligand binding or amino acid mutations at an allosteric site can lead to alterations in either catalysis or binding at another active site or a second binding site. The connection between conformational mobility, allosteric regulation, and catalysis has been discussed recently [4].

In spite of the difficulty in obtaining structural information for each allosteric state in any given protein or receptor, drug discovery with allosteric systems is well underway and bearing fruit. Our purpose is to review the underlaying principles of allosteric proteins and how these may be used to design "allosteric" drugs, and some of the current literature illustrating various allosteric targets for drug discovery.

2. ALLOSTERIC THEORY AND MODELS

Enzymes function as monomers, multidomained protomers or complex oligomeric species. They bind substrates, cofactors, and/or effectors, stabilize transition states to increase the rate of catalysis and release product. Apart from regulation arising at the transcriptional, translational, or posttranslational levels, the

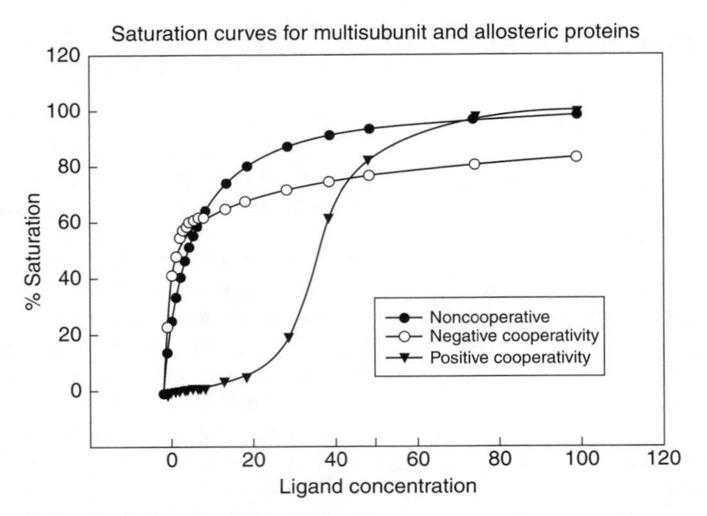

Figure 1. Saturation curves for a protein exhibiting (a) no cooperativity, (b) negative cooperativity, and (c) positive cooperativity.

control of the an enzyme reaction relies on the enzyme's ability to regulate substrate binding or product release (so-called K-type allosteric effects) or the rate of catalysis (so-called V-type allosteric effects) in response to cellular and whole organism changes.

Most proteins exhibit hyperbolic saturation curves, however, a number of important biological processes involve multisubunit proteins whose overall activities are regulated by allosteric effects. These proteins exhibit saturation curves (Fig. 1) that deviate from the normal hyperbolic curve and may involve
effects where the activity increases disproportionately with saturation (a so-called positive cooperativity, giving a sigmoidal saturation curve), or decreases relatively as saturation is increased (a so-called negative cooperativity, giving rise to a saturation curve) that is proportionally too steep below half saturation and less steep, relative to a normal saturation curve, above half saturation. A further deviation from a normal (hyperbolic) saturation curve (Fig. 2a) often seen is the phenomenon of substrate inhibition that also (among other alternatives) can have its root in allosteric interactions. In a Line-Weaver Burk plot, substrate inhibition results in a distinct upturn at high substrate concentrations (Fig. 2b).

Mathematically, cooperativity observed in ligand binding has been described by both the Hill and the Adair equations. The derivations of these equations are modeled on the classic text book example of first ligand binding and then the specifically the Hill and the Adair equations [5,6]. First, consider the simple binding of L [ligand] to P [protein], $P + L \leftrightarrows PL$ with a dissociation constant of K_d given by

$$K_d = \frac{[P][L]}{[PL]} \quad (1)$$

Rearranging Eq. 1, we get

$$[PL] = \frac{[P][L]}{K_d} \quad (2)$$

The fractional saturation, Y, is given as the concentration of bound sites divided by the total number of sites available:

$$Y = \frac{[PL]}{[P]+[PL]} \quad (3)$$

Substituting Eq. 2 in Eq. 3 gives

$$Y = \frac{[L]}{K_d} + [L] \quad (4)$$

A normal saturation curve is observed when Y is plotted as a function of [L].

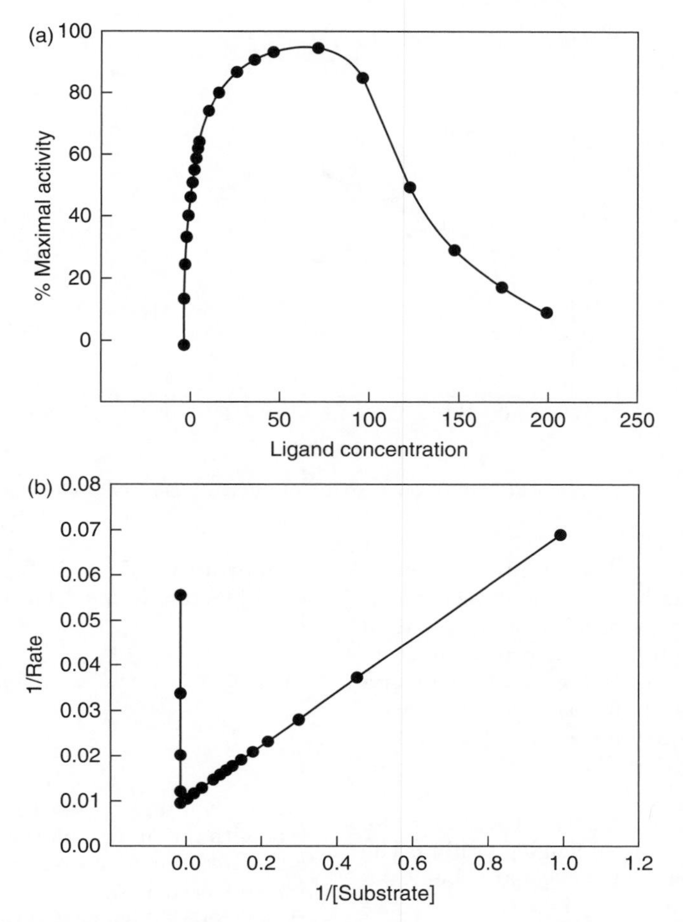

Figure 2. Substrate Inhibition manifested in (a) a saturation plot and (b) a LineWeaver Burk plot of data.

In 1910, Archibald Hill [7] derived an equation similar to Eq. 4 to describe the hemoglobin:O_2 sigmoidal curve of Y versus [L] plots. Hill assumed that a protein has n sites to bind ligand. Upon binding one site, the remaining n-1 sites are immediately occupied. This all or none binding does not allow for accumulation of intermediates and may be represented as $P + nL = PL_n$ with the dissociation constant of L given by

$$K_d = \frac{[P][L]^n}{[PL_n]} \tag{5}$$

The fractional saturation, Y, now becomes

$$Y = \frac{[PL_n]}{[P] + [PL_n]} \tag{6}$$

Substituting Eq 5 for [PL] gives

$$Y = \frac{[L]^n}{K_d + [L]_n} \tag{7}$$

Eq. 7 can be rearranged to give

$$\frac{Y}{1-Y} = \frac{[L]^n}{K_d} \tag{8}$$

where n represents the Hill coefficient. The log of Eq. 8 gives

$$\log \frac{Y}{1-Y} = n \log [L] - \log K_d \tag{9}$$

The plot of log Y/1-Y versus log [L] is linear (Fig. 3) with a slope of n. The all or none binding assumes extreme cooperativity in ligand binding. Under such conditions, n should

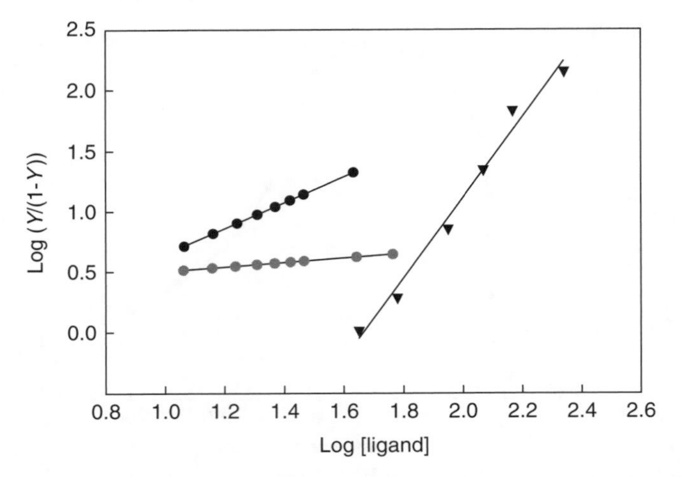

Figure 3. Hill plots of data indicating no cooperativity, negative cooperativity and positive cooperativity.

equal the number of sites within the protein. Few proteins exhibit extreme cooperativity. Therefore, the value for n is usually less than the number of sites but when positive cooperativity occurs, n is greater than 1. In the case of negative cooperativity, n is less than one. An n value of 1 reduces the fractional saturation for the Hill equation (Eq. 7) to that of simple ligand binding (Eq. 4).

The Hill derivation only accounts for two states, liganded and unliganded protein. The observation of partially oxygenated hemoglobin led Gilbert Adair [8] in 1924 to propose that ligand binding could occur in a sequential fashion:

$$\begin{aligned} &\mathrm{P} + \mathrm{L} \leftrightarrows \mathrm{PL} \\ &\mathrm{PL} + \mathrm{L} \leftrightarrows \mathrm{PL}_2 \\ &\mathrm{PL}_2 + \mathrm{L} \leftrightarrows \mathrm{PL}_3 \\ &\mathrm{PL}_{n-1} + \mathrm{L} \leftrightarrows \mathrm{PL}_n \end{aligned}$$

The individual binding steps could be described by a dissociation constant:

$$K_{\mathrm{d}n} = \frac{[\mathrm{PL}_{n-1}][\mathrm{L}]}{[\mathrm{PL}_n]} \qquad (10)$$

The fractional saturation, defined as the concentration L bound divided by total concentration of sites for n sites would give Eq. 11.

$$Y = \frac{[\mathrm{PL}] + 2[\mathrm{PL}_2] + \cdots + [\mathrm{PL}_n]}{n[\mathrm{P}] + [\mathrm{PL}] + [\mathrm{PL}_2] + \cdots + [\mathrm{PL}_n]} \qquad (11)$$

Substituting Eq. 10 in Eq. 11 and factoring out [P] results in the Adair equation:

$$Y = \frac{\left(\frac{[\mathrm{L}]}{K_{\mathrm{d}1}} + \frac{2[\mathrm{L}]^2}{K_{\mathrm{d}1}K_{\mathrm{d}2}} + \cdots + \frac{n[\mathrm{L}]^n}{K_{\mathrm{d}1}K_{\mathrm{d}2}\cdots K_{\mathrm{d}n}}\right)}{n\left(1 + \frac{[\mathrm{L}]}{K_{\mathrm{d}1}} + \frac{[\mathrm{L}]^2}{K_{\mathrm{d}1}K_{\mathrm{d}2}} + \cdots + \frac{[\mathrm{L}]^2}{K_{\mathrm{d}1}K_{\mathrm{d}2}\cdots K_{\mathrm{d}n}}\right)} \qquad (12)$$

An advantage to assessing ligand binding data with the Adair equation is that this derivation does make assumptions about type or presence of cooperativity. Instead, cooperativity is evaluated using $K_{\mathrm{d}n}$, the macroscopic or apparent dissociation constants, and the statistically related microscopic or intrinsic dissociation constants $[K'_{\mathrm{n}}]$, which describe the individual dissociation constants for the ligand binding sites. For a protein containing four binding sites, the relationship between the macroscopic and the microscopic dissociation constants is as follows:

$$K_{\mathrm{d}1} = \frac{K'_1}{4}$$

$$K_{\mathrm{d}2} = 2\frac{K'_2}{3}$$

$$K_{\mathrm{d}3} = 3\frac{K'_3}{2}$$

$$K_{\mathrm{d}4} = 4K'_4$$

No cooperativity is observed when $K'_1 = K'_2$ $K'_3 = K'_4$, the Adair equation reduces to the fractional saturation equation of simple ligand binding. When $K_1' > K_2' > K_3' > K_4'$ positive cooperativity occurs as each ligand binds. If the inverse, $K_1' < K_2' < K_3' < K_4'$, is true, ligand binding proceeds with negative cooperativity.

2.1. The Monod–Wyman–Changeux and Koshland–Nemethy–Filmer/Dalziel–Engel Models

With the mathematical evaluation of cooperativity based upon the Hill and Adair equations, the question remains: how does the enzyme convey ligand binding between sites? Historically, the mechanisms by which proteins convey cooperativity in ligand binding have been described by two models: Monod–Wyman–Changeaux [9] (MWC) and Koshland–Nemethy–Filmer [10]/Dalziel–Engel [11] (KNF/DE). These two models serve as starting points to describe what conformational changes within an enzyme may lead to the observed cooperativity.

2.1.1. Monod–Wyman–Changeaux: The Concerted Model In 1965, Monod, Wyman, and Changeaux proposed the first model to explain the mechanism by which cooperativity may be transmitted within an oligomer. The MWC model begins with two unliganded states, the low affinity, taut state, T, and the high affinity, relaxed state, R, at equilibrium predominated by the T state, refer to Fig. 4. Each complete oligomer may contain either the T state or the R state but not a mixture such that the molecule is symmetric. A ligand molecule binds to the R state, following Le Chatelier's principle, the equilibrium between the T state and the R state is reestablished

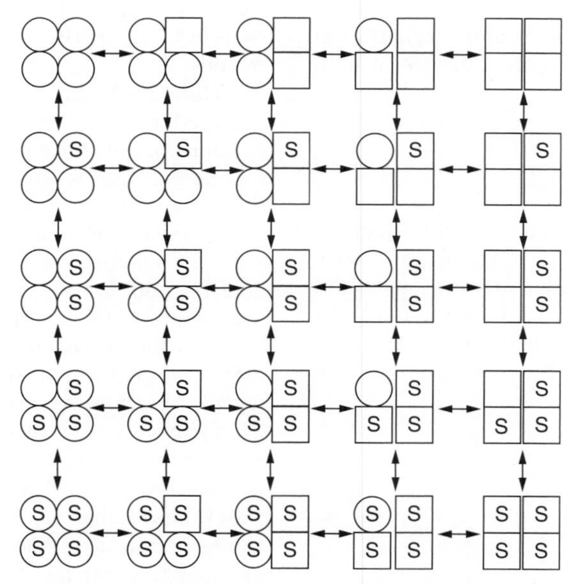

Figure 4. Schematic of the MWC, KNF/DE, and unified mechanistic models of cooperativity. The first and fifth columns represent the nonexclusive MWC model. In this model, an equilibrium between the T state and the R state exists prior to ligand binding. Ligand may bind to either the T state or the R state but will bind preferentially to the R state. Upon ligand binding the equilibrium is reestablished causing a shift from the T to the R state. The newly formed, high-affinity sites are quickly bound by ligand resulting in positive cooperativity. The diagonal represents the KNF/DE sequential model. As ligand binds it induces a conformation change within that subunit which is relayed as either increased or decreased affinity for ligand to surrounding binding sites/subunits. The MWC and KNF/DE models are two restrained examples of a much more general scheme that is represented by the entire figure. Because of the complexity of the general scheme and its mathematical derivations, systems are generally evaluated as following a concerted or sequential pathway.

creating another R state molecule. Ligand will rapidly bind to the newly formed, unoccupied high-affinity sites resulting in positive cooperativity. The MWC model explains positive cooperativity by the transition from low to high-affinity states. Heterotropic effectors, activators or inhibitors, fit into the model by stabilizing/destabilizing the T or R state. The major disadvantage to the MWC model is its inability to describe negative homotropic cooperativity. Because the MWC model begins with a preexisting equilibrium of the T and R state and the binding of ligand stabilizes the R state, the constraints of the model [12] do not allow an increase in the proportion of the T state, negative cooperativity. The model is described mathematically by Eq. 13.

$$Y=\alpha(1+\alpha)^{n-1}+\mathrm{L}\alpha C(1+\alpha C)^{n}/(1+\alpha)^{n}+\mathrm{L}(1+\alpha C)^{n} \quad (13)$$

In Eq. 13, L is the apparent conformational equilibrium constant in the absence of substrate:

$$\mathrm{L}=\frac{\mathrm{T}}{\mathrm{R}} \quad (14)$$

where n is the number of equivalent, independent binding sites for the ligand and corresponds to the number of identical subunits and α and C are constants related to the intrinsic dissociation constants of the two states (K_R and K_T) and the free ligand concentration (F) by

$$C=\frac{K_\mathrm{R}}{K_\mathrm{T}} \quad (15)$$

$$\alpha=\frac{F}{K} \quad (16)$$

2.1.2. Koshland–Nemethy–Filmer and Dalziel–Engel: The Sequential Model A second model to describe cooperativity within systems was proposed independently by both the Koshland and the Dalziel groups. The sequential, unlike the concerted model, does not constrain the oligomer to either the T state or the R state and therefore no preequilibrium of states is present. The sequential model, also known as the induced fit model, hypothesizes that when a ligand binds, a conformational change occurs within its binding site. This change is then transmitted to adjacent protomers inducing either increased or decreased affinity for the binding of the next ligand. The KNF/DE model, because it is not locked into a preexisting equilibrium, can adequately describe both positive and negative homotropic cooperativity. Heterotropic effectors, again, stabilize one form of the enzyme. The sequential model is a mechanistic explanation of the Adair equation and the observed intrinsic dissociation constants.

The MWC and KFN/DE models need not be mutually exclusive. M. Eigen [13] took these two models as limiting cases of a general scheme represented by Fig. 4 as a whole for a tetrameric protein. The mathematical analysis for so many components is complex. Under most circumstances, therefore, cooperativity is generally evaluated under the simplified cases of either the MWC or the KNF/DE models.

While the formal mechanisms of allosteric interactions have usually been invoked in discussions of allosteric proteins there has been increasing interest in more global aspects of allostery. Much of this discussion has focused on the existence of multiple, preexisting pathways for conformational changes and it has been argued that all proteins have such properties and that drug binding at different sites or mutations associated with disease simply shift the resulting ensemble of structures toward some preexisting conformation [14]. This suggests that allosteric properties are a natural result of any dynamic protein [15]. Such a picture of allostery requires that both the underlaying mechanisms of conformational fluctuations and the dynamics of such changes be experimentally demonstrated. For example, the allosteric network involved in the regulation of alpha-isopropylmalate synthase has been investigated using solution-phase H/D exchange monitored by FT-ICR mass spectrometry [16].

The free energy landscapes of proteins can be determined by combining molecular dynamics and NMR techniques [17]. Such approaches have been used to investigate allosteric properties of proteins and have emphasized the

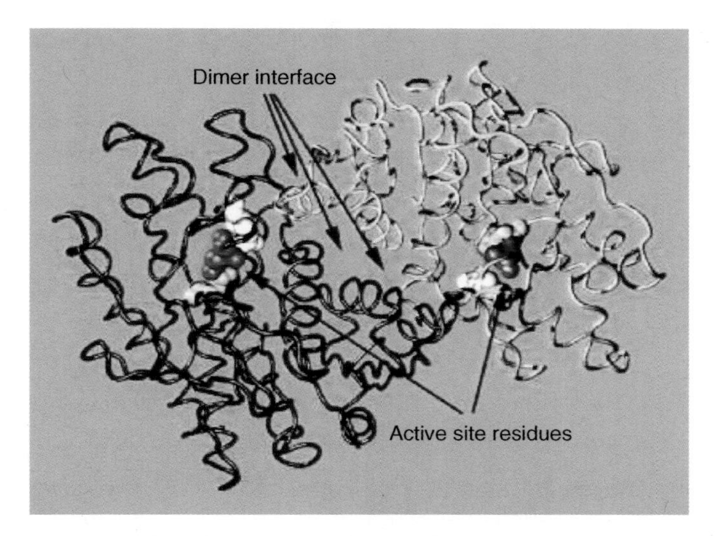

Figure 8 (Chapter 9). The malate dehydrogenase dimer indicating the location of the active sites in each protein plus the dimer interface. Malate dehydrogenase demonstrates substrate inhibition that has been attributed to subunit interactions and allosteric regulation by citrate although the crystal structure of the protein reveals the absence of a separate allosteric site for citrate.

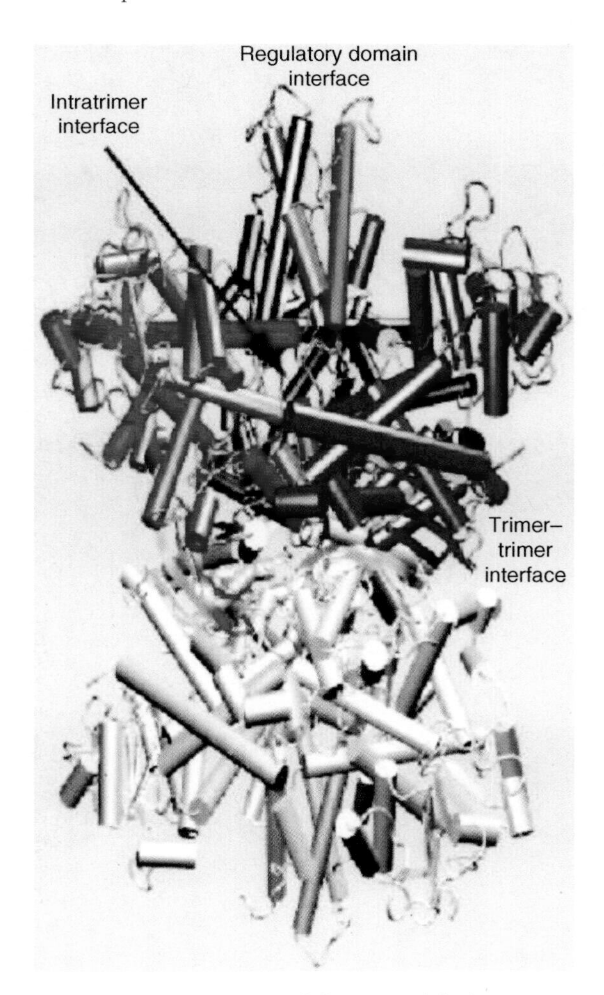

Figure 10 (Chapter 9). The quaternary structure of glutamate dehydrogenase reveals a complex array of subunit interfaces.

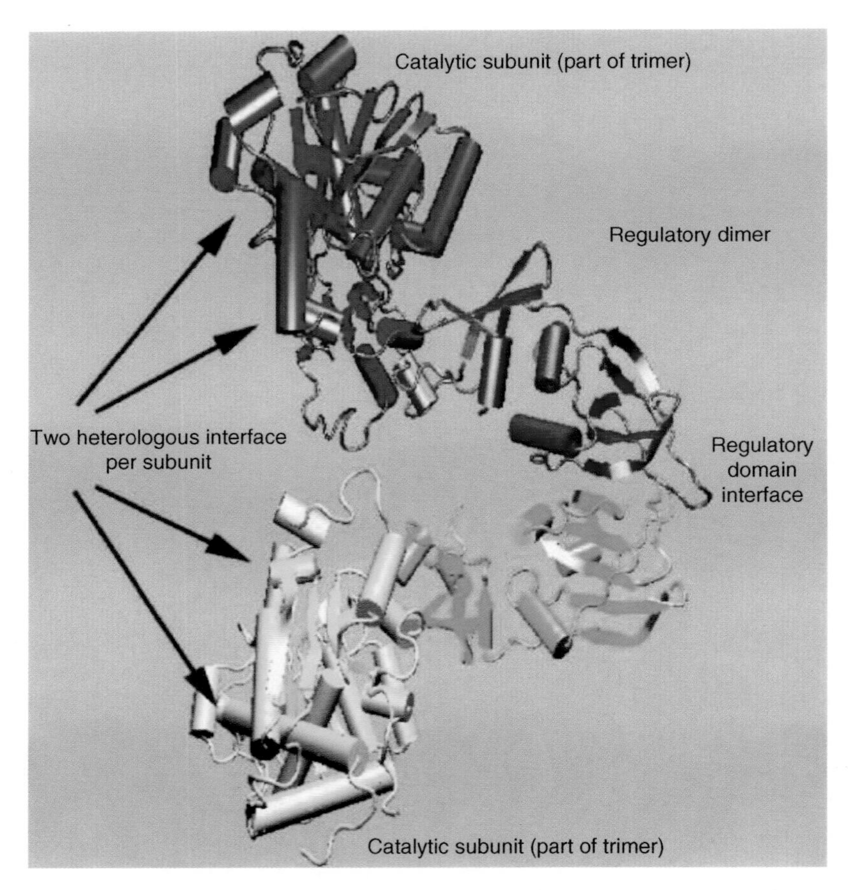

Figure 12 (Chapter 9). Subunit contacts in aspartate transcarbamoylase show heterologous contacts in the catalytic trimers, isologous contacts between regulatory dimers and regulatory subunit–catalytic subunit interactions.

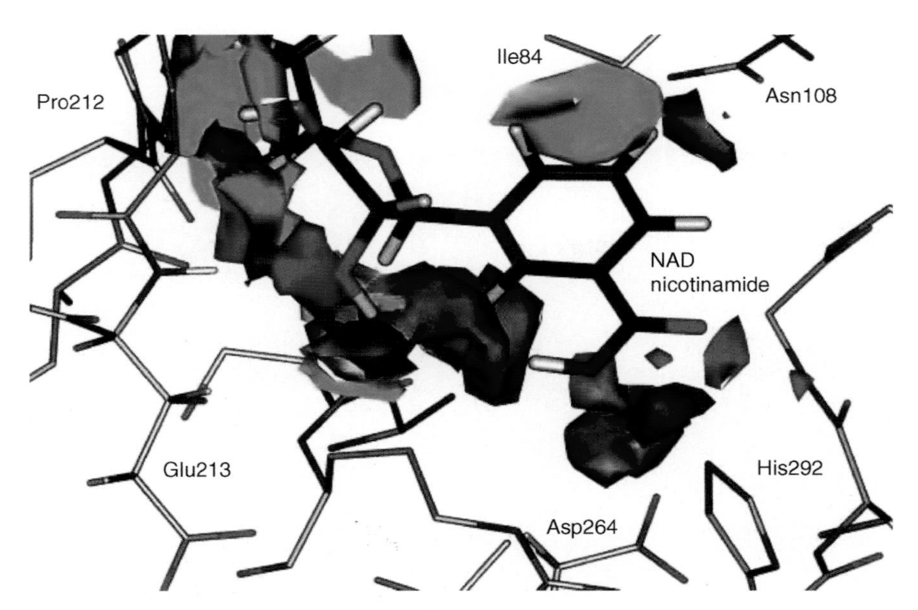

Figure 13 (Chapter 9). HINT analysis of interactions between the cofactor NAD nicotinamide moiety and the surrounding 3-phosphoglycerate dehydrogenase residues. Protein carbons are shown in light gray and NAD carbons are shown in black. Nitrogen atoms are shown in blue and oxygen atoms are shown in red. HINT contours are color coded to represent the different types of noncovalent interactions as follows: blue, favorable hydrogen bonding and acid-base interactions; green, favorable hydrophobic–hydrophobic interactions; and red, unfavorable acid–acid and base–base interactions. Importantly, the volume of the HINT contour represents the magnitude of the interactions.

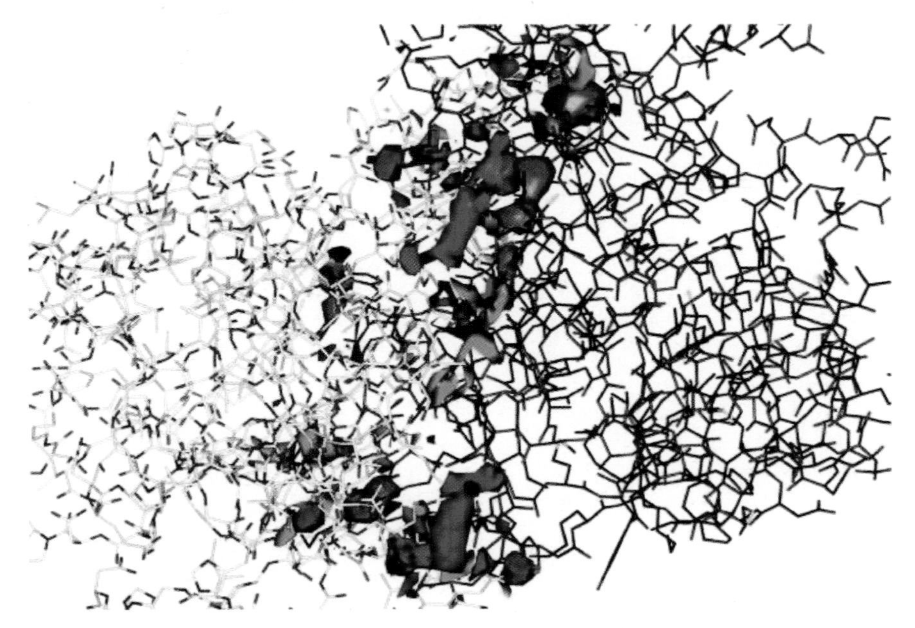

Figure 14 (Chapter 9). HINT Analysis of interactions between residues at the interface between the two subunits of 3-phospoglycerate dehydrogenase. The carbons of the two distinct subunits are colored light and dark gray, respectively. Nitrogen atoms are in blue, oxygen atoms are in red, and sulfur atoms are in yellow. HINT contour maps visually displaying interactions between the residues at the subunit interface are color coded according to the type of interaction: blue represents hydrogen bonding and favorable acid-base interactions, green displays favorable hydrophobic–hydrophobic contacts, red indicates unfavorable base–base and acid–acid interactions, and purple indicates unfavorable hydrophobic–polar interactions. Importantly, the volume of the HINT contour represents the magnitude of the interaction. For the analyzed phosphoglycerate dehydrogenase subunits, HINT indicates that a balance between favorable and unfavorable interactions characterize the interface.

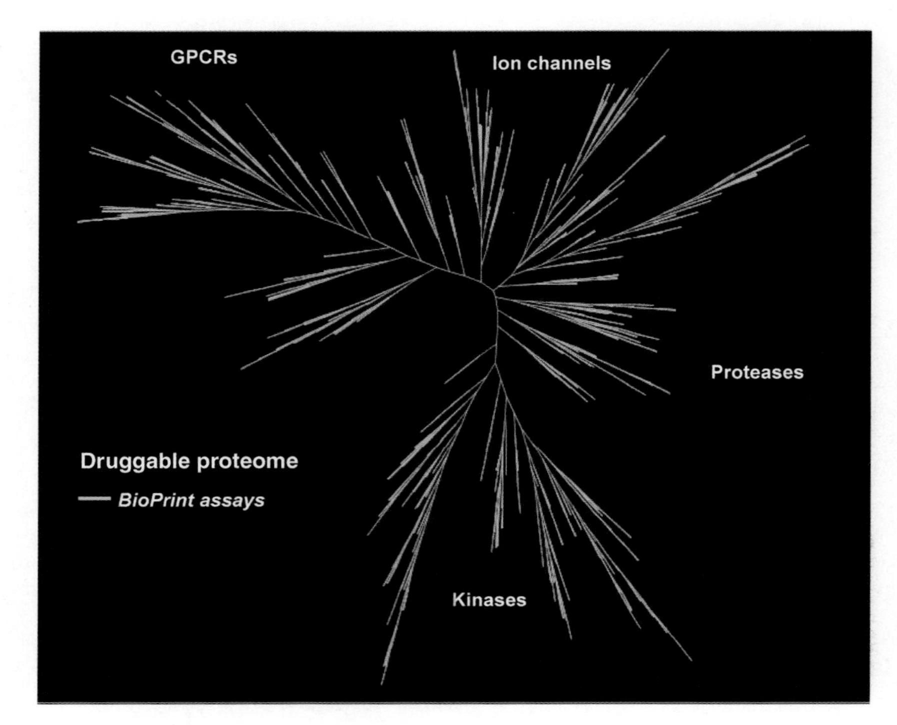

Figure 5 (Chapter 12). An early set of BioPrint assays (cyan) mapped onto the druggable proteome.

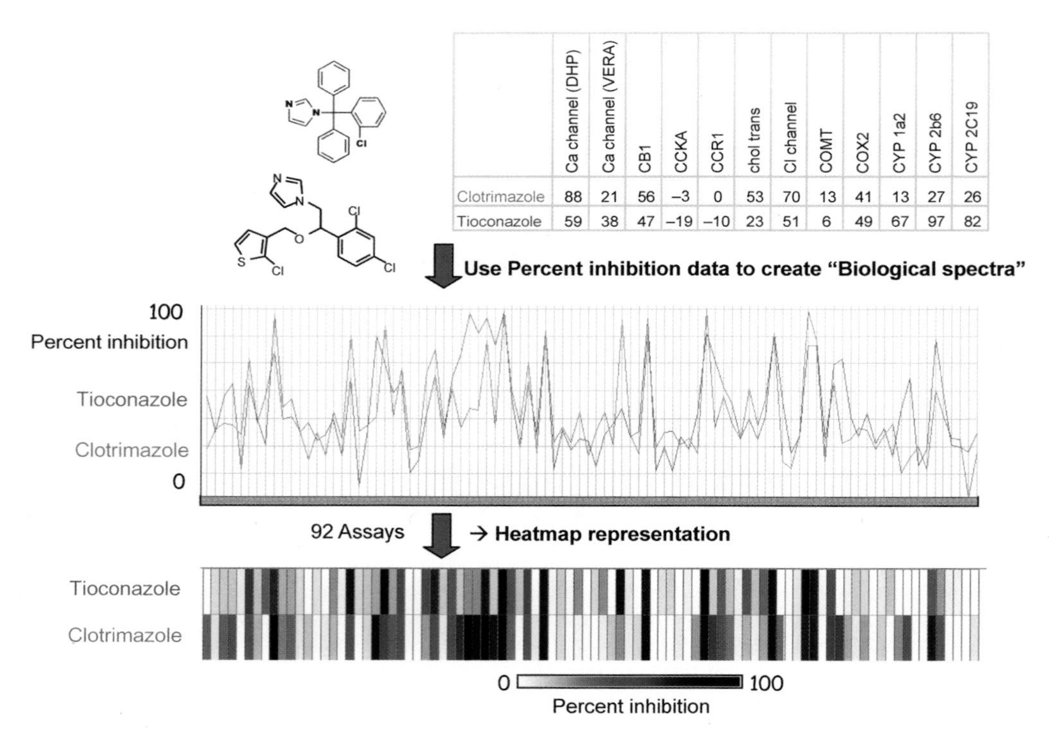

Figure 21 (Chapter 12). Biological spectra approach of Fliri et al., illustrated with two antifungal compounds. A subset of BioPrint profiling data (92 assays) is used (assay percent inhibition data only).

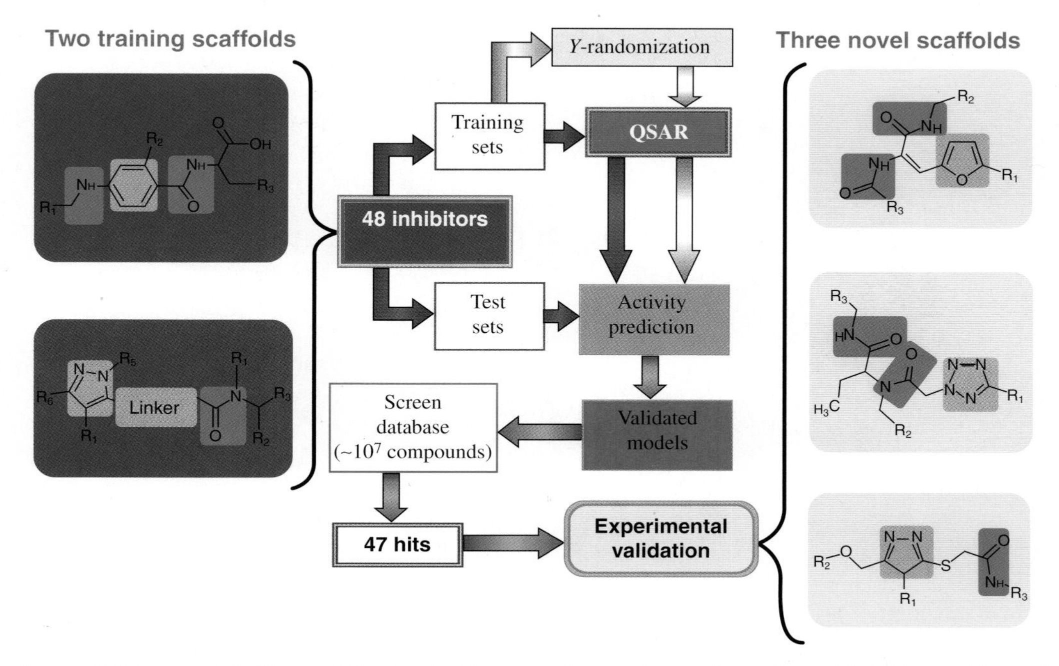

Figure 7 (Chapter 13). The workflow for the discovery of geranylgeranyltransferase-I inhibitors with novel scaffolds by the means of quantitative structure–activity relationship modeling, virtual screening, and experimental validation.

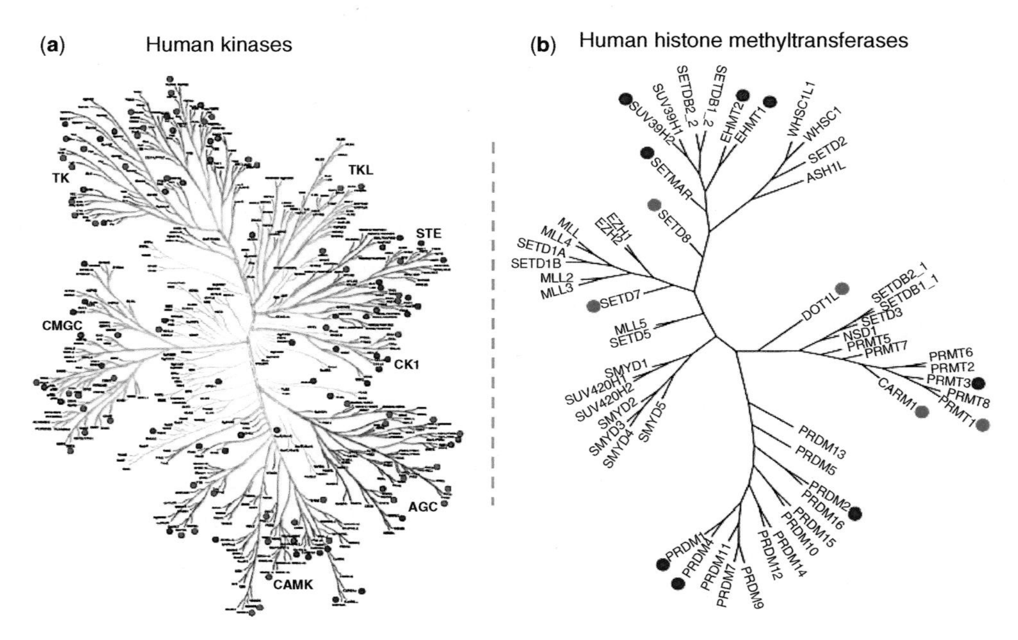

Figure 4 (Chapter 15). Phylogenetic mapping of the structural genomics contribution to the coverage of (a) human kinases and (b) epigenetic methyltransferases. Structures solved in academia, industry, and structural genomics centers are colored red, green, and blue, respectively. (Reproduced with permission from [44], updated version available from http://www.sgc.ox.ac.uk/research/kinases/.)

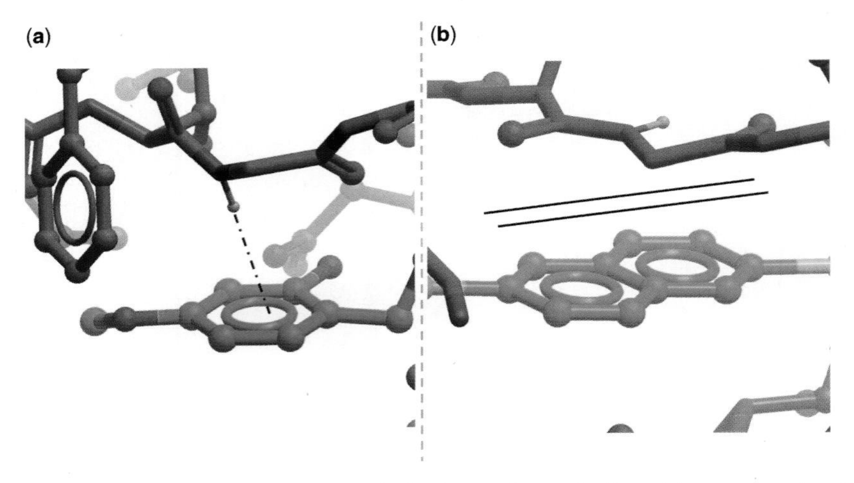

Figure 6 (Chapter 15). Two atypical but regularly observed protein–ligand interactions involving amid groups. Hydrogen bonding ((a) PDB code: 1a8g) or stacking ((b) PDB code: 1ql7) with an aromatic ring. Ligands are in green, protein residues in magenta, oxygens red, nitrogen blue. (Adapted from Ref. [77].)

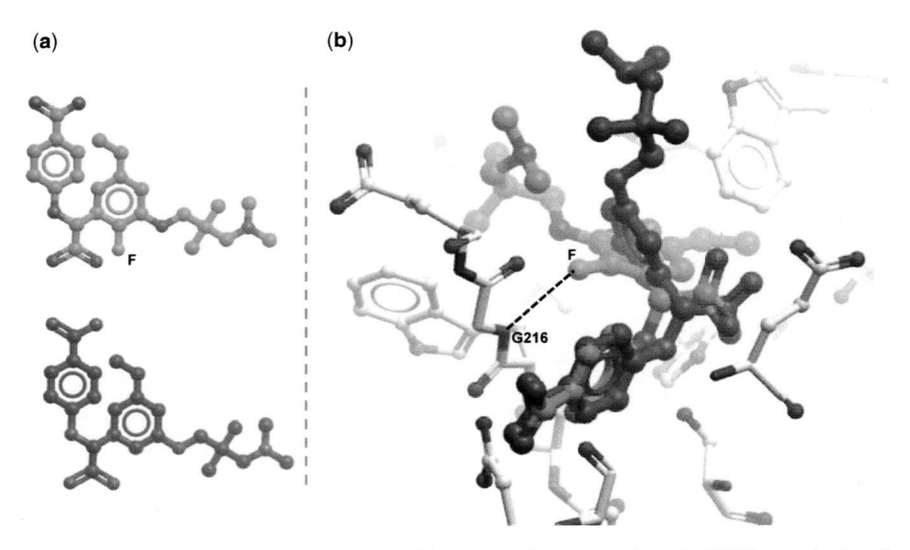

Figure 7 (Chapter 15). Analysis of cocrystallized fluorinated compounds in the PDB reveals that fluorine is regularly involved in dipolar interactions. Replacement of hydrogen (PDB code: 2v3o) by fluorine (PDB code: 2v3h) on a thrombin inhibitor (magenta and green, respectively) results in a dipolar N–H ... F–C interaction that stabilizes an alternate binding conformation. (a) 2D rendition of the inhibitors. (b) Superimposed crystal structures of the complexes. (Adapted from Ref. [78]).

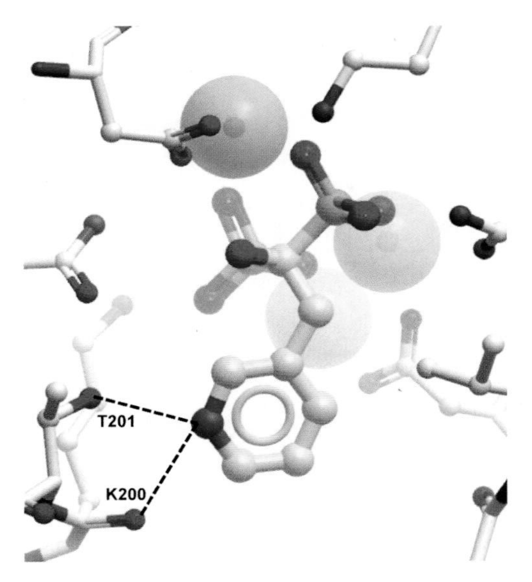

Figure 8 (Chapter 15). Structure of risedronate bound to human FPPS. Three magnesium atoms mediate the interaction between an aspartate-rich region and the phosphonate groups, while the nitrogen hydrogen-bonds to the backbone carbonyl of Lys-200 and the side-chain oxygen of Thr-201(PDB code: 1yv5, [23]).

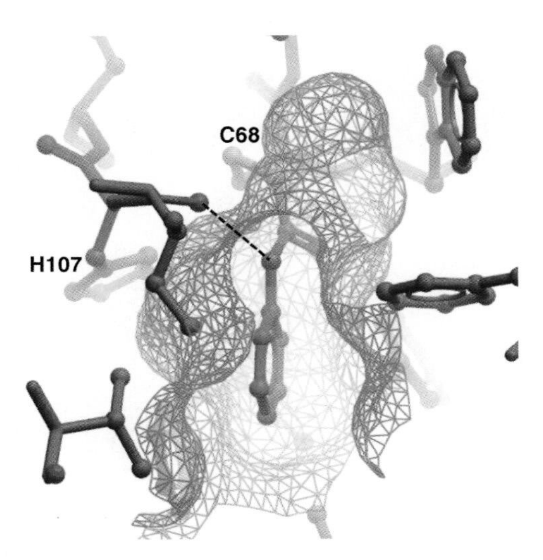

Figure 9 (Chapter 15). Human NAT1 in complex with an irreversible inhibitor. Acetanilide (green) is covalently bond to the catalytic Cys 68, hydrogen bonded to the backbone carbonyl of I106 and is surrounded with hydrophobic, mostly aromatic residues, and with catalytic His 107. The structure of the pocket can be used for metabolic liability prediction of known or prospective compounds ([36], PDB code: 2pqt).

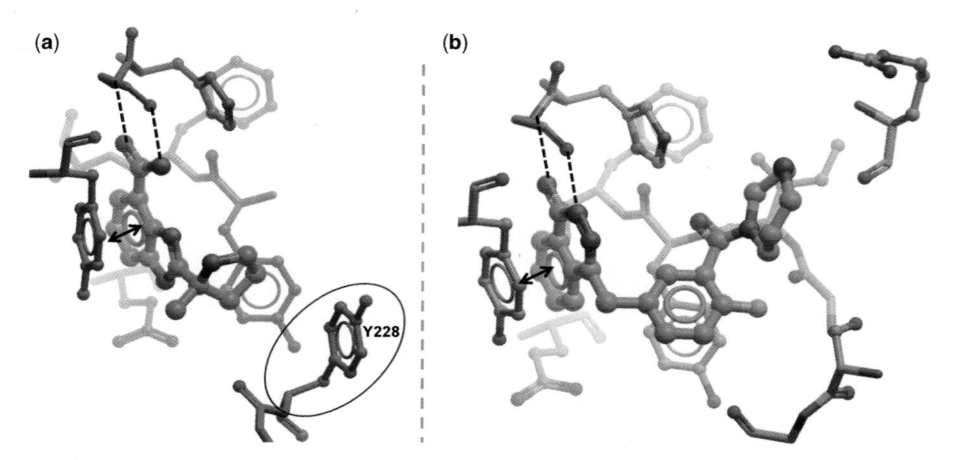

Figure 10 (Chapter 15). Comparison of PARP1 and PARP3 inhibitor binding modes. The structures of human PARP1 ((a)PDB code: 2rd6) and PARP3 ((b)PDB code: 3c49) cocrystallized with different inhibitors (green and cyan, respectively) reveal that residues lining the binding sites are mostly conserved, and so are important interactions, such as a hydrogen bonds with backbone atoms (dashed lines), or pi-stacking of aromatic rings (arrows). Rare, but important differences exist (Tyr 228 in PARP1, Gly in PARP3—highlighted by an oval) that may be used to achieve isoform selectivity.

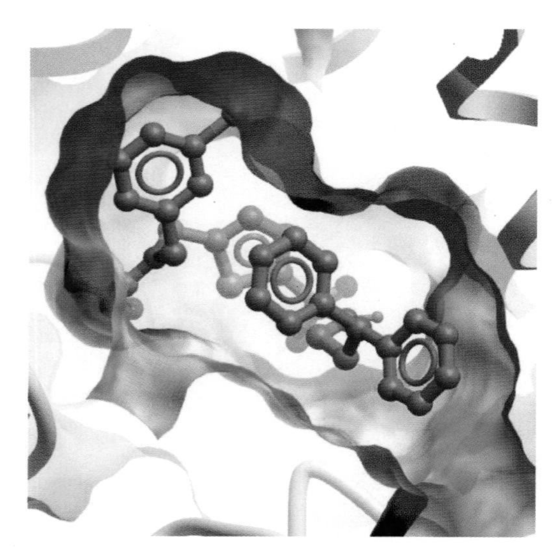

Figure 11 (Chapter 15). Structural mechanism of *Mycobacterium tuberculosis* protein tyrosine phosphatase B inhibition. Understanding the binding mode of a *M. tuberculosis* specific inhibitor (orange) will accelerate the development of improved chemical series ([94], PDB code: 2oz5).

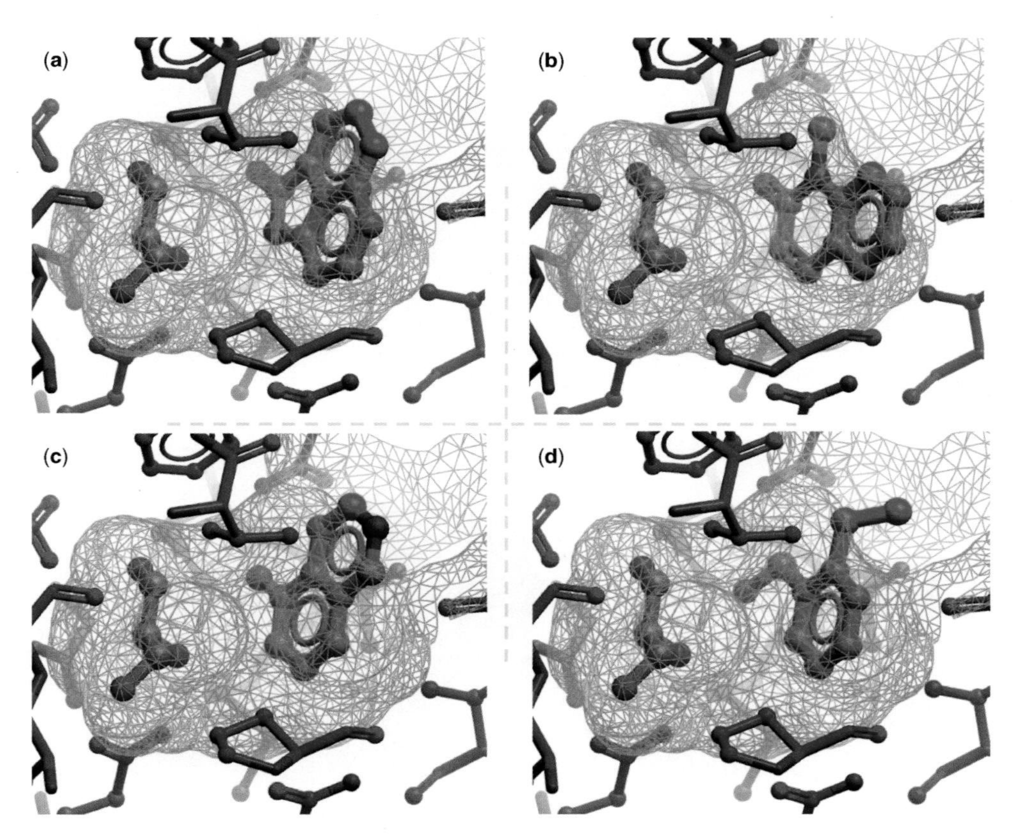

Figure 12 (Chapter 15). Structures of four chemical fragments soaked into the catalytic site of *Trypanosoma brucei* NDRT. The structures clearly indicate a conserved high-efficiency binding mode for ring systems (green), and reveals the presence of a glycerol molecule (orange) occupying another part of the pocket (gray mesh) that may be bused as a complementary chemotype ([99], PDB codes: (a) 2f67, (b) 2f64, (c) 2f2t, and (d) 2f62).

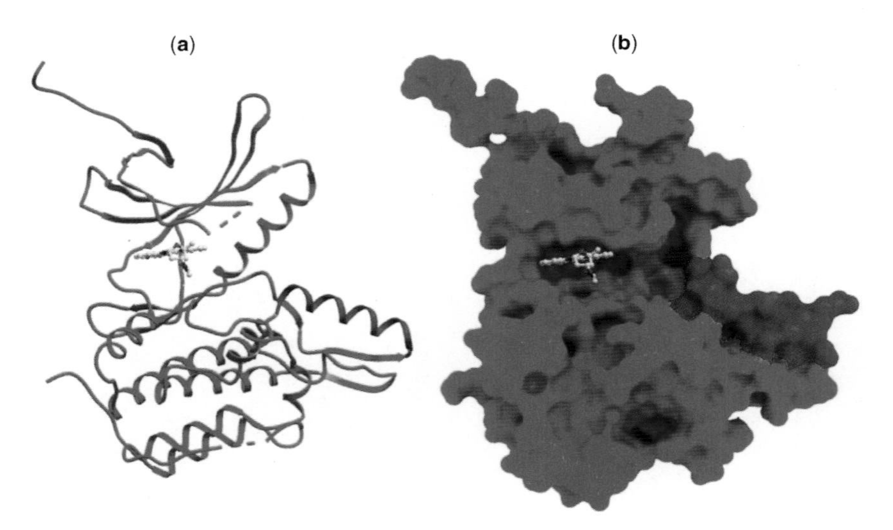

Figure 13 (Chapter 15). The structure of MPSK1 (PDB code: 2buj, (a) ribbon representation, and (b) molecular surface) in complex with staurosporine (sticks) shares only 25% sequence identity with its closest structural neighbor and reveals an alpha-helical insertion (magenta) in its activation segment unseen in any other kinase structure, which may be used to design selective inhibitors [124].

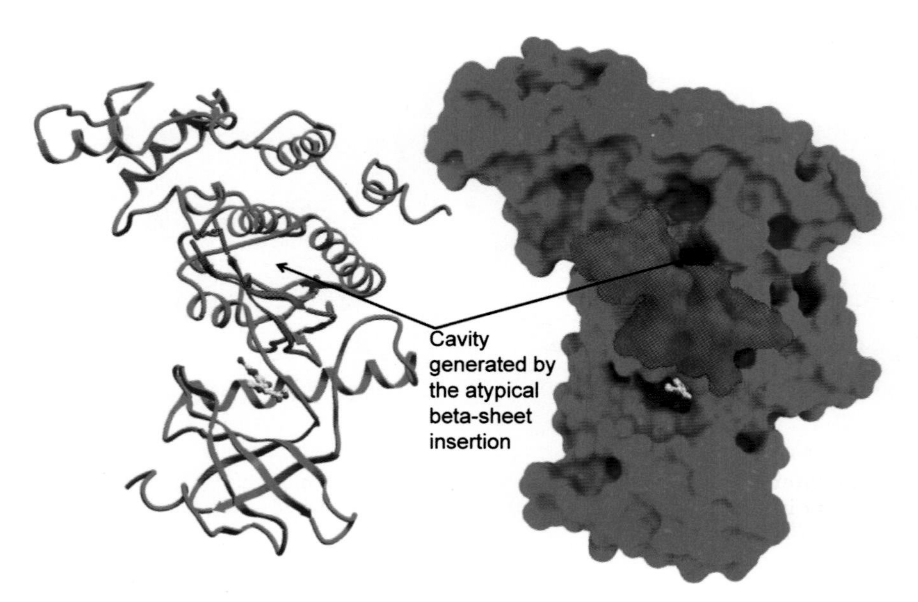

Figure 14 (Chapter 15). The crystal structure of CLK1 (shown, PDB code: 1z57) and CLK3 revealed a unique beta-hairpin insertion (magenta) with unknown function. This subfamily-specific structural motif generates cavities distinct from the ATP site that may be exploited for the generation of synthetic ligands to probe the clinical relevance of the target [125].

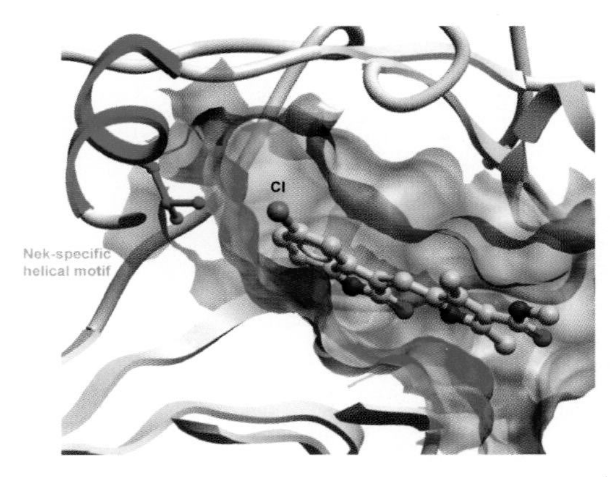

Figure 15 (Chapter 15). The structure of Nek2 uncovered an inhibitory helical motif (cyan) within the activation loop. A chlorine atom (green) of the cocrystallized inhibitor SU11652 makes limited van der Waals contact with a Leu of the Nek-specific helix. A strategy to achieve selective inactive-state inhibition may be to design ligands that better occupy the cavity generated by this atypical structural feature ([31], PDB code: 2jav).

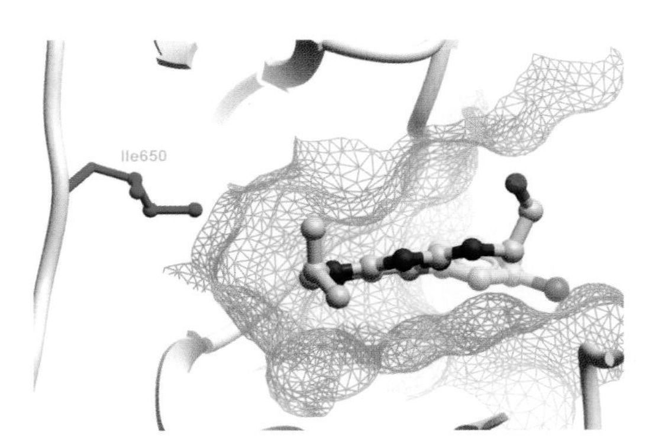

Figure 16 (Chapter 15). The crystal structure of Mer receptor tyrosine kinase (white ribbon) cocrystallized with a nonspecific inhibitor suggests how the presence of a nonconserved isoleucine (cyan) lining the ligand binding pocket (gray mesh) could be exploited to design more selective compounds ([39], PDB code: 3bpr).

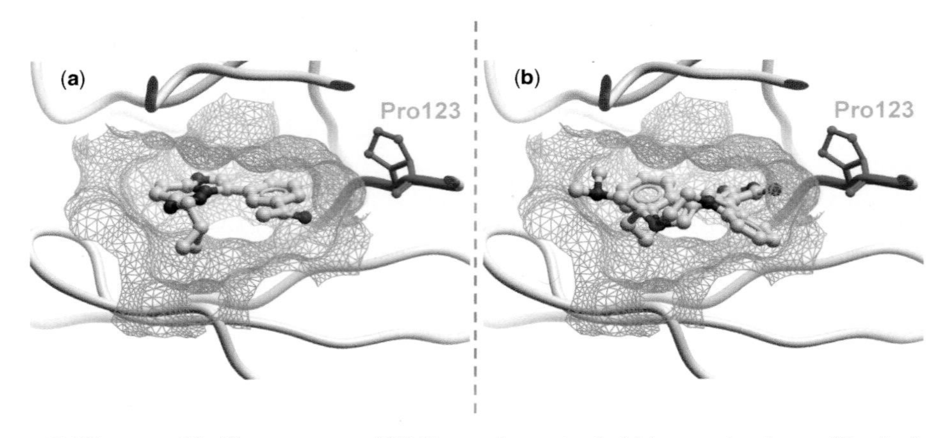

Figure 17 (Chapter 15). The structure of PIM1 reveals an atypical hinge region (cyan ribbon) where the presence of a proline residue (cyan sticks) obliterates one of the two canonical hydrogen-bond opportunities typically exploited by ATP-site inhibitors. (a) A cocrystallized imidazo[1,2-*b*] pyridazine chemotype (colored sticks) illustrates an original binding mode which does not mimic ATP binding, but contacts the binding pocket (gray mesh) mainly at a site opposite to the hinge. This unique binding mechanism results in high selectivity ([26], PDB code: 2c3i). (b) A cocrystallized 200 nM compound in Phase III clinical trial against diabetic complications displays optimal occupation of the binding pocket ([30], PDB code: 2jij).

interdependence of the concept of allostery and enzymatic catalysis [18,19].

Such approaches have been used with CREB binding protein where the KIX domain communicates allosteric information between binding sites resulting in cooperativity between pairs of transcription factors. This involves a redistribution of KIX conformation populations toward a high-energy state where the second binding site is already preformed. This signal transduction involves a network of evolutionarily conserved hydrophobic amino acids [20].

Other approaches have also been used to investigate allosteric phenomena including the engineering of surface sites to create allosteric control of proteins [21], and the generation of pH-dependent protein switches [22].

2.2. Hemoglobin: Classic Example of Allostery

The classic example of allostery and cooperativity is the binding of oxygen to hemoglobin. Hemoglobin (Hb) is a heterotetramer composed of two α- and two β-chains. An α- and a β-chain associate to form a dimer, $\alpha_1\beta_1$ and $\alpha_2\beta_2$, the dimers then associate to form the tetramer. Figure 5 illustrates the topology of both the tetramer and the subunit. Within a subunit the heme is located in a hydrophobic pocket between helices E and F. The coordination of the Fe(II) atom depends upon the oxygenation state. In the deoxy state the Fe(II) is 5-coordinated by the four pyrrole nitrogen atoms and the proximal histidine versus the oxygenated state where the Fe(II) is 6-coordinated with the sixth site occupied by the O_2. The main subunit:subunit interactions in the tetramer occur between the $\alpha_1\beta_1$ and $\alpha_1\beta_2$ interfaces and their symmetry related counterparts. A central cavity resulting from the arrangement of the subunits has been shown to accommodate allosteric effectors such as 2,3-bisphosphoglycerate.

From crystallographic studies, Hb has been shown to exist in at least two structurally distinct states, referred to as the T (deoxy) and R (oxy) states. In the T or constrained state the Fe(II) is ~0.6 Å out of the heme plane toward the proximal histidine. In the β-subunits, Val 62 partially occludes O_2 binding on the distal side of the heme ring. The T state is stabilized by salt bridges between the α_1 and α_2 chains (R141 of the α_1 to D126α_1, CO of V34β_2, and C-terminus of V1α_2) and the β_1- and β_2-chains (Y145β_2 to COV98β_2, H146β_2 to D94, C-terminusβ_2 to K40α_1). Upon oxygenation, the Fe(II) atom moves into the plane of the heme. The proximal His also reorientates and moves with the Fe(II) as does the F helix (of which the His is a part). In order for the F helix to move, a rearrangement at the $\alpha_1\beta_2$ interface is necessary. The contact between the H97 of the β-chain to the T41 of the α-chain must dissociate. H97 then is able to form a contact with T38, one turn down the helix. This movement is much like the knuckles of two hands sliding over one another. The results of these tertiary movements is a gross 15° rotation relative to the $\alpha_1\beta_1$ dimer to the $\alpha_2\beta_2$ dimer and a narrowing of the central solvent filled cavity. Physiologically, the cooperativity of O_2 binding allows Hb to take up and release O_2 over a small range of O_2 pressures comparable to pressure changes between the lungs and the target tissues. For a complete review of hemoglobin, see Ref. [28] and references therein.

The binding of O_2 to hemoglobin can be mathematically explained by both the MWC and the KNF/DE models. Given the crystallographic snapshots of the R and T states and abundant biochemical data, which model best explains the cooperativity in Hb? Both models.

The quaternary transition of Hb between the T state and the R state is consistent with the concerted or MWC model. The two potential interactions between His97 of the β chain and either T41 or T38 of the α chain sterically does not allow intermediary states within the tetramer. Coupled with the constraints of the more rigid $\alpha_1\beta_1$ and $\alpha_2\beta_2$ interfaces, this binary switch (T41 or T38) forces the transition of the entire molecule concomitantly, T → R.

The X-ray structure of human hemoglobin in which only the α subunits are oxygenated offers some evidence for an induced fit model [23]. The partially oxygenated structure resembles the T state with the exception of the α chain Fe(II) atoms which have moved ~0.15 Å closer to the heme plane. Perutz had postulated that the movement of the Fe(II)

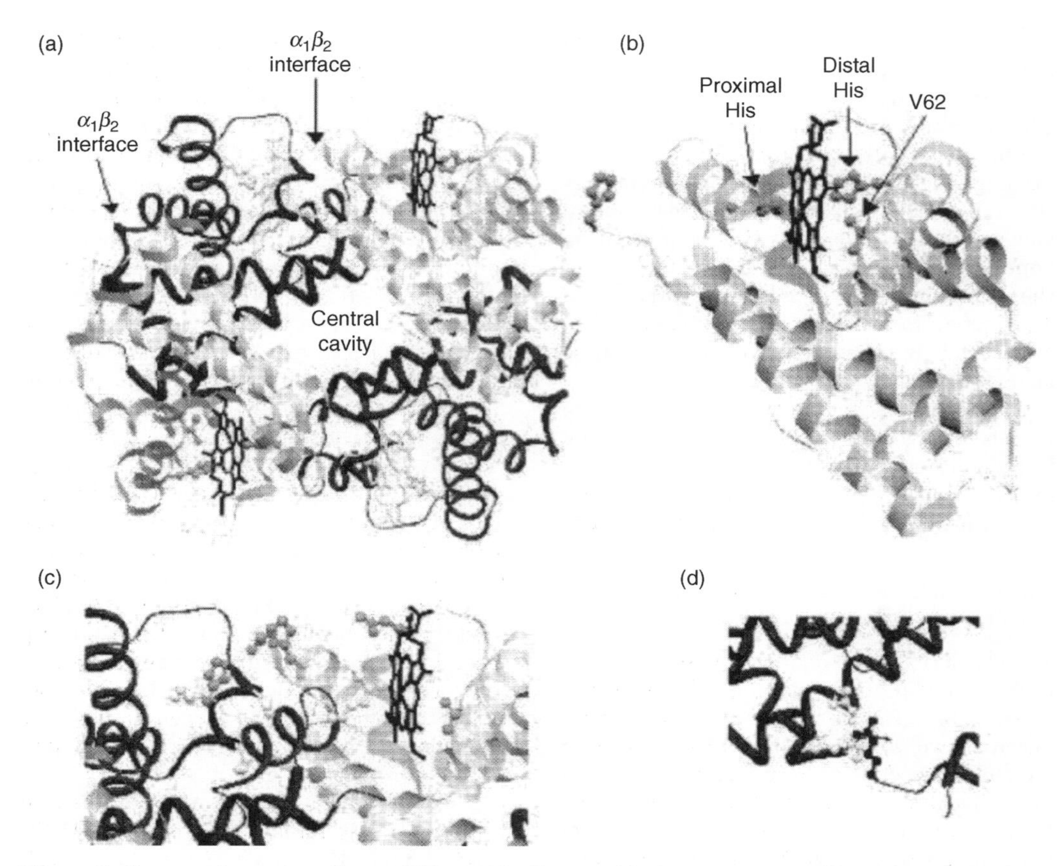

Figure 5. Structural overview of hemoglobin. (a) The T state of the heterotetramer of Hb is composed of two α (black)- and two β (white)-subunits. The packing of the subunits creates a central cavity, the binding site of the allosteric effector, 2,3-bisphosphoglycerate. (b) Both subunits are α-helical. The β-subunit shows the position of the porphyrin ring, located in a hydrophobic pocket between two helices with a proximal and distal histidine capable of coordinating the FeII. Valine 67 of the β subunit in the deoxy state partially blocks O_2 from binding. (c) The two-way switch between the R state and the T state involves His97 of the $\beta 2$ chain and Thr-41 and Thr38 of the $\alpha 1$ chain. In the deoxy state, a salt bridge exists between H97 and T41. When the FeII moves into the plane of the heme upon binding O_2, H92 and helix F must move closer to the heme also. To allow this movement, the H97:T41 bridge is broken and moved down one turn of the helix to T38. A transition between T41 and T38 is unfavored due to steric clashes. A salt bride that stabilizes the T state is also depicted: His146 at the C-terminus of β_2-coordinates its carboxyl to the amino group of Lys-40 in the α_1-chain and the Nε of His146 is coordinated to the COδ of Asp94. With the rearrangement of the H97 when O_2 is bound, the salt bridge between H146 and K40 is broken. (d) Salt bridges between the α-subunits also stabilize the T state. Arg-141 of the α_1-chain forms a hydrogen bond with the amino group of Lys-127 of the α_2-chain through its carboxy terminus. The guanidino group of Arg-141 forms bonds with COδ of Asp126, CO of Val-34 and the C-terminus of Val-1, all in the α_2-subunit. Again, these bonds are disrupted when O_2 binds.

atom into/out of the heme plane in the oxy/deoxy states would exert a tension on the Fe (II) atom proximal His bond. This tension was measured spectroscopically by setting up Hb with NO, which binds with higher affinity than O_2, to pull Fe(II) into the plane of the heme and IHP, which acts as a mimic of BPG and induces the T state, to pull Fe(II) out of the plane. The opposing forces on the Fe(II) if it triggers the transition from $T \rightarrow R$ would exceed the strength of the bonds and they would break. Therefore, the observation of the human Hb intermediary position of the Fe(II) atom suggests that as more O_2 binds to the T state this tension builds within the tertiary structure until the molecule is forced to undergo the T → R.

In part Hb fits both the MWC and the KNF/DE models and underlies the premise that the models provide a starting point for interpreting the mechanistic machinations of a protein but cannot be exclusive of one another or other potential intermediates. The examination of Hb's transition between its "T" and "R" states exemplifies the means by which proteins may transmit and communicate within and between subunits ligand binding events. To summarize Hb utilizes salt bridge formation/disruption at the C terminus, movement of secondary structure (F helix) induced by ligand coordination to a metal and changes in a subunit:subunit interface ($\alpha_1\beta_2$) which, although does not change the nature of the hydrogen bonding pattern, does trigger gross quaternary movement. The knowledge of Hb's T → R transition and other proteins, such as ATCase, form the basis of structural changes that can be used to identify triggering mechanisms in other allosterically regulated and cooperative processes in proteins.

While the above discussion has focused primarily on ligand affinity it should be pointed out that the induced conformational changes or the differences between preexistant states may be reflected in altered catalytic activity not just ligand affinity. This later point illustrates another aspect of allosteric regulation, which can best be described in the concepts of K-type and V-type effects. A K-type effect is one that affects ligand affinity while a V-type effect is one that affects the catalytic activity of the protein. Although the majority of allosteric enzymes are K-type systems, in recent years a number of V-type enzymes have been identified and in many ways may be the more important as potential targets for drug design.

2.3. Features Common to Allosteric Proteins

In summary, allosteric proteins undergo alterations in structure upon binding to ligands. These ligands can be either the native substrate (homotropic) or another effector molecule (heterotropic), and bind at sites spatially distinct (or *allo-steric*) from a given active site. In the case of heterotropic effects, this is usually a binding site quite different from that of the substrates or functional ligands of the protein and may involve a separate subunit or domain of the overall protein. With homotropic effects the spatially separate site is another chemically identical active site in the oligomer. The induced structural changes can be either activating or deactivating, causing an increase or decrease, respectively, in affinity for substrate or catalytic activity. Allosteric proteins are frequently multimeric. This feature usually leads to the phenomenon of cooperativity, whereby protein subunits are not independently acting, but are able to communicate with one another. Substrate binding at one subunit affects the process at neighboring subunits, leading to either positive or negative cooperativity. Experimentally, such effects are usually observed in ligand binding curves or plots of protein activity (or saturation) versus substrate concentration. For proteins not exhibiting binding cooperativity, (the Scatchard plot or Eadie Hofstee Plot) of ν (ligand molecules bound per protein molecule, or rate of reaction for an enzyme) versus ν/[free Ligand] yields a straight line. Cooperativity yields a curved plot (Fig. 6), concave downward for positive interactions and concave upward for negative interactions. In an alternative analysis of the data, where the enzyme velocity is measured, (LineWeaver Burk plot) or the saturation with ligand is determined (Klotz Plot), a double reciprocal representation (Fig. 7) of 1/Rate or 1/[Bound ligand] versus 1/[Substrate] or 1/[Free ligand], that is linear for an enzyme showing no allosteric effects is s concave downward for negative cooperativity or concave upward for positive cooperativity. Such curves better fit the multisite Adair equation introduced earlier or can be described by the Hill equation with terms for the constant of cooperativity, n, also known as the Hill coefficient. This number is less than one for negative cooperativity, equal to one in the noncooperative case, and greater than one for positive cooperativity.

While either ligand binding or kinetic observations are often used to invoke allosteric interactions it is critically important to recognize that such manifestations of cooperative

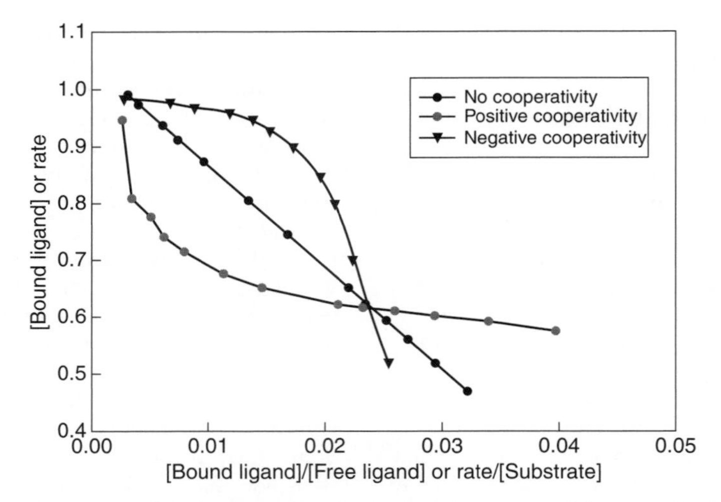

Figure 6. Scatchard or Eadie–Hofstee plots of data indicative of no cooperativity, negative cooperativity, and positive cooperativity.

interactions are not unique. Kinetic plots showing either "positive" or "negative" cooperativity can simply be a manifestation of a complex kinetic mechanism while binding studies indicative or "negative cooperativty" can result from multiple independent species capable of binding the same ligand with different affinities. Sigmoidal saturation curves of binding data however can only be attributed to "positive" cooperative effects. As has been extensively discussed [24–26], a clear demonstration of an allosteric effect requires demonstration of an induced conformational change affecting a distant site (for the Koshland–Dalziel-type models] or the demonstration of a preexistant equilibrium between R and T states of the protein for Monod-type models.

2.4. Structural Basis of Allosteric Transitions

Advances in X-ray crystallographic and spectroscopic techniques have led to much study of the precise structural changes that accompany the transition from T state to R state. One of the most famous structural applications of the MWC theory has been to the molecular description of the allosteric mechanism of the blood oxygen carrier protein, Hb [27,28]. Hemoglobin is a tetramer consisting of four similar heme-containing all α-helical subunits, two α-chains and two β-chains, arranged around a central water cavity. The protein forms an $\alpha_2\beta_2$ tetramer, also referred to as a dimer of $\alpha\beta$ dimers. Hemoglobin's oxygen binding activity displays positive cooperativity; its sigmoidal oxygen-saturation curve is the hallmark feature that distinguishes it from the nonallosteric protein myoglobin, structurally similar to a single hemoglobin subunit. Comparison of crystal structures of deoxy-Hb [29,30] and oxy-Hb [31,32] revealed striking structural changes, both quaternary and tertiary in nature. One $\alpha\beta$ dimer rotates 15° relative to the other, and the two dimers approximate their distance by about 1 Å In addition, the position of the heme iron atom transitions between the out-of-plane position in deoxy-Hb and the in-plane position in oxy-Hb. The iron motion is believed to transmit structural changes to a key "proximal" histidine residue that acts as a trigger to induce subunit motions that induce the breakage of intersubunit salt bridges that maintain the tetramer in the all-T state.

Since the discovery of hemoglobin's stereochemical mechanism, other proteins have been found to undergo similar structural changes upon ligation These include fructose-1,6-bisphosphatase, glucosamine-6-phosphate deaminase, and chorismate mutase [33–35]. There exist still other proteins that undergo more complex $T \rightarrow R$ alterations, encompassing both rigid-body subunit motions and intrasubunit

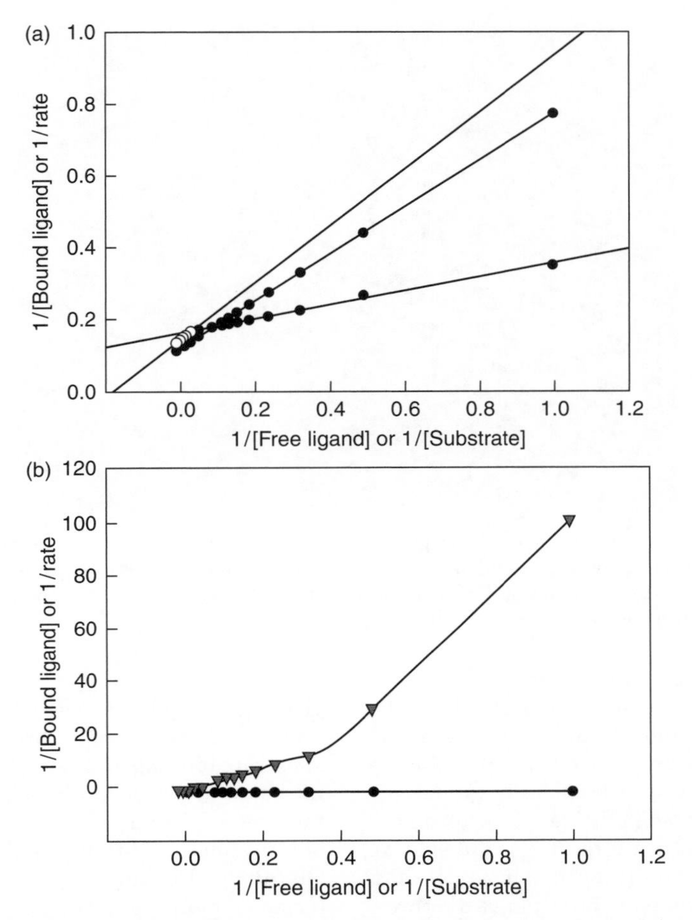

Figure 7. LineWeaver Burk or Klotz plots of data indicative of (a) negative cooperativity where at least two distinct linear regions are seen (as shown here) or (b) positive cooperativity.

domain motions or hinge bending. Recent examples of this more complex allosteric model include pyruvate kinase, hemocyanin, D-3-phosphoglycerate dehydrogenase, and glutamine-5-phospho-1-pyrophosphatase [36–39]. Finally, lactate dehydrogenase represents a transition of intermediate complexity whereby all four subunits undergo rotation, but there are no intrasubunit motions [40].

2.5. The Molecular Mechanisms of Allosteric Changes

It is instructive to consider the various types of subunit structure that have been observed to be associated with allosteric behavior. While such structures are usually discussed in terms of homopolymers (consisting of two or more chemically identical subunits) or heteropolymers (consisting of two or more chemically distinct subunits-one often "catalytic" and the other "regulatory"—see examples later) it is perhaps better to consider the types of interfaces that exist in allosteric proteins particularly when it comes to discussing how potential drugs may be designed to influence protein function at the level of allosteric regulation. It is across these interfaces that the conformational affects associated with allosteric proteins must occur.

Consider first three "homopolymeric" enzymes all shown to involve allosteric effects,

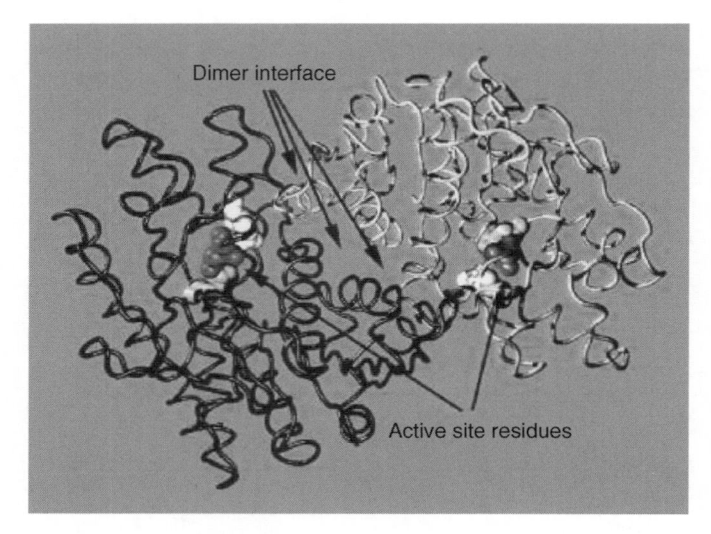

Figure 8. The malate dehydrogenase dimer indicating the location of the active sites in each protein plus the dimer interface. Malate dehydrogenase demonstrates substrate inhibition that has been attributed to subunit interactions and allosteric regulation by citrate although the crystal structure of the protein reveals the absence of a separate allosteric site for citrate. (See color insert.)

malate dehydrogenase, 3-phosphoglycerate dehydrogenase, and glutamate dehydrogenase. Malate dehydrogenase (Fig. 8), a dimer, has a single interface [41], 3-Phosphoglycerate dehydrogenase [42] (Fig. 9), a tetramer, has two quite distinct interfaces, while glutamate dehydrogenase (Fig. 10) a hexamer, has an even more complex situation, with three quite distinct interface regions [43,44]. In a dimer such as MDH, only isologous contacts are possible: the regions of the protein involved are the same on both sides of the interface: the same is true of the tetramer in 3-phosphoglycerate dehydrogenase, although there are clearly two quite distinct types of contacts: both, however, are isologous. In the hexamer structure of glutamate dehydrogenase a more complex situation is in effect. While the interaction between the top and the bottom halves (trimers) of the molecule are clearly isologous, the interactions between subunits in each of the trimers is heterologous, as shown schematically in Fig. 11, otherwise there would be enforced asymmetry in the trimer. Superimpose on these pictures the functional aspects of the various regions of each protein other than malate dehydrogenase. Of the two types of contacts in 3-phosphoglycerate dehydrogenase, one is composed of regions of the protein directly involved in substrate binding and catalysis and the other is composed of regions (domains) of the protein involved in regulator binding, in this case the binding of the V-type regulator, serine. In Glutamate Dehydrogenase the isologous interface involve substrate binding domains while of the two heterologous interfaces, the one in the plane of the trimer also involves substrate binding domains while the other involves regulator binding domains.

This discussion of functional domains within an allosteric protein can be extended to the case of heteropolymeric allosteric enzymes. Consider aspartate transcarbamoylase [45] (Fig. 12). This enzyme consists of catalytic and regulatory subunits as shown and ligand binding must trigger conformational changes that are transmitted through the regulatory subunits to other catalytic subunits

How strong are such interfaces and what governs how the protein will change “shape” during an allosteric conformational change? While the exact details of the overall “strength” of an interaction at an interface depend on the specific nature of the side chains and contacts between the two surfaces

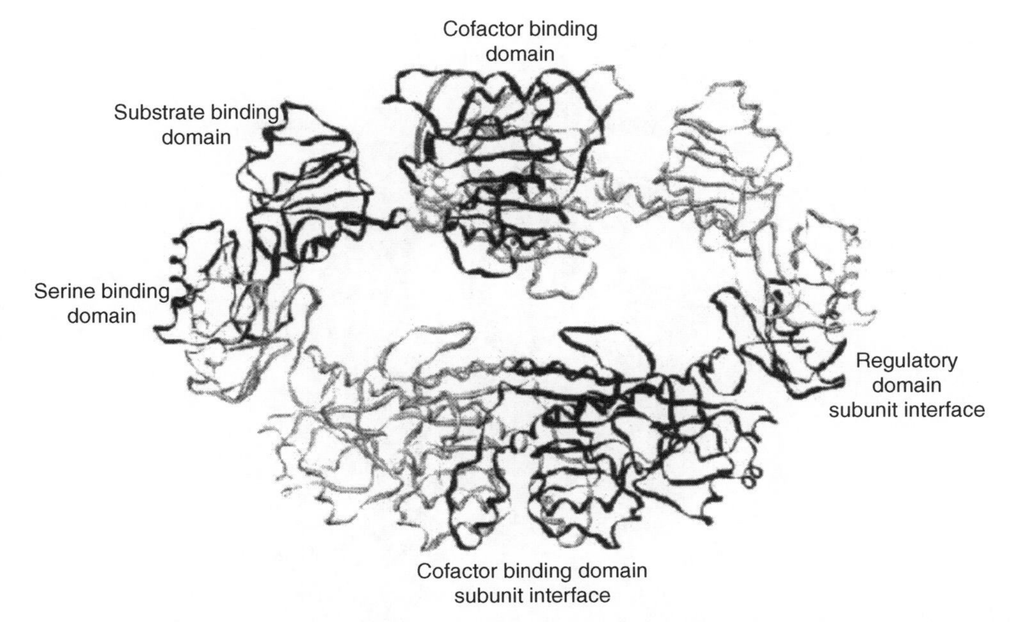

Figure 9. Quaternary structure of 3-phosphoglycerate dehydrogenase. The ribbon diagram represents the homotetramer of PGDH. Two subunits are shown in black versus gray to distinguish each subunit. The NADH and serine are shown for one subunit in van der Waals surfaces. PGDH is unique in that it does not form a compact globular protein but rather a flexible donut shaped moiety. The subunits form a string of three domains connected by hinge regions. The active site is located in a cleft between two of these domains.

(which can be examined in some detail using "HINT" analysis) to a first approximation the area of contact is directly proportional to the strength of the interactions across that interface. If you want to rotate subunits relative to one another, or induce a conformational effect across an interface, these changes are most likely to involve the weakest link of the various interfaces in a given molecule. If one wants to block the induction of an allosteric conformational change (i.e., the R to T transition), one would want to strengthen the "weakest link" in the molecule to the extent that the conformational change induced by ligand binding (whether heterotropic or homotropic) was no longer possible.

Identifying the region of the molecule involved in the direct transmission of allosteric effects is clearly important for understanding how allosteric transitions might be manipulated for therapeutic purposes, particularly if one wished to block an allosteric transition. Understanding the structure–function relationships of the ligands binding to their various sites in an allosteric protein is also critically important. For a homotropic interaction part of the ligand must interact with the active site of the protein and "trigger" the interaction between subunits. In a heterotropic interaction, parts of the regulatory ligand must provide direct binding energy for the binding and part must again act as a trigger for the induced conformational change. In either case, dissecting which parts of the ligand contribute to which types of interaction is critical in the design of an allosteric drug. If the crystal structure of the protein–ligand complex is available, HINT analysis can yield valuable insight into specific contributions to the interaction. For example, in the binding of cofactor to the allosteric protein 3-phosphoglycerate dehydrogenase, HINT analysis, as shown in Fig. 13, indicates regions of both the protein surface and the cofactor that make either strong positive or strong negative contributions to the overall interaction. Using this type of analysis as a starting point can lead to the design of analogs of the cofactor, or mutants of the protein to further dissect contributions to binding and induced conformational changes.

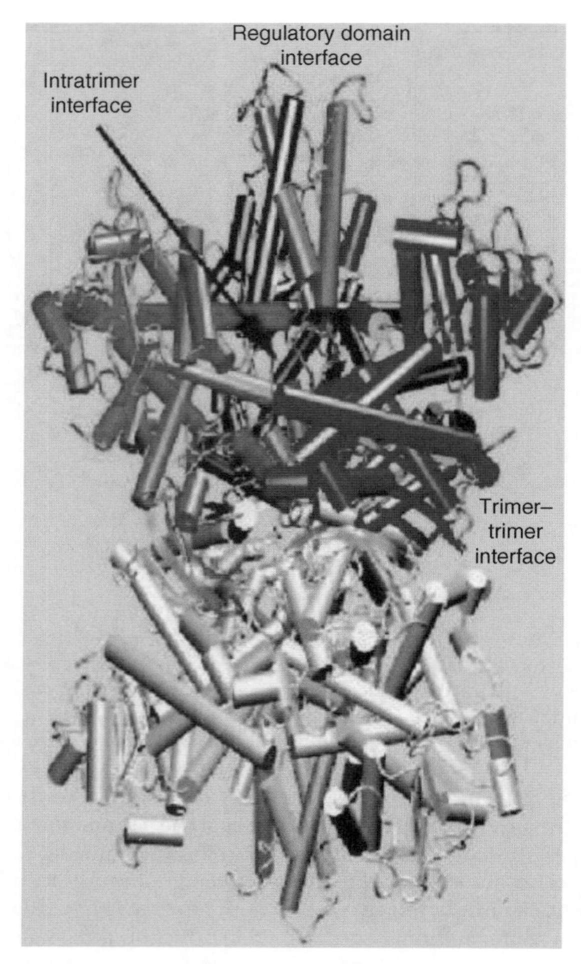

Figure 10. The quaternary structure of glutamate dehydrogenase reveals a complex array of subunit interfaces. (See color insert.)

An understanding of the nature of ligand binding to an allosteric protein at this level is critical to the successful design of an "allosteric" drug (see Section 28.3). Of equal importance is understanding the overall mechanism of transmission of the allosteric effect through the protein conformation to an adjacent subunit. As with the analysis of ligand interactions with a protein discussed above, a combination of computational and experimental approaches will be necessary to fully understand the mechansim of transmission of an allosteric effect through a protein. HINT analysis in addition to being useful for the analysis of ligand–protein interactions can be used to examine potential contributions to subunit interfaces in allosteric proteins. An example of such an analysis is shown in Fig. 14 where the cofactor binding domain subunit interface has been subjected to HINT analysis. Different regions of the subunits contribute either positively or negatively to the overall strength of the interface. Although in Hemoglobin, as discussed earlier much detail is known about how the allosteric changes are triggered, this is not true for most allosteric proteins. A few have had their three-dimensional structures determined in both the so-called R and T states and even this has failed to give significant insight into the nature of the trigger or

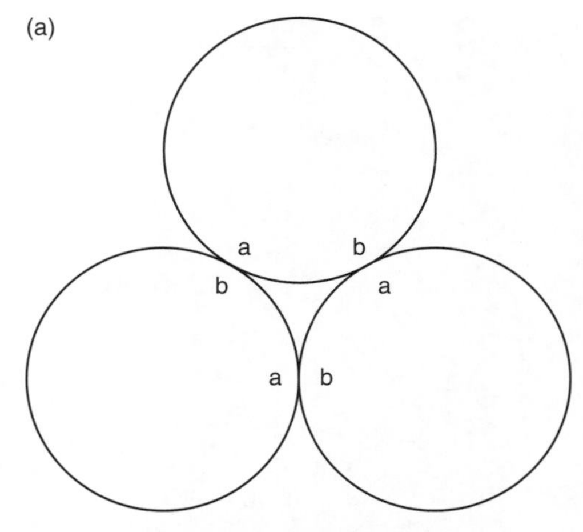

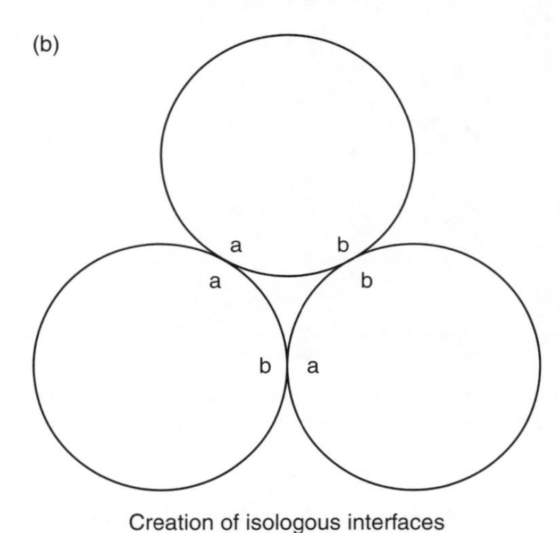

Figure 11. Heterologous interactions between subunits in a trimer.

transmission of allosteric changes. As discussed about computational analysis of crystal structures of oligomeric proteins and of ligand protein complexes can give insight that can lead to tests by direct experimentation. Similar approaches are being developed to examine the ability of ligands to induce conformational changes. These involve docking substrates with the binding sites of the unliganded form and using dynamics calculations to assess how the protein may readjust its conformation. Using snapshots of these dynamic simulations suggestive evidence of how the protein adjusts its conformation can be obtained. If such approaches can be validated by direct experimentation, for example, using site directed mutagenesis to change seemingly important residues in a predictable way, new insights into the triggers and transmission of allosteric changes will be obtained. With detailed information about both the triggers and the transmission of allosteric conformational changes, the rational design of allosteric drugs will become much easier. There are a number of keys to furthering our understanding of allosteric enzymes. It has been cogently argued that a clearer picture of allosteric interactions would result from the recognition that allosteric transitions do not simply involve a ligand induced confromational change but depend upon a comparison of how one ligand affects the binding of a second ligand [46]. Similarly, the global function of the molecule; rather than simply the mechanism of signal transmission must be considered [47].

Whether concerted or sequential models are invoked or a more general view of allostery is taken understanding changes in the interfaces involved in subunit interactions and how they affect either or both of catalysis (V-type effects) or ligand binding (K-type effects) are critical [48].

2.6. The Basis for Allosteric Effectors As Potential Therapeutic Agents

These types of considerations suggest that "allosteric" drugs of various types might be designed. These are schematically illustrated in Fig. 15. For heterotropic effects, a drug might bind but not trigger the necessary allosteric change. For either homotropic or heterotropic effects, a drug might be designed that stabilizes the region of the molecule across which the allosteric change must be transmitted; hence, blocking the required allosteric transition. Finally, if a MWC-type model is in effect yet a third type of "allosteric" drug can be envisaged, one which stabilizes either the R state or the T state as appropriate without directly impacting either the normal allosteric regulator binding or the regions of the molecule involved in the normal conformational change. Such a drug would bind to some

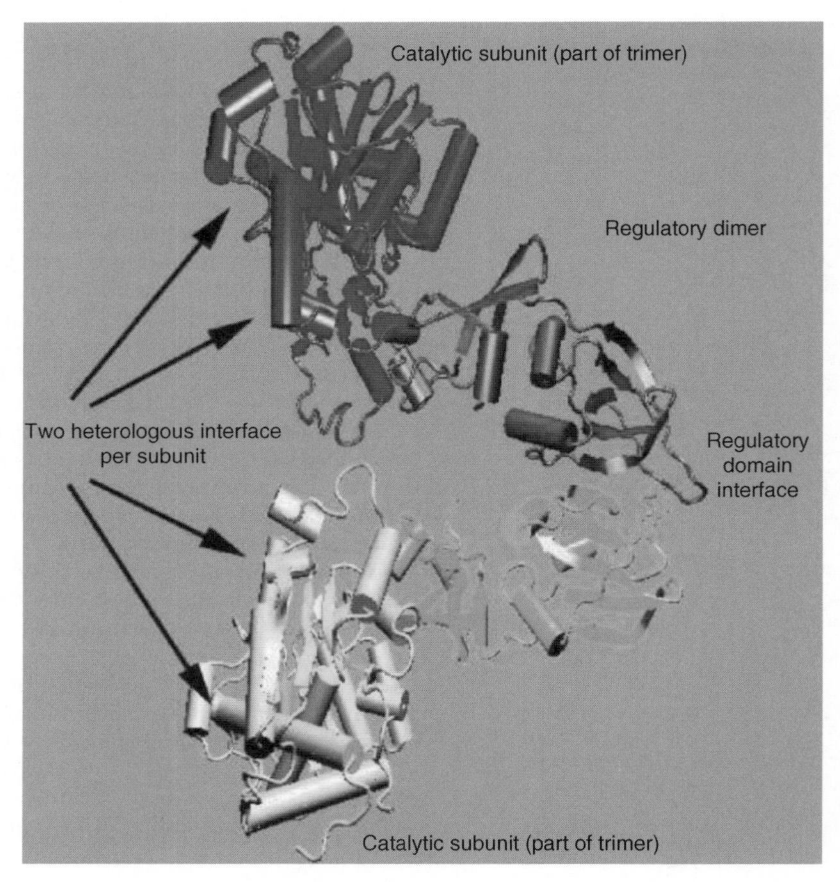

Figure 12. Subunit contacts in aspartate transcarbamoylase show heterologous contacts in the catalytic trimers, isologous contacts between regulatory dimers and regulatory subunit–catalytic subunit interactions. (See color insert.)

aspect of the conformation of either the R state or the T state distinct for that state and displace the equilibrium between R and T toward the appropriate state as shown in Fig. 16.

2.7. Intrinsically Disordered Proteins and "Conformational Changes"

In the past 5 years, perhaps the most important observations with respect to allosteric drug design has been the realization that many proteins, including a large number associated with human diseases, have regions that are intrinsically disordered. In reality this means that such proteins contain regions where there are a number of dynamically interchangeable conformations accessible. The experimental description of such regions is best assessed using NMR techniques [49]. It has been shown that such intrinsically disordered regions are often directly involved in interactions with the physiological target of the protein, whether it be a small ligand or a macromolecular partner. Disorder-to-order transitions allow the adoption of a variety different structures allowing optimization of interactions with a variety of different ligands. The flexibility of intrinsic disordered regions also helps different proteins with disordered regions bind to a common partner with as single binding domain.

In several excellent reviews [50–54], Dunker and colleagues have documented the roles that intrinsically disordered proteins or proteins containing intrinsically disordered regions play in diseases such as diabetes, cardiovascular disease, amyloidoses, neurodegenerative diseases, and cancer.

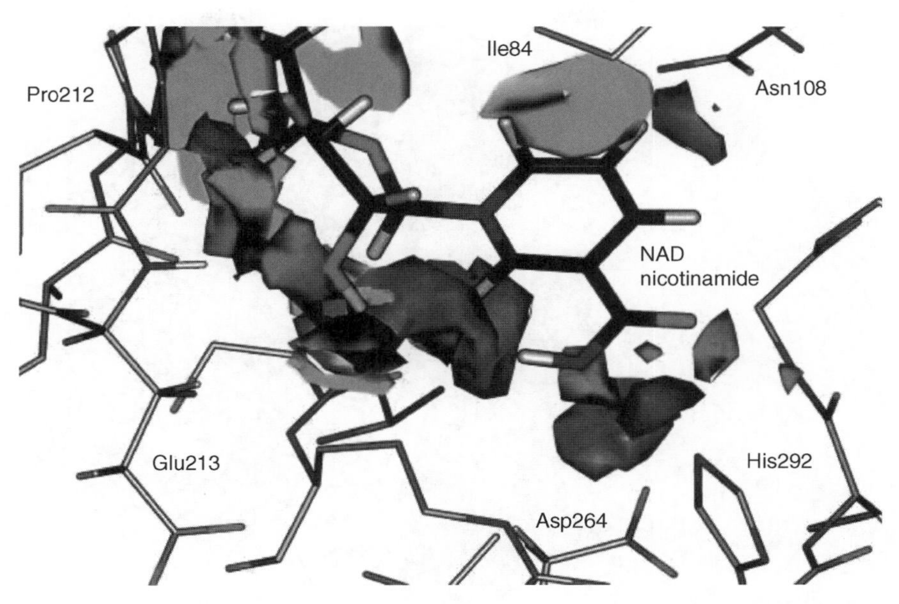

Figure 13. HINT analysis of interactions between the cofactor NAD nicotinamide moiety and the surrounding 3-phosphoglycerate dehydrogenase residues. Protein carbons are shown in light gray and NAD carbons are shown in black. Nitrogen atoms are shown in blue and oxygen atoms are shown in red. HINT contours are color coded to represent the different types of noncovalent interactions as follows: blue, favorable hydrogen bonding and acid-base interactions; green, favorable hydrophobic–hydrophobic interactions; and red, unfavorable acid–acid and base–base interactions. Importantly, the volume of the HINT contour represents the magnitude of the interactions. (See color insert.)

The role that intrinsic disorder plays in both catalysis and ligand binding has been put on a quantitative basis to explain how changes in disorder are related to specificity of promiscuous protein interactions as well as providing insight as to the optimization of protein function [52]. It should be evident that intrinsically disordered proteins, or intrinsically disordered regions in proteins offer the potential for effective allosteric drug design and it is anticipated that this area of research will be the focus of many future efforts as our understanding of the roles that intrinsically disordered proteins or intrinsically disordered regions of protein play in biological function including allostery increases.

3. DISCOVERY OF POTENTIAL ALLOSTERIC SITES IN PROTEINS

With increased interest in allosteric drug design, there is increasing pressure to discover potential allosteric binding sites on proteins. A number of approaches are available including high-throughput screening approaches, structural searches for potential binding sites, and computational approaches. In this section, recent examples in each of these areas will be discussed.

3.1. High-Throughput Screening

Hyperinsulinism/hyperammonemia disorder has been linked with mutations in the regulatory domains of glutamate dehydrogenase that result in loss of allosteric inhibition of GDH by GTP and the concomitant excessive insulin secretion. Studies by Smith and colleagues [55,56] demonstrate that glutamate dehydrogenase may represent a novel drug target for the control a variety of insulin disorders. Subsequent studies demonstrated that glutamate dehydrogenase is inhibited by the green tea polyphenols, epigallocatechin gallate, and epicatechin gallate. High-throughput screening identified several more stable inhibitors, including hexachlorophene,

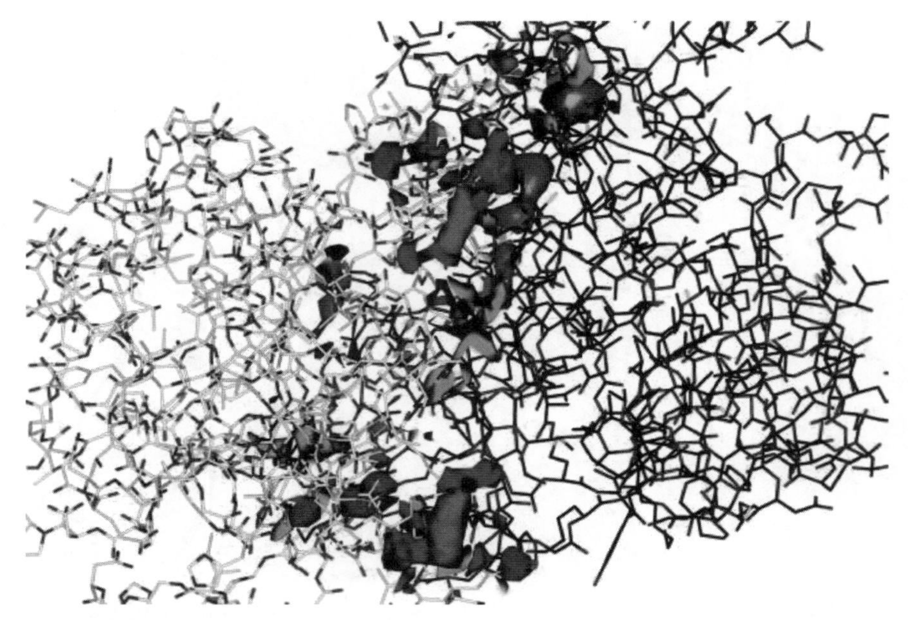

Figure 14. HINT Analysis of interactions between residues at the interface between the two subunits of 3-phospoglycerate dehydrogenase. The carbons of the two distinct subunits are colored light and dark gray, respectively. Nitrogen atoms are in blue, oxygen atoms are in red, and sulfur atoms are in yellow. HINT contour maps visually displaying interactions between the residues at the subunit interface are color coded according to the type of interaction: blue represents hydrogen bonding and favorable acid-base interactions, green displays favorable hydrophobic–hydrophobic contacts, red indicates unfavorable base–base and acid–acid interactions, and purple indicates unfavorable hydrophobic–polar interactions. Importantly, the volume of the HINT contour represents the magnitude of the interaction. For the analyzed phosphoglycerate dehydrogenase subunits, HINT indicates that a balance between favorable and unfavorable interactions characterize the interface. (See color insert.)

GW5074, and bithionol. Although none of the drugs cause conformational changes in the contact residues, all bind to key interface residues involved in ligand induced conformational changes. The drugs appear to inhibit catalysis by inhibiting subunit interactions necessary for normal catalysis.

High-throughput screens have also been used to investigate receptor–agonist interactions [57], and identify allosteric ligands that modify the biological responses to agonists. Pharmacophore screens have also been used [58] to identify a new class of A1AR antagonists that also bind to the Adenosine A1 Receptor allosteric site.

Concomitant with such studies is a need to assess dissociation constants for ligands over a wide range of affinities. A novel approach using capillary electrophoresis [59] has been developed to assess the binding affinities of cyclic nucleotide analogs that function as allosteric regulators of various regulatory proteins.

Finally, it is important to recognize that *in vitro* screens may not be sufficient in some cases to identify ligands that elicit appropriate biological responses. In a recent study [60], *In vivo* Zebrafish chemical screening was used to reveal an inhibitor of the dual specificity phosphatase Dusp6 that has the capability to expand cardiac cell lineages.

3.2. Localization of Binding Sites Using the Multiple Solvent Crystal Structure Method

A ligand binding site consists of a group of subsites each recognizing some specific aspect of the ligand and making a positive interaction

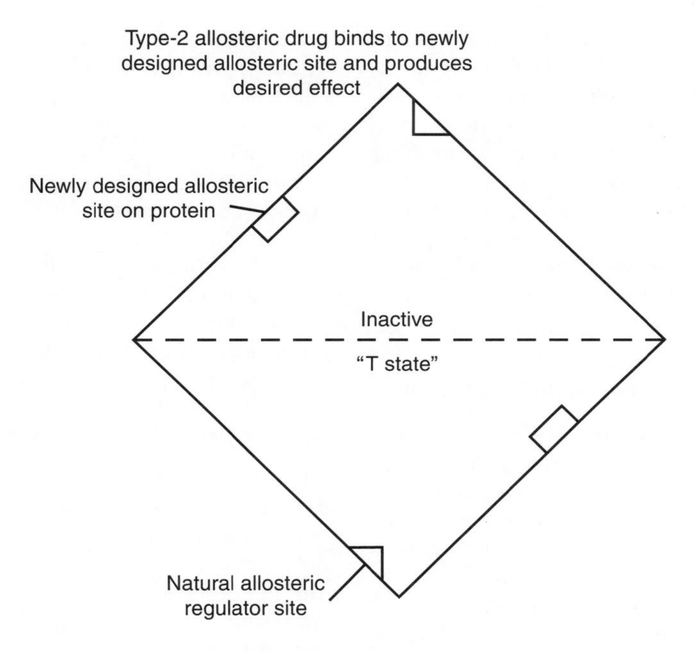

Figure 15. Schematic outline of the basics of allosteric drug design in ligand agonists and antagonists.

with that subsite. This is often counteracted by negative interactions with other parts of the protein surface. The sum of the positive and negative interactions defines both the site and its affinity for a specific ligand. A knowledge of both the positive and the negative interactions contributes significantly to an ability to design a suitable ligand to interact with a defined region of the protein surface. Making use of this concept, Ringe and Mattos [61–64] have developed and utilized an experimental approach to identify unknown sites on proteins. By using a series of organic solvents, each representing some property that could contribute to binding, and determining the three-dimensional structure of a given protein in the presence of each solvent potential subsites involved in binding can be identified. The approach has been validated with elastase [62] and ribonuclease [63] and

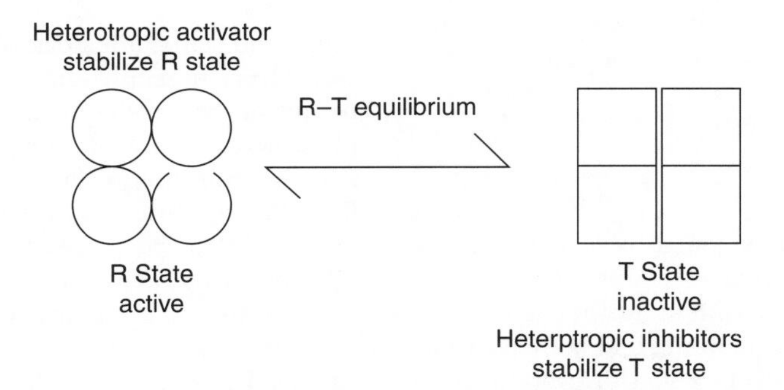

Figure 16. Allosteric drugs could stabilize either the R state or the T state in a Monod–Wyman–Changeux-type allosteric enzyme.

clearly has the potential to aid in the process of rational allosteric drug design. The approach, termed multiple solvent crystal structures (MSCSs) not only can identify potential binding sites and subsites but also allows probing of conformationally flexible "hot" spots, plasticity and hydration simultaneously. The approach provides a powerful complement to guide computational methods in development for binding site determination, ligand docking, and design [64].

3.3. Localization of Binding Sites Using NMR

Similar approaches to the MSCS method discussed above have been developed using NMR [65,66]. Using techniques such as affinity NMR techniques allows detection of binding of small molecules and are valuable tools for rapid screening. NMR observable events that can be used include relaxation, chemical shift perturbations, translational diffusion, and magnetization transfer. These approaches have the advantage of being one-dimensional methods, which increases their utility in high-throughput screening. The approaches also provide insight into the mode of binding.

4. COMPUTATIONAL APPROACHES FOR DISCOVERY OF POTENTIAL ALLOSTERIC SITES IN PROTEINS

With the exponential increase in the number of available structures of potential pharmacological targets, from direct experimentation, from homology modeling and from direct computation of potential structures the availability of validated methods to screen potential ligands computationally is important.

4.1. ROSETTALIGAND

ROSETTALIGAND is an algorithm for protein–small-molecule docking with full side-chain flexibility developed by Baker and his group [67–69], employing a Monte Carlo minimization procedure that simultaneously optimizes the orientation and rigid body position of the small-molecule and the protein side-chain conformations. The energy function comprises van der Waals (vdW) interactions, an implicit solvation model, an explicit orientation hydrogen bonding potential, and an electrostatics model. The approach has been validated using a diverse set of 229 protein–small-molecule complexes. Recently, the program has been enhanced to allow full ligand and receptor flexibility [68]. The modified algorithm has been validated using blind docking of a variety of pharmaceutical compounds [69]. ROSETTALIGAND clearly has the potential to significantly help computational screening to identify potential lead compounds for drug development.

4.2. Molecular Dynamics Approaches

Molecular dynamics simulations of protein–ligand structure can give clues about the conformational space available for optimal ligand interactions and for allosteric proteins can provide clues concerning conformational adjustments during ligand binding. Such approaches have been used to investigate possible allosteric inhibitors for the protein tyrosine phosphatase PTP1B [70]. A 5 ns MD simulation gave potential relative orientations of the alpha7–alpha6–alpha3 helices for protein–allosteric inhibitor complexes. Subsets of the structures were then used to screen small-molecule structure databases for molecules that best matched the allosteric site. Such approaches have the potential to facilitate both identification of potential allosteric drugs and provide insight to the mechanisms of allostery in the target protein.

5. ALLOSTERIC TARGETS FOR DRUG DISCOVERY

Small-molecule allosteric effectors of allosteric proteins represent a growing class of potential therapeutic agents. With continued progress in the elucidation of the three-dimensional structures of both the high- and the low-affinity states of allosteric proteins, opportunities to design state-specific synthetic effectors will become increasingly feasible. As opposed to designing drugs that bind directly to an active site, which can often be limited by the need to generate molecules with receptor affinity that is several fold that of the natural substrate, allosteric effectors offer the unique opportunity of binding to a location that is removed from the substrate binding site, and hence will elicit an effect (e.g., destabilizing an allosteric protein's active state), regardless of the concentration of endogenous substrate present. In addition, allosteric binding sites offer a potential for greater specificity for target proteins [43]. Thus, modulating the activity of allosteric proteins by degree, versus all or none, leaves open the possibility of tailoring drug activity to the severity of the disease state.

A number of allosteric protein targets are outlined below. Many of these targets are covered in greater detail in other volumes and chapters of this edition. The chemical structures of a select number of small-molecule allosteric effectors covered are shown in Fig. 17.

5.1. Transport Proteins

5.1.1. Hemoglobin Hemoglobin is a tetrameric protein composed of two identical alpha subunits and two identical beta subunits. X-ray crystal structures of the low- [30] and high [32]-affinity states of this protein have been solved at high resolution. The discovery of the first nonnaturally occurring hemoglobin allosteric effectors provided a foundation for how to design new effectors that might regulate allosteric processes in a number of important disease states.

Hb effectors of both the low and the high-affinity forms of the protein have been generated and it has been shown that small-molecule allosteric effectors can drive the allosteric mechanism toward either the T state or the R state and in some cases can act bind at the

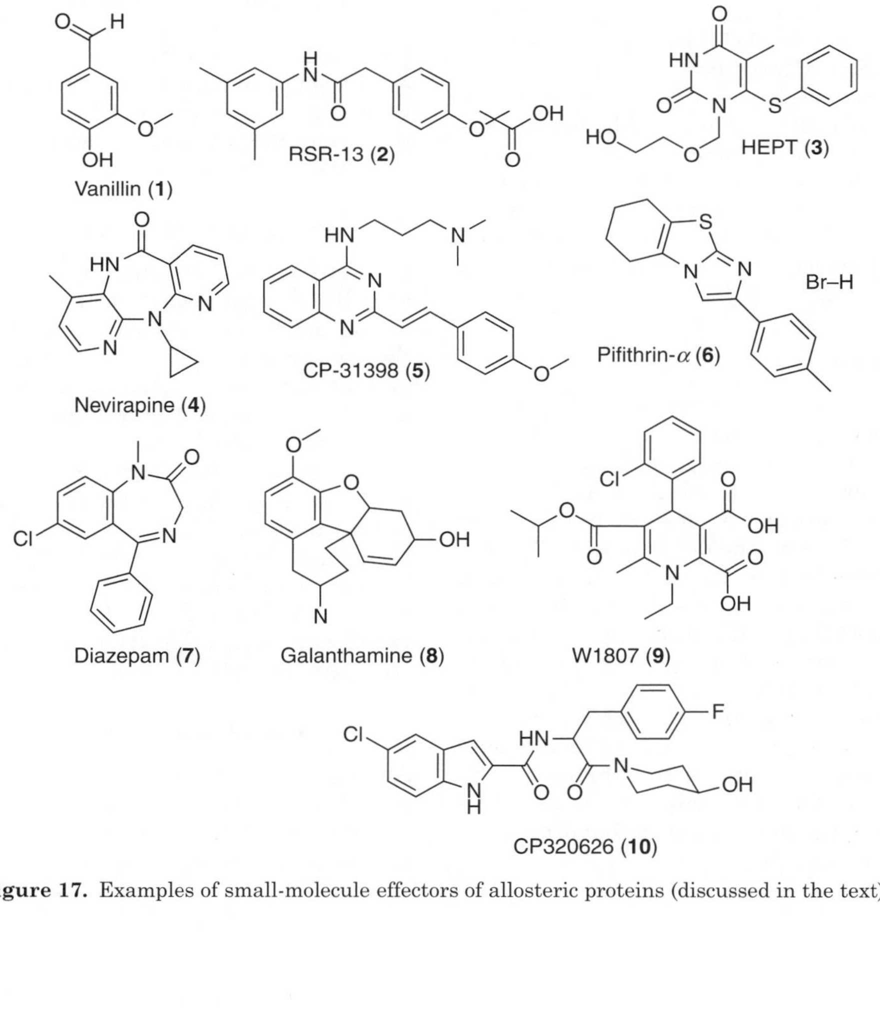

Figure 17. Examples of small-molecule effectors of allosteric proteins (discussed in the text).

same binding site but have opposite allosteric activities [70–73,189–191]. Effectors such as vanillin (**1**) (Fig. 17), which bind to the high-affinity (R) state [70], are potential therapeutics for sickle cell anemia; effectors such as RSR-13 (**2**) (Fig. 17), which bind to the low affinity state [71,72], were under clinical evaluation for the treatment of ischemic conditions. RSR-13, efaproxiral, was evaluated clinically as a radiosensitizing agent to treat a variety of cancers including breast, brain, and lung cancer but failed phase three trials as a radiosensitizing agent to treat breast to brain metastasis [73–75]. myo-Inositol trispyrophosphate (ITPP) is a membrane-permeant molecule that allosterically regulates Hb oxygen binding affinity. ITPP decreases the oxygen binding affinity of Hb, which results in increases of tissue oxygen delivery. ITPP has been shown in normal mice and mice with severe heart failure to increase maximal exercise capacity. ITPP may thus be useful in the therapy of heart failure patients with reduced exercise capacity [76]. In another approach, BW12C (5-[2-formyl-3-hydroxypenoxyl] pentanoic acid) stabilizes oxyhaemoglobin, causing a reversible -shift of the oxygen saturation curve resulting in tissue hypoxia [192] BW12C has been used in conjunction with mitomycin C whose activity is enhanced by hypoxia.

5.2. Kinases

Protein kinases, such as the Src family [77] have been associated with many diseases including cancer and central nervous system disorders. As a result, there has been much interest in the allosteric regulation of kinases [78] and the development of potential therapeutics based upon allosteric protein kinase inhibitors [79–83]. Increased understanding of the mechanisms of allosteric transmission in protein kinases is essential to fully realize the potential of their use as drug targets. Much recent work has focused on the mechanisms of allosteric regulation of both serine/threonine kinases [84,85] and tyrosine kinases [86] including the effects that autophosphorylation may play [87].

5.3. Phosphatases

Phosphatases, such as protein tyrosine phosphatase B1 have been shown to be subject to allosteric regulation [88], and specific small molecules can be designed to bind to the regulatory sight [89]. Given the interest in targeting phosphatases for cancer treatment [90], allosteric drug design targeting phosphatases is likely. These include PTEN phosphatase that is allosterically regulated by phosphatidylinositol 4,5-bisphosphate [91].

In addition to protein phosphatases several small-molecule phosphatases that are allosterically regulated appear to be good candidates for future drug design. Many enzymes involved in metabolic control use allosteric AMP binding sites. Many of these enzymes are potential drug targets for metabolic diseases [92]. Fructose 1,6-bisphosphatase is such an enzyme and libraries of novel allosteric inhibitors have been generated [93]. Interestingly, it has recently been shown that significant gains in potency can be achieved by tethering two inhibitors together such that they can interact with adjacent allosteric sites [94].

5.4. G-Protein-Coupled Receptors

Heterotrimeric G-protein-coupled receptors (GPCRs) are a involved in mediating signaling across membranes and play a critical role in the signal cascades of may physiologically important systems [95]. Allosteric phenomena have been implicated in a number of G-protein-coupled receptors, including the free fatty acid receptors [96], adrenoreceptors [97], PAR-2 receptors [98], and the type-1 cholecystokinin receptor [99]. There are a number of GPRC structures available [100] and has been much discussion about the role that allosteric modulators of G-protein-coupled receptors play in design of future therapeutics [101–104,183,184].

5.5. Protein–Protein Interaction Domains

Any discussion of the regulation of kinases, phosphatases, and G-protein-coupled receptors that does not include the roles played by protein–protein interactions would remiss [105].

The design of small-molecule inhibitors of such interactions offers an attractive alternate approach to modulate signaling pathways without necessarily affecting the activity of individual components [106].

5.6. Ion Channels and Neuroreceptors

Several ion channels and neuroreceptors are allosterically modulated by small molecules [107]. These include calcium channels, sodium channels, GABA receptors, and nicotinic receptors, among others. Cinacalcet, an allosteric enhancer of the Ca^{2+}-receptor is an FDA-approved medicine to treat chronic kidney failure and hyperparathyroidism [185].

Anxiolytic compounds classified as benzodiazepines (e.g., diazepam (**7**)) bind to an allosteric site on GABA receptors [108]. The compound galanthamine (**8**), which is a cholinesterase inhibitor, has been found to modulate nicotinic receptors [109]. Dihydropyridines act upon calcium channels [110].

There has been much work in recent years on the glutamate family of receptors, their allosteric regulation [111–113] and potential as targets for drug design [114,115]. Although less studied that the other glutamate receptors, the kainate receptors also have allosteric properties and are potential targets for drug design [116].

The nicotinic acetylcholine receptors [117] and muscarinic receptors also have been shown to be allosterically regulated and are potential drug targets [118,119].

5.7. Proteases

While most protease inhibitors are active site directed, mutants such as the "flap" mutants of HIV protease [120] suggest that "allosteric" drugs can be developed which inhibit by blocking conformational changes necessary to reveal the active site of some proteases [121].

Many proteases undergo a zymogen activation process, which like the protease involved in steroid signaling [122,123] may be allosterically triggered indicating possible "allosteric" drug development targeted at preventing or promoting zymogen activation as necessary.

Proteases have been a frequent target of small-molecule drug design and the most frequent approach, as illustrated by the plethora of HIV protease inhibitors, has been to target the active site with a substrate analog. Although proteases are rarely though to be under the control of allosteric effects, recent work has indicated that several proteases, including the cysteine protease of Vibrio cholerae [124] and the PDZ proteases such as the degs protease of *E. coli* [125,126] are subject to allosteric control. Since mutations in the human orthologs of the PDZ proteases are associated with a variety of diseases [127–130] it is not unreasonable to think that targeting the allosteric regulation of such proteases for drug design will be feasible.

Although not traditionally thought of as an allosteric site on HIV Protease, attention is still focusing on targeting the "flap" region of the protease for drug design, in effect blocking a required conformational transition necessary for catalysis. Böttcher et al. [131] have developed pyrrolidine based inhibitors targeting the open flap region of HIV protease and obtained a cocrystal structure of one analog, providing a starting point for future drug design.

In a related approach using "Aptazymes" it is apparent that allosteric effects are often involved in the activity of these "enzymes" [132] raising the possibility that allosteric drugs could be developed using naturally occurring or engineered riboswitches for the control of gene expression.

5.8. HIV Proteins

HIV reverse transcriptase (RT) is a heterodimeric protein. The structure of HIV RT has been elucidated by X-ray crystallography in several forms, including the unliganded enzyme [133], in complex with nonnucleoside RT inhibitors (NNTRIs) [134,135], and bound to template primer with [136] and without dNTP substrate [137].

NNRTIs are chemically diverse compounds that are largely hydrophobic [138]. Many NNRTIs fall into the HEPT (**3**), TIBO, a-APA, and nevirapine (**4**) families of compounds [139].

Viral fusion with host cell membranes is mediated by the HIV protein gp41, which upon interaction with a host cell receptor undergoes a conformational change leading to membrane fusion and infection [140,141].

While the normal "allosteric effector" in this system is the host cell receptor a variety of "fusion blocking" drugs have been developed which bind to gp41 [142] and "allosterically" prevent (i.e., stabilize the inactive conformation) the induced conformational change from occurring. Although the early attempts focused on peptide based drugs [143–145] more recent work is developing *N*-carboxyphenylpyrrole ligands that bind to gp41 using the crystal structure to guide design [146] and other nonpeptide small molecules [147]. The HIV coreceptor, CCR5 has also been identified as a potential target for drug design [186]. Maraviroc, a CCR5 antagonist with potent *in vitro* and *in vivo* anti-HIV-1 activity, was recently approved by the FDA for treatment-experienced patients with R5-tropic virus [187,188].

Although not an HIV drug, there is current interest in a variety of other protein target for allosteric drug design including Hepatitis C virus RNA polymerase [148]. It has been demonstrated that conformational changes are required during the progression through the different RNA synthesis steps and hence compounds targeting this site could lock the enzyme at initiation.

5.9. Other Protein Targets

Glycogen phosphorylase (GP) is a dimeric enzyme that catalyzes the degradation of glycogen to glucose. Its structure in both the resting (GPb) [149] and the activated (GPa) forms has been solved in complex with allosteric effectors.

The allosteric effector W1807 (**9**) has been structurally shown to bind to and stabilize the active form (GPa) of the protein [150]. Compound CP320626 (**10**) binds to an allosteric site of the resting (GPb) form of the enzyme and may find therapeutic application as an antidiabetic drug [151]. CP320626 has also been shown to lower cholesterol levels that appear to be mediated by an effect on lanosterol 14a-demethylase (CYP51) [152]. More recent work has used computer modeling to assist design of potent glycogen phosphorylase inhibitors [153].

P53 can exist as multiple-weight aggregates. P53's tetramerization domain has been elucidated by NMR and X-ray crystallographies [154–159]. Researchers at Pfizer have discovered two compounds (CP-31398 (**5**) and CP-25704) that are allosteric activators of P53 mutants [160]. A potential allosteric inhibitor of P53transcription, pifithrin-α (**6**) has also been reported [161]. Recent reports have shown that administration of CP-31398 either alone at high doses or in combination with celecoxib showed enhancement of p53 amounts and significant reduction of colonic tumor poliferation [162]. Pifithrin has been shown to reduce expression of p53 in glial cells [163] suggesting possible Alzheimer's disease treatments by targeting p53-mediated pathways in microglia. MDM2 oncoprotein is involved in the regulation of p53-dependent gene expression and it has recently been shown is allosterically regulated via a mechanism involving the RING finger domain and is the target of anticancer drug design [164].

Adenosine receptors have been shown to be potently inhibited by a variety of imidazo[2,1-*f*]purinones, and although antagonists indicate the possibility that allosteric drug design against adenosine receptors could be used in antiparkinson drug design [165–167]. The adenosine receptor has also been used to define a novel concept—the receptor mosaic concept where allosteric ligands can bias the distribution between various conformers of the protein. Such approaches may assist in allosteric drug development [168].

The vasopressin type-2 receptor is another potential target for allosteric drug design. Current drugs act as competitive inhibitors but recent work suggests that certain peptides can act as negative allosteric inhibitors and hence be lead compounds for allosteric drug development [169].

6. ALLOSTERIC TARGETS FOR DRUG DISCOVERY AND FUTURE TRENDS

Small-molecule allosteric effectors of allosteric proteins represent a growing class of potential therapeutic agents. With continued progress in the elucidation of the three-dimensional structures of both the high- and the low-affinity states of allosteric proteins, together with an emerging understanding of how allosteric effects are transmitted, opportunities to design state specific synthetic effectors will become increasingly feasible. As opposed to designing drugs that bind directly to an active site, which can often be limited by the need to generate molecules with receptor affinity that is several fold that of the natural substrate, allosteric effectors offer the unique opportunity of binding to a location that is removed from the substrate binding site, and hence will elicit an effect (for example, destabilizing an allosteric protein's active state) regardless of the concentration of endogenous substrate present. In addition, allosteric binding sites offer a potential for greater specificity for target proteins. Thus, modulating the activity of allosteric proteins by degree, versus all or none, leaves open the possibility of tailoring drug activity to the severity of the disease state. Although much progress in terms of our understanding of allostery has been made in recent years, the importance of our increasing understanding of the roles that dynamic aspects of protein structure and protein interactions, including the many functions that instrinsically disordered regions of proteins play in function suggests that the coming years will see an ever increasing focus on allosteric drug design. With the reemergence of allostery as a major regulatory mechanism in a wide variety of biological phenomena, especially those involving protein–protein interactions and the ever increasing list of proteins and systems that are subject to allosteric regulation one can speculate that interest in this area will become more "disease" oriented. Already, as documented in this chapter, there has been significant progress in many areas including neurodegenerative diseases (see also Refs [169–173]–) and cancer (see also Refs [174,175]). Current work suggests that future allosteric drugs may be developed for inflammatory diseases [176,177], osteoporosis [178] and trypanosomal diseases [179]. Recent work has also identified porphobilinogen synthase [180] and the integrin family of proteins [181,182] as being subject to allosteric control and hence potential future drug targets.

ACKNOWLEDGMENT

The work of the authors that is included in this review resulted in part from funding from the National Science Foundation: NSF Grant No. MCB 0448905 to J. Ellis Bell.

REFERENCES

1. Fang J, Hsu BY, MacMullen CM, Poncz M, Smith TJ, Stanley CA. Expression, purification and characterization of human glutamate dehydrogenase (GDH) allosteric regulatory mutations. Biochem J 2002;363(Pt 1): 81–87.
2. André I, Strauss CE, Kaplan DB, Bradley P, Baker D. Emergence of symmetry in homooligomeric biological assemblies. Proc Natl Acad Sci USA 2008;105(42): 16148–16152.
3. Whitty A. Cooperativity and biological complexity. Nat Chem Biol 2008;4(8): 435–439.
4. Goodey NM, Benkovic SJ. Allosteric regulation and catalysis emerge via a common route. Nat Chem Biol 2008;4(8): 474–482.
5. Hammes GG, Wu CW. Kinetics of allosteric enzymes. Ann Rev Biophys Bioengin 1974;3:1–33.
6. Bell JE, Bell E. Proteins and Enzymes. Englewood Cliffs, NJ: Prentice-Hall, Inc; 1988.
7. Hill AV. The possible effects of aggregation of the molecules of hemoglobin on its dissociation curve. J Physiol 1910;40:4–11.
8. Adair GS. The hemoglobin system. VI. The oxygen dissociation curve of hemoglobin. J Biol Chem 1925;63:529.
9. Monod J, Wyman J, Changeux JP. On the nature of allosteric transitions: a plausible model. J Mol Biol 1965;12:88–118.
10. KoshlandJr DE, Nemethy G, Filmer D. Comparison of experimental binding data and theoretical models in proteins containing subunits. Biochemistry 1966;51:365–385.
11. Engel PC, Dalziel K. Kinetic studies of glutamate dehydrogenase with glutamate and norvaline as substrates. Coenzyme activation and negative homotropic interactions in allosteric enzymes. Biochem J 1969;115(4): 621–631.
12. Dalziel K, Engel PC. Antagonists homotropic interactions as a possible explanation of coenzyme activation of glutamate dehydrogenase. FEBS Lett 1968;1:349–352.
13. Eigen M. Kinetics of reaction control and information transfer in enzymes and nucleic acids. Nobel Symp 1967;5:333.

14. del Sol A, Tsai CJ, Ma B, Nussinov R. The origin of allosteric functional modulation: multiple pre-existing pathways. Structure 2009;17 (8): 1042–1050.

15. Gunasekaran K, Ma B, Nussinov R. Is allostery an intrinsic property of all dynamic proteins?. Proteins 2004;57(3): 433–443.

16. Frantom PA, Zhang HM, Emmett MR, Marshall AG, Blanchard JS. Mapping of the allosteric network in the regulation of alpha-isopropylmalate synthase from *Mycobacterium tuberculosis* by the feedback inhibitor L-leucine: solution-phase H/D exchange monitored by FT-ICR mass spectrometry. Biochemistry 2009;48(31): 7457–7464.

17. De Simone A, Richter B, Salvatella X, Vendruscolo M. Toward an accurate determination of free energy landscapes in solution states of proteins. J Am Chem Soc 2009;131(11): 3810–3811.

18. Bhattacharyya M, Ghosh A, Hansia P, Vishveshwara S. Allostery and conformational free energy changes in human tryptophanyl-tRNA synthetase from essential dynamics and structure networks. Proteins. 2010; Feb 15: 78(3):506–517.

19. Ghosh A, Vishveshwara S. Variations in clique and community patterns in protein structures during allosteric communication: investigation of dynamically equilibrated structures of methionyl tRNA synthetase complexes. Biochemistry 2008;47(44): 11398–11407.

20. Brüschweiler S, Schanda P, Kloiber K, Brutscher B, Kontaxis G, Konrat R, Tollinger M. Direct observation of the dynamic process underlying allosteric signal transmission. J Am Chem Soc 2009; 131(8), pp 3063–3068.

21. Lee J, Natarajan M, Nashine VC, Socolich M, Vo T, Russ WP, Benkovic SJ, Ranganathan R. Surface sites for engineering allosteric control in proteins. Science 2008;322(5900): 438–442.

22. Sagermann M, Chapleau RR, DeLorimier E, Lei M. Using affinity chromatography to engineer and characterize pH-dependent protein switches. Protein Sci 2009;18(1): 217–212.

23. Luisi B, Shibayama N. Structure of haemoglobin in the deoxy quaternary state with ligand bound at the alpha haems. J Mol Biol 1989;206:723–736.

24. Bell JE, Dalziel K. A conformational transition of the oligomer of glutamate dehydrogenase induced by half-saturation with NAD+ or NADP+. Biochim Biophys Acta 1973;309: 237–242.

25. O'Connell L, Bell ET, Bell JE. 3,4,5,6-Tetrahydrophthalic anhydride modification of glutamate dehydrogenase: the construction and activity of heterohexamers. Arch Biochem Biophys 1987;263:315–322.

26. Alex S, Bell JE. Dual nucleotide specificity of bovine glutamate dehydrogenase. The role of negative co-operativity. Biochem J 1980;191: 299–304.

27. Perutz MF. Stereochemistry of cooperative effects in haemoglobin. Nature 1970;228: 726–739.

28. Perutz MF, Wilkinson AJ, Paoli M, Dodson GG. The stereochemical mechanism of the cooperative effects in hemoglobin revisited. Ann Rev Biophys Biomol Struct 1998;27:1–34.

29. Fermi G. Three-dimensional Fourier synthesis of human deoxyhaemoglobin at 2.5 Å resolution: refinement of the atomic model. J Mol Biol 1975;97:237–256.

30. Fermi G, Perutz MF, Shaanan B, Fourme R. The crystal structure of human deoxyhaemoglobin at 1.74 Å resolution. J Mol Biol 1984;175:159–174.

31. Perutz MF, Muirhead H, Cox JM, Goaman LCG. Three-dimensional Fourier synthesis of horse oxyhaemoglobin at 2.8 Å resolution: the atomic model. Nature 1968;219:131–139.

32. Shaanan B. Structure of human oxyhaemoglobin at 2.1 Å resolution. J Mol Biol 1983;171: 31–59.

33. Zhang Y, Liang J-Y, Huang S, Lipscomb WN. Toward a mechanism for the allosteric transition of pig kidney fructose-1,6-bisphosphatase. J Mol Biol 1994;244:609–624.

34. Oliva G, Fontes MRM, Garratt RC, Altamirano MM, Calcagno ML, Horjales E. Structure and catalytic mechanism of glucosamine-6-phosphate deaminase from *Escherichia coli* at 2.1 Å resolution. Structure 1995;3:1323–1332.

35. Strater N, Hakansson K, Schnappauf G, Braus G, Lipscomb WN. Crystal structure of the T state of allosteric yeast chorismate mutase and comparison with the R state. Proc Natl Acad Sci USA 1996;93:3330–3334.

36. Mattevi A, Valentini G, Rizzi M, Speranza ML, Bolognesi M, Coda A. Crystal structure of *Escherichia coli* pyruvate kinase type I: molecular basis of the allosteric transition. Structure 1995;3:729–741.

37. Hazes B, Magnus KA, Bonaventura C, Bonaventura J, Dauter Z, Kalk KH, Hol WGJ. Crystal structure of deoxygenated *Limulus polyphemus* subunit II hemocyanin at 2.18 Å

resolution: clues for a mechanism for allosteric regulation. Protein Sci 1993;2:597–619.

38. Grant GA, Schuller DJ, Banaszak L. A model for the regulation of D-3-phosphoglycerate dehydrogenase, a V_{max}-type allosteric enzyme. Protein Sci 1996;5:34–41.
39. Smith JL. Enzymes of nucleotide synthesis. Curr Opin Struct Biol 1995;5:752–757.
40. Iwata S, Kamata K, Yoshida S, Minowa T, Ohta T. T and R states in the crystals of bacterial L-lactate dehydrogenase reveal the mechanism for allosteric control. Nat Struct Biol 1994;1:176–185.
41. Hall MD, Levitt DG, Banaszak LJ. Crystal structure of *Escherichia coli* malate dehydrogenase. A complex of the apoenzyme and citrate at 1. 87 Å resolution. J Mol Bio 1992;226:867–882.
42. Schuller DJ, Grant G, Banaszak L. The allosteric ligand site in the Vmax-type cooperative enzyme phosphoglycerate dehydrogenase. Nat Struct Biol 1995;2:69–76.
43. Peterson PE, Smith TJ. The structure of bovine glutamate dehydrogenase provides insights into the mechanism of allostery. Structure 1999;7:769.
44. Smith TJ, Schmidt T, Fang J, Wu J, Siuzdak G, Stanley CA. The structure of Apo human glutamate dehydrogenase details subunit communication and allostery. J Mol Biol 2002;318:777.
45. Williams MK, Stec B, Kantrowitz ER. A single mutation in the regulatory chain of *Escherichia coli* aspartate transcarbamoylase results in an extreme T-state structure. J Mol Biol 1998;281:121.
46. Fenton AW. Allostery: an illustrated definition for the 'second secret of life'. Trends Biochem Sci 2008;33(9): 420–425.
47. Tsai CJ, Del Sol A, Nussinov R. Protein allostery, signal transmission and dynamics: a classification scheme of allosteric mechanisms. Mol Biosyst 2009;5(3): 207–216.
48. Zandany N, Ovadia M, Orr I, Yifrach O. Direct analysis of cooperativity in multisubunit allosteric proteins. Proc Natl Acad Sci USA 2008;105(33): 11697–11702.
49. Wright PE, Dyson HJ. Linking folding and binding. Curr Opin Struct Biol 2009;19(1): 31–38.
50. Uversky VN, Oldfield CJ, Midic U, Xie H, Xue B, Vucetic S, Iakoucheva LM, Obradovic Z, Dunker AK. Unfoldomics of human diseases: linking protein intrinsic disorder with diseases. BMC Genomics. 2009;10(Suppl 1): S7.
51. Dunker AK, Oldfield CJ, Meng J, Romero P, Yang JY, Chen JW, Vacic V, Obradovic Z, Uversky VN. The unfoldomics decade: an update on intrinsically disordered proteins. BMC Genomics. 2008;9(Suppl 2): S1.
52. Liu Jintao, Faederb James R, Camachob Carlos J. Toward a quantitative theory of intrinsically disordered proteins and their function. Proc Natl Acad Sci USA 2009;106(47): 19819–19823.
53. Hazy E, Tompa P. Limitations of induced folding in molecular recognition by intrinsically disordered proteins. Chemphyschem 2009;10 (9–10): 1415–1419.
54. Sickmeier M, Hamilton JA, LeGall T, Vacic V, Cortese MS, Tantos A, Szabo B, Tompa P, Chen J, Uversky VN, Obradovic Z, Dunker AK. DisProt: the database of disordered proteins. Nucleic Acids Res 2007;35(Database issue): D786–D793.
55. Li M, Smith CJ, Walker MT, Smith TJ. Novel inhibitors complexed with glutamate dehydrogenase: allosteric regulation by control of protein dynamics. J Biol Chem 2009;284(34): 22988–3000.
56. Li M, Allen A, Smith TJ. High throughput screening reveals several new classes of glutamate dehydrogenase inhibitors. Biochemistry 2007;46(51): 15089–15102.
57. Terry Kenakin Allosteric agonist modulators. J Recept Signal Transduct Res 27:2007; 247–259.
58. Ferguson GN, Valant C, Horne J, Figler H, Flynn BL, Linden J, Chalmers DK, Sexton PM, Christopoulos A, Scammells PJ. 2-Aminothienopyridazines as novel adenosine A1 receptor allosteric modulators and antagonists. J Med Chem 2008;51(19): 6165–6172.
59. Gavina JM, Mazhab-Jafari MT, Melacini G, Britz-McKibbin P. Label-free assay for thermodynamic analysis of protein–ligand interactions: a multivariate strategy for allosteric ligand screening. Biochemistry 2009;48(2): 223–225.
60. Molina G, Vogt A, Bakan A, Dai W, Queiroz de Oliveira P, Znosko W, Smithgall TE, Bahar I, Lazo JS, Day BW, Tsang M. Zebrafish chemical screening reveals an inhibitor of Dusp6 that expands cardiac cell lineages. Nat Chem Biol 2009;5(9): 680–687.
61. Ringe D. What makes a binding site a binding site? Curr Opin Struct Biol 1995;5:825–829.

62. Mattos C, Ringe D. Locating and characterizing binding sites on proteins. Nat Biotechnol 1996;14(5): 595–599.
63. Dechene M, Wink G, Smith M, Swartz P, Mattos C. Multiple solvent crystal structures of ribonuclease A: an assessment of the method. Proteins 2009;76(4): 861–881.
64. Mattos C, Bellamacina CR, Peisach E, Pereira A, Vitkup D, Petsko GA, Ringe D. Multiple solvent crystal structures: probing binding sites, plasticity and hydration. J Mol Biol 2006;357(5): 1471–1482.
65. Zartler ER, Yan J, Mo H, Kline AD, Shapiro MJ. NMR Methods in ligand–receptor interactions. Curr Top Med Chem 2003;3(1): 25–37.
66. McCoy MA, Wyss DF. Alignment of weakly interacting molecules to protein surfaces using simulations of chemical shift perturbations. J Biomol NMR 2000;18(3): 189–198.
67. Meiler J, Baker D. ROSETTALIGAND: protein–small molecule docking with full side-chain flexibility. Proteins 2006;65(3): 538–548.
68. Davis IW, Baker D. RosettaLigand docking with full ligand and receptor flexibility. J Mol Biol 2009;385(2): 381–392.
69. Davis IW, Raha K, Head MS, Baker D. Blind docking of pharmaceutically relevant compounds using RosettaLigand. Protein Sci 2009;18(9): 1998–2002.
70. Bharatham K, Bharatham N, Kwon YJ, Lee KW. Molecular dynamics simulation study of PTP1B with allosteric inhibitor and its application in receptor based pharmacophore modeling. J Comput Aided Mol Des 2008;22(12): 925–933.
71. DeDecker BS. Allosteric drugs: thinking outside the active-site box. Chem Biol 2000;7: R103–R107.
72. Abraham DJ, Mehanna AS, Wireko FC, Whitney J, Thomas RP, Orringer EP. Blood 1991;77:1334–1341Abraham DJ, Kister J, Joshi GS, Marden MC, Poyart C. J Mol Biol 1995;248:845–855
73. Scott C, Suh J, Stea B, Nabid A, Hackman J. Improved survival, quality of life, and quality-adjusted survival in breast cancer patients treated with efaproxiral (Efaproxyn) plus whole-brain radiation therapy for brain metastases. Am J Clin Oncol 2007;30(6): 580–587.
74. Choy H, Nabid A, Stea B, Scott C, Roa W, Kleinberg L, Ayoub J, Smith C, Souhami L, Hamburg S, Spanos W, Kreisman H, Boyd AP, Cagnoni PJ, Curran WJ. Phase II multicenter study of induction chemotherapy followed by concurrent efaproxiral (RSR13) and thoracic radiotherapy for patients with locally advanced non-small-cell lung cancer. J Clin Oncol 2005;23(25): 5918–5928.
75. Viani GA, Manta GB, Fonseca EC, De Fendi LI, Afonso SL, Stefano EJ. Whole brain radiotherapy with radiosensitizer for brain metastases. J Exp Clin Cancer Res 2009;28:1.
76. Biolo A, Greferath R, Siwik DA, Qin F, Valsky E, Fylaktakidou KC, Pothukanuri S, Duarte CD, Schwarz RP, Lehn JM, Nicolau C, Colucci WS. Enhanced exercise capacity in mice with severe heart failure treated with an allosteric effector of hemoglobin, myo-inositol trispyrophosphate. Proc Natl Acad Sci USA 2009;106 (6): 1926–1929.
77. Ingley E. Src family kinases: regulation of their activities, levels and identification of new pathways. Biochim Biophys Acta 2008 Jan; 1784(1):56–65.
78. Cheng HC, Johnson TM, Mills RD, Chong YP, Chan KC, Culvenor JG. Allosteric networks governing regulation and catalysis of Src-family protein tyrosine kinases: implications for disease-associated kinases. Clin Exp Pharmacol Physiol 2010 Jan; 37(1):93–101.
79. Chico LK, Van Eldik LJ, Watterson DM. Targeting protein kinases in central nervous system disorders. Nat Rev Drug Discov 2009;8 (11): 892–909.
80. Limvorasak S, Posadas EM. Kinase inhibitors in prostate cancer. Anticancer Agents Med Chem 2009;9(10): 1089–1104.
81. Kaminska B, Gozdz A, Zawadzka M, Ellert-Miklaszewska A, Lipko M. MAPK Signal transduction underlying brain inflammation and gliosis as therapeutic target. Anat Rec 2009;292(12): 1902–1913.
82. Zhang J, Yang PL, Gray NS. Targeting cancer with small molecule kinase inhibitors. Nat Rev Cancer 2009;9(1): 28–39.
83. Kirkland LO, McInnes C. Non-ATP competitive protein kinase inhibitors as anti-tumor therapeutics. Biochem Pharmacol 2009;77 (10): 1561–1571.
84. Calleja V, Laguerre M, Parker PJ, Larijani B. Role of a novel PH-kinase domain interface in PKB/Akt regulation: structural mechanism for allosteric inhibition. PLoS Biol 2009;7(1): e17.
85. Calleja V, Laguerre M, Larijani B. 3-D structure and dynamics of protein kinase B-new mechanism for the allosteric regulation of an AGC kinase. J Chem Biol 2009;2(1): 11–25.

86. De Meyts P, Gauguin L, Svendsen AM, Sarhan M, Knudsen L, Nøhr J, Kiselyov VV. Structural basis of allosteric ligand–receptor interactions in the insulin/relaxin peptide family: implications for other receptor tyrosine kinases and G-protein-coupled receptors. Ann NY Acad Sci 2009;1160:45–53.
87. Arias-Palomo E, Recuero-Checa MA, Bustelo XR, Llorca O. Conformational rearrangements upon Syk auto-phosphorylation. Biochim Biophys Acta 2009;1794(8): 1211–1217.
88. Wiesmann C, Barr KJ, Kung J, Zhu J, Erlanson DA, Shen W, Fahr BJ, Zhong M, Taylor L, Randal M, McDowell RS, Hansen SK. Allosteric inhibition of protein tyrosine phosphatase 1B. Nat Struct Mol Biol 2004;11(8): 730–737.
89. Tappan E, Chamberlin AR. Activation of protein phosphatase 1 by a small molecule designed to bind to the enzyme's regulatory site. Chem Biol 2008;15(2): 167–174.
90. Lazo JS, Wipf P. Phosphatases as targets for cancer treatment. Curr Opin Investig Drugs 2009;10(12): 1297–1304.
91. Campbell RB, Liu F, Ross AH. Allosteric activation of PTEN phosphatase by phosphatidylinositol 4,5-bisphosphate. J Biol Chem 2003;278(36): 33617–33620.
92. Erion MD, Dang Q, Reddy MR, Kasibhatla SR, Huang J, Lipscomb WN, van Poelje PD. Structure-guided design of AMP mimics that inhibit fructose-1,6-bisphosphatase with high affinity and specificity. J Am Chem Soc 2007;129(50): 15480–15490.
93. Heng S, Gryncel KR, Kantrowitz ER. A library of novel allosteric inhibitors against fructose 1,6-bisphosphatase. Bioorg Med Chem 2009;17(11): 3916–3922.
94. Hebeisen P, Kuhn B, Kohler P, Gubler M, Huber W, Kitas E, Schott B, Benz J, Joseph C, Ruf A. Allosteric FBPase inhibitors gain 10 (5) times in potency when simultaneously binding two neighboring AMP sites. Bioorg Med Chem Lett 2008;18(16): 4708–4712.
95. Bridges TM, Lindsley CW. G-protein-coupled receptors: from classical modes of modulation to allosteric mechanisms. ACS Chem Biol 2008;3(9): 530–541.
96. Lee T, Schwandner R, Swaminath G, Weiszmann J, Cardozo M, Greenberg J, Jaeckel P, Ge H, Wang Y, Jiao X, Liu J, Kayser F, Tian H, Li Y. Identification and functional characterization of allosteric agonists for the G protein-coupled receptor FFA2. Mol Pharmacol 2008;74(6): 1599–1609.
97. Jiang JL, Peng YP, Qiu YH, Wang JJ. Adrenoreceptor-coupled signal-transduction mechanisms mediating lymphocyte apoptosis induced by endogenous catecholamines. J Neuroimmunol 2009;213(1–2): 100–111.
98. Nhu QM, Shirey K, Teijaro JR, Farber DL, Netzel-Arnett S, Antalis TM, Fasano A, Vogel SN. Novel signaling interactions between proteinase-activated receptor 2 and Toll-like receptors *in vitro* and *in vivo*. Mucosal Immunol 2009; Oct 28. Epub ahead of print.
99. Cawston EE, Miller LJ. Therapeutic potential for novel drugs targeting the type 1 cholecystokinin receptor. British Journal of Pharmacology 2010;159: 1009–1021.
100. Congreve M, Marshall F. The impact of GPCR structures on pharmacology and structure-based drug design. British Journal of Pharmacology 2010; 159: 986–996.
101. Wang L, Martin B, Brenneman R, Luttrell LM, Maudsley S. Allosteric modulators of g protein-coupled receptors: future therapeutics for complex physiological disorders. J Pharmacol Exp Ther 2009;331(2): 340–348.
102. Piñeyro G. Membrane signalling complexes: implications for development of functionally selective ligands modulating heptahelical receptor signalling. Cell Signal 2009;21(2): 179–185.
103. Conn PJ, Christopoulos A, Lindsley CW. Allosteric modulators of GPCRs: a novel approach for the treatment of CNS disorders. Nat Rev Drug Discov 2009;8(1): 41–54.
104. Franco R, Casadó V, Cortés A, Pérez-Capote K, Mallol J, Canela E, Ferré S, Lluis C. Novel pharmacological targets based on receptor heteromers. Brain Res Rev. 2008;58(2): 475–482.
105. Whitty A, Riera TV. New ways to target old receptors. Curr Opin Chem Biol 2008;12(4): 427–433.
106. Arkin MR, Whitty A. The road less traveled: modulating signal transduction enzymes by inhibiting their protein–protein interactions. Curr Opin Chem Biol 2009;13(3): 284–290.
107. Nekouzadeh A, Silva JR, Rudy Y. Modeling subunit cooperativity in opening of tetrameric ion channels. Biophys J 2008;95(7): 3510–3520.
108. Sigel E, Buhr A. Trends Pharmacol Sci 1997;18:425–429.
109. Coyle J, Kershaw P. Biol Psychiatry 2001;49: 289–299.
110. Hockerman GH, Peterson BZ, Sharp E, Tanada TN, Scheuer T, Catterall WA. Proc Natl Acad Sci USA 1997;94:14906–14911.

111. Niswender CM, Lebois EP, Luo Q, Kim K, Muchalski H, Yin H, Conn PJ, Lindsley CW. Positive allosteric modulators of the metabotropic glutamate receptor subtype 4 (mGluR4): Part I. Discovery of pyrazolo[3,4-d]pyrimidines as novel mGluR4 positive allosteric modulators. Bioorg Med Chem Lett 2008;18(20): 5626–5630.
112. Schröder H, Wu DF, Seifert A, Rankovic M, Schulz S, Höllt V, Koch T. Allosteric modulation of metabotropic glutamate receptor 5 affects phosphorylation, internalization, and desensitization of the micro-opioid receptor. Neuropharmacology 2009;56(4): 768–778.
113. Yang CR, Svensson KA. Allosteric modulation of NMDA receptor via elevation of brain glycine and D-serine: the therapeutic potentials for schizophrenia. Pharmacol Ther 2008;120 (3): 317–332.
114. Williams R, Niswender CM, Luo Q, Le U, Conn PJ, Lindsley CW. Positive allosteric modulators of the metabotropic glutamate receptor subtype 4 (mGluR4): Part II. Challenges in hit-to-lead. Bioorg Med Chem Lett 2009; 19(3): 962–966.
115. Johnson KA, Conn PJ, Niswender CM. Glutamate receptors as therapeutic targets for Parkinson's disease. CNS Neurol Disord Drug Targets 2009 Dec; 8(6):475–491.
116. Jane DE, Lodge D, Collingridge GL. Kainate receptors: pharmacology, function and therapeutic potential. Neuropharmacology 2009;56 (1): 90–113.
117. Buccafusco JJ, Beach JW, Terry AV Jr. Desensitization of nicotinic acetylcholine receptors as a strategy for drug development. J Pharmacol Exp Ther 2009;328(2): 364–370.
118. Chan WY, McKinzie DL, Bose S, Mitchell SN, Witkin JM, Thompson RC, Christopoulos A, Lazareno S, Birdsall NJ, Bymaster FP, Felder CC. Allosteric modulation of the muscarinic M4 receptor as an approach to treating schizophrenia. Proc Natl Acad Sci USA 2008;105(31): 10978–10983.
119. Nawaratne V, Leach K, Suratman N, Loiacono RE, Felder CC, Armbruster BN, Roth BL, Sexton PM, Christopoulos A. New insights into the function of M4 muscarinic acetylcholine receptors gained using a novel allosteric modulator and a DREADD (designer receptor exclusively activated by a designer drug). Mol Pharmacol 2008;74(4): 1119–1131.
120. Shao W, Everitt L, Manchester M, Loeb DD, Hutchison CA. Proc Natl Acad Sci USA 1997;94:2243–2248.
121. Rose RB, Craik CS, Stroud RM. Domain flexibility in retroviral proteases: structural implications for drug resistant mutations. Biochemistry 1998;37(8): 2607–2621.
122. Goldstein JL, Brown MS. Science 2001;292: 1310–1312.
123. Matsuda M, Korn BS, Hammer RE, Moon YA, Komuro R, Horton JD, Goldstein JL, Brown MS, Shimomura I. Genes Dev 2001;15: 1206–1216.
124. Lupardus PJ, Shen A, Bogyo M, Garcia KC. Small molecule-induced allosteric activation of the Vibrio cholerae RTX cysteine protease domain. Science 2008;322(5899): 265–268.
125. Sohn J, Grant RA, Sauer RT. Allosteric activation of DegS, a stress sensor PDZ protease. Cell 2007;131(3): 572–583.
126. Sohn J, Sauer RT. OMP peptides modulate the activity of DegS protease by differential binding to active and inactive conformations. Mol Cell 2009;33(1): 64–74.
127. Vande Walle L, Lamkanfi M, Vandenabeele P. The mitochondrial serine protease HtrA2/ Omi: an overview. Cell Death Differ 2008;15 (3): 453–460.
128. Ehrmann M, Clausen T. Proteolysis as a regulatory mechanism. Annu Rev Genet 2004;38:709–724.
129. Wilken C, Kitzing K, Kurzbauer R, Ehrmann M, Clausen T. Crystal structure of the DegS stress sensor: How a PDZ domain recognizes misfolded protein and activates a protease. Cell 2004;117(4): 483–494.
130. Grau S, Baldi A, Bussani R, Tian X, Stefanescu R, Przybylski M, Richards P, Jones SA, Shridhar V, Clausen T, Ehrmann M. Implications of the serine protease HtrA1 in amyloid precursor protein processing. Proc Natl Acad Sci USA 2005;102(17): 6021–6026.
131. Böttcher J, Blum A, Dörr S, Heine A, Diederich WE, Klebe G. Targeting the open-flap conformation of HIV-1 protease with pyrrolidine-based inhibitors. ChemMedChem 2008;3(9): 1337–1344.
132. de Silva C, Walter NG. Leakage and slow allostery limit performance of single drug-sensing aptazyme molecules based on the hammerhead ribozyme. RNA 2009;15(1): 76–84.
133. Hsiou Y, Ding J, Das K, Clark AD, Hughes SH, Arnold E. Structure 1996;4:853–860.
134. Kohlstaedt LA, Wang J, Friedman JM, Rice PA, Steitz TA. Science 1992;256:1783–1790.
135. Das K, Ding J, Hsiou Y, Clark AD Jr, Moereels H, Koymans L, Andries K, Pauwels

R, Janssen PA, Boyer PL. J Mol Biol 1996;264: 1085–1100.
136. Huang H, Chopra R, Verdine GL, Harrison SC. Science 1998;282:1669–1675.
137. Jacobo-Molina A, Ding J, Nanni RG, Clark AD Jr, Lu X, Tantillo C, Williams RL, Kamer G, Ferris AL, Clark P. Proc Natl Acad Sci USA 1993;90:6320–6324.
138. Tachedjian G, Orlova M, Sarafianos SG, Arnold E, Goff SP. Proc Natl Acad Sci USA 2001;98:7188–7193.
139. Tanaka H, Walker RT, Hopkins AL, Ren J, Jones EY, Fujimoto K, Hayashi M, Mlyasaka T, Baba M, Stammers DK, Stuart DI. Antiviral Chem Chemother 1998;9:325–332.
140. Chow YH, Wei OL, Phogat S, Sidorov IA, Fouts TR, Broder CC, Dimitrov DS. Biochemistry 2002;41:7176–7182.
141. Morris K., HIV-1 fusion mechanism could be target for new drugs., Lancet 1997;349:1227.
142. Reeves JD, Lee FH, Miamidian JL, Jabara CB, Juntilla MM, Doms RW. Enfuvirtide resistance mutations: impact on human immunodeficiency virus envelope function, entry inhibitor sensitivity, and virus neutralization. J Virol 2005;79(8): 4991–4999.
143. Alexander L, Zhang S, McAuliffe B, Connors D, Zhou N, Wang T, Agler M, Kadow J, Lin PF. Inhibition of envelope-mediated CD4 + -T-cell depletion by human immunodeficiency virus attachment inhibitors. Antimicrob Agents Chemother 2009;53(11): 4726–4732.
144. He Y, Cheng J, Lu H, Li J, Hu J, Qi Z, Liu Z, Jiang S, Dai Q. Potent HIV fusion inhibitors against Enfuvirtide-resistant HIV-1 strains. Proc Natl Acad Sci USA 2008;105(42): 16332–16337.
145. He Y, Cheng J, Li J, Qi Z, Lu H, Dong M, Jiang S, Dai Q. Identification of a critical motif for the human immunodeficiency virus type 1 (HIV-1) gp41 core structure: implications for designing novel anti-HIV fusion inhibitors. J Virol 2008; Jul 82(13): 6349–6358.
146. Wang Y, Lu H, Zhu Q, Jiang S, Liao Y. Structure-based design, synthesis and biological evaluation of new N-carboxyphenylpyrrole derivatives as HIV fusion inhibitors targeting gp41. Bioorg Med Chem Lett 2010 Jan 1; 20(1): 189–192.
147. Liu S, Wu S, Jiang S. HIV entry inhibitors targeting gp41: from polypeptides to small-molecule compounds. Curr Pharm Des 2007;13(2): 143–162.
148. Betzi S, Eydoux C, Bussetta C, Blemont M, Leyssen P, Debarnot C, Ben-Rahou M, Haiech J, Hibert M, Gueritte F, Grierson DS, Romette JL, Guillemot JC, Neyts J, Alvarez K, Morelli X, Dutartre H, Canard B. Identification of allosteric inhibitors blocking the hepatitis C virus polymerase NS5B in the RNA synthesis initiation step. Antiviral Res 2009;84(1): 48–59.
149. Oikonomakos NG, Skamnaki VT, Tsitsanou KE, Gavalas NG, Johnson LN. Struct Fold Des 2000;8:575–584.
150. Tsitsanou KE, Skamnaki VT, Oikonomakos NG. Biochem Biophys 2000;384:245–254.
151. Oikonomakos NG, Zographos SE, Skamnaki VT, Archontis G. Bioorg Med Chem 2002;10:1313–1319.
152. Deng Q, Lu Z, Bohn J, Ellsworth KP, Myers RW, Geissler WM, Harris G, Willoughby CA, Chapman K, McKeever B, Mosley R. Modeling aided design of potent glycogen phosphorylase inhibitors. J Mol Graph Model 2005;23(5): 457–464.
153. Harwood HJ Jr, Petras SF, Hoover DJ, Mankowski DC, Soliman VF, Sugarman ED, Hulin B, Kwon Y, Gibbs EM, Mayne JT, Treadway JL. Dual-action hypoglycemic and hypocholesterolemic agents that inhibit glycogen phosphorylase and lanosterol demethylase. J Lipid Res 2005;46(3): 547–563.
154. Clore GM, Omichinski JG, Sakaguchi K, Zambrano N, Sakamoto H, Appella E, Gronenborn AM. Science 1994;265:386–391.
155. Lee W, Harvey TS, Yin Y, Yau P, Lichtfield D, Arrowsmith CH. Struct Biol 1994;1:877–890.
156. Jeffrey PD, Gorina S, Pavletich NP. Science 1995;267:1498–1502.
157. Clore GM, Omichinski JG, Sakaguchi K, Zambrano N, Sakamoto H, Appella E, Gronenborn AM. Science 1995;267:1515–1516.
158. Clore GM, Ernst J, Clubb R, Omichinski JG, Poindexter Kennedy WM, Sakaguchi K, Appella E, Gronenborn AM. Struct Biol 1995;2:321–333.
159. Miller M, Lubkowski J, Mohana Rao JK, Danishefsky AT, Omichinski JG, Sakaguchi K, Sakamoto H, Appella E, Gronenborn AM, Clore GM. FEBS Lett 1996;399:166–170.
160. Foster BA, Coffey HA, Morin MJ, Rastinejad F. Science 1999;286:2507–2510.
161. Komarov PG, Komarova EA, Kondratov RV, Christov-Tselkov K, Coon JS, Chernov MV, Gudkov AV. Science 1999;285:1733–1737.
162. Rao CV, Steele VE, Swamy MV, Patlolla JM, Guruswamy S, Kopelovich L. Inhibition of azoxymethane-induced colorectal cancer by

CP-31398, a TP53 modulator, alone or in combination with low doses of celecoxib in male F344 rats. Cancer Res. 2009 Oct 15;69(20): 8175–8182.

163. Davenport CM, Sevastou IG, Hooper C, Pocock JM. Inhibiting p53 pathways in microglia attenuates microglial-evoked neurotoxicity following exposure to Alzheimer peptides. Cancer Res 2009;69(20): 8175–8182Davenport CM, Sevastou IG, Hooper C, Pocock JM. Inhibiting p53pathways in microglia attenuates microglial-evoked neurotoxicity following exposure to Alzheimer peptides. J Neurochem 2010 Jan; 112(2):552–563.

164. Wawrzynow B, Pettersson S, Zylicz A, Bramham J, Worrall E, Hupp TR, Ball KL. A function for the RING finger domain in the allosteric control of MDM2 conformation and activity. J Biol Chem 2009;284(17): 11517–11530.

165. Baraldi PG, Preti D, Tabrizi MA, Romagnoli R, Saponaro G, Baraldi S, Botta M, Bernardini C, Tafi A, Tuccinardi T, Martinelli A, Varani K, Borea PA. Structure–activity relationship studies of a new series of imidazo[2,1-*f*]purinones as potent and selective A(3) adenosine receptor antagonists. Bioorg Med Chem. 2008;16(24): 10281–10294.

166. Francesco AL, Diego G, Susanna G, Giuseppina L, Amina WS, Sergi F, Rafael F, Kjell F. Integrative action of receptor mosaics: relevance of receptor topology and allosteric modulators. J Recept Signal Transduct Res 2008;28(6): 543–565.

167. Pinna A, Tronci E, Schintu N, Simola N, Volpini R, Pontis S, Cristalli G, Morelli M. A new ethyladenine antagonist of adenosine A(2A) receptors: behavioral and biochemical characterization as an antiparkinsonian drug. Neuropharmacology 2010 Mar;58(3):613–623.

168. Rihakova L, Quiniou C, Hamdan FF, Kaul R, Brault S, Hou X, Lahaie I, Sapieha P, Hamel D, Shao Z, Gobeil F Jr, Hardy P, Joyal JS, Nedev H, Duhamel F, Beauregard K, Heveker N, Saragovi HU, Guillon G, Bouvier M, Lubell WD, Chemtob S. VRQ397 (CRAVKY): a novel noncompetitive V2 receptor antagonist. Am J Physiol Regul Integr Comp Physiol 2009;297 (4): R1009–R1018.

169. Buccafusco JJ. Emerging cognitive enhancing drugs. Expert Opin Emerg Drugs 2009; Sep 22. Epub ahead of print.

170. Sattelle DB, Buckingham SD, Akamatsu M, Matsuda K, Pienaar I, Jones AK, Sattelle BM, Almond A, Blundell CD. Comparative pharmacology and computational modelling yield insights into allosteric modulation of human alpha7 nicotinic acetylcholine receptors. Biochem Pharmacol. 2009;Oct 1 78(7): 836–43.

171. Wierońska JM, Pilc A. Metabotropic glutamate receptors in the tripartite synapse as a target for new psychotropic drugs. Neurochem Int 2009;55(1–3): 85–97.

172. Barron SC, McLaughlin JT, See JA, Richards VL, Rosenberg RL. An allosteric modulator of alpha7 nicotinic receptors, *N*-(5-Chloro-2,4-dimethoxyphenyl)-*N'*-(5-methyl-3-isoxazolyl)-urea (PNU-120596), causes conformational changes in the extracellular ligand binding domain similar to those caused by acetylcholine. Mol Pharmacol 2009;76(2): 253–263.

173. Shirey JK, Brady AE, Jones PJ, Davis AA, Bridges TM, Kennedy JP, Jadhav SB, Menon UN, Xiang Z, Watson ML, Christian EP, Doherty JJ, Quirk MC, Snyder DH, Lah JJ, Levey AI, Nicolle MM, Lindsley CW, Conn PJ. A selective allosteric potentiator of the M1 muscarinic acetylcholine receptor increases activity of medial prefrontal cortical neurons and restores impairments in reversal learning. J Neurosci 2009;29(45): 14271–14286.

174. Joseph JD, Wittmann BM, Dwyer MA, Cui H, Dye DA, McDonnell DP, Norris JD. Inhibition of prostate cancer cell growth by second-site androgen receptor antagonists. Proc Natl Acad Sci USA 2009;106(29): 12178–12183.

175. Hamm S, Rath S, Michel S, Baumgartner R. Cancer immunotherapeutic potential of novel small molecule TLR7 and TLR8 agonists. J Immunotoxicol 2009;6(4): 257–265.

176. Ompraba G, Velmurugan D, Louis PA, Rafi ZA. Molecular modeling of the additional inhibitor site located in secretory phospholipase a(2)*. J Biomol Struct Dyn 2010;27(4): 489–500.

177. Wang Z, Wesche H, Stevens T, Walker N, Yeh WC. IRAK-4 inhibitors for inflammation. Curr Top Med Chem 2009;9(8): 724–737.

178. Marie PJ. The calcium-sensing receptor in bone cells: a potential therapeutic target in osteoporosis. Bone 2009; Aug 4. Epub ahead of print.

179. Willert EK, Phillips MA. Regulated expression of an essential allosteric activator of polyamine biosynthesis in African trypanosomes. PLoS Pathog 2008;4(10): e1000183. Epub 2008 Oct 24.

180. Lawrence SH, Ramirez UD, Selwood T, Stith L, Jaffe EK. Allosteric inhibition of human porphobilinogen synthase. J Biol Chem 2009; Oct 7. Epub ahead of print.

181. Miller MW, Basra S, Kulp DW, Billings PC, Choi S, Beavers MP, McCarty OJ, Zou Z, Kahn ML, Bennett JS, DeGrado WF. Small-molecule inhibitors of integrin alpha2beta1 that prevent pathological thrombus formation via an allosteric mechanism. Proc Natl Acad Sci USA 2009;106(3): 719–724.

182. Yuki K, Astrof NS, Bracken C, Yoo R, Silkworth W, Soriano SG, Shimaoka M. The volatile anesthetic isoflurane perturbs conformational activation of integrin LFA-1 by binding to the allosteric regulatory cavity. FASEB J 2008;22(12): 4109–4116.

183. Christopoulos A, May LT, Avlani VA, Sexton PM. G-protein-coupled receptor allosterism: the promise and the problem(s). Biochem Soc Trans 2004;32(Pt 5): 873–877.

184. May LT, Leach K, Sexton PM, Christopoulos A. Allosteric modulation of G protein-coupled receptors. Annu Rev Pharmacol Toxicol 2007;47: 1–51.

185. Torres PU. Cinacalcet HCl: a novel treatment for secondary hyperparathyroidism caused by chronic kidney disease. J Ren Nutr 2006;16(3): 253–258.

186. Kutikuppala P, Rao S. CCR5 inhibitors: Emerging promising HIV therapeutic strategy. Indian J Sex Transm Dis 2009;30:1–9.

187. Lobritz M, Ratcliff A, Marozsan A, Dudley D, Arts E. Different mechanisms of HIV-1 inhibition by Ccr5 agonists/antagonists (Maraviroc, Tak-779 and Psc-Rantes) are linked to different drug resistance mechanisms. Antivir Ther 2009;14(Suppl 1) (A14): Abstract 12.

188. Sayana S, Khanlou H. Maraviroc: a new CCR5 antagonist. Expert Rev Anti Infect Ther 2009;7 (1): 9–19. Review.

189. Abraham DJ, Wireko FC, Randad RS, Poyart C, Kister J, Bohn B. Allosteric modifiers of hemoglobin: 2-[4-[[(3,5-disubstituted anilino) carbonyl]methyl]phenoxy]-2-methylpropionic acid derivatives that lower the oxygen affinity of hemoglobin in red cell suspensions, in whole blood, and *in vivo* in rats. Biochemistry 1992;31:9141–9149.

190. Abraham DJ, Safo MK, Boyiri T, Danso-Danquah RE, Kister J, Poyart C. How allosteric effectors can bind to the same protein residue and produce opposite shifts in the allosteric equilibrium. Biochemistry 1995;34:15006–15020.

191. Boyiri T, Safo MK, Danso-Danquah RE, Kister J, Poyart C, Abraham DJ. Bis-aldehyde allosteric effectors as molecular ratchets and probes. Biochemistry 1995;34:15021–15036.

192. Propper DJ, Levitt NC, O'Byrne K, Braybrooke JP, Talbot DC, Ganesan TS, Thompson CH, Rajagopalan B, Littlewood TJ, Dixon RM, Harris AL. Phase II study of the oxygen saturation curve left shifting agent BW12C in combination with the hypoxia activated drug mitomycin C in advanced colorectal cancer. Br J Cancer 2000;82(11): 1776–1782.

PRINCIPLES OF DRUG METABOLISM

Bernard Testa
Department of Pharmacy, University Hospital Centre, Lausanne, Switzerland

1. INTRODUCTION

Xenobiotic metabolism, which includes drug metabolism, has become a major pharmacological science with particular relevance to biology, therapeutics and toxicology. Drug metabolism is also of great importance in medicinal chemistry because it influences in qualitative, quantitative, and kinetic terms the deactivation, activation, detoxification, and toxification of the vast majority of drugs. As a result, medicinal chemists engaged in drug discovery and development should be able to integrate metabolic considerations into drug design. To do so, however, requires a fair or even good knowledge of xenobiotic metabolism.

This chapter, which is written by a medicinal chemist for medicinal chemists, aims at offering knowledge and understanding rather than encyclopedic information. Readers wanting to go further in the study of xenobiotic metabolism may consult various classical or recent books [1–15], broad reviews, and book chapters [16–27].

1.1. Definitions and Concepts

Drugs are but one category among the many *xenobiotics* (Table 1) that enter the body but have no nutritional or physiological value [14]. The study of the disposition—or fate—of xenobiotics in living systems includes the consideration of their absorption into the organism, how and where they are distributed and stored, the chemical and biochemical transformations they may undergo, and how and by which route(s) they are finally excreted and returned to the environment. As for "metabolism," this word has acquired two meanings, being synonymous with disposition (i.e., the sum of the processes affecting the fate of a chemical substance in the body), and with biotransformation as understood in this chapter.

In pharmacology, one speaks of *pharmacodynamic effects* to indicate what a drug does to the body, and *pharmacokinetic effects* to indicate what the body does to a drug, two aspects of the behavior of drugs that are strongly interdependent. Pharmacokinetic effects will obviously have a decisive influence on the intensity and duration of pharmacodynamic effects, while metabolism will generate new chemical entities (metabolites) that may have distinct pharmacodynamic properties of their own. Conversely, by its own pharmacodynamic effects, a compound may affect the state of the organism (e.g., hemodynamic changes, enzyme activities, etc.) and hence the organism's capacity to handle xenobiotics. Only a systemic approach can help one appreciate the global nature of this interdependence.

1.2. Types of Metabolic Reactions Affecting Xenobiotics

A first discrimination to be made among metabolic reactions is based on the nature of the catalyst. Reactions of xenobiotic metabolism, like other biochemical reactions, are *catalyzed by enzymes*. However, while the vast majority of reactions of xenobiotic metabolism are indeed enzymatic ones, some *nonenzymatic reactions* are also well documented. This is due to the fact that a variety of xenobiotics have been found to be labile enough to react nonenzymatically under biological conditions of pH and temperature [28]. But there is more. In a normal enzymatic reaction, metabolic intermediates exist *en route* to the product (s) and do not leave the catalytic site. However, many exceptions to this rule are known, with the metabolic intermediate leaving the active site and reacting with water, with an endogenous molecule or macromolecule, or with a xenobiotic. Such reactions are also of a nonenzymatic nature but are better designated as *postenzymatic reactions*.

The metabolism of drugs and other xenobiotics is often a biphasic process in which the compound may first undergo a *functionalization reaction* (phase I reaction) of oxidation, reduction, or hydrolysis. This introduces or unveils a functional group such as a hydroxy

Table 1. Major Categories of Xenobiotics (Modified from Ref. [14])

- Drugs
- Food constituents devoid of physiological roles
- Food additives (preservatives, coloring and flavoring agents, antioxidants, etc.)
- Chemicals of leisure, pleasure, or abuse (ethanol, coffee and tobacco constituents, hallucinogens, doping agents, etc.)
- Agrochemicals (fertilizers, insecticides, herbicides, etc.)
- Industrial and technical chemicals (solvents, dyes, monomers, polymers, etc.)
- Pollutants of natural origin (radon, sulfur dioxide, hydrocarbons, etc.)
- Pollutants produced by microbial contamination (e.g., aflatoxins)
- Pollutants produced by physical or chemical transformation of natural compounds (polycyclic aromatic hydrocarbons by burning, Maillard reaction products by heating, etc.)

or amino group suitable for coupling with an endogenous molecule or moiety in a second metabolic step known as a *conjugation reaction* (phase II reaction) [19,24]. In a number of cases, phase I metabolites may be excreted prior to conjugation, while many xenobiotics can be directly conjugated. Furthermore, reactions of functionalization may follow some reactions of conjugation, for example, some conjugates are hydrolyzed and/or oxidized prior to their excretion.

Xenobiotic biotransformation thus produces two types of metabolites, namely functionalization products and conjugates. But with the growth of knowledge, biochemists and pharmacologists have progressively come to recognize the existence of a third class of metabolites, namely xenobiotic–macromolecule adducts, also called macromolecular conjugates. Such peculiar metabolites are formed when a xenobiotic binds covalently to a biological macromolecule, usually following metabolic activation (i.e., postenzymatically). Both functionalization products and conjugates have been found to bind covalently to biological macromolecules, the reaction often being toxicologically relevant.

1.3. Specificities and Selectivities in Xenobiotic Metabolism

The words "selectivity" and "specificity" may not have identical meanings in chemistry and biochemistry. In this chapter, the specificity of an enzyme is taken to mean an ensemble of properties, the description of which makes it possible to specify the enzyme's behavior. In contrast, the term selectivity is applied to metabolic processes, indicating that a given metabolic reaction or pathway is able to select some substrates or products from a larger set. In other words, the selectivity of a metabolic reaction is the detectable expression of the specificity of an enzyme. Such definitions may not be universally accepted, but they have the merit of clarity.

What, then, are the various types of selectivities (or specificities) encountered in xenobiotic metabolism? What characterizes an enzyme from a catalytic viewpoint is first its chemospecificity, that is, its specificity in terms of the type(s) of reaction it catalyzes. When two or more substrates are metabolized at different rates by a single enzyme under identical conditions, substrate selectivity is observed. In such a definition, the nature of the product(s) and their isomeric relationship are not considered. Substrate selectivity is distinct from product selectivity, which is observed when two or more metabolites are formed at different rates by a single enzyme from a single substrate. Thus, substrate-selective reactions discriminate between different compounds, while product-selective reactions discriminate between different groups or positions in a given compound.

The substrates being metabolized at different rates may share various types of relationships. They may be chemically dissimilar or similar (e.g., analogs), in which case the term of substrate selectivity is used in a narrow sense. Alternatively, the substrates may be isomers such as positional isomers (regioisomers) or stereoisomers, resulting in substrate regioselectivity or substrate stereoselectivity. Substrate enantioselectivity is a particular case of the latter (see Section 4.2.1).

Products formed at different rates in product-selective reactions may also share various types of relationships. Thus, they may be analogs, regioisomers, or stereoisomers, resulting in product selectivity (narrow sense), product regioselectivity, or product stereoselectivity (e.g., product enantioselectivity). Note that the product selectivity displayed by two distinct substrates in a given metabolic reaction may be different, in other words the product selectivity may be substrate-selective. The term substrate–product selectivity can be used to describe such complex cases, which are conceivable for any type of selectivity but have been reported mainly for stereoselectivity.

1.4. Pharmacodynamic Consequences of Xenobiotic Metabolism

The major function of xenobiotic metabolism can be seen as the elimination of physiologically useless compounds, some of which may be harmful, witness the tens of thousands of toxins produced by plants. The function of toxin inactivation justifies the designation of detoxification originally given to reactions of xenobiotic metabolism. However, the possible pharmacological consequences of biotransformation are not restricted to detoxification. In the simple case of a xenobiotic having a single metabolite, four possibilities exist [25], namely:

1. Both the xenobiotic and its metabolite are devoid of biological effects (at least in the concentration or dose range investigated); such a situation has no place in medicinal chemistry.
2. Only the xenobiotic elicits biological effects, a situation that in medicinal chemistry is typical of drugs yielding no bioactive metabolite, as seen for example with soft drugs.
3. Both the xenobiotic and its metabolite are biologically active, the two activities being comparable or different either qualitatively or quantitatively.
4. The observed biological activity is due exclusively to the metabolite, a situation that in medicinal chemistry is typical of prodrugs.

When a drug or another xenobiotic is transformed into a toxic metabolite, the reaction is one of *toxification*. Such a metabolite may act or react in a number of ways to elicit a variety of toxic responses at different biological levels. However, it is essential to stress that the occurrence of a reaction of toxification (i.e., toxicity at the molecular level) does not necessarily imply toxicity at the levels of organs and organisms, as discussed later in this chapter.

1.5. Setting the Scene

In drug research and development, metabolism is of pivotal importance due to the interconnectedness between pharmacokinetic and pharmacodynamic processes [14]. *In vitro* metabolic studies are now initiated early during discovery and development to assess the overall rate of oxidative metabolism, to identify the metabolites, and to obtain primary information on the enzymes involved, and to postulate metabolic intermediates. Based on these findings, the metabolites must be synthesized and tested for their own pharmacological and toxicological effects. In preclinical and early clinical studies, many pharmacokinetic data must be obtained and relevant criteria must be satisfied before a drug candidate can enter large-scale clinical trials [29,30]. As a result of these demands, the interest of medicinal chemists for drug metabolism has grown remarkably in recent years.

As will become apparent, the approach followed in this chapter is an analytical one, meaning that the focus is on metabolic reactions, the target groups they affect, and the enzymes by which they are catalyzed. This information provides the foundations of drug metabolism, but it must be complemented by a synthetic view to allow a broader understanding and meaningful predictions. Two steps are required to approach these objectives, namely (a) the elaboration of metabolic schemes where the competitive and sequential reactions undergone by a given drug are ordered and (b) an assessment of the various biological factors that influence such schemes both quantitatively and qualitatively. As an example of a metabolic scheme, Fig. 1 presents the biotransformation of propranolol (**1**) in humans [31]. There are relatively few studies as comprehensive and clinically relevant as this one, which remains as current today as it was

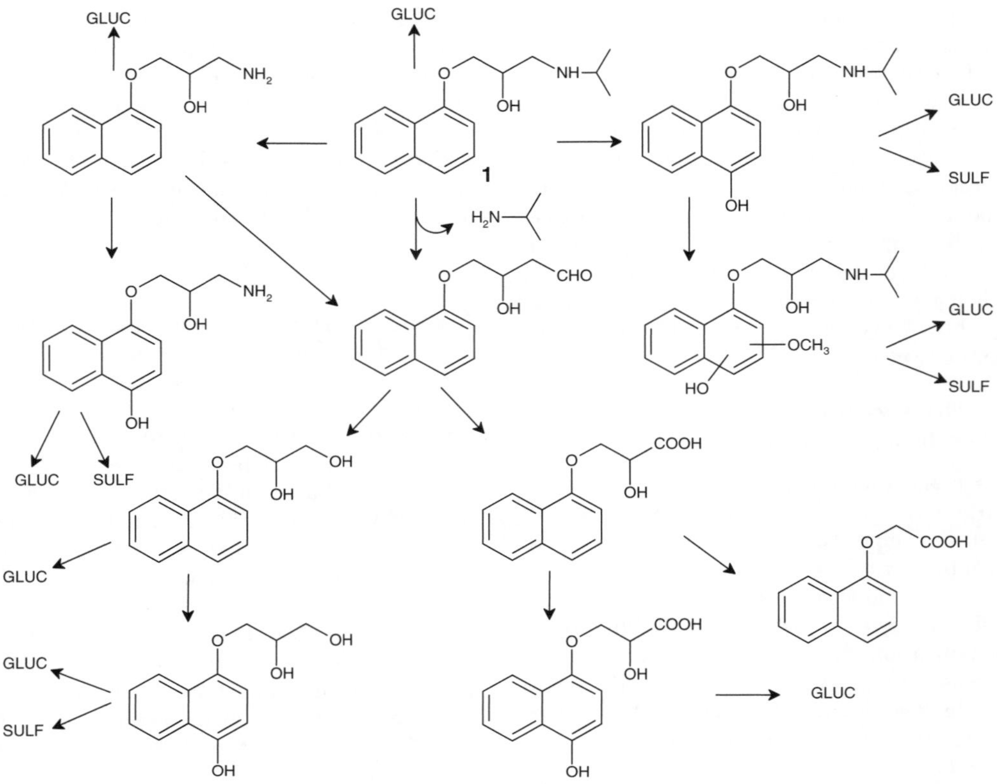

Figure 1. The metabolism of propranolol (1) in humans, accounting for more than 90% of the dose; GLUC = glucuronide(s); SULF = sulfate(s) [31].

when published in 1985. Indeed, over 90% of a dose were accounted for and consisted mainly in products of oxidation and conjugation. The missing 10% may represent other, minor and presumably quite numerous metabolites, for example, those resulting from ring hydroxylation at other positions or from the progressive breakdown of glutathione conjugates.

A large variety of enzymes and metabolic reactions are presented in Sections 2 and 3. As will become clear, some enzymes catalyze only a single type of reaction (e.g., *N*-acetylation), whereas others use a basic catalytic mechanism to attack a variety of moieties and produce different types of metabolites (e.g., cytochromes P450). As an introduction to these enzymes and reactions, we present in Table 2 an estimate of their relative importance in drug metabolism. In this table, the correspondence between the number of substrates and the overall contribution to drug metabolism does not need to be perfect, since some enzymes show a limited capacity (e.g., sulfotransferases) whereas others make a significant contribution to the biotransformation of their substrates (e.g., hydrolases).

2. FUNCTIONALIZATION REACTIONS

2.1. Introduction

Reactions of functionalization comprise oxidations (electron removal, dehydrogenation, and oxygenation), reductions (electron addition, hydrogenation, and removal of oxygen), and hydrations/dehydrations (hydrolysis and addition or removal of water). The reactions of oxidation and reduction are catalyzed by a very large variety of oxidoreductases, while various hydrolases catalyze hydrations. A

Table 2. Estimate of the Relative Contributions of Major Drug-Metabolizing Enzymes[a]

Enzymes	Overall Contribution to Drug Metabolism[b]
Cytochromes P450 (Section 2.2.1)	+ + + +
Dehydrogenases and reductases (Section 2.2.2)	+ + +
Flavin-containing monooxygenases (Section 2.2.2)	+
Hydrolases (Section 2.2.3)	+ +
Methyltransferases (Section 3.2)	+
Sulfotransferases (Section 3.3)	+
Glucuronyltransferases (Section 3.4)	+ + +
N-Acetyltransferases (Section 3.5)	+
Acyl-coenzyme A synthetases (Section 3.6)	+
Glutathione *S*-transferases (Section 3.7)	+ +
Phosphotransferases (Section 3.8)	(+)

[a] (+) very low; + low; + + intermediate; + + + high; + + + + very high.
[b] Including drug metabolites.

large majority of enzymes involved in xenobiotic functionalization are briefly reviewed in Section 2.2 [14]. Metabolic reactions and pathways of functionalization constitute the main body of this section.

2.2. Enzymes Catalyzing Functionalization Reactions

2.2.1. Cytochromes P450 Monooxygenation reactions are of major significance in drug metabolism and are mediated by various enzymes, which differ markedly in their structure and properties. Among these, the most important as far as xenobiotic metabolism is concerned are the cytochromes P450 (EC 1.14.14.1, also 1.14.13), a very large group of enzymes belonging to heme-coupled monooxygenases [4–7,12,15–18,20,22,32–36]. The cytochrome P450 enzymes (CYPs) are encoded by the *CYP* gene superfamily and are classified in families and subfamilies as summarized in Table 3. Cytochrome P450 is the major drug-metabolizing enzyme system, playing a key role in detoxification and toxification, and is of additional significance in medicinal chemistry because several CYP enzymes are drug targets, for example, thromboxane synthase (CYP5) and aromatase (CYP19). The three CYP families mostly involved in xenobiotic metabolism are CYP1 to CYP3, whose relative importance is given in Table 4.

The present section focuses on the metabolic reactions, beginning with the catalytic cycle of cytochrome P450 (Fig. 2). This cycle involves a number of steps that can be simplified as follows:

(a) The enzyme in its ferric (oxidized) form exists in equilibrium between two spin states, a hexacoordinated low-spin form that cannot be reduced, and a pentacoordinated high-spin form. Binding of the substrate to enzyme induces a shift to the reducible high-spin form (reaction a).

(b) A first electron enter the enzyme–substrate complex (reaction b).

(c) The enzyme in its ferrous form has a high affinity for diatomic gases such as CO (a strong inhibitor of cytochrome P450) and dioxygen (reaction c).

(d) Electron transfer from Fe^{2+} to O_2 within the enzyme–substrate–oxygen ternary complex reduces the dioxygen to a bound molecule of superoxide. Its possible liberation in the presence of compounds with good affinity but low reactivity (uncoupling) can be cytotoxic (reaction d).

(e) The normal cycle continues with a second electron entering via either F_{P1} or F_{P2} and reducing the ternary complex (reaction e).

(f) Electron transfer within the ternary complex generates bound peroxide anion (O_2^{2-}).

(g) The bound peroxide anion is split, liberating H_2O (reaction f).

(h) The remaining oxygen atom is an oxene species. This is the reactive form of oxygen that will attack the substrate.

(i) The binary enzyme–product complex dissociates, thereby regenerating the initial state of cytochrome P450 (reaction h).

Oxene is a rather electrophilic species, being neutral but having only six electrons in

Table 3. The Human *CYP* Gene Superfamily: A Table of the Families and Subfamilies of Gene Products [34–37]

Families	Subfamilies (Representative Gene Products)
P450 1 Family (*Aryl hydrocarbon hydroxylases; xenobiotic metabolism; inducible by polycyclic aromatic hydrocarbons*)	
	P450 1A Subfamily (CYP1A1, CYP1A2)
	P450 1B Subfamily (CYP1B1)
P450 2 Family (*Xenobiotic metabolism; constitutive and xenobiotic-inducible*)	
	P450 2A Subfamily (CYP2A6, CYP2A7, CYP2A13)
	P450 2B Subfamily (CYP2B6)
	P450 2C Subfamily (CYP2C8, CYP2C9, CYP2C18, CYP2C19)
	P450 2D Subfamily (CYP2D6)
	P450 2E Subfamily (CYP2E1)
	P450 2F Subfamily (CYP2F1)
	P450 2J Subfamily (CYP2J2)
	P450 2R Subfamily (CYP2R1)
	P450 2S Subfamily (CYP2S1)
	P450 2U Subfamily (CYP2U1)
	P450 2W Subfamily (CYP2W1)
P450 3 Family (*Xenobiotic and steroid metabolism; steroid-inducible*)	
	P450 3A Subfamily (CYP3A4, CYP3A5, CYP3A7, CYP3A43)
P450 4 Family (*Peroxisome proliferator-inducible*)	
	P450 4A Subfamily (CYP4A11, CYP4A20, CYP4A22)
	P450 4B Subfamily (CYP4B1)
	P450 4F Subfamily (CYP4F2, CYP4F3, CYP4F8, CYP4F11, CYP4F12, CYP4F22)
	P450 4V Subfamily (CYP4V2)
	P450 4X Subfamily (CYP4X1)
P450 5 Family	
	P450 5A Subfamily (CYP5A1)
P450 7 Family (*Steroid 7-hydroxylases*)	
	P450 7A Subfamily (CYP7A1)
	P450 7B Subfamily (CYP7B1)
P450 8 Family	
	P450 8A Subfamily (CYP8A1)
	P450 8B Subfamily (CYP8B1)
P450 11 Family (*Mitochondrial steroid hydroxylases*)	
	P450 11A Subfamily (CYP11A1)
	P450 11B Subfamily (CYP11B1, CYP11B2)
P450 17 Family (*Steroid 17β-hydroxylase*)	
	P450 17A Subfamily (CYP17A1)
P450 19 Family (CYP19)	
P450 20 Family (CYP20)	
P450 21 Family (CYP21)	
P450 24 Family (CYP24)	
P450 26 Family	
	P450 26A Subfamily (CYP26A1)
	P450 26B Subfamily (CYP26B1)
	P450 26C Subfamily (CYP26C1)
P450 27 Family (*Mitochondrial steroid hydroxylases*)	
	P450 27A Subfamily (CYP27A1)
	P450 27B Subfamily (CYP27B1)
	P450 27C Subfamily (CYP27C1)
P450 39 Family (CYP39)	
P450 46 Family (CYP46)	
P450 51 Family (CYP51)	

This list reports all 57 known human *CYP* gene products.

Table 4. Levels and Variability of Human CYP Enzymes Involved in Drug Metabolism [16]

CYP	Level of Enzyme in Liver (% of Total)	Variability Range	Percent of Drugs (or Other Xenobiotics) Interacting with Enzyme: As Substrates	As Inhibitors	As Inducers/Activators
1A1			3 (12)	3 (13)	6 (33)
1A2	ca. 13	ca. 40-fold	10 (15)	12 (18)	3 (25)
1B1	<1		1 (10)	1 (7)	1 (9)
2A6	ca. 4	ca. 30- to 100-fold	3 (10)	2 (6)	2 (1)
2B6	<1	ca. 50-fold	4 (9)	3 (5)	13 (no data)
2C	ca. 18	25- to 100-fold	25 (13)	27 (17)	21 (6)
2D6	up to 2.5	>1000-fold	15 (2)	22 (8)	2 (2 activators)
2E1	up to 7	ca. 20-fold	3 (16)	4 (10)	7 (7)
3A4	up to 28	ca. 20-fold	36 (13)	26 (16)	45 (17)

its outer shell. Its detailed reaction mechanisms are beyond the scope of this chapter, but some indications will be given when discussing the various reactions of oxidation catalyzed by cytochromes P450.

2.2.2. Other Oxidoreductases Besides cytochromes P450, other monooxygenases of importance are the flavin-containing monooxygenases (see Table 5), while dopamine β-monooxygenase (EC 1.14.17.1) plays only a minor role. Other oxidoreductases that can

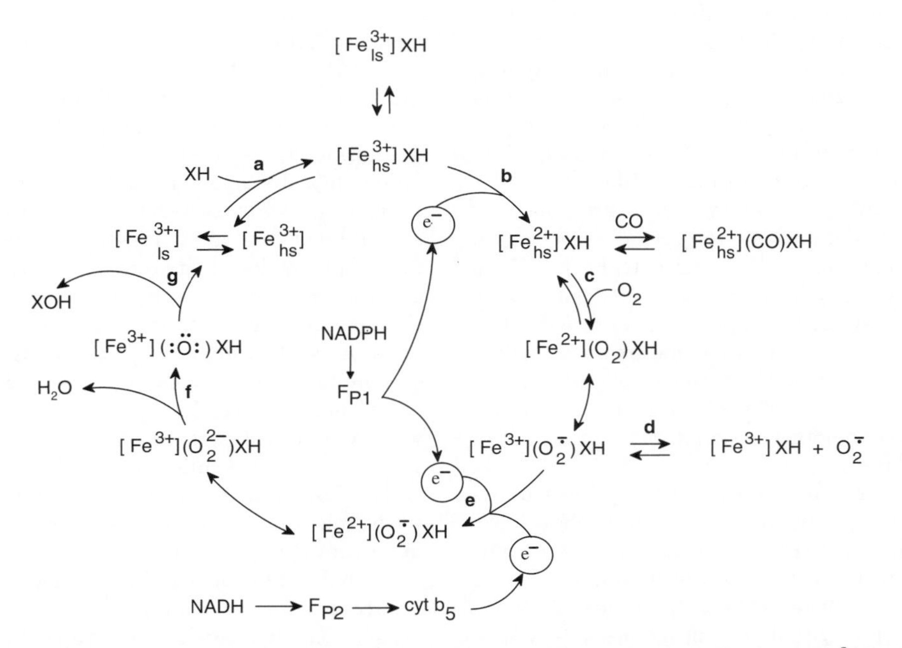

Figure 2. Catalytic cycle of cytochrome P450 associated with monooxygenase reactions. [Fe^{3+}] = ferricytochrome P450; hs = high spin; ls = low spin; [Fe^{2+}] = ferrocytochrome P450; F_{P1} = flavoprotein 1 = NADPH-cytochrome P450 reductase; F_{P2} = NADH-cytochrome b_5 reductase; cyt b5 = cytochrome b_5; XH = substrate (modified from Refs [4,14]).

[1](+) very low; + low; + + intermediate; + + + high; + + + + very high.

Table 5. A Survey of Oxidoreductases Other than Cytochromes P450 Playing a Role in Drug Metabolism [14,22]

Enzymes	EC Numbers	*Gene Root* (or *Gene*) and Major Human Enzymes
Flavin-containing monooxygenases	EC 1.14.13.8	*FMO* (FMO1 to FMO5)
Monoamine oxidases	EC 1.4.3.4	*MAO* (MAO-A and MAO-B)
Copper-containing amine oxidases	EC 1.4.3.6	*AOC* (DAO and SSAO)
Aldehyde oxidase	EC 1.2.3.1	*AOX1* (AO)
Xanthine oxidoreductase	EC 1.17.1.4 and 1.17.3.2	*XOR* (XDH and XO)
Various peroxidases	EC 1.11.1.7 and 1.11.1.8	For example, *EPO* (EPO), *MPO* (MPO) and *TPO* (TPO)
Prostaglandin G/H synthase	EC 1.14.99.1	*PTGS* (COX-1 and COX-2)
Alcohol dehydrogenases	EC 1.1.1.1	*ADH* (ADH1A, 1B and 1C, ADH4, ADH5, ADH6, and ADH7)
Aldehyde dehydrogenases	EC 1.2.1.3 and 1.2.1.5	*ALDH* (e.g., ALDH1A1, 1A2 and 1A3, 1B1, 2, 3A1, 3A2, 3B1, 3B2, 8A1, and 9A1)
Aldo-keto reductases	In EC 1.1.1 and 1.3.1	*AKR* (e.g., ALR1, ALR2, DD1, DD2, DD3, DD4, AKR7A2, 7A3, and 7A4)
Carbonyl reductases	EC 1.1.1.184	*CBR* (CR1, CR3)
Quinone reductases	EC 1.6.5.2 and 1.10.99.2	*NQO* (NQO1 and NGO2)

play a major or less important role in drug metabolism are the two monoamine oxidases that are essentially mitochondrial enzymes, and the broad group of copper-containing amine oxidases. Cytosolic oxidoreductases are the molybdenum hydroxylases, namely aldehyde oxidase and xanthine oxidase.

Various peroxidases are progressively being recognized as important enzymes in drug metabolism. Several cytochrome P450 enzymes have been shown to have peroxidase activity. A variety of peroxidases may oxidize drugs, for example, myeloperoxidase (MPO). Prostaglandin G/H synthase (prostaglandin-endoperoxide synthase) is able to use a number of xenobiotics as cofactors in a reaction of cooxidation.

The large and important ensemble of dehydrogenases/reductases include alcohol dehydrogenases (ADH) that are zinc enzymes found in the cytosol of the mammalian liver and in various extrahepatic tissues. Mammalian liver alcohol dehydrogenases (LADHs) are dimeric enzymes. The human enzymes belong to three different classes: Class I (ADH1), comprising the various isozymes that are homodimers or heterodimers of the alpha (ADH1A, ADH2A, ADH3A), beta and gamma subunits.

Enzymes categorized as aldehyde dehydrogenases (ALDHs) include the NAD^+- and $NAD(P)^+$-dependent enzymes. They exist in multiple forms in the cytosol, mitochondria, and microsomes of various mammalian tissues. Among the aldo-keto reductases, we find aldehyde reductase, alditol dehydrogenase, a number of hydroxysteroid dehydrogenases, and dihydrodiol dehydrogenase. These enzymes are widely distributed in nature and occur in a considerable number of mammalian tissues. Their subcellular location is primarily cytosolic, and for some also mitochondrial. The carbonyl reductases belong to the short-chain dehydrogenases/reductases and are of noteworthy significance in xenobiotic metabolism. There are many similarities, including some marked overlap in substrate specificity, between monomeric, NADPH-dependent aldehyde reductase (AKR1), alditol dehydrogenase (AKR2), and carbonyl reductase (AKR3).

Other reductases that have a role to play in drug metabolism include the important quinone reductase, also known as DT-diaphorase.

2.2.3. Hydrolases Hydrolases constitute a very complex ensemble of enzymes many of which are known or suspected to be involved in xenobiotic metabolism (Table 6). Relevant en-

Table 6. A Survey of Hydrolases Playing a Role in Drug Metabolism [9,14,23]

Classes of Hydrolases	Examples of Enzymes (With Some *Gene Roots* and Human Enzymes)
EC 3.1.1: carboxylic ester hydrolases	EC 3.1.1.1: Carboxylesterase (*CES*) CES1A1, CES2, CES3
	EC 3.1.1.2: Arylesterase (*PON*, see 3.1.8.1)
	EC 3.1.1.8: Cholinesterase (*BCHE*)
EC 3.1.2: thiolester hydrolases	EC 3.1.2.20: Acyl-CoA hydrolase
EC 3.1.3: phosphoric monoester hydrolases	EC 3.1.3.1: Alkaline phosphatase (*ALP*)
	EC 3.1.3.2: Acid phosphatase (*ACP*)
EC 3.1.6: sulfuric ester hydrolases	EC 3.1.6.1: Arylsulfatase
EC 3.1.8: phosphoric triester hydrolases	EC 3.1.8.1: Paraoxonase (*PON*) PON1, PON2, PON3
	EC 3.1.8.2: Diisopropyl-fluorophosphatase
EC 3.2: glycosylases	EC 3.2.1.31: β-Glucuronidase (*GUSB*)
EC 3.3.2: ether hydrolases	EC 3.3.2.9: Microsomal epoxide hydrolase (*EPHX1*) mEH
	EC 3.3.2.10: Soluble epoxide hydrolase (*EPHX2*) sEH
EC 3.4.11: aminopeptidases	EC 3.4.11.1: Leucyl aminopeptidase (*LAP*)
EC 3.4.13 and 3.4.14: peptidases acting on di- and tripeptides	EC 3.4.14.5: Dipeptidyl-peptidase IV (*DPP4*)
EC 3.4.16 to 3.4.18: carboxypeptidases	EC 3.4.16.2: Lysosomal Pro-Xaa carboxypeptidase
	EC 3.4.17.1: Carboxypeptidase A (*CPA*)
EC 3.4.21 to 3.4.25: endopeptidases	EC 3.4.21.1: Chymotrypsin (*CTRB*)
	EC 3.4.24.15: Thimet oligopeptidase (*THOP*)
EC 3.5.1: hydrolases acting on linear amides	EC 3.5.1.4: Amidase
	EC 3.5.1.39: Alkylamidase
EC 3.5.2: hydrolases acting on cyclic amides	EC 3.5.2.1: Barbiturase
	EC 3.5.2.2: Dihydropyrimidinase (*DPYS*)
	EC 3.5.2.6: β-Lactamase

zymes among the serine hydrolases include carboxylesterases, arylesterases, cholinesterase, and a number of serine endopeptidases (EC 3.4.21). The role of arylsulfatases, paraoxonase, β-glucuronidase, and epoxide hydrolases is worth noting. Some metalloendopeptidases (EC 3.4.24) and amidases (EC 3.5.1 and 3.5.2) are also of interest [9,14,19,23].

2.3. Reactions of Carbon Oxidation and Reduction

When examining reactions of carbon oxidation (oxygenations and dehydrogenations) and carbon reduction (hydrogenations), it is convenient from a mechanistic viewpoint to distinguish between sp^3-, sp^2-, and sp-carbon atoms.

2.3.1. sp^3-Carbon Atoms Reactions of oxidation and reduction of sp^3-carbon atoms (plus some subsequent reactions at carbonyl groups) are schematized in Fig. 3 and discussed sequentially below. In the simplest cases, a nonactivated carbon atom in an alkyl group undergoes cytochrome P450-catalyzed hydroxylation. The penultimate position is a preferred site of attack (reaction 1b), but hydroxylation can also occur in the terminal position (reaction 1a) or in other positions in case of steric hindrance or with some specific cytochromes P450. Dehydrogenation by dehydrogenases can then yield a carbonyl derivative (reactions 1c and 1e) that is either an aldehyde or a ketone. Note that reactions 1c and 1e act not only on metabolites, but also on xenobiotic alcohols, and are reversible (i.e., reactions 1d and 1f) since dehydrogenases catalyze the reactions in both directions. And while a ketone is very seldom oxidized further, aldehydes are good substrates for aldehyde dehydrogenases or other enzymes, and lead irreversibly to carboxylic acid metabolites (reaction 1g). A classical example is that of ethanol, which in the body exists in redox equilibrium with acetaldehyde, this metabolite being rapidly and irreversibly oxidized to acetic acid.

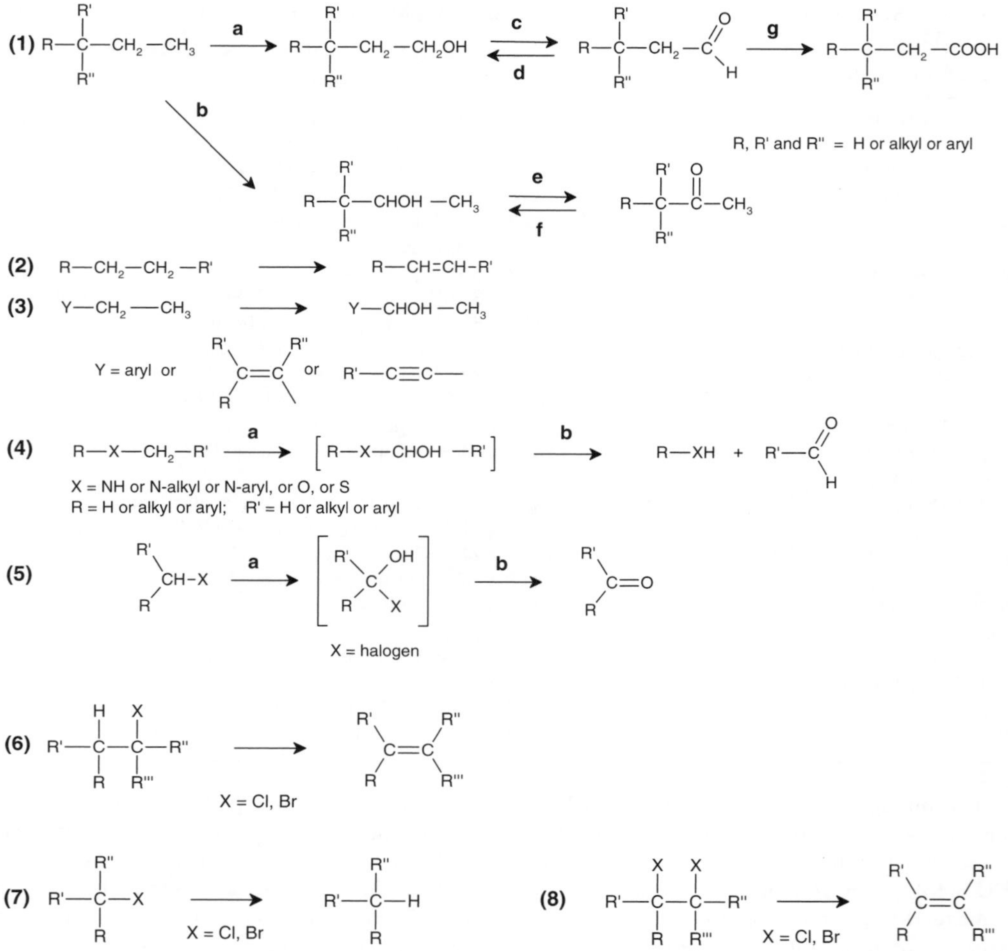

(1) $R{-}C(R')(R''){-}CH_2{-}CH_3$ —a→ $R{-}C(R')(R''){-}CH_2{-}CH_2OH$ ⇄(c/d) $R{-}C(R')(R''){-}CH_2{-}CHO$ —g→ $R{-}C(R')(R''){-}CH_2{-}COOH$

—b→ $R{-}C(R')(R''){-}CHOH{-}CH_3$ ⇄(e/f) $R{-}C(R')(R''){-}CO{-}CH_3$

R, R' and R'' = H or alkyl or aryl

(2) $R{-}CH_2{-}CH_2{-}R'$ ⟶ $R{-}CH{=}CH{-}R'$

(3) $Y{-}CH_2{-}CH_3$ ⟶ $Y{-}CHOH{-}CH_3$

Y = aryl or $(R')(R)C{=}C(R''){-}$ or $R'{-}C{\equiv}C{-}$

(4) $R{-}X{-}CH_2{-}R'$ —a→ $[R{-}X{-}CHOH{-}R']$ —b→ $R{-}XH + R'{-}CHO$

X = NH or N-alkyl or N-aryl, or O, or S
R = H or alkyl or aryl; R' = H or alkyl or aryl

(5) $(R')(R)CH{-}X$ —a→ $[(R')(R)C(OH)(X)]$ —b→ $(R')(R)C{=}O$

X = halogen

(6) $R'{-}C(H)(R){-}C(X)(R'''){-}R''$ ⟶ $(R')(R)C{=}C(R'')(R''')$

X = Cl, Br

(7) $R'{-}C(R'')(R){-}X$ ⟶ $R'{-}C(R'')(R){-}H$

X = Cl, Br

(8) $R'{-}C(X)(R){-}C(X)(R'''){-}R''$ ⟶ $(R')(R)C{=}C(R'')(R''')$

X = Cl, Br

Figure 3. Major functionalization reactions involving an sp^3-carbon in substrate molecules; some reactions at carbonyl groups are also included since they often follows the former. The reactions shown here are mainly oxidations (oxygenations and dehydrogenations) and reductions (hydrogenations), plus some postenzymatic reactions of hydrolytic cleavage.

For a number of substrates, the oxidation of primary and secondary alcohol and of aldehyde groups can also be catalyzed by cytochrome P450. A typical example is the C(10)-demethylation of androgens and analogs catalyzed by aromatase (CYP19).

A special case of carbon oxidation, recognized only recently and probably of underestimated significance, is desaturation of a dimethylene unit by cytochrome P450 to produce an olefinic group (reaction 2). An interesting example is provided by testosterone, which among many cytochrome P450-catalyzed reactions undergoes allylic hydroxylation to 6β-hydroxytestosterone and desaturation to 6,7-dehydrotestosterone [38].

There is a known regioselectivity in cytochrome P450-catalyzed hydroxylations for carbon atoms adjacent (alpha) to an unsaturated system (reaction 3) or a heteroatom such as N, O, or S (reaction 4a). In the former cases, hydroxylation can easily be followed by dehydrogenation (not shown). In the latter cases, however, the hydroxylated metabolite is usually unstable and undergoes a rapid, postenzymatic elimination (reaction 4b). Depending on the substrate, this pathway produces a secondary or primary amine, an alcohol or phenol, or a thiol, while the alkyl group is cleaved as an aldehyde or a ketone. Reactions 4 constitute a very common and frequent pathway as far as drug metabolism is concerned, since it underlies some well-known metabolic reactions of N–C cleavage discussed later. Note that the actual mechanism of such reactions is usually more complex than shown here and may involve intermediate oxidation of the heteroatom.

Aliphatic carbon atoms bearing one or more halogen atoms (mainly chlorine or bromine) can be similarly metabolized by hydroxylation and loss of HX to dehalogenated products (reactions 5a and 5b) (see later). Dehalogenation reactions can also proceed reductively or without change in the state of oxidation. The latter reactions are dehydrohalogenations (usually dehydrochlorination or dehydrobromination) occurring nonenzymatically (reaction 6). Reductive dehalogenations involve replacement of a halogen by a hydrogen (reaction 7), or *vic*-bisdehalogenation (reaction 8). Some radical species formed as intermediates may have toxicological significance.

Reactions 1a, 1b, 3, 4a, and 5a are catalyzed by cytochromes P450. Here, the iron-bound oxene (Section 2.2.1) acts by a mechanism known as "oxygen rebound" whereby a H atom is exchanged for a OH group. In simplified terms, the oxene atom attacks the substrate by cleaving a C–H bond and removing the hydrogen atom (hydrogen radical). This forms an iron-bound $HO^{\bullet}$ species and leaves the substrate as a C-centered radical. In the last step, the iron-bound $HO^{\bullet}$ species is transferred to the substrate.

Halothane (**2**) offers a telling example of the metabolic fate of halogenated compounds of medicinal interest. Indeed, this agent undergoes two major pathways, oxidative dehalogenation leading to trifluoroacetic acid (**3**) and reduction producing a reactive radical (**4**) (Fig. 4).

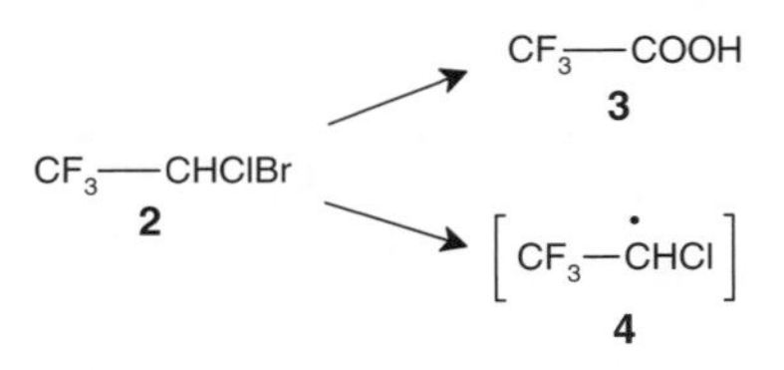

Figure 4. The structure of halothane (**2**) and two of its metabolites, namely trifluoroacetic acid (**3**) produced by oxidation, and a reactive radical (**4**) produced by reduction.

2.3.2. sp^2- and sp-Carbon Atoms Reactions at sp^2-carbons are characterized by their own pathways, catalytic mechanisms, and products (Fig. 5). Thus, the oxidation of aromatic rings generates a variety of (usually stable) metabolites. Their common precursor is often a reactive epoxide (reaction 1a) that can either be hydrolyzed by epoxide hydrolase (reaction 1b) to a dihydrodiol, or rearrange under proton catalysis to a phenol (reaction 1c). The production of a phenol is a very common metabolic reaction for drugs containing one or more aromatic rings. The *para*-position is the preferred position of hydroxylation for unsubstituted phenyl rings, but the regioselectivity of the reaction becomes more complex with substituted phenyl or with other aromatic rings.

Dihydrodiols are seldom observed, as are catechol metabolites produced by their dehydrogenation catalyzed by dihydrodiol dehydrogenase (reaction 1d). It is interesting to note that this reaction restores the aromaticity that had been lost upon epoxide formation. The further oxidation of phenols and phenolic metabolites is also possible, the rate of reaction and the nature of products depending on the ring and on the nature and position of its substituents. Catechols are thus formed by reaction 1e, while hydroquinones are sometimes also produced (reaction 1f).

In some cases, catechols and hydroquinones have been found to undergo further oxidation to quinones (reactions 1g and 1i). Such reactions occur by two single-electron steps and can be either enzymatic or nonenzymatic (i.e., resulting from autoxidation and yielding as byproduct the superoxide anion-radical $O_2^{\bullet}-$). The intermediate is this reaction is a semiquinone. Both quinones and semiquinones are reactive, in particular toward biomolecules, and have been implicated in many toxification reactions. For example, the high toxicity of benzene for bone marrow is believed to be due to the oxidation of catechol and hydroquinone catalyzed by myeloper-oxidase.

The oxidation of diphenols to quinones is reversible (reactions 1h and 1j), a variety of

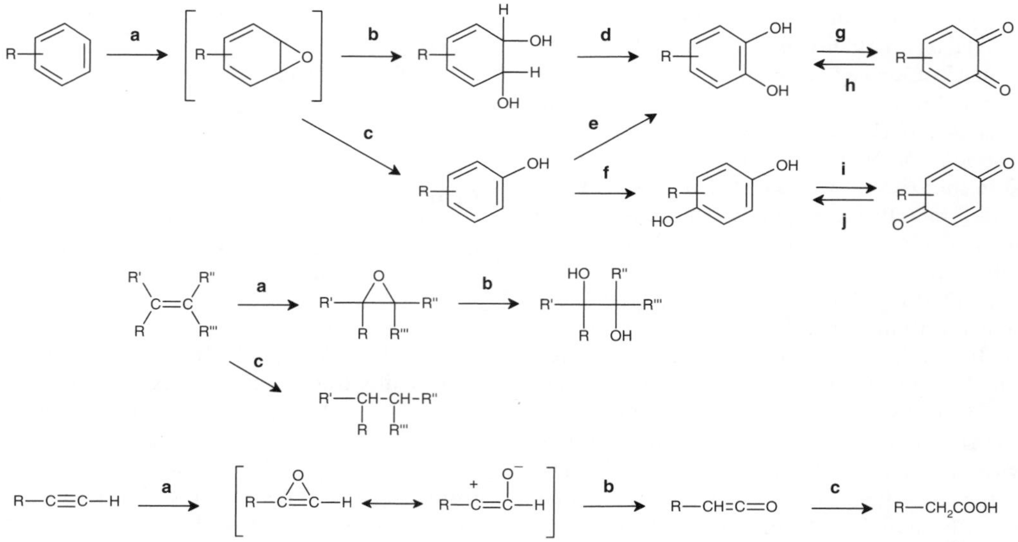

Figure 5. Major functionalization reactions involving an sp^2- or sp-carbon in substrate molecules. These reactions are oxidations (oxygenations and dehydrogenations), reductions (hydrogenations), and hydrations, plus some postenzymatic rearrangements.

cellular reductants being able to mediate the reduction of quinones either by a two-electron mechanism or by two single-electron steps. The two-electron reduction can be catalyzed by carbonyl reductase and quinone reductase, while cytochrome P450 and some flavoproteins act by single-electron transfers. The nonenzymatic reduction of quinones can occur for example in the presence of $O_2^{\bullet}-$ or some thiols such as glutathione.

Olefinic bonds in xenobiotic molecules can also be targets of cytochrome P450-catalyzed epoxidation (reaction 2a). In contrast to arene oxides, the resulting epoxides are fairly stable and can be isolated and characterized. But like arene oxides, they are substrates of epoxide hydrolase to yield dihydrodiols (reaction 2b).

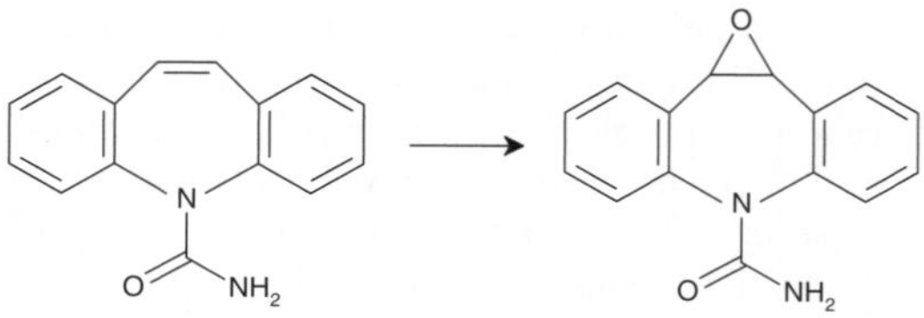

Figure 6. The structure of carbamazepine (**5**) and its 10,11-epoxide metabolite (**6**).

This is exemplified by carbamazepine (**5**), whose 10,11-epoxide (**6**) is a major and pharmacologically active metabolite in humans, and is further metabolized to the inactive dihydrodiol [39] (Fig. 6).

The reduction of olefinic groups (reaction 2c) is documented for a few drugs bearing an α,β-ketoalkene function. The reaction is thought to be catalyzed by various NAD(P)H oxidoreductases.

The few drugs that contain an acetylenic moiety are also targets for cytochrome P450-catalyzed oxidation. Oxygenation of the triple bond (reaction 3a) yields an intermediate that depending on the substrate can react in a number of ways, for example binding covalently to the enzyme, or forming a highly reactive ketene whose hydration produces a substituted acetic acid (reactions 3b and 3c).

2.4. Reactions of Nitrogen Oxidation and Reduction

The main metabolic reactions of oxidation and reduction of nitrogen atoms in organic molecules are summarized in Fig. 7. The functional groups involved are amines and amides and their oxygenated metabolites, as well as 1,4-

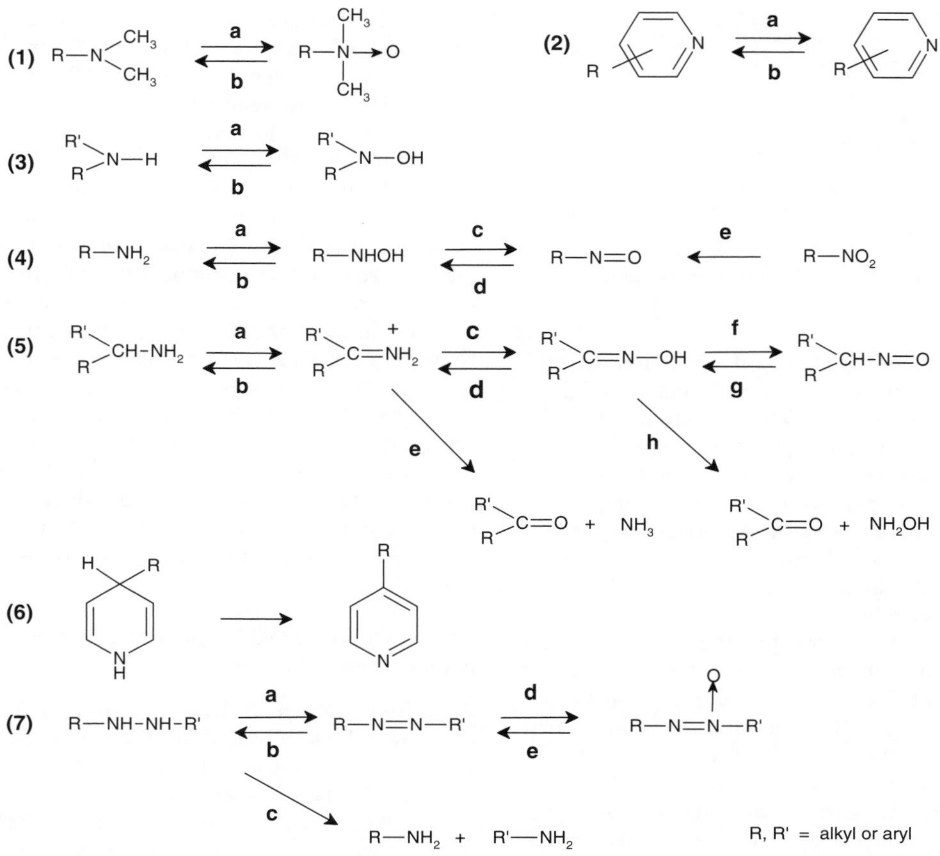

(1) $R{-}N(CH_3)_2 \rightleftharpoons R{-}N(CH_3)_2{\rightarrow}O$ (a, b)

(2) R-pyridine ⇌ R-pyridine N→O (a, b)

(3) $R'(R)N{-}H \rightleftharpoons R'(R)N{-}OH$ (a, b)

(4) $R{-}NH_2 \rightleftharpoons R{-}NHOH \rightleftharpoons R{-}N{=}O \leftarrow R{-}NO_2$ (a, b; c, d; e)

(5) $R'(R)CH{-}NH_2 \rightleftharpoons R'(R)C{=}NH_2^+ \rightleftharpoons R'(R)C{=}N{-}OH \rightleftharpoons R'(R)CH{-}N{=}O$ (a, b; c, d; f, g); e: $\rightarrow R'(R)C{=}O + NH_3$; h: $\rightarrow R'(R)C{=}O + NH_2OH$

(6) 4-R-1,4-dihydropyridine → 4-R-pyridine

(7) $R{-}NH{-}NH{-}R' \rightleftharpoons R{-}N{=}N{-}R' \rightleftharpoons R{-}N{=}N(\rightarrow O){-}R'$ (a, b; d, e); c: $\rightarrow R{-}NH_2 + R'{-}NH_2$

R, R' = alkyl or aryl

Figure 7. Major functionalization reactions involving nitrogen atoms in substrate molecules. The reactions shown here are mainly oxidations (oxygenations and dehydrogenations) and reductions (deoxygenations and hydrogenations).

dihydropyridines, hydrazines, and azo compounds. In many cases, the reactions can be catalyzed by cytochrome P450 and/or flavin-containing monooxygenases. The first oxygenation step in reactions 1–4 and 6 have frequently been observed.

Nitrogen oxygenation is an (apparently) straightforward metabolic reaction of tertiary amines (reaction 1a), be they aliphatic or aromatic. Numerous drugs undergo this reaction, the resulting *N*-oxide metabolite being more polar and hydrophilic than the parent compound. Identical considerations apply to pyridines and analogous aromatic azaheterocycles (reaction 2a). Note that these reactions are reversible, a number of reductases being able to deoxygenate *N*-oxides back to the amine (i.e., reactions 1b and 2b).

Secondary and primary amines also undergo *N*-oxygenation, the first isolable metabolites being hydroxylamines (reactions 3a and 4a, respectively). Again, reversibility is documented (reactions 3b and 4b). These compounds can be aliphatic or aromatic amines, and the same metabolic pathway occurs in secondary and primary amides (i.e., R = acyl), while tertiary amides appear resistant to *N*-oxygenation. The oxidation of secondary amines and amides usually stops at the hydroxylamine/hydroxylamide level, but formation of short-lived nitroxides (not shown) has been reported.

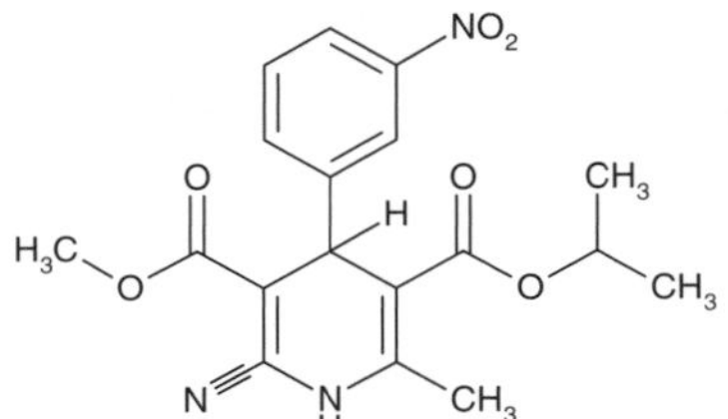

Figure 8. The structure of nivaldipine (7).

As opposed to secondary amines and amides, their primary analogs can be oxidized to nitroso metabolites (reaction 4c), but further oxidation of the latter compounds to nitro compounds does not seem to occur *in vivo*. In contrast, aromatic nitro compounds can be reduced to primary amines via reactions 4e, 4d, and finally 4b. This is the case for numerous chemotherapeutic drugs such as metronidazole.

Note that primary aliphatic amines having a hydrogen on the alpha-carbon can display additional metabolic reactions shown as reactions 5 in Fig. 7. Indeed, *N*-oxidation may also yield imines (reaction 5a), whose degree of oxidation is equivalent to that of hydroxylamines [40]. Imines can be further oxidized to oximes (reaction 5c), which are in equilibrium with their nitroso tautomer (reactions 5f and 5g).

1,4-Dihydropyridines, and particularly calcium channel blockers such as nivaldipine (**7** in Fig. 8), are effectively oxidized by cytochrome P450. The reaction is one of aromatization (reaction 6 in Fig. 7), yielding the corresponding pyridine.

Dinitrogen moieties are also targets of oxidoreductases. Depending on their substituents, hydrazines are oxidized to azo compounds (reaction 7a), some of which can be oxygenated to azoxy compounds (reaction 7d). Another important pathway of hydrazines is their reductive cleavage to primary amines (reaction 7c). Reactions 7a and 7d are reversible, the corresponding reductions (reactions 7b and 7e) being mediated by cytochrome P450 and other reductases. A toxicologically significant pathway thus exists for the reduction of some aromatic azo compounds to potentially toxic primary aromatic amines (reactions 7b and 7c).

2.5. Reactions of Oxidation and Reduction of Sulfur and Other Atoms

A limited number of drugs contain a sulfur atom, usually as a thioether. The major redox reactions occurring at sulfur atoms in organic compounds are summarized in Fig. 9.

Thiol compounds can be oxidized to sulfenic acids (reaction 1a), then to sulfinic acids (reaction 1e), and finally to sulfonic acids (re-

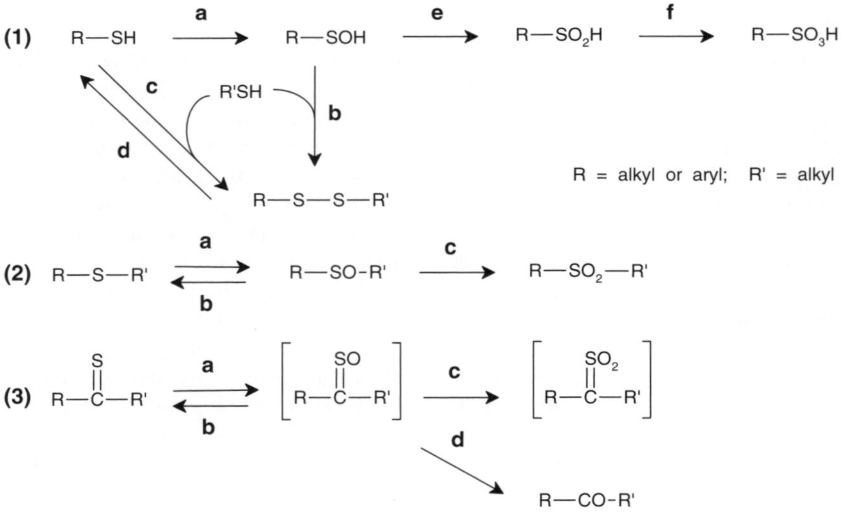

Figure 9. Major reactions of oxidation and reduction involving sulfur atoms in organic compounds.

action 1f). Depending on the substrate, the pathway is mediated by cytochrome P450 and/or flavin-containing monooxygenases. Another route of oxidation of thiols is to disulfides either directly (reaction 1c via thiyl radicals), or by dehydration between a thiol and a sulfenic acid (reaction 1b). However, our understanding of sulfur biochemistry is incomplete, and much remains to be learned. This is particularly true for reductive reactions. While reaction 1c is well-known to be reversible (i.e., reaction 1d), the reversibility of reaction 1a is unclear, while reduction of sulfinic and sulfonic acids appears unlikely.

The metabolism of sulfides (thioethers) is rather straightforward. Besides *S*-dealkylation reactions discussed earlier, these compounds can also be oxygenated by monooxygenases to sulfoxides (reaction 2a) and then to sulfones (reaction 2c). Here, it is known with confidence that reaction 2a is indeed reversible, as documented by many examples of reduction of sulfoxides (reaction 2b, while the reduction of sulfones has never been found to occur.

Thiocarbonyl compounds are also substrates of monooxygenases, forming *S*-monoxides (sulfines, reaction 3a) and then *S*-dioxides (sulfenes, reaction 3c). As a rule, these metabolites cannot be identified as such due to their reactivity. Thus, *S*-monoxides rearrange to the corresponding carbonyl by expelling a sulfur atom (reaction 3d). This reaction is known as oxidative desulfuration and occurs in thioamides and thioureas (e.g., thiopental). As for the *S*-dioxides, they react very rapidly with nucleophiles, and particularly with nucleophilic sites in biological macromolecules. This covalent binding results in the formation of adducts of toxicological significance. Such a mechanism is believed to account for the carcinogenicity of a number of thioamides.

Some other elements besides carbon, nitrogen, and sulfur can undergo metabolic redox reactions. The direct oxidation of oxygen atoms in phenols and alcohols is well documented for some substrates. Thus, the oxidation of secondary alcohols by some peroxidases can yield a hydroperoxide and ultimately a ketone. Some phenols are known to be oxidized by cytochrome P450 to a semiquinone and ultimately to a quinone. A classical example is that of the antiinflammatory drug

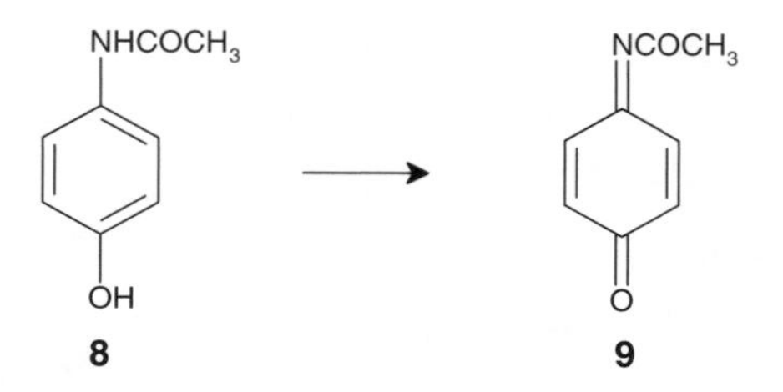

Figure 10. The structure of paracetamol (**8**) and its toxic quinoneimine metabolite (**9**).

paracetamol (**8** in Fig. 10, acetaminophen), a minor fraction of which is oxidized by CYP2E1 to the highly reactive and toxic quinoneimine **9**.

Additional elements of limited significance in medicinal chemistry able to enter redox reactions are silicon, phosphorus, arsenic, and selenium, among others (Fig. 11). Note however that the enzymology and mechanisms of these reactions are insufficiently understood. For example, a few silanes have been shown to yield silanols *in vivo* (reaction 1). The same applies to some phosphines that can be oxygenated to phosphine oxides by monooxygenases (reaction 2).

Arsenicals have received some attention due to their therapeutic significance. Both inorganic and organic arsenic compounds display an As(III)–As(V) redox equilibrium in the body. This is illustrated with the arsine-arsine oxide and arsenoxide–arsonic acid equilibria (reactions 3a and 3b and reactions 4b and 4c, respectively). Another reaction of interest is the oxidation of arseno compounds to arsenoxides (reaction 4a), a reaction of importance in the bioactivation of a number of chemotherapeutic arsenicals.

The biochemistry of organoselenium compounds is of some interest. For example, a few selenols have been seen to be oxidized to selenenic acids (reaction 5a) and then to seleninic acids (reaction 5b).

2.6. Reactions of Oxidative Cleavage

A number of oxidative reactions presented in the previous sections yield metabolic intermediates that readily undergo postenzymatic cleavage of a C–X bond (X being an heteroatom). As briefly mentioned, reactions 4a and

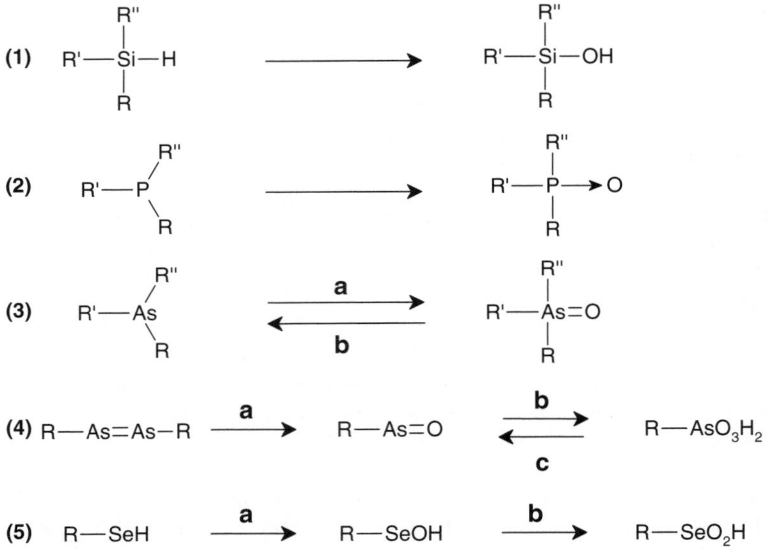

(1) R'—Si(R'')(R)—H ⟶ R'—Si(R'')(R)—OH

(2) R'—P(R'')(R) ⟶ R'—P(R'')(R)→O

(3) R'—As(R'')(R) ⇌(a, b) R'—As(R'')(R)=O

(4) R—As=As—R ⟶(a) R—As=O ⇌(b, c) R—AsO_3H_2

(5) R—SeH ⟶(a) R—SeOH ⟶(b) R—SeO_2H

Figure 11. Some selected reactions of oxidation and reduction involving silicon, phosphorus, arsenic and selenium in xenobiotic compounds.

4b in Fig. 3 represent important metabolic pathways that affect many drugs. When X = N (by far the most frequent case), the metabolic reactions are known as *N*-demethylations, *N*-dealkylations, or deaminations, depending on the moiety being cleaved. This is aptly exemplified by the metabolic fate of fenfluramine (**10** in Fig. 12). This withdrawn drug undergoes *N*-deethylation to norfenfluramine (**11**), an active metabolite, and deamination to (*m*-trifluoromethyl)phenylacetone (**12**), an inactive metabolite that is further oxidized to *m*-trifluoromethylbenzoic acid (**13**).

When X = O or S in reactions 4 (Fig. 3), the metabolic reactions are known as *O*-dealkylations or *S*-dealkylations, respectively. *O*-Demethylations are a typical case of the former reaction. And when X = halogen in reactions

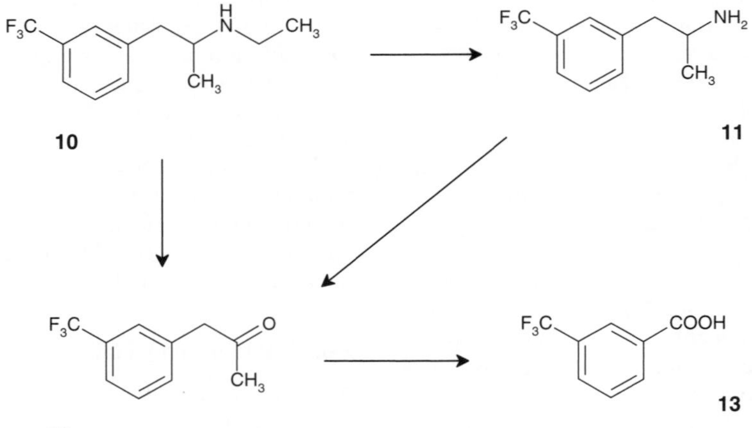

F_3C H N CH_3 CH_3 **10** ⟶ F_3C NH_2 CH_3 **11**

F_3C O CH_3 **12** ⟶ F_3C COOH **13**

Figure 12. Fenfluramine (**10**), norfenfluramine (**11**), (*m*-trifluoromethyl)phenylacetone (**12**), and *m*-trifluoromethylbenzoic acid (**13**).

(1) R—COO—R' ⟶ R—COOH + R'—OH

(2) R—ONO_2 ⟶ R—OH + HNO_3

(3) R—OSO_3H ⟶ R—OH + H_2SO_4

(4) R—CONHR' ⟶ R—COOH + R'—NH_2

Figure 13. Major hydrolysis reactions involving esters (organic and inorganic) and amides.

5a and 5b (Fig. 3), loss of halogen can also occur and is known as oxidative dehalogenation.

The reactions of oxidative C–X cleavage discussed above result from carbon hydroxylation and are catalyzed by cytochrome P450. However, *N*-oxidation reactions followed by hydrolytic C–N cleavage can also be catalyzed by cytochrome P450 (e.g., reactions 5e and 5h in Fig. 7). The sequence of reactions 5a and 5e in Fig. 7 is of particular interest since it is the mechanism by which monoamine oxidase deaminates endogenous and exogenous amines.

2.7. Reactions of Hydration and Hydrolysis

Hydrolases catalyze the addition of a molecule of water to a variety of functional moieties [9,14,23]. Thus, epoxide hydrolase hydrates epoxides to yield *trans*-dihydrodiols (reaction 1b in Fig. 5). This reaction is documented for many arene oxides, in particular metabolites of aromatic compounds, and epoxides of olefins. Here, a molecule of water has been added to the substrate without loss of a molecular fragment, hence the use of the term "hydration" sometimes found in the literature.

Reactions of hydrolytic cleavage (hydrolysis) are shown in Fig. 13. They are frequent for organic esters (reaction 1), inorganic esters such as nitrates (reaction 2) and sulfates (reaction 3), and amides (reaction 4). These reactions are catalyzed by esterases, peptidases, or other enzymes, but nonenzymatic hydrolysis is also known to occur for sufficiently labile compounds under biological conditions of pH and temperature. Acetylsalicylic acid, glycerol trinitrate, and lidocaine are three representative examples of drugs undergoing extensive cleavage of the organic ester, inorganic ester, or amide group, respectively. The reaction is of particular significance in the activation of ester prodrugs.

3. CONJUGATION REACTIONS

3.1. Introduction

As defined in Section 1, conjugation reactions (also infelicitously known as phase II reactions) result in the covalent binding of an endogenous molecule or moiety to the substrate. Such reactions are of critical significance in the metabolism of endogenous compounds, witness the impressive battery of enzymes that have evolved to catalyze them. Conjugation is also of great importance in the biotransformation of xenobiotics, involving parents compounds or metabolites thereof [15,19,24].

Conjugation reactions are characterized by a number of criteria:

1. The substrate is coupled to an endogenous molecule sometimes designated as the endocon ...
2. ... which is usually polar ...
3. ... of medium molecular weight (ca. 100–300 Da) ...
4. ... and carried by a cofactor.
5. The reaction is catalyzed by an enzyme known as a transferase (Table 7).

First and above all, an endogenous molecule (called the endogenous conjugating moiety, and sometimes abbreviated as the "endocon") is coupled to the substrate. This

Table 7. A Survey of Transferases (EC 2) [15,19,24,32]

Methyltransferases (EC 2.1.1)	
S-Adenosyl-L-methionine (SAM)	EC 2.1.1.6: catechol *O*-methyltransferase (*COMT*) EC 2.1.1.1: nicotinamide *N*-methyltransferase (*NNMT*) EC 2.1.1.8: histamine *N*-methyltransferase (*HNMT*) EC 2.1.1.28: noradrenaline *N*-methyltransferase (*PNMT*) EC 2.1.1.49: arylamine *N*-methyltransferase, indolethylamine *N*-methyltransferase (*INMT*) EC 2.1.1.9: Thiol *S*-methyltransferase (TMT) EC 2.1.1.67: Thiopurine *S*-methyltransferase (*TPMT*)
Sulfotransferases (EC 2.8.2) (*SULT*)	
3′-Phosphoadenosine 5′-phosphosulfate (PAPS)	EC 2.8.2.1: aryl sulfotransferase (SULT1A1, 1A2, and 1A3) Thyroid hormone sulfotransferase (SULT1B1) SULT1C1, and SULT1C2 EC 2.8.2.4: estrogen sulfotransferase (SULT1E1) EC 2.8.2.14: alcohol/hydroxysteroid sulfotransferase (SULT2A1) EC 2.8.2.2: hydroxysteroid sulfotransferase (SULT2B1a and 2B1b) EC 2.8.2.15: steroid sulfotransferase EC 2.8.2.18: cortisol sulfotransferase EC 2.8.2.3: amine sulfotransferase (SULT3A1)
UDP-Glucuronosyltransferases (2.4.1.17) (*UGT*)	
Uridine-5′-diphospho-α-D-glucuronic acid (UDPGA)	Subfamily UGT1: UGT1A1, 1A3, 1A4 to 1A10 Subfamily UGT2A: UGT2A1 to 2A3 Subfamily UGT2B: UGT2B4, 2B7, 2B10, 2B11, 2B15, 2B17, 2B28 Subfamily UGT3A: UGT3A1, 3A2 Subfamily UGT8A: UGT8
Acetyltransferases	
Acetylcoenzyme A (AcCoA)	EC 2.3.1.5: *N*-acetyltransferase (*NAT*) NAT1 and NAT2 EC 2.3.1.56: aromatic-hydroxylamine *O*-acetyltransferase EC 2.3.1.118: *N*-hydroxyarylamine *O*-acetyltransferase
Acyl-CoA synthetases	
Coenzyme A (CoA)	EC 6.2.1.1: short-chain fatty acyl-CoA synthetase (*ACSS*) EC 6.2.1.2: medium-chain acyl-CoA synthetase EC 6.2.1.3: long-chain acyl-CoA synthetase (*ACSL*) EC 6.2.1.7: cholate-CoA ligase EC 6.2.1.25: benzoyl-CoA synthetase
Acyltransferases	
Xenobiotic acyl-Coenzyme A	EC 2.3.1.13: glycine *N*-acyltransferase (*GLYAT*) EC 2.3.1.71: glycine *N*-benzoyltransferase EC 2.3.1.14: glutamine *N*-phenylacetyltransferase EC 2.3.1.68: glutamine *N*-acyltransferase EC 2.3.1.65: cholyl-CoA glycine-taurine *N*-acyltransferase (*BAAT*) EC 2.3.1.20: diacylglycerol *O*-acyltransferase EC 2.3.1.22: 2-acylglycerol *O*-acyltransferase EC 2.3.1.26: sterol *O*-acyltransferase (*ACAT*)
Glutathione *S*-transferases (EC 2.5.1.18) (*GST*)	
(Glutathione)	Microsomal GST superfamily (homotrimers): *MGST*: MGST1 to MGST3 Cytoplasmic GST superfamily (homodimers, and a few heterodimers): *GSTA*: Alpha class, GST A1-1, A1-2, A2-2, A3-3, A4-4, A5-5 *GSTK*: Kappa class, GST K1-1 *GSTM*: Mu class, GST M1a-1a, M1a-1b, M1b-1b, M2-2, M3-3, M4-4, M5-5 *GSTO*: Omega class, GST O1-1, O2 *GSTP*: Pi class: GST P1-1 *GSTS*: Sigma class: GST S1 *GSTT*: Theta class, GST T1-1, T2 *GSTZ*: Zeta class, GST Z1-1

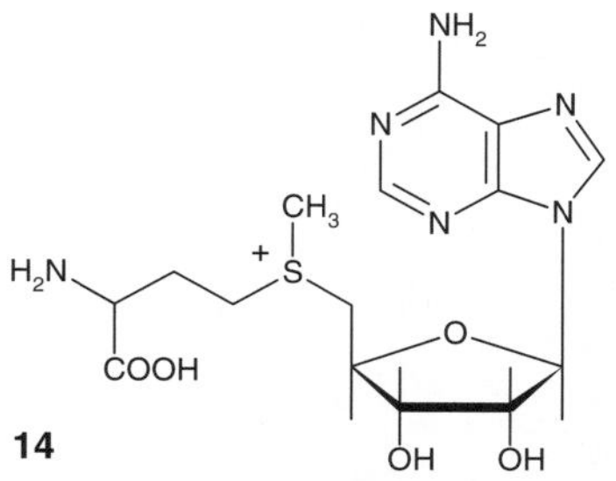

Figure 14. *S*-Adenosyl-L-methionine (**14**).

is the absolute criterion of conjugation reactions. Second, this endogenous molecule or moiety is generally polar (hydrophilic) or even highly polar, but there are exceptions. Third, the size of the endocon is generally in the range 100–300 Da. Fourth, the endogenous conjugating moiety is usually carried by a cofactor, with the chemical bond linking the cofactor and the endocon being a high energy one such that the Gibbs energy released upon its cleavage helps drive the transfer of the endocon to the substrate. Fifth, conjugation reactions are catalyzed by enzymes known as transferases (EC 2) that bind the substrate and the cofactor in such a manner that their close proximity allows the reaction to proceed. The metaphor of transferases being a "nuptial bed" has not escaped some biochemists. It is important from a biochemical and practical viewpoint to note that criteria 2 to 5 considered separately are neither sufficient nor necessary to define conjugations reactions. They are not sufficient, since in hydrogenation reactions (i.e., typical reactions of oxidoreduction) the hydride is also transferred from a cofactor (NADPH or NADH). And they are not necessary, since they all suffer from some important exceptions.

3.2. Methylation

3.2.1. Introduction Reactions of methylation imply the transfer of a methyl group from the cofactor *S*-adenosyl-L-methionine (**14**, SAM). As shown in Fig. 14, the methyl group in SAM is bound to a sulfonium center, giving it a marked electrophilic character and explaining its reactivity. Furthermore, it becomes pharmacokinetically relevant to distinguish methylated metabolites in which the positive charge has been retained or lost as a proton.

A number of methyltransferases are able to methylate small molecules (see Table 7) [15,19,24,41]. Thus, reactions of methylation fulfill only two of the three criteria defined above, since the methyl group is small compared to the substrate. The main enzyme responsible for *O*-methylation is catechol *O*-methyltransferase, which is mainly cytosolic but also exists in membrane-bound form. Several enzymes catalyze reactions of xenobiotic *N*-methylation with different substrate specificities, for example, nicotinamide *N*-methyltransferase, histamine methyltransferase, phenylethanolamine *N*-methyltransferase (noradrenaline *N*-methyltransferase), and the nonspecific arylamine *N*-methyltransferase. Reactions of xenobiotic *S*-methylation are mediated by the membrane-bound thiol methyltransferase and the cytosolic thiopurine methyltransferase.

The above classification of enzymes makes explicit the three types of functionalities undergoing biomethylation, namely hydroxyl (phenolic), amino and thiol groups.

3.2.2. Methylation Reactions Figure 15 summarizes the main methylation reactions seen in drug metabolism [15,19,24,41]. *O*-Methylation is a common reaction of compounds containing a catechol moiety (reaction 1), with a usual regioselectivity for the *meta* position. The substrates can be xenobiotics and particularly drugs, L-DOPA being a classic example. More frequently, however, *O*-methylation occurs as a late event in the metabolism of aryl groups, after they have been oxidized to catechols (reactions 1, Fig. 5). This sequence was seen for example in the metabolism of the antiinflammatory drug diclofenac (**15** in Fig. 16), which in humans yielded 3′-hydroxy-4′-methoxy-diclofenac as a major metabolite with a very long plasma half-life.

Three basic types of *N*-methylation reactions have been recognized (reactions 2–4, Fig. 15). A number of primary and secondary amines (e.g., some phenylethanolamines and tetrahydroisoquinolines) have been shown to be substrates of *N*-methyltransferase (reaction 2). However, such reactions are seldom of significance *in vivo*, presumably due to

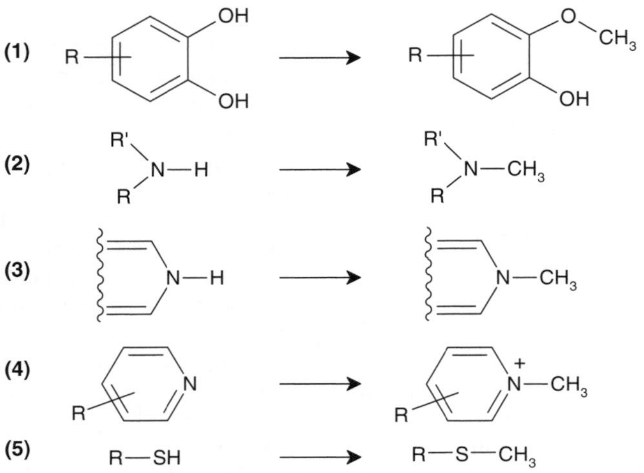

Figure 15. Major methylation reactions involving catechols, various amines, and thiols.

effective oxidative *N*-demethylation. A comparable situation involves the N–H group in an imidazole ring (reaction 3), as exemplified by histamine. A therapeutically relevant example is that of theophylline (**16**) whose *N* (9)-methylation is masked by *N*-demethylation in adult but not newborn humans.

N-Methylation of pyridine-type nitrogen atoms (reaction 4, Fig. 15) appears to be of greater *in vivo* pharmacological significance than reactions 2 and 3, and this for two reasons. First, the resulting metabolites, being quaternary amines, are more stable than tertiary or secondary amines toward *N*-demethylation. And second, these metabolites are also more polar than the parent compounds, in contrast to the products of reactions 2 and 3. Good substrates are nicotinamide (**17**), pyridine and a number of monocyclic and bicyclic derivatives.

S-Methylation of thiol groups (reaction 5) is documented for such drugs as captopril (**18**) and 6-mercaptopurine. Other substrates are metabolites (mainly thiophenols) resulting from the S–C cleavage of (aromatic) glutathione and cysteine conjugates (see below). Once formed, such methylthio metabolites can be further processed to sulfoxides and sulfones

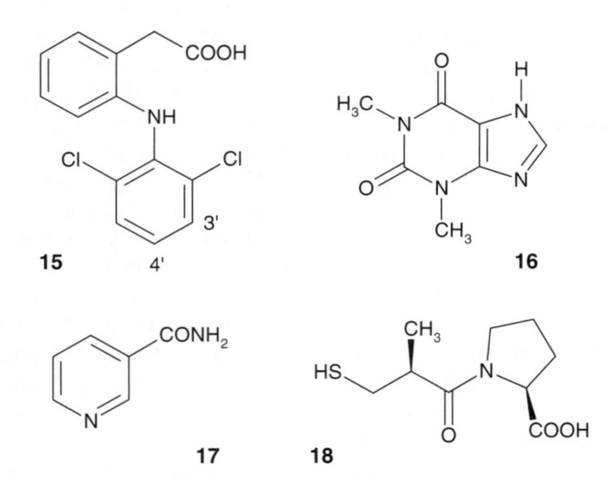

Figure 16. Diclofenac (**15**), theophylline (**16**), nicotinamide (**17**), and captopril (**18**).

before excretion (i.e., reaction 2a and 2c in Fig. 9).

From Fig. 15, it is apparent that methylation reactions can be subdivided into two classes:

(a) Those where the substrate and the product have the same electrical state, a proton in the substrate having been exchanged for a positively charged methyl group (reactions 1–3 and 5);

(b) Those where the product has acquired a positive charge, namely has become a pyridine-type quaternary ammonium (reaction 4).

3.3. Sulfonation

3.3.1. Introduction Sulfonation reactions consist in an SO_3 moiety being transferred from the cofactor 3′-phosphoadenosine 5′-phosphosulfate (**19**, PAPS) (Fig. 17) to the substrate under catalysis by a sulfotransferase [15,19,24,42,43]. The three criteria of conjugation are met in these reactions. Sulfotransferases, which catalyze a variety of physiological reactions, are soluble enzymes (see Table 7). The phenol sulfotransferases include aryl sulfotransferases, thyroid hormone sulfotransferase, SULT1C1 and SULT1C2, and estrogen sulfotransferase. The alcohol sulfotransferases include alcohol/hydroxysteroid sulfotransferase, hydroxysteroid sulfotransferase, steroid sulfotransferase, and cortisol sulfotransferase. There is also an amine sulfotransferase.

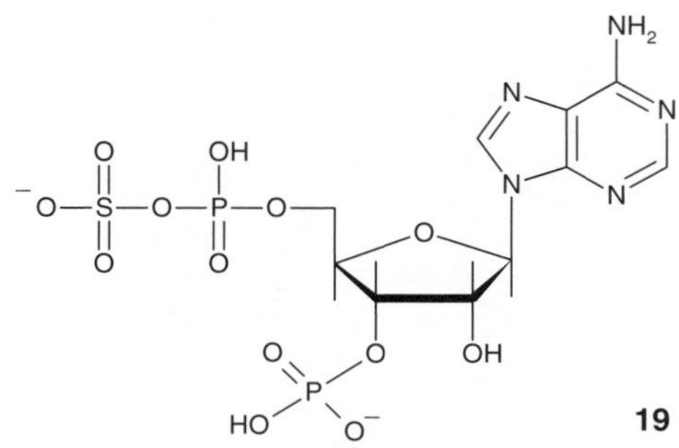

Figure 17. 3′-Phosphoadenosine 5′-phosphosulfate (**19**, PAPS).

The sulfate moiety in PAPS is linked to a phosphate group by an anhydride bridge whose cleavage is exothermic and supplies enthalpy to the reaction. The electrophilic –OH or –NH– site in the substrate will react with the leaving SO3 moiety, forming an ester sulfate or a sulfamate (Fig. 18). Some of these conjugates are unstable under biological conditions and will form electrophilic intermediates of considerable toxicological significance.

3.3.2. Sulfonation Reactions The sulfoconjugation of alcohols (reaction 1 in Fig. 18) leads

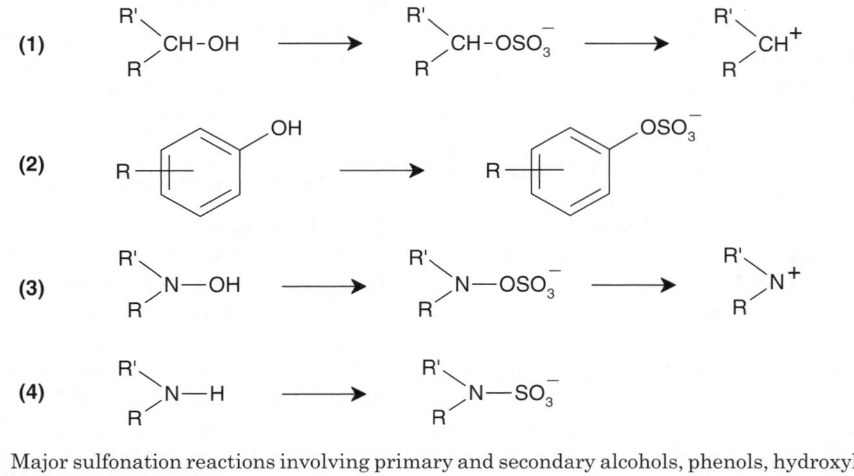

Figure 18. Major sulfonation reactions involving primary and secondary alcohols, phenols, hydroxylamines and hydroxylamides, and amines.

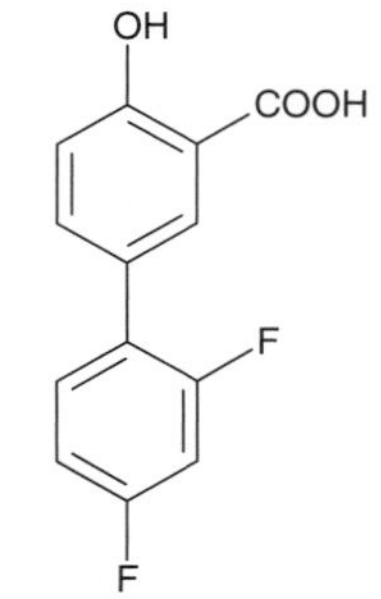

Figure 19. Diflunisal (**20**), minoxidil (**21**), and its *N,O*-sulfate ester (**22**).

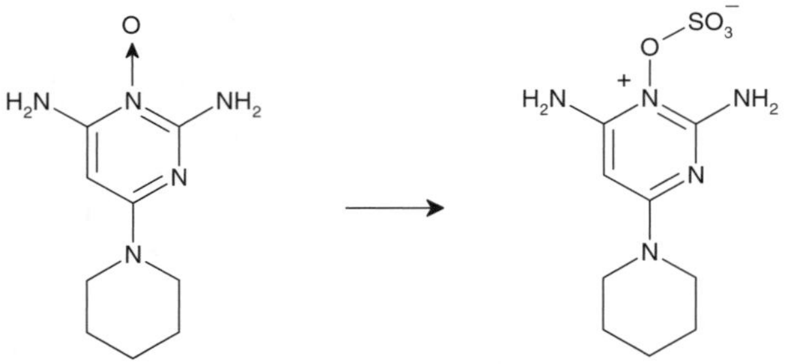

to metabolites of different stabilities. Endogenous hydroxysteroids (i.e., cyclic secondary alcohols) form relatively stable sulfates, while some secondary alcohol metabolites of allylbenzenes (e.g., safrole and estragole) may form genotoxic carbocations. Primary alcohols, for example, methanol and ethanol, can also form sulfates whose alkylating capacity is well known [54]. Similarly, polycyclic hydroxymethylarenes yield reactive sulfates believed to account for their carcinogenicity.

In contrast to alcohols, phenols form chemically stable sulfate esters (reaction 2). The reaction is usually of high affinity (i.e., rapid), but the limited availability of PAPS restricts the amounts of conjugate being produced. Typical drugs undergoing limited sulfonation include paracetamol (**8** in Fig. 10) and diflunisal (**20** in Fig. 19).

Aromatic hydroxylamines and hydroxylamides are good substrates for some sulfotransferases and yield unstable sulfate esters (reaction 3 in Fig. 18). Indeed, heterolytic N–O cleavage produces a highly electrophilic nitrenium ion. This is a mechanism believed to account for part or all of the cytotoxicity of arylamines and arylamides (e.g., phenacetin). In contrast, significantly more stable products are obtained upon *N*-sulfoconjugation of amines (reaction 4). A few primary, secondary, and alicyclic amines are known to yield sulfamates. The significance of these reactions in humans is still poorly understood.

An intriguing and very seldom reaction of conjugation occurs for minoxidil (**21** in Fig. 19), an hypotensive agent also producing hair growth. This drug is an *N*-oxide, and the actual active form responsible for the different therapeutic effects is the *N,O*-sulfate ester (**22**).

3.4. Glucuronidation and Glucosidation

3.4.1. Introduction Glucuronidation is a major and very frequent reaction of conjugation. It involves the transfer to the substrate of a molecule of glucuronic acid from the cofactor uridine-5′-diphospho-α-D-glucuronic acid (**23**, UDPGA) (Fig. 20). The enzymes catalyzing this reaction are known as UDP-glucuronosyltransferases and consist in a number of proteins coded by genes of the *UGT* superfamily (see Table 7). The human UDPGT known to

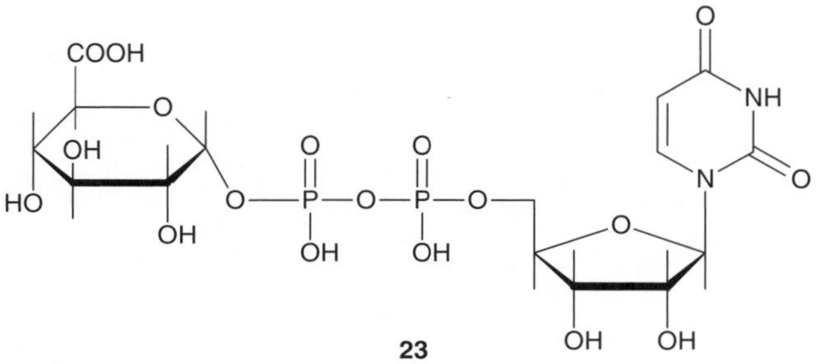

Figure 20. Uridine-5′-diphospho-α-D-glucuronic acid (**23**, UDPGA).

metabolize xenobiotics are the products of main two gene families, *UGT1* and *UGT2*. These enzymes include UGT1A1 (bilirubin UDPGTs) and several UGT1A, as well as numerous phenobarbital-inducible or constitutively expressed UGT2B [15,19,24,44,45].

In addition to glucuronidation, this section briefly mentions glucosidation, a minor metabolic pathway seen for a few drugs. These reactions are also catalyzed by UGT-glucuronosyltransferases.

3.4.2. Glucuronidation Reactions Glucuronic acid exists in UDPGA in the 1α-configuration, but the products of conjugation are β-glucuronides. This is due to the mechanism of the reaction being a nucleophilic substitution with inversion of configuration. Indeed, and as shown in Fig. 21, all functional groups able to undergo glucuronidation are nucleophiles, a common characteristic they share despite their great chemical variety. As a consequence of this diversity, the products of glucuronida-

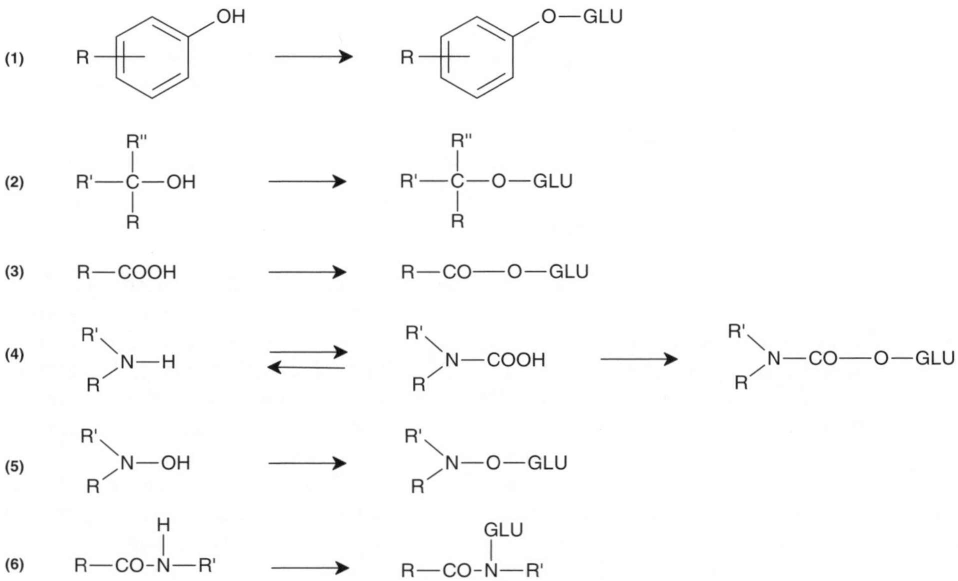

Figure 21. Major glucuronidation reactions involving phenols, alcohols, carboxylic acids, carbamic acids, hydroxylamines and hydroxylamides, carboxamides, sulfonamides, various amines, thiols, dithiocarboxylic acids, and 1,3-dicarbonyl compounds.

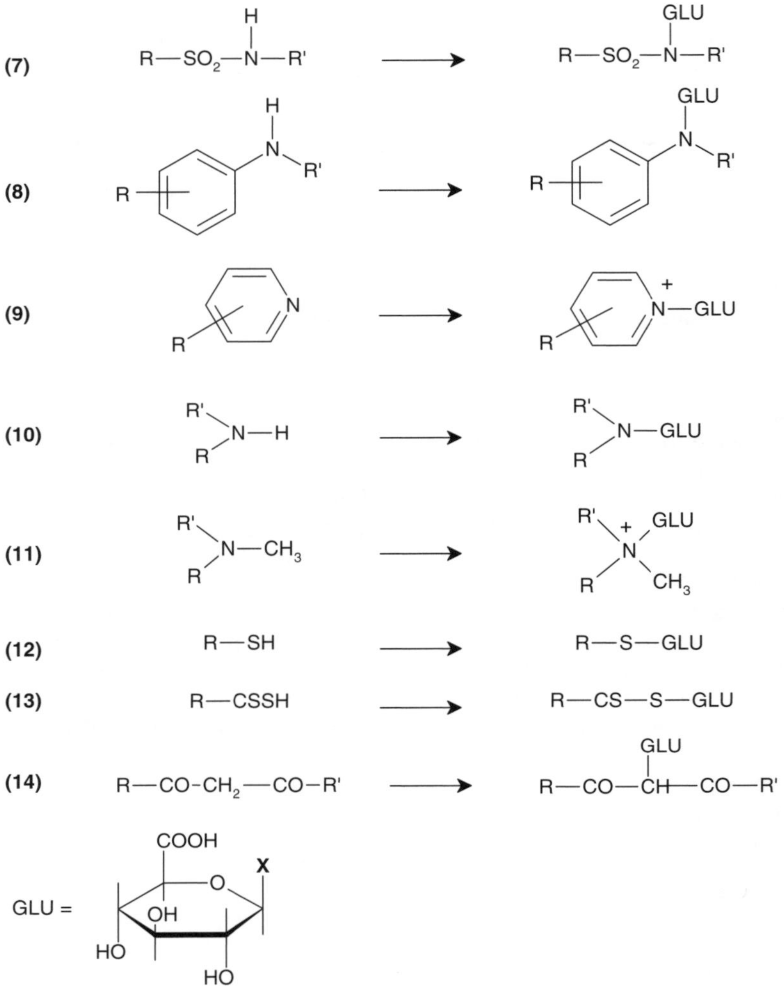

(7) $R-SO_2-N(H)-R'$ ⟶ $R-SO_2-N(GLU)-R'$

(8) R–C₆H₄–NH–R' ⟶ R–C₆H₄–N(GLU)–R'

(9) R–pyridine (N) ⟶ R–pyridinium N^+-GLU

(10) R(R')N–H ⟶ R(R')N–GLU

(11) $R(R')N-CH_3$ ⟶ $R(R')N^+(GLU)CH_3$

(12) R—SH ⟶ R—S—GLU

(13) R—CSSH ⟶ R—CS—S—GLU

(14) $R-CO-CH_2-CO-R'$ ⟶ $R-CO-CH(GLU)-CO-R'$

GLU = (COOH, O, X, OH, HO, HO)

Figure 21. *(Continued).*

tion are classified as *O*-, *N*-, *S*- and *C*-glucuronides.

O-Glucuronidation is shown in reactions 1–5 (Fig. 21). A frequent metabolic reaction of phenolic xenobiotics or metabolites is their glucuronidation to yield polar metabolites excreted in urine and/or bile. *O*-Glucuronidation is often in competition with *O*-sulfonation (see above), with the latter reaction predominating at low doses and the former at high doses. In biochemical terms, glucuronidation is a reaction of low affinity and high capacity, while sulfonation displays high affinity and low capacity. A typical drug undergoing extensive glucuronidation is paracetamol (**8** in Fig. 10). Another major group of substrates are alcohols, be they primary, secondary, or tertiary (reaction 2, Fig. 21). Medicinal examples include chloramphenicol and oxazepam. Another important example is that of morphine, which is conjugated on its phenolic and secondary alcohol groups to form the 3-*O*-glucuronide (a weak opiate antagonist) and the 6-*O*-glucuronide (a strong opiate agonist), respectively.

An important pathway of *O*-glucuronidation is the formation of acyl glucuronides (reaction 3). Substrates are arylacetic acids

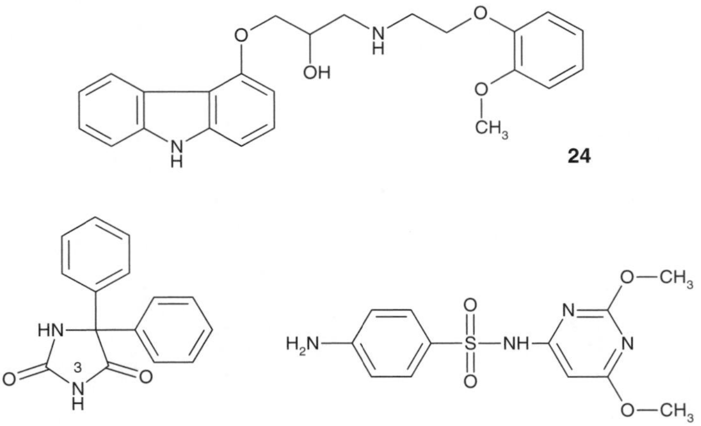

24

25

26

Figure 22. Carvedilol (**24**), phenytoin (**25**), and sulfadimethoxine (**26**).

(e.g., diclofenac, **15** in Fig. 16) and aliphatic acids (e.g., valproic acid). Aromatic acids are seldom substrates, a noteworthy exception being diflunisal (**20** in Fig. 19) that yields both the acyl and the phenolic glucuronides. The significance of acyl glucuronides has long been underestimated perhaps because of analytical difficulties. Indeed, these metabolites are quite reactive, rearranging to positional isomers and binding covalently to plasma and seemingly also tissue proteins [46]. Thus, acyl glucuronide formation cannot be viewed solely as a reaction of inactivation and detoxification.

A special class of acyl glucuronides are the carbamoyl glucuronides (reaction 4 in Fig. 21). A number of primary and secondary amines have been found to yield this type of conjugate, while as expected the intermediate carbamic acids are not stable enough to be characterized. Carvedilol (**24** in Fig. 22) is one drug exemplifying the reaction, in addition to forming an *O*-glucuronide on its alcohol group and a carbazole-*N*-linked glucuronide (see below). Much remains to be understood concerning the chemical and biochemical reactivity of carbamoyl glucuronides.

Hydroxylamines and hydroxylamides may also form *O*-glucuronides (reaction 5, Fig. 21). Thus, a few drugs and a number of aromatic amines are known to be *N*-hydroxylated and then *O*-glucuronidated. The glucuronidation of N–OH groups competes with *O*-sulfation, but the reactivity of *N*-*O*-glucuronides to undergo heterolytic cleavage and form nitrenium ions does not appear to be well characterized.

Second in importance to *O*-glucuronides are the *N*-glucuronides formed by reactions 6–11 in Fig. 21, that is, amides (reactions 6–7), amines of medium basicity (reactions 8 and 9), and basic amines (reactions 10 and 11). The *N*-glucuronidation of carboxamides (reaction 6) is exemplified by carbamazepine (**5** in Fig. 6) and phenytoin (**25** in Fig. 22). In the latter case, *N*-glucuronidation was found to occur at *N*(3). The reaction has special significance for sulfonamides (reaction 7) and particularly antibacterial sulfanilamides such as sulfadimethoxine (**26** in Fig. 22) since it produces highly water-soluble metabolites that show no risk of crystallizing in the kidneys.

N-Glucuronidation of aromatic amines (reaction 8, Fig. 21) has been observed in a few cases only (e.g., conjugation of the carbazole nitrogen in carvedilol (**24**). Similarly, there are a number of observations that pyridine-type nitrogens and primary and secondary basic amines can be *N*-glucuronidated (reactions 9 and 10, respectively). As far as human drug metabolism is concerned, another reaction of significance is the *N*-glucuronidation of lipophilic, basic tertiary amines containing one or two methyl groups (reaction 11). More and more drugs of this type (e.g., antihistamines and neuroleptics), are found to undergo this reaction to a marked extent in

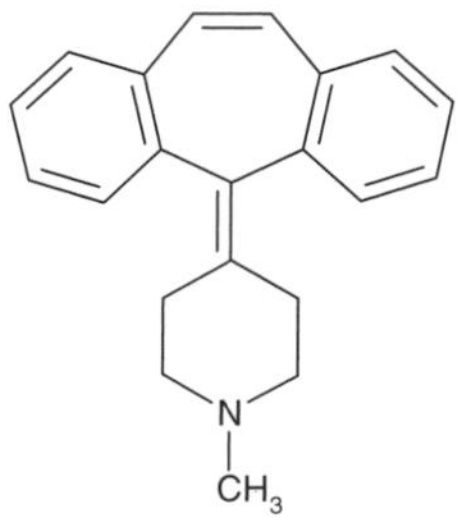

Figure 23. Cyproheptadine (**27**) and sulfinpyrazone (**28**).

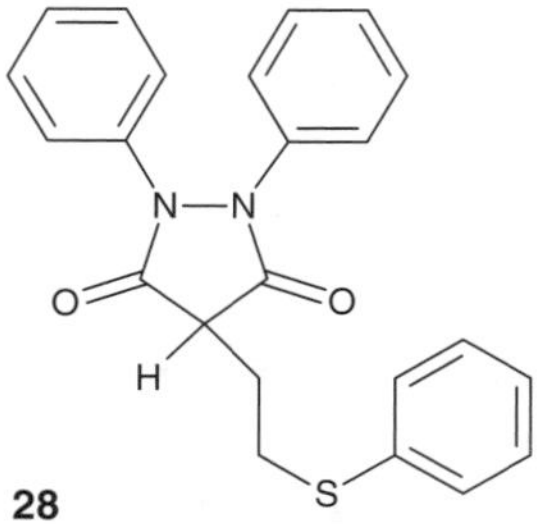

Figure 24. Acetylcoenzyme A (**29**, when R = acetyl).

humans, for example, cyproheptadine (**27** in Fig. 23).

Third in importance are the *S*-glucuronides formed from aliphatic and aromatic thiols (reaction 12 in Fig. 21), and from dithiocarboxylic acids (reaction 13) such as diethyldithiocarbamic acid, a metabolite of disulfiram. As for *C*-glucuronidation (reaction 14), this reaction has been seen in humans for 1,3-dicarbonyl drugs such as phenylbutazone and sulfinpyrazone (**28** in Fig. 23).

3.4.3. Glucosidation Reactions

A few drugs have been observed to be conjugated to glucose in mammals. This is usually a minor pathway in some cases where glucuronidation is possible. An interesting medicinal example is that of some barbiturates such as phenobarbital that yield the *N*-glucoside.

3.5. Acetylation and Acylation

All reactions discussed in this section involve the transfer of an acyl moiety to an acceptor group. In most cases, an acetyl is the acyl moiety being transferred, while the acceptor group may be an amino or hydroxyl function [15,19,24,47].

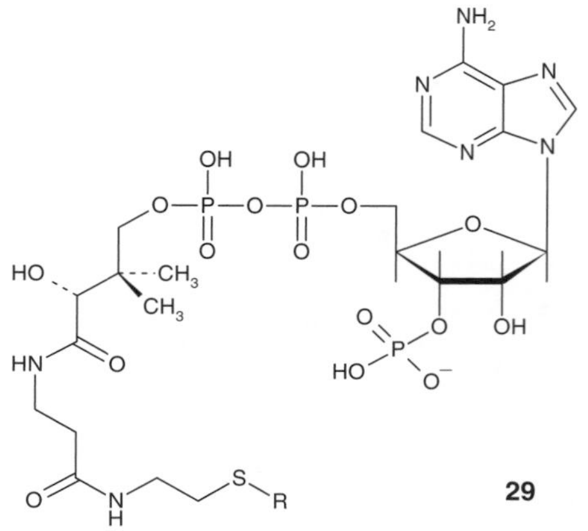

3.5.1. Acetylation Reactions

The major enzyme system catalyzing acetylation reactions is arylamine *N*-acetyltransferase (see Table 7). Two enzymes have been characterized, NAT1 and NAT2, the latter as two closely related isoforms NAT2A and NAT2B whose levels are considerably reduced in the liver of slow acetylators. The cofactor of *N*-acetyltransferase is acetylcoenzyme A (CoA-S-Ac, **29** with R = acetyl) (Fig. 24) where the acetyl moiety is bound by a thioester linkage.

Two other activities, aromatic hydroxylamine *O*-acetyltransferase and *N*-hydroxyarylamine *O*-acetyltransferase, are also involved in the acetylation of aromatic amines and hydroxylamines (see below). Other acetyltransferases exist, for example, diamine *N*-acetyltransferase (putrescine acetyltransferase; EC 2.3.1.57) and aralkylamine *N*-acetyltransferase (serotonin acetyltransferase; EC 2.3.1.87), but their involvement in xenobiotic metabolism does not appear to be documented.

The substrates of acetylation, as schematized in Fig. 25, are mainly amines of medium basicity. Very few basic amines (primary or secondary) of medicinal interest have been reported to form *N*-acetylated metabolites (reaction 1), and when so the yields were low. In contrast, a large variety of primary aromatic amines are *N*-acetylated (reaction 2). Thus, several drugs such as sulfonamides and *para*-aminosalicylic acid (**30**, PAS) (Fig. 26) are

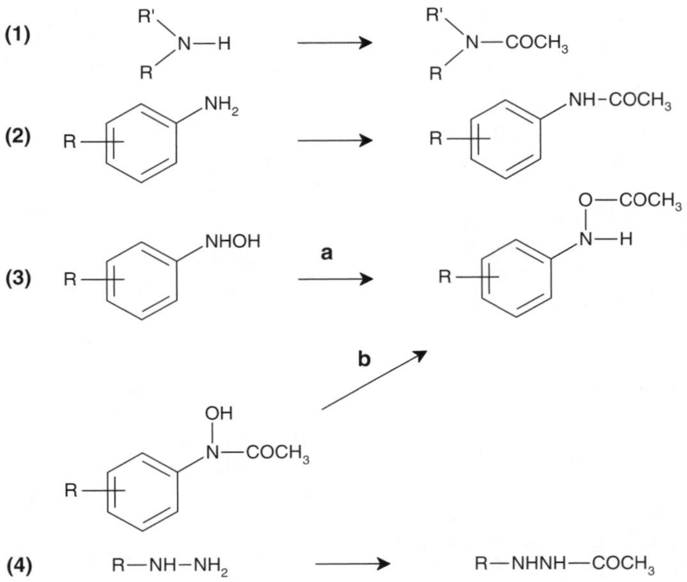

Figure 25. Major acetylation reactions involving aliphatic amines, aromatic amines, arylhydroxylamines, hydrazines and hydrazides.

acetylated to large extents, not to mention various carcinogenic amines such as benzidine.

Arylhydroxylamines can also be acetylated, but the reaction is one of *O*-acetylation (reaction 3a in Fig. 25). This is the reaction formally catalyzed by EC 2.3.1.118 with acetyl-CoA acting as the acetyl donor, the *N*-hydroxy metabolites of a number of arylamines being known substrates. The same conjugates can be formed by intramolecular *N,O*-acetyl transfer, when an arylhydroxamic acid (an *N*-aryl-*N*-hydroxy-acetamide) is substrate of, for example, EC 2.3.1.56 (reaction 3b). In addition, such an arylhydroxamic acid can transfer its acetyl moiety to an acetyltransferase, which can then acetylate an arylamine or an arylhydroxylamine (intermolecular *N,N*- or *N,O*-acetyl transfer).

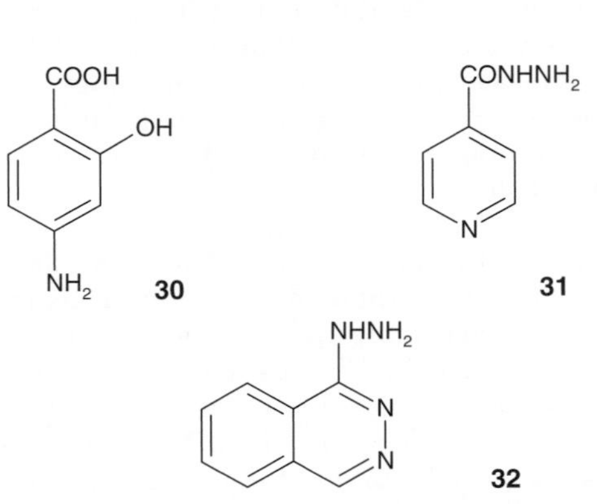

Figure 26. *para*-Aminosalicylic acid (**30**), isoniazid (**31**), and hydralazine (**32**).

Besides amines, other nitrogen-containing functionalities undergo *N*-acetylation, hydrazines, and hydrazides being particularly good substrates (reaction 4, Fig. 25). Medicinal examples include isoniazid (**31** in Fig. 26) and hydralazine (**32**).

3.5.2. Other Acylation Reactions A limited number of studies have shown *N*-formylation to be a genuine route of conjugation for some arylalkylamines and arylamines, and particularly polycyclic aromatic amines. There is evidence to indicate that the reaction is catalyzed by arylformamidase (EC 3.5.1.9) in the presence of *N*-formyl-L-kynurenine.

A very different type of reaction is represented by the conjugation of xenobiotic alcohols with fatty acids, yielding highly lipophilic metabolites accumulating in tissues. Thus, ethanol and haloethanols form esters with, for example, palmitic acid, oleic acid, linoleic acid, and linolenic acid; enzymes catalyzing such reactions are cholesteryl ester synthase (EC 3.1.1.13) and fatty-acyl-ethyl-ester synthase (EC 3.1.1.67) [48,49]. Larger xenobiotics such as tetrahydrocannabinols and codeine are also acylated with fatty acids, possibly by sterol *O*-acyltransferase (EC 2.3.1.26).

3.6. Conjugation with Coenzyme A and Subsequent Reactions

3.6.1. Conjugation with Coenzyme A The reactions described in this section all have in common the fact that they involve xenobiotic carboxylic acids forming an acyl-CoA metabolic intermediate (**29** in Fig. 24, R = xenobiotic acyl moiety). The reaction requires ATP and is catalyzed by various acyl-CoA synthetases of overlapping substrate specificity, for example, short-chain fatty acyl-CoA synthetase, medium-chain acyl-CoA synthetase, long-chain acyl-CoA synthetase, and benzoate-CoA ligase (EC 6.2.1.25) (see Table 7).

The acyl-CoA conjugates thus formed are seldom excreted, but they can be isolated and characterized relatively easily in *in vitro* studies. They may also be hydrolyzed back to the parent acid by thiolester hydrolases (EC 3.1.2). In a number of cases, such conjugates have pharmacodynamic effects and may even represent the active forms of some drugs, for example, hypolipidemic agents. In the present context, the interest of acyl-CoA conjugates is their further transformation by a considerable variety of pathways (Table 8) [15,19,24]. The most significant routes are discussed below.

Table 8. Metabolic Consequences of the Conjugation of Xenobiotic Acids to Coenzyme A (CoA-SH) [15,24]

Depending on structure and other factors, the resulting R-CO-S-CoA intermediate may enter the following metabolic reactions:

- Hydrolysis
- Formation of amino acid conjugates
- Formation of hybrid triglycerides
- Formation of phospholipids
- Formation of cholesteryl esters
- Formation of bile acid esters
- Formation of acyl-carnitines
- Protein acylation
- Unidirectional configurational inversion of arylpropionic acids (profens)
- Dehydrogenation and β-oxidation
- Two-carbon chain elongation

3.6.2. Formation of Amino Acid Conjugates Amino acid conjugation is a major route for a number of small aromatic acids and involves the formation of an amide bond between the xenobiotic acyl-CoA and the amino acid. Glycine is the amino acid most frequently used for conjugation (reaction 1 in Fig. 27), while a few glutamine conjugates (reaction 2) and some taurine conjugates (reaction 3) have been characterized in humans. The enzymes catalyzing these transfer reactions are various *N*-acyltransferases (see Table 7), for example glycine *N*-acyltransferase, glycine *N*-benzoyltransferase, glutamine *N*-phenylacetyltransferase and glutamine *N*-acyltransferase, and cholyl-CoA glycine-taurine *N*-acyltransferase. In addition, other amino acids can be used for conjugation in various animal species, for example, alanine, as well as some dipeptides [3].

The xenobiotic acids undergoing amino acid conjugation are mainly substituted benzoic acids. In humans for example, hippuric acid and salicyluric acid (**33** in Fig. 28) are the major metabolites of benzoic acid and salicylic acid, respectively. Similarly, *m*-trifluoromethylbenzoic acid (**13** in Fig. 12), a major metabolite of fenfluramine (**10**), is excreted as the glycine conjugate. Phenylacetic acid derivatives can yield glycine and glutamine conjugates. Some drugs containing a carboxylic group do form the taurine conjugate as a minor metabolite.

3.6.3. Formation of Hybrid Lipids and Sterol Esters Incorporation of xenobiotic acids into lipids forms highly lipophilic metabolites that may burden the body as long retained residues. In the majority of cases, triacylglycerol analogs (reaction 4 in Fig. 27) or cholesterol esters (reaction 5) are formed. The enzymes catalyzing such reactions are *O*-acyltransferases (see Table 7), including diacylglycerol

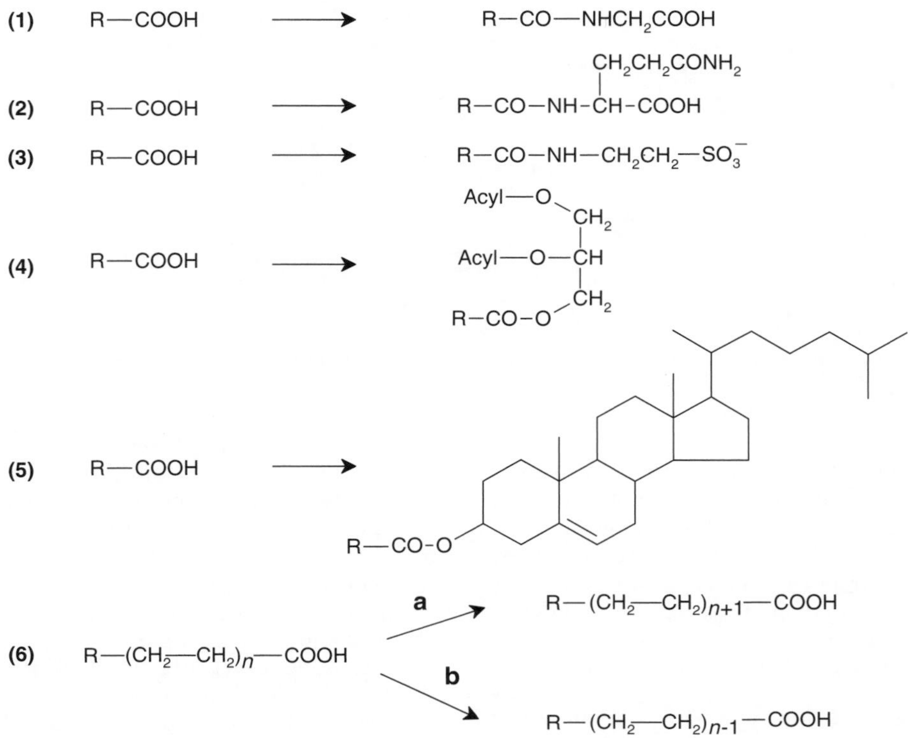

(1) R—COOH ⟶ R—CO—NHCH$_2$COOH

(2) R—COOH ⟶ R—CO—NH—CH(CH$_2$CH$_2$CONH$_2$)—COOH

(3) R—COOH ⟶ R—CO—NH—CH$_2$CH$_2$—SO$_3^-$

(4) R—COOH ⟶ Acyl—O—CH$_2$ / Acyl—O—CH / R—CO—O—CH$_2$

(5) R—COOH ⟶ R—CO-O-(cholesteryl)

(6) R—(CH$_2$—CH$_2$)$_n$—COOH ⟶ a: R—(CH$_2$—CH$_2$)$_{n+1}$—COOH; b: R—(CH$_2$—CH$_2$)$_{n-1}$—COOH

Figure 27. Metabolic reactions involving acyl-CoA intermediates of xenobiotic acids, namely conjugations and two-carbon chain lengthening or shortening. Other products of β-oxidation are shown in Fig. 30.

O-acyltransferase, 2-acylglycerol *O*-acyltransferase, and sterol *O*-acyltransferase (cholesterol acyltransferase). Some phospholipid analogs, as well as some esters to the 3-hydroxy group of biliary acids, have also been characterized [15,24,48–52].

The number of drugs and other xenobiotics that are currently known to form glyceryl or cholesteryl esters is limited, but should increase due to increased awareness of researchers. One telling example is that of ibuprofen (**34** in Fig. 28), a much used antiinflammatory drug whose (*R*)-enantiomer forms hybrid triglycerides detectable in rat liver and adipose tissue.

3.6.4. Configurational Inversion of Arylpropionic Acids

Ibuprofen (**34** in Fig. 28) and other arylpropionic acids (i.e., profens) are chiral drugs existing as the (+)-(*S*) eutomer and the (−)-(*R*) distomer. These compounds undergo an intriguing metabolic reaction such that the (*R*)-enantiomer is converted to the (*S*)-enantiomer, while the reverse reaction is negligible. This unidirectional configurational inversion is thus a reaction of bioactivation, and its mechanism is now reasonably well understood (Fig. 29) [53].

The initial step in the reaction is the substrate stereoselective formation of an acyl-CoA conjugate with the (*R*)-form but not with

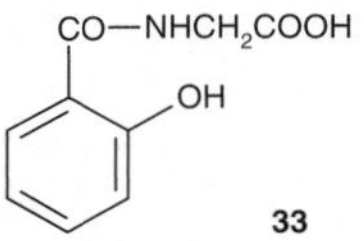

CO—NHCH$_2$COOH, OH — **33**

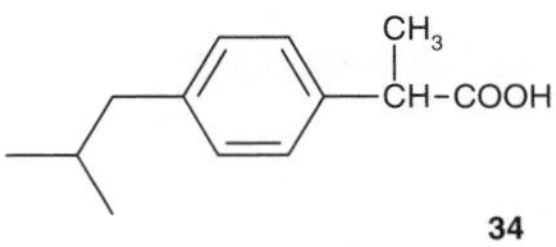

CH$_3$, CH-COOH — **34**

Figure 28. Salicyluric acid (**33**) and ibuprofen (**34**).

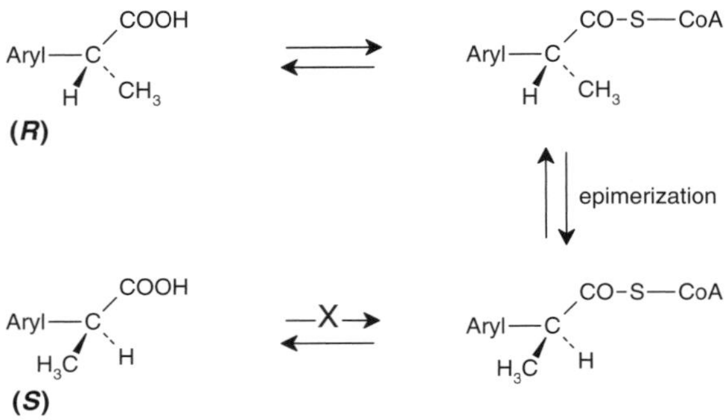

Figure 29. Mechanism of the unidirectional configurational inversion of some profens.

the (*S*)-form. This conjugate then undergoes a reaction of epimerization possibly catalyzed by methylmalonyl-CoA epimerase (EC 5.1.99.1), resulting in a mixture of the (*R*)-profenoyl- and (*S*)-profenoyl-CoA conjugates. The latter can then be hydrolyzed as shown in Fig. 29, or undergo other reactions such as hybrid triglyceride formation (see below).

3.6.5. β-Oxidation and Two-Carbon Chain Elongation In some cases, acyl-CoA conjugates formed from xenobiotic acids can also enter the physiological pathways of fatty acid catabolism or anabolism. A few examples are known of xenobiotic alkanoic and arylalkanoic acids undergoing two-carbon chain elongation, or two-, four- or even six-carbon chain shortening (reactions 6a and 6b in Fig. 27). In addition, intermediate metabolites of β-oxidation may also be seen, as illustrated by valproic acid (**35** in Fig. 30). Approximately 50 metabolites of this drug have been characterized; they are formed by β-oxidation, glucuronidation, and/or cytochrome P450-catalyzed dehydrogenation or oxygenation. Fig. 30 shows the β-oxidation of valproic acid seen in mitochondrial preparations [25,54]. The resulting metabolites have also been found in unconjugated form in the urine of humans or animals dosed with the drug.

3.7. Conjugation and Redox Reactions of Glutathione

3.7.1. Introduction Glutathione (**36** in Fig. 31, GSH) is a thiol-containing tripeptide of major significance in the detoxification and toxification of drugs and other xenobiotics [15,24,55–57]. In the body, it exists in a redox equilibrium between the reduced form (GSH) and an oxidized form (GS-SG). The metabolism of glutathione (i.e., its synthesis, redox equilibrium and degradation) is quite complex and involves a number of enzymes.

Glutathione reacts in a variety of manners. First, the nucleophilic properties of the thiol (or rather thiolate) group make it an effective conjugating agent, as emphasized in this section. Second, and depending on its redox state, glutathione can act as a reducing or oxidizing agent (e.g., reducing quinones, organic nitrates, peroxides and free radicals, or oxidizing superoxide). Another dichotomy exists in the reactions of glutathione, since these can be enzymatic (e.g., conjugations catalyzed by glutathione *S*-transferases, and peroxide reductions catalyzed by glutathione peroxidase) or nonenzymatic (e.g., some conjugations and various redox reactions).

The glutathione *S*-transferases (see Table 7) are multifunctional proteins coded by two gene superfamilies. They can act as enzymes as well as binding proteins. These enzymes are mainly localized in the cytosol as homodimers and heterodimers, but microsomal enzymes also exist [58–60]. The GST A1-2, A2-2 and P1-1 display selenium-independent glutathione peroxidase activity, a property also characterizing the selenium-containing enzyme glutathione peroxidase (EC 1.11.1.9). The GST A1-1 and A1-2 are also known as ligandin when they act as binding or carrier proteins, a property also displayed

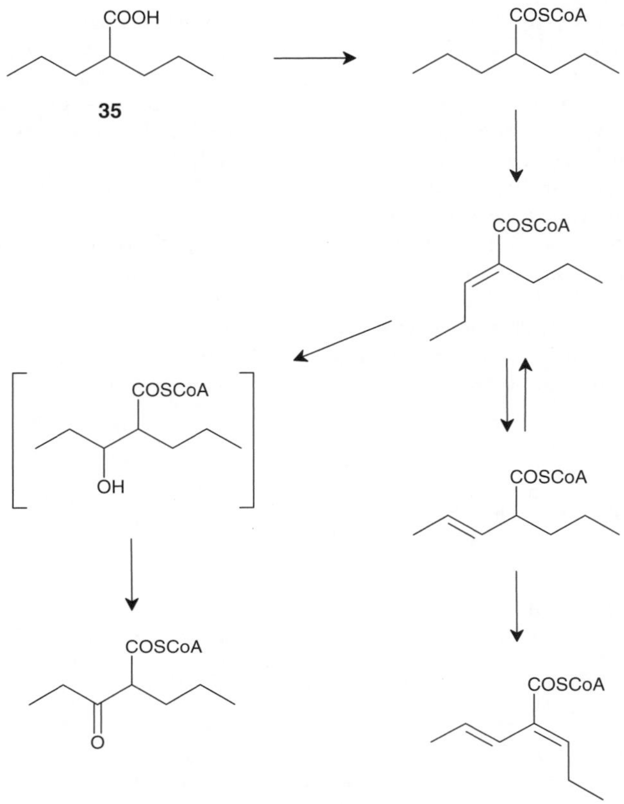

Figure 30. Mitochondrial β-oxidation of valproic acid (35).

by M1a-1a and M1b-1b. In the latter function, these enzymes bind and transport a number of active endogenous compounds (e.g., bilirubin, cholic acid, steroid and thyroid hormones, and hematin), as well as some exogenous dyes and carcinogens.

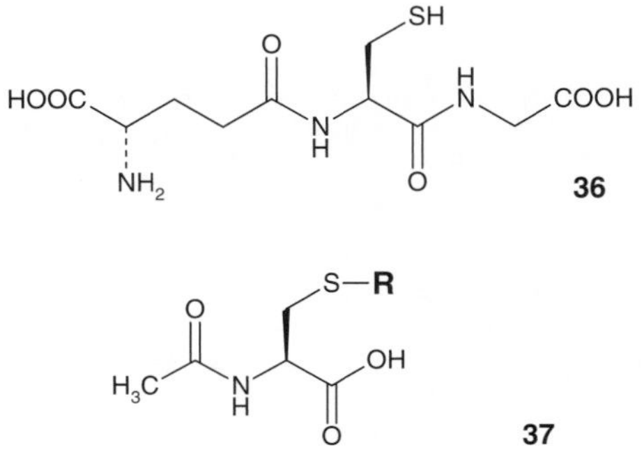

Figure 31. Glutathione (36) and *N*-acetylcysteine conjugates (37).

The nucleophilic character of glutathione is due to its thiol group (pK_a 9.0) in its neutral form and even more to the thiolate form. In fact, an essential component of the catalytic mechanism of glutathione transferase is the marked increase in acidity (pK_a 6–7) experienced by the thiol group upon binding of glutathione to the active site of the enzyme. As a result, GSTs transfer glutathione to a very large variety of electrophilic groups; depending on the nature of the substrate, the reactions can be categorized as nucleophilic additions or nucleophilic additions–eliminations. And with compounds of sufficient reactivity, these reactions can also occur nonenzymatically [61].

Once formed, glutathione conjugates (*R*-SG) are seldom excreted as such (they are best characterized *in vitro* or in the bile of laboratory animals), but usually undergo further biotransformation prior to urinary or fecal excretion. Cleavage of the glutamyl moiety by glutamyl transpeptidase (EC 2.3.2.2), and of the cysteinyl moiety by cysteinylglycine dipeptidase (EC 3.4.13.6) or aminopeptidase M (EC 3.4.11.2), leaves a cysteine conjugate (*R-S*-Cys) that is further *N*-acetylated by cysteine-*S*-conjugate *N*-acetyltransferase (EC 2.3.1.80) to yield an *N*-acetylcysteine conjugate (**37** in Fig. 31, *R-S*-CysAc). The latter type of conjugates are known as mercapturic acids, a name that clearly indicates that they were first characterized in urine. This however does not imply that the degradation of unexcreted glutathione conjugates must stop at this stage, since cysteine conjugates can be substrates of cysteine-*S*-conjugate β-lyase (EC 4.4.1.13) to yield thiols (*R*-SH). These in turn can rearrange as discussed below, or be *S*-methylated and then *S*-oxygenated to yield thiomethyl conjugates (*R-S*-Me), sulfoxides (*R*-SO-Me), and sulfones (*R*-SO_2-Me).

3.7.2. Reactions of Glutathione The major reactions of glutathione, both conjugations and redox reactions, are summarized in Fig. 32. Reactions 1 and 2 are nucleophilic additions and additions-elimination to sp^3-carbons, respectively, while reactions 3–8 are

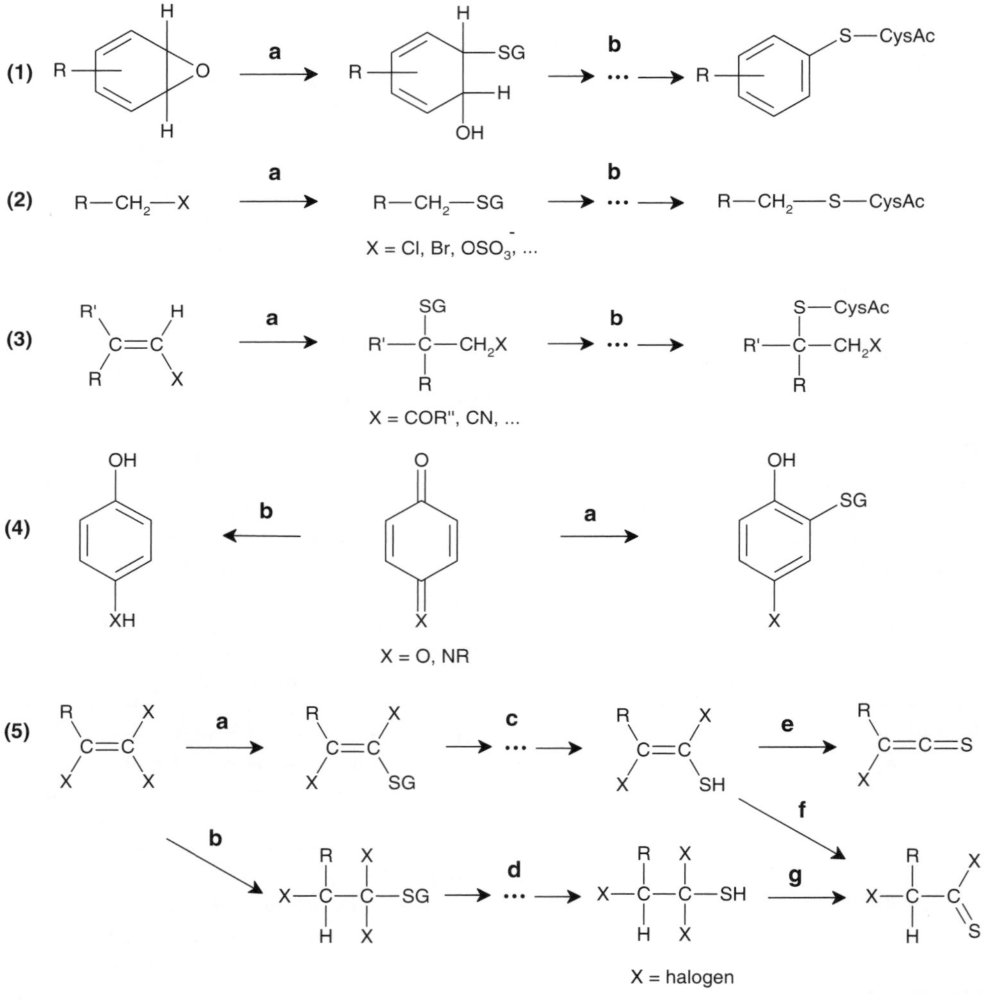

Figure 32. Major reactions of conjugation of glutathione, sometimes accompanied by a redox reaction.

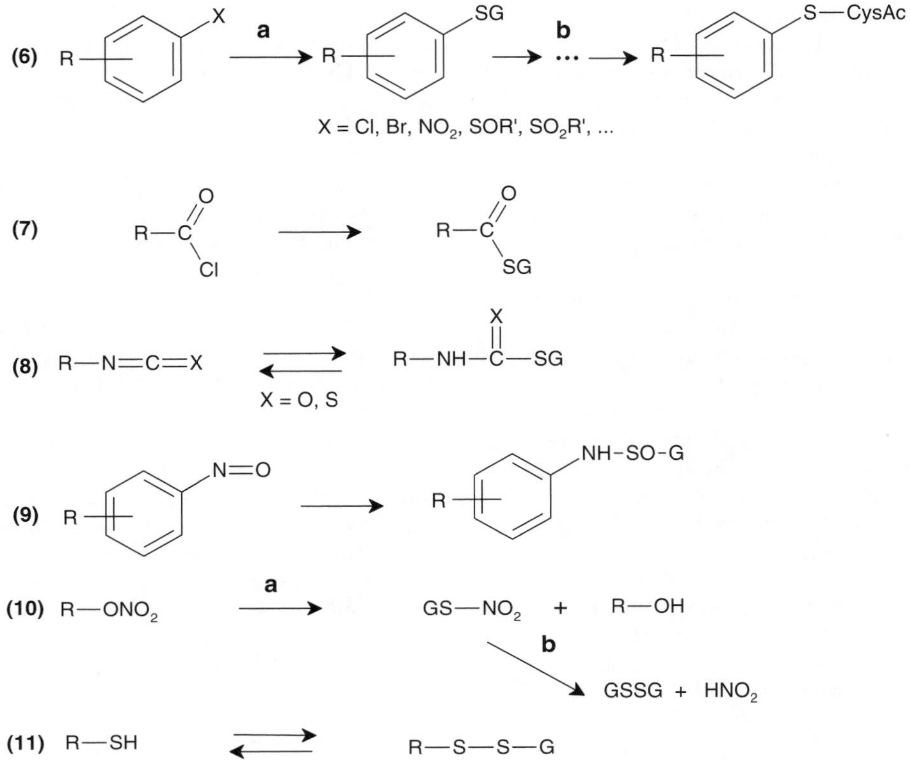

(6) R–C$_6$H$_4$–X —a→ R–C$_6$H$_4$–SG —b→ ··· → R–C$_6$H$_4$–S—CysAc

X = Cl, Br, NO_2, SOR', SO_2R', ...

(7) R—C(=O)—Cl ⟶ R—C(=O)—SG

(8) R—N=C=X ⇄ R—NH—C(=X)—SG

X = O, S

(9) R–C$_6$H$_4$–N=O ⟶ R–C$_6$H$_4$–NH-SO-G

(10) R—ONO_2 —a→ GS—NO_2 + R—OH

—b→ GSSG + HNO_2

(11) R—SH ⇄ R—S—S—G

Figure 32. (*Continued*).

nucleophilic substitutions or additions at sp^2-carbons, sometimes accompanied by a redox reaction. Reactions at nitrogen or sulfur atoms are shown in reactions 9–11.

The first reaction in Fig. 32 is nucleophilic addition to epoxides (reaction 1a) to yield a nonaromatic conjugate. This is followed by several metabolic steps (reaction 1b) leading to an aromatic mercapturic acid. This is a frequent reaction of metabolically produced arene oxides (Fig. 5), as documented for naphthalene and numerous drugs and xenobiotics containing a phenyl moiety. Note that the same reaction can also occur readily for epoxides of olefins (not shown in Fig. 32).

An important pathway of addition-elimination at sp^3-carbons is represented in reaction 2a, followed by the production of mercapturic acids (reaction 2b). Various electron-withdrawing leaving groups (X in reaction 2) may be involved that are either of xenobiotic (e.g., halogens) or metabolic origin (e.g., a sulfate group). Such a reaction occurs for example at the $-CHCl_2$ group of chloramphenicol and at the NCH_2CH_2Cl group of anticancer alkylating agents.

The reactions at sp^2-carbons are quite varied and complex. Addition to activated olefinic groups (e.g., α,β-unsaturated carbonyls) is shown in reaction 3. A typical substrate is acrolein ($CH_2{=}CH{-}CHO$). Quinones (*ortho* and *para*) and quinoneimines react with glutathione by two distinct and competitive routes, namely nucleophilic addition to form a conjugate (reaction 4a), and reduction to the hydroquinone or aminophenol (reaction 4b). A typical example is provided by the toxic quinoneimine metabolite (**9** in Fig. 10) of paracetamol (**8**). Since in most cases quinones and quinoneimines are produced by the bio-oxidation of hydroquinones and aminophenols, respectively, their reduction by GSH can be seen as a futile cycle that consumes reduced glutathione. As for the conjugates produced by reaction 4a, they may undergo reoxidation to *S*-glutathionyl quinones or *S*-glutathionyl quinoneimines of considerable reactivity. These quinone or quinoneimine thioethers are known to undergo further GSH conjugation and reoxidation.

Haloalkenes are a special group of substrates of GS-transferases since they may react with GSH either by substitution (reaction 5a) or by addition (reaction 5b). The formation of mercapturic acids occurs as for other glutathione conjugates, but in this case S–C cleavage of the *S*-cysteinyl conjugates by the renal β-lyase (reactions 5c and 5d) yields thiols of significant toxicity since they rearrange by hydrohalide expulsion to form highly reactive thioketenes (reaction 5e) and/or thioacyl halides (reactions 5f and 5g) [62].

With good leaving groups, nucleophilic aromatic substitution reactions also occur at aromatic rings containing additional electron-withdrawing substituents and/or heteroatoms (reaction 6a). As for the detoxification of acyl halides with glutathione (reaction 7), a good example is provided by phosgene ($O{=}CCl_2$), an extremely toxic metabolite of chloroform that is inactivated to the diglutathionyl conjugate $O{=}C(SG)_2$.

The addition of glutathione to isocyanates and isothiocyanates has received some attention due in particular to its reversible character (reaction 8) [63]. Substrates of the reaction are xenobiotics such as the infamous toxin methyl isocyanate, whose glutathione conjugate behaves as a transport form able to carbamoylate various macromolecules, enzymes and membrane components. The reaction is also of interest from a medicinal viewpoint since anticancer agents such as methylformamide appear to work by undergoing activation to isocyanates and then to the glutathione conjugate.

The reaction of *N*-oxygenated drugs and metabolites with glutathione may also have toxicological and medicinal implications. Thus, the addition of GSH to nitrosoarenes, probably followed by rearrangement, forms sulfinamides (reaction 9) that have been postulated to contribute to the idiosyncratic toxicity of a few drugs such as sulfonamides [64]. As for organic nitrate esters such as nitroglycerine and isosorbide dinitrate, the mechanism of their vasodilating action is now known to result from their reduction to nitric oxide (NO). Thiols, and particularly glutathione, play an important role in this activation. In the first step, a thionitrate is formed (reaction 10a) whose *N*-reduction may occur by more than one route. For example, a GSH-dependent reduction may yield nitrite (reaction 10b) that undergoes further reduction to NO; *S*-nitrosoglutathione (GS-NO) has also been postulated as an intermediate.

The formation of mixed disulfides between GSH and a xenobiotic thiol (reaction 11) has been observed in a few cases, for example, with captopril.

Finally, glutathione and other endogenous thiols (including albumin) are able to inactivate free radicals (e.g., $R^\bullet$, $HO^\bullet$, $HOO^\bullet$, $ROO^\bullet$) and have thus a critical role to play in cellular protection [55–57]. The reactions involved are highly complex and incompletely understood; the simplest are

$$\mathrm{GSH + X^\bullet \rightarrow GS^\bullet + XH}$$

$$\mathrm{GS^\bullet + GS^\bullet \rightarrow GSSG}$$

$$\mathrm{GS^\bullet + O_2 \rightarrow GS{-}OO^\bullet}$$

$$\mathrm{GS{-}OO^\bullet + GSH \rightarrow GS{-}OOH + GS^\bullet}$$

3.8. Other Conjugation Reactions

The above Sections 3.2–3.7 present the most common and important routes of xenobiotic conjugation, but these are not the only ones. A number of other routes have been reported whose importance is at present restricted to a few exogenous substrates, or that have received only limited attention [15,24]. In the present section, two pathways of pharmacodynamic significance will be mentioned, namely phosphorylation and carbonyl conjugation (Fig. 33). In both cases, the xenobiotic substrates belong to narrowly defined chemical classes.

Phosphorylation reactions are of great significance in the processing of endogenous compounds and macromolecules. It is therefore astonishing that relatively few xenobiotics are substrates of phosphotransferases (e.g., EC

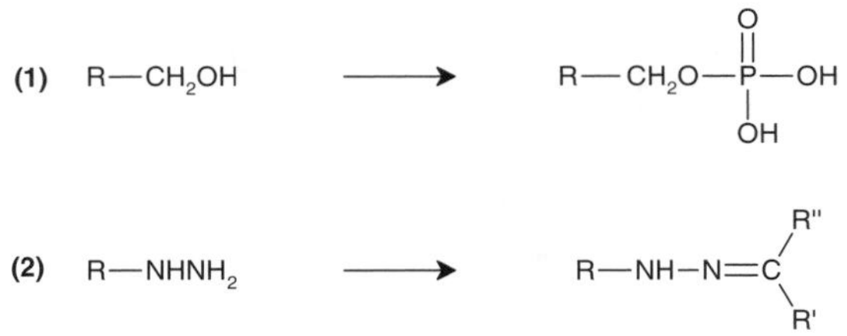

(1) $R-CH_2OH \longrightarrow R-CH_2O-P(=O)(OH)-OH$

(2) $R-NHNH_2 \longrightarrow R-NH-N=C(R'')(R')$

Figure 33. Additional conjugation reactions discussed in Section 3.8, namely formation of phosphate esters and of hydrazones.

2.7.1) to form phosphate esters (reaction 1 in Fig. 33). The phosphorylation of phenol is a curiosity observed by some workers. In contrast, a number of antiviral nucleoside analogs are known to yield the mono-, di- and triphosphates *in vitro* and *in vivo*, for example, zidovudine (**38** in Fig. 34, AZT) and numerous analogs. These conjugates are active forms of the drugs, being in particular incorporated in the DNA of virus-infected cells [65].

The second pathway of conjugation to be presented here is the reaction of hydrazines with endogenous carbonyls (reaction 2) [62]. This reaction occurs nonenzymatically and involves a variety of carbonyl compounds, namely aldehydes (mainly acetaldehyde) and ketones (e.g., acetone, pyruvic acid $CH_3-CO-COOH$, and α-ketoglutaric acid $HOOC-CH_2CH_2-CO-COOH$). The products thus formed are hydrazones, which may be excreted as such or undergo further transformation. Isoniazid (**31** in Fig. 26) and hydralazine (**32** in Fig. 26) are two drugs that form hydrazones in the body. For example, the reaction of hydralazine with acetyldehyde or acetone is a reversible one, meaning that the hydrazones are hydrolyzed under biological conditions of pH and temperature. In addition, the hydrazone of hydralazine with acetaldehyde or pyruvic acid undergoes an irreversible reaction of cyclization to another metabolite, methyltriazolophthalazine (**39** in Fig. 34).

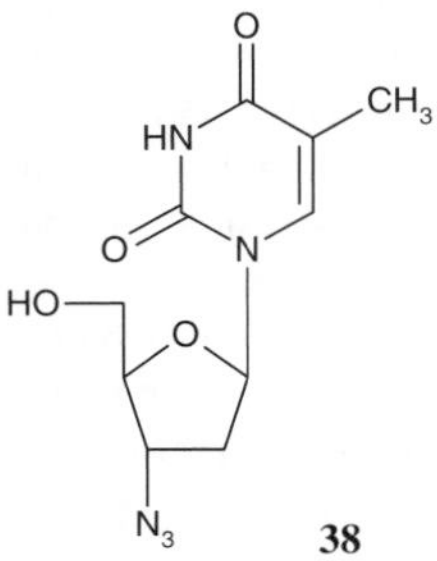

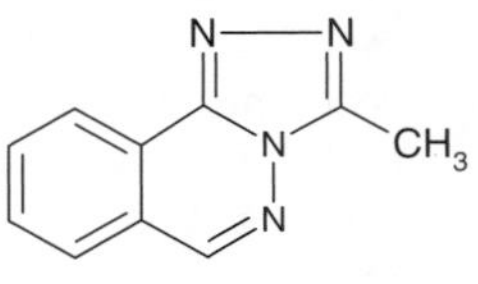

Figure 34. Zidovudine (**38**) and methyltriazolophthalazine (**39**).

4. DRUG METABOLISM AND THE MEDICINAL CHEMIST

Having presented the various reactions of drug metabolism, we now briefly review a few topics of potential interest to medicinal chemists, namely the influence of metabolism on activity and toxicity, the prediction of drug metabolism, and the biological factors that may influence biotransformation.

4.1. Metabolism and Bioactivity

The pharmacological and toxicological consequences of drug metabolism explain to a large extent the immense significance this discipline has acquired. Indeed, such consequences are of utmost relevance in fields such as drug discovery and development, clinical pharmacology and toxicology, and therapeutics.

4.1.1. Pharmacological Activity An important information in any drug's dossier is the activity (or lack thereof) of its metabolites. What should be realized, however, is that "activity" is usually understood to imply the same pharmacological target as the parent molecule [67–69]. Here, we outline various possible combinations, from a drug having no active metabolite to an intrinsically inactive "drug" (i.e., a prodrug) whose therapeutic effects necessitate metabolism to an active metabolite (i.e., bioactivation).

A number of drugs have no active metabolites, for example the 3-hydroxylated

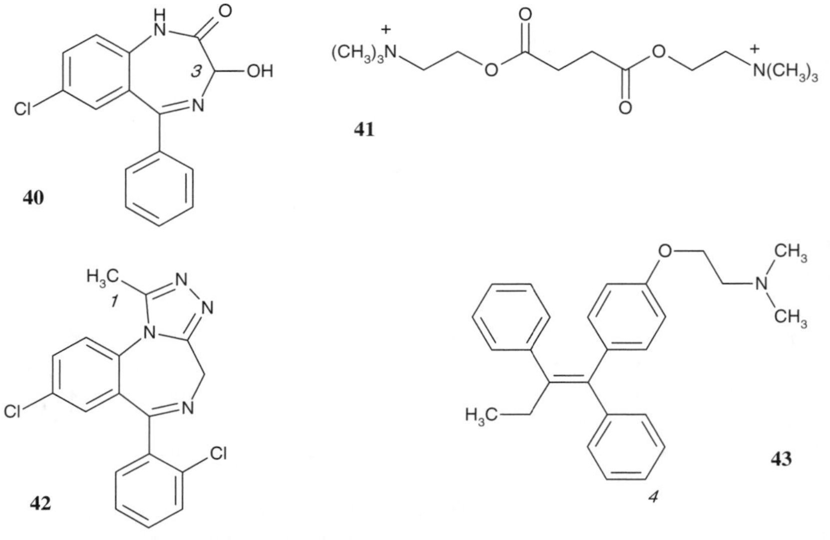

Figure 35. Oxazepam (**40**) and succinylcholine (**41**) as examples of drugs having no active metabolite. In contrast, part of the activity of triazolam (**42**) and tamoxifen (**43**) is due to one or more active metabolites.

benzodiazepines such as lorazepam, oxazepam (**40** in Fig. 35), and temazepam, which undergo *O*-glucuronidation and cleavage reactions. A case of drugs designed to have neither active nor toxic metabolites is that of soft drugs, a concept pioneered and extensively developed by Bodor and collaborators [70–72]. In practical terms, soft drugs (a) are bioisosteres of known drugs, (b) contain a labile bridge (often an ester group), and (c) are cleaved to metabolites known to lack activity and toxicity. In other words, soft drugs bear a close stereoelectronic analogy with the target drugs, but a rapid breakdown to inactive metabolites is programmed into their chemical structure. A typical example is succinylcholine (**41** in Fig. 35), although the discovery of this agent predates by decades the concept and term of soft drugs. In most individuals, this curarimimetic agent is very rapidly hydrolyzed to choline and succinic acid by plasma cholinesterase with a half-life of about four minutes [73].

Drugs having both intrinsic activity and active metabolites may be more numerous than generally believed, but information is not always available. Several benzodiazepines have one or more active metabolite(s). For example, triazolam (**42** in Fig. 35) forms the active 1-hydroxytriazolam, both agents having a short half-life compatible with the use of the drug as a short-acting short-duration hypnotic [74].

A drug whose activity owes much to metabolism is tamoxifen (**43** in Fig. 35) as summarized here. This estrogen receptor antagonist is extensively used for endocrine treatment of breast cancer. Its metabolism is quite complex and leads to at least 13 oxidative metabolites, not to mention conjugation reactions. A pharmacologically significant route is aromatic oxidation to the active 4-hydroxytamoxifen. However, the most important metabolic pathway in qualitative and quantitative terms is *N*-demethylation to the secondary amine *N*-desmethyltamoxifen, a reaction catalyzed mainly by CYP3A4. This metabolite, while reaching high plasma levels, is not as active as 4-hydroxytamoxifen and the second generation metabolite *N*-desmethyl-4-hydroxytamoxifen (known as endoxifen). Indeed, a number of investigations have demonstrated that 4-hydroxytamoxifen and endoxifen are equipotent and many times more active as antiestrogens than the parent drug [75]. When taking plasma levels in consideration [76], it seems that the main contributor to therapeutic activity in patients is endoxifen, and to a lesser extent tamoxifen and 4-hydroxy-tamoxifen.

An extreme case of activation by metabolism is that of prodrugs, namely intrinsically inactive (or poorly active) medicinal compounds whose biotransformation produces the active agent. This case is treated in a separate chapter and will not be considered here.

4.1.2. Toxicity The toxicological consequences of the metabolism of drugs and other xenobiotics can be favorable (i.e., result in detoxification) or unwanted (i.e., result in toxification) [15,25]. The risks of toxification associated with biotransformation [77–81] have now become a major issue in drug discovery and development, where minimizing metabolic toxification is given a high priority [82–89], for example, screening for reactive intermediates and assessing toxicity [90], with metabonomics [91,92] and toxicogenomics [93,94] being increasingly useful tools.

The main types of Averse Drug Reactions (ADRs) are summarized in Table 9 [15,25,96–98]. On-target ADRs result from an exaggerated response caused by drug overdosing or too high levels of an active metabolite; they are predictable in principle and generally dose-dependent, and are labeled as Type A. Off-target ADRs result from the interaction of the drug or a metabolite with a nonintended target such as a receptor or an enzyme. They also are predictable in principle and generally dose-dependent. These two types are pharmacological in nature and fall outside the scope of the present work. ADRs caused by reactive metabolites are the ones of greatest in our context. They involve covalent binding to macromolecules, and/or oxidative stress following the formation of reactive oxygen species (ROSs). These ADRs are predictable (or rationalizable) in terms of the drug's or metabolite's structure, and they are generally dose-dependent. They are often labeled as Type C. Idiosyncratic drug reactions (IDRs) (also known as Type B ADRs) are rare to very rare, unpredictable, and apparently dose-independent. They are also poorly understood, yet appear to be usually due to reactive metabolites.

Leaving aside on-target and off-target interactions, toxic responses are mostly due to reactive metabolites acting either directly on proteins and other macromolecules, or indirectly via reactive oxygen species, reactive nitrogen species (RNSs), and reactive carbonyl species (RCSs) [99–102]. Immunological reactions (not all of which should be classified as idiosyncratic) can be against self (autoimmunity) or against normal, harmless substances (allergy) [103–105]. In some cases, the immunotoxic drug or chemical acts as a hapten (by direct reaction with a protein). In most cases, however, the drug or chemical behaves as a prohapten since it needs metabolic activation to form adducts. In both cases, a hapten-carrier conjugate is formed that may elicit an immune response.

Table 9. Types and Mechanisms of Adverse Drug Reactions (ADRs) [15,25]

On-target ADRs	Predictable in principle and generally dose-dependent. Based on the pharmacology of the drug and its metabolite(s), often an exaggerated response or a response in a nontarget tissue
Off-target ADRs	Predictable in principle and generally dose-dependent. Resulting from the interaction of the drug or a metabolite with a nonintended target
ADRs involving reactive metabolites	Predictable in principle and generally dose-dependent. A major mechanism is covalent binding to macromolecules (adduct formation) resulting in cytotoxic responses, DNA damage, or hypersensitivity and immunological reactions. A distinct (and synergetic) mechanism is the formation of reactive oxygen species and oxidative stress
Idiosyncratic drug reactions (IDRs)	Unpredictable, apparently dose-independent, and rare (<1 in 5000 cases). They might result from a combination of genetic and external factors, but their nature is poorly understood. IDRs include anaphylaxis, blood dyscrasias, hepatotoxicity, and skin reactions

Drug- or chemical-induced cytotoxicity can occur through apoptosis or necrosis, molecular mechanisms being immune injury, adduct formation, oxidative stress, and/or DNA damage. Many organs are targets, but hepatotoxicity is of particular significance due to the high metabolizing capacity of the liver [106–111]. Other organs include the kidneys (nephrotoxicity), lungs and airways (pneumotoxicity), the central nervous system (neurotoxicity), reproductive organs, bone marrow (hematotoxicity), and the skin (sensitization). As for teratogenesis, a number of embryotoxic chemicals are believed to be proteratogens activated to reactive intermediates [112].

A look at toxicophoric groups (also called toxicophores, toxophores, or toxophoric groups, see Table 10) is particularly illustrative of the unity that unlies their chemical diversity [95,113–115]. Indeed, the toxic potential of many toxicophores is explained by their metabolic toxification to electrophilic intermediates or to free radicals. In more detail, the major functionalization reactions that activate toxophoric groups include oxidation to electrophilic intermediates, reduction to free radicals, and autooxidation with oxygen reduction that leads to superoxide, other reactive oxygen species, and reactive nitrogen species. The electrophilic intermediates and the free radicals then react with bio(macro) molecules, mainly proteins and nucleic acids, producing chemical lesions. ROSs also react with unsaturated and mainly polyunsaturated fatty acids in membranes and elsewhere, leading to lipid peroxidation. Of more recent awareness is the fact that some conjugation reactions may also lead to toxic metabolites, namely reactive acyl glucuronides or conjugates with deleterious physicochemical properties [15,25].

However, it would be wrong to conclude from the above that the presence of a toxophoric group necessarily implies toxicity. Reality is far less gloomy, as only potential toxicity is indicated. Given the presence of a toxophoric group in a compound, a number of factors will operate to render the latter either toxic or nontoxic:

(a) The molecular properties of the substrate will increase or decrease its affinity and reactivity toward toxification and detoxification pathways.
(b) Metabolic reactions of toxification are always accompanied by competitive and/or sequential reactions of detoxification that compete with the formation of the toxic metabolite and/or inactivate it. A profusion of biological factors control the relative effectiveness of these competitive and sequential pathways.
(c) The reactivity and half-life of a reactive metabolite control its sites of action and determine whether it will reach sensitive sites.
(d) Dose, rate, and route of entry into the organism are all factors of known significance.

Table 10. Major Toxophoric Groups and Their Metabolic Reactions of Toxification [15,25]

Functionalization reactions
- Some aromatic systems that can be oxidized to epoxides, quinines, or quinonimines (Fig. 5, reaction 1)
- Ethynyl moieties activated by cytochrome P450 (Fig. 5, reaction 3)
- Some halogenated alkyl groups that can undergo reductive dehalogenation (Fig. 3, reaction 7)
- Nitroarenes that can be reduced to nitro anion-radicals, nitrosoarenes, nitroxides, and hydroxylamines (Fig. 7, reaction 4)
- Some aromatic amides that can be activated to nitrenium ions (reaction 3 in Fig. 7, followed by reaction 3 in Fig. 18)
- Some thiocarbonyl derivatives, particularly thioamides, which can be oxidized to *S*,*S*-dioxide (sulfene) metabolites (Fig. 9, reaction 3)
- Thiols that can form mixed disulfides (Fig. 9, reaction 1)

Conjugation reactions
- Some carboxylic acids that can form reactive acylglucuronides (Fig. 21, reaction 3)
- Some carboxylic acids that can form highly lipophilic conjugates (Fig. 27, reactions 4 and 5)

(e) Above all, there exist essential mechanisms of survival value that operate to repair molecular lesions, remove them immunologically, and/or regenerate the lesioned sites.

In conclusion, the presence of a toxophoric group is not a sufficient condition for observable toxicity, a sobering and often underemphasized fact. Nor is it a necessary condition since other mechanisms of toxicity exist, for example the acute toxicity characteristic of many solvents.

4.2. Structure–Metabolism Relationships and the Prediction of Metabolism

Two classes of factors will influence qualitatively (how ? what ?) and quantitatively (how much ? how fast ?) the metabolism a given xenobiotic in a given biological system. The second class consists in the biological factors briefly presented below. The first class of factors are the various molecular properties that will influence a metabolic reaction, most notably (a) global molecular properties such as configuration (e.g., chirality), electronic structure, and lipophilicity and (b) local properties of the target sites such as steric hindrance, electron density, and reactivity. When speaking of a metabolic reaction, however, we should be clear about what is meant. Indeed, enzyme kinetics allows a metabolic reaction to be readily decomposed into a binding and a catalytic phase. In spite of its limitations, Michaelis–Menten analysis offers an informative approach for assessing the binding and catalytic components of a metabolic reaction [116]. Here, the Michaelis constant K_m represents the affinity, with V_{max} being the maximal velocity, k_{cat} the turnover number, and K_m/V_{max} the catalytic efficiency.

4.2.1. Chirality and Drug Metabolism The influence of stereochemical factors in xenobiotic metabolism is a well-known and reviewed phenomenon [117]. Because metabolic reactions can produce more than one response (i.e., several metabolites), two basic types of stereoselectivity are seen, namely substrate stereoselectivity and product stereoselectivity (see Section 1.3). Substrate stereoselectivity occurs when stereoisomers are metabolized differently (in quantitative and/or qualitative terms) and by the "same" biological system under identical conditions. Substrate stereoselectivity is a well-known and abundantly documented phenomenon under *in vivo* and *in vitro* conditions. In fact, it is the rule for many chiral drugs, ranging from practically complete to moderate. A complete absence of substrate enantioselectivity has seldom been seen.

Michaelis–Menten analysis suggests that the molecular mechanism of substrate enantioselectivity can occur in the binding step (different affinities K_m), in the catalytic step (different reactivities, V_{max}) or in both. There are cases in which stereoselective metabolism is of toxicological relevance. This situation can be illustrated with the antidepressant drug mianserin that undergoes substrate stereoselective oxidation. In human liver microsomes, mianserin (**44** in Fig. 36) occurred by aromatic oxidation with a marked preference for its (*S*)-enantiomer, while *N*-demethylation was the major route for the (*R*)-enantiomer. At low drug concentrations, cytotoxicity toward human mononuclear leucocytes was due to (*R*)-mianserin more than to (*S*)-mianserin, and showed a significant correlation with *N*-demethylation [118]. Thus, the toxicity of mianserin seemed associated with *N*-demethylation rather than with aromatic oxidation. The chemical nature of the toxic intermediates was not established.

Product stereoselectivity occurs when stereoisomeric metabolites are generated differently (in quantitative and/or qualitative terms) and from a single substrate with a suitable prochiral center or face. Examples of metabolic pathways producing new centers of chirality in substrate molecules include ketone reduction, reduction of carbon–carbon double bonds, hydroxylation of prochiral methylenes, oxygenation of tertiary amines to *N*-oxides, and oxygenation of sulfides to sulfoxides. Product stereoselectivity may be due to the action of distinct isoenzymes, or it may result from different binding modes of a prochiral substrate to a single isoenzyme. In this case each productive binding mode will bring another of the two enantiotopic or diastereotopic target groups in the vicinity of the

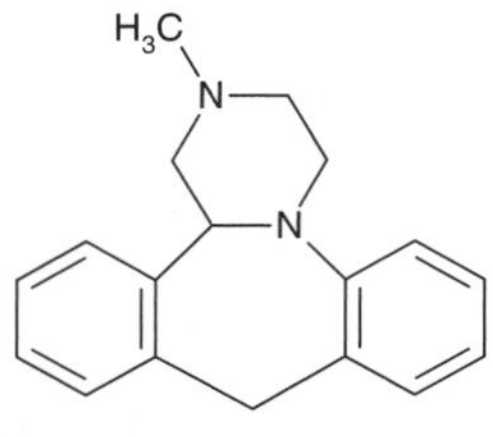

44

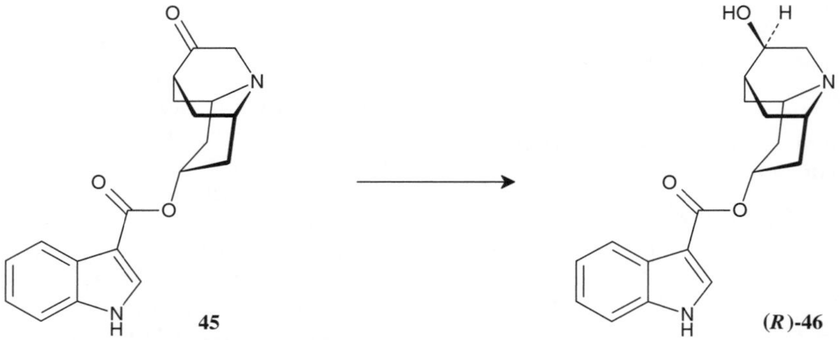

45

(*R*)-46

Figure 36. Mianserine (**44**), dolasetron (**45**), and its active alcohol metabolite (**46**).

catalytic site, resulting in diastereoisomeric enzyme–substrate complexes. Thus, the ratio of products depends on the relative probability of the two binding modes. In addition, the catalytic step, which involves diastereoisomeric transition states, may also influence or control product selectivity.

The potential pharmacological consequence of product-selective metabolism can be illustrated with the antiemetic 5-HT3 antagonist dolansetron (**45** in Fig. 36). This compound is prochiral by virtue of a center of prochirality at the keto group featuring two enantiotopic faces. This drug is reduced rapidly, extensively and stereoselectively in humans to its (*R*)-alcohol (**46** in Fig. 36) [119,120]. The enzymes involved are aldo-keto reductases (AKR1C1, 1C2, and 1C4) and carbonyl reductases. The metabolite proved to be manyfold more active than dolasetron, allowing the latter to be viewed as a prodrug. And significantly, the (*R*)-alcohol is markedly more active than its (*S*)-enantiomer.

It thus appears that substrate and product stereoselectivities are the rule in the metabolism of stereoisomeric and stereotopic drugs. But if the phenomenon is to be expected *per se*, it is not trivial to predict which enantiomer will be the preferred substrate of a given metabolic reaction, or which enantiomeric metabolite of a prochiral drug will predominate. This is due to the fact that the observed stereoselectivity of a given reaction will depend both on molecular properties of the substrate and on enzymatic factors (e.g., the stereoelectronic architecture of the catalytic site in the various isozymes involved). In fact, substrate and product stereoselectivities are determined by the binding mode(s) of the substrate and by the resulting topography of the enzyme–substrate complex.

4.2.2. Relations Between Metabolism and Lipophilicity Comparing the overall metabolism of numerous drugs clearly reveals a global relation with lipophilicity [116]. Indeed, there exist some highly polar xenobiotics known to be essentially resistant to any metabolic reaction, for example, saccharin, disodium cromoglycate, and zanamivir. Furthermore, many *in vivo* metabolic studies have demonstrated a dependence of biotransformation on lipophilicity, suggesting a predominant role for transport and partitioning processes. A particularly illustrative example is offered by β-blockers, where the more lipophilic drugs are extensively if not completely metabolized (e.g., propranolol), whereas the more hydrophilic ones undergo biotransformation for only a fraction of the dose (e.g., atenolol).

This global trend is in line with the Darwinian rationale for xenobiotic metabolism, which is believed to have evolved in an animal–plant "warfare," with herbivores adaptating to the emergence of protective chemicals (e.g., alkaloids) in plants [121]. The exception to the global and direct relation between extent of metabolism and lipophilicity is offered by the vast number of human-made, highly lipophilic polyhalogenated xenobiotics, which now pollute our entire biosphere. Such compounds include polyhalogenated insecticides (e.g., DDT), polyhalogenated biphenyls, and dioxins, which have a strong propensity to accumulate in adipose tissues. In addition, these compounds are highly resistant to biotransformation in animals due in part to their very high lipophilicity, and in part to the steric shielding against enzymatic attack provided by the halo substituents.

When the results of Michaelis–Menten analyses are examined for quantitative structure–metabolism relationships (QSMRs), it is often found that lipophilicity correlates with K_m but not with V_{max} or k_{cat}. This is documented for ester hydrolysis and oxidation of various chemical series by monooxygenases and other oxidases [116]. Depending on the explored property space, the relationships between K_m and lipophilicity are linear or parabolic. Such results indicate that when relations are found between rate of metabolism and lipophilicity, the energy barrier of the reaction is largely similar for all compounds in the series, allowing lipophilicity to become the determining factor. A relevant example is provided by the CYP2D6-catalyzed oxidation of a series of fluorinated propranolol derivatives. Their K_m values spanned a 400-fold range and showed a weak but real correlation with the distribution coefficient. The same was true for the catalytic efficiency k_{cat}/K_m, but only because the k_{cat} values spanned a narrow sixfold range [122].

4.2.3. The Influence of Electronic Factors

Stereoelectronic properties may influence the binding of substrates to enzymatic active sites in the same manner as they influence the binding of ligands to receptors. For example, the K_m values of the above-mentioned propranolol derivatives were highly correlated with their basicity (pK_a), confirming that an ionic bond plays a critical role in the binding of amine substrates to CYP2D6 [122].

Electronic properties are of particular interest in structure–metabolism relationships since they control the cleavage and formation of covalent bonds characteristic of a biotransformation reaction (i.e., the catalytic step). Correlations between electronic parameters and catalytic parameters obtained from *in vitro* studies (e.g., V_{max} or k_{cat}) allow a rationalization of substrate selectivity and some insight into reaction mechanism.

An example of a quantitative SMR study correlating electronic properties and catalytic parameters is provided by the glutathione conjugation of *para*-substituted 1-chloro-2-nitrobenzene derivatives [123]. The values of $\log k_2$ (second order rate constant of the nonenzymatic reaction) and $\log k_{cat}$ (enzymatic reaction catalyzed by various glutathione transferase preparations) were correlated with the Hammett resonance σ-value of the substrates, a measure of their electrophilicity. Regression equations with positive slopes and r^2 values in the range 0.88–0.98 were obtained. These results quantitate the influence of substrate electrophilicity on nucleophilic substitutions mediated by glutathione, be they enzymatic or nonenzymatic.

Quantum mechanical calculations may also shed light on SMRs, revealing correlations between rates of metabolic oxidation and energy barrier in cleavage of the target C–H bond [124].

4.2.4. 3D-QSMRs and Molecular Modeling

When lipophilicity and electronic parameters are used as independent variables and a metabolic parameter (often assessing affinity) as the dependent variable, correlation equations are obtained for rather limited and/or related series of substrates, implying exploration of a relatively narrow structural diversity space. Also, the metabolic parameters are usually

Table 11. A Classification of *In Silico* Methods to Predict Biotransformation [14,21]

(A) "Local" methods

Methods applicable to series of compounds with "narrow" chemical diversity, and/or to biological systems of "low" complexity:

- QSAR: *linear, multilinear, multivariate* (may predict affinities, relative rates, and so on, depending on the physicochemical properties considered)
- Quantum mechanical: *MO methods* (may predict regioselectivity, mechanisms, relative rates, etc.)
- 3D-QSAR: *CoMFA*™, *Catalyst*™, *GRID/GOLPE*™, for example (may predict substrate behavior, relative rates, inhibitor behavior, etc.)
- Molecular modeling and docking (may predict substrate or inhibitor behavior, regioselectivity, etc.)
- Experts systems combining docking, 3D-QSAR and MO: *MetaSite*™, and so on

(B) "Global" methods

Methods applicable to series of compounds with "broad" chemical diversity (*and, in the future, to biological systems of "high" complexity*):

- "Meta"-systems combining (A) docking, 3D-QSAR, MO, and B) a number of enzymes and other functional proteins: *MetaDrug*™, and so on
- Databases: *Metabolite*™, *Metabolism*™, and so on (allow metabolites to be deduced by analogy)
- Rule-based expert systems: *Meta*™, *MetabolExpert*™, *Meteor*™, and so on (may predict metabolites, metabolic trees, reactive/adduct-forming metabolites, relative importance of these metabolites)

obtained from relatively simple biological systems, resulting in QSMR correlations that are typically "local" methods (Table 11) of low extrapolative capacity.

Quantum mechanical methods are also classified as local in Table 11. Here, a word of caution is necessary, since such methods are in principle applicable to any chemical system. However, they cannot handle more than one metabolic reaction or catalytic mechanism at a time, and as such can only predict metabolism in simple biological systems, in contrast to the global methods presented below.

Three-dimensional (3D) methods are also of value in SMRs, namely 3D-QSARs and the molecular modeling of xenobiotic-metabolizing enzymes (Table 11). Indeed, they represent a marked progress in predicting the metabolic behavior (be it as substrates or inhibitors) of novel compounds. An obvious restriction, however, is that each enzyme requires a specific model. 3D-QSMRs methods yield a partial view of the binding/catalytic site of a given enzyme as derived from the 3D molecular fields of a series of substrates or inhibitors (the training set). In other words, they yield a "photographic negative" of such sites, and will allow a quantitative prediction for novel compounds structurally related to the training set. Two popular methods in 3D-QSARs are CoMFA (comparative molecular field analysis) and catalyst. Numerous applications can be found in the literature.

The same is true for the molecular modeling of xenobiotic-metabolizing enzymes, which affords another approach to rationalize and predict drug–enzyme interactions. The methodology of molecular modeling is explained in detail elsewhere in this work and will not be presented here. Suffice it to say that its application to drug metabolism was made possible by the crystallization and X-ray structural determination of the first bacterial cytochromes P450 in the mid-1980s [125], followed a few years ago by human cytochromes P450 [126,127]; the crystal structure of numerous other drug-metabolizing enzymes is also available. As more and more amino acid sequences of drug-metabolizing enzymes become available, their tertiary structure can be modeled by homology superimposition with the experimentally determined template of a closely related protein (homology modeling). Known substrates and inhibitors can then be docked *in silico*. Given the assumptions made in homology modeling and in scoring functions, such models cannot give quantitative affinity predictions. However, they can afford fairly reliable yes/no answers as to the affinity of test set compounds, and in favorable cases may also predict the regioselectivity of metabolic attack (Table 11).

The last approach among "local" systems are expert systems combining (a) 3D-models obtained by molecular modeling and (b) sophisticated QSAR approaches based on multivariate analyses of parameters obtained from molecular interaction fields (MIFs), as found in the MetaSite algorithm [128,129]. MetaSite is a specific system in the sense that it is currently restricted to the major human cytochromes P450. At the end of the procedure the atoms of the substrate are ranked according to their accessibility and reactivity. In other words, MetaSite takes the 3D stereoelectronic structure of both the enzyme and the ligand into account to prioritize the potential target sites of CYP-catalyzed oxidation in the molecule.

4.2.5. "Global" Expert Systems to Predict Biotransformation While medicinal chemists are not usually expected to possess a deep knowledge of the mechanistic and biological factors that influence drug metabolism, they will find it quite useful to have a sufficient understanding of structure–metabolism relationships to be able to predict reasonable metabolic schemes. A qualitative prediction of the biotransformation of a novel xenobiotic should allow (a) the identification of all target groups and sites of metabolic attack, (b) the listing of all possible metabolic reactions able to affect these groups and sites, and (c) the organization of the metabolites into a metabolic tree. The next desirable features would be a warning for potentially reactive/adduct-forming metabolites, and a bridge to a toxicity-predicting algorithm.

Given the available information, there exists an ever increasing interest in expert systems hopefully able to meet the above goals [130–132]. A few systems of this type are now available, for example a system known as MetaDrug combining (a) docking, 3D-QSAR, MO and (b) a number of enzymes and other functional proteins such as transporters [133,134]. Metabolic databases allow reasoning by analogy and can prove quite useful [130]. Rule-based systems are perhaps the most versatile; they include Meta [135,136], MetabolExpert [137,138], and Meteor [139–141]. Such systems will make correct qualitative or even semiquantitative predictions for a number of metabolites, but the risk of false positives must be taken seriously. This is due to the great difficulty of devising efficient filters to remove unlikely metabolites, based on the molecular properties of substrates. Taking biological factors (tissues, animal species, etc.) into account, is still far away.

4.2.6. Biological Factors Influencing Drug Metabolism A variety of physiological and pathological factors influence xenobiotic metabolism and hence the wanted and unwanted activities associated with a drug. This issue complicates significantly the task of medicinal chemists, as optimizing the pharmacokinetic property of a given lead for behavior in laboratory animals often results in clinical failure. While it is not the objective of this chapter to discuss biological factors in any depth, a brief overview will nevertheless be presented for the sake of clarity and to help medicinal chemists prioritize the challenges they face. A structured, comprehensive and amply illustrated discussion of these factors can be found in recent works [15,26,27].

It can be helpful to distinguish between interindividual and intraindividual factors that can influence the capacity of an individual to metabolize drugs (Table 12). The interindividual factors are viewed as remaining constant throughout the life span of an organism and are the expression of its genome. In contrast, the intraindividual factors may vary depending on time (age, even time of day), pathological states, or external factors (nutrition, pollutants, drug treatment).

The most significant interindividual factor to be considered in early drug development is clearly the species-related differences. The use of different species as a surrogate for investigating and predicting metabolism in humans has been and remains an indispensable step in drug development. In parallel, there has been a steady switch during early phases of drug discovery from obtaining metabolism data with animal tissues and whole animals to screening studies with human

Table 12. Biological Factors Affecting Xenobiotic Metabolism [15,26,27]

Interindividual Factors	Intraindividual Factors
Constant for a given organism	*Variable during the lifetime of a given organism*
Animal species	Physiological changes: Age Biological rhythms Pregnancy
Genetic factors (genetic polymorphism)	Pathological changes: Disease Stress
Sex	External influences: Nutrition Enzyme induction by xenobiotics Enzyme inhibition by xenobiotics

cDNA expressed enzymes, cell fractions (microsomes), and cells. Nevertheless, a greater understanding of the molecular, genetic, and physiological differences between humans and animal models is still required for correctly interpreting pharmacological safety studies required by regulatory agencies [142].

Genetic factors may result in polymorphic biotransformations that may cause serious problems at the stage of clinical trials and postmarketing. As much as possible, lead optimization may try to avoid affinity for cytochromes P450 known to exhibit phenotypic differences. After the specific CYP(s) involved in the metabolism of a drug are known, how does this information relate to the overall metabolism of a drug in a diverse population? The ability for an individual to metabolize a drug is dependent on the nature (genotype), location and amount of enzyme present. Studying such factors is the field of pharmacogenetics and pharmacogenomics. At the level of populations, ethnopharmacology investigates the differences in proportions of "normal" and slow metabolizers observed in different populations, a study complicated by the unavoidable influence of external factors such as nutrition and lifestyle cannot be excluded. Comparable arguments are valid for sex-related differences in metabolism, which in humans are known for a limited number of drugs only [143].

Intraindividual factors are numerous and for some difficult to investigate. A major one is age, since differences between the levels of metabolism enzymes for the fetal and neonatal (first four weeks postpartum) liver versus the adult liver have been observed in both animals and humans [144]. The problem is also significant for the growing population of elderly patients, where a marked reduction is usually observed. The origin of such differences is quite complex and often related to other physiological changes such as reduction in liver blood flow and liver size. As the number of older individuals increase in the population, especially those aged 80–90 years, adaptations in drug therapy and posology as related to metabolism become increasingly important.

There is increasing evidence that drug metabolism may undergo some variations due to diurnal, monthly and yearly rhythms. Thus, it is well established that nearly all physiological functions and parameters can vary with the time of day, for example, heart rate, blood pressure, hepatic blood flow, urinary pH, and plasma concentrations of hormones, other signal molecules, glucose, and plasma proteins [145].

A number of diseases may sometimes influence drug metabolism, although no generalization appears in sight. First and above all, diseases of the liver will have a direct effect on enzymatic activities in this organ, for example, cirrhosis, hepatitis, jaundice, and tumors. Diseases of extra-hepatic organs such as the lungs and the heart may also affect hepatic drug metabolism by influencing blood flow, oxygenation, and other vital functions. Renal diseases have a particular significance due to a decrease of the intrinsic xenobiotic-metabolizing activity of the kidneys, and a decreased rate of urinary excretion. Infectious and inflammatory diseases have been associated with differential expression of various CYP enzymes.

Interference with the metabolism of a drug by inhibition of the primary enzymes involved in its metabolism can lead to serious side-effects or therapeutic failure. Inhibitors of the CYP enzymes are considered to be the most problematic, being the primary cause of drug–drug and drug–food interactions [16,146].

A comparable situation is created by enzyme induction, that is, when the amount and activity of a metabolizing enzyme has increased following exposure to a drug or chemical. In contrast to inhibition, induction is generally a slow process caused by an increase in the synthesis of the enzyme. A drug that causes induction may increase its own metabolism (autoinduction) or increase metabolism of another drug. Clinically, drug induction is considered less of a problem than drug inhibition, but there are some marked exceptions. Primary interest has been directed toward the CYP enzymes for which a number of classes of inducing agents have been identified.

5. CONCLUDING REMARKS

Drug discovery and development are becoming more complex by the day, with physicochemical, pharmacokinetic and pharmacodynamic properties being screened and assessed as early and as simultaneously as possible. But the real challenge lies with the resulting deluge of data, which must be stored, analyzed and interpreted. This calls for a successful synergy between humans and algorithmic machines, in other words between human and artificial intelligence. However, making sense of the data, interpreting their analyses, and rationally planning subsequent steps is an entirely different issue. The difference is a qualitative one, and it is the difference between information and knowledge.

This chapter is about both. Our first objective in writing it was obviously to supply useful information, namely structured data as exemplified by the classification of metabolic reactions (Sections 2 and 3).

Information becomes knowledge when it is connected to a context from which it receives meaning. This treatise is about medicinal chemistry, and indeed medicinal chemistry is the context of our chapter and of all others. By discussing the connection between drug metabolism and the medicinal chemistry context (Section 4), we have tried to be true to our second objective that was to present medicinal chemists with meaningful information. Had this chapter been written for a treatise of molecular biology, much of the basic information would have been the same, but the context and the chapter's meaning would have been different. This also tells us that medicinal chemistry itself needs a context to acquire meaning, the context of human welfare to which responsible medicinal chemists are proud to contribute [147].

REFERENCES

1. Testa B, Jenner P. Drug Metabolism. Chemical and Biochemical Aspects. New York: Dekker; 1976.
2. Silverman RB. The Organic Chemistry of Drug Design and Drug Action. San Diego: Academic Press; 1992.
3. Kauffman FC, editor. Conjugation–Deconjugation Reactions in Drug Metabolism and Toxicity. Berlin: Springer Verlag; 1994.
4. Testa B. The Metabolism of Drugs and Other Xenobiotics—Biochemistry of Redox Reactions. London: Academic Press; 1995.
5. Ortiz de Montellano PR, editor. Cytochrome P450. Structure, Mechanism, and Biochemistry. 2nd ed. New York: Plenum Press; 1996.
6. Woolf TF, editor. Handbook of Drug Metabolism. New York: Dekker; 1999.
7. Ioannides C, editor. Enzyme Systems that Metabolise Drugs and Other Xenobiotics. Chichester, UK: John Wiley & Sons; 2002.
8. Rodrigues AD, editor. Drug–Drug Interactions. New York: Dekker; 2002.
9. Testa B, Mayer JM. Hydrolysis in Drug and Prodrug Metabolism—Chemistry, Biochemistry and Enzymology. Zurich/Weinheim: VHCA/Wiley-VCH; 2003.
10. Boullata J, Armenti VT, editors. Handbook of Drug–Nutrient Interactions. Totowa, NJ: Humana Press; 2004.
11. Lash LH, editor. Drug Metabolism and Transport. Totowa, NJ: Humana Press; 2005.
12. Phillips IR, Shephard EA, editors. Cytochrome P450 Protocols. Totowa NJ: Humana Press; 2006.
13. Utrecht JP, Trager WF. Drug Metabolism—Chemical and Enzymatic Aspects. New York: Informa; 2007.

14. Testa B, Krämer SD. The Biochemistry of Drug Metabolism. Principles, Redox Reactions, Hydrolyses Vol. 1. Zurich/Weinheim: VHCA/Wiley-VHC; 2008.
15. Testa B, Krämer SD. The Biochemistry of Drug Metabolism. Conjugation Reactions, Consequences of Metabolism, Influencing Factors, Vol. 2. Zurich/Weinheim: VHCA/Wiley-VHC; 2009.
16. Rendic S. Summary of information on human CYP enzymes: Human P450 metabolism data. Drug Metab Rev 2002;34:83–448.
17. Pelkonen O. Human CYPs: *In vivo* and clinical aspects. Drug Metab Rev 2002;34:37–46.
18. Trager WF. Principles of drug metabolism 1: redox reactions. In: Testa B, van de Waterbeemd H, editors. ADME-Tox Approaches. In: Taylor JB, Triggle DJ, editors. Comprehensive Medicinal Chemistry, Vol. 5, 2nd ed. Oxford: Elsevier; 2007. p 87–132.
19. Testa B. Principles of drug metabolism 2: hydrolysis and conjugation reactions. In: Testa B, van de Waterbeemd H, editors. ADME-Tox Approaches. In: Taylor JB, Triggle DJ, editors. Comprehensive Medicinal Chemistry, Vol. 5, 2nd ed. Oxford: Elsevier; 2007. p 133–166.
20. Totah RA, Rettie AE. Principles of drug metabolism 3: enzymes and tissues. In: Testa B, van de Waterbeemd H, editors. ADME-Tox Approaches. In: Taylor JB, Triggle DJ, editors. Comprehensive Medicinal Chemistry, Vol. 5, 2nd ed. Oxford: Elsevier; 2007. p 167–191.
21. Testa B, Krämer SD. The biochemistry of drug metabolism—an introduction. Part 1: Principles and overview. Chem Biodiver 2006; 3:1053–1101.
22. Testa B, Krämer SD. The biochemistry of drug metabolism—an introduction. Part 2: Redox Reactions and their Enzymes. Chem Biodiver 2007;4:257–405.
23. Testa B, Krämer SD. The biochemistry of drug metabolism—an introduction. Part 3: Reactions of hydrolysis and their enzymes. Chem Biodiver 2007;4:2031–2122.
24. Testa B, Krämer SD. The biochemistry of drug metabolism—an introduction. Part 4: Reactions of Conjugation and their Enzymes. Chem Biodiver 2008;5:2171–2336.
25. Testa B, Krämer SD. The biochemistry of drug metabolism—an introduction. Part 5: Metabolism and Bioactivity. Chem Biodiver 2009;6: 591–684.
26. Krämer SD, Testa B. The biochemistry of drug metabolism—an introduction. Part 6: Inter-individual factors affecting drug metabolism. Chem Biodiver 2008;5:2465–2578.
27. Krämer SD, Testa B. The biochemistry of drug metabolism—an introduction. Part 7: Intra-individual factors affecting drug metabolism. Chem Biodiver 2009;6:1477–1660.
28. Testa B. Nonenzymatic contributions to xenobiotic metabolism. Drug Metab. Rev. 1982;13: 25–50.
29. Balant LP, Roseboom H, Guntert-Remy RM. Pharmacokinetic criteria for drug research and development. In: Testa B, editor. Advances in Drug Research. Vol. 19. London: Academic Press; 1990. p 1–138.
30. Gaviraghi G, Barnaby RJ, Pellegatti M. Pharmacokinetic challenges in lead optimization. In: Testa B, van de Waterbeemd H, Fokers G, Guy R, editors. Pharmacokinetic Optimization in Drug Research—Biological, Physicochemical and Computational Strategies. Zurich/Weinheim; VHCA/Wiley-VCH; 2001. p 3–14.
31. Walle T, Walle UK, Olanoff LS. Quantitative account of propranolol metabolism in urine of normal man. Drug Metab Dispos 1985; 13:204–209.
32. Nomenclature Committee of the International Union of Biochemistry and Molecular Biology (IUBMB). Enzyme Nomenclature. Available at http://www.chem.qmul.ac.uk/iubmb/enzyme/. Accessed 2008 Oct 27.
33. Nelson DR, Zeldin DC, Hoffman SMG, Maltais LJ, Wain HM, Nebert DW. Comparison of cytochrome P450 (CYP) genes from the mouse and human genomes, including nomenclature recommendations for genes, pseudogenes and alternative splice variants. Pharmacogenetics 2004;14:1–18.
34. Nelson DR. Cytochrome P450 Homepage. Available at http://drnelson.utmem.edu/CytochromeP450.html. Accessed 2008 Oct 27.
35. Directory of P450-containing Systems. Available at http://www.icgeb.org/~p450srv/. Accessed 2008 Oct 27.
36. Brenda: The Comprehensive Enzyme Information System. Available at http://www.brenda-enzymes.info/. Accessed 2008 Oct 27.
37. ExPASy Proteomics Server. Available at www.expasy.org. Accessed 2008 Oct 27.
38. Korzekwa KR, Trager WF, Nagata K, Parkinson A, Gillette JR. Isotope effect studies on the mechanism of the cytochrome P-450IIA1-catalyzed formation of delta 6-testosterone from testosterone. Drug Metab Dispos 1990;18:974–979.

39. Rambeck B, May T, Juergens U. Serum concentrations of carbamazepine and its epoxide and diol metabolites in epileptic patients: the influence of dose and comedication. Ther Drug Monitor 1987;9:298–303.
40. Gorrod JW, Raman A. Imines as intermediates in oxidative aralkylamine metabolism. Drug Metab Rev 1989;20:307–339.
41. Weinshilboum RM, Otterness DM, Szumlanski CL. Methylation pharmacogenetics: catechol O-methyltransferase, thiopurine methyltransferase, and histamine N-methyltransferase. Annu Rev Pharmacol Toxicol 1999; 39:19–52.
42. Nagata K, Yamazoe Y. Pharmacogenetics of sulfotransferases. Annu Rev Pharmacol Toxicol 2000;40:159–176.
43. Blanchard RL, Freimuth RR, Buck J, Weinshilboum RM, Coughtrie MW. A proposed nomenclature system for the cytosolic sulfotransferase (SULT) superfamily. Pharmacogenetics 2004;14:199–211.
44. Mackenzie PI, Owens IS, Burchell B, Bock KW, Bairoch A, Bélanger A, Fournel-Gigleux S, Green M, Hum DW, Iyanagi T, Lancet D, Louisot P, Magdalou J, Chowdhury JR, Ritter JK, Schachter H, Tephly TR, Tipton KF, Nebert DW. The UDP glycosyltransferase gene superfamily: recommended nomenclature update based on evolutionary divergence. Pharmacogenetics 1997;7:255–269.
45. Radominska-Pandya A, Czernik PJ, Little JM, Battaglia E, Mackenzie PI. Structural and functional studies of UDP-glucuronosyltransferases. Drug Metab Rev 1999;31:817–899.
46. Spahn-Langguth H, Benet LZ. Acyl glucuronides revisited: is the glucuronidation process a toxification as well as a detoxification mechanism?. Drug Metab Rev 1992;24:5–48.
47. Grant DM, Blum M, Meyer UA. Polymorphisms of N-acetyltransferase genes. Xenobiotica 1992;22:1073–1081.
48. Dodds PF. Xenobiotic lipids: the inclusion of xenobiotic compounds in pathways of lipid synthesis. Prog Lipid Res 1995;34:219–247.
49. Kaphalia BS, Fritz RR, Ansari GAS. Purification and characterization of rat liver microsomal fatty acid ethyl and 2-chloroethyl ester synthase and their relationship with carboxylesterase (pI 6.1). Chem Res Toxicol 1997;10: 211–218.
50. Fears R. Lipophilic xenobiotic conjugates: the pharmacological and toxicological consequences of the participation of drugs and other foreign compounds as substrates in lipid biosynthesis. Prog Lipid Res. 1986;24:177–195.
51. Vickery S, Dodds PF. Incorporation of xenobiotic carboxylic acids into lipids by cultured 3T3-L1 adipocytes. Xenobiotica 2004; 34:1025–1042.
52. Williams K, Day R, Knihinicki R, Duffield A. The stereoselective uptake of ibuprofen enantiomers into adipose tissue. Biochem Pharmacol 1986;35:3403–3405.
53. Mayer JM, Testa B. Pharmacodynamics, pharmacokinetics and toxicity of ibuprofen enantiomers. Drugs of the Future 1997; 22:1347–1366.
54. Bjorge SM, Baillie TA. Studies on the beta-oxidation of valproic acid in rat liver mitochondrial preparations. Drug Metab. Dispos. 1991;19:823–829.
55. Sies H. Glutathione and its role in cellular functions. Free Radic. Biol. Med. 1999;27: 916–921.
56. Dickinson DA, Forman HJ. Cellular glutathione and thiols metabolism. Biochem Pharmacol 2002;64:1019–1026.
57. Pompella A, Visvikis A, Paolicchi A, De Tata V, Casini AF. The changing faces of glutathione, a cellular protagonist. Biochem Pharmacol 2003;66:1499–1503.
58. Hayes JD, Flanagan JU, Jowsey IR. Glutathione transferases. Annu Rev Pharmacol Toxicol 2005;45:51–88.
59. Armstrong RN. Structure, catalytic mechanism, and evolution of the glutathione transferases. Chem Res Toxicol 1997;10:2–18.
60. van der Aar EM, Tan KT, Commandeur JNM, Vermeulen NPE. Strategies to characterize the mechanisms of action of glutathione S-transferases: a review. Drug Metab. Rev. 1998;30:569–643.
61. Ketterer B. The role of nonenzymatic reactions of glutathione in xenobiotic metabolism. Drug Metab. Rev. 1982;13:161–187.
62. Anders MW, Dekant W. Glutathione-dependent bioactivation of haloalkenes. Annu Rev Pharmacol Toxicol 1998;38:501–537.
63. Baillie TA, Slatter JG. Glutathione: a vehicle for the transport of chemically reactive metabolites *in vivo*. Acc Chem Res 1991;24: 264–270.
64. Cribb AE, Miller M, Leeder JS, Hill J, Spielberg SP. Reactions of the nitroso and hydroxylamine metabolites of sulfamethoxazole with reduced glutathione—implications for idiosyncratic toxicity. Drug Metab Dispos 1991;19:900–906.
65. De Clercq E. The discovery of antiviral agents: ten different compounds, ten different stories. Med Res Rev 2008;28:929–953.

66. O'Donnell JP. The reaction of amines with carbonyls: its significance in the nonenzymatic metabolism of xenobiotics. Drug Metab. Rev. 1982;13:123–159.
67. Fura A, Shu YZ, Zhu M, Hanson RL, Roongta V, Humphreys WG. Discovering drugs though biological transformation: role of pharmacologically active metabolites in drug discovery. J Med Chem 2004;47:4339–4351.
68. Caccia S, Garattini S. Formation of active metabolites of psychotropic drugs. An updated review of their significance. Clin Pharmacokin 1990;18:434–459.
69. Garattini S. Active drug metabolites. An overview of their relevance in clinical pharmacokinetics. Clin Pharmacokin 1985;10:216–227.
70. Bodor N, Buchwald P. Soft drug design: general principles and recent applications. Med Res Rev 2000;20:58–101.
71. Bodor N. Novel approaches to the design of safer drugs: soft drugs and site-specific chemical delivery systems. In: Testa B, editor. Advances in Drug Research, Vol. 13. London: Academic Press; 1984. p. 255–331.
72. Bodor N. Soft drugs: principles and methods for the design of safe drugs. Med Res Rev 1984;4:449–469.
73. Durant NN, Katz RL. Suxamethonium. Br J Anaesth 1982;54:195–207.
74. Greenblatt DJ, Divoli M, Abernethy DR, Ochs HR, Shader RI. Clinical pharmacokinetics of the newer benzodiazepines. Clin Pharmacokin 1983;8:233–252.
75. Lim YC, Desta Z, Flockhart DA, Skaar TC. Endoxifen has anti-estrogenic effects in breast cancer cells with potency similar to 4-hydroxytamoxifen. Canc Chemother Pharmacol 2005;55:471–478.
76. Farlanut M, Franceschi L, Pasqual E, Bacchetti S, Poz D, Giorda G, Cagol P. Tamoxifen and its main metabolites serum and tissue concentrations in breast cancer women. Ther Drug Monit. 2007;29:349–352.
77. Josephy PD, Mannervik B. Molecular Toxicology. 2nd ed. Oxford: Oxford University Press; 2006.
78. Anders MW, editor. Bioactivation of Foreign Compounds. London: Academic Press; 1985.
79. Oesch-Bartlomowicz B, Oesch F. Mechanisms of toxification and detoxification that challenge drug candidates and drugs. In: Testa B, van de Waterbeemd H, editors. ADME-Tox Approaches. In: Taylor JB, Triggle DJ, editors. Comprehensive Medicinal Chemistry, Vol. 5, 2nd ed. Oxford: Elsevier; 2007. p. 193–214.
80. Macherey AC, Dansette PM. Biotransformations leading to toxic metabolites: chemical aspects. In: Wermuth CG, editor. The Practice of Medicinal Chemistry. 3rd ed. London: Academic Press; 2008. p 674–696.
81. Liebler DC. The poison within: application of toxicity mechanisms to fundamental disease processes, Chem Toxicol Res 2006;19: 610–613.
82. Baillie TA. Future of toxicology—metabolite activation and drug design: challenges and opportunities in chemical toxicology. Chem Res Toxicol 2006;19:889–893.
83. Doss GA, Baillie TA. Addressing metabolic activation as an integral component of drug design. Drug Metab. Rev. 2006;38:641–649.
84. Kumar S, Kassahun K, Tschirret-Guth RA, Mitra K, Baillie TA. Minimizing metabolic activation during pharmaceutical lead optimization: progress, knowledge gaps and future directions. Curr Opin Drug Discov Devel 2008;11:43–52.
85. Shu YZ, Johnson BM, Yang TJ. Role of biotransformation studies in minimizing metabolism-related liabilities in drug discovery. AAPS J. 2008;10:178–192.
86. Guengerich FP, MacDonald JS. Applying mechanisms of chemical toxicity to predict drug safety. Chem Res Toxicol 2007;20: 344–369.
87. Humphreys WG, Unger SE. Safety assessment of drug metabolites: characterization of chemically stable metabolites. Chem Res Toxicol 2006;19:1564–1569.
88. Hop CECA, Kalgutkar AS, Soglia JR. Importance of early assessment of bioactivation in drug discovery. Annu Rep Med Chem 2006;41: 369–381.
89. Stevens JL. Future of toxicology—mechanisms of toxicity and drug safety: where do we go from here? Chem Res Toxicol 2006;19: 1393–1401.
90. Caldwell GW, Yan Z. Screening for reactive intermediates and toxicity assessment in drug discovery. Curr Opin Drug Discov Devel 2006; 9:47–60.
91. Wilson ID, Nicholson JK. Metabonomics. In: Testa B, van de Waterbeemd H, editors. ADME-Tox Approaches. In: Taylor JB, Triggle DJ, editors. Comprehensive Medicinal Chemistry, Vol. 5. 2nd ed. Oxford: Elsevier; 2007. p. 989–1007.

92. Mortishire-Smith RJ, Skiles GL, Lawrence JW, Spence S, Nicholls AW, Johnson BA, Nicholson JK. Use of metabonomics to identify impaired fatty acid metabolism as the mechanism of drug-induced toxicity. Chem Res Toxicol 2004;17:165–173.
93. Khor TO, Ibrahim S, Kong ANT. Toxicogenomics in drug discovery and drug development: potential applications and future challenges. Pharm Res 2006;23:1659–1664.
94. Ulrich R, Friend SH. Toxicogenomics and drug discovery: will new technologies help us produce better drugs? Nature Rev Drug Discov 2002;1:84–88.
95. Williams DP, Naisbitt DJ. Toxicophores: groups and metabolic routes associated with increased safety risks. Curr Opin Drug Discov. Devel. 2002;5:104–115.
96. Schulz M, Schmoldt A. Therapeutic and toxic blood concentrations of more than 500 drugs. Pharmazie 1997;52:895–911.
97. Pirmohamed M, Park BK. Genetic susceptibility to adverse drug reactions. Trends Pharmacol. Sci. 2001;22:298–305.
98. Park BK. Prediction of adverse drug reactions. Drug Metab Rev 2006;38(S1) 14–15.
99. Darley-Usmar V, Halliwell B. Blood radicals: reactive nitrogen species, reactive oxygen species, transition metal ions, and the vascular system'. Pharm Res 1996;13:649–662.
100. Lancaster JR, Jr. Nitroxidation, nitrosative, and nitrative stress: kinetic predictions of reactive nitrogen species chemistry under biological conditions. Chem Res Toxicol 2006;19:1160–1174.
101. Buettner GR. The pecking order of free radicals and antioxidants: lipid peroxidation, α-toxopherol, and ascorbate. Arch Biochem Biophys 1993;300:535–543.
102. Aldini G, Dalle-Donne I, Maffei Facino R, Milzani A, Carini M. Intervention strategies to inhibit protein carbonylation by lipoperoxidation-derived reactive carbonyls. Med Res Rev 2007;27:817–868.
103. Uetrecht JP. New concepts in immunology relevant to idiosyncratic drug reactions: the "danger hypothesis" and innate immune system. Chem Res Toxicol 1999;12:387–395.
104. Uetrecht JP. Evaluation of which reactive metabolite, if any, is responsible for a specific idiosyncratic reaction. Drug Metab Rev 2006; 38:745–753.
105. Esser C. Immunotoxicology In. Testa B, van de Waterbeemd H, editors. ADME-Tox Approaches. In: Taylor JB, Triggle DJ, editors. Comprehensive Medicinal Chemistry, Vol. 5. 2nd ed. Oxford: Elsevier; 2007. p. 215–229.
106. Gram TE, Okine LK, Gram RA. The metabolism of xenobiotics by certain extrahepatic organs and its relation to toxicity. Annu Rev Pharmacol Toxicol 1986;26:259–291.
107. Park BK, Kitteringham NR, Maggs JL, Pirmohamed M, Williams DP. The role of metabolic activation in drug-induced hepatotoxicity. Annu Rev Pharmacol Toxicol 2005;45: 177–202.
108. Halegoua-De Marzio D, Navarro VJ. Drug-induced hepatotoxicity in humans'. Curr Opin Drug Discov Develop 2008;11:53–59.
109. Welch KD, Wen B, Goodlett DR, Yi EC, Lee H, Reilly TP, Nelson SD, Pohl LR. Proteomic identification of potential susceptibility factors in drug-induced liver disease. Chem Res Toxicol 2005;18:924–933.
110. Hultin-Rosenberg L, Jagannathan J, Nilsson KC, Matis SA, Sjögren N, Huby RDJ, Salter AH, Tugwood JD. Predictive models of hepatotoxicity using gene expression data from primary rat hepatocytes. Xenobiotica 2006;36:1122–1139.
111. Dieterle F, Marrer E, Suzuki E, Grenet O, Cordier A, Vonderscher J. Monitoring kidney safety in drug development: emerging technologies and their implications. Curr Opin Drug Discov Develop 2008;11:60–71.
112. Wells PG, Winn LM. Biochemical toxicology of chemical teratogenesis. Crit Rev Biochem Mol Biol 1996;31:1–40.
113. Masubuchi N, Makino C, Murayama N. Prediction of in vivo potential for metabolic activation of drugs into chemically reactive intermediates: correlation of *in vitro* and *in vivo* generation of reactive intermediates and in vitro glutathione conjugate formation in rats and humans. Chem Res Toxicol 2007;20: 455–464.
114. Kazius J, McGuire R, Bursi R. Derivation and validation of toxicophores for mutagenicity prediction. J Med Chem 2005;48:312–320.
115. Kalgutkar AS, Gardner I, Obach RS, Shaffer CL, Callegari E, Henne KR, Mutlib AE, Dalvie DK, Lee JS, Nakai Y, O'Donnell JP, Boer J, Harriman SP. A comprehensive listing of bioactivation pathways of organic functional groups. Curr Drug Metab 2005;6:161–225.
116. Testa B, Crivori P, Reist M, Carrupt PA. The influence of lipophilicity on the pharmacokinetic behavior of drugs: Concepts and

examples. Perspect Drug Discov. Design 2000;19:179–211.

117. Eichelbaum M, Testa B, Somogyi A, editors. Stereochemical Aspects of Drug Action and Disposition. Berlin: Springer Verlag; 2003.

118. Riley RJ, Lambert C, Kitteringham NR, Park BK. A stereochemical investigation of the cytotoxicity of mianserin metabolites *in vitro*. Br J Clin Pharmacol 1989;27:823–830.

119. Martin HJ, Breyer-Pfaff U, Wsol V, Venz S, Block S, Maser E. Purification and characterization of AKR1B10 from human liver: role in carbonyl reduction of xenobiotics. Drug Metab Dispos 2006;34:464–470.

120. Dow J, Berg C. Stereoselectivity of the carbonyl reduction of dolasetron in rats, dogs and humans. Chirality 1995;7:342–348.

121. Gonzalez FJ, Nebert DW. Evolution of the P450 gene superfamily: animal-plant "warfare", molecular drive and human genetic differences in drug oxidation. Trends Genet. 1990;6:182–186.

122. Upthagrove AL, Nelson WL. Importance of amine pKa and distribution coefficient in the metabolism of fluorinated propranolol derivatives. Preparation, identification of metabolite regioisomers, and metabolism by CYP2D6. Drug Metab. Dispos. 2001;29:1377–1388.

123. Morgenstern R, Lundqvist G, Hancock V, DePierre JW. Studies on the activity and activation of rat liver microsomal glutathione transferase, in particular with a substrate analogue series. J Biol Chem 1988;263:6671–6675.

124. Hines RN, Luo Z, Cresteil T, Ding X, Prough RA, Fitzpatrick JL, Ripp SL, Falkner KC, Ge NL, Levine A, Elferink CJ. Molecular regulation of genes encoding xenobiotic-metabolizing enzymes: mechanisms involving endogenous factors. Drug Metab. Dispos. 2001;29:623–633.

125. Poulos TL, Finkel BC, Howard AJ. High-resolution crystal structure of cytochrome P450cam. J Mol Biol 1987;195:687–700.

126. Williams PA, Cosme J, Ward A, Angove HC, Vinkovic DM, Jhoti H. Crystal structure of human cytochrome P450 2C9 with bound warfarin. Nature 2003;424:464–468.

127. Williams PA, Cosme J, Vinkovic DM, Ward A, Angove HC, Day PJ, Vonrhein C, Tickle IJ, Jhoti H. Crystal structure of human cytochrome P450 3A4 bound to metyrapone and progesterone. Science 2004;305:683–686.

128. Cruciani G, Carosati E, De Boeck B, Ethirajulu K, Mackie C, Howe T, Vianello R. MetaSite: Understanding metabolism in human cytochromes from the perspective of the chemist. J Med Chem 2005;48:6970–6979.

129. Available at http://www.moldiscovery.com/. Accessed 2008 Dec 5.

130. Erhardt PW, editor. Drug Metabolism—Databases and High-Throughput Testing During Drug Design and Development. London: International Union of Pure and Applied Chemistry and Blackwell Science; 1999.

131. Testa B, Cruciani G. Structure–metabolism relations, and the challenge of predicting biotransformation. In: Testa B, van de Waterbeemd H, Folkers G, Guy R, editors. Pharmacokinetic Optimization in Drug Research: Biological, Physicochemical and Computational Strategies. Zurich/Weinheim: Verlag Helvetica Chimica Acta/Wiley-VCH; 2001. p. 65–84.

132. Hawkins DR. Comprehensive expert systems to predict drug metabolism. In: Testa B, van de Waterbeemd H, editors. ADME-Tox Approaches. In: Taylor JB, Triggle DJ, editors. Comprehensive Medicinal Chemistry, Vol. 5. 2nd ed. Oxford: Elsevier; 2007. p. 795–807.

133. Ekins S, Andreyev S, Ryabov A, Kirillov E, Rakhmatulin EA, Bugrim A, Nikolskaya T. Computational prediction of human drug metabolism. Exp Opin Drug Metab Toxicol 2005;1:303–324.

134. Available at http://www.genego.com/metadrug.php/. Accessed 2008 Dec 5.

135. Klopman G, Tu M. META: a program for the prediction of the products of mammal metabolism of xenobiotics. In: Erhardt PW, editor. Drug Metabolism: Databases and High-Throughput Testing During Drug Design and Development. London: International Union of Pure and Applied Chemistry and Blackwell Science; 1999. p. 271–276.

136. Available at http://www.multicase.com/. Accessed 2008 Dec 5.

137. Darvas P, Marokházi S, Kormos P, Kulkarni G, Kalász H, Papp A. MetabolExpert: its use in metabolism research and in combinatorial chemistry. In: Erhardt PW, editor. Drug Metabolism: Databases and High-Throughput Testing During Drug Design and Development. London: International Union of Pure and Applied Chemistry and Blackwell Science; 1999. p. 237–270.

138. Available at http://www.compudrug.com/. Accessed 2008 Dec 5.

139. Langowski JJ, Long A. Computer systems for the prediction of xenobiotic metabolism. Adv Drug Deliv. Rev. 2002;54:407–415.

140. Testa B, Balmat AL, Long A, Judson P. Predicting drug metabolism—an evaluation of the expert system METEOR. Chem Biodiv 2005;2:872–885.
141. Available at http://www.lhasalimited.org/. Accessed 2008 Dec 5.
142. Collins JM. Inter-species differences in drug properties. Chem -Biol Interact 2001;134: 237–242.
143. Tanaka E. Gender-related differences in pharmacokinetics and their clinical significance. J Clin Pharmacol Ther 1999; 24:339–346.
144. Gow PJ, Ghadrial H, Smallwood RA, Morgan DJ, Ching MS. Neonatal hepatic drug elimination. Pharmacol Toxicol 2001;88:3–15.
145. Lemmer B. Relevance for chronopharmacology in practical medicine. Seminars Perinatology 2000;24:280–290.
146. Lin JH, Yu AYH. Interindividual variability in inhibition and induction of cytochrome P450 enzymes. Annu Rev Pharmacol Toxicol 2001;41:535–567.
147. Testa B. Missions and finality of drug research: a personal view. Pharm News 1996; 3:10–12.

PHARMACOPHORES

Yvonne C. Martin
Martin Consulting, Waukegan, IL

1. BACKGROUND

1.1. Definition of Pharmacophore

This discussion uses the IUPAC definition of a pharmacophore [1]:

A pharmacophore is the ensemble of steric and electronic features that is necessary to ensure the optimal supramolecular interactions with a specific biological target structure and to trigger (or to block) its biological response. A pharmacophore does not represent a real molecule or a real association of functional groups, but a purely abstract concept that accounts for the common molecular interaction capacities of a group of compounds towards their target structure. The pharmacophore can be considered as the largest common denominator shared by a set of active molecules. This definition discards a misuse often found in the medicinal chemistry literature which consists of naming as pharmacophores simple chemical functionalities such as guanidines, sulfonamides or dihydroimidazoles (formerly imidazolines), or typical structural skeletons such as flavones, phenothiazines, prostaglandins, or steroids.

Typically pharmacophores are described as distances between points of a certain character such as hydrogen bond donor or acceptor, positive or negative ionizable, aromatic, and hydrophobic. The definition might also include the location of exclusion spheres, that is, points in space that are not occupied active ligands.

More precisely, a pharmacophore consists of a set of geometric objects (points, lines, and planes) and the required relationships between them (distances, angles, and torsion angles). The points may be centered on atoms, on centroids of a set of atoms (for example, a centroid of an aromatic ring), or calculated from other geometric objects in the pharmacophore. For example, the location of an extension point to a protein H-bond acceptor might be calculated from the ligand atoms that define the direction of the hydrogen bond. In a similar vein, lines can be described between two atoms, as the least square line between several atoms, or in relationship to other geometric objects. An elaborate example of a point definition is one described as a certain distance from the plane of an aromatic ring along the line that goes through the centroid of the ring and is perpendicular to the plane of the ring. It is important to recognize that distances are the same in mirror images of structures—hence transforming a pharmacophore into a 3D alignment of molecules requires that one consider both images.

In other words, a pharmacophore is that set of features and distances between them that a molecule must possess in order for it to be active in the biological test of interest. The required features of the pharmacophore are described, not as atomic symbols, but rather by their propensity to engage in intermolecular interactions. For example, Figs 1 and 2 show a potent D1 agonist, our pharmacophore for D1 dopaminergic agonists based on this molecule and others, and the lead molecule designed from the pharmacophore [2].

1.2. History of the Pharmacophore Concept

The concept of a 3D pharmacophores dates to approximately 50 years ago [3,4], but widespread use is much more recent. Although the generation and use of pharmacophores required the development of 3D molecular modeling and graphics [5–8], the facile generation of pharmacophores lagged these developments by a decade. Ligand-based methods require a method to automatically detect the common pharmacophore(s) in a set of structures, a development that occurred in the early 1990s [9,10]. Structure-based pharmacophores depend on having 3D structures of the macromolecular target, information that is becoming much more available. Pharmacophores from either source are frequently used to search 3D databases—computer programs that do so were first available in the late 1980s [11–14]. Practical use of such databases

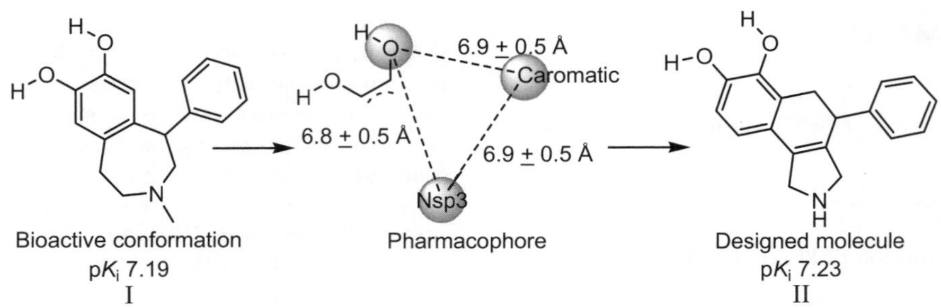

Figure 1. A pharmacophore of the requirements for D1 agonist activity based on the bioactive conformation of an active molecule established by synthesis of conformationally informative analogs. The molecule designed to meet these requirements is shown. Note that the pharmacophore also requires a second phenolic oxygen at an unspecified distance from the basic nitrogen atom [2].

required computer programs that generate one or more 3D structures from a structure diagram: This was accomplished with the program CONCORD in 1988 [15]. Many such programs are now available as are commercial systems for 3D searching.

The power of pharmacophores is exhibited in the fact that programs to generate pharmacophores and to perform 3D searches are widely available, that there are two books [16,17], and a Web site [18], devoted to the topic, and that a search of MedLine in early 2009 found 2730 articles with the pharmacophore keyword. The observation of increasing interest in the topic is shown by the fact that 330 of these articles were published in 2008.

1.3. Uses of Pharmacophores

This chapter will present the use of 3D pharmacophores in drug design. They may be used as queries for a 3D search, to suggest new molecules to synthesize, and to align molecules for 3D QSAR. To illustrate their utility, Figs 3–8 show active compounds discovered using ligand-based pharmacophores [19–24] and Figs 9–14 show compounds discovered using structure-based pharmacophores [25–30]. These examples, selected from the larger set discussed later, highlight the diversity of novel molecules that can be discovered and also illustrate some of the different computer programs used for this purpose.

Pharmacophore are often used as a query for 3D database searching, a form of virtual screening. The compounds that match the pharmacophore would be tested for biological activity with the expectation that most of the compounds would be active. As with any research, the process may be iterative, involving several cycles of defining, searching, and testing.

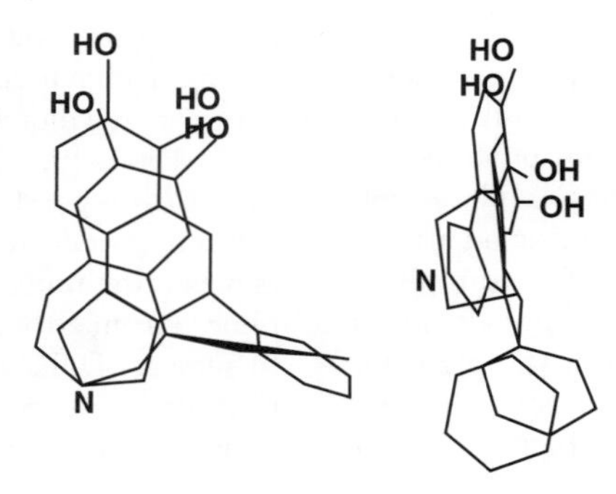

Figure 2. 3D views of the superposition of the bioactive conformation of the D1 agonist lead molecule and the molecule designed to mimic it [2].

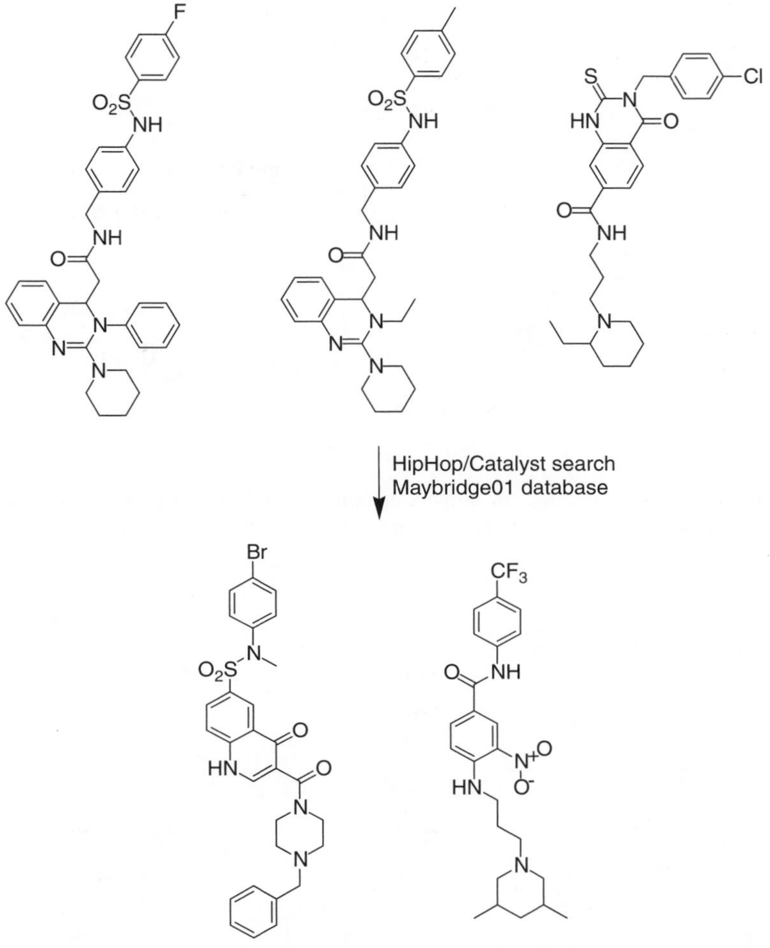

HipHop/Catalyst search
Maybridge01 database

Figure 3. Examples of structures used to define a ligand-based Catalyst pharmacophore for T-type calcium channel blockers and the molecules found with a Catalyst search [19].

A pharmacophore hypothesis can also be used to suggest compounds for synthesis, as shown in Fig. 1. For example, one might use a pharmacophore to design conformationally constrained analogs of more flexible compounds (Fig. 7). Alternatively, one might use a pharmacophore hypothesis to suggest how to swap functional groups from one core molecule to another.

Lastly, pharmacophores are typically used to select the bioactive conformer or to align a set of compounds for a 3D QSAR analysis—in fact, some computational methods combine the two procedures.

1.4. Sources of Pharmacophores

A direct method to identify the 3D pharmacophore in a set of molecules is to examine the interactions that each makes in its 3D structure with the target biomolecule. The consensus interactions form the pharmacophore. This approach has become much more popular as the number of 3D structures of protein–ligand complexes has increased.

In the absence of the structure of the ligands as bound to the target biomolecule, there might be enough known structure–activity relationships that one or a few

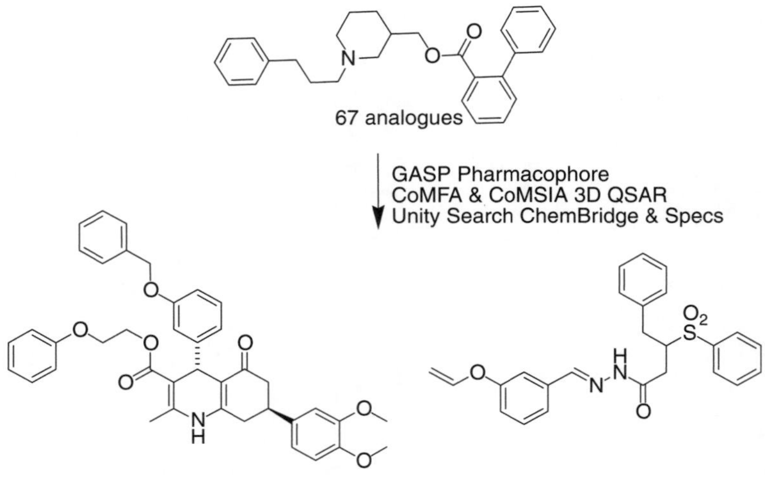

Figure 4. Examples of structures used to define a ligand-based GASP pharmacophore for noncompetitive inhibitors of the nicotinic receptor and the molecules found with a Unity search and subsequent synthesis [20].

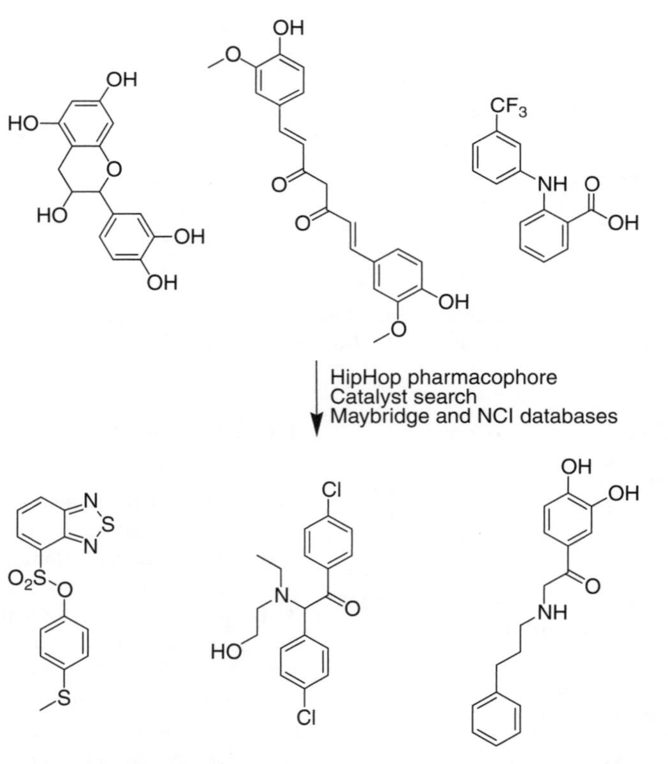

Figure 5. Examples of structures used to define a ligand-based Catalyst HipHop pharmacophore for androgen receptor downregulating agents and the molecules found with a Catalyst search [21].

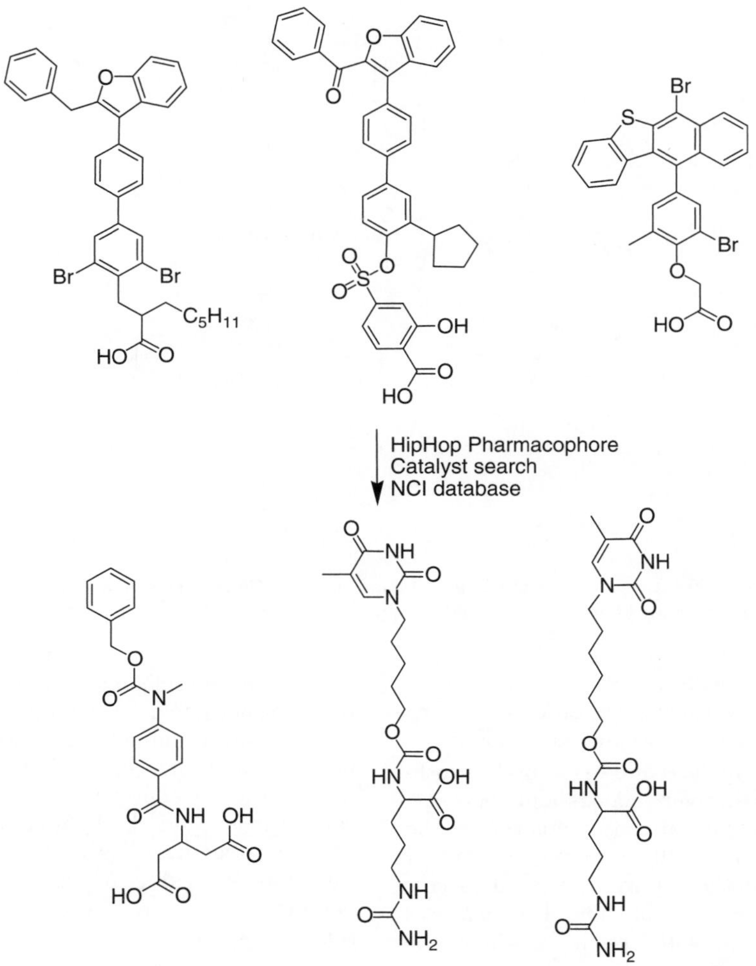

Figure 6. Examples of structures used to define ligand-based Catalyst HipHop pharmacophores for human protein tyrosine phosphatase 1B and the molecules found with a Catalyst search [22].

reasonable pharmacophores may be found with computational methods. However, if this is not the case, the discovery of such 3D pharmacophores can be accomplished with a collaboration of medicinal and computational chemists. Each suggests molecules that will lead to the definition of the pharmacophore—these can be known or newly synthesized compounds. A specific role of the medicinal chemist is to synthesize molecules that probe both the 3D relationships between features that are postulated to be important, but also molecules that probe the necessity for each of the proposed features. The role of the computational chemist is to model the active compounds and to define the limits of the proposed 3D pharmacophore.

2. TECHNIQUES FOR PHARMACOPHORE IDENTIFICATION

There are three principal strategies to establish one or more pharmacophore hypotheses:

(1) Extract the key interacting points from the 3D structures of the ligand–macromolecule complex generated by X-ray crystallography, NMR, or homology

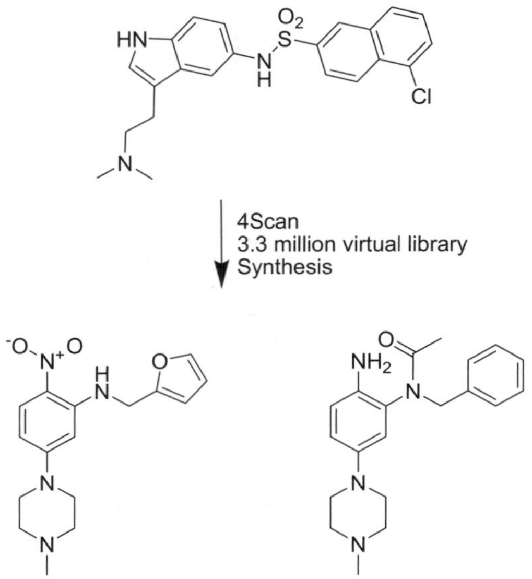

4Scan
3.3 million virtual library
Synthesis

Figure 7. Examples of structures used to define a ligand-based 4SCan pharmacophore for 5-HT_6 ligands and the novel active molecules found in a designed combinatorial library [23].

modeling. These structures may be refined by various computational methods before a pharmacophore is proposed.

(2) Establish the required groups by synthesis of simpler analogs and then prepare conformationally constrained and enantiomerically pure compounds that position the required groups at specific geometric relationships. The distances associated with activity establish the pharmacophore. Separate investigations would establish which groups are required for activity.

(3) Generate and refine many pharmacophores using ligand structure–activity data, molecular, and restricted database searches. The surviving pharmacophores would be further evaluated by their ability to retrieve active compounds from a database or the statistics of a resulting 3D quantitative structure–activity analysis.

Scientists have yet to agree on how high the relative energy the bound conformation of a molecule can be. Although several studies have addressed this problem [31–35], the ultimate answer to this question depends on the accuracy of the calculation used to assess the relative energy. Such calculations are complicated by our incomplete understanding of the behavior and energetics of interaction with water—the low-energy reference state for the calculations—and also by inaccuracies in the protein structures used for the calculations [36–38]. Values of the energies also depend on the force field used to generate the relative energies.

2.1. Pharmacophores from the 3D Structure of Bound Ligand

The 3D structure of a ligand–protein complex provides detailed information on the conformation of the bound ligand and the interactions that it makes with the protein. It also provides information on the properties and shape of the binding site. Both pieces of information are needed to propose a pharmacophore for 3D searching. In spite of the information that the structure of such a complex provides, it is not necessarily straightforward to derive a pharmacophore from this information. There are two sources of this difficulty: ambiguities in the 3D structure of the complex [36,39] and our incomplete understanding of the energetics of noncovalent

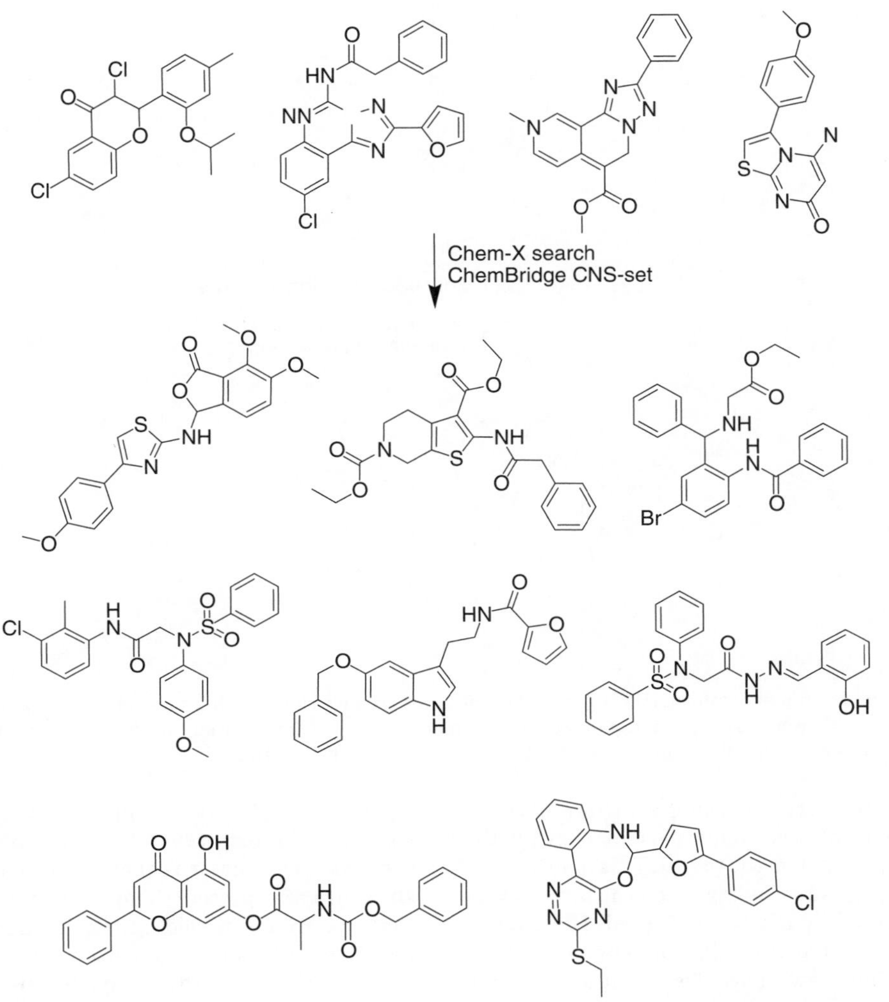

Figure 8. Examples of structures used to define ligand-based ChemX 3-point pharmacophore fingerprints for adenosine receptor ligands and the molecules found with a ChemX search [24].

interactions [40]. Careful investigation of the structure of a complex is necessary before one uses it to propose a pharmacophore. Clearly, if the structure of the complex is derived from homology modeling or even a poorly resolved NMR structure, the coordinates of the protein and ligand atoms are ambiguous. Even a well-resolved crystal structure can present problems: For example, the protein might change conformation slightly to accommodate different ligands [41]. In addition, unless the structure is solved at high enough resolution to identify the locations of hydrogen atoms, in the structure as presented it is possible that the relative positions of the nitrogen and oxygen atoms of primary amides are confused, that the protonation state of carboxylates is incorrect, and that an incorrect tautomer of the ligand is indicated. To address these problems, workers may use a detailed docking program such as AutoDock [42] or molecular dynamics [43–45] to study the fluctuations of the structure of the protein–ligand complex that will be used for the pharmacophore search. Only interactions that persist would be included in the pharmacophore.

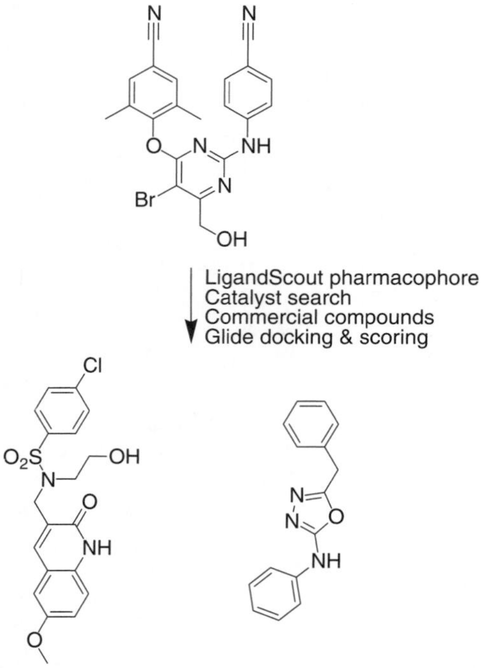

Figure 9. Examples of known HIV-1 nonnucleoside reverse transcriptase inhibitors and those discovered by a Catalyst search based on a LigandScout pharmacophore developed from the X-ray structure of an enzyme–inhibitor complex. The hits were further processed with GLIDE [25].

The first step in using a 3D structure to design is pharmacophore is to identify the binding pocket on the target. This may be done by simply using the region on the protein that is accessed by the bound ligand. However, if there is no structure of a bound ligand or if one hopes that new molecules might access additional sites, then an investigation of the potential binding pockets is necessary [46–48]. The procedure may involve investigating the contours of the protein to identify indentations that are large enough to accommodate a ligand, or it may also involve calculating interaction energies at all regions of the protein surface.

A fundamental obstacle to our ability to forecast the affinity of a ligand for a macromolecule is our incomplete understanding of the energetics of intermolecular interactions [40]. Although this is evidenced most directly in the lack of the ability of various scoring functions to forecast biological affinity [49], even more exact state-of-the-art calculations provide estimates with a root mean square error of 1.3 kcal/mole [50]. This corresponds to approximately ten-fold error in the estimate. The result is that even though the structure of a protein–ligand complex shows that two groups are close to each other, we do not know if this is because there is a strong attraction between the groups (if this adds to the interaction energy) or if the interaction contributes little or nothing to the strength of interaction. Hence, it might not be clear if a particular interaction is important for a pharmacophore. To work around this problem, workers often investigate several alternative pharmacophore hypotheses and select the most reliable one(s) by their ability to distinguish active from inactive molecules in a database of compounds that have been tested for the target activity.

Several computer programs help one identify potential pharmacophores, ideal locations of ligand atoms, from the structure of a protein–ligand complex. One approach uses molecular mechanics force fields of varying sophistication to locate "hot spots" in the

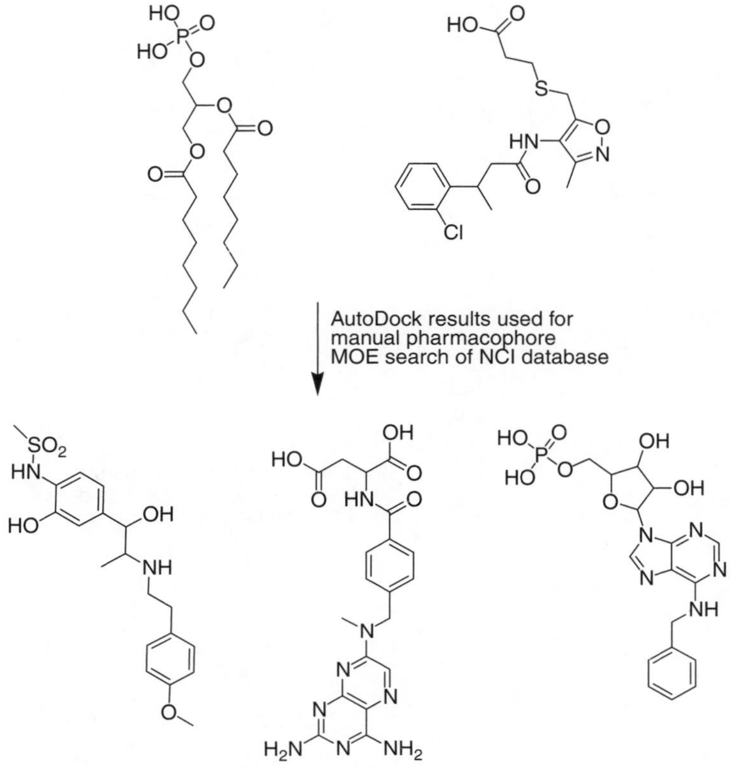

Figure 10. Examples of known LPA3 inhibitors and those discovered by a MOE 3D search. The search was based on a manual pharmacophore developed from a model structure of the receptor–ligand complex optimized with AutoDock. Hits were scored with AutoDock [26].

binding site. This is performed by positioning probe atoms within the active site and recording the ideal type of interaction group and its position of favorable interaction energy [51]. An alternative method uses the knowledge of the preferred distances and angles of atoms involved in intermolecular interactions [52].

Although structures of a complex can also be used to search databases using general docking and scoring methods, often such general methods perform poorly [49]. As a result, many workers add a 3D pharmacophore search to their docking protocol—either the hits from a 3D pharmacophore search are triaged by docking into the active site or *vice versa*.

Pharmacophores based on 3D structures of the target have the advantage that they provide clear boundaries of the binding site. This limits the number of possible matches, especially if the 3D structure is experimentally determined as bound to a ligand. In addition, a structure-based pharmacophore might suggest interacting sites that have not been explored by existing ligands. The disadvantage is that one must have a 3D structure. Although modeled structures can be used, such cases require extensive computational analysis.

2.2. Pharmacophores from the Structure–Activity Relationships of Ligands

Shape from ligand-based pharmacophores is usually established by the union surface of the superimposed molecules. If there are inactive molecules that can match the pharmacophore, then the new regions in space that they occupy would be specific forbidden regions in 3D searching.

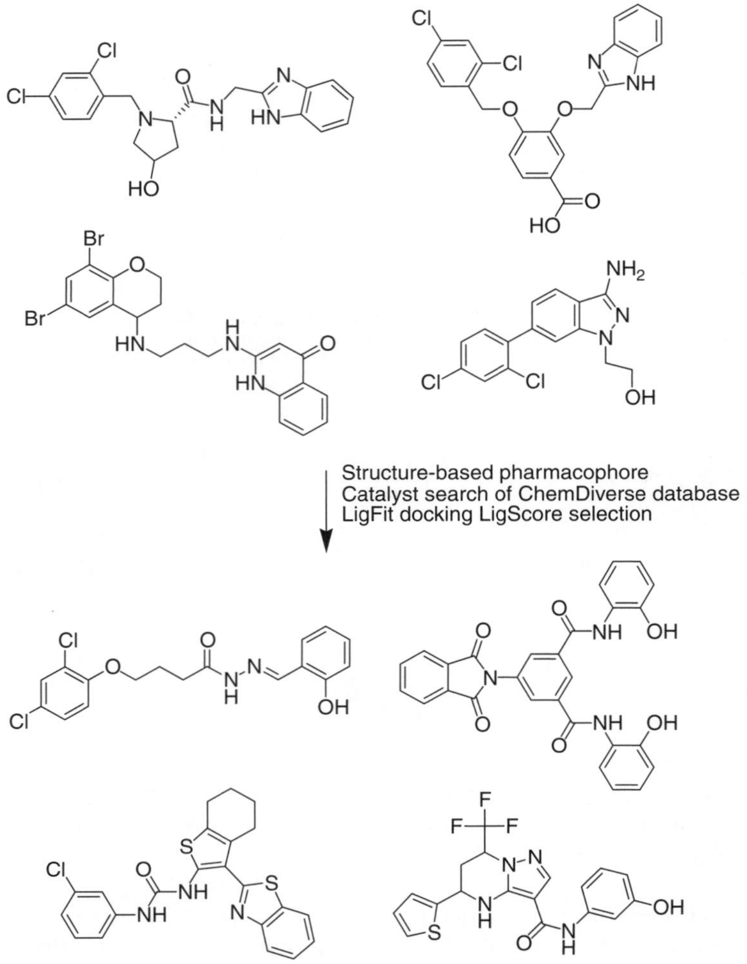

Figure 11. Examples of known inhibitors of Staphylococcus aureus MetRS and those discovered by a Catalyst search of a manual pharmacophore based on the X-ray structures of four enzyme–ligand complexes. Hits were docked with LigFit and scored with LigScore [27].

2.2.1. Ligand Preparation

The 2D Structures of the Molecules Although often biological activity has been measured on a salt, for pharmacophore investigation and 3D searching the counterion of the active molecule is usually ignored. Typically the ligands are entered in their neutral form, even though the bioactive species might be charged, because the algorithms for discovering pharmacophores label the types of interaction a particular atom can participate independently of how the structure is drawn. The algorithms also recognize certain tautomeric possibilities such as that of imidazoles. However, it may be necessary to explicitly include structures for less common tautomeric transformations. In contrast to algorithms that consider interpoint distances and angles as the fit criterion are not sensitive to the enantiomer considered, those consider overlap volume are sensitive to the enantiomer used.

Conformations Strategies that identify pharmacophores from only the structures of the ligands require one or more 3D structures of each. Some methods search for common

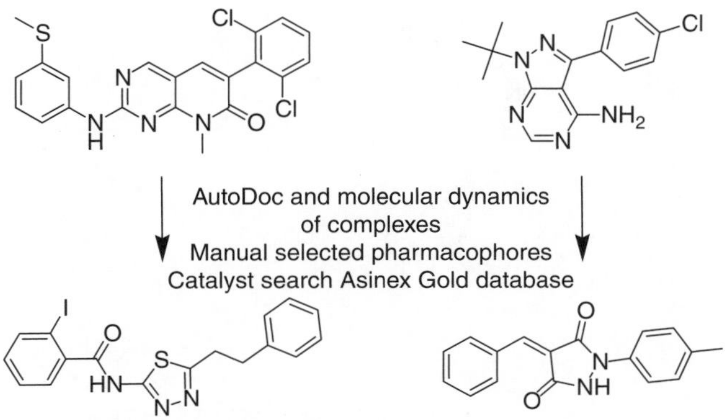

AutoDoc and molecular dynamics
of complexes
Manual selected pharmacophores
Catalyst search Asinex Gold database

Figure 12. Examples of known cSrc tyrosine kinase inhibitors and those found by Catalyst searches of several manual pharmacophores. The pharmacophores were based on an X-ray structure modeled with five inhibitors using AutoDock [28].

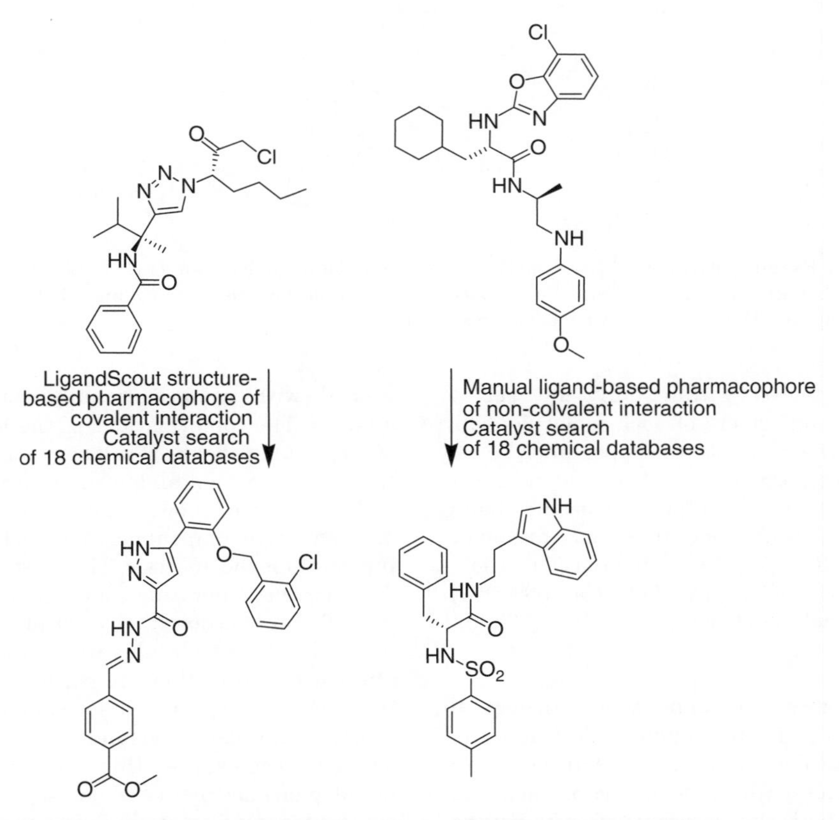

Figure 13. Examples of known cathepsin S inhibitors and those found by Catalyst searching of pharmacophores derived with LigandScout from the X-ray structure of an enzyme–inhibitor complex and also a ligand-based pharmacophore [29].

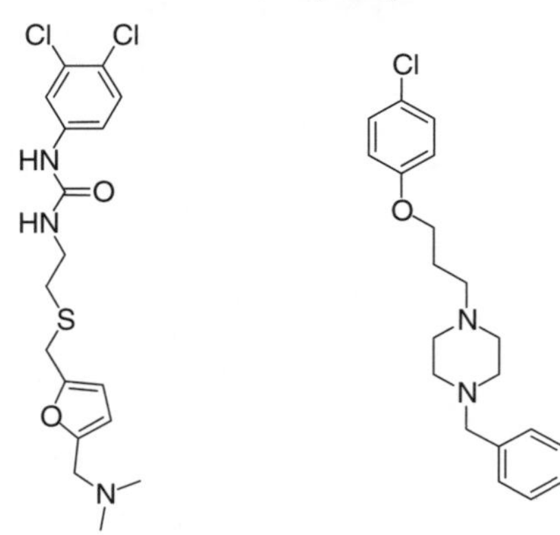

Figure 14. Example of a known histamine H_3 antagonist and antagonists found by a Catalyst search using manual pharmacophores derived from the assumed bioactive conformations of antagonists. Hits were further processed by GOLD docking into a homology model [30].

pharmacophores within only the set of input 3D conformers of each molecule. This presents the user with the opportunity to use the best conformer generation and optimization method for the type of molecules being considered. Many general methods are available: see Ref. [53], Sybyl Search [54], CONFORT [55], CORINA [56], Catalyst [57], Macromodel [58], Omega [59], MOE [60], and Rubicon [61].

2.2.2. Synthesis of Constrained Analogs to Establish the Pharmacophore A traditional method to determine a 3D pharmacophore involves the synthesis of analogs of a molecule to probe both the necessity of certain functional groups and the spatial relationship between them [62]. The spatial relationships are probed with conformationally constrained analogs. The conformations of the molecules of interest may be the result of energy calculations or experimental determination.

If there are known conformationally restricted active compounds, they can be a starting point for the analysis. The strategy would first involve generating all conformers of all possible stereoisomers of each of the active molecules. The molecules would then be examined to see of there is perhaps only one possible 3D arrangement of chemical functionality that can be attained by each of the active molecules. If so, this would be the proposed pharmacophore. If not so, one might synthesize conformationally restricted analogs or compounds that omit certain groups to test each possible 3D arrangement.

A parallel effort would establish which type of group should be at each pharmacophore point. For example, if one of the functional groups is an aliphatic hydroxyl group, this can act as either a hydrogen bond donor or a hydrogen bond acceptor or both. If activity is retained when the hydroxyl is changed into a carbonyl or the hydroxyl proton is replaced with a methyl group would suggest that the oxygen acts as a hydrogen bond acceptor. The problem is that if such compounds are inactive one does not know if this is because the geometry of the compound has changed or because the proton is necessary. Other analogs would be required to test this.

For the pharmacophore shown in Fig. 1, we capitalized on previous investigations on D2 dopaminergic agonists [63]. From this, we knew the required distance between the phenolic oxygen and the basic nitrogen and that affinity and efficacy is increased if the oxygen has a neighboring phenolic oxygen; it is part of a catechol. We designed and synthesized conformationally constrained analogs to probe the location of the phenyl group of I, which provides a 100× increase in affinity [2]. Once the locations were known, it was clear where to add a phenyl group to a screening hit.

The advantage of the manual/synthesis method of defining a pharmacophore is that interesting molecules are tested as the work proceeds. However, one disadvantage is that the necessary molecules may be very difficult to synthesize. Furthermore, if the lead molecules are very conformationally flexible, synthesis and testing of many molecules may be needed to establish a bioactive conformation. Yet another disadvantage is that there might be steric inhibition of binding by the atoms used to hold the presumed pharmacophoric points in the desired geometric relationship. For example, in the case of our study on D1 agonists every molecule used to probe the location of the phenyl group binds only weakly. It was only because for some compounds the corresponding analog that lacked the phenyl binds even more weakly. Subsequent CoMFA analysis suggests that all of these probe molecules occupy a region that is sterically unfavorable for ligands, presumably because it is occupied by the receptor.

2.2.3. Computational Methods It is possible to generate one or more pharmacophore hypotheses if one has at least a few molecules that have been tested in the biological system of interest. The number of possible pharmacophores decreases if at least some of the molecules are conformationally restricted and if some possess only a few of the features likely to be recognized by the target biomolecule.

Some computational methods identify the pharmacophore but provide no quantitative estimate of potency. For such methods one need not include every molecule in the original analysis. Rather, only molecules that provide nonredundant information need to be included. For example, if the data set includes a particular substructure in either a ring or a chain, it is not necessary to include the chain compound in the original analysis. Its presumed bioactive conformation would be modeled once a pharmacophore is proposed from more constrained analogs.

Other computational methods provide many pharmacophore hypotheses and rank each one by a QSAR analysis of the whole set of tested compounds. In such cases, one should set aside a test set that will be used to judge the quality of the models. Various strategies are used to select the test set; the important thing is to be certain that the molecules in the test set are different enough from those of the training set that a true estimate of predictivity can be made.

Proposing Bioactive Conformations of a Set of Molecules Given the Corresponding Atoms These methods are useful if some of the active molecules contain only a few features so that it is fairly obvious which atoms correspond in the various molecules. If there is more than one possible correspondence, one can repeat the search with different criteria. Frequently, these methods will suggest more than one pharmacophore—these can be evaluated by the methods discussed below. As indicated above, one need not include every known ligand in the analysis; rather, only those molecules that provide unique information on the possible pharmacophores are needed. The methods perform best if conformationally constrained analogs are included.

To propose a bioactive conformation the active analog approach uses rigid rotation around single bonds, constrained by energy and by distances between the selected atoms found in the previously searched molecules of the set [7,64]. The second-generation program RECEPTOR uses the same strategy but has many computational enhancements that increase the speed 100× [65,66]. Usually ring conformations are fixed. Because the rotations are systematic, subject to the increment used for the rotations, all of the possible corresponding conformations are searched.

In a different approach, ensemble distance geometry, uses distance geometry to generate random sets of conformations in which the corresponding points are superimposed in all molecules [67,68]. The resulting conformations are usually optimized with a standard molecular mechanics program. Ensemble distance geometry is especially suitable if the set of molecules contains one or more with a flexible ring. It sometimes highlights molecules for which more than one conformation matches a particular pharmacophore. Because the method depends on the random generation of ensembles of molecules, enough solutions must be generated that the user is satisfied that none have been missed.

Lastly, in the Boltzmann jump strategy the corresponding points in the various molecules are tethered with a strong potential while the Monte Carlo molecular mechanics-based search relaxes other constraints on conformational and translational movement [69,70]. By using this method, one must decide on the balance between the potential that tethers the corresponding atoms together with the other sources of energy of the system: If one uses too high a tether energy the molecules will adopt a high energy conformation in order to fit together, whereas if one uses too low a tether energy the corresponding points will not be aligned as closely as one might expect. As with ensemble distance geometry, one must use an independent method to ascertain if the search is complete.

Selecting Bioactive Conformations and Corresponding Atoms of a Set of Conformations of Molecules DISCO [9], HipHop [71], DISCOtech [72], and PharmID [73] discover common pharmacophores within a set of molecules using precalculated three-dimensional structures as input. This gives the user full control over the methods used to generate and refine the conformations as well as the relative energy above which the conformations will not be included. Typically the user would have the option to specify how many points are the minimum acceptable, that certain feature types must be included, and the trade-off between precision of the overlay versus the relative energy of the structures included.

The first three of the above methods use a clique-detection method to find the common distances between points. The 3D structures are encoded as graphs with the nodes the points of a particular property and the edges the distances between the properties.

PharmID first generates a bit-string that encodes each conformation into the two-point pharmacophores that it contains and concatenates all such bit-strings for a particular molecule. It then uses Gibbs sampling of the bit-strings to identify which two-point pharmacophores are common to all of the input molecules. It then uses the corresponding 3D structures and features as input to the final clique-detection analysis [73]. It thus makes the identification of the pharmacophores much faster.

Identifying Bioactive Conformations and Corresponding Atoms of a Set of Molecules GASP [74], uses genetic algorithms to optimize the alignment of pharmacophoric features while minimizing the steric energy and the overlap volume of the included conformers. It generates the conformers by rotation about single bonds. Each GASP run produces one or more pharmacophores that include all molecules with the user-defined relative weighting of pharmacophore point overlap, strain energy, and overlap volume.

GAMMA [75,76], MOGA [77,78], and GALAHAD [79,80] use Pareto optimization to produce pharmacophores that are optimal in different relative weightings. Thus, the user can decide if a pharmacophore with a good match between features is preferable to one that produces a smaller union volume or one that uses lower energy conformations. The later version of MOGA [78] and GALAHAD also include the option to include pharmacophores in which some of the features are not shared by all molecules. If the missing features are present only in potent molecules, they might be important to include in subsequent 3D searches or compound design.

QSAR-Based Proposing Bioactive Conformations and Corresponding Atoms of a Set of Molecules The HypoGen [81] feature of Catalyst is used in situations where there is quantitative structure–activity data on at least 16 compounds. Using precalculated conformations it generates hypotheses that contain a set of 3D features weighted to fit the bioactivity data. In particular, it generates potential pharmacophores from the active molecules and then deletes those that are also present in the inactive compounds. In the optimization phase, it generates regression QSAR models. The ten best models are reported with various estimates of their validity.

Phase [82,83] also identifies pharmacophores and automates generating 3D QSARs. The pharmacophores are identified by a proprietary high-dimensional partitioning algorithm that uses interpoint distances. As a default the pharmacophores are scored based on the superposition RMSD and the alignment of the vectors between the pharmacophore points. The QSAR is developed with a method similar to CoMFA.

Using Encoded Representations of 3D structures The methods above describe pharmacophores as three-dimensional objects. However, the pharmacophore features can also be represented in more condensed form. For example, pharmacophore fingerprints are a long array of 1s and 0s [84]. Each bit in a two-point fingerprint would correspond to a particular pair of feature types at a within a specific distance range. For example, one bit might indicate if the molecule contains a hydrogen-bond donor 3–4 Å from a positive charge and another might indicate if the same features are 6–7 Å apart, and so on. Depending on the granularity of the distance bins and the number of types of features, these arrays can be very long. The length also increases if three- or four-point pharmacophores are encoded. Such arrays are typically used as a filter for 3D pharmacophore searching. However, they can also be used to discover novel compounds that share 3D properties with known actives [84–90].

Autocorrelation vectors are another method to encode 3D information into a linear array. In an autocorrelation vector, each bin again represents a distance but it contains a number that is calculated over all occurrences of a particular pair of property types. By analogy to pharmacophore fingerprint in one type of autocorrelation vector, a particular bin might sum the number of times a hydrogen-bond donor is 3–4 Å from a positive charge in a molecule. However, any atom-based or surface-based property can be encoded into an autocorrelation vector. Although they are not discussed in this chapter, autocorrelation vectors can be used for many of the same purposes as explicit pharmacophores [91–99].

2.2.4. Limitations of Ligand-Based Pharmacophore Methods It is important to remember that a pharmacophore is a hypothesis. Ligand-based pharmacophore hypotheses are usually the simplest explanation of the data, but this is not necessarily the correct one. For example, even with a known 3D structure of a protein, we cannot predict with certainty that a particular ligand will bind in a particular orientation in a protein binding site—the protein might move to accommodate a different orientation of the ligand or there might be more favorable interactions with an unexpected orientation. Pharmacophores based on ligand structure–activity relationships do not have enough information to suggest such possibilities.

2.3. Methods to Evaluate Proposed Pharmacophores

Because pharmacophores are hypotheses, it is important to challenge them with as much information as possible. A good pharmacophore should explain information not used in its derivation such as the bioactive stereoisomer of compounds or why certain similar compounds are inactive.

If there are many known actives, a common validation method involves adding these actives to a database of (presumably) inactive molecules and using the pharmacophore to search the database. If most of the actives but few of the inactives are retrieved, the pharmacophore is expected to be useful for searching for new active molecules.

If the potency of more than a dozen or so compounds is known, then a 3D QSAR that

Table 1. Examples of Novel Compounds Identified from a Ligand-Based Pharmacophore

Title	Reference	Methods
Pharmacophore-based discovery of potential antimalarial agent targeting haem detoxification pathway	[100]	Catalyst Hypogen QSAR, Catalyst search of proposed molecules, synthesis
Discovery of DPP-IV Inhibitors by pharmacophore modeling and QSAR analysis followed by *in silico* screening	[101]	Catalyst Hypogen QSAR model, Catalyst search
Discovery and optimization of a novel series of *N*-arylamide oxadiazoles as potent, highly selective and orally bioavailable cannabinoid receptor 2 CB2 agonists	[102]	FLAME, 3D search, synthesis
Quinolone 3-carboxylic acid pharmacophore: design of second-generation HIV-1 integrase inhibitors	[103]	Catalyst HipHop, Catalyst search
A refined pharmacophore model for HIV-1 integrase inhibitors: optimization of potency in the 1*H*-benzylindole series	[104]	Catalyst HipHop revision of previous pharmacophore, synthesis
3D pharmacophore-based virtual screening of T-type calcium channel blockers	[19]	Catalyst pharmacophore and search
2-Aminothienopyridazines as novel adenosine A1 receptor allosteric modulators and antagonists	[105]	2D pharmacophore search, synthesis
Discovery of a novel class of selective human CB1 inverse agonists	[106]	Manual pharmacophore, Catalyst search
Design and evaluation of a novel series of 2,3-oxidosqualene cyclase inhibitors with low systemic exposure, relationship between pharmacokinetic properties and ocular toxicity	[107]	Manual pharmacophore, synthesis
Identification of natural-product-derived inhibitors of 5-lipoxygenase activity by ligand-based virtual screening	[108]	SpeedCATS 2D autocorrelation topological 2-point pharmacophore searches, CATS 3D and MACCS 2D key searches on hits
8,9-Dihydroxy-2,3,7,11*b*-tetrahydro-1*h*-naph[1,2,3-*de*]isoquinoline—a potent full dopamine D-1 agonist containing a rigid beta-phenyldopamine pharmacophore	[109]	Semi-rigid template, manual design
Three-dimensional database mining identifies a unique chemotype that unites structurally diverse botulinum neurotoxin serotype a inhibitors in a three-zone pharmacophore	[110]	Manual pharmacophore, Catalyst searches
Identification of pharmacophore model, synthesis and biological evaluation of *N*-phenyl-1-arylamide and *N*-phenylbenzenesulfonamide derivatives as BACE 1 inhibitors	[111]	Catalyst ligand-based pharmacophores and synthesis
Design, synthesis, and biological evaluation of 1,3-dioxoisoindoline-5-carboxamide derivatives as T-type calcium channel blockers	[112]	Catalyst HipHop pharmacophore, Catalyst search
New serotonin 5-HT_6 ligands from common feature pharmacophore hypotheses	[113]	Catalyst HipHop pharmacophores, recursive partitioning, similarity searching
Discovery of a quorum-sensing inhibitor of drug-resistant staphylococcal infections by structure-based virtual screening	[114]	Manual pharmacophore of natural peptide ligand, ISIS 3D search

Pharmacophore-based design of sphingosine 1-phosphate-3 receptor antagonists that include a 3,4-dialkoxybenzophenone scaffold	[115]	HipHop Catalyst, Guner-Henry scoring, synthesis
Antitumor compounds based on a natural product consensus pharmacophore	[116]	Manual pharmacophore, synthesis
Screening for inhibitors of tau protein aggregation into alzheimer paired helical filaments: a ligand-based approach results in successful scaffold hopping	[117]	Catalyst Hypogen 3D QSAR models, Catalyst search
Pharmacophore design and database searching for selective monoamine neurotransmitter transporter ligands	[118]	Catalyst Hypogen pharmacophores, Catalyst searches
Discovery of novel CB2 receptor ligands by a pharmacophore-based virtual screening workflow	[119]	Catalyst HipHop, Catalyst search
Analogs of methyllycaconitine as novel noncompetitive inhibitors of nicotinic receptors: pharmacological characterization, computational modeling, and pharmacophore development	[20]	GASP, CoMFA and CoMSIA, Unity, synthesis
Molecular modeling in the discovery of drug leads	[120]	Manual pharmacophore, ChemX searches
Metastin (KiSS-1) mimetics identified from peptide structure–activity relationship-derived pharmacophores and directed small molecule database screening	[121]	Manual pharmacophores based on 3D conformation of the peptide ligand, 3D searches
First Pharmacophore-based identification of androgen Receptor down-regulating agents: discovery of potent anti-prostate cancer agents	[21]	HipHop Catalyst
On designing non-saccharide, allosteric activators of antithrombin	[122]	Manual pharmacophore, design rigid, synthesis
Refinement of histamine H3 ligands pharmacophore model leads to a new class of potent and selective naphthalene inverse agonists	[123]	New feature manually added to previous pharmacophore, synthesis
Discovery of potent cholecystokinin-2 receptor antagonists: elucidation of key pharmacophore elements by X-ray crystallographic and NMR conformational analysis	[124]	Manual pharmacophore, synthesis to probe details
Combining ligand-based pharmacophore modeling, quantitative structure–activity relationship analysis and *in silico* screening for the discovery of new potent hormone sensitive lipase inhibitors	[125]	Catalyst Hypogen QSAR model, Catalyst search
Discovery of new MurF inhibitors via pharmacophore modeling and QSAR analysis followed by *in-silico* screening	[126]	Catalyst Hypogen QSAR model, Catalyst search
Discovery of new potent human protein tyrosine phosphatase inhibitors via pharmacophore and QSAR analysis followed by *in silico* screening	[22]	Catalyst HipHop pharmacophore, Catalyst search
Discovery of 5-HT_6 receptor ligands based on virtual HTS	[23]	4SCan pharmacophore, combinatorial library synthesis
Identification of novel cannabinoid CB1 receptor antagonists by using virtual screening with a pharmacophore model	[127]	Catalyst HipHop pharmacophore, Catalyst search
The utilization of a unified pharmacophore query in the discovery of new antagonists of the adenosine receptor family	[24]	ChemX pharmacophore fingerprints, ChemX search
Discovery of inhibitors for gpiib/iia receptor from chinese herbal drugs database by pharmacophore-based virtual searching	[128]	Catalyst HipHop pharmacophore, Catalyst search
Pharmacophore modeling and *in silico* screening for new KDR kinase inhibitors	[129]	Catalyst HipHop pharmacophore, Catalyst search

Table 2. Examples of Novel Compounds Identified Using Pharmacophores and 3D Structures of the Biological Targets

Title	Reference	Protein Structure	Strategy
Ligand based virtual screening and biological evaluation of inhibitors of chorismate mutase (Rv1885c) from Mycobacterium tuberculosis H37Rv	[130]	X-ray	GASP pharmacophore, Unity search, FlexX docking, and scoring
Structure-based pharmacophore identification of new chemical scaffolds as non-nucleoside reverse transcriptase inhibitors	[25]	X-ray	LigandScout pharmacophore, Catalyst search
A refined pharmacophore identifies potent 4-amino-7-chloroquinoline-based inhibitors of the botulinum neurotoxin serotype A metalloprotease	[131]	X-ray, molecular dynamics	HINT-based pharmacophore, Catalyst docking and scoring
Developing a dynamic pharmacophore model for HIV-1 integrase	[43]	X-ray, molecular dynamics	Molecular dynamics docking of probe molecules and pharmacophore from conserved interactions
Novel inhibitors of anthrax edema factor	[132]	X-ray	HINT docked fragments, Unity search, and AutoDock docking and scoring
Docking ligands into flexible and solvated macromolecules 2 Development and application of fitted 15 to the virtual screening of potential HCV polymerase inhibitors	[133]	X-ray	Fitted 1.5 docking using manual pharmacophore as part of the docking
Pharmacophore modelling and virtual screening for identification of new aurora-A kinase inhibitors	[134]	X-ray	Catalyst HypoGen pharmacophores and search, docking
Identification of *in vitro* inhibitors of *Mycobacterium tuberculosis* lysine ε-aminotransferase by pharmacophore mapping and three-dimensional flexible searches	[135]	X-ray	AutoDock-based manual pharmacophore f + D21rom known inhibitors, Unity search, hits redocked
Computational discovery of novel low micromolar human pregnane X Receptor antagonists	[136]	X-ray	Catalyst HipHop pharmacophore, GOLD docking and scoring
Identification of non-lipid LPA3 antagonists by virtual screening	[26]	X-ray	MOE and AutoDock pharmacophore, MOE search, AutoDock scoring
Identification of novel inhibitors of methionyl-tRNA synthetase Met	[27]	X-ray	Manual pharmacophore from structure, Catalyst search, LigandFit scoring
Computational studies to discover a new NR2B/NMDA receptor antagonist and evaluation of pharmacological profile	[137]	model	Catalyst HipHop pharmacophore model, Gold docking confirmation of pharmacophore, Catalyst search
Identification of a potent, selective, and orally active leukotriene A4 hydrolase inhibitor with anti-inflammatory activity	[138]	X-ray	Manual pharmacophore from structure, Catalyst search
Computer based design, synthesis and biological evaluation of novel indole derivatives as HCV NS3-4A serine protease inhibitors	[139]	X-ray	Catalyst pharmacophore and search, Molsoft ICM docking and scoring
Identification of *Plasmodium falciparum* spermidine synthase active site binders through structure-based virtual screening	[140]	X-ray	Manual pharmacophores based on structures, Phase search, Glide docking and scoring

Novel GSK-3β inhibitors from sequential virtual screening	[141]	X-ray	HipHop ligand-based pharmacophore, Catalyst search, LigFit docking and scoring
Discovery of substituted sulfonamides and thiazolidin-4-one derivatives as agonists of human constitutive androstane receptor	[142]	X-ray	MOLCAD and GRID-based manual pharmacophore from homology model, Unity search, GOLD docking and scoring
Discovery of a highly active ligand of human pregnane X receptor: a case study from pharmacophore modeling and virtual screening to "*in vivo*" biological activity	[143]	X-ray	Manual 2D pharmacophore from three ligand complexes, ISIS search, Surflex docking and scoring
Definition of new pharmacophores for nonpeptide antagonists of human urotensin-II. Comparison with the 3D-structure of human urotensin-II and URP	[144]	NMR of peptide ligand	Catalyst HipHop pharmacophore, NMR of natural ligand to refine pharmacophore
Virtual screening application of a model of full-length HIV-1 integrase complexed with viral DNA	[145]	Model	Catalyst HipHop and structure-based focusing pharmacophores, Catalyst search, Glide docking
Optimization of the pharmacophore model for 5-HT_{7R} antagonism. Design and synthesis of new naphtholactam and naphthosultam derivatives.	[146]	Model	Catalyst pharmacophore, docking to model, synthesis
Novel aldosterone synthase inhibitors with extended carbocyclic skeleton by a combined ligand-based and structure-based drug design approach	[147]	Model	GALAHAD pharmacophore, manual compound design, FlexX-Pharm docking
Pharmacophore modeling and molecular docking led to the discovery of inhibitors of human immunodeficiency virus-1 replication targeting the human cellular aspartic acid-glutamic acid-alanine-aspartic acid box polypeptide 3	[148]	X-ray	Manual structure-based pharmacophore, 3D database search, Gold docking and scoring of hits
A Combination of docking/dynamics simulations and pharmacophoric modeling to discover new dual c-Src/Abl kinase inhibitors	[28]	X-ray	Molecular dynamics and AutoDock manual structure-based pharmacophores, Catalyst searches
N-(Thiazol-2-yl)-2-thiophene carboxamide derivatives as Abl inhibitors identified by a pharmacophore-based database screening of commercially available compounds	[149]	X-ray	Ligand and structure-based pharmacophores, 3D database searches, Gold docking and scoring
Structure–activity study in the class of 6-3′-hydroxyphenyl naphthalenes leading to an optimization of a pharmacophore model for 17beta-hydroxysteroid dehydrogenase type 1 17beta-HSD1 inhibitors	[150]	X-ray	Manual ligand and structure-based pharmacophore, synthesis
Discovery of novel PPAR ligands by a virtual screening approach based on pharmacophore modeling, 3D shape, and electrostatic similarity screening	[151]	X-ray	HipHop ligand-based and LigandScout structure-based pharmacophores, Catalyst search, EON shape and electrostatic similarity scoring

(*Continued*)

Table 2 (*continued*)

Title	Reference	Protein Structure	Strategy
Discovery of novel cathepsin S inhibitors by pharmacophore-based virtual high-throughput screening	[29]	X-ray	Catalyst and LigandScout pharmacophores, Catalyst searching
Generation of a homology model of the human histamine H3 receptor for ligand docking and pharmacophore-based screening	[30]	Homology model	Molecular dynamics, GOLD docking: Catalyst ligand-based manual pharmacophores, Catalyst search
Discovery of nonsteroidal 17beta-hydroxysteroid dehydrogenase 1 inhibitors by pharmacophore-based screening of virtual compound libraries	[152]	X-ray	SPROUT and LigandScout structure-based pharmacophores, Catalyst searches GOLD docking and scoring
Discovery of novel agonists and antagonists of the free fatty acid receptor 1 FFAR1 using virtual screening	[153]	Model	2D similarity search in MOE, 3D pharmacophore from GLIDE docking, 3D search with Unity, GLIDE search and scoring
4-Phenylaminopyrrolopyrimidines: potent and selective, ATP site directed inhibitors of the EGF-receptor protein tyrosine kinase	[154]	Model	Manual pharmacophore, manual design
Discovery of multitarget inhibitors by combining molecular docking with common pharmacophore matching	[155]	X-ray	Pharmacophore from bound structure by Pocket v2, DOCK search, hits filtered by Pscore for fitting to pharmacophores, AutoDock on hits, further filter with Pscore
Structure-based virtual screening for identification of novel 11beta- HSD1 inhibitors	[156]	X-ray	LigandScout pharmacophore, Catalyst search, Glide docking and scoring
Discovery of novel inhibitors of 11β-hydroxysteroid dehydrogenase type 1 by docking and pharmacophore modeling	[157]	X-ray	DOCK and Glide structure-based searches, Catalyst HipHop models and searches on the DOCK hits.
The discovery of novel vascular endothelial growth factor receptor tyrosine kinases inhibitors: pharmacophore modeling, virtual screening and docking studies	[158]	X-ray	Catalyst HipHop pharmacophores, CatShape from bound ligand, Catalyst search, LigFit docking and scoring
Structural analysis of the contacts anchoring moenomycin to peptidoglycan glycosyltransferases and implications for antibiotic design	[159]	X-ray	Manual analysis and design
Pharmacophore guided discovery of small-molecule human apurinic/apyrimidinic endonuclease 1 inhibitors	[160]	X-ray	Manual structure-based pharmacophore, Catalyst searches, Gold docking and scoring

makes use of the pharmacophores should be significant.

3. EXAMPLES OF SUCCESSES

Table 1 lists studies up to the beginning of 2009 that led to novel structures using ligand-based pharmacophores [19–24,100–129] and Table 2 lists the examples using structure-based pharmacophores [25–30,43,117,121,130–160]. The examples have been published since the 41 ligand-based examples renewed in 2005 [161]. Although not every compound that matches a pharmacophore is active, there is always substantial enrichment compared to random screening.

Table 1 demonstrates that ligand-based pharmacophores do lead to novel compounds. The lack of a structure of the target does not hinder discovery of novel compounds.

One could question whether using a pharmacophore improves results compared to raw docking. In a direct comparison, Muthas and colleagues found increased enrichments if after docking one rejects compounds that do not match a structure-based pharmacophore [162]. Table 2 shows that another common strategy is to perform a structure-based pharmacophore 3D search first and filter the hits by docking to the protein. Because 3D searching is fast compared with docking, with this strategy only a small portion of the database is docked, which saves time [163]. Others have developed docking methods that incorporate pharmacophores into the algorithm [164,165].

4. CONCLUSIONS

The examples enumerated in the tables demonstrate that the use of pharmacophores facilitates the discovery of novel active compounds. Several studies used a pharmacophore to design novel compounds that were synthesized and found active. Other studies used one or more pharmacophores as the query for a 3D database search or to triage compounds identified with docking to a binding site.

There are many different ways to identify and to use pharmacophores. The method to be used for a particular study will depend on the goals of the research and the information and software available. Pharmacophores derived from either the analysis of a ligand–macromolecule complex or from the analysis of structure–activity relationships of ligands are equally productive. In addition, it appears that identification of novel active molecules in a screening database is enhanced if one uses quantitative structure–activity relationships derived from ligands to postprocess hits from virtual screening based on a macromolecular binding site.

In summary, the concept of pharmacophores has provided a powerful tool for the medicinal chemist. It provides a 3D framework to complement the more traditional 2D view of structure–activity relationships.

REFERENCES

1. Wermuth CG, Ganellin CR, Lindberg P, Mitscher LA. Glossary of Terms Used in Medicinal Chemistry. IUPAC, 2008, http://www.chem.qmul.ac.uk/iupac/medchem/.
2. Martin YC, Kebabian JW, MacKenzie R, Schoenleber R.In: Silipo C, Vittoria A, editors. QSAR: Rational Approaches on the Design of Bioactive Compounds. Amsterdam: Elsevier; 1991. p 469–482.
3. Kier LB. Molecular Orbital Theory in Drug Research. New York: Academic Press; 1971.
4. Gund P, Wipke WT, Langridge R. Computers in Chemical Research, Education, and Technology. Vol. 3. Elsevier, Amsterdam; 1974. p 5–21.
5. Langridge R, Ferrin TE, Kuntz ID, Connolly ML. Science 1981;211:661–667.
6. Blaney JM, Jorgensen EC, Connolly ML, Ferrin TE, Langridge R, Oatley SJ, Burridge JM, Blake CCF. J Med Chem 1982;25:785–790.
7. Marshall GR, Barry CD, Bosshard HE, Dammkoehler RA, Dunn DA.In: Olson EC, Christoffersen RE, editors. Computer-Assisted Drug Design. Washington, DC: American Chemical Society; 1979. p 205–226.
8. Sufrin JR, Dunn DA, Marshall GR. Mol Pharmacol 1981;19:307–313.
9. Martin YC, Bures MG, Danaher EA, DeLazzer J, Lico I, Pavlik PA. J Comput Aided Mol Des 1993;7:83–102.
10. Golender VE, Vorpagel ER.In: Kubinyi H, editor. 3D QSAR in Drug Design. Theory Methods and Applications. Leiden: ESCOM; 1993. p 137–149.

11. Brint AT, Willett P. J Mol Graph 1987;5: 49–56.
12. Jakes SE, Watts N, Willett P, Bawden D, Fisher JD. J Mol Graph 1987;5:41–48.
13. Van Drie JH, Weininger D, Martin YC. J Comput Aided Mol Des 1989;3:225–251.
14. Sheridan RP, Nilakantan R, Rusinko A, Bauman N, Haraki K, Venkataraghavan R. J Chem Inf Comput Sci 1989;29:255–260.
15. CONCORD, Tripos Associates, Inc., 1699 S. Hanley Road, Suite 303, St. Louis, MO.
16. Güner OF. Pharmacophore Perception, Development, and Use in Drug Design. La Jolla, CA: Internatonal University Line; 1999. p. 537.
17. Langer T, Hoffmann RD.In: Mannhold R, Kubinyi H, Folkers G, editors. Methods and Principles in Medicinal Chemistry. Weinheim: Wiley-VCH; 2006. p 375.
18. van Drie JH.Pharmacophores http://pharmacophore.org/ 2009.
19. Doddareddy MR, Choo H, Cho YS, Rhim H, Koh HY, Lee J-H, Jeong S-W, Pae AN. Bioorg Med Chem 2006;15:1091–1105.
20. McKay DB, Chang C, González-Cestari TF, McKay SB, El-Hajj RA, Bryant DL, Zhu MX, Swaan PW, Arason KM, Pulipaka AB, Orac CM. Bergmeier SC. Mol Pharmacol 2007;71: 1288–1297.
21. Purushottamachar P, Khandelwal A, Chopra P, Maheshwari N, Gediya LK, Vasaitis TS, Bruno RD, Clement OO, Njar VCO. Bioorg Med Chem 2007;15:3413–3421.
22. Taha MO, Bustanji Y, Al-Bakri AG, Yousef A-M, Zalloum WA, Al-Masri IM, Atallaha N. J Mol Graph Model 2006;25:870–884.
23. Tasler S, Kraus J, Wuzik A, Müller O, Aschenbrenner A, Cubero E, Pascual R, Quintana-Ruiz J-R, Dordal A, Mercè R, Codony X. Bioorg Med Chem Lett 2007;17:6224–6229.
24. Webb TR, Melman N, Lvovskiy D, Ji X-d, Jacobson KA. Bioorg Med Chem Lett 2000;10:31–34.
25. Barreca ML, De Luca L, Iraci N, Rao A, Ferro S, Maga G, Chimirri A. J Chem Inf Model 2007;47:557–562.
26. Fells JI, Tsukahara R, Fujiwara Y, Liu J, Perygin DH, Osborne DA, Tigyi G, Parrill AL. Bioorg Med Chem 2008;16:6207–6217.
27. Finn J, Stidham M, Hilgers M, Kedar CG. Bioorg Med Chem Lett 2008;18:3932–3937.
28. Manetti F, Locatelli GA, Maga G, Schenone S, Modugno M, Forli S, Corelli F, Botta M. J Med Chem 2006;49:3278–3286.
29. Markt P, McGoohan C, Walker B, Kirchmair J, Feldmann C, De MG, Spitzer G, Distinto S, Schuster D, Wolber G, Laggner C, Langer T. J Chem Inf Model 2008;48:1693–1705.
30. Schlegel B, Laggner C, Meier R, Langer T, Schnell D, Seifert R, Stark H, Höltje HD, Sippl W. J Comput Aided Mol Des 2007;21: 437–453.
31. Nicklaus MC, Wang S, Driscoll JS, Milne GWA. Bioorg Med Chem 1995;3:411–428.
32. Boström J, Norrby PO, Liljefors T. J Comput Aided Mol Des 1998;12:383–396.
33. Vieth M, Hirst JD, Brooks CL 3rd. J Comput Aided Mol Des 1998;12:563–572.
34. Perola E, Charifson PS. J Med Chem 2004;47:2499–2510.
35. Butler KT, Luque FJ, Barril X. J Comput Chem 2008;30:601–609.
36. Rhodes G. Crystallography Made Crystal Clear: A Guide for Users of Macromolecular Models. 3rd ed. Oxford: Elsevier; 2006.
37. Steuber H, Zentgraf M, Gerlach C, Sotriffer CA, Heine A, Klebe G. J Mol Biol 2006;363: 174–187.
38. Rauh D, Klebe G, Stubbs MT. J Mol Biol 2004;335:1325–1341.
39. Acharya KR, Lloyd MD. Trends Pharmacol Sci 2005;26:10–14.
40. Gilson MK, Zhou H-X. Annu Rev Biophys Biomol Struct 2007;36:21–42.
41. Antel J, Weber A, Sotriffer CA, Klebe G, Carbonic Anhydrase-Hs inhibitors and Activators CRC Press, Boca Raton; 2004. p 45–65.
42. AutoDock,The Scripps Research Institute. http://autodock.scripps.edu/.
43. Carlson HA, Masukawa KM, Rubins K, Bushman FD, Jorgensen WL, Lins RD, Briggs JM, McCammon JA. J Med Chem 2000;43: 2100–2114.
44. Carlson HA, McCammon JA. Mol Pharmacol 2000;57:213–218.
45. Meagher KL, Carlson HA. J Am Chem Soc 2004;126:13276–13281.
46. Sotriffer C, Klebe G. Farmaco 2002;57: 243–251.
47. Campbell SJ, Gold ND, Jackson RM, Westhead DR. Curr Opin Struct Biol 2003;13:389–395.
48. Jones S, Thornton JM. Curr Opin Chem Biol 2004;8:3–7.
49. Warren GL, Andrews CW, Capelli AM, Clarke B, La Londe J, Lambert MH, Lindvall M, Nevins N, Semus SF, Senger S, Tedesco G, Wall ID, Woolven JM, Peishoff CE, Head MS. J Med Chem 2006;49:5912–5931.

50. Almlof M, Carlsson J, Aqvist J. J Chem Theory Comput 2007;3:2162–2175.
51. GRID, Molecular Discovery Ltd., Oxford.
52. Böhm HJ. J Comput Aided Mol Des 1992;6: 61–78.
53. Agrafiotis DK, Gibbs AC, Zhu F, Izrailev S, Martin E. J Chem Inf Model 2007;47: 1067–1086.
54. Tripos, Inc., St. Louis, MO.
55. CONFORT, Tripos, Inc., St. Louis, MO. http://tripos.com/data/SYBYL/confort_072505.pdf.
56. CORINA, Molecular Networks GmbH Computerchemie, Erlangen, www.mol-net.com.
57. Catalyst in Discovery Studio, Accelrys, San Diego, CA. http://accelrys.com/products/datasheets/ds-pharmacophore-0308.pdf.
58. Macromodel, Schrödinger, LLC, New York. http://www.schrodinger.com/ProductDescription.php?mID=6&sID=8&cID=0.
59. Omega, OpenEye Scientific Software, 3600 Cerrillos Rd., Suite 1107, Santa Fe, NM. www.eyesopen.com/products/applications/omega.html.
60. Moe, Chemical Computing Group, Montreal. http://www.chemcomp.com/software-chem.htm.
61. Rubicon, Daylight Chemical Information Systems, Inc., Aliso Viejo, CA. http://www.daylight.com/dayhtml/doc/man/man1/rubicon.html.
62. Wermuth CG.In: Langer T, Hoffman RD, editors. Pharmacophores and Pharmacophore Searches. Weinheim: Wiley-VCH; 2006. p 3–13.
63. Seeman P, Watanabe M, Grigoriadis D, Tedesco JL, George SR, Svensson U, Lars J, Nilsson G, Neumeyer JL. Mol Pharmacol 1985;28: 391–399.
64. Beusen DD, Marshall GR.In: Güner O, editor. Pharmacophore Perception, Development, and Use in Drug Design. La Jolla, CA: International University Line; 1999. p 21–45.
65. Dammkoehler RA, Karasek SF, Shands EF, Marshall GR. J Comput Aided Mol Des 1995;9:491–499.
66. Beusen DD, Shands E, Karasek SF, Marshall GR, Dammkoehler RA. J Mol Struct 1996; 370:2–3.
67. Sheridan RP, Nilakantan R, Dixon JS, Venkataraghavan R. J Med Chem 1986;29:899–906.
68. DGEOM. Distance Geometry; QCPE 590, Quantum Chemistry Program Exchange, Indiana University, Bloomington, IN.
69. Barakat MT, Dean PM. J Comput Aided Mol Des 1990;4:295–316.
70. Barakat MT, Dean PM. J Comput Aided Mol Des 1991;5:107–117.
71. Clement OO, Mehl AT.In: Güner OF, editor. Pharmacophore Perception, Development, and Use in Drug Design. Vol. 1.La Jolla, CA: International University Line; 1999. p 69–84.
72. DISCOtech, Tripos, Inc., 2009. http://www.tripos.com/data/SYBYL/DISCOTech_072505.pdf.
73. Feng J, Sanil A, Young SS. J Chem Inf Model 2006;46:1352–1359.
74. Jones G, Willett P, Glen RC.In: Güner OF, editor. Pharmacophore Perception, Development, and Use in Drug Design. La Jolla CA: International University Line; 1999. p 85–107.
75. Handschuh S, Wagener M, Gasteiger J. J Chem Inf Comput Sci 1998;38:220–232.
76. Handschuh S, Gasteiger J. J Mol Model 2000;6:358–378.
77. Cottrell SJ, Gillet VJ, Taylor R, Wilton DJ. J Comput Aided Mol Des 2004;18:665–682.
78. Cottrell SJ, Gillet VJ, Taylor R. J Comput Aided Mol Des 2006;20:735–749.
79. GALAHAD, Tripos, St, Louis, MO. http://www.tripos.com/data/SYBYL/GALAHAD_9-7-05.pdf.
80. Richmond NJ, Abrams CA, Wolohan PRN, Abrahamian E, Willett P, Clark RD. J Comput Aided Mol Des 2006;20:567–587.
81. Li H, Sutter J, Hoffman R.In: Güner OF, editor Pharmacophore Perception, Development, and Use in Drug Design. Vol. 1.La Jolla, CA: International University Line; 1999. p 171–189.
82. Phase, Schrödinger, 120 W 45th St., New York.
83. Dixon SL, Smondyrev AM, Knoll EH, Rao SN, Shaw DE, Friesner RA. J Comput Aided Mol Des 2006;20:647–671.
84. Mason JS, Good AC, Martin EJ. Curr Pharm Des 2001;7:567–597.
85. Nair PC, Sobhia ME. J Chem Inf Model 2008;48:1891–1902.
86. Askjaer S, Langgård M. J Chem Inf Model 2008;48:476–488.
87. Jenkins JL, Glick M, Davies JW. J Med Chem 2004;47:6144–6159.
88. Good AC, Cho SJ, Mason JS. J Comput Aided Mol Des 2004;18:523–527.
89. Makara GM. J Med Chem 2001;44: 3563–3571.

90. Nettles JH, Jenkins JL, Williams C, Clark AM, Bender A, Deng Z, Davies JW, Glick M. J Mol Graph Model 2007;26:622–633.
91. Wagener M, Sadowski J, Gasteiger J. J Am Chem Soc 1995;117:7769–7775.
92. Sadowski J, Wagener M, Gasteiger J. Angew Chem Int Ed 1996;34:23–24.
93. Devillers J. Analusis 1999;27:23–29.
94. Pastor M, Cruciani G, McLay I, Pickett S, Clementi S. J Med Chem 2000;43: 3233–3243.
95. Breneman CM, Sundling CM, Sukumar N, Shen LL, Katt WP, Embrechts MJ. J Comput Aided Mol Des 2003;17:231–240.
96. Hollas B. J Math Chem 2003;33:91–101.
97. Moro S, Bacilieri M, Cacciari B, Spalluto G. J Med Chem 2005;48:5698–5704.
98. Rhodes N, Clark DE, Willett P. J Chem Inf Model 2006;46:615–619.
99. ADRIANA, Molecular Networks, Erlangen, http://www.mol-net.com.
100. Acharya BN, Saraswat D, Kaushik MP. Eur J Med Chem 2008;43:2840–2852.
101. Al-masri IM, Mohammad MK, Taha MO. ChemMedChem 2008;3:1763–1779.
102. Cheng Y, Albrecht BK, Brown J, Buchanan JL, Buckner WH, DiMauro EF, Emkey R, Fremeau RT Jr, Harmange JC, Hoffman BJ, Huang L, Huang M, Lee JH, Lin FF, Martin MW, Nguyen HQ, Patel VF, Tomlinson SA, White RD, Xia X, Hitchcock SA. J Med Chem 2008;51:5019–5034.
103. Dayam R, Al-Mawsawi LQ, Zawahir Z, Witvrouw M, Debyser Z, Neamati N. J Med Chem 2008;51:1136–1144.
104. De Luca L, Barreca ML, Ferro S, Iraci N, Michiels M, Christ F, Debyser Z, Witvrouw M, Chimirri A. Bioorg Med Chem Lett 2008;18:2891–2895.
105. Ferguson GN, Valant C, Horne J, Figler H, Flynn BL, Linden J, Chalmers DK, Sexton PM, Christopoulos A, Scammells PJ. J Med Chem 2008;51:6165–6172.
106. Foloppe N, Allen NH, Bentley CH, Brooks TD, Kennett G, Knight AR, Leonardi S, Misra A, Monck NJT, Sellwood DM. Bioorg Med Chem Lett 2008;18:1199–1206.
107. Fouchet MH, Donche F, Martin C, Bouillot A, Junot C, Boullay AB, Potvain F, Magny SD, Coste H, Walker M, Issandou M, Dodic N. Bioorg Med Chem 2008;16:6218–6232.
108. Franke L, Oliver S, Müller-Kuhrt L, Hoernig C, Fischer L, George S, Tanrikulu Y, Schneider P, Werz O, Steinhilber D, Schneider G. J Med Chem 2007;50:2640–2646.
109. Ghosh D, Snyder SE, Watts VJ, Mailman RB, Nichols DE. J Med Chem 1996;39:549–555.
110. Hermone AR, Burnett JC, Nuss JE, Tressler LE, Nguyen TL, Solaja BA, Vennerstrom JL, Schmidt JJ, Wipf P, Bavari S, Gussio R. ChemMedChem 122008 1905–1912.
111. Huang W, Yu H, Sheng R, Li J, Hu Y. Bioorg Med Chem 2008;16:10190–10197.
112. Kim HS, Kim Y, Doddareddy MR, Seo SH, Rhim H, Tae J, Pae AN, Choo H, Cho YS. Bioorg Med Chem Lett 2006;17:476–481.
113. Kim H-J, Doddareddy MR, Choo H, Cho YS, No KT, Park W K, Pae AN. J Chem Inf Model 2008;48:197–206.
114. Kiran MD, Adikesavan NV, Cirioni O, Giacometti A, Silvestri C, Scalise G, Ghiselli R, Saba V, Orlando F, Shoham M, Balaban N. Mol Pharmacol 2008;73:1578–1586.
115. Koide Y, Uemoto K, Hasegawa T, Sada T, Murakami A, Takasugi H, Sakurai A, Mochizuki N, Takahashi A, Nishida A. J Med Chem 2007; 442–454.
116. Lagisetti C, Pourpak A, Jiang Q, Cui X, Goronga T, Morris SW, Webb TR. J Med Chem 2008;51:6220–6224.
117. Larbig G, Pickhardt M, Lloyd DG, Schmidt B, Mandelkow E. Curr Alzheimer Res 2007;4.
118. Macdougall IJA, Griffith R. J Mol Graphics Modell 2008;26:1113–1124.
119. Markt P, Feldmann C, Rollinger JM, Raduner S, Schuster D, Kirchmair J, Distinto S, Spitzer GM, Wolber G, Laggner C, Altmann K-H, Langer T, Gertsch J. J Med Chem 2009;52: 369–378.
120. Milne GW, Wang S, Nicklaus MC. J Chem Inf Comput Sci 1996;36:726–730.
121. Orsini MJ, Klein MA, Beavers MP, Connolly PJ, Middleton SA, Mayo KH. J Med Chem 2007;50:462–471.
122. Raghuraman A, Liang A, Krishnasamy C, Lauck T, Gunnarsson GT, Desai UR. Eur J Med Chem 2009;44:2626–2631.
123. Roche O, Nettekoven M, Vifian W, Sarmiento RMR. Bioorg Med Chem Lett 2008;18: 4377–4379.
124. Rosen MD, Hack MD, Allison BD, Phuong VK, Woods CR, Morton MF, Prendergast CE, Barrett TD, Schubert C, Li L, Wu X, Wu J, Freedman JM, Shankley NP, Rabinowitz MH. Bioorg Med Chem 2008;16: 3917–3925.

125. Taha MO, Dahabiyeh LA, Bustanji Y, Zalloum H, Saleh S. J Med Chem 2008;51: 6478–6494.
126. Taha MO, Atallah N, Al BAG, Paradis BC, Zalloum H, Younis KS, Levesque RC. Bioorg Med Chem 2008;16:1218–1235.
127. Wang H, Duffy RA, Boykow GC, Chackalamannil S, Madison VS. J Med Chem 2008;51:2439–2446.
128. Xu-Dong L, Xiao-Jie X, Juan H. Acta Phys Chim Sin 2008;24:307–312.
129. Yu H, Wang Z, Zhang L, Zhang J, Huang Q. Bioorg Med Chem Lett 2007;17:2126–2133.
130. Agrawal H, Kumar A, Bal NC, Siddiqi MI, Arora A. Bioorg Med Chem Lett 2007;17: 3053–3058.
131. Burnett JC, Opsenica D, Sriraghavan K, Panchal RG, Ruthel G, Hermone AR, Nguyen TL, Kenny TA, Lane DJ, McGrath CF, Schmidt JJ, Vennerstrom JL, Gussio R, Šolaja BA, Bavari S. J Med Chem 2007;50:2127–2136.
132. Chen D, Misra M, Sower L, Peterson JW, Kellogg GE, Schein CH. Bioorg Med Chem 2008;16:7225–7233.
133. Corbeil CR, Englebienne P, Yannopoulos CG, Chan L, Das SK, Bilimoria D, L'Heureux L, Moitessier N. J Chem Inf Model 2008;48: 902–909.
134. Deng XQ, Wang HY, Zhao YL, Xiang ML, Jiang PD, Cao ZX, Zheng YZ, Luo SD, Yu LT, Wei YQ, Yang SY. Chem Biol Drug Des 2008;71: 533–539.
135. Dube D, Tripathi SM, Ramachandran R. Med Chem Res 2008;17:182–188.
136. Ekins S, Kholodovych V, Ai N, Sinz M, Gal J, Gera L, Welsh WJ, Bachmann K, Mani S. Mol Pharmacol 2008;74:662–672.
137. Gitto R, De LL, Ferro S, Occhiuto F, Samperi S, De SG, Russo E, Ciranna L, Costa L, Chimirri A. ChemMedChem 2008;3: 1539–1548.
138. Grice CA, Tays KL, Savall BM, Wei J, Butler CR, Axe FU, Bembenek SD, Fourie AM, Dunford PJ, Lundeen K, Coles F, Xue X, Riley JP, Williams KN, Karlsson L, Edwards JP. J Med Chem 2008;51:4150–4169.
139. Ismail NSM, El Dine RS, Hattori M, Takahashi K, Ihara M. Bioorg Med Chem 2008;16: 7877–7887.
140. Jacobsson M, Gäredal M, Schultz J, Karlén A. J Med Chem 2008;51:2777–2786.
141. Kim H-J, Choo H, Cho YS, No KT, Pae AN. Bioorg Med Chem 2008;16:636–643.
142. Küblbeck J, Jyrkkärinne J, Poso A, Turpeinen M, Sippl W, Honkakoski P, Windshügel B. Biochem Pharmacol 2008;76:1288–1297.
143. Lemaire G, Benod C, Nahoum V, Pillon A, Boussioux A-M, Guichou J-F, Subra G, Pascussi J-M, Bourguet W, Chavanieu A, Balaguer P. Mol Pharmacol 2007;72:572–581.
144. Lescot E, Santos JS-dO Dubessy C, Oulyadi H, Lesnard A, Vaudry H, Bureau R, Rault S. J Chem Inf Model 2007;47:602–612.
145. Liao C, Karkia RG, Marchand C, Pommier Y, Nicklaus MC. Bioorg Med Chem Lett 2007;17: 5361–5365.
146. Lopez-Rodriguez ML, Porras E, Morcillo MJ, Benhamu B, Soto LJ, Lavandera JL, Ramos JA, Olivella M, Campillo M, Pardo L. J Med Chem 2003;46:5638–5650.
147. Lucas S, Heim R, Negri M, Antes I, Ries C, Schewe KE, Bisi A, Gobbi S, Hartmann RW. J Med Chem 2008;51:6138–6149.
148. Maga G, Falchi F, Garbelli A, Belfiore A, Witvrouw M, Manetti F, Botta M. J Med Chem 2008;51:6635–6638.
149. Manetti F, Falchi F, Crespan E, Schenone S, Maga G, Botta M. Bioorg Med Chem Lett 2008;18:4328–4331.
150. Marchais OS, Frotscher M, Ziegler E, Werth R, Kruchten P, Messinger J, Thole H, Hartmann RW. Mol Cell Endocrinol. 2009;301:205–211.
151. Markt P, Petersen RK, Flindt EN, Kristiansen K, Kirchmair J, Spitzer G, Distinto S, Schuster D, Wolber G, Laggner C, Langer T. J Med Chem 2008;51:6303–6317.
152. Schuster D, Nashev LG, Kirchmair J, Laggner C, Wolber G, Langer T, Odermatt A. J Med Chem 2008;51:4188–4199.
153. Tikhonova IG, Sum CS, Neumann S, Engel S, Raaka BM, Costanzi S, Gershengorn MC. J Med Chem 2008;51:625–633.
154. Traxler PM, Furet P, Mett H, Buchdunger E, Meyer T, Lydon N. J Med Chem 1996;39: 2285–2292.
155. Wei D, Jiang X, Zhou L, Chen J, Chen Z, He C, Yang K, Liu Y, Pei J, Lai L. J Med Chem 2008;51:7882–7888.
156. Yang H, Shen Y, Chen J, Jiang Q, Leng Y, Shen J. Eur J Med Chem. 2008.
157. Yang H, Dou W, Lou J, Leng Y, Shen J. Bioorg Med Chem Lett 2008;18:1340–1345.
158. Yu H, Wang Z, Zhang L, Zhang J, Huang Q. Chem Biol Drug Des 2007;69:204–211.
159. Yuan Y, Fuse S, Ostash B, Sliz P, Kahne D, Walker S. ACS Chem Biol 2008;3:429–436.

160. Zawahir Z, Dayam R, Deng J, Pereira C, Neamati N. J Med Chem 2009;52:20–32.
161. Martin YC.In: Mason JS, editor. Comprehensive Medicinal Chemistry II. Vol. 4.Oxford: Elsevier; 2007. p 515–536.
162. Muthas D, Sabnis YA, Lundborg M, Karlén A. J Mol Graph Model 2008;26:1237–1251.
163. Kontoyianni M, Madhav P, Suchanek E, Seibel W. Curr Med Chem 2008;15:107–116.
164. Hindle SA, Rarey M, Buning C, Lengauer T. J Comput Aided Mol Des 2002;16:129–149.
165. Yang JM, Shen TW. Proteins 2005;59: 205–220.

USE OF BIOLOGICAL FINGERPRINTS VERSUS STRUCTURE/CHEMOTYPES TO DESCRIBE MOLECULES

JONATHAN S. MASON[1,2]
[1] Heptares Therapeutics, Welwyn Garden City, Hertfordshire, UK
[2] Lundbeck Research, Valby, Denmark

1. INTRODUCTION

There have been significant efforts to characterize molecules by fingerprints based on their chemical structure. These range from 2D structure-based methods (e.g., substructures, atom paths, and circular connectivity) that only represent the underlying structure that gives rise to the properties recognized by a biological target to 3D pharmacophores or molecular interaction fields (MIFs) that much better represent how the protein binding sites would "see" a molecule.

However, all of these have many limitations. 3D structure-based approaches are more appealing in representing structures closer to how a biological target would see them, but suffer as the bioactive conformation is not known for all (or often any!) of the potential protein targets of a particular molecule. Thus, an ensemble of conformations needs to be evaluated, that may or may not include the bioactive ones, and will contain "noise" from biologically irrelevant conformations. For molecules that interact at multiple targets, the bioactive conformations may differ from target to target.

A more relevant descriptor for a molecule in biological space would be a fingerprint based on *in vitro* binding affinities for a diverse range of biological targets. Many names for this description of a molecule by a broad/diverse set of experimental bioactivity data have been coined, including "biological fingerprints" and "biological profiles" [1–4]; "biological spectra/biospectra" [5–7]; "bioactivity spectra" [8] and "affinity fingerprints" [9,10]; and "chemical genomic profiles" [11] and "chemical–genetic fingerprints" [12] have been used for related fingerprints. Such *in vitro* fingerprints need currently to be experimentally determined to be fully effective, as reliable bioactivity models are not available for many targets, and for when new "chemotypes" are profiled that are outside the predictive capability of a model (although with the use of 3D pharmacophoric descriptors there is less dependence on chemical structure). Large-scale biological data generation and integration enables the development of *in silico* models for a subset of relevant targets [13], but these will currently only produce a "partial" biological fingerprint that may miss a key activity or new selectivity. *In silico* models are discussed in a later section.

At a minimum, binding affinities from a large and diverse panel of biological targets are needed, consistently generated, together with reference data for known drugs etc.; optionally additional functional data that distinguishes agonists from antagonists, and so on can be included. Such a major and costly undertaking has indeed been done by the BioPrint® initiative from Cerep [14–19], with support from several major Pharma companies, and this chapter will focus on results using biological fingerprints from this project, as applied to the differentiation of structures/chemotypes; see Fig. 1 for a "heatmap" of dugs and related compounds versus BioPrint assays. Another goal of such projects is to investigate if *in vivo* effects, such as adverse drug reactions (ADRs) can be associated with/predicted from these *in vitro* profiles, and investigations into this have been published based on both BioPrint and the analysis of large data sets generated in some large pharmaceutical companies.

An important concept in the use of broad, but nevertheless limited, pharmacological profiles is that the actual assays, or combinations of the assays, may be surrogates for a far larger set of targets of relevance. This principle was used in the affinity fingerprint method [9,10], discussed in Stanton and Cao [2], in which a small "diverse" and "orthogonal" set of protein targets and compounds are used to model the activity of a new protein. The broad biological profile can also be used as a type of spectral analysis of the molecule (for example, using percent inhibition values, so that all assays have a continuous numerical value) as described by Fliri et al. [5–7] (see later section).

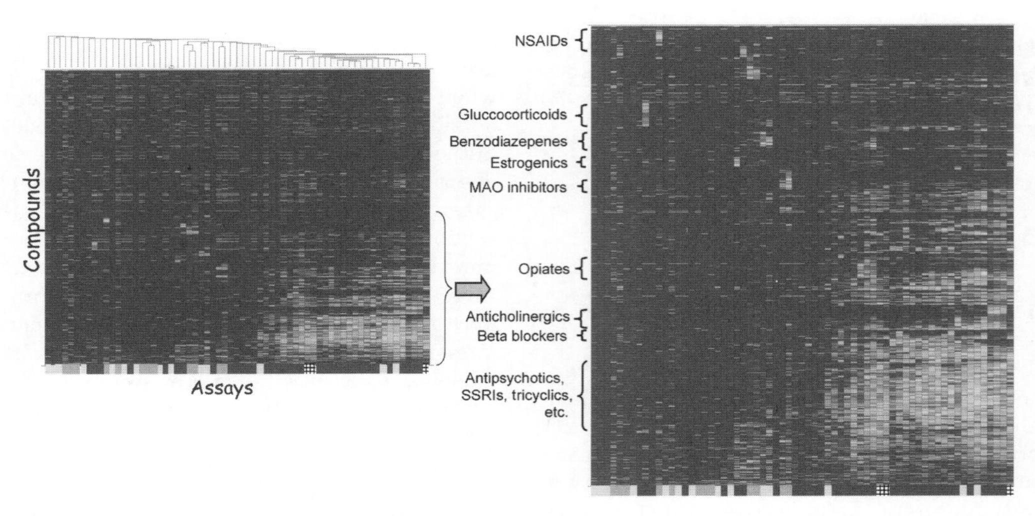

Figure 1. Compounds (~1800 drugs and related compounds) versus assays (65 pharmacological from BioPrint database, pIC_{50}). The rows contain the biological fingerprint of a compound as a heatmap of the biological assay data (x-axis, coded to show the target type: light gray: ion channel; medium gray: enzyme; dark gray: receptor; grid: transporter) with red: most active and blue-green: inactive. Hierarchical clustering has been performed on both axes: compounds by their fingerprint of biological activities and targets by the fingerprint of the activities of the same set of compounds for each target (1-Pearson correlation coefficient as the distance metric and a single linkage inclusion method). Shown on the right is an enlargement showing the data-rich compounds and some therapeutic areas that the drug compounds tended to cluster into. (This figure is available in full color at http://mrw.interscience.wiley.com/emrw/9780471266945/home.)

In this chapter, the focus is on using the biological fingerprints based on dose–response data (binding IC_{50} values from Cerep BioPrint) as a tool to aid/improve decision making in drug discovery. These fingerprints are full-matrix in that an IC_{50} determination was attempted for any compound with >30% inhibition at 10 μM (50% cutoff used for some project compound examples), with a "no activity" value actually meaning that and thus being usable in an analysis, versus the more normal case of it being due to missing data. The results for the differentiation of hit/lead often use a subset of these assays, the 70–100 pharmacological assays that provide the most signal. It was found that a decision on the prioritization of future work could normally be made using these reduced but information-rich biological fingerprints; this saves time and money. However, for the analysis of key tool/reference compounds a full BioPrint profile was used and this is highly recommended as unexpected off-target activities are found and cannot often be predicted. Examples are described that indicate that such a description for a compound can be more relevant for decision making than one based on the 2D structure and "chemotype." The early selection of the "best" leads and the exclusion of unsuitable tool compounds provide both time and efficiency improvements to the drug discovery process.

2. CHEMOTYPES: A BIOLOGICAL VIEWPOINT

Drug molecules are very commonly represented as 2D structures, often grouped and analyzed by "scaffold" and substructure [20]. Unfortunately, such a representation and derived fingerprints, and so on are not relevant to how a protein target sees a molecule. A pharmacophoric description (hydrogen bond acceptor/donor, lipophilic, etc.) is better, particularly if calculated from a 3D structure rather than estimated from 2D connectivity. The use of MIFs, such as those generated by the well established program GRID [21–23], as illustrated in Fig. 2, provide a powerful *in silico* descriptor in which the properties of the molecule are projected into "receptor

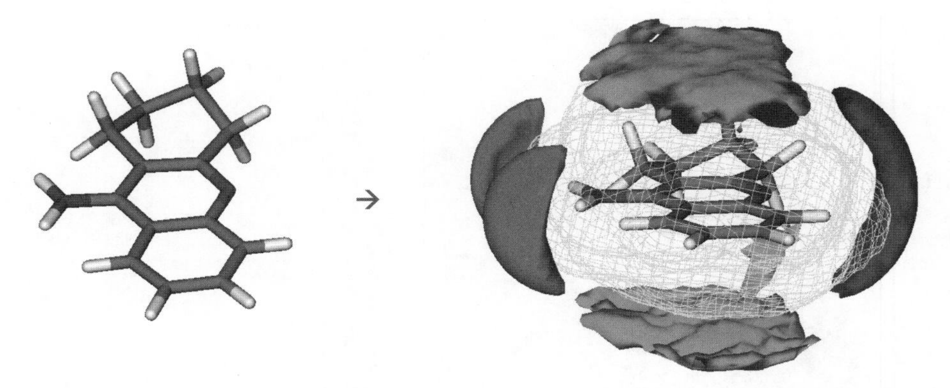

Figure 2. GRID analysis of a molecule leading to MIFs, from which pharmacophoric hotspots can be derived. The favorable region for a hydrogen bond acceptor probe (carbonyl group) is shown in red (due to the ligand donor group), for an donor probe in blue and a hydrophobic probe in green. (This figure is available in full color at http://mrw.interscience.wiley.com/emrw/9780471266945/home.)

space" and analyzed in this "protein space," as in the FLAP method [24,25]. All approaches based on 3D structures need to address the conformation problem, in that the bioactive conformation of a particular molecule is not known for all (or any) target(s), being possibly different for different proteins. Generating descriptors from conformations not relevant to binding adds noise, and failing to sample a bioactive conformation means potentially missing key descriptors. Thus, similarity studies based on 3D descriptors may fail to find compounds that ranked high by simpler 2D approaches, and thus have lower enrichment rates, but the devil is in the details and 3D methods will often find very interesting "scaffold hopping" style compounds not found by the 2D structure-based methods [26].

However, none of these approaches describe the molecule directly in terms of how it interacts with a broad range of biological systems, which is likely to be more relevant to the *in vivo* situation. When considering the suitability of a candidate for clinical trials, a profile based on how the molecule interacts with a diverse range of relevant biological targets provides a unique way to describe and differentiate compounds. This "biological fingerprint," such as used in BioPrint, describes the molecule now in terms of binding affinities to a diverse set of biological targets. Such a description is useful at all levels from hit/lead identification to clinical candidate selection and prioritization. Selectivity issues clearly are highlighted directly by such an approach (and in contrast to gene profiling, etc., the individual data points of binding to a target are directly usable by the medicinal chemist), but interestingly it is found by such analyses that "similar" compounds in terms of 2D structure, can quite often have (very) different profiles of binding (albeit at a relatively low level on many but not on all targets) across diverse targets. As chemists tend to make compounds where some key pharmacophoric components for activity are kept constant, similar compounds by 2D methods will tend to have similar primary activities. This basic concept in medicinal chemistry that similar compounds have similar activities can be misleading, as it is based on biased data sets (many analogs with key pharmacophoric elements retained), and normally only a few activities related to the primary one are considered. Primary potent activities are more likely to be similar in similar compounds, although when the bias in data sets is considered, the earlier claim that compounds with a 2D (Daylight fingerprint [27]) Tanimoto similarity >0.85 will likely have similar activities has more recently [28] been modified to that there is only a 30% chance of a compound that is >0.85 (Tanimoto on Daylight fingerprints) similar to an active is itself active.

Figure 3 shows that when a diverse set of biological targets is considered (using Cerep BioPrint, see below), with a binary indication

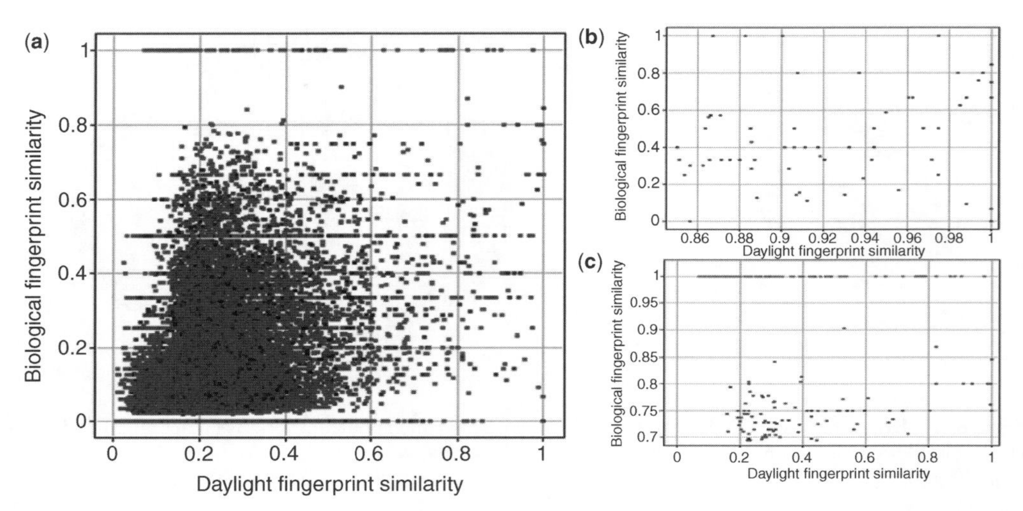

Figure 3. (a) Similarity in broad biological space versus chemical structure space for 347 drugs from the BioPrint dataset. Daylight structural fingerprints are used for structural similarity (x-axis) and BioPrint biological activity fingerprints (154 assays) where active is defined as an $IC_{50} < 100\,\mu M$ (y-axis), with pairwise Tanimoto distance (0–1, 1: identical) shown; points are color coded by MW blue (200) to black (600). Enlarged views are shown in (b) of the region where Daylight fingerprint similarity is >0.85 and in (c) of the region where biological activity similarity is >0.7. (This figure is available in full color at http://mrw.interscience.wiley.com/emrw/9780471266945/home.)

of binding or not (and a cutoff at $10\,\mu M\ IC_{50}$ or >50% at $10\,\mu M$, etc.), there is no apparent correlation between 2D similarity and broad biological similarity. The main plot shows the comparison of similarity of biological fingerprints created from a panel of 154 assays from the BioPrint database (measured by pairwise Tanimoto distance for 347 drugs with MW 200–600; 60031 points) with Daylight structural fingerprints [2]. The overall scatter plot of the whole data set in Fig. 3 demonstrates, at least for these similarity measures, the poor correlation ($R^2 = 0.13$) of general biological and chemical (structural) similarity. In Fig. 3b, a more detailed view of the region where the structural similarity >0.85 is shown. Neighborhood behavior would be expected to be stronger here, but the lack of correlation is still observed, although similarity of "primary" activity will be better. The low correlation coefficient of 0.05 supports the finding of Martin et al. noted above [28]. A focus on the region of high biological fingerprint similarity (Tanimoto > 0.7), Fig. 3c, shows that correlation is still poor. When similarity between profiles is nonbinary and takes the level of binding into account then there is an improvement in the similarity correlation, particularly if a broad profile of related targets such as kinases is used. However, correlation between structural and biological similarity soon fades away when the compounds are not very similar (e.g., a Daylight Tanimoto < 0.85).

There is no simple answer to the question of whether chemical/structural similarity or similarity from structurally derived descriptors infers biological similarity as there is an almost limitless number of possible definitions of chemical similarity, that can use 2D or 3D structures. Horvath and Jeandenans [29–31] investigated which chemical/structural fingerprints best correlate with biological similarity, studying many descriptors such as 2D topological, shape, 3- and 4-point pharmacophores and fuzzy bipolar pharmacophores, and many similarity metrics. 3D descriptors, in particular the "fuzzy pharmacophores" [32] (counts of the number of feature pairs—hydrophobes, aromatic groups, hydrogen bond donors/acceptors, and positive/negative charges—with binned distances), were encouragingly found to have improved performance in replicating the biological similarities calculated from a smaller panel of 42 targets and to be able to correlate best with the broad biological similarity. Conformational space differences seemed to cause poorer results with the more specific 3- and 4-point pharmacophore fingerprints. A recent study from Steffen et al. [33] at Astra Zeneca using a much more recent and larger version of BioPrint with 146 assays confirmed that fingerprint methods which describe global features of a molecule such as pharmacophore patterns and physicochemical properties are likely to be better suited to describe similarity of biological activity profiles than purely structural fingerprint methods. The authors suggest that the usage of these fingerprint methods could increase the probability of finding molecules with a similar biological activity profile but yet a different chemical structure; this had been the experience of this author with 4-point pharmacophore fingerprints [34–37].

Thus small changes to structure, such as only minor changes to the "decoration" on a core scaffold as well as minor scaffold changes, can cause large changes in broad biological profile, and thus of potential off-target effects, and so on in the clinic. If the targets used to generate the biological profile are considered in combination, they can be surrogates for many other activities, that is, binding to another target may be represented by a combination of activities on other targets or may be much stronger for a related target (e.g., with kinases). The data indicates strongly that it is not necessary to condemn a whole "series" because of one clinical failure etc., the series can be mined for compounds with different/better profiles (e.g., BioPrint) and those used for clinical evaluation. This in itself could lead to a large efficiency increase and enable faster delivery of better drugs to meet medical needs.

A lot of information on how drugs, attrited compounds and medicinal chemistry project compounds has been generated through the BioPrint initiative from Cerep [14–20], with several major pharma partners. Lessons learnt from these analyses will be focused on in this chapter, as they are the first-hand experience of the author. As much of the data described in this chapter is from this Cerep BioPrint profiling, this will first be described.

3. THE CEREP BIOPRINT DATABASE

The commercially available BioPrint [14–20] database package consists of a large database and a set of tools to access both the data and the models generated from the data. The database contains structural information, *in vitro* and *in vivo* data on most marketed pharmaceuticals and a variety of other reference compounds; Fig. 4 illustrates the general concept. The full matrix *in vitro* data that are generated consist of panels of pharmacology and early ADME assays, with consistently generated data (over time and over different compound sets) that enable analyses that could not be effectively done using data with many structure–activity "holes" or inconsistent data. Figure 1 shows a heatmap (red: most active and blue-green: inactive) of about 2000 drugs and related compounds versus 70 pharmacological assays from BioPrint, in which hierarchical clustering has been performed on both axes: compounds by their fingerprint of biological activities and targets by the fingerprint of the activities of the same set

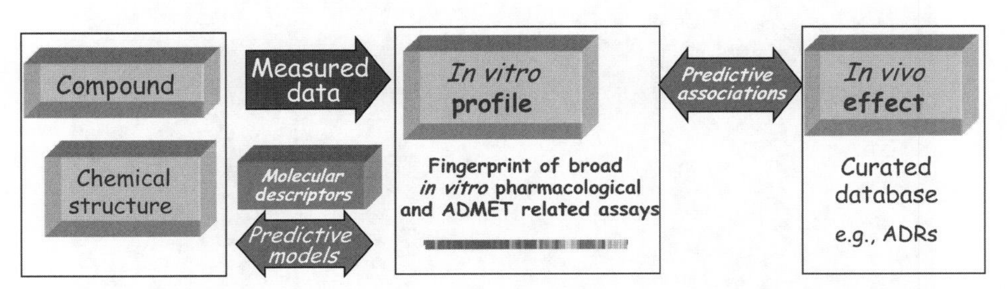

Figure 4. BioPrint approach. The *in vitro* data are all measured in a consistent manner with full dose–response for any activity >30% at 10 μM; the *in vivo* data are curated from available data, supplemented by custom measured data from collaborators. (This figure is available in full color at http://mrw.interscience.wiley.com/emrw/9780471266945/home.)

of compounds for each target. The polypharmacology of many drugs can clearly be seen; some therapeutic areas that the drug compounds tended to cluster into are also indicated.

The *in vivo* data consist of ADR data extracted from drug labels, mechanisms of action, associated therapeutic areas, PK data, and route of administration data. The data in BioPrint provide direct information on a compound, but importantly also a context to interpret and cluster new data and can also be used to develop predictive models etc.

The BioPrint database is continually growing, having around 150 pharmacological and 30 pharmaceutical/ADME assays, covering most launched drugs, pharmacological reference compounds and some drugs that have failed during development. The database that has the power of being full matrix (IC_{50} values were generated if percent inhibition at 10 μM was >30%), has been built over more than 10 years through collaborations with major pharmaceutical companies. The pharmacological part of the BioPrint Profile is designed to address relevant biological diversity. Assays have been selected primarily for their scientific interest but due consideration was also given to robustness and the consistency/quality of the data; criteria included phylogenetic analysis, coverage of relevant therapeutic areas, the concept of the "druggable" proteome [38] and various constraints including technical issues. The BioPrint profile matches approximately a therapeutic target classes distribution with about 67% receptors (with GPCRs most represented), 24% enzymes, 13% ion channels, 3% transporters, and 2% nuclear receptors. Figure 5 shows a distribution of assays in 2005 across the druggable proteome (larger diversity of assays now). The BioPrint project began in 1997, with Bristol-Myers-Squibb as the first partner. As one of the goals was to try and predict attrition and *in vivo* toxicity, assays were added to a core initial set with a focus on assays that could be associated with adverse effects and this concept has been retained as other partners (e.g., Pfizer) got involved and further developed the database.

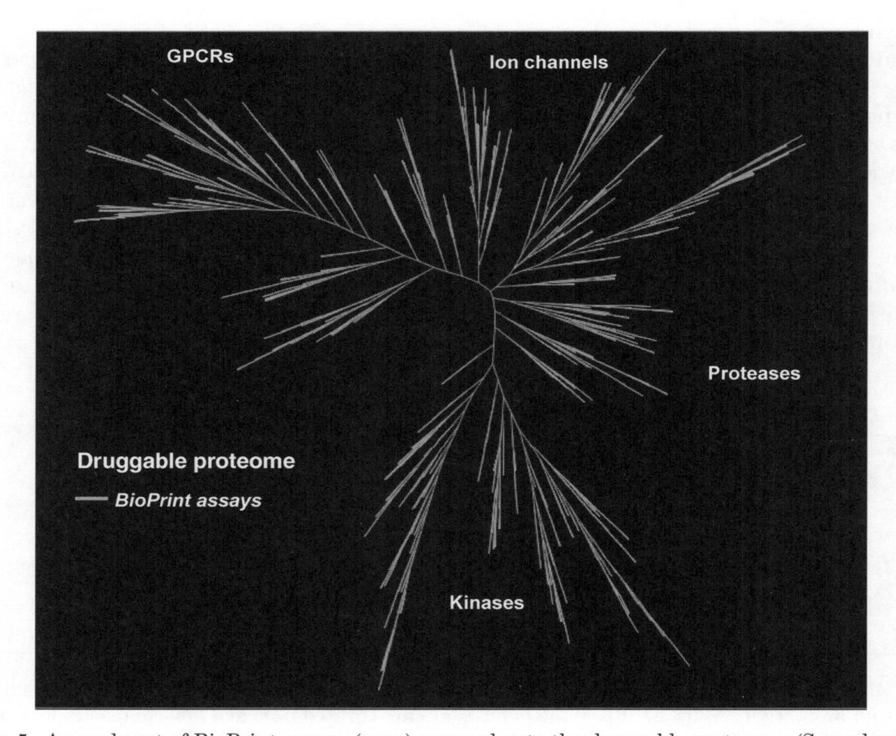

Figure 5. An early set of BioPrint assays (cyan) mapped onto the druggable proteome. (See color insert.)

Redundancy in the assay data could be an issue as data from several receptor subtypes are included, and an analysis was performed to verify that there were no obvious correlations. A subset of 80 assays and 1600 compounds was analyzed using the KEM approach [39,40] developed by Ariana [41], a systematic rule-based method that identifies all relations of the type A → B, A and B → C and not D, and so on. The analysis showed that no such noncontradicted relations where found, even using very large activity bins, showing that there is signal in all the assay data, including from related assays.

4. CHOOSING THE BEST HIT/LEAD COMPOUND

This is an important and common task in medicinal chemistry, and if the best choice can be made upfront then significant time/cost savings can be made; as resources often do not allow all possibilities to be pursued, in a worst case scenario the "best" starting point, that could lead to a clinical candidate, may never be pursued. A major challenge in the drug discovery process is the selection of development candidates that have the best chance of survival (and of course efficacy). The attrition risk should ideally be minimized, but at a minimum the risk should be orthogonalized as much as possible for multiple candidates, to avoid multiple compounds attriting for the same unexpected cause. It is thus very important to prioritize which leads are of most interest, that have the best chance of becoming a suitable development candidates' with the most desired profile of biological and ADME related properties, including differentiation from other compounds (in-house or competitor). An early identification of the best leads is critical, and one component of the issue can be stated that it is better to identify the best lead series than to try and find the best candidate from a possibly suboptimal series, that can happen at late stages of lead optimization, where it is very difficult to make major changes to the lead series chemistry.

An early pilot for BioPrint at Pfizer enabled a parallel approach whereby all the four potent hit/lead series for a project that were identified from HTS were pursued by medicinal chemistry and measurement and analyses of the BioPrint profiles performed. This study, described below, clearly illustrates the power of early biological profiling and that a "better" starting point can be critical to project success. Only the series prioritized by BioPrint profiling could be developed to a clinical candidate; the other compounds had key selectivity issues that could not resolved. This result, reinforced with other similar findings at other sites, led Pfizer to embark on a larger initiative with Cerep to use BioPrint profiling for hit/lead compounds from all therapeutic areas; continued successes lead to a larger renewal package in terms of project compound profiling. Only a few examples are described below, but many examples were found where decisions could be clearly made from the biological profile (using the 70–90 assays with the highest hit-rate) that were not evident from a chemical structure-based analysis. Many insights into unexpected activities (outside of the target class) were obtained

The early BioPrint pilot study at Pfizer [42,43] is illustrated in Fig. 6, where for a project four potent hit/lead series were identified from HTS; these are shown together with a reference compound in clinical development at the time, Duloxetine (see Fig. 7). All the compounds had the desired primary dual target activities ("polypharmacology"), but the BioPrint analysis showed clearly that they had quite different off-target activities in their overall biological profiles (fingerprints). Many of the activities were significant, and one particular important selectivity target could thus be identified early and monitored. Interestingly, all four series were pursued by medicinal chemistry, but only the compound that was highlighted by the BioPrint profile clustering as a much cleaner and differentiated starting point could be optimized to a clinical candidate. Activity on the troublesome selectivity target that affected the other series could not removed while retaining other desired activities/properties. The significant differences in the off-target activities of the different series was neither predictable from a simple structural analysis nor expected.

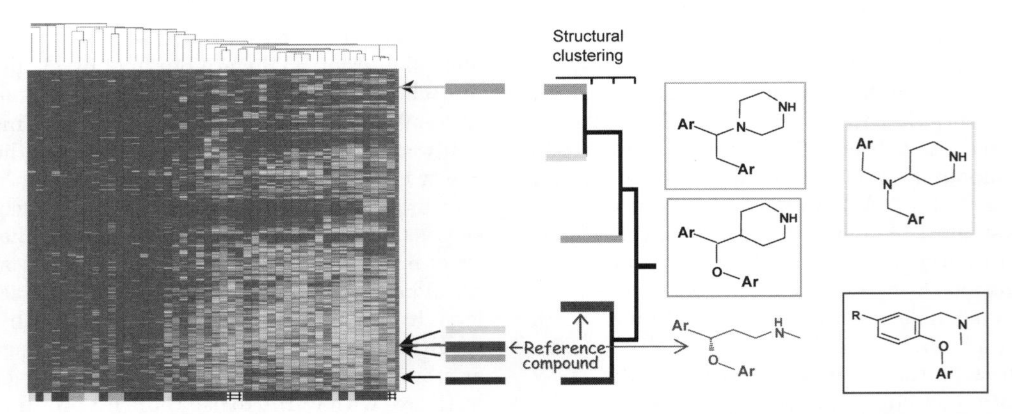

Figure 6. On the left is shown a clustering of four hit/lead compounds (colored bars/boxes) with the BioPrint drug compound set and a clinical reference compound Duloxetine (shown in blue, see Fig. 7 for full structure) based on their biological fingerprints (BioPrint). All compounds had similar potent low nanomolar "primary" activities. The rows show the partial heatmap of activity for a compound (46 assays on the x-axis, coded to show the target type: light gray: ion channel; medium gray: enzyme; dark gray: receptor; grid: transporter). Activity is color coded from red (very active) to blue-green (inactive). In the center for comparison is shown a structure-based clustering using Daylight fingerprints of the compounds. (This figure is available in full color at http://mrw.interscience.wiley.com/emrw/9780471266945/home.)

The key elements of the chemical structures are shown in Fig. 6 to illustrate how a "chemotype"-based analysis would not obviously lead to the best starting point being highlighted and chosen for optimization. The structures are relatively similar in terms of key features (basic and aromatic groups) and the most interesting compound, the piperazine compound (highlighted in orange), is not obvious from a structural viewpoint. The piperidine (highlighted in yellow) has similar nonselectivity such as the reference compound (blue), and a typical clustering that may normally have been done, using a standard 2D structural fingerprint, the Daylight [28] fingerprint, is shown in the center of Fig. 6. Such an analysis is commonly used to reduce compound lists and to select representatives for further analysis, and from this data it is clear that this could lead to wrong selections and most critically miss the optimal "clean" series ("orange" piperazine compound in Fig. 6). If the piperidine compound (yellow) was selected to represent this part of structural "diversity," a series with similar off-target issues to the reference compound and substantial polypharmacology would be pursued, and the most interesting piperazine compound (orange) missed. It is thus possible that the suboptimal piperidine compounds (yellow and green) compounds could be chosen to represent a structurally diverse selection for follow-up work, yet both have similar profiles to the reference (blue) compound, so this approach would fail. The bis-aromatic ether compound (highlighted in dark blue) clusters in 2D with the reference compound, but this compound actually has a somewhat differentiated biological profile compared to the reference compound and the piperidine compounds (highlighted in yellow and green). The small nitrogen atom shift between the piperazine (orange) and amino-piperidine (yellow) compounds produces the most differentiated compound in biological space, moving into a space previously more occupied by opiates. This selection hypothesis was fully validated, with all

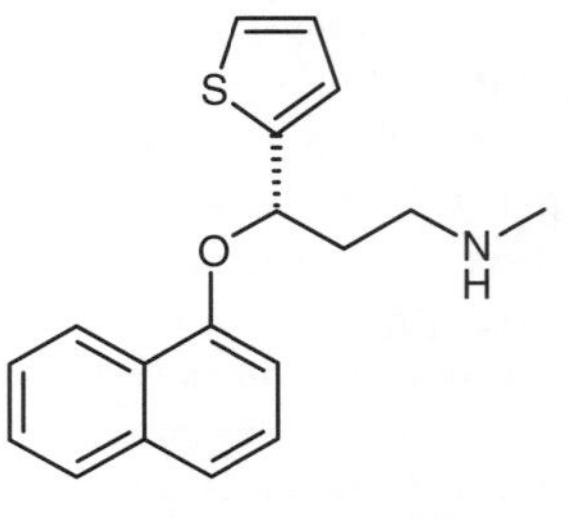

Figure 7. Structure of Duloxetine.

the lead series being followed up with further chemistry, armed with the knowledge of key selectivity assays, and only the piperazine compound (orange) moving into clinical development (staying relatively clean); the "SAR out of the selectivity problem" approach failed for all the series with close BioPrint profiles to the reference compound.

BioPrint biological profiling was applied systematically across different therapeutic area projects to the analysis of hit, leads and tool compounds to enable better data-driven decisions/prioritization. The data surprisingly usually enabled clear differentiation of different hits/leads, with few ambiguous situations. The clustering by the biological profile together with other compounds for which the BioPrint profile had been measured (both drug and company project compounds) aided this and providing further insights. It was found that the off-target activity difference for compounds of the same "chemotype" can be quite dramatic, changing from quite clean to promiscuous. Figure 8 shows an example for a pair of compounds with 2–4 nM activity, both with a distinctive "chemotype" of a bicyclic polyheteroaromatic core linked to a cyclic base and two substituents. It would make a big difference for *in vivo* studies etc. as to which was selected, with (a) on the left being quite promiscuous (binding at <1 μM to 31 pharmacological off-targets), whereas the other (b) on the right being relatively clean, binding to only two pharmacological off-targets and CYP3A4 at <1 μM [43]. The simple change in the substituents of alkoxy to alkyl and pyrimidine to phenyl causes this dramatic effect, which can be associated with both a pharmacophoric change and a $c\log p$ change (see Section 6 and Ref. [44]). The situation is more complex than a simple lipophilicity/$c\log p$ difference, as a compound in which one of the nitrogens is moved from the pyrimidine substituent (→ pyridine) to the core bicyclic ring gives a compound with a similar $c\log p$ (2.2) but increased pharmacological promiscuity (2 → 10 <1 μM pharmacological off-targets) and ADME issues (1 → 3 <1 μM CYP inhibition + Pgp efflux). Clearly representing the "chemotype" associated with the core structure in biological space is not a simple matter and a biological versus chemical fingerprint is needed.

5. TOOL COMPOUNDS AND TARGET VALIDATION

Obtaining broad biological profiling data, such as the Cerep BioPrint *in vitro* pharmacology data, is an important and often critical step in the selection of a "tool" compound that is to be used *in vivo* to validate if a desired biological effect is obtained through a hypothesized mechanism. Analysis of many "tool" compounds with BioPrint has shown many reported "selective" compounds are only selective on the some related assays used, and that they can have significant activity on other targets. Such activity could actually be responsible for the desired biological effect—that is, a wrong decision to pursue a target for a particular

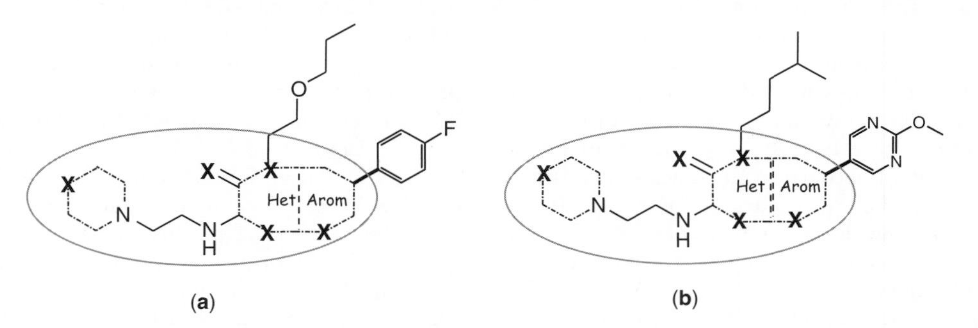

Figure 8. Two compounds with similar potent primary activities ((a) $IC_{50} = 4$ nM and (b) $IC_{50} = 2$ nM). Compound (a) [$c\log p = 5$] is found by BioPrint profiling to be promiscuous, with <1 μM activities on 31 off-targets, whereas compound (b) [$c\log p = 2$], of a similar "chemotype," is much cleaner with <1 μM activities for only 2 off-targets and CYP3A4. (This figure is available in full color at http://mrw.interscience.wiley.com/emrw/9780471266945/home.)

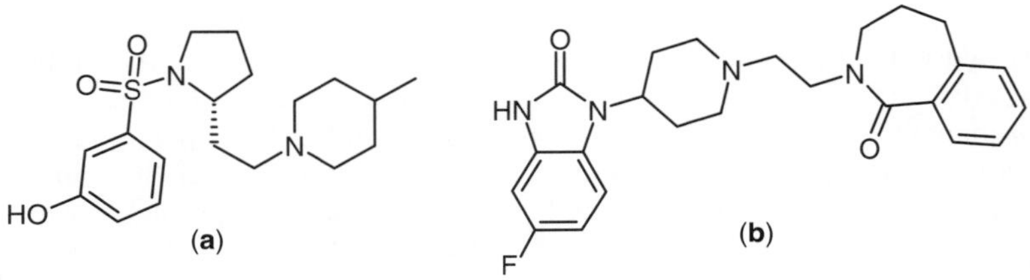

Figure 9. Two potential tool compounds reported to be selective $5HT_7$ inhibitors. (a) SB-269970 is selective (>80× in BioPrint profiling) whereas (b) SB-691673 has poor selectivity and usefulness as a tool compound being <3× selective against five targets from BioPrint profiling.

indication could be made, which may not be discovered until significant wasted work had been undertaken.

For example, full biological profiling of several published $5HT_7$ antagonist revealed that one of them, SB-269970, was more than 80× selective in all the BioPrint assays and would be a good *in vivo* tool compound to evaluate a mechanistic hypothesis, whereas another compound SB-691673 was less than 3× selective against five targets. This would have been a poor choice to evaluate the hypothesis, as one of the "off-target" activities is known to cause the desired *in vivo* effect, thus a wrong target validation could have been made. The structures are shown in Fig. 9; the "clean" compound was not active in the *in vivo* assay, thus no further time was wasted pursuing that hypothesis.

In a related manner, if a side effect is seen *in vivo* with several different compounds, then the broad biological profile can be used to see if the various compounds, even if structurally diverse, share a common off-target effect(s) that could cause the undesired effect or if it is most likely to be mechanism dependant. A recent presentation from Bell et al. [45] illustrates this. They used BioPrint broad biological profiling to validate that structurally different and relatively selective compounds with potent activity on the mechanism under consideration, α_{2A} adrenoceptor inhibition, had different broad pharmacological profiles, and were able to conclude that there was mechanism-based undesirable CV effects for selective α_{2A} inhibitors. Historically a nonsubtype selective α_2 antagonist (RS79948) had been shown to increase bladder capacity in animal models and a selective α_{2A} antagonist that was CNS penetration impaired had been shown to increase bladder capacity in the rat, whereas selective CNS penetrant α_{2B} and α_{2C} antagonists had been shown to be inactive in incontinence models. A functional HTS on the Pfizer screening file identified series without the typical α pharmacophore, with an early focus on selectivity, PK, CNS permeability, and the Cerep BioPrint profile. Several new series were developed by this approach to compounds suitable for *in vivo* evaluation. They used *in silico* Bayesian models derived from a large database of *in vitro* activities [13] as part of the HTS triage process to help remove false positives; they used a "conservative" approach of excluding hits with >3 nonaminergic GPCR target predicted activities and including if >0 activities predicted for aminergic GPCR targets. With early profiling data to aid the process, they were able to design selective α_{2A} antagonists despite poor literature precedence. An aryl pyridine series had significant polypharmacology ever present, variable between different compounds (see Fig. 10), and the BioPrint broad profiling data enabled a focus for the optimization work to deliver the most selective tool for *in vivo* evaluation. The α selectivity and polypharmacology of the oxadiazole carboxamide series was improved with the addition of a basic amine (see Fig. 11), an unexpected result. Further optimization lead to a neutral oxadiazole with the only other significant activity being on α_{2C}, producing the cleanest compound. The *in vivo* evaluation of the two structurally and biologically differentiated compounds, that shared primary activity on α_2a but not secondary pharmacology (with the neutral oxadiazole having a much cleaner *in vitro* pharmacological profile, see Fig. 12) unfortunately showed that the undesirable

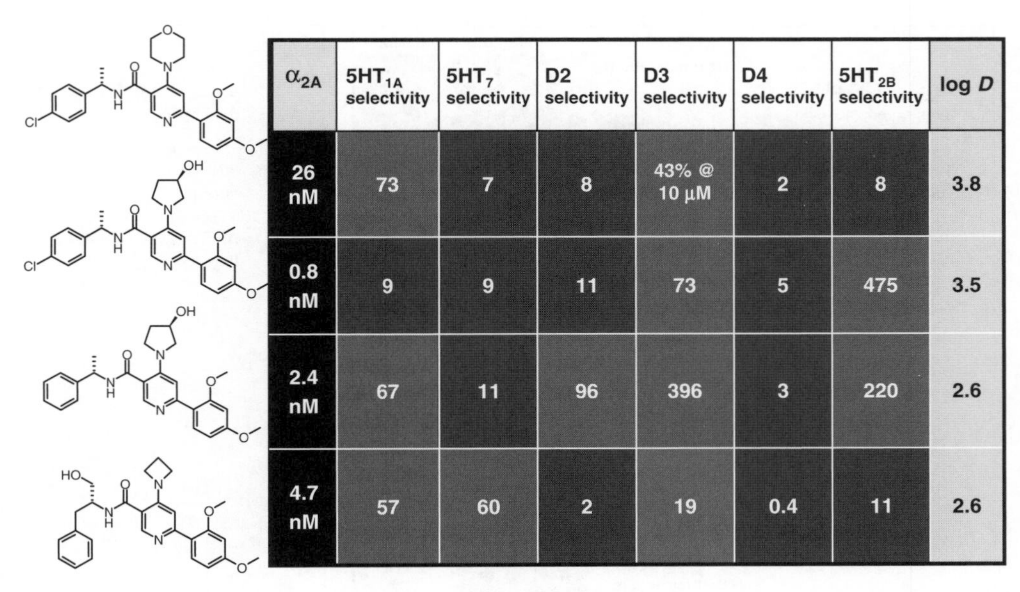

α_{2A}	$5HT_{1A}$ selectivity	$5HT_7$ selectivity	D2 selectivity	D3 selectivity	D4 selectivity	$5HT_{2B}$ selectivity	log *D*
26 nM	73	7	8	43% @ 10 μM	2	8	3.8
0.8 nM	9	9	11	73	5	475	3.5
2.4 nM	67	11	96	396	3	220	2.6
4.7 nM	57	60	2	19	0.4	11	2.6

Figure 10. Variable polypharmacology (from BioPrint profiling) of the aryl pyridine α_{2A} antagonist series. The selectivity relative to the α_{2A} activity is shown for three serotonin and three dopamine targets. The selectivity relative to the α_{2A} activity is shown as a color code: green for good (>25×), orange for moderate (>12×), and red for poor (<12) selectivity. (This figure is available in full color at http://mrw.interscience.wiley.com/emrw/9780471266945/home.)

CV effect was mechanism-related to α_{2A} antagonism and that further work was unlikely to yield a CV "clean" compound.

6. POLYPHARMACOLOGY OF DRUGS

The issue of "Can we rationally design promiscuous drugs?" was discussed by Hopkins et al. [3] and different views of the data linking structural properties and activity on multiple targets ("promiscuity") were discussed. An analysis of the BioPrint data set showed that compounds that are active at <1 μM on more than 10 targets generally are more lipophilic, with a $c\log p > \sim 3$ (see Fig. 13). Another analysis based on in house screening and HTS

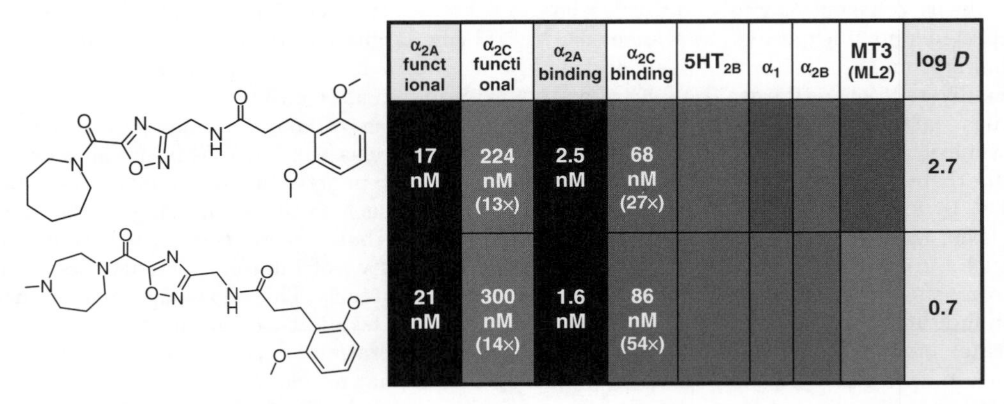

α_{2A} functional	α_{2C} functional	α_{2A} binding	α_{2C} binding	$5HT_{2B}$	α_1	α_{2B}	MT3 (ML2)	log *D*
17 nM	224 nM (13×)	2.5 nM	68 nM (27×)					2.7
21 nM	300 nM (14×)	1.6 nM	86 nM (54×)					0.7

Figure 11. Variable polypharmacology (from BioPrint profiling) of the α_{2A} antagonist oxadiazole carboxamides. Surprisingly, introduction of a basic center improved the α_{2A} selectivity (and overall polypharmacology profile). The selectivity relative to the α_{2A} activity is shown as a color code: green for good (>25×), orange for moderate (>12×) and red for poor (<12) selectivity. (This figure is available in full color at http://mrw.interscience.wiley.com/emrw/9780471266945/home.)

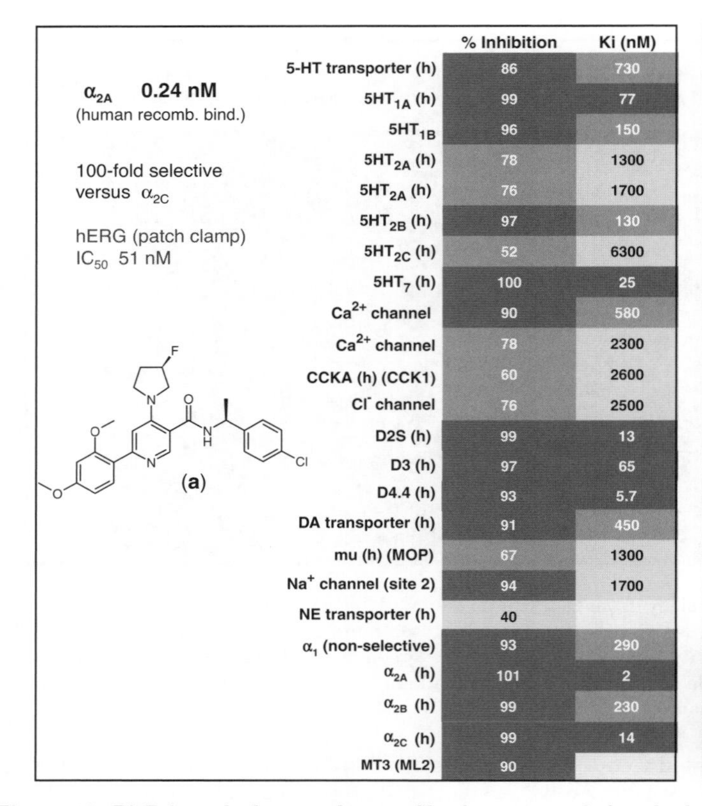

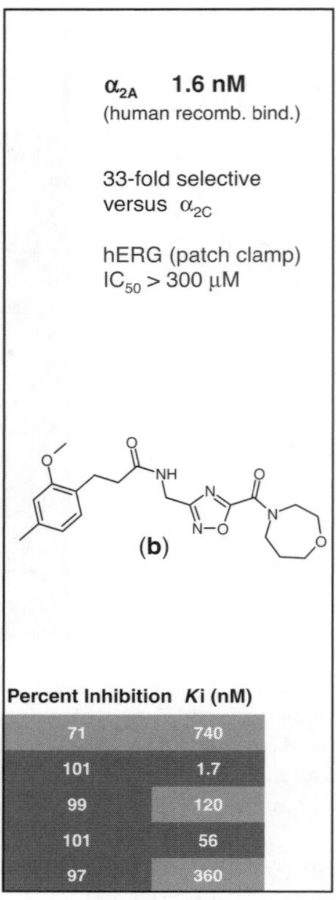

Figure 12. BioPrint polypharmacology profiles for compounds from both α_{2A} series. The neutral oxadiazole compound (b) had a clean off-target profile, in comparison to the basic pyridine compound (a). (This figure is available in full color at http://mrw.interscience.wiley.com/emrw/9780471266945/home.)

data showed that smaller (and thus simpler in terms of 3D pharmacophoric patterns) compounds hit more assays.

The polypharmacology of many drugs, both expected and unexpected, was a key early finding from the BioPrint analysis of drugs. The diversity of "off-target" binding of many drugs (at levels sometimes close to the "primary" activity) is an interesting insight into their potential pharmacological activities. In Fig. 1, the BioPrint data for drugs (y-axis) against assays (x-axis) is shown, in which the data have been hierarchically clustered using both sets of data; color coding of red indicates high binding affinity, yellow indicates medium binding affinity, and green-blue indicates low binding affinity. It was noted at this stage that many drugs of a particular therapeutic class tend to cluster together, although through use of such analyses it was found that compounds with a desired profile could be identified that clustered into a different part of space (see, for example, Fig. 6). A histogram of the number of different targets (assays) that a set of drugs (1388) in the BioPrint database were found to be active against is shown in Fig. 14 (using a cutoff of 50% inhibition at 10 μM).

Analyses of candidate compounds for newer target classes such as kinases showed that many potent activities are found at unrelated targets, such as the aminergic GPCRs, highlighting that selectivity issues beyond the known issue of kinase selectivity should be evaluated early. This knowledge enabled such issues to be addressed at an earlier stage for next generation compounds. Knowledge of the broad pharmacology is important even for drugs for which polypharmacology is part of their action/efficacy as activity on all the targets may not be desirable, and be responsible for undesired side effects etc. Thus, differen-

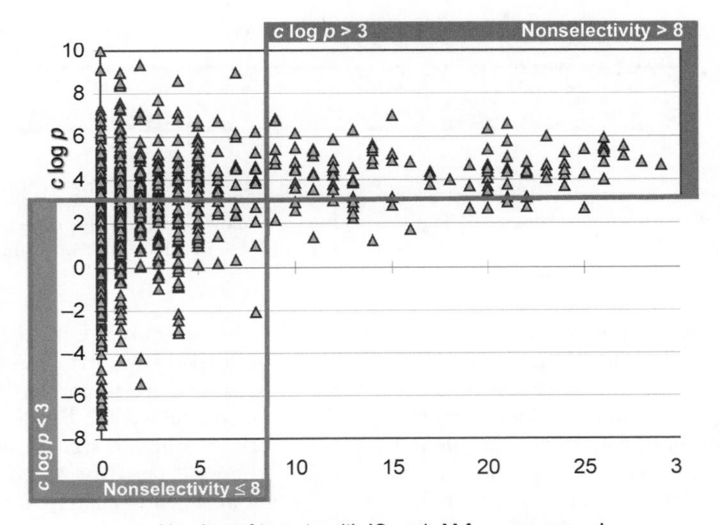

Figure 13. "Promiscuity"/lack of selectivity as defined by the number of targets hit (x-axis) for 1098 drugs (triangles) profiled in the BioPrint assay panel (with <1 μM IC_{50} taken as active) versus $c\log p$ (hydrophobicity, y-axis). The more promiscuous compounds (>8 targets hits) tend to have a $c\log p > 3$. (This figure is available in full color at http://mrw.interscience.wiley.com/emrw/9780471266945/home.)

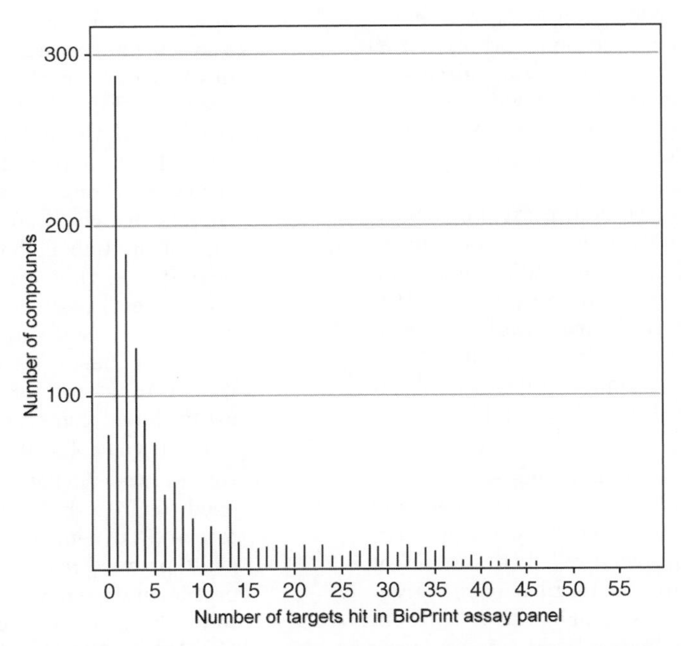

Figure 14. Histogram of the number of targets hit (with 50% inhibition at 10 μM taken as active) for 1388 drugs profiled in BioPrint assay panel. The x-axis is the number of targets hit and the y-axis is the number of compounds with that activity.

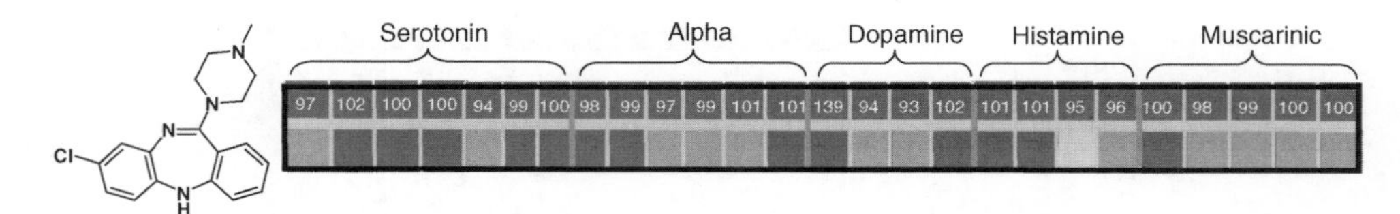

Figure 15. Biological fingerprint (BioPrint) for clozapine, showing the results for assays with a percent inhibition >90% at 10 μM (upper line, red). The IC_{50} values are shown with a color code on the lower line (red < 100 nM, orange < 1 μM, yellow < 5 μM). (This figure is available in full color at http://mrw.interscience.wiley.com/emrw/9780471266945/home.)

tiation by this broad profile could be a useful approach for multiple clinical candidates. Figure 15 shows the polypharmacology of clozapine, with the partial BioPrint biological fingerprint showing only assays with a percent inhibition >90% at 10 μM on the upper heatmap (all red) and the related IC_{50} values below color coded with red most active: red (<100 nM); note the clearer differentiation.

7. EVALUATION OF THE USEFULNESS OF A CHEMOTYPE APPROACH THROUGH THE ANALYSIS OF DRUGS AND ATTRITED COMPOUNDS FOR SPECIFIC TARGETS

As part of an investigation as to whether a "chemotype"-based decision process was scientifically relevant for drug discovery, groups of drugs and withdrawn/attrited compounds for some specific targets were analyzed using the Cerep BioPrint biological profiles. The results demonstrated that "successful" versus "failed" compounds could be similar "chemotypes," and that similar compounds could have quite different biological profiles in terms of broad pharmacology (binding). Similarly a structurally different compound may actually have similar off-target effects, for example, due to similar "decoration" on a different "scaffold."

(i) An analysis of six HMGCoA inhibitor drugs, including a discontinued and a withdrawn drug showed that small changes in structure lead to significant changes in the profiles of off-target binding. This represents a class in which the chemical structures are very similar, yet through a broad profiling analysis differences in off-target effects can be identified; these are mostly at a relatively low level, but the differences could be an important indicator of *in vivo* effects. The BioPrint profiles and the compound structures are shown in Fig. 16. The diversity of the polypharmacology is clearly seen, with a few strong off-target activities (red). This analysis also shows that it would not be correct to condemn a whole series because of one bad clinical result, as very 2D similar compounds (i.e., same "chemotype") are successful drugs.

(ii) An analysis of eight compounds developed for activity on a serotonin (5HT) receptor also showed the quite different profiles of similar compounds with the same primary target. The biological fingerprints for the compounds are shown in Fig. 17. The diversity and richness of the polypharmacology of most of the compounds is evident (red: strong binding). Three attrited compounds (toxicity issues) are shown at the bottom (purple box). The targeted activity, together with binding for related targets is shown in the blue box. There is clearly significant difference in binding to these related receptors, that may be of importance, but if this data are excluded there remains significant differences in the off-target profiles.

(iii) An analysis of a set compounds developed against the same phosphodiesterase (PDE) enzyme target. The targeted primary activity was part of the profile, and is shown in a black box on the right of the profile (last column of Fig. 18). Included in the analysis are several compounds that had attrited during development. The profiles show

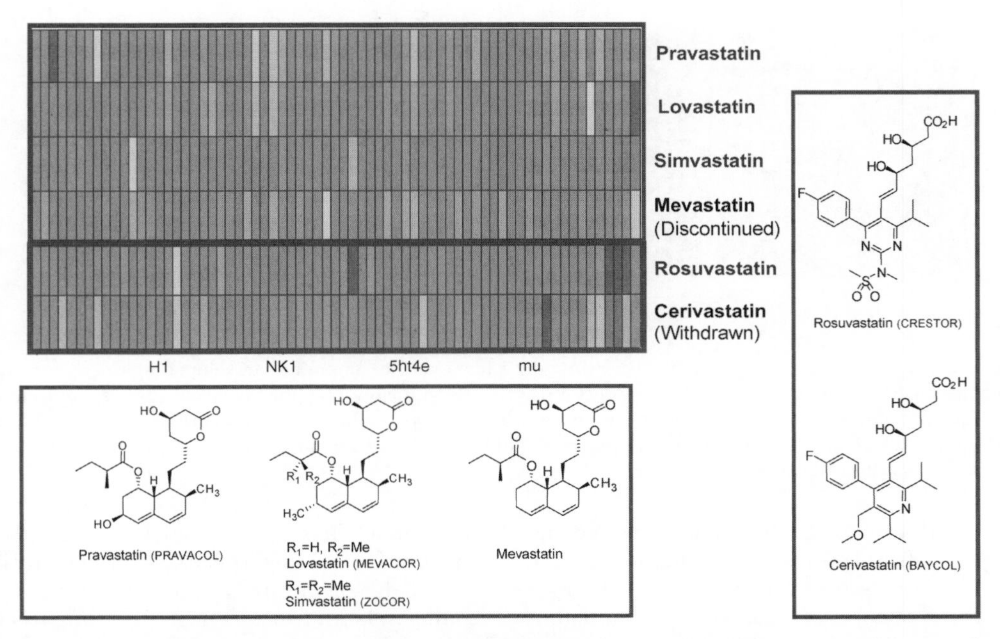

Figure 16. A biological profile (partial BioPrint) of six HMGCoA inhibitors. The compounds are on the y-axis and assays on the x-axis with activity color coded from red (very active) to green (inactive). The two main structural classes are indicated by blue and purple boxes. (This figure is available in full color at http://mrw.interscience.wiley.com/emrw/9780471266945/home.)

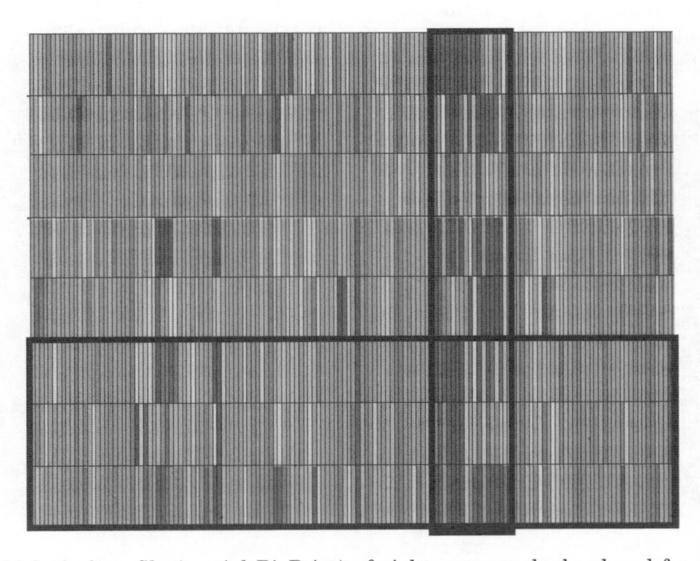

Figure 17. The biological profile (partial BioPrint) of eight compounds developed for activity against a serotonin receptor, showing the polypharmacology for both related and other targets. The compounds are on the y-axis and assays on the x-axis with activity color coded from red (very active) to green (inactive). The three compounds that attrited for some type of toxicity issue are shown at the bottom in a purple box. The assays for serotonin receptors and transporters are highlighted with a blue box. (This figure is available in full color at http://mrw.interscience.wiley.com/emrw/9780471266945/home.)

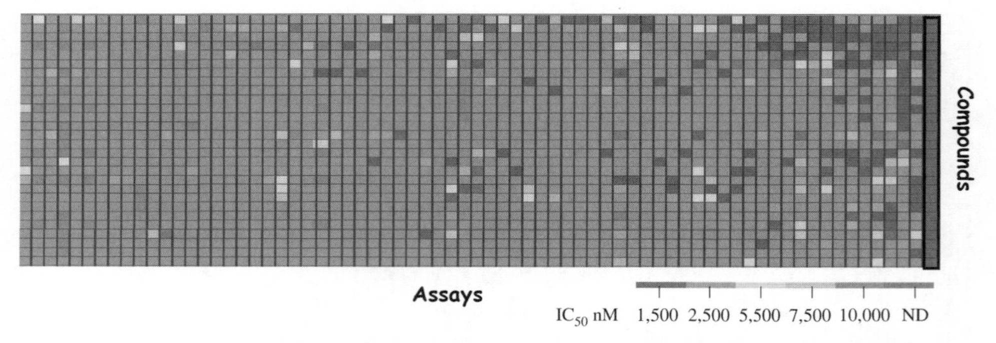

Figure 18. Biological profile (partial BioPrint) data for a set of compounds developed against the same enzyme (PDE) target, illustrating the varying (and at times significant) polypharmacology in these candidate-quality compounds. The compounds are on the y-axis and assays on the x-axis with activity color coded from red (active) to green (inactive). (This figure is available in full color at http://mrw.interscience.wiley.com/emrw/9780471266945/home.)

that many of the compounds had significant off-target binding on non-PDE targets (red squares), much of which was not expected, but also that potent yet broad pharmacologically "clean" compounds could be developed.

8. ANALYSIS OF ATTRITED COMPOUNDS

Several sets of compounds that include successful marketed drugs and attrited compounds, that failed in clinical development or post-launch have been analyzed using BioPrint biological fingerprints. A large analysis included more than 130 preclinical attrited compounds known to Pfizer. By comparison with reference drug set it was shown that showed that for many of the attrited compounds there was a richness of off-target pharmacology and CYP inhibition. Fifteen assays that were hit multiple times by these compounds, compared to results for the same assays for the general drug set, are shown in Fig. 19. Several of these assays are often hit

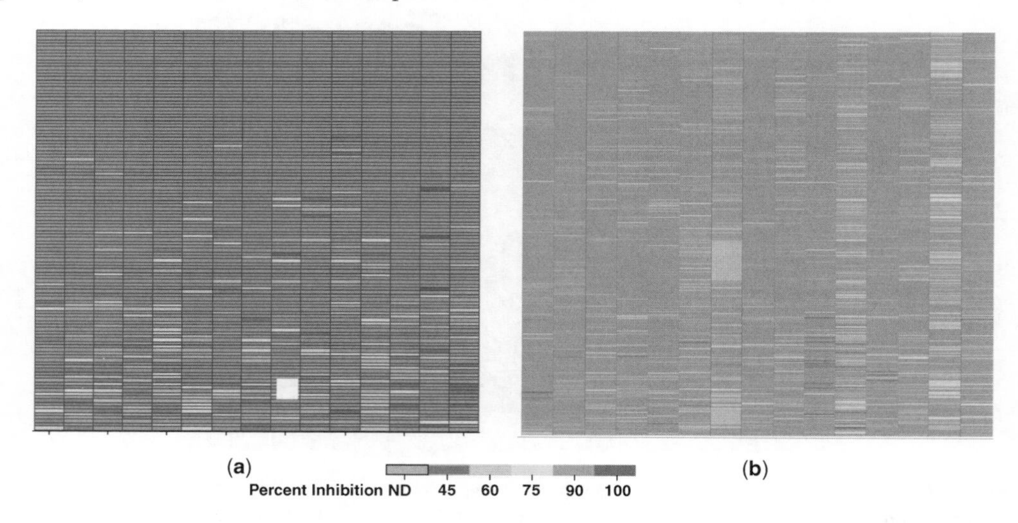

Figure 19. (a) Analysis of >130 attrited compounds using the BioPrint assays showing the 15 targets hit by more than 2 attrited compounds (>50% inhibition at 10 μM). (b) Activity of the BioPrint drug compound set against the same 15 assays. Activity is color coded from red (very active) to green (inactive). (This figure is available in full color at http://mrw.interscience.wiley.com/emrw/9780471266945/home.)

together, but not always the same ones; it is hoped that through further analyses with larger data sets and incorporation of functional and exposure data that profiles with a higher risk can be delineated. The failure of the chemotype approach to risk management, that compounds of the same/similar chemotype will have similar risks of attrition and that the most structurally different chemotypes offer the best approach to minimize attrition risk, was also evident from the structure-profile analyses, as described above. The failure of 2D structure-based methods to be able to effectively differentiate compounds was shown through an analysis of the compounds developed for the three different targets described above, together with a set of fluoroquinolone antibiotics, based on a principal components analysis (PCA) of 2D structure-based property fingerprints. This showed that successful and attrited compounds were not differentiated; a plot is shown in Fig. 20. Clustering/differentiation of compounds for a particular therapeutic activity was obtained, but those that failed in development (gray compound) were not differentiated from those that had not attrited (at that point of time). As these were all compounds that had met typical large pharma quality criteria, obvious structural alerts would not be expected to be present.

9. OTHER APPLICATIONS OF PHARMACOLOGICAL PROFILING DATA IN DRUG DISCOVERY: USE OF BROAD PROFILING DATA TO ADDRESS SAFETY ISSUES AND SELECTIVITY/PROMISCUITY

This chapter focuses on how broad pharmacological profiling data ("biological fingerprints") provides a different and powerful way of describing and differentiation of compounds/"chemotypes" that is medicinal chemistry relevant. Other uses of pharmacological/biological profiling data have been reported, focused more on promiscuity and/or safety issues.

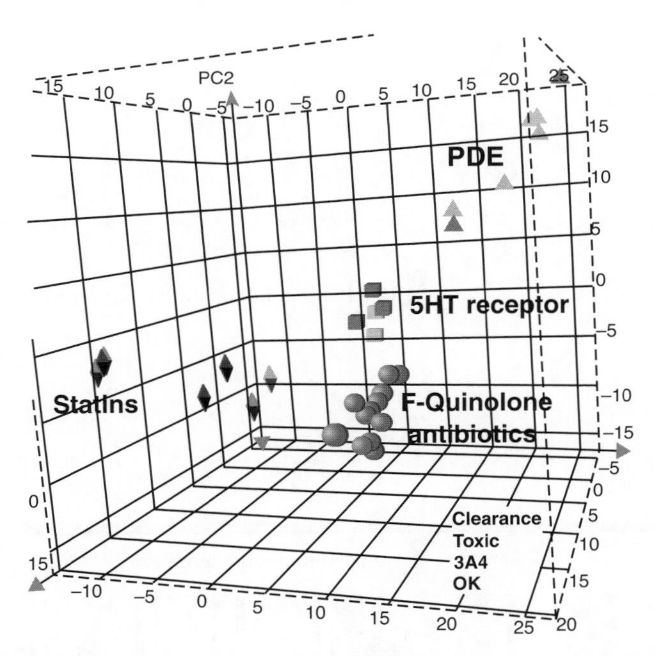

Figure 20. Plot of both failed and "successful" (at least until the time the plot was made) compounds developed for different target classes, based on a PCA of 2D structure-based atom pair property fingerprints. Compounds are coded according to their target class (diamond: statin; square: 5HT receptor; triangle: PDE; circle: F-quinoline antibiotics.) and clinical status at that time (gray: ok; yellow: clearance issue; red: CYP3A4 inhibition issue; purple: toxicity issue). (This figure is available in full color at http://mrw.interscience.wiley.com/emrw/9780471266945/home.)

Fliri et al. have published several interesting papers on the use of BioPrint data [5–7]. They used the complete percentage inhibition data set (versus the IC_{50} data used in the hit and tool evaluation studies described earlier), treating the data as "biological spectra" or "biospectra" of compounds. This approach has the advantage of no apparent "missing data," as even with a 30% inhibition at 10 μM cutoff to try and generate an IC_{50}, many compound-assay cells will have no numerical IC_{50}, only a >entry). The disadvantage is that the refined dose–response data (IC_{50}) are not used. Both approaches have yielded useful results, and the group using this percent inhibition approach was able to address early safety issues (e.g., to find more effectively a new series to avoid muscle toxicity issues) as well as to prioritize compounds to avoid certain ADRs. This new "biospectra" approach (see Fig. 21) to understand the "proteome interaction potential" for small molecules allows for a new approach for the grouping of compounds and their potential properties.

Mignon and coworkers [1,17,18] have used the BioPrint profile of individual hits, and the profile similarity to known compounds to analyze potential ADR liabilities. They find BioPrint to be particularly useful in placing new drug candidates in the context of drugs and related compounds, for which much *in vivo* data are often known. Thus, a the BioPrint profile of a candidate is generated using the same assays as the BioPrint compounds and the resulting profile is analyzed in two ways: either the binding activities within the profile are analyzed and assessed for potential ADR liabilities (using the extensive collections of ADR associations in BioPrint) or the entire profile is used to identify compounds with similar profiles. In the latter approach, potential ADR liabilities are assessed based on those of the similar compounds identified. Pharmacokinetic (PK) data are also used to

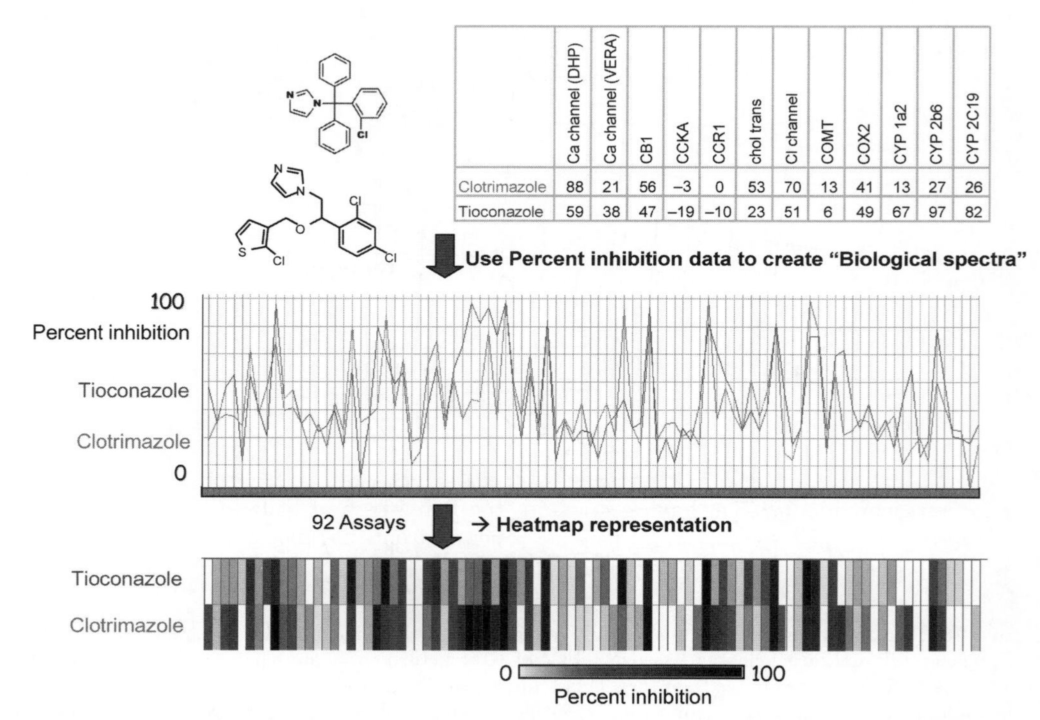

	Ca channel (DHP)	Ca channel (VERA)	CB1	CCKA	CCR1	chol trans	Cl channel	COMT	COX2	CYP 1a2	CYP 2b6	CYP 2C19
Clotrimazole	88	21	56	−3	0	53	70	13	41	13	27	26
Tioconazole	59	38	47	−19	−10	23	51	6	49	67	97	82

Figure 21. Biological spectra approach of Fliri et al., illustrated with two antifungal compounds. A subset of BioPrint profiling data (92 assays) is used (assay percent inhibition data only). (See color insert.)

confirm that the strength of the *in vitro* hit is consistent with *in vivo* exposure levels. The profile similarity approach was applied to Duloxetine, the reference compound in the leas selection example, before it was added to the database. Several potential ADRs were predicted based on the nearest neighbors, including dry mouth that was reported during clinical trials that could be associated with $5HT_7$ activity.

The analysis of pharmacology data and the prediction of adverse drug reactions and off-target effects, including from chemical structure, has been reported in several interesting and informative papers from groups at Novartis [46–49]. They have also published with a focus on pharmacological promiscuity and properties [50,51] and evaluated classification methods for data sets of molecules according to their chemical structure for their biological relevance, including rule-based, scaffold-oriented classification methods and clustering based on molecular descriptors [52]. An analysis of the BioPrint database was used to identify pharmacological targets with a positive association to drug-induced seizure from groups at AstraZeneca [53]. They looked for statistically significant associations between targets and clinical reports of seizure/convulsion, assessed using the "assay by ADR" function of BioPrint with the search term "convulsion" for each target. They also obtained *in vitro* binding profiles of 266 compounds in their seizure set, with compounds considered to be active in a given assay if their percent inhibition was 50% or greater. These data were then analyzed to determine which binding activities were statistically more highly represented in the seizure associated compounds (18 GPCRs, 2 ligand-gated ion channels, 4 voltage-gated ion channels, and 3 transporters were identified as having a possible link).

Leeson [44] at AstraZeneca used the BioPrint data set to look for promiscuity–property relations, and in the very informative paper relations similar to Fig. 13 were found, together with many other interesting analyses of structure–activity data. Bamborough et al. [54] at GSK have published on kinome space and selectivity among kinase targets, showing that compounds often exhibited some off-target kinase activity that could not be predicted from similarity in binding-site amino acids.

10. *IN SILICO* APPROACHES

Full-matrix data sets such as BioPrint provides a very consistent and complete data set for the generation of predictive models. Simple binary models could be generated for about 50% of the assays using various modeling approaches, including Bayesian models and the FCPF6 circular fingerprint (Scitegic Pipeline Pilot software). When the BioPrint data was combined with published activity data (e.g., in *Journal of Medicinal Chemistry*) and Pfizer in-house screening data (that includes compounds/data from many other companies that have been "incorporated" into Pfizer) a unique set of 617,694 data points on 238,655 compounds covering 698 targets was obtained [13]. Bayesian models and human polypharmacology interaction networks, that represent relationships between proteins in chemical space (two proteins are deemed interacting in chemical space, that is, joined by an edge, if both bind one or more compounds within a defined difference in binding energy threshold) were generated from this data set [13]; the Bayesian models were used in the α_{2A} example above. Bender et al. [55] have used "Bayes affinity fingerprints" to improve retrieval rates in virtual screening and to define orthogonal bioactivity space, discussing when are multitarget drugs a feasible concept.

The clustering of biological fingerprints with those of other compounds is an important way to differentiate compounds, but is compound set and descriptor dependent. Taking the four hits and Duloxetine reference compound from the Choosing the Best Hit/Lead Compound section 4 and reclustering with other active compounds used as part of the chemotype analysis work described (PDEs, etc.), showed that for this case a change of descriptor to the Scitegic FCFP6 circular fingerprints enable the "clean" compound to be separated (see Fig. 22), although unfortunately this is not a universal solution.

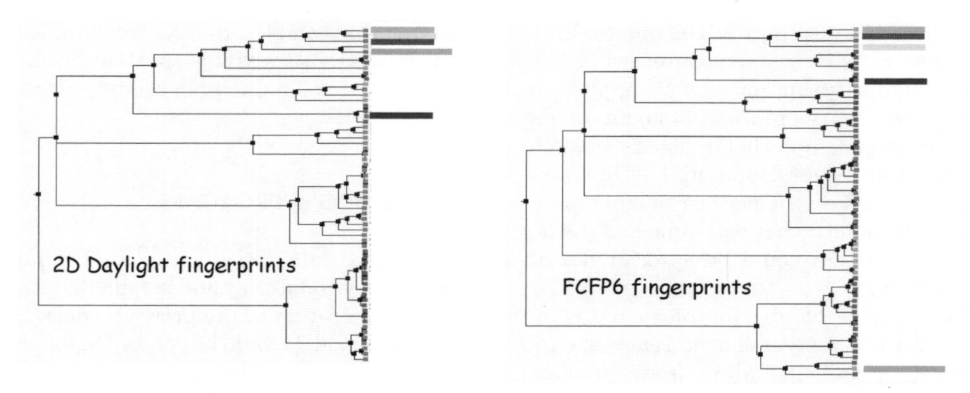

Figure 22. Comparison of the structure-based clustering of the four hit/lead compounds and the clinical reference compound described in the section 4 on choosing the best hit/lead compound, using two different fingerprints. Included in the clustering this time are the compounds from the evaluation of the usefulness of chemotype studies described above. The color coding (bars) for the compounds is the same as for figure 6. (This figure is available in full color at http://mrw.interscience.wiley.com/emrw/9780471266945/home.)

The use of BioPrint data for ADMET prediction and in particular for CYP2D6 inhibition has been published [56], and data showing the many inhibitors are also substrates generated, that can cause variability in data and derived predictive models [1].

Mestres and coworkers have published several interesting papers on chemogenomic approaches and *in silico* pharmacology using large structure–activity data sets [57–60]. Sutherland et al. have published recently on the use of chemical fragments for understanding target space and activity prediction [61].

11. FUTURE

Polypharmacology can be both a desired and undesired (leading to safety issues, etc.) property of a compound. The increased knowledge around drugs and attrited compounds of activities beyond the desired primary pharmacology is an important step forward. The potential of a "biological profile" or "biological fingerprint" to characterize a medicinal chemistry compound is being increasingly explored as both experimental broad profiling approaches such as Cerep BioPrint are used more and can now embrace functional screening at the same time/cost as binding assays, and *in silico* approaches become more effective with the broader availability of experimental data, enabling more predictive models to be developed. Thus, more biological profiling/fingerprinting data of relevance can be generated for project compounds, candidates, drugs, and attrited compounds and this should lead to improved predictions for *in vivo* effects, including ADRs of compounds. Having *in vitro* data generated in a consistent fashion year after year is a challenge, and is one of the advantages of initiatives such as Cerep BioPrint. Other initiatives are bringing together more *in vivo* data, both human and animal, generated for compounds and together with good *in vitro* data sets enable powerful analyses to seek [structure and *in vitro*]–[*in vivo*] associations.

Describing and differentiating compounds by biological fingerprints impacts the drug discovery process at all stages, from "validating" the tool compounds that are used for target validation, through more effective selection and optimization of hits/leads to clinical candidate selection and differentiation. At this important stage, biases based on structural similarity or dissimilarity to known compounds ("good" and "bad") can be put aside and an early evaluation of whether a compound is likely to have similar biological effects made on a more rational basis, the broad *in vitro* profile. This is still a step away from the critical *in vivo* profile, but has a relevance and potential far greater than structure-based analysis. Data can be generated quickly using small sample quantities, with relatively low costs. It is therefore expected that the positive impact of this approach on the drug discovery process will lead to a wider use in medicinal chemistry, using both experimental and *in silico* data.

ABBREVIATIONS

5HT	serotonin receptor
ADME(T)	absorption distribution metabolism excretion (toxicity)
ADR	adverse drug reaction
BioPrint	Cerep BioPrint®
CNS	central nervous system
CV	cardiovascular
CYP	cytochrome P450
GPCR	G-protein coupled receptor
H2	histamine H2 receptor
$c \log p$	calculated logarithm of the partition coefficient, p (the ratio of concentration of neutral species in octanol divided the concentration of neutral species in water)
HTS	high-throughput screening
MIF	molecular interaction field
NHR	nuclear hormone receptor
PCA	principal components analysis
PDE	phospodiesterase
Pgp	P-glycoprotein transporter
PK	pharmacokinetic

REFERENCES

1. Mason JS, Migeon J, Dupuis P, Otto-Bruc A. Use of broad biological profiling as a relevant descriptor to describe and differentiate compounds: structure–*in vitro*–*in vivo* (safety) relationships. Antitargets: Prediction and Prevention of Drug Side Effects. Vol.38. Methods and Principles in Medicinal Chemistry. Wiley-VCH; Verlag GmbH 2008. p 23–52.
2. Stanton R, Cao Q. Biological fingerprints. In: Mason JS, editor. Taylor JB, Triggle DJ, editors-in-chief. Comprehensive Medicinal Chemistry II. Vol. 4. Oxford: Elsevier; 2007. p 807–818.
3. Hopkins AL, Mason JS, Overington JP. Can we rationally design promiscuous drugs? Curr Opin Struct Biol 2006;16:127–136.
4. Bender A, Young DW, Jenkins JL, Serrano M, Mikhailov D, Clemons PA, Davies JW. Chemogenomic data analysis: prediction of small-molecule targets and the advent of biological fingerprints. Comb Chem High Throughput Screen 2007;10(8): 719–731.
5. Fliri AF, Loging WT, Thadeio P, Volkmann RA. Biological spectra analysis: linking biological activity profiles to molecular structure. Proc Natl Acad Sci USA 2005;102:261–266.
6. Fliri AF, Loging WT, Thadeio P, Volkmann RA. Analysis of drug-induced effect patterns linking structure and side effects of medicines. Nat Chem Biol 2005;1:389–397.
7. Fliri AF, Loging WT, Thadeio P, Volkmann RA. Biospectra analysis: model proteome characterizations for linking molecular structure and biological response. J Med Chem 2005;48: 6918–6925.
8. Filimonov DA, Poroikov VV, Karaicheva EI, Exp Clin Pharmacol 1995;58:56–62.
9. Kauvar LM, Higgins DL, Villar HO, Sportsman JR, Engqvist-Goldstein A, Bukar R, Bauer KE, Dilley H, Rocke DM. Predicting ligand binding to proteins by affinity fingerprinting. Chem Biol 1995;2:107–118.
10. Dixon SL, Villar HO. Bioactive diversity and screening library selection via affinity fingerprinting. J Chem Inf Comput Sci 1998;38(6): 1192–1203.
11. Haggarty SJ, Clemons PA, Schreiber SL. Chemical genomic profiling of biological networks using graph theory and combinations of small molecule perturbations. J Am Chem Soc 2003;125:10543–10545.
12. Kim YK, Arai MA, Arai T, Lamenzo JO, Dean EF 3rd, Patterson N, Clemons PA, Schreiber SL. Relationship of stereochemical and skeletal diversity to cellular measurement space. J Am Chem Soc 2004;126:14740–14745.
13. Paolini GV, Shapland RHB, van Hoorn WP, Mason JS, Hopkins AL. Global mapping of pharmacological space. Nat Biotechnol 2006;24: 805–815.
14. Jean T, Chapelain B. Method of identification of leads or active compounds. CEREP, 1999, International publication number WO-09915894.
15. Krejsa CM, Horvath D, Rogalski SL, Penzotti JE, Mao B, Barbosa F, Migeon JC. Predicting

ADME properties and side effects: the BioPrint approach. Curr Opin Drug Discov Dev 2003; 6:471–480.

16. Froloff N, Hamon V, Dupuis P, Otto-Bruc A, Mao B, Merrick S, Migeon J. Construction of a homogeneous and informative *in vitro* profiling database for anticipating the clinical effects of drugs. In: Jacoby E, editor. Chemogenomics Knowledge-Based Approaches to Drug Discovery. London: Imperial College Press; 2006. p 175–206.
17. Krejsa CM, Horvath D, Rogalski SL, Penzotti JE, Mao B, Barbosa F, Migeon JC. Predicting ADME properties and side effects: the BioPrint approach. Curr Opin Drug Discov Dev 2003;6: 471–480.
18. Armstrong D, Migeon J, Rolf MG, Bowes J, Crawford M, Valentin J-P. Secondary pharmacodynamic studies and *in vitro* pharmacological profiling. In: Shayne Cox Gad. Preclinical Development Handbook: Toxicology. John Wiley & Sons Inc.; 2008. p 581–609.
19. http://www.cerep.fr/cerep/users/pages/productsservices/bioprintservices.asp & http://www.chemaxon.com/forum/vberenz_cerep_2006_v2-download1572.pdf.
20. Goodford PJ. A computational procedure for determining energetically favorable binding sites on biologically important macromolecules. J Med Chem 1985;28(7):849–857.
21. Merlot C, Domine D, Cleva C, Church DJ. Drug Discov Today 2003;8(13):594–602.
22. Cruciani G, editor. Molecular Interaction Fields: Applications in Drug Discovery and ADME Prediction. Wiley-VCH; 2005. p 328.
23. Fox T. Protein selectivity studies using GRID-MIFs. Molecular Interaction Fields. Vol. 27. Methods and Principles in Medicinal Chemistry. 2006. p 45–82.
24. Baroni M, Cruciani G, Sciabola S, Perruccio F, Mason JS. A common reference framework for analyzing/comparing proteins and ligands. Fingerprints for ligands and proteins (FLAP): theory and application. J Chem Inf Model 2007; 47(2):279–294.
25. Perruccio F, Mason JS, Sciabola S, Baroni M. FLAP: 4-point pharmacophore fingerprints from GRID. Molecular Interaction Fields. Vol. 27. Methods and Principles in Medicinal Chemistry. Wiley-VCH Verlag GmbH; 2006. p 83–102.
26. Zhang Q, Muegge I. Scaffold hopping through virtual screening using 2D and 3D similarity descriptors: ranking, voting, and consensus scoring. J Med Chem 2006;49(5):1536–1548.
27. Daylight fingerprints. Daylight Chemical Information Systems, Inc., Irvine, CA. www.daylight.com.
28. Martin YC, Kofron JL, Traphagen LM. Do structurally similar molecules have similar biological activity? J Med Chem 2002;45:4350–4358.
29. Horvath D, Jeandenans C. Neighborhood behavior of *in silico* structural spaces with respect to *in vitro* activity spaces: a novel understanding of the molecular similarity principle in the context of multiple receptor binding profiles. J Chem Inf Comput Sci 2003;43:680–690.
30. Horvath D, Jeandenans C. Neighborhood behavior of *in silico* structural spaces with respect to *in vitro* activity spaces: a benchmark for neighborhood behavior assessment of different *in silico* similarity metrics. J Chem Inf Comput Sci 2003;43:691–698.
31. Horvath D. ComPharm: automated comparative analysis of pharmacophoric patterns and derived QASAR approaches, novel tools in high throughput drug discovery. A proof of concept study applied to farnesyl protein transferase inhibitor design. In: Diudea editor. QSPAR/QASAR Studies by Molecular Descriptors. Nova Science Publishers; 2001. p 395–439.
32. Horvath D. Throughput conformational sampling and fuzzy similarity metrics: a novel approach to similarity searching and focused combinatorial library design and its role in the drug discovery laboratory. In: Ghose A, Viswandhan V, editors. Principles, Software Tools and Applications. New York: Marcel Dekker; 2001. p 429–472.
33. Steffen A, Kogej T, Tyrchan C, Engkvist O. Comparison of molecular fingerprint methods on the basis of biological profile data. J Chem Inf Model 2009;49(2):338–347.
34. Mason JS, Morize I, Menard PR, Cheney DL, Hulme C, Labaudiniere RF. New 4-point pharmacophore method for molecular similarity and diversity applications: overview of the method and applications, including a novel approach to the design of combinatorial libraries containing privileged substructures. J Med Chem 1999; 42(17):3251–3264.
35. Good AC, Mason JS, Pickett SD. Pharmacophore pattern application in virtual screening, library design and QSAR. Virtual Screening for Bioactive Molecules. Vol.10. Methods and Principles in Medicinal Chemistry. 2000. p 131–159.
36. Mason JS, Cheney DL. Ligand–receptor 3-D similarity studies using multiple 4-point pharmacophores. Pacific Symposium on Biocomputing'99,

Mauna Lani, Hawaii, January 4–9, 1999. p 456–467.

37. Mason JS, Cheney DL. Library design and virtual screening using multiple 4-point pharmacophore fingerprints. Pacific Symposium on Biocomputing 2000, Honolulu, January 4–9, 2000; 576–587.
38. Hopkins AL, Groom CR. The druggable genome. Nat. Rev. Drug Discov. 2002;1:727–730.
39. Sallantin J, Dartnell C, Afshar M. A pragmatic logic of scientific discovery. In: Lavrac N, Todrovski L, Jantke JP, editors. Discovery Science. Vol. 4265. Lecture Notes in Computer Science. Berlin, Heidelberg: Springer Verlag; 2006. p 231–242.
40. Afshar M, Lanoue A, Sallantin J. Multiobjective/multicriteria optimization and decision support in drug discovery. In: Mason JS, editor. Taylor JB, Triggle DJ, editors-in-chief. Comprehensive Medicinal Chemistry II. Vol. 4. Oxford: Elsevier; 2007. p 767–772.
41. Ariana Pharmaceuticals, Paris. www.ariana-pharma.com.
42. Mason JS, Mills JE, Barker C, Loesel J, Yeap K, Snarey M. Higher-throughput approaches to property and biological profiling, including the use of 3-D pharmacophore fingerprints and applications to virtual screening and target class-focused library design. In: Abstracts of Papers, 225th ACS National Meeting, New Orleans, LA, March 23–27, 2003. COMP-343.
43. Mason JS. Understanding leads and chemotypes from a biological viewpoint: chemogenomic, biological profiling and data mining approaches. In: Book of Abstracts of First European Conference on Chemistry for Life Sciences: Understanding the Chemical Mechanisms of Life, Rimini, Italy, October 4–8, 2005.
44. Leeson PD, Springthorpe B. The influence of drug-like concepts on decision-making in medicinal chemistry. Nat Rev Drug Discov 2007; 6(11):881–890.
45. Bell A, Andrews M, Brown A, van Hoorn W, Lewthwaite R. Trials and tribulations: the identification of selective. CNS penetrant alpha-2a antagonists. East of England Medicinal Chemistry Symposium, Hatfield, UK, April 30, 2009.
46. Scheiber J, Jenkins JL, Sukuru SCK, Bender A, Mikhailov D, Milik M, Azzaoui K, Whitebread S, Hamon J, Urban L, Glick M, Davies JW. Mapping adverse drug reactions in chemical space. J Med Chem 2009;52(9):3103–3107.
47. Bender A, Scheiber J, Glick M, Davies JW, Azzaoui K, Hamon J, Urban L, Whitebread S, Jenkins JL. Analysis of pharmacology data and the prediction of adverse drug reactions and off-target effects from chemical structure. ChemMedChem 2007;2(6):861–873.
48. Whitebread S, Hamon J, Bojanic D, Urban L. Keynote review: *in vitro* safety pharmacology profiling: an essential tool for successful drug development. Drug Discov Today 2005;10: 1421–1433.
49. Scheiber J, Chen B, Milik M, Sukuru SCK, Bender A, Mikhailov D, Whitebread S, Hamon J, Azzaoui K, Urban L, Glick M, Davies JW, Jenkins JL. Gaining insight into off-target mediated effects of drug candidates with a comprehensive systems chemical biology analysis. J Chem Inf Model 2009;49(2):308–317.
50. Azzaoui K, Hamon J, Faller B, Whitebread S, Jacoby E, Bender A, Jenkins JL, Urban L. Modeling promiscuity based on *in vitro* safety pharmacology profiling data. ChemMedChem 2007;2:874–880.
51. Faller B, Wang J, Zimmerlin A, Bell L, Hamon J, Whitebread S, Azzaoui K, Bojanic D, Urban L. High-throughput *in vitro* profiling assays: lessons learnt from experiences at Novartis. Expert Opin Drug Metab Toxicol 2006;2(6): 823–833.
52. Schuffenhauer A, Brown N, Ertl P, Jenkins JL, Selzer P, Hamon J. Clustering and rule-based classifications of chemical structures evaluated in the biological activity space. J Chem Inf Model 2007;47(2):325–336.
53. Easter A, Bell ME, Damewood JR Jr, Redfern WS, Valentin J-P, Winter MJ, Fonck C, Bialecki RA. Approaches to seizure risk assessment in preclinical drug discovery. Drug Discov Today 2009;14(17/18):876–884.
54. Bamborough P, Drewry D, Harper G, Smith GK, Schneider K. Assessment of chemical coverage of kinome space and its implications for kinase drug discovery. J Med Chem 2008;51(24): 7898–7914.
55. Bender A, Jenkins JL, Glick M, Deng Z, Nettles JH, Davies JW. "Bayes affinity fingerprints" improve retrieval rates in virtual screening and define orthogonal bioactivity space: when are multitarget drugs a feasible concept? J Chem Inf Model 2006;46(6):2445–2456.
56. Gozalbes R, Barbosa F, Froloff N, Horvath D. The BioPrint approach for the evaluation of ADMET properties: application to the prediction of cytochrome P450 2D6 inhibition. In: Testa B, Krämer SD, Wunderli-Allenspach H, Folkers G, editors. Pharmacokinetic Profiling in Drug Research: Biological, Physicochemical

and Computational Strategies. Zürich: HCA; Weinheim: Wiley-VCH; 2006. p 395–415.

57. Mestres J, Martin-Couce L, Gregori-Puigjane E, Cases M, Boyer S. Ligand-based approach to *in silico* pharmacology: nuclear receptor profiling. J Chem Inf Model 2006;46(6):2725–2736.
58. Cases M, Mestres J. A chemogenomic approach to drug discovery: focus on cardiovascular diseases. Drug Discov Today 2009;14(9/10): 479–485.
59. Mestres J. Mapping the chemogenomic space. Chemogenomics 2006; 39–57.
60. Cases M, Garcia-Serna R, Hettne K, Weeber M, van der Lei J, Boyer S, Mestres J. Chemical and biological profiling of an annotated compound library directed to the nuclear receptor family. Curr Top Med Chem. 2005;5(8):763–772.
61. Sutherland JJ, Higgs RE, Watson I, Vieth M. Chemical fragments as foundations for understanding target space and activity prediction. J Med Chem 2008;51(9):2689–2700.

RECENT ADVANCES IN DEVELOPMENT, VALIDATION, AND EXPLOITATION OF QSAR MODELS

ALEXANDER TROPSHA
Laboratory for Molecular Modeling, School of Pharmacy, University of North Carolina, Chapel Hill, NC

1. INTRODUCTION

The 15th EuroQSAR meeting that took place in Istanbul in 2004 was officially dedicated to the 40th anniversary of QSAR modeling linking the origin of the field to the seminal (and one of the most cited in the field) paper published by Hansch and Fujita in 1964 [1]. Needless to say, that celebratory symposium only confirmed that QSAR modeling field has long reached maturity and it continues to expand as indicated by the strong upward trend in the rate of QSAR publications (Fig. 1).

In its more than 45 years of rich history of methodology developments and applications, the QSAR modeling field has evolved and transformed dramatically. Initially, the QSAR models were relatively simple employing a small number of easily interpretable physical chemical descriptors (such as octanol–water distribution coefficient, log p, molar refractivity, or dipole moment) and relatively unsophisticated statistical data modeling methodologies such as (multiple) linear regression. QSAR modeling was viewed solely as a tool for lead optimization, that is, it was employed to elucidate the relationship between structure and activity in relatively small congeneric compound series (summarized in the BioBite database [2]) and predict relatively small structural modifications leading to the enhanced activity. These simple methodologies were nevertheless successful in enabling several documented cases of drug (or at least viable drug candidate) discovery as discussed in the preface to the first volume of the famous book series titled *Reviews in Computational Chemistry* [3].

The field has changed dramatically in the last 20 years fueled by the dramatic changes in size, complexity, and availability of experimental data sets of biologically active compounds (summarized in a recent review [4]). These changes have been coincidental with the advances in chemometrics resulting in a significant increase in the number of chemical descriptors as well as broader reliance on machine learning and advanced statistical modeling techniques for QSAR studies. The dramatic change in the content and complexity of QSAR models shifted the field away from the original simple and easily interpretable linear models built with simple descriptors toward complex multiparametric and mostly nonlinear approaches. The "old" QSAR models employed very small number of molecular descriptors and were built for small congeneric data sets allowing their straightforward interpretation and even incremental extrapolation to predict the (incremental) change in the target bioactivity due to small modifications in the single scaffold. In the "New World," we deal with much larger and substantially more chemically diverse data sets characterized by hundreds of chemical descriptors. Hence, the model fitness (which on the surface is more difficult to achieve but in reality, with the increase in the number of descriptors the chance correlation is even more probable) cannot be merely equated to its external predictive power as was pointed out and illustrated in one of our earlier publications [5].

We would argue strongly that many modern QSAR modeling approaches are much closer in spirit to such subdisciplines of computational science as data mining and knowledge discovery in databases than to physical organic chemistry where the field originated. Unfortunately, this critical shift in the complexity and meaning of modern QSAR models developed for modern data sets was not recognized and captured by many users of the method. Ignorance of the paradigm shift has led to many studies that confused model fitness with its predictive power, lacking proper model validation, and misinterpreted correlation as causation as discussed in a recent critical review [6] where the author even posed a question as to whether QSAR is dead or alive. The unfortunate abundance of poorly executed QSAR studies also led to a recent editorial published by the leading Cheminformatics *Journal of Chemical Information and*

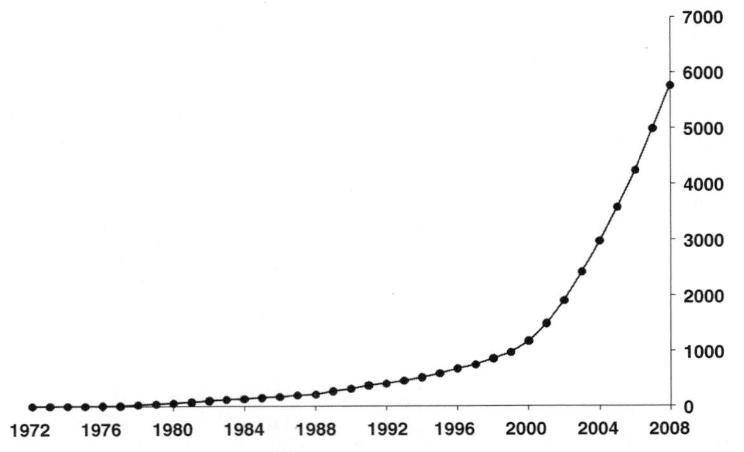

Figure 1. The growth of publications related to QSAR modeling based on the data in PubMed (Figure courtesy of Prof. A. Cherkasov, University of British Columbia).

Modeling (also reproduced by the *Journal of Medicinal Chemistry*) that introduced severe limitations on the quality of QSAR papers to be considered acceptable [7]. In the same vein, at the 37th Joint Meeting of Chemicals Committee and Working Party on Chemicals, Pesticides and Biotechnology, held in Paris in November 17–19 2004, the OECD (Organization for Economic Co-operation and Development) adopted the following five principles that valid (Q)SAR models should support to allow their use in regulatory assessment of chemical safety: (i) a defined endpoint, (ii) an unambiguous algorithm, (iii) a defined domain of applicability, (iv) appropriate measures of goodness-of-fit, robustness, and predictivity, (v) a mechanistic interpretation, if possible (see Ref. [8]).

These recent major steps taken by leading professional journals and a scientific council within an intergovernmental organization highlight both the importance of the QSAR field and the need to present a deeper analysis of the complexity and challenges faced by the QSAR modeling field today. Thus, some of the main focus of this chapter include the discussion of the best practices for model development, validation, and exploitation that should be followed by QSAR modelers. The previous and ongoing basic and applied research in the area of QSAR modeling suggests that there is still a substantial room for further advancement of the field via innovative methodologies and important applications. We believe strongly that many examples of low impact QSAR research have been due to frequent exploration of data sets of limited size with little attention paid to model validation. This limitation has led to models having questionable "mechanistic" explanatory power but perhaps little if any forecasting ability outside of the training sets used for model development. We believe that the latter ability along with the capabilities of QSAR models to explore chemically diverse data sets with complex biological properties should be viewed as the chief focus of QSAR studies. This focus requires the reevaluation of the success criteria for the modeling as well as the development of novel chemical data mining algorithms and model validation approaches. In fact, we think that the most interesting era in QSAR modeling is just beginning with the rapid growth of the experimental publicly available SAR data space due to such projects as PubChem [9].

Naturally, any QSAR investigation includes the following four stages: (i) data collection and preparation, (ii) model building (that implies the calculation of chemical descriptors and the application of statistical and machine learning approaches to establish an empirical relationship between the values of descriptors and the target property), (iii) model validation, and (iv) model exploitation (which includes model interpretation and prospective use for discovering novel compounds with the desired property). This chapter examines the recent developments in QSAR modeling field that concern all major components of QSAR modeling. To set the stage for discussing the recent and emerging developments in the field, we start from a few introductory paragraphs presenting the general outline of QSAR modeling. We then discuss recent trends in the context of aforementioned four components of the QSAR modeling process illustrating these trends with examples where appropriate. We focus our discussion on what we consider best practices for model development implemented as part of the data analytical modeling workflow developed in our laboratory. This workflow incorporates modules for combinatorial QSAR model development (i.e., using all possible binary combinations of available descriptor sets and statistical data modeling techniques), rigorous model validation, and virtual screening of available chemical databases to identify novel biologically active compounds. We present examples of studies where the application of rigorously validated QSAR models for virtual screening afforded computational hits that were confirmed by subsequent experimental investigations. We stress that the emerging focus of QSAR modeling on target property forecasting brings it forward as predictive, as opposed to evaluative, modeling approach that is suitable not only for traditional lead optimization but also for lead discovery.

1.1. The General QSAR Modeling Framework

Modern QSAR modeling is a very complex and complicated field requiring deep understanding and thorough practicing to develop robust models. Multiple types of chemical descriptors and numerous statistical model development approaches can be found in specialized literature and so need not be discussed in this chapter. Instead, we shall present several unifying concepts that underlie practically any QSAR methodology. As a matter of fact, this general framework of model development has not really changed as compared to the birth time of the field since we are still talking about calculating descriptors and relating them to specific target activities or properties of compounds. What has changed is the complexity of models, that is, the number and diversity of available chemical descriptors and the sophistication of techniques used to relate descriptors and activity. Thus, to reflect on recent trends, we shall first describe the basics of the QSAR modeling framework.

Any QSAR method can be generally defined as an application of mathematical and statistical methods to the problem of finding empirical relationships (QSAR models) of the form $P_i = \hat{k}(D_1, D_2, \ldots D_n)$, where P_i is biological activities (or other properties of interest) of molecules, $D_1, D_2, \ldots, D_\mathrm{n}$ is calculated (or, sometimes, experimentally measured) structural properties (molecular descriptors) of compounds, and $\hat{k}$ is some empirically established mathematical transformation that should be applied to descriptors to calculate the target property (e.g., activity) values for all molecules. The relationship between values of descriptors D and target properties P can be linear or nonlinear. The example of the former relationship is given by multiple linear regression (MLR) common to the Hansch QSAR approach [1], where target property can be predicted directly from the descriptor values. On the contrary, nearest neighbor QSAR methods (e.g., see Ref. [10]) serve as examples of nonlinear techniques where descriptor values are used in characterizing chemical similarities between molecules, which are then used to infer compound activity. The goal of QSAR modeling is to establish a trend in the descriptor values, which parallels the trend in biological activity. In essence, all QSAR approaches rely, directly or indirectly, on a simple similarity principle, which for a long time has provided a foundation for the experimental medicinal chemistry: compounds with similar structures are expected to have similar biological activities.

The differences in various QSAR methodologies can be understood in terms of the types of target property values, descriptors, and optimization algorithms used to relate descriptors to the target properties. Target properties (regarded as dependent variables in statistical data modeling sense) can be of three general types: continuous (i.e., real values covering certain range, e.g., IC_{50} values, or binding constants); categorical related, or rank-based (e.g., classes of rank-ordered target properties covering certain range of values, e.g., classes of metabolic stability such as unstable, moderately stable, or stable); and categorical unrelated (i.e., classes of target properties that do not relate to each other in any continuum, e.g., different pharmacological classes of compounds). As simple as it appears, understanding this classification is actually very important since the choice of descriptor types and modeling techniques as well as model accuracy metrics is often dictated by the type of the target properties. Thus, in general, the latter two types require classification modeling approaches whereas the former type of the target properties allows the use of (multi)linear regression type modeling. The corresponding methods of data analysis are referred to as either classification or continuous property QSAR.

1.2. Brief Summary of Recent Trends

Modern QSAR approaches are characterized by the use of multiple descriptors of chemical structure combined with the application of both linear and nonlinear optimization approaches, and a growing emphasis on rigorous model validation to afford robust and predictive models (see separate discussion in Section 3.2). Many QSAR approaches have been developed during the past few decades (e.g., see recent reviews [11,12]). The major differences between various approaches are due to structural parameters (descriptors) used to characterize molecules and the mathematical approaches used to establish a correlation between descriptor values and biological activity. Most of the modeling techniques assume a linear relationship between the molecular descriptors and a target property, which may be an adequate methodology for many data sets. However, the advances in combinatorial chemistry and high-throughput screening technologies have resulted in the explosive growth of the amount of structural and biological data making the problem of developing robust QSAR models more challenging. This progress has provided an impetus for the development of fast, nonlinear QSAR methods that can capture structure–activity relationships for large and complex data. New nonlinear methods of multivariate analysis such as different types of artificial neural networks [13–16], generalized linear models [14,17–19], classification and regression trees [17,20–23], random forests [24–26], MARS(multivariate adaptive regression splines) [26,27], support vector machines [28–31], and some other methods have become routine tools in QSAR studies. Interesting examples of applications have been reported for all types of the above methods. In some cases, the comparisons between different techniques as applied to the same data set have been made but, in general, there appears to be no universal QSAR approach that produces the best QSAR models for any data sets.

As mentioned above, the most important recent developments in the field concur with a substantial increase in the size and the endpoint complexity of experimental data sets available for the analysis and an increased application of QSAR models as virtual screening tools to discover biologically active molecules in chemical databases and/or virtual chemical libraries [32]. The latter focus differs substantially from the traditional emphasis on developing so-called explanatory QSAR models characterized by high statistical significance but only as applied to training sets of molecules with known chemical structure and biological activity. Figure 2 shows a general schema for the development and application of QSAR models for hit discovery by the means of virtual screening. Examples of the prospective use of QSAR models as virtual screening tools are given in Section 4.1.

1.3. Why Models May Fail

As is true perhaps for any computational field, QSAR modeling has been both blessed and, sometimes, cursed in the literature. Our group was one of the first groups emphasizing the importance of statistical validation of QSAR models [5]. As we pointed out and demonstrated with examples, the high accuracy of model fitness characterized with leave-one-out cross validated R^2 (q^2) is not indicative of the high external predictive power. Thus, the exclusive reliance on training set modeling

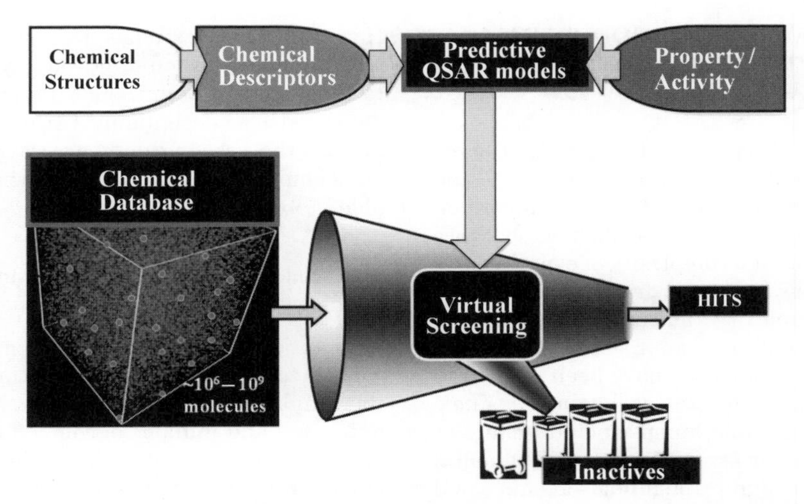

Figure 2. A general framework for QSAR model development and application to virtual screening. (This figure is available in full color at http://mrw.interscience.wiley.com/emrw/9780471266945/home.)

without any external validation is one of the reasons why many models cannot be considered reliable. Another important paper examined the reasons behind the failure of *in silico* ADME/Tox models [33] linking the frequent failures to the inappropriate use of models, false expectations, or procedures used to develop models. In a brief but very important editorial note, G. Maggiora [34] outlined limitations and some reasons for failures of QSAR modeling that relate to the so-called "activity cliffs," which are known cases when a small change in chemical structure leads to dramatic changes in the target activity. Such cases are indeed difficult to foresee and hard to capture and explain using QSAR models since the models work best in reflecting relatively smooth trends in structure–activity correlations. Addressing the activity cliffs problem is indeed a hard problem in QSAR modeling and in some cases a source of poor predictions. A summary of various reasons leading to erroneous QSAR models was given in a recent critical overview of the field [6]. Another recent important paper listed as many as 21 possible sources of error when developing QSAR models [35] and provided some recipes as to how at least some common errors could be avoided.

In most cases, the authors concerned with the quality and practical utility of QSAR models looked deeply into possible sources of errors or offered approaches to improve the robustness of models. On the other hand, the author of the negative opinion letter published in early 2008 [36] made an unfortunate attempt to equate the fraction of papers not paying enough attention to the statistical quality of models with the entire field. As we discuss in this chapter, it is critically important to avoid the oversimplification of the QSAR modeling process and employ statistically robust approaches for both model development and validation. Authors ignoring the complexity of the problem or those paying insufficient attention to model validation do end up developing and in some cases, publishing models that could not be regarded reliable as we illustrate in subsequent sections of the chapter. Conversely, the criticism of the field should be balanced and based on the thorough analysis of possible sources of error rather than equating the entire field to one large error as the aforementioned opinion letter [36] did. Thus, in this chapter, we make a fair attempt to balance the substantiated criticism of published QSAR literature versus objective challenges facing (but not dooming!) the field (such as activity cliffs) and we emphasize the importance of developing and practicing rigorous approaches to both model development and validation.

2. THE COMPLEXITY OF MODERN DATA SETS AND THE EMERGING NEED FOR DATA CURATION

In the early days of QSAR modeling, the experimental data sets were relatively small and chemically congeneric and the techniques employed were relatively unsophisticated. Since then, the size and complexity of experimental data sets has increased dramatically, and so had the complexity and challenges of respective data analytical approaches. Traditionally, QSAR approaches have been applied to modeling data sets tested against a single target, for example, in specific enzymatic or receptor binding assays. Recent experimental advances in high-throughput screening and multitarget testing of compound libraries have led to the establishment of data sets of biologically active compounds (often publicly available) that we shall define below as complex.

The increase in size and complexity of data sets used for QSAR modeling raises an important issue of data accuracy. Indeed, modelers typically "grab" the data on chemical structures and their biological activities without necessarily spending time to curate this data before embarking on model development. However, recent studies [37] suggest that QSAR modelers should indeed pay attention to the quality of both chemical and biological data used in model building. In this section, we discuss the growing databases available for model building and then address the unavoidable problem of data curation.

2.1. Modern Data Sets for QSAR Model Building

As mentioned above, modern data sets have evolved to incorporate as many as hundreds of thousands of compounds in target bioactivity databases and millions of compounds in chemical databases. Table 1 lists several sources of currently existing and constantly growing databases that contain bioactivity information for many organic molecules; the table also lists several large chemistry databases that can be used for virtual screening in the context of the schema in Fig. 2. Further discussion of many such databases emphasizing the need of their integration to advance the science of cheminformatics and computational chemical genomics can be found in a recent review [4].

Generally speaking, one could distinguish two types of complex databases: first, those that include collections of many cases when a

Table 1. Selected Sources of Data for QSAR Model Building and Virtual Screening

Structures and bioactivities:
- PubChem: http://pubchem.ncbi.nlm.nih.gov/
- NCI: http://dtp.nci.nih.gov/docs/dtp_search.html
- WOMBAT: http://sunsetmolecular.com/
- BINDING DB: http://www.bindingdb.org/bind/index.jsp

Metabolites:
- http://www.hmdb.ca/

Drugs and clinical candidates:
- NLM's Dailymed: http://dailymed.nlm.nih.gov/
- DrugBank: http://drugbank.ca/
- FDA: http://www.accessdata.fda.gov/scripts/cder/drugsatfda/
- WHO Essential Drugs: http://www.who.int/medicines/publications/essentialmedicines/en/

Toxicology data:
- NIEHS: http://ntp.niehs.nih.gov/ntpweb/
- EPA ToxCast: http://www.epa.gov/ncct/toxcast/
- EPA DSS-Tox: http://www.epa.gov/ncct/dsstox/index.html

Chemical structures:
- PubChem: http://pubchem.ncbi.nlm.nih.gov/
- ChemSpider: http://www.chemspider.com/
- ZINC: http://zinc.docking.org/
- e-Molecules: http://www.emolecules.com/

large number of molecules were tested against a single target and, second, those that contain data on a series of compounds tested concurrently in multiple assays. The first type is typically represented by the activity or "property" data sets (e.g., pharmacokinetics data, or solubility, or toxicity) when the property is naturally measured across many molecules. Examples of data collections of the first type are provided by commercial databases WOMBAT and WOMBAT-PK [38]. For instance, version 2009.1 of WOMBAT contains 295,435 entries (242,485 unique SMILES), representing 1966 unique targets, captured from 14,367 papers published in medicinal chemistry journals between 1975 and 2008. Approximately 61% of these papers are from the ACS journal, *Journal of Medicinal Chemistry*; another 30.3% of the papers are from the Elsevier journal, *Bioorganic and Medicinal Chemistry Letters*. Each bioactive molecule has indexed target and bioassay protocol information, with links to the original publication as well as computed chemical descriptors. WOMBAT-PK 2009 contains 1230 entries (1230 unique SMILES), totaling over 13,000 clinical PK measurements. WOMBAT-PK 2009 drugs are indexed from multiple literature sources [39,40]; FDA Approved Drug Products [41]; peer-reviewed literature, etc.; 1085 drugs and 36 active metabolites have drug target annotations on 618 targets; an additional 231 drugs are annotated for antitargets [42]. Several physicochemical property measurements (e.g., water solubility at neutral pH, $\log d_{7.4}$; octanol–water distribution coefficient, $\log p$; $\mathrm{p}K_\mathrm{a}$; water solubility) are also included.

Arguably the largest single collection of toxicity data sets is *DSSTox* (http://www.epa.gov/nheerl/dsstox/About.html), which includes data such as tumor target site incidence and TD50 potencies for 1354 chemical substances tested in rats and mouse, 80 chemical substances tested in hamsters, 5 chemicals tested in dogs, and 27 chemical substances tested in nonhuman primates; data reviewed and compiled from literature and NTP studies. EPAFHM (EPA Fathead Minnow Aquatic Toxicity Database) includes acute toxicities of 617 chemicals tested in a common assay, with mode-of-action assessment and confirmatory measures. In addition, a large collection of single target property data sets is available from http://www.cheminformatics.org/datasets/. The EPA ToxCast™ project initiated in 2007 (http://www.epa.gov/ncct/toxcast/) promises to afford the largest collection of environmental chemicals (and potentially, some drug molecules as well) tested against a large panel of *in vitro* assays in an unprecedented effort to dissect the molecular mechanisms of toxicity and establish rapid and relatively inexpensive approaches for toxicity testing. The ToxCast database currently contains data for over 300 well-characterized chemicals (primarily pesticides) profiled in high-throughput screening (HTS) experiments against over 400 endpoints.

The databases of the second type are rapidly emerging. The NIH's Molecular Libraries Roadmap Initiative [43] laid out a strategy plan to house information on the biological activities of small molecules (in PubChem [9]) and transform them into chemical probes to perturb specific biological pathways. Currently, PubChem contains more than 25.5 million unique structures for the Compound database derived from over 60.7 million records in the PubChem Substance database, with links to bioassay description, literature, references, and assay data for each entry. BioAssay Database provides searchable descriptions of nearly 2000 bioassays, including descriptions of the conditions and readouts specific to a screening protocol. It integrated a vast array of resources including the 60 human tumor cell lines data from molecular targets databases of DTP/NCI and 1478 MLPCN (Molecular Libraries Probe Production Centers Network) related assays. It is especially useful when chemical information is needed for specific targets, cell lines or diseases.

It should be pointed out that the Substance database sourced data information from a multitude of major databases, for example, Binding Database, ChemBank, NCI/DTP, KEGG, SMID and ZINC. The Binding Database is a public database of measured binding affinities for biomolecules, containing experimental data of 21,143 binders to 244 biological targets [44]. ChemBank is a suite of informatics tools and databases created by the

Broad Institute, aimed at promoting the development of chemical genetics [45].

The Developmental Therapeutics Program (DTP) of the NCI has collected 127,000 compounds in both 2D and 3D formats that are freely available. They were generally screened for evidence of the ability to inhibit the growth of 60 human tumor cell lines over the past forty years. KEGG (Kyoto Encyclopedia of Genes and Genomes) is an informatics resource for biological systems [46]. It includes four constituent databases, categorized as building blocks in the genomic space (KEGG GENES, 1,720,795 genes), the chemical space (KEGG LIGAND, 14,238 compounds), wiring diagrams of interaction networks and reaction networks (KEGG PATHWAY, 42,314 pathways) and KEGG BRITE, 5642 hierarchical classifications. The Small Molecule Interaction Database (SMID) [47] is a database of protein domain–small molecule interactions by using structural data from the Protein Data Bank (PDB). SMID is essentially a "listing" of all small molecules (5117 records) that have been shown to bind to any given conserved protein domain (3508 records), including total of 274,917 interactions.

As part of the NIMH Psychoactive Drug Screening Program, PDSP K_i Database (http://pdsp.med.unc.edu/indexR.html) currently contains 47458 K_i values, embracing 749 types of receptors and 6935 test ligands. The majority of the receptors are GPCRs (549 types), along with various enzymes, ion channel and transporters, thus the largest database of its kind in the public domain. As the common observations in GPCR–ligand interactions, small molecules can bind to multiple set of GPCRs with high affinities creating interesting opportunities for developing QSAR models that could predict the receptor selectivity of a drug. Another GPCR–ligand database (GLIDA) is a unique database tailored for GPCR-related chemical genomic research [48]. Currently, 3738 entries of GPCRs are searchable together with 649 ligand entries and 1989 GPCR–ligand pair entries.

Finally, there are interesting examples of chemogenomics databases that capture the effects of chemicals on gene expression. CEBS Microarray Database, available from the National Center for Toxicogenomics at NIEHS (http://www.niehs.nih.gov/cebs-df/incebs.cfm), provides an integrated solution for searching, analyzing, and interpreting data from several microarray platforms. This is the largest publicly available collection of toxicogenomic data for diverse chemicals including data on toxicogenomic profiles for over 100 chemicals provided by Johnson & Johnson.

In most cases, the biological endpoint (e.g., any toxicity) is very complex with many possible underlying biological mechanisms. Naturally, the complex data sets call for the development of more sophisticated computational tools and corresponding models that place particular emphasis on statistical model validation and external predictive power rather than mechanistic interpretation. Some of these techniques are discussed below in Sections 3.2 and 3.3.

2.2. Data Quality and Data Curation: Impact on Model Quality

Molecular modelers and cheminformaticians typically analyze data generated by other scientists doing experimental research. Consequently, when it comes to the experimental data accuracy the modelers are always at the mercy of data providers who may, for different reasons provide (partially) erroneous data. Practically, any QSAR study entails calculation of chemical descriptors that are expected to accurately reflect intricate details of underlying chemical structures. Obviously, any error in the structure translates either into inability to calculate descriptors for erroneous chemical records or into erroneous descriptors making the subsequent data models valid for only a fraction of formally available data or, what is even worse, making models inaccurate. Since both the amount of data and the number of data models as well as the body of related QSAR publications (Fig. 1) continue to grow it becomes increasingly important to address the issue of data quality that inherently effectuates the quality of models.

Surprisingly, the investigations into how the primary data quality influences the quality of cheminformatics models have been almost absent in the published literature. It appears that cheminformaticians and molecular modelers tend to take published chemical and biological data at their face value and launch calculations without carefully examining the accuracy of the data records. Perhaps, it is indeed difficult to verify the results of biological assays although it is well known that numerical values of bioactivity for similar compounds measured in similar assays frequently disagree between different laboratories. However, there should be inherently much less disagreement concerning the exact representation of the chemical structure of compounds in the databases except for arguably difficult issues such as tautomers.

How significant is the problem of accurate structure representation as it concerns the exploratory cheminformatics research? As mentioned above, there appears to be almost no published studies on the subject. Southan et al. [49] mentioned briefly some procedures used to determine the number of unique chemical structure in a chemical database. Recent benchmarking investigations by a large group of collaborators from six laboratories [50] have clearly demonstrated that the type of chemical descriptors has much greater influence on the prediction performances of QSAR models than the nature of model optimization techniques. These findings suggest that having erroneous structures represented by erroneous descriptors should have a detrimental effect on the model performance. Indeed, in a recent seminal publication [37] the authors clearly pointed out the importance of chemical data curation in the context of QSAR modeling. They have discussed several illustrative examples of incorrect structures generated from either correct or incorrect SMILEs using commercial software. They also discussed the error rates in several known databases and evaluated the consequences of both random and systematic errors on the prediction performances of the derivative QSAR models. The main conclusions of their study were that small structural errors within a data set could lead to significant losses of predictive abilities of QSAR models. The authors further demonstrated that manual curation of structural data leads to substantial increase in the model predictivity. This conclusion becomes especially important in light of a study [51] showing that about 10% of data sets used in QSAR studies published in *Journal of Medicinal Chemistry* and in *QSAR and Combinatorial Science* contain errors with respect to their chemical structure and/ or biological activity.

Both the common sense and the recent investigations described above indicate that chemical record curation should be viewed as a separate and perhaps critical component of cheminformatics research. By comparison, the community of protein X-ray crystallographers has long recognized the importance of structural data curation; indeed, the PDB team includes a large group of curators whose major job is to process and validate primary data submitted to the PDB by crystallographers [52]. Furthermore, NIH recently awarded a significant Center grant to a group of scientists from the University of Michigan (http://csardock.org/) where one of the major tasks is to curate primary data on protein–ligand complexes deposited to the PDB. Conversely, to the best of our knowledge the largest publicly funded cheminformatics project, that is, PubChem, is considered as a data repository, that is, no special effort is dedicated to the curation of structural information deposited to PubChem by various contributors. Chemical data curation has been addressed whenever possible by the publicly available Chemspider project [53]; however, until now most effort has focused on data collection and database expansion. Thus, it is critical that scientists who build models using data derived from current databases or extracted from publications dedicate their own effort to the task of data curation.

Although there are obvious and compelling reasons to believe that chemical data curation should be given a lot of attention it is also obvious that for the most part the basic steps to curate a data set of compounds have been either considered trivial or ignored by the experts in the field. There is an apparent trend within the community of QSAR modelers to develop and follow the standardized guidelines for developing statistically robust and externally predictive QSAR models [54]. It appears timely to emphasize the importance of and develop best practices for the data preparation prior to initiating the modeling process because it is merely senseless to launch massive cheminformatics or molecular modeling investigations if the underlying chemical structures are not correct.

Our group is in the process of developing a standard curation protocol that should be helpful in the preprocessing of any chemical data set. We specifically focus on chemical structure curation procedures and do not cover highly relevant but special topic of name to structure conversion, which is often used to create some chemical databases; several publications have addressed this important topic [55–58]. Indeed, many data curation tasks can be accomplished automatically using existing software packages many of which are free for academic investigators. Figure 3 summarizes key steps of chemical structure curation including the removal of inorganics, counterions and mixtures; structure cleaning; ring aromatization; normalization of specific chemotypes; curation of tautomeric forms, and the deletion of duplicates. Table 2 provides references for the available software that helps speeding up the time-consuming, but essential, curation procedures.

3. MODEL DEVELOPMENT AND VALIDATION: INTEGRATED PREDICTIVE QSAR MODELING WORKFLOW

As mentioned in the Introduction, in the beginning of the QSAR modeling field both the data set and the models were relatively simple. The chief challenge was to find a way of fitting small number of data points (no more than two dozen) to a small number of physical chemical descriptors. Thus, the goal was to build models capable of explaining the observed trends in bioactivity using simple physical properties of compounds.

With the increased complexity of the data sets and substantial expansion in the number of descriptors used in advanced QSAR modeling techniques (especially, CoMFA [77]), we have begun to pay attention to the issue of model validation. Leave-one-out (LOO) cross-validation was introduced as part of CoMFA to ensure at least the internal statistical significance of the models developed for the training set. However, as we and others have demonstrated [5],

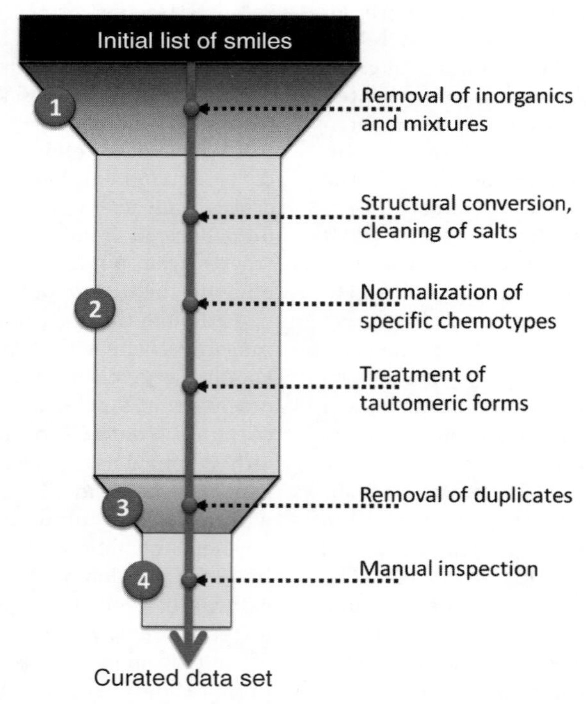

Figure 3. Major steps for chemical data curation. (This figure is available in full color at http://mrw.interscience.wiley.com/emrw/9780471266945/home.)

Table 2. Summary of Major Procedures and Corresponding Relevant Software for Every Step of the Chemical Data Curation Process

Procedures	Software	Availability
Inorganics removal	ChemAxon/Standardizer	Free for academia [59]
Structure normalization	ChemAxon/Standardizer	Free for academia [60]
(fragment removal,	OpenBabel	Free [61]
structural curation, salt	Molecular Networks/CHECK,TAUTOMER	Commercial [62]
neutralization)	ISIDA/Duplicates	Free for academia [63]
Duplicate removal	HitQSAR	Free for academia [64]
	CCG/MOE	Commercial [65]
SDF management/viewer	ISIDA/EdiSDF	Free [66]
File format converter	Hyleos/ChemFileBrowser	Free [67]
	OpenBabel	Free [68]
	ChemAxon/MarvinView	Free for Academia [69]
	CambridgeSoft/ChemOffice	Commercial [70]
	Schrödinger/Canvas	Commercial [71]
	ACD/ChemFolder	Commercial [72]
	Symix Cheminformatics	Commercial [73]
	CCG/MOE	Commercial [74]
	Accelrys/Accord	Commercial [75]
	Tripos/Benchware Pantheon	Commercial [76]

the LOO cross-validated R^2 (q^2) was established to be insufficient parameter to ensure the external predictive power of QSAR models. This realization led us to establish an integrated workflow for QSAR model development and validation (reviewed in Ref. [54]).

Our integrated strategy is outlined in Fig. 4. It describes the predictive QSAR modeling workflow focused on delivering validated models and ultimately, computational hits to be confirmed by the experimental validation. We start by randomly selecting a fraction of compounds (typically, 10–15%) as an external validation set. The remaining compounds are then divided rationally (using the Sphere Exclusion protocol implemented in our laboratory [78]) into multiple training and test sets that are used for model development and validation, respectively, using criteria discussed in more detail below. As described below, we employ multiple QSAR techniques based on the combinatorial exploration of all possible pairs of descriptor sets coupled with various statistical data mining techniques (combi-QSAR) and select models characterized by high accuracy in predicting both training and test sets data. An ensemble of validated models is used for consensus prediction of the external validation set. The critical step of the external validation is the use of applicability domains. If external validation demonstrates the significant predictive power of the models we use all such models for virtual screening of available chemical databases (e.g., ZINC [79]) to identify putative active compounds and work with collaborators who could validate such hits experimentally. The detailed description of this approach and examples of its application can be found in several recent papers and reviews (e.g., see Refs [12,32,80]).

We note that our approach shifts the emphasis on ensuring good (best) statistics for the model that fits known experimental data toward generating testable hypothesis about purported bioactive compounds. Thus, the output of the modeling has exactly the same format as the input, that is, chemical structures and (predicted) activities making model interpretation and utilization completely seamless for medicinal chemists.

In the following sections, we discuss the essential components of the workflow. We believe that the overall model development and validation schema (Fig. 4) should be generally followed by QSAR practitioners irrespective of specific QSAR modeling approaches employed in their studies.

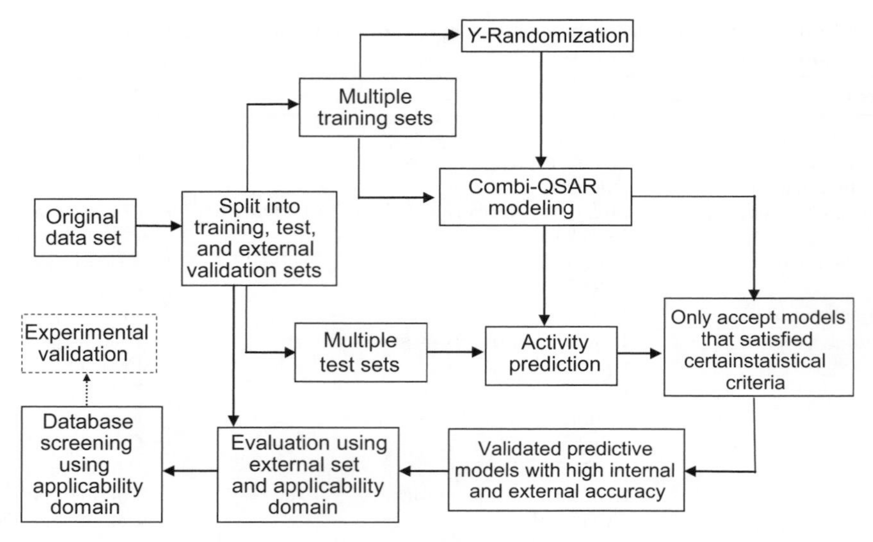

Figure 4. Integrated schema for predictive QSAR model development and validation.

3.1. Combinatorial QSAR Modeling

Our experience suggests that there is no universal QSAR method that is guaranteed to give the best results for any data set. QSAR is a highly experimental area of statistical data modeling where it is impossible to decide *a priori* as to which particular QSAR modeling method will prove most successful. To achieve QSAR models of the highest internal, and most importantly, external accuracy, the combi-QSAR approach (shown in Fig. 5) explores all possible binary combinations of various descriptor types and optimization methods along with external model validation. Each combination of descriptor sets and optimization techniques is likely to capture certain unique aspects of the structure–activity relationship. Since our ultimate goal is to use the resulting models as reliable activity (property)

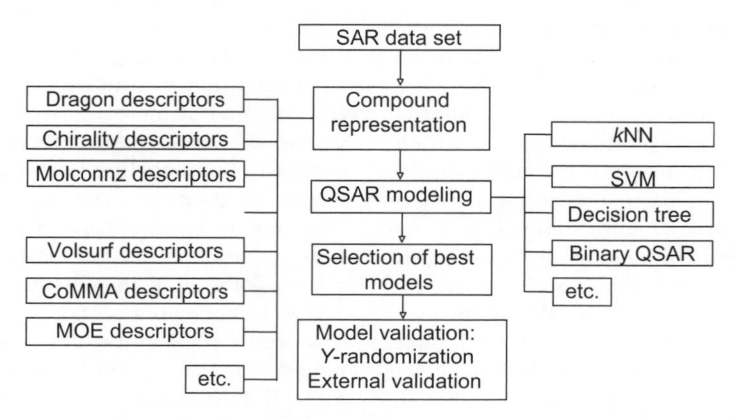

Figure 5. Flowchart of the combinatorial QSAR methodology. All descriptor sets and methods currently implemented in our laboratory are listed for illustration but practically any combination of descriptors and modeling techniques could be added.

predictors, application of different combinations of modeling techniques and descriptor sets will increase our chances for achieving successful models.

An important question that always concerns QSAR modelers is the somewhat awkwardly (see below) defined dimensionality of QSAR models. This issue came to light with the arrival of CoMFA [77] that was dubbed as 3D QSAR to reflect the fact that CoMFA descriptors were calculated from 3D structure of molecules taking into account their shape and surface charge distribution. Consequently, the original Hansch analysis that employed only constitutional molecular descriptors (such as molecular weight or log *p*) could be referred to as 1D QSAR whereas models employing chemical graph descriptors (such as molecular connectivity or E-state indices [81,82]) would be called 2D QSAR. This line of QSAR terminology has received an interesting extension when A. Hopfinger introduced the so-called 4D QSAR [83] that took into account multiple conformations of the ligands aligned in 3D (unlike CoMFA that employs a single "pharmacophoric" conformation for each ligand). Recently, J. Polanski [84] went as far as to propose a separate systematic classification of *m*-dimensional QSAR models where *m* refers to "the static ligand representation (3D), multiple ligand representation (4D), ligand-based virtual or pseudo-receptor models (5D), multiple solvation scenarios (6D), and real receptor or target-based receptor model data (7D)."

Although there have been indeed interesting QSAR studies that gave rise to specific examples of *m*D QSAR reviewed in Ref. [84] we still find this terminology awkward. In our view, practically any QSAR model (with possible exception of simple linear models) employs multiple descriptors of underlying molecules and hence any models are *m*-dimensional by default. The differences between models should be described due to the type and sources of chemical descriptors. We would agree that the original classification of QSAR models based on the dimensionality of physical space in which compounds are considered (i.e., 1D for constitutional descriptors, 2D for chemical graphs, or 2D connectivity based descriptors, and 3D for compounds described by their coordinated in 3D space), has indeed some basis in physical reality. In this regard, a time-dependent QSAR model (if it could possibly exist) could be in principle called 4D QSAR. However, all other examples of *m*D QSAR models discussed in Ref. [84] should really belong to the class of 3D QSAR augmented by additional special types of chemical descriptors.

A more important question as to what type of descriptors should be used to achieve QSAR models of the highest statistical significance and external predictive power may be paraphrased from a famous retort of Prince Hamlet as "2D, or not 2D?" Obviously, a major benefit of 2D compared to 3D chemometric methods is that the former neither require a conformational search nor structural alignment of molecules to derive models. Accordingly, 2D methods are more easily automated and adapted to the task of database searching, or virtual screening [85]. Conversely, 3D descriptors should more accurately describe chemical structures that are inherently three dimensional. Perhaps, because of the complexity of QSAR modeling and a large diversity of available optimization approaches and descriptor sets, the issue of 2D versus 3D QSAR comparison has not been addressed systematically using large collections of both methods and compound sets. However, there have been individual reports addressing this issue for individual data sets. For instance, Brown and Martin [86] concluded that 2D descriptors have been shown to be superior to 3D descriptors in database mining. On the contrary, Hillebrecht and Klebe came to the conclusion [87] that for their data set the 3D QSAR CoMSIA models outperformed 2D QSAR models in virtual screening. Perhaps the most extensive comparative QSAR modeling studies to date was conducted by Gedeck et al. [88] where as many as almost 17,000 QSAR models were generated for nearly 1000 industrial data sets comprising approximately 143,000 compounds. Because of the scale of the study, only one of QSAR method used was 3D, and it was not found to be the best. However, the overall conclusion reached by the authors was that "none of the descriptors is best for all data sets; it is therefore necessary to test in each individual case, which descriptor produces the best model". We completely agree with this conclusion and believe that the statement above reflects most accurately the *status quo* of QSAR modeling and provides support for practicing the combi-QSAR modeling approach.

We believe that multiple alternative QSAR models should be developed (as opposed to a single model using some favorite QSAR method) for each data set to identify all successful approaches (i.e., those that achieve certain level of internal and external accuracy) in the context of the given data set. Since QSAR modeling is relatively fast, these alternative models could be explored simultaneously when making predictions for external data sets. The consensus predictions of biological activity for novel test set compounds on the basis of several QSAR models, especially when they converge, are more reliable and provide better justification for the experimental exploration of hits as we and others have demonstrated in several recent publications [50,88–91].

3.2. Critical Importance of Model Validation

It should sound almost axiomatic that validation should be a natural part of any QSAR modeling study. Indeed, what is the (ultimate) purpose of any modeling approach such as QSAR, if not developing models with a significant external predictive power? Unfortunately, as we and others have indicated in many publications (e.g., see Refs [5,80,92]), the field of QSAR modeling has been plagued with insufficient attention paid to the issue of external validation. Most practitioners have merely presumed that internally cross-validated models built from available training set data should be externally predictive. As mentioned in the Introduction, the large number of QSAR publications exploring small to medium size data sets to produce models with little statistical significance led to the editorial published by *Journal of Chemical Information and Modeling* 2 years ago [7] that explicitly discouraged researchers from submitting the "introspective" QSAR/QSPR publications and requested that "evidence that any reported QSAR/QSPR model has been properly validated using data not in the training set must be provided." We and others have demonstrated that the training set statistics using most common internal validation techniques such as leave-one-out or even leave-many-out cross-validation approaches is insufficient and the statistical figures of merit of such models serve as misleading indicators of the external predictive power of QSAR models [5,80].

In our highly cited publication, "Beware of q^2!" [5], we have demonstrated the insufficiency of the training set statistics for developing externally predictive QSAR models and formulated the main principles of model validation. At the time of that publication (in 2002), the majority of papers on QSAR analysis ignored any model validation except for cross-validation performed during model development. Despite earlier observations of several authors [93–95] warning that high cross-validated correlation coefficient R^2 (q^2) is the necessary, but not the sufficient condition for the model to have high predictive power, many authors continued to consider q^2 as the only parameter characterizing the predictive power of QSAR models. In Ref. [5], we have shown that the predictive power of QSAR models can be claimed only if the model was successfully applied for prediction of the external test set compounds, which were not used in model development. We have demonstrated that the majority of the models with high q^2 values have poor predictive power when applied for prediction of compounds in the external test set. In another publication [80], the importance of rigorous validation was again emphasized as a crucial, integral component of model development. Several examples of published QSPR models with high-fitted accuracy for the training sets, which failed rigorous validation tests, have been considered. We presented a set of simple guidelines for developing validated and predictive QSPR models and discussed several validation strategies such as the randomization of the response variable (Y-randomization) external validation using rational division of a data set into training and test sets. We highlighted the need to establish the domain of model applicability in the chemical space to flag molecules for which predictions may be unreliable, and discussed some algorithms that can be used for this purpose. We advocated the broad use of these guidelines in the development of predictive QSPR models [78,80,96].

To be more specific, QSAR modelers must always distinguish between model fitness to the training set data (that can be characterized by q^2) versus its external predictivity (which can be characterized by the external q^2 or external R^2 generated for the data that was never used in either model development or model selection). In our critical publications [5,80], we have recommended a set of statistical criteria that must be satisfied by an externally predictive model. For continuous QSAR, we employ q^2 for the training set and the following parameters to characterize the model accuracy for the test set: (i) correlation coefficient R between the predicted and the observed activities; (ii) coefficients of determination [97] (predicted versus observed activities R_0^2, and observed versus predicted activities $R_0'^2$ for regressions through the origin); (iii) slopes k and k' of regression lines through the origin of predicted versus observed activities and observed vs. predicted activities, respectively. We consider a QSAR model predictive if the following conditions are satisfied:

$$q^2 > 0.5 \quad (1)$$

$$R^2 > 0.6 \quad (2)$$

$$\frac{(R^2 - R_0^2)}{R^2} < 0.1 \quad (3)$$

$$0.85 \le k \le 1.15 \quad \text{or} \quad \frac{(R^2 - R_0'^2)}{R^2} < 0.1 \quad (4)$$

and

$$0.85 \le k' \le 1.15 \quad \text{or} \quad |R_0^2 - R_0'^2| < 0.3 \quad (5)$$

where q^2 is the cross-validated correlation coefficient calculated for the training set, but all other criteria are calculated for the test set.

Validation of QSAR models remains one of the most critical problems of QSAR. Recently, we have extended our requirements for the validation of multiple QSAR models selected by acceptable statistics criteria of prediction of the test set [98]. Additional studies in this critical component of QSAR modeling should establish reliable and commonly accepted "good practices" for model development.

3.3. Applicability Domains of QSAR Models

One of the most important problems in QSAR analysis is establishing the models' domain of applicability in the chemistry space. In the absence of the applicability domain, each model can formally predict the activity of any compound, even when its chemical structure is completely different from those included in the training set. Thus, the absence of the model applicability domain as a mandatory component of any QSAR model would lead to the unjustified extrapolation of the model in the chemistry space and, as a result, a high likelihood of inaccurate predictions. In our research, we have always paid particular attention to this issue [80,99–105]. Thus, the need for establishing the applicability domain for every model adds another critical degree of complexity to the model building process. It should be noted that when variable selection QSAR approaches are used, the applicability domains should be defined both in terms of global similarity between the training set and external compounds and in the context of each individual model.

The applicability domain problem has been addressed by many researchers. Mandel [106] introduced the so-called Effective Prediction Domain that was based on the ranges of descriptors included in the regression equation. Afantitis et al. [107] built a multiple linear regression model for a data set of apoptotic agents. They defined the applicability domain for each compound as a leverage defined as a corresponding diagonal element of the hat matrix. In fact, it is a method for detecting possible leverage outliers. If, for some compound, leverage is higher than $3K/N$, where K is the number of descriptors and N is the number of compounds, the compound is an outlier. To use this approach, for each external compound it would be necessary to recalculate the leverage. Netzeva et al. [108] and Saliner et al. [109] defined the applicability domain by ranges of descriptors. This definition of the applicability domain has a significant drawback, because domain has a shape of hyper- parallelepiped in the high-dimensional descriptor space encapsulating nonuniformly distributed points representing compounds in a data set. For this reason, the so defined applicability domain is likely to be too large and not reflective of the specific data set A similar definition of the applicability domain was proposed by Tong et al. [110]. The authors built QSAR models for two data sets of estrogen receptor ligands using the Decision Forest method and studied the dependence of the model predictive power versus the applicability domain threshold. The prediction accuracy within the domain is defined as a ratio of the number of correct predictions to the total number of compounds in the domain. The accuracy was changing from about 90% for the initial applicability domain to about 50% when the applicability domain increased by 30%. Interestingly, for one of the data sets the prediction accuracy was increasing until the domain was extended by about 20%. Another important aspect of this study was that the authors defined the confidence level of prediction. The probability that a compound belongs to a certain class was defined as the percentage of active compounds in the leaf node that the compound belongs to. The authors found (as expected), that the confidence level correlated with the prediction accuracy.

In Ref. [111], a lazy learning *k*NN-like method was applied for the prediction of rodent carcinogenicity and Salmonella mutagenicity. The applicability domain was defined by a so-called confidence index. A compound was assigned to one of the two classes by a weighted majority vote of its nearest neighbors. The confidence index is the weighted majority quote divided by the number of nearest neighbors. If the absolute value of the confidence index is low (<0.05) a compound is said to be out of the applicability domain. This definition of the applicability domain captures the areas in the descriptor space where compounds of both classes are close to each other, and possibly mixed. In this area, the precise and accurate prediction of a compound's class is impossible. A Tanimoto-like coefficient is used as a similarity measure. Nearest neighbors are defined by the value of this coefficient higher than 0.3, which limits the possibility of overextrapolation.

In most of our QSAR studies, we have defined the applicability domain as the distance cutoff value $D_{\mathrm{cutoff}} = \langle D \rangle + Zs$, where Z is a similarity threshold parameter defined by a user, and $\langle D \rangle$ and s are the average and standard deviation of all Euclidian distances in the multidimensional descriptor space between each compound and its nearest neighbors for all compounds in the training set (e.g., see Ref. [78]). This definition of the applicability domain has several major drawbacks that we continue to address in our ongoing studies: (i) Currently, applicability domain is direction-independent in the descriptor space. We shall consider the directions in the descriptor space in which the distribution of representative points has smaller spread as less important than those that have higher spread. Thus, the applicability domain will be represented as a multidimensional ellipsoid in the principal component space. (ii) Too strict definition of the applicability domain: if a compound is outside of the model applicability domain, we currently do not predict its activity. Naturally, we shall establish the lower and upper bounds for the applicability domain. (iii) Finally, it seems reasonable to introduce a confidence level of prediction, which will depend on the distance of the compound under prediction from its nearest neighbor of the training set. These considerations provide just a few examples that illustrate the importance of ongoing research in this area of QSAR modeling. Not surprisingly, the model applicability domain was the subject of a special symposium organized at a recent 235th meeting of the American Chemical Society in New Orleans, LO.

3.4. Ensembles of QSAR Models and Consensus Prediction

The application of combi-QSAR strategy and model acceptability thresholds (cf. Fig. 4 and Eqs) inherently leads to a library of models that satisfy both internal (training set) and external (test set) accuracy criteria. It is always appealing (and in fact, unfortunately done by many authors) to attempt to identify the "best" model, for example, the one that has the highest external predictive power. It should be noted, however, that such model would not be characterized, in general, by the highest internal accuracy (i.e., q^2) as was convincingly shown in several earlier publications [5,78]. Note that the "best" model in this case is chosen based on the knowledge of the test set data, that is, the entire modeling set (including both training and test sets) is used to make the call. By the logical extension of the "beware of q^2" principle [5], one should not expect that this "best" model will actually perform best on the completely independent external set; in fact, an example based on our recent publication [50] that will be discussed below demonstrates explicitly that this is not the case.

Our experience with combi-QSAR modeling suggests that making the judgment call on any basis to select the "best" model is in fact unnecessary if not entirely wrong. Instead, we recommend using the entire ensemble of acceptable models to make consensus prediction for the external data set. The following example illustrates that this approach ensures the highest accuracy of the external prediction; it also illustrates the complexity and power of modern QSAR modeling approaches and highlights the importance of collaborative and consensual model development.

In this recent study [50], the combi-QSAR approach was applied to a diverse series of organic compounds tested for aquatic toxicity in *Tetrahymena pyriformis*. The unique aspect of this research was that it was conducted in collaboration between six academic groups specializing in cheminformatics and computational toxicology. The common goals for our virtual collaboratory were to explore the relative strengths of various QSAR approaches in their ability to develop robust and externally predictive models of this particular toxicity endpoint. We have endeavored to develop the most statistically robust, validated, and externally predictive QSAR models of aquatic toxicity. Each group relied on its own QSAR modeling approaches to develop toxicity models using the same modeling set, and we agreed to evaluate the realistic model performance using the same external validation set that was not available to any of the groups during the study.

Different groups have employed different techniques and (sometimes) different statistical parameters to evaluate the performance of models developed independently for the modeling set (described below). To harmonize the results of this study the same standard parameters were chosen to describe each model's performance as applied to the modeling and external test set predictions. Thus, we have employed Q^2_{abs} (squared leave-one-out cross-validation correlation coefficient) for the modeling set, R^2_{abs} (frequently described as coefficient of determination) for the external validations sets, and MAE (mean absolute error) for the linear correlation between predicted (Y_{pred}) and experimental (Y_{exp}) data (here, $Y = \mathrm{pIGC}_{50}$); these parameters were defined as follows:

$$Q^2_{abs} = 1 - \frac{\sum_Y (Y_{exp} - Y_{LOO})^2}{\sum_Y (Y_{exp} - <Y>_{exp})^2} \quad (6)$$

$$R^2_{abs} = 1 - \frac{\sum_Y (Y_{exp} - Y_{pred})^2}{\sum_Y (Y_{exp} - <Y>_{pred})^2} \quad (7)$$

$$\mathrm{MAE} = \frac{\sum_Y Y - Y_{pred}}{n} \quad (8)$$

Many other statistical characteristics can be used to evaluate model performance; however, we restricted ourselves to these three parameters that provide minimal but sufficient information concerning any model's ability to reproduce the trends both in experimental data for the test sets and in the mean accuracy of predicting all experimental values. The models were considered acceptable if R^2_{abs} exceeded 0.5.

The objective of this study from methodological prospective was to explore the suitability of different QSAR modeling tools for the analysis of a data set with an important toxicological endpoint. Typically, such data sets are analyzed with one (or several) modeling technique(s), with a great emphasis on the (high value of) statistical parameters of the training set models. In this study, we went well beyond the modeling studies reported in the original publications in several respects. First, we have compiled all reported data on chemical toxicity against

T. pyriformis in a single large data set and attempted to develop global QSAR models for the entire set. Second, we have employed multiple QSAR modeling techniques thanks to the engagement of six collaborating groups. Third, we have focused on defining model performance criteria not only using training set data but also, most importantly, using external validation sets that were not used in model development in any way (unlike any common cross-validation procedure) [112]. This focus afforded us the opportunity to evaluate and compare all models using simple and objective universal criteria of external predictive accuracy, which in our opinion is the most important single figure of merit for a QSAR model that is of practical significance for experimental toxicologists. Fourth, we have explored the significance of applicability domains and the power of consensus modeling in maximizing the accuracy of external predictivity of our models.

Of the 15 QSAR approaches used in this study, 9 implemented method-specific applicability domains. For the most part, all models succeeded in achieving reasonable accuracy of external prediction especially when using the AD. It then appeared natural to bring all models together to explore the power of consensus prediction. Thus, the consensus model was constructed by averaging all available predicted values taking into account the applicability domain of each individual model. In this case, we could use only 9 of 15 models that had the AD defined. Since each model had its unique way of defining the AD, each external compound could be found within the AD of anywhere between one and nine models so for averaging we only used models covering the compound. The advantage of this data treatment was that the overall coverage of the prediction (i.e., the fraction of compounds in the external set that is within the AD) was still high because it was rare to have an external compound outside of the ADs of all available models. The results shown for the external set (Fig. 6) clearly indicate that the consensus model had absolutely best performance compared to any individual model.

Indeed, the consensus model achieved the highest external prediction accuracy characterized by $R^2_{abs} = 0.7$ with MAE of 0.32 at 100% coverage. Several individual models could be identified with higher R^2_{abs} and lower MAE (e.g., Models 3 or 11) but in this case the coverage is really low, only about 70%. Conversely, models that did not use the AD criteria (i.e., 6, 8, and 12–15) have achieved 100% coverage but their accuracy parameters were found to be low, with R^2_{abs} never exceeding 0.65 with MAE of 0.4. This observation suggests

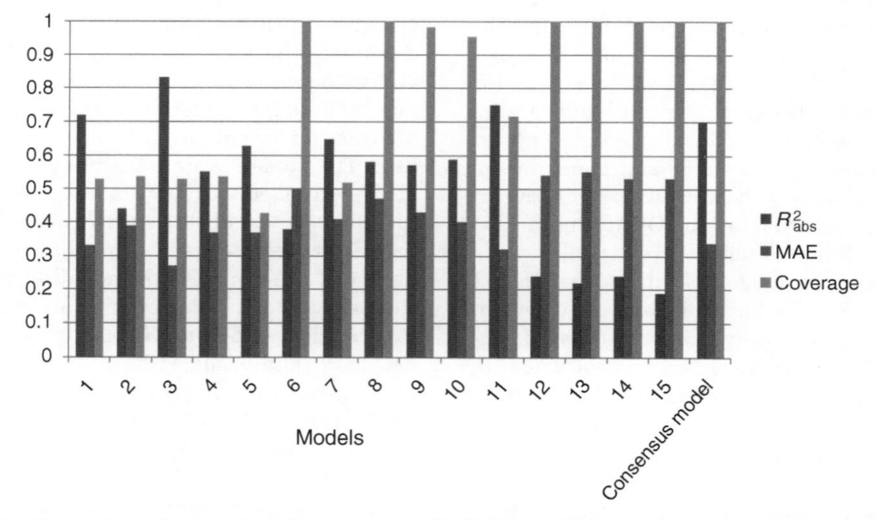

Figure 6. Comparison of external predictive power of individual QSAR models versus consensus models. See text for the definition of R^2_{abs}, MAE, and coverage. (This figure is available in full color at http://mrw.interscience.wiley.com/emrw/9780471266945/home.)

that consensus models afford both high space coverage and high accuracy of prediction unlike any of the individual models.

In summary, this study presents an example of fruitful international collaboration between researchers that use different techniques and approaches but share general principles of QSAR model development and validation. However, the most significant single result of our studies is the demonstrated best performance of the consensus modeling approach when all models are used concurrently and predictions from individual models are averaged. We have shown that both the predictive accuracy and the coverage of the final consensus QSAR models were better as compared to these parameters for individual models. The consensus models appeared robust as it was found to be insensitive to both incorporating individual models with low prediction accuracy and the inclusion or exclusion of the AD. Another important result of this study is the power of addressing complex problems in QSAR modeling by forming a virtual collaboratory of independent research groups leading to the formulation and empirical testing of best modeling practices. This latter endeavor is especially critical in light of the growing interest of regulatory agencies to developing most reliable and predictive models for environmental risk assessment [113] and placing such models in public domain.

3.5. QSAR Modeling Using Hybrid Chemical Biological Descriptors

As discussed above, traditional QSAR modeling aims to establish rigorous correlations between calculated chemical descriptors for a set of compounds and their experimentally studied biological activities. However, in some cases, additional physicochemical properties, such as water partition coefficient ($\log p$) [114], water solubility [115], and melting point [116] were used successfully to augment computed chemical descriptors and improve the predictive power of QSAR models. These studies suggest that using hybrid descriptor sets in QSAR modeling could prove beneficial.

The recent public availability of HTS data for large sets of chemical agents (due to projects such as PubChem [9] as discussed in Section 2.1) offers an attractive avenue for exploring its utility in hybrid descriptor-based QSAR modeling. In a recent study [117], we tested this hypothesis for a data set of chemical carcinogens tested in six cell viability assays [118]. We felt that the HTS data represent attractive and potentially mechanistically relevant *in vitro* "biological" descriptors for modeling the adverse health effects *in vivo*. Our study tested a hypothesis that improved QSAR predictions can be developed using a combination of chemical and biological descriptors (i.e., hybrid descriptors) of environmental chemicals.

To this end, we have developed k-nearest neighbor (kNN) QSAR [10] models of carcinogenicity using HTS data. Initially, we evaluated if a correlation exists between the HTS assay results and the *in vivo* rodent carcinogenic potency of chemicals. We found that the compounds classified by HTS as "actives" in at least one cell line were likely to be rodent carcinogens (sensitivity 77%); however, HTS "inactives" were far less informative (specificity 46%). Using chemical descriptors only, kNN QSAR modeling resulted in 62.3% prediction accuracy for rodent carcinogenicity. Importantly, the prediction accuracy of the model was significantly improved (to 72.7%) when chemical descriptors were augmented by HTS data, which were regarded as biological descriptors.

Thus, our studies suggested that combining HTS profiles (i.e., biological descriptors) with conventional chemical descriptors could considerably improve the predictive power of computational approaches in toxicology. Nowadays, the HTS studies of both drug candidates and environmental chemicals are highly popular. Therefore, we believe that the modeling approach discussed in this section that employs both chemical and biological descriptors has a promise of delivering models with increased statistical significance and biological relevance. With sufficient improvements in resulting model predictive performance, *in vitro* HTS bioassays, coupled with traditional chemical structure-based descriptors, may be ultimately helpful in prioritizing or even partially replacing *in vivo* toxicity testing.

4. MODEL EXPLOITATION

Arguably, the most important component of QSAR modeling workflow (Fig. 4) is the prospective use of models as opposed to a frequent belief that first and foremost the QSAR models should be "mechanistically" interpretable. Notice that mechanistic interpretation (if possible (!)) is listed as the last, fifth principle of the OECD guidelines cited in the beginning of this chapter, following other guidelines that place strong emphasis on model external validation. Some authors (e.g., see Ref. [119]) indeed prefer descriptors which are mechanistically interpretable. On the other hand, Estrada and Patlewicz [120] argued that in many cases a biological response is a result of a multitude of different processes, some of which can be even not known, and it's *a posteriori* mechanistic interpretation is difficult if not impossible. The authors suggested an alternative approach where a biological system is considered as a black box, when considering several possible mechanisms would be more productive. At the same time, some variables included in the model can describe several different mechanisms simultaneously, for example, $\log p$, so in many cases it makes no sense to suggest that the use of this descriptor in QSAR models affords any mechanistic interpretation (see also Ref. [121]). We would add those descriptors that give better models in terms of their predictive power are actually preferable. As we mentioned above, we consider building predictive models as the main goal of QSAR analysis. Of course, the interpretation of the model is also important, and if possible, it should be done. However, in many cases it is impossible, even when models with high predictive power have been obtained (for instance, in one study the best models were found to be those built with molecular connectivity indices but these models were disregarded by the authors for the lack of mechanistic interpretability [119]).

We believe that mechanistic interpretation of the externally validated QSAR model is an important *a posteriori* exercise that should be done after the model has been internally and externally validated, and descriptors that afford models with the highest predictive power should be always used preferentially. In this section, we consider cases when validated QSAR models can be exploited as virtual screening tools to prioritize compounds for experimental investigations as well as comment on perhaps yet uncommon but exciting use of models to curate experimental (but possibly erroneous) biological data.

4.1. Validated and Predictive QSAR Models as Virtual Screening Tools

The development of truly validated and predictive QSAR models affords their growing application in chemical data mining and combinatorial library design [122,123]. In one of the earliest studies, three-dimensional (3D) stereoelectronic pharmacophore based on QSAR modeling was used to search the National Cancer Institute Repository of Small Molecules [124] to find new leads for inhibiting HIV type 1 reverse transcriptase at the nonnucleoside binding site [125]. In another early investigation, QSAR models were used for virtual screening to discover novel dopamine transported inhibitors [126].

Recent studies have shown that QSAR models could be used successfully as virtual screening tools [127] to discover compounds with the desired biological activity in chemical databases or virtual libraries [30,32,85,103,128,129]. We shall present one example to illustrate the use of QSAR models as virtual screening tools for lead identification when novel scaffolds were actually discovered.

In this recent study [130], we employed the QSAR modeling workflow (Fig. 4) to discover novel Geranylgeranyltransferase type I (GGTase-I) inhibitors. Geranylgeranylation is critical to the function of several proteins including Rho, Rap1, Rac, Cdc42, and G-protein gamma subunits. GGTase-I inhibitors (GGTIs) have therapeutic potential to treat inflammation, multiple sclerosis, atherosclerosis, and many other diseases. Following our standard QSAR modeling workflow, we have developed and rigorously validated models for 48 GGTIs using variable selection k-nearest neighbor [10] and automated lazy learning [104], and genetic algorithm-partial least square [123] QSAR methods. The QSAR models were employed for virtual screening of 9.5 million commercially

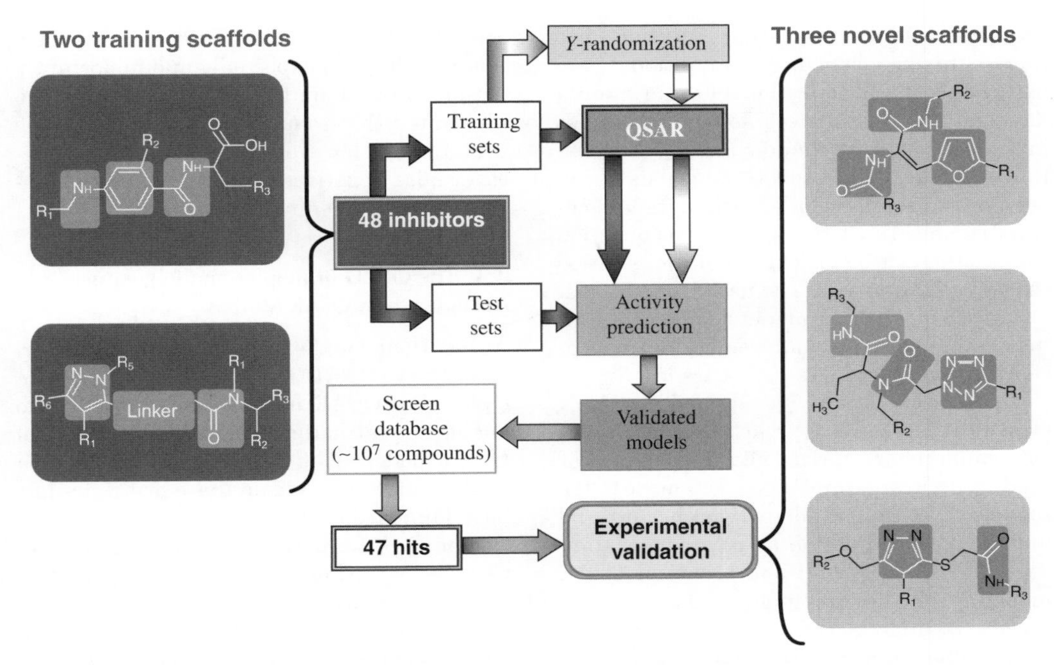

Figure 7. The workflow for the discovery of geranylgeranyltransferase-I inhibitors with novel scaffolds by the means of quantitative structure–activity relationship modeling, virtual screening, and experimental validation. (See color insert.)

available chemicals yielding 47 diverse computational hits. Seven of these compounds with novel scaffolds and high predicted GGTase-I inhibitory activities were tested *in vitro,* and all were found to be *bona fide* and selective micromolar inhibitors.

Figure 7 shows the simplified model development and application workflow. We should emphasize that QSAR models have been traditionally viewed as lead optimization tools capable of predicting compounds with chemical structure similar to the structure of molecules used for the training set. However, this study clearly indicates (Fig. 7) that with enough attention given to the model development process and using chemical descriptors characterizing whole molecules (as opposed to, e.g., chemical fragments) it is indeed possible to discover compounds with novel chemical scaffolds. Furthermore, in our study we have additionally demonstrated that these novel hits could not be identified using traditional chemical similarity search [130].

In the aforementioned publications as well as in several other studies, we have relied on the predictive QSAR modeling workflow (Fig. 4) with the emphasis on model validation and collaborated with colleagues who tested our predictions experimentally. The latter, experimental validation is indeed the most critical test of actual utility of QSAR models. The discovery of novel bioactive chemical entities is the primary goal of computational drug discovery, and the development of validated and predictive QSAR models is critical to achieve this goal.

4.2. Application of QSAR Models for Experimental Data Curation

In section 2.2 we have addressed an important issue of experimental data quality and its influence on the quality of QSAR models. Obviously, the model quality depends on the accuracy of both chemical and biological data. We have discussed the need and technical opportunities for chemical data curation indicating that several protocols and software do exist to this effect (Fig 3 and Table 2). However, it appears even more challenging to address the issue of biological data curation since there are no obvious computational approaches that could question the experimental data quality. Nevertheless, as we shall discuss in this section of the chapter, one may suggest that rigorous, validated QSAR models developed on more reliable biological data and curated chemical structures may in fact question the accuracy of biological annotations in external validation sets. The use of QSAR models to curate biological data represents an interesting and potentially, emerging application that we would like to discuss briefly.

In a recent study, we applied the *k*NN classification QSAR approach to a data set of compounds characterized either as binders or binding decoys of AmpC beta-lactamase [131]. Models were subjected to rigorous internal and external validation as part of our standard workflow (Fig. 4) and a special QSAR modeling scheme was employed that took into account the imbalanced ratio of true inhibitors to nonbinders (1:4) in this data set. Several hundred predictive models were obtained with correct classification rate (CCR) for both training and test sets as high as 0.90 or higher. The prediction accuracy was as high as 100% (CCR = 1.00) for the external validation set composed of 10 compounds (5 true binders and 5 decoys) selected randomly from the original data set. For an additional external set of 50 known nonbinders, we have achieved the CCR of 0.87 using very conservative model applicability domain threshold.

The validated binary *k*NN QSAR models were further employed for mining the AmpC screening data set (69,653 compounds) available from PubChem. Surprisingly, the consensus prediction of 64 compounds identified as screening hits in the AmpC PubChem assay disagreed with their annotation in PubChem but was in agreement with the results of secondary assays obtained independently [132]. At the same time, 15 compounds were identified as potential binders contrary to their annotation in PubChem. Five of them were tested experimentally and showed inhibitory activities in millimolar range with the highest binding constant K_i of 135 μM. Thus, our studies suggested that validated QSAR models could indeed identify promising hits by virtual screening of molecular libraries. Furthermore, perhaps unexpectedly and importantly in the context of this discussion, this study also illustrated that robust QSAR models cold be used to question true positives as well as recover false negatives resulting from the high-throughput screening campaigns.

4.3. The OECD Principles and Regulatory Acceptance of QSAR Models

As mentioned in the beginning of this chapter, in 2004, the OECD accepted five guiding principles that QSAR modelers should follow to enable the ultimate regulatory acceptance of QSAR models. The development of the open QSAR software based on these principles has been launched to "facilitate practical application of (Q)SAR approaches in regulatory contexts by governments and industry and to improve their regulatory acceptance" [8].

Since then, most authors publishing in environmental QSAR field tend to include a statement that their models fully comply with the OECD principles (e.g., see Refs [109,133–135]). For instance, two aspects of QSAR modeling outlined in the OECD principles were considered by Estrada and Patlewicz [120]. The first aspect concerns the theoretical approaches used in chemistry in general, and in QSAR in particular, that is, which method should be selected for theoretical studies: more sophisticated and complex or more simple. The authors criticized the common belief that applying more sophisticated methods should always lead to significantly better results. They considered an example of polycyclic aromatic hydrocarbons (PAHs) the toxicity of which is believed to depend on the energy gap between HOMO and LUMO values. The authors showed that a simple Hückel molecular orbital theory gives practically the same values of HOMO and LUMO as the sophisticated *ab initio* methods yet the calculations are 10^4–10^7 times faster. They reached the conclusion that if a more simple method is capable of giving results better or similar to those of more sophisticated method, one should naturally use a more simple method!

Unfortunately, even today in spite of many published recommendations on proper model validation, highly visible *Journal of Complementary and Integrated Medicine* editorial [7], and growing proliferation and popularity of OECD principles (e.g., see Ref. [136]) one can still see publications even in reputable professional journals that fail to present meaningful QSAR models. An example is provided by a QSAR study of a data set of 197 nitroarenes tested in the Ames genotoxicity assay, which was published as recently as in 2008 [137]. On the surface, the paper appeared to follow major guidelines for developing externally validated models and even referenced several important publications on the validation methodologies but in practice its quality and study design was surprisingly low. The authors attempted to develop comparative QSAR models of Ames genotoxicity using CoMFA, CoMSIA, HQSAR, and GFA (genetic function approximation) approaches. Multiple aspects of the study are highly questionable. Thus, the data in the form of the logarithm of the number of revertants ($\log R$) in the Ames test using *Salmonella typhimurium* strain TA98 were collected from four various sources and in cases when several results were available for the same compounds they were simply averaged. For instance, $\log R$ values for 9-nitrophenanthrene were -0.3 and 2.25, indicating a dramatic difference in reported data between different publications.

Further, the authors formally followed the OECD guidelines for model validation and removed ~30% of compounds to be used as external validation set; however, the results of this validation were not discussed and only predicted values were reported. A quick analysis of the reported data showed that the models were not predictive at all: R^2_{ext} obtained from the reported predictions for the external test set compounds was between 0 and 0.28; Q^2_{ext} ranged between -0.70 and 0.2 with the standard error of prediction (SEP) of 1.5–2. These statistical parameters of the reported "model" made the eight pages long discussion of the model if not the entire study quite unnecessary. This recent example serves as a perfect illustration that the wrong usage of low-quality data will most certainly result in worthless models. Thus, attention should be given not only to the knowledge of OECD principles for model development and validation but also, more importantly, to deep understanding of the nature of each principle as they relate to the rigor with which every QSAR model should be developed.

The regulatory acceptance of QSAR models is probably one of the highest goals that the modelers should establish for themselves. Undoubtedly, the use of QSAR models as part of the regulatory process by federal agencies such as FDA or EPA would significantly reduce the cost, animal use, and time needed for allowing pharmaceutical or industrial products reach the marketplace. To achieve this level of trust, models must be developed with the highest level of rigor paying special attention to external validation. We sincerely expect that as QSAR methodologies and software tools continue to evolve we will indeed begin to see the proliferation of more reliable model building protocols and reliable specific endpoint models within the regulatory agencies.

5. CONCLUSIONS

Following more than 45 years of methodological developments and applications, the field of QSAR modeling remains at the forefront of computational chemistry research. In fact, the need for novel methods and innovative applications is even greater today than in previous decades due to the exponential growth of bioactivity and chemical databases in the public domain. Federally funded projects such as the NIH Molecular Libraries Roadmap Initiative [43] contributing the results of screening assays in PubChem [9] or DSSTox [138] with its collection of toxicity testing results as well as many other experimental data generating projects have made available many publicly available databases of biologically tested compounds as discussed in Section 2.2.

This growth creates new challenges for QSAR modeling such as development of novel tools and approaches for the analysis of complex biomolecular data sets including the emerging problems of chemical and biological data curation. The diversity of relevant databases including data not only on traditional biological assays of chemical compounds but also on gene or protein expression or metabolite profiles modulated by chemicals may require new data analytical approaches integrating cheminformatics and bioinformatics methods. The novel approaches should address new challenges in drug discovery such as designing compounds with polypharmacological mechanisms of action.

Methodological developments and application studies discussed in this chapter have established that QSAR models could be used successfully as virtual screening tools to discover compounds with the desired biological activity in chemical databases or virtual libraries. The discovery of novel bioactive chemical entities is the primary goal of computational drug discovery, and the development of validated and predictive QSAR models is critical to achieving this goal. Additional challenges to improve the accuracy and reliability of QSAR models are poised by a growing interest of regulatory agencies in employing the models for chemical risk assessment. We conclude that QSAR modeling remains a mature but definitely live and constantly growing and exciting research discipline addressing a strong need in widely accessible and reliable computational approaches and specific endpoint predictors.

ACKNOWLEDGMENTS

The author appreciates funding from the National Institutes of Health (grants R01-GM66940, R21GM076059, and P20HG003898) that enabled many of the recent studies described in this chapter. The author is also grateful to all members of his laboratory who have been involved in many important research projects that helped shaping the author's views of modern QSAR modeling, but especially to Drs. Golbraikh, Fourches, Wang, and Zhu.

REFERENCES

1. Hansch C, Fujita T. ρ–σ–π Analysis. A method for the correlation of biological activity and chemical structure. J Amer Chem Soc 1964;86:1616–1626.
2. Hansch C, Kurup A, Garg R, Gao H. Chem-bioinformatics and QSAR: a review of QSAR lacking positive hydrophobic terms. Chem Rev 2001;101(3): 619–672.
3. Boyd D. Successes of computer-assisted molecular design. In: Boyd D, Lipkowitz KB,editors. Reviews in Computational Chemistry. New York: VCH Publishers; 1990. 355–371.
4. Oprea T, Tropsha A. Target, chemical and bioactivity databases: integration is key. Drug Discov Today 2006;3:357–365.
5. Golbraikh A, Tropsha A. Beware of q2! J Mol Graph Model 2002;20(4): 269–276.
6. Doweyko AM. QSAR: dead or alive? J Comput Aided Mol Des 2008;22(2): 81–89.
7. Jorgensen WL. QSAR/QSPR and proprietary data. J Chem Inf Model 2006;46:937.
8. OECD (Q)SAR Project. http://www.oecd.org/document/23/0,3343,en_2649_34379_33957015_1_1_1_1,00.html. 2009.
9. PubChem. http://pubchem.ncbi.nlm.nih.gov/. 2009.
10. Zheng W, Tropsha A. Novel variable selection quantitative structure–property relationship approach based on the *k*-nearest-neighbor principle. J Chem Inf Comput Sci 2000;40(1): 185–194.
11. Tropsha A, Recent trends in quantitative structure–activity relationships. In: Abraham D,editor. Burger's Medicinal Chemistry and Drug Discovery. New York: John Wiley & Sons, Inc.; 2003. p 49–77.
12. Tropsha A. Predictive QSAR (quantitative structure–activity relationships) modeling. In: Martin YC,edtior. Comprehensive Medicinal Chemistry II. Elsevier; 2006. 113–126.
13. Papa E, Villa F, Gramatica P. Statistically validated QSARs, based on theoretical descriptors, for modeling aquatic toxicity of organic chemicals in *Pimephales promelas* (fathead minnow). J Chem Inf Model 2005;45(5): 1256–1266.
14. Tetko IV. Neural network studies. 4. Introduction to associative neural networks. J Chem Inf Comput Sci 2002;42(3): 717–728.
15. Zupan J, Novic M, Gasteiger J. Neural networks with counter-propagation learning-strategy used for modeling. Chemometr Intell Lab Syst 1995;27(2): 175–187.
16. Devillers J. Strengths and weaknesses of the back propagation neural network in QSAR and QSPR studies. In: Devillers J,editor. Genetic Algorithms in Molecular Modeling. Academic Press; 1996. 1–24.

17. Engels MFM, Wouters L, Verbeeck R, Vanhoof G. Outlier mining in high throughput screening experiments. J Biomol Screen 2002;7(4): 341–351.
18. Schuurmann G, Aptula AO, Kuhne R, Ebert RU. Stepwise discrimination between four modes of toxic action of phenols in the *Tetrahymena pyriformis* assay. Chem Res Toxicol 2003;16(8): 974–987.
19. Xue Y, Li H, Ung CY, Yap CW, Chen YZ. Classification of a diverse set of *Tetrahymena pyriformis* toxicity chemical compounds from molecular descriptors by statistical learning methods. Chem Res Toxicol 2006;19(8): 1030–1039.
20. Breiman L, Friedman JH, Olshen RA, Stone CJ. Classification and Regression Trees. Wadsworth; 1984.
21. Deconinck E, Hancock T, Coomans D, Massart DL, Vander Heyden Y. Classification of drugs in absorption classes using the classification and regression trees (CART) methodology. J Pharm Biomed Anal, 2005;39(1–2): 91–103.
22. MOE. http://www.chemcomp.com/fdept/prodinfo.htm#Cheminformatics. 2005.
23. Put R, Perrin C, Questier F, Coomans D, Massart DL, Vander Heyden YV. Classification and regression tree analysis for molecular descriptor selection and retention prediction in chromatographic quantitative structure–retention relationship studies. J Chromatogr A 2003;988(2): 261–276.
24. Breiman L. Random forests. Mach Learn 2001;45(1): 5–32.
25. Svetnik V, Liaw A, Tong C, Culberson JC, Sheridan RP, Feuston BP. Random forest: a classification and regression tool for compound classification and QSAR modeling. J Chemical Inf Comput Sci 2003;43(6): 1947–1958.
26. Put R, Xu QS, Massart DL, Heyden YV. Multivariate adaptive regression splines (MARS) in chromatographic quantitative structure–retention relationship studies. J Chromatogr A 2004;1055(1–2): 11–19.
27. Friedman JH. Multivariate adaptive regression splines. Ann Stat 1991;19(1): 1–67.
28. Vapnik VN. The Nature of Statistical Learning Theory. New York: Springer-Verlag; 1995.
29. Aires-de-Sousa J, Gasteiger J. Prediction of enantiomeric excess in a combinatorial library of catalytic enantioselective reactions. J Comb Chem 2005;7(2): 298–301.
30. Oloff S, Mailman RB, Tropsha A. Application of validated QSAR models of D1 dopaminergic antagonists for database mining. J Med Chem 2005;48(23): 7322–7332.
31. Chohan KK, Paine SW, Waters NJ. Quantitative structure–activity relationships in drug metabolism. Curr Top Med Chem 2006;6(15): 1569–1578.
32. Tropsha A. Application of predictive QSAR models to database mining. In: Oprea T,editor. Cheminformatics in Drug Discovery. Wiley-VCH; 2005. 437–455.
33. Stouch TR, Kenyon JR, Johnson SR, Chen XQ, Doweyko A, Li Y. *In silico* ADME/Tox: why models fail. J Comput Aided Mol Des 2003;17(2–4): 83–92.
34. Maggiora GM. On outliers and activity cliffs: why QSAR often disappoints. J Chem Inf Model 2006;46(4): 1535.
35. Dearden JC, Cronin MT, Kaiser KL. How not to develop a quantitative structure–activity or structure–property relationship (QSAR/QSPR). SAR QSAR Environ Res 2009;20 (3–4): 241–266.
36. Johnson SR. The trouble with QSAR (or how I learned to stop worrying and embrace fallacy). J Chem Inf Model 2008;48(1): 25–26.
37. Young D, Martin D, Venkatapathy R, Harten P. Are the chemical structures in your QSAR correct? QSAR Comb Sci 2008;27(11–12): 1337–1345.
38. Olah M, Rad R, Ostopovici L, Bora A, Hadaruga N, Hadaruga D, et al. WOMBAT and WOMBAT-PK: bioactivity databases for lead and drug discovery. In: Schreiber SL, Kapoor TM, Weiss G,editors. Chemical Biology: From Small Molecules to Systems Biology and Drug Design. Wiley-VCH; 2007. 760–786.
39. Brunton L, Lazo J, Parker K.Goodman & Gilman's The Pharmacological Basis of Therapeutics. 2005.
40. Physicians' Desk Reference. 63rd ed. PDR; 2009.
41. US. Food, Drug Administration FDA Approved Drug Products. 2009.
42. Vaz R, Klabunde T. Antitargets: Prediction and Prevention of Drug Side Effects. Methods and Principles in Medicinal Chemistry. Wiley-VCH; 2008.
43. Austin CP, Brady LS, Insel TR, Collins FS. NIH molecular libraries initiative. Science 2004;306(5699): 1138–1139.
44. Chen X, Liu M, Gilson MK. BindingDB: a Web-accessible molecular recognition database. Comb Chem High Throughput Screen 2001;4 (8): 719–725.

45. Strausberg RL, Schreiber SL. From knowing to controlling: a path from genomics to drugs using small molecule probes. Science 2003;300(5617): 294–295.
46. Kanehisa M, Goto S, Hattori M, Aoki-Kinoshita KF, Itoh M, Kawashima S, et al. From genomics to chemical genomics: new developments in KEGG. Nucleic Acids Res 2006;34 (Database issue): D354–D357.
47. Snyder KA, Feldman HJ, Dumontier M, Salama JJ, Hogue CW. Domain-based small molecule binding site annotation. BMC Bioinformatics 2006;7:152.
48. Okuno Y, Yang J, Taneishi K, Yabuuchi H, Tsujimoto G. GLIDA: GPCR–ligand database for chemical genomic drug discovery. Nucleic Acids Res 2006;34(Database issue): D673–D677.
49. Southan C, Varkonyi P, Muresan S. Complementarity between public and commercial databases: new opportunities in medicinal chemistry informatics. Curr Top Med Chem 2007;7 (15): 1502–1508.
50. Zhu H, Tropsha A, Fourches D, Varnek A, Papa E, Gramatica P, et al. Combinatorial QSAR Modeling of Chemical Toxicants Tested against *Tetrahymena pyriformis*. J Chem Inf Model 2008;48(4): 766–784.
51. Oprea T, Olah M, Ostopovici L, Rad R, Mracec M. On the propagation of Errors in the QSAR Literature. Euro QSAR 2002, Bournemouth, UK, Poster. 2002.
52. Dutta S, Burkhardt K, Young J, Swaminathan GJ, Matsuura T, Henrick K, et al. Data deposition and annotation at the worldwide Protein Data Bank. Mol Biotechnol 2009;42(1): 1–13.
53. ChemSpider. ChemSpider. http://www.chemspider.com. 2009.
54. Tropsha A, Golbraikh A. Predictive QSAR modeling workflow, model applicability domains, and virtual screening. Curr Pharm Des 2007;13:3494–3504.
55. Garfield E. An algorithm for translating chemical name to chemical formula. Essays of an Information Scientist, Vol. 7, ISI Press, Philadelphia, PA; 1984. p 441–513.
56. Brecher J. From chemical name to structure: finding a noodle in the haystack. CAS/IUPAC Conference on Chemical Identifiers and XML for Chemistry, Columbus, OH. 2002.
57. Brecher J. Name=Struct: a practical approach to the sorry state of real-life chemical nomenclature. J Chem Inf Comput Sci 1999;39: 943–950.
58. Olah M, Rad R, Ostopovici L, Bora A, Hadaruga N, Hadaruga D, et al. WOMBAT and WOMBAT-PK: bioactivity databases for lead and drug discovery. Chemical Biology: From Small Molecules to Systems Biology and Drug Design. Weinheim: Wiley-VCH; 2007. p 760–786.
59. Standardizer. ChemAxon JChem. http://www.chemaxon.com. 2009.
60. Standardizer. ChemAxon JChem. http://www.chemaxon.com. 2009.
61. OpenBabel. Openbabel org. 2009.
62. Molecular Networks GmbH. http://www.molecular-networks.com/products. 2009.
63. ISIDA. Laboratoire d'Infochimie, Louis Pasteur University, Strasbourg, France (infochim u-strasbg fr). 2009.
64. Kuz'min VE, Artemenko AG, Muratov, EN. Hierarchical QSAR technology based on the Simplex representation of molecular structure. J Comput Aided Mol Des 2008;22(6–7): 403–421.
65. MOE. Chemical Computing Group. http://www.chemcomp.com/software.htm. 2009.
66. ISIDA. Laboratoire d'Infochimie, Louis Pasteur University, Strasbourg, France (infochim u-strasbg fr). 2009.
67. Hyleos. http://www.hyleos.net/. 2009.
68. OpenBabel. Openbabel org. 2009.
69. ChemAxon/Marvin. http://www.chemaxon.com. 2009.
70. CambridgeSoft. http://www.cambridgesoft.com/. 2009.
71. Schrodinger. http://www.schrodinger.com/. 2009.
72. ACDLabs. Advanced Chemistry Development. http://www.acdlabs.com. 2009.
73. Symyx. http://www.symyx.com/. 2009.
74. MOE. Chemical Computing Group. http://www.chemcomp.com/software.htm. 2009.
75. Accelrys. http://accelrys.com/. 2009.
76. Tripos. http://tripos com/. 2009.
77. Cramer RD III, Patterson DE, Bunce JD. Comparative molecular field analysis (CoMFA). 1. Effect of shape on binding of steroids to carrier proteins. J Am Chem Soc 1988;110:5959–5967. 1988.
78. Golbraikh A, Shen M, Xiao Z, Xiao YD, Lee KH, Tropsha A. Rational selection of training and test sets for the development of validated QSAR models. J Comput Aided Mol Des 2003;17(2–4): 241–253.
79. Irwin JJ, Shoichet BK. ZINC: a free database of commercially available compounds for virtual

screening. J Chem Inf Model 2005;45(1): 177–182.

80. Tropsha A, Gramatica P, Gombar VK. The Importance of Being Earnest: Validation is the Absolute Essential for Successful Application and Interpretation of QSPR Models. Quant Struct Act Relat Comb Sci 2003;22:69–77.
81. Kier LB, Hall LH. Molecular Connectivity in Chemistry and Drug Research. New York: Academic Press; 1976.
82. Kier LB, Hall LH. Molecular Structure Description: TheElectrotopological State. New York: Academic Press; 1999.
83. Klein CD, Hopfinger AJ. Pharmacological activity and membrane interactions of antiarrhythmics: 4D-QSAR/QSPR analysis. Pharm Res 1998;15(2): 303–311.
84. Polanski J. Receptor dependent multidimensional QSAR for modeling drug–receptor interactions. Curr Med Chem 2009;16(25): 3243–3257.
85. Tropsha A, Zheng W. Identification of the descriptor pharmacophores using variable selection QSAR: applications to database mining. Curr Pharm Des 2001;7(7): 599–612.
86. Brown RD, Martin YC. The information content of 2D and 3D structural descriptors relevant to ligand–receptor binding. J Chem Inf Comp Sci 1997;37(1): 1–9.
87. Hillebrecht A, Klebe G. Use of 3D QSAR models for database screening: a feasibility study. J Chem Inf Model 2008;48(2): 384–396.
88. Gedeck P, Rohde B, Bartels C. QSAR: how good is it in practice? Comparison of descriptor sets on an unbiased cross section of corporate data sets. J Chem Inf Model 2006;46(5): 1924–1936.
89. Wang XS, Tang H, Golbraikh A, Tropsha A. Combinatorial QSAR modeling of specificity and subtype selectivity of ligands binding to serotonin receptors $5HT_{1E}$ and $5HT_{1F}$. J Chem Inf Model 2008;48(5): 997–1013.
90. de Cerqueira LP, Golbraikh A, Oloff S, Xiao Y, Tropsha A. Combinatorial QSAR modeling of P-glycoprotein substrates. J Chem Inf Model 2006;46(3): 1245–1254.
91. Kovatcheva A, Golbraikh A, Oloff S, Xiao YD, Zheng W, Wolschann P, et al. Combinatorial QSAR of ambergris fragrance compounds. J Chem Inf Comput Sci 2004;44(2): 582–595.
92. Kubinyi H, Hamprecht FA, Mietzner T. Three-dimensional quantitative similarity–activity relationships (3D QSiAR) from SEAL similarity matrices. J Med Chem 1998;41(14): 2553–2564.
93. Novellino E, Fattorusso C, Greco G. Use of comparative molecular field analysis and cluster analysis in series design. Pharm Acta Helv 1995;70:149–154.
94. Norinder U. Single and domain made variable selection in 3D QSAR applications. J Chemomet 1996;10:95–105.
95. Tropsha A, Cho SJ. Cross-validated R2-guided region selection for CoMFA studies. In: Kubinyi H, Folkers G, Martin YC,editors. 3D QSAR in Drug Design. Dordrecht, The Netherlands: Kluwer Academic Publishers; 1998. 57–69.
96. Golbraikh A, Tropsha A. Predictive QSAR modeling based on diversity sampling of experimental datasets for the training and test set selection. J Comput Aided Mol Des 2002;16 (5–6): 357–369.
97. Sachs L. Handbook of Statistics. Springer; 1984.
98. Zhang S, Golbraikh A, Tropsha A. Development of quantitative structure–binding affinity relationship models based on novel geometrical chemical descriptors of the protein–ligand interfaces. J Med Chem 2006;49(9): 2713–2724.
99. Golbraikh A, Bonchev D, Tropsha A. Novel chirality descriptors derived from molecular topology. J Chem Inf Comput Sci 2001;41(1): 147–158.
100. Kovatcheva A, Buchbauer G, Golbraikh A, Wolschann P. QSAR modeling of alpha-campholenic derivatives with sandalwood odor. J Chem Inf Comput Sci 2003;43(1): 259–266.
101. Shen M, Xiao Y, Golbraikh A, Gombar VK, Tropsha A. Development and validation of *k*-nearest-neighbor QSPR models of metabolic stability of drug candidates. J Med Chem 2003;46(14): 3013–3020.
102. Shen M, LeTiran A, Xiao Y, Golbraikh A, Kohn H, Tropsha A. Quantitative structure--activity relationship analysis of functionalized amino acid anticonvulsant agents using *k* nearest neighbor and simulated annealing PLS methods. J Med Chem 2002;45(13): 2811–2823.
103. Shen M, Beguin C, Golbraikh A, Stables JP, Kohn H, Tropsha A. Application of predictive QSAR models to database mining: identification and experimental validation of novel anticonvulsant compounds. J Med Chem 2004; 47(9): 2356–2364.
104. Zhang S, Golbraikh A, Oloff S, Kohn H, Tropsha A. A novel automated lazy learning QSAR (ALL-QSAR) approach: method development,

applications, and virtual screening of chemical databases using validated ALL-QSAR models. J Chem Inf Model 2006;46(5): 1984–1995.

105. Golbraikh A, Shen M, Xiao Z, Xiao YD, Lee KH, Tropsha A. Rational selection of training and test sets for the development of validated QSAR models. J Comput Aided Mol Des 2003;17(2–4): 241–253.

106. Mandel J. Use of the singular value decomposition in regression-analysis. Am Stat 1982; 36(1): 15–24.

107. Afantitis A, Melagraki G, Sarimveis H, Koutentis PA, Markopoulos J, Igglessi-Markopoulou O. A novel QSAR model for predicting induction of apoptosis by 4-aryl-4*H*-chromenes. Bioorg Med Chem 2006;14(19): 6686–6694.

108. Netzeva TI, Gallegos SA, Worth AP. Comparison of the applicability domain of a quantitative structure–activity relationship for estrogenicity with a large chemical inventory. Environ Toxicol Chem 2006;25(5): 1223–1230.

109. Saliner AG, Netzeva TI, Worth AP. Prediction of estrogenicity: validation of a classification model. SAR QSAR Environ Res 2006;17(2): 195–223.

110. Tong W, Xie Q, Hong H, Shi L, Fang H, Perkins R. Assessment of prediction confidence and domain extrapolation of two structure–activity relationship models for predicting estrogen receptor binding activity. Environ Health Perspect 2004;112(12): 1249–1254.

111. Helma C. Lazy structure–activity relationships (lazar) for the prediction of rodent carcinogenicity and Salmonella mutagenicity. Mol Divers 2006;10(2): 147–158.

112. Gramatica P. Principles of QSAR models validation: internal and external. QSAR Comb Sci 2007;26(5): 694–701.

113. Yang C, Richard AM, Cross KP. The art of data mining the minefields of toxicity databases to link chemistry to biology. Curr Comput Aided Drug Des 2006;2:135–150.

114. Klopman G, Zhu H, Ecker G, Chiba P. MCASE study of the multidrug resistance reversal activity of propafenone analogs. J Comput Aided Mol Des 2003;17(5–6): 291–297.

115. Stoner CL, Gifford E, Stankovic C, Lepsy CS, Brodfuehrer J, Prasad JVNV, et al. Implementation of an ADME enabling selection and visualization tool for drug discovery. J Pharm Sci 2004;93(5): 1131–1141.

116. Mayer P, Reichenberg F. Can highly hydrophobic organic substances cause aquatic baseline toxicity and can they contribute to mixture toxicity? Environ Toxicol Chem 2006;25(10): 2639–2644.

117. Zhu H, Rusyn I, Richard A, Tropsha A. Use of cell viability assay data improves the prediction accuracy of conventional quantitative structure–activity relationship models of animal carcinogenicity. Environ Health Perspect 2008;116(4): 506–513.

118. Xia M, Huang R, Witt KL, Southall N, Fostel J, Cho MH, et al. Compound cytotoxicity profiling using quantitative high-throughput screening. Environ Health Perspect 2008;116 (3): 284–291.

119. Moss GP, Cronin MTD. Quantitative structure–permeability relationships for percutaneous absorption: re-analysis of steroid data. Int J Pharm 2002;238(1–2): 105–109.

120. Estrada E, Patlewicz G. On the usefulness of graph-theoretic descriptors in predicting theoretical parameters. Phototoxicity of polycyclic aromatic hydrocarbons (PAHs). Croatica Chemica Acta 2004;77(1–2): 203–211.

121. Leo AJ, Hansch C. Role of hydrophobic effects in mechanistic QSAR. Perspect Drug Discov Des 1999;17:1–25.

122. Tropsha A, Cho SJ, Zheng W."New tricks for an old dog": development and application of novel QSAR methods for rational design of combinatorial chemical libraries and database mining. In: Parrill AL, Reddy MR, editors. American Chemical Society: Washington, DC. Rational Drug Design: Novel Methodology and Practical Applications. 1999. p 198–211.

123. Cho SJ, Zheng W, Tropsha A. Rational combinatorial library design. 2. Rational design of targeted combinatorial peptide libraries using chemical similarity probe and the inverse QSAR approaches. J Chem Inf Comput Sci 1998;38(2): 259–268.

124. NCI. http://dtp.nci.nih.gov/docs/3d_database/structural_information/smiles_strings.html. 2007.

125. Gussio R, Pattabiraman N, Kellogg GE, Zaharevitz DW. Use of 3D QSAR methodology for data mining the National Cancer Institute Repository of Small Molecules: application to HIV-1 reverse transcriptase inhibition. Methods 1998;14(3): 255–263.

126. Hoffman BT, Kopajtic T, Katz JL, Newman AH. 2D QSAR modeling and preliminary database searching for dopamine transporter inhibitors using genetic algorithm variable

selection of Molconn Z descriptors. J Med Chem 2000;43(22): 4151–4159.

127. Varnek A, Tropsha A. Cheminformatics Approaches to Virtual Screening. London: RSC; 2008.

128. Zhang S, Wei L, Bastow K, Zheng W, Brossi A, Lee KH, et al. Antitumor Agents 252. Application of validated QSAR models to database mining: discovery of novel tylophorine derivatives as potential anticancer agents. J Comput Aided Mol Des 2007;21(1–3): 97–112.

129. Hsieh JH, Wang XS, Teotico D, Golbraikh A, Tropsha A. Differentiation of AmpC beta-lactamase binders vs. decoys using classification *k*NN QSAR modeling and application of the QSAR classifier to virtual screening. J Comput Aided Mol Des 2008;22(9):593–609.

130. Peterson YK, Wang XS, Casey PJ, Tropsha A. Discovery of geranylgeranyltransferase-I inhibitors with novel scaffolds by the means of quantitative structure–activity relationship modeling, virtual screening, and experimental validation. J Med Chem 2009;52(14): 4210–4220.

131. Hsieh JH, Wang XS, Teotico D, Golbraikh A, Tropsha A. Differentiation of AmpC beta-lactamase binders vs. decoys using classification *k*NN QSAR modeling and application of the QSAR classifier to virtual screening. J Comput Aided Mol Des 2008;22(9): 593–609.

132. Babaoglu K, Simeonov A, Irwin JJ, Nelson ME, Feng B, Thomas CJ, et al. Comprehensive mechanistic analysis of hits from high-throughput and docking screens against beta-lactamase. J Med Chem 2008;51(8): 2502–2511.

133. Pavan M, Netzeva TI, Worth AP. Validation of a QSAR model for acute toxicity. SAR QSAR Environ Res 2006;17(2): 147–171.

134. Vracko M, Bandelj V, Barbieri P, Benfenati E, Chaudhry Q, Cronin M, et al. Validation of counter propagation neural network models for predictive toxicology according to the OECD principles: a case study. SAR QSAR Environ Res 2006;17(3): 265–284.

135. Roberts DW, Aptula AO, Patlewicz G. Mechanistic applicability domains for non-animal based prediction of toxicological endpoints. QSAR analysis of the Schiff base applicability domain for skin sensitization. Chem Res Toxicol 2006;19(9): 1228–1233.

136. Fjodorova N, Novich M, Vrachko M, Smirnov V, Kharchevnikova N, Zholdakova Z, et al. Directions in QSAR modeling for regulatory uses in OECD member countries, EU and in Russia. J Environ Sci Health C Environ Carcinog Ecotoxicol Rev 2008;26(2): 201–236.

137. Nair PC, Sobhia ME. Comparative QSTR studies for predicting mutagenicity of nitro compounds. J Mol Graph Model 2008;26(6): 916–934.

138. DSSTox. http://www.epa.gov/nheerl/dsstox/About.html. 2008.

THE APPLICATION OF RECOMBINANT DNA TECHNOLOGY IN MEDICINAL CHEMISTRY AND DRUG DISCOVERY

Soumitra Basu[1]
Adegboyega Oyelere[2]
[1] Kent State University, Kent, OH
[2] Georgia Tech, Atlanta, GA

1. INTRODUCTION

Discovering new drugs has never been a simple matter. From ancient times to the beginning of the nineteenth century, treatment for illness or disease was based mainly on folklore and traditional curative methods derived from plants and other natural sources. The isolation and chemical characterization of the principal components of some of these traditional medicines, mainly alkaloids and the likes, spawned the development of the modern pharmaceutical industry and the production of drugs in mass quantities. Consequently, the industry has undergone profound changes within the last millennium. As the companion chapters of this volume describe, the emphasis has changed from isolation of active constituents to creation of new, potent chemical entities. This evolution from folklore to science is responsible for the thousands of pharmaceuticals available worldwide at present [1].

1.1. Chemistry-Driven Drug Discovery

The exacting process of discovering new chemical entities that are safe and effective drugs has itself undergone many changes, each of which was prompted by the introduction of some new technology [2,3]. In the 1920s, the first efforts at understanding why and how morphine works in terms of its chemical structure were initiated. During the 1940s, challenges for mass production of medicinally valuable natural products, like the penicillins, were conquered. By the late 1950s, advances in synthetic organic chemistry enabled the generation of multitudes of novel structures for broad testing into the major focus of the modern pharmaceutical industry. Although serendipitous at best, this approach yielded many valuable compounds, most notably the benzodiazepine tranquilizers chlordiazepoxide and diazepam [4]. Even with these successful compounds, however, the process of drug discovery amounted to little more than evaluating available chemical entities in animal models suggestive of human disease.

By the mid-1960s, medicinal chemistry had clearly become the cornerstone technology of modern drug discovery. Systematic development of structure–activity relationships (SARs), even to the point at which predictions about activity might be made, became the hallmark of new drug discovery. Even then, however, an understanding of the actions of drugs at the molecular level was often lacking. Receptors and enzymes were still considered as functional "black boxes" whose structures and functions were poorly understood. The first successful attempts at actually designing a drug to work at a particular molecular target happened nearly simultaneously in the 1970s, with the discovery of cimetidine, a selective H_2-antagonist for the treatment of ulcers [5], and captopril, an angiotensin-converting enzyme inhibitor for hypertension [6]. The success of these two drugs sparked a realignment of chemistry-driven pharmaceutical research. Since then, the art of rational drug design has undergone an explosive evolution, making use of sophisticated computational and structural methodology to help in the effort [7]. During the 1980s, mechanism-targeted design and screening combined to produce a number of novel chemical entities. Two landmark examples include the natural product HMG-CoA reductase inhibitor lovastatin for the treatment of hypercholesteremia [8] and the antihypertensive angiotensin II receptor antagonist losartan, synthetically optimized from a chemical library screening lead [9].

There is little doubt that the task of discovering new therapeutic agents that work potently, specifically, and with minimum side effects has become increasingly important and coincidentally more difficult. Advances in medical research that have provided new clues to the previously obscured etiologies of diseases have revealed new opportunities for therapeutic intervention. Much progress were recorded in the 1990s and the current dispensation that the idea of personalized medicine is no more far fetched but a reality. This has forced the science of medicinal chemistry, once founded almost solely in near-blind synthesis and screening for *in vivo* effects, to become keenly aware of biochemical mechanisms as an intimate part of the development process. Even with these major advances in the medicinal and pharmaceutical sciences, more fundamental questions remain: What determines a useful biological property? And how is it measured in the discovery process? The answers can determine the success or failure of any drug discovery program, since both the observation of a useful biological property in a novel molecule and the optimization of structure–activity relationships associated with ultimate clinical candidate selection have rightfully relied heavily on practices, and sometimes prejudices, founded in decades of empirical success [10]. Although the task of drug development has now been refined into a process without major unidentified obstacles, the challenge to bring the discovery of novel compounds to a comparable state of maturity remains. As in the past, another research avenue synergistic with existing discovery technologies is necessary.

1.2. The Advent of Recombinant DNA Technology

The evolution of recombinant DNA (rDNA) technology, from scientific innovation to pharmaceutical discovery process, has occurred in parallel with the development of contemporary medicinal chemistry [11–14]. The traditional products of biotechnology research share few of the traits characteristic of traditional pharmaceuticals. These biotechnologically derived therapeutics are large extracellular proteins, peptides, glycoconjugates, and oligonucleotides destined to be, with few exceptions, injectables for use in either chronic replacement therapies or in acute or near-term chronic situations for the treatment of life-threatening indications [15–17]. Many of these products also satisfy urgent and previously unfulfilled therapeutic needs. However, their dissimilarity to traditional medicinal agents does not end there. Unlike most low molecular weight pharmaceuticals, these biopharmaceuticals were developed not because of the novelty of their structures, but because of the novelty of their actions. Their discovery hinged on the recognition of a useful biological activity, its subsequent association with an effector protein, and the genetic identification, expression, and production of the effector by the application of recombinant DNA technology [18,19].

If modulation of biochemical processes by a low molecular weight compound has been the traditional goal of medicinal chemistry, then association of a biological effect with a distinct protein and its identification and production have been considered the domain of molecular genetics. The application of recombinant DNA technology to the identification of proteins and other macromolecules as drugs or drug targets and their production in meaningful quantity as products or discovery tools, respectively, provide an answer to at least one of the persistent problems of new lead discovery. Because a comprehensive review of the genetic engineering of important proteins is well beyond the scope of this volume, this chapter will instead highlight some novel examples of advances in recombinant DNA technology, with respect to both exciting new pharmaceuticals and potential applications of recombinantly produced proteins, be they enzymes, receptors, or hormones, to the more traditional processes of drug discovery.

2. NEW THERAPEUTICS FROM RECOMBINANT DNA TECHNOLOGY

The traditional role of the pharmaceutical industry, that is, synthesis of new chemical entities as therapeutic agents, was suddenly expanded by the introduction of the first biotechnologically derived products in the 1980s. The approval of recombinant human insulin in 1982 broke important ground for products produced by genetic engineering [20]. In 1985 another milestone was achieved when Genentech became the first biotechnology company to be granted approval to market a recombinant product, human growth hormone. These events set an entire industry into motion, to produce not only natural proteins for the treatment of deficiency-associated diseases but also true therapeutics for both acute and chronic care.

Industry estimates show that the upward trend in biotechnologically derived drugs will continue well into the current millennium. There are over 100 companies involved in research and development of rDNA products and over 400 products generated by biotechnology are estimated to be somewhere in the development pipeline or approval process [21–23]. A comprehensive list of FDA approved biotechnology drugs is available on the Web at http://www.biopharma.com/approvals1.html; http://www.fda.gov/BiologicsBloodVaccines/ucm121134.htm. The variety of products—from hormones and enzymes to receptors, vaccines, and monoclonal antibodies (mAbs)—seeks to treat a broad range of clinical indications thought untreatable just three decades ago. Yet despite this period of phenomenal growth for recombinant DNA-derived therapeutics, the promise of biotechnology, once touted to be limitless, has instead become more realistically defined to include not only the actual recombinant products and the difficulties inherent in their production but also many spin-off technologies, including diagnostics and genetically defined drug discovery tools [24–26].

One particular area of traditional pharmaceutical research in which recombinant DNA technology has made a profound impact has been in the engineering of antibiotic-producing organisms [27–33]. Always an important source of new bioactive compounds, especially antibiotics [34,35], fermentation procedures can be directly improved by strain optimization techniques, including genetic recombination and cloning. More exciting is the possibility of producing hybrid antibiotics that combine desirable features of one or more individual compounds for improved potency, bioavailability, or specificity [36]. The art of finding new natural product-based lead compounds by screening fermentation broths, plant sources, and marine organisms by using genetically engineered reagents is becoming of special importance as more of the relevant targets identified by molecular biology operate in obscure or even unknown modes. The structural diversity provided by natural products combined with the ability to test molecular biology-driven biochemical hypotheses has already become an important route for the discovery of new therapeutics [37–43].

3. PROTEIN ENGINEERING AND SITE-DIRECTED MUTAGENESIS

Rapid developments in the technique of site-directed mutagenesis have created the ability to change essentially any amino acid, or even substitute or delete whole domains, in any protein, with the goal of designing and constructing new proteins with novel binding, clearance, or catalytic activities [44–46]. The concomitant changes in protein folding and tertiary structure, protein physiology, binding affinities (for a receptor or hormone), binding specificities (either for substrate or receptor), or catalytic activity (for enzyme active site mutants) are all effects that are measurable against the "wild-type" parent, assuming that expression of the gene and subsequent proper folding have successfully occurred. Several surprising observations have been made during the short period that this technology has been available: amino acid substitutions lead, in general, to highly localized changes in protein structure with few global changes in overall folding; substitutions of residues not involved in internal hydrophobic contacts are extremely well accommodated, leading to few unsynthesizable mutants; and proteins seem extremely tolerant of domain substitution, even among unrelated proteins, allowing often even crude first attempts at producing chimeric proteins to be successful. The implications of this technology for the discovery of new pharmaceuticals lie in two areas: second-generation protein therapeutics and site- or domain-specific mutant proteins for structure–function investigations.

Throughout this chapter, amino acids are denoted by their standard one-letter codes; site-specific mutations are represented by the code for the wild-type amino acid, the residue number, and the code for the replacement amino acid [44].

3.1. Second-Generation Protein Therapeutics

The cloning, expression, and manufacture of proteins as therapeutics involve the same problems encountered in the development and successful clinical approval of any drug. Potency, efficacy, bioavailability, metabolism, and pharmaceutical formulation challenges presented by the natural protein suggest that second-generation products might be engineered to alleviate the particular problem at hand, producing desired therapeutic improvements. The parent proteins to which this technology has been applied extend across the range of recombinant products already approved and those in advanced stage of clinical evaluation [47,48]. As an example, for tissue-type plasminogen activator (t-PA), one of the most studied recombinant products [49–51], four properties functioning in concert (i.e., substrate specificity, fibrin affinity, stimulation of t-PA activity by fibrin and fibrinogen, and sensitivity of the enzyme to inhibition by plasminogen activator inhibitors (PAIs)) are responsible for the localization and potentiation of the lytic reaction at a clot surface and are readily analyzed using molecular variants [52]. A consensus structure combining the major domains of t-PA has been predicted based on the significant sequence homology with other serum proteins and serine proteases. The complexity of this structure is reflected in its functional multiplicity: efficient production of plasmin by cleavage of the R560-V561 bond of plasminogen, very low binding to plasminogen in the absence of fibrin, moderately high affinity for fibrin, increase in the efficiency of plasminogen activation by 500-fold in the presence of fibrin, rapid inactivation by PAI-1, and rapid hepatic elimination by receptor-mediated endocytosis [53]. BM 06.022, a recombinantly engineered t-PA deletion mutant (t-PA del (V4-E175)), made up of the Kringle 2 and protease domains, has been reported to have the same plasminogenolytic activity but a lower fibrin affinity compared with wild-type t-PA [54]. Another variant of t-PA (T103N, KHRR 296–299 AAAA) was demonstrated to have the combined desirable properties of decreased plasma clearance, increased fibrin specificity, resistance to PAI-1, and *in vivo* increased potency and decreased systemic activation of plasminogen when administered by bolus dose [55].

Although the systematic changes exemplified by t-PA site-directed mutagenesis studies are the rDNA equivalents of medicinal chemistry (multiple analog synthesis for structure–activity relationship development), more recent applications of this technology bear a less straightforward resemblance to medicinal chemistry-driven drug discovery paradigms [56–59]. However, these same recombinant techniques can be used to combine domains from different proteins to produce chimeric constructs that incorporate multiple desired properties into a single final product or reagent. For instance, in an effort to overcome the short plasma half-life associated with soluble CD4, chimeric molecules termed *immunoadhesins* (Fig. 1) have been recombinantly constructed from the gp120-specific domains of CD4 and the effector domains of various immunoglobulin classes [60,61]. In addition to dramatically improved pharmacokinetics, these chimeric constructs incorporate functions such as Fc receptor binding, protein A binding, complement fixation, and placental transfer, all of which are imparted by the Fc portion of immunoglobulins. Dimeric constructs from human (CD4-2γl and CD4-4γl) and mouse (CD4-Mγ2a) IgG and a pentameric chimera (CD4-Mμ) from mouse IgM exhibit evidence of retained gpl20 binding and anti-HIV infectivity activity. Both CD4-2γl and CD4-4-γl show significantly increased plasma half-lives of 6.7 and 48 h, respectively, compared with 0.25 h for rCD4. Furthermore, the immunoadhesin CD4-2γl (CD4-IgG) mediates antibody-dependent, cell-mediated cytotoxicity (ADCC) toward HIV-infected cells and is efficiently transferred across the placenta of primates [62]. These early studies have inspired the investigation and identification of other chimeric gpl20 binding proteins as potential therapeutic agents against HIV [63–67].

3.1.1. Drug Efficacy and Personalized Medicine

It is becoming clear that genetic variations play critical roles in patients' response to certain medications. Differential expression

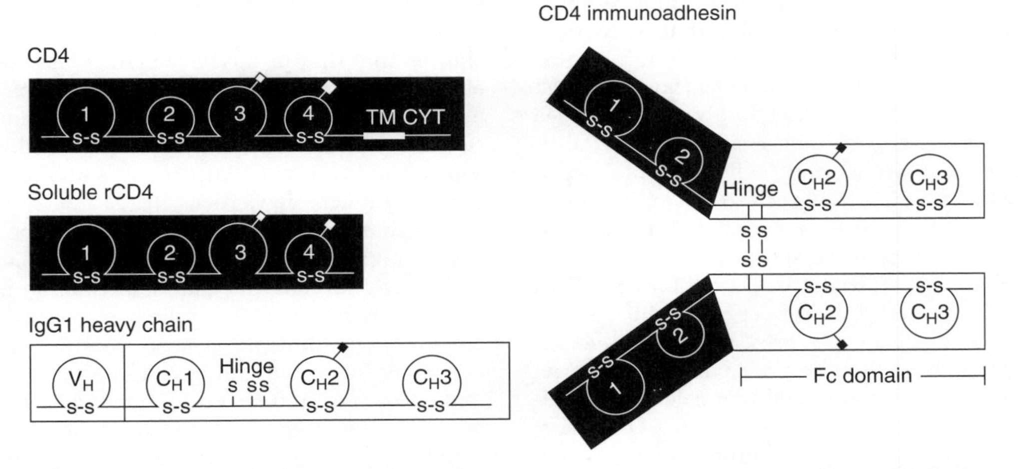

Figure 1. Structure of CD4 immunoadhesin, soluble rCD4, and the parent human CD4 and IgG1 heavy-chain molecules. CD4- and IgG1-derived sequences are indicated by shaded and unshaded regions, respectively. The immunoglobulin like domains are numbered 1–4, TM is the transmembrane domain, and CYT is the cytoplasmic domain. Soluble CD4 is truncated after P368 of the mature CD4 polypeptide. The variable (V_H) and constant (C_H1, Hinge, C_H2, and C_H3) regions of IgG1 heavy chains are shown. Disulfide bonds are indicated by S–S. CD4 immunoadhesin consists of residues 1–180 of the mature CD4 protein fused to IgG1 sequences, beginning at D216, which is the first residue in the IgG1 hinge after the cysteine residue involved in heavy–light chain bonding. The CD4 immunoadhesins shown, which lacks a C_H1 domain, was derived from a C_H1-containing CD immunoadhesin by oligonucleotide-directed deletional mutagenesis, expressed in Chinese hamster ovary cells and purified to >99% purity using protein A-sepharose chromatography [42].

of drug targets and or metabolic enzymes has been shown to lead to differences in efficacy and toxicity profiles of drugs in section of population that harbors this genetic variation [68–73]. Molecular biology and its associated techniques feature prominently in bringing to birth an interdisciplinary field, pharmacogenetics (now more broadly defined as pharmacogenomics), which promises to unravel how genetic make up and variation thereof affect human response to medication [74,75]. It is a widely held view that advances in pharmacogenomics will revolutionize drug dispensation and drug discovery processes. When fully realized, the gains of pharmacogenomics will positively impact all stages of drug discovery processes in numerous ways including: (i) identification of new and novel therapeutic targets; (ii) an increased understanding of the molecular "uniqueness" of diseases; (iii) genetic tagging of diseases with the consequence of developing designer medications that best combat an ailment; and (iv) efficient design of clinical trials, with a better chance of success, aided by a genetic prescreening of candidates. Medications that were judged ineffective by traditional validation methods in a random patient population may be found beneficial to a population of patients having an overexpressed "susceptibility gene(s)."

A notable success story in new target identification and validation is seen in the introduction of a new class of nonstereoidal anti-inflammatory drugs (NSAIDs) by Merck Frosst (Vioxx) and Monsanto-Pfizer (Celebrex). Until recently, the onset of inflammation and pain has been linked to one cyclooxygenase enzyme, COX. Clinically useful NSAIDs such as aspirin, diclofenac, and ibuprofen exhibit their anti-inflammatory and antipyretic activity by inhibiting COX. A prolonged use of most NSAIDs results in gastrointestinal (GI) toxicity [76], which may be debilitating enough to require hospitalization in many patients. Advances in the understanding of the biology of COX, championed by elegant biochemical and recombinant DNA studies, revealed that it exists in three isoforms: constitutive COX-1, inducible COX-2, and COX-3, a splice variant of COX-1 (Fig. 2) [77,78]. COX-1 is always expressed and mediates the synthesis of prostagladins that regulate normal cell functions, while COX-2 is active only at the onset of inflammation [79,80]. However, the precise role of COX-3 in inflammatory response is still a subject of debate [81]. Early NSAIDs inhibit both COX isoforms, thereby interfering with the production of the protective prostagladin products of COX-1 in addition to their pain relieve activity. It was hypothesized that selective inhibitors of inflammation associated COX-2 may produce a better drug profile and possibly avoid many side effects caused by nonselective NSAIDs, most especially GI tract toxicity [82]. The discovery of COX-2 gene stimulated intensive studies aimed at verifying this hypothesis [83,84]. The clinical and commercial success of the first-generation COX-2-specific inhibitors validated COX-2 as a new target for anti-inflammation and antipyretic therapy. Furthermore, the safety profiles of these drugs showed that they do not have many toxic side effects, such as gastrotoxicity and platelet aggregation, commonly associated with traditional nonselective NSAIDs [85–88]. Sadly, Vioxx has been withdrawn due to unanticipated cardiovascular risks.

Alternatively, recombinant DNA technology is making it possible to decipher the roles of disease subtypes and genetic variability in disease progression and response to drugs. New markers that aid disease classification are being identified and explored as targets to develop new and novel therapeutic strategies to better treat or manage diseases. The insights gained from such studies are beginning to yield genetically engineered pharmaceuticals for treating various human disease conditions including diabetes, multiple sclerosis, rheumatoid arthritis, cancer, and viral diseases. For example, an overexpression of epidermal growth factor receptor-2 (HER-2) is

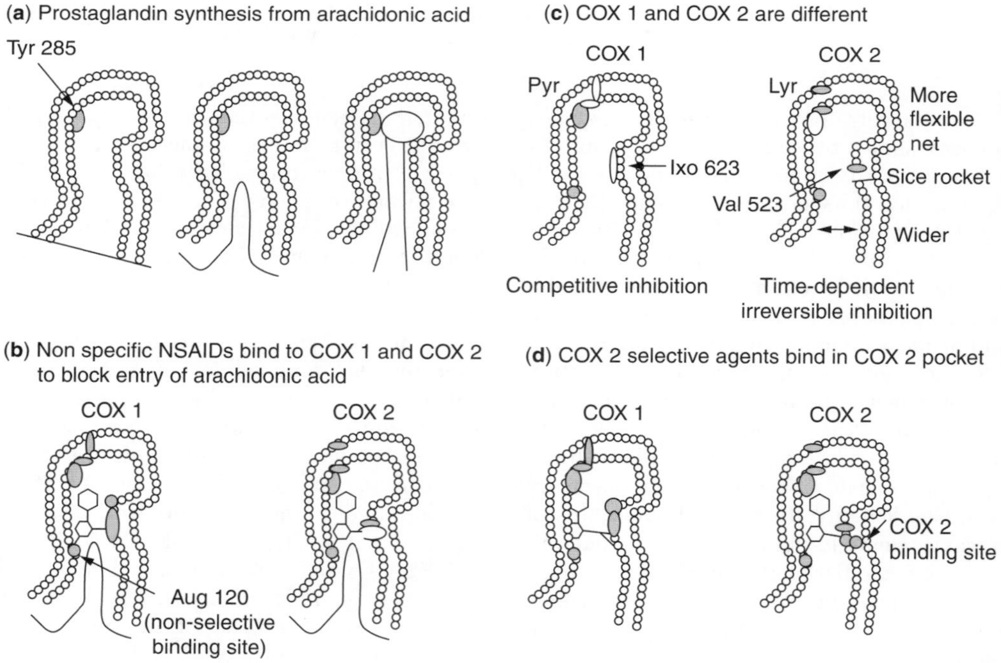

Figure 2. Prostaglandin synthesis and inhibition in COX-1 and COX-2. (a) Initial stages of prostaglandin synthesis. (b) Binding stages of standard NSAIDs to arginine 120 to inhibit prostaglandin synthesis by direct blockade of cyclooxygenase channel. (c) Differences between COX-1 and COX-2. (d) Specific blockade of COX-2.

known to occur in about 25–30% of women with breast cancer [89]. Herceptin, a humanized anti-HER-2 monoclonal antibody, was introduced by Genentech to treat metastatic breast cancer and it is proven beneficial in patients with metastatic breast cancer in which HER-2 is overexpressed [90–95].

Although the use of herceptin in breast cancer treatment is frequently cited as an ideal example of the feasibility of personalized medicine, several examples of personalized medicines are now in clinical use with many more undergoing clinical trials [75]. The discoveries of these agents are all facilitated by the application of the principles of pharmacogenomics. Nevertheless, The fulfillment of the many promises of pharmacogenomics is strongly hinged on the identification of unique markers that correlate genetic makeup to drug response. Molecular biology is and will continue to play frontline roles in the identification of these genetic markers. Public and private efforts are currently ongoing in identifying such informative markers, details of which are beyond the scope of this chapter.

3.2. Antibody-Based Therapeutics

Antibody therapeutics can potentially treat diseases that can be as diverse as autoimmune disorders to cancer and infectious diseases. Antibodies are currently rated as an important and growing class of biotherapeutics. Other than vaccines monoclonal antibodies currently in development outnumbered all other classes of therapeutics. Recombination technology plays a key role in the development and commercialization of therapeutic antibodies. In fact, eight out of nine antibody products available in the US market are recombinant products [96–99].

During the early days of mAb use their therapeutic applications were limited due to immunogenicity in humans as they were murine antibodies that induced human antimouse antibodies, leading to allergic reactions and reduced efficacy [100]. After the discovery of murine antibodies in 1975, the next generations of antibodies were chimeric in nature with 66% human and 34% mouse produced via genetic engineering. During the 1980s and early 1990s, complementarity-determining region (CDR) grafting and veneering techniques were established, which reduced the mouse portion of the sequence to less than 10%. Lately, genetically engineered transgenic animals and plants that can be used for production of humanized antibodies have been developed [100–104].

3.3. Epitope Mapping

Site-directed mutagenesis technology has also been applied to one of the most challenging problems in structural biochemistry: the nature of the protein–protein interaction. While numerous examples of models of enzyme–ligand complexes have been developed based on active-site modifications, site-directed mutagenesis is being extended to the formidable problem of defining the essential elements of a protein–protein (e.g., a protein substrate to a protease or a hormone to its receptor) binding epitope.

An early example of a systematic search for a binding epitope is seen in the work used to define the human growth hormone-somatogenic receptor interaction [105,106]. First, using a technique termed homolog-scanning mutagenesis, segments of sequences (7–30 ammo acids in length) from homologous proteins known not to bind to the hGH receptor or to hGH-sensitive mAb were systematically substituted throughout the hGH structure, by using a working model based on the three-dimensional folding pattern found by X-ray crystallographic analysis of the highly homologous porcine growth hormone [107]. Using an ELISA (*E*nzyme-*L*inked *I*mmunosorbent *A*ssay)-based binding assay, which measures the affinity of the mutant hGH for its recombinantly derived receptor, researchers discovered that swap mutations that disrupted binding were found to map within close proximity on the three-dimensional model, even though the residues changed within each subset were usually distant in the primary sequence [108]. By this analysis, three discontinuous polypeptide determinants (the loop between residues 54 and 74, the central portion of helix 4 to the C-terminus, and to a lesser extent, the amino-terminal region of helix 1) were identified as being important for binding to the receptor.

A second technique, termed alanine-scanning mutagenesis was then applied. Single alanine mutations (62 in total) were introduced at every residue within the regions implicated in receptor recognition. The alanine scan revealed a cluster of a dozen large side chains that when mutated to alanine exhibited more than a fourfold decrease in binding affinity. Many of the residues that constitute the hGH binding epitope for its receptor are altered in close homologs, such as placental lactogen and the prolactins. The overall correct folding of the mutant proteins was determined by cross-reactivity with a single set of conformationally sensitive mAb reagents. Using the receptor binding determinants identified in these studies, a variant of human prolactin (hPRL) was engineered, to contain eight mutations. This variant had an association constant for the hGH receptor that was increased by more than 10,000-fold [109].

Finally, biophysical studies, including calorimetry, size-exclusion chromatography, fluorescence quenching binding assay [110], and X-ray crystallography [111] revealed the presence of two overlapping binding epitopes (Fig. 3) on growth hormone, through which it causes dimerization of two membrane-bound receptors to induce its effect. The crystal structure confirmed both the 1:2 hormone-to-receptor complex structure and the interface residues identified by the scanning mutagenesis mapping technique. These results gave an early indication that the homolog and alanine-scanning mutagenesis techniques should be generally useful starting points in helping to identity amino acid residues important to any protein–protein interaction [112] and that these techniques have potential to provide essential information for rational drug design.

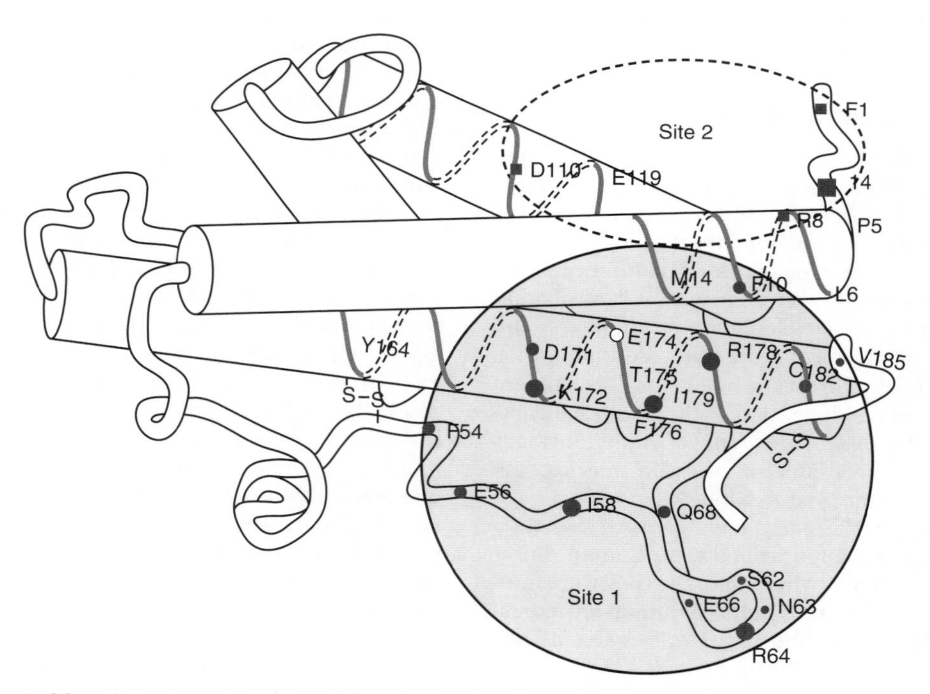

Figure 3. Map of alanine substitution in hGH disrupt binding of hGHbp at either Site 1 or Site 2. The two sites are generally delineated by the large shaded circles. Residues for which alanine mutations reduce Site 2 binding are shown: ▪, 2- to 4-fold; ▪, 4- to 10-fold; ■, 10- to 50-fold; and ■, >50-fold. Sites where alanine mutations in Site 1 cause changes in binding affinity for the hGHbp using an immunoprecipitation assay are shown: •, 2- to 4-fold reduction; •, 4- to 10-fold reduction; ●, >10-fold reduction; and ●, 4-fold increase.

3.4. Future Directions

In an intriguing example of what might be termed *reverse* small molecule design, randomized mutagenesis techniques also have been directly applied to the ever-growing problem of antibiotic resistance. Bacterial resistance to increasingly complex antibiotics has become widespread, severely limiting the useful therapeutic lifetime of most marketed antimicrobial agents [113–115]. Using a mutagenesis technique that randomizes the DNA sequence of a short stretch (3–6 codons) of a gene, followed by determination of the percentage of functional mutants expressed from the randomized gene, localization of the regions of the protein critical to either structure or function can be determined. Application of this technique to TEM-1 β-lactamase (the enzyme responsible for bacterial resistance to β-lactam antibiotics such as penicillins and cephalosporins) over a 66-codon stretch revealed that the enzyme is extremely tolerant of amino acid substitutions: 44% of all mutants function at some level, and 20% function at the level of the wild-type enzyme [116]. The region identified as most sensitive to substitution are located either in the active site or in buried positions that likely contribute to the core structure of the protein (Fig. 4). Such a library of functional mutant β-lactamases could, in theory, be used to simulate multiple next generations of natural mutations, but at an accelerated pace. Screening of new synthetic β-lactams against such a mutant library might then be used to discover compounds with the potential for increased useful therapeutic lifetime. The art of rDNA site-directed mutagenesis, although advancing rapidly, is still limited to the repertoire of the 20 natural amino acids encoded by DNA. To effect more subtle changes in proteins, such as increased or decreased acidity, nucleophilicity, or hydrogen-bonding characteristics without dramatically altering the size of the residue and without affecting the overall tertiary structure, it has been proposed that site-directed mutagenesis using unnatural amino acids might offer the needed advantages. In the past, such changes were accomplished semisynthetically on chemically

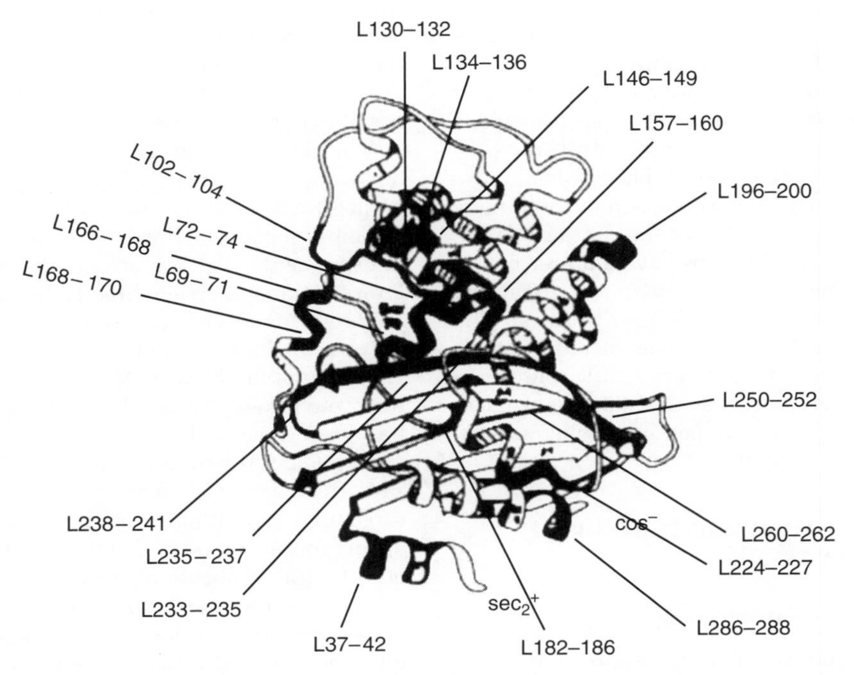

Figure 4. Position of random libraries on a ribbon diagram of the homologous *S. aureus* β-lactamase. Dark regions correspond to the position of random libraries. Lines point to the position of individual libraries.

reactive residues such as Cys. However, methodology for carrying out such mutations recombinantly has been successfully used. The key requirements for successful site-specific incorporation of (un)natural amino acids include (1) generation of an amber (TAG) "blank" codon in the gene of interest at the position of the desired mutation, (2) identification of a suppressor tRNA that can efficiently translate the amber message but that is not a substrate for any endogenous aminoacyl-tRNA synthetases, (3) development of a method for the efficient acylation of the $tRNA_{CCA}$ with novel amino acids, and (4) availability of a suitable *in vitro* protein synthesis system to which a plasmid bearing the mutant gene or corresponding mRNA and the acylated $tRNA_{CCA}$ can be added. These requirements are individual hurdles that have been crossed in the course of development of this technology, although there is a room for further development. Advances in nucleic acids synthesis have ensured a rapid access to aminoacylated CCA for semisynthesis of tRNAs bearing amino acids of interest [117–119]. Furthermore, a recent development of wholly *in vitro* translation method promises to dramatically increase the array of unnatural amino acids that can be substituted into proteins [120]. An early successful demonstration of this methodology involved replacement of F66 with three phenylalanine analogs in RTEM β-lactamase and subsequent determination of the kinetic constants k_{cat} and K_m of the mutants [121]. Subsequent applications have centered on the critical issue of the introduction of unnatural amino acid replacements into proteins. The artificial residues can probe effects on stability and folding governed by subtle changes in hydrophobicity and residue side-chain packing to a degree not possible using the 20 natural amino acids [122,123].

4. GENETICALLY ENGINEERED DRUG DISCOVERY TOOLS

4.1. Reagents for Screening

An increasingly important application of recombinant technology lies not in new protein drug product discovery per se but in the ability to provide cloned and expressed proteins as reagents for medicinal chemistry investigations. The common practice of *in vitro* screening for enzyme activity or receptor binding using animal tissue homogenates (nonhuman, and therefore nontarget) has begun to give way to the use of solid-phase or whole-cell binding assays based on recombinantly produced and isolated or cell surface expressed reagent quantities of the relevant target protein [124,125].

The ability to carry out large-scale, high flux screening of chemical, natural product, and recombinantly or synthetically derived diversity libraries (Nisbet, 1986 #128; Hylands, 1991 #129; Williams, 1989 #130; Waterman, 1990 #131; Moos, 1993 #57; Fellows, 1992 #146; Krebs, 1986 #55; Geysen, 1984 #54; Fodor, 1991 #53; Lam, 1991 #52; Scott, 1992 #51; Houghten, 1991 #50; Simon, 1992 #49; Wolff, 1990 #48; Marks, 1992 #47; Bock, 1992 #46; Ellington, 1990 #44; Tuerk, 1990 #45; Brenner, 1992 #43) also critically depends on reagent availability and consistency. The inherent differences in these potential sources of drug design information, especially from large combinatorially generated libraries requires that assay variations be reduced to the absolute minimum to ensure the ability to analyze data consistently from possibly millions of assay points.

The discovery of the HIV Tat inhibitor Ro 5-3335 and its eventual development as the analog Ro 24-7429 are recent successes from screening chemical libraries using recombinant reagents. Tat is a strong positive regulator of HIV expression directed by the HIV-1 long terminal repeat (LTR) and as such constitutes an important and unique target for HIV regulation, because the *tat trans*-activator protein (one of the HIV-1 gene products) has been clearly demonstrated to regulate expression of the complete genome [126]. Assays to detect inhibitors of *tat* function by screening [127,128], presented immediate opportunities to control a key step in the HIV-1 viral replication process. In this instance, to screen for Tat inhibitors, two plasmids were cotransfected into COS cells: a gene for either Tat or the reporter gene for secreted alkaline phosphatase (SeAP) was put under the control of the HIV-1 LTR promoter [129].

Because Tat is necessary for HIV expression and SeAP expression is under the control of the HIV LTR, an inhibitor of Tat would necessarily lower the apparent alkaline phosphatase activity. This assay was standardized and used in high flux screening to identify structure (**3**), which was then subjected to medicinal chemistry optimization to produce structure (**4**) as the ultimate clinical candidate [130]. Such rapid lead discovery highlights the continued importance of highly directed screening to drug discovery, now better enabled by use of recombinant reagents and techniques.

Even more to the point of human pharmaceutical discovery and design, however, is the issue of species and/or tissue specificities. Sometimes the differences between tissue isolates and recombinant reagent are small; more frequently, however, the sequence homologies and even functional characteristics can vary greatly, providing a distinct advantage in favor of the recombinant protein. When the possibility of achieving subtype specificity, because of either tissue distribution or differential gene expression, determines a particular isoenzyme as a target for selective drug action, it is of obvious importance to be able to test for the desired specificity. Polymerase chain reaction (PCR), an enzymatic method for the *in vitro* amplification of specific DNA fragments, has revolutionized the search for receptor and enzyme subspecies, making whole families of target proteins available for comparative studies [131]. Classic cloning requires knowledge of at least a partial sequence for low stringency screening. This method is unlikely to detect cDNAs corresponding to genes expressed at low levels in the tissue from which the library was constructed. In contrast, the PCR technique can uncover and amplify sequences present in low copy number in the mRNA and offers a greater likelihood of obtaining useful, full-length clones. The selective amplification afforded by PCR can also be used to identify subspecies present in tissue in especially short supply, offering yet another advantage over classic methods. PCR has also been applied to the generation of recombinant diversity libraries of DNA [132], RNA [133,134], and novel chemical diversity "tagged" for detection and amplification [135].

4.2. Combinatorial Biosynthesis and Microbe Reengineering

Many clinically important pharmaceuticals and initial drug candidates are derived from natural sources such as microbes and plants [136]. In most cases, the structural complexity of these drugs precludes chemical synthesis as a practical approach to commercially produce them. This consequently contributes to the dearth of derivatives of these compounds for evaluation as potential drug candidates. Also, slow generation time and low quantities of the drugs from their natural producers are usually major obstacles to contend with.

The tools of recombinant DNA technology are now being elegantly applied in pharmaceutical industries to overcome these and other problems. Several examples of biosynthetic pathway engineering, designed to enhance the production of known compounds or generate novel products in microbes, plants and animals have been recently reported (for recent reviews, see Refs [137–141]. Particularly, major strides have been made in carotenoid and macrocyclic polyketides production [141,142]. Natural product producing organisms have been engineered to produce enhanced levels of the desired compound compared to the wild type [143]. Driven by the realization of the slow growth rate and difficulties in genetic manipulation of natural source-organisms, researchers are making progress in introduction of non-native biosynthetic pathways into genetically amenable organisms [141]. For example, the biosynthetic pathways of 6-deoxyerthronolide B (6-dEB), the macrocylic-aglycon portion of antibiotic erythromycin, has been successfully engineered into *E. coli* [144]. The engineered *E. coli* strain was reported to produce 6-dEB in yields comparable to the high-producing mutant of *Saccharopolyspora erythraea*, the source-organism of 6-dEB [144].

Genetic manipulation of gene clusters within the biosynthetic pathways of natural products has been used to rationally design new and novel products [145]. Similarly, combinatorial genetic approach has been used to dramatically increase the repertoire of pharmaceutically important metabolites produced by natural producing-organisms [146]. Alterations of the erythromycin polyketide synthase genes have been recently demonstrated to generate a mini-library of more than 50 potential pharmaceutically useful macrolides (Fig. 5) [147]. The structural richness produced by this combinatorial biosynthesis approach is of the sort that will be most tasking to even the best of organic chemist. Potential applications of such combinatorial

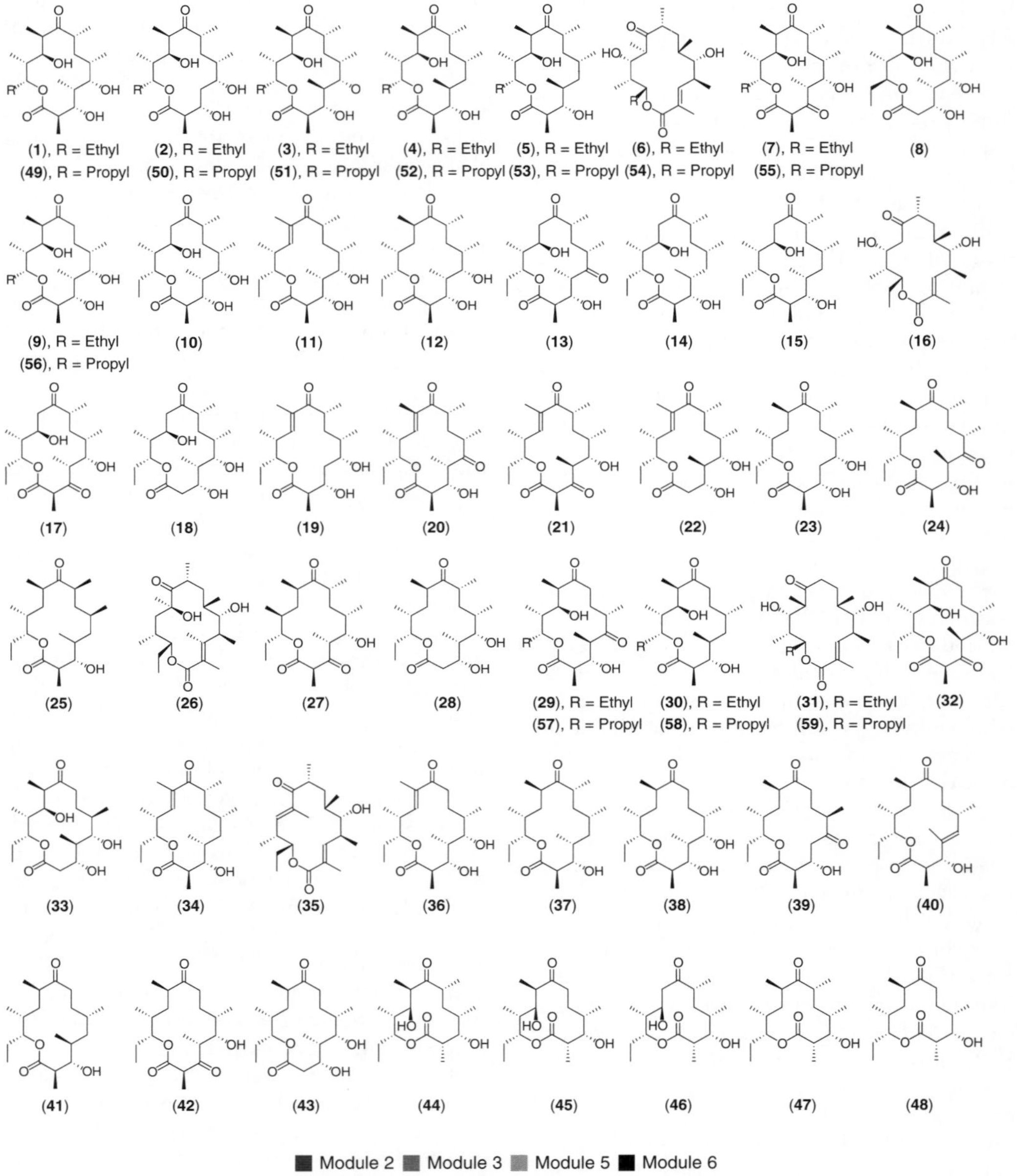

(1), R = Ethyl (49), R = Propyl; (2), R = Ethyl (50), R = Propyl; (3), R = Ethyl (51), R = Propyl; (4), R = Ethyl (52), R = Propyl; (5), R = Ethyl (53), R = Propyl; (6), R = Ethyl (54), R = Propyl; (7), R = Ethyl (55), R = Propyl; (8)

(9), R = Ethyl (56), R = Propyl; (10); (11); (12); (13); (14); (15); (16)

(17); (18); (19); (20); (21); (22); (23); (24)

(25); (26); (27); (28); (29), R = Ethyl (57), R = Propyl; (30), R = Ethyl (58), R = Propyl; (31), R = Ethyl (59), R = Propyl; (32)

(33); (34); (35); (36); (37); (38); (39); (40)

(41); (42); (43); (44); (45); (46); (47); (48)

Module 2 Module 3 Module 5 Module 6

Figure 5. DEBS combinatorial library. Colors indicate the location of the engineered carbon(s) resulting from catalytic domain substitutions in module 2 (red), module 5 (green), module 6 (blue), or modules 1, 3, or 4 (yellow). (This figure is available in full color at http://mrw.interscience.wiley.com/emrw/9780471266945/home.)

biosynthesis methods include lead generation and optimization of existing pharmaceuticals [148].

4.3. SELEX or *In Vitro* Evolution

In vitro selection (or SELEX), first reported in 1990 has developed into a powerful technology for drug discovery, diagnostics, structure–function studies and for creating molecules with novel catalytic properties [133,134,149–153]. SELEX involves the selection of RNA or DNA molecules from random sequence pools that can bind to small or large molecule or perform a chosen function. The small or large molecule binders are called aptamers and they often bind with dissociation constants in the picomolar range. A scheme of SELEX is depicted in Fig. 6. PCR and subcloning techniques play a major role in SELEX.

4.3.1. Aptamers as Drugs The success of aptamers as potential drugs relies on finding solutions to many of the similar issues faced by conventional pharmaceutical agents. The aptamers must have high affinity and specificity to their targets, must be biologically stable and must reach the target molecule. Aptamers generated by SELEX have been shown to have tight binding and a high degree of specificity. For example, an aptamer designed to bind the vascular endothelial growth factor (VEGF) protein has a dissociation constant of about 100 pM [154]. Aptamers have been shown to distinguish between protein kinase C (PKC) isozymes that are 96% identical, reflecting the high level specificity that can be attained by these molecules [155]. The issue of biological stability of aptamers has also been addressed. By incorporating chemically modified nucleotide analogs such as 2′-amine or 2′-fluoro modified pyrimidines can impart stability to nucleases when incorporated into aptamers and increase affinity for the target [154]. These analogs are conveniently enough substrates of T7 RNA polymerase and AMV reverse transcriptase, two of the essential enzymes utilized during the aptamer selection process. Key advantages of aptamers are its apparent lack of immunogenicity and stability that are similar to small molecule drugs, while possessing large enough surface area that allows it to make multiple chemical interactions, which provides it with high degree of binding specificity and tight binding that are reminiscent of antibody-target interactions. However, limitation in terms of cellular uptake and *in vivo* half-life are significant drawbacks as are true for most of oligonucleotide-based technologies.

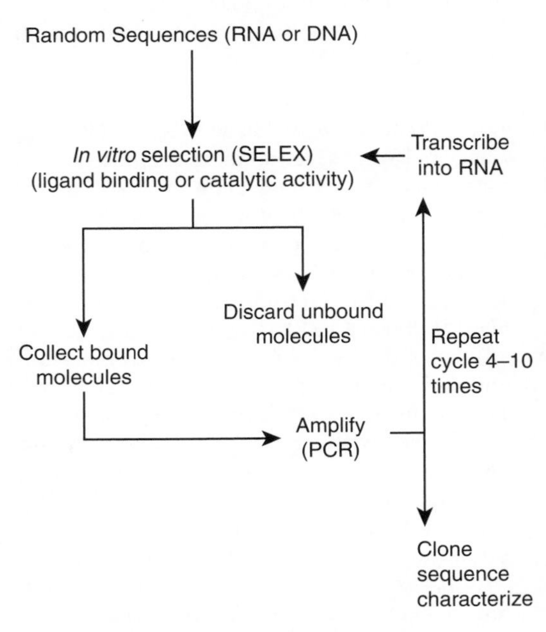

Figure 6. Schematic of SELEX.

The aptamer technology has come a long way as far as realization of therapeutics potential is concerned. It is remarkable that in 15 years since the inception of the technology, an aptamer-based drug Macugen (pegaptanib sodium injection) is approved in 2004 for the treatment neovascular age-related macular degeneration (AMD). This aptamer is designed to bind with high degree of specificity to the isoform 165 of vascular endothelial growth factor protein. Currently, several aptamer molecules are in various stages of clinical trial.

4.3.2. Aptamers as Diagnostics Aptamers rival antibodies in terms of affinity and specificity. Aptamers being significantly smaller and simpler molecules than antibodies make them attractive for diagnostic use. In fact some of them surpass the discriminating abilities of antibodies. For example, an RNA aptamer was generated against theophylline with a K_d of 400 nM, and it showed >10,000 weaker binding to caffeine, which was an order of magnitude higher discrimination than offered by antibodies against theophylline [156].

In vitro evolution or SELEX has proven to be a powerful technology that may have a variety if applications in the development of therapeutics and diagnostics. Chemically modified nucleotides have addressed the concerns about biological stability, but issues related to delivery and mass production still remain as concerns.

4.4. Small Interfering RNA

Small interfering RNA (short interfering RNA, silencing RNA, siRNA) is a new class of about 22 nucleotides double-stranded RNA molecules that are responsible for modulation of gene expression posttranscriptionally, and are endogenous to both plants and animals including humans [157–160]. The basic mechanism involves processing of dsRNAs into siRNAs, about 22 nucleotides in length, by the RNase enzyme Dicer. These siRNAs are then incorporated into a silencing complex called RISC (RNA-induced silencing complex), which in turn identifies complementary sequences in mRNA and silences them. Because, RNA interference is considered to be a robust and natural cellular event, its potential therapeutic applications are enormous. However, as true with all oligonucleotide-based therapeutics delivery of siRNAs is a major challenge, as without modification or help from delivery vehicle their cellular uptake is poor. Recombinant DNA technology can play a significant role in addressing this problem. Although, synthetic siRNAs are commonly used, for viral delivery or when a plasmid encoding a particular siRNA are used for gene silencing recombinant technologies are used.

4.4.1. Vector-Mediated Delivery of siRNA

Gene silencing for therapeutic applications can utilize synthetic siRNAs. However, oligonucleotides-based siRNA delivery suffers from several limitations, such as, high cost of manufacture, nonspecific toxicity caused by transfecting agents, and transient suppression of gene expression. DNA vector-mediated delivery of siRNA with the help of recombinant technology can overcome the problems [161,162]. Viral vector-based system that can scan perform genome-scale loss of function has been reported [163]. There are many reports of delivery of siRNA via viral vector-based systems, although such strategies also suffer from many drawbacks.

4.5. Phage Display

With the identification of new therapeutic targets it has become imperative to identify newer ligands that can bind to them, and one way to identify such molecules is by creating a random library and selecting the members with the desired properties. Coupling recombinant DNA technology with phage biology, the phage display technique has revolutionized the identification of novel peptides that can be used for therapeutic purposes as well as for structural studies such as epitope mapping, identification of critical amino acids responsible for protein–protein interactions, and so on. It also provides an opportunity to physically link genotype with the phenotype that has allowed linking DNA encoding novel functions to be selected directly from complex libraries [164,165].

4.5.1. Preparation of Phage Display Libraries

Phage M13 and other members of the filamentous phage family have been used as expression vectors in which foreign gene products are fused to the phage coat proteins and are displayed on the surface of the phage particle. The probability of finding a ligand in a random peptide library (RPL) is proportional to the affinity of the ligand for the selector molecule and its frequency of occurrence in the library. The frequency can be enhanced by constructing libraries with large sequence diversity and flexibility of the insert, which can be long and unconstrained with the goal of making multiple contacts with the selector molecule. But the libraries that produce higher affinity ligands are created by deliberately introducing predefined structural features or structural constraints like disulfide bridges.

Furthermore, it has been established that the chances of finding a good ligand increases by screening more libraries. Various protein and peptides including complex peptide libraries have been displayed on phage. The phage particles that display the desired molecules are mainly isolated by binding to immobilized target molecules or by affinity chromatography. A wide variety of selectors have been used that ranges from organs in whole animals to cell surface proteins, studies with enzyme, antibody or receptor proteins.

4.5.2. Phage Display Selections Against Purified Proteins Peptides that will selectively bind to purified cell surface receptor proteins or other therapeutically relevant proteins can be isolated from phage display libraries. One of the classes of therapeutically relevant targets is enzymes. Small molecules are traditionally good binders of active site clefts or allosteric regulatory sites that are often buried within the enzyme structure. Phage display libraries can also effectively identify enzyme inhibitors. Recently, Hyde-DerRuyscher et al. could obtain peptide inhibitors against six of the seven different enzyme classes that they tested [166]. Additionally, they demonstrated that the isolated peptides can also be useful for identifying small molecules inhibitors of the target enzyme in high-throughput screens.

Selective inhibition of individual proteases of the coagulation cascade may have enormous therapeutic value but is extremely difficult. Dennis et al. isolated a peptide that noncompetitively inhibits the activity of serine protease factor VIIa (FVIIa), a key regulator of the coagulation cascade with a high degree of specificity and potency [167]. Extensive structure–function characterization showed that the peptide binds to a previously unknown "exosite" that is distinct from the active site. Apparently the inhibition was via an allosteric mechanism.

The advantage of a peptide-based enzyme inhibitor is that it can theoretically sample the entire exposed surface of the enzyme with more contact points than a small molecule inhibitor. They can therefore act as both drug molecules and as reagents in drug discovery.

4.5.3. Cell-Specific Targeting Peptides that can home in on specific cell types have enormous potential both in basic research and in therapeutic applications [168]. Therapeutic applications include identification of specific cell surface proteins in diseased cells, as diagnostic agents, as gene delivery agents.

The selection of cell targeting peptides may involve targeting a known cell surface molecule or screening whole cells without *a priori* knowledge of the chemical nature of the cell surface. The latter approach can lead to identification of hitherto unknown cell surface receptors or target molecules.

Selection from a phage display library of peptides that will bind to a known target has its advantage in that specific ligands can be identified in the presence of a complex milieu of biomolecules. Sometimes this may also pose problem during selection because of the chemical complexity of available targets leading to nonspecific binding.

One of the first peptides to be isolated from a selection based on whole cells is an antagonist for the unknown plasminogen activator receptor. This selection was carried out on transfected COS-7 and SF-9 insect cells that overexpressed the receptor. Another example is isolation of ligands for the thrombin receptor. Interestingly, the selection was executed on human platelet cells that naturally express the thrombin receptor. To identify the specific ligand for the receptor a known agonist for the thrombin receptor was used to elute the bound phages. The selection led to the identification of two peptides one of which acted as an agonist with the ability to activate the thrombin receptor. The other peptide could immunoprecipitate the thrombin receptor form membrane extract and promoted aggregation of platelets, thereby acting as a true antagonist.

4.5.4. *In Vivo* Phage Display In a pioneering study, Pasqualini and Ruoslahti demonstrated for the first time an *in vivo* phage display can specifically target organs [169]. Two pools of peptide-phage libraries were injected into mice via the tail vein. The mice were sacrificed and the two targeted organs, brain and kidney, were isolated. The isolated tissues were homogenized and the bound phages were reamplified in *E. coli*. Several dominant peptide motifs were identified. The isolated phages were injected back into the mice and the selectivity of the phages was measured. The brain-targeting phage accumulated in the brain tissue 4–9 times better compared to the kidney-targeted phage. A synthetic peptide inhibited the uptake of the corresponding phage in the brain showing the specificity of the isolated peptide.

In another exciting study of *in vivo* selection of phage display libraries was used to isolate peptides that home exclusively to tumor blood vessels [170]. Phage libraries were injected into the circulation of nude mice bearing breast carcinoma xenografts. Three main peptide motifs were identified that targeted the phages into the tumors. One of the motifs, CDCRGDCFC (embedded RGD motif), homed into several tumor types (including carcinoma, sarcoma, and melanoma) in a highly selective manner and the targeting was specifically inhibited by the cognate peptide. By conjugating doxorubicin a common chemotherapeutic agent, to the identified peptides, the efficacy of doxorubicin was increased and the toxicity was markedly decreased.

Phage display has also been effectively used to identify peptides that will achieve gene delivery by targeting cell surface receptors to augment receptor-mediated gene delivery. This has tremendous potential for gene therapy where a major hurdle is delivering the gene [171,172].

The above examples demonstrate vividly the power of recombinant technology for developing techniques and agents that directly advances medicinal chemistry.

4.6. Reagents for Structural Biology Studies

In combination with molecular genetics, structural biology also has used physical techniques—nuclear magnetic resonance (NMR) spectroscopy and X-ray crystallography—to its advantage in the study of proteins as drug targets, models for new drugs, and discovery tools [173]. These two techniques can be used independently, or in concert, to determine the complete three-dimensional structure of proteins. Recent advances in NMR techniques, especially multidimensional heteronuclear studies, offer dramatic improvements in spectral resolution and interpretation [174,175]. Identification of differences in the results from comparative studies on the same protein can reveal important structural or dynamic information [176], possibly relevant to the design of synthetic ligands or inhibitors. Inclusion of such structural biology results into the more traditional synthesis-driven discovery paradigm has become a recognized and important component of drug design [177,178].

The variety of studies undertaken using these structural biology techniques spans the range of proteins of interest, from enzymes and hormones to receptors and antibodies. Recombinantly produced reagents (accessible as either purified, soluble proteins, or cell surface expressed, functional enzymes and receptors) with potential application to drug discovery fall into a number of general categories: enzymes (with catalytic function), receptors (with signal transduction function), and binding proteins (with cellular adhesion properties). Rather than exhaustively catalog further examples, the next sections will highlight instances in which combinations of directed specific assays and structural biology studies have aided in nonprotein drug discovery.

4.7. Enzymes as Drug Targets

A large number of enzymes have been cloned and expressed in useful quantities for biochemical characterization. The advent of rational drug design paradigms, in particular the methodology surrounding mechanism-based enzyme inhibition [179] and the market success of various enzyme inhibitors such as captopril, have made enzymes of all types more reasonable laboratory tools and therefore accessible targets for medicinal chemistry efforts. Many enzymes linked to pathologies or known to regulate important biochemical pathways have been cloned for subspecies differentiation and/or access to human isotypes. Also, important advances have been made in the molecular biology and target validation of various classes of potential target enzymes including protein kinase C [180,181], cyclooxygenase (COX) isozymes, phosphodiesterase [182–184] and the phospholipase A_2 [185–189] families.

The rational basis of enzyme–inhibitor interactions, especially to predict or explain specificity, is among the most intensely active areas of structural biology. One of the most studied therapeutic targets is dihydrofolate reductase (DHFR), an enzyme essential for growth and replication at the cellular level. Inhibitors of DHFR, most notably the antifolates methotrexate (MTX) and trimethoprim (TMP), are used extensively in the treatment of neoplastic and infectious disorders. Some of the observed species selectivities for these inhibitors have been explained in terms of distinctive structural differences at the binding sites of the chicken and *E. coli* enzymes [190], but some of the conclusions made based on the enzyme–inhibitor binding interaction have been challenged by a crystal structure of human recombinant DHFR complexed with folate, the natural substrate [191]. Comparisons of the conformations of the conserved human and mouse DHFR side chains revealed differences in packing, most noticeably the orientation of F31. Site-directed mutagenesis studies confirmed the importance of this observation. The mutant F31L (human F to *E. coli* L mutation) gave equivalent K_i values for inhibition by TMP, but gave a 10-fold increase in K_m for dihydrofolate [192]. Similar results were found for the F31S mutant, for which there was also a 10-fold increase in K_m for dihydrofolate and a 100-fold increase in K_d for MTX. The F34S mutant, however, showed greater differences: a threefold reduction in K_m for NADPH (nicotinamide adenine dinucleotide phosphate, reduced form), a 24-fold increase in K_m for dihydrofolate, a 3-fold reduction in k_{cat} and an 80,000-fold increase in K_d for MTX, suggesting that phenylalanines 31 and 34 make different contributions to ligand binding and catalysis in human DHFR [193]. These results helped to pinpoint major differences among DHFRs of various species and thus suggest ways to design new and more species-specific inhibitors that would preferentially target pathogen versus host DHFR. Such compounds would be expected to be more potent chemotherapeutics exhibiting less toxicity in humans. The design and refinement of inhibitors of *E. coli* thymidilate synthetase such as (**5**) attests to the viability and potential cost-effectiveness of this rational design approach [194–197].

In contrast to the DHFR investigations for which the goal is refinement, problems in *de novo* design of inhibitors require more fundamental help, specifically the availability of the target enzyme in quantity for screening. The ability of rDNA technology to expedite access to quantities of a specific enzyme in a situation in which some indication of specificity would eventually be required of the final inhibitor is no where more evident than in the case of the retroviral aspartic HIV-1 protease (HIV-1 PR) [198]. From among the multitude of potential points of intervention into viral replication of the HIV-1 genome, this enzyme was identified as a viable target for anti-AIDS drugs because mutation of the active site aspartic acid (D25) effectively prevents processing of retroviral polyprotein, producing immature, noninfective virions. In addition to the residues DTG at positions 25 to 27, mutations within the sequence GRD/N (positions 86 to 88 in HIV-1 PR)—a highly conserved domain in the retroviral proteases but not present in cellular aspartic proteases—were found to be completely devoid of proteolytic activity, potentially pinpointing a site critical for design of specific inhibitors capable of recognizing the viral, but not the host proteases.

The search for important tertiary structural differences between HIV-1 PR and known eukaryotic proteases began by determination of the X-ray crystal structure (Fig. 7) of recombinantly expressed material at 3 Å resolution [199]. Subsequent crystallographic studies on both synthetic (at 2.8 Å) and recombinantly expressed (at 2.7 Å) material helped locate side chains and resolved some ambiguities in the dimer interface region [200,201]. From this information, a model of the substrate binding site was proposed [202]. Far more useful for inhibitor design purposes, complexes of four structurally distinct inhibitors bound to HIV-1 PR were solved [203–206], from which a generalized closest contact map (Fig. 8) was developed [198]. With the functional role and tertiary structure of the protease determined, additional studies with both recombinant and synthetic material have

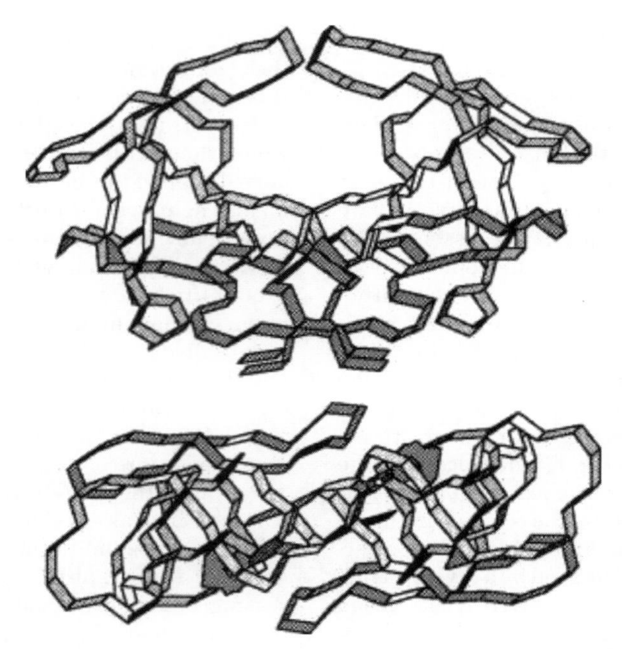

Figure 7. The structure of native HIV-1 protease drawn as a ribbon connecting the positions of α-carbons. The upper structure, in which the pseudo-twofold axis relating one monomer to the other is vertical and the plane of the page, represents a view along the substrate binding cleft. The lower structure is a top view, with the pseudo-twofold axis perpendicular to the page.

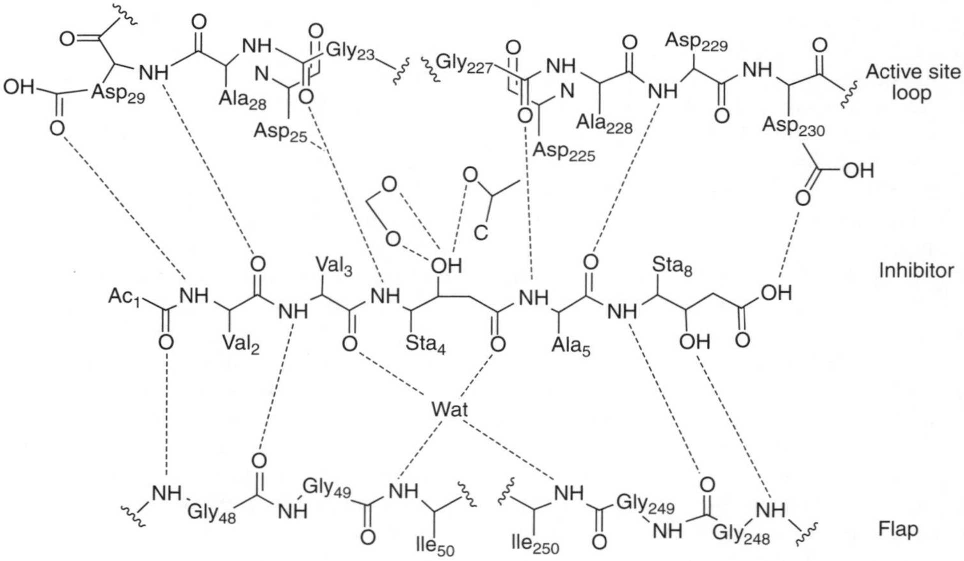

Figure 8. Hydrogen bonds between a prototypical aspartic protease inhibitor (acetylpepstatin) and HIV-1 protease. The residues are labeled at the C-β position (C-α for glycine). The residues labeled 25–50 are from monomer A, those labeled 225–250 are from monomer B, and those labeled 1–6 are with acetyl pepstatin.

yielded automated robotics assays for screening of chemical libraries, fermentation broths, and designed inhibitors using HIV-1 PR cleavage of synthetic pseudo-substrates. Peptide sequences derived from specific retroviral polyprotein substrates and inhibition by pepstatin and other renin inhibitors identified (S/T)P_3P_2(Y/F)P as a consensus cleavage site for HIV-1 PR. One such inhibitor, SGN(Fψ[CH_2N]P)IVQ, has been used as an affinity reagent for large-scale purification of recombinant HIV-1 PR [207], while Ac-TI(nLψ[CH_2NHl-nL]QR-NH_2 was used in the cocrystallization studies mentioned above. From among the large numbers of peptides identified as HIV-1 PR inhibitors, only a limited number have been shown to inhibit effectively viral proteolytic processing and syncytia formation in chronically infected T-cell cultures [208,209]. The most advanced peptidomimetic compound is structure (**6**), which both inhibits HIV-1 PR and exhibits effective and noncytotoxic antiviral activity in chronically infected cells at nanomolar concentrations [210]. However, as with other peptidomimetic structures such as inhibitors of another aspartyl protease, renin [211], their transformation into potential drugs will require additional synthetic work. The short interval from identification of the enzyme as a target from among the possibilities presented by the HIV-l genome to accessing material for assay and structural purposes has obviously hastened the determination of the viability of HIV-1 PR inhibitors as AIDS therapeutics and also has provided an excellent example of structurally driven rational drug design.

The search for a common mechanism of action of the immunosuppressive drugs cyclosporin (CsA) (**7**) and FK-506 (**8**) highlights another possibility, in which the drug was discovered by screening in cellular or *in vivo* models, but the exact mechanism or site of action is unknown [212]. Using the active molecules, cyclophilin (CyP) and the FK binding protein (FKBP) were identified as the specific receptors for cyclosporin and FK-506, respectively. These binding proteins were discovered to be distinct and inhibitor-specific *cis–trans* peptidyl-prolyl isomerases that catalyze the slow *cis–trans* isomerization of proline peptide bonds in oligopeptides and accelerate the rate-limiting steps in the folding of some proteins [213,214]. The biochemical mechanism of inhibition was first proposed to involve a specific covalent adduct between inhibitor and its rotamase [214], but the hypothesis was soon challenged by evidence that showed that the binding interactions are peptide-sequence specific [215] for both proteins. Using recombinant CyP as the standard, four cysteine-to-alanine mutants (C52A, C62A, C115A, and C161A) were shown to retain full affinity for CsA and equivalent rotamase catalytic activity, indicating that the cysteines play no essential role in binding or catalysis [216]. In the case of FKBP, NMR studies of [8,9-^{13}C]FK-506 bound to recombinant FKBP, wherein the likely mechanism of inhibition also is noncovalent, suggesting that the α-ketoamide of FK-506 serves as an effective surrogate for the twisted amide of a bound peptide substrate [217]. Numerous further NMR, X-ray crystallographic, and computational modeling studies have been carried out on both CsA/CyP and FK-506/FKBP complexes to attempt to fully determine a structural basis for activity [218,219].

The exact signal transduction mechanism that triggers the immunosuppressive response in T cells, however, remained unknown until an elegant set of experiments identified calcineurin, a calcium- and calmodulin-dependent serine/threonine phosphatase, and a complex of calcineurin with calmodulin as the binding targets of the immunophilin–drug complexes [220]. The immunosuppressant, displayed by the immunophilin protein, then effectively functions as the critical element that binds the pentapartite complex together, causing inhibition of the phosphatase (Fig. 9). The complex seems also to exert two subsequent effects: halting DNA translation in T-lymphocyte nuclei and inhibition of IgE-induced mast cell degranulation. These *in vitro* observations must necessarily be confirmed by further *in vivo* work, but the elucidation of these molecular-level mechanisms will allow the development of more highly tailored and potent immunosuppressive agents [221]. Toward this goal, experiments that sought to probe the relationship between the immunosuppressive effect and the common

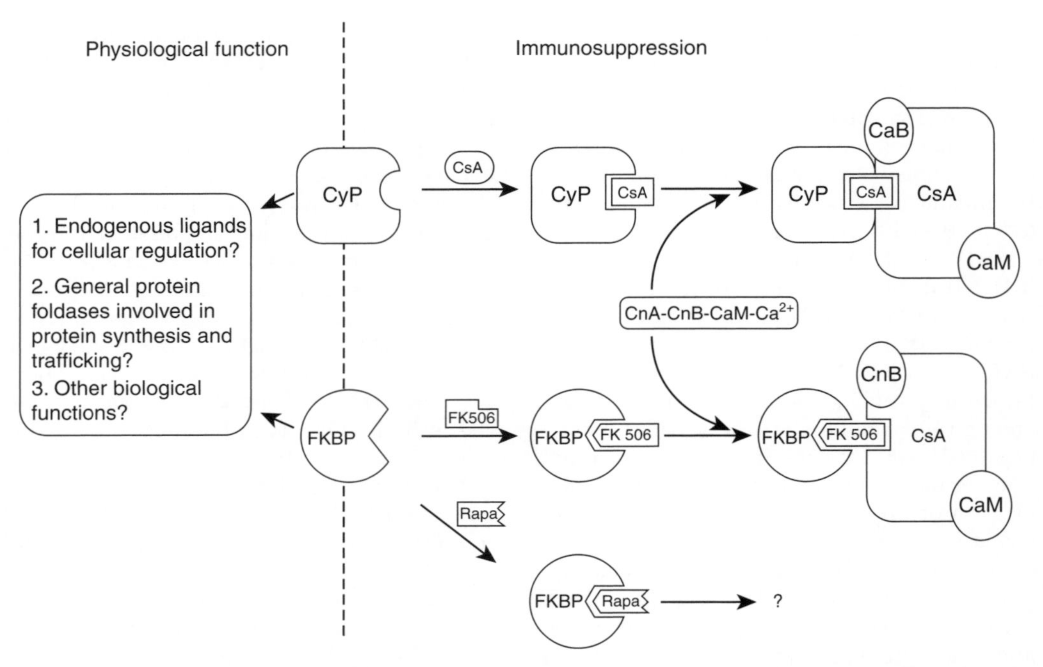

Figure 9. Schematic representation of immunosuppressant–immunophilin complex interactions. CyPs bind CsA to form a complex in which both components undergo change in structure. This complex binds to, and inhibits, calcineurin (CnA, A subunit; CnB, B subunit; CaM, camodulin) in a calcium-dependent manner. FK506 binding protein complexes with FK506 or rapamycin (Rapa). FKBP-FK506 also binds calcineurin. The target of FKBP-rapamycin is unknown but is presumed to be different from cacineurin.

side effects, such as hypertension and nephrotoxicity of CsA was designed by Russo et al. [222]. It was identified that the immunosuppressive effect of CsA is unconnected with its hypertensive effect; a vital information that may be of use in design of CsA derivatives devoid of these side effects.

4.8. Receptors as Drug Targets

Even more so than with enzymes, molecular genetics has been primarily responsible for the identification of functional receptor subtypes. Success in the case of cimetidine, a selective H_2-receptor antagonist, made it clear that the design of specific ligands for at least some receptors is a task amenable to medicinal chemistry. The classic tissue binding pharmacological methods that made distinctions on the basis of ligand selectivity have been supplemented, and in most cases supplanted, by further subtyping made possible by cross-hybridization cloning using the known receptor genes. Any studies that profile the *in vitro* receptor subtype specificity of compounds can theoretically help identify potential *in vivo* side effects of the compounds if the association of subtype to effect is known or suspected. At this point, just the indication of a more specific profile with fewer side effects is enough to help choose one compound over another for preclinical development.

One of the earliest examples of the role of molecular biology in discerning receptor subtype roles was the case of the muscarinic cholinergic receptors (MAChRs). The two subtypes M_1 and M_2 had been defined pharmacologically by their affinity, or lack thereof, for pirenzepine and were later confirmed by molecular cloning to be distinct gene products *m1* and *m2*, respectively [223–225]. Three additional muscarinic receptor genes (*m3* to *m5*) were subsequently isolated [226–228]. From this work, a subtype-specific heterologous stable expression system in Chinese hamster ovary (CHO) cells suitable for screening potential subtype-specific ligands was developed. From this assay, pirenzepine, previously thought to bind to *M1* only, was found to have only a 50-fold reduced affinity for *M2*, and an almost equivalent (to *M1*) binding affinity for *M3* and *M4*, suggesting that studies using pirenzepine on tissue homogenate have failed to distinguish adequately among the subtypes [229]. Similar breakthroughs have been realized across the rest of the family of signal transduction G-protein-coupled receptors, because in addition to the muscarinics, the primary structures of the adrenergic (α_1, α_2, β_1, and β_2), serotonergic, tachykinin, and rhodopsin receptors have been determined [230–232]. All of these display the now-familiar homology pattern of seven membrane-spanning domains packed into antiparallel helical bundles (Fig. 10). The exceedingly high homology among the large family of G-protein-coupled receptors also has allowed the development of three-dimensional models of the proteins to aid in drug refinement [233]. For example, mutagenesis studies on the β_2-adrenergic receptor have localized the intracellular domains involved in (1) the coupling of the receptor to G-proteins [234]; (2) homologous desensitization by β-adrenergic receptor kinase (β-ARK) [235], itself cloned and a possible target for downregulation inhibitors [236]; (3) heterologous desensitization bv cAMP-dependent protein kinase [237]; and (4) an

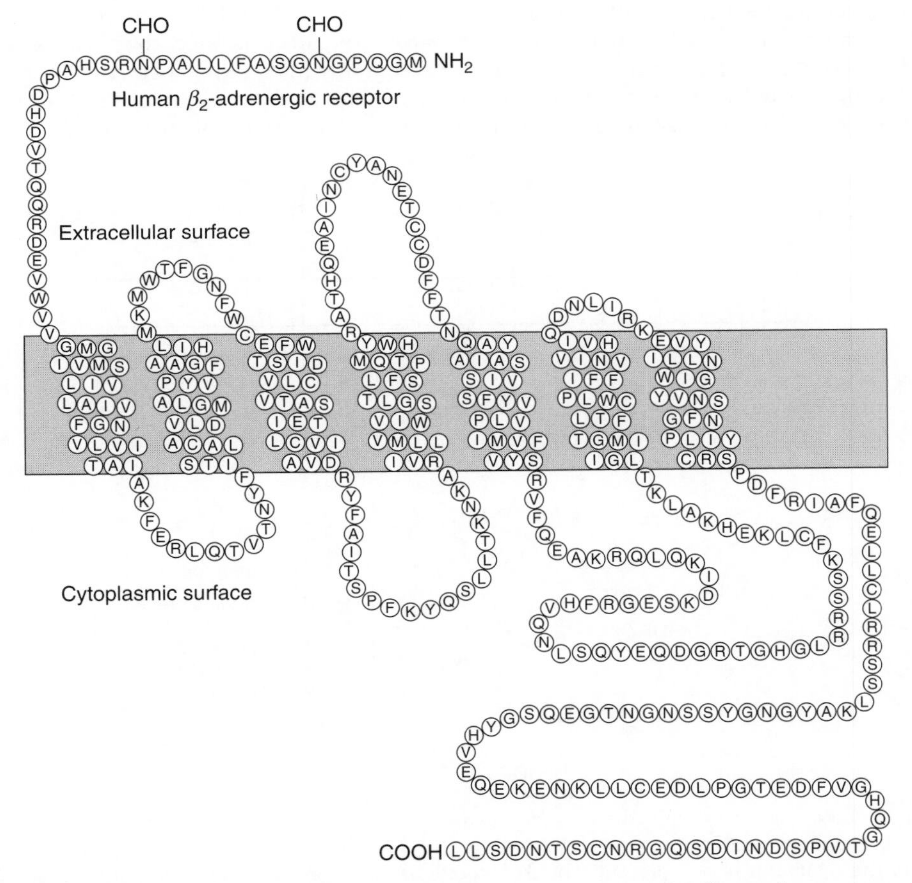

Figure 10. Topographical representation of primary sequence of human β_2-adrenergic receptor, a typical G-protein-coupled receptor. The receptor protein is illustrated as possessing seven hydrophobic regions each capable of spanning the plasma membrane, thus creating intracellular and extracellular loops as well as an extracellular amino terminus and a cytoplasmic carboxyl terminal region.

extracellular domain with conserved cysteine residues implicated in agonist ligand binding [238]. A chimeric muscarinic cholinergic: β-adrenergic receptor engineered to activate adenylyl cyclase (a second messenger system not coupled to MAChR agonism) also has helped identify which intracellular loops may be involved in direct G-protein interactions [239]. The diverse signal transduction functional roles of the many G-proteins to which these receptors are coupled (Fig. 11) also makes them viable drug targets [240].

The complicated biochemical pharmacology of natriuretic peptides, the regulatory system that acts to balance the renin–angiotensin–aldosterone system [241], has been significantly clarified by the cloning of three receptor subtypes, which revealed the functional characteristics of a new paradigm for second messenger signal transduction via guanylate cyclase (GC). The α-atrial natriuretic peptide (α-ANP) receptor (NPA-R) and the brain natriuretic peptide (BNP) receptor (NPB-R) contain both protein kinase and guanylate cyclase domains, as determined by both sequence homologies and catalytic activities, while the clearance receptor (ANP-C) completely lacks the necessary intracellular domains for signal transduction via the guanylate cyclase pathway [242]. This system defines the first example of a cell surface receptor that enzymatically synthesizes a diffusible second messenger system in response to hormonal stimulation [243] (Fig. 12). Data from experiments performed with C-ANP_{4-23} indicates that the clearance receptor (NPC-R) may be coupled to the adenylate cyclase/cAMP signal transduction system through an inhibitory guanine nucleotide regulatory protein [244]. Because the NPs have differential, but not absolute, affinities for their corresponding receptors [245] and because both agonism [246] and antagonism [247] of the GC activity have been demonstrated *in vitro* using

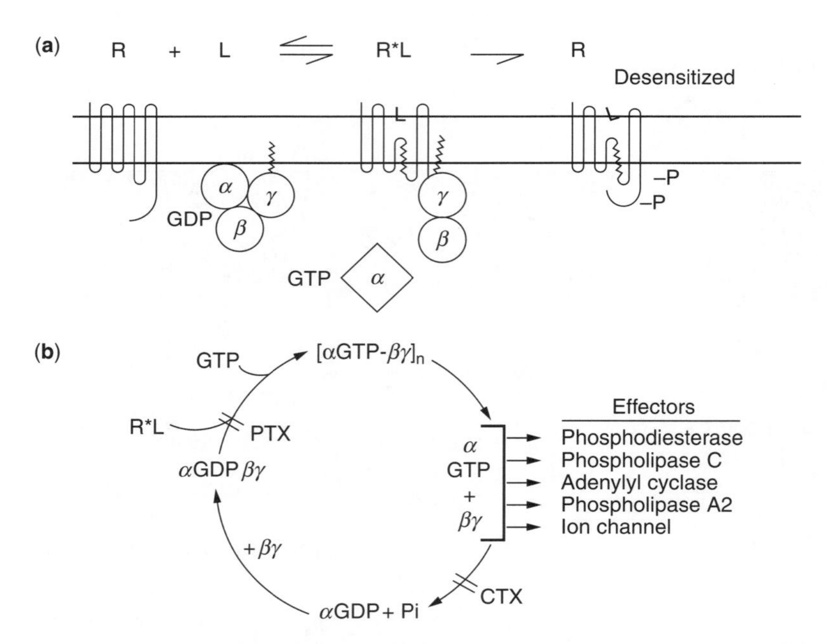

Figure 11. Receptor G-protein-mediated signal transduction. (a) Receptor (R) associates with a specific ligand (L), stabilizing an activated form of the receptor (R^*), which can catalyze the exchange of GTP for GDP bound to the α-subunit of a G-protein. The βγ-heterodimer may remain associated with the membrane through a 20-carbon isoprenyl modification of the γ-subunit. The receptor is desensitized by specific phosphorylation (-P). (b) The G-protein cycle. Pertussis toxin (PTX) blocks the catalysis of GTP exchange by receptor. Activated α-subunits (aGTP) and βγ-heterodimers can interact with different effectors (E). Cholera toxin (CTX) blocks the GTPase activity of some α-subunits, fixing them in an activated form.

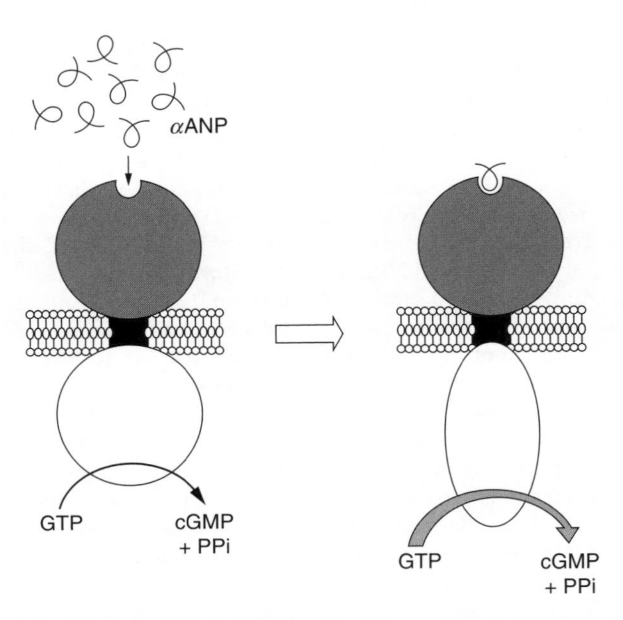

Figure 12. Model for ANP-A and ANP-B receptor function. The unoccupied ANP-A receptor is shown on the left with a basal rate of cGMP synthesis (indicated by a thin arrow). The effect of ligand binding to the amino-terminal extracellular domain is shown on the right. Proposed allosteric modulation of guanylate cyclase by α-ANP is schematically illustrated by a change in shape of the intracellular domain and a thicker arrow to denote an increase in guanylate cyclase-specific activity with greater production of the second messenger cGMP.

ANP analogs, it may be possible to discriminate among the receptor GCs to obtain more subtle structure–activity information for the design of selective NP analogs. Homology between the NP receptors and another guanylate cyclase firmly identified the latter as the elusive heat-stable enterotoxin receptor St (a)-R [248], which aided in the identification of both the biochemically isolated guanylin [249] and the cloned proguanylin [250] versions of the endogenous natural ligand. This system is presumed to play a role in water retention through cGMP modulation of the CFTR chloride ion channel [251].

The number of receptors of biological significance cloned and expressed for further study continues to grow at an exponential rate [252]. These include epidermal growth factor (EGFR), insulin (INSR), insulin-like growth factor-1 (IGF-1R), platelet-derived growth factor (PDGFR) receptors and related tyrosine kinases [253], tumor necrosis factor receptors 1 and 2 [254], subtypes of the $GABA_A$-benzodiazepine receptor complex [255–257], human γ-interferon receptor [258], inositol 1,4,5-triphosphate (IP3) binding protein P400 [259], kianate-subtype glutamate receptor [260], follicle-stimulating hormone receptor [261], multiple members of the steroid (ER, PR, AR, GR, MR), thyroid hormone (TRα and β), and retinoid (RARα, β and γ, and RXRα) receptor superfamily of nuclear transcriptional factors [262–265], multiple members of the interleukin cytokine receptor family [266], and subtypes of the glutamate [267] and adenosine [268] receptor families. The importance of access to human cloned receptors continues to be underscored as receptor binding plays an increasingly critical role in modern drug discovery [269].

To make the case for using cloned human receptors for drug discovery even stronger, dramatic evidence that minor amino acid sequence variations inherent in species variability can produce profoundly different pharmacological effects was recently provided in two instances. In the comparison of rodent versus human analogs of the 5-hydroxytryptamine receptor subtype 5-HT_{1B}, the natural receptors were found to bind 5-HT identically, but they differed profoundly in their affinities for many serotonergic drugs. These striking differences could be reversed by change of a single transmembrane domain residue (T355N), which effectively rendered the two receptors pharmacologically identical [270]. A similar study comparing human to chicken and hamster progesterone receptors showed that the steroidal abortifacient RU486 (**9**) shows antagonist activity in humans but not in the two other species, because of the presence of a glycine at position 575. Both the chicken and hamster receptors have a cysteine at this position, and replacement by glycine (C575G) generated a mutant receptor that could bind RU486. Likewise, mutation of the human receptor (G575C) abrogated RU486 binding [271]. These and many similar findings emphasize the critical importance of the availability of human clones proteins as potential targets for drug action and as sources of structural information in the discovery of potent and selective therapeutic agents.

4.9. Cellular Adhesion Proteins

The understanding of the molecular processes that govern cell localization in various pathological conditions has been significantly expanded because of the cloning and expression of some of the major cellular adhesion proteins, especially in the integrin family, a highly related and widely expressed group of αβ-heterodimeric membrane proteins [272]. The interaction in the antigen-receptor cross-linking adhesion of T-cells mediated by the intercellular adhesion molecule (ICAM-1) and the lymphocyte function-associated molecule (LFA-1) was clarified by the cloning and expression of ICAM-1, the major cell surface receptor for rhinovirus [273]. A soluble form of human ICAM-1 effectively inhibits rhinovirus infection at nanomolar concentrations [274]. As the pivotal interaction in the adhesion of leukocytes to activated endothelium and tissue components exposed during injury (Fig. 13), inhibition of ICAM-l/LFA-1 binding represents a potential prime intervention point for new anti-inflammatory drugs.

Other members of the integrin family have been also successfully cloned [275]. The availability of the individual members of this heterodimer superfamily—characterized by gross similarities in structure (Fig. 13), function, and in some cases avidity for RGD-containing peptides—will allow their individual roles in specific disease pathophysiologies to be ascertained and provide the means to develop integrin-specific antagonists for a multitude of utilities. Of importance to antithrombotic drug discovery efforts, the integrin $\alpha_{IIb}\beta3$ also known as $GPII_bIII_a$ the platelet fibrinogen receptor [276], was successfully expressed as the functional heterodimer, showing that prior association of the endogenous subunits is necessary to produce the cell surface complexes [277]. Surface expression of $GPII_bIII_a$ is the common endpoint in platelet

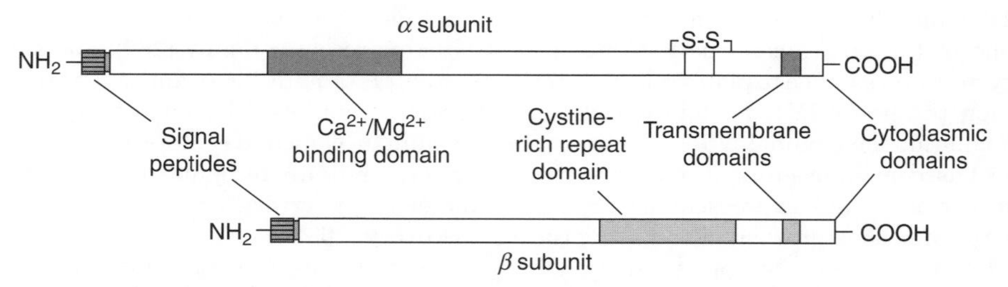

Figure 13. General polypeptide structure of integrins. The α-subunit of the integrins is translated from a single mRNA, and in some cases, it is processed into two polypeptides that remain disulfide bonded to one another. The α-subunit and the β-subunit contain typical transmembrane domain that is thought to traverse the cell membrane and bring COOH-termini of the subunits into the cytoplasmic side of the membrane. The α-subunit contains a series of short-sequence elements homologous to known calcium binding sites in other proteins; the β-subunit is tightly folded by numerous intrachain disulfide bonds.

activation, initiating the platelet–platelet cross-linking via fibrinogen that is responsible for thrombus formation. A number of compounds, including disintegrin snake venoms, RGD peptides and organic mimetics, have been shown to inhibit thrombus formation and platelet aggregation in animal and human clinical trials [278,279].

Lymphocyte and neutrophil trafficking, the first step in the development of an inflammatory response, is known to occur via specific cell surface receptors and ligands, which match the inflammatory cell to the right target. Different receptors and ligands are expressed at different time points during inflammatory processes, from seconds to hours. These protein recognition signals—previously termed homing receptors (HR) and now uniformly called selectins—are membrane-bound proteins, which target circulating lymphocytes to specialized targets provide one such opportunity for intervention [280,281]. Molecular cloning of the murine HR now called L-selectin revealed that the receptor contains a lectin (carbohydrate binding) domain that is responsible for the binding event [282]. The selectin family consists of three cell surface receptors that share this affinity for carbohydrate ligands, specifically the tetrasaccharide sialyl Lewis X (**10**). The carbohydrates are likely displayed at multiple *O*-glycosylation sites on mucin-like glycoproteins such as Spg50, a novel endothelial ligand for L-selectin discovered by a combination of biochemical isolation, sequencing, and cloning techniques [280]. The carbohydrate itself, however, offers a viable starting point for drug design as a structural lead for both carbohydrate analogs and noncarbohydrates [283,284]. Molecular modeling of E-selectin based on antibody mapping and homology to mannose binding protein [285] also suggests drug design possibilities based on proposed analogous structural interactions of the selectin with structure (**10**) [286]. Inhibition of selectin-mediated cellular trafficking at specific times might help break a spiraling acute inflammatory cycle, allowing control of acute inflammatory processes, such as shock and adult respiratory distress syndrome (ARDS), and helping in the management of integrin-mediated inflammatory processes that follow [287].

5. FUTURE PROSPECTS

Molecular genetics is only now beginning to identify new targets for drug action. For example, regulation of inducible or tissue-specific gene expression has been an obvious but elusive target for pharmacological intervention [288–290]. The tools to monitor such events are now available, as in the case of the low-density lipoprotein receptor, for which tissue-specific upregulation of receptor population may successfully compete with other cholesterol-lowering agents [291,292]. In parallel, another genetic marker for atherosclerotic disease, lipoprotein(a), is a target for selective expression downregulation [293]. An even more direct method to interfere with gene expression is selectively to bind the gene using sequence-specific recognition elements that prohibit transcription. Antisense oligonucleotides are the first sequence-directed molecules designed to inhibit protein expression at the level of translation of mRNA into the undesired protein [294]. The first FDA approved drug using antisense technology is Vitravene, introduced by Isis pharmaceuticals in 1998 for local treatment of CMV retinitis in AIDS patients [295]. Genasense, another antisense-based drug, designed to block the production of Bcl-2 protein, a protein overexpressed in most cancer cells, is in late-stage clinical trials as an adjunct cancer treating drug [296]. The ability to test for inhibition of gene expression or to measure the effects of species-specific agents against relevant pharmacological targets in animal models has also been advanced by the development and use of transgenic [39] or engineered gene knockout animals, such as the CFTR-defective mouse for cystic fibrosis [297], in screening and evaluation procedures.

The power of molecular genetics to provide unique and valuable tools for drug discovery is beginning to be exploited. The prospects for uncovering the molecular etiology of a disease state or for gaining access to a disease-relevant target enzyme or receptor are already being realized. The recent successful mapping of human genome [298,299] further promises better chances of rationally intervening in disease states at previously inaccessible or unknown points.

The development of recombinant DNA technology into a fully integrated component of the drug discovery process is inevitable [300,301]. In 1987, Kornberg remarked that "the two cultures, chemistry and biology, [are] growing further apart even as they discover more common ground" [302]. However, the broad area of drug development might qualify as one such meeting place for medicinal chemistry and molecular biology where the trend is reversing. The application of genetic engineering techniques to biochemical and pharmacological problems will facilitate the discovery of novel therapeutics with potent and selective actions. There is little doubt among medicinal chemists that effective collaboration between chemistry and biology is not only needed but is actually growing in importance in drug design [303,304]. The eventual extent of the impact that molecular biology will have on the drug discovery process is, and will be for some time, unknown. However, the reality of recombinant protein therapeutics offers the assurance that this same technology, in conjunction with structural biology, computer-assisted molecular modeling, computational analysis, and medicinal chemistry [305], will help make possible better therapies for those diseases already controllable and new therapies for diseases never before treatable.

REFERENCES

1. Liebenau J. In: Hansch C, Sammes PG, Taylor JB,editors. Comprehensive Medicinal Chemistry. Vol. 1. Oxford: Pergamon Press; 1990. 81–98.
2. Burger A. In: Hansch C, Sammes PG, Taylor JB,editors. Comprehensive Medicinal Chemistry. Vol. 1. Oxford: Pergamon Press; 1990. p 1–5.
3. Sneader W. In: Hansch C, Sammes PG, Taylor JB,editors. Comprehensive Medicinal Chemistry. Vol. 1. Oxford: Pergamon Press; 1990.
4. Sterbach LH. Prog Drug Res 1978;22:229–266.
5. Ganellin CR. In: Bindra JS, Lednicer D,editors. Chronicles of Drug Discovery. Vol. 1.New York: John Wiley & Sons Inc.; 1982. p 1–38.
6. Ondetti MA, Cushman DW. J Med Chem 1981;24:355–361.
7. Seydel JK. In: Mutschler E, Winterfeldt E, editors. Trends in Medicinal Chemistry. New York: VCH Verlagsgesellschaft; 1987. p 83–103.
8. Henwood JM, Heel RC. Drugs 1986;36: 429–454.
9. Duncia JV, Carini DJ, Chiu AT, Johnson AL, Price WA, Wong PC, Wexler RR, Timmermans PBMWM. Med Res Rev 1992;12:149–191.
10. Freter KR. Pharm Res 1988;5:397–400.
11. Lowe JA, Hobart PM. Annu Rep Med Chem 1983;18:307–316.
12. Venuti MC. Annu Rep Med Chem 1990;25: 201–211.
13. Harvey A. Trends Pharmacol Sci 1991;12: 317–319.
14. Venuti MC. In: Friedmann T,editor. Molecular Genetic Medicine. Vol. 1.New York: Academic Press, Inc.; 1991. p 133–168.
15. Szkrybalo W. Pharm Res 1987;4:361–363.
16. Cometta S. Arzneim Forsch Drug Res 1989;39:929–934.
17. Roush W. Antisense aims for a renaissance. Science 1997; 1192–1193.
18. Swetly P. In: Martin YC, Kutter E, Austel V, editors. Modern Drug Research: Path to Better and Safer Drugs, Medicinal Research Series. Vol. 12.New York: Marcel Dekker, Inc.; 1989. p 217–241.
19. Adams SP. In: Hansch C, Sammes PG, Taylor JB,editors. Comprehensive Medicinal Chemistry. Vol. 1.Oxford: Pergamon Press; 1990. p 409–454.
20. Johnson IS. Science 1983;219:632–637.
21. Pavlou AK, Reichert JM. Recombinant protein therapeutics—success rates, market trends and values to 2010. Nat Biotechnol 2004;22: 1513–1519.
22. Reichert JM. Trends in US approvals: new biopharmaceuticals and vaccines. Trends Biotechnol 2006;24:293–298.
23. Tsuji K, Tsutani K. Approval of new biopharmaceuticals 1999-2006: comparison of the US, EU and Japan situations. Eur J Pharmaceut Biopharmaceut 2008;68:496–502.
24. Hentschel C. In: Copsey DN, Delnatte SYJ, editors. Genetically Engineered Human Therapeutic Drugs. New York: Stockton Press; 1988. p 3–6.
25. Gordon SL. In: Copsey DN, Delnatte SYJ, editors. Genetically Engineered Human Therapeutic Drugs. New York: Stockton Press; 1988. p l37–146.
26. Bost PGB, Jouanneau A. In: Hansch C, Sammes PG, Taylor JB,editors. Comprehen-

sive Medicinal Chemistry. Vol. 1.Oxford: Pergamon Press; 1990. p 455–479.

27. Hutchinson CR. Med Res Rev 1988;8:557–567.
28. Hutchinson CR, Borell CW, Otten SL, Stutzmann-Engwall KJ, Wang Y. J Med Chem 1989;32:929–937.
29. Katz L, Hutchinson CR. Annu Rep Med Chem 1992;27:129–138.
30. Oksman-Caldentey K-M, Inzé D. Plant cell factories in the post-genomic era: new ways to produce designer secondary metabolites. Trends Plant Sci 2004;9:433–440.
31. Olano C, Lombó F, Méndez C, Salas JA. Improving production of bioactive secondary metabolites in actinomycetes by metabolic engineering. Metab Eng 2008;10:281–292.
32. Sato F, Yamada Y. Engineering formation of medicinal compounds in cell cultures. Adv Plant Biochem Mol Biol 2008;1:311–345.
33. Lee SY, Kim HU, Park JH, Park JM, Kim TY. Metabolic engineering of microorganisms: general strategies and drug production. Drug Discov Today 2009;14:78–88.
34. Nisbet LJ, Westley JW. Annu Rep Med Chem 1986;21:149–157.
35. Hylands PJ, Nisbet LJ. Annu Rep Med Chem 1991;26:259–269.
36. Katz L, Donadio S. Polyketide synthesis: prospects for hybrid antibiotics. Annu Rev Microbiol 1993;47:875–912.
37. Williams DH, Stone MJ, Hauck PR, Rahman SK. J Nat Prod 1989;52:1189–1208.
38. Waterman PG. J Nat Prod 1990;53:13–22.
39. Coombes JD. New Drugs from Natural Sources. London: IBC Technical Services; 1992.
40. Cordell GA. Biodiversity and drug discovery—a symbiotic relationship. Phytochemistry 2000;55:463–480.
41. Ortholand J-Y, Ganesan A. Natural products and combinatorial chemistry: back to the future. Curr Opin Chem Biol 2004;8:271–280.
42. Rishton GM. Natural products as a robust source of new drugs and drug leads: past successes and present day issues. Am J Cardiol 2008;101(Suppl 1): S43–S49.
43. Harvey AL. Natural products in drug discovery. Drug Discov Today 2008;13:894–901.
44. Knowles JR. Science 1987;236:1252–1258.
45. Kaiser ET. Angew Chem Inc Ed Engl 1988;27: 913–922.
46. Brannigan JA, Wilkinson AJ. Protein engineering 20 years on. Nat Rev Mol Cell Biol 2002;3:964–970.
47. Livingston DJ. Annu Rep Med Chem 1989;24:213–221.
48. Kubinyi H. Strategies and recent technologies in drug discovery. Pharmazie 1995;50: 647–662.
49. Ross MJ, Grossbard EB, Hotchkiss A, Higgins D, Andersen S. Annu Rep Med Chem 1988;23:111–120.
50. Haber E, Quertermous T, Matsueda GR, Runge MS. Science 1989;243:51–56.
51. Higgins DL, Bennett WF. Annu Rev Pharmacol Toxicol 1990;30:91–121.
52. Haigwood NL, Mullenbach GT, Moore GK, DesJardin LE, Tabrizi A, Brown-Shimer SL, Stauss H, Stohr HA, Paques EP. Prot Eng 1989;2:631–620.
53. Krause J, Tanswell P. Arzneim-Forsch Drug Res 1989;39:632–637.
54. Kohnert U, Rudolph R, Verheijen JH, Jacoline E, Weening-Verhoeff D, Stern A, Opitz U, Martin U, Lill H, Prinz H, et al. Prot Eng 1992;5:93–100.
55. Refino CJ, Paoni NF, Keyt BA, Pater CS, Badillo JM, Wurm FM, Ogez J, Bennett WF. Thromb Haemostas 1993;70:313–319.
56. Lee J, Lee HJ, Shin MK, Ryu W-S. Versatile PCR-mediated insertion or deletion mutagenesis. BioTechniques 2004;36:398–400.
57. Ko J-K, Ma J. A rapid and efficient PCR-based mutagenesis method applicable to cell physiology study. Am J Physiol Cell Physiol 2005;288: C1273–C1278.
58. Harris CJ, Stevens AP. Chemogenomics: structuring the drug discovery process to gene families. Drug Discov Today 2006;11: 880–888.
59. Triggle DJ. Drug discovery and delivery in the 21st century. Med Princ Pract 2007;16:1–14.
60. Capon DJ, Chamow SM, Mordenti J, Marsters SA, Gregory T, Mitsuya H, Byrn RA, Lucas C, Wurm FM, Groopman JE, et al. Nature 1989;337:525–531.
61. Traunecker A, Schneider J, Kiefer H, Karjalainen K. Nature 1989;339:68–70.
62. Byrn RA, Mordenti J, Lucas C, Smith D, Marsters SA, Johnson JS, Cossum P, Chamow SM, Wurm FM, Gregory T, et al. Nature 1990;334: 667–670.
63. Jacobson JM, Lowy I, Fletcher CV, et al. Single dose safety, pharmacology, and antiviral activity of the human immunodeficiency virus (HIV) type 1 entry inhibitor PRO 542 in HIV-infected adults. J Infect Dis 2000;182: 326–329.

64. Kilby JM, Eron JJ. Novel therapies based on mechanisms of HIV-1 cell entry. NEJM 2003;348:2228–2238.
65. Vu JR, Fouts T, Bobb K, Burns J, McDermott B, Israel DI, Godfrey K, DeVico A. AIDS Res Hum Retroviruses 2006;22:477–490.
66. McFadden K, Cocklin S, Gopi H, Baxter S, Ajith S, Mahmood N, Shattock R, Chaiken I. Proteins 2007;67:617–629.
67. Kaushik-Basu N, Basu A, Harris D. Peptide inhibition of HIV-1: current status and future potential. Biodrugs 2008;22:161–175.
68. Rininger JA, DiPippo VA, Gould-Rothberg BE. Drug Discov Today 2000;5:560–568.
69. Gunther EC, Stone DJ, Gerwien RW, Bento P, Heyes MP. Prediction of clinical drug efficacy by classification of drug-induced genomic expression profiles *in vitro*. Proc Natl Acad Sci USA 2003;100:9608–9613.
70. Cattaneo D, Perico N, Remuzzi G. From pharmacokinetics to pharmacogenomics: a new approach to tailor immunosuppressive therapy. Am J Transplant 2004;4:299–310.
71. Lewin DA, Weiner MP. Molecular biomarkers in drug development. Drug Discovery Today 2004;9:976–983.
72. Grant SFA, Hakonarson H. Recent development in pharmacogenomics: from candidate genes to genome-wide association studies. Expert Rev Mol Diagn 2007;7:371–393.
73. Gottlieb B. Human genome variation and pharmacogenetics. Hum Mutat 2008;29: 453–455.
74. Henry CM. Chem Eng News 2001;79:37–42.
75. Issa AM. Personalized medicine and the practice of medicine in the 21st century. McGill J Med 2007;10:53–57.
76. Fosslien E. Ann Clin Lab Sci 1998;28:67–81.
77. Willoughby DA, Moore AR, Colville-Nash PR. COX-1, COX-2, and COX-3 and the future treatment of chronic inflammatory disease. Lancet 2000;355:646–648.
78. Hersh EV, Lally ET, Moore PA. Update on cyclooxygenase inhibitors: has a third COX isoform entered the fray. Curr Med Res Opin 2005;21:1217–1226.
79. Vane JR, Bakhle YS, Botting RM. Pharmacol Toxicol. 1998;38:97–120.
80. Portanova J, Zhang Y, Anderson GD. J Exp Med 1996;184:883–891.
81. Kam PCA, So A. COX-3: Uncertainties and controversies. Curr Anaesth Crit Care 2009;20:50–53.
82. Vane J. Nature 1994;367:215–216.
83. Holtzman MJ, Turk J, Sharnick LP. J Biol Chem 1992;267:21438–21445.
84. Hla T, Nelson K. Proc Natl Acad Sci USA 1992;89:7384–7388.
85. Simon LS, Weaver AL, Graham DY. JAMA 1999;282:1921–1928.
86. Simon LS. Arthritis Rheum. 1998;41: 1591–1602.
87. Langman MJ, Jensen DM, Watson DJ. JAMA 1999;282:1929–1933.
88. Stichtenoth DO, Frolich JC. The second generation of COX-2 inhibitors: what advantages do the newest offer? Drugs 2003 2003;63: 33–45.
89. Hynes NE, Stern DF. Biochim Biophys Acta 1994;1198:165–185.
90. Slamon DJ. N Engl J Med 2001;344:783–792.
91. Harris M. Monoclonal antibodies as therapeutic agents for cancer. Lancet Oncol 2004;5: 292–302.
92. Adams GP, Weiner LM. Monoclonal antibody therapy of cancer. Nat Biotechnol 2005;23:1147–1157.
93. Valabrega G, Montemurro F, Aglietta M. Trastuzumab: mechanism of action, resistance and future perspectives in HER2-overexpressing breast cancer. Ann Oncol 2007;18:977–984.
94. Whenham ND, Hondt V, Piccart MJ. HER2-positive breast cancer: from trastuzumab to innovatory anti-HER2 strategies. Clin Breast Cancer 2008;8:38–49.
95. Ross JS, Slodkowska EA, Symmans WF, Pusztai L, Ravdin PM, Hortobagyi GN. The HER-2 receptor and breast cancer: ten years of targeted anti-HER-2 therapy and personalized medicine. Oncologist 2009;14:320–368.
96. Dillman RO. Monoclonal antibodies in the treatment of malignancy: basic concepts and recent developments. Cancer Invest 2001;19: 833–841.
97. Drewe E, Powell RJ. Clinically useful monoclonal antibodies in treatment. J Clin Pathol 2002;55:81–85.
98. Guillemard V, Saragovi HU. Novel approaches for targeted cancer therapy. Curr Cancer Drug Targets 2004;4:313–326.
99. Chari RV. Targeted cancer therapy: conferring specificity to cytotoxic drugs. Acc Chem Res 2008;41:98–107.
100. Chadd HE, Chamow SM. Therapeutic antibody expression technology. Curr Opin Biotech 2001;12:188–194.

101. Ma JK-C, Drake PMW, Christou P. The production of recombinant pharmaceutical proteins in plants. Nat Rev Genet 2003;4: 794–805.
102. Zhu L, van de Lavoir M-C, Albanese J, Beenhouwer DO, Cardarelli PM, Cuison S, Deng DF, Deshpande S, Diamond JH, Green L. Production of human monoclonal antibody in eggs of chimeric chickens. Nat Biotechnol 2005;23:1159–1170.
103. Birch JR, Racher AJ. Antibody production. Adv Drug Delivery Rev 2006;58:671–685.
104. Steinmeyer DE, McCormick EL. The art of antibody process development. Drug Discov Today 2008;13:613–618.
105. Cunningham BC, Jhurani P, Ng P, Wells JA. Science 1989;243:1330–1336.
106. Cunningham BC, Wells JA. Science 1989;244:1081–1085.
107. Abdel-Meguid SS, Shieh HS, Smith WW, Dayringer HE, Violand BN, Bentle LA. Proc Natl Acad Sci USA 1987;84:6434–6437.
108. Fuh G, Mulkerrin MG, Bass S, McFarland N, Brochier M, Bourell JH, Light DR, Wells JA. J Biol Chem 1990;265:3111–3115.
109. Cunningham BC, Henner DJ, Wells JA. Science 1990;247:1461–1465.
110. Cunningham BC, Ultsch M, de Vos AM, Mulkerrin MG, Klausner KR, Wells JA. Science 1991;254:821–825.
111. de Vos AM, Ultsch M, Kossiakoff AA. Science 1992;255:306–312.
112. Wells JA, Cunningham BC, Fuh G, Lowman HB, Bass SH, Mulkerrin MG, Ultsch M, de Vos AM. Recent Prog Hormone Res 1992;48:253–275.
113. Nau HC. Science 1992;257:1064–1073.
114. Morens DM, Folkers GK, Fauci AS. The challenge of emerging and re-emerging infectious diseases. Nature 2004;430:242–249.
115. Spellberg B, Guidos R, Gilbert D, Bradley J, Boucher HW, Scheld WM, Bartlett JG, Edwards J. The epidemic of antibiotic-resistant infections: a call to action for the medical community from the infectious diseases society of America. Clin Infect Dis Chicago 2008;46: 155–164.
116. Palzkill T, Botstein D. Prot Struct Funct Genet 1992;14:29–44.
117. Ellman J, Mendel D, Anthony-Cahill S, Noren CJ, Schultz PG. Biosynthetic method for introducing unnatural amino acids site-specifically into proteins. Methods Enzymol. 1991;202:301–336.
118. Oliver JS, Oyelere A. Aminoacylation of nucleosides with FMOC amino acid fluorides. J. Org. Chem. 1996;61:4168–4171.
119. Oliver JS, Oyelere A. Efficient preparation of aminoacylated dinucleoside phosphates with N-FMOC amino acid fluorides. Tetrahedron Lett. 1997;38:4005–4008.
120. Shimizu Y, Inoue A, Tomari Y, Suzuki T, Yokogawa T, Nishikawa K, Ueda T. Cell-free translation reconstituted with purified components. Nat Biotech 2001;19:751–755.
121. Noren CJ, Anthony-Cahill SJ, Griffith MC, Schultz PG. Science 1989;244:182–188.
122. Ellman JA, Mendel D, Schultz PG. Science 1992;255:197–200.
123. Mendel D, Ellman JA, Chang Z, Veenstra DL, Kollman PA, Schultz PG. Science 1992;256: 1798–1802.
124. Burch RM, Kyle DJ. Pharm Res 1991;8: 141–147.
125. Walkinshaw MD. Med Res Rev 1992;12: 317–372.
126. Greene WC. Annu Rev Immunol 1990;8: 453–475.
127. Batcheler LT, Strehl LL, Neubauer RH, Petteway SR Jr., Ferguson BQ. AIDS Res Hum Retroviruses 1989;5:275–278.
128. Hasler JM, Weighous TF, Pitts TW, Evans DB, Sharma SK, Tarpley WG. AIDS Res Hum Retroviruses 1989;5:507–516.
129. Hsu M-C, Schutt AD, Holly M-C, Slice LW, Sherman MI, Richrnan DD, Potash MJ, Volsky DJ. Biochem Soc Trans 1992;20:525–531.
130. Steinmetz M, IBC USA Conferences; Southborough MA; 1992.
131. Erlich HA, Gelfand D, Sninsky JJ. Science 1991;252:1643–1651.
132. Bock LC, Griffin LC, Latham JA, Vermaas EH, Toole JJ. Nature 1992;355:564–566.
133. Ellington AD, Szostak JW. Nature 1990;346: 818–822.
134. Tuerk C, Gold L. Science 1990;249:505–510.
135. Brenner S, Lerner RA. Proc Natl Acad Sci USA 1992;89:5381–5383.
136. Marderosian AH,Stickley, G. F. Co., Philadelphia. 1988.
137. Chartrain M, Salmon PM, Robinson DK, Buckland BC. Curr Opin Biotech 2000;11:209–214.
138. Neilsen J. Appl Microbiol Biotech 2001;55: 263–283.
139. Rohlin L, Oh MK, Liao JC. Curr Opin Microbiol 2001;4:330–335.

140. Baltz RH. Trends Microbiol. 1998;6:76–83.
141. Schmidt-Dannert C. Curr Opin Biotech 2000;11:255–261.
142. Kealey JT, Liu L, Santi DV, Betlach MC, Barr PJ. Proc Natl Acad Sci USA 1998;95:505–509.
143. Albrecht M, Misawa N, Sandmann G. Biotech Lett 1999;21:791–795.
144. Pfeifer BA, Admiraal SJ, Gramajo H, Cane DE, Khosla C. Science 2001;291:1790–1792.
145. McDaniel R, Ebert-Khosla S, Hopwood DA, Khosla C. Nature 1995;375:549–554.
146. Xue Q, Ashley G, Hutchinson CR, Santi DV. Proc Natl Acad Sci USA 1999;96:11740–11745.
147. McDaniel R, Thamchaipenet A, Gustafsson C, Fu H, Betlach M, Ashley G. Proc Natl Acad Sci USA 1999;96:1846–1851.
148. Cane DE, Walsh CT, Khosla C. Science 1998;282:63–68.
149. Osborne SE, Matsumura I, Ellington AD. Aptamers as therapeutic and diagnostic reagents: problems and prospects. Curr Opin Chem Biol. 1997;1:5–9.
150. Famulok M, Jenne A. Oligonucleotide libraries—variatio delectat. Curr Opin Chem Biol 1998;2:320–327.
151. Jayasena SD. Aptamers: an emerging class of molecules that rival antibodies in diagnostics. Clin Chem 1999;45:1628–1650.
152. Patel DJ, Suri AK. Structure, recognition and discrimination in RNA aptamer complexes with cofactors, amino acids, drugs and aminoglycoside antibiotics. J Biotechnol 2000;74: 39–60.
153. Sun S. Technology evaluation: SELEX. Curr Opin Mol Ther 2000;2:100–105.
154. Green LS, Jellinek D, Bell C, Beebe LA, Feistner BD, Gill SC, Jucker FM, Janjic N. Chem Biol 1995;2:683–695.
155. Conrad R, Keranen LM, Ellington AD, Newton AC. Isozyme-specific inhibition of protein kinase C by RNA aptamers. J Biol Chem 1994;269:32051–32054.
156. Jenison RD, Gill SC, Pardi A, Polisky B. Science 1994;263:1425–1429.
157. Hamilton AJ, Baulcombe DC. A species of small antisense RNA in posttranscriptional gene silencing in plants. Science 1999;286: 950–952.
158. Elbashir SM, Harborth J, Lendeckel W, Yalcin A, Weber K, Tuschl T. Duplexes of 21-nucleotide RNAs mediate RNA interference in cultured mammalian cells. Nature 2001;411: 494–498.
159. Zamore PD, Tuschl T, Sharp PA, Bartel DP. RNAi Cell 2000;101:25–33.
160. Hannon GJ, Rossi JJ. Unlocking the potential of the human genome with RNA interference. Nature 2004;431:371–378.
161. Grimm D. Small silencing RNAs: state-of-the-art. Adv Drug Delivery Rev 2009;61:672–703.
162. Castanotto D, Rossi JJ. The promises and pitfalls of RNA-interference-based therapeutics. Nature 2009;457:426–433.
163. Root DE, Hacohen N, Hacohen WC, Lander ES, Sabatini DM. Genome-scale loss-of-function screening with a lentiviral RNAi library. Nat Methods 2006;3:715–719.
164. Sidhu SS. Phage display in pharmaceutical biotechnology. Curr Opin Biotech 2000;11: 610–616.
165. Hoess RH. Protein design and phage display. Chem Rev 2001;101:3205–3218.
166. Hyde-DeRuyscher R, Paige LA, Christensen DJ, Hyde-DeRuyscher N, Lim A, Fredericks ZL, Kranz J, Gallant P, Zhang J, Rocklage SM, et al. Detection of small-molecule enzyme inhibitors with peptides isolated from phage-displayed combinatorial peptide libraries. Chem Biol 2000;7:17–25.
167. Dennis MS, Roberge M, Quan C, Lazarus RA. Selection and characterization of a new class of peptide exosite inhibitors of coagulation factor VIIa. Biochemistry 2001;40:9513–9521.
168. Brown KC. New approaches for cell specific targeting: identification of cell selective peptides from combinatorial libraries. Curr Opin Chem Biol 2000;4:16–21.
169. Pasqualini R, Rouslahti E. Nature 1996;380: 364–367.
170. Arap W, Pasqualini R, Rouslahti E. Cancer treatment by targeted drug delivery to tumor vasculature in a mouse model. Science 1998;279:377–380.
171. Larocca D, Witte A, Johnson W, Pierce GF, Baird A. Targeting bacteriophage to mammalian cell surface receptors for gene delivery. Hum Gene Ther 1998;9:2393–2399.
172. Larocca D, Burg MA, Jensen-Pergakes K, Ravey EP, Gonzalez AM, Baird A. Evolving phage display vectors for cell targeted gene delivery. Curr Pharm Biotechnol 2002;3:45–57.
173. Erickson JW, Fesik SW. Annu Rep Med Chem 1992;27:271–289.
174. Clore GM, Gronenborn AM. Science 1991;252: 1390–1399.
175. Fesik SW. J Med Chem 1991;34:2937–2945.

176. Shaanan B, Gronenborn AM, Cohen GH, Gilliland GL, Veerapandian B, Davies DR, Clore GM. Science 1992;257:961–964.
177. Kuntz ID. Science 1992;257:1078–1082.
178. Mason JS, Good AC, Martin EJ. 3-D pharmacophores in drug discovery. Curr Pharm Des 2001;7:567–597.
179. Rando RR. Pharmacol Rev 1984;36:111–142.
180. Parker PJ, Coussens L, Totty N, Rhee L, Young S, Chen E, Stabel S, Waterfield MD, Ullrich A. Science 1986;233:853–859.
181. Nishizuka Y. Science 1992;258:607–614.
182. Nicholson CD, Challiss RAJ, Shahid M. Trends Pharmacol Sci 1991;12:19–27.
183. Crocker IC, Townley RG. Drugs Today 1999;35:519–535.
184. Stief CG. Drugs Today 2000;36:93–99.
185. Kramer RM, Hession C, Johansen B, Hayes G, McGray P, Chow EP, Tizard R, Pepinsky RB. J. Biol. Chem. 1989;264:5678–5775.
186. Seilhamer JL, Pruzanski W, Vadas P, Plant S, Miller JA, Kloss J, Johnson LK. J Biol Chem 1989;264:5335–5338.
187. Lehr M. Expert Opin Ther Patents 2001;11: 1123–1136.
188. Capper EA, Marshall LA. Mammalian phospholipases A(2): mediators of inflammation, proliferation and apoptosis. Prog Lipid Res 2001;40:167–197.
189. Murakami M, Kudo I. Phospholipase a(2). J Biochem 2002;131:285–292.
190. Matthews DA, Bolin JT, Burridge JM, Filman DJ, Volz W, Kaufman BT, Beddell CR, Champness JN, Stammers DK, Kraut J. J Biol Chem 1985;260:381–391.
191. Oefner C, D'Arcy A, Winkler F. Eur J Biochem 1988;174:377–385.
192. Prendergast NJ, Appleman JR, Delchamp TJ, Blakley RL, Freisham JH. Biochemistry 1989;28:4645–4650.
193. Schweitzer BI, Srirnatkandata S, Gritsman H, Sheridan R, Venkataraghavan R, Bertino J. J Biol Chem 1989;264:20786–20795.
194. Appelt K, Backquet RJ, Bartlett DA, Booth CLJ, Freer ST, Fuhry MAM, Gehring JR, Herrmann SM, Howland EF, Janson CA, et al. J Med Chem 1991;34:1925–1934.
195. Jones TR, Varney MD, Webber SE, Lewis KK, Marzoni GP, Palmer CL, Kathardekar V, Welsh KM, Webber S, Matthews DA, et al. Structure-based design of lipophilic quinazoline inhibitors of thymidylate synthase. J Med Chem 1996;39:904–917.
196. Stout TJ, Tondi D, Rinaldi M, Barlocco D, Pecorari P, Santi DV, Kuntz ID, Stroud RM, Shoichet BK, Costi MP. Structure-based design of inhibitors specific for bacterial thymidylate synthase. Biochemistry 1999;38: 1607–1617.
197. Klebe G. Recent developments in structure-based drug design. J Mol Med 2000;78: 269–268.
198. Huff JR. J Med Chem 1991;34:2305–2314.
199. Navia MA, Fitzgerald PMD, McKeever BM, Leu C-T, Heimbach JC, Herber WK, Sigal IS, Drake PL, Springer JP. Nature 1989;337: 615–620.
200. Wlodawer A, Miller M, Jaskolski M, Sathyanarayana BK, Baldwin E, Weber IT, Selk LM, Clawson L, Schneider J, Kent SBH. Science 1989;245:616–621.
201. Lapatto P, Blundell T, Hemmings A, Overington J, Wilderspin A, Wood S, Merson JR, Whittle PJ, Danley DE, Geoghegan KF, et al. Nature 1989;342:299–302.
202. Weber IT, Miller M, Jaskolski M, Leis J, Skalka AM, Wlodawer A. Science 1989;143: 928–931.
203. Miller M, Schneider J, Sathyanarayana BK, Toth MV, Marshall GR, Clawson L, Selk L, Kent SBH, Wlodawer A. Science 1989;246: 1149–1152.
204. Fitzgerald PMD, McKeever BM, VanMiddelsworth J, Springer JP, Heimbach JC, Leu CT, Herber WK, Dixon RAF, Darke PL. J Biol Chem 1990;265:14209–14219.
205. Erickson J, Neidhart DJ, VanDrie J, Kempf DJ, Wang XC, Norbeck DW, Plattner JJ, Rittenhouse JW, Turon M, Wideburg N, et al. Science 1990;249:527–533.
206. Swain AL, Miller MM, Green J, Rich DH, Schneider J, Kent SBH, Wlodawer A. Proc Natl Acad Sci USA 1990;87:8805–8809.
207. Heimbach JC, Garsky VM, Michaelson SR, Dixon RAF, Sigal IS, Darke FL. Biochem Biophys Res Commun 1989;164:955–960.
208. Meek TD, Lambert DM, Dreyer GB, Carr TJ, Tomaszek TA, Moore ML, Strickler JE, Debouck C, Hyland LJ, Matthews TJ, et al. Nature 1990;343:90–92.
209. McQuade TJ, Tomasseili AG, Liu L, Karacostas V, Moss B, Sawyer TK, Heinrikson RL, Tarpley WG. Science 1990;247:454–456.
210. Roberts NA, Martin JA, Kinchington D, Broadhurst AV, Craig JC, Duncan IB, Galpin SA, Handa BK, Kay J, Krohn A, et al. Science 1990;248:358–361.

211. Greenlee W. Med Res Rev 1990;10:173–276.
212. Schreiber SL. Science 1991;251:283–287.
213. Takahashi N, Hayano T, Suzuki M. Nature 1989;337:473–475.
214. Fischer G, Wittmann-Liebold B, Lang K, Kiefhaber T, Schmid FX. Nature 1989;337: 476–478.
215. Harrison RK, Stein RL. Biochemistry 1990;29:3813–3816.
216. Liu J, Albers MW, Chen C-M, Schreiber SL, Walsh CT. Proc Natl Acad Sci USA 1990;87: 2304–2308.
217. Rosen MK, Standaert S, Galat A, Nakatsuka M, Schreiber SL. Science 1990;248:863–866.
218. Wuthrich K, von Freberg B, Weber C, Wider G, Traber R, Widmer H, Braun W. Science 1991;254:953–954.
219. Gallion S, Ringe D. Prot Eng 1992;5:391–397.
220. Liu J, Farmer JD Jr., Lane WS, Friedmann J, Weissman I, Schreiber SL. Cell 1991;66: 807–815.
221. McKeon F. Cell 1991;66:823–826.
222. Russo AL, Passaquin AC, Andre P, Skutella M, Ruegg UT. Effect of cyclosporin A and analogues on cytosolic calcium and vasoconstriction: possible lack of relationship to immunosuppressive activity. Br J Pharmacol 1996;118: 885–892.
223. Sokolovsky M. Adv Drug Res 1989;18: 431–509.
224. Mei L, Roeske WR, Yamamura HI. Life Sci 1989;45:1831–1851.
225. Hulme EC, Birdsall NJM, Buckley NJ. Annu Rev Pharmacol Toxicol 1990;30:633–673.
226. Peralta EG, Ashkenazi A, Winslow JW, Smith DH, Ramachandran J, Capon DJ. EMBO J 1987;6:3923–3929.
227. Bonner TI, Buckley NJ, Young AC, Brann MR. Science 1987;237:527–532.
228. Bonner TI, Young AC, Brann MR, Buckley NJ. Neuron 1988;1:403–410.
229. Peralta EG, Winslow JW, Ashkenazi A, Smith DH, Ramachandran J, Capon DJ. Trends Pharmacol Sci 1988;9(Suppl): 6–11.
230. Gilman AG. Annn Rev Biochem 1987;56: 615–649.
231. Birnbaumer L. Annu Rev Pharmacol Toxicol 1990;30:675–705.
232. Dohlman HG, Caron MG, Lefkowitz RJ. Biochemistry 1987;26:2657–2664.
233. Humblet C, Mirzadegan T. Annu Rep Med Chem 1992;27:291–300.
234. O'Dowd BF, Hnatowich M, Regan JW, Leader WM, Caron MG, Lefkowitz RJ. J Biol Chem 1988;263:15985–15992.
235. Benovic JL, DeBlasi A, Stone WC, Caron MG, Lefkowitz RL. Science 1989;246:235–240.
236. Lohse MJ, Lefkowitz RJ, Caron MG, Benovic JL. Proc Natl Acad Sci USA 1989;86: 3011–3015.
237. Clark RB, Friedman J, Dixon RAF, Strader CD. Mol Pharmacol 1989;36:343–348.
238. Fraser CM. J Biol Chem 1989;264:9266–9270.
239. Wong SK-F, Parker EM, Ross EM. J Biol Chem 1990;265:6219–6224.
240. Simon MI, Strathmann MP, Gautam N. Science 1991;252:802–808.
241. Bovy P. Med Res Rev 1990;10:115–142.
242. Schultz S, Yuen PST, Garbers DL. Trends Pharmacol Sci 1991;12:166–120.
243. Lowe DG, Chang M-S, Hellmiss R, Cheo E, Singh S, Garbers DG, Goeddel DV. EMBO J 1989;8:1377–1384.
244. Anand-Srivastava MB, Sairam MR, Cantin M. J Biol Chem 1990;265:8566–8572.
245. Chang M-S, Lowe DG, Lewis M, Hellmiss R, Chen E, Goeddel DV. Nature 1989;341:68–72.
246. Bovy PR, O'Neal JM, Ollins GM, Patton DR, Mehta PP, McMahon EG, Palomo M, Schuh J, Blehm D. J Biol Chem 1989;264:20309–20313.
247. Kambayashi Y, Nakajima S, Ueda M, Inouye K. FEBS Lett 1989;248:28–34.
248. Schultz S, Green CK, Yuen PST, Garbers DL. Cell 1990;63:941–948.
249. Currie MG, Fok KF, Kato J, Moore RJ, Hamra FK, Duffin KL, Smith CE. Proc Natl Acad Sci USA 1992;89:947–951.
250. deSauvage FJ, Horuk R, Bennett G, Quan C, Burnier JP, Goeddel DV. J Biol Chem 1992;267:6429–6482.
251. Chao AC, deSauvage FJ, Dong YJ, Wagner JA, Goeddel DV, Gardner P. EMBO J 1994;13: 1065–1072.
252. Abbott A. Trends Pharmacol Sci 1992;13(TiPS Receptor Nomenclature Suppl): 169.
253. Ullrich A, Schlessinger J. Cell 1990;61: 203–212.
254. Tartaglia L, Goeddel DV. Immunol Today 1992;13:151–153.
255. Sprengel R, Werner P, Seeburg PH, Mukhin AG, Santi MR, Grayson DR, Guidotti A, Krueger KE. J Biol Chem 1989;264:20415–20421.
256. Pritchett DB, Luddens H, Seeburg PH. Science 1989;245:1389–1392.

257. Olsen RW, Tobin AJ. FASEB J 1990;4:1469–1480.
258. Jung V, Jones C, Kumar CS, Stefanos S, O'Connell S, Peska S. J Biol Chem 1990;265: 1827–1830.
259. Furuichi T, Yoshikawa S, Miyawaki A, Wada K, Maeda N, Mikoshiba K. Nature 1989;342:32–38.
260. Hollmann M, O'Shea-Greenfield A, Rogers SW, Heinemann S. Nature 1989;342:43–48.
261. Sprengel R, Braun T, Nikolics K, Segaloff DL, Seeberg P. Mol Endocrinol 1990;4:525–530.
262. Evans RM. Science 1988;240:889–895.
263. Godowski PJ, Picard D. Biochem Pharmacol 1989;38:3135–3143.
264. Power RF, Conneely OM, O'Malley BW. Trends Pharmacol Sci 1992;13:318–323.
265. McDonnell DP, Clevenger B, Dana S, Santiso-Mere D, Tzukerman MT, Gleeson MA. J Clin Pharmacol 1993;33:1165–1172.
266. Kishimoto T, Akira S, Taga T. Science 1992;258:593–597.
267. Nakanishi S. Science 1992;258:597–603.
268. van Galen PJM, Stiles GL, Michaels G, Johnson KA. Med Res Rev 1992;12:423–471.
269. Williams M. Med Res Rev 1991;11:147–184.
270. Oksenberg D, Marsters SA, O'Dowd BF, Jin H, Havlik S, Peroutka SJ, Ashkenazi A. Nature 1992;360:161–163.
271. Benhamou B, Garcia T, Lerouge T, Vergezac A, Gofflo D, Bigogne C, Chambon P, Gronemeyer H. Science 1992;255:206–209.
272. Hynes RO. Cell 1992;69:11–25.
273. Springer TA. Nature 1990;346:425–434.
274. Marlin SD, Staunton DE, Springer TE, Stratowa C, Sommergruber W, Merluzzi V J. Nature 1990;344:70–72.
275. Ruoslahti E, Pierchbaucher MD. Science 1987;238:491–497.
276. Phillips DR, Charo IF, Scarborough RM. Cell 1991;65:359–362.
277. Bodary SC, Napier MA, McLean JW. J Biol Chem 1989;264:18859–18862.
278. Jakubowski JA, Smith GF, Sall DJ. Annu Rep Med Chem 1993;27:99–108.
279. Blackburn BK, Gadek TR. Annu Rep Med Chem 1993;28:79–88.
280. Lasky L. Science 1992;258:964–969.
281. Yednock TA, Rosen SD. Adv Immunol 1989;44:313–378.
282. Lasky LA, Singer MS, Yednock TA, Dowbenko D, Fennie C, Rodriguez H, Nguyen T, Stachel S, Rosen SD. Cell 1989;56:1045–1055.
283. Karlsson KA. Trends Phamacol Sci 1991;12: 265–272.
284. Musser JH. Annu Rep Med Chem 1992;27: 301–310.
285. Weis WI, Drickamer K, Hendrickson WA. Nature 1992;360:127–134.
286. Erbe DV, Wolitzky BA, Presta LG, Norton CR, Ramos RJ, Burns DK, Rumberger JM, Rao BNN, Foxall C, Brandley BK, et al. J Cell Biol 1992;119:215–227.
287. Osborn L. Cell 1990;62:3–6.
288. Maniatis T, Goodbourn S, Fischer JA. Science 1987;236:1237–1245.
289. Frankel AD, Kim PS. Cell 1991;65:717–719.
290. Shea RG, Milligan JF. Annu Rep Med Chem 1992;27:311–320.
291. Catapano AL. Pharmacol Ther 1989;43: 187–219.
292. Schneider WJ. Biochim Biophys Acta 1989;988:303–317.
293. Utermann G. Science 1989;246:904–910.
294. Matteucci MD, Bischofberger N. Annu Rep Med Chem 1991;26:287–296.
295. Jabs DA, Griffiths PD. Fomivirsen for the treatment of cytomegalovirus retinitis. Am J Ophthalmol 2002;133:552–556.
296. Banerjee D. Curr Opin Mol Ther 1999;1:404–408.
297. Snouwaert JN, Brigman KK, Latour AM, Malouf NN, Bouchere RC, Smithies O, Koller BH. Science 1992;257:1083–1088.
298. Consortium IHGS. Nature 2001;409:860–921.
299. Venter JC, et al. Science 2001;291:1304–1351.
300. Hurley LH. J. Med. Chem. 1987;30:7A–8A.
301. Zeelen FJ. Trends Pharmcol Sci 1989;10:472.
302. Kornberg A. Biochemistry 1987;26:6888–6891.
303. Hirschmann RH. Angew Chem Int Ed Engl 1991;30:1278–1301.
304. Schreiber SL. Chem Eng News 1992;70:22–32.
305. Knight P. Biotechnology 1990;8:105–107.

STRUCTURAL GENOMICS, ITS APPLICATION IN CHEMISTRY, BIOLOGY, AND DRUG DISCOVERY

MATTHIEU SCHAPIRA[1,2]
[1] The Structural Genomics Consortium, University of Toronto, Toronto, Canada
[2] Department of Pharmacology, University of Toronto, Toronto, Canada

1. INTRODUCTION

Structural knowledge of target proteins has been used to design small-molecule ligands that fit in the active or allosteric site for more than 30 years in the pharmaceutical industry and academia [1], and structural biology contributed to the development of numerous marketed drugs across a variety of therapeutic areas in the past decade, such as the HIV protease inhibitor Nelfinavir, the neuraminidase inhibitor Tamiflu, the kinase inhibitor imatinib, or the antithrombin ximelagatran, to name a few (see Ref. [2] for review). The exponential growth of computing power, and increased accuracy and efficiency of docking algorithms allows for rapid virtual screening of libraries of a million compounds or more against experimental structures to select focused collections that can be tested experimentally [3,4]; cocrystal structures of leads and rational design of optimized derivatives can effectively accelerate the progression of chemical series to the clinic.

A simple and probably accurate conclusion is that more experimental protein structures will generate more opportunities to accelerate drug development. Structural genomics efforts, which are significantly accelerating the production of protein structures, should have a direct impact on drug discovery. What are the goals and strategies that are defining the main structural genomics centers? What is their contribution to the structural coverage of current or potential therapeutic targets? What is their current and expected impact on drug discovery? We will review the current trends and future landscape of structural genomics based on current statistics and published case studies.

2. STRATEGY, IMPLEMENTATION, AND PRODUCTION OF STRUCTURAL GENOMICS CENTERS

A combination of technological breakthroughs in the 1990s made large-scale approaches to structural biology tractable. More efficient engineering and expression of recombinant proteins opened the way to high-throughput crystallization (see Ref. [5] for review). Increased accessibility to robust synchrotron radiation sources made high-resolution data sets more accessible [6]. More robust experimental and computational approaches facilitated crystal structure determination [7,8]. The scientific community realized the potential of a large scale, systematic approach to structural biology, and structural genomics pilot projects were created to set up technology platforms and establish proofs of concept [9–11]. These evolved into more mature programs, the largest of which are the Protein Structure Initiative (PSI) in the United States, the RIKEN Structural Genomics/Proteomics Initiative (RSGI) in Japan, and the Structural Genomics Consortium (SGC), operating out of laboratories in Canada, Sweden, and the United Kingdom.

2.1. Strategies of Structural Genomics Centers

Complete sequencing of the human genome and of the genomes of human pathogens has significantly increased the spectrum of protein targets for therapeutic intervention, as well as the number of genes that may be suitable targets but have not been validated as such. Structural genomics efforts around the world aim at characterizing poorly understood proteins, validating potential new targets, and accessing a deeper knowledge of well-known or emerging target gene classes as structural systems. Different strategies are defining the 30 structural genomics centers that contributed at least one structure to the Protein Data Bank (PDB), the central repository for high-resolution protein structures. We will review here the three main trends, which significantly differ in their expected short-term impact on drug discovery.

2.1.1. Coverage of the Structure Space Structure and function of biological systems are intimately linked. The structure of a protein can help understand its function, and the central aim of structural genomics is to predict the structure of each protein from its sequence. Such prediction is not within reach *ab initio*, although progress is being made for small protein domains [12], and it is generally accepted that reliable computer models of a protein structure can be derived from the experimental structure of a protein homolog if the sequence identity is greater than 40% [13].

Based on these observations, the PSI, part of the RIKEN project, and other smaller scale structural genomics centers are populating the PDB with structures of proteins for which no close homolog was structurally characterized. The goal is to cover the structure space of all existing protein domain 3D folds with a diverse and representative set of experimental structures. The rest of the structure space could be computationally derived by homology [11,14].

Since 2000, approximately 179,000 protein structures were made computationally accessible by homology to experimental structures generated by structural genomics centers, representing 27% of the novel modeling leverage derived from all PDB depositions [14]. In the past 3 years, 50% of structures representing the first of a protein sequence family deposited in the PDB originated from the PSI or RIKEN [15]. These efforts should clearly contribute to our understanding of protein functions and biological systems. A corollary is that novel disease associated biological networks and therapeutic targets are expected to be identified, even though these important outcomes are not the central goal of structural genomics centers that adopted a blanket approach to the protein structure universe. Importantly for the drug discovery community, the accuracy of computational models of protein structures derived from homologs that only share 40% sequence identity is insufficient for structure-based drug design.

2.1.2. Genome Centric Approach Some structural genomics efforts are focused on the genome of a specific organism, or class of organisms. The goal is to better understand critical protein networks, identify and characterize promising therapeutic targets, and provide a structural framework to accelerate the development of drugs against specific diseases.

For instance, the TB structural genomics consortium (www.tbgenomics.org) has solved 192 structures of proteins from *Mycobacterium tuberculosis* (as of November 2008) [16,17]. Genes encoded in the complete genome sequences of a laboratory strain (H37Rv) and a clinical strain (CDC1551) of *M. tuberculosis* are subjected to bioinformatics, gene expression profiling and comparative genomics analyses, to prioritize potential drug targets, which are then crystallized in individual laboratories from 31 organizations out of 11 countries, coordinated by the consortium. Functional studies and identification of small-molecule ligands by virtual screening are also integrated into the pipeline for preliminary target validation [17]. Other structural genomics efforts focusing on mycobacteria in Europe are associated with SPinE (structural proteomics in Europe—www.spineurope.org) and X-MTB (www.xmtb.org).

While its primary focus is on human proteins, the structural genomics consortium also solved, as of November 2008, 100 structures of proteins from protozoan phylum Apicomplexa parasites, which are responsible for diseases such as malaria, toxoplasmosis, and cryptosporidiosis (www.thesgc.org). More than 400 distinct *Plasmodium falciparum* target genes were chosen representing different cellular classes, along with select orthologs from four other *Plasmodium* species as well as *Cryptosporidium parvum* and *Toxoplasma gondii*. A standardized production platform using *Escherichia coli* proved successful for genome-scale crystallography of apicomplexan proteins [18]. The SGC also aims at delivering to the scientific community detailed information on structural determinants of inhibitor binding, and seven structures were solved cocrystallized with a small-molecule compound.

The Medical Structural Genomics of Pathogenic Protozoa (MSGPP) solved, as of November 2008, 66 structures, 14 of which bound to a small-molecule ligand from 9 organisms responsible for malaria, Chagas' disease, sleeping sickness, leishmaniasis, amebiasis, giardiasis, toxoplasmosis, and cryptosporidiosis (www.msgpp.org). Here again, the goal is to impact target validation and early stage discovery, as illustrated by the implementation of "fragment cocktail crystallography" for medical structural genomics [19].

The "Viral Enzymes Involved in Replication" (VIZIER) project is an interesting variation on the genome centric theme of structural genomics (http://www.vizier-europe.org/). This effort set out to produce the structure of as many replicases as possible from life-threatening RNA viruses that break out in a regular, yet unpredictable manner. The idea is to raise the level of scientific preparedness, which should translate into faster development of drugs against new viruses, as they emerge. The project coordinates a pipeline spanning from viral genomics—to identify and prioritize targets—to structure-based precompetitive lead discovery [20]. As a proof of concept, VIZIER solved the structure of 13 targets from flaviviruses, a genus that includes a large number of human pathogens.

2.1.3. Protein Family-Based Coverage of the Human Genome The SGC and RIKEN are the two main centers focusing on families of human proteins with validated or emerging therapeutic relevance. The cellular activity of a protein relies first on its spatial and temporal expression level, and second on its ability to recognize specific substrates, interaction partners, or small-molecule modulators such as hormones, xenobiotics, or drugs. Structural biology analyzes at the atomic level the structural determinants for the latter. Different proteins belonging to the same family and sharing the same fold are expressed differentially in a pathogenic context. Molecular tools that can specifically inhibit—or in some cases activate—an individual target, such as chemical probes (i.e., a preclinical synthetic ligands) and drugs are important to dissect and block the contribution of a protein to the disease state. While structural models of proteins can be derived computationally from crystallized homologs, only experimental structures are sufficiently accurate to design compounds that selectively bind one member of a protein family, but not the other.

Solving the structure of many different members of a protein family results in (i) increased technical expertise and efficiency—which construct should be cloned, what are the optimum buffer conditions, purification, and crystallization protocols—and (ii) deeper scientific knowledge—what is the structural mechanism, which conformation should be targeted, and which critical structural motifs should be targeted to achieve selectivity.

For instance, the SGC set out not only to build expertise in a portfolio of therapeutically relevant gene classes approved by industrial funders, such as kinases, acetyltransferases, or poly(ADP-ribose)polymerase to name a few, but also to develop and characterize synthetic molecular probes that selectively bind individual proteins and help prioritize the most promising therapeutic targets. To reach this goal, a significant number of proteins are cocrystallized with small-molecule inhibitors (Table 1) to capture induced-fit conformational states and map interaction field potential of ligand binding pockets—all data that can significantly accelerate subsequent discovery programs, as discussed later.

The percent of structures solved by major structural genomics centers in the presence of a synthetic small-molecule inhibitor (buffer, cofactors, and natural substrates are excluded) is a metric that reflects which of these protein-centric efforts are progressing toward the chemistry space (Fig. 1). Although we believe that they accurately illustrate different trends and goals of structural genomics centers, we acknowledge that the numbers are not absolutely accurate due to the subjective assignment of a compound as substrate, cofactor, buffer, or real inhibitor. Clearly a few organizations are focusing more than others on the characterization of protein–ligand interaction. For instance, 11%, 5.5% and 10% of structures deposited, respectively, by SPinE, the TBSGC, and the SGC are in complex with small-molecule inhibitors (Fig. 1, Table 1).

Table 1. Structures in Complex with Drug-Like Inhibitors Released by the SGC as of November 2008

PDB Code	Inhibitor PDB Chain ID and Name	Reference(s)
1xws	BI1 3-{1-[3-(Dimethylamino)propyl]-1*H*-indol-3-yl}-4-(1*H*-indol-3-yl)-1*H*-pyrrole-2,5-dione	[21,22]
1yb1	AE2 Aetiocholanolone	
1yv5	RIS 1-Hydroxy-2-(3-pyridinyl)ethylidene bis-phosphonic acid	[23]
1z57	DBQ Debromohymenialdisine	
1zw5	ZOL Zoledronic acid	[23]
1zx1	CB1 5-(Aziridin-1-yl)-2,4-dinitrobenzamide	[24]
2bel	CBO Carbenoxolone	[25]
2bik	BI1 3-{1-[3-(Dimethylamino)propyl]-1*H*-indol-3-yl}-4-(1*H*-indol-3-yl)-1*H*-pyrrole-2,5-dione	[21,22]
2bil	BI1 3-{1-[3-(Dimethylamino)propyl]-1*H*-indol-3-yl}-4-(1*H*-indol-3-yl)-1*H*-pyrrole-2,5-dione	[21,22]
2buj	STU Staurosporine	
2bzh	HB1 Ruthenium-pyridocarbazole-1	
2bzi	DW2 Ru-pyridocarbazole-2	
2bzj	ME3 Ruthenium-pyridocarbazole-3	
2c3i	IYZ 1-(3-{6-[(Cyclopropylmethyl)amino]imidazo[1,2-*b*]pyridazin-3-yl}phenyl)ethanone	[26]
2c47	5ID (2*R*,3*R*,4*S*,5*R*)-2-(4-Amino-5-iodo-7*H*-pyrrolo[2,3-*d*]pyrimidin-7-yl)-5-(hydroxymethyl)tetrahydrofuran-3,4-diol	
2cdz	23D N2-[(1*R*,2*S*)-2-Aminocyclohexyl]-N6-(3-chlorophenyl)-9-ethyl-9*H*-purine-2,6-diamine	[27]
2chl	DKI 5-Amino-3-{[4-(aminosulfonyl)phenyl]amino}-*N*-(2,6-difluorophenyl)-1*H*-1,2,4-triazole-1-carbothioamide	
2clq	STU Staurosporine	[28]
2cmw	OLP 2-(2-Hydroxyethylamino)-6-(3-chloroanilino)-9-isopropylpurine	
2f57	23D N2-[(1*R*,2*S*)-2-Aminocyclohexyl]-N6-(3-chlorophenyl)-9-ethyl-9*H*-purine-2,6-diamine	[27]
2gwh	PCI Pentachlorophenol	[29]
2her	ZOL Zoledronic acid	[18]
2iwi	HB1 Ruthenium-pyridocarbazole-1	
2izr	BRK {(2*Z*)-4-Amino-2-[(4-methoxyphenyl)imino]-2,3-dihydro-1,3-thiazol-5-yl}(4-methoxyphenyl)methanone	
2izs	BRQ {[4-Amino-2-(3-chloroanilino)-1,3-thiazol-5-yl](4-fluorophenyl)methanone	
2izt	23D N2-[(1*R*,2*S*)-2-Aminocyclohexyl]-N6-(3-chlorophenyl)-9-ethyl-9*H*-purine-2,6-diamine	
2izu	P01 2-({6-[(3-Chlorophenyl)amino]-9-isopropyl-9*H*-purin-2-yl}amino)-3-methylbutan-1-ol	
2j2i	LY4 (9*R*)-9-[(Dimethylamino)methyl]-6,7,10,11-tetrahydro-9*H*,18*H*-5,21:12,17-dimethenodibenzo[*E*,*K*]pyrrolo[3,4-*H*][1,4,13]oxadiazacyclohexadecine-18,20-dione	[30]
2j51	DKI 5-Amino-3-{[4-(aminosulfonyl)phenyl]amino}-*N*-(2,6-difluorophenyl)-1*H*-1,2,4-triazole-1-carbothioamide	
2j7t	274 (3*Z*)-*N*-(3-Chlorophenyl)-3-({3,5-dimethyl-4-[(4-methylpiperazin-1-yl)carbonyl]-1*H*-pyrrol-2-yl}methylene)-*N*-methyl-2-oxoindoline-5-sulfonamide	
2j90	iza 2-tert-butyl-9-fluoro-3,6-dihydro-7*h*-benz[*h*]-imidaz[4,5-f]isoquinoline-7-one	

Table 1. (*Continued*)

PDB Code	Inhibitor PDB Chain ID and Name	Reference(s)
2jam	J60 5-[(*E*)-(5-Chloro-2-oxo-1,2-dihydro-3*H*-indol-3-ylidene) methyl]-*N*-[2-(diethylamino)ethyl]-2,4-dimethyl-1*H*-pyrrole-3-carboxamide	
2jav	5Z5 5-[(*Z*)-(5-Chloro-2-oxo-1,2-dihydro-3*H*-indol-3-ylidene) methyl]-*N*-(diethylamino)ethyl]-2,4-dimethyl-1*H*-pyrrole-3-carboxamide	[31]
2jc6	QPP *N*-(5-Methyl-1*H*-pyrazol-3-yl)-2-phenylquinazolin-4-amine	
2jfl	DKI 5-Amino-3-{[4-(aminosulfonyl)phenyl]amino}-*N*-(2,6-difluoro-phenyl)-1*H*-1,2,4-triazole-1-carbothioamide	
2nyr	SVR 8,8′-[Carbonylbis[imino-3,1-phenylenecarbonylimino(4-methyl-3,1-phenylene)carbonylimino]]bis-1,3,5-naphthalenetri-sulfonic acid	[32]
2o1o	RIS 1-Hydroxy-2-(3-pyridinyl)ethylidene bis-phosphonic acid	[33]
2pa9	GAB 3-Aminobenzoic acid	
2p0e	TIZ (1*R*)-1-[4-(Aminocarbonyl)-1,3-thiazol-2-yl]-1,4-anhydro-D-ribitol	[34]
2pqo	SHH Octanedioic acid hydroxyamide phenylamide	[35]
2pqt	2Bromoacetanilide	[36]
2pqp	TSN Trichostatin A	[35]
2qis	RIS 1-Hydroxy-2-(3-pyridinyl)ethylidene bis-phosphonic acid	[23]
2qkr	IXM (*Z*)-1*H*,1′*H*-[2,3′]Biindolylidene-3,2′-dione-3-oxime	
2uv2	GVD [4-({4-[(5-Cyclopropyl-1*H*-pyrazol-3-yl)amino]quinazolin-2-yl}imino)cyclohexa-2,5-dien-1-yl]acetonitrile	
2v7o	DRN Bisindolylmaleimide IX	
2vag	V25 Ethyl 3-[(*E*)-2-amino-1-cyanoethenyl]-6,7-dichloro-1-methyl-1*H*-indole-2-carboxylate	
2vd5	BI8 3-[1-(3-Aminopropyl)-1*H*-indol-3-yl]-4-(1-methyl-1*H*-indol-3-yl)-1*H*-pyrrole-2,5-dione	
2vd7	PD2 pyridine-2,4-dicarboxylic acid	[37]
2vn9	GVD [4-({4-[(5-Cyclopropyl-1*H*-pyrazol-3-yl)amino]quinazolin-2-YL}imino)cyclohexa-2,5-dien-1-yl]acetonitrile	
2vuw	5ID (2*I*,3*R*,4*S*,5*R*)-2-(4-Amino-5-iodo-7*H*-pyrrolo[2,3-*d*]pyrimidin-7-yl)-5-(hydroxymethyl)tetrahydrofuran-3,4-diol	
2vx3	D15 *N*-(5-{[(2*S*)-4-Amino-2-(3-chlorophenyl)butanoyl]amino}-1*H*-indazol-3-yl)benzamide	
2vz6	FEF (2*Z*,3*E*)-2,3′-Biindole-2′,3(1*H*,1′*H*)-dione 3-{*O*-[(3*R*)-3,4-dihy-droxybutyl]oxime}	
3b7l	M0N (1-Hydroxy-2-imidazo[1,2-*a*]pyridin-3-ylethane-1,1-diyl)bis (phosphonic acid)	
3bhh	5CP [4-({4-[(5-Cyclopropyl-1*H*-pyrazol-3-yl)amino]-6-(methylami-no)pyrimidin-2-yl}amino)phenyl]acetonitrile	
3bhy	7CP (4*R*)-7,8-Dichloro-1′,9-dimethyl-1-oxo-1,2,4,9-tetrahydrospiro [beta-carboline-3,4′-piperidine]-4-carbonitrile	
3bkb	STU Staurosporine	[38]
3bpr	OLP 2-(2-Hydroxyethylamino)-6-(3-chloroanilino)-9-isopropylpurine	[39]
3bpt	QUE 3,5,7,3′,4′-Pentahydroxyflavone	
3bqr	4RB 4-(6-{[(1*R*)-1-(Hydroxymethyl)propyl]amino}imidazo[1,2-*b*] pyridazin-3-yl)benzoic acid	
3c49	KU8 4-[3-(1,4-Diazepan-1-ylcarbonyl)-4-fluorobenzyl]phthalazin-1 (2*H*)-one	
3c4h	DRL 2-Methyl-3,5,7,8-tetrahydro-4*H*-thiopyrano[4,3-*d*]pyrimidin-4-one	
3cbl	STU Staurosporine	[38]

Table 1. (*Continued*)

PDB Code	Inhibitor PDB Chain ID and Name	Reference(s)
3cd3	STU Staurosporine	[38]
3ce0	P34 N~2~,N~2~-Dimethyl-N~1~-(6-oxo-5,6-dihydrophenanthridin-2-YL)glycinamide	
3ckl	STL Resveratrol	
3cp6	RSX (4A*S*,7A*R*)-Octahydro-1*H*-cyclopenta[*B*]pyridine-6,6-diylbis (phosphonic acid)	
3cxw	MB0 (4*R*)-7,8-Dichloro-1′,9-dimethyl-1-oxo-1,2,4,9-tetrahydrospiro[beta-carboline-3,4′-piperidine]-4-carbonitrile	
3cy2	MB9 (4*R*)-7-Chloro-9-methyl-1-oxo-1,2,4,9-tetrahydrospiro[beta-carboline-3,4′-piperidine]-4-carbonitrile	
3cy3	JN5 (2*S*)-1,3-Benzothiazol-2-yl{2-[(2-pyridin-3-ylethyl)amino]pyrimidin-4-yl}ethanenitrile	
3da2	4MD *N*-(4-Chlorobenzyl)-*N*-methylbenzene-1,4-disulfonamide	
3dko	IHZ 5-[(2-Methyl-5-{[3-(trifluoromethyl)phenyl]carbamoyl}phenyl) amino]pyridine-3-carboxamide	
3dzq	IFC *N*-[2-Methyl-5-({[3-(4-methyl-1*H*-imidazol-1-yl)-5-(trifluoromethyl)phenyl]carbonyl}amino)phenyl]isoxazole-5-carboxamide	
3e7v	DZO 3-(3-Aminophenyl)-*N*-(3-chlorophenyl)pyrazolo[1,5-*a*]pyrimidin-5-amine	
3eb0	DRK 3-({[(3*S*)-3,4-Dihydroxybutyl]oxy}amino)-1*H*,2′*H*-2,3′-biindol-2′-one	
3ez3	ZOL Zoledronic acid	
3f2n	IZZ (2*S*)-2-{[3-(3-Aminophenyl)imidazo[1,2-*b*]pyridazin-6-yl]amino}-3-methylbutan-1-ol	
3f2r	HC6 (2*S*,2′*S*)-2,2′-Biphenyl-4,4′-diylbis(2-hydroxy-4,4-dimethylmorpholin-4-ium)	
3f2s	HC6 (2*S*,2′*S*)-2,2′-Biphenyl-4,4′-diylbis(2-hydroxy-4,4-dimethylmorpholin-4-ium)	
3f3z	DRK 3-({[(3*S*)-3,4-Dihydroxybutyl]oxy}amino)-1*H*,2′*H*-2,3′-biindol-2′-one	
3feg	HC7 (2*S*)-2-[4′-({dimethyl[2-(phosphonooxy)ethyl]ammonio}acetyl)biphenyl-4-yl]-2-hydroxy-4,4-dimethylmorpholin-4-ium	

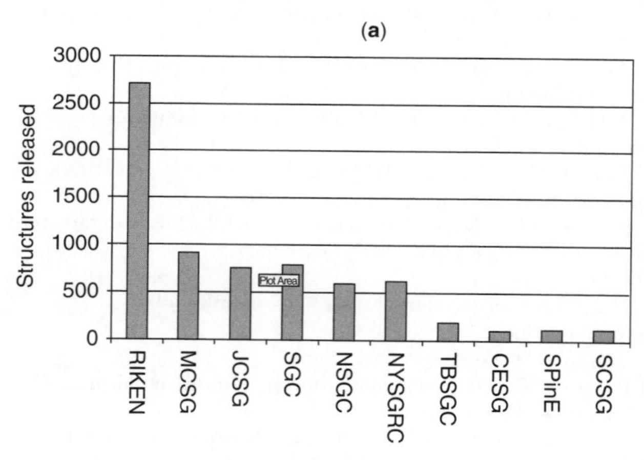

Figure 1. Relative volume of structural genomics structures addressing ligand chemistry. (a) Number of structures released in the PDB by structural genomics centers as of November 2008. (b) Number of structures cocrystallized with drug-like inhibitors (buffer, substrate and cofactor molecules were removed upon visual inspection). Only centers with at least 100 released structures are shown. (This figure is available in full color at http://mrw.interscience.wiley.com/emrw/9780471266945/home.)

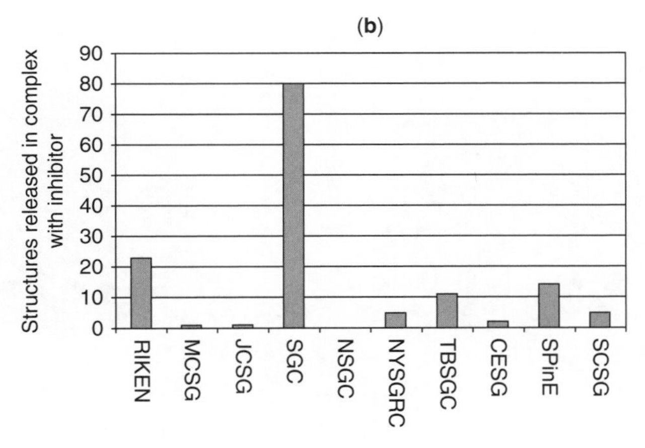

Figure 1. *(Continued)*

2.2. Contribution to PDB

2.2.1. Contribution to the Coverage of the Structure Space While the number of protein structures available from the PDB has grown exponentially since 1995 (3826 structures in 1995, 13,612 in 2000, 34,221 in 2005, 54,466 in 2008), the fraction of non-redundant protein structures started to decrease in 1995 [40]. For instance, 34.5% of all enzyme structures in the PDB covered only 1% of enzymes in 2005 [41]. The trend was inverted in 2003, when the PSI, which aims at covering the structure space, started to deliver. While the impact on our understanding of the relationship between structure and function is generally accepted, the relevance to drug discovery is not clear. One line of thinking is that structural characterization may highlight the druggable potential of today's poorly understood protein, suggesting its relevance as a therapeutic target. Another rational is that in depth understanding of mechanisms ruling protein structure space may translate into improved technologies to modulate protein activity and accelerate drug development. The observation that packing defects called dehydrons present at the surface of protein structures can be used to design selective inhibitors (discussed below) illustrates how physical phenomena learnt from systematic observation of protein structures may have direct application in drug design [42].

2.2.2. Contribution to the Coverage of the Human Proteome According to the UniProt database, the human genome is composed of approximately 20,500 genes, encoding around 11,000,000 amino acids [43]. In 2000, experimental structures were available for approximately 1% of the residues in the human proteome. This number raised to 6.9% in 2008, with structural genomics centers contributing to 50% of the coverage increase from 2005 to 2008, and 25% of the total coverage (Fig. 2, [44])

It is clear from Fig. 2 that the SGC and RIKEN are the two structural genomics centers contributing massively to the structural characterization of human proteins. It should be noted here that these institutes are focused on a number of specific gene classes, generally more relevant to the drug discovery industry. For instance, the SGC has contributed more than 50% of all new structures of kinase catalytic domains solved between 2005 and 2008 (Fig. 3, see Refs [44,45] for review).

The rational is that kinases constitute a biologically and structurally validated drug target family. However, while validated drug targets are over represented in the PDB (e.g., no less than 24 records can be found in the PDB for ABL tyrosine kinase, the target of imatinib (Gleevec)), few structures are available for poorly characterized branches of the kinome phylogenetic tree (Fig. 4a). The structures of underrepresented kinases solved by structural genomics centers can be used to design selective chemical probe inhibitors,

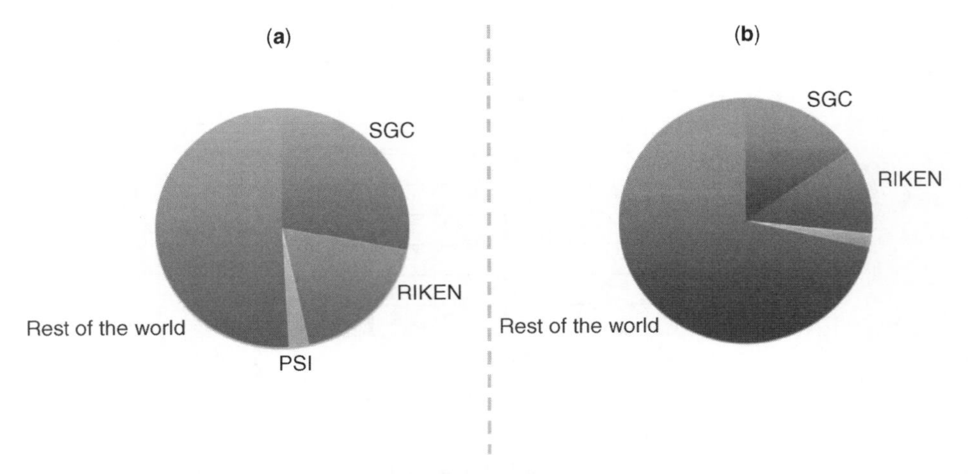

Figure 2. Contribution of structural genomics centers to the PDB coverage of the human proteome (percent residues). (a) From July 2005 to September 2008. (b) All time. (Adapted from Ref. [44].) (This figure is available in full color at http://mrw.interscience.wiley.com/emrw/9780471266945/home.)

which will constitute important tools to better understand the biology of these structurally druggable proteins, and potentially validate some as therapeutically relevant drug targets [30] (see discussion below). Kinases solved at the SGC for which closest structures shared only 25% sequence identity at the time of their PDB deposition include a "never in mitosis" (NIMA) family member Nek2 [31], ASK1 [46], and haspin (PDB code: 2vuw).

Histone methyltransferases constitute another family where structural genomics centers contributed significantly to PDB coverage. These proteins methylate lysines and

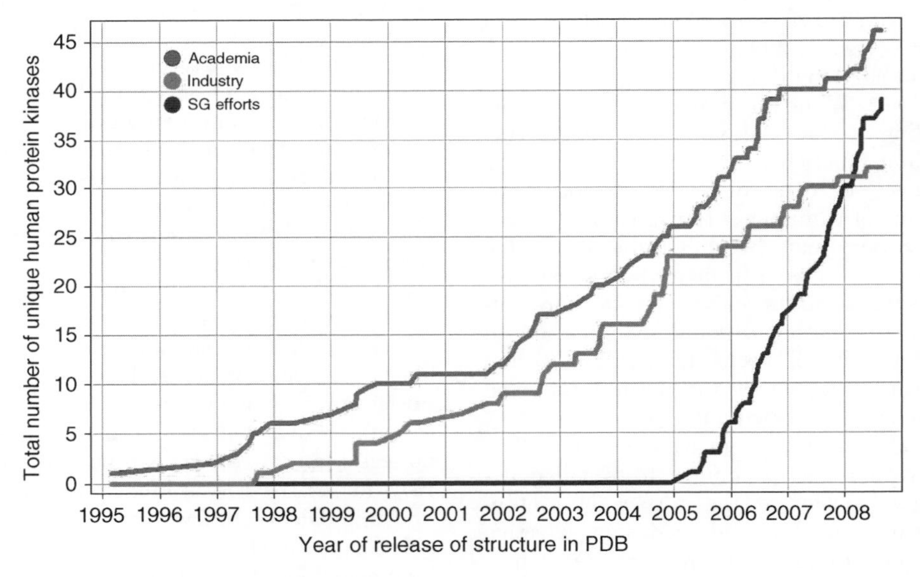

Figure 3. Contribution of academia, industry, and structural genomics centers to the structural characterization of human kinase catalytic domains. (Reproduced with permission from [44], updated version available from http://www.sgc.ox.ac.uk/research/kinases/.) (This figure is available in full color at http://mrw.interscience.wiley.com/emrw/9780471266945/home.)

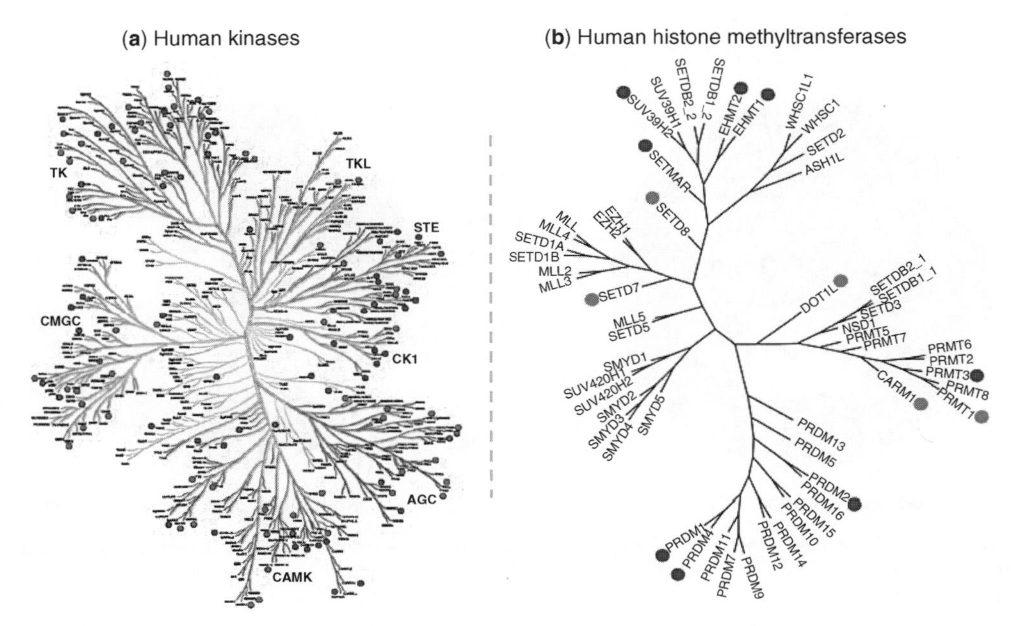

Figure 4. Phylogenetic mapping of the structural genomics contribution to the coverage of (a) human kinases and (b) epigenetic methyltransferases. Structures solved in academia, industry, and structural genomics centers are colored red, green, and blue, respectively. (Reproduced with permission from [44], updated version available from http://www.sgc.ox.ac.uk/research/kinases/.) (See color insert.)

arginines on histone tails and are writers of the epigenetic code (see Ref. [47] for review). Several members of this protein family are implicated in cancer [48], such as EZH2, which plays a critical role in stem cell renewal, PRDMs, involved in cell growth and apoptosis, and MLL (mixed lineage leukemia) proteins, important for normal hematopoiesis. Based on mounting literature in the epigenetic field suggesting that histone methyltransferases constitute an emerging class of drug targets, and on recent evidence that these proteins are structurally druggable [49], structural genomics centers recently allocated some effort on this protein family, and already delivered 8 of the 12 human histone methyltransferases structures available (Fig. 4b). These and future structures will allow a systematic analysis of the mechanism of methylation state selectivity, peptide substrate specificity, and will guide the design of synthetic inhibitors that may serve as chemical tools for target validation and drug design.

Several other human gene families with validated or potential therapeutic relevance are systematically targeted by structural genomics initiatives. These include protein tyrosine phosphatases, proteins involved in G-protein signaling or histone acetyltransferases to name a few (see Ref. [44] for review).

2.2.3. Contribution to the Coverage of Clinical Drug Targets Another metric of interest is the extent to which structural genomics centers are contributing to the structural coverage of relevant drug targets. The problem is that the definition of a relevant drug target is not clear. Is it a functionally validated target, which in the industry, probably means that a small-molecule inhibitor with clear mechanism of action has reached the clinic, or is at least potent *in vivo*? Does it include novel targets with promising cellular or *in vivo* data, but no advanced compound? Does it exclude targets with no obvious binding site?

To stay away from such ambiguities, a metric recently used was the number of structures of clinical drug targets released in the

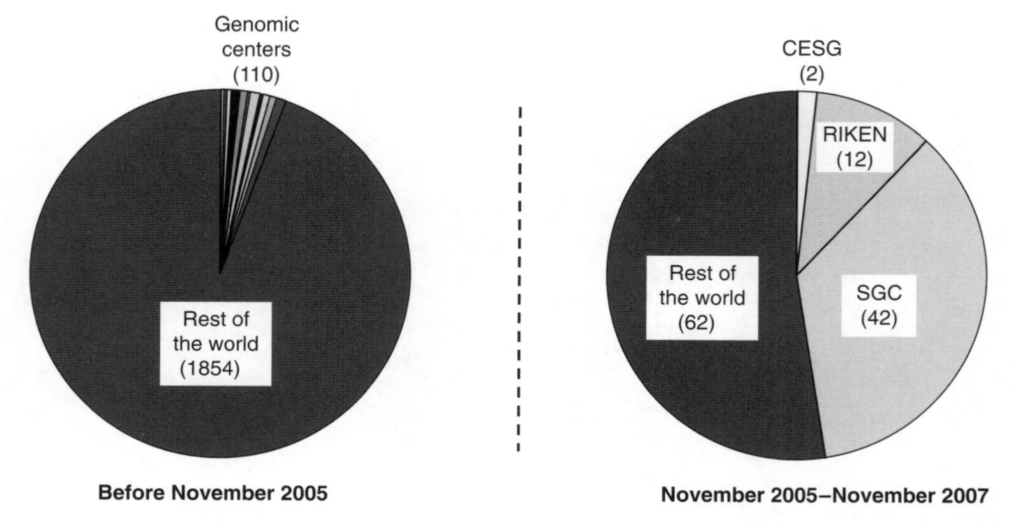

Figure 5. Structural genomics contribution to the structural coverage of approved drug targets. (Adapted from Ref. [50].) (This figure is available in full color at http://mrw.interscience.wiley.com/emrw/9780471266945/home.)

PDB, where only structures of previously unsolved proteins were counted [50]. From November 2005 to November 2007, structural genomics centers contributed to approximately 47% of the structural coverage of approved drugs, with the SGC, and RIKEN bringing 35% and 10% of the total PDB contribution (Fig. 5). These figures are in stark contrast with the situation in November 2005, when structural genomics had only contributed to 5%. Clearly, this metric is once again subject to debate, as follow-on structures of a same target in different conformational states, or in complex with different ligands—ignored here—are also important for structure based drug design. Yet, Fig. 5 illustrates what impact structural genomics is having on our shared understanding of drug target structures. Public dissemination of undisclosed structures that populate file systems in the pharmaceutical industry would probably resonate with even greater impact, but such discussion is beyond the scope of this review.

2.2.4. Emerging Efforts in Membrane Proteins

Forty percent of FDA approved drugs were targeting either G-protein coupled receptors (GPCRs) or ion channels in 2006. Overall, 60% of drugs targets are cell surface proteins, while these represent only 22% of the human genome [51]. This bias of existing drugs toward membrane proteins is in stark contrast with structures available in the PDB: as of November 2008, the structure of 178 unique membrane proteins (including paralogs from different organisms) were available in the PDB [52]. Obviously, this does not reflect a lack of interest by the structural biology community for this important class of proteins, but the presence of well-known technical challenges at each step of the structure determination pipeline, from protein expression to data collection.

Solving the structure of a new membrane protein remains to this day a rare scientific feat, achieved by expert teams that have usually spent five, ten or more years on the subject, which is widely acclaimed and has significant impact in the scientific community (e.g., [53–55]).

Recognizing that solving membrane proteins requires both a high degree of expertise and a very systematic approach, the Center for Structures of Membrane Proteins (SMP, http://csmp.ucsf.edu/) in California is focused on this class of proteins, and deposited 10 structures in the PDB, as of November 2008. The SGC is also allocating part of its effort on membrane proteins, with two structures solved to date [56]. These successes remain sporadic, and the future will tell whether structural genomics operations can make a significant breakthrough in the structural coverage of this class of proteins important for drug discovery.

3. CURRENT AND FUTURE IMPACT OF STRUCTURAL GENOMICS IN DRUG DISCOVERY

3.1. Impact of Technology Development in Drug Discovery

Proof of principle for a technologically and economically sound approach to large-scale structural biology was initially established by a number of pilot projects that paved the way for more mature programs [9–11]. Systematic methods and customized instrumentation were developed toward more efficient protein production and crystallization.

Methods developed for the industrialization of the protein crystallization and structure determination pipeline at structural genomics centers [57,58,59], can directly be applied in the drug discovery industry, both in big pharma [60,61] and in niche biotechs focused on high-throughput crystallography, such as Astex Therapeutics, Structural Genomix, acquired by Eli Lilly in 2008, or Syrrx, acquired by Takeda in 2005 [62–64]. Collective analysis of the expression and purification of more than 10,000 different proteins carried out in different structural genomics centers resulted in a decision-making tree that can be applied to any protein expression/purification effort [65].

Robotic systems developed out of necessity by and/or for the structural genomics community are now used in the drug discovery industry. Examples include the Cartesian Microsys and the Technology Partnership Mosquito II liquid dispensing systems, capable of screening thousands of crystallization parameters in nanolitre droplets, the Minstrel III crystal recognition system (Robodesign, USA), which combines high-density plate storage and imaging, and crystal mounting robots, such as Bruker-Nonius from Bruker AXS (see Refs [63,66] for review).

3.2. Dissemination of Data, Methods, and Reagents

Visualization and interpretation of protein structures is not trivial to nonspecialists, and some structural genomics centers such as the SGC are distributing annotated 3D animations of their structures that can be inspected interactively from their website [67]. Probably as important as the structures themselves are materials and methods used to solve structures. While docking of compounds to structures delivered by genomics centers can be a of value in a drug discovery program, nothing replaces actual cocrystal structures to understand with atomic precision how a lead interacts with its target, and design follow-up compounds. Scientists, mostly in the industry, need to be able to reproduce crystals obtained by genomics centers, and soak them with their proprietary compounds to solve complex structures that will guide development. To facilitate this technology transfer, the SGC publishes on its website for each deposited structure a detailed description of the successful protocol and makes corresponding DNA constructs available to the scientific community upon request. The SGC also initiated a program to develop small-molecule chemical probes that can be used for target validation, and is pledging to keep all resulting chemistry available upon request, and unencumbered from intellectual property claims, to foster scientific progress and shared knowledge, all relevant to drug discovery [68].

3.3. Impact of Delivered Structures in Drug Discovery

3.3.1. Increased Opportunities for Structure-Based Drug Design More structures of therapeutically relevant targets directly translate into more opportunities for structure based drug design. First, a new structure published in the PDB can be used to screen virtually large (one million or more) compound libraries to produce focused collections enriched in hits that can be rapidly screened [3,4]. Second, redundant structures in complex with different ligands, depicting different conformational states, as well as structures of close homologs within a proteins family, provide a depth of understanding which can translate into more informed design of ligands presenting improved specificity profiles (see discussion below). Third, reagents and protocols made available by structural genomics centers can be used to cocrystallize primary hits or leads to guide development programs.

Though generally reliable, the accuracy of experimental structures should not be taken for granted. In 2002, it was estimated that up to 3% of residues in the PDB were not accurately fitted, mostly because of flipped histidine rings or side-chain amide groups [69]. As technology progresses, it is reasonable to assume that this number would be lower today, and software exists that automatically checks the integrity of hydrogen bond networks in a structure (see Ref. [40] for review).

3.3.2. Development of New In Silico Technologies New technologies are being developed to interrogate and analyze the wealth of structural data, which are growing exponentially. One example is the ability to screen thousands of binding sites to identity putative targets for a compound of interest. This approach identified five possible targets for a series of 1,3,5-triazepan-2,6-diones by docking the compounds to 2000 binding sites from the PDB; and one of them, phospholipase A2, was subsequently confirmed experimentally [70]. This virtual profiling technology should become more relevant and accurate as the number of binding sites grows in the PDB, and as accuracy of docking algorithms improves.

Another emerging technology is to scan the PDB for binding sites sharing similar structural chemistry (general shape, hydrophobicity profile, hydrogen bonding network) against a pocket of interest. This may be useful to predict off-target liabilities, identify scaffolds cocrystallized to sites promiscuous to the query pocket, and guide the design of focused libraries, or inspire lead optimization [71–73]. Though promising, these technologies still need to demonstrate that they can have a significant impact in drug discovery.

A valuable service freely available on the web is PROCOGNATE, a database of cognate ligands for domains in enzyme structures as described by CATH, SCOP, and Pfam [74]. All compounds cocrystallized to a specific protein fold can be easily retrieved, and be used to guide or inspire rational development of new molecules.

3.3.3. Extracting from PDB Universal Rules for Medicinal Chemists Universal rules defining protein–ligand interactions, such as preferred orientation of hydrogen bond relative to a carbonyl acceptor, are used on a daily basis by medicinal and computational chemists to optimize chemical series. A recent analysis by Roche scientists shows that out of 2308 basic scaffolds present in drugs that reached the clinic, 420 are present in complex with a receptor in the PDB [75]. Expectedly, the cores most observed in drugs are better covered structurally: while more than 80% of fragments present in at least 10 drugs are described in a cocrystallized state, only 10% of scaffolds present in only 1 drug are in the PDB. As the number of cocrystal structures—contributed in a part by structural genomics-grows, meaningful statistical analysis of the stereochemistry of cocrystallized ligands is used in the drug discovery industry and in academia to identify preferred binding pauses of basic scaffolds, commonly observed interaction types that were previously thought atypical, or torsional energy levels sustained by induced fit. We will present here a few examples.

A Boehringer Ingelheim team systematically compared the calculated potential energy landscape of free ligands with conformations bound to proteins, and could draw several conclusions important for drug design: (1) a good overall correlation was observed between calculated energy minima and bound states, showing that in most cases the potential energy is a good predictor for the most likely bound conformation; (2) ligands bound to a protein sustain on average 0.6 kcal/mol of strain energy per rotatable bond; (3) induced fit conformations imposing more than 4 kcal/mol of torsional strain are extremely unlikely [76].

Another analysis conducted at Novartis examined the environment of amide groups in 3275 cocrystal structures with a resolution better than 2.5 Å, and revealed general rules that can be used in drug discovery: (1) most of the C=O and NH groups at the protein–ligand interface are engaged in hydrogen bonds, and among the small percentage of outliers, C=O

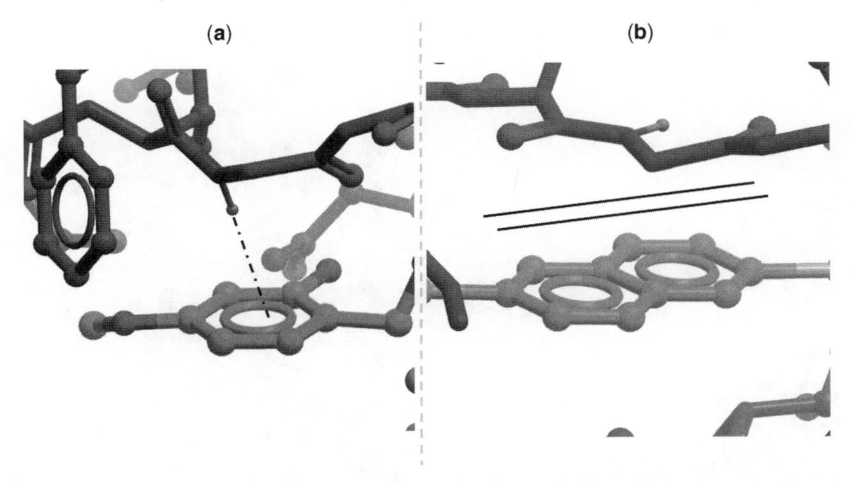

Figure 6. Two atypical but regularly observed protein–ligand interactions involving amid groups. Hydrogen bonding ((a) PDB code: 1a8g) or stacking ((b) PDB code: 1ql7) with an aromatic ring. Ligands are in green, protein residues in magenta, oxygens red, nitrogen blue. (Adapted from Ref. [77].) (See color insert.)

are more often solvated or embedded in a hydrophobic environment than NH; (2) NH have a significant propensity to interact with aromatic rings, either by acting as a hydrogen bond donor or, more often, by stacking between the amide and the aromatic planes (Fig. 6) [77].

In another example with direct lessons for medicinal chemists, a group working in the pharmaceutical industry explored the specific influence of carbon–fluorine single bonds—a function widely used to avoid pharmacokinetic liabilities—on protein–ligand interaction [78]. A systematic dissection of interaction patterns observed in the PDB showed that fluorine (1) interacts favorably with peptidic NH and C=O moieties, (2) undergoes dipolar interactions with side-chain amides of glutamines and asparagines, (3) favors positively polarized environments (Fig. 7).

A very surprising result recently came from the comparison of the physicochemical properties, binding affinity, cocrystal structure and binding sites of enzyme and nonenzyme inhibitors listed in Binding-MOAD, a database of more than 10,000 protein–ligand crystal structures extracted from the PDB [79]. The authors observed that high-affinity ligands of enzymes are larger, while there is no correlation between size and affinity for nonenzyme inhibitors. A corollary was that nonenzyme inhibitors have overall a higher binding efficiency, which is not intrinsic to the ligand, but is due to differences in the general amino acid composition of enzyme versus nonenzyme ligand binding site [80].

Focused analysis of data present in the PDB can not only provide rules that should be accounted for when designing ligands but also provide a rational for the failure of some compounds to comply with a well-established SAR: Astrazenecca scientists systematically analyzed cocrystal structures of ligands grouped within chemical series, and could show that, even if the pose is conserved between two analogs, minor modifications in the ligand can generate receptor side-chain (or even backbone) movements that corrupt structure–activity relationship [81].

These examples illustrate how the wealth of data on protein–ligand interaction present in the PDB can be analyzed to address questions of relevance to drug design. As illustrated and discussed above, some structural genomics centers are directing their effort toward delivering more cocrystal structures. These efforts should translate in a better understanding of the structural chemistry of protein–ligand interaction, and will add to the intellectual tool kit that is used on a daily basis by seasoned medicinal chemists.

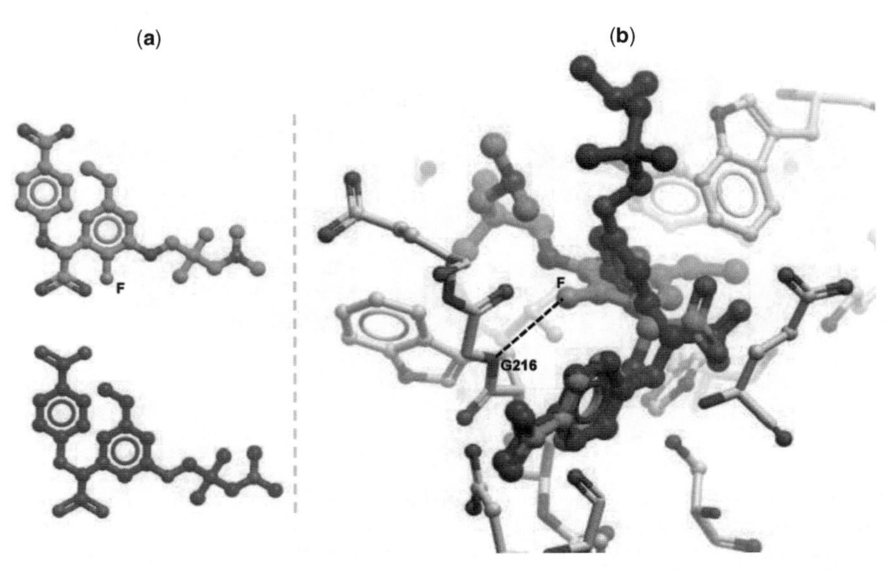

Figure 7. Analysis of cocrystallized fluorinated compounds in the PDB reveals that fluorine is regularly involved in dipolar interactions. Replacement of hydrogen (PDB code: 2v3o) by fluorine (PDB code: 2v3h) on a thrombin inhibitor (magenta and green, respectively) results in a dipolar N–H ... F–C interaction that stabilizes an alternate binding conformation. (a) 2D rendition of the inhibitors. (b) Superimposed crystal structures of the complexes. (Adapted from Ref. [78]). (See color insert.)

3.4. Case Studies

Data contributed by structural genomics centers, obviously, not only are useful for statistical analysis but also can help directly in the development of inhibitors for target validation and drug discovery.

3.4.1. Nitrogen-Containing Bisphosphonates

Nitrogen-containing bisphosphonates (BPs), such as risedronate or alendronate, are the drugs of choice against osteoporosis and other diseases characterized by excessive bone resorption. Inhibition of farnesyl pyrophosphate synthase (FPPS) by BPs reduces prenylation of small GTPase proteins, which mediates the antiresorptive response. FPPS orthologs from parasitic protozoans are also potential targets against Chagas' disease, cryptosporidiosis and leishmaniasis [82]. The SGC solved in 2005 the first structure of human FPPS in complex with risedronate and zoledronate [23], two of the leading BPs drugs, used in the treatment of osteoporosis, Paget's disease, hypercalcemia, and osteolysis associated with multiple myeloma and metastatic cancers [83]. The structures and accompanying biochemical studies showed that magnesium-dependent binding of BPs occurs at the geranyl pyrophosphate/dimethylallyl pyrophosphate site, while positioning the nitrogen in a proposed carbocation binding site increases affinity (Fig. 8). The structures confirmed that the slow-tight binding mechanism of BPs to FPPS is due to isomerization of the enzyme–inhibitor complex, which necessitates a significant conformational rearrangement of the enzyme, and produces more potent inhibition. The structures also suggest that enhanced potency might be achieved by decreasing the entropic penalty through targeted reduction of flexibility, and by increasing binding enthalpy through additional branching groups that would extend further into the geranyl pyrophosphate/dimethylallyl pyrophosphate site, toward Tyr-204 and Thr-167 [23].

As a follow-up study, a structure–activity relationship among the clinical BPs and analogs was establish to further investigate the time-dependent mechanism of inhibition, and compare potency with *in vivo* efficacy in bone

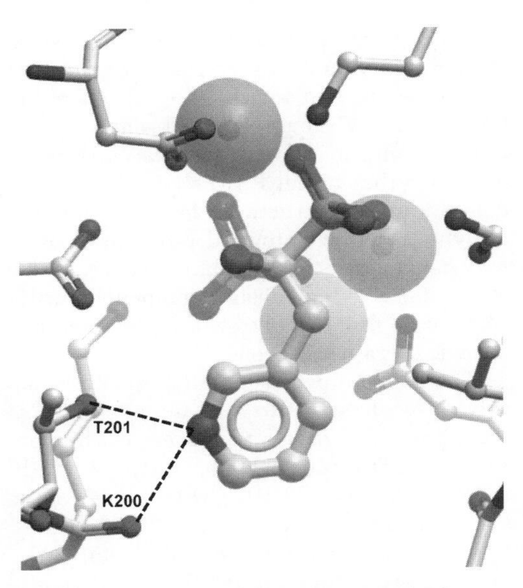

Figure 8. Structure of risedronate bound to human FPPS. Three magnesium atoms mediate the interaction between an aspartate-rich region and the phosphonate groups, while the nitrogen hydrogen-bonds to the backbone carbonyl of Lys-200 and the side-chain oxygen of Thr-201(PDB code: 1yv5, [23]) (See color insert.)

resorption models. In the light of the cocrystal structures, relative contributions of the phosphonate and hydroxyl moieties interacting with the aspartate-rich region of the enzyme, and contribution of the spatial positioning of the critical nitrogen were assessed biochemically and *in vivo* [84].

The SGC also solved the crystal structure of the *C. parvum* FDPS ortholog in complex with risedronate and zoledronate (PDB code: 2o1o and 2her, respectively—unpublished), which may help in the development of leads that inhibit the protozoan and not the human target, toward new treatments against cryptosporidiosis, a neglected condition associated with significant morbidity in the developing world [85].

3.4.2. Structural Mechanism of Aromatic Amine and Hydrazine Drugs Acetylation *N*-acetyl transferases are enzymes transferring an acetyl group from acetyl-coA to the substrate nitrogen of drugs, endogenous ligands, or peptides. Of 29 human *N*-acetyltransferases, the structures of 8 are available, 7 of which were contributed by structural genomics centers (PDB codes: 2pfr, 2pqt, 1z4r, 2o28, 2fxf, 2bei, and 2ob0). The SGC alone contributed six of these, including the two important drug metabolic enzymes NAT1 and NAT2 [36]. The hepatic arylamine *N*-acetyltransferases NAT1 and NAT2 constitute a major route in the metabolism and detoxification of various drugs and xenobiotics, and participate in the bioactivation of known carcinogens. NAT2 is also the site of a common genetic polymorphism of clinical relevance [86]. These enzymes influence both the pharmacokinetic properties of aromatic amine and hydrazine drugs in the clinic, and the balance between detoxification and metabolic activation of various carcinogens [87].

The SGC solved the first crystal structures of human NAT1 and NAT2, including a cocrystal structure of NAT1 in complex with an irreversible inhibitor (Fig. 9), and of NAT2 with the cofactor CoA [36]. Novel features, which could not have been derived from previous procaryote NAT structures, were observed, and structural determinants for sub-

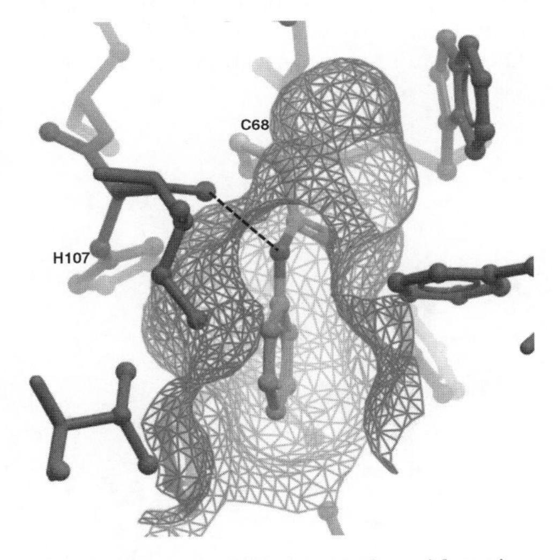

Figure 9. Human NAT1 in complex with an irreversible inhibitor. Acetanilide (green) is covalently bond to the catalytic Cys 68, hydrogen bonded to the backbone carbonyl of I106 and is surrounded with hydrophobic, mostly aromatic residues, and with catalytic His 107. The structure of the pocket can be used for metabolic liability prediction of known or prospective compounds ([36], PDB code: 2pqt). (See color insert.)

strate selectivity were defined. A site-directed NAT1 mutant was also solved, and defective functions of naturally occurring NAT1 and NAT2 genetic variants were interpreted.

These structures may be valuable to rationalize observed pharmacokinetics of aromatic amine and hydrazine drugs in the clinic. Crystallization methods available to the public, and docking studies can be used for future development of prodrugs NAT substrates and of optimized compounds that evade metabolic liabilities.

3.4.3. Structural Determinants for Inhibition of Poly(ADP)Ribosylation Poly(ADP-ribose) polymerases (PARPs) catalyze the ADP-ribosylation of DNA, histones, and various DNA repair enzymes in response to DNA damage. PARPs constitute an emerging class of therapeutic targets. PARP inhibitors potentiate DNA-damaging agents, such as alkylators, platinums, topoisomerase inhibitors or radiation, and several investigational drugs are in advanced oncology clinical trials (see Refs [88,89] for review). Additionally, activation of PARP has been implicated in the pathogenesis of stroke, myocardial ischemia, diabetes, diabetes-associated cardiovascular dysfunction, shock, traumatic central nervous system injury, arthritis, colitis, allergic encephalomyelitis, and various other forms of inflammation. Various chemical classes with different pharmacological actions have been described that target PARPs [90]. Better understanding the structural determinants underlying selectivity toward different PARPs may help dissect the polypharmacology of the PARP family, and design compounds with selectivity profiles better targeted against specific therapeutic areas.

The SGC launched a PARP program and solved seven structures of human PARPs, including the first structures of human PARP3 (PDB code: 2pa9), PARP12 (PDB code: 2pqf), and PARP15 (PDB code: 3blj); the first structure of the PARP domain of human tankyrase 1 (PDB code: 2rf5, [91]); and four structures of PARP3 (PDB codes: 2pa9, 3c4h, 3c49, 3ce0) cocrystallized with small-molecule inhibitors. These structures (Fig. 10), in addition with 2 PARP1 structures cocrystallized with inhibitors (PDB codes: 2rcw and 2rd6) lay the foundation for a pharmacophoric model of PARP inhibition that may accelerate the discovery of chemical probes useful for target validation (see

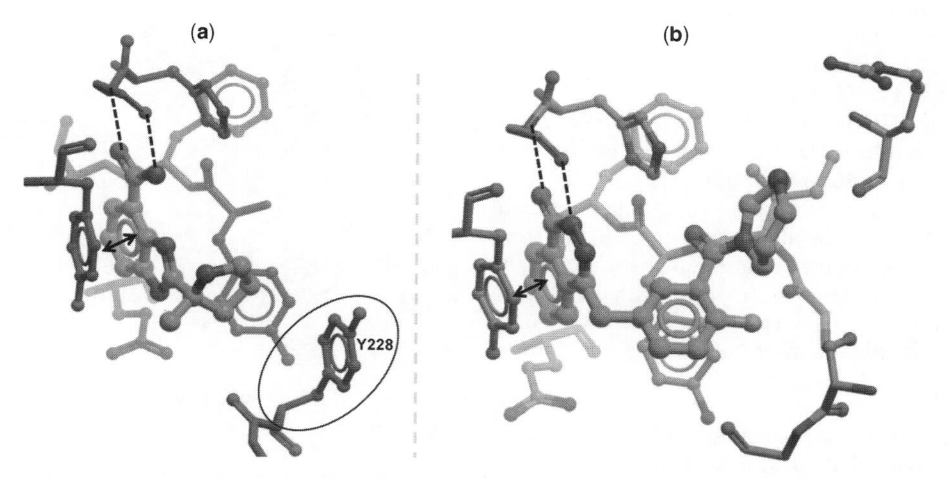

Figure 10. Comparison of PARP1 and PARP3 inhibitor binding modes. The structures of human PARP1 ((a) PDB code: 2rd6) and PARP3 ((b)PDB code: 3c49) cocrystallized with different inhibitors (green and cyan, respectively) reveal that residues lining the binding sites are mostly conserved, and so are important interactions, such as a hydrogen bonds with backbone atoms (dashed lines), or pi-stacking of aromatic rings (arrows). Rare, but important differences exist (Tyr 228 in PARP1, Gly in PARP3—highlighted by an oval) that may be used to achieve isoform selectivity. (See color insert.)

SGC's epigenetics program below), and the development of novel investigational drugs with enhanced pharmacology.

3.4.4. Structure-Based Development of Antituberculosis Drugs Tuberculosis kills approximately two million people worldwide each year. Emergence of drug resistance and the persistent state of the bacterium, which necessitates long antibiotic treatment, are major issues that need to be addressed with the validation of new drug targets and the development of new classes of drugs. The TBSGC has solved 199 structures of known or potential tuberculosis targets, as of November 2008, 10% of which cocrystallized to small-molecule inhibitors (Fig. 1). This effort should have a significant impact in the development of future drugs [92]. Structures delivered by the TBSGC are already used in promising discovery programs.

One virulence strategy of pathogenic bacteria, including *Mycobacterium tuberculosis*, is to produce and secrete protein tyrosine phosphatases to attenuate host immune defenses [93]. Protein tyrosine phosphatase B (PtpB) is a key virulence factor in *M. tuberculosis*. The TBSGC solved the first crystal structure of this potential target in 2005 [94] and follow-up work resulted in the discovery and cocrystallization of a submicromolar inhibitor of PtpB with over 60-fold selectivity over a limited panel of human PTPs (PTP1B, Pac1, Glepp1, and PTPH1) (Fig. 11, [94]). Both the structure of the complex and the method are valuable to dock and cocrystallize novel chemical classes that may have improved development potential, such as an isoxazole series reaching 220 nM Ki value and good selectivity versus human isoforms recently identified by the same research group [95].

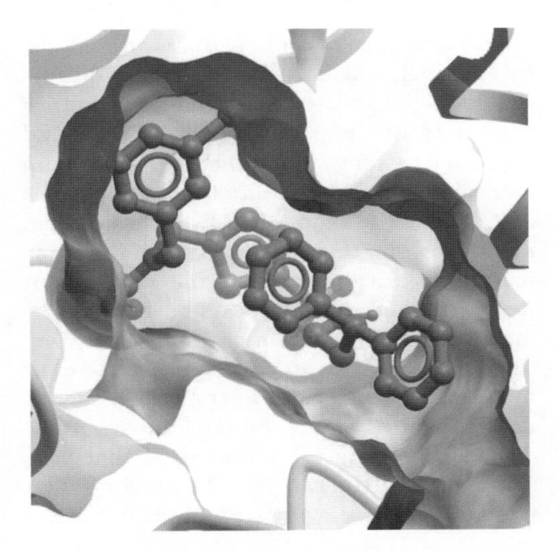

Figure 11. Structural mechanism of *Mycobacterium tuberculosis* protein tyrosine phosphatase B inhibition. Understanding the binding mode of a *M. tuberculosis* specific inhibitor (orange) will accelerate the development of improved chemical series ([94], PDB code: 2oz5). (See color insert.)

Another example is the structure of the enoyl-(acyl-carrier-protein) reductase (InhA), the target of the preferred antitubercular drug isoniazid, solved by the TBSGC [96]. Isoniazid requires conversion to an activated form of the drug by catalase-peroxidase (KatG) [97], associated with increased liability toward resistance mechanisms. The TBSGC structure shows that the activated form of the drug is covalently attached to the nicotinamide ring of nicotinamide adenine dinucleotide bound within the active site of InhA. This structure can be used for the development of novel classes of *M. tuberculosis* InhA inhibitors with improved structural mechanism of action. This avenue is currently actively pursued within a drug discovery program managed by the public–private partnership global alliance for TB drug development (TB Alliance, 2008 Annual Report).

3.4.5. Fragment Cocrystallography Approach Against a Sleeping Sickness Target *Trypanosoma brucei*, the causative agent of sleeping sickness, is a serious threat in sub-Saharan Africa, with more than 60 million people at risk. If untreated, the disease is fatal. The parasite lacks enzymes necessary for *de novo* purine biosynthesis, and uses nucleoside 2-deoxyribosyltransferase (NDRT), an enzyme absent in humans, to recycle nucleosides from the host [98].

The MSGPP solved the structure of apo-NDRT and soaked cocktails of small-molecule fragments for crystallographic screening: 69 crystals were examined by crystallography, probing 304 different compounds in 31 chemically diverse cocktails. Most cocktails did not affect the resolution of soaked crystals. Interestingly, soaks that resulted in diffraction worse than 3 Å were rescreened with a shorter incubation time (down to 10 s), and diffracted better than 2.5 Å. Four ligands were identified in the active site ([99], PDB codes: 2f2t, 2f62, 2f64, and 2f67). All active site fragments position their ring system in the same place, next to a glycerol molecule, also present in the pocket, revealing a pharmacophoric model for NDRT inhibition, and suggesting an optimization strategy where a glycerol or mimetic could be bridged to one of the identified scaffolds (Fig. 12). Equally important for drug discovery is the observation that very short soaks of 10 s can be sufficient for a fragment to reach the active site. The method validated here may be applied with more fragile crystals of unrelated targets.

These few examples published by academic groups illustrate how structures produced by structural genomics centers are already used to characterize known drugs and discover new ligands. Research often originating in academia must reach the pharmaceutical industry before it can impact medicine. It is reasonable to hope that some of the precompetitive discovery efforts relying on structural genomics data raise sufficient interest in the private sector, and trigger successful drug discovery programs. In this regard, chemogenomics approaches may leverage the contribution of structural genomics and significantly increase its impact.

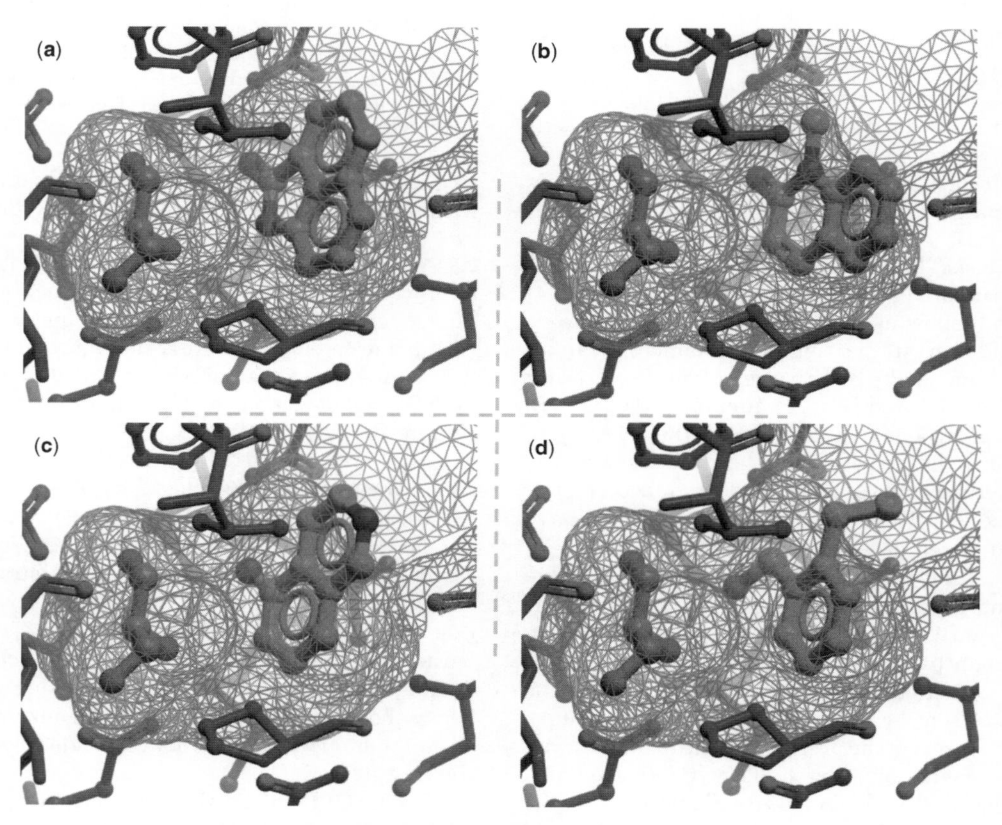

Figure 12. Structures of four chemical fragments soaked into the catalytic site of *Trypanosoma brucei* NDRT. The structures clearly indicate a conserved high-efficiency binding mode for ring systems (green), and reveals the presence of a glycerol molecule (orange) occupying another part of the pocket (gray mesh) that may be bused as a complementary chemotype ([99], PDB codes: (a) 2f67, (b) 2f64, (c) 2f2t, and (d) 2f62). (See color insert.)

3.5. Structural Genomics and Chemogenomics

3.5.1. Chemogenomics: A New Paradigm for Drug Discovery While the goal of drug discovery programs until recently was to identify compounds as potent and selective as possible against one specific target, the one disease/one gene/one drug paradigm is challenged in today's drug discovery industry [100]. Current clinical attrition rates are not supporting the simple concept, prevailing for 20 years, that increased drug selectivity translates into reduced toxicity.

One example is the development of histone deacetylase (HDAC) inhibitors, in oncology. There are 18 known human HDACs, divided into 4 groups. HDACs remove acetyl marks from histone, thereby contributing to epigenetic signaling, and have many other protein substrates involved in regulation of gene expression, cell proliferation, and cell death. Several HDAC inhibitors have been identified with acute selectivity and potency against the isoforms that are believed to be the best targets, and 14 were in clinical trial in 2007 [101]. The only molecule that successfully achieved FDA approval is SAHA, a comparatively weak compound that inhibits all HDACs but NAD+-dependent class III enzymes [102]. The wide spectrum of action and suspected lack of redundancy of HDACs is still not fully understood and needs to be dissected with chemical probes that each inhibit one individual isoform. The prevailing theory is that SAHA's pan-HDAC activity mediates its clinical efficacy, and that its acceptable safety profile relies on it mediocre potency.

Many lessons can be learnt from this and other examples: (1) target monoselectivity is not necessarily the way to safety, (2) *in vitro* potency is not necessarily driving clinical efficacy, and (3) understanding the biology of each target of a gene family is necessary for a knowledge-based approach to pharmacology and drug design.

If the term polypharmacology—knowledge-based approach to the design of compounds hitting a select set of targets—has gained in popularity over the past 2 years, the concept surfaced in the industry while the human genome project was nearing completion. The idea was—and still is—to "translate the information from genome-sequencing efforts into knowledge that will aid in the discovery of drugs" [103]. Systematic exploration of a gene class with chemical tools each inhibiting one individual target, also referred to as chemogenomics coverage of a family, is increasingly considered as a critical path toward target validation and modern drug discovery [104–106]. Integrating sequence, structure, and SAR data over a protein family can also facilitate cross-SAR exploitation, identify preferred selectivity profiles and reveal entry point chemistry for drug development [105].

Glaxo-Wellcome scientists pioneered the systematic approach to the development of chemical probes to dissect function and disease association, using the nuclear hormone receptors as a model protein family. Several family members were targets of known drugs, and it was anticipated that many others may be valid drug targets as well [107]. However, no ligand was available for three quarters of the 48 nuclear hormone receptors. A reverse endocrinology program was launched to identify natural and synthetic ligands of orphan receptors, which resulted in the discovery of unanticipated signaling pathways and the validation of new targets [108]. Importantly, availability of crystal structures of the targets in their unliganded states, and cocrystallized to agonists or antagonists played an important role in the chemogenomics coverage of the nuclear hormone receptor family [109]. Virtual screening methods were validated against the family [110]. All precompetitive chemical probes developed were made available to the public through commercial suppliers, and many were used extensively to dissect the biology of some family members. For instance, use of compound GW0742, a peroxisome proliferator activated receptor delta agonist, was reported in 392 papers and cited in 7212, while figures are 250 and 4482, respectively, for compound GW4064, a farnesyl X receptor ligand (data compiled from Google Scholar, October 5, 2007—Tim Willson, personal communication). The wealth of biological characterization of nuclear hormone receptor genes that resulted from the availability of these selective chemical probes was critical for the identification of the best clinical intervention points, and guided drug discovery programs in the pharmaceutical industry.

The drug discovery industry as a whole has since fully recognized the importance of chemogenomics for target validation and for the development of drugs with improved polypharmacology [105,106,111]. While not a necessity, understanding the structural determinants for isoform selectivity adds depth to the chemogenomics coverage of a gene class, and can accelerate the systematic discovery of chemical probes and subsequent development of drugs against validated targets.

3.5.2. Contribution of Protein Structures to the Chemogenomics Understanding of Clinical Kinase Inhibitors Kinases constitute a prominent target class with a diverse array of compounds in development and in the clinic [112,113]. We will see here how extensive structural analysis, essentially focused on today's validated kinase targets, in different conformational states and complexed to a variety of ligands, put some rationale on the observed selectivity of clinical compounds, and what are the lessons for maximizing the impact of structural genomics in drug discovery.

Most kinase inhibitors bind at the ATP site, which is relatively well conserved across the family. Systematic profiling revealed that (1) few compounds are selective and (2) kinase inhibitors cover a wide spectrum of specificity profiles [30,114]. Interestingly, clustering kinases based on the profile of compounds that bind their ATP site produces different results than clustering based on sequence similarity, indicating that sequence alone cannot be a good predictor of ligand selectivity [115]. For instance, the anticancer drug imatinib is active against Bcr-ABL, inactive against the SRC subfamily that is close to Bcr-ABL on the phylogenetic dendogram of tyrosine kinases, and active against CKIT and PDGFR tyrosine kinases, evolutionarily remote from Bcr-ABL [114,116]. However, phylogenetic classification based on a structurally more relevant set of residues—that is, residues lining cocrystallized imatinib—brings CKIT and PDGFR in proximity to Bcr-ABL, while SRC appears more dissimilar [117]. Clearly, structure data can reconcile sequence similarity and compound selectivity profile, and help establish cross-family SAR.

Structural studies also revealed that the catalytic domain of kinases can adopt multiple conformations. In particular, active and inactive states—often referred to as DFG-in and DFG-out, respectively—are associated with very different inhibitor binding site topologies, and two kinases presenting almost identical DFG-in ATP sites can have diverging DFG-out pockets, amenable to the development of selective inhibitors [118]. This conformational selection mechanism underlies the relative selectivity of the three anticancer drugs dasatinib, erlotinib, and lapatinib: dasatinib can recognize multiple conformational states of multiple targets and is promiscuous; erlotinib is a rather selective inhibitor of EGFR, but recognizes both DFG-in and DFG-out conformations, which makes it more prone to off-target activity; lapatinib is a highly selective DFG-out EGFR inhibitor with no known off-target effect [119]. The availability of multiple kinase structures complexed to a diversity of ligands in varied conformational states was necessary to understand the conformational mechanism of ligand specificity, and knowledge-based approaches can be derived to rapidly design novel compounds with improved selectivity profiles [120].

A last example illustrates how crystal structures were used to develop a strategy toward the development of selective cKIT inhibitors. cKIT is the primary target against gastriontestinal stromal tumors (GISTs). While imatinib was developed as a Bcr-ABL inhibitor, it is also potent against PDGFR and cKIT. It is believed that Bcr-ABL inhibition mediates the observed cardiotoxicity of the drug. Additionally, JNK, another kinase, blocks mitochondrial depolarization pathway, and JNK inhibitors have cardioprotective effect. cKIT/JNK selective compounds would have a better therapeutic index against GISTs. Structural alignment of Bcr-Abl and cKIT cocrystallized with imatinib, and of JNK in complex with a pyrazoloquinolone revealed a dehydron (structure packing defect that can be exploited to achieve selectivity [42]) specific to cKIT and JNK, which was used to rationally design a cKIT selective inhibitor with reduced

Bcr-Abl inhibition, and enhanced JNK inhibition, currently in clinical trial [121]. Extensive structural coverage of the clinically validated space of the kinome in complex with inhibitors guided development of a promising drug.

Extending this chemogenomics driven structural coverage to less well-understood kinases, or other gene families, will shift the discovery process to tomorrow's drug targets. This is where structural genomics can impact drug discovery most.

3.5.3. The SGC Kinase Program The human kinome is composed of 518 genes [113]. We saw above that extensive structural characterization of clinically validated kinases was critical to our profound understanding of their structural chemistry and is used to design novel drugs. A number of phylogenetic subclasses of the human kinome are comparatively poorly characterized, and the clinical relevance of the majority of kinases remains unknown. Selective inhibitors are necessary for therapeutic validation. Crystal structures are the fastest way to selective inhibitors. The strategy defining the kinase program of the SGC is to solve the structure of kinases representative of poorly characterized phylogenetic classes, and actively contribute to the development of selective chemicals that can be used to probe the biology of these kinases, validate novel targets, and eventually reduce attrition in drug discovery [122]. Examples illustrating new structural features, some of which may be exploited to achieve ligand specificity, follow.

The canonical activation segment of protein kinases is composed of a magnesium-binding DFG motif, a short beta-sheet, and the activation and P + 1 loops contributing to substrate recognition in their active conformation [123]. The SGC solved the structure of MPSK1 (myristoylated and palmitoylated serine/threonine kinase 1), a member of the NAK family (Numb-associated kinases) that shared only 25% sequence identity with its closest structural neighbor Aurora A, cocrystallized with the pan-inhibitor staurosporine. The structure revealed an atypical activation segment, comprising a beta-sheet and a large alpha-helical insertion, which had not been observed in any other kinase structure. ([124], Fig. 13). Sequence comparison suggested that this novel topology is shared by other members of the NAK family. The authors further linked the atypical structural motif discovered to the recruitment of specific substrates [124]. Whether this unique structural feature can be

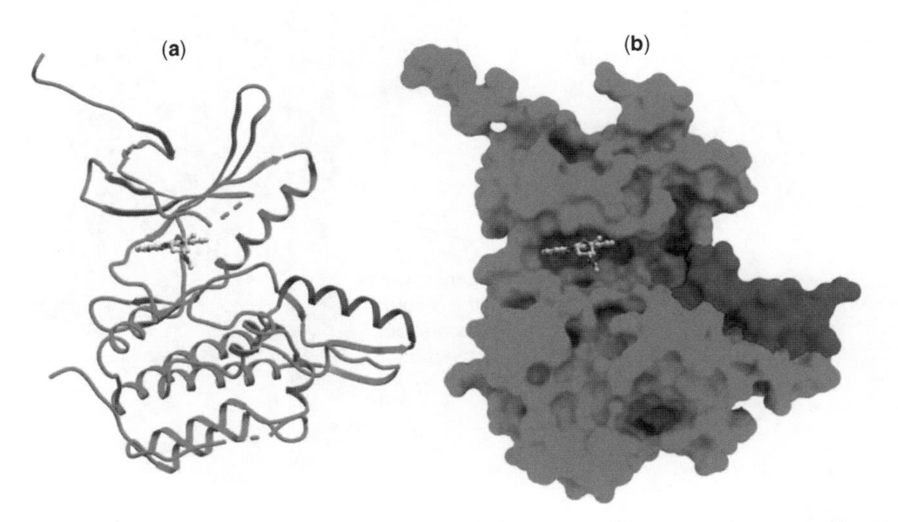

Figure 13. The structure of MPSK1 (PDB code: 2buj, (a) ribbon representation, and (b) molecular surface) in complex with staurosporine (sticks) shares only 25% sequence identity with its closest structural neighbor and reveals an alpha-helical insertion (magenta) in its activation segment unseen in any other kinase structure, which may be used to design selective inhibitors [124]. (See color insert.)

used to design selective ligands targeting the substrate binding remains to be investigated.

Another novel structural feature was uncovered by the crystal structures of CLK1 and CLK3. The Clk family consists of four dual-specificity kinases, capable of phosphorylating both Ser/Thr and Tyr residues. The structures of CLK1 and CLK3, both in their apo form and in complex with active site inhibitors, revealed the presence of a long beta-hairpin segment, located opposite the hinge region, with so far unknown function (PDB codes: 1z57, 2eu9, 2exe, and 2vag [125]). This insertion generates unique cavities that may be exploited to design nonactive site ligands useful to probe the biology and therapeutic relevance of these proteins (Fig. 14).

One more example is the structure of the dimeric Ser/Thr kinase Nek2 in complex with the pyrrole-indolinone inhibitor SU11652 ([31], PDB code: 2jav). Nek2 is a member of the NIMA6-related kinase (or Nek) family of serine/threonine protein kinases. It regulates centrosome architecture [126], and aberrant regulation of Nek2 activity can lead to aneuploid defects characteristic of cancer cells. Nek2 is upregulated in a number of human cancers and downregulation can inhibit cell proliferation [127]. Nek kinases represent 2% of the human kinome. The Nek2 crystal structure was the first representative of the family. Interestingly, the highly conserved DFG motif that lies at the N-terminal end of the T-loop and typically contributes to the coordination of a divalent metal ion adopts an unusual helical topology that results in an autoinhibitory state of the enzyme (Fig. 15) [31]. This DFGLARIL helical motif is conserved within Nek kinases, suggesting that the inhibitory mechanism revealed by this structure is shared with other enzymes.

The SGC structure of the kinase domain of Mer was the first from the Mer/Axl/Sky subfamily of receptor tyrosine kinases ([39], PDB code: 3bpr). A number of reports suggest that Mer is a promising oncology target: Mer is involved in cellular transformation of B-lymphocytes, its expression is elevated in adenomas ([128], lymphomas [129], and lymphoblastic leukemia [130]; transgenic mice overexpressing Mer develop leukemia/lymphoma [131]. On the other hand, Mer

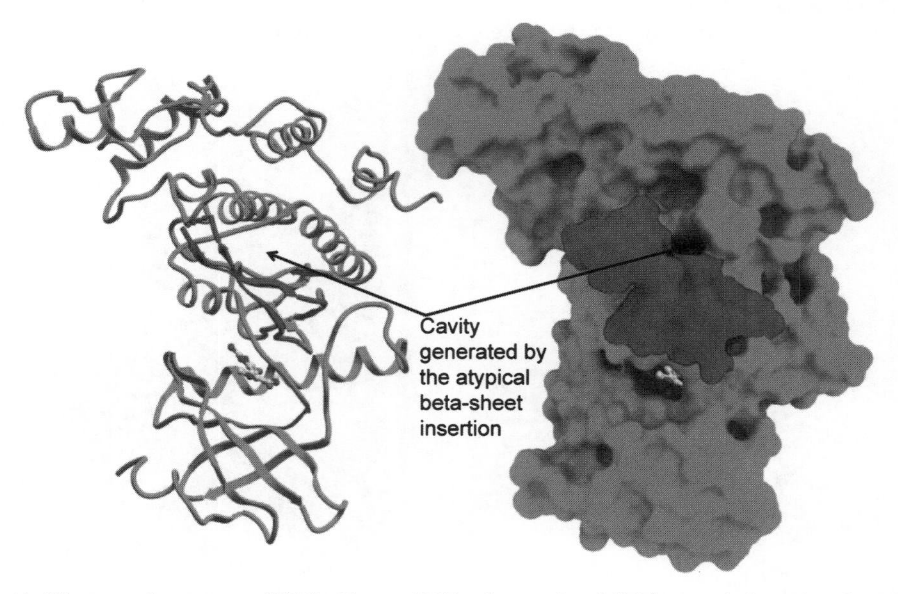

Figure 14. The crystal structure of CLK1 (shown, PDB code: 1z57) and CLK3 revealed a unique beta-hairpin insertion (magenta) with unknown function. This subfamily-specific structural motif generates cavities distinct from the ATP site that may be exploited for the generation of synthetic ligands to probe the clinical relevance of the target [125]. (See color insert.)

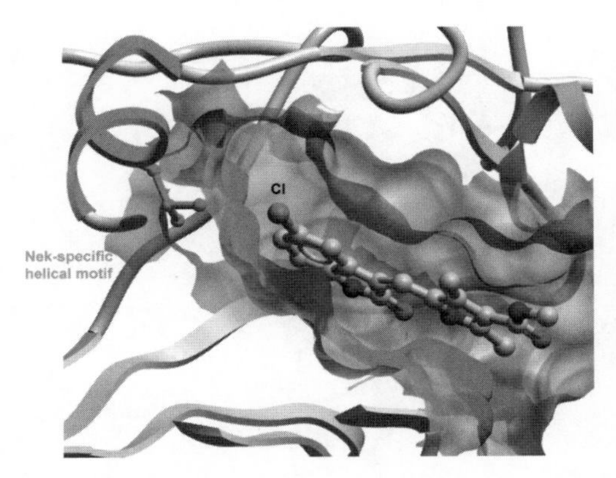

Figure 15. The structure of Nek2 uncovered an inhibitory helical motif (cyan) within the activation loop. A chlorine atom (green) of the cocrystallized inhibitor SU11652 makes limited van der Waals contact with a Leu of the Nek-specific helix. A strategy to achieve selective inactive-state inhibition may be to design ligands that better occupy the cavity generated by this atypical structural feature ([31], PDB code: 2jav). (See color insert.)

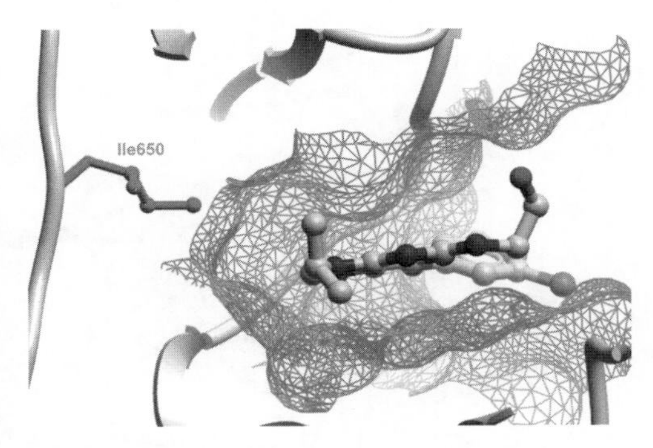

Figure 16. The crystal structure of Mer receptor tyrosine kinase (white ribbon) cocrystallized with a nonspecific inhibitor suggests how the presence of a nonconserved isoleucine (cyan) lining the ligand binding pocket (gray mesh) could be exploited to design more selective compounds ([39], PDB code: 3bpr). (See color insert.)

mediates phagocytic processes in retinal pigment epithelial cells, and mutations in Mer result in retinitis pigmentosa in patients [132], suggesting that Mer inhibition may mediate unacceptable side effect. RNAi or knockout experiments do not mimic the effect of a small-molecule drug, and cannot be used to estimate whether Mer represents a valid intervention point. *In vivo* efficacy and toxicology assessment of a potent and selective small-molecule Mer inhibitor is necessary to either invalidate the target or generate sufficiently promising data for promotion to preclinical discovery in the private sector.

The Mer structure reveals a so-called "DFG-Asp-in/alphaC-Glu-out" conformational state previously observed in the structures of Abl, SRC, EGFR, Chk, and Met kinases [39]. Interestingly, the isopropyl group of the cocrystallized inhibitor points toward Ile650, one of the few binding pocket residues not conserved among Axl/Mer/Sky kinases (Fig. 16). This structural feature may be exploited to develop selective chemical probes for target validation.

As a last example, the PIM Ser/Thr kinase family is composed of three members. PIM1 and PIM2 are overexpressed in leukemia and lymphoma [133,134], and activation of PIM1 and PIM2 mediates malignant transformation by oncogenic tyrosine kinases such as FLT3 internal tandem duplications, expressed in 25% of human acute myelogenous leukemia [135–137]. While a widely conserved mechanism of inhibitor binding at the kinase ATP site is to engage two hydrogen bonds with backbone atoms of the hinge region, the SGC structure of PIM1 revealed a unique hinge region characterized by the insertion of an additional residue and the presence of a proline, thereby lacking the second canonical hydrogen bond donor ([21,22], PDB code: 2bil). A first cocrystal structure showed how a kinase inhibitor can interact with the hinge region in a classic ATP-mimetic fashion [21,22]. In subsequent work, cocrystallization of an imidazo-[1,2-*b*] pyridazine inhibitor revealed an atypical binding mode, which translated in excellent binding selectivity, with over 100-fold specificity for PIM1 against PIM2 and cross-reactivity with only one other kinase out of a panel of 50 ([26], PDB code: 2c3i, Fig. 17a).

Finally, the same team conducted the systematic profiling of a library of 156 kinase inhibitors against a panel of 60 human Ser/Thr kinases, and showed that LY333'531, a PKCbeta inhibitor in phase III clinical trials for the treatment of diabetic complications, was able to inhibit PIM1 with and IC50 of 200 nM. The authors went on with a crystal structure showing that LY333'531 forms one

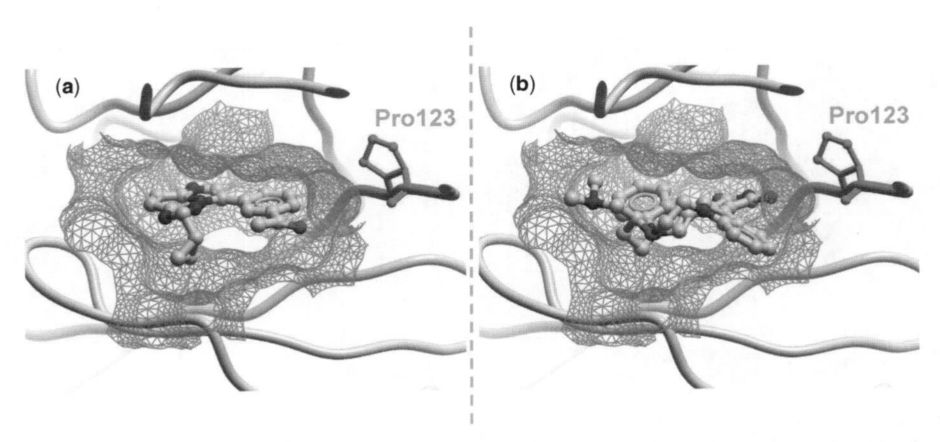

Figure 17. The structure of PIM1 reveals an atypical hinge region (cyan ribbon) where the presence of a proline residue (cyan sticks) obliterates one of the two canonical hydrogen-bond opportunities typically exploited by ATP-site inhibitors. (a) A cocrystallized imidazo[1,2-*b*] pyridazine chemotype (colored sticks) illustrates an original binding mode which does not mimic ATP binding, but contacts the binding pocket (gray mesh) mainly at a site opposite to the hinge. This unique binding mechanism results in high selectivity ([26], PDB code: 2c3i). (b) A cocrystallized 200 nM compound in Phase III clinical trial against diabetic complications displays optimal occupation of the binding pocket ([30], PDB code: 2jij). (See color insert.)

hydrogen bond to the hinge region, and occupies optimally the ATP pocket ([30], PDB code: 2j2i. Fig. 17b). LY333'531 reduced the growth of leukemic blast cells from five patients diagnosed with acute myeloid leukemia, which suggests an avenue to repurpose this clinical compound for cancer therapy [30].

These examples illustrate how structural genomics coverage of undercharacterized phylogenetic branches of the human kinome can reveal novel structural features, which can be exploited toward the development of selective and potent chemical probes, for target validation and prioritization of original discovery programs in the pharmaceutical industry.

3.5.4. Epigenetics: An Attractive Area for Potential Targets The field of epigenetics focuses on the study of covalent and noncovalent modifications of DNA and histone proteins, and how these control chromatin structure and gene expression [138]. This area of biology has gained considerable attention in recent years, as new fundamental principles surface, unknown biological mechanisms unfold and novel genes are linked to disease states. As a simple metric, the number of articles published in year 2008 responding to the keyword "epigenetics" was more than 10 times the value in year 2000 (14,189 in 2008 versus 1274 in 2000), while the ratio is less than five when using the buzzword "kinase."

The pharmaceutical industry is well aware that proteins writing, reading, or deleting epigenetics marks and regulating downstream signaling events likely constitute a trove of new potential intervention points for a diverse array of diseases [139–142] just like proteins controlling phosphorylation mediated signaling (kinases and phosphatases) represent an important fraction of today's validated therapeutic targets. For instance, and as we saw earlier, histone deacetylases, which remove acetyl marks from histones and other proteins, represent an important class of new epigenetics drug targets: 17 HDAC inhibitors were in clinical trial in 2007 and one of them, SAHA, was recently approved for the treatment of cutaneous T-cell lymphoma [102].

Considering the vast array of genes that control epigenetic signaling, the drug discovery industry needs to find ways to efficiently prioritize the most promising targets for therapeutic intervention. The rapid development of selective chemical probes to test the clinical relevance of specific genes, and their cocrystallization to dissect the structural mechanism of specificity, will guide drug discovery teams and accelerate the development of

Table 2. SGC Contribution, as of December 2008, to the Structural Coverage of Genes Involved in Writing, Reading, or Deleting Epigenetics Marks

Domain Family		Typical Substrate Class	Targets	Solved by SGC
Histone lysine demethylase		Histone/protein K(me)*n*	25	2
Bromodomain		Histone/protein K(ac)	55	13
ROYAL	Tudor domain	Histone Rme2s	26	1
	Tandem tudor	Histone K(me)2/3		2
	Chromodomain	Histone/protein K(me)3	31	8
	MBT repeat domain	Histone K(me)3	11	5
PHD finger		Histone K(me)*n*	78	1
Acetyltransferase		Histone/protein K	21	10
Histone Deacetylase		Histone/protein K	18	2
Methyltransferase		Histone/proteins K and R	54	9
Poly(ADP-ribose) polymerase		Histone carboxy	16	3

The last column lists the number of targets for which a previously unknown structure of the indicated domain was solved.
ROYAL = royal family of epigenetics marks reading modules; K = lysine; R = arginine; me = methyl; ac = acetyl.

tomorrows' drugs (Trump R, GSK, Epigenetics in health and disease—from biology to medicine. Oxford, November 13–14, 2008). To that end, the SGC, the NIH Chemical Genome Center and GSK launched in 2008 a large-scale partnership to develop and characterize chemical probes against epigenetics targets. The SGC already significantly contributed to the systematic structural coverage of the epigenetics landscape (Table 2, Fig. 4, [37,143,144]), and is beginning to deliver cocrystallized structures of novel inhibitors [145]. It is expected that this and other similar structural genomics efforts will significantly impact drug discovery.

4. CONCLUSION

The applications of structural genomics in chemistry, biology, and drug discovery are past, present, and future. The technology development that was necessary to build efficient protein production, purification, and crystallization platforms has already benefited laboratories outside structural genomics settings. Structural genomics centers have made major contributions to the structure coverage of the human proteome, as well as proteomes from human pathogens, such as the causative agents of tuberculosis and malaria. This raises the collective scientific knowledge necessary to understand biological mechanisms in health and disease. The increased volume of cocrystal structures teaches statistically significant rules for preferred conformations and interactions of overrepresented cocrystallized chemotypes. A number of such rules have already been published by medicinal chemists in the pharmaceutical industry. Systematic efforts focused on validated and emerging gene classes are mapping mechanistic determinants for receptor specificity and dissecting the structural chemistry of selective inhibition, thereby contributing to the chemogenomics coverage of new potential intervention points, uncovered by the human genome sequence, and accelerating target validation and drug discovery. Finally, a longer term and more fundamental goal defined at the outset of structural genomics initiatives remains the systematic blanketing of the structure space through experimental characterization of select representatives and computational modeling of structural neighbors.

ACKNOWLEDGMENTS

I wish to thank Yong Zhao for collecting the data necessary to generate Figure 1, and Aled Edwards, Cheryl Arrowsmith, and Stefan Knapp for their comments on the manuscript. The author is affiliated with the Structural Genomics Consortium, a registered charity (No. 1097737) that receives funds from the Canadian Institutes for Health Research, the Canadian Foundation for Innovation, Genome

Canada through the Ontario Genomics Institute, GlaxoSmithKline, Karolinska Institutet, the Knut and Alice Wallenberg Foundation, the Ontario Innovation Trust, the Ontario Ministry for Research and Innovation, Merck & Co., Inc., the Novartis Research Foundation, the Swedish Agency for Innovation Systems, the Swedish Foundation for Strategic Research, and the Wellcome Trust.

ABBREVIATIONS

CESG	Center for Eukaryotic Structural Genomics
GPCR	G-protein coupled receptor
HDAC	histone deacetylase
JCSG	Joint Center for Structural Genomics
MCSG	Midwest Center for Structural Genomics
MSGPP	Medical Structural Genomics of Pathogenic Protozoa
NDRT	Nucleoside 2-deoxyribosyltransferase
NSGC	Northeast Structural Genomics Consortium
NYSGRC	New York SGX Research Center for Structural Genomics
PARP	poly(ADP-ribose) polymerase
PDB	Protein Data Bank
PSI	Protein Structure Initiative
SAR	structure–activity relationship
SGC	Structural Genomics Consortium
SMP	Center for Structures of Membrane Proteins
SPinE	Structural Proteomics in Europe
TBSGC	Tuberculosis Structural Genomics Consortium
VIZIER	Viral Enzymes Involved in Replication

REFERENCES

1. Beddell CR, Goodford PJ, et al. Compounds designed to fit a site of known structure in human haemoglobin. Br J Pharmacol 1976;57(2): 201–209.
2. Congreve M, Murray CW, et al. Structural biology and drug discovery. Drug Discov Today 2005;10(13): 895–907.
3. Abagyan R, Totrov M. High-throughput docking for lead generation. Curr Opin Chem Biol 2001;5(4): 375–382.
4. Huang N, Shoichet BK, et al. Benchmarking sets for molecular docking. J Med Chem 2006;49(23): 6789–6801.
5. Gileadi O, Burgess-Brown NA, et al. High throughput production of recombinant human proteins for crystallography. Methods Mol Biol 2008;426:221–246.
6. Moffat K, Ren Z. Synchrotron radiation applications to macromolecular crystallography. Curr Opin Struct Biol 1997;7(5): 689–696.
7. Hendrickson WA. Synchrotron crystallography. Trends Biochem Sci 2000;25(12): 637–643.
8. Lamzin VS, Perrakis A. Current state of automated crystallographic data analysis. Nat Struct Biol 2000;7(Suppl): 978–981.
9. Zarembinski TI, Hung LW, et al. Structure-based assignment of the biochemical function of a hypothetical protein: a test case of structural genomics. Proc Natl Acad Sci USA 1998;95(26): 15189–15193.
10. Heinemann U, Frevert J, et al. An integrated approach to structural genomics. Prog Biophys Mol Biol 2000;73(5): 347–362.
11. Yokoyama S, Hirota H, et al. Structural genomics projects in Japan. Nat Struct Biol 2000;7 (Suppl): 943–945.
12. Bradley P, Misura KM, et al. Toward high-resolution de novo structure prediction for small proteins. Science 2005;309(5742): 1868–1871.
13. Vitkup D, Melamud E, et al. Completeness in structural genomics. Nat Struct Biol 2001;8(6): 559–566.
14. Burley SK, Joachimiak A, et al. Contributions to the NIH-NIGMS Protein Structure Initiative from the PSI Production Centers. Structure 2008;16(1): 5–11.
15. Levitt M. Growth of novel protein structural data. Proc Natl Acad Sci USA 2007;104(9): 3183–3188.
16. Goulding CW, Apostol M, et al. The TB Structural Genomics Consortium: providing a structural foundation for drug discovery. Curr Drug Targets Infect Disord 2002;2(2): 121–141.
17. Murillo AC, Li HY, et al. High throughput crystallography of TB drug targets. Infect Disord Drug Targets 2007;7(2): 127–139.

18. Vedadi M, Lew J, et al. Genome-scale protein expression and structural biology of *Plasmodium falciparum* and related apicomplexan organisms. Mol Biochem Parasitol 2007;151 (1): 100–110.
19. Fan E, Baker D, et al. Structural genomics of pathogenic protozoa: an overview. Methods Mol Biol 2008;426:497–513.
20. Coutard B, Gorbalenya AE, et al. The VIZIER project: preparedness against pathogenic RNA viruses. Antiviral Res 2008;78(1): 37–46.
21. Bullock AN, Debreczeni J, et al. Structure and substrate specificity of the Pim-1 kinase. J Biol Chem 2005;280(50): 41675–41682.
22. Bullock AN, Debreczeni JE, et al. Structural basis of inhibitor specificity of the human protooncogene proviral insertion site in moloney murine leukemia virus (PIM-1) kinase. J Med Chem 2005;48(24): 7604–7614.
23. Kavanagh KL, Guo K, et al. The molecular mechanism of nitrogen-containing bisphosphonates as antiosteoporosis drugs. Proc Natl Acad Sci USA 2006;103(20): 7829–7834.
24. Ludwig C, Michiels PJ, et al. SALMON: solvent accessibility, ligand binding, and mapping of ligand orientation by NMR spectroscopy. J Med Chem 2008;51(1): 1–3.
25. Hult M, Shafqat N, et al. Active site variability of type 1 11beta-hydroxysteroid dehydrogenase revealed by selective inhibitors and cross-species comparisons. Mol Cell Endocrinol 2006;248(1–2): 26–33.
26. Pogacic V, Bullock AN, et al. Structural analysis identifies imidazo[1,2-*b*]pyridazines as PIM kinase inhibitors with *in vitro* antileukemic activity. Cancer Res 2007;67(14): 6916–6924.
27. Eswaran J, Lee WH, et al. Crystal Structures of the p21-activated kinases PAK4, PAK5, and PAK6 reveal catalytic domain plasticity of active group II PAKs. Structure 2007;15(2): 201–213.
28. Bunkoczi G, Salah E, Filippakopoulos P, Fedorov O, Müller S, Sobott F, Parker SA, Zhang H, Min W, Turk BE, Knapp S Structural and functional characterization of the human protein kinase ASK1 Structure. 2007Oct;15 (10):1215–26.
29. Allali-Hassani A, Pan PW, et al. Structural and chemical profiling of the human cytosolic sulfotransferases. PLoS Biol 2007;5(5): e97.
30. Fedorov O, Marsden B, et al. A systematic interaction map of validated kinase inhibitors with Ser/Thr kinases. Proc Natl Acad Sci USA 2007;104(51): 20523–20528.
31. Rellos P, Ivins FJ, et al. Structure and regulation of the human Nek2 centrosomal kinase. J Biol Chem 2007;282(9): 6833–6842.
32. Schuetz A, Min J, et al. Structural basis of inhibition of the human NAD+-dependent deacetylase SIRT5 by suramin. Structure 2007;15(3): 377–389.
33. Artz JD, Dunford JE, et al. Targeting a uniquely nonspecific prenyl synthase with bisphosphonates to combat cryptosporidiosis. Chem Biol 2008;15(12): 1296–1306.
34. Tempel W, Rabeh WM, et al. Nicotinamide riboside kinase structures reveal new pathways to NAD+. PLoS Biol 2007;5(10): e263.
35. Schuetz A, Min J, et al. Human HDAC7 harbors a class IIa histone deacetylase-specific zinc binding motif and cryptic deacetylase activity. J Biol Chem 2008;283(17): 11355–11363.
36. Wu H, Dombrovsky L, et al. Structural basis of substrate-binding specificity of human arylamine *N*-acetyltransferases. J Biol Chem 2007;282(41): 30189–30197.
37. Ng SS, Kavanagh KL, et al. Crystal structures of histone demethylase JMJD2A reveal basis for substrate specificity. Nature 2007;448 (7149): 87–91.
38. Filippakopoulos P, Kofler M, et al. Structural coupling of SH2-kinase domains links Fes and Abl substrate recognition and kinase activation. Cell 2008;134(5): 793–803.
39. Huang X, Finerty P Jr, et al. Structural insights into the inhibited states of the Mer receptor tyrosine kinase. J Struct Biol 2008.
40. Kirchmair J, Markt P, et al. The Protein Data Bank (PDB), its related services and software tools as key components for *in silico* guided drug discovery. J Med Chem 2008.
41. Mestres J. Representativity of target families in the Protein Data Bank: impact for family-directed structure-based drug discovery. Drug Discov Today 2005;10(23–24): 1629–1637.
42. Fernandez A. Incomplete protein packing as a selectivity filter in drug design. Structure 2005;13(12): 1829–1836.
43. Boutet E, Lieberherr D, et al. UniProtKB/Swiss-Prot. Methods Mol Biol 2007;406: 89–112.
44. Edwards A. Large-scale structural biology of the human proteome. Annu Rev Biochem 2009; 78.
45. Marsden BD, Knapp S. Doing more than just the structure–structural genomics in kinase drug discovery. Curr Opin Chem Biol 2008;12(1): 40–45.

46. Bunkoczi G, Salah E, et al. Structural and functional characterization of the human protein kinase ASK1. Structure 2007;15(10): 1215–1226.
47. Kouzarides T. Chromatin modifications and their function. Cell 2007;128(4): 693–705.
48. Schneider R, Bannister AJ, et al. Unsafe SETs: histone lysine methyltransferases and cancer. Trends Biochem Sci 2002;27(8): 396–402.
49. Kubicek S, O'Sullivan RJ, et al. Reversal of H3K9me2 by a small-molecule inhibitor for the G9a histone methyltransferase. Mol Cell 2007;25(3): 473–481.
50. Weigelt J, McBroom-Cerajewski LD, et al. Structural genomics and drug discovery: all in the family. Curr Opin Chem Biol 2008;12(1): 32–39.
51. Overington JP, Al-Lazikani B, et al. How many drug targets are there? Nat Rev Drug Discov 2006;5(12): 993–996.
52. White SH. The progress of membrane protein structure determination. Protein Sci 2004;13 (7): 1948–1949.
53. Zhou Y, Morais-Cabral JH, et al. Chemistry of ion coordination and hydration revealed by a $K+$ channel-Fab complex at 2.0 Å resolution. Nature 2001;414(6859): 43–48.
54. Huang Y, Lemieux MJ, et al. Structure and mechanism of the glycerol-3-phosphate transporter from *Escherichia coli*. Science 2003;301 (5633): 616–620.
55. Jaakola VP, Griffith MT, et al. The 2.6 angstrom crystal structure of a human A2A adenosine receptor bound to an antagonist. Science 2008;322(5905): 1211–1217.
56. Lunin VV, Dobrovetsky E, et al. Crystal structure of the CorA Mg^{2+} transporter. Nature 2006;440(7085): 833–837.
57. Vedadi M, Niesen FH, et al. Chemical screening methods to identify ligands that promote protein stability, protein crystallization, and structure determination. Proc Natl Acad Sci USA 2006;103(43): 15835–15840.
58. Dong A, Xu X, et al. In situ proteolysis for protein crystallization and structure determination. Nat Methods 2007;4(12): 1019–1021.
59. Gileadi O, Knapp S, et al. The scientific impact of the Structural Genomics Consortium: a protein family and ligand-centered approach to medically-relevant human proteins. J Struct Funct Genomics 2007;8(2–3): 107–119.
60. Scapin G. Structural biology and drug discovery. Curr Pharm Des 2006;12(17): 2087–2097.
61. Skarzynski T, Thorpe J. Industrial perspective on X-ray data collection and analysis. Acta Crystallogr D Biol Crystallogr 2006;62(Pt 1): 102–107.
62. Hosfield D, Palan J, et al. A fully integrated protein crystallization platform for small-molecule drug discovery. J Struct Biol 2003;142(1): 207–217.
63. Blundell TL, Patel S. High-throughput X-ray crystallography for drug discovery. Curr Opin Pharmacol 2004;4(5): 490–496.
64. Gao X, Bain K, et al. High-throughput limited proteolysis/mass spectrometry for protein domain elucidation. J Struct Funct Genomics 2005;6(2–3): 129–134.
65. Graslund S, Nordlund P, et al. Protein production and purification. Nat Methods 2008;5(2): 135–146.
66. Manjasetty BA, Turnbull AP, et al. Automated technologies and novel techniques to accelerate protein crystallography for structural genomics. Proteomics 2008;8(4): 612–625.
67. Abagyan R, Lee WH, et al. Disseminating structural genomics data to the public: from a data dump to an animated story. Trends Biochem Sci 2006;31(2): 76–78.
68. Edwards A. Open-source science to enable drug discovery. Drug Discov Today 2008;13 (17–18): 731–733.
69. Badger J, Hendle J. Reliable quality-control methods for protein crystal structures. Acta Crystallogr D Biol Crystallogr 2002;58(Pt 2): 284–291.
70. Muller P, Lena G, et al. *In silico*-guided target identification of a scaffold-focused library: 1,3,5-triazepan-2,6-diones as novel phospholipase A2 inhibitors. J Med Chem 2006;49(23): 6768–6778.
71. Jambon M, Imberty A, et al. A new bioinformatic approach to detect common 3D sites in protein structures. Proteins 2003;52(2): 137–145.
72. Powers R, Copeland JC, et al. Comparison of protein active site structures for functional annotation of proteins and drug design. Proteins 2006;65(1): 124–135.
73. Najmanovich R, Kurbatova N, et al. Detection of 3D atomic similarities and their use in the discrimination of small molecule protein-binding sites. Bioinformatics 2008;24(16): i105–i111.
74. Bashton M, Nobeli I, et al. PROCOGNATE: a cognate ligand domain mapping for enzymes.

Nucleic Acids Res 2008;36(Database issue): D618–D622.
75. Brameld KA, Kuhn B, et al. Small molecule conformational preferences derived from crystal structure data. A medicinal chemistry focused analysis. J Chem Inf Model 2008;48(1): 1–24.
76. Hao MH, Haq O, et al. Torsion angle preference and energetics of small-molecule ligands bound to proteins. J Chem Inf Model 2007;47 (6): 2242–2252.
77. Cotesta S, Stahl M. The environment of amide groups in protein–ligand complexes: H-bonds and beyond. J Mol Model 2006;12 (4): 436–444.
78. Muller K, Faeh C, et al. Fluorine in pharmaceuticals: looking beyond intuition. Science 2007;317(5846): 1881–1886.
79. Benson ML, Smith RD, et al. Binding MOAD, a high-quality protein–ligand database. Nucleic Acids Res 2008;36(Database issue): D674–D678.
80. Carlson HA, Smith RD, et al. Differences between high- and low-affinity complexes of enzymes and nonenzymes. J Med Chem 2008;51 (20): 6432–6441.
81. Bostrom J, Hogner A, et al. Do structurally similar ligands bind in a similar fashion? J Med Chem 2006;49(23): 6716–6725.
82. Yardley V, Khan AA, et al. *In vivo* activities of farnesyl pyrophosphate synthase inhibitors against *Leishmania donovani* and *Toxoplasma gondii*. Antimicrob Agents Chemother 2002;46 (3): 929–931.
83. Rodan GA, Reszka AA. Bisphosphonate mechanism of action. Curr Mol Med 2002;2(6): 571–577.
84. Dunford JE, Kwaasi AA, et al. Structure–activity relationships among the nitrogen containing bisphosphonates in clinical use and other analogues: time-dependent inhibition of human farnesyl pyrophosphate synthase. J Med Chem 2008;51(7): 2187–2195.
85. Ricci KA, Girosi F, et al. Reducing stunting among children: the potential contribution of diagnostics. Nature 2006;444(Suppl 1): 29–38.
86. Blum M, Grant DM, et al. Human arylamine *N*-acetyltransferase genes: isolation, chromosomal localization, and functional expression. DNA Cell Biol 1990;9(3): 193–203.
87. Hein DW. Molecular genetics and function of NAT1 and NAT2: role in aromatic amine metabolism and carcinogenesis. Mutat Res 2002;506–507,65–77.
88. Ratnam K, Low JA. Current development of clinical inhibitors of poly(ADP-ribose) polymerase in oncology. Clin Cancer Res 2007;13 (5): 1383–1388.
89. Rodon J, Iniesta MD, et al. Development of PARP inhibitors in oncology. Expert Opin Investig Drugs 2009;18(1): 31–43.
90. Southan GJ, Szabo C. Poly(ADP-ribose) polymerase inhibitors. Curr Med Chem 2003;10(4): 321–340.
91. Lehtio L, Collins R, et al. Zinc binding catalytic domain of human tankyrase 1. J Mol Biol 2008;379(1): 136–145.
92. Arcus VL, Lott JS, et al. The potential impact of structural genomics on tuberculosis drug discovery. Drug Discov Today 2006;11(1–2): 28–34.
93. DeVinney R, Steele-Mortimer O, et al. Phosphatases and kinases delivered to the host cell by bacterial pathogens. Trends Microbiol 2000;8(1): 29–33.
94. Grundner C, Ng HL, et al. *Mycobacterium tuberculosis* protein tyrosine phosphatase PtpB structure reveals a diverged fold and a buried active site. Structure 2005;13(11): 1625–1634.
95. Soellner MB, Rawls KA, et al. Fragment-based substrate activity screening method for the identification of potent inhibitors of the *Mycobacterium tuberculosis* phosphatase PtpB. J Am Chem Soc 2007;129(31): 9613–9615.
96. Rozwarski DA, Grant GA, Barton DH, Jacobs WR Jr, Sacchettini JC. Modification of the NADH of the isoniazid target (InhA) from Mycobacterium tuberculosis Science. 1998Jan 2;279(5347):98–102.
97. Sacchettini JC, Blanchard JS. The structure and function of the isoniazid target in *M. tuberculosis*. Res Microbiol 1996;147(1–2): 36–43.
98. Barrett MP, Burchmore RJ, et al. The trypanosomiases. Lancet 2003;362(9394): 1469–1480.
99. Bosch J, Robien MA, et al. Using fragment cocktail crystallography to assist inhibitor design of *Trypanosoma brucei* nucleoside 2-deoxyribosyltransferase. J Med Chem 2006;49(20): 5939–5946.
100. Hopkins AL. Network pharmacology: the next paradigm in drug discovery. Nat Chem Biol 2008;4(11): 682–690.

101. Xu WS, Parmigiani RB, et al. Histone deacetylase inhibitors: molecular mechanisms of action. Oncogene 2007;26(37): 5541–5552.
102. Marks PA. Discovery and development of SAHA as an anticancer agent. Oncogene 2007;26(9): 1351–136.
103. Frye SV. Structure–activity relationship homology (SARAH): a conceptual framework for drug discovery in the genomic era. Chem Biol 1999;6(1): R3–R7.
104. Jacoby E, Schuffenhauer A, et al. Chemogenomics knowledge-based strategies in drug discovery. Drug News Perspect 2003;16(2): 93–102.
105. Harris CJ, Stevens AP. Chemogenomics: structuring the drug discovery process to gene families. Drug Discov Today 2006;11(19–20): 880–888.
106. Jacoby E. Chemogenomics: drug discovery's panacea? Mol Biosyst 2006;2(5): 218–220.
107. Ribeiro RC, Kushner PJ, et al. The nuclear hormone receptor gene superfamily. Annu Rev Med 1995;46:443–453.
108. Kliewer SA, Lehmann JM, et al. Orphan nuclear receptors: shifting endocrinology into reverse. Science 1999;284(5415): 757–760.
109. Willson TM, Moore JT. Genomics versus orphan nuclear receptors: a half-time report. Mol Endocrinol 2002;16(6): 1135–1144.
110. Schapira M, Abagyan R, et al. Nuclear hormone receptor targeted virtual screening. J Med Chem 2003;46(14): 3045–3059.
111. Klabunde T. Chemogenomic approaches to drug discovery: similar receptors bind similar ligands. Br J Pharmacol 2007;152(1): 5–7.
112. Cohen P. Protein kinases: the major drug targets of the twenty-first century? Nat Rev Drug Discov 2002;1(4): 309–315.
113. Manning G, Whyte DB, et al. The protein kinase complement of the human genome. Science 2002;298(5600): 1912–1934.
114. Karaman MW, Herrgard S, et al. A quantitative analysis of kinase inhibitor selectivity. Nat Biotechnol 2008;26(1): 127–132.
115. Vieth M, Higgs RE, et al. Kinomics-structural biology and chemogenomics of kinase inhibitors and targets. Biochim Biophys Acta 2004;1697(1–2): 243–257.
116. Deininger M, Buchdunger E, et al. The development of imatinib as a therapeutic agent for chronic myeloid leukemia. Blood 2005;105(7): 2640–2653.
117. Verkhivker GM. Imprint of evolutionary conservation and protein structure variation on the binding function of protein tyrosine kinases. Bioinformatics 2006;22(15): 1846–1854.
118. Liu Y, Gray NS. Rational design of inhibitors that bind to inactive kinase conformations. Nat Chem Biol 2006;2(7): 358–364.
119. Verkhivker GM. Exploring sequence–structure relationships in the tyrosine kinome space: functional classification of the binding specificity mechanisms for cancer therapeutics. Bioinformatics 2007;23(15): 1919–1926.
120. Kufareva I, Abagyan R. Type-II kinase inhibitor docking, screening, and profiling using modified structures of active kinase states. J Med Chem 2008;51(24): 7921–7932.
121. Fernandez A, Sanguino A, et al. An anticancer C-Kit kinase inhibitor is reengineered to make it more active and less cardiotoxic. J Clin Invest 2007;117(12): 4044–4054.
122. Fedorov O, Sundstrom M, et al. Insights for the development of specific kinase inhibitors by targeted structural genomics. Drug Discov Today 2007;12(9–10): 365–372.
123. Nolen B, Taylor S, et al. Regulation of protein kinases; controlling activity through activation segment conformation. Mol Cell 2004;15 (5): 661–675.
124. Eswaran J, Bernad A, et al. Structure of the human protein kinase MPSK1 reveals an atypical activation loop architecture. Structure 2008;16(1): 115–124.
125. Bullock AN, Das S, Debreczeni JE, Rellos P, Fedorov O, Niesen FH, Guo K, Papagrigoriou E, Amos AL, Cho S, Turk BE, Ghosh G, Knapp S. Kinase domain insertions define distinct roles of CLK kinases in SR protein phosphorylation. Structure. 2009;17(3):352–362
126. Fry AM. The Nek2 protein kinase: a novel regulator of centrosome structure. Oncogene 2002;21(40): 6184–6194.
127. Hayward DG, Fry AM. Nek2 kinase in chromosome instability and cancer. Cancer Lett 2006;237(2): 155–166.
128. Evans CO, Young AN, et al. Novel patterns of gene expression in pituitary adenomas identified by complementary deoxyribonucleic acid microarrays and quantitative reverse transcription-polymerase chain reaction. J Clin Endocrinol Metab 2001;86(7): 3097–3107.
129. Ek S, Hogerkorp CM, et al. Mantle cell lymphomas express a distinct genetic signature affecting lymphocyte trafficking and growth regulation as compared with subpopulations of normal human B cells. Cancer Res 2002;62 (15): 4398–4405.

130. Graham DK, Salzberg DB, et al. Ectopic expression of the proto-oncogene Mer in pediatric T-cell acute lymphoblastic leukemia. Clin Cancer Res 2006;12(9): 2662–2669.
131. Keating AK, Salzberg DB, et al. Lymphoblastic leukemia/lymphoma in mice overexpressing the Mer (MerTK) receptor tyrosine kinase. Oncogene 2006;25(45): 6092–6100.
132. Gal A, Li Y, et al. Mutations in MERTK, the human orthologue of the RCS rat retinal dystrophy gene, cause retinitis pigmentosa. Nat Genet 2000;26(3): 270–271.
133. Amson R, Sigaux F, et al. The human protooncogene product p33pim is expressed during fetal hematopoiesis and in diverse leukemias. Proc Natl Acad Sci USA 1989;86(22): 8857–8861.
134. Cohen AM, Grinblat B, et al. Increased expression of the hPim-2 gene in human chronic lymphocytic leukemia and non-Hodgkin lymphoma. Leuk Lymphoma 2004;45(5): 951–955.
135. Mizuki M, Schwable J, et al. Suppression of myeloid transcription factors and induction of STAT response genes by AML-specific Flt3 mutations. Blood 2003;101(8): 3164–3173.
136. Kim KT, Baird K, et al. Pim-1 is up-regulated by constitutively activated FLT3 and plays a role in FLT3-mediated cell survival. Blood 2005;105(4): 1759–1767.
137. Adam M, Pogacic V, et al. Targeting PIM kinases impairs survival of hematopoietic cells transformed by kinase inhibitor-sensitive and kinase inhibitor-resistant forms of Fms-like tyrosine kinase 3 and BCR/ABL. Cancer Res 2006;66(7): 3828–3835.
138. Goldberg AD, Allis CD, et al. Epigenetics: a landscape takes shape. Cell 2007;128(4): 635–638.
139. Jones PA, Baylin SB. The epigenomics of cancer. Cell 2007;128(4): 683–692.
140. Iacobuzio-Donahue CA. Epigenetic changes in cancer. Annu Rev Pathol 2008.
141. Jiang Y, Langley B, et al. Epigenetics in the nervous system. J Neurosci 2008;28(46): 11753–11759.
142. Szyf M. Epigenetics, DNA methylation, and chromatin modifying drugs. Annu Rev Pharmacol Toxicol 2008.
143. Min J, Allali-Hassani A, et al. L3MBTL1 recognition of mono- and dimethylated histones. Nat Struct Mol Biol 2007;14(12): 1229–1230.
144. Avvakumov GV, Walker JR, et al. Structural basis for recognition of hemi-methylated DNA by the SRA domain of human UHRF1. Nature 2008;455(7214): 822–825.
145. Rose NR, Ng SS, et al. Inhibitor scaffolds for 2-oxoglutarate-dependent histone lysine demethylases. J Med Chem 2008.

INDEX